W.H. ORTTUNG $25.00

molecular structure
and dynamics

molecular structure and dynamics

W. H. FLYGARE

School of Chemical Sciences
University of Illinois, Urbana

Prentice-Hall, Inc., Englewood Cliffs, New Jersey 07632

Library of Congress Cataloging in Publication Data

Flygare, W H (date)
Molecular structure and dynamics.

Includes bibliographical references and index.
1. Molecular theory. 2. Quantum chemistry.
I. Title.
QD461.F58 541'.28 77-16786
ISBN 0-13-599753-4

Printed in the United States of America

10 9 8 7 6 5 4 3 2 1

Prentice-Hall International, Inc., *London*
Prentice-Hall of Australia Pty. Limited, *Sydney*
Prentice-Hall of Canada, Ltd., *Toronto*
Prentice-Hall of India Private Limited, *New Delhi*
Prentice-Hall of Japan, Inc., *Tokyo*
Prentice-Hall of Southeast Asia Pte. Ltd., *Singapore*
Whitehall Books Limited, *Wellington, New Zealand*

contents

guide to major figures

guide to major tables

preface

This book evolved through several years of teaching a nine-month combined advanced undergraduate–introductory graduate course to students in the Chemistry, Physics, Astronomy, and Electrical Engineering Departments at the University of Illinois. The students who take the course have a wide diversity of backgrounds and degrees of mathematical sophistication. The text starts at an elementary but basic level and then proceeds, largely by example, to the more advanced topics that are of current interest in studying the structure and dynamics of molecules.

The approach of the text is from the perspective of an experimental scientist, and a balance is attempted between laboratory measurements and the theoretical models to explain the measurements. The intent is to develop a unity in approach toward a wide range of subjects in molecular structure and dynamics in a single text. In this way, the student may begin to understand the interplay between theory and experiment and the relations between the various methods of studying molecular structure and dynamics.

Chapter 1 examines the wave nature of electromagnetic radiation and the elastic scattering of electromagnetic fields by atoms and molecules. The particle nature of electromagnetic radiation is illustrated by examining blackbody radiation, the photoelectric effect, and the Compton effect. In Chapter 2 we outline some important classical equations of motion. We begin with Newton's second law and demonstrate the center of mass (c.m.) separation of potential and kinetic energies and follow with a discussion of molecular rotation. Newton's second law is also used to describe the interaction of a charged particle with an electric field. We then examine molecular vibrations, the use of generalized coordinates and Lagrange's equations, and molecular displacement coordinates. We end this chapter with a discussion of Hamilton's formalism including conservation theorems, and we derive the Hamiltonian for the interaction of an electromagnetic field with a charged particle.

Chapter 3 lays out the formalism needed to use quantum mechanical equations to understand the structure and dynamics of atoms and molecules. The early quantum theory of atoms established (for bound particles) the concept of discrete states, the quantization of angular momentum, and the Bohr frequency rule. Free particles can be described by a wave packet developed, by the superposition principle, from a free particle wavefunction. Following these preliminaries, the postulates of quantum mechanics are developed. Hermitian operators, state functions, stationary states, and the matrix formalism are all emphasized. Time-independent perturbation and variational theories are examined in connection with the matrix representation of the Hamiltonian operator. Finally, time-dependent perturbation theory is developed, and the semiclassical theory of the interaction of electromagnetic radiation and matter is examined. Chapter 4 applies the quantum theory including perturbation, matrix, and variational techniques to the solution of a number of stationary state model problems that include the particle-in-a-ring, internal rotation, the particle-in-a-box, the hydrogen atom, spin-orbit interactions, ESR, NMR, the rigid rotor, and the harmonic oscillator. Extensive problems with applications to current spectroscopic observations on atomic and molecular systems are given.

Chapter 5 develops the many-electron theories for atoms. The Hartree–Fock SCF single determinant method is developed and extended to configuration interaction (CI) corrections leading finally to the exact wavefunction for the atom. The use of Slater-type orbitals as an approximation to the Hartree–Fock atomic orbitals is discussed in detail, and several observables arising from one-electron operators are examined with these atomic orbitals. Various types of atomic orbital expansion are reviewed. The electron theory of the spin-orbit and nuclear spin–electron spin coupling in atoms is also given. An "atoms-in-molecules" approach to transition metal complexes involving crystal field theory is then developed and applied also to nuclear quadrupole interactions. The chapter ends with a view of solid-state properties in terms of atoms. Chapter 6 gives the many-electron *ab initio* theory of molecules. Both the LCAO-MO-SCF single determinant Hartree–Fock method and the LCAO-MO-SCF-CI technique are examined. Various stages of sophistication in the basis sets in the single determinant method are examined as well as various techniques of applying CI improvements. Several applications in molecular structure are reviewed. Approximate LCAO-MO-SCF theories are also developed with applications to chemical structural problems. The last half of Chapter 6 uses perturbation theory on the many-electron molecular functions to develop the theories of the static electric and magnetic field interactions with electrons in molecules.

Molecular magnetic susceptibilities are related to molecular g-values and molecular quadrupole moments. Nuclear magnetic shielding in molecules is related to the spin-rotation interaction and chemical shifts. Electric polarizabilities are also related to the electronic structure of molecules.

In Chapter 7, radiation absorption processes, Doppler–Lorentz convolution, and correlation functions are discussed in reference to molecular spectroscopy. The magnetic and electric Bloch equations are derived and molecular relaxation (T_2 and T_1) processes are examined. Following a discussion of double-resonance saturation effects, we examine the transient solutions to the Bloch equations. Transient nutation and free induction decay processes, Fourier transform spectroscopy, and multiple pulse experiments are all considered. Optical interferometers, lasers, and picosecond spectroscopy are also examined.

Chapter 8 gives a more advanced treatment of the nonresonant scattering of electromagnetic radiation (including X rays) from atoms and molecules in gases and liquids. Starting with the kinetic theory of gases, the single-molecule Rayleigh and Raman scattering processes are developed and the measurements of rotational and vibrational relaxation times are examined. This is followed by an examination of the static and dynamic information obtained in the rotationally quenched liquid state where vibrational relaxation times and rotational correlation times are measured. Hydrodynamic methods of describing fluctuations in the liquid state are also discussed. The methods of X-ray scattering to determine molecular and liquid structure are followed by an examination of the methods of studying the hydrodynamic motions of macromolecules in solution by light scattering. Chapter 9 develops the theories necessary for understanding particle-particle scattering experiments. Following a discussion of classical scattering, wave packets and their scattering properties are examined. The concepts of plane waves and partial waves are developed and the extraction of interparticle interaction potentials from the scattering data is developed. The chapter ends with a discussion of the determination of molecular structure from the elastic scattering of high energy electrons (electron diffraction).

The Appendix deserves the special attention of the serious student. It begins with a discussion of notational conventions and their units, and conversion factors are developed between the SI and cgs-Gaussian systems. Several useful mathematical relations and techniques are summarized. These include vector and tensor algebra, Fourier transform methods, Dirac delta functions, the solution of several important differential equations, and a discussion of numerical methods. Methods of using group theory in the examination of molecular structure are also summarized in the Appendix. The Appendix ends with some comments on references.

I have found the sequence presented here to be satisfactory for a nine-month course. It is difficult to cover all the material by the traditional lecture style of deriving equations and illustrating experiments on a blackboard while the students take notes. Using this technique, the instructor then has the choice of omitting those topics he finds peripheral to the course. On the other hand, by using handwritten viewgraphs prepared before the lecture, most of the material can be covered in a two-semester or three-quarter sequence. Of course, a great deal is asked of the students through the wide range of material covered, the many references to more detailed and extensive treatments, and the extensive problems at the end of the chapters. The student should obtain a good handbook such as H. B. Dwight, *Tables of Integrals and Other Mathematical Data*, 4th ed. (Macmillan Co., New York, 1961) to aid in solving the problems as well as working in the text. The problems are designed both to illustrate the principles through experimental numbers and to branch out to new applications that will sometimes couple strongly with the current literature. Many of the problems will require the use of a digital computer, a facility available to most students. The problems requiring machine computation are listed with an asterisk (*). Many useful programs (in Fortran language), including matrix diagonalization, function orthogonalization, spectroscopic applications, molecular structure calculations, *ab initio* LCAO-MO calculations, approximate LCAO-MO methods, and least-squares methods are available from the Quantum Chemistry Program Exchange (QCPE), Chemistry Department, Indiana University, Bloomington, Indiana, 47401. After going through the material in this text, the student should be fairly well prepared for advanced work in some field of molecular structure and dynamics. If the student's main interests lie elsewhere, he or she will at least have an appreciation for this field of study from a variety of viewpoints at a fairly sophisticated level.

This material can also be used for shorter courses. A semester or quarter course on introductory quantum chemistry as illustrated by spectroscopy might use Sections 1.1, 2.2, 3.3, 3.4, 3.5, 3.6, 3.7, all of Chapter 4, and Appendix E as the basis of the course. A semester or quarter course in the electronic structure of molecules might use parts of Chapter 3, parts of Chapter 5, and the first seven sections of Chapter 6. A semester or quarter course on modern molecular spectroscopy might use Sections 1.1, 2.2, Chapter 3, parts of Chapter 4, and all of Chapter 7. A special topics course on electric and magnetic interactions in molecules might use Sections 2.2, 3.7, 5.5, 5.6, 5.7, 5.8, 6.8, 6.9, and 6.10 as a basis for the course. Chapters 8 and 9 can be used as the basis for special topics courses in electromagnetic or particle scattering, respectively. If the text is used for one of the shorter courses suggested here, the student should find the remaining parts of the text useful in studying other topics.

Finally, I hope the book will be of use to those instructors, scientists, and engineers who are no longer students but who are interested in a somewhat integrated view of the field of molecular structure and dynamics.

I am indebted to those many people who have reviewed parts of or all of the manuscript and who have offered comments and suggestions for improvement. These people include T. Balle, R. A. Bernheim, M. T. Bowers, W. Buxton, M. Coltrin, L. Duda, T. H. Dunning, B. Fujimoto, T. D. Gierke, L. B. Harding, R. T. Hofmann, S. Kurtz, B. Mahan, R. A. Marcus, K. J. Miller, J. Pochan, W. P. Reinhardt, T. G. Schmalz, R. L. Shoemaker, K. Smith, and H. L. Voss. I am also indebted to R. Flygare, K. McCormick, D. Myers, and N. Way for the excellent secretarial and editorial help that I've had with this manuscript.

W. H. Flygare

molecular structure and dynamics

1

electromagnetic radiation and photons

1.1 THE ELECTROMAGNETIC WAVE THEORY OF RADIATION

Prior to Maxwell, the disciplines of optics and electromagnetism were developed independently. Contributions to an understanding of optics were made by Newton, Huygens, Young, Fresnel, and others [1]†. Light and the observed phenomena of dispersion, refraction, reflection, diffraction, polarization, and interference could be explained on the basis of a wavelike character. Contributions to an understanding of electromagnetism were made by Oersted, Faraday, Ampere, and others [2].

In 1846 Faraday discovered the first connection between optical and electromagnetic phenomena now called the *Faraday effect*. When plane-polarized radiation passes through certain materials exposed to a magnetic field that is parallel to the propagation direction of the radiation, the plane of polarization is rotated. The

† Numbers in square brackets correspond to the numbered sources found in the Reference section on p. 37 at the end of the chapter.

degree of rotation is dependent on the nature of the medium and is proportional to the magnitude of the field. After this discovery, there was much speculation on the relation between the seemingly unrelated subjects of optics and electromagnetism. Faraday sought the answer to this problem and it seemed predestined that the solution would evolve around him and his work. It was not long before Maxwell, guided by Faraday's work, forged the union of optics and electromagnetism.

Maxwell's equations put the wave theory of electromagnetic interaction on a firm foundation. Explanations of common optical phenomena such as dispersion, refraction, and other properties were put into quantitative predictive terms with Maxwell's wave equations. Also, many observations of the interaction of electromagnetic radiation with matter on a microscopic level such as nonresonant light scattering (Section 1.2) and resonant atomic and molecular emission and absorption of radiation (Section 2.2) could be explained on the basis of the wave theory.

It is important to gain a firm understanding of the classical equations for the propagation of electromagnetic energy because most descriptions of the interaction of radiation and matter use a semiclassical approach. The *semiclassical approach* assumes classical radiation interacting with a quantum mechanical system. Maxwell's equations are given here in both cgs and SI units [1], [3]:

cgs	*SI*	
$\nabla \times \boldsymbol{E} + \dfrac{1}{c}\dfrac{\partial \boldsymbol{B}}{\partial t} = 0$	$\nabla \times \boldsymbol{E} + \dfrac{\partial \boldsymbol{B}}{\partial t} = 0$	(1-1)
$\nabla \times \boldsymbol{H} - \dfrac{1}{c}\dfrac{\partial \boldsymbol{D}}{\partial t} = \dfrac{4\pi}{c}\boldsymbol{J}$	$\nabla \times \boldsymbol{H} - \dfrac{\partial \boldsymbol{D}}{\partial t} = \boldsymbol{J}$	(1-2)
$\nabla \cdot \boldsymbol{D} = 4\pi\bar{\rho}$	$\nabla \cdot \boldsymbol{D} = \bar{\rho}$	(1-3)
$\nabla \cdot \boldsymbol{B} = 0$	$\nabla \cdot \boldsymbol{B} = 0$	(1-4)
$\boldsymbol{D} = \boldsymbol{\varepsilon} \cdot \boldsymbol{E}, \qquad \boldsymbol{B} = \boldsymbol{\mu}_p \cdot \boldsymbol{H}$	$\boldsymbol{D} = \varepsilon_0 \boldsymbol{\varepsilon} \cdot \boldsymbol{E}, \qquad \boldsymbol{B} = \mu_0 \boldsymbol{\mu}_p \cdot \boldsymbol{H}$	(1-5)

E and ***H*** are the electric and magnetic field vectors, respectively. ***D*** is the *electric induction* or *displacement field* and $\boldsymbol{\varepsilon}$ is the *relative electric permittivity* (*dielectric constant*) of the medium. ***B*** is the *magnetic induction* or *magnetic flux density* and $\boldsymbol{\mu}_p$ is the *relative magnetic permeability* of the medium. $\boldsymbol{\varepsilon}$ and $\boldsymbol{\mu}_p$ are, in general, dyadics [4] and the elements have no dimensions as these are relative quantities. For an isotropic medium, the scalar ε and μ_p are used. $\varepsilon = \mu_p = 1.0$ for a vacuum and ε ranges from 1.0 (vacuum) to 80 (H_2O) for various substances. μ_p is very near unity for most substances. If μ_p is less than 1.0, the substance is diamagnetic; if μ_p is larger than 1.0, the substance is paramagnetic. μ_p and ε are always positive.

$$\varepsilon_0 = \frac{10^7}{4\pi c^2}\mathrm{C}^2 \cdot \mathrm{N}^{-1} \cdot \mathrm{m}^{-2} = 8.854187818 \times 10^{-12}\ \mathrm{s}^4 \cdot \mathrm{A}^2 \cdot \mathrm{kg}^{-1} \cdot \mathrm{m}^{-3}\ (\mathrm{F} \cdot \mathrm{m}^{-1})$$

is the *permittivity of a vacuum* and

$$\mu_0 = 4\pi \times 10^{-7}\ \mathrm{kg} \cdot \mathrm{m} \cdot \mathrm{s}^{-2} \cdot \mathrm{A}^{-2}\ (\mathrm{H} \cdot \mathrm{m}^{-1})$$

is the *permeability of free space*. It is evident that $(1/\varepsilon_0 \mu_0)^{1/2} = c\ (\mathrm{m} \cdot \mathrm{s}^{-1})$. $\bar{\rho}$ is a scalar representing the *electric charge density* (charge · volume^{-1}) and ***J*** is the *electric* *current density* (charge · area^{-1} · t^{-1}), which is equal to the velocity of the charge distribution times the charge density: $\boldsymbol{J} = \bar{\rho}\boldsymbol{v}$. The proportionality between $\boldsymbol{J}$ and $\boldsymbol{E}$ is called the *conductivity*, $\boldsymbol{\sigma}$, a dyadic,

$$\boldsymbol{J} = \boldsymbol{\sigma} \cdot \boldsymbol{E}, \tag{1-6}$$

which is appropriate for both sets of units. Returning to Eqs.(1-1) to (1-5), we note that $\boldsymbol{\nabla}$ is the vector operator given by

$$\boldsymbol{\nabla} = \hat{\boldsymbol{i}}\frac{\partial}{\partial x} + \hat{\boldsymbol{j}}\frac{\partial}{\partial y} + \hat{\boldsymbol{k}}\frac{\partial}{\partial z}, \tag{1-7}$$

where $\hat{\boldsymbol{i}}$, $\hat{\boldsymbol{j}}$, and $\hat{\boldsymbol{k}}$ are the Cartesian unit vectors. $\boldsymbol{\nabla} \times$ is called the *curl*, $\boldsymbol{\nabla}\cdot$ is called the *divergence*, and $\boldsymbol{\nabla}$ is called the *grad*.

$$\nabla^2 \quad \text{or} \quad \boldsymbol{\nabla} \cdot \boldsymbol{\nabla} = \nabla^2 = \frac{\partial^2}{\partial x^2} + \frac{\partial^2}{\partial y^2} + \frac{\partial^2}{\partial z^2} \tag{1-8}$$

is called the *Laplacian operator*.

The physical interpretations of Maxwell's equations are

1. Equation (1-1) is Faraday's law of electromagnetic induction showing that a time-dependent magnetic flux density, $\boldsymbol{B}$, gives rise to an electric field, $\boldsymbol{E}$, in a direction perpendicular to the original magnetic field.
2. Equation (1-2) is the Ampere–Oersted law showing that a magnetic field will exist near a current density, $\boldsymbol{J}$. The equation also contains Maxwell's assumption of a displacement field, $\boldsymbol{D}$, which is necessary to propagate electromagnetic energy through space. In the absence of a current density, $\boldsymbol{J}$, a time-dependent displacement field will still lead to a magnetic field. Equations (1-1) and (1-5) show that the time-dependent magnetic field produces an electric field, thereby completing the cycle for propagation.
3. Equation (1-3) is the Coulomb law in electrostatics.
4. Equation (1-4) embodies the nonexistence of a magnetic entity analogous to electric charge.

Equations (1-1) to (1-4) represent the great synthesis of macroscopic electric and magnetic phenomena and their interdependence. Equations (1-5) and (1-6) show the response of a material medium to electric and magnetic fields.

The classical continuity expression for electric charge can be obtained from Maxwell's equations by taking the divergence ($\boldsymbol{\nabla}\cdot$) of Eq.(1-2) [4] giving

$$\text{cgs:} \quad \boldsymbol{\nabla} \cdot (\boldsymbol{\nabla} \times \boldsymbol{H}) - \left(\frac{\boldsymbol{\nabla}}{c}\right) \cdot \left(\frac{\partial \boldsymbol{D}}{\partial t}\right) - \frac{4\pi \boldsymbol{\nabla} \cdot \boldsymbol{J}}{c} = 0, \qquad \text{SI:} \quad \boldsymbol{\nabla} \cdot (\boldsymbol{\nabla} \times \boldsymbol{H}) - \boldsymbol{\nabla} \cdot \frac{\partial \boldsymbol{D}}{\partial t} - \boldsymbol{\nabla} \cdot \boldsymbol{J} = 0. \tag{1-9}$$

Noting that

$$\boldsymbol{\nabla} \cdot [\boldsymbol{\nabla} \times \boldsymbol{H}] = 0 \tag{1-10}$$

by expansion [see Eq.(B-41)] and taking the time derivative of Eq.(1-3) gives

$$\begin{array}{cc} cgs & SI \\ \nabla \cdot \left(\frac{\partial \boldsymbol{D}}{\partial t}\right) = 4\pi \frac{\partial \bar{\rho}}{\partial t}, & \nabla \cdot \left(\frac{\partial \boldsymbol{D}}{\partial t}\right) = \frac{\partial \bar{\rho}}{\partial t}. \end{array} \tag{1-11}$$

Comparing Eqs.(1-9) and (1-11) gives

$$\nabla \cdot \boldsymbol{J} = -\frac{\partial \bar{\rho}}{\partial t}, \tag{1-12}$$

which is valid for either set of units. Equation (1-12) is the classical hydrodynamic equation of continuity. The divergence of a vector $\boldsymbol{J}$, $(\nabla \cdot \boldsymbol{J})$, is interpreted at each point as the rate per unit volume at which the current, $\boldsymbol{J}$, is flowing from that point. Thus, according to Eq.(1-12), the divergence of the current density at a certain point in space is equal to the decrease in the charge density in time.

In treatment of electromagnetic radiation, it is frequently convenient to express the magnetic and electric field vectors in terms of an electromagnetic *vector potential*, $\boldsymbol{A}(\boldsymbol{r}, t)$, and an electromagnetic *scalar potential*, $\varphi(\boldsymbol{r}, t)$. This transformation from $\boldsymbol{E}$ and $\boldsymbol{H}$ to $\boldsymbol{A}$ and φ reduces the equations in six components (E_x, E_y, E_z, H_x, H_y, and H_z) to equations in four components (A_x, A_y, A_z, and φ). The scalar potential, φ, is defined for time-independent (static) fields when Eq.(1-1) reduces to $\nabla \times \boldsymbol{E} = 0$. Under these conditions we can write

$$\boldsymbol{E} = -\nabla \varphi, \tag{1-13}$$

as $\nabla \times \boldsymbol{E} = -\nabla \times \nabla \varphi = 0$. Now, if the medium is isotropic and $\boldsymbol{\varepsilon}$ reduces to a scalar ε [see discussion following Eq.(B-37)], we can substitute Eq.(1-5) into Eq.(1-3) and use Eq.(1-13) to give

$$\begin{array}{cc} cgs & SI \\ \nabla^2 \varphi = -\frac{4\pi}{\varepsilon} \bar{\rho} & \nabla^2 \varphi = -\frac{1}{\varepsilon \varepsilon_0} \bar{\rho}, \end{array} \tag{1-14}$$

which is *Poisson's equation* for electrostatics (scalar ε). In regions of space (or in materials) where the charge density, $\bar{\rho}$, is zero, Poisson's equation reduces to *Laplace's equation*:

$$\nabla^2 \varphi = 0. \tag{1-15}$$

In the case of time-dependent (dynamic) fields, we must also include the vector potential, $\boldsymbol{A}$, in a relationship with $\boldsymbol{E}$. We construct $\boldsymbol{E}$ according to

$$\begin{array}{cc} cgs & SI \\ \boldsymbol{E} = -\frac{1}{c}\frac{\partial \boldsymbol{A}}{\partial t} - \nabla \varphi & \boldsymbol{E} = -\frac{\partial \boldsymbol{A}}{\partial t} - \nabla \varphi. \end{array} \tag{1-16}$$

Substituting Eq.(1-16) into Eq.(1-1) gives

$$\frac{\partial \nabla \times A}{\partial t} = \frac{\partial B}{\partial t} \tag{1-17}$$

for both sets of units. This equation is satisfied if

$$\nabla \times A = B, \tag{1-18}$$

which defines the magnetic induction in terms of the vector potential. Thus, Eqs.(1-16) and (1-18) satisfy Maxwell's equations and define the vector and scalar potentials A and φ. The vector and scalar potentials in an isotropic medium are related by the usual classical equation of flow in Eq.(1-12) by using the *Lorentz convention* given by

cgs	*SI*	
$\nabla \cdot A + \frac{\varepsilon \mu_p}{c} \frac{\partial \varphi}{\partial t} = 0$	$\nabla \cdot A + \varepsilon \mu_p \varepsilon_0 \mu_0 \frac{\partial \varphi}{\partial t} = 0.$	(1-19)

The Lorentz convention is a convenient choice for A and φ that is consistent with Eqs.(1-16), (1-18), and Maxwell's equations. The choice of A and φ is not unique, however, and any choice within the limits set by the following equations is valid:

$$\begin{aligned} A &\longrightarrow A + \nabla f = A' \\ \varphi &\longrightarrow \varphi - \frac{1}{c}\frac{\partial f}{\partial t} = \varphi', \end{aligned} \tag{1-20}$$

where f is a scalar. The choice of f is called a choice of *gauge*. Thus, Eq.(1-19) defines a convenient gauge [5]. Other gauges may be used, when convenient, within the limitations above.

Substituting Eqs.(1-16), (1-18), and (1-19) into Maxwell's equations for an isotropic medium ($\boldsymbol{\varepsilon} = \varepsilon$ and $\boldsymbol{\mu}_p = \mu_p$) gives the characteristic wave equations for the scalar and vector potentials (where $\varepsilon_0 \mu_0 = 1/c^2$):

cgs	*SI*	
$-\nabla^2 \varphi + \frac{\varepsilon \mu_p}{c^2} \frac{\partial^2 \varphi}{\partial t^2} = \frac{4\pi \bar{\rho}}{\varepsilon}$	$-\nabla^2 \varphi + \frac{\varepsilon \mu_p}{c^2} \frac{\partial^2 \varphi}{\partial t^2} = \frac{\bar{\rho}}{\varepsilon \varepsilon_0}$	(1-21)
$-\nabla^2 A + \frac{\varepsilon \mu_p}{c^2} \frac{\partial^2 A}{\partial t^2} = \frac{4\pi \mu_p}{c} J$	$-\nabla^2 A + \frac{\varepsilon \mu_p}{c^2} \frac{\partial^2 A}{\partial t^2} = \mu_p \mu_0 J.$	(1-22)

For instance, Eq.(1-21) is obtained from Eqs.(1-3), (1-5), and (1-16) by

cgs	*SI*	
$\nabla \cdot E = -\frac{1}{c}\frac{\partial}{\partial t} \nabla \cdot A - \nabla^2 \varphi = \frac{4\pi \bar{\rho}}{\varepsilon}$	$\nabla \cdot E = -\frac{\partial}{\partial t} \nabla \cdot A - \nabla^2 \varphi = \frac{\bar{\rho}}{\varepsilon \varepsilon_0}.$	(1-23)

From Eq.(1-19) we can write

$$\text{cgs} \qquad\qquad \text{SI}$$

$$-\frac{1}{c}\frac{\partial}{\partial t}(\nabla \cdot \boldsymbol{A}) = \frac{\varepsilon\mu_p}{c^2}\frac{\partial^2 \varphi}{\partial t^2} \qquad -\frac{\partial}{\partial t}\nabla \cdot \boldsymbol{A} = \varepsilon\mu_p\varepsilon_0\mu_0\frac{\partial^2 \varphi}{\partial t^2}.$$

Comparing this equation with Eq.(1-23) gives Eq.(1-21). Equation (1-22) is also easily derived.

We shall now examine the propagation of electromagnetic radiation in a neutral ($\bar{\rho} = 0$) and nonconducting ($\boldsymbol{J} = 0$) medium where Eq.(1-22) reduces to

$$\nabla^2 \boldsymbol{A} = \frac{\varepsilon\mu_p}{c^2}\frac{\partial^2 \boldsymbol{A}}{\partial t^2}. \tag{1-24}$$

We choose the plane traveling wave solution to this equation given by

$$\boldsymbol{A} = \boldsymbol{A}_0 \exp[i(\boldsymbol{k}\cdot\boldsymbol{r} - \omega t)] = \boldsymbol{A}_0[\cos(\boldsymbol{k}\cdot\boldsymbol{r} - \omega t) + i\sin(\boldsymbol{k}\cdot\boldsymbol{r} - \omega t)], \tag{1-25}$$

where $\boldsymbol{r}$ is the vector distance from the origin of the coordinate system to the point of measurement, $\boldsymbol{k}$ is called the *propagation vector* with units of inverse length, ω is the *angular frequency* in units of rad $\cdot$ s^{-1}, and t is the time. The frequency in units of hertz (Hz) is given by $\nu = \omega/2\pi$. Substituting Eq.(1-25) into Eq.(1-24) gives

$$-k^2\boldsymbol{A} + \frac{\omega^2\varepsilon\mu_p}{c^2}\boldsymbol{A} = 0$$

$$k = (\varepsilon\mu_p)^{1/2}\left(\frac{\omega}{c}\right) = (\varepsilon\mu_p)^{1/2}\left(\frac{2\pi\nu}{c}\right) = (\varepsilon\mu_p)^{1/2}\left(\frac{2\pi}{\lambda_0}\right). \tag{1-26}$$

λ_0 is the *vacuum wavelength* of the electromagnetic radiation. The equation above relating the magnitude of the propagation vector to the vacuum wavelength is called the *dispersion relation.* Equation (1-26) shows that the radiation propagates with a speed of $c/(\varepsilon\mu_p)^{1/2}$. In a nonmagnetic medium where $\mu_p = 1.0$, as it is in most materials, the radiation speed is $c/(\varepsilon)^{1/2}$. In the case of optical radiation, the square root of the relative dielectric constant, ε, is called the *refractive index*, n:

$$n = (\varepsilon)^{1/2}, \qquad k = \frac{2\pi n}{\lambda_0}. \tag{1-27}$$

Of course, in a vacuum, $n = 1.0$ and $k = 2\pi/\lambda_0$ and the radiation propagates with a speed c.

Substituting the plane wave solution for $\boldsymbol{A}$ in Eq.(1-25) into Eq.(1-18) gives the value of the magnetic induction vector [6]:

$$\begin{aligned}\boldsymbol{B} = \nabla \times \boldsymbol{A} &= \nabla \times \boldsymbol{A}_0 \exp[i(\boldsymbol{k}\cdot\boldsymbol{r} - \omega t)] = -\boldsymbol{A}_0 \times \nabla \exp[i(\boldsymbol{k}\cdot\boldsymbol{r} - \omega t)] \\ &= -\boldsymbol{A}_0 \times [\nabla(i\boldsymbol{k}\cdot\boldsymbol{r})]\exp[i(\boldsymbol{k}\cdot\boldsymbol{r} - \omega t)] = -i\boldsymbol{A}_0 \times \boldsymbol{k}\exp[i(\boldsymbol{k}\cdot\boldsymbol{r} - \omega t)] \\ &= \boldsymbol{B}_0 \exp[i(\boldsymbol{k}\cdot\boldsymbol{r} - \omega t)] = \boldsymbol{B}_0[\cos(\boldsymbol{k}\cdot\boldsymbol{r} - \omega t) + i\sin(\boldsymbol{k}\cdot\boldsymbol{r} - \omega t)].\end{aligned} \tag{1-28}$$

Similarly, the electric vector may be obtained by substituting Eq.(1-25) into Eq.(1-16). We mentioned, however, that the choice of $\boldsymbol{A}$ and φ in Eqs.(1-16) and (1-18) is not unique. Accordingly, when $\boldsymbol{J} = 0$ and $\bar{\rho} = 0$, we can choose $\varphi = 0$ with no loss in generality [7]. Substituting $\varphi = 0$ and Eq.(1-25) into Eq.(1-16) gives

cgs	*SI*
$\boldsymbol{E} = -\dfrac{1}{c}\dfrac{\partial}{\partial t}\boldsymbol{A}$	$\boldsymbol{E} = -\dfrac{\partial}{\partial t}\boldsymbol{A}$
$= -\dfrac{\omega}{c}\{-i\boldsymbol{A}_0 \exp[i(\boldsymbol{k}\cdot\boldsymbol{r} - \omega t)]\}$	$= -\omega\{-i\boldsymbol{A}_0 \exp[i(\boldsymbol{k}\cdot\boldsymbol{r} - \omega t)]\}$
$= \boldsymbol{E}_0 \exp[i(\boldsymbol{k}\cdot\boldsymbol{r} - \omega t)]$	$= \boldsymbol{E}_0 \exp[i(\boldsymbol{k}\cdot\boldsymbol{r} - \omega t)]$

$$\boldsymbol{E} = \boldsymbol{E}_0[\cos(\boldsymbol{k}\cdot\boldsymbol{r} - \omega t) + i\sin(\boldsymbol{k}\cdot\boldsymbol{r} - \omega t)]. \qquad (1\text{-}29)$$

It is clear from Eqs.(1-28) and (1-29) that the electric field direction is parallel to the direction of the vector potential and the direction of the magnetic field is perpendicular to both the vector potential (and therefore the parallel electric field) and the propagation vector direction. Thus, $\boldsymbol{E}$ and $\boldsymbol{B}$ oscillate in phase, at the same frequency, and both field directions in free space are perpendicular to each other and the direction of propagation. Equation (1-26) shows that the speed of propagation of electromagnetic waves in a vacuum is equal to c. In a neutral nonconducting medium the speed of propagation is $c/(\varepsilon\mu_p)^{1/2}$.

Now, returning to Eqs.(1-28) and (1-29) we note that the fields are complex. We shall be interested in only the real or measurable fields. In a vacuum where k is real, the real fields are given by the cosine terms in Eqs.(1-28) and (1-29). In a medium, however, k may be complex and we shall have to examine the complex forms of Eqs.(1-28) and (1-29) carefully to extract the real parts in order to make predictions about a measurement.

The results above on the electric and magnetic fields could also have been obtained directly from Maxwell's equations. Straightforward manipulations of Eqs.(1-1) to (1-5) for a neutral ($\bar{\rho} = 0$) nonconducting ($\boldsymbol{J} = 0$) medium give (see Problem 1-2)

$$\nabla^2\boldsymbol{E} = \frac{\varepsilon\mu_p}{c^2}\frac{\partial^2\boldsymbol{E}}{\partial t^2}, \qquad \nabla^2\boldsymbol{B} = \frac{\varepsilon\mu_p}{c^2}\frac{\partial^2\boldsymbol{B}}{\partial t^2}. \qquad (1\text{-}30)$$

The plane wave solutions to these equations are the same as in Eqs.(1-28) and (1-29).

Now, if we specify propagation in a vacuum, $k = \omega/c$ is real and the real parts of Eqs.(1-28) and (1-29) for propagation along the y axis are given by

$$B_x = B_0\cos(ky - \omega t), \qquad E_z = E_0\cos(ky - \omega t), \qquad (1\text{-}31)$$

where the perpendicular values of $\boldsymbol{B}_0$ and $\boldsymbol{E}_0$ are written along the x and z axes, respectively. These fields (shifted by $\pi/2$) are shown in Fig. 1-1 where a spatial projection is shown at one instant in time. If time is allowed to advance, the entire sinusoidal pattern propagates along the y axis from $-y$ to $+y$. This situation is called

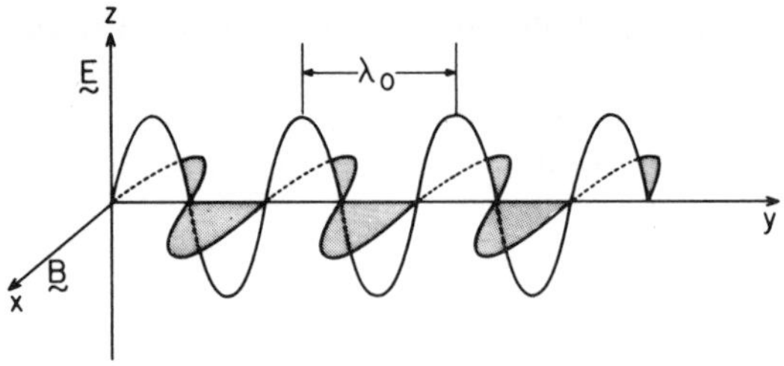

Figure 1-1 A schematic representation of plane wave electromagnetic radiation projected along the y axis at one instant in time [see Eqs.(1-31)]. The magnetic induction, $\boldsymbol{B}$, is in the xy plane and the electric field, $\boldsymbol{E}$, is in the yz plane. In a vacuum, the frequency is $\nu = c/\lambda_0$, where λ_0 is the vacuum wavelength of the radiation. In a medium with refractive index n, the speed of light is given by $\nu\lambda_0/n = c/n$.

a *plane traveling wave.* A *plane standing wave* is produced by placing boundary conditions along the y axis to fix the sinusoidal pattern in space for all time leading to an equation of $E = E_0 \sin ky \sin \omega t$ compared to $\sin(ky - \omega t)$ for the traveling wave.

The general features of the polarization of the radiation can be described by using the right-handed coordinate system in Fig. 1-1 for the projection of the radiation along the y axis from $-y$ to $+y$. The electric field can be decomposed into components along the z and x axes. The field vector in the xz plane is given by

$$\mathbf{E} = \hat{\boldsymbol{i}}E_x + \hat{\boldsymbol{k}}E_z$$

$$E_x = E_x^0 \cos(ky - \omega t + \alpha_x), \qquad E_z = E_z^0 \cos(ky - \omega t + \alpha_z)$$

$$\alpha = \alpha_x - \alpha_z, \tag{1-32}$$

where $\hat{\boldsymbol{i}}$ and $\hat{\boldsymbol{k}}$ are unit vectors along the x and z axes, respectively, as described previously. *Plane polarized* radiation is obtained when the phase factor $\alpha = 0$ or π and $E_x^0 = E_z^0$. When $E_x^0 = E_z^0$ and $\alpha = 0$, E_x and E_z are in phase and the plane-polarized radiation bisects the angle between the x and z axes and oscillates between the (x, z) and $(-x, -z)$ quadrants. If $E_x^0 = E_z^0$ and $\alpha = \pi$, E_x and E_z are out of phase by π and the plane polarized radiation bisects the angle between the x and $-z$ axes and oscillates between the $(x, -z)$ and $(-x, z)$ quadrants. Of course, plane polarized radiation polarized along the z axis (in the yz plane) can be achieved by letting $E_x^0 = 0$ as shown in Fig. 1-1.

Elliptically polarized radiation can be obtained from E_x and E_z in Eq.(1-32) when $\alpha = \pm\pi/2$ as given by

$$E_x = E_x^0 \cos(ky - \omega t)$$

$$E_z = E_z^0 \cos\left(ky - \omega t \pm \frac{\pi}{2}\right) = \pm E_z^0 \sin(ky - \omega t)$$

$$\mathbf{E}_\pm = \hat{\boldsymbol{i}}E_x \pm \hat{\boldsymbol{k}}E_z = \hat{\boldsymbol{i}}E_x^0 \cos(ky - \omega t) \pm \hat{\boldsymbol{k}}E_z^0 \sin(ky - \omega t).$$

Now, if $E_x^0 = E_z^0 = \mathscr{E}$, we have *circular polarized* radiation given by

$$\boldsymbol{E}_\pm = \mathscr{E}[\hat{\boldsymbol{i}} \cos (ky - \omega t) \pm \hat{\boldsymbol{k}} \sin (ky - \omega t)]. \tag{1-33}$$

The field rotates clockwise or counterclockwise about the y axis when viewed looking back toward the source of the radiation. When $\alpha = +\pi/2$ leading to $\boldsymbol{E}_+$, the field appears to rotate counterclockwise about y. $\boldsymbol{E}_+$ and $\boldsymbol{E}_-$ indicate the right (counterclockwise) and left (clockwise) polarized radiation.

differs from Ditchburn p. 363

The flow of electromagnetic energy can also be determined from Maxwell's equations. Multiplying Eqs.(1-1) by $\boldsymbol{H}\cdot$ and Eqs.(1-2) by $-\boldsymbol{E}\cdot$, adding the two equations, and using Eqs.(1-5) gives

See Born & Wolf; pp 27-9 Ditchburn (& B&W) are "traditional" Flygare is "Natural"

cgs

$$\boldsymbol{H}\cdot(\nabla \times \boldsymbol{E}) - \boldsymbol{E}\cdot(\nabla \times \boldsymbol{H}) = -\frac{1}{c}\left[\mu_p \boldsymbol{H}\cdot\frac{\partial \boldsymbol{H}}{\partial t} + \varepsilon \boldsymbol{E}\cdot\frac{\partial \boldsymbol{E}}{\partial t}\right] - \frac{4\pi}{c}\boldsymbol{E}\cdot\boldsymbol{J},$$

SI

$$\boldsymbol{H}\cdot(\nabla \times \boldsymbol{E}) - \boldsymbol{E}\cdot(\nabla \times \boldsymbol{H}) = -\mu_0\mu_p \boldsymbol{H}\cdot\frac{\partial \boldsymbol{H}}{\partial t} + \varepsilon_0 \varepsilon \boldsymbol{E}\cdot\frac{\partial \boldsymbol{E}}{\partial t} - \boldsymbol{E}\cdot\boldsymbol{J},$$

which can be written as [8]

cgs

$$\nabla\cdot(\boldsymbol{E} \times \boldsymbol{H}) = -\frac{1}{2c}\frac{\partial}{\partial t}(\mu_p H^2 + \varepsilon E^2) - \frac{4\pi}{c}\boldsymbol{E}\cdot\boldsymbol{J},$$

SI

$$\nabla\cdot(\boldsymbol{E} \times \boldsymbol{H}) = -\frac{1}{2}\frac{\partial}{\partial t}(\mu_0\mu_p H^2 + \varepsilon_0\varepsilon E^2) - \boldsymbol{E}\cdot\boldsymbol{J}.$$

Multiplying the *cgs* equation by $c/4\pi$ gives

cgs	*SI*
$\nabla\cdot\boldsymbol{S} = -\frac{\partial}{\partial t}\rho - \boldsymbol{E}\cdot\boldsymbol{J}$	$\nabla\cdot\boldsymbol{S} = -\frac{\partial}{\partial t}\rho - \boldsymbol{E}\cdot\boldsymbol{J}$
$\boldsymbol{S} = \frac{c}{4\pi}\boldsymbol{E}\times\boldsymbol{H}$	$\boldsymbol{S} = \boldsymbol{E}\times\boldsymbol{H}$
$\rho = \frac{1}{8\pi}(\mu_p H^2 + \varepsilon E^2)$	$\rho = \frac{1}{2}(\mu_0\mu_p H^2 + \varepsilon_0\varepsilon E^2)$. (1-34)

ρ is the *energy density* (in a plane wave) or energy per unit volume. $\boldsymbol{S}$ (*the Poynting vector*) is the *energy current density* (energy flux) or energy flowing through a unit cross section per unit time in a plane wave. The Poynting vector is parallel to the propagation vector, $\boldsymbol{k}$, or the direction of flow of the electromagnetic energy. When $\boldsymbol{J} = 0$ and $\varepsilon = \mu_p = 1.0$, Eq.(1-34) is the hydrodynamic continuity equation for electromagnetic energy, which is similar to Eq.(1-12) for charge density and current. The gradient of the electromagnetic energy flux at a point leads to a first-order decrease in electromagnetic energy density at that point.

The *radiation intensity*, I, is equal to the radiation energy density, ρ, times the speed of light. Intensity has units of energy · area^{-1} · t^{-1},

$$I = \rho c. \tag{1-35}$$

The radiation intensity is also easily related to the radiation power, P, by (power has units of energy·t^{-1} and 1 W = 1 J·s^{-1} = 10^7 erg·s^{-1})

$$P = I\bar{A}, \tag{1-36}$$

where $\bar{A}$ is the cross-sectional area being irradiated. For example, a monochromatic microwave oscillator entering a waveguide with a 1-cm^2 cross-sectional area with a *power* of 1 W has an *intensity* of 1 W · cm^{-2}. The *energy density* of the 1-W oscillator in the waveguide with a 1-cm^2 cross-sectional area is

$$\rho = \frac{I}{c} = \frac{P}{\bar{A}c} = 3.333 \times 10^{-4} \text{ erg} \cdot \text{cm}^{-3} \text{ (or dyne} \cdot \text{cm}^{-2}) = 3.29 \times 10^{-10} \text{ atm}$$

$$= 3.333 \times 10^{-5} \text{ J} \cdot \text{m}^{-3} = 3.333 \times 10^{-5} \text{ Pa.}$$

We have also recognized that ρ or energy density has units of pressure (dyne · cm^{-2} or J · m^{-3} = Pa) and we have shown the radiation pressure of the 1-W oscillator where 1 dyne · cm^{-2} = 0.987×10^{-6} atm or 1 atm = 101325 Pa.

Given an energy density or radiation intensity it is interesting to compute the magnitudes of the electric and magnetic field strengths produced by the radiation. Equation (1-34) relates the energy density to the field strengths. In a vacuum where $\varepsilon = \mu_p = 1.0$, k is real and we can write values of B^2 and E^2 from the real parts of Eqs.(1-28) and (1-29). Starting with the magnetic induction we have

$$B^2 = k^2 A_0^2 \cos^2 (\boldsymbol{k} \cdot \boldsymbol{r} - \omega t) = B_0^2 \cos^2 (\boldsymbol{k} \cdot \boldsymbol{r} - \omega t).$$

We now take the long time average of B^2 indicated by angular brackets as

$$\langle B^2 \rangle = \lim_{T \to \infty} \frac{1}{T} \int_0^T B^2 \, dt = k^2 A_0^2 \lim_{T \to \infty} \frac{1}{T} \int_0^{\omega T} \frac{\cos^2 \omega t}{\omega} \, d(\omega t)$$

$$= k^2 A_0^2 \lim_{T \to \infty} \frac{1}{\omega T} \left(\tfrac{1}{2}\omega t + \tfrac{1}{4} \sin 2\omega t\right)\Big|_0^{\omega T} = \tfrac{1}{2} k^2 A_0^2 = \tfrac{1}{2} B_0^2. \tag{1-37}$$

A similar analysis of E^2 from Eqs.(1-29) gives

cgs	*SI*
$\langle E^2 \rangle = \frac{1}{2} k^2 A_0^2 = \frac{1}{2} E_0^2$	$\langle E^2 \rangle = \frac{1}{2} \omega^2 A_0^2 = \frac{1}{2} E_0^2.$

Comparing these values of $\langle E^2 \rangle$ with $\langle B^2 \rangle$ above shows the following relations for electromagnetic radiation in a vacuum:

cgs	*SI*
	$\mathbf{B} = \mu_0 \mathbf{H}$
$\langle E^2 \rangle = \langle B^2 \rangle = \langle H^2 \rangle$	$\langle E^2 \rangle = c^2 \langle B^2 \rangle = \dfrac{\mu_0}{\varepsilon_0} \langle H^2 \rangle$
$E_0^2 = B_0^2 = H_0^2$	$E_0^2 = c^2 B_0^2 = \dfrac{\mu_0}{\varepsilon_0} H_0^2.$

We can now write the appropriate expressions for the energy density and resultant field strengths of the electromagnetic radiation in a vacuum:

cgs

$$\rho = \frac{1}{8\pi}(\langle H^2 \rangle + \langle E^2 \rangle)$$

$$= \frac{1}{4\pi}\langle H^2 \rangle = \frac{1}{4\pi}\langle E^2 \rangle$$

$$= \frac{1}{8\pi} E_0^2 = \frac{1}{8\pi} H_0^2$$

$$E_0 = B_0 = H_0 = (8\pi\rho)^{1/2}$$

$$\rho = 3.333 \times 10^{-4}\ \text{erg} \cdot \text{cm}^{-3}$$

$$E_0 = 0.09152\ \text{SV} \cdot \text{cm}^{-1}$$

$$B_0 = H_0 = 0.09152\ \text{G (gauss)}$$

SI

$$\rho = \tfrac{1}{2}(\mu_0 \langle H^2 \rangle + \varepsilon_0 \langle E^2 \rangle)$$

$$= \varepsilon_0 c^2 \langle B^2 \rangle = \mu_0 \langle H^2 \rangle = \varepsilon_0 \langle E^2 \rangle$$

$$= \tfrac{1}{2}\varepsilon_0 c^2 B_0^2 = \tfrac{1}{2}\mu_0 A_0^2 = \tfrac{1}{2}\varepsilon_0 E_0^2$$

$$E_0 = \left(\frac{2\rho}{\varepsilon_0}\right)^{1/2}$$

$$B_0 = \frac{1}{c}\left(\frac{2\rho}{\varepsilon_0}\right) = \frac{1}{c} E_0$$

$$\rho = 3.333 \times 10^{-5}\ \text{J} \cdot \text{m}^{-3}$$

$$E_0 = 2.744 \times 10^3\ \text{V} \cdot \text{m}^{-1}$$

$$B_0 = 9.152 \times 10^{-6}\ \text{T (tesla)}$$

$$H_0 = \frac{B_0}{\mu_0} = 7.283\ \text{T} \cdot \text{m} \cdot \text{H}^{-1}. \qquad (1\text{-}38)$$

It is evident from these results that $1\ \text{V} \cdot \text{m}^{-1} = c^{-1} \times 10^6\ \text{SV} \cdot \text{cm}^{-1}$ (c is in cgs units in this equation) and $1\ \text{G} = 10^{-4}\ \text{T}$, which agrees with the conversion factors in Table A-1. In SI units the ratio $E_0/H_0 = c\mu_0 = 3.76730314 \times 10^2\ \text{H} \cdot \text{s}^{-1}\ (\text{V} \cdot \text{A}^{-1}) \cong 377\ \text{V} \cdot \text{A}^{-1}$ is sometimes called the impedance of free space, where the quantity in units of $\text{V} \cdot \text{A}^{-1}$ is called the *ohm* (see Table A-1).

Induced Electric and Magnetic Moments

We now give a description of the response of a system to external electric or magnetic fields. We consider electric field effects first.

The field-induced electric dipole moment in an atom or molecule is proportional to the field.

$$\boldsymbol{D}_{\text{ind}} = \boldsymbol{\alpha} \cdot \boldsymbol{E}. \tag{1-39}$$

$\boldsymbol{D}_{\text{ind}}$ is the induced moment (units of charge times length), $\boldsymbol{\alpha}$ is called the *electric polarizability* dyadic, and $\boldsymbol{E}$ is the external inducing electric field [4]. In an isotropic medium, Eq.(1-39) reduces to [see discussion following Eq.(B-32)]

$$\boldsymbol{D}_{\text{ind}} = \alpha \boldsymbol{E}. \tag{1-40}$$

Next we define the *polarization*, $\boldsymbol{P}$, of the isotropic medium as the induced moment per unit volume, given by

$$\boldsymbol{P} = \rho_0 \boldsymbol{D}_{\text{ind}} = \rho_0 \alpha \boldsymbol{E}, \tag{1-41}$$

where ρ_0 is the number density, $\boldsymbol{D}_{\text{ind}}$ is the induced dipole moment of a single particle, and α is the single-particle scalar polarizability.

The electric induction, $\boldsymbol{D}$, is given in terms of the electric field, $\boldsymbol{E}$, and polarization, $\boldsymbol{P}$, according to [9]

cgs	*SI*	
$\boldsymbol{D} = \boldsymbol{E} + 4\pi \boldsymbol{P}$	$\boldsymbol{D} = \varepsilon_0 \boldsymbol{E} + \boldsymbol{P}.$	(1-42)

Using Eqs.(1-5) and (1-41) gives the polarization and scalar dielectric constant ε of

cgs	*SI*	
$\boldsymbol{P} = \dfrac{(\varepsilon - 1)}{4\pi} \boldsymbol{E}$	$\boldsymbol{P} = (\varepsilon - 1)\varepsilon_0 \boldsymbol{E}$	
$\varepsilon = 1 + 4\pi \rho_0 \alpha$	$\varepsilon = 1 + \dfrac{\rho_0 \alpha}{\varepsilon_0}.$	(1-43)

It is evident that the units of the electric dipole moment are SC · cm (cgs) and C · m(SI) and the units of electric polarizability are cm^3 (cgs) and $\text{C} \cdot \text{m}^2 \cdot \text{V}^{-1}$(SI). The conversion factors are in Table A-1.

We can also make a parallel set of definitions for the response of a medium to a magnetic field [9]. The field induced magnetic dipole moment in an atom or molecule is proportional to the magnetic field according to

$$\boldsymbol{\mu}_{\text{ind}} = \boldsymbol{\chi} \cdot \boldsymbol{H}. \tag{1-44}$$

$\boldsymbol{\mu}_{ind}$ is the induced moment, $\boldsymbol{\chi}$ is called the *magnetic susceptibility* dyadic, and $\boldsymbol{H}$ is the magnetic field. In an isotropic medium, Eq.(1-44) reduces to [see discussion following Eq.(B-37)]

$$\boldsymbol{\mu}_{ind} = \chi \boldsymbol{H}. \tag{1-45}$$

Next we define the macroscopic magnetic moment density or *magnetization* of the isotropic medium, $\boldsymbol{M}$, as the induced moment per unit volume,

$$\boldsymbol{M} = \rho_0 \boldsymbol{\mu}_{ind} = \rho_0 \chi \boldsymbol{H}. \tag{1-46}$$

The magnetic induction, $\boldsymbol{B}$, is related to the magnetic field, $\boldsymbol{H}$, and magnetization, $\boldsymbol{M}$, according to [9]

$$\begin{array}{cc} cgs & SI \\ \boldsymbol{B} = \boldsymbol{H} + 4\pi \boldsymbol{M} & \boldsymbol{B} = \mu_0(\boldsymbol{H} + \boldsymbol{M}). \end{array} \tag{1-47}$$

Using Eq.(1-5) and (1-46) gives the macroscopic moment and scalar magnetic permeability, μ_p, of

$$\begin{array}{cc} cgs & SI \\ \boldsymbol{M} = \dfrac{\mu_p - 1}{4\pi} \boldsymbol{H} & \boldsymbol{M} = (\mu_p - 1)\boldsymbol{H} \\ \mu_p = 1 + 4\pi \rho_0 \chi & \mu_p = 1 + \rho_0 \chi \end{array} \tag{1-48}$$

It is evident that the units of the magnetic dipole moment are $\mathrm{erg \cdot G^{-1}}$ (cgs) and $\mathrm{J \cdot T^{-1}}$ (SI). The units of magnetic susceptibility are $\mathrm{erg \cdot G^{-2}}$ (cgs) and $\mathrm{J \cdot T^{-2}}$ (SI). The conversion factors are in Table A-1.

The analogy between the electric and magnetic properties is evident from these expressions:

Electric		*Magnetic*
$\boldsymbol{E}$	$\longleftrightarrow$	$\boldsymbol{H}$
$\boldsymbol{D}$	$\longleftrightarrow$	$\boldsymbol{B}$
$\boldsymbol{\alpha}$	$\longleftrightarrow$	$\boldsymbol{\chi}$
$\boldsymbol{P}$	$\longleftrightarrow$	$\boldsymbol{M}$

We shall have occasion to use the equations above throughout the text.

The combination of the sciences of optics and electromagnetism into one concise theory via Maxwell's equations was one of the high points of science during the last third of the nineteenth century. In 1887, H. Hertz first demonstrated that electromagnetic waves could be transmitted and received through space [10].

In 1897, the discovery and characterization of the electron [11] led to a further strengthening of the electromagnetic wave theory of radiation. H. A. Lorentz incorporated the electron into the electromagnetic theory by rewriting Maxwell's equations for application at a microscopic level [12]. This extrapolation to the microscopic region of electrons in atoms and molecules was attempted by merely replacing $\bar{\rho}$ and $\boldsymbol{J}$ in Maxwell's equations with ρ^* and $\rho^*\boldsymbol{v}$. ρ^* is the charge density of the electron and $\boldsymbol{v}$ is the velocity of the electron at the point at which the electron density is taken. Lorentz also added an additional equation for the force, $\boldsymbol{F}$, experienced by a particle with charge e in an electromagnetic field, which is [5]

cgs	*SI*	
$\boldsymbol{F} = e\left[\boldsymbol{E} + \frac{1}{c}(\boldsymbol{v} \times \boldsymbol{B})\right]$	$\boldsymbol{F} = e(\boldsymbol{E} + \boldsymbol{v} \times \boldsymbol{B}).$	(1-49)

We shall discuss Lorentz's classical electron theory in Section 2.2.

1.2 LIGHT SCATTERING

One of the more successful applications of the wave theory of light is found in the explanation of the *nonresonant elastic scattering* of electromagnetic radiation. Two rather distinct classes of light scattering include scattering from *uncorrelated centers* and scattering from *correlated centers*. Uncorrelated systems include atoms in the upper atmosphere and dilute solute molecules in an inert solvent. In this case, the fields scattered from different atoms have random phases and the total scattering intensity is the sum of intensities from each uncorrelated atom. Scattering from correlated centers can lead to constructive and destructive interference (diffraction), which is described by the Bragg law. Two centers are correlated if (for instance) they are connected together by a rigid bond. Beginning with a description of scattering from a single atom (uncorrelated scatterer) we shall derive the Rayleigh formula and show the λ^{-4} dependence in the scattering intensity. Then we shall derive the Bragg equation for scattering from atoms that are correlated in molecules or in crystals.

In a description of Rayleigh light scattering from a single stationary particle, we start with incident plane-polarized monochromatic electromagnetic radiation propagating along the y axis (Fig. 1-1) with the electric vector in the zy plane given by the real part of Eq.(1-29) as given in Eq.(1-31):

$$E_z = E_0 \cos(ky - \omega t). \tag{1-50}$$

The coordinate system and scattering geometry are shown in Fig. 1-2: r is the distance from the source to the stationary scattering center. $\boldsymbol{R}$ is the vector from the stationary scattering center to the observation point, p, and θ_s is the scattering angle. The intensity of the incident radiation is given from Eqs.(1-34) and (1-35) by (in a vacuum where $\mu_p = \varepsilon = 1$)

cgs	*SI*	
$I = \frac{c}{8\pi}(E^2 + H^2) = \frac{c}{4\pi}E^2$	$I = \frac{c}{2}(\mu_0 H^2 + \varepsilon_0 E^2) = c\varepsilon_0 E^2.$	(1-51)

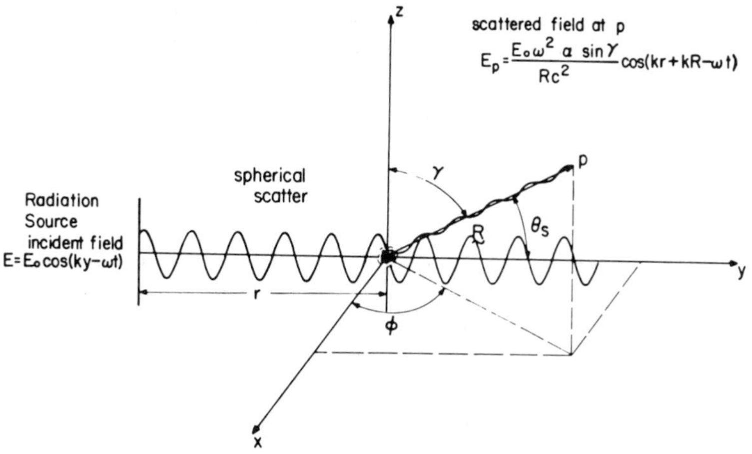

Figure 1-2 Scattering diagram showing the zy polarized incident light propagating along the y axis and then scattering into the spherical polar angle γ and the scattering angle θ_s along the $\boldsymbol{R}$ vector. E_p is polarized and in the zR plane for spherical scatterers. $\boldsymbol{E}_p$ is parallel to $\boldsymbol{E}_0$ when $\gamma = \pi/2$. The spherical polar azimuthal angle is also shown.

Substituting Eq.(1-50) for the incident field and taking the time average gives the incident intensity of

$$\textit{cgs} \qquad\qquad \textit{SI}$$

$$I_0 = \frac{c}{8\pi} E_0^2 \qquad I_0 = \tfrac{1}{2} c \varepsilon_0 E_0^2. \tag{1-52}$$

where the time average of $\cos^2 (ky - \omega t) = \frac{1}{2}$ is used [see Eq.(1-37)].

The incident field in Eq.(1-50) induces a dipole moment, $\boldsymbol{D}_{\text{ind}}$, in the scattering center that we assume is spherical and therefore isotropic. Thus, the induced moment is proportional to the incident field, as shown in Eq.(1-40). We shall also assume that the scattering center is small relative to the radiation wavelength, λ_0.

Consider now the scalar potential at p in Fig. 1-2 due to static dipole moment at the scattering center. Referring to Fig. 1-3, the dipole moment, $D = eqd$, is shown by the separation of two charges, eq and $-eq$ where q is the fraction of charge and d is the distance between the separated charges. ℓ_+ is the distance from the scattering origin

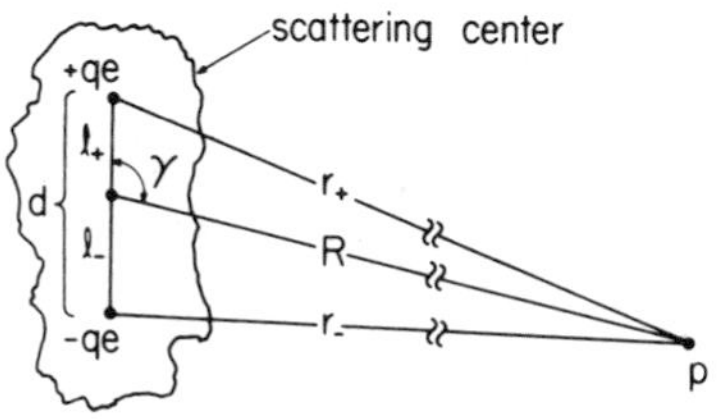

Figure 1-3 Diagram used to compute an electric field at p due to a dipolar charge distribution in the scattering center. $D = dqe$ is the dipole moment and γ is the angle defined in Fig. 1-2. The distance from the scattering center to the detector is normally much larger than the size of the scatterer; $R \gg \ell$.

to the positive charge and ℓ_- is the distance from the origin to the negative charge. The scalar potential at point p in Fig. 1-3 is given by

$$\text{cgs} \qquad\qquad \text{SI}$$

$$\varphi_p = qe\left(\frac{1}{r_+} - \frac{1}{r_-}\right) \qquad \varphi_p = \frac{qe}{4\pi\varepsilon_0}\left(\frac{1}{r_+} - \frac{1}{r_-}\right). \tag{1-53}$$

Now, we refer to Fig. 1-3 and let $\boldsymbol{\ell}_+$ and $\boldsymbol{R}$ be vectors originating at the center of the scatterer and let $\boldsymbol{r}_+$ be a vector originating from $+qe$. According to these definitions we can write $\boldsymbol{r}_+ = \boldsymbol{R} - \boldsymbol{\ell}_+$. Squaring $\boldsymbol{r}_+$ gives

$$\boldsymbol{r}_+ \cdot \boldsymbol{r}_+ = r_+^2 = (\boldsymbol{R} - \boldsymbol{\ell}_+)\cdot(\boldsymbol{R} - \boldsymbol{\ell}_+) = R^2 + \ell_+^2 - 2\boldsymbol{R}\cdot\boldsymbol{\ell}_+ = R^2 + \ell_+^2 - 2R\ell_+ \cos\gamma,$$

where γ is the angle between $\boldsymbol{R}$ and $\boldsymbol{\ell}_+$. Using this and a similar expression for r_-^2 gives

$$r_- = [\ell_-^2 + R^2 - 2\ell_- R\cos(\pi - \gamma)]^{1/2} = R\left[1 + \frac{2\ell_-}{R}\cos\gamma + \left(\frac{\ell_-}{R}\right)^2\right]^{1/2}$$

$$r_+ = (\ell_+^2 + R^2 - 2\ell_+ R\cos\gamma)^{1/2} = R\left[1 - \frac{2\ell_+}{R}\cos\gamma + \left(\frac{\ell_+}{R}\right)^2\right]^{1/2}. \tag{1-54}$$

Now, if $\ell_+ \ll R$ and $\ell_- \ll R$, $1/r_-$ and $1/r_+$ can be expanded in a power series to give

$$\frac{1}{r_-} = \frac{1}{R}\left(1 - \frac{\ell_-}{R}\cos\gamma + \cdots\right)$$

$$\frac{1}{r_+} = \frac{1}{R}\left(1 + \frac{\ell_+}{R}\cos\gamma + \cdots\right). \tag{1-55}$$

Truncating and substituting into Eq.(1-53) gives the potential at the distance R from the origin of the dipole moment,

$$\text{cgs} \qquad\qquad \text{SI}$$

$$\varphi_p \cong qe\left(\frac{\ell_+ + \ell_-}{R^2}\right)\cos\gamma = \frac{qed}{R^2}\cos\gamma \qquad \varphi_p \cong \frac{qed\cos\gamma}{4\pi\varepsilon_0 R^2} = \frac{\boldsymbol{D}\cdot\boldsymbol{R}}{4\pi\varepsilon_0 R^3} = \frac{Dz}{4\pi\varepsilon_0 R^3},$$

$$= \frac{D\cos\gamma}{R^2} = \frac{\boldsymbol{D}\cdot\boldsymbol{R}}{R^3} = \frac{Dz}{R^3}, \tag{1-56}$$

where qed is the electric dipole moment of the scattering center and z is the projection of $\boldsymbol{R}$ on the z axis.

The electric field at p in Fig. 1-3 due to a stationary dipole is obtained from Eq.(1-13) and φ_p in Eq.(1-56) for the static field where $\boldsymbol{\nabla}$ in Cartesian coordinates is given in Eq.(1-7). Thus, the static field at p is given by ($R^2 = x^2 + y^2 + z^2$)

$$\boldsymbol{E}_p = -\boldsymbol{\nabla}\varphi_p = -\left(\hat{\boldsymbol{i}}\frac{\partial}{\partial x} + \hat{\boldsymbol{j}}\frac{\partial}{\partial y} + \hat{\boldsymbol{k}}\frac{\partial}{\partial z}\right)\varphi_p = \hat{\boldsymbol{i}}E_{px} + \hat{\boldsymbol{j}}E_{py} + \hat{\boldsymbol{k}}E_{pz}$$

cgs	SI
$E_{px} = -\frac{\partial}{\partial x}\left(\frac{Dz}{R^3}\right) = \frac{3xzD}{R^5}$	$E_{px} = -\frac{\partial}{\partial x}\left(\frac{Dz}{4\pi\varepsilon_0 R^3}\right) = \frac{3xzD}{4\pi\varepsilon_0 R^5}$
$E_{py} = -\frac{\partial}{\partial y}\left(\frac{Dz}{R^3}\right) = \frac{3yzD}{R^5}$	$E_{py} = -\frac{\partial}{\partial y}\left(\frac{Dz}{4\pi\varepsilon_0 R^3}\right) = \frac{3yzD}{4\pi\varepsilon_0 R^5}$
$E_{pz} = -\frac{\partial}{\partial z}\left(\frac{Dz}{R^3}\right) = 3\frac{z^2D}{R^5} - \frac{D}{R^3}$	$E_{pz} = -\frac{\partial}{\partial z}\left(\frac{Dz}{4\pi\varepsilon_0 R^3}\right) = \frac{1}{4\pi\varepsilon_0}\left(\frac{3z^2D}{R^5} - \frac{D}{R^3}\right). \quad (1\text{-}57)$

As a simple example, the electric field in the xy plane 10 Å from an HCl molecule ($D = 1.8 \times 10^{-18}$ SC · cm) is $E_p \cong 5.4 \times 10^5$ V · cm^{-1}.

Now, if D is an oscillating dipole, we also need the vector potential at p and with Eq.(1-16) we can obtain the time-dependent field at p (see Fig. 1-3). The resultant real field far from the scatterer ($R \gg \lambda_0$) is given by [5,8] (see Fig. 1-3)

cgs	SI
$\boldsymbol{E}_p = -\frac{\omega^2 \alpha \sin\gamma}{Rc^2} \times \hat{\gamma} E_0 \cos(ky + kR - \omega t)$	$\boldsymbol{E}_p = -\frac{\omega^2 \alpha \sin\gamma}{16\pi^2\varepsilon_0^2 Rc^2} \times \hat{\gamma} E_0 \cos(ky + kR - \omega t) \quad (1\text{-}58)$

$\hat{\gamma}$ is the unit vector perpendicular to $\boldsymbol{R}$. When $\gamma = \pi/2$, $\hat{\gamma}$ is parallel to $\boldsymbol{E}_0$; $\hat{\gamma}E_0 = -\boldsymbol{E}_0$.

Finally, substituting Eq.(1-58) into Eq.(1-51) gives the intensity of the scattered radiation for a single uncorrelated scatterer,

cgs	SI
$I = I_0\left[\frac{\omega^4\alpha^2 \sin^2\gamma}{R^2c^4}\right],$	$I = I_0\left[\frac{\omega^4\alpha^2 \sin^2\gamma}{16\pi^2\varepsilon_0^2 R^2c^4}\right].$

The time average value of $\cos^2(ky + kR - \omega t) = \frac{1}{2}$ has been used and I_0 is given in Eq.(1-52) as the intensity of the zy plane polarized incident radiation. Rearranging gives (remember that $\omega/c = 2\pi/\lambda_0$)

cgs	SI
$\frac{I}{I_0} = \frac{\omega^4\alpha^2 \sin^2\gamma}{R^2c^4} = \frac{16\pi^4\alpha^2 \sin^2\gamma}{R^2\lambda_0^4},$	$\frac{I}{I_0} = \frac{\pi^2\alpha^2 \sin^2\gamma}{\varepsilon_0^2 R^2\lambda_0^4}. \quad (1\text{-}59)$

In Problem 1-5 we show that $\sin^2\gamma$ is replaced by $(\frac{1}{2})(1 + \cos^2\theta_s)$ for unpolarized incident light. These results show the $1/\lambda_0^4$ wavelength dependence of the scattered radiation as first pointed out by Rayleigh [13]. Therein lies the reason for the apparent blue color of the sky at midday and the red-orange color at sunset as viewed from the surface of the earth. Electromagnetic blackbody radiation from the sun ($T \cong 5000$ K, see Fig. 1-7) is scattered in the upper atmosphere. The shorter wavelengths are scattered more intensely. During midday the observer on Earth looks away from the sun at the blue light that is scattered more intensely than the red light in the upper atmosphere. The blue sky polarization is examined in Problem 1-6. At sunset the observer looks directly at the sun across the horizon and the blue light from the sun is scattered away as it passes through the atmosphere (dust) leaving a red-orange hue.

We can estimate I/I_0 for the scattering of a monochromatic He–Ne laser source of electromagnetic waves ($\lambda = 6328$ Å $= 632.8$ nm) from a single neon atom ($\alpha \approx 10^{-24}$ cm^3). The ratio of I/I_0 per unit atom at a point 1 m and $\gamma = \pi/2$ from the scattering center is $I/I_0 \cong 9.72 \times 10^{-33}$. The total intensity at a detector is obtained by multiplying this result by I_0, the solid angle of detector acceptance, $d\Omega = \sin\gamma\, d\gamma\, d\phi$, the number density, ρ_0, and the illuminated volume containing the scatterers that is focused onto a detector, V_s. Thus, in a dilute gas or dilute solute in solution where each scattering center is independent, the Rayleigh scattering of z-polarized radiation at small angles γ (see Fig. 1-2) would be negligible. At $\gamma \cong \pi/2$ (in the xy plane) the scattering intensity will be maximized. It is also evident that the scattering intensity from polarized incident radiation is independent of the scattering angle θ_s in Fig. 1-2.

In the dilute gas considered above, we can also relate the particle polarizability, α, to the refractive index, the concentration, and the molecular weight of the scatterer. Comparing Eq.(1-43) with Eq.(1-27) we write the refractive index as

cgs | *SI*

$$n = [\varepsilon]^{1/2} = (1 + 4\pi\rho_0\alpha)^{1/2} \cong 1 + 2\pi\rho_0\alpha, \qquad n = [\varepsilon]^{1/2} = \left(1 + \frac{\rho_0\alpha}{\varepsilon_0}\right)^{1/2} \cong 1 + \frac{\rho_0\alpha}{2\varepsilon_0}, \tag{1-60}$$

where the last step assumes $4\pi\rho_0\alpha \ll 1$ or the dilute gas limit. Remembering that the refractive index and ε are unity in a vacuum, we can write another expression for the refractive index of a dilute gas by expanding in terms of the concentration, C (C is in units of mass per volume), giving

$$n = 1.0 + \left(\frac{\partial n}{\partial C}\right)_0 C + \cdots,$$

where the zero subscript indicates $\partial n/\partial C$ extrapolated to zero concentration. Comparing this result with Eq.(1-60) gives an expression relating the polarizability to the concentration:

cgs | *SI*

$$\alpha = \frac{1}{2\pi\rho_0}\left(\frac{\partial n}{\partial C}\right)_0 C = \frac{M}{2\pi}\left(\frac{\partial n}{\partial C}\right)_0, \qquad \alpha = \frac{2\varepsilon_0}{\rho_0}\left(\frac{\partial n}{\partial C}\right)_0 C = 2M\varepsilon_0\left(\frac{\partial n}{\partial C}\right)_0,$$

where $C/\rho_0 = M$ is used in the last step, M being the mass of the scatterer. Substituting this result into Eq.(1-59) gives

$$\frac{I}{I_0} = \frac{4\pi^2 M^2 \sin^2\gamma}{R^2\lambda_0^4}\left[\left(\frac{\partial n}{\partial C}\right)_0\right]^2, \tag{1-61}$$

for the single-particle scattering ratio. The refractive index of the dilute gas, n, and the change in n with concentration, $(\partial n/\partial C)$, can be measured and extrapolated to $C \longrightarrow 0$ to give $(\partial n/\partial C)_0$. The experimentally observed scattering ratio is obtained by multiplying the result above for I/I_0 by the number of scatterers, $N = \rho_0 V_s$ (where ρ_0 is the scatterer number density and V_s is the scattering volume), and the solid angle defining the entrance aperture of the detector, $d\Omega = \sin\gamma\, d\gamma\, d\phi \ll 4\pi$,

$$\left(\frac{I}{I_0}\right)_{\text{exp}} = \frac{4\pi^2 M C V_s \sin^2\gamma\, d\Omega}{R^2 \lambda_0^4} \left[\left(\frac{\partial n}{\partial C}\right)_0\right]^2, \tag{1-62}$$

where we have again used $\rho_0 = C/M$. If $d\Omega$ is not small relative to 4π, the appropriate integral over the angles for the detection aperture is easily written. Equation (1-62) was written for polarized incident radiation. The corresponding equation for unpolarized incident radiation is given in Eq.(1-87).

With a few modifications we can use the development above to describe the light scattered from a dilute solution of scatterers. The refractive index of the solution can be written as

$$n = n_0 + \left(\frac{\partial n}{\partial C}\right)_0 C + \cdots,$$

where n_0 is the solvent refractive index and n is the solution refractive index. Squaring and dropping higher-order terms gives

$$n^2 = n_0^2 + 2n_0\left(\frac{\partial n}{\partial C}\right)_0 C.$$

Comparing this result with

cgs	*SI*
$n^2 = n_0^2 + 4\pi\rho_0\alpha$	$n^2 = n_0^2 + \dfrac{\rho_0\alpha}{2\varepsilon_0}$

from Eq.(1-60) gives

cgs	*SI*
$\alpha = \dfrac{n_0}{2\pi\rho_0}\left(\dfrac{\partial n}{\partial C}\right)_0 C$	$\alpha = \dfrac{2\varepsilon_0 n_0}{\rho_0}\left(\dfrac{\partial n}{\partial C}\right)_0 C.$

Substituting into Eq.(1-59) and multiplying by $\rho_0 V_s\, d\Omega$ as described before gives the experimental scattering ratio (the Rayleigh ratio):

$$\left(\frac{I}{I_0}\right)_{\text{exp}} = \frac{4\pi^2 M C V_s n_0^2 \sin^2\gamma\, d\Omega}{R^2 \lambda_0^4} \left[\left(\frac{\partial n}{\partial C}\right)_0\right]^2, \tag{1-63}$$

Equation (1-63) is identical to Eq.(1-62) except for the n_0^2 factor above where n_0 is the refractive index of the solvent and n is the refractive index of the solution. For unpolarized incident radiation, we replace $\sin^2 \gamma$ with $\frac{1}{2}(\cos^2 \theta_s + 1)$, where θ_s is the scattering angle (see Problem 1-5). The molecular weights of many molecules have been measured using these developments in light scattering [13]. The success of these methods illustrates the utility of the wave picture of light in describing nonresonant elastic scattering from uncorrelated systems.

1.3 INTERFERENCE AND DIFFRACTION

We shall now discuss interference which arises by scattering from two centers which are fixed (correlated) in space with respect to each other as shown in Fig. 1-4. Referring to Fig. 1-4, the field at the detector will be a sum of the two scattered fields given by [see Eq.(1-58) and let $\gamma = \pi/2$ for convenience, use cgs units]

$$E_{p_1} = -\frac{\omega^2 \alpha}{R_1 c^2} E_0 \cos\left[k(r_1 + R_1) - \omega t\right]$$

$$E_{p_2} = -\frac{\omega^2 \alpha}{R_2 c^2} E_0 \cos\left[k(r_2 + R_2) - \omega t\right].$$

R_1 and R_2, which appear before the cosine term, affect only the amplitude of E_{p_1} and E_{p_2}. The phases of the scattered fields at the detector are determined by the $k(r_1 + R_1)$ and $k(r_2 + R_2)$ terms. The difference between these phase factors must be some multiple of 2π, for constructive interference between fields E_{p_1} and E_{p_2}:

$$k(r_2 + R_2 - r_1 - R_1) = m(2\pi), \qquad m = 0, \pm 1, \pm 2, \ldots$$

A direct determination of the difference in path lengths of the light scattered from 1 and 2, respectively, in Fig. 1-4 $(r_2 + R_2 - r_1 - R_1)$, can be obtained in general from the incident and scattering unit vectors, $\boldsymbol{U}_0$ and $\boldsymbol{U}_s$ defined along the axes of incident and scattered radiation, respectively. r_i and R_i are the magnitudes of $\boldsymbol{r}_i$ and $\boldsymbol{R}_i$. We now write

$$r_2 - r_1 = \boldsymbol{U}_0 \cdot \boldsymbol{r}_{12}$$

$$R_2 - R_1 = -\boldsymbol{U}_s \cdot \boldsymbol{r}_{12}$$

$$(r_2 - r_1) + (R_2 - R_1) = \boldsymbol{r}_{12} \cdot (\boldsymbol{U}_0 - \boldsymbol{U}_s).$$

Multiplying by $k = 2\pi n/\lambda_0$, where n is the refractive index of the medium, and comparing with $k(r_2 + R_2 - r_1 - R_1) = 2\pi m$ gives

$$2\pi m = k(r_2 - r_1 + R_2 - R_1) = \boldsymbol{r}_{12} \cdot \left[\boldsymbol{U}_0\left(\frac{2\pi n}{\lambda_0}\right) - \boldsymbol{U}_s\left(\frac{2\pi n}{\lambda_0}\right)\right]$$

$$= \boldsymbol{r}_{12} \cdot (\boldsymbol{k}_0 - \boldsymbol{k}_s) = \boldsymbol{r}_{12} \cdot \boldsymbol{K}, \tag{1-64}$$

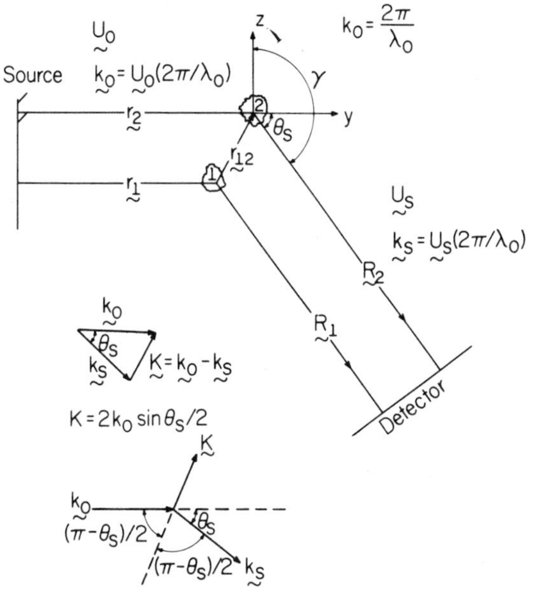

Figure 1-4 Scattering from two centers represented by 1 and 2 and the resultant constructive interference at the detector. γ is the angle between the z axis and the plane of observation and θ_s is the scattering angle. If $k_0 = k_s$, the magnitude of the scattering vector is $K = 2k_0 \sin(\theta_s/2) = (4\pi/\lambda_0)\sin(\theta_s/2)$ in a vacuum and $K = (2\pi n/\lambda_0)\sin(\theta_s/2)$ in a medium with refractive index n. It is also evident from the lower diagrams that $\boldsymbol{K}$ bisects the angle between the incident and scattered light if $k_0 = k_s$.

where $\boldsymbol{K} = \boldsymbol{k}_0 - \boldsymbol{k}_s$ defines the *scattering vector*. The magnitude of the electric field scattered from j, with a phase of $\boldsymbol{r}_j \cdot \boldsymbol{K}$ relative to the field scattered from i, can be written as (cgs units)

$$E_j(t) = \frac{\omega^2 \sin\gamma}{Rc^2} E_0 \alpha_j \cos(\boldsymbol{K}\cdot\boldsymbol{r}_j - \omega t),$$

where $\boldsymbol{r}_j$ is the vector from the ith to the jth scattering center and we assume that $r_j \ll R_j = R$ (see Fig. 1-4). If we refer all $\boldsymbol{r}_j$ to a common origin, the total field scattered into the detector is a sum over all j scattering centers given by

$$E(t) = \frac{\omega^2 \sin\gamma}{Rc^2} E_0 \sum_j \alpha_j \cos(\boldsymbol{K}\cdot\boldsymbol{r}_j - \omega t). \tag{1-65}$$

Noting that θ_s is the angle between $\boldsymbol{k}_0$ and $\boldsymbol{k}_s$ as shown in the inset of Fig. 1-4, we can write

$$\boldsymbol{K}\cdot\boldsymbol{K} = K^2 = (\boldsymbol{k}_0 - \boldsymbol{k}_s)\cdot(\boldsymbol{k}_0 - \boldsymbol{k}_s) = k_0^2 + k_s^2 - 2\boldsymbol{k}_0\cdot\boldsymbol{k}_s = k_0^2 + k_s^2 - 2k_0 k_s \cos\theta_s.$$

Now, assuming elastic scattering where $k_0 = k_s$ gives

$$K = 2k_0 \sin\left(\frac{\theta_s}{2}\right) = \frac{4\pi n}{\lambda_0} \sin\left(\frac{\theta_s}{2}\right). \tag{1-66}$$

It is also evident from this discussion and Fig. 1-4 that $\boldsymbol{K}$ bisects the angle between the incident and scattered light for elastic scattering when $k_0 = k_s$.

We can derive the Bragg law for scattering from a periodic lattice from the results above. In Bragg scattering from a surface perpendicular to $\boldsymbol{r}_{12}$, the incident radiation and scattered radiation have the same angle with respect to the surface and $\boldsymbol{r}_{12}$ is parallel to $\boldsymbol{K}$. Thus, Eq.(1-64) can be written as $\boldsymbol{r}_{12} \cdot \boldsymbol{K} = r_{12} K = 2\pi m$. Now, according to Eq.(1-66), if $k_0 = k_s$, $K = 2k_0 \sin(\theta_s/2)$, which leads to $2r_{12} k_0 \sin(\theta_s/2) = 2\pi m$ or

$$2r_{12} \sin\left(\frac{\theta_s}{2}\right) = m\lambda_0,$$

which is the *Bragg law* where $\theta_s/2$ is the Bragg angle or the equivalent angle between either $\boldsymbol{k}_0$ or $\boldsymbol{k}_s$ and the surface plane or the plane perpendicular to $\boldsymbol{r}_{12}$. Equation (1-64) describes the condition for constructive interference between two correlated scattering centers for any type of wave motion and the special case of Bragg scattering is given in Eq.(1-66). For instance, if r_{12} is an interatomic distance, we think of X-ray, electron, or neutron diffraction where the wavelength of the X ray or particle is of the order of an internuclear distance. The scattering mechanism is different for X rays, electrons, and neutrons, however, and the intensity ratio given in Eq.(1-61) is no longer appropriate (see Chapter 9).

According to Eq.(1-66) the scattering angle depends on the distances between correlated centers. For instance, in the case of optical radiation with $\lambda = 5145$ Å, the distance between scattering centers at $\theta_s = \pi/2$ is $r_{12} = 3.66 \times 10^{-7}$ nm and in the two limits $\theta_s = \pi$, $r_{12} \longrightarrow m\lambda_0/2$, and $\theta_s = 0$, $r_{12} \longrightarrow \infty$.

Constructive interference is illustrated in Fig. 1-5 for the plane wave stimulation of two spherical scatterers. Lines drawn from the scattering region to the detector screen pass through the points where the wavefronts from the two scatterers coincide; these are the points of constructive interference where the electric fields from the two scatterers are in phase. The sine of the angle between two adjacent points of constructive interference, as shown in Fig. 1-5, is given by

$$\sin\theta = \frac{\ell}{R} \cong \theta,$$

when $\ell \ll R$. Using Eqs.(1-64) and (1-66) for adjacent points of constructive interference on the detector screen gives (assume small θ)

$$\lambda = 2d \sin(\theta/2) \cos\eta \cong d\theta \cos\eta,$$

where d is the distance between the scatterers and η is the angle between the line between the two scatterers and the scattering vector, $\boldsymbol{K}$. Combining these two equations gives the distance ℓ between points of constructive interference on the screen:

$$\ell \cong \frac{\lambda R}{d \cos\eta}.$$

It should be evident from Fig. 1-5 that the diffraction pattern at the detector screen is independent of the phase of the incoming plane waves. The important point is that

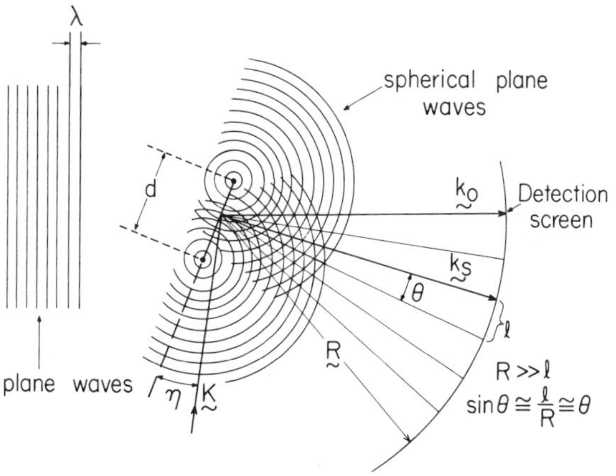

$\lambda = 2d \sin(\theta/2) \cos\eta$

$\ell = \dfrac{\lambda R}{d \cos\eta}$

Figure 1-5 Diagram showing an incident plane wave stimulating two spherical scatterers whose scattered fields then demonstrate constructive and destructive interference at the detector screen where $d \ll R$. Lines are drawn through points of overlap or points of constructive interference in the spherical scattered waves. The two spherical scatterers are separated by distance d and $\ell = \lambda R/(d \cos \eta)$ is the spacing between points of constructive interference on the detector screen.

the phase difference between the fields from two different scatterers is required to be independent of time.

The discussions above on the correlated scattering of monochromatic radiation leads naturally to a discussion of the use of a diffraction grating for spatially dispersing a spectrum of frequencies [14] as shown in Fig. 1-6. The source of radiation emits from a point source with a spectrum, $I(\nu)$, and is collimated onto a planar diffraction grating with parallel slits spaced a distance d apart. The forward scattered light is shown refocused to a point (zeroth order). Also shown is the first-order beam of constructive interference focused at an angle θ from the forward scattered light that occurs when the grating plane distances are λ and 2λ, respectively. The second order will scatter at another angle θ' when the corresponding grating-plane distances are 2λ and 4λ, respectively. Further orders are obtained by extensions of these arguments. The relation between θ_m for the mth order, λ, and d is given by Eq.(1-64) where the angle between the diffraction grating and the $\boldsymbol{K}$ vector is always $\theta/2$ if the diffraction grating is perpendicular to the direction of propagation of the incident radiation. Thus, according to Eq.(1-64) we write

$$2\pi m = \boldsymbol{d} \cdot \boldsymbol{K} = dK \cos\left(\frac{\theta}{2}\right) = \frac{4\pi d}{\lambda} \sin\left(\frac{\theta}{2}\right) \cos\left(\frac{\theta}{2}\right) = \frac{2\pi d}{\lambda} \sin\theta,$$

which gives

$$\sin\theta_m = \frac{m\lambda}{d}, \qquad m = 0, 1, 2, 3, \ldots = \text{order}. \tag{1-67}$$

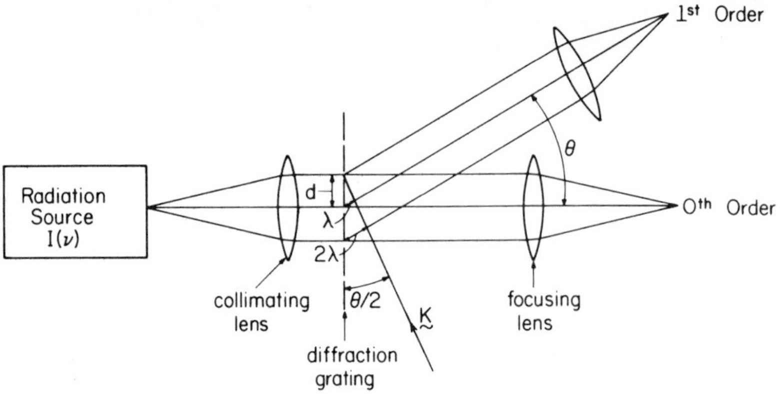

Figure 1-6 Schematic showing the zero-order and first-order images of the source of the radiation after passing through the diffraction grating. The angles between the orders are given by Eq.(1-67). Images for a single radiation wavelength are shown. If the source is composed of a band of frequencies, with distribution, $I(\nu)$, the various frequencies in the band will be spatially separated according to Eq.(1-67). Spatial collimation at the mth-order image will lead to separation of the frequency components in the source. Higher dispersion is achieved at higher orders.

It is evident that the dispersion is inversely proportional to d and that the dispersion increases with the order of diffraction, m. A monochromatic source of radiation will give diffraction maxima at each of the $m = 0, 1, 2, \ldots$ orders (the $m = 0$ and $m = 1$ orders are shown in Fig. 1-6). In summary, a band or spectrum of frequencies will be dispersed in space at the first and higher orders of diffraction, the dispersion being inversely proportional to d. Higher dispersion of a band of frequencies is also achieved at the higher orders of diffraction and of course higher fidelity in the higher-order images is achieved with increased numbers of slits in the diffraction grating.

1.4 BLACKBODY RADIATION

Just as it appeared that the wave theory of light was on a firm foundation, a series of experiments were performed that could not be explained on the basis of wave theory. The first break with the wave theory of light came with Planck's explanation of the frequency and temperature dependence of thermal radiation from materials. At constant temperature, the spectrum of the thermal radiation has a single maximum in frequency and the intensity falls to zero on each side of the maximum. As the temperature of the source increases, the intensity and frequency of the single maximum peak frequency of the electromagnetic radiation increases. Thermal radiation is caused by oscillations in the material that give rise to the emission of electromagnetic radiation. In order to attempt to explain the temperature and frequency dependence of thermal radiation, we note that the energy density per unit frequency of a system of oscillators is given by [15]

$$\rho(\nu) = \frac{8\pi\nu^2}{c^3}\,\bar{\varepsilon}, \tag{1-68}$$

where $\rho(\nu)$ is the energy density per unit frequency interval, c is the velocity of light, and $\bar{\varepsilon}$ is the mean energy of the oscillator emitting the frequency ν. The mean energy of an oscillator is obtained (according to classical theories) by starting with Boltzmann's theorem. In a state of equilibrium, the relative probability of a system being in the state having energy ε is $\exp(-\varepsilon/kT)$, where $k = 1.38066 \times 10^{-16}\ \text{erg}\cdot\text{K}^{-1}$ is *Boltzmann's constant* [16]. Thus, the number of systems per unit volume in state ε_1 (denote by N_1) is related to the number of systems per unit volume in state ε_2 (denote by N_2) by

$$\frac{N_1}{N_2} = \exp - \left(\frac{\varepsilon_1 - \varepsilon_2}{kT}\right).$$

The mean value of ε, which we denote by $\bar{\varepsilon}$, is given by the classical equation

$$\bar{\varepsilon} = \frac{\sum_{i=0}^{\infty} \varepsilon_i \exp(-\varepsilon_i/kT)}{\sum_{i=0}^{\infty} \exp(-\varepsilon_i/kT)}, \tag{1-69}$$

where ε_i is multiplied by the probability, $\exp(-\varepsilon_i/kT)$ and summed over all possible ε_i. The denominator is the normalization factor to ensure that a certainty has a probability of unity. Substituting $\beta = 1/kT$, we have

$$\bar{\varepsilon} = \frac{\sum_{i=0}^{\infty} \varepsilon_i \exp(-\beta\varepsilon_i)}{\sum_{i=0}^{\infty} \exp(-\beta\varepsilon_i)} = -\frac{d}{d\beta} \ln \sum_{i=0}^{\infty} \exp(-\beta\varepsilon_i).$$

Now, if the ε_i are continuous, the summation can be replaced with an integral giving

$$\bar{\varepsilon} = -\frac{d}{d\beta} \ln\left[\int_0^{\infty} \exp(-\beta\varepsilon)\, d\varepsilon\right] = -\frac{d}{d\beta} \ln\left(\frac{1}{\beta}\right) = \frac{1}{\beta} = kT.$$

Thus, the classical mean energy of an oscillator is kT. Substituting this result into Eq.(1-68) gives

$$\rho(\nu) = \frac{8\pi\nu^2}{c^3} kT, \tag{1-70}$$

which is the Rayleigh–Jeans radiation law and predicts increasing energy density (per unit frequency interval) with the square of ν at any temperature. This leads to the famous ultraviolet catastrophe that predicts divergent energy emission at high frequency. As the experimental data show a falloff in radiation intensity at both high and low frequencies, the classical description is obviously wrong.

Planck gave a correct explanation of the spectrum of thermal radiation by making a break with the classical concepts of an oscillator. Planck suggested that the energy of the oscillators in the medium comprised a discrete set of values instead of the classical continuous distribution [17]. This assumption changes the expression for the mean energy, $\bar{\varepsilon}$. Planck assumed that the radiating electromagnetic field originated from harmonic oscillators and that the energy emitted was proportional to integral multiples of the fundamental vibrational frequency. Assuming the first excited state energy ε_0 is proportional to the fundamental frequency, ν, we can write

$$\varepsilon_0 = h\nu, \tag{1-71}$$

where h is the proportionality constant called *Planck's constant*. According to Eq.(1-71) the units of h must be energy $\cdot\, t$. The higher energy states increase in integral multiples of ε_0 according to $0, \varepsilon_0, 2\varepsilon_0, 3\varepsilon_0, 4\varepsilon_0, \ldots, n\varepsilon_0$. Using the expression in Eq.(1-69) gives the mean energy of oscillation as ($\beta = 1/kT$)

$$\bar{\varepsilon} = \frac{\sum_{n=0}^{\infty} n\varepsilon_0 \exp(-n\varepsilon_0\beta)}{\sum_{n=0}^{\infty} \exp(-n\varepsilon_0\beta)} = -\frac{d}{d\beta}\ln\left[\sum_{n=0}^{\infty}\exp(-\beta n\varepsilon_0)\right] = \frac{\varepsilon_0\exp(-\beta\varepsilon_0)}{1-\exp(-\beta\varepsilon_0)}. \tag{1-72}$$

Substituting Eqs.(1-72) and (1-71) into Eq.(1-68) and rearranging gives the Planck distribution curve (energy density per unit frequency). The intensity per unit frequency interval is obtained from Eq.(1-35):

$$\rho(\nu) = \frac{8\pi h\nu^3}{c^3}\left[\frac{1}{\exp(h\nu/kT)-1}\right], \qquad I(\nu) = c\rho(\nu). \tag{1-73}$$

$I(\nu)$ is in units of energy per unit area, which is equivalent to intensity per unit frequency interval. The integral of $I(\nu)$ over a band of frequencies, $\nu_a \longrightarrow \nu_b$, gives the intensity in units of energy per unit area per unit time:

$$I = \int_{\nu_a}^{\nu_b} I(\nu)\,d\nu = c\int_{\nu_a}^{\nu_b}\rho(\nu)\,d\nu. \tag{1-74}$$

These equations are very convenient for calculating the powers and intensities from a thermal source. We note, however, that the energy densities and intensities given above are for the integral over the full solid angle of the thermal source. A more appropriate expression would be an intensity per unit frequency interval per unit solid angle, which is called the *brightness*, $B(\nu)$. Thus, the brightness and total intensities are given by

$$B(\nu) = \frac{1}{4\pi}I(\nu) = \frac{2h\nu^3}{c^2}\left[\frac{1}{\exp(h\nu/kT)-1}\right]$$

$$I = \int_{\nu_a}^{\nu_b} I(\nu)\,d\nu = \int_0^{2\pi}\int_0^{\pi}\int_{\nu_a}^{\nu_b} B(\nu)\,d\nu\,\sin\theta\,d\theta\,d\phi, \tag{1-75}$$

where θ and ϕ are the normal spherical polar angles. The brightness of thermal sources is plotted in Fig. 1-7 as a function of frequency for a variety of temperatures. The frequency scale is designed to coincide with the arrangement in Fig. 1-8, which shows the electromagnetic spectrum.

Comparing the Planck distribution functions in Eqs.(1-73) to (1-75) to the experimental data gave the first value of h of about 6×10^{-34} J · s. Thus, with this extraordinary suggestion that electromagnetic radiation was emitted and absorbed in discrete quanta, one could explain the temperature dependence of the electromagnetic emissions from a blackbody. The low frequency $h\nu \ll kT$ limit of $\rho(\nu)$ in Eq.(1-73) gives the Rayleigh–Jeans expression in Eq.(1-70). In order to compare the

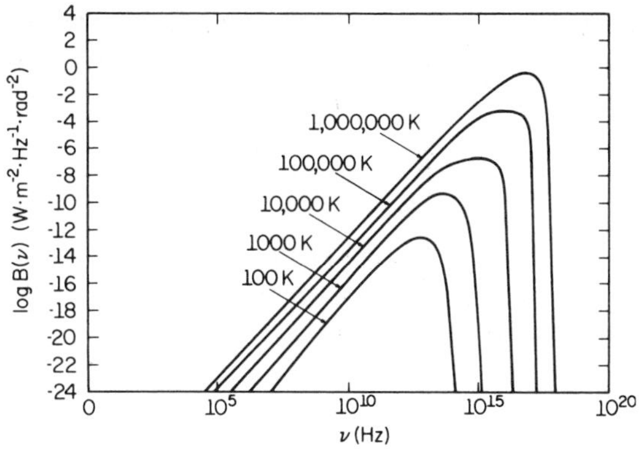

Figure 1-7 log $B(\nu)$ as a function of ν for several temperatures in degrees Kelvin. The brightness is given in units of watts per square meter per unit frequency per unit solid angle ($W \cdot m^{-2} \cdot Hz^{-1} \cdot rad^{-2}$). This figure is adapted from J. D. Kraus, *Radio Astronomy* (McGraw-Hill Book Co., New York, 1966).

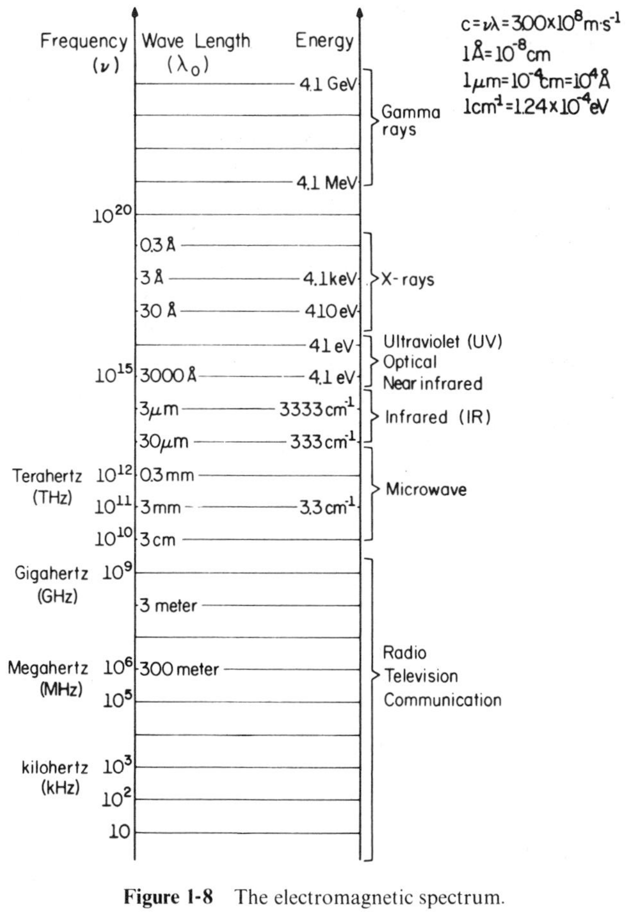

Figure 1-8 The electromagnetic spectrum.

results above with the energy density derived from electromagnetic theory in Eq.(1-34), we must integrate $\rho(\nu)$ in Eq.(1-73) over the appropriate band of frequencies. Thus, the energy density from Eq.(1-73) is

$$\rho = \int_{\nu_a}^{\nu_b} \rho(\nu)\, d\nu \cong \rho_\nu \Delta\nu,$$

where $\Delta\nu$ is a narrow band in frequencies.

The photon flux, J, or photons per unit area per unit time is equal to the radiation intensity divided by the energy of one photon, $h\nu$,

$$J = \frac{I}{h\nu} = \frac{\rho c}{h\nu}. \tag{1-76}$$

A 1-W monochromatic microwave oscillator with frequency $\nu = 10^{10}$ Hz and cross section of 1 cm^2 has a photon flux of $J = 1.51 \times 10^{23}$ photons $\cdot$ cm$^{-2} \cdot$ s^{-1}.

1.5 THE PHOTOELECTRIC EFFECT, COMPTON SCATTERING, AND PHOTOELECTRON SPECTROSCOPY

Evidence for the particle nature of light also came from Einstein's successful explanation of the photoelectric effect as shown in Fig. 1-9. Several years earlier Hertz observed that when certain crystals were irradiated with light in the optical or ultraviolet region of the electromagnetic spectrum, electrons were emitted by the crystal. By using a prism or grating as a dispersive element for the radiation, the energy of the emitted electron and resultant electron current could be observed as a function of the intensity and frequency of the radiation. It was found experimentally that the energy of the emitted electrons was independent of the *intensity* of the incident radiation but was proportional to the *frequency* of the incident radiation. Above the threshold frequency ν', however, the intensity of the emitted electrons is proportional to the intensity of the incident radiation.

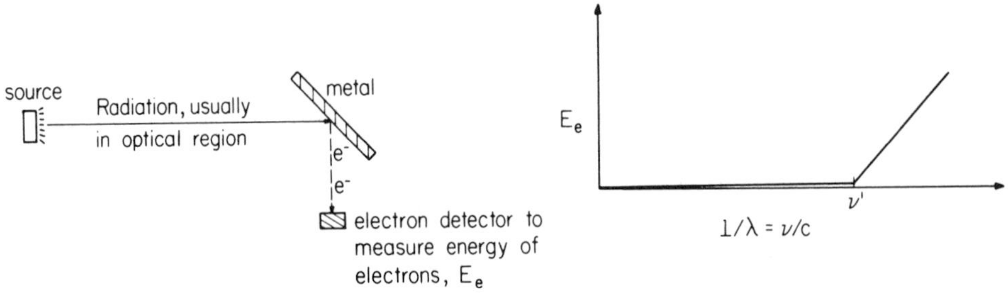

Figure 1-9 The photoelectric effect where optical radiation is directed toward a solid causing electrons to be ejected. The E_e, $1/\lambda$ plot indicates the threshold electromagnetic frequency, ν', necessary to expel an electron. E_e is the energy of the emitted electrons.

If light were transmitted by a wavelike mechanism, it would continuously give its energy to the metal until an electron had enough energy to escape from the crystal. According to the classical theory, the energy of the escaping electron should be proportional to the intensity of the incident radiation, which contradicts the findings of the experiment.

Einstein assumed that the incident radiation exhibited particle-like behavior by possessing a kinetic energy $h\nu$ and that the particle gave some of its kinetic energy to the electron during a collision [18]. The experimental relationship between the energy of the electron, E_e and the light particle energy, $h\nu$, could be explained by

$$E_e = h\nu - E_B, \tag{1-77}$$

where h is Planck's constant and E_B is the binding energy or energy necessary for the electron to escape from the material after it has collided with a particle of light. Particles of light with energy $h\nu$ are called *photons*. These arguments brought back the long-dormant notions of the particle nature of light. The photoelectric effect is, of course, employed in many useful devices that use photodiodes and photomultipliers to convert photon energy to electrical energy (see Problem 1-7).

Another demonstration of the particle nature of light is given by the Compton effect [19], which arises from the apparent collision between a photon and an electron, demonstrating that light (X rays) can exchange linear momentum during a collision. Compton scattering, which is shown in Fig. 1-10, is normally observed with higher energy radiation (X rays) than the optical radiation used in the photoelectric effect. As a result, both the electron and the electromagnetic radiation are present for observation after collision with the material. In the photoelectric effect, the radiation is lost in the material medium.

Compton scattering of X rays can be explained by postulating that light has a particle-like nature that is demonstrated by a simple collision between the light particle (photon) and the electron. It is important to emphasize that the X rays used in this experiment are about 10,000 times more energetic (10^5 eV) than the energy that holds electrons in the vicinity of their respective atomic nuclei (10 eV).

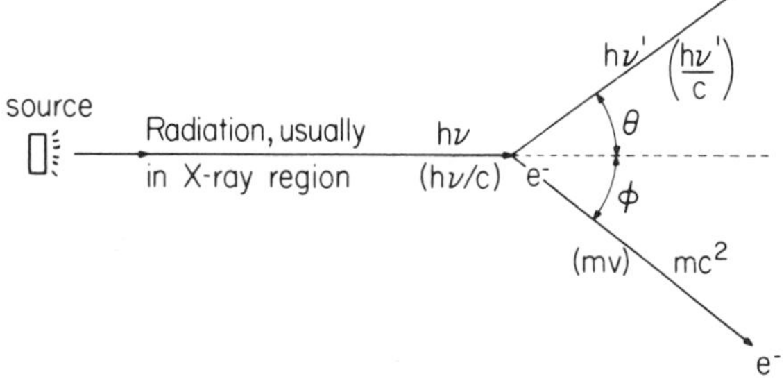

Figure 1-10 The Compton effect where a high energy photon collides with an electron and the photon is scattered into an angle θ and the electron is scattered into an angle ϕ. The energies in each region are listed above the lines and the momenta are listed below the lines in parentheses.

Consider the collision of a photon and electron in Fig. 1-10 and the conservation of energy and momentum. Assuming the electron is originally at rest, it has an energy equal to m_0c^2 and momentum equal to zero where m_0 is the rest mass. After the collision, the electron has an energy mc^2 and momentum mv where m is the mass of the electron at velocity, v. The photon has energy $h\nu$ and we now assume that the photon also possesses a momentum $h\nu/c$ before the collision. The energy and momentum of the scattered photon are $h\nu'$ and $h\nu'/c$, respectively, where ν' indicates a different frequency than the incident radiation, ν. Considering the conservation of energy and momentum and Fig. 1-10 gives the following equations:

Conservation of energy

$$h\nu + m_0c^2 = h\nu' + mc^2. \tag{1-78}$$

Conservation of momentum

The momentum parallel to the incident radiation gives

$$\frac{h\nu}{c} = \frac{h\nu'}{c}\cos\theta + mv\cos\phi, \tag{1-79}$$

and the momentum perpendicular to the incident radiation gives

$$0 = \frac{h\nu'}{c}\sin\theta - mv\sin\phi, \tag{1-80}$$

where the angles ϕ and θ are defined in Fig. 1-10. Thus, we have three equations from Fig. 1-10 and the assumption that the light acts as a particle with energy $h\nu$ and momentum $h\nu/c$. The unknowns in Eqs.(1-78), (1-79), and (1-80) are m, ν', ϕ, v, and θ. It is a simple matter to manipulate Eqs.(1-78) to (1-80) to exclude the value of ϕ to yield a relationship among m, v, ν, ν', and θ to compare with the experimental results. Rearranging Eqs.(1-79) and (1-80) and squaring gives

$$m^2v^2\cos^2\phi = \frac{h^2\nu^2}{c^2} + \frac{h^2\nu'^2}{c^2}\cos^2\theta - \frac{2h^2\nu\nu'}{c^2}\cos\theta$$

$$m^2v^2\sin^2\phi = \frac{h^2\nu'^2}{c^2}\sin^2\theta.$$

Summing and rearranging cancels the ϕ dependence,

$$m^2v^2c^2 = h^2(\nu^2 + \nu'^2 - 2\nu\nu'\cos\theta). \tag{1-81}$$

Rearranging and squaring Eq.(1-78) gives

$$m^2c^4 = [h(\nu - \nu') + m_0c^2]^2 = h^2(\nu^2 + \nu'^2 - 2\nu\nu') + m_0^2c^4 + 2h(\nu - \nu')m_0c^2.$$

Subtracting this result from Eq.(1-81) and rearranging gives

$$c^2[m^2(v^2 - c^2) + m_0^2c^2] = 2h^2\nu\nu'(1 - \cos\theta) - 2h(\nu - \nu')m_0c^2. \tag{1-82}$$

The term in the brackets on the left-hand side of this equation contains the Lorentz–Einstein relation between the mass of a particle moving at constant velocity, m, and the rest mass, m_0, given by

$$m = \frac{m_0}{[1 - (v^2/c^2)]^{1/2}}. \tag{1-83}$$

Thus, the left-hand side is zero, giving

$$h^2 \nu\nu'(1 - \cos\theta) = h(\nu - \nu')m_0 c^2.$$

Rearranging gives

$$(1 - \cos\theta) = \frac{m_0 c^2}{h}\left(\frac{1}{\nu'} - \frac{1}{\nu}\right) = \frac{m_0 c}{h}(\lambda' - \lambda). \tag{1-84}$$

Now define, for an electron, the *Compton wavelength*, λ_C (see Table A-3),

$$\lambda_C = \frac{h}{m_0 c} = 2.4263089 \times 10^{-12}\ \text{m}.$$

Finally, the change in X-ray wavelength as a function of θ is given by

$$(\lambda' - \lambda) = \Delta\lambda = \lambda_C(1 - \cos\theta), \tag{1-85}$$

which agrees with the experimental results.

The theory above with the result in Eq.(1-85) assumes that the scattering electron is at rest. Of course, if the electron is in motion, the Compton shift will depend on the initial momentum of the scattering electron. Returning to the experimental arrangement in Fig. 1-10 we note that if the electron is imbedded in a molecule or solid medium, there will be a distribution of X-ray wavelengths observed at any angle θ due to the velocity distribution in the scattering electrons. This Compton profile or distribution of wavelengths in the scattered X rays can be related back to the momentum distribution of the electrons in the scattering medium. In many cases, knowledge of the momentum distribution of electrons in molecules or solids can be very useful in examining chemical bonding and electron structure from the momentum viewpoint [20].

The recently revived field of photoelectron spectroscopy is a careful use of the photoelectric effect to obtain significant information on the nature of electrons in molecules by making accurate measurements of the energies of the photon-ejected electrons. We use Eq.(1-77) to describe the energy of the electron, $E_e = h\nu - E_B$, where the photon energy, $h\nu$, varies from 10 to 10,000 eV and E_B is the electron's binding energy in the atom or molecule. Photoelectron spectroscopy [21] and X-ray photoelectron spectroscopy [22] can use a variety of sources of radiation. An example of X-ray photoelectron spectroscopy is shown in Fig. 1-11 where the count rate is plotted as a function of the energy of the ejected electron. The ejected electrons were initially in the 1*s* or inner electron shell in each of the four carbon atoms in the molecule shown in the figure. Differences in the binding energies of the inner 1*s* electrons are due to the different valence structures of the molecular bonding electrons in the various carbon atoms. E_B values for *F* and O atoms are far removed on the scale of energies shown in Fig. 1-11. The small differences in E_B from the same atom in

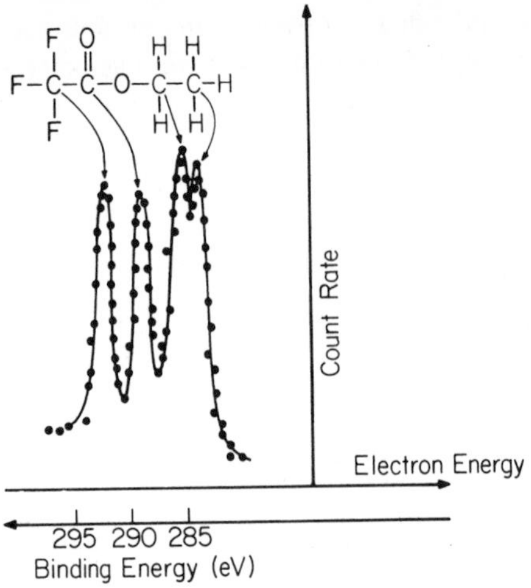

Figure 1-11 X-Ray photoelectron spectrum of an organic molecule. The count rate is plotted as a function of the electron energy of the ejected electrons. The original location of each inner shell electron on the carbon atoms is identified with the arrows. The different binding energies of the carbon atoms 1s inner electrons are clearly evident. The data are adapted from reference 24.

different chemical environments (as shown in Fig. 1-11) are called *chemical shifts* and analysis of these chemical shifts can contribute to the identification of molecular structure [23]. ESCA (electron spectroscopy for chemical analysis) is the acronym used to describe these broad ranges of photoelectron spectroscopic studies [24].

In summary, despite the success of the electromagnetic wave theory of light at the turn of the century, Planck, Einstein, and Compton have shown that electromagnetic radiation exhibits particle-like behavior in some experiments. Planck showed that the thermal radiation being emitted from a material arose from oscillators with discrete, not continuous, energy levels. Einstein showed that light exhibited the features of kinetic energy when colliding with an electron. Compton showed that light also exhibited the features of linear momentum when colliding with an electron. Apparently light exhibits a wave-particle duality [25].

PROBLEMS

The answer is Eq.(3-181)

1. Derive an expression for the energy density in terms of the vector potential.

2. Use Maxwell's equations to derive the wave equations for the electric and magnetic fields in a neutral nonconducting medium where $\bar{\rho} = \boldsymbol{J} = 0$ [see Eqs.(1-30)].

3. What are the magnitudes of the electric and magnetic fields in a monochromatic signal from a 100-mW, 6328 Å (=632.8 nm) helium-neon laser that has a beam cross section of 1 cm^2? What is the photon flux? What is the radiation pressure? What are the magnitudes of the electric and magnetic fields of a corresponding 100-mW microwave oscillator at 30 GHz (=30 × 10^9 Hz) with the same 1-cm^2 cross section? What is the photon flux and corresponding radiation pressure for the microwaves?

4. Following Eq.(1-36), we calculated a radiation pressure of 3.29×10^{-10} atm for an electromagnetic source of energy with an intensity on the detector of $1\ \mathrm{W \cdot cm^{-2}}$. Show that the corresponding pressure variations 7 m for a riveting machine (or 7 m from a passing subway train) with a noise level of 100 dB is also near 3×10^{-10} atm. Consider a cool (273 K) dry morning at sea level where the velocity of sound is $3.32 \times 10^4\ \mathrm{cm \cdot s^{-1}}$. Remember that the decibel level of a signal or noise is defined by

$$\text{decibel (dB)} = 10 \log \frac{I_{\text{signal (or noise)}}}{I_{\text{reference}}}, \tag{1-86}$$

where $I_{\text{reference}} = 10^{-16}\ \mathrm{W \cdot cm^{-2}}$, which is the approximate lowest level of human audible detection for sound at 1000 Hz.

5. Equation (1-63) gives the Rayleigh ratio for the scattering of plane-polarized incident light from a solution of macromolecules.
   (a) Show that the corresponding ratio in the case of completely unpolarized incident light is obtained by replacing $\sin^2 \gamma$ with $\frac{1}{2}(1 + \cos^2 \theta_s)$ to give

$$\left(\frac{I}{I_0}\right)_{\text{exp}} = \frac{4\pi^2 M C n_0^2 V_s\, d\Omega}{R^2 \lambda_0^4} \left[\left(\frac{\partial n}{\partial C}\right)_0\right]^2 \tfrac{1}{2}(1 + \cos^2 \theta_s), \tag{1-87}$$

   where θ_s is the angle between the direction of the incident light and the line of observation (the scattering angle). Unpolarized incident radiation can be described by a linear combination of yz and xy plane-polarized components (see Fig. 1-2).
   (b) Integrate through the appropriate solid angle to evaluate the total light scattered in the case of both plane-polarized and unpolarized incident light. If you were designing an experiment to measure the total I/I_0, would you use polarized or unpolarized incident light in order to maximize the signal-to-noise ratio?

6. Light scattered from the sun in the upper atmosphere (blue sky) is strongly polarized. Use Fig. 1-2 and the concepts of scattering from spherical particles to predict the blue sky polarization.

7. A photoelectric tube uses the photoelectric effect to convert an optical photon to an electron. The efficiency of the conversion is β and $\beta = 0.1$ for typical photoelectric tubes. The signal is amplified, resulting in a total electronic gain of G where typical gains are 10^6. Consider a photoelectric tube with $\beta = 0.1$ and $G = 10^6$ that has a dark current (current with no photons hitting the photoelectric surface) of 10 pA (1 pA = 1 picoamp = 10^{-12} amp).
   (a) What is the minimum number of photons per second that can be observed by this photoelectric cell?
   (b) Assuming $\lambda = 514.5$ nm radiation, what is the minimum observable power that can be detected with this photoelectric tube?

8. (a) Consider the light scattered from a solution of bovine serum albumin (BSA) molecules that have a molecular weight of 67,000 u (atomic mass units). Using 1 W of plane-polarized incident radiation at $\lambda_0 = 514.5$ nm from an Ar^+ laser, calculate the minimum observable concentration of BSA in a water solution that will produce enough scattered light to be observable using the photoelectric tube described in Problem 1-7. Assume a 1 cm^3 scattering volume and a 1 mm^2 collection aperture for the detector, and use $\gamma = \pi/2$ (see Fig. 1-2), $(dn/dC)_0 = 0.2 \text{ ml} \cdot \text{g}^{-1}$, and $R = 1$ m. Use the resultant concentration, C, to calculate the number density of BSA molecules in the solution.

(b) Consider now the effects on the intensity of light scattering in the BSA solution due to contamination with dust particles. Assuming spherical dust particles that have 50 μm $= 5 \times 10^{-3}$ cm diameters with mass densities of 2.5 $\text{g} \cdot \text{cm}^{-3}$, calculate the number of dust particles that will give rise to scattered light in excess of the light scattered from a 1% by weight aqueous solution of BSA.

9. Using 1 W of plane-polarized incident radiation at $\lambda_0 = 514.5$ nm from an Ar^+ ion laser, calculate the minimum number of Kr atoms that will scatter enough light to be observable with the phototube described in Problems 1-7 and 1-8. Use $\alpha = 5 \times 10^{-24}$ cm^3 for the Kr atom polarizability, $\sin^2 \gamma = 1$ where γ is defined in Fig. 1-2, and $R = 1$ m.

10. Consider the accurate measurement of the wavelengths of standing sound waves in a gas by the diffraction of optical radiation from the sonic standing wave pattern. Figure 1-12 illustrates an experiment where an ultrasonic transducer at frequency ν_s generates the energy to produce a sound wave with velocity v_s in the gas. The boundaries of the container set up stationary standing longitudinal sound waves in the medium that have a wavelength given by

$$\lambda_s = \frac{v_s}{\nu_s}. \tag{1-88}$$

Thus, ν_s is generated with a transducer and λ_s can be measured by the diffraction pattern in the scattered light to give a measurement of the velocity of sound in the medium.

(a) Referring to Fig. 1-12, show that the wavelength of the propagating sound wave is related to the radiation wavelength, λ_0, and scattering angle, θ, by (assume unity for the refractive index of the gas)

$$\sin \theta_m = \frac{\lambda_0}{\lambda_s} m, \tag{1-89}$$

where m is an integer; $m = 0, 1, 2, 3$. Under normal conditions where $\lambda_s \gg \lambda_0$, the low angle limit in Eq.(1-89) can be used giving $\theta_m = m\lambda_0/\lambda_s$. Thus, points of constructive interference are observed at θ_m values equal to $m\lambda_0/\lambda_s$ according to $m = 0, 1, 2, \ldots$, the order of diffraction. In the low angle limit, the distance of separation of the high intensity points on the

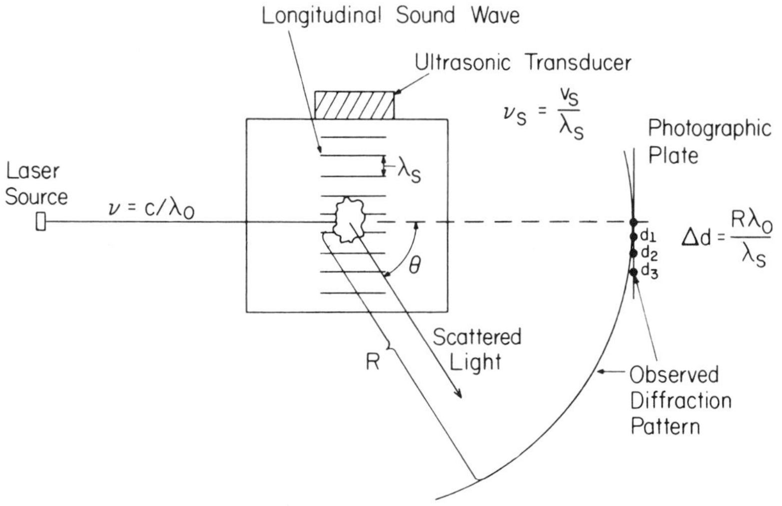

Figure 1-12 Diagram showing the measurement of the wavelength, λ_s, of a sound wave by scattering optical radiation in the gas being modulated by a sound wave. A diffraction pattern appears on the photographic plate with maxima in intensity appearing at distances $d_m = R_m\lambda/\lambda_s$ where m is the diffraction order and $\lambda_s \gg \lambda_0$. The direction of sound propagation is perpendicular to the direction of the light propagation and the refractive index of the gas is assumed to be equal to unity.

photographic plate is given by ($\sin\theta_m = d_m/R = \theta_m)d_m = R(m\lambda_0/\lambda_s)$. The distance between the exposed spots on the photographic plate is $\Delta d = R\lambda_0/\lambda_s$.

(b) If the frequency of the sound wave is 1 MHz and the wavelength of the argon ion laser radiation is 514.5 nm, successive maxima in light intensity are observed at $\theta_n = 2^\circ$ and $\theta_{n+1} = 2^\circ 9'$ at the photographic plate. What is the velocity of the sound wave?

(c) What is the spacing between the maxima on a photographic plate for the angles in part b if the detector is 1 m from the scattering region? Does normal film have adequate resolution to record these diffraction patterns?

In conclusion, this method is an accurate technique for measuring sound velocities in a gas by measuring the wavelength of the sound being excited at a fixed frequency. We note that the velocity of sound in a perfect gas is equal to $(C_p kT/C_v M)^{1/2}$ where C_p and C_v are the heat capacities at constant pressure and constant volume, respectively, M is the mass of the gas molecule, k is Boltzmann's constant, and T is the temperature.

11. The electronically excited sodium atom emits yellow-orange light that is actually two lines at 16,956.18 cm^{-1} and 16,974.38 cm^{-1}. This emission is called the *sodium doublet line*. Design a diffraction grating that will allow the resolution of this doublet in the first order. The degree of resolution should be defined and then the diffraction grating designed around the definition.

12. Consider one mole of HCl molecules at room temperature under the influence of a thermal radiator at $T = 2000$ K with an emission cross section of 1 cm^2. Assume that the HCl molecules will absorb completely a band of frequencies 100 MHz wide at the fundamental vibrational frequency of 8.65×10^{13} Hz from the total frequencies emitted by the blackbody. The statement above assumes that only a single vibration-rotation transition at 8.65×10^{13} Hz is active, which has an approximate Lorentzian line shape with half width at half height of about 50 MHz.

(a) What is the rate of energy increase in the HCl molecules? That is, how much power are the molecules absorbing?

(b) Assuming that the vibrational excitation is all converted to translational energy and that this energy is not lost by wall collisions, how much will the temperature of the HCl gas rise after 10 minutes of irradiation?

(c) The calculations above are appropriate for a Nernst glower, which is the source used for many infrared spectrographs. Carbon arcs may give blackbody temperatures up to 8000 K. Repeat the computations above for $T = 8000$ K.

(d) Repeat the steps above for the $J = 0 \longrightarrow J = 1$ rotational transition at 20 cm^{-1}. Use the same line widths.

13. The Stefan–Boltzmann law states that the total intensity, I, irradiated from a blackbody is proportional to T^4 (total of all frequencies),

$$I = \sigma T^4, \tag{1-90}$$

where $\sigma = 5.67 \times 10^{-8}\ \mathrm{J \cdot m^{-2} \cdot s^{-1} \cdot K^{-1}}$ is the proportionality constant. Show that this relation is consistent with Planck's radiation formula [Eq.(1-73)] and derive the value of σ from Planck's law [26].

14. Show that the Rayleigh–Jeans radiation law [Eq.(1-70)] can be obtained from the Planck radiation law [Eq.(1-73)] for high temperatures or low frequencies ($kT \gg h\nu$).

15. The fireball of an atomic bomb reaches a temperature of 10^6 K. What is the wavelength or frequency of the most intense electromagnetic radiation emitted from the fireball? If 5 mg were converted to energy and all the energy were emitted as electromagnetic energy at the peak of the radiation curve, how many photons of what energy would be emitted?

16. Consider a CO_2 laser that operates at a single vibration-rotation transition at 3×10^{13} Hz with a full width at half height of 10 Hz. The power output of a continuous wave CO_2 laser at the single vibration-rotation transition mentioned above is typically on the order of 10 W. What is the equivalent temperature of a blackbody source of 1-cm^2 cross section to give the same power output per unit frequency as the CO_2 laser?

17. (a) Assuming the temperature at the surface of the earth is 300 K, where does the blackbody radiation peak occur (in frequency) and what consequences does this background radiation have as far as electromagnetic communication on the earth?

(b) The background radiation temperature of outer space is about 3.2 K. At what frequency and wavelength does the blackbody radiation peak occur for this temperature and what consequences does this have in radio astronomy?

18. A typical klystron or backward wave oscillator that is used in microwave spectroscopy delivers 0.1 W of power at a given microwave frequency (1–100 GHz, see Fig. 1-8) with a bandwidth of 100 kHz.

(a) What is the corresponding output power of a 1-cm^2, 2000 K thermal source at 30,000 MHz? Assume a bandwidth for the blackbody radiation that is equal to the klystron width of 100 kHz.

(b) Do you think a blackbody source (see part a) would be a suitable replacement for a typical klystron or other microwave oscillator used in rotational spectroscopy?

19. Use the concepts presented on Compton scattering to compute the change in momentum of an electron scattering from a stationary neutron. Assume that the electron has an initial energy of 0.1 MeV.

20. It is well known that an electron-positron pair annihilates [27] to give two photons that have sufficient energy to be called *gamma rays.*

$$e^- + e^+ = 2\gamma. \tag{1-91}$$

Assuming negligible initial kinetic energy of the e^- and e^+, what is the frequency of this gamma "ray" and what is the energy of this gamma photon?

21. X-Ray generators operate by exposing a metal surface with high velocity electrons. What is the minimum wavelength of X rays produced by electrons excited to a potential difference of 90,000 V?

22. If an X ray of 0.03 Å wavelength collided with a deuterium nucleus, what would be the wavelength of the X rays observed at right angles to the incident radiation?

23. In principle, the frequency of radiation confined in a reflecting box with no losses can be increased or decreased by changing the dimensions of the box. Consider a box of volume 1 cm^3 with 10^{20} photons of monochromatic $\nu = 10^{11}$ Hz radiation.

(a) What is the photon pressure and energy density in the box? Now decrease the volume of the box by one half and assume no loss of photons.

(b) What is the new photon pressure and what is the radiation frequency?

(c) Suggest the design of a piston engine using blackbody radiation from the sun for fuel. Describe the basic cycles of the engine.

REFERENCES

[1] M. Born and E. Wolf, *Principles of Optics* (Pergamon Press, Oxford, 1964).

[2] An interesting historical background is provided in D. C. Mattis, *The Theory of Magnetism* (Harper & Row, New York, 1965).

[3] R. BECKER, *Electromagnetic Fields and Interactions*, Vol. I and II (Blaisdell Publishing Co., New York, 1964). A discussion of units is given in Appendix A. We shall give parallel cgs and SI unit expressions throughout this chapter. If the equation is the same for both sets of units, only a single result is given.

[4] Vectors and dyadics are reviewed in Appendix B.1.

[5] J. D. JACKSON, *Classical Electrodynamics* (John Wiley & Sons, Inc., New York, 1962).

[6] The reader may wish to verify the identity $(\boldsymbol{A}_0 \times \nabla)(\boldsymbol{k} \cdot \boldsymbol{r}) = \boldsymbol{A}_0 \times \boldsymbol{k} = -\boldsymbol{k} \times \boldsymbol{A}_0$.

[7] L. I. SCHIFF, *Quantum Mechanics*, 3rd ed. (McGraw-Hill Book Co., New York, 1968), p. 399.

[8] P. LORRAIN and D. R. CORSON, *Electromagnetic Fields and Waves* (W. H. Freeman and Co., San Francisco, 2nd ed., 1970). The reader should prove the identity $\nabla \cdot (\boldsymbol{E} \times \boldsymbol{H}) = -\boldsymbol{E} \cdot (\nabla \times \boldsymbol{H}) + \boldsymbol{H} \cdot (\nabla \times \boldsymbol{E})$.

[9] L. D. LANDAU and E. M. LIFSHITZ, *Electrodynamics of Continuous Media* (Pergamon Press, New York, 1960).

[10] H. HERTZ, *Ann. Physik* **31**, 983 (1887).

[11] J. J. THOMSON, *Phil. Mag.* **44**, 293 (1897).

[12] H. A. LORENTZ, *The Theory of Electrons*, 2nd ed. (Teubner, Leipzig, 1916; reprinted by Dover, New York, 1952).

[13] See historical summary in M. Kerker, *The Scattering of Light* (Academic Press, New York, 1969).

[14] J. F. JAMES and R. S. STERNBERG, *The Design of Optical Spectrometers* (Chapman and Hall, Ltd., London, 1969).

[15] M. BORN, *Atomic Physics* (Hafner Publishing Co., New York, 1956). This book gives a good description of blackbody radiation, the photoelectric effect, and the Compton effect.

[16] S. GOLDEN, *Theoretical Physical Chemistry* (Addison Wesley Pub. Co., Reading, Mass., 1961).

[17] M. PLANCK, *Ann. d. Physik* **4**, 553 (1901).

[18] A. EINSTEIN, *Ann. d. Physik* **17**, 145 (1905).

[19] A. H. COMPTON, *Phys. Rev.* **22**, 409 (1923); P. P. Debye, *Z. Phys.* **24**, 161 (1923); A. B. Arons and M. B. Peppard, *Am. J. Phys.* **33**, 367 (1965).

[20] I. R. EPSTEIN, *Accts. Chem. Res.* **6**, 145 (1973).

[21] W. C. PRICE, *Adv. in Atom. and Mol. Phys.*, Ed. by D. R. Bates and B. Bederson, **10**, 131 (Academic Press, New York, 1974).

[22] J. M. HOLLANDER and W. C. JOLLY, *Accts. Chem. Res.* **3**, 193 (1970).

[23] D. W. TURNER *et al.*, *Molecular Photoelectron Spectroscopy* (Wiley, London, 1970).

[24] K. SIEGBAHN *et al.*, *ESCA Applied to Free Molecules* (American Elsevier, New York, 1969); see also D. A. Shirley, *Adv. in Chem. Phys.*, Ed. by I. Prigogine and S. A. Rice, **23**, 85 (John Wiley & Sons, New York, 1973).

[25] H. MARGENAU, *The Nature of Physical Reality* (McGraw-Hill Book Co., New York, 1950). M. BORN, *The Restless Universe* (Dover Pub., Inc., New York, 1951). M. JAMMER, *The Philosophy of Quantum Mechanics* (John Wiley & Sons, New York, 1974).

[26] C. Kittel, *Thermal Physics* (John Wiley & Sons, Inc., New York, 1969).

[27] See chemical applications in J. A. MERIGAN, J. H. GREEN, and S.-J. TAO, "Positronium Annihilation," *Physical Methods of Chemistry*, Ed. by A. Weissbarger and B. Rossiter (Wiley-Interscience, New York, 1972).

2

the nature of matter

2.1 CLASSICAL MECHANICS AND ROTATIONAL KINETIC ENERGY

This study of classical mechanics begins with Newton's equations of motion. The observed continuous nature of macroscopic motion in space and time can be explained by applying integral and differential calculus to the solution of dynamic problems. The principal ideas embodied in the determinalistic formulation of Newtonian and subsequent forms of classical mechanics are threefold [1,2].†

1. The system in question is observationally accessible at all times and observations can be executed without disturbing the system.
2. By defining the state of a system in time, all the mechanical variables of the system at that particular time can be determined.
3. Combining the specification of the state in statement 2 with the laws of mechanics allows a complete description of the system as a function of time.

† Numbers in square brackets correspond to the numbered sources found in the Reference section on p. 67 at the end of the chapter.

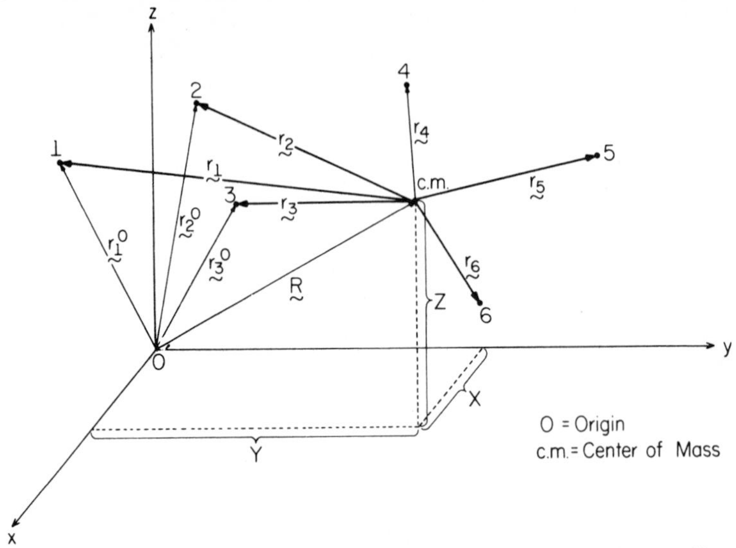

Figure 2-1 Cartesian coordinate axis system. 0 is the origin of the coordinate system chosen by convenience to be the laboratory frame and c.m. is the center of mass based in the many-particle frame. The $\boldsymbol{r}_i^0$ and $\boldsymbol{r}_i$ vectors originate at the arbitrary origin and c.m., respectively, and terminate at the point masses designated by 1, 2, ... $\boldsymbol{R}$ is the vector from the arbitrary origin to the c.m. and X, Y, and Z are the Cartesian components of $\boldsymbol{R}$.

The Cartesian coordinate system for a many-particle system is shown in Fig. 2-1. There are two coordinate frames of reference that will be important: the arbitrary origin at 0 in the laboratory frame where all observations take place and the center of mass (c.m.) frame fixed in a well-defined point in the many-particle system.

We begin our discussion of classical mechanics with Newton's second law for the external force on the kth particle, $\boldsymbol{F}_k^{\text{ex}}$, with respect to an origin in the laboratory frame:

$$\boldsymbol{F}_k^{\text{ex}} = m_k \frac{d\boldsymbol{v}_k^0}{dt} = \frac{d}{dt} m_k \boldsymbol{v}_k^0 = \frac{d}{dt}\boldsymbol{p}_k^0 = \dot{\boldsymbol{p}}_k^0, \tag{2-1}$$

where m_k, $\boldsymbol{v}_k^0$, $\boldsymbol{p}_k^0$, $\dot{\boldsymbol{p}}_k^0$ are the *mass, velocity, linear momentum* ($\boldsymbol{p}_k^0 = m_k \boldsymbol{v}_k^0$), and time rate of change of the linear momentum for the kth particle with respect to the laboratory frame. From Eq.(2-1) we can state that the linear momentum is conserved in time if the force on a particle is zero (Newton's first law).

The *angular momentum*, $\boldsymbol{J}_k^0$, is defined by

$$\boldsymbol{J}_k^0 = \boldsymbol{r}_k^0 \times \boldsymbol{p}_k^0. \tag{2-2}$$

A conservation law also exists for the angular momentum. Defining a *torque*, $\boldsymbol{Y}_k^0$ as $\boldsymbol{r}_k^0 \times \boldsymbol{F}_k^0$ and using Eq.(2-1) gives

$$\boldsymbol{Y}_k^0 = \boldsymbol{r}_k^0 \times \boldsymbol{F}_k^0 = \boldsymbol{r}_k^0 \times \frac{d}{dt} m_k \boldsymbol{v}_k^0 = \frac{d}{dt} m_k \boldsymbol{r}_k^0 \times \boldsymbol{v}_k^0$$

$$= \frac{d}{dt}(\boldsymbol{r}_k^0 \times \boldsymbol{p}_k^0) = \frac{d}{dt}\boldsymbol{J}_k^0 = \dot{\boldsymbol{J}}_k^0. \tag{2-3}$$

Thus, if the torque is zero, the angular momentum is conserved in time.

The conservation concept for mechanical energy is obtained by stating that a *conservative force* is given by the negative gradient of the potential energy, V:

$$\boldsymbol{F}_k^0 = -\boldsymbol{\nabla}_k V. \tag{2-4}$$

The work done by an external device in moving the kth particle from position a to position b is given by

$$E_k(a \longrightarrow b) = \int_a^b \boldsymbol{F}_k^{\text{ex}} \cdot d\boldsymbol{r}. \tag{2-5}$$

Using this and Eq.(2-4) gives

$$E_k(a \longrightarrow b) = \int_a^b -\boldsymbol{\nabla} V_k \cdot d\boldsymbol{r}_k = -V_{kb} + V_{ka}. \tag{2-6}$$

If we substitute $\boldsymbol{F}_k^0 = \dot{\boldsymbol{p}}_k^0$ from Eq.(2-1) into Eq.(2-5), we obtain

$$E_k(a \longrightarrow b) = \int_a^b m_k \frac{d\boldsymbol{v}_i^0}{dt} \cdot d\boldsymbol{r}_k. \tag{2-7}$$

Substituting $d\boldsymbol{r} = \boldsymbol{v}\, dt$ and $(d\boldsymbol{v}/dt)\cdot\boldsymbol{v} = \frac{1}{2}(dv^2/dt)$ gives

$$E_k(a \longrightarrow b) = \int_a^b m_k \frac{d\boldsymbol{v}_k^0}{dt}\cdot\boldsymbol{v}_k^0\, dt = \frac{m_k}{2}\int_a^b \frac{d(v_k^2)}{dt}\, dt = \frac{m_k}{2}\int_a^b d(v_k^2) = T_{kb} - T_{ka}, \tag{2-8}$$

where T is the kinetic energy. The right-hand sides of Eqs.(2-6) and (2-8) are equal, which shows the equivalence of the total energy $(T_k + V_k)$ at points a and b. Thus, $T_k + V_k$ is conserved in a motion that is in response to a conservative force.

We shall now consider in more detail a system of n particles. The total mass, M, is obtained by summing over all n particles as given by

$$M = \sum_{k=1}^{n} m_k. \tag{2-9}$$

The vector from the laboratory frame origin to the *center of mass* (c.m.) origin, $\boldsymbol{R}$, is given by (see Fig. 2-1)

$$\boldsymbol{R} = \frac{1}{M}\sum_k m_k \boldsymbol{r}_k^0, \tag{2-10}$$

where we omit the sum limits from $k = 1$ to $k = n$ (for convenience). The linear momentum of the c.m. is given by

$$\boldsymbol{p}^0 = M\frac{d}{dt}\boldsymbol{R} = \sum_k m_k \frac{d\boldsymbol{r}_k^0}{dt} = \sum_k m_k \boldsymbol{v}_k^0. \tag{2-11}$$

In the case of a many-particle system we must distinguish between internal and external forces. Let $\boldsymbol{F}_k^{\text{ex}}$ indicate the external force on the kth particle as in Eq.(2-1) and let $\boldsymbol{F}_{kj}$ indicate the internal force on the kth particle due to the jth particle $(F_{kk} = 0)$. Newton's second law for the kth particle is then written by

$$\sum_{j=1}^{n} \boldsymbol{F}_{kj} + \boldsymbol{F}_k^{\text{ex}} = \dot{\boldsymbol{p}}_k^0. \tag{2-12}$$

Summing over all k particles gives

$$\sum_{\substack{j,k\\ j\neq k}} \boldsymbol{F}_{kj} + \sum_k \boldsymbol{F}_k^{\text{ex}} = \sum_k \dot{\boldsymbol{p}}_k^0 = \frac{d^2}{dt^2}\sum_k m_k \boldsymbol{r}_k^0. \tag{2-13}$$

$\sum_k \boldsymbol{F}_k^{\text{ex}} = \boldsymbol{F}^{\text{ex}}$ is the total external force. The first term on the left-hand side of Eq.(2-13) goes to zero as $F_{kj} = -F_{jk}$ due to Newton's third law of action and reaction, which states that the forces exerted between two particles on each other are equal and opposite and lie along a line joining the particles. Thus, using this and also substituting Eq.(2-10) into Eq.(2-13) gives

$$\boldsymbol{F}^{\text{ex}} = \sum_k \boldsymbol{F}_k^{\text{ex}} = M\frac{d^2\boldsymbol{R}}{dt^2}, \tag{2-14}$$

which shows that the c.m. moves as though the total external force were acting on the total mass of the system located at the c.m. Thus, *internal* forces have no effect on the motion of the center of mass. Substituting Eq.(2-11) into Eq.(2-14) gives the conservation law for the total linear momentum of a system of particles:

$$\boldsymbol{F}^{\text{ex}} = M\frac{d}{dt}\boldsymbol{v} = \frac{d}{dt}\boldsymbol{p}^0 = \dot{\boldsymbol{p}}^0. \tag{2-15}$$

If the total external force is zero, the linear momentum is conserved. We can also show that an analogous conservation law for the total angular momentum exists; if the total torque on the system is zero, the total angular momentum is conserved (see Problem 2-1).

The total angular momentum with respect to the laboratory frame is given by [see Eq.(2-2)]

$$\boldsymbol{J}^0 = \sum_k \boldsymbol{r}_k^0 \times \boldsymbol{p}_k^0 = \sum_k m_k \boldsymbol{r}_k^0 \times \boldsymbol{v}_k^0. \tag{2-16}$$

Now, according to Fig. 2-1, $\boldsymbol{r}_k^0 = \boldsymbol{R} + \boldsymbol{r}_k$. Taking the first time derivative gives $\boldsymbol{v}_k^0 = \boldsymbol{v} + \boldsymbol{v}_k$. Substituting into Eq.(2-16) gives

$$\begin{aligned}\boldsymbol{J}^0 &= \sum_k m_k(\boldsymbol{R} + \boldsymbol{r}_k) \times (\boldsymbol{v} + \boldsymbol{v}_k)\\ &= \boldsymbol{R} \times \boldsymbol{v}\sum_k m_k + \boldsymbol{R} \times \sum_k m_k \boldsymbol{v}_k - \boldsymbol{v} \times \sum_k m_k \boldsymbol{r}_k + \sum_k m_k \boldsymbol{r}_k \times \boldsymbol{v}_k.\end{aligned}$$

The second and third terms vanish as both contain $\sum_k m_k \boldsymbol{r}_k$ ($d/dt\ \sum_k m_k \boldsymbol{r}_k$ in the second term) and $\boldsymbol{r}_k$ are the c.m. coordinates, which leads to

$$\boldsymbol{J}^0 = M\boldsymbol{R} \times \boldsymbol{v} + \sum_k m_k \boldsymbol{r}_k \times \boldsymbol{v}_k = M\boldsymbol{R} \times \boldsymbol{v} + \sum_k \boldsymbol{J}_k = M\boldsymbol{R} \times \boldsymbol{v} + \boldsymbol{J}. \tag{2-17}$$

Thus, the total angular momentum as determined in the laboratory frame is a sum of a c.m. contribution, $M\boldsymbol{R} \times \boldsymbol{v}$, as well as a total internal component, $\boldsymbol{J}$, in the c.m. frame. If the c.m. is at rest, $\boldsymbol{J}^0 = \boldsymbol{J}$.

If all particles in the many-particle system are rigidly connected together, the angular momentum $\boldsymbol{J}$, in the molecular frame takes on the features of a *rigid rotor*. In a rigid rotor, the velocity vector $\boldsymbol{v}_k$ of the kth particle in the c.m. coordinate system is related to the *angular velocity*, $\boldsymbol{\omega}$, of the rotating particles by

$$\boldsymbol{v}_k = \boldsymbol{\omega} \times \boldsymbol{r}_k, \tag{2-18}$$

where $\boldsymbol{\omega}$ has its origin at the c.m. and does not carry an index, as the angular velocity of each particle is identical in a rigid system. Substituting Eq.(2-18) into Eq.(2-17) for $\boldsymbol{J}$ gives

$$\boldsymbol{J} = \sum_k m_k(\boldsymbol{r}_k \times \boldsymbol{v}_k) = \sum_k m_k[\boldsymbol{r}_k \times (\boldsymbol{\omega} \times \boldsymbol{r}_k)]. \tag{2-19}$$

Using $\boldsymbol{A} \times (\boldsymbol{B} \times \boldsymbol{C}) = \boldsymbol{B}(\boldsymbol{A} \cdot \boldsymbol{C}) - \boldsymbol{C}(\boldsymbol{A} \cdot \boldsymbol{B})$ in Eq.(2-19) gives (see Appendix B.1)

$$\boldsymbol{J} = \sum_k m_k[\boldsymbol{\omega}(\boldsymbol{r}_k \cdot \boldsymbol{r}_k) - \boldsymbol{r}_k(\boldsymbol{r}_k \cdot \boldsymbol{\omega})] = \sum_k m_k(r_k^2\mathbf{1} - \boldsymbol{r}_k\boldsymbol{r}_k) \cdot \boldsymbol{\omega} = \mathbf{I} \cdot \boldsymbol{\omega}$$

$$= \begin{pmatrix} \mathsf{I}_{xx} & \mathsf{I}_{xy} & \mathsf{I}_{xz} \\ \mathsf{I}_{yx} & \mathsf{I}_{yy} & \mathsf{I}_{yz} \\ \mathsf{I}_{zx} & \mathsf{I}_{zy} & \mathsf{I}_{zz} \end{pmatrix} \begin{pmatrix} \omega_x \\ \omega_y \\ \omega_z \end{pmatrix}. \tag{2-20}$$

$\mathbf{1}$ is the unit dyadic and $\mathbf{I}$ is the *moment of inertia dyadic*. The relations among vectors, dyadics, and the matrix expression given here are described in Appendix B.1. $\boldsymbol{J} = \mathbf{I} \cdot \boldsymbol{\omega}$ is obtained by the row ($\mathbf{I}$), column ($\boldsymbol{\omega}$) matrix multiplication. The components in the $\mathbf{I}$ matrix are easily determined from the expressions above and the results in the c.m. coordinate system are (see Fig. 2-1)

$$\mathsf{I}_{xx} = \sum_k m_k(y_k^2 + z_k^2), \qquad \mathsf{I}_{yx} = \mathsf{I}_{xy} = -\sum_k m_k x_k y_k, \tag{2-21}$$

and cyclic permutations for I_{yy}, I_{zz}, $\mathsf{I}_{zx} = \mathsf{I}_{xz}$, and $\mathsf{I}_{zy} = \mathsf{I}_{yz}$. If, on the other hand, the moment of inertia tensor is computed with respect to an arbitrary origin, 0 (see Fig. 2-1), the moment of inertia components are the same as in Eq.(2-21) with x_k^0 replacing x_k. The moments in the c.m. system are related to those at an arbitrary origin by

$$\mathsf{I}_{xx} = \sum_k m_k[(y_k^0)^2 + (z_k^0)^2] - M(Y^2 + Z^2), \qquad \mathsf{I}_{xy} = -\sum_k m_k x_k^0 y_k^0 + MXY, \tag{2-22}$$

and cyclic permutations as described above. X, Y, and Z are the Cartesian coordinates from the arbitrary origin, 0, to the center of mass as shown in Fig. 2-1.

The total kinetic energy of the system is given by

$$T = \frac{1}{2}\sum_k m_k \boldsymbol{v}_k^0 \cdot \boldsymbol{v}_k^0. \tag{2-23}$$

Using $\boldsymbol{v}_k^0 = \boldsymbol{v} + \boldsymbol{v}_k$ from the first time derivative of $\boldsymbol{r}_k^0 = \boldsymbol{R} + \boldsymbol{r}_k$ from Fig. 2-1 gives

$$T = \frac{1}{2}\sum_k m_k(\boldsymbol{v} + \boldsymbol{v}_k) \cdot (\boldsymbol{v} + \boldsymbol{v}_k) = \frac{1}{2}v^2 \sum_k m_k + \frac{1}{2}\sum_k m_k v_k^2 + \boldsymbol{v} \cdot \sum_k m_k \boldsymbol{v}_k. \tag{2-24}$$

The last term is zero as

$$\sum_k m_k \boldsymbol{v}_k = \frac{d}{dt}\sum_k m_k \boldsymbol{r}_k$$

and $\boldsymbol{r}_k$ are c.m. coordinates. Thus,

$$T = \frac{1}{2}Mv^2 + \frac{1}{2}\sum_k m_k v_k^2 = \frac{1}{2}Mv^2 + T_{\text{c.m.}}. \tag{2-25}$$

Thus, the total kinetic energy is a sum of two parts: The first part is the kinetic energy of the total mass at the c.m. and the second part, $T_{\text{c.m.}}$, is the internal kinetic energy in the c.m. frame. In the case of a rigid system of particles, we can use $\boldsymbol{v}_k$ in Eq.(2-18) for the kinetic energy with respect to the c.m. frame to give

$$T_{\text{c.m.}} = \frac{1}{2}\sum_k m_k \boldsymbol{v}_k \cdot \boldsymbol{v}_k = \frac{1}{2}\sum_k m_k \boldsymbol{v}_k \cdot (\boldsymbol{\omega} \times \boldsymbol{r}_k) = \frac{1}{2}\boldsymbol{\omega} \cdot \sum_k m_k(\boldsymbol{r}_k \times \boldsymbol{v}_k) = \frac{1}{2}\boldsymbol{\omega} \cdot \boldsymbol{J}, \tag{2-26}$$

where Eq.(2-17) is used in the last step. Substituting Eq.(2-20) gives (see Appendix B.1)

$$T_{\text{c.m.}} = \tfrac{1}{2}\boldsymbol{\omega}\cdot \mathsf{I} \cdot \boldsymbol{\omega} = \tfrac{1}{2}(\omega_x \omega_y \omega_z)\begin{pmatrix} \mathsf{I}_{xx} & \mathsf{I}_{xy} & \mathsf{I}_{xz} \\ \mathsf{I}_{yx} & \mathsf{I}_{yy} & \mathsf{I}_{yz} \\ \mathsf{I}_{zx} & \mathsf{I}_{zy} & \mathsf{I}_{zz} \end{pmatrix}\begin{pmatrix} \omega_x \\ \omega_y \\ \omega_z \end{pmatrix}. \tag{2-27}$$

We can now apply these results to the calculation of the classical kinetic energy of a molecular rigid rotor. In order to compute I for a molecule we start by choosing an arbitrary origin and then use Eq.(2-10) to find the c.m. Atomic masses are used for m_k. We shall examine the independent contributions to the moments due to electrons and nuclei in Chapter 6 and show that to a very good approximation molecular moments of inertia are determined by using atomic masses (see Table A-5) centered at the nuclear positions. After finding the c.m., the moments are computed using Eqs.(2-22). From this point it is convenient to find an axis system fixed in the molecule (c.m. frame) where I is diagonal. The axis system in which I is diagonal is called the *principal* inertial axis system. We use the *a*, *b*, and *c* axes to designate the principal inertial axis system and we conventionally choose

$$\mathsf{I}_{aa} \le \mathsf{I}_{bb} \le \mathsf{I}_{cc}, \tag{2-28}$$

and classify molecules according to the following types:

$$\begin{aligned} \mathsf{I}_{aa} = 0, \quad \mathsf{I}_{bb} &= \mathsf{I}_{cc}; && \text{linear} \\ \mathsf{I}_{aa} < \mathsf{I}_{bb} &= \mathsf{I}_{cc}; && \text{prolate symmetric} \\ \mathsf{I}_{aa} = \mathsf{I}_{bb} &< \mathsf{I}_{cc}; && \text{oblate symmetric} \\ \mathsf{I}_{aa} = \mathsf{I}_{bb} &= \mathsf{I}_{cc}; && \text{spherical} \\ \mathsf{I}_{aa} < \mathsf{I}_{bb} &< \mathsf{I}_{cc}; && \text{asymmetric} \end{aligned} \tag{2-29}$$

These various types of molecules, classified according to the diagonal elements in the principal inertial tensor, are illustrated in Fig. 2-2.

The rotational kinetic energy reduces to a very convenient form in the principal inertial axis system. According to Eq.(2-27), we can write

$$T_{\text{c.m.}} = \tfrac{1}{2}(\omega_a^2 \mathsf{I}_{aa} + \omega_b^2 \mathsf{I}_{bb} + \omega_c^2 \mathsf{I}_{cc}). \tag{2-30}$$

We can also write this expression in terms of the angular momentum in Eq.(2-20). If I is diagonal (the principal inertial axis system),

$$\boldsymbol{J} \cdot \boldsymbol{J} = \boldsymbol{\omega} \cdot \mathsf{I} \cdot \mathsf{I} \cdot \boldsymbol{\omega} = \mathsf{I}_{aa}^2 \omega_a^2 + \mathsf{I}_{bb}^2 \omega_b^2 + \mathsf{I}_{cc}^2 \omega_c^2 = J_a^2 + J_b^2 + J_c^2. \tag{2-31}$$

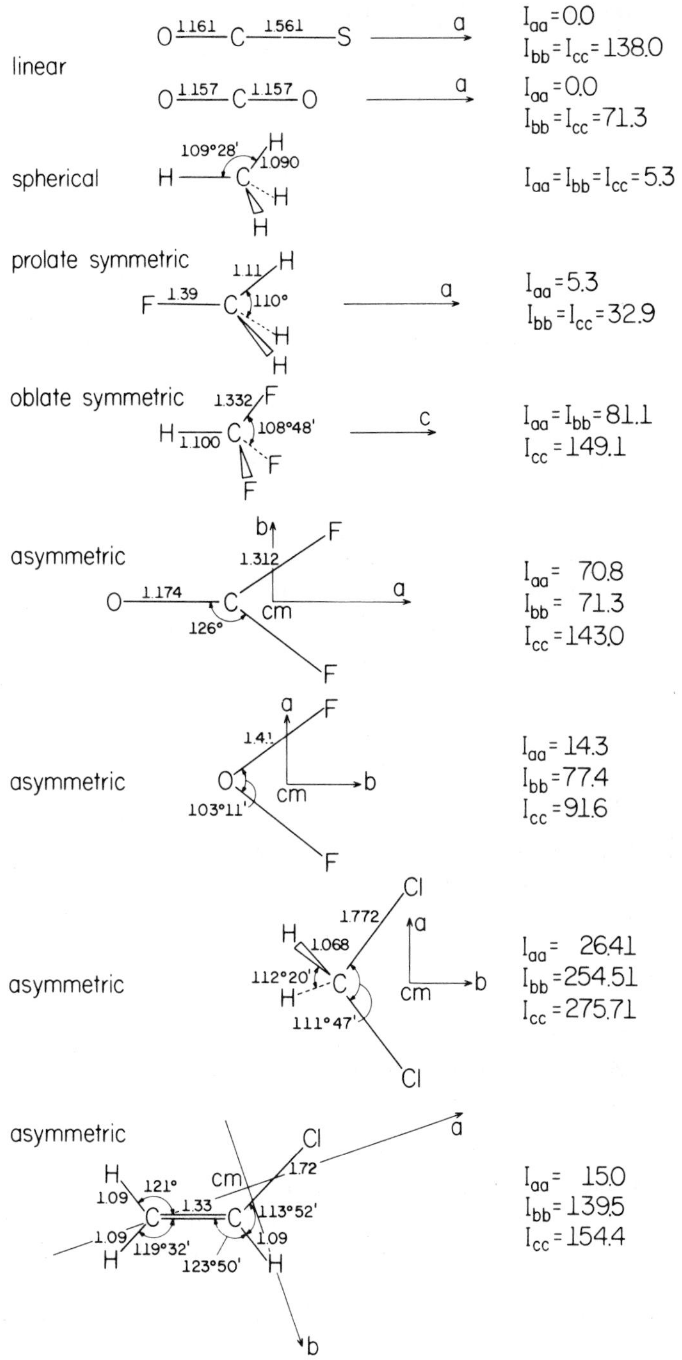

Figure 2-2 Moments of inertia of several molecules in the principal inertial axis system (a, b, c) for the common isotopes (^{16}O, ^{12}C, ^{32}S, ^{19}F, and ^{35}Cl). The bond distances are in units of $Å = 10^{-8}$ cm. The units of the moments are 10^{-40} g · cm^2, or 10^{-47} kg · m^2.

Thus, we can rewrite Eq.(2-30) as

$$T_{\text{c.m.}} = \frac{J_a^2}{2\mathsf{I}_{aa}} + \frac{J_b^2}{2\mathsf{I}_{bb}} + \frac{J_c^2}{2\mathsf{I}_{cc}}. \tag{2-32}$$

Now, consider the kinetic energy of a linear molecule where $\mathsf{I}_{aa} = 0$, $\mathsf{I}_{bb} = \mathsf{I}_{cc}$, and the energy is given by [Eq.(2-30)]

$$T_{\text{c.m.}} = \tfrac{1}{2}(\omega_b^2 + \omega_c^2)\mathsf{I}_{bb} = \tfrac{1}{2}\omega^2\mathsf{I}_{bb}. \tag{2-33}$$

Assuming the equipartition principle for kinetic energy that gives $\frac{1}{2}kT$ for each rotational degree of freedom, where k is Boltzmann's constant and T is the temperature, the kinetic energy of rotation for a single linear molecule at a temperature of 300 K is 4.14×10^{-21} J. Comparing this energy along with the value of $\mathsf{I} = 138.0 \times 10^{-47}\ \text{kg}\cdot\text{m}^2$ from Fig. 2-2 gives $\omega \cong 2.45 \times 10^{12}\ \text{rad}\cdot\text{s}^{-1}$ and $\nu = \omega/2\pi = 3.90 \times 10^{11}$ Hz. Thus, under normal equilibrium conditions of equipartition of rotational and translational kinetic energy, the OCS molecule will rotate with a frequency of about 4×10^{11} Hz.

It is also of interest to compute the component of linear velocity on the outer atoms due to the rotational motion at $T = 300$ K. Assuming an average of 10^{-8} cm from the c.m. to the outer atoms in OCS, and using Eq.(2-18), gives a linear velocity of $v = \omega r = 2.5 \times 10^4\ \text{cm}\cdot\text{s}^{-1}$. This outer atom linear velocity is comparable with the average translational speed of the c.m. of OCS, which is easily computed from standard kinetic theory [3] to give

$$v = \left(\frac{8kT}{\pi M}\right)^{1/2} \cong 3.2 \times 10^4\ \text{cm}\cdot\text{s}^{-1}.$$

It will be interesting to return to this classical concept of continuous rotational motion after studying the quantized picture for rotation provided by quantum mechanics.

In summary, we note that the classical view of rotation allows continuous values of angular velocity, $\boldsymbol{\omega}$, angular momentum, $\boldsymbol{J}$, and rotational energy.

2.2 THE CLASSICAL INTERACTION OF A CHARGED PARTICLE WITH AN ELECTRIC FIELD

The classical interaction of a charged particle with an electric field is an important problem that can be successfully applied to many physical systems with Newton's equations of motion. We shall consider both static and dynamic fields here. We start by examining the motion of a free particle with mass M and charge q in the presence of a static electric field along the x axis, E_x. This model is appropriate for an ion in solution, for instance.

Under the influence of a static field, E_x, the particle with charge q will experience a force given by Eq.(1-49):

$$F_x = qE_x. \tag{2-34}$$

This electrostatic driving force will lead to a particle drift in the medium with a velocity $\dot{x}$ along the x axis. It is reasonable to assume that the drifting particle in a viscous medium will experience a frictional drag in the fluid arising from a force that is proportional to the velocity:

$$F_x = -f\dot{x}, \tag{2-35}$$

where f is the frictional force constant. Combining Eqs.(2-34) and (2-35) and using Newton's second law in Eq.(2-1) gives an equation of motion for the particle:

$$M\ddot{x} = qE_x - f\dot{x}. \tag{2-36}$$

Rearranging gives

$$\ddot{x} + \frac{\dot{x}}{\tau} = \frac{q}{M} E_x, \tag{2-37}$$

where $\tau = M/f$ is the relaxation time for the particle.

We now turn on the external field, E_x, at $t = 0$ and solve Eq.(2-37) for the velocity of the initially stationary particle giving

$$\dot{x}(t) = \dot{x}(t = \infty)\left[1 - \exp\left(-\frac{t}{\tau}\right)\right], \tag{2-38}$$

where $\dot{x}(t = \infty)$ is the drift velocity at $t = \infty$. According to this result, $\dot{x}(t = 0) = 0$ and grows to $\dot{x}(t = \infty)$ with a $1 - \exp(-t/\tau)$ dependence. Substituting Eq.(2-38) into Eq.(2-37) gives $\dot{x}(t = \infty) = (\tau q/M)E_x$, which leads to a definition of the *mobility* as the drift velocity at $t = \infty$ divided by the value of the static field,

$$\mu = \frac{\dot{x}(\infty)}{E_x}. \tag{2-39}$$

Typical mobilities for ions in water solutions at $T = 300$ K are H^+ (36.3×10^{-8} $m^2 \cdot s^{-1} \cdot V^{-1}$), Na^+ (5.2×10^{-8} $m^2 \cdot s^{-1} \cdot V^{-1}$), Ba^{2+} (6.6×10^{-8} $m^2 \cdot s^{-1} \cdot V^{-1}$), Cl^- (7.2×10^{-8} $m^2 \cdot s^{-1} \cdot V^{-1}$), and SO_4^{2-} (8.3×10^{-8} $m^2 \cdot s^{-1} \cdot V^{-1}$). Thus, at a typical laboratory field of $100\ V \cdot cm^{-1} = 10^4\ V \cdot m^{-1}$, the drift velocity of a Na^+ ion in a water solution is equal to $v = 5.2 \times 10^{-4}\ m \cdot s^{-1} = 5.2 \times 10^{-2}\ cm \cdot s^{-1}$.

The steady-state (long-time average) *conductivity*, σ, of the solution of charged particles is obtained as the proportionality between the current density, J(charge $A^{-1} \cdot t^{-1}$), and the applied field, E_x, as shown in Eq.(1-6),

$$J = \sigma E_x = \rho_0 q \dot{x}(\infty), \tag{2-40}$$

where ρ_0 is the charge carrier number density. Thus,

$$\sigma = \rho_0 q\left[\frac{\dot{x}(\infty)}{E_x}\right] = \rho_0 q\mu = \rho_0 q^2 \frac{\tau}{M}. \tag{2-41}$$

where μ is the mobility given in Eq.(2-39). The *resistivity* of the solution is the inverse of the conductivity. In the case of electrons (in a metal, for instance), Eq.(2-41) can also be written in the form

$$\sigma = \rho_0 e^2 \frac{\tau}{m}. \tag{2-42}$$

We shall return to this expression for the electron conductivity in Chapter 5 where we discuss the electronic structure of metals.

Dynamic Fields and Bound Particles

Consider now the effects of a time-dependent field on the motion of a charged particle in a medium. The case of an unbound particle is considered above and the time-dependent field will add no new features except that the velocity, $\dot{x}(t)$, will follow the time dependence of the field and therefore alternate its direction with the field. A more interesting case is provided by the bound particle. Using this model we assume that the charged particle is bound to an equilibrium position in the medium at $x = 0$. Now, if the particle moves (or is driven) from its equilibrium position on the x axis, it will experience a restoring force that is proportional to its extension, x, from equilibrium. This restoring force is called the *Hooke's law force* (see Section 2.3, especially Fig. 2-3). This force is written as

$$F = -kx, \tag{2-43}$$

where k is the force constant. Adding this force to the results in Eq.(2-36), which were considered previously, gives

$$M\ddot{x} = qE_x - f\dot{x} - kx. \tag{2-44}$$

We use a traveling plane wave for the dynamic field as given in Eq.(1-29). For radiation propagating along the y axis and polarized along the x axis, this field is given by

$$E_x(t) = E_0 \exp\,[i(ky - \omega t)]. \tag{2-45}$$

We assume that the displacement has the same spatial and temporal dependence, but not necessarily the same phase as the inducing field, giving

$$x(t) = x_0 \exp\,[i(ky - \omega t)] \tag{2-46}$$

where x_0 can be complex to represent the amplitude and phase shift. Substituting Eq.(2-46) into Eq.(2-44) gives

$$x(t) = \frac{(q/M)E_x(t)}{\omega_0^2 - \omega^2 - (i\omega/\tau)}, \tag{2-47}$$

where $\omega_0^2 = k/M$ and $\tau = M/f$. $\omega_0 = (k/M)^{1/2}$ is the harmonic oscillator angular frequency determined from Eq.(2-44) when $E_x = 0 = f$ giving $\ddot{x} = -(k/M)x$, which has a solution of $x = x_0 \cos \omega_0 t$ with $\omega_0^2 = k/M$ as shown in detail from Eqs.(2-57) to (2-61). Returning to Eq.(2-47) we note that the amplitude of the displacement, x_0, is not in phase with the amplitude of the inducing field, E_x, and we can rewrite Eq.(2-47) as

$$x(t) = BE_x(t) \exp(i\theta), \tag{2-48}$$

where B is a real constant and θ is the phase factor. Comparing Eqs.(2-48) and (2-47) gives

$$\tan\theta = \frac{\omega/\tau}{\omega^2 - \omega_0^2}. \tag{2-49}$$

Thus, if $\omega < \omega_0$, the phase of $E_x(t)$ leads $x(t)$, but if $\omega_0 < \omega$, the phase of $x(t)$ leads $E_x(t)$. The measurement of the phase of the response leads to a measure of the relaxation time, τ.

The displacement of the charged bound particle under the influence of $E_x(t)$ as described above can be measured by relating the displacement to the observable dielectric constant or conductivity. In order to make this connection to the dielectric constant, we must return to Eq.(1-43). We can relate the polarizability in Eq.(1-43) to the displacement by noting that the induced dipole moment of the spherical charged particle in the field is given by Eq.(1-40):

$$D_x(t) = \alpha E_x(t) = qx(t). \tag{2-50}$$

Thus, $\alpha = qx(t)/E_x(t)$. Substituting into Eq.(1-43) gives

$$\begin{array}{cc} cgs & SI \\ \varepsilon = 1 + 4\pi\rho_0 q\dfrac{x(t)}{E_x(t)} & \varepsilon = 1 + \dfrac{\rho_0 q x(t)}{\varepsilon_0 E_x(t)}. \end{array} \tag{2-51}$$

Substituting the displacement in Eq.(2-47) into this result gives real, $\varepsilon_r(\omega)$, and imaginary, $\varepsilon_i(\omega)$, components in the dielectric constant. The result is

$$\begin{aligned} \varepsilon(\omega) &= 1 + \frac{\omega_p^2}{\omega_0^2 - \omega^2 - (i\omega/\tau)} = \varepsilon_r(\omega) + i\varepsilon_i(\omega) \\ &= 1 + \frac{\omega_p^2(\omega_0^2 - \omega^2)}{(\omega_0^2 - \omega^2)^2 + (\omega^2/\tau^2)} + i\frac{\omega_p^2(\omega/\tau)}{(\omega_0^2 - \omega^2)^2 + (\omega^2/\tau^2)} \\ \omega_p^2 &= \frac{4\pi q^2}{M}\rho_0 \text{ (cgs)} = \frac{q^2}{\varepsilon_0 M}\rho_0 \text{ (SI)}. \end{aligned} \tag{2-52}$$

ω_p is called the *plasma frequency*. Equation (2-52) should describe the frequency dependence of the dielectric response of a bound charged particle subjected to a viscous medium. Of course, for a free particle, $\omega_0 = 0$, and the curves are then shifted appropriately. This latter case of $\omega_0 = 0$ is appropriate for the dielectric response of free electrons in a metal or a gaseous plasma of electrons or other charged particles (see Problem 2-10). The classical expression for $\varepsilon(\omega)$ above is used extensively in interpreting systems of charged particles that behave classically. $\varepsilon(\omega)$ is obtained by measuring the displacement, $\boldsymbol{D}$, produced by an external field, $\boldsymbol{E}$, as shown in Eq.(1-5); $\boldsymbol{D} = \boldsymbol{\varepsilon} \cdot \boldsymbol{E}$. The frequency dependence of the conductivity for the bound forced oscillator can also be examined as in Problem 2-12.

Before proceeding we note that $\varepsilon_r(\omega)$ and $\varepsilon_i(\omega)$ are related by the Kramers–Kronig relations [4]:

$$\varepsilon_r(\omega') = 1 + \frac{1}{\pi}\lim_{\delta\to 0}\left[\int_{-\infty}^{\omega'-\delta}\frac{\varepsilon_i(\omega)}{\omega'-\omega}\,d\omega' + \int_{\omega'+\delta}^{\infty}\frac{\varepsilon_i(\omega)}{\omega'-\omega}\,d\omega'\right]$$

$$\varepsilon_i(\omega') = -\frac{1}{\pi}\lim_{\delta\to 0}\left[\int_{-\infty}^{\omega'-\delta}\frac{\varepsilon_r(\omega)-1}{\omega'-\omega}\,d\omega' + \int_{\omega'+\delta}^{\infty}\frac{\varepsilon_r(\omega)-1}{\omega'-\omega}\,d\omega'\right].$$

Thus, if either $\varepsilon_r(\omega)$ or $\varepsilon_i(\omega)$ can be measured, the other can be computed by the Kramers–Kronig relation. This is generally true for any property that is written like $\varepsilon = \varepsilon_r + i\varepsilon_i$, as a sum of real and imaginary parts.

The Lorentz Electron Theory of Matter

Before the advent of the quantum theory and indeed before the discovery of the planetary atom, Lorentz [5] and others attempted to apply the previous models to explain the resonant interactions between electromagnetic radiation and electrons in atoms and molecules. The model of electrons in atoms and molecules involves a harmonically bound electron where losses are rationalized by the friction arguments given previously. The radiation field then leads to the electron displacement given in Eq.(2-47). We shall now relate the absorption coefficient in a radiation absorption experiment to the imaginary part of the refractive index that can be evaluated with the displacement in Eq.(2-47).

First we relate the displacement to the refractive index in the case where $\varepsilon - 1 \ll 1$. Substituting $\alpha = qx(t)/E_x(t)$ from Eq.(2-50) into Eq.(1-60) gives

cgs	*SI*	
$n = 1 + 2\pi\rho_0 q\dfrac{x(t)}{E_x(t)}$	$n = 1 + \dfrac{\rho_0 q x(t)}{2\varepsilon_0 E_x(t)}.$	(2-53)

Substituting the displacement in Eq.(2-47) into this equation gives

$$n = n_r + in_i = 1 + \frac{\omega_p^2}{2}\left[\frac{1}{\omega_0^2 - \omega^2 - (i\omega/\tau)}\right]$$

$$= 1 + \frac{\frac{1}{2}\omega_p^2(\omega_0^2-\omega^2)}{(\omega_0^2-\omega^2)^2 + (\omega^2/\tau^2)} + i\frac{\frac{1}{2}\omega_p^2(\omega/\tau)}{(\omega_0^2-\omega^2)^2 + (\omega^2/\tau^2)}, \tag{2-54}$$

where n_r and n_i are the real and imaginary components of the refractive index. ω_p is the plasma frequency defined in Eq.(2-52).

Next we relate n_i to the absorption experiment in spectroscopy. In the case of a neutral ($\bar{\rho} = 0$), nonconducting ($\boldsymbol{J} = 0$), nonmagnetic ($\mu_p = 1.0$), but dielectric ($\varepsilon \neq 1.0$) material, the electric real field is given in Eq.(1-29) with the dispersion being discussed in Eq.(1-27) where $k = n(2\pi/\lambda_0)$. Now, as the refractive index is complex, $n = n_r + in_i$, the real part of the electric field will not be a simple cosine as shown (for a vacuum) in Eq.(1-29). Substituting $n = n_r + in_i$ into $k = (\omega/c)n$ and substituting this into Eq.(1-29) for the electric field propagating along the y axis (from $-y$ to $+y$) gives

$$E = E_0 \exp\left\{i\omega\left[\frac{y}{c}(n_r + in_i) - \omega t\right]\right\} = E_0 \exp\left(-\frac{\omega}{c} n_i y\right) \exp\left[i\left(\frac{\omega}{c} n_r y - \omega t\right)\right],$$

which has a real component given by

$$E = E_0 \exp\left(-\frac{\omega}{c} n_i y\right) \cos\left(\frac{\omega}{c} n_r y - \omega t\right).$$

Thus, the field is attenuated as it passes through the medium with a nonzero imaginary refractive index, n_i: The imaginary part of the refractive index leads to absorption of the radiation by the medium. The intensity of the radiation obtained from the real field is given by

cgs	*SI*
$I(y) = \frac{c}{4\pi}\langle E^2 \rangle$	$I(y) = c\varepsilon_0 \langle E^2 \rangle$
$= \frac{c}{8\pi} E_0^2 \exp\left(-\frac{2\omega}{c} n_i y\right)$	$= \frac{c\varepsilon_0}{2} E_0^2 \exp\left(-\frac{2\omega}{c} n_i y\right)$
$= I(0) \exp(-\gamma y)$	$= I(0) \exp(-\gamma y)$

$$\gamma = \frac{2\omega n_i}{c}, \qquad (2\text{-}55)$$

where the angular brackets indicate the usual time average. $I(0)$ is the incident intensity at the edge of the medium ($y = 0$) with refractive index n. γ is the absorption coefficient that is proportional to the imaginary part of the refractive index. Substituting $n_i(\omega)$ from Eq.(2-54) into Eq.(2-55) gives the absorption coefficient for the bound electron:

$$\gamma(\omega) = \frac{1}{\tau c}\left[\frac{\omega_p^2 \omega^2}{(\omega_0^2 - \omega^2)^2 + (\omega^2/\tau^2)}\right].$$

Now, $\gamma(\omega)$ is large only near resonance where $\omega_0 \approx \omega$, allowing us to write

$$\omega_0^2 \cong \omega^2, \qquad \omega^2 - \omega_0^2 = (\omega + \omega_0)(\omega - \omega_0) \cong 2\omega(\omega - \omega_0).$$

Substituting into the expression for $\gamma(\omega)$ above gives

$$\gamma(\omega) \cong \frac{\pi\omega_p^2}{2c}\left\{\frac{1}{\pi}\left[\frac{(1/2\tau)}{(\omega_0 - \omega)^2 + (1/2\tau)^2}\right]\right\} = \frac{\pi\omega_p^2}{2c}\mathscr{L}(\omega - \omega_0), \qquad (2\text{-}56)$$

where $\mathscr{L}(\omega - \omega_0)$ is the normalized Lorentzian $[\int_0^{+\infty} \mathscr{L}(\omega - \omega_0)\, d\omega = 1.0]$ centered at ω_0 with half width at half height given by $\Delta\omega = \frac{1}{2}\tau$ (see Fig. 7-1). Equation (2-56) for the resonant absorption coefficient for a weakly absorbing bound electron in an atom or molecule with relaxation time τ was a high point of the Lorentz theory of matter. This theory was capable of explaining the observed discrete spectra and their observed Lorentzian line shapes. An atom or molecule could continuously absorb electromagnetic radiation by exciting harmonic oscillations in the electrons. Emission of radiation in an atom or molecule could also be excited by external radiation of the correct frequency. The sharp frequency dependence of atomic and molecular absorptions and emissions was explained by Lorentz by the concept of harmonic oscillation where electrons in a particular atom or molecule will oscillate more favorably at some frequencies than others. By using the simple models above, Lorentz was able to explain most of the known resonant electromagnetic interactions observed up to the early 1900's. The Lorentz theory could not explain all the finer details of atomic spectra, however, and it became clear that this electron theory of matter either was incomplete or was founded on false postulates.

The most serious shortcoming of the classical electron theory of matter became evident in 1911 by interpreting the experimental results of alpha-particle scattering from metal foils (see Problem 9-2 for a more complete discussion and references). By examining data obtained by scattering alpha-particles (helium nuclei) from a thin gold foil, Rutherford discovered that an atom consists of a positive center having a diameter of about 10^{-13} cm and an outer shell of electronic charge having a diameter of about 10^{-8} cm. It appeared that atoms were analogous to the solar system with the positive heavy nucleus at the center of a revolving system of lighter electrons. It appeared that electrons in atoms and molecules could not be described as harmonic oscillators and furthermore if the electrons are orbiting the nucleus, they are in a state of constant acceleration that, according to the classical theory, should give rise to electromagnetic radiation. Thus, the electron should give off radiation, thereby losing energy and finally it would cascade into the nucleus. These dilemmas could not be answered by the Lorentz classical harmonic electron theory of matter and the subsequent successes of the quantum theory have replaced the more intuitive (but erroneous) basis of understanding matter that was given by Lorentz and his school of thought.

2.3 VIBRATIONAL MOTION AND LAGRANGE'S EQUATIONS

In this section we shall examine simple one-dimensional displacements and harmonic oscillation. Following that, we define a set of generalized coordinates and write Lagrange's equations, which are independent of the coordinate system.

Consider the motion of a mass m suspended on a massless spring from a solid wall as shown in Fig. 2-3. The equation of motion for this simple system is given in Eq.(2-12) where the external force is zero and the value of $\dot{p}_k^0$ in the laboratory frame is equal to $\dot{p}_k$, the corresponding value in the c.m. frame where the c.m. is at the surface of the infinitely heavy wall. The result is

$$F = \dot{p} = \frac{d}{dt}mv = m\frac{d^2}{dt^2}x, \tag{2-57}$$

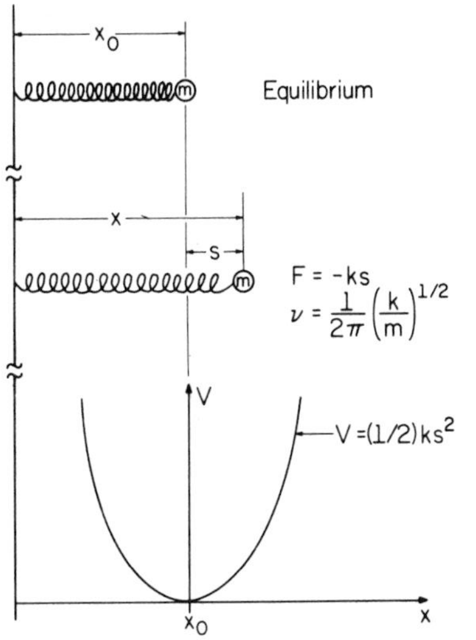

Figure 2-3 A mass represented by m suspended from a fixed wall with a massless spring. The restoring force of the extended spring is described by Hooke's law leading to a $V = \frac{1}{2}ks^2$ potential function, which is also shown. The classical vibrational frequency, ν, is also given.

where the x axis contains the particle as shown in Fig. 2-3. The force at equilibrium ($x = x_0$) is zero. As the particle moves a distance s off equilibrium, however, there will be a restoring force that is proportional to the displacement (Hooke's law) and given according to Fig. 2-3, by $F = -ks$ where k is the force constant. The resultant equation of motion is

$$-ks = m\frac{d^2}{dt^2}s. \tag{2-58}$$

The potential energy curve is also shown in Fig. 2-3 as obtained from

$$V = -\int \boldsymbol{F} \cdot d\boldsymbol{s} = \tfrac{1}{2}ks^2 + \text{constant}. \tag{2-59}$$

The potential energy can only be determined to within a constant. Returning to Eq.(2-58) we note that a sinusoidal function,

$$s = s_0 \cos(\omega t + \alpha), \tag{2-60}$$

is a solution where s_0 is a constant amplitude and α is an arbitrary phase factor. Substituting this harmonic solution in Eq.(2-60) into Eq.(2-58) gives

$$\omega^2 = \frac{k}{m} \qquad \nu = \frac{1}{2\pi}\left(\frac{k}{m}\right)^{1/2}, \tag{2-61}$$

where ν is the vibrational frequency of the oscillator. Equation (2-61) relates the force constant k to the fundamental vibrational frequency (see Fig. 2-3). The case of HCl is very similar to the small mass suspended by a spring from a heavy wall. If we replace the mass m in Eq.(2-61) with the reduced mass, however, we obtain the vibrational frequency for the *relative* H—Cl vibrational motion. The reduced mass is (we use ^{35}Cl here)

$$\mu = \frac{m_{\mathrm{H}} m_{\mathrm{Cl}}}{m_{\mathrm{H}} + m_{\mathrm{Cl}}} = 1.615 \times 10^{-24} \text{ g}.$$

In this case where $m_{\mathrm{Cl}} > m_{\mathrm{H}}$, $\mu \cong m_{\mathrm{H}}$. Using the observed HCl vibrational frequency of 8.65×10^{13} Hz and μ in place of m in Eq.(2-61) gives

$$k = (2\pi\nu)^2\mu = 4.77 \times 10^5 \text{ dyne} \cdot \text{cm}^{-1} = 4.77 \times 10^2 \text{ N} \cdot \text{m}^{-1},$$

as the force constant for the HCl band.

It is difficult to extend these techniques above to a polyatomic molecule because of the constraints imposed by the molecular structure. A better approach is obtained by writing Lagrange's equations in terms of generalized coordinates, which are then independent of the coordinate system. We can then conveniently use displacement coordinates to describe the vibrations.

We start by returning to the total kinetic energy of the system of particles in the c.m. frame from Eq.(2-25):

$$T_{\text{c.m.}} = \frac{1}{2}\sum_k m_k \boldsymbol{v}_k \cdot \boldsymbol{v}_k.$$

The first derivative with respect to the v_x component of the ith particle gives

$$\frac{\partial T}{\partial v_{ix}} = m_i v_{ix}, \qquad \frac{d}{dt}\left(\frac{\partial T}{\partial v_{ix}}\right) = \frac{d}{dt}(m_i v_{ix}) = \dot{p}_{ix}. \tag{2-62}$$

In the absence of external forces, we assume that the internal forces are conservative and that the force on the ith particle is given by

$$-\boldsymbol{\nabla}_i V = \dot{\boldsymbol{p}}_i = \frac{d}{dt} m\boldsymbol{v}_i.$$

Using Eq.(2-62), this equation can be rewritten to give

$$\frac{\partial V}{\partial x_i} + \frac{d}{dt}\left(\frac{\partial T}{\partial v_{ix}}\right) = 0, \tag{2-63}$$

with similar equations for y and z.

Another way of writing Eq.(2-63) was introduced by Lagrange by defining a function $L = T - V$, now called the *Lagrangian*, which depends explicitly on the three velocity components in T and the three position coordinates in V for each particle. Thus, from

$$\frac{\partial T}{\partial v_x} = \frac{\partial L}{\partial v_x} \qquad \frac{\partial V}{\partial x} = -\frac{\partial L}{\partial x}, \tag{2-64}$$

we can write

$$\frac{d}{dt}\left(\frac{\partial L}{\partial v_{ix}}\right) - \frac{\partial L}{\partial x_i} = 0, \tag{2-65}$$

with similar expressions for the y and z coordinates.

It is also easy to show that

$$\frac{\partial L}{\partial v_x} = p_x. \tag{2-66}$$

It is clear that the corresponding expression for the angular velocity, $\boldsymbol{\omega}$, and angular momentum, $\boldsymbol{J}$, is given by

$$\frac{\partial L}{\partial \omega_x} = J_x. \tag{2-67}$$

The important step forward in the Lagrange formulation is that Eq.(2-65) is independent of the choice of coordinates. Each of the Cartesian coordinates can be reexpressed in terms of generalized coordinates, q_i, and their time derivatives, $\dot{q}_i$. The three Cartesian coordinates for each of the n particles are functions of the generalized coordinates that take into account restrictive forces or constraints in the system. If there are r constraints, we can write

$$\begin{aligned} x_1 &= x_1(q_1, q_2, \ldots, q_{3n-r}) \\ y_1 &= y_1(q_1, q_2, \ldots, q_{3n-r}) \\ z_1 &= z_1(q_1, q_2, \ldots, q_{3n-r}) \\ x_2 &= x_2(q_1, q_2, \ldots, q_{3n-r}) \\ &\vdots \\ z_n &= z_n(q_1, q_2, \ldots, q_{3n-r}). \end{aligned} \tag{2-68}$$

If the constraints are time-dependent, then the Cartesian coordinates are also time-dependent. The transformation from Cartesian coordinates to the generalized coordinates gives Lagrange's generalized equations [2]:

$$\frac{d}{dt}\left(\frac{\partial L}{\partial \dot{q}_i}\right) - \frac{\partial L}{\partial q_i} = 0, \tag{2-69}$$

where the generalized coordinate, q_i, depends on the $3n - r$ coordinates of the n particles. We now define the generalized momentum:

$$p_i = \frac{\partial L}{\partial \dot{q}_i}. \tag{2-70}$$

Lagrange's equation can now be rewritten as

$$\dot{p}_i = \frac{\partial L}{\partial q_i}, \tag{2-71}$$

which implies a conservation theorem. If the Lagrangian, L, is independent of the generalized coordinate describing the particle, q_i, the generalized momentum, p_i, is conserved in time. The kinetic energy in the c.m. frame can be written in terms of the generalized coordinates according to

$$T = \frac{1}{2}\sum_i m_i v_i^2 = \frac{1}{2}\sum_i m_i\left(\frac{dr_i}{dt}\right)^2 = \frac{1}{2}\sum_i m_i\left(\sum_j \frac{\partial r_i}{\partial q_j}\frac{dq_j}{dt}\right)^2,$$

which can be rewritten in terms of velocities as

$$T = \frac{1}{2}\sum_i m_i \sum_{j,k}\left(\frac{\partial \mathbf{r}_i}{\partial q_j}\cdot\frac{\partial \mathbf{r}_i}{\partial q_k}\right)\dot{q}_j\dot{q}_k = \frac{1}{2}\sum_{j,k} t_{jk}\dot{q}_j\dot{q}_k,$$

$$t_{jk} = \sum_i m_i\left(\frac{\partial \mathbf{r}_i}{\partial q_j}\cdot\frac{\partial \mathbf{r}_i}{\partial q_k}\right). \tag{2-72}$$

Next we define the *mass-weighted* displacement coordinates in terms of the displacements from equilibrium:

$$\begin{aligned} S_1 &= (m_1)^{1/2}\Delta x_1, & S_2 &= (m_1)^{1/2}\Delta y_1, & S_3 &= (m_1)^{1/2}\Delta z_1 \\ S_4 &= (m_2)^{1/2}\Delta x_2, & S_5 &= (m_2)^{1/2}\Delta y_2, & &\ldots, \end{aligned} \tag{2-73}$$

where Δx_i, Δy_i, and Δz_i are the displacements of the ith particle (atom in a molecule) from the equilibrium structure. The S_i coordinates reduce t_{ij} in Eq.(2-72) to $t_{ij} = \sum_i \delta_{ij}$. Substituting into the kinetic energy expression gives

$$T = \frac{1}{2}\sum_{i=1}^{3n}\left(\frac{dS_i}{dt}\right)^2 = \frac{1}{2}\sum_{i=1}^{3n}\dot{S}_i^2. \tag{2-74}$$

The potential energy can also be defined in terms of the mass-weighted displacement coordinates where we consider a general potential function $V(S_1, S_2, \ldots, S_{3n})$. For small displacements, the potential energy function can be expanded about the equilibrium configuration ($S_i = 0$) giving

$$\begin{aligned} V(S_1, S_2, \ldots, S_{3n}) = V &= V_0 + \sum_{i=1}^{3n}\left(\frac{\partial V}{\partial S_i}\right)_0 S_i + \frac{1}{2}\sum_{i,j=1}^{3n}\left(\frac{\partial^2 V}{\partial S_i\,\partial S_j}\right)_0 S_i S_j \\ &\quad + \frac{1}{6}\sum_{i,j,k=1}^{3n}\left(\frac{\partial^3 V}{\partial S_i\,\partial S_j\,\partial S_k}\right)_0 S_i S_j S_k + \cdots \\ &= V_0 + \sum_i V_i S_i + \frac{1}{2}\sum_{i,j} V_{ij} S_i S_j + \frac{1}{6}\sum_{i,j,k} V_{ijk} S_i S_j S_k + \cdots \end{aligned} \tag{2-75}$$

All derivatives are evaluated at equilibrium ($S_i = 0$) and V_0 is a constant that represents the potential energy of the system at equilibrium. We can reference all our results to V_0 or in effect set the constant V_0 equal to zero. If the potential energy is at a minimum at equilibrium ($S_i = 0$), then the net force is zero, $V_i = (\partial V/\partial S_i)_0 = 0$. V_{ij} is the harmonic, V_{ijk} is the cubic, and V_{ijkl} is the quartic contribution to the potential function. Assuming most vibrators are reasonably well-behaved harmonic oscillators, Eq.(2-75) can be approximated by including only the first nonzero term:

$$V(S_1, S_2, \ldots, S_{3n}) = V \cong \frac{1}{2}\sum_{i,j} V_{ij} S_i S_j, \tag{2-76}$$

which is analogous to the quadratic potential based on an analogy with Hooke's law. Substituting the Lagrangian function, $L = T - V$, from Eqs.(2-76) and (2-74) into Lagrange's equation of motion in Eq.(2-69), gives

$$\ddot{S}_k + \sum_i V_{ki} S_i = 0, \tag{2-77}$$

where we have used $V_{ki} = V_{ik}$. A harmonic solution to Eq.(2-77) is obtained by use of the sinusoidal function,

$$S_k = a_k \cos(\omega t + \alpha), \tag{2-78}$$

where $\omega = 2\pi\nu$ is the angular frequency and α is an arbitrary phase factor. Substituting Eq.(2-78) into Eq.(2-77) and canceling $\cos(\omega t + \alpha)$ gives

$$-a_k\omega^2 + \sum_{i=1}^{3n} V_{ki} a_i = \sum_{i=1}^{3n} (V_{ki} - \omega^2 \delta_{ki}) a_i = 0. \tag{2-79}$$

This set of $3n$ linear homogeneous equations has a nontrivial solution when the determinant of the coefficients of a_i is zero:

$$\begin{vmatrix} V_{11} - \omega^2 & V_{12} & \cdots \\ V_{21} & V_{22} - \omega^2 & \cdots \\ \vdots & & \ddots \end{vmatrix} = 0. \tag{2-80}$$

The $3n$ roots of this secular equation (the solution of the polynomial in ω^2) give the vibrational frequencies. Three of the roots will be zero, representing the translational motion that does not exist in the c.m. frame. In a linear molecule, two of the roots will be zero for the two rotational degrees of freedom. There are only two rotational degrees of freedom in a linear molecule because it only takes two angles to describe any orientation of the molecule with respect to a laboratory-fixed, reference axis system. Three angles are necessary to describe the orientation of a nonlinear molecule. Thus, the degrees of freedom for molecules break down as follows:

	Translational	Rotational	Vibrational
linear	3	2	$3n - 5$
nonlinear	3	3	$3n - 6$

The five (or six) translational and rotational degrees of freedom lead to zero values for the roots in Eq.(2-80) and the remaining $3n - 5$ or $3n - 6$ roots of the secular equation give the vibrational frequencies for the system. Equation (2-78) gives the coordinates that describe the vibration where the a_k are given by substituting the ω^2 roots from Eq.(2-80) into Eq.(2-79). Thus, for each of the $3n - 6$ (or $3n - 5$) values of ω, the corresponding S_k vibrates at frequency $\nu = \omega/2\pi$ with an amplitude, a_k. The $3n - 6$ (or $3n - 5$) independent modes of vibrational motion are called *normal modes of vibration*. The normal coordinates can be written as a linear combination of Cartesian displacement (from equilibrium) coordinates that describe the physical motion. Normal modes of motion and their observed vibrational frequencies for diatomic and linear and nonlinear triatomic molecules are shown in Fig. 2-4.

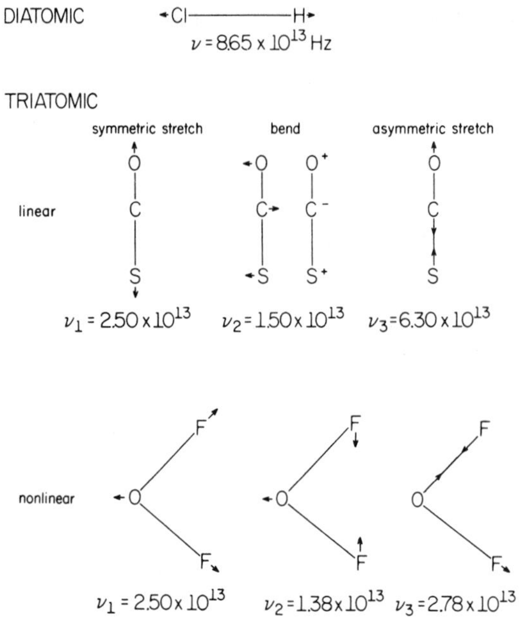

Figure 2-4 Normal modes of motion in diatomic and in linear and nonlinear triatomic molecules. In a diatomic molecule there is one vibrational degree of freedom and two rotational degrees of freedom. In the linear triatomic molecule there are four modes of vibrational motion and two rotational degrees of freedom. In the nonlinear triatomic molecule there are three modes of vibration and three rotational degrees of freedom. The observed vibrational frequencies are also shown.

2.4 HAMILTON'S EQUATIONS

We shall now generalize the classical equations of motion by developing Hamiltonian mechanics, which leads to the most evident link with quantum mechanics.

Equation (2-69) gives Lagrange's equation in terms of the generalized velocities and coordinates. Writing a new expression called the *Hamiltonian function* in terms of p_i, $\dot{q}_i$, and the Lagrangian, L, as

$$H = \sum_i p_i \dot{q}_i - L(\dot{q}_i, q_i), \tag{2-81}$$

and differentiating this equation with respect to q_i gives

$$\frac{\partial H}{\partial q_i} = p_i\left(\frac{\partial \dot{q}_i}{\partial q_i}\right) - \left(\frac{\partial L}{\partial \dot{q}_i}\right)\left(\frac{\partial \dot{q}_i}{\partial q_i}\right) - \frac{\partial L}{\partial q_i}.$$

L is an explicit function of only $\dot{q}_i$ and q_i. Substituting Eq.(2-70) into this expression shows that the first and second terms cancel giving

$$\frac{\partial H}{\partial q_i} = -\frac{\partial L}{\partial q_i} = -\dot{p}_i, \tag{2-82}$$

where Eq.(2-71) has been used in the last step. Differentiating H in Eq.(2-81) with respect to p_i and remembering that L is not an explicit function of the generalized momentum gives

$$\frac{\partial H}{\partial p_i} = \dot{q}_i. \tag{2-83}$$

Equations (2-82) and (2-83) give the conservation theorems in Hamilton's formulation. If the Hamiltonian function is independent of coordinate (particle position), q_i, the generalized momentum is conserved in time [Eq.(2-82)]. If the Hamiltonian is independent of the generalized momentum, p_i, the coordinate describing the particle, q_i, is conserved in time [Eq.(2-83)]. The generalized momentum and coordinate describing the particle are called conjugate variables. Equations (2-82) and (2-83) give Hamilton's $6n$ first-order differential equations, which have replaced Lagrange's $3n$ second-order differential equations.

We shall now show that the Hamiltonian is the total energy. From Eq.(2-72) we can write

$$\frac{\partial T}{\partial \dot{q}_i} = \sum_k t_{ik}\dot{q}_k.$$

Multiplying both sides of this expression by $\sum_i \dot{q}_i$ gives

$$\sum_i \dot{q}_i \frac{\partial T}{\partial \dot{q}_i} = \sum_{k,\, i} t_{ik}\dot{q}_i\dot{q}_k = 2T. \tag{2-84}$$

Returning to Eq.(2-70) and using $L = T - V$, we can write

$$p_i = \frac{\partial L}{\partial \dot{q}_i} = \frac{\partial T}{\partial \dot{q}_i} \tag{2-85}$$

if V is independent of velocity. Substituting Eq.(2-85) into the Hamiltonian in Eq.(2-81) where $L = T - V$ gives

$$H = \sum_i \frac{\partial T}{\partial \dot{q}_i}\dot{q}_i - T + V.$$

Substituting Eq.(2-84) into this result gives

$$H = 2T - T + V = T + V. \tag{2-86}$$

Thus, for a conservative system in which the potential energy is independent of the velocity, the Hamiltonian is the total energy.

Electromagnetic Field-Charged Particle Interaction

We shall now derive the Hamiltonian for the interaction of a free particle with mass m and charge e with an electromagnetic field.

We begin with Eq.(2-81) for the Hamiltonian for a single particle:

$$H = p_x\dot{x} + p_y\dot{y} + p_z\dot{z} - L.$$

Substituting Eq.(2-70) gives

$$H = \dot{x}\frac{\partial L}{\partial \dot{x}} + \dot{y}\frac{\partial L}{\partial \dot{y}} + \dot{z}\frac{\partial L}{\partial \dot{z}} - L. \tag{2-87}$$

Our task is now to write the Hamiltonian as a function of momenta and coordinates.

Starting with the Lorentz force in Eq.(1-49) for a particle with charge e in an electric and magnetic field, it is convenient to introduce the vector and scalar potentials defined in Eqs.(1-16) and (1-18) for the electric and magnetic fields into Eq.(1-49) to give (cgs units are used in this derivation)

$$\boldsymbol{F} = \frac{e}{c}\,\boldsymbol{v} \times (\nabla \times \boldsymbol{A}) - \frac{e}{c}\frac{\partial \boldsymbol{A}}{\partial t} - e\,\nabla\varphi, \tag{2-88}$$

which has Cartesian components written as

$$F_x = \frac{d}{dt}m\dot{x} = \frac{e}{c}\left(\dot{y}\frac{\partial A_y}{\partial x} + \dot{z}\frac{\partial A_z}{\partial x}\right) - \frac{e}{c}\left(\dot{y}\frac{\partial A_x}{\partial y} + \dot{z}\frac{\partial A_x}{\partial z} + \frac{\partial A_x}{\partial t}\right) - e\frac{\partial \varphi}{\partial x}. \tag{2-89}$$

F_y and F_z are obtained by cyclic permutations. Remembering that $\boldsymbol{A}$ is a function of x, y, z, and t, we can write

$$\frac{dA_x}{dt} = \frac{\partial A_x}{\partial t} + \dot{x}\frac{\partial A_x}{\partial x} + \dot{y}\frac{\partial A_x}{\partial y} + \dot{z}\frac{\partial A_x}{\partial z}. \tag{2-90}$$

Rearranging and substituting into the second bracket of Eq.(2-89) gives

$$\frac{d}{dt}m\dot{x} = \frac{e}{c}\left(\dot{x}\frac{\partial A_x}{\partial x} + \dot{y}\frac{\partial A_y}{\partial x} + \dot{z}\frac{\partial A_z}{\partial x}\right) - \frac{e}{c}\frac{dA_x}{dt} - e\frac{\partial \varphi}{\partial x}. \tag{2-91}$$

We can now show that a Lagrangian given by

$$L = \tfrac{1}{2}m\dot{x}^2 + \tfrac{1}{2}m\dot{y}^2 + \tfrac{1}{2}m\dot{z}^2 + \frac{e}{c}\,\boldsymbol{v}\cdot\boldsymbol{A} - e\varphi, \tag{2-92}$$

where φ is a velocity-independent static potential, can give the force expression in Eq.(2-91). Substituting Eq.(2-92) into Lagrange's equation in Eq.(2-69) for a single particle gives an equation identical to Eq.(2-91).

Continuing now with the derivation, we can substitute Eq.(2-92) into Eq.(2-87) to give

$$H = \frac{m}{2}(\dot{x}^2 + \dot{y}^2 + \dot{z}^2) + e\varphi = \frac{1}{2m}[(m\dot{x})^2 + (m\dot{y})^2 + (m\dot{z})^2] + e\varphi. \tag{2-93}$$

The values of $m\dot{x}$, $m\dot{y}$, and $m\dot{z}$ are obtained by substituting Eq.(2-92) into Eq.(2-70) and rearranging to give

$$m\dot{x} = p_x - \frac{e}{c}A_x, \qquad m\dot{y} = p_y - \frac{e}{c}A_y, \qquad m\dot{z} = p_z - \frac{e}{c}A_z. \tag{2-94}$$

Thus, the linear momentum of the particle in the electromagnetic field depends on the vector potential that describes the field. Substituting Eq.(2-94) into Eq.(2-93) gives the final result (in cgs units)

$$H = \frac{1}{2m}\left[\left(p_x - \frac{e}{c}A_x\right)^2 + \left(p_y - \frac{e}{c}A_y\right)^2 + \left(p_z - \frac{e}{c}A_z\right)^2\right] + e\varphi, \qquad (2\text{-}95)$$

which is the classical Hamiltonian for the interaction of a free particle with charge e and mass m with an electromagnetic field described by the vector potential. For an electron, e must be replaced with $-e$ in Eq.(2-95).

Time Dependence of a Dynamical Variable, G

We can also investigate the time-dependence of any function, G, which is an explicit function of the generalized coordinates, the generalized momenta, and the time:

$$\frac{dG}{dt} = \frac{\partial G}{\partial t} + \sum_j \left(\frac{\partial G}{\partial q_j}\frac{\partial q_j}{\partial t} + \frac{\partial G}{\partial p_j}\frac{\partial p_j}{\partial t}\right) = \frac{\partial G}{\partial t} + \sum_j \left(\frac{\partial G}{\partial q_j}\dot{q}_j + \frac{\partial G}{\partial p_j}\dot{p}_j\right). \qquad (2\text{-}96)$$

From Eqs.(2-82) and (2-83) we can write

$$\frac{dG}{dt} = \frac{\partial G}{\partial t} + \sum_j \left(\frac{\partial G}{\partial q_j}\frac{\partial H}{\partial p_j} - \frac{\partial G}{\partial p_j}\frac{\partial H}{\partial q_j}\right) = \frac{\partial G}{\partial t} + i\mathscr{L}G = \frac{\partial G}{\partial t} + \{G, H\}, \qquad (2\text{-}97)$$

where $\{G, H\}$ is called the *Poisson bracket* of the two dynamical quantities G and H. The $\mathscr{L}$ operator,

$$\mathscr{L} = \frac{1}{i}\sum_j \left(\frac{\partial H}{\partial p_j}\frac{\partial}{\partial q_j} - \frac{\partial H}{\partial q_j}\frac{\partial}{\partial p_j}\right), \qquad (2\text{-}98)$$

is called the *Liouville operator*. If G in Eq.(2-97) is not explicitly dependent on time, the solution to the differential equation is given by

$$G(t) = \exp(i\mathscr{L}t)G(0),$$

where $G(0)$ and $G(t)$ represent the appropriate property of the system at times 0 and t, respectively, and the $\exp(i\mathscr{L}t)$ operator (the propagator) generates $G(t)$ from $G(0)$.

Returning to Eq.(2-97) we note that if the Poisson bracket of G and the Hamiltonian is zero and G is not explicitly dependent on time, the value of G is conserved in time. We shall return to a discussion of Eq.(2-97) in Section 3.6 after we have introduced the concepts of quantum mechanics.

The Virial Theorem

We end our formal discussion with a derivation of the virial theorem, which has been of historical and practical importance in both classical and quantum mechanics. Consider a function G that is the product of linear momenta and coordinates summed over i particles given by $G = \sum_i p_i q_i$. As there is no explicit time-dependence in $\sum_i p_i q_i$, we can use Eq.(2-97) to give

$$\frac{d}{dt}\sum_i p_i q_i = \sum_i (p_i \dot{q}_i + q_i \dot{p}_i). \tag{2-99}$$

We now take the long-time average of both sides of this equation, indicated by angular brackets according to

$$\begin{aligned}\left\langle \frac{d}{dt}\sum_i p_i q_i \right\rangle &= \lim_{T\to\infty}\frac{1}{T}\int_0^T \frac{d}{dt}\left(\sum_i p_i q_i\right)dt = \lim_{T\to\infty}\frac{1}{T}\int_0^T d\left(\sum_i p_i q_i\right)\\ &= \lim_{T\to\infty}\frac{1}{T}\left[\sum_i (p_i q_i)_T - \sum_i (p_i q_i)_0\right].\end{aligned} \tag{2-100}$$

Now, if the trajectories of the particles are bounded, both p_i and q_i will be bounded and, therefore, finite. Thus, the expression above goes to zero as $T \longrightarrow \infty$, giving from Eq.(2-99)

$$\left\langle \sum_i (p_i \dot{q}_i + q_i \dot{p}_i) \right\rangle = 0. \tag{2-101}$$

Rearranging, and remembering that $\sum_i p_i \dot{q}_i = 2T$, gives

$$\langle 2T \rangle = -\left\langle \sum_i q_i \dot{p}_i \right\rangle = -\left\langle \sum_i q_i F_i \right\rangle.$$

We have used Eq.(2-1) in the last step where F_i is the force. In Cartesian coordinates, this equation leads to

$$\langle 2T \rangle = -\left\langle \sum_i (x_i F_{xi} + y_i F_{yi} + z_i F_{zi}) \right\rangle = -\left\langle \sum_i \boldsymbol{r}_i \cdot \boldsymbol{F} \right\rangle. \tag{2-102}$$

Remembering that $\boldsymbol{F} = -\nabla V$ for a conservative force, we can show that Eq.(2-102) reduces further for a potential given by $V = Cr^n$ where C and n are constants. Differentiating gives the force

$$\boldsymbol{F} = -\nabla V = -nCr^{n-1}\left(\frac{\boldsymbol{r}}{r}\right).$$

Substituting into Eq.(2-102) gives

$$\langle 2T \rangle = \left\langle \sum_i \boldsymbol{r}_i \cdot (nCr_i^{n-1})\frac{\boldsymbol{r}_i}{r_i} \right\rangle = \left\langle n \sum_i Cr_i^n \right\rangle = n\langle V \rangle. \tag{2-103}$$

In the case of the single-mode harmonic oscillator considered in Section 2.3, $n = 2$ and $\langle T \rangle = \langle V \rangle$. If the oscillator has a total energy equal to kT, then $\langle T \rangle + \langle V \rangle = kT$ and $\langle T \rangle = \langle V \rangle = kT/2$ from the virial theorem. In the case of an electrostatic potential between two charges placed a distance r apart, $n = -1$ and Eq.(2-103) reduces to

$$\langle 2T \rangle = -\langle V \rangle. \tag{2-104}$$

We shall return to this important result in later chapters.

PROBLEMS

1. Show that the total angular momentum of a many-particle system is conserved in time if the total external torque is zero.

2. Use the concepts of potential and kinetic energies to compute the escape velocities of a molecule or any object from the earth and moon. The potential energy is obtained from the law for gravitational force. Compare these escape velocities with the velocities of molecules such as N_2 and H_2O at thermal equilibrium on the surfaces of the earth and moon and explain why the moon has no atmosphere.

3. Substituting Eq.(2-18) into Eq.(2-1) for the force on an orbiting particle and remembering that $\boldsymbol{\omega}$ is perpendicular to $\boldsymbol{r}$ ($v = \omega r$) gives mv^2/r. A free particle with charge e and mass m in a magnetic field is held into a stable orbit by a centripetal force equal to evH/c from Eq.(1-49). Thus, the angular frequency of the orbiting particle is given by

cgs	SI
$\frac{mv^2}{r} = \frac{e}{c} vH$	$\frac{mv^2}{r} = evB$
$\frac{v}{r} = \omega = \frac{eH}{mc}$	$\frac{v}{r} = \omega = \frac{eB}{m},$ (2-105)

where $\nu = \omega/2\pi$ is the cyclotron frequency for the free particle.

(a) What is the cyclotron frequency of an electron at $H = 1000$ G?
(b) What is the velocity attained by the electron at a radius of 1 mm?
(c) What is the resultant kinetic energy of an electron attained at the 1-mm, 1000-G conditions?
(d) Calculate the cyclotron frequency of a free electron in the earth's ionosphere where the magnetic field is $H \cong 0.5$ G.

The radioastronomy "window" through our atmosphere on earth extends from about 3 MHz (above the cyclotron frequency of the free electron in the ionosphere in the earth's magnetic field) to about 30 GHz (far infrared, 1 cm^{-1}) [6]. One of the major sources of radio energy that passes through the earth's "window" is synchrotron radiation from ionized particles traveling at relativistic velocities. The synchrotron frequency for an electron is obtained from the cyclotron frequency by correcting the mass for the relativistic velocities, $m = m_0/[1 - (v^2/c^2)]^{1/2}$, where m_0 is the electron's rest mass. The resultant synchrotron frequency is

cgs	SI
$\omega_s = \frac{eH}{cm_0}[1 - (v^2/c^2)]^{1/2},$	$\omega_s = \frac{eB}{m_0}[1 - (v^2/c^2)]^{1/2},$ (2-106)

which reduces to the cyclotron frequency in Eq.(2-105) as $v \to 0$.

(e) Ion cyclotron resonance (ICR) involves the absorption of radio frequency energy by an ion in a magnetic field at its cyclotron frequency [7]. What is the resonance frequency of a $^{35}Cl^-$ ion at 10,000 G? What is the corresponding radius of that $^{35}Cl^-$ ion (in an orbit perpendicular to the field) at its thermal linear velocity at 300 K?

(f) Consider now the problem of an orbiting satellite around the earth. The mechanical force, $F = mv^2/r$, is equated to the gravitational force for a stable orbit. Show that the condition for a stable orbit is independent of the mass of the orbiting object. What is the stable radius for a circular orbit for an object traveling with a velocity of 2000 $m \cdot s^{-1}$?

4. Consider Fig. 2-2 and the molecular structure of $CH_2{=}CH^{35}Cl$.

(a) Calculate the moment of inertia tensor with the CH_2 carbon atom as origin in an axis system with the x axis in the molecular plane along the C=C bond and the y axis also in the molecular plane.

(b) Use the method described in Appendix B.1 to diagonalize the **I** tensor in part a.

5. Consider the mechanical force on the HCl diatomic molecule due to rotational motion. Assume the molecule's center of mass is at the Cl nucleus and that the classical rotational frequency is high enough to maintain the thermal rotational kinetic energy of kT.

(a) What is the force on the hydrogen atom at the rotational frequency above?

(b) Using the HCl force constant from Section 2.3, compute the approximate increase in the bond length due to the rotational frequency above.

6. Consider the vibration of the HCl molecule where we assume for simplicity that the Cl atom is stationary and the H atom vibrates with the known HCl force constant.

(a) Compute the force on the hydrogen atom when the HCl internuclear distance is extended to a length that is 0.1 Å longer than the equilibrium length.

(b) What is the value of the electric field (from the Lorentz force) which will lead to the same force on the proton as experienced in the HCl vibrator which is extended 0.1 Å beyond equilibrium as described in part a?

7. Consider the collision of a moving atom with a vibrating and rotating HCl molecule whose c.m. is at rest. Assume the atom has a normal room temperature c.m. velocity of 300 $m \cdot s^{-1}$. The collision begins to take place when the atom is about 10 Å from the molecular center of mass. It is of interest to examine the dynamical behavior of the molecule during the approach of the atom.

(a) How many cycles of vibration take place in the molecule during the time taken for the atom to travel 1 Å?

(b) Assuming translational and rotational thermal equilibrium, how many cycles of rotation take place in the molecule during the time taken for the atom to travel 1 Å?

8. Consider the magnetic deflection of an electron emitting from a hot cathode as shown in Fig. 2-5. The electron has a circular trajectory within the uniform magnetic field of $H = 1000$ G. Neglecting fringe effects at the edge of the magnet, calculate the velocity of the electrons that strike the detector plate at an angle of 5°.

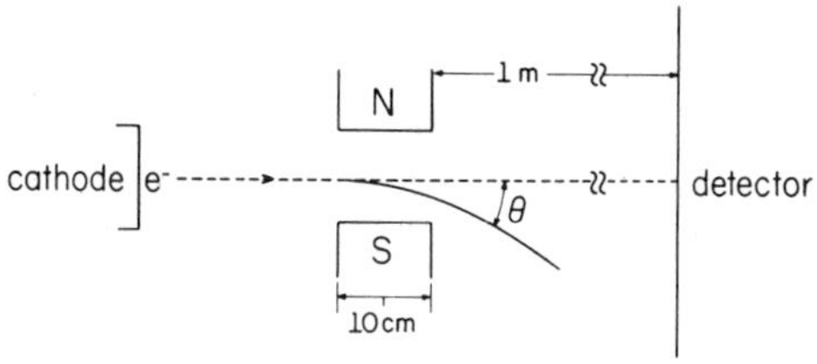

Figure 2-5 Diagram showing the magnetic deflection of an electron.

9. Repeat the derivation from Eq.(2-88) to the result in Eq.(2-95) using SI units.

10. We shall now apply the concept of the refractive index to describe the propagation of electromagnetic radiation in a free collisionless electron gas or plasma of ions. In the case of a collisionless free electron gas, both $\tau \longrightarrow \infty$ and $\omega_0 \longrightarrow 0$ and the dielectric constant in Eq.(2-52) reduces to

$$\lim_{\substack{\tau \to \infty \\ \omega_0 = 0}} \varepsilon = 1 - \frac{\omega_p^2}{\omega^2}, \qquad (2\text{-}107)$$

and the refractive index is

$$n = (\varepsilon)^{1/2} = [1 - (\omega_p^2/\omega^2)]^{1/2}. \qquad (2\text{-}108)$$

If $\omega_p > \omega$, n is pure imaginary leading to an absorption of the radiation [see Eq.(2-55)] and if $\omega > \omega_p$, the refractive index is real and the radiation transmits through the plasma. Thus, frequencies at and below ω_p are not propagated through the ionized plasma region; the plasma acts as a high-pass filter.

(a) Calculate the plasma frequency, ω_p, for electrons in our ionosphere where $\rho_0 \cong 10^4$ electrons · cm^{-3} (typically). How does this plasma-frequency cutoff of radio energy compare to the cyclotron frequency cutoff frequency of electrons in the ionosphere as computed in Problem 2-3?

(b) We can also use this model to estimate the high-frequency cutoff for reflection from metal surfaces [8]. Assuming that the conduction electrons near a metal surface are free, calculate the plasma frequencies for the conduction electrons in Cu and Al. Which metal would be the better reflector for the ultraviolet $\lambda = 1720$ Å radiation in the excited atom xenon laser ($Xe_2^* \rightleftarrows Xe + Xe^*$)?

11. Use the concepts in Section 2.2 to derive an expression for the polarizability of an electron in a molecule.

(a) Show that in the low-frequency limit the polarizability is given by (let $\tau = \infty$)

$$\text{cgs} \qquad\qquad\qquad\qquad \text{SI}$$

$$\lim_{\omega \ll \omega_0} \alpha = \frac{e^2}{m\omega_0^2} = \frac{\hbar^2 e^2}{m(h\nu_0)^2} \qquad \lim_{\omega \ll \omega_0} \alpha = \frac{e^2}{4\pi\varepsilon_0 m\omega_0^2} = \frac{\hbar^2 e^2}{4\pi\varepsilon_0 m(h\nu_0)^2}. \qquad (2\text{-}109)$$

Now, if we estimate $h\nu_0$ in this equation by the ionization energy, E, and sum over all n electrons in an atom, we have the atom polarizability:

$$\text{cgs} \qquad\qquad\qquad\qquad \text{SI}$$

$$\lim_{\omega \ll \omega_0} \alpha = \frac{\hbar^2 e^2}{m} \sum_{i=1}^{n} \frac{1}{[E(i)]^2} \qquad \lim_{\omega \ll \omega_0} \alpha = \frac{\hbar^2 e^2}{4\pi\varepsilon_0 m} \sum_{i=1}^{n} \frac{1}{[E(i)]^2}. \qquad (2\text{-}110)$$

(b) Estimate α for H, He, Ne, and Ar (to 5%) according to Eq.(2-110) by looking up the ionization potentials. Compare your calculated values of α with the experimentally known values.

(c) Show that the high-frequency polarizability of an electron in an atom is given by (let $\tau = \infty$)

$$\lim_{\omega \gg \omega_0} \alpha = -\frac{e^2}{m\omega^2} \text{ (cgs)} = -\frac{e^2}{4\pi\varepsilon_0 m\omega^2} \text{ (SI)}.$$

(d) Compute the high-frequency polarizability of an Ar atom for X rays with $\lambda = 1$ Å.

(e) What is the ratio of X-ray scattering intensities from equal concentrations of electrons and protons?

12. Develop equations for the frequency dependence of the conductivity of the forced damped oscillator from the development in Section 2.2. Show in detail the values of the real and imaginary values of the conductivity and show how these values are related to the corresponding real and imaginary components of the dielectric constant given in Eq.(2-52).

REFERENCES

[1] R. C. Tolman, *The Principles of Statistical Mechanics* (Oxford University Press, Oxford, England, 1938).

[2] H. Goldstein, *Classical Mechanics* (Addison-Wesley Pub. Co., Reading, Mass., 1950).

[3] S. M. Blinder, *Advanced Physical Chemistry* (The Macmillan Co., Toronto, Canada, 1969).

[4] L. D. Landau and E. M. Lifshitz, *Electrodynamics of Continuous Media* (Pergamon Press, New York, 1960).

[5] H. A. LORENTZ, *The Theory of Electrons*, 2nd ed. (Teubner, Leipzig, 1916; reprinted by Dover, New York, 1962); T. HIROSIGE, *Hist. Stud. Phys. Sci.*, Ed. by R. McCormmach, **1**, 151 (University of Pennsylvania Press, Philadelphia, 1969); R. MCCORMMACH, *ibid.* **2**, 41 (University of Pennsylvania Press, Philadelphia, 1969).

[6] J. D. KRAUS, Jr., *Radio Astronomy* (McGraw-Hill Book Co., New York, 1966).

[7] J. D. BALDESCHWEILER and S. S. WOODGATE, *Accts. Chem. Res.* **4**, 114 (1971); G. A. GRAY, *Adv. Chem. Phys.*, Ed. by I. Prigogine and S. A. Rice, **19**, 141 (John Wiley & Sons, Inc., New York, 1971).

[8] H. EHRENREICH, *IEEE Spectrum* **2**, 162 (1965).

3

quantum theory

3.1 EARLY QUANTUM THEORY OF ATOMS

Building on the Rutherford experimental determination of the planetary atom and the Planck–Einstein photon concept, Bohr and others attempted to find a viable theory of atomic structure [1]. This pre-Schrödinger–Heisenberg era of atomic structure from 1910 to 1920 established, among other things, that energy levels could be restricted to discrete values. In the case of the hydrogen-like atom, this meant that the orbiting electron in Rutherford's planetary atom would not emit electromagnetic radiation and cascade into the nucleus; instead, the electron and proton could achieve a stable state of relative motion that did not lose or gain energy. Bohr assumed that the atom gains or loses energy by changing from one discrete state to another [1].†

† Numbers in square brackets correspond to the numbered sources found in the Reference section on p. 118 at the end of the chapter.

According to the discussion in Problem 2-3, the centripetal force on an electron orbiting about a stationary proton, $-mv^2/r$, is equal to the electrostatic force given by

$$\begin{array}{cc} cgs & SI \\ F = -\dfrac{Ze^2}{r^2} & F = -\dfrac{Ze^2}{4\pi\varepsilon_0 r^2}, \end{array} \tag{3-1}$$

where Ze is the nuclear charge. Thus, the radius of the orbiting electron is given by

$$\begin{array}{cc} cgs & SI \\ r = \dfrac{Ze^2}{mv^2} & r = \dfrac{Ze^2}{4\pi\varepsilon_0 mv^2}. \end{array} \tag{3-2}$$

Bohr assumed that the electronic angular momentum in the hydrogen atom was quantized according to (we use L for the electronic angular momentum)

$$L = mrv = n\hbar. \tag{3-3}$$

Rearranging to give

$$mv^2 = \frac{n^2\hbar^2}{r^2 m}, \tag{3-4}$$

and substituting this result into Eq.(3-2) gives

$$\begin{array}{cc} cgs & SI \\ r = \dfrac{n^2\hbar^2}{Ze^2 m} & r = 4\pi\varepsilon_0\left(\dfrac{n^2\hbar^2}{Ze^2 m}\right). \end{array} \tag{3-5}$$

The value of r in Eq.(3-5) for $n = 1$ and $Z = 1$ is called the *Bohr radius.*

$$\begin{array}{cc} cgs & SI \\ r_{n=1} = a_0 = \dfrac{\hbar^2}{e^2 m} & r_{n=1} = a_0 = 4\pi\varepsilon_0\left(\dfrac{\hbar^2}{e^2 m}\right) \\ = 0.52918 \times 10^{-8}\ \text{cm} & = 0.52918 \times 10^{-10}\ \text{m}, \end{array} \tag{3-6}$$

which is the radius of the electron in the lowest state in an atom with $Z = 1$ and an infinitely heavy nucleus (stationary nucleus).

The total energy is a sum of kinetic and potential energies given by

$$\begin{array}{cc} cgs & SI \\ E = \dfrac{mv^2}{2} - \dfrac{Ze^2}{r} & E = \dfrac{mv^2}{2} - \dfrac{Ze^2}{4\pi\varepsilon_0 r}. \end{array}$$

Substituting from Eq.(3-2) gives

$$\text{cgs} \qquad E = \frac{Ze^2}{2r} - \frac{Ze^2}{r} = -\frac{Ze^2}{2r} \qquad\qquad \text{SI} \qquad E = -\frac{Ze^2}{8\pi\varepsilon_0 r}.$$

Substituting r from Eq.(3-5) gives the energy levels of the hydrogen-like atom:

cgs

$$E_n = -\frac{e^4 m Z^2}{2n^2\hbar^2} = -\frac{Z^2e^2}{2a_0}\left(\frac{1}{n^2}\right) = -R_\infty\left(\frac{Z^2}{n^2}\right) = -13.6075\left(\frac{Z^2}{n^2}\right)\text{ eV}$$

SI

$$E_n = -\frac{e^4 m Z^2}{(4\pi\varepsilon_0)^2 2n^2\hbar^2} = -\frac{Z^2e^2}{2a_0^2}\left(\frac{1}{n^2}\right) = -R_\infty\left(\frac{Z^2}{n^2}\right), \qquad (3\text{-}7)$$

where R_∞ is the infinitely heavy nucleus Rydberg constant (see Appendix A.1 for numerical values). Equation (3-7) shows that the energy levels in the hydrogen-like atom depend on the square of the integer $n = 1, 2, 3, \ldots$. In a real hydrogen-like atom the nucleus has a finite mass and all the previous equations must be derived in the c.m. coordinate system. Doing this gives the same result as in Eq.(3-7) except that the electron mass m is replaced by the reduced mass, $\mu = Mm/(M + m)$, where M is the nuclear mass, to give the energy level differences:

$$E_{n_2} - E_{n_1} = -\frac{e^4\mu Z^2}{2\hbar^2}\left(\frac{1}{n_2^2} - \frac{1}{n_1^2}\right)(\text{cgs}) = -\frac{e^4\mu_0 Z^2}{(4\pi\varepsilon_0)^2 2\hbar^2}\left(\frac{1}{n_2^2} - \frac{1}{n_1^2}\right)(\text{SI}). \qquad (3\text{-}8)$$

The Bohr frequency rule relates the energy difference between the discrete states described above to the frequency of the transition:

$$\nu = \frac{E_2 - E_1}{h}. \qquad (3\text{-}9)$$

This important equation combined Bohr's discrete state assumption and the Planck–Einstein frequency expression $h\nu$ for the energy of a photon. Equation (3-8) agrees with the experimental low resolution zero field electronic spectra of the hydrogen-like atom.

The early quantum theory of atomic structure as described above is incomplete, however. These early theories could not give a correct description of the Stark effect (perturbations on the atomic system by an external laboratory electric field), the Zeeman effect (perturbations on the atomic system by an external magnetic field), or the relative intensities of electromagnetic absorption in the hydrogen atom. The early quantum theories also had major difficulties in dealing with many-electron atoms.

The reason for the failure in the Lorentz and Bohr theories of matter involved the extrapolation of classical physical theory into the atomic and molecular domain.

In classical Newtonian physics, an observer can perform his observation of any system without affecting or perturbing that system. It is now clear, however, that a physical scientist cannot observe the atomic or molecular world without perturbing the system he wishes to observe. This confuses the direct cause-effect concept of classical physics. In the microscopic realm we shall see later that a cause leads to a most probable effect rather than to a specific outcome [2].

Consider the observation of the position of an electron with electromagnetic radiation. A well-known principle of image formation is that the resolving power of an instrument is dependent on the wavelength of the radiation and the physical size of the object to be viewed. The wavelength of the radiation must be less than the size of the object to be viewed in order for the reflected light to carry information about the shape of the object. An atom has a radius of about 10^{-10} m. Nuclei, according to the Rutherford scattering experiments, have radii on the order of 10^{-14} to 10^{-15} m, and electrons have a classical radius of $e^2/mc^2 = 2.82 \times 10^{-15}$ m (Chapter 9). Thus, it is clear that both the electron and nucleus are quite small and in order to observe the position of the electron with electromagnetic radiation we must use radiation with wavelengths on the order of 10^{-15} m. Wavelengths on the order of 10^{-15} m are in the X-ray region of the spectrum, however, and according to Compton's analysis in Section 1.5 the photon-electron interaction will impart an appreciable momentum to the electron, thereby changing its position. Thus, we cannot observe the electron without disturbing its position [3].

A quantum mechanical description of nature resolves some of the dilemmas above and leads to a more consistent description of electrons and nuclei.

3.2 WAVE-PARTICLE DUALITY, THE WAVE EQUATION, WAVE PACKETS, AND UNCERTAINTY

Early attempts to obtain a satisfactory description of small particles and their interactions involved the wave theories of matter. Previously we discussed the particle behavior of electromagnetic radiation where the momentum of a photon was given by [see Fig. 1-10 and Eq.(1-79)]

$$p = \frac{h\nu}{c} = \frac{h}{\lambda}. \tag{3-10}$$

De Broglie suggested that this equation also applied to a description of particles [4]. This apparent wave character of particles was verified experimentally by Davisson and Germer [5] who observed electron diffraction by scattering electrons from the surface of a nickel crystal. The observed electron interference pattern was very similar to the X-ray diffraction patterns that were attributed to the wave character of the X rays (see Section 1.3). By a correct description of electron diffraction similar to the X-ray diffraction discussed earlier, Davisson and Germer found that Eq.(3-10) indeed gave an accurate description of the wavelength of the electron that appeared to be diffracted by the crystal. Today, electron diffraction is an important tool for the determination of molecular and surface structure (see Chapter 9). Heavier particles can also be described by Eq.(3-10); for example, neutron diffraction is also a useful tool for the determination of molecular structures and intramolecular interactions [6]. Helium atoms have also been diffracted off crystals.

The Wave Equation

We shall now attempt to write a quantum mechanical wave equation describing the wave properties of a particle by incorporating the features of Eq.(3-10). Consider first the wave equation for the electric field for electromagnetic radiation propagating in a nonconducting ($\boldsymbol{J} = 0$) and neutral ($\bar{\rho} = 0$) medium. Under these conditions we take the curl of Eq.(1-1) and substitute from Eq.(1-2) to give the wave equation for the electric field $[1/(\varepsilon_0 \mu_0)^{1/2} = c]$:

$$\nabla^2 \boldsymbol{E} = \frac{\varepsilon \mu_p}{c^2} \frac{\partial^2 \boldsymbol{E}}{\partial t^2}. \tag{3-11}$$

A complex sinusoidal solution is given by

$$\boldsymbol{E} = \boldsymbol{E}_0 \exp\left[i(\boldsymbol{k} \cdot \boldsymbol{r} - \omega t)\right] \tag{3-12}$$

as shown in Eq.(1-29) where the magnitude of k is defined by the dispersion relation in Eq.(1-26). A similar expression can be obtained for the magnetic field. The corresponding energy density and intensity of electromagnetic radiation are discussed in Eqs.(1-34) and (1-35). The energy densities and intensities are proportional to the squares of the fields.

We shall now define a wavefunction for a particle in a manner similar to the classical expression above for a wave according to

$$\Psi(\boldsymbol{r}, t) = \psi^0 \exp\left[i(\boldsymbol{k} \cdot \boldsymbol{r} - \omega t)\right], \tag{3-13}$$

where ψ^0 is the amplitude. Later (Section 3.6) we shall interpret the probability density of the particle described by the wavefunction $\Psi(\boldsymbol{r}, t)$ as the square of the function, again by analogy to the classical wave interpretation as described above. Remembering that the magnitude of the wave vector, k, is equal to $2\pi/\lambda$, we can return to Eq.(3-10) to write a dispersion relation for the particle:

$$k = \frac{2\pi}{\lambda} = \frac{2\pi}{h} p = \frac{p}{\hbar}. \tag{3-14}$$

The angular frequency, ω, is related to the energy or momentum by the Planck–Einstein relation given by

$$\omega = 2\pi\nu = \frac{E}{\hbar} = \frac{p^2}{2m\hbar}. \tag{3-15}$$

Substituting Eqs.(3-14) and (3-15) into Eq.(3-13) gives

$$\Psi(\boldsymbol{r}, t) = \psi^0 \exp\left[\frac{i}{\hbar}(\boldsymbol{p} \cdot \boldsymbol{r} - Et)\right]. \tag{3-16}$$

We have now introduced the wave character of particles directly into the wavefunction, which implies a different form for the wave equation than the standard form given in Eq.(3-11). Taking the first derivative of $\Psi(\mathbf{r}, t)$ with respect to time and applying the Laplacian [∇^2, see Eq.(1-8)] to $\Psi(\mathbf{r}, t)$ gives

$$\frac{\partial \Psi(\mathbf{r}, t)}{\partial t} = -\frac{iE}{\hbar}\Psi(\mathbf{r}, t) \tag{3-17}$$

$$\nabla^2 \Psi(\mathbf{r}, t) = -\frac{p^2}{\hbar^2}\Psi(\mathbf{r}, t). \tag{3-18}$$

Multiplying Eq.(3-17) by $\hbar/i$, multiplying Eq.(3-18) by $\hbar^2/2m$, and subtracting the resultant equations gives

$$\frac{\hbar}{i}\frac{\partial \Psi(\mathbf{r}, t)}{\partial t} - \frac{\hbar^2 \nabla^2}{2m}\Psi(\mathbf{r}, t) = \left(-E + \frac{p^2}{2m}\right)\Psi(\mathbf{r}, t) = 0, \tag{3-19}$$

where the last step follows from $E = p^2/2m$ for a free particle giving

$$i\hbar \frac{\partial \Psi(\mathbf{r}, t)}{\partial t} = -\frac{\hbar^2 \nabla^2}{2m}\Psi(\mathbf{r}, t). \tag{3-20}$$

Equation (3-20) is a wave equation for a free particle with Eq.(3-16) as the solution. We note from Eq.(3-18) that the classical kinetic energy, $T = p^2/2m$, is replaced by an operator, $-\hbar^2\nabla^2/2m$, in Eq.(3-20). In the presence of forces, Eq.(3-20) is replaced by

$$i\hbar \frac{\partial \Psi(\mathbf{r}, t)}{\partial t} = \left(-\frac{\hbar^2 \nabla^2}{2m} + V\right)\Psi(\mathbf{r}, t) = \mathscr{H}\Psi(\mathbf{r}, t), \tag{3-21}$$

where V is the potential energy arising from the forces and $\mathscr{H}$ is the Hamiltonian operator. Equation (3-21) was first derived by Schrödinger [7] and is called the *Schrödinger equation.* The form of the wave equation is considerably different from the corresponding classical equation in Eq. (3-11).

Superposition and Wave Packets

The wave function in Eq.(3-16), $\Psi_i(\mathbf{r}, t)$, describes a state for the particle with momentum, $\mathbf{p}$, and position, $\mathbf{r}$, at time t. It is also evident that a linear combination of $\Psi_i(\mathbf{r}, t) = \Psi_i$ defines a new state of the system:

$$\Psi = C_1\Psi_1 + C_2\Psi_2,$$

as Ψ satisfies Eq.(3-20) for all C_1 and C_2. The momentum of the particle in state Ψ above is no longer obvious, however, as the value must lie somewhere between $\mathbf{p}_1$ (from Ψ_1) and $\mathbf{p}_2$ (from Ψ_1). This principle of *superposition of states* is inherent to quantum mechanics and arises from the linear nature of Schrödinger's equation.

We shall now examine some features of a *wave packet* by taking a superposition of momentum states, or linear combination of k states, for the free particle traveling along the z axis: From Eq.(3-16) where $p = k\hbar$ we write

$$\Psi(z, t) = \sum_j f_j \exp\left[i\left(k_j z - \frac{E_j}{\hbar}t\right)\right], \tag{3-22}$$

where

$$E_j = \frac{p_j^2}{2m} = \frac{k_j^2\hbar^2}{2m}. \tag{3-23}$$

If the momentum and energy values are continuous, we can replace the summation by an integral over k giving

$$\Psi(z, t) = \int_{-\infty}^{+\infty} f(k) \exp\left[i\left(kz - \frac{E}{\hbar}t\right)\right] dk = \int_{-\infty}^{+\infty} f(k) \exp\left[i\left(kz - \frac{k^2\hbar t}{2m}\right)\right] dk. \tag{3-24}$$

Using the integral form for $\Psi(z, t)$ in Eq.(3-24), we shall now develop a very simple k-pulse model for the wave packet. In this k-pulse model, we assume that $f(k) = f_0$, a constant, from $-\Delta k$ to $+\Delta k$ centered at k_0 and $f(k) = 0$ at all other values of k. This is similar to assigning the particle a velocity, $v_0 (v_0 = p_0/m = \hbar k_0/m)$ with an uncertainty in velocity given by $2\hbar\Delta k/m$. We shall examine the wavefunction at $t = 0$ using $\Psi(z, t = 0)$ in Eq.(3-24) and the k-pulse, which gives

$$\begin{aligned}\Psi(z, 0) &= f_0 \int_{k_0 - \Delta k}^{k_0 + \Delta k} \exp(ikz)\, dk = f_0 \int_{k_0 - \Delta k}^{k_0 + \Delta k} (\cos kz + i \sin kz)\, dk \\ &= \frac{2f_0}{z} \cos k_0 z \sin \Delta kz + \frac{2if_0}{z} \sin k_0 z \sin \Delta kz.\end{aligned} \tag{3-25}$$

If we assume that $\Delta k < k_0$, we note that the real part of $\Psi(z, 0)$ is a sinusoidal function in k_0, which is amplitude-modulated by the slower varying sinusoidal function in Δk_0. The amplitude modulation drops off to zero outside the $\Delta z = \pm\pi/\Delta k$ boundaries with the $1/z$ factor in Eq.(3-25). Thus, $\Psi(z, 0)$ is localized with the major amplitude lying between the boundaries defined by the amplitude modulation in Δk. The probability per unit length for the existence of the particle along the z axis is given by (see Section 3.6)

$$\Psi(z, 0)\Psi^*(z, 0) = \frac{4f_0^2}{z^2} \sin^2 \Delta kz = 4f_0^2(\Delta k)^2 \frac{\sin^2 \Delta kz}{(\Delta kz)^2}.$$

This type of curve is also examined in Section 3.7 (see Fig. 3-1). At 0.405 times the maximum, $\Delta k \Delta z = 1.0$, where Δz is the full width along the z axis. Thus, the full width, Δz, at the approximate half height of $\Psi(z, 0)\Psi^*(z, 0)$ along the z axis satisfies

$$\Delta z \cong \frac{1}{\Delta k} = \frac{\hbar}{\Delta p}, \qquad \Delta z \Delta p \cong \hbar. \tag{3-26}$$

Thus, the superposition of states and the form of the wave equation in Eq.(3-22) leads to the uncertainty relation between the position and momentum of the particle. If the momentum of a particle is known with an uncertainty of Δp, the position of the particle can be known with the uncertainty no less than $\hbar/\Delta p$. If the momentum or velocity of the free particle is defined very precisely, then its position has a large uncertainty. If the position of the free particle is well defined, then the velocity (momentum) will be very uncertain. It should be evident that the uncertainty relation between

Δz and Δp arises from the nature of the quantum mechanical wave equation [see Section 3.6 for further interpretation of $\Psi(\mathbf{r}, t)$ and $\Psi(\mathbf{r}, t)\Psi^*(\mathbf{r}, t)$].

We shall sharpen up our definition of the uncertainty in Section 9.2 as we examine in more detail the nature of a Gaussian wave packet; its time dependence and its transmission and reflection characteristics through a barrier and over a well.

As a final point, we note that $\Psi(z, t)$ and $f(k)$ in Eq.(3-24) are Fourier transform pairs. According to Eqs.(B-44) and (B-45) in the Appendix, $f(k)$ in Eq.(3-24) is given by (we set $t = 0$ for this illustration)

$$f(k) = \frac{1}{2\pi} \int_{-\infty}^{+\infty} \Psi(z, 0) \exp(-ikz)\, dk, \tag{3-27}$$

which is called the momentum representation for the state of the system that is described in the coordinate representation by $\Psi(z, 0)$.

Before attempting to interpret the wave function any further, we shall examine some features of operators, functions, and matrix representations.

3.3 THE POSTULATES OF QUANTUM MECHANICS

In this section we shall develop the axiomatic basis of quantum mechanics and describe the convenient use of the matrix formalism to solve problems. We begin by examining the features of linear operators and functions. Operators play a key role in quantum mechanics as they are the link between dynamical variables in macroscopic classical mechanics and the observables in microscopic atomic and molecular systems.

A linear operator A is defined by its operation on two functions $\Phi_1(x)$ and $\Phi_2(x)$ according to (we use the coordinate representation for the functions throughout this section)

$$A[\Phi_1(x) + \Phi_2(x)] = A\Phi_1(x) + A\Phi_2(x).$$

For instance, let $A = d/dx$, $\Phi_1(x) = \sin x$, and $\Phi_2(x) = \cos x$, which gives

$$\frac{d}{dx}[\sin x + \cos x] = \frac{d}{dx}\sin x + \frac{d}{dx}\cos x = \cos x - \sin x.$$

The square, for instance, is not a linear operator. Let A be the square operator, giving

$$A(\Phi_1 + \Phi_2) = (\Phi_1 + \Phi_2)^2 = \Phi_1^2 + \Phi_2^2 + 2\Phi_1\Phi_2.$$

Two operators A and B are said to commute if

$$AB\Phi_1(x) = BA\Phi_1(x). \tag{3-28}$$

It is evident that x and d/dx do not commute from

$$x\frac{d}{dx}\Phi_1(x) = x\frac{d\Phi_1(x)}{dx}, \qquad \frac{d}{dx}x\Phi_1(x) = \Phi_1(x) + x\frac{d\Phi_1(x)}{dx},$$

where the $(d/dx)x\Phi(x)$ differentiation is done by parts. Rewriting Eq.(3-28) gives

$$AB\Phi_1(x) - BA\Phi_1(x) = 0, \qquad (AB - BA)\Phi_1(x) = 0$$

$$[A, B]\Phi_1(x) = 0, \qquad [A, B] = AB - BA,$$

where $[A, B]$ is called the *commutator* for the two operators A and B. Of course, an operator does commute with itself and we can write $AAA\Phi_1(x) = A^3\Phi_1(x)$. Thus, any operator will commute with any power of itself according to $A^mA^n = A^nA^m$.

If A is an operator and $A\Phi_j(x)$ is equal to some constant times $\Phi_j(x)$, then $\Phi_j(x)$ is called an *eigenfunction* of A with *eigenvalue* a_j given according to

$$A\Phi_j(x) = a_j\Phi_j(x). \tag{3-29}$$

$\sin xt$ is an eigenfunction of the d^2/dx^2 operator with eigenvalue $-t^2$: $d^2(\sin xt)/dx^2 = -t^2 \sin xt$. Most problems in quantum mechanics involve solving eigenvalue equations where additional boundary conditions are imposed on the function by the nature of the physical system. One important restriction that will become evident later for bound state systems is that the functions should be square integrable and therefore normalizable,

$$\int_{-\infty}^{+\infty} \Phi_j^*(x)\Phi_j(x)\, dx = 1.0, \tag{3-30}$$

where we use the absolute value square as the functions may be complex; $\Phi_j^*(x)$ is the *complex conjugate* of $\Phi_j(x)$ taken by changing all $i = \sqrt{-1}$ in $\Phi_j(x)$ to $-i$. In addition, $\Phi_j(x)$ must be *continuous* and *single-valued*. If Φ_j is a function of an angle α, then a continuous function satisfies $\Phi_j(\alpha) = \Phi_j(\alpha + 2\pi)$. If a function $\Phi_j(\boldsymbol{r})$ describes the state of a particle with coordinates $\boldsymbol{r}$ and if the same state of the system is described by $\Phi_j(\boldsymbol{r}')$ with coordinates $\boldsymbol{r}'$, then a single-valued function requires that $\Phi_j(\boldsymbol{r}) = \Phi_j(\boldsymbol{r}')$. Equation (3-30) also indicates that Φ_j can only be determined to within a phase factor whose absolute value squared is unity such as $\exp(i\alpha)$ where α is some undetermined number.

We also limit most of our discussion in this section to functions, Φ_j, which yield discrete (as opposed to continuous) eigenvalues. We shall now discuss briefly functions that have continuous eigenvalues. The plane wave in Eq.(3-16), as described in Section 3.2, is an example of a function that has continuous eigenvalues and, therefore, Eq.(3-30) will not lead to a proper normalization. The $t = 0$ form of Eq.(3-16) for propagation down the z axis is given by

$$\Phi_k(z) = \Phi^0 \exp(ikz),$$

where Φ^0 is an amplitude or normalization factor. A special kind of normalization can be introduced for these types of functions. Due to the continuous values of the momentum, the superposition state is given in Eq.(3-24) as an integral instead of a summation:

$$\psi(z) = \int_{-\infty}^{+\infty} f(k)\Phi_k(z)\, dk = \Phi^0 \int_{-\infty}^{+\infty} f(k) \exp(ikz)\, dk.$$

Multiplying $\psi(z)$ by $\Phi_{k'}^*(z)$ and integrating over z gives

$$\begin{aligned}\int_{-\infty}^{+\infty} \Phi_{k'}^*(z)\psi(z)\,dz &= \int_{-\infty}^{+\infty}\int_{-\infty}^{+\infty} f(k)\Phi_k(z)\Phi_{k'}^*(z)\,dk\,dz \\ &= (\Phi^0)^2 \int_{-\infty}^{+\infty}\int_{-\infty}^{+\infty} f(k)\exp(ikz)\exp(-ik'z)\,dk\,dz \\ &= 2\pi(\Phi^0)^2 \int_{-\infty}^{+\infty} f(k)\delta(k-k')\,dk = 2\pi(\Phi^0)^2 f(k'),\end{aligned}$$

where the delta function from Eq.(B-67) has been used; it requires that

$$\int_{-\infty}^{+\infty} \Phi_{k'}^*(z)\Phi_k(z)\,dz = 2\pi(\Phi^0)^2\delta(k-k').$$

Thus, $\Phi_k(z)$ is orthogonal to within a delta function restriction in k and the normalization constant is given within that restriction by $\Phi^0 = (2\pi)^{-1/2}$.

We now return to our discussion of functions that have discrete eigenvalues to examine the eigenfunctions of Hermitian operators.

Hermitian Operators

An operator A is Hermitian if

turnover rule (see Pilar, pp 71-73)

$$\int \Phi_i^* A\Phi_j\,dV = \int \Phi_j A^*\Phi_i^*\,dV = \int (A^*\Phi_i^*)\Phi_j\,dV, \tag{3-31}$$

matrix definition (see Schiff p. 124)

where the brackets indicate that the A operation is contained within the brackets and $\int dV$ indicates the integral over the space defined by the functions Φ_i and Φ_j. Hermitian operators play a key role in quantum mechanics because they form the basis of the first postulate of quantum mechanics:

POSTULATE I The position and momentum variables in classical mechanics are replaced with Hermitian operators in quantum mechanics.

The correspondence is shown in Table 3-1.

Table 3-1 Classical dynamical variables and the corresponding quantum mechanical operators (Cartesian coordinate values are given here).

Classical Quantity	Quantum Mechanical Form of Operator
coordinate; x, y, z	x, y, z
linear momentum; p_x, p_y, p_z	$-i\hbar\frac{\partial}{\partial x}, -i\hbar\frac{\partial}{\partial y}, -i\hbar\frac{\partial}{\partial z}$
energy; E	$\mathscr{H}$

As the Hamiltonian is an explicit function of coordinates and momentum, it represents the energy operator. For instance, the correspondence is normally written as

$$H = T + V = \sum_j \frac{1}{2m_j} p_j^2 + V(r_1, r_2, \ldots) \longrightarrow \mathscr{H} = -\hbar^2 \sum_j \frac{1}{m_j} \nabla_j^2 + V(r_1, r_2, \ldots), \tag{3-32}$$

where the linear momentum operators from Table 3-1 are used and r_j represents the coordinate of the jth particle. Operators for other observable properties can be constructed from the Hermitian operators in Table 3-1.

Hermitian operators have three important features:

1. Their eigenvalues are real.
2. The eigenfunctions form an orthogonal set.
3. The eigenfunctions form a complete set.

Let Φ_j be an eigenfunction of a Hermitian operator A with eigenvalue a_j as shown in Eq.(3-29). Multiplying by Φ_j^*, integrating over the space defined by Φ_j, and using Eq.(3-30) gives

$$\int \Phi_j^* A \Phi_j \, dV = \int \Phi_j^* a_j \Phi_j \, dV = a_j \int \Phi_j^* \Phi_j \, dV = a_j. \tag{3-33}$$

Now, taking the complex conjugate of Eq.(3-29) gives

$$A^* \Phi_j^* = a_j^* \Phi_j^*.$$

Multiplying by Φ_j and integrating again gives

$$\int \Phi_j A^* \Phi_j^* \, dV = a_j^*. \tag{3-34}$$

Comparing the left-hand sides of Eqs.(3-33) and (3-34) with Eq.(3-31) for $i = j$ shows that $a_j = a_j^*$ or that the eigenvalues of Hermitian operators are real.

We shall now show that the eigenfunctions of Hermitian operators form an *orthogonal set*. Returning to Eq.(3-31), we can write

$$\int \Phi_i^* A \Phi_j \, dV = a_j \int \Phi_i^* \Phi_j \, dV = \int (A^* \Phi_i^*) \Phi_j \, dV = a_i \int \Phi_i^* \Phi_j \, dV$$

$$(a_j - a_i) \int \Phi_i^* \Phi_j \, dV = 0.$$

According to this result, if $a_j - a_i \neq 0$, then $\int \Phi_i^* \Phi_j \, dV = 0$. Thus, the eigenfunctions of the Hermitian operator A are orthogonal if the eigenvalues are unequal. If $a_j - a_i = 0$, the eigenfunctions give *degenerate* eigenvalues. If $a_j = a_i$, however, then any linear combination of Φ_j and Φ_i is also an eigenfunction of A as shown by (b is a numerical constant)

$$A(\Phi_i + b\Phi_j) = a_i \Phi_i + a_j b \Phi_j = a_i(\Phi_i + b\Phi_j) = a_j(\Phi_i + b\Phi_j). \tag{3-35}$$

As long as any linear combination of Φ_i and Φ_j is also an eigenfunction of A, we can form an orthogonal set given by

$$\Phi_m = \Phi_i, \qquad \Phi_n = \Phi_i + b\Phi_j. \tag{3-36}$$

Now, Φ_m is orthogonal to Φ_n if the constant b is adjusted accordingly (assume Φ_i and Φ_j are normalized):

$$\int \Phi_m^* \Phi_n \, dV = 0 = \int \Phi_i^*(\Phi_i + b\Phi_j)\, dV = 1 + b \int \Phi_i^* \Phi_j \, dV$$

$$b = -\frac{1}{\int \Phi_i^* \Phi_j \, dV}. \tag{3-37}$$

The technique of orthogonalization above is discussed in detail in Appendix D.2 and the results given here can be easily extended to higher degrees of degeneracy. In summary, the eigenfunctions of Hermitian operators are or can be made orthonormal.

The eigenfunctions of Hermitian operators are also a *complete set of functions.* A linear combination of a complete set of functions $\Phi_1, \Phi_2, \ldots,$ can be constructed to give another function ψ, which satisfies the same restrictions as the Φ set. ψ is given by

$$\psi = \sum_{j=1}^{\infty} \mathrm{b}_j \Phi_j, \tag{3-38}$$

where b_j are constant coefficients. We shall not prove here that the eigenfunctions of Hermitian operators form a complete set as this lengthy proof is given elsewhere [8]. Requiring ψ to be normalized, however, leads to

$$\int \psi^* \psi \, dV = 1.0 = \sum_{i,j} \mathrm{b}_i^* \mathrm{b}_j \int \Phi_i^* \Phi_j \, dV. \tag{3-39}$$

Now, as the functions Φ_i form a complete orthonormal set, we can write

$$\int \Phi_i^* \Phi_j \, dV = \delta_{ij} \begin{cases} = 1 & \text{if } i = j \\ = 0 & \text{if } i \neq j \end{cases}, \tag{3-40}$$

where δ_{ij} is called the *Kronecker delta.* Substituting into Eq.(3-39) gives

$$\int \psi^* \psi \, dV = \sum_{i,j} \mathrm{b}_i^* \mathrm{b}_j \delta_{ij} = \sum_i |\mathrm{b}_i|^2, \tag{3-41}$$

which is sometimes referred to as the Parseval relation. Any of the coefficients b_i in Eq.(3-38) can be obtained by multiplying both sides of Eq.(3-38) by Φ_i^* and integrating to give

$$\int \Phi_i^* \psi \, dV = \sum_j \mathrm{b}_j \int \Phi_i^* \Phi_j \, dV = \sum_j \mathrm{b}_j \delta_{ij} = \mathrm{b}_i. \tag{3-42}$$

There is a striking analogy between the development of orthonormal functions given here with theories of linear vector space and indeed it is common to develop quantum mechanics around the existing vector space theories [9,10,11,12]. We shall draw often upon this strong analogy in this text and we now focus attention to the common features. Referring to the expansion in Eq.(3-38), the complete set of orthonormal functions, Φ_i, are analogous to a set of unit vectors in an infinite dimension vector space. Φ_i are pairwise orthogonal and normalized to unit length. The function ψ is analogous to a vector in the infinite dimension space with the coefficients $b_j = \int \Phi_j^* \psi \, dV$ being the components along the unit base vectors (denoted by Φ_j). We write the components of the vector ψ as a column matrix denoted by

b is a vector that represents ψ; ψ is not a vector here.

$$\mathbf{b} = \begin{pmatrix} b_1 \\ b_2 \\ b_3 \\ \vdots \end{pmatrix}. \tag{3-43}$$

Thus, the vector ψ, which is given by a product of the unit vectors Φ_i with the components b_i, is written as

$$\psi = \sum_{j=1}^{\infty} b_j \Phi_j = (\Phi_1 \Phi_2 \Phi_3 \cdots) \begin{pmatrix} b_1 \\ b_2 \\ b_3 \\ \vdots \end{pmatrix}, \tag{3-44}$$

which is a row matrix containing the unit vectors Φ_i times the column matrix containing the vector components b_i. We use the conventional row-column multiplication order when multiplying the two matrices together [13]. We also note that the basis set Φ_i or the unit vectors, by analogy to vector space, are written as row matrices;

$$\mathbf{\Phi} = (\Phi_1 \Phi_2 \Phi_3 \cdots). \tag{3-45}$$

We now note that ψ in Eq.(3-44) is only one of many possible linear combinations of $\mathbf{\Phi}$. Thus, we denote the kth function ψ by ψ_k and rewrite Eq.(3-44) to give

$$\psi_k = \mathbf{\Phi b}_k = (\Phi_1 \Phi_2 \cdots) \begin{pmatrix} b_{1k} \\ b_{2k} \\ b_{3k} \\ \vdots \end{pmatrix} = \sum_j \Phi_j b_{jk}, \tag{3-46}$$

where $\mathbf{b}_k$ is a column matrix representing the kth set of vector components b_{1k}, $b_{2k}, b_{3k}, \ldots$. We then write the set of unit vectors ψ_k by analogy to the unit vectors Φ_j according to

$$(\psi_1 \psi_2 \psi_3 \cdots) = (\Phi_1 \Phi_2 \Phi_3 \cdots) \begin{pmatrix} b_{11} & b_{12} & b_{13} & \cdots \\ b_{21} & b_{22} & b_{23} & \\ b_{31} & b_{32} & b_{33} & \\ \vdots & & & \ddots \end{pmatrix} \tag{3-47}$$

$$\boldsymbol{\psi} = \mathbf{\Phi b},$$

ψ is a vector here!

where $\mathbf{b}$ is now a matrix with the number of rows keyed to the dimension of $\mathbf{\Phi}$ and the number of columns keyed to the dimension of $\boldsymbol{\psi}$. Of course, in order for Eq.(3-47) to be valid, the basis sets must have dimensions of infinity. Thus, we have defined our convention in relating basis functions and vectors with their corresponding matrices. The basis sets (analogous to unit vectors in vector space) are written as row matrices and the expansion coefficients (analogous to the vector components) are written as column matrices [13].

Further analogy between vectors and the eigenfunctions of Hermitian operators is seen by noting that the integral in Eq.(3-40) is analogous to a scalar vector product between unit vectors. Starting with $\boldsymbol{\psi}$ in Eq.(3-47) as a row matrix with elements given by the unit vectors, $\psi_1, \psi_2, \ldots$, we define the *transpose* of the row matrix as

$$\tilde{\boldsymbol{\psi}} = \begin{pmatrix} \psi_1 \\ \psi_2 \\ \psi_3 \\ \vdots \end{pmatrix}. \tag{3-48}$$

The transpose interchanges rows and columns and, therefore, converts the row to a column. Taking the complex conjugate of each element in $\tilde{\psi}$ gives the *adjoint matrix*,

$$\tilde{\boldsymbol{\psi}}^* = \boldsymbol{\psi}^\dagger = \begin{pmatrix} \psi_1^* \\ \psi_2^* \\ \psi_3^* \\ \vdots \end{pmatrix}. \tag{3-49}$$

Thus, the scalar product of all unit vectors in $\boldsymbol{\psi}$ is given by

$$\int \boldsymbol{\psi}^\dagger \boldsymbol{\psi}\, dV = \begin{pmatrix} \int \psi_1^* \psi_1\, dV & \int \psi_1^* \psi_2\, dV & \cdots \\ \int \psi_2^* \psi_1\, dV & \int \psi_2^* \psi_2\, dV & \\ \vdots & & \ddots \end{pmatrix} = \mathbf{1}, \tag{3-50}$$

where $\mathbf{1}$ is the unit matrix with elements equal to the Kronecker delta [Eq.(3-40)] with unit on the diagonal and zero on the off-diagonal. Now from Eq.(3-47) we write the adjoint of $\boldsymbol{\psi}$ to give

$$\boldsymbol{\psi}^\dagger = \mathbf{b}^\dagger \boldsymbol{\Phi}^\dagger, \tag{3-51}$$

where we note that the transpose of the product of two matrices is given by the product in reverse order of the transposed individual matrices [13]. Substituting Eq.(3-51) into Eq.(3-50) gives

$$\mathbf{1} = \mathbf{b}^\dagger \left(\int \boldsymbol{\Phi}^\dagger \boldsymbol{\Phi}\, dV \right) \mathbf{b} = \mathbf{b}^\dagger \mathbf{1} \mathbf{b} = \mathbf{b}^\dagger \mathbf{b}, \tag{3-52}$$

where we note that $\mathbf{b}^\dagger \mathbf{b}$ gives a unit matrix. In this special case $\mathbf{b}$ is called a *unitary matrix*, and the transformation $\mathbf{b}$ between two orthonormal basis sets [as in Eq.(3-47)] is called a *unitary tranformation*.

We refer to $\int \Phi_i^* A \Phi_j \, dV = \mathrm{A}_{ij}^{\Phi}$ as a matrix element of the operator A in the $\mathbf{\Phi}$ basis. Thus, there exists a *matrix representation* of A defined by the matrix elements in $\mathbf{A}^{\Phi}$ according to

$$\mathbf{A}^{\Phi} = \begin{pmatrix} \mathrm{A}_{11}^{\Phi} & \mathrm{A}_{12}^{\Phi} & \mathrm{A}_{13}^{\Phi} & \cdots \\ \mathrm{A}_{21}^{\Phi} & \mathrm{A}_{22}^{\Phi} & \mathrm{A}_{23}^{\Phi} & \\ \mathrm{A}_{31}^{\Phi} & \mathrm{A}_{32}^{\Phi} & \mathrm{A}_{33}^{\Phi} & \\ \vdots & & & \ddots \end{pmatrix} = \int \mathbf{\Phi}^{\dagger} A \mathbf{\Phi} \, dV = \begin{pmatrix} \int \Phi_1^* A \Phi_1 \, dV & \int \Phi_1^* A \Phi_2 \, dV & \cdots \\ \int \Phi_2^* A \Phi_1 \, dV & \int \Phi_2^* A \Phi_2 \, dV & \\ \vdots & & \ddots \end{pmatrix}, \tag{3-53}$$

where $\mathbf{\Phi}^{\dagger}$ and $\mathbf{\Phi}$ are the column and row forms given in Eqs.(3-45) and (3-49), for instance. Now, according to Eq.(3-31), if A is a Hermitian operator, the $\mathbf{A}$ matrix is equal to its own adjoint:

$$\mathbf{A} = \mathbf{A}^{\dagger}. \tag{3-54}$$

We can also write the matrix representation of the operator A in the $\boldsymbol{\psi}$ basis according to

$$\mathbf{A}^{\psi} = \int \boldsymbol{\psi}^{\dagger} A \boldsymbol{\psi} \, dV. \tag{3-55}$$

The relation between the matrix representation of A in the two different basis sets is obtained by substituting $\boldsymbol{\psi} = \mathbf{\Phi b}$ and $\boldsymbol{\psi}^{\dagger} = \mathbf{b}^{\dagger}\mathbf{\Phi}^{\dagger}$ from Eqs.(3-47) and (3-51) into Eq.(3-55) to give

$$\mathbf{A}^{\psi} = \int \boldsymbol{\psi}^{\dagger} A \boldsymbol{\psi} \, dV = \mathbf{b}^{\dagger}\left(\int \mathbf{\Phi}^{\dagger} A \mathbf{\Phi} \, dV \right)\mathbf{b} = \mathbf{b}^{\dagger}\mathbf{A}^{\Phi}\mathbf{b}. \tag{3-56}$$

Thus, the unitary transformation $\mathbf{b}^{\dagger}$ transforms the matrix representation of A in the $\mathbf{\Phi}$ basis into the matrix representation of A in the $\boldsymbol{\psi}$ basis. Postmultiplying Eq.(3-56) by $\mathbf{b}^{\dagger}$, premultiplying by $\mathbf{b}$, and using Eq.(3-52) gives

$$\mathbf{A}^{\Phi} = \mathbf{b}\mathbf{A}^{\psi}\mathbf{b}^{\dagger}. \tag{3-57}$$

We note in Problem 3-14 that the sum of diagonal elements (called the *trace* of the matrix) is invariant to a unitary transformation. If $\boldsymbol{\psi}$ contains elements that are all eigenfunctions of the A operator, then $\mathbf{A}^{\psi}$ is a diagonal matrix with the eigenvalues along the diagonal.

We shall now show that commuting operators can share, simultaneously, the same eigenfunctions. Consider two commuting operators A and B and their matrix representation in the $\mathbf{\Phi}$ basis:

$$AB = BA \qquad \mathbf{A}^{\Phi}\mathbf{B}^{\Phi} = \mathbf{B}^{\Phi}\mathbf{A}^{\Phi}.$$

The ijth matrix elements in $\mathbf{A}^{\Phi}\mathbf{B}^{\Phi}$ must equal the corresponding ijth matrix element in $\mathbf{B}^{\Phi}\mathbf{A}^{\Phi}$ giving (dropping the Φ superscript)

$$\sum_k \mathrm{A}_{ik}\mathrm{B}_{kj} = \sum_k \mathrm{B}_{ik}\mathrm{A}_{kj}.$$

Now, if $\mathbf{A}^{\Phi}$ is diagonal ($\mathbf{\Phi}$ contains the eigenfunctions for A), $\mathrm{A}_{ik} = \mathrm{A}_{ik}\delta_{ik}$ giving

$$\mathrm{A}_{ii}\mathrm{B}_{ij} = \mathrm{B}_{ij}\mathrm{A}_{jj} \qquad \mathrm{B}_{ij}(\mathrm{A}_{ii} - \mathrm{A}_{jj}) = 0.$$

Thus, if $\mathrm{A}_{ii} \neq \mathrm{A}_{jj}$, $\mathrm{B}_{ij} = 0$. Thus, there are no off-diagonal elements in $\mathbf{B}^{\Phi}$ between functions Φ_i and Φ_j that have different eigenvalues. Now, if $\mathrm{A}_{ii} = \mathrm{A}_{jj}$, any linear combination of Φ_i and Φ_j is also an eigenfunction of A [see Eq.(3-35) and associated discussion]. Thus, linear combinations of Φ_i and Φ_j can be found that will also be eigenfunctions of B. We prove this by writing equal dimension subblocks of the $\mathbf{B}^{\Phi}$ and $\mathbf{A}^{\Phi}$ matrices, call them $\mathbf{B}'$ and $\mathbf{A}'$, which are the subblocks associated with the degeneracies in the $\mathbf{A}^{\Phi}$ matrix discussed above. Now, all elements in the diagonal $\mathbf{A}'$ matrix are equal and any unitary transformation on $\mathbf{A}'$ still gives $\mathbf{A}'$, $\mathbf{U}^{\dagger}\mathbf{A}'\mathbf{U} = \mathbf{A}'$ (see Problem 3-12). Thus, we can require $\mathbf{U}$ to diagonalize $\mathbf{B}'$, $\mathbf{U}^{\dagger}\mathbf{B}'\mathbf{U}^{\dagger} = \mathbf{B}'_{\lambda}$ without changing $\mathbf{A}'$. Thus, we can always find simultaneous eigenfunctions of two commuting operators. We now return to our discussion of Hermitian operators and their matrix representations to summarize the important properties: *Eigenfunctions of Hermitian operators can be normalized and made to form a complete orthonormal set of functions. The analogy between these eigenfunctions and vectors in an infinite dimension vector space will be of use to us in quantum mechanics.*

We now need to relate the quantum mechanical Hermitian operators to observables through a description of the system by a set of functions $\Phi_1, \Phi_2, \ldots$. First we return to Schrödinger's equation in Eq.(3-21), however, and state a second postulate.

POSTULATE II The wave function, $\Psi(\mathbf{r}, t)$, describing the state of a system is obtained from a solution of Schrödinger's equation given by

$$i\hbar \frac{\partial \Psi(\mathbf{r}, t)}{\partial t} = \mathscr{H}\Psi(\mathbf{r}, t), \tag{3-58}$$

where $\mathscr{H}$ is the Hamiltonian operator. If the Hamiltonian operator is time-independent, $\Psi(\mathbf{r}, t)$ is separable into time-independent and time-dependent parts:

$$\Psi(\mathbf{r}, t) = f(t)\psi(\mathbf{r}), \tag{3-59}$$

where we use the capital Ψ to include both the time and spatial dependence and the lowercase ψ to denote only the spatial dependence. Substituting Eq.(3-59) into Eq.(3-58) and dividing the result by $\Psi(\mathbf{r}, t) = f(t)\psi(\mathbf{r})$ gives

$$\frac{1}{\psi(\mathbf{r})}\mathscr{H}(\mathbf{r})\psi(\mathbf{r}) = \frac{i\hbar}{f(t)}\frac{\partial}{\partial t}f(t).$$

Both sides of this equation are equal for all $\mathbf{r}$ and t if both sides are equal to a constant energy, E, giving

$$\frac{df(t)}{dt} = -\frac{iE}{\hbar}f(t)$$

with solution

$$f(t) = \exp\left(-\frac{iE}{\hbar}t\right). \tag{3-60}$$

The remaining time-independent Schrödinger equation is given by

$$\mathscr{H}(\mathbf{r})\psi(\mathbf{r}) = E\psi(\mathbf{r}). \tag{3-61}$$

Thus, $\psi(\mathbf{r})$ is the eigenfunction of the Hermitian operator $\mathscr{H}(\mathbf{r})$ with eigenvalue E. The eigenfunctions of the time-independent Hamiltonian are called *stationary states*. If we acknowledge several stationary states in a system, the jth gives

$$\mathscr{H}(\mathbf{r})\psi_j(\mathbf{r}) = E_j\psi_j(\mathbf{r}), \tag{3-62}$$

where E_j is the energy eigenvalue for the jth stationary state, $\psi_j(\mathbf{r})$, in the system.

In the following discussion we shall refer often to the state of the system. A stationary state is defined above. In general, we refer to the state of a system as a time-independent state, but it need not be a stationary state. For instance, if we apply our previous discussion of superposition [following Eq.(3-21)] to Eq.(3-62), the linear nature of Schrödinger's equation requires that any linear combination of stationary states also be a solution to the time-independent differential equation. For instance, use $\psi = \psi_i + b\psi_j$ to give (b is a constant)

$$\mathscr{H}(\mathbf{r})\psi = \mathscr{H}(\mathbf{r})(\psi_i + b\psi_j) = E_i\psi_i + E_jb\psi_j. \tag{3-63}$$

Thus, $\mathscr{H}(\mathbf{r})\psi$ gives a new function $E_i\psi_i + E_jb\psi_j$ that is not, in general, a constant times ψ. If $E_i = E_j$, then $\psi = \psi_i + b\psi_j$ is still an eigenfunction of $\mathscr{H}(\mathbf{r})$ for all values of b. In general, however, ψ is not an eigenfunction of $\mathscr{H}(\mathbf{r})$ and is therefore not a stationary state, yet, $\psi = \psi_i + b\psi_j$ is a perfectly good solution to Schrödinger's linear equation. Thus, we conclude that the system can be described as being in one of an infinite number of states. The next postulate examines the nature of observables with respect to the possible states available to the system.

POSTULATE III The only values which a measurement on an observable with an operator Q *can yield are the eigenvalues,* Q_i, *of the equation*

$$Q\Phi_i = Q_i\Phi_i. \tag{3-64}$$

If the system happens to be in or is prepared to be in one of the orthonormal eigenstates Φ_i before the measurement, then the observed result will be the eigenvalue Q_i. In general, however, the system will be in a state described by a superposition of the states in $\mathbf{\Phi}$. Call the state of the system χ_j, given by [see Eq.(3-46)] a superposition of the Φ_i states,

$$\chi_j = \sum_i \Phi_i \mathrm{b}_{ij}, \tag{3-65}$$

where the components in $\boldsymbol{\chi} = (\chi_1\chi_2\chi_3 \cdots)$ also form a complete orthonormal set. Under these circumstances we need Postulate IV to predict the results of an experiment.

POSTULATE IV If a system is in a state χ_j, *the average value of a series of measurements on the observable with operator* Q *is given by*

$$Q_{jj}^{\chi} = \int \chi_j^* \mathrm{Q} \chi_j \, dV, \tag{3-66}$$

where we again emphasize that χ_j is one component in the orthonormal set $\boldsymbol{\chi}$. Now according to Postulate III, only eigenvalues (or differences in eigenvalues) can be measured. Thus, we substitute Eq.(3-65) into Eq.(3-66) to give the results of a sequence of measurements in terms of the eigenvalues:

$$\mathrm{Q}_{jj}^{\chi} = \sum_i \mathrm{b}_{ij}^* \sum_k \mathrm{b}_{kj} \int \Phi_i^* Q \Phi_k \, dV = \sum_{i,k} \mathrm{b}_{ij}^* \mathrm{b}_{kj} \mathrm{Q}_k \delta_{ik}, \tag{3-67}$$

where Q_k is the eigenvalue of Φ_k, as in Eq.(3-64). The Kronecker delta reduces the double sum to a single sum given by

$$\mathrm{Q}_{jj}^{\chi} = \sum_k \mathrm{b}_{kj}^* \mathrm{b}_{kj} Q_k = \sum_k |\mathrm{b}_{kj}|^2 Q_k. \tag{3-68}$$

Thus, the coefficients $|\mathrm{b}_{kj}|^2$ give the probability of one measurement yielding the eigenvalue Q_k. A sequence of measurements of the observable corresponding to Q in a system in a state χ_j will give different eigenvalues Q_k during each measurement; the probability of measuring Q_k on any single measurement will be directly proportional to the weighting factor $|\mathrm{b}_{kj}|^2$.

We now mention another method of predicting the results of a sequence of measurements on an observable Q on a system in a state χ_j which involves the *density matrix*. Assuming that Q is not the Hamiltonian operator, we expand χ_j in Eq.(3-66) in terms of the stationary states $\boldsymbol{\psi}$ [the eigenfunctions of the Hamiltonian operator where $\mathscr{H}\psi_i = E_i\psi_i$ from Eq.(3-62), for instance]. Substituting

$$\chi_j = \sum_k \psi_k \mathrm{a}_{kj} \tag{3-69}$$

into Eq.(3-66) gives

$$\mathrm{Q}_{jj}^{\chi} = \sum_{i,m} \mathrm{a}_{ij}^* \mathrm{Q}_{im}^{\psi} \mathrm{a}_{mj} = \mathrm{Tr}\ \boldsymbol{\rho}^j \mathbf{Q}^{\psi}$$

$$\boldsymbol{\rho}^j = \begin{pmatrix} \mathrm{a}_{1j}^*\mathrm{a}_{1j} & \mathrm{a}_{1j}^*\mathrm{a}_{2j} & \mathrm{a}_{1j}^*\mathrm{a}_{3j} & \cdots \\ \mathrm{a}_{2j}^*\mathrm{a}_{1j} & \mathrm{a}_{2j}^*\mathrm{a}_{2j} & \mathrm{a}_{2j}^*\mathrm{a}_{3j} & \\ \mathrm{a}_{3j}^*\mathrm{a}_{1j} & \mathrm{a}_{3j}^*\mathrm{a}_{2j} & \mathrm{a}_{3j}^*\mathrm{a}_{3j} & \\ \vdots & & & \ddots \end{pmatrix}, \tag{3-70}$$

where $\boldsymbol{\rho}^j$ is the density matrix associated with the jth state, χ_j, and the expansion in Eq.(3-69). Tr indicates the trace or diagonal sum of the $\boldsymbol{\rho}^j\mathbf{Q}^{\psi}$ matrix. If $\boldsymbol{\rho}^j$ is a diagonal matrix, the diagonal values are related to the thermal equilibrium probabilities for the stationary states. If $\mathbf{Q}^{\psi}$ is diagonal, then $\boldsymbol{\psi}$ are the eigenfunctions of Q and Eq.(3-70) reduces to

$$\mathrm{Q}_{jj}^{\chi} = \mathrm{Tr}\ \boldsymbol{\rho}^j\mathbf{Q}^{\psi} = \sum_{i,m} \mathrm{a}_{ij}^* \mathrm{Q}_{im}^{\psi} \delta_{im} \mathrm{a}_{mj} = \sum_m |\mathrm{a}_{mj}|^2 \mathrm{Q}_{mm}^{\psi} = \sum_m |\mathrm{a}_{mj}|^2 Q_m.$$

Thus, the observed value of Q_{jj}^{χ} will be the $|\mathrm{a}_{mj}|^2$ weighted sum of eigenvalues $\mathrm{Q}_{mm} = Q_m$ as shown above.

The major efforts in Chapters 4, 5, and 6 involve the determination of stationary states of various atomic and molecular systems. The stationary states are the eigenfunctions of the time-dependent Hamiltonian operator and the eigenvalues give the discrete energy states of the system. Even though these energy states are not always observed directly, they form a convenient frame of reference and we shall now consider some of the interpretations of observables in reference to stationary states. We begin by rewriting Eq.(3-62):

$$\mathscr{H}(\mathbf{r})\psi_j(\mathbf{r}) = E_j\psi_j(\mathbf{r}), \tag{3-71}$$

where $\psi_j(r)$ are the stationary states. Now according to our previous discussion of superposition and the postulates of quantum mechanics, we note that a free molecule that has stationary states, ψ_j, need not be in a stationary state but, in general, is in some state that will be a linear combination of stationary states, given by

$$\Phi_i = \sum_k \psi_k \mathrm{a}_{ki}. \tag{3-72}$$

It is possible to prepare a molecule, atom, or other system in a stationary state prior to the observation on the system. For instance, a unidirectional beam of molecules can be passed through electrostatic fields that will select a single stationary state, the molecules in all other states being deflected or defocused from the beam. Thus, a system can be prepared in a stationary state but, in general, the system will be in a superposition state described by Eq.(3-72). In the case of atoms or molecules in thermal equilibrium with the external world, the weighting factors for the probability of the existence of the stationary state ψ_k in the state Φ_i in Eq.(3-72) given by $(\mathrm{a}_{ki})^2$ will be proportional to the Boltzmann factor for the kth stationary state.

In Chapters 4, 5, and 6 we concentrate on the stationary states and energy eigenvalues as spectroscopic measurements are capable of measuring energy differences between stationary states. The measurement is not direct in that the radiation field probes the dipole moment of the system and the dipole moment is found to oscillate at a frequency given by the stationary state energy difference divided by Planck's constant (the Bohr frequency rule). Thus, we shall focus a great deal of attention on the determination of the stationary states of atomic, molecular, and condensed phase systems.

The task is to solve Eq.(3-71) for systems of interest to obtain the eigenvalues and stationary state eigenfunctions. Multiplying both sides of Eq.(3-71) by ψ_i^* and integrating gives

$$\int \psi_i^* \mathscr{H}(\mathbf{r})\psi_j \, dV = E_j \int \psi_i^*\psi_j \, dV = E_j\delta_{kj}$$

$$\int \boldsymbol{\psi}^\dagger \mathscr{H} \boldsymbol{\psi} \, dV = \mathscr{H}^\psi = \mathbf{E},$$

where $\mathscr{H}^\psi$ is a diagonal matrix with eigenvalues E_j. Many times Eq.(3-71) is not readily solvable and $\boldsymbol{\psi}$ cannot be determined directly. Under these circumstances $\boldsymbol{\psi}$ can be expanded in a known orthonormal set $\boldsymbol{\Phi}$ by

$$\boldsymbol{\psi} = \boldsymbol{\Phi}\mathbf{a} \tag{3-73}$$

giving

$$\mathscr{H}^\psi = \mathbf{a}^\dagger \mathscr{H}^\Phi \mathbf{a} = \mathbf{E}, \tag{3-74}$$

where $\mathscr{H}^{\Phi}$ is the matrix representation of $\mathscr{H}$ in the $\mathbf{\Phi}$ basis. Thus, the problem has been reduced to calculating the matrix elements in the $\mathscr{H}^{\Phi}$ matrix in the known basis $\mathbf{\Phi}$ and then diagonalizing the $\mathscr{H}^{\Phi}$ matrix to give the stationary state eigenvalues. As $\mathscr{H}$ is normally a real Hermitian operator, $\mathscr{H}^{\Phi}$ is a symmetric matrix that leads to straightforward numerical diagonalization on a digital computer. In summary, the problem of solving quantum mechanical problems is reduced to an appropriate choice of initial functions, $\mathbf{\Phi}$; calculation of the elements in $\mathscr{H}^{\Phi}$; and subsequent diagonalization of $\mathscr{H}^{\Phi}$ to yield $\mathbf{E}$ and $\boldsymbol{\psi}$. The widespread availability of high-speed digital computers makes simple work of diagonalizing a symmetric matrix that arises from the real Hermitian Hamiltonian operator.

Uniqueness

Before proceeding, the uniqueness of the $\mathbf{E}$ matrix in Eq.(3-74) will be examined. Assume that $\mathbf{a}$ is not unique and that a new unitary matrix, $\mathbf{b}$, defined by $\boldsymbol{\psi}' = \mathbf{\Phi b}$, will also solve Eq.(3-74). That is, assume two solutions to the matrix form of the Schrödinger equation:

$$\mathbf{E} = \mathscr{H}^{\psi} = \mathbf{a}^{\dagger}\mathscr{H}^{\Phi}\mathbf{a}, \qquad \boldsymbol{\varepsilon} = \mathscr{H}^{\psi'} = \mathbf{b}^{\dagger}\mathscr{H}^{\Phi}\mathbf{b}. \tag{3-75}$$

As both $\mathbf{a}$ and $\mathbf{b}$ are unitary, they are related by another unitary matrix, $\mathbf{u}$, according to

$$\mathbf{a} = \mathbf{bu}, \qquad \mathbf{b} = \mathbf{au}^{\dagger}, \qquad \mathbf{a}^{\dagger}\mathbf{a} = \mathbf{1} = \mathbf{u}^{\dagger}\mathbf{b}^{\dagger}\mathbf{bu} = \mathbf{u}^{\dagger}\mathbf{u}. \tag{3-76}$$

From Eq.(3-75) we can write

$$\mathscr{H}^{\Phi} = \mathbf{aEa}^{\dagger} = \mathbf{b}\boldsymbol{\varepsilon}\mathbf{b}^{\dagger}. \tag{3-77}$$

Substituting $\mathbf{b} = \mathbf{au}^{\dagger}$ into $\mathbf{b}\boldsymbol{\varepsilon}\mathbf{b}^{\dagger}$ gives

$$\mathscr{H}^{\Phi} = \mathbf{b}\boldsymbol{\varepsilon}\mathbf{b}^{\dagger} = \mathbf{au}^{\dagger}\boldsymbol{\varepsilon}\mathbf{ua}^{\dagger}. \tag{3-78}$$

Comparing Eqs.(3-77) and (3-78) gives

$$\mathbf{E} = \mathbf{u}^{\dagger}\boldsymbol{\varepsilon}\mathbf{u} \qquad \mathbf{uE} = \boldsymbol{\varepsilon}\mathbf{u}$$

$$\sum_{j} \mathrm{u}_{ij}\mathrm{E}_{jm}\delta_{jm} = \sum_{k} \varepsilon_{ik}\delta_{ik}\mathrm{u}_{km} \qquad \mathrm{u}_{im}(\mathrm{E}_{mm} - \varepsilon_{ii}) = 0. \tag{3-79}$$

According to these results, if the $\mathbf{u}$ matrix is diagonal and nonzero ($\mathrm{u}_{ij} = \mathrm{u}_{ij}\delta_{ij}$), $\boldsymbol{\varepsilon} = \mathbf{E}$. Each diagonal element in $\mathbf{u}$ may contain an arbitrary phase factor, $e^{i\alpha}$, however, as we require [in Eq.(3-76)] only that $\mathbf{u}^{\dagger}\mathbf{u} = \mathbf{1}$. Thus, this arbitrary phase in $\mathbf{u}$ will result in multiplying the corresponding eigenfunctions by a constant whose absolute value squared is unity. On the other hand, if $\mathrm{u}_{im} \neq 0$, $\mathrm{E}_{mm} = \varepsilon_{ii}$. In this case, $\mathbf{u}$ is not diagonal and the diagonal elements in $\boldsymbol{\varepsilon}$ and $\mathbf{E}$ have been permuted with respect to each other. This permutation of the diagonal elements results from a permutation of the ordering in $\mathbf{\Phi}$ caused by $\mathbf{u}$ as shown by substituting $\mathbf{a} = \mathbf{bu}$ into $\boldsymbol{\psi} = \mathbf{\Phi a}$ to give

$$\boldsymbol{\psi} = \mathbf{\Phi a} = \mathbf{\Phi bu} = \boldsymbol{\psi}'\mathbf{u}. \tag{3-80}$$

Thus, the ordering in the resultant eigenfunctions undergo the same permutations as the eigenvalues. The result is that the unitary matrix that diagonalizes $\mathscr{H}^{\Phi}$, to yield the eigenvalues and eigenfunctions, is unique within possible permutations of the eigenvalues and associated eigenfunctions or the multiplication of the eigenfunctions by arbitrary phase factors.

Returning to the solution of Eq.(3-74), it is important to choose an initial known basis set, $\mathbf{\Phi}$, that represents the system fairly well and nearly approximates ψ. If $\mathbf{\Phi}$ is a good approximation to ψ, the near-diagonal $\mathscr{H}^{\Phi}$ matrix can be safely truncated before diagonalization. A convenient method of obtaining a good original known basis set, $\mathbf{\Phi}$, is to separate the Hamiltonian artificially into two parts, one of which solves Schrödinger's equation exactly, giving $\mathbf{\Phi}$. The remaining component in the Hamiltonian then acts as a perturbation on the original basis yielding the actual eigenfunction, ψ, by the expansion, $\psi = \mathbf{\Phi a}$. Let the Hamiltonian be

$$\mathscr{H} = \mathscr{H}_0 + \mathscr{H}_1, \tag{3-81}$$

where $\mathscr{H}$ satisfies Eq.(3-74) and $\mathscr{H}_0$ satisfies

$$\mathscr{H}_0^{\Phi} = \boldsymbol{\varepsilon}. \tag{3-82}$$

$\boldsymbol{\varepsilon}$ is a diagonal matrix in the known $\mathbf{\Phi}$ basis yielding the eigenvalues of the $\mathscr{H}_0$ operator. If the effects of $\mathscr{H}_0$ are larger than $\mathscr{H}_1$, $\mathbf{\Phi}$ in Eq.(3-82) will be a fairly accurate description of the system. Returning to Eq.(3-71) with Eq.(3-81), we write

$$\mathscr{H}\psi = (\mathscr{H}_0 + \mathscr{H}_1)\psi, \tag{3-83}$$

where ψ will approach $\mathbf{\Phi}$ as the effects of $\mathscr{H}_1$ become negligible relative to the effects of $\mathscr{H}_0$. The perturbation on the original $\mathbf{\Phi}$ basis due to $\mathscr{H}_1$ can be investigated by expanding the eigenfunctions, ψ, in Eq.(3-83), in terms of the known basis, $\mathbf{\Phi}$, in Eq.(3-82). Substituting $\psi = \mathbf{\Phi a}$ into Eq.(3-83) gives

$$\mathscr{H}\psi = \mathscr{H}\mathbf{\Phi a} = (\mathscr{H}_0 + \mathscr{H}_1)\mathbf{\Phi a}. \tag{3-84}$$

Applying $\int \psi^{\dagger}\, dV = \int \mathbf{a}^{\dagger}\mathbf{\Phi}^{\dagger}\, dV$ to Eq.(3-84) gives

$$\mathscr{H}^{\psi} = \mathbf{E} = \int \mathbf{a}^{\dagger}\mathbf{\Phi}^{\dagger}(\mathscr{H}_0 + \mathscr{H}_1)\mathbf{\Phi a}\, dV = \mathbf{a}^{\dagger}(\mathscr{H}_0^{\Phi} + \mathscr{H}_1^{\Phi})\mathbf{a} = \mathbf{a}^{\dagger}(\boldsymbol{\varepsilon} + \mathscr{H}_1^{\Phi})\mathbf{a}. \tag{3-85}$$

The matrix to be diagonalized by the unitary transformation $\mathbf{a}$ is the sum of the diagonal $\mathscr{H}_0^{\Phi}$ matrix and the perturbation $\mathscr{H}_1^{\Phi}$ matrix:

$\mathbf{a}^{\dagger}(\mathscr{H}_0^{\Phi} + \mathscr{H}_1^{\Phi})\mathbf{a}$

$$= \mathbf{a}^{\dagger}\begin{pmatrix} (\mathscr{H}_0)_{11}^{\Phi} + (\mathscr{H}_1)_{11}^{\Phi} & (\mathscr{H}_1)_{12}^{\Phi} & (\mathscr{H}_1)_{13}^{\Phi} & \cdots \\ (\mathscr{H}_1)_{21}^{\Phi} & (\mathscr{H}_0)_{22}^{\Phi} + (\mathscr{H}_1)_{22}^{\Phi} & (\mathscr{H}_1)_{23}^{\Phi} & \\ (\mathscr{H}_1)_{31}^{\Phi} & (\mathscr{H}_1)_{32}^{\Phi} & (\mathscr{H}_0)_{33}^{\Phi} + (\mathscr{H}_1)_{33}^{\Phi} & \\ & \vdots & & \ddots \end{pmatrix}\mathbf{a} = \mathbf{E}. \tag{3-86}$$

Equation (3-86) summarizes the solution of the stationary state problem. The Hamiltonian, $\mathscr{H} = \mathscr{H}_0 + \mathscr{H}_1$, is defined; the matrix representation of $\mathscr{H}$ in the basis of $\mathscr{H}_0$ eigenfunctions, $\mathbf{\Phi}$, is constructed; and then the $\mathscr{H}^{\Phi}$ matrix is diagonalized to give the eigenvalues, $\mathbf{E}$. We shall now examine the solution of the eigenvalue problem by perturbation theory.

3.4 TIME-INDEPENDENT PERTURBATION THEORY

In order to solve the Schrödinger equation using perturbation theory, Eq.(3-71) for the jth state is written as

$$\mathscr{H}\psi_j = (\mathscr{H}_0 + \lambda\mathscr{H}_1)\psi_j = E_j\psi_j, \tag{3-87}$$

where λ is a scalar parameter. The zero-order equation from $\mathscr{H}_0$ is given by

$$\mathscr{H}_0\Phi_j = \varepsilon_j\Phi_j. \tag{3-88}$$

We shall now examine *nondegenerate perturbation theory*, which requires that $\varepsilon_i \neq \varepsilon_j$. The energy and associated eigenfunction in Eq.(3-87) can both be expanded in a power series in λ:

$$\begin{aligned} \psi_j &= \Phi_j + \lambda\psi_j^{(1)} + \lambda^2\psi_j^{(2)} + \cdots \\ E_j &= \varepsilon_j + \lambda E_j^{(1)} + \lambda^2 E_j^{(2)} + \cdots. \end{aligned} \tag{3-89}$$

$E_j^{(1)}$ and $\psi_j^{(1)}$ are the first-order corrections to the unperturbed energy and eigenfunction, respectively. $E_j^{(2)}$ and $\psi_j^{(2)}$ are the second-order corrections to the energy and eigenfunction, respectively. Substituting Eqs.(3-89) into Eq.(3-87) and rearranging gives

$$\begin{aligned} &\mathscr{H}_0\Phi_j + \lambda[\mathscr{H}_0\psi_j^{(1)} + \mathscr{H}_1\Phi_j] + \lambda^2[\mathscr{H}_0\psi_j^{(2)} + \mathscr{H}_1\psi_j^{(1)}] + \cdots \\ &\qquad = \varepsilon_j\Phi_j + \lambda[E_j^{(1)}\Phi_j + \varepsilon_j\psi_j^{(1)}] + \lambda^2[\varepsilon_j\psi_j^{(2)} + E_j^{(1)}\psi_j^{(1)} + E_j^{(2)}\Phi_j] + \cdots, \end{aligned} \tag{3-90}$$

which, in general, can be written as

$$\mathscr{H}_0 \sum_{n=0}^{\infty} \lambda^n\psi_j^{(n)} + \mathscr{H}_1 \sum_{n=0}^{\infty} \lambda^{n+1}\psi_j^{(n)} = \sum_{n=0}^{\infty} \lambda^n \sum_{m=0}^{n} E_j^{(m)}\psi_j^{(n-m)}. \tag{3-91}$$

Equation (3-91) is valid for all values of λ when the coefficients of $\lambda^n (n = 0, 1, 2, \ldots)$ are equal giving the following equations:

$$\begin{aligned} &\lambda^0, \quad \mathscr{H}_0\Phi_j = \varepsilon_j\Phi_j \\ &\lambda^1, \quad (\mathscr{H}_0 - \varepsilon_j)\psi_j^{(1)} + \mathscr{H}_1\Phi_j = E_j^{(1)}\Phi_j \\ &\lambda^2, \quad (\mathscr{H}_0 - \varepsilon_j)\psi_j^{(2)} + \mathscr{H}_1\psi_j^{(1)} = E_j^{(1)}\psi_j^{(1)} + E_j^{(2)}\Phi_j \\ &\quad\vdots \\ &\lambda^n, \quad (\mathscr{H}_0 - \varepsilon_j)\psi_j^{(n)} + \mathscr{H}_1\psi_j^{(n-1)} = \sum_{m=1}^{n-1} E_j^{(m)}\psi_j^{(n-m)} + E_j^{(n)}\Phi_j, \qquad n \neq 0. \end{aligned} \tag{3-92}$$

The $n = 0$ equation in Eqs.(3-92) is the zero-order equation in Eq.(3-88). The first-order energy, $E_j^{(1)}$, is obtained from the $n = 1$ equation by multiplying by Φ_j^* and integrating, giving

$$\int \Phi_j^*(\mathscr{H}_0 - \varepsilon_j)\psi_j^{(1)}\,dV + \int \Phi_j^*\mathscr{H}_1\Phi_j\,dV = \int \Phi_j^* E_j^{(1)}\Phi_j\,dV. \tag{3-93}$$

Now, $\mathscr{H}_0$ is a Hermitian operator and we can write

$$\int \Phi_j^* \mathscr{H}_0 \psi_j^{(1)}\, dV = \int (\mathscr{H}_0^* \Phi_j^*) \psi_j^{(1)}\, dV = \varepsilon_j \int \Phi_j^* \psi_j^{(1)}\, dV$$

and Eq.(3-93) reduces to

$$E_j^{(1)} = \int \Phi_j^* \mathscr{H}_1 \Phi_j\, dV, \tag{3-94}$$

for the first-order energy. Further manipulation of Eqs.(3-92) gives the higher-order energies:

$$\begin{aligned}
E_j^{(2)} &= \int \Phi_j^* [\mathscr{H}_1 - E_j^{(1)}] \psi_j^{(1)}\, dV \\
E_j^{(3)} &= \int \Phi_j^* [\mathscr{H}_1 - E_j^{(1)}] \psi_j^{(2)}\, dV - E_j^{(2)} \int \Phi_j^* \psi_j^{(1)}\, dV \\
&\vdots \\
E_j^{(n)} &= \int \Phi_j^* \mathscr{H}_1 \psi_j^{(n-1)}\, dV - \sum_{m=1}^{n-1} E_j^{(m)} \int \Phi_j^* \psi_j^{(n-m)}\, dV.
\end{aligned} \tag{3-95}$$

We also require that the final functions in Eqs.(3-95) be normalized giving

$$\begin{aligned}
\int \psi_j^* \psi_j\, dV = 1 = \int \Phi_j^* \Phi_j\, dV &+ \lambda\left[\int \Phi_j^* \psi_j^{(1)}\, dV + \int \psi_j^{(1)*} \Phi_j\, dV\right] \\
&+ \lambda^2\left[\int \Phi_j^* \psi_j^{(2)}\, dV + \int \psi_j^{(2)*} \Phi_j\, dV + \int \psi^{(1)*} \psi^{(1)}\, dV\right] + \cdots.
\end{aligned} \tag{3-96}$$

Now, as Φ_j is also orthonormal, $\int \Phi_j^* \Phi_j\, dV = 1.0$ and all remaining terms in the equation must sum to zero. As all terms in $\lambda, \lambda^2, \ldots$ must sum to zero for all values of λ, each term within the square brackets must individually be zero, giving

$$\begin{aligned}
&\lambda^1, \quad \int \Phi_j^* \psi_j^{(1)}\, dV + \int \psi_j^{(1)*} \Phi_j\, dV = 0 \\
&\lambda^2, \quad 2 \int \Phi_j^* \psi_j^{(2)}\, dV + \int \psi_j^{(1)*} \psi_j^{(1)}\, dV = 0 \\
&\quad \vdots
\end{aligned} \tag{3-97}$$

We shall now use expansion methods in $\psi^{(n)}$ to obtain the perturbation solutions in terms of the zero-order complete set, $\mathbf{\Phi}$. Consider the second equation in Eqs.(3-92). The first-order correction to the wavefunction, $\psi_j^{(1)}$, can be expanded in terms of the unperturbed functions according to

$$\psi_j^{(1)} = \sum_k \Phi_k a_{kj}. \tag{3-98}$$

Substituting into the second equation in Eqs.(3-92) gives

$$(\mathscr{H}_0 - \varepsilon_j) \sum_k \Phi_k a_{kj} = [E_j^{(1)} - \mathscr{H}_1]\Phi_j. \tag{3-99}$$

By applying $\int \Phi_m^* \, dV$ to this equation, we obtain

$$\sum_k a_{kj}(\varepsilon_k - \varepsilon_j)\delta_{mk} = E_j^{(1)}\delta_{mj} - (\mathscr{H}_1)_{mj}^{\Phi}.$$

Using the Kronecker delta in the summation gives

$$a_{mj}(\varepsilon_m - \varepsilon_j) = E_j^{(1)}\delta_{mj} - (\mathscr{H}_1)_{mj}^{\Phi}. \tag{3-100}$$

There are two solutions to this equation:

1. $m = j$, which gives

$$E_j^{(1)} = (\mathscr{H}_1)_{jj}^{\Phi}. \tag{3-101}$$

2. $m \neq j$, which gives

$$a'_{mj} = \frac{(\mathscr{H}_1)_{mj}^{\Phi}}{\varepsilon_j - \varepsilon_m}, \qquad a_{jj} = 0, \tag{3-102}$$

   where the prime indicates exclusion of the $m = j$ term for nondegenerate perturbation theory. Comparing Eq.(3-93) with the λ part of Eq.(3-97) shows that the real part of a_{jj} is zero. The imaginary part of a_{jj} can be arbitrarily set equal to zero by adjusting the phase of $\psi_j^{(1)}$.

Equation (3-101) gives the first-order correction to the energy in Eq.(3-89) and Eq.(3-102) gives the expansion coefficients in Eq.(3-98) defining the first-order correction to the function.

Continuing with the third equation in Eqs.(3-92), we can expand the second-order correction to the wavefunction in terms of the unperturbed function:

$$\psi_j^{(2)} = \sum_i \Phi_i b_{ij}. \tag{3-103}$$

Substituting Eqs.(3-98) and (3-103) into the third equation of Eqs.(3-92) gives

$$\sum_i b_{ij}(\mathscr{H}_0 - \varepsilon_j)\Phi_i = \sum_k a_{kj}[E_j^{(1)} - \mathscr{H}_1]\Phi_k + E_j^{(2)}\Phi_j, \tag{3-104}$$

and applying $\int \Phi_s^* \, dV$ to this result leads to

$$\sum_i b_{ij}(\varepsilon_i - \varepsilon_j)\delta_{si} = \sum_k a_{kj}E_j^{(1)}\delta_{sk} - \sum_k a_{kj}(\mathscr{H}_1)_{sk}^{\Phi} + E_j^{(2)}\delta_{sj}. \tag{3-105}$$

Using the Kronecker deltas in the summations gives

$$b_{sj}(\varepsilon_s - \varepsilon_j) = a_{sj}E_j^{(1)} - \sum_k a_{kj}(\mathscr{H}_1)_{sk}^{\Phi} + E_j^{(2)}\delta_{sj}. \tag{3-106}$$

There are two solutions to this equation:

1. $s = j$, which gives

$$a_{jj}E_j^{(1)} - \sum_k a_{kj}(\mathscr{H}_1)_{jk}^{\Phi} + E_j^{(2)} = 0. \tag{3-107}$$

   Remember that $a_{jj} = 0$ from Eq.(3-102) gives

$$E_j^{(2)} = \sum_k a_{kj}(\mathscr{H}_1)_{jk}^{\Phi}. \tag{3-108}$$

Substituting the a_{kj} coefficient from Eq.(3-102) gives the second-order correction to the energy:

$$E_j^{(2)} = \sum_k{}' \frac{(\mathscr{H}_1)^{\Phi}_{jk}(\mathscr{H}_1)^{\Phi}_{kj}}{\varepsilon_j - \varepsilon_k} = \sum_k{}' \frac{|(\mathscr{H}_1)^{\Phi}_{jk}|^2}{\varepsilon_j - \varepsilon_k}. \tag{3-109}$$

2. $s \neq j$, giving

$$b_{sj}(\varepsilon_s - \varepsilon_j) = a_{sj}E_j^{(1)} - \sum_k a_{kj}(\mathscr{H}_1)^{\Phi}_{sk}. \tag{3-110}$$

Substituting the a_{sj} coefficients from Eq.(3-102) and rearranging gives the coefficients in Eq.(3-103),

$$b_{sj} = -\frac{(\mathscr{H}_1)^{\Phi}_{sj}(\mathscr{H}_1)^{\Phi}_{jj}}{(\varepsilon_s - \varepsilon_j)^2} - \sum_k{}' \frac{(\mathscr{H}_1)^{\Phi}_{sk}(\mathscr{H}_1)^{\Phi}_{kj}}{(\varepsilon_s - \varepsilon_j)(\varepsilon_j - \varepsilon_k)}, \tag{3-111}$$

which is valid only for nondegenerate systems with $s \neq j$ and $j \neq k$. b_{jj} can be found by normalization of the corrected wavefunction to the desired order.

Equation (3-109) gives the second-order correction to the energy in Eq.(3-89); Eq.(3-111) gives the expansion coefficients in Eq.(3-103) for the second-order correction to the function. This process could be continued to give the higher-order perturbation terms in the same straightforward manner.

The energy of the jth state of a system corrected to third order is easily obtained:

$$E_j = \varepsilon_j + (\mathscr{H}_1)^{\Phi}_{jj} + \sum_k{}' \frac{(\mathscr{H}_1)^{\Phi}_{jk}(\mathscr{H}_1)^{\Phi}_{kj}}{\varepsilon_j - \varepsilon_k}$$

$$+ \left\{ \sum_{l,m}{}' \frac{(\mathscr{H}_1)^{\Phi}_{jl}(\mathscr{H}_1)^{\Phi}_{lm}(\mathscr{H}_1)^{\Phi}_{mj}}{(\varepsilon_j - \varepsilon_l)(\varepsilon_j - \varepsilon_m)} - \sum_k{}' \frac{(\mathscr{H}_1)^{\Phi}_{jk}(\mathscr{H}_1)^{\Phi}_{kj}(\mathscr{H}_1)^{\Phi}_{jj}}{(\varepsilon_j - \varepsilon_k)^2} \right\} + \cdots. \tag{3-112}$$

The first term, ε_j, is the unperturbed energy from Eq.(3-88); the second term is the first-order correction; the third term is the second-order correction; and the fourth term is the third-order correction. As the perturbation theory expansion of the energy in Eq.(3-112) gives the same eigenvalues, E_j, as the matrix diagonalization in Eq.(3-86), the following conclusions involving the two methods may be made:

1. The first-order perturbation correction to the energy of the jth level is merely the diagonal matrix element of the perturbation Hamiltonian, $(\mathscr{H}_1)^{\Phi}_{jj}$, in the zero-order basis, $\mathbf{\Phi}$.
2. The second-order correction to the energy of the jth state involves a sum over all k off-diagonal elements coupling all other originally unperturbed or zero-order states in the system.
3. The third-order correction to the energy of the jth state involves a correction acknowledging the second-order perturbation on all other states involved in the summation in statement 2. The summation there involves coupling between unperturbed states. These states were actually also perturbed and the subsequent error introduced in statement 2 is corrected in the third-order term.

If the spacings between the unperturbed energies, ε_j, are large or the off-diagonal elements, $(\mathscr{H}_1)^{\Phi}_{ji}$ are small, the series in Eq.(3-112) will converge rapidly. The corresponding concept of a perturbation in Eq.(3-86) is also evident in the character of the off-diagonal elements in the matrix. If $\varepsilon_1 < \varepsilon_2 < \varepsilon_3 < \cdots < \varepsilon_n$, only the off-diagonal elements near the diagonal elements will affect the resultant eigenvalues significantly. Similarly, if the off-diagonal elements are smaller than the unperturbed energy, ε_j, the effect of the off-diagonal elements would be small. As $\boldsymbol{\psi} = \boldsymbol{\Phi}\mathbf{a}$ is an infinite expansion, the Hamiltonian matrix in Eq.(3-86) is also infinite. Truncation of the infinite matrix in Eq.(3-86) is equivalent to truncating the perturbation summation in Eq.(3-112). In order to deal with finite Hamiltonian matrices, $\boldsymbol{\Phi}$ must be chosen to yield the $\mathscr{H}^{\Phi}$ matrix as diagonal as possible so that the levels of interest have virtually no coupling with the parts of the matrix that have been truncated.

The function in Eq.(3-89) is expanded in a manner similar to the energy giving, through second order,

$$\psi_j = \Phi_j + \sum_k{}' \frac{\Phi_k(\mathscr{H}_1)^{\Phi}_{kj}}{\varepsilon_j - \varepsilon_k} + \sum_n{}' \left\{ - \frac{\Phi_n(\mathscr{H}_1)^{\Phi}_{nj}(\mathscr{H}_1)^{\Phi}_{jj}}{(\varepsilon_n - \varepsilon_j)^2} - \sum_k{}' \frac{\Phi_n(\mathscr{H}_1)^{\Phi}_{nk}(\mathscr{H}_1)^{\Phi}_{kj}}{(\varepsilon_n - \varepsilon_j)(\varepsilon_j - \varepsilon_k)} \right\}. \qquad (3\text{-}113)$$

The expansion coefficients for each Φ_n in Eq.(3-113) are related to the $\mathbf{a}$ matrix in $\boldsymbol{\psi} = \boldsymbol{\Phi}\mathbf{a}$. $\mathbf{a}^{\dagger}\mathbf{a} = \mathbf{1}$, however, which requires $\boldsymbol{\psi}$ to be orthonormal as are the original $\boldsymbol{\Phi}$. The unnormalized ψ_j component in Eq.(3-113) can be normalized to an order square in the correction coefficients by adding

$$-\frac{1}{2}\sum_k{}' \frac{|(\mathscr{H}_1)_{kj}|^2}{(\varepsilon_j - \varepsilon_k)^2}\Phi_j$$

to Eq.(3-113), which is equivalent to defining the b_{jj} term mentioned after Eq.(3-111) [14]. If there are degeneracies in the matrix, then the expansions in Eqs.(3-112) and (3-113) are not valid and the matrix expression in Eq.(3-86), or the equivalent form of degenerate perturbation theory, must be used.

Degenerate Perturbation Theory

Consider a set of zero-order degenerate levels with energy ε and Schrödinger equations $\mathscr{H}_0\Phi_1 = \varepsilon\Phi_1$, $\mathscr{H}_0\Phi_2 = \varepsilon\Phi_2, \ldots$, where we call $\Phi_1, \Phi_2, \ldots$, the zero-order basis set. As a result of the degeneracy, any linear combination of the Φ_i is also an eigenfunction of the zero-order Hamiltonian, $\mathscr{H}_0$. If there are n degenerate levels, however, there are n linearly independent orthonormal functions, $\Phi'_1, \Phi'_2, \ldots, \Phi'_n$, which are obtained by taking linear combinations of the original set $\Phi_1, \Phi_2, \ldots, \Phi_n$:

$$\Phi'_j = \sum_{i=1}^{n} c_{ji}\Phi_i. \qquad (3\text{-}114)$$

Φ'_i are *linearly independent* if $\sum_{j=1}^{n} b_j\Phi'_j = 0$, and only if all b_j are zero. We now use the orthonormal linearly independent set Φ'_j as the zero-order functions and return to the standard perturbation expansions and expressions from Eqs.(3-88) to (3-92) where we now use Φ'_j in place in Φ_j to give equations similar to Eqs.(3-92):

$$\lambda^0, \quad \mathscr{H}_0\Phi'_j = \varepsilon_j\Phi'_j$$

$$\lambda^1, \quad (\mathscr{H}_0 - \varepsilon_j)\psi_j^{(1)} + \mathscr{H}_1\Phi'_j = E_j^{(1)}\Phi'_j$$

$$\vdots$$

with the same meanings as before. The λ^0 equation gives no new information as the ε_j are the degenerate eigenvalues. Substituting Eq.(3-114) into the λ^1 equation above gives

$$(\mathscr{H}_0 - \varepsilon_j)\psi_j^{(1)} + \sum_{i=1}^{n} c_{ji}\mathscr{H}_1\Phi_i = E_j^{(1)}\sum_{i=1}^{n} c_{ji}\Phi_i.$$

Now substituting the expansion $\psi_j^{(1)}$ in Eq.(3-98) into this expression gives

$$(\mathscr{H}_0 - \varepsilon_j)\sum_k \Phi_k a_{kj} + \sum_{i=1}^{n} c_{ji}\mathscr{H}_1\Phi_i = E_j^{(1)}\sum_{i=1}^{n} c_{ji}\Phi_i. \qquad (3\text{-}115)$$

If $k = 0$ to n for the n degenerate levels, $(\mathscr{H}_0 - \varepsilon_j)\Phi_k a_{kj} = 0$ giving

$$\sum_{i=1}^{n} c_{ji}\mathscr{H}_1\Phi_i = E_j^{(1)}\sum_{i=1}^{n} c_{ji}\Phi_i.$$

Multiplying by Φ_j^* and integrating gives

$$\sum_{i=1}^{n} c_{ji}(\mathscr{H}_1)_{ji} - E_j^{(1)}c_{jj} = 0.$$

The trivial solution to this equation is obtained by setting all $c_{ji} = 0$. The nontrivial solution is obtained by setting the determinant of the coefficients of the c_{ij} equal to zero:

$$\text{Det}\,[\mathscr{H}_1 - E^{(1)}\mathbf{1}] = 0, \qquad (3\text{-}116)$$

where "Det" indicates the *determinant*. $\mathscr{H}_1$ is an $n \times n$ matrix in the $\mathbf{\Phi}$ original basis, $\mathbf{1}$ is the $n \times n$ unit matrix and $E^{(1)}$ are the roots of the secular equation (n roots). Thus, the first-order corrections to the energy are the roots of the secular equation in Eq.(3-116).

The solution obtained here in Eq.(3-116) is also evident from our previous work on the matrix formulation as summarized in Eq.(3-86). Consider an $n \times n$ block of $\mathscr{H}^{\Phi} = \mathscr{H}_0^{\Phi} + \mathscr{H}_1^{\Phi}$ in Eq.(3-86) that contains degenerate or equal values of $(\mathscr{H}_0)_{ii}$ from $i = 1$ to n. Under these circumstances we can separate all the equal values of $(\mathscr{H}_0)_{ii}$ ($i = 1, n$) and write

$$\mathbf{c}^\dagger\mathscr{H}\mathbf{c} = \mathscr{H}_0 + \mathbf{c}^\dagger\mathscr{H}_1\mathbf{c}$$

and we need only diagonalize $\mathscr{H}_1$ to give the first-order corrections [see Problem 3-12, which shows that $\mathbf{c}^\dagger(\mathscr{H}_0 + \mathscr{H}_1)\mathbf{c} = \mathscr{H}_0 + \mathbf{c}^\dagger\mathscr{H}_1\mathbf{c} = \mathbf{E} = \boldsymbol{\varepsilon} + \mathbf{E}^{(1)}$ only if all elements on the diagonal in $\mathscr{H}_0$ are equal],

$$\mathbf{c}^\dagger\mathscr{H}\mathbf{c} = \mathbf{E}^{(1)}.$$

An alternate solution to this equation is the secular equation in Eq.(3-116) where the roots $E^{(1)}$ are the diagonal elements in $\mathbf{E}^{(1)}$. The $n \times n$ matrix above contains the c_{ij} coefficients in Eq.(3-114). It is now evident that $\mathbf{E}^{(1)}$ contains the complete set of corrections due to $\mathscr{H}_1$; there are no higher-order corrections within the $n \times n$ degenerate block.

Now, if the $n \times n$ block of originally degenerate eigenvalues interacts with levels outside the degenerate block, the coefficients a_{sj} $(s > n)$ in Eq.(3-115) can be obtained by multiplying Eq.(3-115) by $\Phi_s^*(s > n)$ and integrating to give

$$(\varepsilon_s - \varepsilon_j)a_{sj} + \sum_{i=1}^{n} c_{ji}(\mathscr{H}_1)_{si} = E_j^{(1)} \sum_{i=1}^{n} c_{ji}\delta_{si}.$$

As $s > i$, the right-hand side is zero giving

$$a_{sj} = \frac{\sum_{i=1}^{n} c_{ji}(\mathscr{H}_1)_{si}}{\varepsilon_s - \varepsilon_j}, \tag{3-117}$$

for the first-order corrections to the function in Eq.(3-98) for couplings between the jth state in the degenerate $n \times n$ block with the sth state outside the degenerate block. The energy correct to second order is given by

$$E_j = E_j^{(1)} + \sum_{s>n} \frac{|(\mathscr{H}_1)_{js}|^2}{\varepsilon_s - \varepsilon_j}, \tag{3-118}$$

where $E_j^{(1)}$ is obtained from Eq.(3-116) or the matrix equivalent and $s > n$ indicates s states outside the $n \times n$ original block of zero-order degenerate eigenvalues. We see again in this later case the analogy with the direct diagonalization of the $\mathscr{H} = \mathscr{H}_0 + \mathscr{H}_1$ matrix where now $\mathscr{H}_0$ contains both degenerate blocks as well as diagonal values at different energies.

3.5 THE VARIATION METHOD FOR ANY STATE AND THE SECULAR EQUATION

The variation theorem is usually expressed for the lowest energy level in the system, E_1, by

$$\frac{\int \Phi_1^* \mathscr{H} \Phi_1 \, dV}{\int \Phi_1^* \Phi_1 \, dV} \geq E_1 = \int \psi_1^* \mathscr{H} \psi_1 \, dV, \tag{3-119}$$

where Φ_1 is a general unnormalized trial function that approximates the eigenfunction, ψ_1, for the ground state of the system. In order to prove this theorem, we expand Φ_1 in terms of the complete orthonormal basis, ψ (which is an eigenfunction of $\mathscr{H}$), according to

$$\Phi_1 = \sum_k \psi_k a_{k1},$$

and substitute into Eq.(3-119) to give

$$\frac{\sum_k a_{k1}^* \sum_j a_{j1} \int \psi_k^* \mathscr{H} \psi_j \, dV}{\sum_k a_{k1}^* \sum_j a_{j1} \int \psi_k^* \psi_j \, dV} = \frac{\sum_{k,j} a_{k1}^* a_{j1} E_j \delta_{kj}}{\sum_{k,j} a_{k1}^* a_{j1} \delta_{kj}} = \frac{\sum_k |a_{k1}|^2 E_k}{\sum_k |a_{k1}|^2} \geq E_1. \quad (3\text{-}120)$$

As E_1 was defined as the lowest energy level of the system, all $E_k \geq E_1$. Thus, as the total fractions of E_k values from the $|a_{k1}|^2/\sum_k |a_{k1}|^2$ coefficients in Eq.(3-120) must equal unity, the left side of Eq.(3-120) is always larger or equal to E_1. If $\mathbf{a}$ in $\mathbf{\Phi} = \boldsymbol{\psi}\mathbf{a}$ is a unitary matrix, the denominator in Eq.(3-120) goes to unity and the inequality is perhaps more readily obvious. The equality in Eq.(3-120) holds only when $\Phi_1 = \psi_1$. Equation (3-119) gives a systematic method of testing one's intuition in choosing trial functions that approximate the actual ground state function for a system.

The variational theorem is also valid for all excited states, E_j, if the trial function for the jth state, Φ_j, is orthogonal to all eigenfunctions with energy levels lower than E_j,

$$\frac{\int \Phi_j^* \mathscr{H} \Phi_j \, dV}{\int \Phi_j^* \Phi_j \, dV} \geq E_j = \int \psi_j^* \mathscr{H} \psi_j \, dV. \quad (3\text{-}121)$$

Expanding Φ_j in terms of the eigenfunctions, $\boldsymbol{\psi}$,

$$\Phi_j = \sum_k \psi_k a_{kj},$$

and substituting into Eq.(3-121) gives

$$\frac{\sum_{k=1} a_{kj} \int \Phi_j^* \mathscr{H} \psi_k \, dV}{\sum_{k=1} a_{kj} \int \Phi_j^* \psi_k \, dV} \geq E_j.$$

If we require Φ_j to be orthogonal to all eigenfunctions with eigenvalues below E_j, ψ_k where $k = 1, 2, 3, \ldots, j-1$, the summations start with $k = j$ and this equation reduces to

$$\frac{\sum_{k=j} a_{kj} \int \Phi_j^* \mathscr{H} \psi_k \, dV}{\sum_{k=j} a_{kj} \int \Phi_j^* \psi_k \, dV} = \frac{\sum_{k=j} \sum_{n=1} a_{kj} a_{nj}^* \int \psi_k^* \mathscr{H} \psi_n \, dV}{\sum_{k=j} \sum_{n=1} a_{kj} a_{nj}^* \int \psi_k^* \psi_n \, dV} = \frac{\sum_{k=j} |a_{kj}|^2 E_k}{\sum_{k=j} |a_{kj}|^2} \geq E_j. \quad (3\text{-}122)$$

As all E_k are larger than or equal to E_j and the total fractional amount of all E_k must be equal to unity, the inequality in Eq.(3-121) involving excited states must be valid. Equation (3-121) provides a method of testing or finding excited state functions for any excited state E_j, providing, of course, the orthogonality conditions on the Φ_j functions are met. The orthogonality requirements may be obtained from the symmetric or Schmidt methods (see Appendix D). Thus, approximations to the eigenstates of a system can be obtained by Eq.(3-119), for the ground state, and Eq.(3-121) for each successive excited state by requiring all excited state trial functions to be orthogonal to the previously determined lower energy approximations to the eigenstates.

The Secular Equation

We shall now discuss the popular and convenient method of minimizing the energy in Eq.(3-119) by the variation in expansion coefficients. The principle is to expand the trial function, Φ_1, in Eq.(3-119) in terms of a known basis set $\boldsymbol{\chi}$ with the expansion coefficients **b**. Next, the energy is minimized with respect to the coefficients in the **b** matrix that leads to the secular equation. Expanding Φ_j in terms of a general nonorthonormal basis $\boldsymbol{\chi}$,

$$\Phi_j = \sum_k \chi_k \mathrm{b}_{kj},$$

and substituting this result into Eq.(3-119) gives

$$\frac{\sum_k \mathrm{b}_{kj}^* \sum_i \mathrm{b}_{ij} \int \chi_k^* \mathscr{H} \chi_i \, dV}{\sum_{k,i} \mathrm{b}_{kj}^* \mathrm{b}_{ij} \int \chi_k^* \chi_i \, dV} = \varepsilon_j \geq E_j. \tag{3-123}$$

The summations over k and i are over all functions in the $\boldsymbol{\chi}$ basis and ε_j is the calculated energy. The left side quantity in Eq.(3-123) can be minimized with respect to the b_{kj} coefficients to obtain the lowest calculated energy with a given basis set of $\boldsymbol{\chi}$. Expanding Eq.(3-123) in more detail for the jth state gives

$$\begin{aligned}\varepsilon_j &= \frac{\int (\chi_1 \mathrm{b}_{1j} + \chi_2 \mathrm{b}_{2j} + \cdots)^* \mathscr{H} (\chi_1 \mathrm{b}_{1j} + \chi_2 \mathrm{b}_{2j} + \cdots) \, dV}{\int (\chi_1 \mathrm{b}_{1j} + \chi_2 \mathrm{b}_{2j} + \cdots)^* (\chi_1 \mathrm{b}_{1j} + \chi_2 \mathrm{b}_{2j} + \cdots) \, dV} \\ &= \frac{H_{11}^\chi \mathrm{b}_{1j}^2 + 2H_{12}^\chi \mathrm{b}_{1j}\mathrm{b}_{2j} + 2H_{13}^\chi \mathrm{b}_{1j}\mathrm{b}_{3j} + \cdots}{\mathrm{S}_{11}^\chi \mathrm{b}_{1j}^2 + 2\mathrm{S}_{12}^\chi \mathrm{b}_{1j}\mathrm{b}_{2j} + 2\mathrm{S}_{13}^\chi \mathrm{b}_{1j}\mathrm{b}_{3j} + \cdots},\end{aligned}$$

where $\mathrm{S}_{ij}^\chi = \int \chi_i^* \chi_j \, dV$ represents the matrix elements in the overlap matrix in the $\boldsymbol{\chi}$ basis and we have assumed that **b** is a real matrix. The equation in ε_j above is then successively minimized with respect to all the $b_{mj}(m = 1, 2, \ldots)$ coefficients giving the following series of linear equations:

$$\begin{aligned}\frac{\partial \varepsilon_j}{\partial \mathrm{b}_{1j}} &= 0 = (H_{11}^\chi - \mathrm{S}_{11}^\chi \varepsilon_j)\mathrm{b}_{1j} + (H_{12}^\chi - \mathrm{S}_{12}^\chi \varepsilon_j)\mathrm{b}_{2j} + \cdots \\ \frac{\partial \varepsilon_j}{\partial \mathrm{b}_{2j}} &= 0 = (H_{21}^\chi - \mathrm{S}_{21}^\chi \varepsilon_j)\mathrm{b}_{1j} + (H_{22}^\chi - \mathrm{S}_{22}^\chi \varepsilon_j)\mathrm{b}_{2j} + \cdots. \\ &\vdots\end{aligned} \tag{3-124}$$

We can now rewrite Eq.(3-124) in matrix form to give

$$\mathscr{H}^\chi \mathbf{b}_j - \mathbf{S}^\chi \mathbf{b}_j \varepsilon_j = \mathbf{0}_j$$

$$\begin{pmatrix} \mathscr{H}_{11}^\chi & \mathscr{H}_{12}^\chi & \cdots \\ \mathscr{H}_{21}^\chi & \mathscr{H}_{22}^\chi & \\ \vdots & & \ddots \end{pmatrix} \begin{pmatrix} \mathrm{b}_{1j} \\ \mathrm{b}_{2j} \\ \vdots \end{pmatrix} - \begin{pmatrix} \mathrm{S}_{11}^\chi & \mathrm{S}_{12}^\chi & \cdots \\ \mathrm{S}_{21}^\chi & \mathrm{S}_{22}^\chi & \\ \vdots & & \ddots \end{pmatrix} \begin{pmatrix} \mathrm{b}_{1j}\varepsilon_j \\ \mathrm{b}_{2j}\varepsilon_j \\ \vdots \end{pmatrix} = \begin{pmatrix} 0_j \\ 0_j \\ \vdots \end{pmatrix}. \tag{3-125}$$

The nontrivial solution of Eq.(3-125) is obtained by the associated secular equation

$$\mathrm{Det}\,(\mathscr{H}^\chi - \mathbf{S}^\chi \varepsilon_j) = \mathbf{0}. \tag{3-126}$$

After the roots are obtained, the associated **b** matrix can be obtained from successive solutions of Eq.(3-125) for the columns in the **b** matrix.

Returning to the matrix formulation in Eq.(3-125), we note that the equation is valid for each column labeled by j. Therefore, we can write the more complete matrix statement as

$$\mathscr{H}^\chi \mathbf{b} - \mathbf{S}^\chi \mathbf{b} \boldsymbol{\varepsilon} = \mathbf{0}, \tag{3-127}$$

where all columns $\mathbf{b}_j$ are now included. Multiplying by $\mathbf{b}^\dagger$ gives

$$\mathbf{b}^\dagger \mathscr{H}^\chi \mathbf{b} = \mathbf{b}^\dagger \mathbf{S}^\chi \mathbf{b} \boldsymbol{\varepsilon}. \tag{3-128}$$

Thus, the nonunitary $\mathbf{b}$ matrix must simultaneously diagonalize the $\mathscr{H}^\chi$ matrix and transform the overlap matrix $\mathbf{S}^\chi$ to unity in order to yield the eigenvalues, $\boldsymbol{\varepsilon}$. Consider the nonunitary transformation of the nonorthogonal χ to an orthonormal basis $\boldsymbol{\Phi}$ given by

$$\boldsymbol{\Phi} = \chi \mathbf{c}. \tag{3-129}$$

$\mathbf{c}$ is, of course, nonunitary and nonunique. $\boldsymbol{\Phi}$ is orthonormal but not necessarily an eigenfunction of the Hamiltonian. Continuing gives

$$\int \boldsymbol{\Phi}^\dagger \boldsymbol{\Phi} = \mathbf{c}^\dagger \mathbf{S}^\chi \mathbf{c} = \mathbf{1} \tag{3-130}$$

$$\mathscr{H}^\Phi = \mathbf{c}^\dagger \mathscr{H}^\chi \mathbf{c}. \tag{3-131}$$

We can now diagonalize the symmetric $\mathscr{H}^\Phi$ matrix with a unitary transformation $\mathbf{a}$ to give the eigenvalues:

$$\mathbf{a}^\dagger \mathscr{H}^\Phi \mathbf{a} = \boldsymbol{\varepsilon} = \mathscr{H}^\psi. \tag{3-132}$$

Thus, the eigenfunctions are given in terms of the original nonorthonormal basis χ by

$$\boldsymbol{\psi} = \boldsymbol{\Phi}\mathbf{a} = \chi \mathbf{c}\mathbf{a} = \chi \mathbf{b}. \tag{3-133}$$

The method of symmetric orthogonalization is most generally used where

$$\mathbf{c} = \mathbf{S}^{-1/2}. \tag{3-134}$$

$\mathbf{S}^{-1/2}$ is obtained by diagonalizing the overlap matrix $\mathbf{S}^\chi$ with a unitary transformation, $\mathbf{u}$, taking the inverse square root of the result, and transforming this final result back to the original basis with $\mathbf{u}$. The details of this symmetric orthogonalization method as well as other methods of generating the $\mathbf{c}$ matrix in Eq.(3-129) are given in Appendix D [15].

3.6 PROBABILITY AND TIME-DEPENDENCE

In Section 3.3 we discussed four postulates of quantum mechanics in terms of operators, functions, and their interpretation regarding observations. We shall now discuss more directly the interpretation of the function, $\Psi(\mathbf{r}, t)$, which is obtained from a solution of the time-dependent Schrödinger equation in Eq.(3-58). The probability per unit volume or *probability density*, $P(\mathbf{r}, t)$, corresponding to the system described by Ψ is given by

$$P(\mathbf{r}, t) = \Psi^*(\mathbf{r}, t)\Psi(\mathbf{r}, t). \tag{3-135}$$

Thus, a particle in quantum mechanics is no longer described as a mathematical point but is described by the probability of it being observed at a given location at a given time [16]. In order to determine the probability of finding the particle within a certain volume element at a given time, $\Psi^*(\mathbf{r}, t)\Psi(\mathbf{r}, t)$ is integrated over the appropriate volume element. The integration of $\Psi^*\Psi$ over all regions of space described by Ψ is unity corresponding to a certainty of finding the system somewhere in the region described by Ψ at any time t.

Consider now the change in $P(\mathbf{r}, t)$ with time for a general time-dependent system,

$$\frac{\partial P(\mathbf{r}, t)}{\partial t} = \Psi^* \frac{\partial \Psi}{\partial t} + \frac{\partial \Psi^*}{\partial t} \Psi. \tag{3-136}$$

Substituting the time-dependent function from Eq.(3-58) into this equation gives

$$\frac{\partial P(\mathbf{r}, t)}{\partial t} = \frac{1}{i\hbar} \{\Psi^* \mathscr{H}(\mathbf{r}, t)\Psi - [\mathscr{H}^*(\mathbf{r}, t)\Psi^*]\Psi\}. \tag{3-137}$$

The conservation of probability may be obtained from Eq.(3-137) by integrating over the spatial coordinates, dV,

$$\int \frac{\partial}{\partial t} P(\mathbf{r}, t)\, dV = \frac{\partial}{\partial t} \int \Psi^*\Psi\, dV = \frac{1}{i\hbar}\left[\int \Psi^* \mathscr{H}\Psi\, dV - \int (\mathscr{H}^*\Psi^*)\Psi\, dV\right] = 0. \tag{3-138}$$

The conservation of probability follows directly from the Hermitian nature of $\mathscr{H}$.

The probability density reduces to an expression that can be given a physical interpretation if the Hamiltonian operator takes the time-independent form:

$$\mathscr{H} = T + V = -\sum_j \frac{\hbar^2}{2m_j} \nabla_j^2 + V, \tag{3-139}$$

which gives

$$\begin{aligned} \frac{\partial}{\partial t} P(\mathbf{r}, t) &= -\frac{\hbar}{2i} \sum_j \frac{1}{m_j} [\Psi^* \nabla_j^2 \Psi - (\nabla_j^2 \Psi^*)\Psi] \\ &= \frac{i\hbar}{2} \sum_j \frac{1}{m_j} \nabla_j \cdot [\Psi^* \nabla_j \Psi - (\nabla_j \Psi^*)\Psi]. \end{aligned} \tag{3-140}$$

Substituting the expression for the linear momentum from Table 3-1 gives

$$\frac{\partial}{\partial t} P(\mathbf{r}, t) = -\sum_j \nabla_j \cdot \left\{\frac{1}{2m_j} [\Psi^* \boldsymbol{p}_j \Psi + (\boldsymbol{p}_j^* \Psi^*)\Psi]\right\}. \tag{3-141}$$

If $\Psi(\mathbf{r}, t)$ describes electrons and if we multiply both sides of Eq.(3-141) by the electron charge, $-e$, we have the quantum mechanical analog of the electrical current flow as derived from Maxwell's equations [see Eq.(1-12)].

Quantum Mechanics

$$-\frac{\partial (eP)}{\partial t} = \frac{1}{2} \sum_j \nabla_j \cdot \left\{\frac{1}{m_j} [\Psi^* e\boldsymbol{p}_j \Psi + (e\boldsymbol{p}_j \Psi)^* \Psi]\right\} = \sum_j \nabla_j \cdot \boldsymbol{J}_j. \tag{3-142}$$

The sum over j is over all electrons described by Ψ, and $\boldsymbol{J}_j$ is the current density of the jth electron.

Maxwell

$$-\frac{\partial \bar{\rho}}{\partial t} = \nabla \cdot \boldsymbol{J}. \tag{3-143}$$

Thus, the analogy is given by

$$\text{charge density} = \bar{\rho} = eP = e\Psi^*\Psi$$

$$\text{current density} = \boldsymbol{J}_j = \frac{1}{2m_j}\,[\Psi^* e\boldsymbol{p}_j\Psi + (e\boldsymbol{p}_j\Psi)^*\Psi].$$

Time-Dependence of an Operator

Consider now the matrix element of a general time-dependent operator, $G(\boldsymbol{r}, t)$, in the $\Psi(\boldsymbol{r}, t)$ basis as given by

$$G^{\Psi} = \int \Psi(\boldsymbol{r}, t)^* G(\boldsymbol{r}, t)\Psi(\boldsymbol{r}, t)\, dV. \tag{3-144}$$

The total derivative with respect to time where $\boldsymbol{r}$ is not an explicit function of time gives

$$\frac{d}{dt}G^{\Psi} = \int \frac{\partial \Psi^*}{\partial t} G\Psi\, dV + \int \Psi^* \frac{\partial G}{\partial t}\Psi\, dV + \int \Psi^* G \frac{\partial \Psi}{\partial t}\, dV.$$

Substituting $\partial\Psi^*/\partial t$ and $\partial\Psi/\partial t$ from Eq.(3-58) gives

$$\begin{aligned} \frac{d}{dt}G^{\Psi} &= \frac{i}{\hbar}\int \mathscr{H}^*\Psi^* G\Psi\, dV + \int \Psi^* \frac{\partial G}{\partial t}\Psi\, dV - \frac{i}{\hbar}\int \Psi^* G\mathscr{H}\Psi\, dV \\ &= \frac{i}{\hbar}\int \Psi^*[\mathscr{H}, G]\Psi\, dV + \int \Psi^*\frac{\partial G}{\partial t}\Psi\, dV = \frac{i}{\hbar}[\mathscr{H}, G]^{\Psi} + \left(\frac{\partial G}{\partial t}\right)^{\Psi}, \end{aligned} \tag{3-145}$$

where $\mathscr{H}$ is Hermitian, $[\mathscr{H}, G]$ is the commutator of $\mathscr{H}$ and G,

$$[\mathscr{H}, G]^{\Psi} = \int \Psi^*(\boldsymbol{r}, t)[\mathscr{H}, G]\Psi(\boldsymbol{r}, t)\, dV,$$

and

$$\left(\frac{\partial G}{\partial t}\right)^{\Psi} = \int \Psi^*(\boldsymbol{r}, t)\left(\frac{\partial G}{\partial t}\right)\Psi(\boldsymbol{r}, t)\, dV.$$

Equation (3-145) is the quantum mechanical analog of Eq.(2-97) in classical mechanics. If G is not an explicit function of time, Eq.(3-145) reduces to

$$\frac{dG^{\Psi}}{dt} = \frac{i}{\hbar}[\mathscr{H}, G]^{\Psi}. \tag{3-146}$$

Equation (3-146) shows that if G commutes with $\mathscr{H}$ and if G is not explicitly dependent on time, then G^Ψ is time-independent. If $G = \mathscr{H}$, and $\mathscr{H}$ is time-independent, then Eq.(3-146) shows that $\partial \mathscr{H}^\Psi/\partial t = 0$, which is the quantum mechanical equivalent of the classical law for the conservation of energy.

As both G^Ψ and $[\mathscr{H}, G]^\Psi$ are integrals over Ψ, Eq.(3-146) is a single dimension matrix equation. Furthermore, the operators must obey the same equations of motion as their matrix representations, which leads to

$$\frac{dG}{dt} = \frac{i}{\hbar}[\mathscr{H}, G] \tag{3-147}$$

for the G and $\mathscr{H}$ operators. This equation, which is sometimes called Heisenberg's equation of motion, has the following formal solution for $G(t)$:

$$G(t) = \exp\left(\frac{i\mathscr{H}t}{\hbar}\right)G(0)\exp\left(\frac{-i\mathscr{H}t}{\hbar}\right). \tag{3-148}$$

$G(0)$ is the value of the G operator at $t = 0$. Equation (3-148) describes the time evolution of an operator in the Heisenberg representation of quantum mechanics [11]. The Heisenberg representation gives a form of quantum mechanics that is more analogous to the corresponding classical expressions for the equations of motion. Of course, if $\mathscr{H}$ and G commute, then $dG/dt = 0$ and G in Heisenberg's representation is also time-independent [17].

Stationary States

Returning now to Eq.(3-146), we require $\mathscr{H}$ to be time-independent. Now assuming that the system is in a stationary state with $\Psi(\boldsymbol{r}, t)$ given by

$$\Psi(\boldsymbol{r}, t) = \psi_j(\boldsymbol{r})\exp\left(-\frac{iE_jt}{\hbar}\right),$$

where $\mathscr{H}\psi_j = E_j\psi_j$, we substitute into Eq.(3-146) to give

$$\begin{aligned}\frac{dG^\Psi}{dt} &= \frac{i}{\hbar}\int \exp\left(\frac{iE_jt}{\hbar}\right)\psi_j^*(\boldsymbol{r})[\mathscr{H}, G]\exp\left(-\frac{iE_jt}{\hbar}\right)\psi_j(\boldsymbol{r})\,dV \\ &= \frac{i}{\hbar}[\mathscr{H}, G]_{jj} = \frac{i}{\hbar}\sum_k(\mathscr{H}_{jk}\delta_{jk}\mathrm{G}_{kj} - \mathrm{G}_{jk}\mathscr{H}_{kj}\delta_{kj}) = 0.\end{aligned} \tag{3-149}$$

This equation shows that if G and $\mathscr{H}$ are both time-independent and the system is in a stationary state, G^Ψ is time-independent.

In summary, the analysis above shows that if G is not explicitly dependent on time and if G commutes with $\mathscr{H}$, the average value of G in any state is conserved in time. If *both* G and $\mathscr{H}$ are independent of time, however, then the average value of G is conserved in time in a stationary state regardless of the commutation properties of G and $\mathscr{H}$.

Momentum-Position Relations

We shall now consider several more examples for the G operator in Eq.(3-145) and a Hamiltonian given by $\mathscr{H}(\boldsymbol{r}, t) = -(\hbar^2/2m)\nabla^2 + V$.

1. $G = \boldsymbol{p}$ (*linear momentum*)

Using the time-independent Hamiltonian $\mathscr{H}(\boldsymbol{r}) = -(\hbar^2/2m)\nabla^2 + V$ gives the commutator

$$[\mathscr{H}, \boldsymbol{p}] = V\boldsymbol{p} - \boldsymbol{p}V = -i\hbar(V\boldsymbol{\nabla} - \boldsymbol{\nabla}V) = i\hbar\boldsymbol{\nabla}V.$$

Substituting into Eq.(3-146) gives

$$\frac{d\boldsymbol{p}^{\Psi}}{dt} = -\int \Psi^*\boldsymbol{\nabla}V\Psi\, dV = -(\boldsymbol{\nabla}V)^{\Psi} = \boldsymbol{F}^{\Psi}, \tag{3-150}$$

where the total force on all particles, $\boldsymbol{F} = -\boldsymbol{\nabla}V$ from Eq.(2-4) is also used.

2. $G = \boldsymbol{r}$ (*position coordinate*)

The commutator of $\boldsymbol{r}$ with $\mathscr{H} = -(\hbar^2/2m)\nabla^2 + V$ is given by

$$[\mathscr{H}, \boldsymbol{r}] = -\hbar^2\left(\frac{\nabla^2}{2m}\right)\boldsymbol{r} + \boldsymbol{r}\hbar^2\left(\frac{\nabla^2}{2m}\right) = -\frac{\hbar^2}{m}\boldsymbol{\nabla}.$$

Substituting into Eq.(3-146) gives

$$\frac{d\boldsymbol{r}^{\Psi}}{dt} = -i\hbar\int \Psi^*\left(\frac{\boldsymbol{\nabla}}{m}\right)\Psi\, dV = \int \Psi^*\left(\frac{\boldsymbol{p}}{m}\right)\Psi\, dV = \frac{1}{m}\boldsymbol{p}^{\Psi} = -\frac{i\hbar}{m}\boldsymbol{\nabla}^{\Psi}. \tag{3-151}$$

Now if the Hamiltonian is time-independent and the system is in a stationary state, we can write one of the matrix elements for the left-hand side of Eq.(3-151) according to

$$\frac{d}{dt}\left[\exp\left(\frac{iE_j t}{\hbar}\right)\left(\int \psi_j^*\boldsymbol{r}\psi_k\, dV\right)\exp\left(-\frac{iE_k t}{\hbar}\right)\right] = \frac{d}{dt}\exp(i\omega_{jk}t)\boldsymbol{r}_{jk} = i\omega_{jk}\exp(i\omega_{jk}t)\boldsymbol{r}_{jk},$$

where the Bohr frequency rule, $\hbar\omega_{jk} = E_j - E_k$, has also been used. The jkth matrix element on the right-hand side of Eq.(3-151) can be written as

$$-\frac{i\hbar}{m}\left[\exp\left(\frac{iE_j t}{\hbar}\right)\left(\int \psi_j^*\boldsymbol{\nabla}\psi_k\, dV\right)\exp\left(-\frac{iE_k t}{\hbar}\right)\right] = -\frac{i\hbar}{m}\exp(i\omega_{jk}t)\boldsymbol{\nabla}_{jk}.$$

Equating these two results gives

$$\omega_{jk}\boldsymbol{r}_{jk} = -\frac{\hbar}{m}\boldsymbol{\nabla}_{jk}. \tag{3-152}$$

Returning to Eq.(3-151) and differentiating with respect to time and using Eq.(3-150) gives

$$\frac{d^2\boldsymbol{r}^{\Psi}}{dt^2} = \frac{1}{m}\frac{d}{dt}\boldsymbol{p}^{\Psi} = \frac{1}{m}\boldsymbol{F}^{\Psi},$$

which is Ehrenfest's theorem, the quantum mechanical analog of Newton's second law.

3. $G = \boldsymbol{r} \cdot \boldsymbol{p}$

The commutator of $\boldsymbol{r} \cdot \boldsymbol{p}$ with $\mathscr{H} = -(\hbar^2/2m)\nabla^2 + V$ is given by

$$[H, \boldsymbol{r} \cdot \boldsymbol{p}] = \frac{\hbar}{i}(2T - \boldsymbol{r} \cdot \nabla V).$$

According to our previous discussion [Eq.(3-149)], $[\mathscr{H}, \boldsymbol{r} \cdot \boldsymbol{p}]_{jj} = 0$ in a stationary state, $\psi_j(\boldsymbol{r})$. Thus,

$$\frac{\hbar}{i} \int \psi_j^* [H, \boldsymbol{r} \cdot \boldsymbol{p}] \psi_j \, dV = \frac{\hbar}{i} \int \psi_j^* (2T - \boldsymbol{r} \cdot \nabla V) \psi_j \, dV = 0$$

$$2\mathrm{T}_{jj} = (\boldsymbol{r} \cdot \nabla V)_{jj}, \tag{3-153}$$

which is the quantum mechanical virial theorem for a stationary state. If $V = Cr^n$, where n is any number and C is a constant, Eq.(3-153) reduces to

$$2\mathrm{T}_{jj} = n\mathrm{V}_{jj}.$$

$n = -1$ for an electrostatic potential between two charges a distance r apart, which reduces Eq.(3-153) to

$$2\mathrm{T}_{jj} = -\mathrm{V}_{jj}, \tag{3-154}$$

which is the quantum mechanical analog of Eq.(2-104) in classical mechanics. The derivation for a single particle above is easily extended to many particles.

3.7 TIME-DEPENDENT PERTURBATION THEORY AND THE SEMICLASSICAL INTERACTION BETWEEN RADIATION AND MATTER

When the Hamiltonian operator for a system contains time-dependence, the total Hamiltonian can normally be written as a sum of time-independent and time-dependent parts:

$$\mathscr{H}(\boldsymbol{r}, t) = \mathscr{H}_0(\boldsymbol{r}) + \mathscr{H}_1(\boldsymbol{r}, t). \tag{3-155}$$

The method of solution of the time-dependent Schrödinger equation in Eq.(3-58) is to first solve

$$\mathscr{H}_0 \boldsymbol{\psi}(\boldsymbol{r}) = \boldsymbol{\psi}(\boldsymbol{r})\boldsymbol{E} \tag{3-156}$$

for the stationary state energies, $\boldsymbol{E}$, and the eigenfunctions, $\boldsymbol{\psi}(\boldsymbol{r})$, as shown in Eqs.(3-58) to (3-62). We then write the complete time-dependent wavefunction, $\Psi(\boldsymbol{r}, t)$, as a linear combination of stationary states, $\psi(\boldsymbol{r})$, with time-dependent coefficients, $\mathrm{b}_i(t)$:

$$\Psi(\boldsymbol{r}, t) = \sum_i \psi_i(\boldsymbol{r})\mathrm{b}_i(t) = \boldsymbol{\psi}\mathbf{b}, \tag{3-157}$$

where $\mathbf{b}$ is a column matrix. The coefficients, $b_i(t)$, will now be related back to the time-dependent part of the Hamiltonian, $\mathscr{H}(\mathbf{r}, t)$, in Eq.(3-155). Substituting Eq.(3-157) and $\mathscr{H} = \mathscr{H}_0 + \mathscr{H}_1$ from Eq.(3-155) into Eq.(3-58) gives

$$(\mathscr{H}_0 + \mathscr{H}_1)\boldsymbol{\psi}\mathbf{b} = i\hbar\boldsymbol{\psi}\frac{\partial}{\partial t}\mathbf{b}. \tag{3-158}$$

Substituting Eq.(3-156) where $\boldsymbol{E}$ contains the stationary state eigenvalues, multiplying by $\boldsymbol{\psi}^\dagger$, and integrating gives

$$\frac{\partial}{\partial t}\mathbf{b} = \frac{1}{i\hbar}[\mathbf{E}\mathbf{b} + \mathscr{H}_1^{\psi}\mathbf{b}] \qquad \frac{\partial}{\partial t}b_j(t) = \frac{1}{i\hbar}[E_j b_j(t) + \sum_l (\mathscr{H}_1)_{jl} b_l(t)],$$

where

$$(\mathscr{H}_1)_{jl} = \int \psi_j^*(\mathbf{r})\mathscr{H}_1\psi_l(\mathbf{r})\,dV.$$

We now substitute

$$b_j(t) = \exp\left(-\frac{iE_jt}{\hbar}\right)c_j(t) \tag{3-159}$$

to give

$$-\frac{iE_j}{\hbar}\exp\left(-\frac{iE_jt}{\hbar}\right)c_j(t) + \exp\left(-\frac{iE_jt}{\hbar}\right)\frac{\partial c_j(t)}{\partial t}$$

$$= \frac{1}{i\hbar}\left[E_j\exp\left(-\frac{iE_jt}{\hbar}\right)c_j(t) + \sum_l(\mathscr{H}_1)_{jl}\exp\left(-\frac{iE_lt}{\hbar}\right)c_l(t)\right].$$

Canceling the first terms on each side of the equation and postmultiplying the result by $\exp(+iE_jt/\hbar)$ [$c_j(t)$ is a function only of t and we replace the partial with the total derivative] gives

$$\frac{dc_j(t)}{dt} = \frac{1}{i\hbar}\sum_l \exp\left(\frac{iE_jt}{\hbar}\right)(\mathscr{H}_1)_{jl}\exp\left(-\frac{iE_lt}{\hbar}\right)c_l(t) = \frac{1}{i\hbar}\sum_l(\mathscr{H}_1)_{jl}\exp(i\omega_{jl}t)c_l(t), \tag{3-160}$$

where we have also used the Bohr frequency rule,

$$\omega_{jl} = \frac{E_j - E_l}{\hbar}. \tag{3-161}$$

Now, expanding Eq.(3-157) [use also Eq.(3-159)] and (3-160) we write

$$\Psi(\mathbf{r}, t) = \psi_1\exp\left(-\frac{iE_1t}{\hbar}\right)c_1(t) + \psi_2\exp\left(-\frac{iE_2t}{\hbar}\right)c_2(t) + \cdots$$
$$+ \psi_k\exp\left(-\frac{iE_kt}{\hbar}\right)c_k(t) + \cdots$$

$$\frac{dc_j(t)}{dt} = \frac{1}{i\hbar}[(\mathscr{H}_1)_{j1}\exp(i\omega_{j1}t)c_1(t) + (\mathscr{H}_1)_{j2}\exp(i\omega_{j2}t)c_2(t) + \cdots$$
$$+ (\mathscr{H}_1)_{jk}\exp(i\omega_{jk}t)c_k(t) + \cdots]. \tag{3-162}$$

The complex square of the $b_j(t)$ or $c_j(t)$ coefficients gives the probability of finding the system in the jth stationary state described by $\psi_j(\boldsymbol{r})$. At time $t = 0$, $\mathscr{H}_1 = 0$ and the expansion of $\Psi(\boldsymbol{r}, t)$ in Eq.(3-162) is limited to a single term, say the kth, and thus $c_k(t = 0) = 1.0$ and all other $c_l(0)$ are zero. In the presence of the perturbation, however, $\mathscr{H}_1(t) \neq 0$ and all other $c_l(t)$ are nonzero. Thus, $c_j(t)$ for the new state is initially zero and grows in time according to Eq.(3-162). As a first approximation, we examine the growth of $c_j(t)$ over a very small time period where $c_k(t) \cong 1.0$ and all other $c_l(t)$, including $c_j(t)$, are approximately equal to zero. The first-order evaluation of $c_j(t)$ is then given by

$$\frac{dc_j(t)}{dt} \cong \frac{1}{i\hbar}(\mathscr{H}_1)_{jk} \exp(i\omega_{jk}t). \tag{3-163}$$

If another state, say the mth, is involved, the first-order growth is given by

$$\frac{dc_m(t)}{dt} \cong \frac{1}{i\hbar}(\mathscr{H}_1)_{mk} \exp(i\omega_{mk}t).$$

Now, there will also be second-order contributions to $dc_j(t)/dt$ and $dc_m(t)/dt$ according to Eq.(3-162), which are written for the mth state as

$$\frac{dc_m(t)}{dt} \cong \frac{1}{i\hbar}(\mathscr{H}_1)_{mk} \exp(i\omega_{mk}t) + \frac{1}{i\hbar}\sum_j (\mathscr{H}_1)_{mj} c_j(t) \exp(i\omega_{mj}t). \tag{3-164}$$

The $c_j(t)$ are obtained in Eq.(3-163) and substituted into Eq.(3-164) to give $c_m(t)$ correct to second order. This process can be repeated for higher orders of correction.

Semiclassical Interactions

In Sections 2.2 and 2.4 we discussed the classical interaction of radiation and matter. In this section we shall now use the time-dependent perturbation theory above to develop a semiclassical theory for the interaction of radiation and matter that describes the interaction between classical radiation with a quantum mechanical system. The Hamiltonian describing the interaction of electromagnetic fields with a particle of mass m and charge e is given in Eq.(2-95). Using the operator form for the linear momentum gives (cgs units are used here; see results of Problem 2-9 for the SI expression)

$$\begin{aligned}\mathscr{H} &= \frac{1}{2m}\left[\left(-i\hbar\frac{\partial}{\partial x} - \frac{e}{c}A_x\right)^2 + \left(-i\hbar\frac{\partial}{\partial y} - \frac{e}{c}A_y\right)^2 + \left(-i\hbar\frac{\partial}{\partial z} - \frac{e}{c}A_z\right)^2\right] + e\varphi \\ &= \frac{1}{2m}\left(-\hbar^2\nabla^2 + \frac{i\hbar e}{c}\nabla\cdot\boldsymbol{A} + \frac{i\hbar e}{c}\boldsymbol{A}\cdot\nabla + \frac{e^2}{c^2}A^2\right) + e\varphi.\end{aligned} \tag{3-165}$$

∇ is an operator and we rewrite the second term in this expression as

$$\nabla\cdot\boldsymbol{A} = (\nabla\cdot\boldsymbol{A}) + \boldsymbol{A}\cdot\nabla, \tag{3-166}$$

where the parentheses indicate the limits of the differentiation. We now choose $\varphi = 0$ and $(\nabla\cdot\boldsymbol{A}) = 0$ from Eq.(1-19) for an electromagnetic wave [see discussion following Eq.(1-28)] and we also add the potential energy of the electron, V. Substituting these results into Eq.(3-165) gives

$$\mathscr{H} = \underbrace{-\frac{\hbar^2}{2m}\nabla^2 + V}_{\mathscr{H}_0} + \underbrace{\frac{i\hbar e}{mc}\boldsymbol{A}\cdot\boldsymbol{\nabla}}_{\mathscr{H}_1} + \underbrace{\frac{e^2}{2mc^2}A^2}_{\mathscr{H}_2} \qquad (3\text{-}167)$$

for the final result. Substituting $\mathscr{H}_1$ from this equation into Eq.(3-163) for an electron (use negative sign on e) gives the first-order growth of the jth state from a system that is initially in the kth state:

$$\frac{dc_j(t)}{dt} = -\frac{e}{mc}(\boldsymbol{A}\cdot\boldsymbol{\nabla})_{jk}\exp(i\omega_{jk}t). \qquad (3\text{-}168)$$

Substituting the real part of the vector potential from Eq.(1-25),

$$\boldsymbol{A} = \boldsymbol{A}_0\cos(\boldsymbol{k}\cdot\boldsymbol{r} - \omega t) = \frac{\boldsymbol{A}_0}{2}\{\exp[i(\boldsymbol{k}\cdot\boldsymbol{r} - \omega t)] + \exp[-i(\boldsymbol{k}\cdot\boldsymbol{r} - \omega t)]\}$$

into Eq.(3-168) gives

$$\frac{dc_j(t)}{dt} = -\frac{e}{2mc}\exp[i(\omega_{jk} - \omega)t][\exp(i\boldsymbol{k}\cdot\boldsymbol{r})\boldsymbol{A}_0\cdot\boldsymbol{\nabla}]_{jk}$$

$$-\frac{e}{2mc}\exp[i(\omega_{jk} + \omega)t][\exp(-i\boldsymbol{k}\cdot\boldsymbol{r})\boldsymbol{A}_0\cdot\boldsymbol{\nabla}]_{jk}. \qquad (3\text{-}169)$$

We now expand $\exp(i\boldsymbol{k}\cdot\boldsymbol{r})$ for small $\boldsymbol{k}\cdot\boldsymbol{r}$ giving

$$\exp(i\boldsymbol{k}\cdot\boldsymbol{r}) = 1 + i\boldsymbol{k}\cdot\boldsymbol{r} + \tfrac{1}{2}(i\boldsymbol{k}\cdot\boldsymbol{r})^2 + \cdots. \qquad (3\text{-}170)$$

As $k = \omega/c = 2\pi/\lambda$ from Eq.(1-26), Eq.(3-170) will converge rapidly if the magnitude of $\boldsymbol{r}$ is less than the wavelength, λ, of the electromagnetic radiation. $\boldsymbol{r}$ is the vector distance from the system center of mass to the point of interaction as we have noted previously that the internal vibrational and rotational coordinates can be separated from the translational coordinates of the center of mass.

As the size of atoms and molecules indicates diameters on the order of 1 to 100 Å, the values of λ must be larger than 100 Å in order to assure convergence of the expansion in Eq.(3-170). For electromagnetic radiation in the X-ray region of the spectrum with wavelengths shorter than about 30 Å, the analysis above breaks down. This short wavelength electromagnetic radiation exhibits a particle-like behavior as shown in the Compton effect (see Fig. 1-10). For wavelengths in the optical and ultraviolet regions, 8000 $\longrightarrow$ 1000 Å, and longer wavelengths in the infrared and microwave regions of the electromagnetic spectrum, however, the summation in Eq.(3-170) is approximated quite well by the first term. We shall return to the effects of the $i\boldsymbol{k}\cdot\boldsymbol{r}$ term later in this section. The first-term approximation to $\exp(i\boldsymbol{k}\cdot\boldsymbol{r})$, when substituted into Eq.(3-169) gives

$$\frac{dc_j(t)}{dt} \cong -\frac{e}{2mc}\exp[i(\omega_{jk} - \omega)t](\boldsymbol{A}_0\cdot\boldsymbol{\nabla})_{jk} - \frac{e}{2mc}\exp[i(\omega_{jk} + \omega)t](\boldsymbol{A}_0\cdot\boldsymbol{\nabla})_{jk}.$$

$$(3\text{-}171)$$

Substituting $\mathbf{V}_{jk} = (m\omega_{jk}/\hbar e)(-e\mathbf{r})_{jk}$, or

$$(\mathbf{A}_0 \cdot \mathbf{V})_{jk} = \frac{m\omega_{jk}}{\hbar e}[\mathbf{A}_0 \cdot (-e\mathbf{r})]_{jk} = \frac{m\omega_{jk}}{\hbar e}(\mathbf{A}_0 \cdot \mathbf{D})_{jk} \tag{3-172}$$

from Eq.(3-152) into Eq.(3-171) gives

$$\frac{dc_j(t)}{dt} = -\frac{\omega_{jk}}{2\hbar c}\exp[i(\omega_{jk} - \omega)t](\mathbf{A}_0 \cdot \mathbf{D})_{jk} - \frac{\omega_{jk}}{2\hbar c}\exp[i(\omega_{jk} + \omega)t](\mathbf{A}_0 \cdot \mathbf{D})_{jk}, \tag{3-173}$$

where we have used the definition of the electric dipole moment of $\mathbf{D} = -e\mathbf{r}$. We can now obtain the value of $c_j(t)$ at time t by integrating and remembering that $c_j(t = 0) = 0$, giving

$$c_j(t) = -\frac{\omega_{jk}}{2i\hbar c}(\mathbf{A}_0 \cdot \mathbf{D})_{jk}\left\{\frac{\exp[i(\omega_{jk} - \omega)t] - 1}{(\omega_{jk} - \omega)}\right\}$$
$$- \frac{\omega_{jk}}{2i\hbar c}(\mathbf{A}_0 \cdot \mathbf{D})_{jk}\left\{\frac{\exp[i(\omega_{jk} + \omega)t] - 1}{(\omega_{jk} + \omega)}\right\}. \tag{3-174}$$

We can also use $\mathbf{E}_0 = (i\omega/c)\mathbf{A}_0$ from Eq.(1-29), to give

$$c_j(t) = \left(\frac{\omega_{jk}}{2\omega\hbar}\right)(\mathbf{E}_0 \cdot \mathbf{D})_{jk}\left\{\frac{\exp[i(\omega_{jk} - \omega)t] - 1}{(\omega_{jk} - \omega)}\right\}$$
$$+ \left(\frac{\omega_{jk}}{2\omega\hbar}\right)(\mathbf{E}_0 \cdot \mathbf{D})_{jk}\left\{\frac{\exp[i(\omega_{jk} + \omega)t] - 1}{(\omega_{jk} + \omega)}\right\}. \tag{3-175}$$

We note that if $E_k < E_j$, ω_{jk} is positive and the first terms in Eqs.(3-174) and (3-175) are much larger when $\omega \longrightarrow \omega_{jk}$ (near resonance) than the $\omega_{jk} + \omega$ second terms. If $E_k > E_j$, however, the $\omega_{jk} + \omega$ contribution in the second terms in $c_j(t)$ in Eqs.(3-174) and (3-175) will be dominant. Thus, the $\omega_{jk} - \omega$ terms are appropriate for absorption of electromagnetic energy and the $\omega_{jk} + \omega$ terms are appropriate for stimulated emission of electromagnetic energy. In the case of our example of $k \longrightarrow j$ absorption, we shall consider further only the first terms in Eqs.(3-174) and (3-175). The probability for the existence of the jth state at a later time, t, is the absolute value square of $c_j(t)$ given by

$$P_{jk}(\omega, t) = |c_j(t)|^2 = \frac{\omega_{jk}^2[(\mathbf{A}_0 \cdot \mathbf{D})_{jk}]^2 t^2}{4\hbar^2 c^2}\left\{\frac{\sin^2[(t/2)(\omega_{jk} - \omega)]}{[(t/2)(\omega_{jk} - \omega)]^2}\right\}. \tag{3-176}$$

$P_{jk}(\omega, t)$ (which is dimensionless) is the probability for the existence of the jth state or the probability of a $k \longrightarrow j$ transition at time t and at frequency ω. Of course, we can also write $P_{jk}(\nu, t)$ by using $\omega = 2\pi\nu$. $P_{jk}(\nu, t)$ is plotted in Fig. 3-1 as a function of ν for two values of time. We note from Fig. 3-1 that the maximum in the $P_{jk}(\nu, t)$ curve

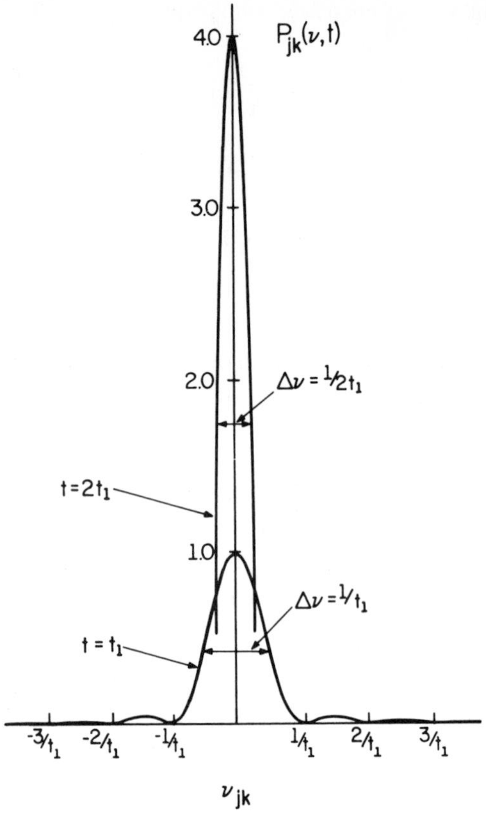

Figure 3-1 The probability curve for a $k \longrightarrow j$ transition in the absence of relaxation mechanisms for two different times t_1 and $2t_1$. The peak height increases with t^2 and the full width at $0.405P_{max}$ decreases according to $\Delta\nu \cong 1/t$. $P_{jk}(\nu, t)$ is from Eq.(3-176) when $\omega = 2\pi\nu$.

increases with the square of time and the full width decreases in time according to $\Delta\nu = 1/t$. The maximum is obtained by taking the limit as $\omega \longrightarrow \omega_{jk}$ giving

$$\lim_{(\omega_{jk}-\omega)\to 0} P_{jk}(\omega, t) = \frac{\omega_{jk}^2[(\boldsymbol{A}_0 \cdot \boldsymbol{D})_{jk}]^2}{4\hbar^2 c^2} t^2 = P_{max}.$$

It is evident from these results that the full width $\Delta\nu = 1/t$ at $0.405P_{max}$.

In certain molecular beam experiments, $P_{jk}(\nu, t)$ can be observed directly [18]. In these experiments a molecular (or atom) beam is formed and a single state k is selected by electrical focusing. The beam of molecules in the kth state then passes through the region that contains the time-dependent electric or magnetic (electromagnetic) field. Those molecules that are excited to the jth state are then focused onto a detector. Thus, only the excited molecules (state j) are detected and in the absence of relaxation processes the frequency dependence of the detected molecules is given by $P_{jk}(\omega, t)$, where t is the molecular transit time in the electromagnetic field. This is an unusual experiment, however, as most experiments measure the absorption coefficient or radiation power loss in a static gas, liquid, or solid. In this case, the time derivative of $P_{jk}(\nu, t)$ is measured as discussed in Chapter 7 when we treat absorption spectroscopy in more detail.

The probability per unit time or the rate constant for the transition is obtained by integrating $P_{jk}(\nu, t)$ over all frequencies. If A_0 is constant over a small range of frequencies near ν_{jk}, it can be factored out of the integral. We can now use

$$\int_{-\infty}^{+\infty} \frac{\sin^2 X}{X^2}\, dX = \pi,$$

where $X = \pi(\nu_{jk} - \nu)t$ and $dX = -\pi t\, d\nu$ where ν has limits from 0 to ∞. It is evident that as $\nu \longrightarrow 0$, $X \longrightarrow \pi\nu_{jk}t$ and as $\nu \longrightarrow \infty$, $X \longrightarrow -\infty$. Thus, in the long-time limit where $\pi\nu_{jk}t$ is a large number, we are justified in the change of variables to integrate X from $+\infty \longrightarrow -\infty$ as ν goes from the limits of $0 \longrightarrow \infty$. Thus, the total probability per unit time or the rate constant for the transition is

$$k_{jk} = \int_{-\infty}^{+\infty} P_{jk}(\nu, t)\, d\nu = \frac{\omega_{jk}^2[(\boldsymbol{A}_0 \cdot \mathbf{D})_{jk}]^2 t}{4\hbar^2 c^2}, \qquad (3\text{-}177)$$

where k_{jk} has the appropriate units for the first-order rate constant of s^{-1}. The results above show that the peak height in an absorption process increases with t^2, but the integrated transition probability or the total rate constant increases only with the first power of t. It is also evident from Eq.(3-177) that the rate constant for the reverse process is equal to the rate constant for the forward process:

$$k_{jk}(j \longleftarrow k) = k_{kj}(k \longleftarrow j).$$

Selection Rules, Plane Polarized and Circularly Polarized Radiation

In Chapter 4 we shall examine the selection rules for several systems that are interacting with both plane polarized and circularly polarized radiation. Plane polarized and circularly polarized radiation are both discussed in Section 1.1 beginning with Eq.(1-32). We now need to examine the nature of the $(\boldsymbol{A} \cdot \nabla)_{jk}$ matrix element in Eq.(3-168) for plane polarized and circularly polarized radiation.

For plane polarized radiation we choose z as the axis of polarization and y as the propagation axis to write

$$A_x = 0 \qquad A_y = 0 \qquad A_z = A_0 \cos(ky - \omega t),$$

which gives

$$\boldsymbol{A} \cdot \nabla = A_0 \cos(ky - \omega t) \frac{\partial}{\partial z}.$$

Following this operator from Eqs.(3-169) to (3-176) leads to

$$(\boldsymbol{A}_0 \cdot \mathbf{D})_{jk} = (A_0 D_z)_{jk},$$

where D_z is the electric dipole component along the laboratory-fixed z axis. Now, if the dipole moment of a cylindrically symmetric molecule (or system) is free to rotate, we write

$$(\boldsymbol{A}_0 \cdot \boldsymbol{D})_{jk} = A_0[D_a \cos\theta]_{jk},$$

where θ is the angle between the a dipole axis (cylindrical axis) in the molecule and the z axis in the laboratory-fixed system and we have also assumed that $\boldsymbol{A}_0$ is constant over the space described by the integral. If there are no angular variables in the eigenfunctions, ψ_j and ψ_k, and A_0 is a constant over the space described by ψ, we can write

$$(\boldsymbol{A}_0 \cdot \boldsymbol{D})_{jk} = A_0(D_a)_{jk} \cos\theta \qquad [(\boldsymbol{A}_0 \cdot \boldsymbol{D})_{jk}]^2 = A_0^2(D_a)_{jk}^2 \cos^2\theta.$$

Now, if the atom or molecule carrying D_a is free to rotate in three dimensions, we average this expression over the spherical polar solid angle, $\sin\theta\, d\theta\, d\phi$, to give

$$\frac{\int_0^{2\pi}\int_0^{\pi} A_0^2 D_{jk}^2 \cos^2\theta \sin\theta\, d\theta\, d\phi}{\int_0^{2\pi}\int_0^{\pi} \sin\theta\, d\theta\, d\phi} = \frac{1}{3} A_0^2 D_{jk}^2. \tag{3-178}$$

In the case of circularly polarized radiation we switch axis systems for convenience and let the radiation propagate along the z axis and define circularly polarized radiation as arising from the circularly polarized vector potential $\boldsymbol{A}_\pm$ in the xy plane according to [see Eq.(1-33)]

$$\boldsymbol{A} = \boldsymbol{A}_\pm = A_0[\hat{\boldsymbol{i}} \cos(kz - \omega t) \pm \hat{\boldsymbol{j}} \sin(kz - \omega t)],$$

where $A_z = 0$. The interpretation of right and left circular polarization is given following Eq.(1-33). Now, the circularly polarized vector potential will couple with the gradient according to

$$\boldsymbol{A} \cdot \boldsymbol{\nabla} = \boldsymbol{A}_\pm \cdot \boldsymbol{\nabla} = A_0(\hat{\boldsymbol{i}} \cos a \pm \hat{\boldsymbol{j}} \sin a) \cdot \left(\hat{\boldsymbol{i}} \frac{\partial}{\partial x} + \hat{\boldsymbol{j}} \frac{\partial}{\partial y} + \hat{\boldsymbol{k}} \frac{\partial}{\partial z}\right)$$

$$= A_0\left(\cos a \frac{\partial}{\partial x} \pm \sin a \frac{\partial}{\partial y}\right)$$

$$a = kz - \omega t.$$

Expanding the sines and cosines gives

$$\boldsymbol{A}_\pm \cdot \boldsymbol{\nabla} = \frac{A_0}{2}\left\{[\exp(ia) + \exp(-ia)] \frac{\partial}{\partial x} \pm i[\exp(ia) - \exp(-ia)] \frac{\partial}{\partial y}\right\}$$

$$= \frac{A_0}{2}\left[\exp(ia)\left(\frac{\partial}{\partial x} \pm i \frac{\partial}{\partial y}\right) + \exp(-ia)\left(\frac{\partial}{\partial x} \mp i \frac{\partial}{\partial y}\right)\right]. \tag{3-179}$$

Using this operator for circularly polarized radiation in place of $\boldsymbol{A} \cdot \boldsymbol{\nabla}$ in Eq.(3-168) and following through to Eq.(3-174) shows that the $\exp(ia)$ terms above lead to absorption, the $\exp(-ia)$ terms lead to emission, and the $(\boldsymbol{A}_0 \cdot \boldsymbol{D})_{jk}$ terms in Eq.(3-168) are replaced with $A_0(D_x \pm iD_y)$. Thus, we note that the $D_x + iD_y$ dipole operator couples with right circularly polarized radiation and $D_x - iD_y$ couples with left circularly polarized radiation.

Radiation Rate Constants

We now return to the case of plane-polarized radiation and randomly oriented dipoles and substitute Eq.(3-178) into Eq.(3-177) to give the rate constant for the randomly oriented dipoles where D_{jk} contains no angular dependence.

$$k_{jk} = \frac{\omega_{jk}^2 A_0^2 D_{jk}^2 t}{12\hbar^2 c^2}. \tag{3-180}$$

A_0^2 can be related to the Planck radiation formula for the energy density of a thermal source. The time average of the squares of the electric and magnetic fields, $\langle E^2 \rangle$ and $\langle H^2 \rangle$, are related to the magnitude of the vector potential as shown following Eq.(1-36). The values of A_0^2 can also be related to the energy density giving

$$A_0^2 = 8\pi \frac{c^2}{\omega^2} \rho. \tag{3-181}$$

The energy density per unit frequency interval of electromagnetic radiation emitted from a thermal source is given by the Planck radiation law in Eq.(1-73). Multiplying $\rho(\nu)$ in Eq.(1-73) by $\Delta\nu$, which is the frequency band being absorbed by the sample, and substituting this result into Eq.(3-181) gives

$$A_0^2 = \frac{16\nu h}{c}\left[\frac{1}{\exp(h\nu/kT) - 1}\right]\Delta\nu. \tag{3-182}$$

Substituting Eq.(3-182) into Eq.(3-180) gives the rate constant:

$$k_{jk} = \frac{4\omega_{jk}^3 D_{jk}^2 t\Delta\nu}{3\hbar c^3}\left[\frac{1}{\exp(\hbar\omega_{jk}/kT) - 1}\right]. \tag{3-183}$$

According to Eq.(3-176) the width of frequencies absorbed by the system, $\Delta\nu$, is approximately equal to the full width at half height of the absorption curve. Thus, we use $\Delta\nu t \cong 1$, which gives the following expression for the rate constant for the $k \longrightarrow j$ transition under the influence of the thermal source:

$$k_{jk} = \frac{4\omega_{jk}^3 D_{jk}^2}{3\hbar c^3}\left[\frac{1}{\exp(\hbar\omega_{jk}/kT) - 1}\right]. \tag{3-184}$$

We shall now evaluate the rate constant, k_{jk}, for several examples including typical electronic, vibrational, and rotational transitions. According to Fig. 1-8 we can establish approximate values of ω_{jk} for the transitions as follows: electronic (6 eV transition), $\omega_E = 2\pi\nu_E \cong 10^{16}$ rad · s^{-1}; vibrational (520 cm^{-1} transition), $\omega_V \cong 10^{14}$ rad · s^{-1}; and rotational (16 GHz transition), $\omega_R \cong 10^{11}$ rad · s^{-1}. Furthermore, we shall assume a thermal source of radiation at 4000 K and a dipole moment matrix element of $D_{jk} \cong 10^{-18}$ SC · cm. A 4000 K thermal source is somewhere between a glo-bar temperature of 1000–2000 K and a carbon arc temperature of 6000–8000 K. The value of $D_{jk} \cong 10^{-18}$ SC · cm is typical for an atomic or molecular electronic transition. Similarly, $D_{jk} \cong 10^{-18}$ SC · cm is also appropriate for vibrational and rotational transitions, as we shall note later. Using these estimates allows an estimation of k_{ik} for each type of transition ($T = 4000$ K, $D_{jk} = 10^{-18}$

SC · cm) giving $k_{jk} \cong 10^{-1}\ s^{-1}$, $10^2\ s^{-1}$, and $10^6\ s^{-1}$ for the electronic, vibrational, and rotational transitions, respectively.

Consider now a large number of molecules in contact with a radiation source at temperature T that causes the $j \longleftrightarrow k$ transition according to the $k_{jk} = k_{kj}$ rates discussed above. We indicate the number of molecules per unit volume in the kth and jth states as N_k and N_j, respectively. The time rate of change of the number of molecules per unit volume in the kth state is given by the first-order rate law:

$$\frac{dN_k}{dt} = -k_{jk}N_k + k_{kj}N_j. \tag{3-185}$$

Thus, in the presence of the radiation, N_k *decreases* in time because of $k \longrightarrow j$ transitions and the value of N_k *increases* because of $j \longrightarrow k$ transitions. The time rate of change for the number density of molecules in the jth state is given by

$$\frac{dN_j}{dt} = k_{jk}N_k - k_{kj}N_j. \tag{3-186}$$

Subtracting Eq.(3-186) from Eq.(3-185) and remembering that $k_{jk} = k_{kj}$ for the electromagnetic induced rates gives $d\Delta N/\Delta N = -2k_{jk}dt$, where $\Delta N = N_k - N_j$. Solving for $\Delta N(t)$ gives

$$\Delta N(t) = \Delta N_0 \exp(-2k_{jk}t), \tag{3-187}$$

where ΔN_0 is the initial value of $N_k - N_j$ at $t = 0$ and $\Delta N(t)$ is the value of $N_k - N_j$ at t.

We shall now assume that, in the absence of the electromagnetic perturbation, the two-level system is in thermal equilibrium with the relative populations being governed by the Boltzmann equation:

$$\frac{N_j}{N_k} = \exp\left[-\frac{E_j - E_k}{kT}\right], \tag{3-188}$$

where we have also assumed that the kth and jth energy levels have equal degeneracies. It is now evident from Eq.(3-187) that the electromagnetic perturbation which causes transitions from the lower kth to the upper jth levels will drive the system to populations which are in thermal equilibrium with the radiation source. According to Eq.(3-187), the time necessary to reduce the original value (before the radiation) of ΔN_0 to $1/e$ of its initial value is only $1/2k_{jk}$. We can now compare our previous estimates of k_{jk} to note that the $1/2k_{jk}$ times for electronic, vibrational, and rotational transitions are approximately 5 s, 5×10^{-3} s, and 5×10^{-7} s, respectively. Thus, in the absence of any effects to return the system to equilibrium, all energy levels would *saturate* in a very short time and no further absorption of radiation energy would be observed as the probability for absorption (k_{jk}) is equal to the corresponding probability for emission (k_{kj}). Of course, we know from a good deal of experience that it is relatively difficult to drive a system of molecular energy levels to the point of saturation. Thus, nature must provide a number of relaxation processes to return the molecular energy levels to thermal equilibrium. These relaxation processes are normally termed *radiative* (spontaneous emission) and *nonradiative* (collisional deactivation by energy transfer).

Spontaneous Emission

Experimentally we observe that in the absence of the collisional deactivation of an excited system by the environment, atoms and molecules proceed to equilibrium by spontaneous emission from the excited state. We can understand spontaneous emission by making an equilibrium argument. In the presence of electromagnetic radiation from a thermal source, the number of systems (per unit volume) making a transition from the lower to the upper state $k \longrightarrow j$ is $N_k k_{jk}$ (per unit time) and the number making a transition from the upper to the lower state $j \longrightarrow k$ is $N_j(k_{kj} + A_{kj})$. $k_{jk} = k_{kj}$ is the total rate constant or probability per unit time for an electromagnetic transition as shown in Eq.(3-184). A_{kj} is the rate constant or probability per unit time of a spontaneous emission from the upper to the lower state. At equilibrium we have

$$N_k k_{jk} = N_j(k_{jk} + A_{kj}). \tag{3-189}$$

Rearranging, using Boltzmann's equation in Eq.(3-188), and solving for A_{kj} gives

$$A_{kj} = k_{jk}\left[\exp\left(\frac{h\nu_{jk}}{kT}\right) - 1\right]. \tag{3-190}$$

Assuming that the two-level system is in equilibrium with a blackbody source of radiation, we can substitute k_{jk} from Eq.(3-184) to give

$$A_{kj} = \frac{4\omega_{jk}^3 D_{jk}^2}{3\hbar c^3}. \tag{3-191}$$

It is clear from Eqs.(3-190) and (3-191) that when $h\nu_{jk} \gg kT$, $A_{kj} \gg k_{jk}$ and spontaneous emission is a very important process [19]. Typical rates of spontaneous emission for electronic, vibrational, and rotational transitions are ($D_{jk} = 10^{-13}$ SC · cm) electronic ($\omega_{jk} = 10^{16}$), $A_{kj} \cong 4 \times 10^7\ \mathrm{s}^{-1}$; vibrational ($\omega_{jk} = 10^{14}$), $A_{kj} = 40\ \mathrm{s}^{-1}$; and rotational ($\omega_{jk} = 10^{11}$), $A_{kj} = 4 \times 10^{-8}\ \mathrm{s}^{-1}$. In the case of electronic energy level spacings, spontaneous emission occurs at a rate much faster than induced absorption or emission (from a thermal ($T \cong 8000$ K) source of radiation), $A_{kj} > k_{jk} = k_{kj}$. In the case of vibrational energy levels, $A_{kj} \cong k_{jk}$. In the case of rotational energy levels, the rate of spontaneous emission is negligible relative to the induced rate, $A_{kj} \ll k_{jk}$.

Using a first-order rate law to compute the rate of decay of molecules in the excited jth state gives

$$\frac{dN_j}{dt} = -A_{kj}N_j \qquad N_j(t) = N_j(t = 0)\exp(-A_{jk}t). \tag{3-192}$$

The time taken for $1/e$ of the initial number of molecules in the jth state $[N_j(t = 0)]$ to decay is called the *lifetime* or *relaxation time* of the state, τ_j, and it is clear from Eq.(3-192) that

$$\tau_j = \frac{1}{A_{jk}}. \tag{3-193}$$

Collisional Relaxation

Collisions can also maintain or drive a system to equilibrium. Consider the average molecular speed, v, from kinetic theory, which is

$$v = \left(\frac{8kT}{\pi M}\right)^{1/2}. \tag{3-194}$$

For carbon monoxide (CO) at 273°K, we have $v = 4.50 \times 10^2\ \mathrm{m \cdot s^{-1}}$. Further derivations from kinetic theory give the mean collision time, t_c, as the mean free path, $\lambda = [(2)^{1/2}\pi\rho_0 d^2]^{-1}$ divided by the molecular speed, v. Thus,

$$t_c = \frac{\lambda}{v} = \frac{1}{(2)^{1/2}\pi\rho_0 d^2 v}, \tag{3-195}$$

where ρ_0 is the molecular number density and d is the effective molecular diameter. For CO at standard conditions of $p = 1$ atm, $T = 273$°K, and assuming $d = 2 \times 10^{-10}$ m, we have $t_c = 4.7 \times 10^{-10}$ s. Comparing this t_c with the typical spontaneous emission lifetimes given earlier shows that the time between collisions in CO at 1 atm is short relative to the natural lifetimes. Thus, if each collision in time t_c causes an energy exchange that drives the system toward equilibrium, the collisional relaxation time will be competitive with the spontaneous emission mechanism for optical transitions and it will be the dominant relaxation mechanism in the case of vibrational and rotational transitions. The mechanisms of collisional energy transfer are much more efficient between the lower energy rotational energy levels, however, than for vibrational and electronic energy levels. Thus, the collisional relaxation rates have the following trend in the rotational-vibrational-electronic series: $k_R > k_V > k_E$. These compare to the reverse trend of $A_R < A_V < A_E$, as shown previously for the rate of spontaneous emission. Thus, the spontaneous emission relaxation is normally the dominant relaxation mechanism for retaining thermal equilibrium in the case of electronic energy spacings and the collisional relaxation mechanisms normally become more important for vibrational and rotational energy levels.

Magnetic Dipole and Electric Quadrupole Interactions

All the discussions above involved the electric dipole mechanism of interaction with the radiation field. This result stemmed from truncating Eq.(3-170) at the first term. The higher terms in the exp $(i\boldsymbol{k} \cdot \boldsymbol{r})$ expansion in Eq.(3-170) for the plane wave solution to Maxwell's equations for electromagnetic radiation yield higher-order transition mechanisms [20]. The second, or $i\boldsymbol{k} \cdot \boldsymbol{r}$ term in Eq.(3-170), leads to the *electric quadrupole* and *magnetic dipole* transition probabilities, which we consider now.

Substituting $i\boldsymbol{k} \cdot \boldsymbol{r}$ in Eq.(3-170) in place of exp $(i\boldsymbol{k} \cdot \boldsymbol{r})$ in the first term of Eq.(3-169) gives the following differential equation involving the growth of the jth state from the initial kth state:

$$\frac{dc_j(t)}{dt} = -\frac{e}{mc} \exp\,[i(\omega_{jk} - \omega)t] \int \psi_j^*(i\boldsymbol{k} \cdot \boldsymbol{r})(\boldsymbol{A}_0 \cdot \boldsymbol{\nabla})\psi_k\, dV. \tag{3-196}$$

We can rewrite the form of the operator as

$$\begin{aligned}(\boldsymbol{k}\cdot\boldsymbol{r})(\boldsymbol{A}\cdot\nabla) &= (k_x x + k_y y + k_z z)(A_x\nabla_x + A_y\nabla_y + A_z\nabla_z)\\ &= \tfrac{1}{2}(\boldsymbol{k}\times\boldsymbol{A})\cdot(\boldsymbol{r}\times\nabla) + \frac{1}{2}\sum_{\substack{\alpha,\beta=x\\ y\\ z}} k_\alpha A_\beta(\alpha\nabla_\beta + \beta\nabla_\alpha).\end{aligned} \tag{3-197}$$

The first term in this expression leads to the magnetic dipole moment mechanism of interaction and the second term leads to the electric quadrupole moment term. Starting with the first term we substitute the linear momentum operator $\boldsymbol{p} = -i\hbar\nabla$ to give

$$\tfrac{1}{2}(\boldsymbol{k}\times\boldsymbol{A}_0)\cdot(\boldsymbol{r}\times\nabla) = -\frac{1}{2i\hbar}(\boldsymbol{k}\times\boldsymbol{A}_0)\cdot(\boldsymbol{r}\times\boldsymbol{p}) = -\frac{1}{2i\hbar}(\boldsymbol{k}\times\boldsymbol{A}_0)\cdot\boldsymbol{L},$$

where $\boldsymbol{L}$ is the angular momentum operator. We now note that the magnetic dipole operator is proportional to the angular momentum and is given by $\boldsymbol{\mu} = -(\mu_B/\hbar)\boldsymbol{L}$, where μ_B is the Bohr magneton [see Eq.(4-20)]. Substituting gives

$$\tfrac{1}{2}(\boldsymbol{k}\times\boldsymbol{A}_0)\cdot(\boldsymbol{r}\times\nabla) = -\frac{1}{2i\mu_B}(\boldsymbol{k}\times\boldsymbol{A}_0)\cdot\boldsymbol{\mu}.$$

Substituting this result back into Eq.(3-196) and following steps identical to those followed from Eqs.(3-170) to (3-184) shows all the same results except that the electric dipole term $|(\boldsymbol{A}_0\cdot\boldsymbol{D})_{jk}|^2$ is replaced by the magnetic dipole term $|(\boldsymbol{A}_0\cdot\boldsymbol{\mu})_{jk}|^2$ where the magnetic dipole moment matrix element is given by

$$(\boldsymbol{A}_0\cdot\boldsymbol{\mu})_{jk} = \int \psi_j^* \boldsymbol{A}_0\cdot\boldsymbol{\mu}\psi_k\, dV. \tag{3-198}$$

Returning to the second term in Eq.(3-186), we consider now a single term $k_x A_y(x\nabla_y + y\nabla_x)$ that will lead to electric quadrupole transitions. Substituting into Eq.(3-196) gives

$$\frac{dc_j}{dt} = -\frac{ie}{mc}\exp\,[i(\omega_{jk}-\omega)t]k_x A_y \int \psi_j^*\left(x\frac{\partial}{\partial y} + y\frac{\partial}{\partial x}\right)\psi_k\, dV. \tag{3-199}$$

We can evaluate the integral with the help of Eq.(3-145). When $G = xy$ and $\mathscr{H} = -(\hbar^2\nabla^2/2m) + V$ with commutator

$$[\mathscr{H}, xy] = -\frac{\hbar^2}{m}\left(x\frac{\partial}{\partial y} + y\frac{\partial}{\partial x}\right),$$

Eq.(3-145) leads to

$$\frac{d}{dt}\exp\,(i\omega_{jk}t)\int \psi_j^*(xy)\psi_k\, dV = -\frac{i\hbar}{m}\exp\,(i\omega_{jk}t)\int \psi_j^*\left(x\frac{\partial}{\partial y} + y\frac{\partial}{\partial x}\right)\psi_k\, dV.$$

Differentiating, substituting the result into Eq.(3-199), and following the theory through as before leads to expressions similar to those obtained earlier where the components of the electric dipole moment $\boldsymbol{D}_{jk}$ are replaced by the following components of the electric quadrupole moment:

$$\frac{\omega_{jk}}{c} Q(xy)_{jk} = \frac{\omega_{jk}}{c} \int \psi_j^*(exy)\psi_k \, dV. \tag{3-200}$$

Substituting a few numbers shows that for atoms and molecules the strongest transitions will arise from electric dipole transitions, the next strongest will arise from magnetic dipole transitions, and the weakest transitions will arise from the electric quadrupole-type transitions.

Selection Rules

In summary, D_{jk}, μ_{jk}, and $(\omega_{jk}/c)Q(xy)_{jk}$ represent the electric dipole, magnetic dipole, and electric quadrupole transitions, respectively, and they are listed in decreasing order of importance. A transition is *allowed* if its matrix element over the interaction operator (electric or magnetic dipole or electric quadrupole) is nonzero. The transition is *forbidden* if the matrix element is zero. If the value of D_{jk} is zero, the transition is not allowed by the electric dipole mechanism but may be allowed by either the magnetic dipole or the electric quadrupole mechanism. These latter two mechanisms generally lead to a much weaker transition probability.

In the following three chapters we shall consider selection rules for plane polarized and circularly polarized radiation. The general approach is to write down the Hamiltonian and transition moment matrices $\mathcal{H}^{\Phi}$ and $\boldsymbol{T}^{\Phi}$ in a convenient basis, $\boldsymbol{\Phi}$. $\mathcal{H}^{\Phi}$ is then diagonalized by a unitary transformation, $\boldsymbol{u}$, and $\boldsymbol{T}^{\Phi}$ is transformed to the same basis by $\boldsymbol{u}^{\dagger}\boldsymbol{T}^{\Phi}\boldsymbol{u} = \boldsymbol{M}$. The elements in $\boldsymbol{M}$ are then squared to obtain the probability for the transition.

We shall give a detailed discussion of the measurement of the transition moments by spectroscopy in Chapter 7, along with the concepts of relaxation processes, transition line shapes, and correlation functions.

PROBLEMS

1. Calculate the Bohr radius for the Be^{3+} ion.
2. At what atomic number would the Bohr radius be enclosed in a spherical nucleus if the nucleus had a diameter of 10^{-12} cm? What would happen to the electron under these conditions?
3. According to the Bohr theory, what is the instantaneous linear velocity of the electron in the first Bohr orbit in the hydrogen atom?
4. One of the main sources of electromagnetic energy from interstellar sources is emission from hydrogen atoms arising from a combination of an electron and a proton $[e^- + p^+ = \mathrm{H}\ (\text{excited})]$. The radiation is emitted when an excited state of the hydrogen atom falls to the next lower state.

(a) Use the Bohr theory and Eq.(3-8) to compute which hydrogen atom emission ($\Delta n = -1$) will occur near a formaldehyde rotational transition at $\nu = 4829.7$ MHz. If the line widths are on the order of 1 MHz, would you expect any confusion in identifying the formaldehyde and hydrogen atom spectra? The best known value of the constant in Eq.(3-8) is $(1/hc)(e^4\mu/2\hbar^2) = 1.09677576(3) \times 10^5\ \text{cm}^{-1}$.

(b) What is the radius of the electron in the $n = 100$ state of the hydrogen atom? What is the instantaneous electric dipole moment of H with the electron in the $n = 100$ state?

5. In Section 1.5 we defined the Compton wavelength of an electron. In Section 3.2 we discussed the de Broglie wavelength of a particle. What are the similarities and differences in the Compton and de Broglie definitions for the wavelength of an electron?

6. Using the angular momentum definition in Eq.(2-2), compute the commutators for $[x, J_x]$, $[x, J_y]$, $[x, J_z]$, $[\mathbf{r} \cdot \mathbf{p}, J_x]$, $[J^2, \mathscr{H}]$, $[J_x, \mathscr{H}]$, $[J_y, \mathscr{H}]$, and $[J_z, \mathscr{H}]$. Use $\mathscr{H} = -(\hbar/2m)\nabla^2 + V(r)$ for the Hamiltonian.

7. Show that the x, p_x, and J_x operators are Hermitian.

8. Show that if A is a Hermitian operator, A^2 is also Hermitian. Show that the diagonal elements of the matrix representation of the square of a Hermitian operator are real positive.

9. Consider the following functions:

$$\psi_1 = 1 \qquad \psi_2 = x \qquad \psi_3 = x^2 + bx + c.$$

(a) Show that ψ_1 and ψ_2 are orthogonal over the interval $(-1, 1)$.

(b) Find b and c such that ψ_3 is orthogonal to ψ_1 and ψ_2 over the interval $(-1, 1)$.

(c) Normalize the functions above to obtain the orthonormal set.

(d) Expand $f(x) = 2x^2 + 3x - 6$ in terms of the orthonormal set above. Some orthogonalization methods are discussed in Appendix D.1.

10. Prove that the diagonal sums of the matrices **BA** and **AB** are equal.

11. Prove that the diagonal sum of the product of three matrices **ABC** is invariant to cyclic permutations.

12. Consider a symmetric matrix **A** that is diagonalized by a unitary transformation **u** to give diagonal values $\boldsymbol{\lambda}$,

$$\mathbf{u}^\dagger \mathbf{A} \mathbf{u} = \boldsymbol{\lambda}.$$

Prove that $\mathbf{u}^\dagger(\mathbf{c} + \mathbf{A})\mathbf{u} = \mathbf{c} + \mathbf{u}^\dagger\mathbf{A}\mathbf{u}$ only if **c** is diagonal and all elements in **c** are identical.

13. **A** and **B** are defined by

$$\mathbf{A} = \begin{pmatrix} -2.0 & 0.5 & 0 \\ 0.5 & -3.0 & 0 \\ 0 & 0 & 1.0 \end{pmatrix} \qquad \mathbf{B} = \begin{pmatrix} 1.0 & 0 & 0 \\ 0 & 2.0 & 0 \\ 0 & 0 & 3.0 \end{pmatrix}.$$

(a) Diagonalize $\mathbf{A}$ with $\mathbf{u}$ to give $\boldsymbol{\lambda}_A$ and show the $\mathbf{u}$ and $\boldsymbol{\lambda}_A$ matrices.

$$\mathbf{u}^\dagger \mathbf{A} \mathbf{u} = \boldsymbol{\lambda}_A$$

(b) Next, diagonalize $\mathbf{A} + \mathbf{B}$

$$\mathbf{T}^\dagger(\mathbf{A} + \mathbf{B})\mathbf{T} = \boldsymbol{\lambda}$$

It should be evident that $\boldsymbol{\lambda}_A + \mathbf{B} \neq \boldsymbol{\lambda}$.

14. Prove that the sum of diagonal elements of a square matrix (called the *trace* of the matrix) is invariant under a unitary transformation.

15. Consider a Hamiltonian that is the sum of a zero-order term, $\mathscr{H}_0$, plus a smaller perturbation $\mathscr{H}_1$. Consider a two-state system in the absence of the perturbation described by $\mathscr{H}_0 \mathbf{\Phi} = \mathbf{\Phi}\mathbf{E}$ and $\mathbf{\Phi} = (\Phi_0 \Phi_1)$. In the presence of the perturbation, the unnormalized ground-state wavefunction corrected to first order is

$$\psi_0 = \Phi_0 + \frac{(\mathscr{H}_1)_{10}^{\Phi}}{E_0 - E_1} \Phi_1. \tag{3-201}$$

Show that the (variational) energy computed with ψ_0 includes energy contributions to second order in perturbation theory. In other words, a wavefunction, correct to first order, leads to an energy that is correct at least to second order.

REFERENCES

[1] N. BOHR, *Phil. Mag.* **26**, 1 (1913). Of course, major events such as the Bohr atom are complicated developments and many people are involved. An interesting account of the concepts leading to the Bohr atom is given in J. L. Heilbron and T. S. Kuhn, *Historical Studies of the Physical Sciences*, Ed. by R. McCormmach, Vol. 1, 211 (University of Pennsylvania Press, Philadelphia, 1969).

[2] The modification or rejection of metaphysical causality by quantum mechanical ideas is discussed in detail in M. JAMMER, *The Conceptual Development of Quantum Mechanics* (McGraw-Hill, New York, 1966). Causality and quantum theory and its relation to other intellectual movements in Germany from 1918–1927 is examined in P. FORMAN, *Historical Studies of the Physical Sciences*, Vol. 3, Ed. by R. McCormmach (University of Pennsylvania Press, Philadelphia, 1971), p. 1.

[3] The general philosophy of indeterminancy in nature goes much further than the simple discussion given here. In fact, the Copenhagen school of thought would argue that the indeterminancy exists in the absence of the measurement. Further discussion of the indeterminancy of nature is given in M. JAMMER, *The Philosophy of Quantum Mechanics* (John Wiley & Sons, New York, 1974), which also gives an excellent historical perspective on the arguments. See also B. S. DEWITT and R. N. GRAHAM, *Am. J. Phys.* **39**, 724 (1971).

[4] L. DE BROGLIE, *Nature*, **112**, 540 (1923) and *Phil. Mag.* **47**, 446 (1924).

[5] C. DAVISSON and L. H. GERMER, *Nature* **119**, 558 (1927) and *Phys. Rev.* **30**, 705 (1927).

[6] J. M. HASTINGS and W. C. HAMILTON, "Neutron Scattering," *Physical Methods in Chemistry*, Ed. by A. Weissberger and B. Rossiter (Wiley-Interscience, New York, 1972). See summary of applications to biologically important structures in B. P. SCHOENBORN, *Chemical and Engineering News*, 31, (January 24, 1977).

[7] E. Schrödinger, *Ann. d. Phys.* **79**, 361, 489 (1926).

[8] H. Margenau and G. M. Murphy, *The Mathematics of Physics and Chemistry* (D. Van Nostrand Co., Inc., New York, 1956).

[9] P. A. M. Dirac, *The Principles of Quantum Mechanics*, 4th ed. (Oxford at the Clarendon Press, 1957).

[10] L. D. Landau and E. M. Lifshitz, *Quantum Mechanics* (Pergamon Press, New York, 1974).

[11] A. Messiah, *Quantum Mechanics*, Vol. I (North Holland Publishing Co., Amsterdam, 1961).

[12] *Quantum Theory I. Elements*, Ed. by D. R. Bates (Academic Press, New York, 1961).

[13] See Appendix B.1 for more details on matrices, vectors, and the analogies being drawn upon here.

[14] Additional details involving perturbation theory, including higher-order terms, are found in A. Dalgarno, *Quantum Theory I. Elements*, Ed. by D. R. Bates (Academic Press, New York, 1961).

[15] See more details on the variation method in S. T. Epstein, *The Variation Method in Quantum Chemistry* (Academic Press, New York, 1974).

[16] There is still considerable controversy in this definition as discussed in M. Jammer, *The Philosophy of Quantum Mechanics* (Wiley-Interscience, New York, 1974). See also L. E. Ballentine, *Rev. Mod. Phys.* **42**, 358 (1970).

[17] More details, including a discussion of time inversion symmetry, are given in S. M. Blinder, *Foundations of Quantum Dynamics* (Academic Press, New York, 1974).

[18] N. F. Ramsey, *Molecular Beams* (Oxford University Press, Oxford, 1963) and T. R. Dyke *et al.*, *J. Chem. Phys.* **57**, 2277 (1972).

[19] R. Loudon, *The Quantum Theory of Light* (Clarendon Press, Oxford, 1973).

[20] L. R. B. Elton, *Introductory Nuclear Theory* (Interscience Publishers, Inc., New York, 1959).

4

model quantum mechanical problems

4.1 PARTICLE-IN-A-RING

The particle-in-a-ring is a good starting point for a discussion of model problems because many of the features that arise in more sophisticated systems are also evident in this example. Consider a particle of mass M constrained to a planar, circular orbit (see Fig. 4-1) with constant r, but variable ϕ. The energy levels can be obtained by applying de Broglie's relation between the wavelength and linear momentum. According to de Broglie's wave concept for particles, the particle will exhibit wavelike properties with an associated wavelength λ, which must be a multiple of the circumference of the ring, given by

$$m\lambda = 2\pi r, \qquad m = 0, \pm 1, \pm 2, \ldots, \tag{4-1}$$

in order that the wave not interfere destructively with itself. Using this and Eq.(3-10) gives the quantized momentum of the particle,

$$p = \frac{h}{\lambda} = \frac{hm}{2\pi r} = \frac{\hbar m}{r}. \tag{4-2}$$

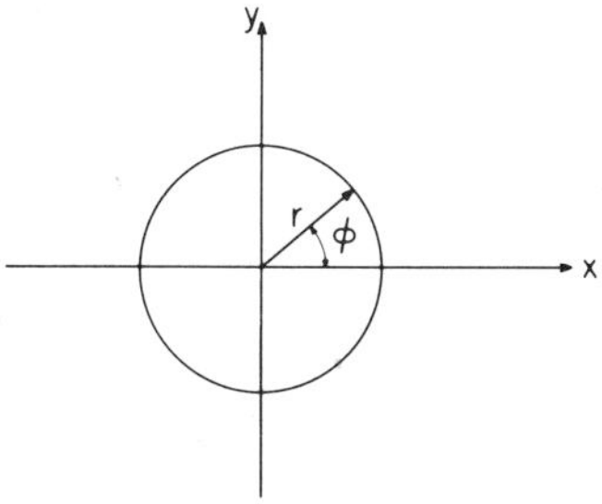

Figure 4-1 Coordinate system for the particle-in-a-ring

Substituting the momentum into the kinetic energy expression for the particle and adding the constant potential energy gives the energy of the particle-in-a-ring according to de Broglie:

$$E = T + V = \frac{p^2}{2M} + V = \frac{\hbar^2 m^2}{2Mr^2} + V = \frac{\hbar^2 m^2}{2\mathrm{I}_{zz}} + V, \tag{4-3}$$

where $\mathrm{I}_{zz} = Mr^2$ is the moment of inertia for the particle-in-a-ring and V is the constant potential energy for a constant radius, r.

We shall now write the classical Hamiltonian, convert it to the operator form, and find the eigenfunctions that describe the stationary states for the Hamiltonian operator. Referring to Fig. 4-1 and Sections 2.1 and 2.4, we can write the Hamiltonian for the particle-in-a-ring as

$$H = T + V = \frac{\mathrm{I}_{zz}\omega^2}{2} + V = \frac{J_z^2}{2\mathrm{I}_{zz}} + V, \tag{4-4}$$

where ω is the angular frequency and J_z is the angular momentum. Using the definition of the angular momentum in terms of the linear momentum from Eq.(2-2) and converting the Hamiltonian to operator form from Table 3-1 gives

$$\mathscr{H} = -\frac{\hbar^2}{2\mathrm{I}_{zz}}\left(x\frac{\partial}{\partial y} - y\frac{\partial}{\partial x}\right)^2 + V. \tag{4-5}$$

It is convenient to rewrite Eq.(4-5) in terms of polar coordinates where the transformations are (see Fig. 4-1)

$$\begin{aligned} x &= r\cos\phi \qquad & r &= (x^2 + y^2)^{1/2} \\ y &= r\sin\phi \qquad & \tan\phi &= \frac{y}{x}. \end{aligned} \tag{4-6}$$

Using the chain rule in differentiation gives

$$\begin{aligned} \frac{\partial}{\partial y} &= \frac{\partial r}{\partial y}\frac{\partial}{\partial r} + \frac{\partial \phi}{\partial y}\frac{\partial}{\partial \phi} = \sin\phi\frac{\partial}{\partial r} + \frac{\cos\phi}{r}\frac{\partial}{\partial \phi} \\ \frac{\partial}{\partial x} &= \frac{\partial r}{\partial x}\frac{\partial}{\partial r} + \frac{\partial \phi}{\partial x}\frac{\partial}{\partial \phi} = \cos\phi\frac{\partial}{\partial r} - \frac{\sin\phi}{r}\frac{\partial}{\partial \phi}. \end{aligned} \tag{4-7}$$

Substituting the results in Eqs.(4-6) and (4-7) into Eq.(4-5) gives

$$\mathscr{H} = -\frac{\hbar^2}{2I_{zz}}\frac{\partial^2}{\partial\phi^2} + V, \tag{4-8}$$

and Schrödinger's time-independent equation becomes

$$-\frac{\hbar^2}{2I_{zz}}\frac{d^2\psi}{d\phi^2} = (E - V)\psi, \tag{4-9}$$

where we remember that V is a constant. The requirement that ψ be single-valued, twice continuously differentiable, and square integrable places restrictions on the form of ψ similar to the restrictions discussed in the de Broglie picture. In order for ψ to be single-valued and continuous, we require continuity around the ring, which is satisfied if

$$\psi(\phi) = \psi(\phi + 2\pi). \tag{4-10}$$

Sinusoidal functions will satisfy Eqs.(4-10) and (4-9). One form for a satisfactory normalized function is

$$\psi_m = \left(\frac{1}{2\pi}\right)^{1/2} \exp(im\phi), \tag{4-11}$$

where m is a number that is restricted to values of 0, ± 1, ± 2, ± 3, ... as in Eq.(4-1). All other nonintegral values of m lead to ψ_m functions that are either not continuous or not single-valued. Thus, the values of m are fixed by the boundary conditions of the problem. Substituting ψ_m in Eq.(4-11) into Eq.(4-9) leads to

$$E_m = \frac{\hbar^2 m^2}{2I_{zz}} + V, \tag{4-12}$$

which is equivalent to the energy given in Eq.(4-3) by de Broglie's relation. An energy level diagram for the particle-in-a-ring is shown in Fig. 4-2. The orthonormal character of the eigenfunctions in Eq.(4-11) is evident from

$$\int_0^{2\pi} \psi_m^* \psi_{m'}\, d\phi = \left(\frac{1}{2\pi}\right) \int_0^{2\pi} \exp[-i\phi(m - m')]\, d\phi = \delta_{mm'}. \tag{4-13}$$

The proof of this identity is left to the reader.

In the case of an electron in a benzene ring, $r \cong 1.4 \times 10^{-8}$ cm, giving $I_{zz} \cong 1.78 \times 10^{-43}$ g $\cdot$ cm^2 and $\hbar^2/2I_{zz} \cong 3.11 \times 10^{-12}$ erg $= 1.94$ eV. If we now place the six delocalized electrons of the benzene ring (one for each carbon atom) into the lowest particle-in-a-ring levels in Fig. 4-2, we have two electrons in each state as shown by the arrows. Note that at $H_z = 0$ all levels above $m = 0$ are doubly degenerate and can accommodate two electron pairs. The electrons pair in each state due to the Pauli principle, which we shall discuss in Chapter 5. The lowest energy transition is between the $m = 1$ and $m = 2$ states and occurs at $3(\hbar^2/2I_{zz}) = 5.83$ eV, which is the correct order of magnitude for the lowest experimentally observed electronic transition in benzene at 4.7 eV. The electronic transition energies decrease with increasing molecular size due to the inverse I_{zz} dependence. This trend is roughly exhibited by the

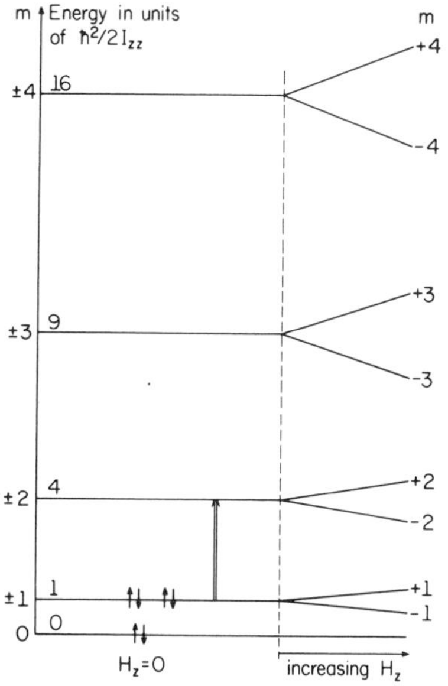

Figure 4-2 The energy levels relative to the constant potential energy for the free particle-in-a-ring at $H_z = 0$ and as a function of increasing magnetic fields, H_z. Only the linear field Zeeman effect is shown. The six arrows in the lowest two levels indicate the delocalized electrons in the benzene ring and the larger vertical double arrow indicates a possible electronic transition.

lowest energy electronic absorption in progressively larger ring systems. The extensions and application of the free electron theory (V = constant) has been of use in correlating trends in electronic spectra with molecular size [1]†.

All $\pm m$ states in the particle-in-a-ring at zero external fields are degenerate due to the m^2 dependence shown in Eq.(4-12). A linear combination of eigenfunctions with degenerate levels (eigenvalues) is also an eigenfunction with the same eigenvalue [see Eq.(3-35) and associated discussion]. We are therefore free to take linear combinations of the $\pm m$ particle-in-a-ring states to yield another set of eigenfunctions of the Hamiltonian operator in Eq.(4-8). The sum of $+m$ and $-m$ eigenfunctions gives

$$\psi_m^+ = \left(\frac{1}{2}\right)^{1/2}(\psi_m + \psi_{-m}) = \frac{1}{2}\left(\frac{1}{\pi}\right)^{1/2}[\exp(im\phi) + \exp(-im\phi)]$$

$$\psi_m^+ = \left(\frac{1}{\pi}\right)^{1/2}\cos m\phi = \psi_{-m}^+, \tag{4-14}$$

where $(\frac{1}{2})^{1/2}$ is necessary to normalize the linear combination.

We can also obtain the imaginary eigenfunctions of the Hamiltonian,

$$\psi_m^- = \left(\frac{1}{2}\right)^{1/2}(\psi_m - \psi_{-m}) = \frac{1}{2}\left(\frac{1}{\pi}\right)^{1/2}[\exp(im\phi) - \exp(-im\phi)]$$

$$= i\left(\frac{1}{\pi}\right)^{1/2}\sin(m\phi) = -\psi_{-m}^-. \tag{4-15}$$

ψ_m^- is an odd function of $m\phi$.

† Numbers in square brackets correspond to the numbered sources found in the Reference section on p. 263 at the end of the chapter.

In Section 3.3 [following Eq.(3-57)], we noted that commuting operators may share simultaneous eigenfunctions. It is evident that the $L_z = -i\hbar\, \partial/\partial\phi$ and $(L_z)^j$ (j is a positive integer) operators both commute with each other and the Hamiltonian. It is also easily shown that the function in Eq.(4-11) is a simultaneous eigenfunction of the $\mathscr{H}$, L_z, and $(L_z)^j$ operators. The Hamiltonian eigenfunctions given in Eqs.(4-14) and (4-15), however, are not eigenfunctions of the L_z and L_z^2 operators. Thus, we emphasize again that two commuting operators *may* possess identical eigenfunctions. Other less general functions can be found that are eigenfunctions of only one of the two commuting operators. If such a set is used as a starting basis, however, the arguments developed following Eq.(3-57) can be used to find simultaneous eigenfunctions for both operators by forming appropriate linear combinations of basis functions.

Static Magnetic Field Perturbation

We shall now discuss the effects of a magnetic field along the z axis (see Fig. 4-1) in a particle-in-a-ring. The classical expression for the interaction of a particle of mass m and charge e with an electromagnetic field is given in Eq.(3-167).

The general forms of $\boldsymbol{A}$ and A^2 in the case of static magnetic fields are also needed here and in later applications. Remembering that in a vacuum where $\boldsymbol{B} = \boldsymbol{H} = \boldsymbol{\nabla} \times \boldsymbol{A}$ in cgs units, we can write the following expressions for a uniform magnetic field ($\partial H/\partial x = \partial H/\partial y = \partial H/\partial z = 0$):

$$\boldsymbol{A} = -\frac{1}{2}\boldsymbol{r} \times \boldsymbol{H} \tag{4-16}$$

$$A^2 = \boldsymbol{A} \cdot \boldsymbol{A} = \tfrac{1}{4}(\boldsymbol{r} \times \boldsymbol{H}) \cdot (\boldsymbol{r} \times \boldsymbol{H}) = \tfrac{1}{4}\boldsymbol{H} \cdot [\boldsymbol{r} \times (\boldsymbol{H} \times \boldsymbol{r})] = \tfrac{1}{4}\boldsymbol{H} \cdot (r^2\mathbf{1} - \boldsymbol{rr}) \cdot \boldsymbol{H}, \tag{4-17}$$

where we have used $(\boldsymbol{A} \times \boldsymbol{B}) \cdot (\boldsymbol{C} \times \boldsymbol{D}) = \boldsymbol{C} \cdot [\boldsymbol{D} \times (\boldsymbol{A} \times \boldsymbol{B})]$ and $\boldsymbol{A} \times (\boldsymbol{B} \times \boldsymbol{C}) = \boldsymbol{B}(\boldsymbol{A} \cdot \boldsymbol{C}) - \boldsymbol{C}(\boldsymbol{A} \cdot \boldsymbol{B})$ in the second equation (see Appendix B.1).

In the case of a magnetic field along the z axis in the particle-in-a-ring example, $H_x = H_y = 0, H_z \neq 0$, and the electric field $\boldsymbol{E} = 0$. Thus, in this simple case of a single static uniform magnetic field component, the vector potential components take on the simple form given by

$$A_x = -\frac{y}{2}H_z \qquad A_y = \frac{x}{2}H_z \qquad A_z = 0. \tag{4-18}$$

These equations for the planar magnetic field H_z are easily verified by Eq.(1-18), $\boldsymbol{H} = \boldsymbol{B} = \boldsymbol{\nabla} \times \boldsymbol{A}$ in cgs units. Substituting Eqs.(4-18) into Eq.(3-167) for an electron (use negative sign on e) gives

$$\mathscr{H} = -\frac{\hbar^2}{2m}\nabla^2 + V - \frac{i\hbar e H_z}{2mc}\left(x\frac{\partial}{\partial y} - y\frac{\partial}{\partial x}\right) + \frac{e^2H_z^2}{8mc^2}(x^2 + y^2), \tag{4-19}$$

which describes the interaction of an electron with negative charge e and mass m in a static magnetic field, H_z. The Bohr magneton is defined as the magnitude of the standard magnetic moment for an electron with charge e and mass m given by

$$\mu_B = \frac{e\hbar}{2mc} = 0.927408 \times 10^{-20} \text{ erg} \cdot \text{G}^{-1} = 0.927408 \times 10^{-23} \text{ J} \cdot \text{T}^{-1}$$

$$\frac{\mu_B}{h} = 1.39961 \text{ MHz} \cdot \text{G}^{-1}, \qquad \frac{\mu_B}{\hbar} = 8.79403 \times 10^6 \text{ Hz} \cdot \text{G}^{-1}. \tag{4-20}$$

Substituting the Bohr magneton for an electron and $L_z = -i\hbar(x\, \partial/\partial y - y\, \partial/\partial x) = -i\hbar\, \partial/\partial\phi$ into Eq.(4-19) gives the Hamiltonian for an electron of

$$\mathscr{H} = \underbrace{-\frac{\hbar^2\nabla^2}{2m} + V}_{\mathscr{H}_0} + \underbrace{\frac{\mu_B H_z}{\hbar} L_z}_{\mathscr{H}_1} + \underbrace{\frac{e^2 H_z^2}{8mc^2}(x^2 + y^2)}_{\mathscr{H}_2}. \tag{4-21}$$

We define the magnetic moment for the electron as the negative first derivative of the energy with respect to the magnetic field [see Eq.(4-213)] to give

$$\boldsymbol{\mu} = -\frac{\mu_B}{\hbar}\mathbf{L}, \tag{4-22}$$

which gives $\mathscr{H}_1 = -\boldsymbol{\mu} \cdot \boldsymbol{H}$ as the linear field perturbation in Eq.(4-21). In the case of a negatively charged particle, the magnetic moment and angular momentum are antiparallel. In the case of a positively charged particle, the angular momentum and magnetic moment are parallel.

Returning now to Eq.(4-21) for a particle-in-a-ring, we note that L_z commutes with $\mathscr{H}_0$ and the original zero-order solution in Eq.(4-11) is also an eigenfunction of $\mathscr{H}_0 + \mathscr{H}_1$ in Eq.(4-21) with an eigenvalue given by

$$(\mathscr{H}_0 + \mathscr{H}_1)\psi_m = \left(-\frac{\hbar^2}{2I_{zz}}\frac{d^2}{d\phi^2} + \frac{\mu_B H_z}{i}\frac{d}{d\phi} + V\right)\psi_m = \psi_m E_m$$

$$E_m = \frac{\hbar^2}{2I_{zz}} m^2 + \mu_B H_z m + V. \tag{4-23}$$

Thus, the m degeneracy is broken in the presence of the external magnetic field, H_z, as shown on the right side of Fig. 4-2. The energy difference between a set of $\pm m$ levels is given from Eq.(4-23):

$$\Delta E = (E_{+m} - E_{-m}) = 2m\mu_B H_z. \tag{4-24}$$

If we choose a typical laboratory field of $H_z = 5000$ G and the Bohr magneton values given in Eq.(4-20), the energy spacings for an electron are $\Delta E = m(0.9274 \times 10^{-23}\text{ J})$ with frequencies given by $\nu = \Delta E/h = m(1.3996 \times 10^4 \text{ MHz})$. These spacings are several orders of magnitude lower in energy than the spacings between the different $|m|$ states in our previous benzene example.

We now return to Eq.(4-21) and a discussion of $\mathscr{H}_2$. We have already pointed out that $\mathscr{H}_1$ commutes with $\mathscr{H}_0$. We now note that $\mathscr{H}_2$ is a constant for fixed $r = (x^2 + y^2)^{1/2}$ and it therefore commutes with $\mathscr{H}_0$ and, of course, $\mathscr{H}_2$ also commutes with $\mathscr{H}_0 + \mathscr{H}_1$. Thus, the original eigenfunctions of $\mathscr{H}_0$ and $\mathscr{H}_0 + \mathscr{H}_1$ are also valid eigenfunctions for $\mathscr{H}_0 + \mathscr{H}_1 + \mathscr{H}_2$ with eigenvalues

$$E_m = \frac{\hbar^2}{2I_{zz}} m^2 + V + \mu_B m H_z + \frac{\mu_B^2 I_{zz} H_z^2}{2\hbar^2}, \tag{4-25}$$

where we have used

$$\frac{e^2}{4mc^2}(x^2 + y^2) = \frac{e^2}{4m^2c^2} I_{zz} = \frac{\mu_B^2}{\hbar^2} I_{zz} \tag{4-26}$$

for constant $r^2 = x^2 + y^2$ [m is the electron mass in Eq.(4-26) and m is the quantum number in Eq.(4-25)]. The term in H_z^2 is not dependent on the quantum number m. Therefore, all energy levels in the particle-in-a-ring (see Fig. 4-2) will be increased by a constant value of $\mu_B^2 I_{zz} H_z^2/2\hbar^2$ in the presence of the external field.

In our previous example of the electron in the benzene ring ($r = 1.4 \times 10^{-8}$ cm and $\hbar^2/2I_{zz} = 3.12 \times 10^{-12}$ erg), we can evaluate the various contributions to the energy in Eq.(4-25) at fields of 10,000 G. The result for $m = 1$ is

$$E = \underbrace{\frac{\hbar^2}{2I_{zz}}} + V + \underbrace{\mu_B H_z} + \underbrace{\frac{e^2 H_z^2}{8mc^2}(x^2 + y^2)}$$

$$E(m = 1) = (3.1 \times 10^{-12} + V) + 9.3 \times 10^{-17} + 6.9 \times 10^{-22} \tag{4-27}$$

in cgs units. It is clear that the H_z^2 term contributes only a small correction at fields on the order of 10^4 G. The last term in Eq.(4-25) contains the magnetic susceptibility χ_{zz} for the electron in the ring, which is defined as the negative value of the second derivative of the energy with respect to the magnetic field [see Eq.(4-213)]:

$$\frac{\partial^2 E}{\partial H_z^2} = -\chi_{zz} = \frac{\mu_B^2 I_{zz}}{\hbar^2}. \tag{4-28}$$

In the case of our previous example for a single electron in the benzene ring with $I_{zz} = 1.78 \times 10^{-43}$ g · cm^2, we have $\chi_{zz} = -1.38 \times 10^{-29}$ erg · G^{-2} (cm^3), where we note that the cgs units of magnetic susceptibility are volume. For six free electrons in the benzene ring, we have $\chi_{zz} = -8.28 \times 10^{-29}$ cm^3. The corresponding ring magnetic susceptibility for a mole of benzene molecules is obtained by multiplying by Avogadro's number, giving $\chi_{zz} = -49.9 \times 10^{-6}$ cm^3 · mol^{-1}. This result should give an estimate of the contribution of the six free electrons to the magnetic susceptibility perpendicular to the benzene molecular ring. In order to compare this result with experiment, we must subtract the contributions to the magnetic susceptibilities of the remaining electrons in the benzene molecule. A fairly reliable method of canceling local effects or effects not due to the six free electrons is obtained by comparing the value of χ_{zz} with the experimental magnetic susceptibility anisotropy given by $\chi_{zz} - \frac{1}{2}(\chi_{xx} + \chi_{yy})$ where χ_{xx} and χ_{yy} are the in-plane susceptibilities. The ring contribution to χ_{xx} and χ_{yy} should be zero and the anisotropy should give an estimate of the non-local or free electron contribution. The experimental values of $\chi_{zz} - \frac{1}{2}(\chi_{xx} + \chi_{yy})$ are listed in Table 4-1 for benzene and several other small molecules which apparently

Table 4-1 Experimental magnetic susceptibility anisotropies in some small ring compounds from D. H. Sutter and W. H. Flygare, *J. Am. Chem. Soc.* **61**, 6895 (1969). The z axis is perpendicular to the ring.

Molecule	$\chi_{zz} - \frac{1}{2}(\chi_{xx} + \chi_{yy})$ $(10^{-6}\ \mathrm{cm}^3 \cdot \mathrm{mol}^{-1})$	Number of Free Electrons
	−59.7	6
	−50.1	5
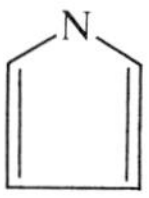	−42.4	5
	−38.7	5
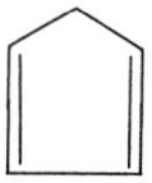	−34.3	5
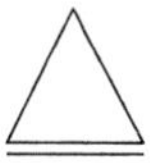	−17.0	3
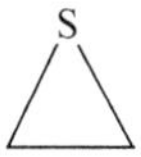	−15.4	3
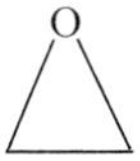	−9.4	3

contain free electrons which can circulate in the ring under the influence of the magnetic field perpendicular to the ring. The values for $\chi_{zz} - \frac{1}{2}(\chi_{xx} + \chi_{yy})$ in Table 4-1 correlate roughly with r^2 and the number of free electrons in the ring. Further details of the molecular magnetic susceptibility are given in Section 6.8.

As a final point, we note that the last term in the energy expression in Eq.(4-25) can be rewritten as

$$E(H_z^2) = \frac{1}{2}\left(\frac{eH_z}{2mc}\right)^2 I_{zz}. \qquad (4\text{-}29)$$

Referring back to our classical discussion of the energy of a particle-in-a-ring, we note that $E = \omega^2 I_{zz}/2$, where ω is the angular frequency. Thus, according to Eq.(4-29) and the classical expression, the quadratic field term in the energy induces an angular frequency that can be obtained by equating the energies,

$$\frac{1}{2}\left(\frac{eH_z}{2mc}\right)^2 I_{zz} = \frac{1}{2}\bar{\omega}^2 I_{zz}, \qquad \bar{\omega} = \frac{eH_z}{2mc}. \tag{4-30}$$

This induced angular frequency is denoted by $\bar{\omega}$, and the field-induced frequency, $\bar{\nu} = \bar{\omega}/2\pi$, is called the *Larmor frequency* [2]. Equation (4-30) for the Larmor frequency is similar to the cyclotron frequency equation [Eq.(2-105)], which is obtained by equating the Lorentz force for the particle in a magnetic field to the classical mechanical force (see Problem 2-3). However, the Larmor frequency is one-half the cyclotron frequency.

4.2 INTERNAL ROTATION

It is also appropriate to apply the particle-in-a-ring model to the relative motion between two methyl groups in ethane or between CH_3 and the COH group in acetaldehyde. In the case of *free* relative rotation of the two groups with respect to each other, the appropriate wavefunctions describing the motion would be the orthonormal particle-in-a-ring functions given in Eq.(4-11) where ϕ is the internal rotation angle. The energy is given in Eq.(4-12) where I_{zz} is the *relative* moment of inertia between the two groups (z is the symmetry axis of the methyl group). For instance, in CH_3COOH,

$$I(CH_3COOH) = \frac{I_m I_0}{I_m + I_0} \approx I_m. \tag{4-31}$$

I_m is the methyl group moment of inertia and I_0 is the moment of inertia of the COOH group where both I_m and I_0 are along the C—C axis in the molecule. In the case of ethane,

$$I(CH_3CH_3) = \frac{I_m I_m}{I_m + I_m} = \frac{I_m}{2}. \tag{4-32}$$

It is easy to show that the moment of inertia along the C_3 symmetry axis of a symmetric methyl group is approximately (see Fig. 2-2) $I_m = 5.3 \times 10^{-40}\ \text{g}\cdot\text{cm}^2$. Thus $I(CH_3CH_3) = I = 2.65 \times 10^{-40}\ \text{g}\cdot\text{cm}^2$ and we have

$$\frac{\hbar^2}{2I} = 2.10 \times 10^{-15}\ \text{erg}, \qquad \frac{1}{hc}\left(\frac{\hbar^2}{2I}\right) = 10.6\ \text{cm}^{-1}. \tag{4-33}$$

If the CH_3 groups in ethane were free rotors, the transitions between the states would be in multiples of 10.6 cm^{-1} beginning in the far infrared region of the electromagnetic spectrum.

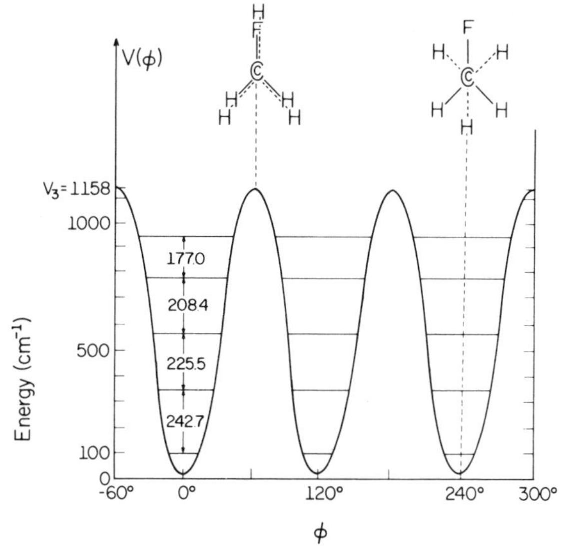

Figure 4-3 The potential function for the hindered internal rotation in CH_3CH_2F. The energy differences in cm^{-1} are from G. Sage and W. Klemperer, *J. Chem. Phys.* **39**, 371 (1963). The value of V_3 given in the figure is accurate to about 1%. The value of $V_6 \cong -5\ cm^{-1}$ was also determined from the observed energy spacings.

It is well known that the free rotor model above is incomplete for most molecules as there is generally a barrier that hinders the internal rotation. Consider ethyl fluoride, CH_3CH_2F, which has a dominant threefold barrier (as in CH_3CH_3) hindering internal rotation and the lower energy configuration is the staggered form (see Fig. 4-3). The observed energy spacings are much larger than predicted by the free rotor model; these observations indicate a high barrier. An appropriate form for the potential function for the hindered internal rotation is

$$V = \frac{V_3}{2}(1 - \cos 3\phi), \tag{4-34}$$

where V_3 is the energy difference between the maxima and minima in the potential function (see Fig. 4-3). The complete Hamiltonian describing the hindered internal rotor is

$$\mathscr{H} = -\frac{\hbar^2}{2I}\frac{d^2}{d\phi^2} + \frac{V_3}{2}(1 - \cos 3\phi) = \mathscr{H}_0 + \mathscr{H}_1. \tag{4-35}$$

The zero-order function is given in Eq.(4-11) and the nonzero values of the matrix elements of $\mathscr{H}_0 = -(\hbar^2/2I)\, d^2/d\phi^2$ in this basis are

$$(\mathscr{H}_0)_{m,m} = m^2 \frac{\hbar^2}{2I}. \tag{4-36}$$

The matrix elements of the perturbation Hamiltonian, $\mathscr{H}_1$, are given by

$$
\begin{aligned}
(\mathscr{H}_1)_{m,m'} &= \int_0^{2\pi} \psi_m^* \left[\frac{V_3}{2}(1 - \cos 3\phi) \right] \psi_{m'}\, d\phi \\
&= \frac{V_3}{2}\delta_{m,m'} - \frac{V_3}{4\pi}\int_0^{2\pi} \exp\left[-i(m - m')\phi\right] \cos 3\phi\, d\phi \\
&= \frac{V_3}{2}\delta_{m,m'} - \frac{V_3}{4}\delta_{m',m\pm 3}.
\end{aligned}
\tag{4-37}
$$

Thus, the complete Hamiltonian matrix in the $\psi_m = (1/2\pi)^{1/2} \exp(im\phi)$ basis is given by a combination of Eqs.(4-36) and (4-37) to yield

$$
\mathscr{H} = \mathscr{H}_0 + \mathscr{H}_1 =
$$

$$
\begin{array}{c|ccccccccc|}
m & 0 & 1 & -1 & 2 & -2 & 3 & -3 & 4 & -4 \\
\hline
0 & \frac{V_3}{2} & 0 & 0 & 0 & 0 & -\frac{V_3}{4} & -\frac{V_3}{4} & 0 & 0 \\
1 & 0 & \frac{\hbar^2}{2I}+\frac{V_3}{2} & 0 & 0 & -\frac{V_3}{4} & 0 & 0 & -\frac{V_3}{4} & 0 \\
-1 & 0 & 0 & \frac{\hbar^2}{2I}+\frac{V_3}{2} & -\frac{V_3}{4} & 0 & 0 & 0 & 0 & -\frac{V_3}{4} \\
2 & 0 & 0 & -\frac{V_3}{4} & 4\left(\frac{\hbar^2}{2I}\right)+\frac{V_3}{2} & 0 & 0 & 0 & 0 & 0 \\
-2 & 0 & -\frac{V_3}{4} & 0 & 0 & 4\left(\frac{\hbar^2}{2I}\right)+\frac{V_3}{2} & 0 & 0 & 0 & 0 \;\cdots \\
3 & -\frac{V_3}{4} & 0 & 0 & 0 & 0 & 9\left(\frac{\hbar^2}{2I}\right)+\frac{V_3}{2} & 0 & 0 & 0 \\
-3 & -\frac{V_3}{4} & 0 & 0 & 0 & 0 & 0 & 9\left(\frac{\hbar^2}{2I}\right)+\frac{V_3}{2} & 0 & 0 \\
4 & 0 & -\frac{V_3}{4} & 0 & 0 & 0 & 0 & 0 & 16\left(\frac{\hbar^2}{2I}\right)+\frac{V_3}{2} & 0 \\
\vdots & & & & & \vdots & & & & \ddots
\end{array}
\tag{4-38}
$$

where the rows and columns are ordered $m = 0, 1, -1, 2, -2, 3, -3, \ldots$ from left to right and top to bottom, respectively. The energy of the ground state ($m = 0$) correct to second order in perturbation theory is given by

$$
E_{m=0} = \frac{V_3}{2} - \frac{IV_3^2}{36\hbar^2}.
\tag{4-39}
$$

It is not readily obvious from Eq.(4-38) which originally degenerate m levels are split under the influence of the perturbation. Diagonalization of $(\mathscr{H}_0 + \mathscr{H}_1)$ in Eq.(4-38) according to Eq.(3-86) by a computer is a simple matter, however, and shows that the m degeneracy of the $m = \pm 3$, $m = \pm 6$, $m = \pm 9, \ldots$ levels has been broken but the other m states remain degenerate. A plot of internal rotation energy levels $[(1/hc)(\hbar^2/2\text{I}) = 17\ \text{cm}^{-1}]$ for $V_3 = 100\ \text{cm}^{-1}$ and $V_3 = 200\ \text{cm}^{-1}$ is shown in Fig. 4-4, which can be compared to the free rotor diagram that is shown in Fig. 4-2. A detailed analysis as a function of V_3 by computer diagonalization gives the correlation diagram in Fig. 4-5 with $(1/hc)(\hbar^2/2\text{I}) = 17\ \text{cm}^{-1}$, which is appropriate for the methyl alcohol system, CH_3OH, where I is the reduced moment of the methyl protons and the alcohol proton about the CO bond.

Several important features appear in Figs. 4-4 and 4-5. In Fig. 4-4, if $V_3 = 100\ \text{cm}^{-1}$, the $m = 0$ and ± 1 levels are below the barrier. At $V_3 = 200\ \text{cm}^{-1}$, the $m = 0$, ± 1, and ± 2 levels are all below the barrier. If the system is in a state below the barrier, it is classically locked in that configuration and does not possess enough energy to go to another equivalent configuration on the other side of the barrier. Quantum mechanically, the system can "tunnel" through the barrier from one configuration to another. The classical tunneling frequency in the case of the methyl group internal rotation is obtained from the Bohr frequency rule and the following arguments. In the original barrier-free case with the energy levels being given by $\hbar^2 m^2/2\text{I}$, one can meaningfully talk about the rotational frequency of the methyl group, which is $\nu_m = (1/h)(\hbar^2 m^2/2\text{I}) = m^2 \times 5.1 \times 10^{11}$ Hz in the case of CH_3OH. The values of $\pm m$ indicate that the methyl group can rotate clockwise or counterclockwise with equal angular frequency. In the presence of a finite barrier, the concept of the angular velocity of the methyl group is not clear as the rotation is hindered even

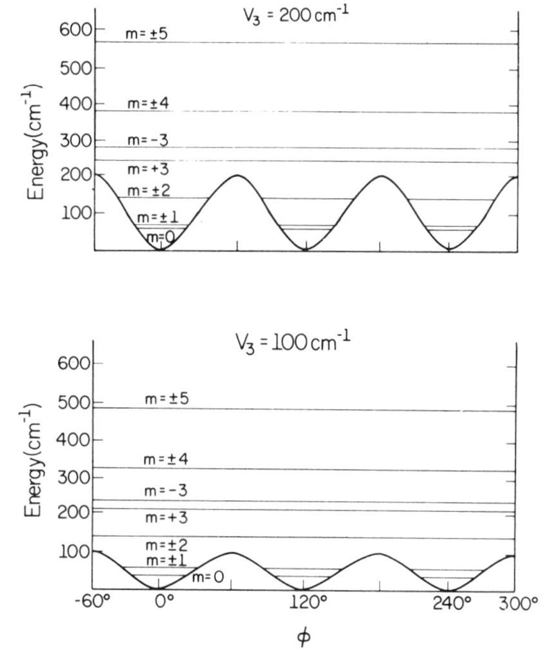

Figure 4-4 Internal rotation energy levels for two barriers hindering the rotation of $V_3 = 100\ \text{cm}^{-1}$ and $V_3 = 200\ \text{cm}^{-1}$ for $(1/hc)(\hbar^2/2\text{I}) = 17\ \text{cm}^{-1}$.

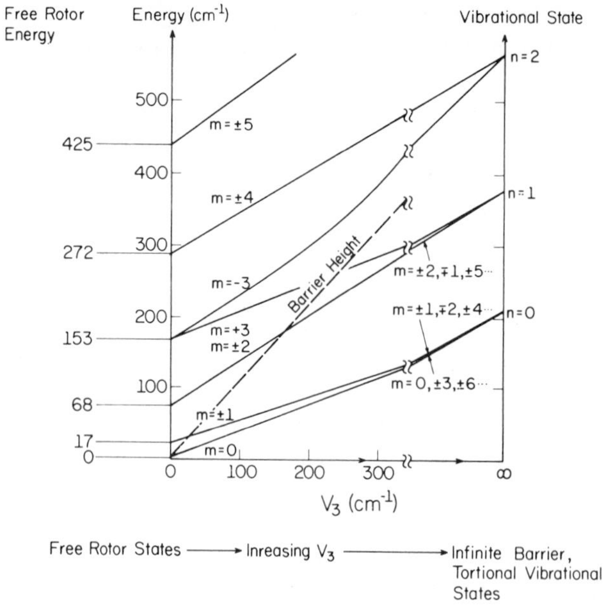

Figure 4-5 Correlation diagram between the free rotor $[V_3 = 0$ and $(1/hc)(\hbar^2/2I) = 17\ \text{cm}^{-1}]$ and high barrier $(V_3 = \infty)$ energy levels. There are three equivalent conformations at $V_3 = \infty$ leading to triple degeneracy and to equally spaced vibrational levels appropriate to a harmonic potential (see Section 4.12) where the quantum number n is used. The free rotor energy levels are also shown in Fig. 4-2. The m levels on the high barrier side indicate the mixed states.

if the state has an energy that is above the barrier. Even though the effective rotational energy and resultant frequency will decrease, however, the total energy will increase due to the addition of the potential energy in the Hamiltonian.

As the barrier increases, the molecular internal rotation becomes highly hindered and takes on a motion familiar as vibrational motion in the states well below the top of the potential function. This is evident from Fig. 4-3; expanding the potential function in Eq.(4-34) about $\phi = 0$ for small amplitude motions gives

$$V(\phi) = \frac{V_3}{2}(1 - \cos 3\phi) = \frac{V_3}{2}\left[1 - \left(1 - \frac{(3\phi)^2}{2}\right) + \cdots\right] \cong \frac{9V_3}{4}\phi^2, \quad (4\text{-}40)$$

where ϕ is the torsional angle described previously. In the high barrier limit, we interpret $9V_3/2$ as a reduced force constant for a torsional vibration by analogy with $V = \frac{1}{2}ks^2$ for the harmonic oscillator. This can be seen better by referring again to Fig. 4-1 where we can write, for small deviations from equilibrium, $\sin\phi = y/r = \phi - (\phi^3/6) + \cdots \cong \phi$. Substituting into Eq.(4-41) gives

$$V(\phi) = \left(\frac{9V_3}{4r^2}\right)y^2, \quad (4\text{-}41)$$

where y is the linear displacement and $9V_3/2r^2$ is the force constant [compare with the harmonic oscillator potential of $V(s) = \frac{1}{2}ks^2$] for the torsional oscillation. Choosing a reasonable value of $r^2 = 10^{-16}$ cm^2, the value of V_3 in Eq.(4-40) to produce a normal bending force constant of about 0.3×10^5 dyne · cm^{-1} is $V_3 = 0.667 \times 10^{-12}$ erg = 3356 cm^{-1}. Thus, a reasonably high barrier of 3356 cm^{-1} will lead to a force constant for torsional motion (in the valley of the potential) of around 0.3×10^5 dyne · cm^{-1}. The lower vibrational levels, all being below the top of the potential, will experience largely harmonic motion about an equilibrium configuration and there will be three equivalent locked positions for this harmonic motion as shown on the right side of Fig. 4-5 and in Fig. 4-3. Thus, in the high barrier case, where levels have coalesced into groups of three, we interpret the singly degenerate states, $m = 0, +3, -3, +6, -6, \ldots$, as arising from the vibrational energy and we interpret the doubly degenerate pairs as arising from the vibrational plus rotational energy components. The $\pm$ degeneracy is still maintained for equivalent clockwise and counterclockwise motion. The energy difference between the $m = 0$ and $m = \pm 1$ states represents the remaining rotational energy in the high barrier case. Using the Bohr frequency relation gives the rotational frequency or tunneling frequency [3].

Other periodic barrier functions may also be appropriate such as a periodic Gaussian; however, sinusoidal potentials have been very successful in describing internal rotation. In general, the potential function for internal rotation is written as

$$\mathscr{H}_1 = V(\phi) = \sum_{n=1}^{\infty} \frac{V_n}{2}(1 - \cos n\phi). \tag{4-42}$$

In practice, the dominant component in this summation is usually the lowest component possessing the correct symmetry. The first nonzero term in Eq.(4-42) for nitromethane or toluene is the sixfold term. In 2-fluorotoluene we are back to a dominant threefold barrier. In 2,6-difluorotoluene, however, the sixfold term is again dominant. After the dominant term in Eq.(4-42) is identified by symmetry, higher-order terms can be added as long as they possess the correct symmetry. For instance, in ethane or propylene, the dominant term is a threefold term but the potential function includes higher-order terms that are multiples of $n = 3$ in Eq.(4-42). The relative signs of V_3 and V_6 are valuable in identifying the shape of the potential. If a small V_6 is

Table 4-2 Methyl group barriers to internal rotation in several molecules from W. H. Flygare, *Ann. Rev. Phys. Chem.* **18**, 325 (1967) and H. D. Rudolph, *Ann. Rev. Phys. Chem.* **21**, 73 (1970) and other sources (Refs. [4] and [5]).

Molecule	V_3 (cal · mol^{-1})	V_6 (cal · mol^{-1})
$CH_3—CH_3$	2930	—
$CH_3—SiH_3$	1655	—
$CH_3—GeH_3$	1240	—
$CH_3—CH_2F$	3306	−14
$CH_3—CH_2Cl$	3560	—
$CH_3—CHF_2$	3200	—
$CH_3—CF_3$	3500	—
CH_3NO_2	0	6
CH_3BF_2	0	14
CH_3COH	1150	—
$CH_3CH{=}CH_2$	2000	−37
$CH_3CF{=}CH_2$	2440	—

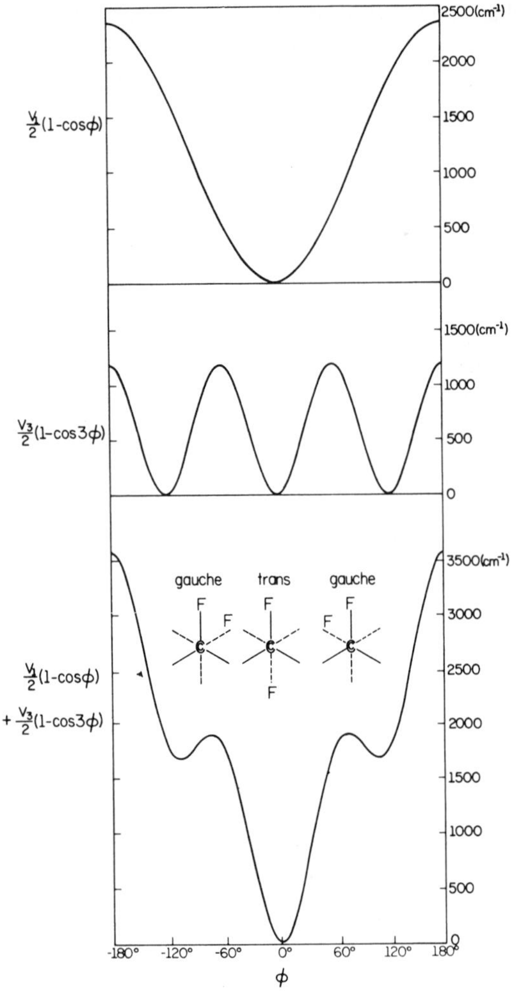

Figure 4-6 Potential function showing the stable *trans* and *gauche* forms of 1,2-difluoroethane.

positive relative to a larger positive, V_3, the total potential will appear to be sharper at the base and broader at the top. The reverse is true for a small negative V_6 relative to a positive V_3. The matrix representation for the potential energy in the free rotor basis is easily constructed by a generalization of Eq.(4-37) to give

$$(\mathscr{H}_1)_{m,m'} = \frac{V_n}{2}\,\delta_{m,m'} - \frac{V_n}{4}\,\delta_{m',m\pm n} + \frac{V_{2n}}{2}\,\delta_{m,m'} - \frac{V_{2n}}{4}\,\delta_{m',m\pm 2n} - \cdots. \qquad (4\text{-}43)$$

In many cases of chemical interest, especially with heavier rotors, the $V_n/4$ off-diagonal terms become much larger than $(\hbar^2/2I)m^2$ for values of m up to a convenient matrix order of 50 to 100. Thus, a high barrier approach is more appropriate.

A few barriers for methyl group rotations are shown in Table 4-2. The rapid convergence in Eq.(4-42) is evident from some of these examples.

Figure 4-6 demonstrates the summation of positive V_1 and V_3 terms for the potential function in 1,2-difluoroethane. Twofold barriers are evident in the 1,2-disubstituted ethylenes. The barriers about double bonds are much larger (40,000 $\text{cal} \cdot \text{mol}^{-1}$), however, than the single-bond barriers [4]. Several additional examples of the use of Eq.(4-42) in determining internal rotation in different molecules are given in the problems at the end of the chapter [5].

The particle-in-a-box is another important model problem and should be examined (see Problem 4-5) by the student before proceeding.

4.3 NONRELATIVISTIC DESCRIPTION OF THE HYDROGEN-LIKE ATOM

One electron in the central field of a single point charge nucleus in the absence of external perturbations leads to a Schrödinger equation that can be solved exactly, yielding the stationary states of the hydrogen-like atomic system. These results can then be used as a model for the distribution of electrons in many-electron atoms. The solution of the time-independent Schrödinger equation for the hydrogen-like atom is historically and conceptually one of the more important developments in our understanding of atoms and molecules. The Hamiltonian for a point charge nucleus of atomic number Z, charge Ze, and mass M, and a single electron with mass m and charge $-e$ is

$$\mathscr{H} = T_n + T_e + V_{en} = -\frac{\hbar^2}{2M}\nabla_n^2 - \frac{\hbar^2}{2m}\nabla_e^2 - \frac{Ze^2}{[(x_n - x_e)^2 + (y_n - y_e)^2 + (z_n - z_e)^2]^{1/2}}, \tag{4-44}$$

where x_n, y_n, and z_n and x_e, y_e, and z_e are the Cartesian coordinates of the nucleus and electron, respectively, with respect to an arbitrary origin. Equation (4-44) is simplified by transforming to a spherical polar coordinate system with the center of mass (c.m.) as the origin. The coordinate system is shown in Fig. 4-7 and the transformation to the c.m. has been discussed in Chapter 2 and is illustrated in Fig. 2-1. The coordinates of the center of mass in Fig. 4-7 are defined from the arbitrary origin by X, Y, and Z [see Eq.(2-10)] where $R^2 = X^2 + Y^2 + Z^2$ and $\boldsymbol{R}$ is the vector distance from the arbitrary origin to the c.m. The electronic and nuclear Cartesian coordinates in Fig. 4-7 $(x_e, y_e, z_e, x_n, y_n, z_n)$ can be expressed in terms of the spherical polar coordinates as follows:

$$x_n = X - \frac{\mu}{M} r \sin\theta\cos\phi \qquad x_e = X + \frac{\mu}{m} r \sin\theta\cos\phi$$

$$y_n = Y - \frac{\mu}{M} r \sin\theta\sin\phi \qquad y_e = Y + \frac{\mu}{m} r \sin\theta\sin\phi$$

$$z_n = Z - \frac{\mu}{M} r \cos\theta \qquad z_e = Z + \frac{\mu}{m} r \cos\theta$$

$$r = [(x_n - x_e)^2 + (y_n - y_e)^2 + (z_n - z_e)^2]^{1/2}. \tag{4-45}$$

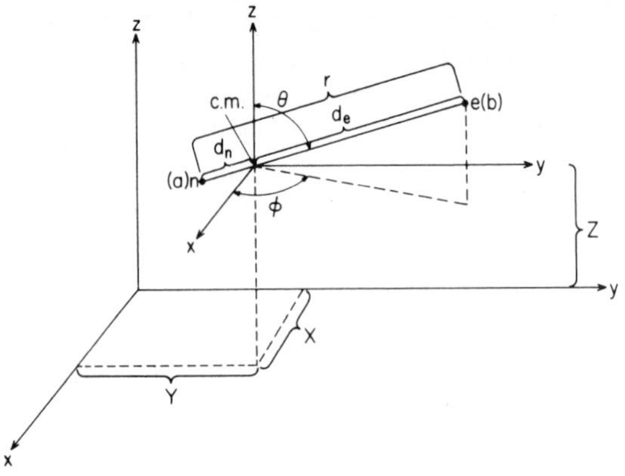

Figure 4-7 Coordinate system for a two-particle system; either an electron and proton or atom (a) and atom (b). The c.m. Cartesian coordinates are shown along with the spherical polar coordinates (θ is the polar angle and ϕ is the azimuthal angle).

$\mu = Mm/(M + m)$ is the reduced mass of the two-body system. Making these substitutions into Eq.(4-44) reduces the potential energy, V_{en}, to $-Ze^2/r$, which is the familiar central field potential. The transformation of the nuclear and electronic Laplacian operators, ∇_n^2 and ∇_e^2, to the spherical polar coordinates by use of the transformations above is a straightforward extension of Eqs.(4-6) and (4-7) to include the polar angle θ. The result gives the following Hamiltonian:

$$\mathscr{H} = -\frac{\hbar^2}{2(M+m)}\left(\frac{\partial^2}{\partial X^2} + \frac{\partial^2}{\partial Y^2} + \frac{\partial^2}{\partial Z^2}\right)$$
$$-\frac{\hbar^2}{2\mu}\left\{\frac{1}{r^2}\frac{\partial}{\partial r}\left(r^2\frac{\partial}{\partial r}\right) + \frac{1}{r^2 \sin\theta}\frac{\partial}{\partial\theta}\left(\sin\theta\frac{\partial}{\partial\theta}\right) + \frac{1}{r^2\sin^2\theta}\frac{\partial^2}{\partial\phi^2}\right\} - \frac{Ze^2}{r}. \quad (4\text{-}46)$$

The first term is the kinetic energy of a particle with mass $(M + m)$ representing the translational motion of the c.m. of the atom with respect to the arbitrary origin. The second term represents the internal motion of the electron and associated nucleus. The time-independent Schrödinger equation for the hydrogen atom is

$$\left[-\frac{\hbar^2}{2(M+m)}\nabla^2_{\text{c.m.}} - \frac{\hbar^2}{2\mu}\nabla^2_{r\theta\phi} - \frac{Ze^2}{r}\right]\Lambda(\boldsymbol{R}, r, \theta, \phi)$$
$$= [\mathscr{H}_t(\boldsymbol{R}) + \mathscr{H}_{in}(r, \theta, \phi)]\Lambda(\boldsymbol{R}, r, \theta, \phi) = E^{\wedge}\Lambda(\boldsymbol{R}, r, \theta, \phi)$$
$$\mathscr{H}_t(\boldsymbol{R}) = -\frac{\hbar^2}{2(M+m)}\nabla^2_{\text{c.m.}}, \qquad \mathscr{H}_{in}(r, \theta, \phi) = -\frac{\hbar^2}{2\mu}\nabla^2_{r\theta\phi} - \frac{Ze^2}{r}$$
$$\nabla^2_{r\theta\phi} = \frac{1}{r^2}\frac{\partial}{\partial r}\left(r^2\frac{\partial}{\partial r}\right) + \frac{1}{r^2\sin\theta}\frac{\partial}{\partial\theta}\left(\sin\theta\frac{\partial}{\partial\theta}\right) + \frac{1}{r^2\sin^2\theta}\frac{\partial^2}{\partial\phi^2}$$
$$\nabla^2_{\text{c.m.}} = \frac{\partial^2}{\partial X^2} + \frac{\partial^2}{\partial Y^2} + \frac{\partial^2}{\partial Z^2}. \quad (4\text{-}47)$$

$\mathscr{H}_t(\boldsymbol{R})$ and $\mathscr{H}_{in}(r, \theta, \phi)$ are the translational and internal Hamiltonians, respectively, and $\nabla^2_{r\theta\phi}$ and $\nabla^2_{\text{c.m.}}$ are the Laplacian operators for the internal r, θ, and ϕ and c.m. coordinates, respectively. Equation (4-47) is a partial differential equation in variables $\boldsymbol{R}$ to the c.m. and internal c.m. coordinates r, θ, and ϕ. Λ is a function of $\boldsymbol{R}$, r, θ, and ϕ. We can separate the $\boldsymbol{R}$ and r, θ, ϕ variables by writing $\Lambda(\boldsymbol{R}, r, \theta, \phi)$ as a product of functions in separated variables according to

$$\Lambda(\boldsymbol{R}, r, \theta, \phi) = \Gamma(\boldsymbol{R})\psi(r, \theta, \phi). \tag{4-48}$$

Substituting into Eq.(4-47), dividing by $\Gamma\psi$, and rearranging gives

$$\frac{1}{\Gamma}\mathscr{H}_t\Gamma + \frac{1}{\psi}\mathscr{H}_{in}\psi = E^{\Lambda}. \tag{4-49}$$

This equation is valid for all values of $\boldsymbol{R}$, r, θ, and ϕ only if both terms left of the equals sign are constants, call them ε and E, respectively, giving

$$\mathscr{H}_t\Gamma = \varepsilon\Gamma, \qquad \mathscr{H}_{in}\psi = E\psi, \qquad E^{\Lambda} = \varepsilon + E. \tag{4-50}$$

The wave equation for a free particle of mass $m + M$ from Eqs.(4-50) is given by

$$\mathscr{H}_t\Gamma(\boldsymbol{R}) = -\frac{\hbar^2\nabla^2_{\text{c.m.}}}{2(m + M)}\Gamma(\boldsymbol{R}) = \varepsilon\Gamma(\boldsymbol{R}). \tag{4-51}$$

A solution to this wave equation is the plane wave solution as discussed in Section 3.2,

$$\Gamma(\boldsymbol{R}) = \exp(-i\boldsymbol{k}\cdot\boldsymbol{R}) = \exp\left(-\frac{i}{\hbar}\boldsymbol{p}\cdot\boldsymbol{R}\right), \tag{4-52}$$

which gives a continuous energy of

$$\varepsilon = \frac{p^2}{2(m + M)}, \tag{4-53}$$

where p is the linear momentum of the c.m. If, alternatively, the hydrogen atom were confined by the walls of a cube, we would have the particle-in-a-box (see Problem 4-5) energy levels for the translational energies of the atom. The separation of internal motion and the translational motion of the center of mass is quite general and is easily extended to many-particle systems.

Equations (4-50) also contain the Schrödinger equation for the internal energy of the hydrogen atom given by

$$\left(-\frac{\hbar^2}{2\mu}\nabla^2_{r\theta\phi} - \frac{Ze^2}{r}\right)\psi(r, \theta, \phi) = \psi(r, \theta, \phi)E, \tag{4-54}$$

where $\nabla^2_{r\theta\phi}$ is defined in Eqs.(4-47). A separation of radial and angular variables in Eq.(4-54) can be obtained by using

$$\psi(r, \theta, \phi) = \eta(r)Y(\theta, \phi). \tag{4-55}$$

Substituting into Eq.(4-54), dividing by ηY, and rearranging gives

$$\frac{1}{\eta}\left\{\frac{\partial}{\partial r}\left(r^2\frac{\partial}{\partial r}\right) + \frac{2\mu Ze^2 r}{\hbar^2} + \frac{2\mu}{\hbar^2}r^2E\right\}\eta = -\frac{1}{Y}\left[\frac{1}{\sin\theta}\frac{\partial}{\partial\theta}\left(\sin\theta\frac{\partial}{\partial\theta}\right) + \frac{1}{\sin^2\theta}\frac{\partial^2}{\partial\phi^2}\right]Y. \tag{4-56}$$

This equation is satisfied for all values of r, θ, and ϕ only if both sides of the equals sign are equal to a constant, λ. Thus, we obtain two differential equations from Eq.(4-56), one in the variable r and the other in the variables θ and ϕ:

$$\left[\frac{\partial}{\partial r}\left(r^2 \frac{\partial}{\partial r}\right) + \frac{2\mu Ze^2 r}{\hbar^2} + \frac{2\mu r^2}{\hbar^2} E\right]\eta(r) = \lambda\eta(r), \tag{4-57}$$

$$\left[\frac{1}{\sin\theta}\frac{\partial}{\partial\theta}\left(\sin\theta\frac{\partial}{\partial\theta}\right) + \frac{1}{\sin^2\theta}\frac{\partial^2}{\partial\phi^2}\right]Y(\theta, \phi) = -\lambda Y(\theta, \phi). \tag{4-58}$$

We shall examine the angular equation involving $Y(\theta, \phi)$ in Section 4.4 and we shall give a solution to the radial equation in Section 4.6.

4.4 THE ANGULAR SOLUTION AND SPHERICAL HARMONICS

We start by separating the θ and ϕ variables in Eq.(4-58):

$$Y(\theta, \phi) = \Phi(\phi)\Theta(\theta). \tag{4-59}$$

Substituting into Eq.(4-58), multiplying by $\sin^2\theta$, dividing by $\Phi\Theta$, and rearranging gives

$$\frac{1}{\Theta}\left[\sin\theta\frac{\partial}{\partial\theta}\left(\sin\theta\frac{\partial}{\partial\theta}\right) + \lambda\sin^2\theta\right]\Theta = -\frac{1}{\Phi}\left(\frac{\partial^2}{\partial\phi^2}\right)\Phi. \tag{4-60}$$

Both sides of this equation are equal for all values of θ and ϕ when both sides are equal to a constant, α. This leads to two separate differential equations in θ and ϕ, respectively:

$$\left[\sin\theta\frac{\partial}{\partial\theta}\left(\sin\theta\frac{\partial}{\partial\theta}\right) + \lambda\sin^2\theta\right]\Theta = \alpha\Theta, \tag{4-61}$$

$$\frac{\partial^2}{\partial\phi^2}\Phi = -\alpha\Phi. \tag{4-62}$$

Equation (4-62) is identical to the equation considered previously in the case of a particle-in-a-ring; with the same boundary conditions, we obtain the solution

$$\Phi_m(\phi) = \left(\frac{1}{2\pi}\right)^{1/2}\exp(im\phi), \tag{4-63}$$

where $\alpha = m^2$ with m being restricted to the values of $m = 0, \pm 1, \pm 2, \ldots$ by the boundary conditions on ϕ.

Substituting $\alpha = m^2$ into Eq.(4-61) gives

$$\left[\sin\theta\frac{\partial}{\partial\theta}\left(\sin\theta\frac{\partial}{\partial\theta}\right) + \lambda\sin^2\theta\right]\Theta = m^2\Theta. \tag{4-64}$$

The solution to this equation is an eigenvalue problem that is well known [6]. In Appendix C.1 we develop a series expansion solution to Eq.(4-64) where truncation of the series requires that $\lambda = l(l+1)$ where l is limited to $l = 0, 1, 2, 3, \ldots$. Furthermore, for each l, m is limited to $m = l, l-1, l-2, \ldots, 0, -1, -2, \ldots, -l$. The

Table 4-3 The first few normalized spherical harmonics, $Y_{lm}(\theta, \phi)$, from Eq.(4-66).

$$Y_{00} = \frac{1}{2(\pi)^{1/2}}$$

$$Y_{10} = \frac{1}{2}\left(\frac{3}{\pi}\right)^{1/2} \cos\theta$$

$$Y_{1\pm1} = \mp\frac{1}{2}\left(\frac{3}{2\pi}\right)^{1/2} \sin\theta \exp(\pm i\phi)$$

$$Y_{20} = \frac{1}{4}\left(\frac{5}{\pi}\right)^{1/2} (3\cos^2\theta - 1)$$

$$Y_{2\pm1} = \mp\frac{1}{2}\left(\frac{15}{2\pi}\right)^{1/2} \sin\theta\cos\theta \exp(\pm i\phi)$$

$$Y_{2\pm2} = \frac{1}{4}\left(\frac{15}{2\pi}\right)^{1/2} \sin^2\theta \exp(\pm 2i\phi)$$

$$Y_{30} = \frac{1}{4}\left(\frac{7}{\pi}\right)^{1/2} (5\cos^3\theta - 3\cos\theta)$$

$$Y_{3\pm1} = \mp\frac{1}{8}\left(\frac{21}{\pi}\right)^{1/2} (5\cos^2\theta - 1)\sin\theta \exp(\pm i\phi)$$

$$Y_{3\pm2} = \frac{1}{4}\left(\frac{105}{2\pi}\right)^{1/2} \sin^2\theta\cos\theta \exp(\pm 2i\phi)$$

$$Y_{3\pm3} = \mp\frac{1}{8}\left(\frac{35}{\pi}\right)^{1/2} \sin^3\theta \exp(\pm 3i\phi)$$

corresponding eigenfunctions under the conditions above are the (normalized) associated Legendre polynomials in argument $\cos\theta$ given by

$$\Theta_l^{|m|}(\theta) = \left[\frac{(2l+1)(l-|m|)!}{2(l+|m|)!}\right]^{1/2} P_l^{|m|}(\cos\theta). \tag{4-65}$$

The $P_l^{|m|}$ functions are defined in Appendix C.1. Substituting Eqs.(4-63) and (4-65) into Eq.(4-59) gives

$$Y_{lm}(\theta, \phi) = \varepsilon\Phi_m(\phi)\Theta_l^{|m|}(\theta) = \varepsilon\left[\frac{(2l+1)(l-|m|)!}{4\pi(l+|m|)!}\right]^{1/2} P_l^{|m|}(\cos\theta)\exp(im\phi). \tag{4-66}$$

$Y_{lm}(\theta, \phi)$ are the normalized spherical harmonics where an arbitrary phase factor of $\varepsilon = (-1)^m$ for $m > 0$ and $\varepsilon = 1$ for $m \le 0$ has been added according to the more standard convention [7]. Several of the l manifolds in the Y_{lm} functions are listed in Table 4-3. The higher-order terms in $Y_{lm}(\theta, \phi)$ can be obtained by the following convenient generation formula (times the phase factor):

$$Y_{lm}(\theta, \phi) = \frac{1}{2^l l!}\left[\frac{(2l+1)(l-|m|)!}{4\pi(l+|m|)!}\right]^{1/2} \sin^{|m|}\theta \frac{d^{l+|m|}(\sin^{2l}\theta)}{(d\cos\theta)^{l+|m|}} \exp(im\phi). \tag{4-67}$$

4.5 ANGULAR MOMENTUM

The separation of variables in Eq.(4-55) assures us that all the angular dependence of the hydrogen-like system is contained in $Y_{lm}(\theta, \phi)$, the spherical harmonics. We shall now show that Y_{lm} is an eigenfunction of the square of the total angular momentum of the system.

The angular momentum operators are given by

$$\mathbf{L} = \mathbf{r} \times \mathbf{p}, \qquad L_x = yp_z - zp_y \longrightarrow \frac{\hbar}{i}\left(y\frac{\partial}{\partial z} - z\frac{\partial}{\partial y}\right),$$

$$L_y = zp_x - xp_z \longrightarrow \frac{\hbar}{i}\left(z\frac{\partial}{\partial x} - x\frac{\partial}{\partial z}\right), \tag{4-68}$$

$$L_z = xp_y - yp_x \longrightarrow \frac{\hbar}{i}\left(x\frac{\partial}{\partial y} - y\frac{\partial}{\partial x}\right),$$

$$L^2 = \mathbf{L}\cdot\mathbf{L} = L_x^2 + L_y^2 + L_z^2,$$

where the linear momentum operators from Table 3-1 have been substituted in place of p_x, p_y, and p_z. The angular momentum operators in Eq.(4-68) can be converted to spherical polar coordinates by familiar methods using the following relationships between the Cartesian coordinates and spherical polar coordinates (see Fig. 4-7):

$$x = r\sin\theta\cos\phi \qquad r = (x^2 + y^2 + z^2)^{1/2}$$

$$y = r\sin\theta\sin\phi \qquad \theta = \arccos\left[\frac{z}{(x^2 + y^2 + z^2)^{1/2}}\right]$$

$$z = r\cos\theta \qquad \phi = \arctan\left(\frac{y}{x}\right).$$

Making these transformations by using

$$\frac{\partial}{\partial z} = \frac{\partial r}{\partial z}\frac{\partial}{\partial r} + \frac{\partial\theta}{\partial z}\frac{\partial}{\partial\theta} + \frac{\partial\phi}{\partial z}\frac{\partial}{\partial\phi},$$

and so forth, gives

$$L_x = \frac{\hbar}{i}\left(-\sin\phi\frac{\partial}{\partial\theta} - \cot\theta\cos\phi\frac{\partial}{\partial\phi}\right), \qquad L_y = \frac{\hbar}{i}\left[\cos\phi\frac{\partial}{\partial\theta} - \cot\theta\sin\phi\frac{\partial}{\partial\phi}\right]$$

$$L_z = \frac{\hbar}{i}\frac{\partial}{\partial\phi}, \qquad L^2 = -\hbar^2\left[\frac{1}{\sin\theta}\frac{\partial}{\partial\theta}\left(\sin\theta\frac{\partial}{\partial\theta}\right) + \frac{1}{\sin^2\theta}\frac{\partial^2}{\partial\phi^2}\right]. \tag{4-69}$$

Comparing the L^2 operator with Eqs.(4-60) and (4-59) and the following development shows that

$$L^2 Y_{lm}(\theta, \phi) = -\hbar^2\left[\frac{1}{\sin\theta}\frac{\partial}{\partial\theta}\left(\sin\theta\frac{\partial}{\partial\theta}\right) + \frac{1}{\sin^2\theta}\frac{\partial^2}{\partial\phi^2}\right]Y = \hbar^2\lambda Y_{lm}(\theta, \phi). \tag{4-70}$$

The eigenvalues of L^2 in the $\boldsymbol{Y}$ basis are, therefore, $\hbar^2 l(l+1)$ from the definition of λ following Eq.(4-64). The spherical harmonics are also eigenfunctions of the L_z operator with eigenvalues equal to $m\hbar$:

$$L_z Y_{lm}(\theta, \phi) = \frac{\hbar}{i}\frac{\partial}{\partial\phi} Y_{lm}(\theta, \phi) = m\hbar Y_{lm}(\theta, \phi). \tag{4-71}$$

It is clear from Eqs.(4-70) and (4-71) that L_z and L^2 must necessarily commute as they have identical eigenfunctions. $Y_{lm}(\theta, \phi)$ is not an eigenfunction of L_x and L_y and, correspondingly, L_x and L_y do not commute with L_z (L_x and L_y do, however, commute with L^2). The commutation relationships of the angular momentum operators are easily obtained from the definitions in Eqs.(4-68) or (4-69) and the results are

$$\begin{aligned}
&[L_x, L^2] = 0 \qquad [L_x, L_y] = L_x L_y - L_y L_x = i\hbar L_z \\
&[L_y, L^2] = 0 \qquad [L_y, L_z] = i\hbar L_x \\
&[L_z, L^2] = 0 \qquad [L_z, L_x] = i\hbar L_y.
\end{aligned} \tag{4-72}$$

The corresponding matrix representation of the angular momentum operators in the spherical harmonic basis are easily determined:

$$\int Y^*_{lm} L^2 Y^*_{l'm'}\, dV \quad \text{and} \quad \int Y^*_{lm} L_z Y_{l'm'}\, dV$$

lead to diagonal matrices as discussed before and

$$\int Y^*_{lm} L_x Y^*_{l'm'}\, dV \quad \text{and} \quad \int Y^*_{lm} L_y Y_{l'm'}\, dV$$

lead to matrices that are diagonal only in l. The angular momentum matrices for $l = 1, m = -1, 0, 1$ are given here. These matrix elements can be verified by actual integration where $\int dV$ is an abbreviation for

$$\int_0^{2\pi}\int_0^{\pi} \sin\theta\, d\theta\, d\phi.$$

$$\mathbf{L}^2 = \begin{pmatrix} \int Y^*_{11} L^2 Y_{11}\, dV & \int Y^*_{11} L^2 Y_{10}\, dV & \int Y^*_{11} L^2 Y_{1-1}\, dV \\ \int Y^*_{10} L^2 Y_{11}\, dV & \int Y^*_{10} L^2 Y_{10}\, dV & \int Y^*_{10} L^2 Y_{1-1}\, dV \\ \int Y^*_{1-1} L^2 Y_{11}\, dV & \int Y^*_{1-1} L^2 Y_{10}\, dV & \int Y^*_{1-1} L^2 Y_{1-1}\, dV \end{pmatrix} = 2\hbar^2 \begin{pmatrix} 1 & 0 & 0 \\ 0 & 1 & 0 \\ 0 & 0 & 1 \end{pmatrix}$$

$$\mathbf{L}_z = \hbar\begin{pmatrix} 1 & 0 & 0 \\ 0 & 0 & 0 \\ 0 & 0 & -1 \end{pmatrix}; \quad \mathbf{L}_x = \frac{\hbar}{(2)^{1/2}}\begin{pmatrix} 0 & 1 & 0 \\ 1 & 0 & 1 \\ 0 & 1 & 0 \end{pmatrix}; \quad \mathbf{L}_y = \frac{i\hbar}{(2)^{1/2}}\begin{pmatrix} 0 & -1 & 0 \\ 1 & 0 & -1 \\ 0 & 1 & 0 \end{pmatrix}. \tag{4-73}$$

Generalization of these results gives the following nonzero matrix elements for the angular momentum operators in the spherical harmonic basis:

$$\int Y_{lm}^* L^2 Y_{lm}\, dV = \int_0^{2\pi} \int_0^{\pi} Y_{lm}^*(\theta, \phi) L^2 Y_{lm}(\theta, \phi) \sin\theta \, d\theta \, d\phi$$

$$= \langle lm | L^2 | lm \rangle = \hbar^2 l(l+1)$$

$$\int Y_{lm}^* L_x Y_{lm\pm 1}\, dV = \langle lm | L_x | lm \pm 1 \rangle = \frac{\hbar}{2} [l(l+1) - m(m \pm 1)]^{1/2}$$

$$\langle lm | L_y | lm \pm 1 \rangle = \pm \frac{i\hbar}{2} [l(l+1) - m(m \pm 1)]^{1/2}$$

$$\langle lm | L_z | lm \rangle = \hbar m, \tag{4-74}$$

where all results can be verified by direct integration. These matrix representations for L_x, L_y, and L_z conform with the phase choices of most conventional textbooks that require the phase factor ε in Eq.(4-66) for the spherical harmonics. The convenient Dirac [8] angular bracket notation is also introduced here for the first time where we note that $\langle lm | l'm' \rangle = \delta_{ll'} \delta_{mm'}$. The bracket notation is convenient when the functions contain several quantum numbers.

It is easy to show from the equations above that the L_x, L_y, and L_z operators are Hermitian. According to Eq.(3-54), an operator is Hermitian if its matrix representation is equal to its transposed complex conjugate (adjoint). According to Eq.(4-73), $\mathbf{L}_x = \mathbf{L}_x^\dagger$, $\mathbf{L}_y = \mathbf{L}_y^\dagger$, and $\mathbf{L}_z = \mathbf{L}_z^\dagger$. The raising and lowering angular momentum operators, $L_\pm = L_x \pm iL_y$, and their relation to L_x, L_y, and L_z are considered in Problem 4-10.

4.6 SOLUTION OF THE RADIAL EQUATION, HYDROGEN-LIKE ATOM ENERGIES, AND THE RADIAL DISTRIBUTION

The radial part of Schrödinger's equation for the hydrogen-like atom is given in Eq.(4-57) where $\lambda = l(l+1)$ from the series solution of Eq.(4-64). A connection can be established with Laguerre's differential equation by making the following substitutions in Eq.(4-57) for negative energies:

$$E = -\frac{\mu e^4 Z^2}{2\hbar^2 n^2} \tag{4-75}$$

and

$$r = \frac{n\hbar^2}{2\mu e^2 Z} x. \tag{4-76}$$

We shall show that n is an integer that is defined below according to the solution of the differential equation obtained by substitution and truncation of the series expansion for the Laguerre polynomial. Substituting $\lambda = l(l+1)$ and Eqs.(4-75) and (4-76) into Eq.(4-57) gives

$$\frac{d^2\eta(x)}{dx^2} + \frac{2}{x}\frac{d\eta(x)}{dx} + \left[-\frac{1}{4} + \frac{n}{x} - \frac{l(l+1)}{x^2}\right]\eta(x) = 0. \tag{4-77}$$

First we examine the asymptotic solutions to Eq.(4-77).

1. $x \longrightarrow \infty$

   Equation (4-77) reduces to

$$\frac{d^2\eta}{dx^2} - \frac{1}{4}\eta = 0, \tag{4-78}$$

which has a solution of

$$\eta = \exp\left(\frac{-x}{2}\right). \tag{4-79}$$

2. $x \longrightarrow 0$

   Rearranging Eq.(4-77) gives

$$\frac{d^2}{dx^2}(x\eta) + \left[\frac{n}{x} - \frac{1}{4} - \frac{l(l+1)}{x^2}\right](x\eta) = 0. \tag{4-80}$$

This equation is convenient to study the asymptotic behavior as $x \longrightarrow 0$, which gives

$$\frac{d^2}{dx^2}(x\eta) - \frac{l(l+1)}{x^2}(x\eta) = 0. \tag{4-81}$$

The solution to this equation, which is well behaved at $x = 0$, is

$$x\eta = x^{l+1}; \qquad \eta = x^l. \tag{4-82}$$

Thus, as $x \longrightarrow \infty$, $\eta = \exp(-x/2)$, and as $x \longrightarrow 0$, $\eta = x^l$. The solution to Eq.(4-77) for all regions of x is then given by

$$\eta(x) = U(x)x^l \exp\left(\frac{-x}{2}\right), \tag{4-83}$$

where $U(x)$ is a polynomial. Substituting this result into Eq.(4-77) gives (after considerable algebra) a recognizable differential equation:

$$x\frac{d^2U}{dx^2} + (2l + 2 - x)\frac{dU}{dx} + (n - l - 1)U = 0. \tag{4-84}$$

We solve this differential equation by a series expansion method in Appendix C.2 where we show that $U(x)$ are the associated Laguerre functions if $n \geq l + 1$, where $l = 0, 1, 2, 3, \ldots$ and $n = 1, 2, 3, 4, \ldots$. The limits on n and l arise from a truncation in the series expansion for $U(x)$. The Laguerre polynomials are given by

$$L_{n+l}(x) = \exp(x)\frac{d^{n+l}}{dx^{n+l}}[x^{n+l}\exp(-x)]. \tag{4-85}$$

Table 4-4 Hydrogen-like radial functions, $\eta_{nl}(r)$, from Eq.(4-87). Z is the atomic number and a is given in Eq.(4-88).

$$\eta_{10} = 2\left(\frac{Z}{a}\right)^{3/2} \exp\left(-\frac{Zr}{a}\right)$$

$$\eta_{20} = \frac{1}{2(2)^{1/2}}\left(\frac{Z}{a}\right)^{3/2}\left(2 - \frac{Zr}{a}\right)\exp\left(-\frac{Zr}{2a}\right)$$

$$\eta_{21} = \frac{1}{2(6)^{1/2}}\left(\frac{Z}{a}\right)^{3/2}\left(\frac{Zr}{a}\right)\exp\left(-\frac{Zr}{2a}\right)$$

$$\eta_{30} = \frac{2}{81(3)^{1/2}}\left(\frac{Z}{a}\right)^{3/2}\left[27 - \frac{18Zr}{a} + 2\left(\frac{Zr}{a}\right)^2\right]\exp\left(-\frac{Zr}{3a}\right)$$

$$\eta_{31} = \frac{4}{81(6)^{1/2}}\left(\frac{Z}{a}\right)^{3/2}\left[6\frac{Zr}{a} - \left(\frac{Zr}{a}\right)^2\right]\exp\left(-\frac{Zr}{3a}\right)$$

$$\eta_{32} = \frac{4}{81(30)^{1/2}}\left(\frac{Z}{a}\right)^{3/2}\left(\frac{Zr}{a}\right)^2\exp\left(-\frac{Zr}{3a}\right)$$

The associated Laguerre polynomials, which satisfy Eq.(4-84), are obtained by taking the $(2l + 1)$th derivative of $L_{n+l}(x)$ with respect to x:

$$U(x) = L_{n+l}^{2l+1}(x) = \frac{d^{2l+1}}{dx^{2l+1}} L_{n+l}(x). \tag{4-86}$$

Thus, the radial functions can be obtained by substituting Eq.(4-86) into Eq.(4-83). The final normalized radial function is given by

$$\eta_{nl}(r) = -\left[\left(\frac{2Z}{na}\right)^3 \frac{(n-l-1)!}{2n[(n+l)!]^3}\right]^{1/2}\left(\frac{2Zr}{na}\right)^l \exp\left(-\frac{Zr}{na}\right)L_{n+l}^{2l+1}\left(\frac{2Zr}{na}\right), \tag{4-87}$$

where $l = 0, 1, 2, 3, \ldots, n = 1, 2, 3, \ldots;\ n \geq l + 1$; and a is the Bohr radius for the one-electron atom or ion with reduced mass μ,

$$a = \frac{\hbar^2}{\mu e^2}. \tag{4-88}$$

We reserve the notation of a_0 for the Bohr radius for an infinitely heavy nucleus where $\mu \longrightarrow m$ [see Eq.(3-6)]. The first few radial functions are given in Table 4-4.

In summary, the product of $Y_{lm}(\theta, \phi)$ in Eq.(4-66) with the radial function $\eta_{nl}(r)$ in Eq.(4-87) gives the hydrogen-like eigenfunctions. The associated eigenvalues for negative energies are given in Eq.(4-75).

The difference between any two energy levels in the hydrogen-like atom is also obtained from Eq.(4-75) and the result is identical to Bohr's result in Eq.(3-8).

The *average* electron-nuclear distance, $\langle r \rangle$, for hydrogen-like atoms is given by [9]

$$\langle r \rangle_{nlm} = \int_0^{2\pi}\int_0^{\pi}\int_0^{\infty} \psi^*_{nlm}[r]\psi_{nlm} r^2\, dr \sin\theta\, d\theta\, d\phi$$

$$\langle r \rangle_{nl} = \int_0^{\infty} \eta_{nl}(r)[r]\eta_{nl}(r) r^2\, dr = \frac{n^2 a}{Z}\left[1 + \frac{1}{2}\left(1 - \frac{l(l+1)}{n^2}\right)\right], \tag{4-89}$$

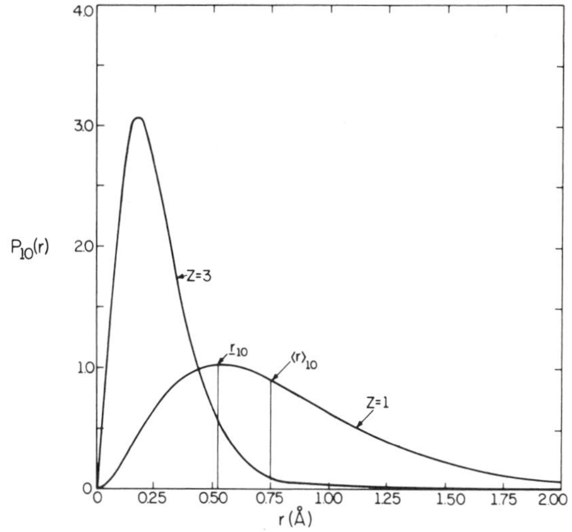

Figure 4-8 The $P_{10}(r)$ functions defined in Eq.(4-90) for the $Z = 1$ and $Z = 3$ hydrogen-like atoms. $\langle r \rangle$ is the average value and $\underline{r}$ is the most probable value of r in the $Z = 1$ case.

where $\eta_{nl}(r)$ are given in Table 4-4 and Eq.(4-87). $\langle r \rangle_{nl}$ for the $n = 1, l = 0$ function is equal to $(\frac{3}{2})a$ when $Z = 1$.

As ψ_{nlm} is not an eigenfunction of r, the average value of r given in Eq.(4-89) will not necessarily be equal to the most probable value of r. The most probable value of r is obtained from Eq.(3-135) as discussed in Section 3.6. The probability of finding the electron within a small region between r and $r + \Delta r$ as Δr goes to zero is

$$P_{nl}(r) = \lim_{\Delta r \to 0} \left[\int_0^{2\pi} \int_0^{\pi} \int_r^{r+\Delta r} \psi_{nlm}^* \psi_{nlm} r^2 \, dr \sin\theta \, d\theta \, d\phi \right] = \eta_{nl}(r)\eta_{nl}(r)r^2. \quad (4\text{-}90)$$

$P_{nl}(r)$ is the radial distribution function for the hydrogen-like atom (probability per unit length). A plot of $P_{nl}(r)$ for the $1s$ hydrogen-like atom is shown in Fig. 4-8 for $Z = 1$ and $Z = 3$.

The most probable radial position, $\underline{r}_{nl}$, is obtained by determining the maximum of $P_{nl}(r)$ in Eq.(4-90) with respect to r. Consider $\underline{r}_{nl}$ (the most probable value) for the $1s$ function ($n = 1, l = 0, m = 0$). We start with

$$P_{10}(r) = [\eta_{10}(r)]^2 r^2 = 4\left(\frac{Z}{a}\right)^3 r^2 \exp\left(-\frac{2Zr}{a}\right)$$

$$\frac{dP_{10}(r)}{dr} = -8\left(\frac{Z}{a}\right)^4 r^2 \exp\left(-\frac{2Zr}{a}\right) + 8\left(\frac{Z}{a}\right)^3 r \exp\left(-\frac{2Zr}{a}\right).$$

Setting $dP_{10}(r)/dr = 0$ and solving for r gives the most probable value of r in the ground state ($n = 1, l = 0$) of

$$\underline{r}_{10} = \frac{a}{Z}. \quad (4\text{-}91)$$

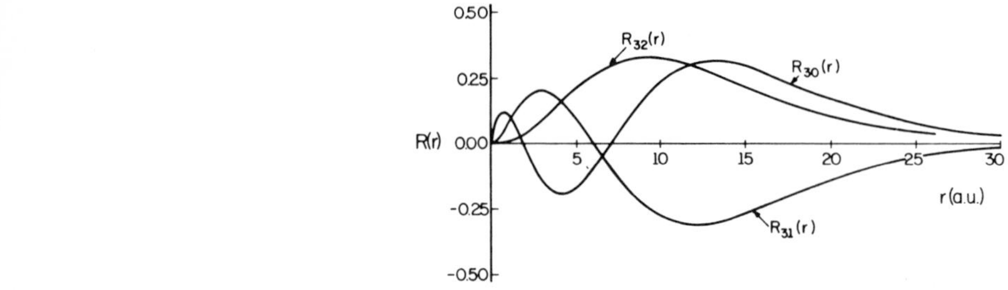

Figure 4-9 The radial amplitude functions for the $n = 3$, $l = 2$, 1, and 0 hydrogen-like functions when $Z = 1$ in atomic units (a.u.).

This value is different from the average value of r in the same state, which is $3a/2Z$, as shown in Fig. 4-8.

It is sometimes convenient to write the radial functions in Table 4-4 in atomic units of length or in units of $a_0 = \hbar^2/me^2$, the Bohr radius for an infinitely heavy nucleus. The radial functions in Table 4-4 can be written in atomic units (a.u.) by setting $a = 1.0$. $P_{10}(r)$ in atomic units for the 1s function of the hydrogen-like atom is

$$P_{10}(r) = 4Z^3r^2 \exp(-2Zr). \tag{4-92}$$

Returning to Eq.(4-90) for the radial probability density, $P_{nl}(r)$, we can write

$$P_{nl}(r) = [r\eta_{nl}(r)][r\eta_{nl}(r)] = [R_{nl}(r)]^2, \tag{4-93}$$

where $R_{nl}(r)$ is defined as the radial amplitude function.

It is evident from Table 4-4 that the number of nodes in the radial function is equal to $n - l - 1$. Thus, the $\eta_{30}(r)$ function has two nodes, the $\eta_{31}(r)$ function has one node, and the $\eta_{32}(r)$ function has zero nodes. The radial amplitude functions, $r\eta(r)$, for the $n = 3$ radial functions are written as

$$R_{30}(r) = r\eta_{30}(r), \qquad R_{31}(r) = r\eta_{31}(r), \qquad R_{32}(r) = r\eta_{32}(r). \tag{4-94}$$

These functions are shown in Fig. 4-9 in atomic units. It will be interesting to compare the radial functions in Fig. 4-9 with the Slater and self-consistent field atomic radial functions which are discussed in the treatment of many-electron atoms in Chapter 5.

It is also necessary to investigate the solution of the radial equation in Eq.(4-57) for positive values of the energy. If we substitute

$$\lambda = l(l+1), \qquad E_k = \frac{\mu e^4 Z^2}{2\hbar^2 k^2}, \quad \text{and} \quad r = \frac{k\hbar^2}{2\mu e^2 Z}x \tag{4-95}$$

into Eq.(4-57), we obtain the radial equation for positive energies,

$$\frac{d^2\eta(x)}{dx^2} + \frac{2}{x}\frac{d\eta(x)}{dx} + \left[\frac{1}{4} + \frac{k}{x} - \frac{l(l+1)}{x^2}\right]\eta(x) = 0. \tag{4-96}$$

Arguments identical with those given previously on the asymptotic character of the functions suggest a solution of the following type:

$$\eta_{kl}(x) = U(x)x^l \exp\left(\pm\frac{ix}{2}\right). \tag{4-97}$$

Substituting this solution into Eq.(4-96) yields a differential equation similar to Eq.(4-84) which is solvable in a polynomial $U(x)$ which does not restrict k to integers. Thus, the positive energies span a continuous array of levels in the hydrogen-like atom. In order for the hydrogen-like eigenfunctions to be a complete set of functions, the continuum solutions must also be included. We now return to a discussion of the discrete states.

4.7 NOTATION, LINEAR TRANSFORMATIONS, HYBRID ORBITALS, AND SPHERICAL HARMONIC SELECTION RULES

The values of n, the principal quantum in the hydrogen-like eigenfunction, $\psi_{nlm}(r, \theta, \phi)$, designate the shell structure of an atom with the following notation:

$$\begin{array}{rcccccccc} n = & 1 & 2 & 3 & 4 & 5 & 6 & \cdots \\ \text{shell} = & K & L & M & N & O & P & \cdots \end{array} \tag{4-98}$$

The common method of designating the angular dependence of the eigenfunctions with various values of l within each value of n is as follows:

$$\begin{array}{rcccccccc} l = & 0 & 1 & 2 & 3 & 4 & 5 & \cdots \\ \textit{state} = & S & P & D & F & G & H & \cdots \end{array} \tag{4-99}$$

Thus, a D state of the hydrogen-like atom, where $n \geq 3$, signifies the $l = 2$ spherical harmonic $Y_{2m}(\theta, \phi)$, which has a possible $(2l + 1) = 5$ states within each given l state. Table 4-5 summarizes the lower energy stationary states of the hydrogen-like atom.

Each ψ_{nlm} is called an *orbital* and lowercase letters from Eq.(4-99) are used to designate the orbital. An s-orbital has $l = 0$, a p-orbital has $l = 1$ with three values of $m(m = 0, \pm 1)$, a d-orbital has $l = 2$ $(m = 0, \pm 1, \pm 2)$, and so on.

Table 4-5 Energy eigenstates of the hydrogen-like atom.

Shell	n	l	m	Orbital, ψ_{nlm}	Spectroscopic Notation	Energy
	∞					0
M	3	2	± 2	$\psi_{32\pm 2}$	D	$-(\mu e^4 Z^2/18\hbar^2)$
	3	2	± 1	$\psi_{32\pm 1}$		
	3	2	0	ψ_{320}		
	3	1	± 1	$\psi_{31\pm 1}$	P	
	3	1	0	ψ_{310}		
	3	0	0	ψ_{300}	S	
L	2	1	± 1	$\psi_{21\pm 1}$	P	$-(\mu e^4 Z^2/8\hbar^2)$
	2	1	0	ψ_{210}		
	2	0	0	ψ_{200}	S	
K	1	0	0	ψ_{100}	S	$-(\mu e^4 Z^2/2\hbar^2) = -(e^2/2a)$

The angular dependence of the hydrogen-like functions, as described by the spherical harmonics, can be transformed from the complex basis to a real basis within each l manifold ($l, l-1, l-2, \ldots, 0, -1, -2, \ldots, -l$) of degenerate eigenvalues. Using the spherical harmonics in Table 4-3 gives

1. $l = 0$

$$s = Y_{00} = \frac{1}{2(\pi)^{1/2}}, \tag{4-100}$$

2. $l = 1$

$$(p_x \;\; p_y \;\; p_z) = (Y_{1-1} \;\; Y_{11} \;\; Y_{10})\left(\frac{1}{2}\right)^{1/2}\begin{pmatrix} +1 & i & 0 \\ -1 & i & 0 \\ 0 & 0 & (2)^{1/2} \end{pmatrix}$$

$$p_x = \frac{1}{2}\left(\frac{3}{\pi}\right)^{1/2} \sin\theta \cos\phi$$

$$p_y = \frac{1}{2}\left(\frac{3}{\pi}\right)^{1/2} \sin\theta \sin\phi$$

$$p_z = \frac{1}{2}\left(\frac{3}{\pi}\right)^{1/2} \cos\theta, \tag{4-101}$$

3. $l = 2$

$$(d_{xy} \; d_{x^2-y^2} \; d_{xz} \; d_{yz} \; d_{z^2}) = (Y_{2-2} \; Y_{22} \; Y_{2-1} \; Y_{21} \; Y_{20})\left(\frac{1}{2}\right)^{1/2}\begin{pmatrix} 1 & i & 0 & 0 & 0 \\ 1 & -i & 0 & 0 & 0 \\ 0 & 0 & +1 & i & 0 \\ 0 & 0 & -1 & i & 0 \\ 0 & 0 & 0 & 0 & (2)^{1/2} \end{pmatrix}$$

$$d_{xy} = \frac{1}{4}\left(\frac{15}{\pi}\right)^{1/2} \sin^2\theta \cos 2\phi$$

$$d_{x^2-y^2} = \frac{1}{4}\left(\frac{15}{\pi}\right)^{1/2} \sin^2\theta \sin 2\phi$$

$$d_{xz} = \frac{1}{2}\left(\frac{15}{\pi}\right)^{1/2} \sin\theta \cos\theta \cos\phi$$

$$d_{yz} = \frac{1}{2}\left(\frac{15}{\pi}\right)^{1/2} \sin\theta \cos\theta \sin\phi$$

$$d_{z^2} = \frac{1}{4}\left(\frac{5}{\pi}\right)^{1/2} (3\cos^2\theta - 1). \tag{4-102}$$

The transformation matrices are, of course, unitary. The real p and d functions are illustrated in Fig. 5-9. The real $f(l = 3)$, $g(l = 4)$, and $h(l = 5)$ functions can be obtained in the same manner. The real $s, p, d, \ldots$ functions are also eigenfunctions of the hydrogen-like Hamiltonian operator as all linear combinations of the spherical harmonics involve degenerate energy states. The $s, p, d, \ldots$ functions are not, however, all eigenfunctions of the L_z operator. The complete notation for hydrogen-like functions (or orbitals) includes n with $ns, np, nd, \ldots$ where n is defined by the radial eigenfunctions and $l = s, p, d, \ldots$ from the angular eigenfunctions. Thus, a $5s$ function corresponds to $\psi_{nlm} = \psi_{500}$ (see Table 4-5). A few of the real hydrogen-like orbitals are listed below. These results are obtained by multiplying the results in Eqs.(4-100) and (4-101) by the radial functions in Table 4-4.

$$
\begin{aligned}
\psi_{1s} &= \left(\frac{1}{\pi}\right)^{1/2}\left(\frac{Z}{a}\right)^{3/2} \exp\left(-\frac{Zr}{a}\right) \\
\psi_{2s} &= \frac{1}{8}\left(\frac{2}{\pi}\right)^{1/2}\left(\frac{Z}{a}\right)^{3/2}\left(2-\frac{Zr}{a}\right) \exp\left(-\frac{Zr}{2a}\right) \\
\psi_{2p_x} &= \frac{1}{8}\left(\frac{2}{\pi}\right)^{1/2}\left(\frac{Z}{a}\right)^{5/2} r \exp\left(-\frac{Zr}{2a}\right) \sin\theta\cos\phi \\
\psi_{2p_y} &= \frac{1}{8}\left(\frac{2}{\pi}\right)^{1/2}\left(\frac{Z}{a}\right)^{5/2} r \exp\left(-\frac{Zr}{2a}\right) \sin\theta\sin\phi \\
\psi_{2p_z} &= \frac{1}{8}\left(\frac{2}{\pi}\right)^{1/2}\left(\frac{Z}{a}\right)^{5/2} r \exp\left(-\frac{Zr}{2a}\right) \cos\theta
\end{aligned}
\tag{4-103}
$$

The solution of the hydrogen-like Schrödinger equation does not limit the linear combinations to l manifolds as all l within a given n have eigenfunctions that lead to degenerate eigenvalues. Thus, linear combinations of $s, p, d, \ldots$ functions within each n shell also yield energy eigenfunctions. These combinations of spherical harmonics with different l-values are called *hybrid orbitals* [10]. A few orbitals are given below.

nsp hybrid

$$
(\alpha_1 \quad \alpha_2) = (ns \quad np_x)\left(\frac{1}{2}\right)^{1/2}\begin{pmatrix} 1 & 1 \\ 1 & -1 \end{pmatrix} \tag{4-104}
$$

nsp^2 *hybrid*

$$
(\alpha_1 \quad \alpha_2 \quad \alpha_3) = (ns \quad np_x \quad np_y)\begin{pmatrix} (\frac{1}{3})^{1/2} & (\frac{1}{3})^{1/2} & (\frac{1}{3})^{1/2} \\ -(\frac{2}{3})^{1/2} & (\frac{1}{6})^{1/2} & (\frac{1}{6})^{1/2} \\ 0 & (\frac{1}{2})^{1/2} & -(\frac{1}{2})^{1/2} \end{pmatrix} \tag{4-105}
$$

nsp^3 *hybrid*

$$
(\alpha_1 \quad \alpha_2 \quad \alpha_3 \quad \alpha_4) = (ns \quad np_x \quad np_y \quad np_z)\left(\frac{1}{4}\right)^{1/2}\begin{pmatrix} 1 & 1 & 1 & 1 \\ 1 & 1 & -1 & -1 \\ 1 & -1 & -1 & 1 \\ 1 & -1 & 1 & -1 \end{pmatrix} \tag{4-106}
$$

Each of the transformation matrices is unitary. Hybrid orbitals can also be formed from higher l combinations by the proper unitary transformations (see a group theoretical discussion in Appendix E.5). The α_n functions in Eqs.(4-104) to (4-106) are easily related back to the spherical harmonics. Consider the nsp^3 hybrids given by

nsp^3 hybrid

$$(\alpha_1 \quad \alpha_2 \quad \alpha_3 \quad \alpha_4) = (\eta_{n0} Y_{00} \quad \eta_{n1} Y_{1-1} \quad \eta_{n1} Y_{11} \quad \eta_{n1} Y_{10})$$

$$\times \begin{pmatrix} 1 & 0 & 0 & 0 \\ 0 & +\dfrac{1}{(2)^{1/2}} & \dfrac{i}{(2)^{1/2}} & 0 \\ 0 & -\dfrac{1}{(2)^{1/2}} & \dfrac{i}{(2)^{1/2}} & 0 \\ 0 & 0 & 0 & 1 \end{pmatrix} \left(\frac{1}{4}\right)^{1/2} \begin{pmatrix} 1 & 1 & 1 & 1 \\ 1 & 1 & -1 & -1 \\ 1 & -1 & -1 & 1 \\ 1 & -1 & 1 & -1 \end{pmatrix}. \tag{4-107}$$

All three sets of functions ψ_{nlm}; $ns, np_x, np_y, \ldots$; and $\alpha_1, \alpha_2, \alpha_3, \ldots$ in Eqs.(4-104) and (4-105) are eigenfunctions of the Hamiltonian having identical eigenvalues (for any given n). Thus, an electron in the N shell, for instance, will have the same energy in any of the possible nine eigenstates of this shell. The different eigenstates do have different spatial dependence as demonstrated in the ns, np_x, np_y, and np_z atomic orbitals and the subsequent hybrid orbitals.

Selection Rules

The selection rules for possible transitions caused by the interaction of electromagnetic radiation and the hydrogen atom will now be derived according to the electric and magnetic dipole operators where we examine the interaction with plane polarized and circular polarized radiation.

In the case of the electric dipole operator, we have

$$D_x = ex = er \sin\theta \cos\phi, \qquad D_y = ey = er \sin\theta \sin\phi, \qquad D_z = ez = er\cos\theta. \tag{4-108}$$

As r is common for the three operators, we write

$$\int \psi^*_{n'l'm'}[erD(\theta, \phi)]\psi_{nlm}\, dV = e\int_0^\infty \eta^*_{n'l'}(r) r \eta_{nl}(r) r^2\, dr \int_0^{2\pi}\int_0^\pi Y^*_{l'm'} D(\theta, \phi) Y_{lm} \sin\theta\, d\theta\, d\phi \tag{4-109}$$

for the hydrogen-like functions. $D(\theta, \phi) = \sin\theta\cos\phi$, $\sin\theta\sin\phi$, and $\cos\theta$ for D_x, D_y, and D_z, respectively. As

$$e\int_0^\infty \eta^*_{n'l'}(r) r \eta_{nl}(r) r^2\, dr \neq 0 \tag{4-110}$$

for all combinations of n' and l', there are no selection rules on n for the electric dipole transitions in the hydrogen-like atom. The selection rules for l are determined from the matrix element over the spherical harmonics. We now rewrite the angular parts in Eq.(4-108) in terms of the spherical harmonics in Table 4-3 giving

$$D_z = 2\left(\frac{\pi}{3}\right)^{1/2} erY_{10}(\theta, \phi), \qquad D_x \pm iD_y = D_\pm = e(x \pm iy) = \mp 2\left(\frac{2\pi}{3}\right)^{1/2} erY_{1\pm 1}(\theta, \phi), \tag{4-111}$$

where D_z interacts with plane polarized radiation and $D_\pm$ interacts with the circular polarized radiation as demonstrated in Eq.(3-179) and the associated discussion.

The selection rules on l and m are obtained from

$$(D_z)_{l'm',lm} = C_1\left[2\left(\frac{\pi}{3}\right)^{1/2}\right]\int_0^{2\pi}\int_0^\pi Y^*_{l'm'}[Y_{10}]Y_{lm}\sin\theta\, d\theta\, d\phi \tag{4-112}$$

$$(D_\pm)_{l'm',lm} = C_1\left[2\left(\frac{2\pi}{3}\right)^{1/2}\right]\int_0^{2\pi}\int_0^\pi Y^*_{l'm'}[Y_{1\pm 1}]Y_{lm}\sin\theta\, d\theta\, d\phi. \tag{4-113}$$

C_1 is the value of the radial operator matrix element in Eq.(4-110). A close look at the integrals over three spherical harmonics (see Table 4-3) shows that the sum of m-values must be zero and the values of l must satisfy a triangular relationship [11]. That is,

$$\int_0^{2\pi}\int_0^\pi Y^*_{l_1m_1}Y_{l_2m_2}Y_{l_3m_3}\sin\theta\, d\theta\, d\phi \neq 0 \tag{4-114}$$

only if $l_1 + l_2 + l_3$ is even and

$$-m_1 + m_2 + m_3 = 0$$

$$l_1 + l_2 \geq l_3, \qquad l_1 + l_3 \geq l_2, \qquad l_2 + l_3 \geq l_1. \tag{4-115}$$

By simple substitution and examination of the integrals above, it is easy to show that plane polarized electric fields in electromagnetic radiation of proper frequency may excite the atom through the D_z operator leading to $\Delta l = \pm 1, \Delta m = 0$ transitions and circularly polarized radiation may excite the atom through the $D_\pm$ operator leading to $\Delta l = \pm 1, \Delta m = \pm 1$ transitions.

Table 4-6 The selection rules on the spherical harmonics, Y_{lm}. C_1 is the radial integral defined in Eq.(4-110) for the hydrogen-like atom. C_1 is the permanent electric dipole moment in the case of a rotating linear molecule ($l \longrightarrow J$ and $m \longrightarrow M$). μ in the right-hand column is the corresponding magnetic moment for the system in question; $\mu = -\mu_B$ for electronic spin or orbital angular momentum ($m \longrightarrow M_S \longrightarrow M_L$), and $\mu = g_i\mu_0$ for nuclear angular momentum ($l \longrightarrow I$, $m \longrightarrow M_I$).

				Transition Moment
Electric dipole	D_z	$\Delta l = \pm 1$	$\Delta m = 0$	$2C_1\left(\frac{\pi}{3}\right)^{1/2}(\langle lm \vert Y_{10} \vert l \pm 1, m\rangle)$
	$D_\pm$	$\Delta l = \pm 1$	$\Delta m = \pm 1$	$2C_1\left(\frac{2\pi}{3}\right)^{1/2}(\langle lm \vert Y_{1\pm 1} \vert l \pm 1, m \pm 1\rangle)$
Magnetic dipole	μ_z	$\Delta l = 0$	$\Delta m = 0$	μm
	$\mu_\pm$	$\Delta l = 0$	$\Delta m = \pm 1$	$\mu[l(l+1) - m(m \pm 1)]^{1/2}$

Possible magnetic dipole transitions may be analyzed similarly. The magnetic dipole operator contains no r-dependence, however, and the resultant selection rules on n are $\Delta n = 0$. The selection rules over the spherical harmonics give

$$(\mu_z)_{l'm',lm} = -\int_0^{2\pi}\int_0^{\pi} Y^*_{l'm'}\left(\frac{\mu_B}{\hbar}L_z\right)Y_{lm}\sin\theta\, d\theta\, d\phi = -\mu_B m \delta_{m',m}\delta_{l',l}$$

$$(\mu_\pm)_{l'm',lm} = -\frac{\mu_B}{\hbar}\int_0^{2\pi}\int_0^{\pi} Y^*_{l'm'}(L_x \pm iL_y)Y_{lm}\sin\theta\, d\theta\, d\phi$$

$$= -\mu_B[l(l+1) - m(m \mp 1)]^{1/2}\delta_{l',l}\delta_{m',m\pm 1}. \tag{4-116}$$

If the magnetic field vector in the electromagnetic radiation is plane polarized, no transitions are allowed as $\Delta l = 0$ and $\Delta m = 0$. If the magnetic field is circularly polarized, the selection rules are $\Delta l = 0$ and $\Delta m = \pm 1$.

Thus, the electric dipole transitions give $\Delta l = \pm 1$ transitions but the magnetic dipole transitions give only transitions between m states. The results above, which are summarized in Table 4-6, give the general selection rules for spherical harmonics. μ is used in the magnetic dipole selection rules in Table 4-6 in a general sense to denote the magnetic dipole moment of the interacting species, which could be an electron, a nucleus, or a molecular magnetic moment.

The selection rules in Table 4-6 are generally applicable for electronic transitions in any free atom system that has Y_{lm} eigenfunctions. In the case of electronic transitions in lower symmetry systems, such as diatomic and polyatomic molecules, the selection rules for electric dipole and magnetic dipole transitions must be evaluated using the appropriate symmetry of the molecule.

Of course, a consideration of the selection rules in the hydrogen atom must also include the spin-orbit interaction as described in Section 4-8.

4.8 ELECTRON SPIN AND THE SPIN-ORBIT INTERACTION

It is well known that the previous description of the hydrogen-like atom is not complete. In addition to the actual finite size of the nucleus causing nuclear-electronic interactions (which will be treated later), the electron experiences further perturbations that can be accounted for by assuming that the electron has an intrinsic spin angular momentum and resultant spin magnetic moment. Experimental evidence for the existence of electron spin came from at least two main lines of work.

The first evidence for the existence of an intrinsic magnetic moment of the electron came from an analysis of the Zeeman effect in the hydrogen atom. We can easily generalize Eq.(4-21) to apply to the hydrogen atom in the presence of a magnetic field H_z along the z axis, giving

$$\mathscr{H} = \mathscr{H}_0 + \frac{\mu_B H_Z}{\hbar} L_z + \frac{e^2 H_z^2}{8mc^2}(x^2 + y^2) + \frac{e^2 Z^2 H_z^2}{8Mc^2}(x_n^2 + y_n^2). \qquad (4\text{-}117)$$

The first term is the zero-field hydrogen-like Hamiltonian in Eq.(4-46), the second term is the magnetic dipole interaction, and the third and fourth terms represent the susceptibility terms for both the electron and nucleus with coordinates from the atomic center of mass. In the hydrogen-like atom, $x_n \ll x$ and $M \gg m$, which allows us to ignore the last term in Eq.(4-117). It is evident that $\mathscr{H}_0$ and $\mu_B H_z L_z/\hbar$ commute and the linear field term in Eq.(4-117) leads to an energy perturbation of $E(H_z) = \mu_B H_Z m$, where m is now the quantum number as in Table 4-5. The observation of the Zeeman effect is considerably more complex than allowed by Eq.(4-117) and the analysis given above. Thus, additional magnetic interactions were evident that led to the postulate of an electron spin magnetic moment.

The second line of evidence that leads to the necessity of electron spin was the observation of space quantization in an $L = 0$ atomic state. Space quantization, as demonstrated by the Stern–Gerlach experiment [12], is the result of a force on a beam of neutral atoms passing through a magnetic field gradient. Consider a beam of neutral atoms traveling in the y direction through a magnetic field gradient along the z axis. The potential energy of the atom in the field is $V = -\mu_z H_z$, where μ_z is the magnetic dipole moment of the atom. The force on the atom along the z axis is obtained from Eq.(2-4);

$$F_z = -\frac{\partial V}{\partial z} = \mu_z \frac{\partial H_z}{\partial z}. \qquad (4\text{-}118)$$

Thus, the atom with magnetic dipole moment μ_z experiences a force perpendicular to the beam direction and parallel to the magnetic field gradient, $\partial H_z/\partial z$. According to Eq.(4-22), the magnetic moment is related to the orbital angular momentum by $\mu_z = -(\mu_B/\hbar)L_z$ and the average value of μ_z for the hydrogen atom is $\langle\mu_z\rangle = -\mu_B m$. The average force on the free hydrogen atom in a field gradient is then given by

$$\langle F_z \rangle = -m\mu_B\left(\frac{\partial H_z}{\partial z}\right). \qquad (4\text{-}119)$$

The result of this force is to separate the atoms in the beam spatially along the z axis (perpendicular to the axis of beam flow). A detection screen should observe

$2l + 1$ spots along the x axis corresponding to the m states. Furthermore, if $l = 0$ in the hydrogen atom, no space quantization is possible according to the arguments above. The Stern–Gerlach experiment demonstrated space quantization, however, in an atom in an $l = 0$ or S state that pointed to the existence of intrinsic electronic angular momentum. This experimental determination of the intrinsic electron magnetic moment gave rise to the assumption of electron spin and its associated angular momentum.

The usual approach to the inclusion of electron spin in the hydrogen model is to assume spin angular momentum properties similar to the electronic angular momentum properties discussed in Section 4.5. Thus, in analogy to Eq.(4-69) we can assume the following spin angular momentum operators exist: $\boldsymbol{S}$, S_x, S_y, and S_z. It is further assumed that the same commutation relations exist for both spin and orbital angular momentum. Furthermore, we shall assume that all components of the spin angular momentum commute with all components of the orbital angular momentum. These assumptions are summarized by

$$\begin{aligned}
&[S_x, S^2] = 0 \qquad [S_x, S_y] = i\hbar S_z \\
&[S_y, S^2] = 0 \qquad [S_y, S_z] = i\hbar S_x \\
&[S_z, S^2] = 0 \qquad [S_z, S_x] = i\hbar S_y \\
&[S_g, L_{g'}] = 0; \qquad [S_g, L^2] = 0 \qquad g = x, y, z \\
&[S^2, L^2] = 0; \qquad [S^2, L_g] = 0 \qquad g' = x, y, z.
\end{aligned} \tag{4-120}$$

The eigenfunctions of S^2 and S_z for integral values of the spin will be analogous to the spatial spherical harmonics. We shall call the new spin eigenfunctions the spin spherical harmonics, Λ_{SM_S}. The eigenvalues of S^2 and S_z are

$$S^2\Lambda_{SM_S} = \hbar^2 S(S+1)\Lambda_{SM_S}, \qquad S_z\Lambda_{SM_S} = \hbar M_S\Lambda_{SM_S}. \tag{4-121}$$

The matrix elements of S_x, S_y, S_z, and S^2 can be obtained from the results in Eq.(4-74) where both integral and half-integral values of S and M_S are allowed. The single electron has a spin of $\frac{1}{2}$ leading to $\hbar^2(\frac{3}{4})$ and $\pm\hbar/2$ for eigenvalues in Eq.(4-121). A spin $\frac{1}{2}$ basis function is denoted by $\boldsymbol{\Lambda} = (\alpha \;\; \beta) = (\Lambda_{1/2\,1/2} \;\; \Lambda_{1/2\,-1/2})$

$$\begin{aligned}
S^2\alpha &= \hbar^2\left(\frac{3}{4}\right)\alpha; \qquad S_z\alpha = \left(\frac{\hbar}{2}\right)\alpha \\
S^2\beta &= \hbar^2\left(\frac{3}{4}\right)\beta; \qquad S_z\beta = -\left(\frac{\hbar}{2}\right)\beta.
\end{aligned} \tag{4-122}$$

The corresponding spin matrices for the hydrogen-like atom (a single electron) are

$$\mathbf{S}^2 = \int \boldsymbol{\Lambda} S^2 \boldsymbol{\Lambda}\, dV = \begin{pmatrix} \int \alpha^* S^2\alpha\, dV & \int \alpha^* S^2\beta\, dV \\ \int \beta^* S^2\alpha\, dV & \int \beta^* S^2\beta\, dV \end{pmatrix} = \hbar^2\begin{pmatrix} \frac{3}{4} & 0 \\ 0 & \frac{3}{4} \end{pmatrix}$$

$$\mathbf{S}_x = \frac{\hbar}{2}\begin{pmatrix} 0 & 1 \\ 1 & 0 \end{pmatrix}; \qquad \mathbf{S}_y = \frac{\hbar}{2}\begin{pmatrix} 0 & -i \\ i & 0 \end{pmatrix}; \qquad \mathbf{S}_z = \frac{\hbar}{2}\begin{pmatrix} 1 & 0 \\ 0 & -1 \end{pmatrix}; \tag{4-123}$$

where the integral is over the spin coordinates denoted by dV_S. Equation (4-123) is the $l = \frac{1}{2}$ analog of the results for $l = 1$ in Eq.(4-73). Other half-integral and integral matrix elements of the spin operators above can be obtained with Eqs.(4-74). The $\mathbf{S}_x$, $\mathbf{S}_y$, and $\mathbf{S}_z$ spin matrices for $S = \frac{1}{2}$ in Eq.(4-123) are usually referred to as the Pauli spin matrices and obey the same commutation relations as the respective spin operators.

The Zeeman Effect of a Free Electron

We shall now consider the Zeeman effect of a free electron or a bound electron with no orbital angular momentum and no nuclear magnetic interactions. We have noted previously that the magnetic moment is proportional to the angular momentum. Therefore, the electron spin magnetic moment is proportional to the spin angular momentum, $\boldsymbol{S}$. The proportionality constant for orbital angular momentum is given in Eq.(4-22) by $-\mu_B/\hbar$. Thus, by analogy with the orbital moment, we write the spin magnetic moment as

$$\boldsymbol{\mu}_S = -\frac{g_e\mu_B}{\hbar}\,\mathbf{S}, \tag{4-124}$$

where g_e is a constant (electron g-value) that depends on the nature of the electrons; $g_e = 2.002319$ for a free electron (see Appendix A).

Consider the matrix representation of the Hamiltonian for the free electron interacting with an external magnetic field along a space-fixed z axis. The Hamiltonian is

$$\mathscr{H} = -\mu_z H_z = \frac{g_e\mu_B}{\hbar} H_z S_z. \tag{4-125}$$

The appropriate basis for the computation of the matrix representation of this Hamiltonian is given in Eq.(4-121) as Λ_{SM_S}. As H_z does not operate on Λ_{SM_S} and Λ_{SM_S} is an eigenfunction of S_z, the Hamiltonian matrix is diagonal and the diagonal elements are given by

$$\int \Lambda_{SM_S}^*[\mathscr{H}]\Lambda_{SM_S}\,dV_S = \frac{g_e\mu_B}{\hbar} H_z\langle SM_S|S_z|SM_S\rangle = g_e\mu_B H_z M_S. \tag{4-126}$$

As $M_S = \pm\frac{1}{2}$, the splitting between the states is given by

$$E_{M_S=\frac{1}{2}} - E_{M_S=-\frac{1}{2}} = h\nu = g_e\mu_B H_z, \tag{4-127}$$

where the Bohr frequency rule is used. Thus, when $H_z = 10{,}000$ G, we have $h\nu = 1.857 \times 10^{-23}$ J or $\nu = 28{,}025$ MHz. The energy spacings are about the same order of magnitude as expected for the orbital moment Zeeman effect (see Section 4.1). Thus, in the case of independent (noninteracting) nonzero orbital and spin moments, the linear Zeeman effect will be described by a sum of the energies in Eq.(4-126) and those obtained from the Hamiltonian in Eq.(4-117).

The hydrogen-like eigenfunction, including the electronic spin information, will be the cross product between $\boldsymbol{\Lambda}$ and ψ given by

$$\boldsymbol{\Psi}(r, \theta, \phi) \times \boldsymbol{\Lambda} = \boldsymbol{\Psi} \times (\alpha\, \beta). \tag{4-128}$$

This function may now be used to investigate the atomic Zeeman effect and the coupling between the orbital angular momentum, $\boldsymbol{L}$, and the spin angular momentum, $\boldsymbol{S}$, in a hydrogen-like atom.

The Spin-Orbit Interaction

The spin-orbit interaction in the hydrogen-like atom is the interaction of the intrinsic spin electronic magnetic moment with the magnetic field due to the orbital motion of the electron. The Hamiltonian is

$$\mathscr{H}_1 = -\boldsymbol{\mu}_S \cdot \boldsymbol{H}_i = \frac{g_e \mu_B}{\hbar} \boldsymbol{S} \cdot \boldsymbol{H}_i, \tag{4-129}$$

where $\boldsymbol{\mu}_S$ is from Eq.(4-124). The internal magnetic field, $\boldsymbol{H}_i$, experienced by the spinning electron with a velocity $\boldsymbol{v}$ at a distance $\boldsymbol{r}$ from the nuclear origin can be obtained from electrodynamic considerations [13]. The internal magnetic field is given by $\boldsymbol{H}_i = -(\boldsymbol{v}/c) \times \boldsymbol{E}$. The electric force $e\boldsymbol{E}$ can be approximated by the negative gradient of the spherically symmetric potential energy, $V(r)$, given by

$$e\boldsymbol{E} = -\frac{\boldsymbol{r}}{r}\frac{dV(r)}{dr}.$$

Thus, the internal magnetic field is given by

$$\boldsymbol{H}_i = \frac{1}{erc}\frac{dV(r)}{dr}\boldsymbol{r} \times \boldsymbol{v} = \frac{1}{emrc}\frac{dV(r)}{dr}\boldsymbol{L}. \tag{4-130}$$

In the case of a hydrogen-like atom, where $V(r) = -Ze^2/r$, we obtain

$$\boldsymbol{H}_i = \frac{Ze}{mcr^3}\boldsymbol{L}. \tag{4-131}$$

It might be instructive to estimate the magnetic field at an electron in the hydrogen atom due to its motion through the central field potential. From Eq.(4-131) we can write $H \cong (v/c)(e/r^2)$ and substitute $r \cong 10^{-8}$ cm and $v \cong 10^8\ \text{cm} \cdot \text{s}^{-1}$ to give a magnetic field of from 10^4 to 10^5 G. Thus, the internal magnetic field is of the same order as a standard laboratory magnetic field. The energy of a magnetic dipole moment of one Bohr magneton in a field of 10^4 G is $E = \mu_B H \cong 10^{-23}$ J. Thus, we are justified in using perturbation theory. Substituting Eq.(4-131) for the internal magnetic field into (4-129) gives

$$\mathscr{H}_1 = -\boldsymbol{\mu}_S \cdot \boldsymbol{H}_i = \left(\frac{g_e \mu_B}{\hbar}\boldsymbol{S}\right) \cdot \left(\frac{Ze}{mcr^3}\boldsymbol{L}\right) = \frac{g_e Ze^2}{2c^2 m^2 r^3}\boldsymbol{S} \cdot \boldsymbol{L}. \tag{4-132}$$

This equation is in error by a factor of $\frac{1}{2}$. The extra factor of $\frac{1}{2}$ is found by deriving the spin-orbit interaction using relativistically invariant classical expressions [14], which gives

No. should be 2

$$\mathscr{H}_1 = \frac{g_e Z e^2}{4c^2 m^2 r^3} \mathbf{S} \cdot \mathbf{L}. \tag{4-133}$$

We shall also consider the atomic Zeeman effect that is the interaction of an external magnetic field, $\mathbf{H}$, with the total electronic magnetic moment, $\boldsymbol{\mu}$,

$$\mathscr{H}_2 = -\boldsymbol{\mu} \cdot \mathbf{H} = -(\boldsymbol{\mu}_S + \boldsymbol{\mu}_L) \cdot \mathbf{H} = \frac{\mu_B}{\hbar}(g_e \mathbf{S} + \mathbf{L}) \cdot \mathbf{H}, \tag{4-134}$$

where $\mu_S = -(g_e \mu_B/\hbar)\mathbf{S}$ from Eq.(4-124) and $\mu_L = -(\mu_B/\hbar)\mathbf{L}$ from Eq.(4-22).

The zero-order Hamiltonian for the internal motion of the hydrogen-like atom is given in Schrödinger's equation in Eq.(4-54). Substituting $\nabla_{r\theta\phi}$ from Eq.(4-47) and L^2 from Eq.(4-69) gives

$$\mathscr{H}_0 = -\frac{\hbar^2}{2\mu}\left[\frac{1}{r^2}\frac{\partial}{\partial r}\left(r^2 \frac{\partial}{\partial r}\right) - \frac{L^2}{r^2 \hbar^2}\right] - \frac{Ze^2}{r}. \tag{4-135}$$

The total Hamiltonian is a sum of $\mathscr{H}_0$ in Eq.(4-135), $\mathscr{H}_1$ in Eq.(4-133), and $\mathscr{H}_2$ in Eq.(4-134), giving

$$\mathscr{H} = \underbrace{-\frac{\hbar^2}{2\mu}\left[\frac{1}{r^2}\frac{\partial}{\partial r}\left(r^2 \frac{\partial}{\partial r}\right) - \frac{1}{r^2 \hbar^2} L^2\right] - \frac{Ze^2}{r}}_{\mathscr{H}_0} + \underbrace{\frac{g_e Z e^2}{4c^2 m^2 r^3} \mathbf{S} \cdot \mathbf{L}}_{\mathscr{H}_1} + \underbrace{\frac{\mu_B}{\hbar}(g_e S_z + L_z) H_z}_{\mathscr{H}_2}, \tag{4-136}$$

where we have employed a planar static magnetic field along the z axis, H_z.

There are two important limits for examining Eq.(4-136): the *uncoupled* and *coupled* bases of representation.

Uncoupled Basis: ψ_{L, M_L, S, M_S}

When $\mathscr{H}_1$ in Eq.(4-136) is negligible, the simple uncoupled product of the orbital and spin functions in Eq.(4-128) is an eigenfunction of $\mathscr{H}_0 + \mathscr{H}_2$. This is evident because $\mathscr{H}_0 + \mathscr{H}_2$, L^2, L_z, S^2, and S_z all commute and the eigenfunctions of L^2, L_z, S^2, and S_z are also eigenfunctions of $\mathscr{H}_0 + \mathscr{H}_2$. The eigenvalues of $\mathscr{H}_0 + \mathscr{H}_2$ in the uncoupled basis for the ground state of the hydrogen atom are obtained from

$$\begin{aligned} E(n, L, M_L, S, M_S) &= E_n + \frac{\mu_B}{\hbar} H_Z \langle L, M_L, S, M_S | g_e S_z + L_z | L, M_L, S, M_S \rangle \\ &= E_n + \mu_B H_z (g_e M_S + M_L), \end{aligned} \tag{4-137}$$

where E_n is the zero-field energy of the hydrogen atom for state n. The effects of $\mathscr{H}_1$ in the uncoupled basis can be treated as a perturbation whose first-order contribution is given by

$$E^{(1)}(n, L, M_L, S, M_S) = \frac{g_e Ze^2}{4c^2m^2} \int_0^\infty \eta_{nL}^*(r)\left(\frac{1}{r^3}\right)\eta_{nL}(r)r^2\,dr$$
$$\times \langle L, M_L, S, M_S | S_x L_x + S_y L_y + S_z L_z | L, M_L, S, M_S \rangle. \tag{4-138}$$

Evaluation of the integral over the r coordinate is a straightforward process for any state defined by n and L. Average values of r^i in the hydrogen-like atom where $i = 1, 2, 3, 4, -1, -2, -3$, and -4 are found elsewhere [15] and the result for $i = -3$ is given by

$$\int_0^\infty \eta_{nL}^*\left(\frac{1}{r^3}\right)\eta_{nL} r^2\,dr = \frac{1}{a^3}\left[\frac{Z^3}{n^3L(L+1)(L+\frac{1}{2})}\right], \tag{4-139}$$

where $a = \hbar^2/\mu e$ is the Bohr radius of the hydrogen atom. Equation (4-139) is not valid for an s-type electron where the integral above diverges. As $L = 0$ for an s electron, however, the spin-orbit coupling is zero. Adding the result in Eq.(4-139) to the nonzero contribution to the $\langle L, M_L, S, M_S|$ matrix element in Eq.(4-138) gives $E^{(1)}(n, L, M_L, S, M_S)$, which, when combined with Eq.(4-137), gives the first-order energy in the uncoupled basis [there are no diagonal matrix elements in S_x, S_y, L_x, or L_y as shown in Eq.(4-74)],

$$E(n, L, M_L, S, M_S) = E_n + \mu_B H_z(g_e M_S + M_L) + \frac{2\Delta_n M_L M_S}{L(L+1)(L+\frac{1}{2})}$$
$$\Delta_n = \frac{g_e Z^4 e^2 \hbar^2}{8c^2m^2a^3n^3}. \tag{4-140}$$

Coupled Basis: ψ_{L,S,J,M_J}

When $\mathscr{H}_2$ in Eq.(4-136) is negligible, the remaining $\mathscr{H}_0 + \mathscr{H}_1$ parts are most appropriately examined in the coupled basis. In Section 2.1 we noted that in a conservative system with zero external force, the torque on the system is zero giving rise to a total angular momentum that is conserved in time. Thus, the total angular momentum $\boldsymbol{J}$, which is a sum of the orbital and spin components, is conserved in time giving

$$\boldsymbol{J} = \boldsymbol{L} + \boldsymbol{S}. \tag{4-141}$$

It is easy to show that all components of $\boldsymbol{J}$ commute with $\mathscr{H}_0 + \mathscr{H}_1$. All components of $\boldsymbol{L}$ and $\boldsymbol{S}$ (and therefore the components of $\boldsymbol{J} = \boldsymbol{L} + \boldsymbol{S}$) commute with $\mathscr{H}_0$. This is made clear by noticing that there are no spin operators in $\mathscr{H}_0$ and recalling the commutation of all components of $\boldsymbol{L}$ with L^2 as discussed in Section 4.5. Furthermore, it is easy to show that the components of $\boldsymbol{J}$ commute with the $\boldsymbol{S} \cdot \boldsymbol{L}$ operator, an exercise we leave for the reader. Thus, all components of $\boldsymbol{J}$ commute with $\mathscr{H}_0 + \mathscr{H}_1$; however, all components of $\boldsymbol{J}$ do not commute with all components of $\boldsymbol{L}$ and $\boldsymbol{S}$. A detailed application of the commutation relations in Eq.(4-120) shows that

S^2, L^2, J^2, and J_z commute with each other and with $\mathscr{H}_0 + \mathscr{H}_1$. Furthermore, it is evident that these five mutually commuting operators may share, simultaneously, the same eigenfunction. The eigenfunction of the S^2, L^2, J^2, J_z, and $\mathscr{H}_0 + \mathscr{H}_1$ operators is called the *coupled basis* and is denoted by ψ_{L,S,J,M_J}.

The angular momentum eigenvalues of the S^2, L^2, J^2, and J_z operators in the coupled basis are $\hbar^2 S(S+1)$, $\hbar^2 L(L+1)$, $\hbar^2 J(J+1)$, and $\hbar M_J$, respectively. The eigenvalues of $\mathscr{H}_0 + \mathscr{H}_1$ are obtained by substituting

$$\boldsymbol{S} \cdot \boldsymbol{L} = \tfrac{1}{2}(J^2 - L^2 - S^2) \tag{4-142}$$

from $J^2 = \boldsymbol{J} \cdot \boldsymbol{J} = (\boldsymbol{L} + \boldsymbol{S}) \cdot (\boldsymbol{L} + \boldsymbol{S})$ into $\mathscr{H}_1$, which gives the following eigenvalue for $\mathscr{H}_0 + \mathscr{H}_1$ in the coupled basis:

$$\begin{aligned} E(n, L, S, J, M_J) &= \int \psi^*_{S,L,J,M_J}[\mathscr{H}_0 + \mathscr{H}_1]\psi_{S,L,J,M_J}\, dV \\ &= E_n + \frac{g_e Z e^2}{8c^2 m^2}\hbar^2[J(J+1) - L(L+1) - S(S+1)] \\ &\quad \times \int_0^\infty \eta^*_{nL}(r)\left(\frac{1}{r^3}\right)\eta_{nL}(r) r^2\, dr. \end{aligned} \tag{4-143}$$

Substituting Eq.(4-139) and using Δ_n in Eq.(4-140) gives the final result when $\mathscr{H}_2 = 0$ (zero field):

$$E(n, L, S, J, M_J) = E_n + \frac{\Delta_n[J(J+1) - S(S+1) - L(L+1)]}{L(L+1)(L+\frac{1}{2})}. \tag{4-144}$$

The notation used to describe the states is the quite natural form of ${}^{(2S+1)}L_J$. In the hydrogen-like atom we have the states $n = 1$, ${}^2S_{1/2}$; $n = 2$, ${}^2S_{1/2}$, ${}^2P_{1/2}$, ${}^2P_{3/2}$; $n = 3$, ${}^2S_{1/2}$, ${}^2P_{1/2}$, ${}^2P_{3/2}$, ${}^2D_{3/2}$, ${}^2D_{5/2}$; and so on.

Using $S = \frac{1}{2}$ and Eq.(4-144) gives two states for each value of J [we drop the M_J and use $E(n, L, S, J)$ notation],

$$\begin{aligned} E(n, L, \tfrac{1}{2}, L + \tfrac{1}{2}) &= -\frac{\mu e^4 Z^2}{2\hbar^2 n^2} + \frac{\Delta_n}{(L+1)(L+\frac{1}{2})} \\ E(n, L, \tfrac{1}{2}, L - \tfrac{1}{2}) &= -\frac{\mu e^4 Z^2}{2\hbar^2 n^2} - \frac{\Delta_n}{L(L+\frac{1}{2})}. \end{aligned} \tag{4-145}$$

The difference between the $J = L + \frac{1}{2}$ and $J = L - \frac{1}{2}$ states can be predicted accurately with this simple theory:

$$E(n, L, \tfrac{1}{2}, L + \tfrac{1}{2}) - E(n, L, \tfrac{1}{2}, L - \tfrac{1}{2}) = \frac{2\Delta_n}{L(L+1)}. \tag{4-146}$$

A few calculated splittings for the hydrogen atom (replace m with μ, the reduced mass in Δ_n) are obtained from Eq.(4-146) and the results are shown in Table 4-7. It is clear from the comparison between the calculated and experimental results that we have only achieved a partial description of the true interactions in the hydrogen atom. For instance, the $n^2S_{1/2} - n^2P_{1/2}$ and $n^2P_{3/2} - n^2D_{3/2}$ splittings are not predicted correctly. These additional refinements on the description of the energy

Table 4-7 Comparison of the experimental spin-orbit doublet splittings in the hydrogen atom with the values predicted from Eq.(4-146). The experimental values are obtained from C. E. Moore, *Atomic Energy Levels* (National Bureau of Standards Circular 467, 1949). The value of $e^2/2a$ is obtained from the experimental difference between the $1^2S_{1/2}$ and $2^2S_{1/2}$ levels in hydrogen. All energies are in cm^{-1} and referenced to the $1^2S_{1/2}$ state.

n	Spectroscopic State	*Theory* Splitting	*Experiment* Levels	Splitting
1	$^2S_{1/2}$	0	0	0
2	$^2P_{1/2}$	0.793	82258.907	0.035
2	$^2S_{1/2}$		82258.942	0.365
2	$^2P_{3/2}$	0.365	82259.272	
3	$^2S_{1/2}$		97492.208	0.010
3	$^2P_{1/2}$	0.108	97492.198	0.108
3	$^2P_{3/2}$	−0.058	97492.306	0.000
3	$^2D_{3/2}$	0.036	97492.306	0.036
3	$^2D_{5/2}$		97492.342	

of the hydrogen-like atom arise from both relativistic effects (see Problem 4-11 for a calculation of the relativistic effects by perturbation theory) and the Lamb shift [14].

Quite a different picture of the hydrogen-like atom is obtained by solving the full relativistic Schrödinger equation as shown by Dirac [8]. The results of the relativistic treatment include an energy contribution identical to what we have called here the spin-orbit interaction. The relativistic result is obtained without postulating electron spin, however, which questions the "existence" of spin; yet there is no doubt that the free electron experiences a Zeeman effect that exhibits a property analogous to a system which has an intrinsic spin magnetic moment. Thus, the magnetic properties of the electron are purely relativistic. We shall not give the complete relativistic treatment here as the magnetic moment development given above is satisfactory in explaining the experimental data.

Next we apply a magnetic field and consider the Zeeman effect, $\mathscr{H}_2$, as a small perturbation on the result given in Eq.(4-144). The first-order matrix elements involving $\mathscr{H}_2$ in the coupled basis are

$$E^{(1)}(S, L, J, M_J) = \langle S, L, J, M_J | \mathscr{H}_2 | S, L, J, M_J \rangle$$
$$= H_z \left\langle S, L, J, M_J \left| \frac{\mu_B}{\hbar}(g_e S_z + L_z) \right| S, L, J, M_J \right\rangle.$$

As ψ_{S,L,J,M_J} is an eigenfunction of neither S_z nor L_z, no simple evaluation of the matrix element above is possible. We can substitute $J_z = S_z + L_z$, giving

$$\langle S, L, J, M_J | \mathscr{H}_2 | S, L, J, M_J \rangle = H_z \left\langle S, L, J, M_J \left| \frac{\mu_B}{\hbar}[J_z + (g_e - 1)S_z] \right| S, L, J, M_J \right\rangle$$
$$= \mu_B M_J H_z + \frac{\mu_B H_z}{\hbar}(g_e - 1)$$
$$\times \langle S, L, J, M_J | S_z | S, L, J, M_J \rangle, \qquad (4\text{-}147)$$

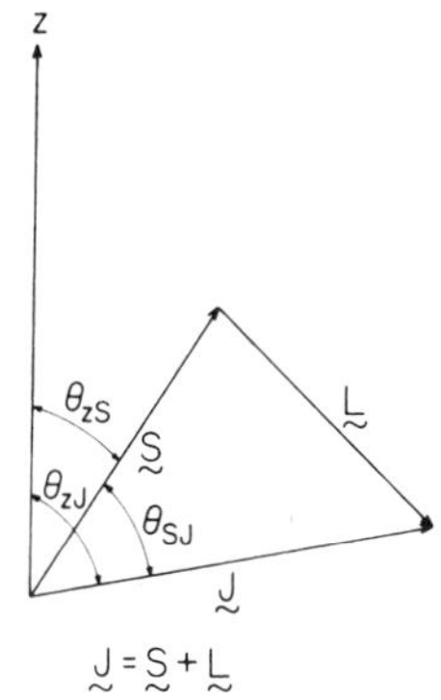

Figure 4-10 Diagram showing the orientation of the ***S***, ***L***, and ***J*** angular momentum vectors with respect to the space-fixed z axis. ***S***, ***J***, and the z axis are all in the same plane.

where the fact that ψ_{S,L,J,M_J} is an eigenfunction of J_z has been used in the first term. We are left with the matrix element of S_z in the coupled basis. An evaluation of $\langle S, L, J, M_J|S_z|S, L, J, M_J\rangle$ is possible by referring to the vector coupling of $\boldsymbol{S} + \boldsymbol{L}$, as shown in Fig. 4-10.

The magnetic moments, $\boldsymbol{\mu}_S$, $\boldsymbol{\mu}_L$, and $\boldsymbol{\mu}_J$, are parallel to and oriented in directions opposing the corresponding angular momentum vectors. The z axis is the laboratory field axis. From the diagram (Fig. 4-10), we can write

$$S_z = S\cos\theta_{zS}, \qquad J_z = J\cos\theta_{zJ}, \tag{4-148}$$

where S and J are the magnitudes of $\boldsymbol{S}$ and $\boldsymbol{J}$, respectively. Rewriting S_z gives

$$S_z = S\cos\theta_{zS} = S\cos(\theta_{zJ} - \theta_{SJ}) = S(\sin\theta_{zJ}\sin\theta_{SJ} + \cos\theta_{zJ}\cos\theta_{SJ}). \tag{4-149}$$

The sine terms in Eq.(4-149) lead to off-diagonal terms in J. This is easily shown with an equation similar to Eq.(4-148):

$$J_x = J\sin\theta_{Jz}, \tag{4-150}$$

which leads to off-diagonal elements in J in the coupled basis as ψ_{S,L,J,M_J} is not an eigenfunction of J_x. Thus, the diagonal matrix element from Eq.(4-147) is given by

$$\langle S, L, J, M_J|\mathscr{H}_2|S, L, J, M_J\rangle$$
$$= H_z\left[\mu_B M_J + \frac{\mu_B}{\hbar}(g_e - 1)\langle S, L, J, M_J|(S\cos\theta_{zJ}\cos\theta_{SJ})|S, L, J, M_J\rangle\right]. \tag{4-151}$$

From Eq.(4-148), we write $\cos\theta_{zJ} = J_z/J$. Similarly, we can evaluate $\cos\theta_{SJ}$ from Fig. 4-10 and Eq.(4-142) from which we write $\boldsymbol{S}\cdot\boldsymbol{J} = SJ\cos\theta_{SJ} = (J^2 + S^2 - L^2)/2$. Substituting these values of $\cos\theta_{zJ}$ and $\cos\theta_{SJ}$ into Eq.(4-151) gives

$$\langle S, L, J, M_J|\mathscr{H}_2|S, L, J, M_J\rangle$$
$$= H_z\mu_B M_J + \frac{\mu_B H_z}{\hbar}(g_e - 1)\left\langle S, L, J, M_J\left|\frac{(J^2 + S^2 - L^2)J_z}{2J^2}\right|S, L, J, M_J\right\rangle$$
$$= H_z\mu_B M_J + H_z\mu_B M_J(g_e - 1)\left\langle S, L, J, M_J\left|\frac{(J^2 + S^2 - L^2)}{2J^2}\right|S, L, J, M_J\right\rangle$$
$$= H_z\mu_B M_J\left\{1 + \left[\frac{J(J+1) + S(S+1) - L(L+1)}{2J(J+1)}\right](g_e - 1)\right\} = H_z\mu_B M_J g_L. \tag{4-152}$$

g_L is called the *Lande* g-factor for the electron. There are two interesting choices for the values of S, L, and J.

1. $S = 0, L = J$

In this case, $g_L = 1$ and the first-order energy becomes $H_z \mu_B M_L$, a result readily obtained from Eq.(4-117) in the absence of electron spin.

2. $L = 0, S = J$

In this case, $g_L = g_e$ and the first-order energy becomes $g_e H_z \mu_B M_S$ in agreement with Eq.(4-126), which is the case for a free electron.

The final energy of the hydrogen atom in the coupled basis is a combination of Eqs.(4-144) and (4-152) giving

$$E(n, L, S, J, M_J) = E_n + \frac{\Delta_n[J(J+1) - S(S+1) - L(L+1)]}{L(L+1)(L+\frac{1}{2})} + H_z \mu_B g_L M_J. \tag{4-153}$$

Δ_n is defined in Eq.(4-140). We see from this equation that the J states, which were originally $2J + 1$ degenerate, are now split into the M_J components.

In summary, we have the low field-coupled result in Eq.(4-153) and the high field-uncoupled result in Eq.(4-140). Limiting cases are illustrated in the correlation diagram in Fig. 4-11 for the $n = 2$, $L = 1$, $S = \frac{1}{2}$ hydrogen-like atom. In the case of

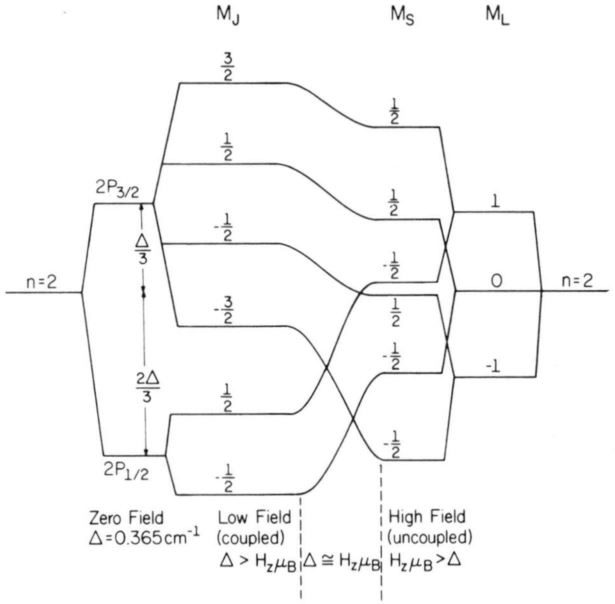

Figure 4-11 Correlation diagram between the low and high field limits for the $^2P_{3/2}$ and $^2P_{1/2}$ states of the hydrogen atom. The magnetic field increases from left to right and the limiting expressions for the energy are in Eqs.(4-140) and (4-153). Δ_n in Eq.(4-153) is equal to 0.365 cm^{-1} for the hydrogen atom when $n = 2$.

the $n = 2$, $^2P_{3/2}$, and $^2P_{1/2}$ states of the hydrogen atom, $\Delta = 0.365\ \text{cm}^{-1}$, as shown in Table 4-7. This value is about the same order of magnitude energy as is produced by the Zeeman effect at about 10,000 G. Thus, the idealized high field limit shown in Fig. 4-11 is difficult to achieve in the case of the hydrogen atom. This conclusion was also evident earlier following Eq.(4-131) where the internal magnetic field at the electron was estimated.

We have attempted to illustrate an application of perturbation and matrix theory that involves the coupling and uncoupling of two angular momenta. In addition to the spin-orbit coupling and atomic Zeeman effect discussed here, nuclear-electronic interactions can further couple the nuclear angular momentum, $\boldsymbol{I}$, to the electronic spin and orbital angular momentum. We shall discuss these effects in Section 4-9.

4.9 THE ELECTRONIC AND NUCLEAR ZEEMAN EFFECT IN THE ABSENCE OF ORBITAL ELECTRONIC ANGULAR MOMENTUM: ESR

Electron spin resonance (ESR) involves the absorption of microwave energy (1000–30,000 MHz) by a transition between the electronic spin energy levels under the influence of an external magnetic field. Most molecules that have been studied by the electronic spin magnetic resonance method have zero orbital angular momentum ($\boldsymbol{L} = 0$) [16]. In addition, most ESR studies are done in condensed phases, which quenches the rotational angular momentum. In the crystalline solid state, the electron spin interactions considered here will experience considerable spatial anisotropy. We consider only liquids or glassy solids (random orientations) and, therefore, only isotropic interactions. Anisotropic interactions in ESR are treated in Problem 5-19. If we consider molecules with $\boldsymbol{L} = 0$, the perturbation terms in Eq.(4-136) reduce to the single term given by

$$\mathscr{H}_2 = \frac{g_e \mu_B}{\hbar} \boldsymbol{S} \cdot \boldsymbol{H}. \tag{4-154}$$

We also consider the effect of the presence of nuclei with spin angular momentum $\boldsymbol{I}_i$ where the subscript i indicates the ith nucleus. The ith nucleus with angular momentum $\boldsymbol{I}_i$ has a magnetic moment that is written in analogy with the electronic orbital and spin moments [Eqs.(4-22) and (4-124)] as

$$\boldsymbol{\mu}_i = \frac{g_i \mu_0}{\hbar} \boldsymbol{I}_i = \gamma_i \boldsymbol{I}_i$$

$$\mu_0 = \frac{\hbar e}{2m_p c} = 5.05082 \times 10^{-24}\ \text{erg} \cdot \text{G}^{-1} = 5.05082 \times 10^{-27}\ \text{J} \cdot \text{T}^{-1}, \tag{4-155}$$

Table 4-8 Nuclear g-values and nuclear spin in units of $\hbar$ for some common nuclei. The nuclear magnetic moment is defined in Eq.(4-155).

Atomic Number	Nucleus	g-Value	Nuclear Spin, $I(\hbar)$
1	^{1}H	5.5854	$\frac{1}{2}$
1	^{2}H	0.8574	1
3	^{6}Li	0.8219	1
3	^{7}Li	2.1707	$\frac{3}{2}$
5	^{10}B	0.6001	3
5	^{11}B	1.7920	$\frac{3}{2}$
6	^{13}C	1.4042	$\frac{1}{2}$
7	^{14}N	0.4036	1
7	^{15}N	−0.5660	$\frac{1}{2}$
8	^{17}O	−0.7572	$\frac{5}{2}$
9	^{19}F	5.2546	$\frac{1}{2}$
11	^{23}Na	1.4774	$\frac{3}{2}$
14	^{29}Si	−1.1094	$\frac{1}{2}$
15	^{31}P	2.2610	$\frac{1}{2}$

where μ_0 is the nuclear magneton, m_p is the proton mass, g_i is the nuclear g-value, and $\gamma_i = g_i\mu_0/\hbar$ is the proportionality constant between the moment and the angular momentum called the *magnetogyric ratio.* The values of several nuclear g-values are listed in Table 4-8. Note that the signs of the nuclear g-values are both positive and negative.

By analogy with our previous discussion on spin-orbit interactions, we shall now write those components of the Hamiltonian describing a system with electron spin $\boldsymbol{S}(\boldsymbol{L} = 0)$ and a number of nuclei with angular momenta $\boldsymbol{I}_i$ that are interacting with each other and with the external magnetic field:

$$\mathscr{H} = \frac{g_e\mu_B}{\hbar}\boldsymbol{S}\cdot\boldsymbol{H} - \frac{\mu_0}{\hbar}\sum_i g_i\boldsymbol{I}_i\cdot\boldsymbol{H}(1-\sigma_i) + h\sum_i\left(\frac{a_i}{\hbar^2}\right)\boldsymbol{I}_i\cdot\boldsymbol{S} + h\sum_{i>j}\left(\frac{J_{ij}}{\hbar^2}\right)\boldsymbol{I}_i\cdot\boldsymbol{I}_j. \tag{4-156}$$

a_i is the nuclear spin-electron spin coupling constant (the electronic theory of this coupling is given in Section 5.6) in units of hertz. J_{ij} is the electron-coupled nuclear spin–nuclear spin coupling constant in units of hertz where the sum over i and j $(i > j)$ includes all internuclear interactions (the direct nuclear spin–nuclear spin coupling is ignored here; see Problem 4-39). The $\boldsymbol{I}_i\cdot\boldsymbol{H}(1-\sigma_i)$ term represents the interaction of the nuclear magnetic moment with the external field where σ_i is the nuclear magnetic shielding caused by the electrons. σ_i, a_i, and J_{ij} are all of significant interest in studying the electronic structure of atoms and molecules.

Even though we shall be dealing primarily with the uncoupled basis sets in the following work, it is instructive to compute the states involved in the coupled basis.

Consider a three-nuclear spin system with $I_1 = I_2 = I_3 = \frac{1}{2}$, and $S = 1$. The total states in a strongly coupled scheme (zero field) are given by successive addition of the angular momentum to give the total angular momentum, $\boldsymbol{F}$. $\boldsymbol{F}$ is the vector sum of the total nuclear, $\boldsymbol{I}$, and total electronic, $\boldsymbol{J} = \boldsymbol{L} + \boldsymbol{S}$, components. When $\boldsymbol{L} = 0$, we have

$$\boldsymbol{I}_1 + \boldsymbol{I}_2 = \boldsymbol{I}'; \qquad 0 \qquad\qquad 1$$

$$\boldsymbol{I}' + \boldsymbol{I}_3 = \boldsymbol{I}; \qquad \frac{1}{2} \qquad\qquad \frac{1}{2} \text{ or } \frac{3}{2}$$

$$\boldsymbol{I} + \boldsymbol{S} = \boldsymbol{F}; \qquad \frac{1}{2} \text{ or } \frac{3}{2}; \qquad \frac{1}{2} \text{ or } \frac{3}{2}; \frac{5}{2} \text{ or } \frac{3}{2} \text{ or } \frac{1}{2}. \tag{4-157}$$

The seven final F states each have $2F + 1$ degenerate M_F levels for a total of 24 levels. We note that the sum of levels in the example above is given by

$$(2I_1 + 1)(2I_2 + 1)(2I_3 + 1)(2S + 1).$$

This method of determining the total number of states is easily generalized.

In correlating the states from zero field to high field, the $M_F = \sum_i M_{I_i} + M_S$ equality is very useful and arises from the fact that the uncoupled basis function is an eigenfunction of the z components of all individual angular momenta and the coupled basis function is an eigenfunction of F_z with eigenvalue M_F. We shall now proceed with several examples of the use of the Hamiltonian in Eq.(4-156) to determine the energy levels and corresponding transitions between the states.

As a first example, consider a single nucleus system with $S = \frac{1}{2}$ and $I = \frac{1}{2}$ such as the hydrogen atom (with $\boldsymbol{L} = 0$) or the acetylene radical $\mathrm{H{-}^{12}C{\equiv}^{12}C\cdot}$, where the dot indicates the unpaired electron. In this case, the Hamiltonian in Eq.(4-156) reduces to the following with a static laboratory H_z magnetic field:

$$\mathscr{H} = \frac{g_e \mu_B}{\hbar} S_z H_z - \frac{\mu_0}{\hbar} g_i (1 - \sigma) I_z H_z + h\left(\frac{a}{\hbar^2}\right)(I_x S_x + I_y S_y + I_z S_z). \tag{4-158}$$

The uncoupled basis set for this system of an electron spin, $\boldsymbol{S}$, and a nuclear spin, $\boldsymbol{I}$, is given by a product of the electron and nuclear spin functions, $\psi_{S,M_S,I,M_I} = \psi_{S,M_S}\psi_{I,M_I}$, which is an eigenfunction of S^2, S_z, I^2, and I_z. The matrix elements of the Hamiltonian in Eq.(4-158) with the uncoupled basis are easy to obtain from our previous work by remembering that all nonzero matrix elements are of the form given in Eqs.(4-74). The results are

Diagonal

$$\langle S, M_S, I, M_I | \mathscr{H} | S, M_S, I, M_I \rangle = g_e \mu_B H_z M_S - \mu_0 g_i H_z M_I (1 - \sigma) + ha M_S M_I$$

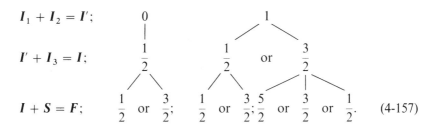

Off-Diagonal

$$\left\langle S, M_S, I, M_I \left| h\left(\frac{a}{\hbar^2}\right)(I_y S_y + I_x S_x) \right| S, M_S \pm 1, I, M_I \pm 1 \right\rangle$$

$$\langle S, M_S, I, M_I | \mathscr{H} | S, M_S + 1, I, M_I + 1 \rangle = \langle S, M_S, I, M_I | \mathscr{H} | S, M_S - 1, I, M_I - 1 \rangle = 0$$

$$\langle S, M_S, I, M_I | \mathscr{H} | S, M_S - 1, I, M_I + 1 \rangle = \frac{ha}{2} [S(S+1) - M_S(M_S - 1)]^{1/2} \times [I(I+1) - M_I(M_I + 1)]^{1/2}$$

$$\langle S, M_S, I, M_I | \mathscr{H} | S, M_S + 1, I, M_I - 1 \rangle = \frac{ha}{2} [S(S+1) - M_S(M_S + 1)]^{1/2} \times [I(I+1) - M_I(M_I - 1)]^{1/2}. \quad (4\text{-}159)$$

Note that only simultaneous $\Delta M_S = \pm 1$ and $\Delta M_I = \mp 1$ off-diagonal matrix elements are nonzero. The matrix representation of the Hamiltonian in Eq.(4-158), for this simple example in the uncoupled basis ($S = \frac{1}{2}, I = \frac{1}{2}$), is given by

M_S \ M_I	M_S: $\frac{1}{2}$, M_I: $\frac{1}{2}$	M_S: $\frac{1}{2}$, M_I: $-\frac{1}{2}$	M_S: $-\frac{1}{2}$, M_I: $\frac{1}{2}$	M_S: $-\frac{1}{2}$, M_I: $-\frac{1}{2}$
$\frac{1}{2}$, $\frac{1}{2}$	A	0	0	0
$\frac{1}{2}$, $-\frac{1}{2}$	0	B	$\frac{ha}{2}$	0
$-\frac{1}{2}$, $\frac{1}{2}$	0	$\frac{ha}{2}$	C	0
$-\frac{1}{2}$, $-\frac{1}{2}$	0	0	0	D

$$\left.\begin{aligned} A &= \frac{g_e \mu_B}{2} H_z - \tfrac{1}{2}\xi + \frac{ha}{4} \\ B &= \frac{g_e \mu_B}{2} H_z + \tfrac{1}{2}\xi - \frac{ha}{4} \\ C &= -\frac{g_e \mu_B}{2} H_z - \tfrac{1}{2}\xi - \frac{ha}{4} \\ D &= -\frac{g_e \mu_B}{2} H_z + \tfrac{1}{2}\xi + \frac{ha}{4} \end{aligned}\right\} \xrightarrow[\text{field}]{\text{high}} \begin{aligned} &= E_1 \\ &= E_2 \\ &= E_3 \\ &= E_4 \end{aligned} \quad (4\text{-}160)$$

$$\xi = \mu_0 g_i (1 - \sigma) H_z.$$

The selection rules and relative intensities for the magnetic dipole allowed transitions are computed from the total magnetic dipole moment operator $\boldsymbol{\mu} = \boldsymbol{\mu}_S + \boldsymbol{\mu}_i = -(g_e \mu_B/\hbar)\boldsymbol{S} + (g_i \mu_0/\hbar)\boldsymbol{I}_i$. The $\mu_{\pm}$ matrix involving circularly polarized radiation is easily set up in the uncoupled basis (see Table 4-6) using the same ordering as shown in the Hamiltonian matrix in Eq.(4-160). The result is

$$\boldsymbol{\mu}_+ + \boldsymbol{\mu}_- = \boldsymbol{\mu}_\pm = \begin{pmatrix} 0 & g_i\mu_0 & -g_e\mu_B & 0 \\ g_i\mu_0 & 0 & 0 & -g_e\mu_B \\ -g_e\mu_B & 0 & 0 & g_i\mu_0 \\ 0 & -g_e\mu_B & g_i\mu_0 & 0 \end{pmatrix}, \qquad (4\text{-}161)$$

where μ_+ fills the matrix elements above the diagonal and μ_- fills the matrix elements below the diagonal.

We can now diagonalize $\mathscr{H}$ in Eq.(4-160), $\mathbf{b}^\dagger \mathscr{H} \mathbf{b} = \mathbf{E}$, at any value of the field by the following simple rotation [see Eqs.(B-36) and (B-37)]:

$$\mathbf{b} = \begin{pmatrix} 1 & 0 & 0 & 0 \\ 0 & \cos\theta & -\sin\theta & 0 \\ 0 & \sin\theta & \cos\theta & 0 \\ 0 & 0 & 0 & 1 \end{pmatrix}. \qquad (4\text{-}162)$$

By transforming $\mathscr{H}$ in Eq.(4-160) to the diagonal form, simple arithmetic will show that [see Eq.(B-37)]

$$\tan 2\theta = \frac{ha}{B - C} = \frac{ha}{[g_e\mu_B + g_i\mu_0(1-\sigma)]H_z}, \qquad (4\text{-}163)$$

where B and C are defined in Eq.(4-160). These methods give a diagonalized energy matrix with nonzero elements given by

$$\begin{aligned} E_1 &= A \\ E_2 &= B\cos^2\theta + ha\cos\theta\sin\theta + C\sin^2\theta \\ E_3 &= B\sin^2\theta - ha\cos\theta\sin\theta + C\cos^2\theta \\ E_4 &= D. \end{aligned} \qquad (4\text{-}164)$$

If we now apply the transformation in Eq.(4-162) that diagonalizes $\mathscr{H}$ to the $\boldsymbol{\mu}_\pm$ matrix given in Eq.(4-161), we obtain $\mathbf{b}^\dagger\boldsymbol{\mu}_\pm\mathbf{b} = \mathbf{T}$. The elements in the $\mathbf{T}$ matrix are obtained by simple multiplication and the resultant nonzero transition probabilities are given by the squares of the elements in $\mathbf{T}$.

$$\begin{aligned} E_1 &\longleftrightarrow E_2 \quad & \mathrm{T}_{12}^2 = \mathrm{T}_{21}^2 &= (g_i\mu_0\cos\theta - g_e\mu_B\sin\theta)^2 \\ E_1 &\longleftrightarrow E_3 \quad & \mathrm{T}_{13}^2 = \mathrm{T}_{31}^2 &= (-g_e\mu_B\cos\theta - g_i\mu_0\sin\theta)^2 \\ E_2 &\longleftrightarrow E_4 \quad & \mathrm{T}_{24}^2 = \mathrm{T}_{42}^2 &= (-g_e\mu_B\cos\theta + g_i\mu_0\sin\theta)^2 \\ E_3 &\longleftrightarrow E_4 \quad & \mathrm{T}_{34}^2 = \mathrm{T}_{43}^2 &= (g_i\mu_0\cos\theta + g_e\mu_B\sin\theta)^2. \end{aligned} \qquad (4\text{-}165)$$

In the high field limit, $\theta \longrightarrow 0$ according to Eq.(4-163) and the high field energy levels are equal to the diagonal values in Eq.(4-160) as shown on the right-hand side of Fig. 4-12. We note that in the high field case, the electronic M_S splittings are on the order of $m_p/m = \mu_B/\mu_0 = 1836$ times larger than the corresponding nuclear splittings. Typical electronic spin splittings at 10,000 G are $g_e\mu_B H_z/h = 20$ to 30×10^9 Hz and

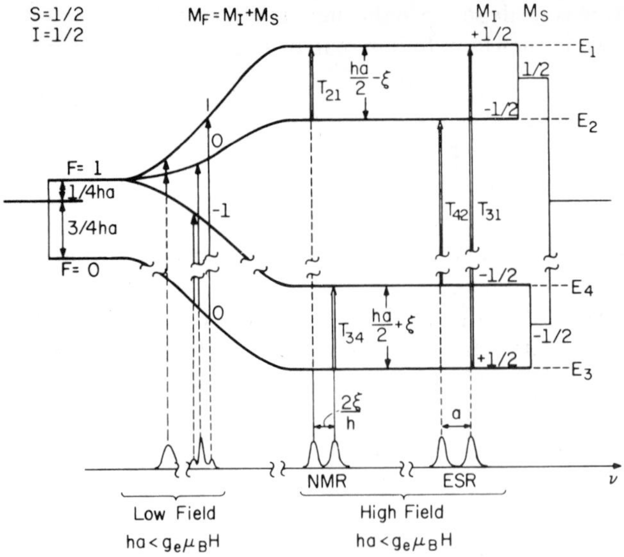

Figure 4-12 Energy level correlation diagram between the high and low fields for an $L = 0$, $S = \frac{1}{2}$, and $I = \frac{1}{2}$ system. The energy levels, NMR, and ESR transitions are discussed in the text. The horizontal breaking lines indicate that the lower two levels are considerably lower in energy because the $4 \longrightarrow 2$ and $3 \longrightarrow 1$ ESR transition energies are about $\mu_B/\mu_0 = 1836$ times as large as the corresponding $2 \longrightarrow 1$ and $3 \longrightarrow 4$ NMR transition energies in the high field limit. We have assumed that $\xi = \mu_0 g_i(1 - \sigma)H_z$ is positive and smaller than a positive ha for the ordering of the energy levels where the high field energies are noted in Eq.(4-160). Other assumptions lead to similar schemes with the $g_e\mu_B H_z$ term dominating the pattern.

typical nuclear splittings are $g_i\mu_0 H_z/h \cong 10$ to 50×10^6 Hz depending on the nuclear g-value. In most cases, $ha \ll g_e\mu_B H_z$ can be obtained with laboratory fields for organic molecules of interest in chemistry. Thus, at relatively high fields, the matrix representation of the Hamiltonian in Eq.(4-158) as given in Eq.(4-160) is nearly diagonal in the uncoupled basis. A large number of the published analyses of $\mathbf{S} + \mathbf{I}$ systems in molecular free radicals in high fields are interpreted successfully by ignoring the off-diagonal elements in M_S and M_I.

The transition matrix in the high field limit where $\theta \longrightarrow 0$ shows relative intensities given by $\mathrm{T}_{12}^2 = \mathrm{T}_{43}^2 = (g_i\mu_0)^2$ for the nuclear magnetic resonance (NMR) transitions and $\mathrm{T}_{13}^2 = \mathrm{T}_{24}^2 = (g_e\mu_B)^2$ for the ESR transitions. The high field limit is the most convenient region for interpreting the chemical composition of a free radical because an ESR doublet is observed with a splitting of (see Fig. 4-12)

$$\frac{1}{h}[(E_1 - E_3) - (E_2 - E_4)] = a, \tag{4-166}$$

which gives directly the nuclear spin–electron spin coupling constant. A similar analysis of the predicted NMR spectrum also indicates that a doublet will appear with a separation $2\xi/h$ as shown in Fig. 4-12.

We can also examine the zero-field limit of Eq.(4-160) where $\mathscr{H}$ reduces to

$$\mathscr{H} = ha\begin{pmatrix} \frac{1}{4} & 0 & 0 & 0 \\ 0 & -\frac{1}{4} & \frac{1}{2} & 0 \\ 0 & \frac{1}{2} & -\frac{1}{4} & 0 \\ 0 & 0 & 0 & \frac{1}{4} \end{pmatrix}. \tag{4-167}$$

$\mathscr{H}$ can be easily diagonalized by using Eqs.(4-162) and (4-163) where $\tan 2\theta = \infty$ and $\theta = \pi/4$, which gives

$$\mathbf{b}^{\dagger}\mathscr{H}\mathbf{b} = ha\begin{pmatrix} \frac{1}{4} & 0 & 0 & 0 \\ 0 & \frac{1}{4} & 0 & 0 \\ 0 & 0 & -\frac{3}{4} & 0 \\ 0 & 0 & 0 & \frac{1}{4} \end{pmatrix}. \tag{4-168}$$

The results obtained here for the zero-field case (three triply degenerate levels at $E = \frac{1}{4}ha$ and a singly degenerate level at $E = -\frac{3}{4}ha$) could have been obtained directly from evaluating the matrix elements of the zero-field Hamiltonian [Eq.(4-158)] in the coupled basis. The first-order energy is [see Eqs.(4-141) through (4-144)]

$$\begin{aligned} E(S, I, F) &= \left\langle S, I, F, M_F \middle| h\left(\frac{a}{\hbar^2}\right)\boldsymbol{I}\cdot\boldsymbol{S} \middle| S, I, F, M \right\rangle \\ &= \left\langle S, I, F, M_F \middle| \frac{ha}{2\hbar^2}(F^2 - I^2 - S^2) \middle| S, I, F, M_F \right\rangle \\ &= \frac{ha}{2}[F(F+1) - I(I+1) - S(S+1)]. \end{aligned} \tag{4-169}$$

Thus, the two F states have the following energies: $E(\frac{1}{2}, \frac{1}{2}, 0) = -\frac{3}{4}ha$ (singlet) and $E(\frac{1}{2}, \frac{1}{2}, 1) = \frac{1}{4}ha$ (triplet), which agrees with the result in Eq.(4-168).

We note from Eq.(4-165) that at zero fields where $\theta \longrightarrow \pi/4$, the squared transition moments are $T_{12}^2 = T_{24}^2 = \frac{1}{2}(g_i\mu_0 - g_e\mu_B)^2$ and $T_{13}^2 = T_{34}^2 = \frac{1}{2}(g_i\mu_0 + g_e\mu_B)^2$ leading to an observable $E(F = 0) \longrightarrow E(F = 1)$ zero-field transition. Figure 4-12 shows a correlation diagram between the low and high field cases.

The analysis above is appropriate for the ground state of the free hydrogen atom ($L = 0$, ${}^2S_{1/2}$ state) and the transition between the zero-field $F = 0$ and $F = 1$ levels is the familiar signal monitored by radioastronomers at a wavelength of 21 cm. Precise intermediate field work in the hydrogen atom has given $a = 1420.40573 \times 10^6$ Hz and $g_e = 2.002296$ [17]. The slightly different result in Appendix A for g_e arises from slight adjustments of fundamental constants.

Next, we shall consider the high field case of $S = \frac{1}{2}$ with two identical nuclear spins of $I_1 = \frac{1}{2}$ and $I_2 = \frac{1}{2}$ as in the hydrogen molecular ion H_2^+. In this case, the Hamiltonian is

$$\begin{aligned} \mathscr{H} &= \frac{g_e\mu_B}{\hbar}H_zS_z - \frac{\mu_0 g_i(1-\sigma)}{\hbar}H_z(I_{1z} + I_{2z}) \\ &\quad + h\left(\frac{a}{\hbar^2}\right)(I_{1x}S_x + I_{1y}S_y + I_{1z}S_z + I_{2x}S_x + I_{2y}S_y + I_{2z}S_z). \end{aligned} \tag{4-170}$$

Table 4-9 The diagonal Hamiltonian matrix elements for $S = \frac{1}{2}, I_1 = I_2 = \frac{1}{2}$, and $\xi = \mu_0 g_i(1 - \sigma)H_z$. a is the nuclear spin-electron spin coupling constant.

M_S	M_{I_1}	M_{I_2}	Diagonal Elements	
$\frac{1}{2}$	$\frac{1}{2}$	$\frac{1}{2}$	$\frac{1}{2}(g_e\mu_B H_z) - \xi + (ha/2)$	E_1
$\frac{1}{2}$	$\frac{1}{2}$	$-\frac{1}{2}$	$\frac{1}{2}(g_e\mu_B H_z) + 0 + 0$	E_2
$\frac{1}{2}$	$-\frac{1}{2}$	$\frac{1}{2}$	$\frac{1}{2}(g_e\mu_B H_z) + 0 + 0$	E_2
$\frac{1}{2}$	$-\frac{1}{2}$	$-\frac{1}{2}$	$\frac{1}{2}(g_e\mu_B H_z) + \xi - (ha/2)$	E_3
$-\frac{1}{2}$	$\frac{1}{2}$	$\frac{1}{2}$	$-\frac{1}{2}(g_e\mu_B H_z) - \xi - (ha/2)$	E_4
$-\frac{1}{2}$	$\frac{1}{2}$	$-\frac{1}{2}$	$-\frac{1}{2}(g_e\mu_B H_z) + 0 + 0$	E_5
$-\frac{1}{2}$	$-\frac{1}{2}$	$\frac{1}{2}$	$-\frac{1}{2}(g_e\mu_B H_z) + 0 + 0$	E_5
$-\frac{1}{2}$	$-\frac{1}{2}$	$-\frac{1}{2}$	$-\frac{1}{2}(g_e\mu_B H_z) + \xi + (ha/2)$	E_6

The small effects due to nuclear spin–nuclear spin coupling are not usually observable in ESR due to the relatively rapid electron spin relaxation and relatively broad ESR line widths. These effects are observed in high resolution NMR (see Section 4.10). The diagonal elements of the Hamiltonian above in the coupled basis are listed in Table 4-9. The number of diagonal elements is equal to the order of the matrix in the uncoupled basis which is given by $(2S + 1)(2I_1 + 1)(2I_2 + 1)$. In this case, the order is 8.

The number of levels and values of the total angular momentum in the coupled basis are

$$\mathbf{I}_1 + \mathbf{I}_2 = \mathbf{I} \longrightarrow 0 \text{ and } 1 \qquad \mathbf{I} + \mathbf{S} = \mathbf{F} \longrightarrow \begin{cases} \frac{1}{2} & \text{when } I = 0 \\ \left.\begin{matrix} \frac{1}{2} \\ \frac{3}{2} \end{matrix}\right\} & \text{when } I = 1. \end{cases} \tag{4-171}$$

A correlation diagram between the low field F states and the high field states listed in Table 4-9 is shown in Fig. 4-13. The transition matrix for this $S = \frac{1}{2}$, $I_1 = \frac{1}{2}$, and $I_2 = \frac{1}{2}$ system is easily determined leading to the NMR and ESR transitions as shown where the ESR spectrum is a triplet with the intensity ratio 1:2:1 and symmetric splitting, a. The NMR spectrum is a simple doublet with a splitting equal to $2\xi/h$.

It is now relatively easy to extend this discussion of nuclear and electronic spin coupling to the methyl radical, $H_3^{12}C\cdot$, which has three equivalent protons and one unpaired electron where $S = \frac{1}{2}$ and $I_1 = I_2 = I_3 = \frac{1}{2}$. The order of the matrix representation of the appropriate Hamiltonian is

$$(2S + 1)(2I_1 + 1)(2I_2 + 1)(2I_3 + 1) = 2^4 = 16.$$

The diagonal elements are easily obtained and a high field energy level diagram is shown in Fig. 4-14 where the NMR transition is still a simple doublet with splitting equal again to $2\xi/h$. The ESR spectrum will appear as a quartet with relative intensities in a 1:3:3:1 ratio with splitting equal to a. The actual ESR spectra of the methyl radical is shown in the lower part of Fig. 4-14 [18].

The ESR multiplet spectra in the $S = \frac{1}{2}$ systems shown in Fig. 4-12 ($I_1 = \frac{1}{2}$), Fig. 4-13 ($I_1 = I_2 = \frac{1}{2}$), and Fig. 4-14 ($I_1 = I_2 = I_3 = \frac{1}{2}$) follow a systematic

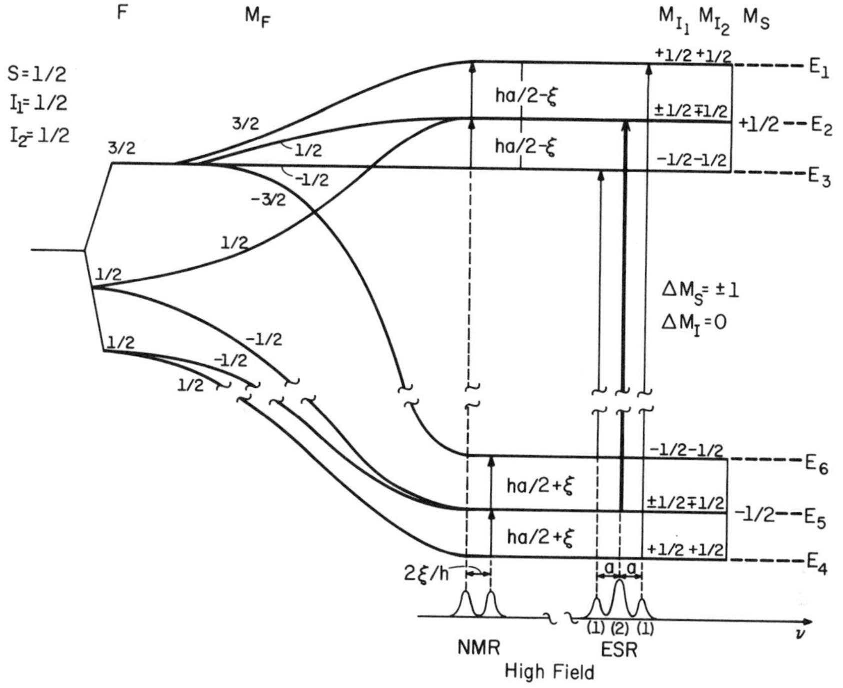

Figure 4-13 Energy level correlation diagram between the high and low fields for a $S = \frac{1}{2}$, $I_1 = \frac{1}{2}$, and $I_2 = \frac{1}{2}$ system. The high field energy levels are listed in Table 4-9 and we have assumed that ha is positive and larger than ξ to order the energy levels. The corresponding predicted NMR ($\Delta M_S = 0$, $\Delta M_I = \pm 1$) and ESR ($\Delta M_S = \pm 1$, $\Delta M_I = 0$) transitions are shown. The heavy line between E_2 and E_5 includes two transitions.

progression that is related to the binomial coefficients. The relative intensities of the components of the multiplet are determined by the M_I degeneracies within each of the two $M_S = \pm\frac{1}{2}$ manifolds in a $S = \frac{1}{2}$ system. The relative ordering of degeneracies and the pattern for the relative intensities of the multiplets are summarized in Table 4-10. These results are easily verified by example.

It is apparent from the analysis above that the observation of $\Delta M_S = \pm 1$ transitions in molecules by ESR can be a definite aid in identifying and characterizing organic and inorganic free radicals. Of course, the spin coupling constants are also of fundamental interest in interpreting the molecular electronic structure. The analysis of paramagnetic ions in the solid state or complexes in a liquid is similar to the work above on organic free radicals.

ESR spectra are normally observed with the sample placed in a fixed resonant microwave cavity with cavity frequency ν_c. The spectrum is obtained by sweeping the field, which varies the electron spin energy levels, until coincidence with the microwave cavity frequency is achieved. The coincidence, $(E_1 - E_2)/h = \nu_c$, leads to the resonance and the resultant absorption or dispersion curve (see Section 2.5) as a function of the magnetic field. In the high field limit, $g_e \mu_B H_z \gg ha$, the relative spacings in the observed ESR multiplet structure will be identical for different

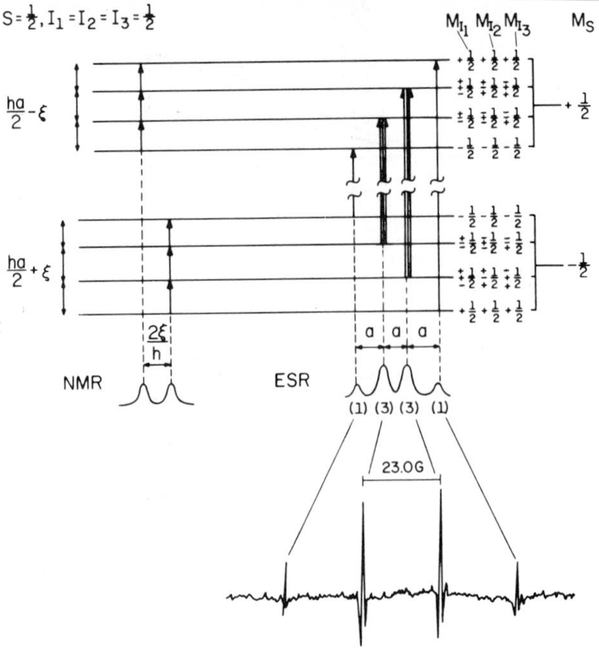

Figure 4-14 Energy levels for the high field limit ($\mu_B H_z \gg ha$) for $S = \frac{1}{2}$ and three equivalent nuclei, $I_1 = I_2 = I_3 = \frac{1}{2}$, where ha is positive and larger than ξ to order the energy levels. The predicted NMR ($\Delta M_S = 0, \Delta M_I = \pm 1$) and ESR ($\Delta M_S = \pm 1$ and $\Delta M_I = 0$) spectra are shown. In the case of the methyl radical, $a = 64.5$ MHz, which is obtained from the experimental ESR spectra which is also shown. The spectrum is from Ref. [18]. The splitting in gauss (G) can be converted to megahertz by multiplying by $g_e \mu_B / h = 2.80247 \text{ MHz} \cdot \text{G}^{-1}$.

Table 4-10 Binomial coefficients and their relation to the total nuclear states in each of the $M_S = \frac{1}{2}$ or $-\frac{1}{2}$ states. The ESR multiplet also follows the same pattern of relative intensities as given by the binomial patterns.

Number of Equivalent $I = \frac{1}{2}$ Nuclei	Total Number of States	Relative Ordering of Intensities
0	1	1
1	2	1 1
2	4	1 2 1
3	8	1 3 3 1
4	16	1 4 6 4 1
5	32	1 5 10 10 5 1
6	64	1 6 15 20 15 6 1

microwave cavity frequencies, ν_c. Thus, one reason for using higher field, higher-frequency ESR spectrometers is to assure a first-order (high field limit) spectra and a simple interpretation as given in this section. In the high field limit, the difference between molecular absorptions in gauss can be converted to megahertz by multiplying by $g_e \mu_B / h = 2.80247$ MHz/G.

Using the high field limit and the relative intensities given by the binomial series in Table 4-10, we can progressively construct spectra in molecules by the technique of equivalent groups of protons. For instance, the ethyl radical, $H_3C—\dot{C}H_2$, may have the two combinations of $\Delta M_S = \pm 1$ spectra illustrated in Fig. 4-15. Figure 4-15(a) illustrates an ESR spectrum when $a(CH_2) \gg a(CH_3)$ and Fig. 4-15(b) shows the case when $a(CH_2) \ll a(CH_3)$. The actual experimental ESR spectrum of $CH_3—\dot{C}H_2$ is shown in Fig. 4-15(c). The experimental spectrum correlates best with $a(CH_3) > a(CH_2)$, which indicates that the unpaired electron with $S = \frac{1}{2}$ couples

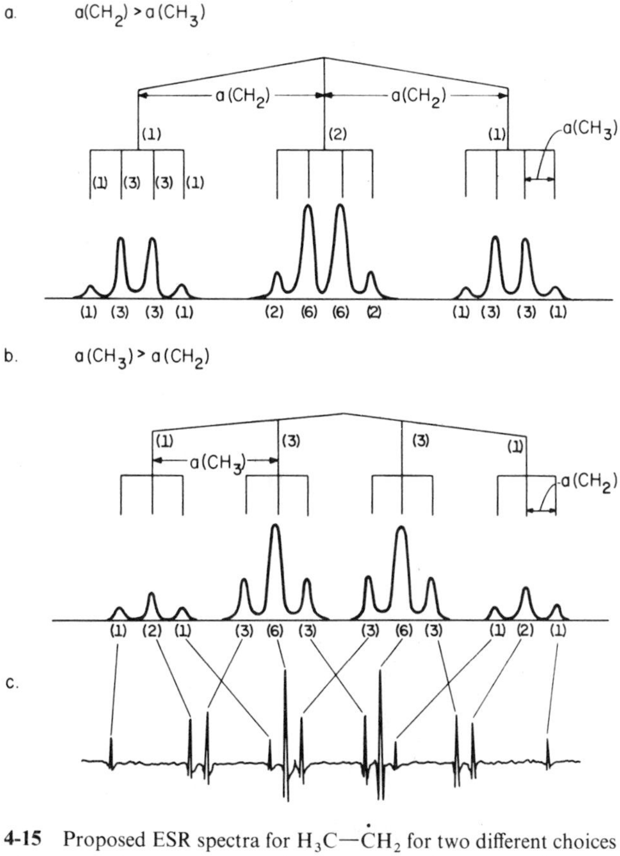

Figure 4-15 Proposed ESR spectra for $H_3C—\dot{C}H_2$ for two different choices for the two nuclear spin-electron spin coupling constants. (a) Predicted spectra when $a(CH_2) \gg a(CH_3)$. The predicted relative intensities are in parentheses. (b) Predicted spectra when $a(CH_2) \ll a(CH_3)$. (c) Experimental spectra from Ref. [18], which shows that $a(CH_2) < a(CH_3)$ and substantial mixing of the two simple pictures is clearly evident. The experimental results can be fit with $a(CH_2) = 62.7$ MHz and $a(CH_3) = 75.2$ MHz.

somewhat more strongly with the protons that are one carbon atom removed than with the apparent nearest protons. This effect is also apparent in the observed ESR spectra of the cyclopentyl radical that is shown in Fig. 4-16. The molecule has one α proton, four equivalent β protons, and four equivalent γ protons. The α, β, and γ notation follows the standard practice of labeling the carbon atoms successively from the apparent point of the unpaired electron (α). The ESR spectra in Fig. 4-16 shows that $a(H_\beta) > a(H_\alpha) \gg a(H_\gamma)$ in the cyclopentyl radical. The spin coupling constants in a number of representative organic radicals are listed in Table 4-11 [19]. It is interesting to note from Table 4-11 that the values of a for the unsaturated rings are considerably smaller than the saturated systems.

The discussion above centered around $S = \frac{1}{2}$ electronic systems in the presence of protons. The results can be extended easily to $S = 1$ systems as well as including the deuterium, ^{13}C, ^{19}F, ^{15}N, and other magnetic nuclei [see Problem 4-15 and Ref. [20]].

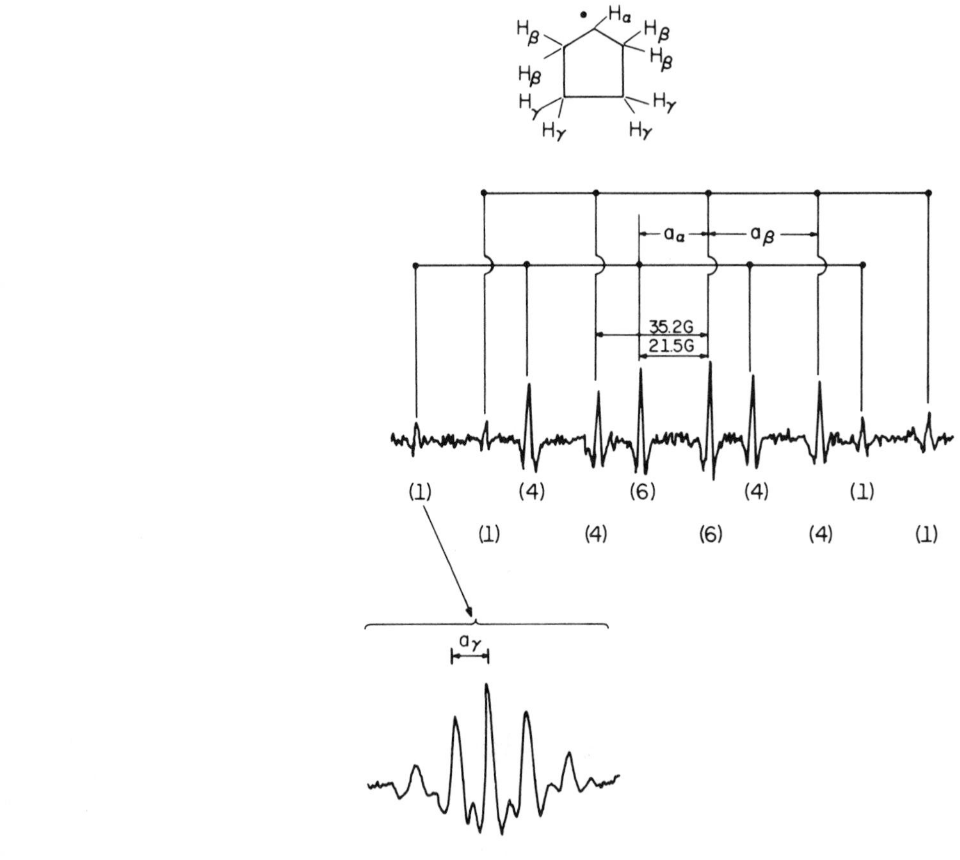

Figure 4-16 The ESR spectra of the cyclopentyl radical from Ref. [18]. The upper experimental trace shows two typical four-spin quintuplets with 1:4:6:4:1 intensity ratios (shown in parentheses) that are separated by the H_α coupling. An experimental blowup of the leftmost transition on the experimental trace shows the γ proton coupling.

Table 4-11 Proton nuclear spin-electron spin coupling constants for some representative organic free radicals that have $S = \frac{1}{2}$ electronic states with the free electronic g-value. The results are adapted from Refs. [18] and [19]. The units of a are megahertz and α, β, and γ follow the $\cdot C_\alpha - C_\beta - C_\gamma$ standard notation.

Radical	a_α (MHz)	a_β (MHz)	a_γ (MHz)
$\cdot CH_3$	64.6	—	—
$\cdot CH_2 CH_3$	62.7	75.3	—
$\cdot CH_2 CH_2 CH_3$	61.9	93.0	1.1
$\cdot CH(CH_3)_2$	61.9	69.1	—
$\begin{matrix} CH_3 \\ \vert \\ \cdot CCH_2CH_3 \\ \vert \\ H \end{matrix}$	61.1	68.7 (CH_3) 78.2 (CH_2)	— —
$\cdot C(CH_3)_3$	—	63.7	—
cyclobutyl	59.4	102.7	3.1
cyclopentyl	60.2	98.5	1.5
cyclohexyl	59.3	128.8	2.0
benzene negative ion	10.5	—	—

As mentioned at the beginning of this section, we have only considered isotropic interactions as observed primarily in the liquid state. Orientational effects are examined in Problem 4-39 and anisotropic ESR interactions are examined in Problem 5-19.

4.10 THE NUCLEAR ZEEMAN EFFECT IN THE ABSENCE OF ELECTRONIC ANGULAR MOMENTUM: NMR

We shall now consider the pure nuclear Zeeman effect and the corresponding nuclear magnetic resonance (NMR) transitions that are observable in most diamagnetic molecules (molecules with zero total electronic angular momentum) where the Hamiltonian in Eq.(4-156) reduces to

$$\mathscr{H} = -\frac{\mu_0}{\hbar} \sum_i g_i(1 - \sigma_i)\boldsymbol{I}_i \cdot \boldsymbol{H} + h \sum_{i > j} \frac{J_{ij}}{\hbar^2} \boldsymbol{I}_i \cdot \boldsymbol{I}_j. \tag{4-172}$$

The nuclear g-values, g_i, are listed in Table 4-8 for a number of common nuclei. J_{ij} is the nuclear spin-nuclear spin coupling constant [21] in units of hertz where typical values range from 0 to 500 Hz. The nuclear magnetic shielding denoted by σ_i originates from the field-induced electron currents in the molecule containing the nucleus with g-value, g_i. There are two field components at the nucleus; the direct external field, $\boldsymbol{H}$, and a smaller field, $\sigma_i \boldsymbol{H}$, that arises from the field-induced currents. If the nuclear magnetic shielding is *diamagnetic*, σ_i is positive and the net field at the nucleus, $(1 - \sigma_i)\boldsymbol{H}$, is smaller than the external field, $\boldsymbol{H}$. If the nuclear magnetic shielding is *paramagnetic*, σ_i is negative and the net field at the nucleus is larger than the external field. A detailed theory of magnetic shielding is given in Section 6.9.

Consider first a two-spin system with $I_1 = I_2 = \frac{1}{2}$ where the Hamiltonian is

$$\mathscr{H} = -\frac{\xi_1}{\hbar} I_{1z} - \frac{\xi_2}{\hbar} I_{2z} + h\left(\frac{J_{12}}{\hbar^2}\right)(I_{1x}I_{2x} + I_{1y}I_{2y} + I_{1z}I_{2z})$$

$$\xi_1 = \mu_0 g_1 H_z(1 - \sigma_1), \qquad \xi_2 = \mu_0 g_2 H_z(1 - \sigma_2). \tag{4-173}$$

The matrix elements are easily obtained in the uncoupled $\psi_{I_1 M_{I_1} I_2 M_{I_2}}$ basis giving

$$\begin{array}{cc|cccc}
 & M_{I_1} & \frac{1}{2} & \frac{1}{2} & -\frac{1}{2} & -\frac{1}{2} \\
M_{I_1}\ M_{I_2} & M_{I_2} & \frac{1}{2} & -\frac{1}{2} & \frac{1}{2} & -\frac{1}{2} \\
\hline
\frac{1}{2} & \frac{1}{2} & -\frac{\xi_1}{2} - \frac{\xi_2}{2} + \frac{hJ_{12}}{4} & 0 & 0 & 0 \\
\frac{1}{2} & -\frac{1}{2} & 0 & -\frac{\xi_1}{2} + \frac{\xi_2}{2} - \frac{hJ_{12}}{4} & \frac{hJ_{12}}{2} & 0 \\
-\frac{1}{2} & \frac{1}{2} & 0 & \frac{hJ_{12}}{2} & +\frac{\xi_1}{2} - \frac{\xi_2}{2} - \frac{J_{12}}{4} & 0 \\
-\frac{1}{2} & -\frac{1}{2} & 0 & 0 & 0 & \frac{\xi_1}{2} + \frac{\xi_2}{2} + \frac{hJ_{12}}{4}
\end{array}$$

$$\mathscr{H} = \begin{pmatrix} -\frac{\xi_1}{2} - \frac{\xi_2}{2} + \frac{hJ_{12}}{4} & 0 & 0 & 0 \\ 0 & -\frac{\xi_1}{2} + \frac{\xi_2}{2} - \frac{hJ_{12}}{4} & \frac{hJ_{12}}{2} & 0 \\ 0 & \frac{hJ_{12}}{2} & +\frac{\xi_1}{2} - \frac{\xi_2}{2} - \frac{J_{12}}{4} & 0 \\ 0 & 0 & 0 & \frac{\xi_1}{2} + \frac{\xi_2}{2} + \frac{hJ_{12}}{4} \end{pmatrix}. \tag{4-174}$$

The transition dipole matrix for the nuclear magnetic dipole transitions in the two-spin system is needed in order to predict the NMR intensities under a variety of conditions. The magnetic dipole operator for the two-spin system is $\boldsymbol{\mu} = \boldsymbol{\mu}_1 + \boldsymbol{\mu}_2 = g_1(\mu_0/\hbar)\boldsymbol{I}_1 + g_2(\mu_0/\hbar)\boldsymbol{I}_2$ and the matrix representation of the $\mu_{\pm}$ magnetic dipole operator, in the same order as the matrix representation of the Hamiltonian in Eq.(4-174), is given by

$$\boldsymbol{\mu}_+ + \boldsymbol{\mu}_- = \boldsymbol{\mu}_{\pm} = \mu_0 \begin{pmatrix} 0 & g_2 & g_1 & 0 \\ g_2 & 0 & 0 & g_1 \\ g_1 & 0 & 0 & g_2 \\ 0 & g_1 & g_2 & 0 \end{pmatrix}, \tag{4-175}$$

where $\boldsymbol{\mu}_+$ fills the matrix elements above the diagonal and $\boldsymbol{\mu}_-$ fills the matrix elements below the diagonal.

In the case of a simple first-order spectrum that is defined by the condition

$$\frac{J_{12}}{2} \ll \frac{(\xi_1 - \xi_2)}{h}, \tag{4-176}$$

we can use the diagonal elements in $\mathscr{H}$ to describe the energy levels, and the transition moments are obtained directly from the elements in $\boldsymbol{\mu}_{\pm}$ as shown in Eq.(4-175).

In order to satisfy Eq.(4-176), ξ_1 and ξ_2 must normally arise from different nuclei. Consider, for instance, fluoroacetylene, F—C≡C—H, where $I_F = I_H = \frac{1}{2}$. The values of ξ_H and ξ_F can be estimated by ignoring the values of σ_H and σ_F as the shieldings are normally on the order of 10^{-4} and 10^{-3} for hydrogen and fluorine atoms, respectively. Thus, the values of ξ_H/h and ξ_F/h at 10,000 G are (see Table 4-8 for the g-values)

$$\frac{\xi_H}{h} = \frac{\mu_0 g_H H_z(1 - \sigma_H)}{h} \cong 42.6 \text{ MHz}, \qquad \frac{\xi_F}{h} = \frac{\mu_0 g_F H_z(1 - \sigma_F)}{h} \cong 40.1 \text{ MHz}. \tag{4-177}$$

It is evident that Eq.(4-176) is satisfied for values of J well above the normal J_{HF} values of up to 100 Hz. The resultant first-order spectra predicted for F—C≡C—H is shown in Fig. 4-17 where the transitions are identified from Eqs.(4-174) and (4-175). Spectra arising from two different magnetic nuclei as in Fig. 4-17 are called AX spectra.

A somewhat different situation arises when $J_{12}/2 \cong (\xi_1 - \xi_2)/h$, which would normally arise from two identical nuclei in different chemical environments (an AB spectrum). Under these circumstances, we can use the rotational transformation in

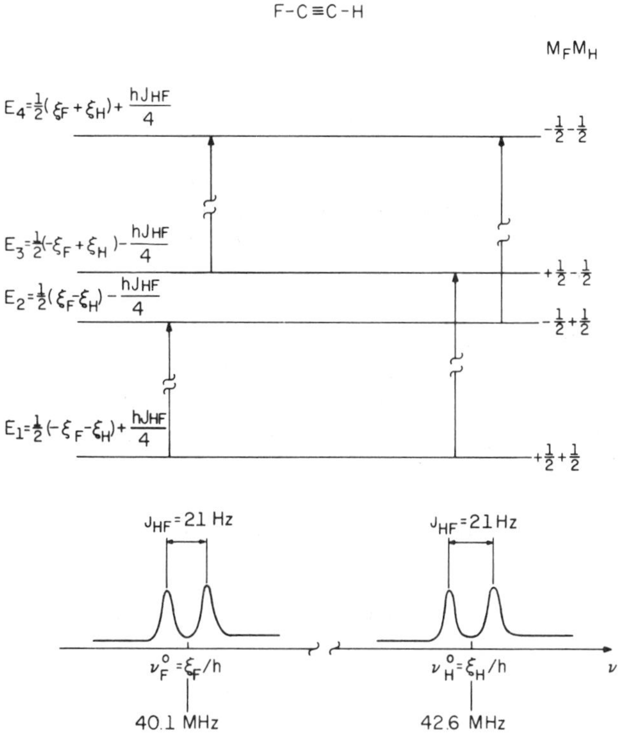

Figure 4-17 Energy level diagram showing $^{19}F(I = \frac{1}{2})$ and $H(I = \frac{1}{2})$ NMR transitions in F—C≡C—H at 10,000 G. The first-order AX spectrum is evident from Eqs.(4-174) and (4-175) where $|(\xi_F - \xi_H)| \gg hJ$. The value of $J_{HF} = 21$ Hz is from W. H. Middleton and W. H. Sharkev. *J. Am. Chem. Soc.* **81**, 803 (1959).

Eq.(4-162) to diagonalize $\mathscr{H}$ by $\mathbf{b}^{\dagger}\mathscr{H}\mathbf{b} = \mathbf{E}$ and transform $\boldsymbol{\mu}_{\pm}$ by $\mathbf{b}^{\dagger}\boldsymbol{\mu}_{\pm}\mathbf{b} = \mathbf{T}$. The diagonal elements in $\mathbf{E}$ and the resultant nonzero elements in $\mathbf{T}$ are given below:

$$E_1 = a, \qquad E_2 = b\cos^2\theta + hJ\cos\theta\sin\theta + c\sin^2\theta$$

$$E_3 = b\sin^2\theta - hJ\cos\theta\sin\theta + c\cos^2\theta, \qquad E_4 = d$$

$$a = \frac{(-\xi_1 - \xi_2)}{2} + \frac{hJ}{4}, \qquad b = \frac{(-\xi_1 + \xi_2)}{2} - \frac{hJ}{4}$$

$$c = \frac{(+\xi_1 - \xi_2)}{2} - \frac{hJ}{4}, \qquad d = \frac{(\xi_1 + \xi_2)}{2} + \frac{hJ}{4}$$

$$E_1 \longleftrightarrow E_2 \qquad T_{12}^2 = [g_2\cos\theta + g_1\sin\theta]^2\mu_0^2$$

$$E_1 \longleftrightarrow E_3 \qquad T_{13}^2 = [g_1\cos\theta - g_2\sin\theta]^2\mu_0^2$$

$$E_2 \longleftrightarrow E_4 \qquad T_{24}^2 = [g_1\cos\theta + g_2\sin\theta]^2\mu_0^2$$

$$E_3 \longleftrightarrow E_4 \qquad T_{34}^2 = [g_2\cos\theta - g_1\sin\theta]^2\mu_0^2$$

$$\tan 2\theta = \frac{hJ}{\xi_2 - \xi_1}, \tag{4-178}$$

where ξ_1 and ξ_2 are defined in Eq.(4-173). The results in Eq.(4-178) are plotted for a fixed field as a function of J and $|\xi_1 - \xi_2|$ in Fig. 4-18. Note from Eq.(4-178) that the relative intensities are in equivalent pairs. We also note that if $\xi_1 - \xi_2 = 0$, a single transition will appear at $\nu = \xi_1/h = \xi_2/h$. A specific example of an AB spectrum for two protons in different chemical environments is shown in Fig. 4-19 for 2-bromo-5-chlorothiophene. The Br and Cl nuclei do not couple with the protons due to their nuclear quadrupole moment relaxation processes. The observed spectra for 2-bromo-5-chlorothiophene in Fig. 4-19 corresponds closely in pattern and relative intensity to the calculated $J = 4.0$ Hz spectra shown in Fig. 4-18. The 7100-G AB spectra of 2-bromo-5-chlorothiophene can be forced to a more nearly first-order spectrum by increasing the magnetic field (J_{12} is independent of the field). Increasing the magnetic field increases the value of $|\xi_1 - \xi_2|/h$ leading ultimately to the first-order condition described in Eq.(4-176). The effect of increasing the magnetic field can be seen from Fig. 4-18 by going from the bottom to the top of the diagram. Thus, the spectra shown in Fig. 4-19 will converge at high fields to a quartet spectra with a peak splitting of $|(\xi_1 - \xi_2)|/h$ between two equivalent sets of doublets split by J_{12}.

The more complex case of three nuclei of spin $\frac{1}{2}$ gives a matrix of order $2^3 = 8$. The Hamiltonian is given by

$$\mathscr{H} = -\left(\frac{\xi_1}{\hbar}\right)I_{1z} - \left(\frac{\xi_2}{\hbar}\right)I_{2z} - \left(\frac{\xi_3}{\hbar}\right)I_{3z} + h\left(\frac{J_{12}}{\hbar^2}\right)\boldsymbol{I}_1 \cdot \boldsymbol{I}_2$$
$$+ h\left(\frac{J_{13}}{\hbar^2}\right)\boldsymbol{I}_1 \cdot \boldsymbol{I}_3 + h\left(\frac{J_{23}}{\hbar^2}\right)I_2 \cdot I_3, \tag{4-179}$$

where $\xi_i = \mu_0 g_i(1 - \sigma_i)H_z$ as before. The resultant Hamiltonian matrix in the uncoupled basis is easily set up, as before, as well as the transition dipole matrix. Features similar to those discussed above are observed. In the case of equivalent nuclei in identical chemical environments, only a single absorption curve is observed

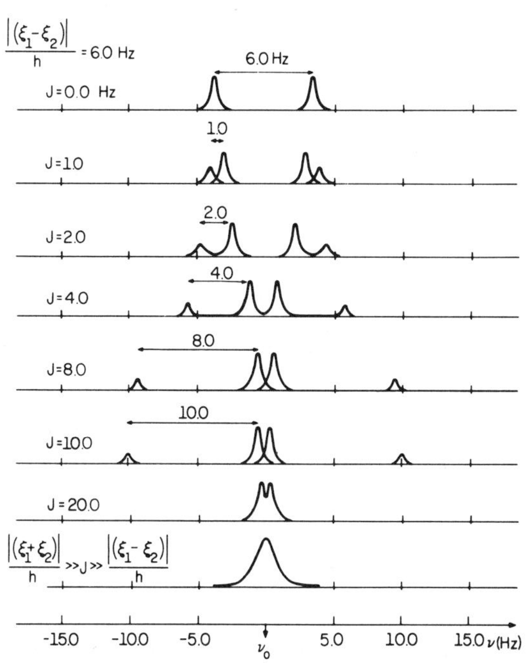

Figure 4-18 Predicted NMR spectra of a two-spin system where the nuclei are identical but in different chemical environments that is determined by $|(\xi_1 - \xi_2)| =$ 6.0 Hz. The spectra are shown for J_{12} varying from $0 \longrightarrow \infty$. Note that if $J_{12} = 0$, the splitting between the two nuclear resonances is equal to $|(\xi_1 - \xi_2)|/h$. If $J_{12} \gg |(\xi_1 - \xi_2)|/h$, a single resonance is observed at ν_0.

and J cannot be measured. This is a general observation. For instance, the proton–proton J cannot be measured in CH_4 because each proton is equivalent. J_{HD} can be measured in CH_3D, however, and J_{HH} in CH_4 and J_{HD} in CH_3D should be related by the ratio of the nuclear g-values.

We shall now illustrate the first-order building-up principle in the proton and fluorine nuclear magnetic resonance spectra in the following series:

$$CH_4 \longleftrightarrow CH_3F \longleftrightarrow CH_2F_2 \longleftrightarrow CHF_3 \longleftrightarrow CF_4.$$

The first-order ^{19}F and ^{1}H NMR spectra of these molecules is shown in Fig. 4-20. The general patterns are obtained from the binomial coefficients from Table 4-10 as discussed in Section 4.9. For instance, the methyl radical $\cdot CH_3$ includes an electron

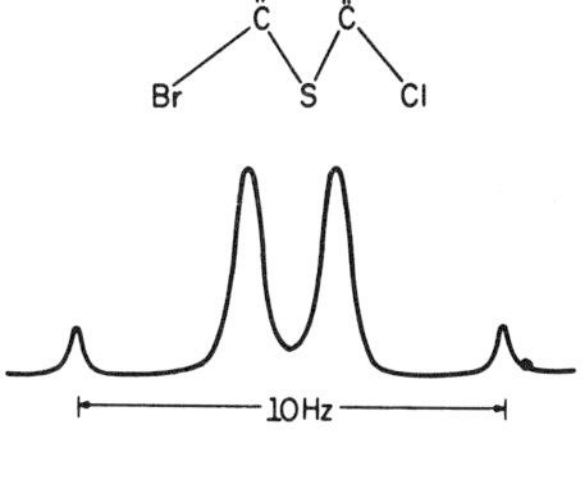

Figure 4-19 The proton NMR spectrum of 2-bromo-5-chlorothiophene which has two protons in different chemical environments. The spectrum is from W. A. Anderson, *Phys. Rev.* **102**, 151 (1956) and the experimentally determined values of $|(\xi_1 - \xi_2)|/h$ and J_{12} are 4.7 ± 0.2 Hz and 3.9 ± 0.2 Hz, respectively. The spectrum was recorded by sweeping the field at a fixed frequency of about 30.5 MHz, which corresponds to a field near 7160 G.

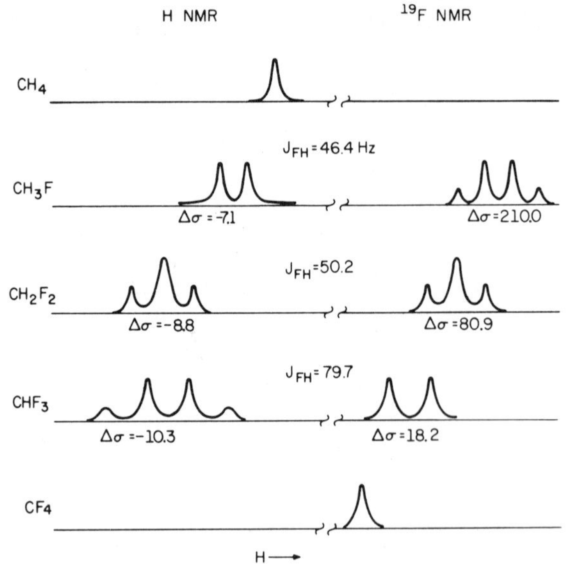

Figure 4-20 The ^{19}F and ^{1}H NMR spectra of the CH_4, CH_3F, CH_2F_2, CHF_3, and CF_4 series of molecules from L. H. Meyer and H. S. Gutowsky, *J. Phys. Chem.* **57**, 481 (1953) and S. G. Frankiss, *J. Phys. Chem.* **67**, 752 (1963). The resonances are shown in the high field limit at fixed frequencies and increasing field. The values of J_{FH} increase with increasing number of F atoms. Both the ^{19}F and ^{1}H NMR resonance fields increase with increased ^{1}H substitution. The chemical shifts in parts per million (ppm) relative to the respective CF_4 and CH_4 references are shown under the spectra [see Eq.(4-182) for the chemical shift in terms of the resonant fields].

with spin $\frac{1}{2}$ interacting with three identical protons with spin $\frac{1}{2}$, which leads to the 1:3:3:1 quartet, as shown in Fig. 4-14. Formally we can replace the electron with spin $\frac{1}{2}$ with the ^{19}F nucleus with spin $\frac{1}{2}$ in CFH_3 and we can expect the ^{19}F NMR to appear as a 1:3:3:1 quartet with splitting equal to J_{HF} as shown in Fig. 4-20. Similarly, the three equivalent protons in CFH_3 are all coupled to the single ^{19}F nucleus leading to the doublet ^{1}H NMR spectra. Both the ^{1}H and ^{19}F NMR spectra in CH_2F_2 are expected to be symmetric 1:2:1 triplets with splittings equal to J_{HF}. Both the ^{19}F—^{1}H spin–spin coupling constants and the *chemical shifts* are shown in Fig. 4-20. The chemical shifts arise from the difference between the magnetic shielding of a nucleus in different chemical environments. The chemical shift of ^{19}F in CHF_3 relative to ^{19}F in CF_4 is the difference in the magnetic shielding at the ^{19}F nucleus in the two molecules.

$$\Delta\sigma = \sigma(\mathrm{CHF_3}) - \sigma(\mathrm{CF_4}). \tag{4-180}$$

$\Delta\sigma$ can be obtained experimentally by observing the NMR spectra at constant field (varying ν) or constant frequency (varying the field). At constant frequency, ν_0, the field is swept through the resonances (see Fig. 4-20). ν_0 is related to the chemical shift through

$$\nu_0 = \frac{\mu_0}{h} g_F H_z(\mathrm{CHF_3})[1 - \sigma_F(\mathrm{CHF_3})] = \frac{\mu_0}{h} g_F H_z(\mathrm{CF_4})[1 - \sigma_F(\mathrm{CF_4})]. \tag{4-181}$$

$H(CHF_3)$ and $H(CF_4)$ are the ^{19}F resonant fields in the two molecules given by

$$H_z(CHF_3) = \frac{h\nu_0}{\mu_0 g_F[1 - \sigma_F(CHF_3)]} \cong \frac{h\nu_0[1 + \sigma_F(CHF_3)]}{\mu_0 g_F}$$

$$H_z(CF_4) = \frac{h\nu_0}{\mu_0 g_F[1 - \sigma_F(CF_4)]} \cong \frac{h\nu_0[1 + \sigma_F(CF_4)]}{\mu_0 g_F}.$$

Thus, the chemical shift is given in terms of the resonant fields by

$$\Delta\sigma = \sigma_F(CHF_3) - \sigma_F(CF_4) \cong \frac{H_z(CHF_3) - H_z(CF_4)}{H_z(CF_4)},$$

or, in general,

$$\Delta\sigma = \sigma - \sigma_r \cong \frac{H - H_r}{H_r}, \qquad (4\text{-}182)$$

where σ_r is the reference shielding, H_r is the resonance field of the reference molecule, and H and σ are the resonance field and shielding of the molecule of interest.

The chemical shift can also be obtained at fixed field by varying the frequency. It is easy to show that

$$\Delta\sigma \cong \frac{\nu(CF_4) - \nu(CHF_3)}{\nu(CF_4)} = \frac{\nu_r - \nu}{\nu_r}. \qquad (4\text{-}183)$$

The proton chemical shifts for a number of molecules and molecular systems relative to tetramethyl silane (TMS) are listed in Fig. 4-21. TMS is used as a reference for proton magnetic resonance chemical shifts because the resonance fields for protons in most organic compounds lie at lower fields than TMS. Following the usual convention we have set the TMS magnetic shielding arbitrarily to 10 ppm. A better

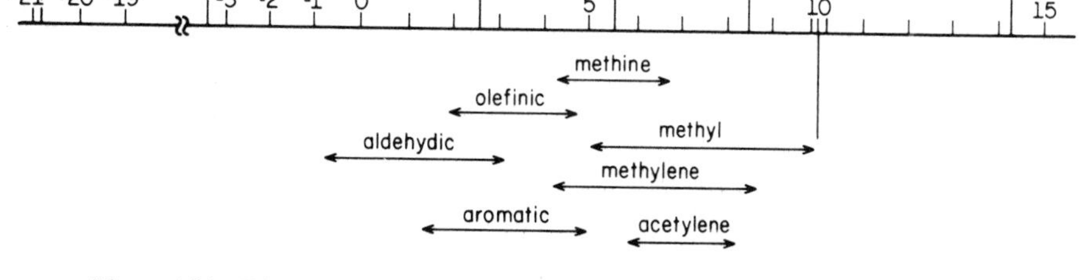

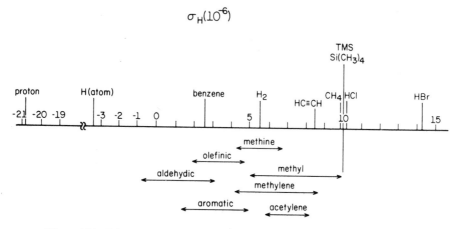

Figure 4-21 Diagram showing the proton chemical shifts of many compounds relative to an arbitrary origin. The origin or reference is the proton magnetic shielding in $Si(CH_3)_4$, which is arbitrarily (for convenience) set at $\sigma_H[Si(CH_3)_4] = 10.0 \times 10^{-6}$ or 10 ppm.

reference would be the bare proton shown at $\sigma = -20.9 \times 10^{-6}$ relative to TMS at $\sigma = 10.0 \times 10^{-6}$. We shall discuss the absolute shielding scales in Section 6.9.

The J_{HF} coupling constants in CFH_3, CF_2H_2, and CF_3H are also shown in Fig. 4-20. The J_{HF} coupling is also evident in HC≡CF in Fig. 4-17 and evidence for J_{HH} in 2-bromo-5-chlorothiophene is shown in Fig. 4-19. Additional values of J_{HH} and J_{HF} for other molecules are listed in Table 4-12.

As a final example, consider the apparent first-order proton spectra for CH_3CHF_2 shown in Fig. 4-22. According to Fig. 4-21, we would expect the upper field methyl proton NMR spectra to be characteristic of a combination of a two-spin ^{19}F and single-spin proton coupling. According to Table 4-12, $J_{HF} > J_{HH}$. Correspondingly, we would expect the methyl group proton NMR spectra to be a 1:2:1 two-spin triplet with splitting equal to J_{HF} and a further splitting of each of the lines in the triplet due

Table 4-12 Typical proton and fluorine spin–spin coupling constants for a number of molecules from J. W. Emsley, J. Feeney, and L. H. Sutcliffe, *High Resolution Nuclear Magnetic Resonance Spectroscopy* (Pergamon Press, Oxford, 1965). See also the continuing periodical series entitled *Progress in Nuclear Magnetic Resonance Spectroscopy*. Ed. by J. W. Emsley, J. Feeney, and L. H. Sutcliffe.

Molecule or Fragment		$X = H$ J_{HH} (Hz)	$X = F$ J_{HF} (Hz)
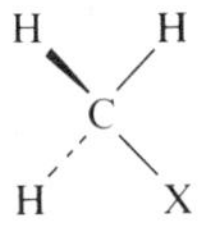		10–15	40–80
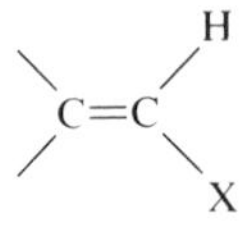		0–5	40–100
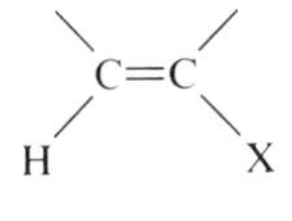		6–15	1–8
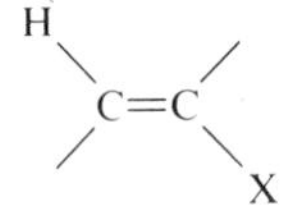		11–18	12–45
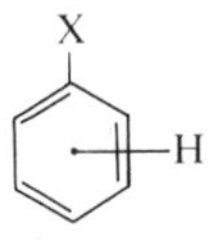	*ortho*	7–10	6–10
	meta	2–3	5–6
	para	1	0
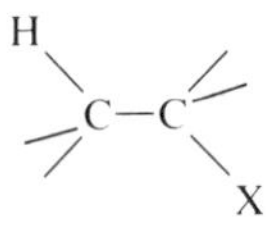		2–9	10–25

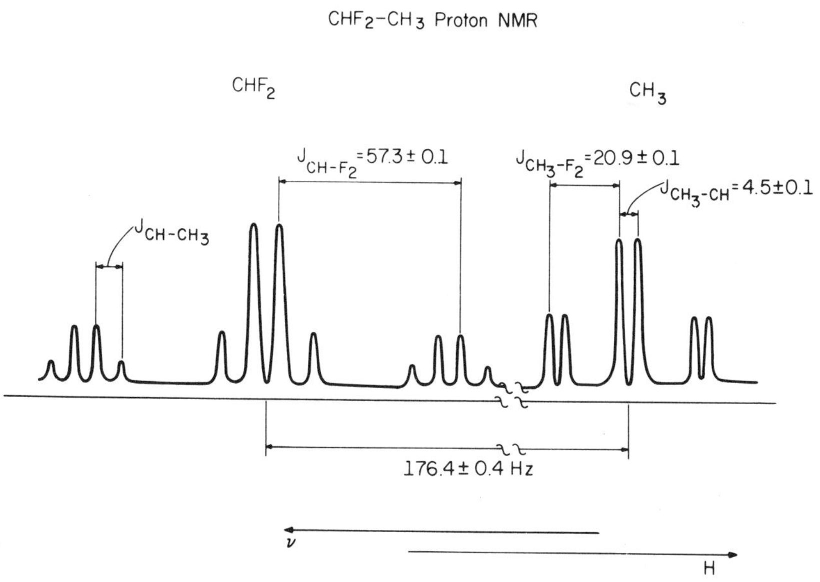

Figure 4-22 Proton NMR spectra of $CHF_2—CH_3$ from G. W. Flynn and J. D. Baldeschwieler, *J. Chem. Phys.* **37**, 2907 (1962) taken at 40 MHz. The spin–spin coupling constants are shown in units of hertz and the splitting between CHF_2 and CH_3 protons leads to a chemical shift of $\sigma_{CH_3} - \sigma_{CHF_2} = 176.4/(40 \times 10^6) = 4.4$ ppm.

to the smaller J_{HH} coupling constant. The resultant six-line methyl group NMR spectra is evident at high fields in Fig. 4-22 with $J_{HF} = 20.9$ and $J_{HH} = 4.5$.

The proton in the CHF_2 group of CH_3CHF_2 appears at lower field in Fig. 4-22. In this case, we would expect a dominant triplet due to J_{HF} coupling and a smaller J_{HH} splitting, which leads to 1:3:3:1 quartets. The splitting between the CH_3 and CHF_2 proton resonances leads to a chemical shift of $\sigma = 4.4 \times 10^{-6}$ as shown.

Many other nuclei in addition to the proton and ^{19}F nuclei considered here are eligible for NMR. Some of the additional nuclei are listed in Table 4-8 and the more commonly studied nuclei include ^{13}C and ^{31}P. Several problems at the end of this section illustrate the NMR spectra, chemical shifts, and spin–spin coupling constants of these other nuclei. The relative intensities within each NMR multiplet are shown in the figures in this section. However, the relative intensities between the spectra arising from different nuclei have not been considered.

The relative sensitivities of equal numbers of different nuclei depend on the details of the observing method. In NMR, this normally involves an external oscillator driven into a circuit including capacitance, inductance, and resistance. The sample is placed in the inductive coil of the circuit and the inductance will change when the nuclei in the sample absorb the radio-frequency energy (see Section 7.4). This leads to a measurable electromotive force due to the nuclei that can be observed as a voltage change in the circuit. A detailed analysis of the relative sensitivities for different nuclei in this type of circuit gives the results in Table 4-13.

Table 4-13 NMR relative sensitivities for equal numbers of different nuclei relative to the proton at a magnetic field of 10,000 G according to H. S. Gutowsky, "Analytical Application of Nuclear Magnetic Resonance" in *Physical Methods in Chemical Analysis*, Vol. III, Ed. by W. G. Berl (Academic Press, New York, 1956).

Nucleus	Resonance Frequency (MHz) at 10,000 G	Relative Intensity
^{1}H	42.58	1.000
^{2}H	6.54	2.45×10^{-2}
^{13}C	10.71	3.17×10^{-2}
^{19}F	40.06	8.58×10^{-1}
^{31}P	17.24	1.04×10^{-1}

4.11 THE RIGID ROTOR

In this section we examine the rigid rotation (see Section 2.1) of linear molecules and we calculate the effects of electric and magnetic fields. We also examine the rigid rotation of nonlinear molecules.

The Linear Rigid Rotor

The Hamiltonian for a rigid diatomic molecule can be obtained from our previous work on the hydrogen-like atom by replacing the nucleus with atom a and the electron with atom b in Fig. 4-7. Rewriting the internal part of Eq.(4-46) gives

$$\mathscr{H} = -\frac{\hbar^2}{2\mu r^2}\frac{\partial}{\partial r}\left(r^2\frac{\partial}{\partial r}\right) - \frac{\hbar^2}{2\mu r^2 \sin^2\theta}\left[\sin\theta\frac{\partial}{\partial\theta}\left(\sin\theta\frac{\partial}{\partial\theta}\right) + \frac{\partial^2}{\partial\phi^2}\right] - \frac{Ze^2}{r}. \quad (4\text{-}184)$$

As r is a constant for a rigid linear molecule, only the second term in Eq.(4-184) will be included in the rotational Hamiltonian for a diatomic molecule. The Schrödinger equation is

$$-\frac{\hbar^2}{2\mu r^2}\left[\frac{1}{\sin\theta}\frac{\partial}{\partial\theta}\left(\sin\theta\frac{\partial}{\partial\theta}\right) + \frac{1}{\sin^2\theta}\frac{\partial^2}{\partial\phi^2}\right]\psi(\theta,\phi) = \psi(\theta,\phi)E$$

$$\frac{1}{2\mu r^2}J^2\psi(\theta,\phi) = \psi(\theta,\phi)E, \quad (4\text{-}185)$$

where we have also substituted J^2, the rotational angular momentum, from Eq.(4-69). The reduced mass is $\mu = M_a M_b/(M_a + M_b)$, where M_a and M_b are the masses of *atoms* (a) and (b) in the diatomic molecule. It is easy to show that μr^2 is the moment of inertia for the two-body system. We can write the moment of inertia for the two-body system in Fig. 4-7 as $I = M_a d_a^2 + M_b d_b^2$, where d_a and d_b are the distances from the center of mass to the a and b atoms. Substituting $d_a = M_b/(M_a + M_b)r$ and $d_b = M_a/(M_a + M_b)r$ gives

$$\mathsf{I} = \frac{M_a^2 M_b r^2}{(M_a + M_b)^2} + \frac{M_a M_b^2 r^2}{(M_a + M_b)^2} = \mu r^2. \tag{4-186}$$

This equation can easily be generalized to a linear molecule, with more than two atoms, giving the c.m. moment of inertia, I, perpendicular to the internuclear axis as

$$\mathsf{I} = \frac{\sum_{i<j} M_i M_j r_{ij}^2}{\sum_i M_i}, \tag{4-187}$$

where r_{ij} is the distance along the internuclear line between the ith and jth atoms and $i < j$ indicates an independent sum over i and j (all atoms in the linear molecule) where $i < j$. Substituting these results into Eq.(4-185) gives the rotational Schrödinger equation for a linear molecule,

$$\mathscr{H} Y_{JM}(\theta, \phi) = \frac{J^2}{2\mathsf{I}} Y_{JM}(\theta, \phi) = \frac{\hbar^2}{2\mathsf{I}} J(J + 1) Y_{JM}(\theta, \phi), \tag{4-188}$$

where the eigenfunctions are the spherical harmonics as listed in Table 4-3.

In Section 2.1 we discussed the classical kinetic rotational frequency of $\nu = 3.90 \times 10^{11}$ Hz for the OCS molecule when the rotational temperature was in equilibrium with the translational temperature of 300 K. If we interpret each rotational state given in Eq.(4-188) as having a rotational frequency $\nu = E_J/h$, we can estimate which J quantum state has the equivalent classical frequency of $\nu = 3.90 \times 10^{11}$ Hz by $\nu = E_J/h = (\hbar^2/2\mathsf{I}h)J(J + 1) = (\hbar/4\pi\mathsf{I})J(J + 1)$. Using $\mathsf{I} = 138.0 \times 10^{-40}$ g · cm^2 for $^{16}O^{12}C^{32}S$ from Fig. 2-2 gives $J(J + 1) \cong 64$, which leads to $J \cong 8$. Thus, according to the results above, if the rotational and translational states are in thermal equilibrium at $T = 300$ K, the $J = 8$ rotational state has approximately the rotational frequency that is apparent in a classical rotor with the same temperature. As we increase the moment of inertia, the corresponding value of J increases. As the J states approach degeneracy, the classical and quantum mechanical systems become equivalent. According to the Bohr frequency rule, a given $J \longrightarrow J + 1$ rotational transition implies a change in rotational frequency of

$$\frac{E(J + 1, M) - E(J, M)}{h} = \nu_{J+1,J} = \frac{\hbar^2}{h\mathsf{I}_{bb}}(J + 1) = \frac{\hbar}{2\pi\mathsf{I}_{bb}}(J + 1) = 2B(J + 1). \tag{4-189}$$

The rotational constant, B, in units of hertz, is defined by

$$B = \frac{\hbar}{4\pi\mathsf{I}_{bb}}, \tag{4-190}$$

where I_{bb} denotes the moment about an axis perpendicular to the interatomic line in the linear molecule. The rotational constants for some of the isotopic species in OCS are listed in Table 4-14 [22].

Table 4-14 The rotational constants and natural isotopic abundance of several of the isotopic combinations in OCS.

Isotopic Species	B (MHz)	Natural Abundance, %
$^{16}O^{12}C^{32}S$	6,081.49	94.00
$^{16}O^{13}C^{32}S$	6,061.89	1.00
$^{17}O^{12}C^{32}S$	5,883.67	0.04
$^{18}O^{12}C^{32}S$	5,704.83	0.20
$^{16}O^{12}C^{34}S$	5,932.82	4.00
$^{16}O^{12}C^{33}S$	6,004.91	0.72
$^{16}O^{13}C^{34}S$	5,911.73	0.04

The $J = 0 \longrightarrow J = 1$ ground vibrational state rotational spectra of OCS is shown in Fig. 4-23 where the relative intensities of the various isotopic species are illustrated. As we shall note later, the $J = 0 \longrightarrow J = 1$ rotational transitions for molecules in the excited vibrational states will also be present and some of these vibrational satellites will be as intense as the weaker isotopic transitions. A major problem in microwave spectroscopy is the differentiation of weak isotopic transitions from the rotational spectra of excited vibration states of the main isotopic species. As line widths are

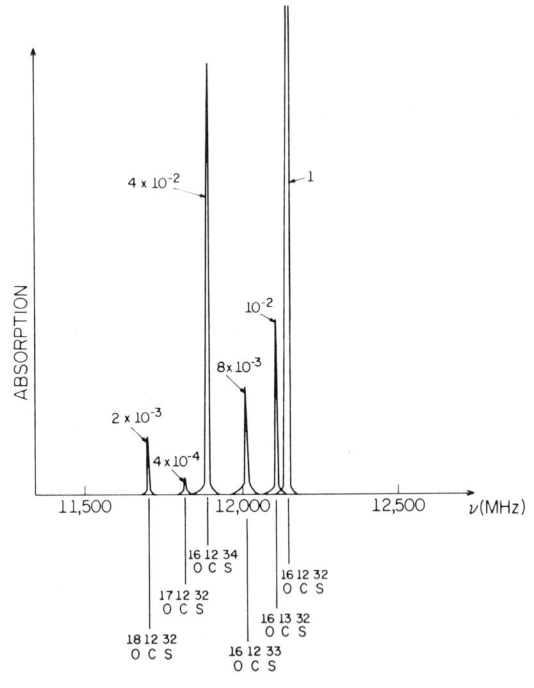

Figure 4-23 A schematic showing the relative intensities of some of the naturally occurring isotopic species of OCS as observed by the relative intensities of the $J = 0 \longrightarrow J = 1$ transitions. See Table 4-14 for natural abundance percentages.

Table 4-15 The rotational transitions of HCN and DCN from C. A. Burrus and W. Gordy, *Phys. Rev.* **101**, 599 (1956). The numbers are in units of megahertz.

J	$\longrightarrow$	$J+1$	$H^{12}C^{14}N$	$D^{12}C^{14}N$
0	$\longrightarrow$	1	88,631.62	72,414.61
1	$\longrightarrow$	2	177,260.99	144,827.86
2	$\longrightarrow$	3	265,886.18	217,238.40
3	$\longrightarrow$	4		286,644.67

normally less than 1 MHz, the different isotopic species are easily resolved if the sensitivity is sufficiently high. Using the data on the OCS isotopes above allows a determination of the moments of inertia for the several isotopes and a determination of the two bond distances.

As another example, the observed rotational spectra of two isotopic species of HCN are listed in Table 4-15. The experimental values of the rotational constants $(\hbar/4\pi I_{bb})$ from Table 4-15 are $H^{12}C^{14}N$, $B = 44{,}315.99$ MHz and $D^{12}C^{14}N$, $B = 36{,}207.42$ MHz. The resulting moments of inertia are (using $\hbar/4\pi = 8.39215 \times 10^{-35}$ $g \cdot cm^2 \cdot MHz$) $I_{bb}(H^{12}C^{14}N) = 18.937 \times 10^{-40}$ $g \cdot cm^2$ and $I_{bb}(D^{12}C^{14}N) = 23.178 \times 10^{-40}$ $g \cdot cm^2$. These numbers with Eq.(4-187) can be used to solve for the internuclear distances giving $r_{CH} = 1.066 \times 10^{-8}$ cm and $r_{CN} = 1.157 \times 10^{-8}$ cm. The observed rotational constants and structures of several additional linear molecules are listed in Table 4-16.

Static Electric and Magnetic Fields

We shall now examine the effects of electric and magnetic fields on the rotational energy levels of linear molecules. We start by expanding the energy of a system (molecule) in an external field in a power series around the zero field. The same series

Table 4-16 The rotational constants and molecular structure of several linear molecules where $\hbar/4\pi = 8.39215 \times 10^{-35}$ $g \cdot cm^2 \cdot MHz$. The internuclear distances (in angstrom units) are given from left to right. The electric dipole moments are also listed.

Molecule	B (MHz)	r_{12} (Å)	r_{23}	r_{34}	r_{45}	Electric Dipole Moment (10^{-18} SC · cm) (Absolute Values)
$^{12}C—^{16}O$	57,897.5	1.128				0.10
$^{6}Li—^{127}I$	15,381.5	2.392				6.25
$^{133}Cs—^{35}Cl$	2,161.2	2.906				10.40
$H—^{12}C—^{14}N$	44,315.8	1.066	1.157			3.00
$^{16}O—^{12}C—^{32}S$	6,081.5	1.161	1.561			0.71
$^{14}N—^{14}N—^{16}O$	12,561.6	1.126	1.191			0.17
$^{19}F—^{12}C—^{14}N$	10,554.8	1.262	1.159			2.17
$^{19}F—^{12}C{\equiv}^{12}C—H$	9,706.2	1.279	1.198	1.053		0.73
$H—^{12}C{\equiv}^{12}C—^{12}C{\equiv}^{14}N$	4,549.1	1.058	1.205	1.378	1.159	3.60

expansion is appropriate for either *electric* or *magnetic* fields. The classical energy of a molecule, W, in the presence of an electric field (E) can be expressed by a Taylor series expansion about the zero field,

$$W = W^0 + \sum_\alpha \left(\frac{\partial W}{\partial E_\alpha}\right)_{E_\alpha=0} E_\alpha + \frac{1}{2}\sum_{\alpha,\beta}\left(\frac{\partial^2 W}{\partial E_\alpha\,\partial E_\beta}\right)_{\substack{E_\alpha=0\\E_\beta=0}} E_\alpha E_\beta + \cdots, \tag{4-191}$$

where W^0 is the zero-field energy and the sums in α and β are over the three Cartesian laboratory axes, x, y, and z. The negative first and second derivatives of the energy with respect to the electric field are called the *electric dipole moment* and *electric polarizability*, respectively:

Electric dipole moment

$$D_\alpha = -\left(\frac{\partial W}{\partial E_\alpha}\right)_0 \tag{4-192}$$

Electric polarizability

$$\alpha_{\alpha\beta} = -\left(\frac{\partial^2 W}{\partial E_\alpha\,\partial E_\beta}\right)_{\substack{0\\0}}. \tag{4-193}$$

Substituting gives

$$\begin{aligned} W &= W^0 - \sum_\alpha D_\alpha E_\alpha - \frac{1}{2}\sum_{\alpha,\beta} E_\alpha \alpha_{\alpha\beta} E_\beta + \cdots \\ &= W^0 - \mathbf{D}\cdot\mathbf{E} - \frac{1}{2}\mathbf{E}\cdot\boldsymbol{\alpha}\cdot\mathbf{E} + \cdots. \end{aligned} \tag{4-194}$$

$\boldsymbol{D}$ is the electric dipole moment vector (the values of the electric dipole moments of several linear molecules are listed in Table 4-16). $\boldsymbol{\alpha}$ is a symmetric polarizability tensor that has the same tensorial form as the moment of inertia tensor. A detailed theory of electric polarizability is given in Section 6.10. To obtain the discrete rotational states in the presence of a field, we consider W in Eq.(4-194) as the Hamiltonian operator for the system and we then solve Schrödinger's equation.

Consider first the $-\boldsymbol{D}\cdot\boldsymbol{E}$ term in Eq.(4-194) as a perturbation, $\mathscr{H}_1$. Both $\boldsymbol{D}$ and $\boldsymbol{E}$ are defined in the laboratory-fixed x, y, and z axis system. The laboratory axes projections of $\boldsymbol{D}$ are related to the components of D in the principal inertial axes a, b, and c (molecular rotating axis system) by the direction cosine transformation that we write as [see Eqs.(B-27) and (B-28)]

$$\begin{pmatrix} D_x \\ D_y \\ D_z \end{pmatrix} = \begin{pmatrix} C_{xa} & C_{xb} & C_{xc} \\ C_{ya} & C_{yb} & C_{yc} \\ C_{za} & C_{zb} & C_{zc} \end{pmatrix}\begin{pmatrix} D_a \\ D_b \\ D_c \end{pmatrix} = \mathbf{C}\begin{pmatrix} D_a \\ D_b \\ D_c \end{pmatrix}$$

$$(D_x \quad D_y \quad D_z) = (D_a \quad D_b \quad D_c)\begin{pmatrix} C_{ax} & C_{ay} & C_{az} \\ C_{bx} & C_{by} & C_{bz} \\ C_{cx} & C_{cy} & C_{cz} \end{pmatrix} = (D_a \quad D_b \quad D_c)\tilde{\mathbf{C}}. \tag{4-195}$$

C_{za} is the cosine of the angle between the a and z axes with similar definitions for all other elements in the $\mathbf{C}$ matrix.

In summary, we write the linear field perturbation as

$$\mathscr{H}_1 = -\boldsymbol{D}(abc)\tilde{\mathbf{C}}\cdot\boldsymbol{E} = -(D_a \quad D_b \quad D_c)\begin{pmatrix} C_{ax} & C_{ay} & C_{az} \\ C_{bx} & C_{by} & C_{bz} \\ C_{cx} & C_{cy} & C_{cz} \end{pmatrix}\begin{pmatrix} E_x \\ E_y \\ E_z \end{pmatrix}. \tag{4-196}$$

In the case of a linear molecule with the nuclei along the a axis, $D_b = D_c = 0.0$. Thus, if only a single E_z electric field is present, the Hamiltonian is

$$\mathscr{H}_1 = -D_a E_z C_{az} = -D_a E_z \cos\theta_{az} = -D_a E_z \cos\theta, \tag{4-197}$$

where θ_{az}, the angle between the a molecular-fixed axis and the z laboratory-fixed axis is equal to θ, the spherical polar angle (see Fig. 4-7). The negative sign always appears in the dipole external-field interactions from the classical consideration that a dipole in a field is in the highest energy state when the dipole vector ($+\longrightarrow-$ or $N\longrightarrow S$) opposes the external-field vector. This leads to $\theta = \pi$ and $\cos\theta = -1$ giving DE. The system is at lowest energy when the dipole and the field are aligned. This gives $\theta = 0$ and $\cos\theta = 1$ leading to $-DE$ and lower energy than the opposed vector value of DE.

We now need the matrix representation of the perturbation Hamiltonian, $\mathscr{H}_1$, in the linear molecule spherical harmonic basis,

$$\begin{aligned}\langle J, M|\mathscr{H}_1|J', M'\rangle &= \int_0^{2\pi}\int_0^{\pi} Y_{JM}^* \mathscr{H}_1 Y_{J'M'} \sin\theta\, d\theta\, d\phi \\ &= -D_a E_z \langle J, M|\cos\theta|J', M'\rangle. \end{aligned} \tag{4-198}$$

It is easy to show that these perturbation matrix elements are diagonal in M_J by noting that there is no ϕ-dependence in $\mathscr{H}_1$. By direct substitution from Table 4-3 for the spherical harmonics we can write

$$\begin{aligned}\langle 0, 0|\mathscr{H}_1|1, 0\rangle &= -2\left(\frac{\pi}{3}\right)^{1/2} D_a E_z \int_0^{2\pi}\int_0^{\pi} Y_{00}^* Y_{10} Y_{10} \sin\theta\, d\theta\, d\phi \\ &= -\left(\frac{1}{3}\right)^{1/2} D_a E_z \\ \langle 1, 0|\mathscr{H}_1|2, 0\rangle &= -\frac{2}{(15)^{1/2}} D_a E_z \\ &\vdots \\ \langle J, M|\mathscr{H}_1|J+1, M\rangle &= -2\left(\frac{\pi}{3}\right)^{1/2} D_a E_z \langle J, M|Y_{10}|J+1, M\rangle \\ &= -\frac{[(J+1)^2 - M^2]^{1/2}}{[(2J+1)(2J+3)]^{1/2}} D_a E_z. \end{aligned} \tag{4-199}$$

The matrix representation of the complete Hamiltonian in the spherical harmonic basis is given by

$$\mathscr{H}^{Y} = \mathscr{H}_0^{Y} + \mathscr{H}_1^{Y} =$$

$$
\begin{array}{cc|c|ccc|ccccc}
 & J & 0 & & 1 & & & & 2 & & \\
 & M & 0 & -1 & 0 & 1 & -2 & -1 & 0 & 1 & 2 \\
J & M & & & & & & & & & \\
\hline
0 & 0 & 0 & 0 & -\frac{D_a E_z}{(3)^{1/2}} & 0 & 0 & 0 & 0 & 0 & 0 \quad \cdots \\
1 & -1 & 0 & \frac{\hbar^2}{I} & 0 & 0 & 0 & -\frac{D_a E_z}{(5)^{1/2}} & 0 & 0 & 0 \\
1 & 0 & -\frac{D_a E_z}{(3)^{1/2}} & 0 & \frac{\hbar^2}{I} & 0 & 0 & 0 & -\frac{2D_a E_z}{(15)^{1/2}} & 0 & 0 \\
1 & 1 & 0 & 0 & 0 & \frac{\hbar^2}{I} & 0 & 0 & 0 & -\frac{D_a E_z}{(5)^{1/2}} & 0 \\
2 & -2 & 0 & 0 & 0 & 0 & \frac{3\hbar^2}{I} & 0 & 0 & 0 & 0 \\
2 & -1 & 0 & -\frac{D_a E_z}{(5)^{1/2}} & 0 & 0 & 0 & \frac{3\hbar^2}{I} & 0 & 0 & 0 \\
2 & 0 & 0 & 0 & -\frac{2D_a E_z}{(15)^{1/2}} & 0 & 0 & 0 & \frac{3\hbar^2}{I} & 0 & 0 \quad \cdots \\
2 & 1 & 0 & 0 & 0 & -\frac{D_a E_z}{(5)^{1/2}} & 0 & 0 & 0 & \frac{3\hbar^2}{I} & 0 \\
2 & 2 & 0 & 0 & 0 & 0 & 0 & 0 & 0 & 0 & \frac{3\hbar^2}{I} \\
 & & & \vdots & & \vdots & & & & & \ddots
\end{array}
\tag{4-200}
$$

The eigenvalues can be obtained from Eq.(4-200) by matrix diagonalization. We can also extract the eigenvalues by perturbation theory, however, where we note from the matrix above that there is no first-order correction. Using perturbation theory we can write the energy of the lowest two levels, correct to second order, as

$$E(J = 0, M = 0) = \frac{|\langle 0, 0|\mathscr{H}_1|1, 0\rangle|^2}{0.0 - (\hbar^2/I)} = -\frac{D_a^2 E_z^2}{3\hbar^2/I}$$

$$E(J = 1, M = 0) = \frac{\hbar^2}{I} + \frac{|\langle 0, 0|\mathscr{H}_1|1, 0\rangle|^2}{\hbar^2/I} + \frac{|\langle 1, 0|\mathscr{H}_1|2, 0\rangle|^2}{(\hbar^2/I) - (3\hbar^2/I)} = \frac{\hbar^2}{I} + \frac{3D_a^2 E_z^2}{15\hbar^2/I}. \tag{4-201}$$

The difference between the $M = 0$ energies of the lowest two states is

$$E(J = 1, M = 0) - E(J = 0, M = 0) = h\nu = \frac{\hbar^2}{I} + \frac{8D_a^2E_z^2 I}{15\hbar^2},$$

and the frequency of the transition is

$$\nu(J = 0, M = 0 \longrightarrow J = 1, M = 0) = 2B + \frac{4D_a^2E_z^2}{15h^2B}, \tag{4-202}$$

where the rotational constant B is in units of hertz (Hz), as defined in Eq.(4-190). Referring to our previous example of HCN in Table 4-15, we can compute the magnitude of the E_z^2 term for the $J = 0, M = 0 \longrightarrow J = 1, M = 0$ transition. Using the electric dipole moment in HCN of $D_a = 3.0 \times 10^{-18}$ SC · cm, the rotational constant of $B = 44{,}315.99$ MHz, and a typical laboratory electric field of $E = 3000$ V · cm^{-1} = 10 SV · cm^{-1} gives $\nu(J = 0, M = 0 \longrightarrow J = 1, M = 0) = (88{,}631.6 + 123.4)$ MHz, where 88,631.6 is the zero-field frequency. Half widths at half height in normal low-pressure conditions for microwave transitions near 90,000 MHz are about 1.0 MHz, so an E-field shift in frequency of 123 MHz is easily measured. The electric field perturbation above, sometimes referred to as the Stark effect, has been used extensively to measure the square of the molecular electric dipole moments in linear molecules; some representative results are listed in Table 4-16.

Returning to Eq.(4-200), we note that there are no diagonal elements in the $\mathscr{H}_1^Y$ matrix and the second-order terms have M^2-dependence. Thus, the M degeneracy is not broken due to the Stark effect. It is also evident that only $\Delta J = \pm 1$, $\Delta M = 0$ matrix elements are nonzero, which couples each (J, M) state to the above $(J + 1, M)$ and below $(J - 1, M)$ states. General second-order perturbation expressions are easily derived for the Stark effect on any $J \neq 0$ level in a linear molecule giving

$$E_{J,M}^{(2)} = \frac{D_a^2E^2}{2hBJ(J + 1)}\left[\frac{J(J + 1) - 3M^2}{(2J - 1)(2J + 3)}\right]. \tag{4-203}$$

We now return to the second term in Eq.(4-194) involving the molecular polarizability tensor, $\boldsymbol{\alpha}$. We use again the direction cosine matrix in Eq.(4-195) to relate the laboratory space-fixed polarizability tensor $\boldsymbol{\alpha}_l$ (xyz axes) to the molecular-fixed tensor $\boldsymbol{\alpha}_m$ (abc axes) as shown in Eq.(B-35),

$$\mathscr{H}_2 = -\tfrac{1}{2}\mathbf{E} \cdot \boldsymbol{\alpha}_l \cdot \mathbf{E} = -\tfrac{1}{2}\mathbf{E} \cdot \mathbf{C}\boldsymbol{\alpha}_m\tilde{\mathbf{C}} \cdot \mathbf{E}. \tag{4-204}$$

The polarizability tensor for a linear molecule in the principal inertial axis system is diagonal with values α_{aa} and $\alpha_{bb} = \alpha_{cc}$ along the diagonal. Using this and a single nonzero laboratory field, E_z, gives

$$\mathscr{H}_2 = -\frac{E_z^2}{2}(\mathrm{C}_{az}^2\alpha_{aa} + \mathrm{C}_{bz}^2\alpha_{bb} + \mathrm{C}_{cz}^2\alpha_{cc}) = -\frac{E_z^2}{2}(\mathrm{C}_{az}^2\alpha_{aa} + \alpha_{bb}(\mathrm{C}_{bz}^2 + \mathrm{C}_{cz}^2)]. \tag{4-205}$$

We can now relate C_{az}^2, C_{bz}^2, and C_{cz}^2 to the spherical polar coordinates with Fig. 4-24. An ABC linear molecule has its a axis passing through the nuclei. As long as $\alpha_{bb} = \alpha_{cc}$, the value of $\mathscr{H}_2$ in Eq.(4-205) will be independent of orientation of the c and b axes about the a axis. Thus, we rotate the molecular b axis into the az plane and the c axis is now in the xy plane with no loss in generality. As a result, we have $\mathrm{C}_{az} = \cos\theta$,

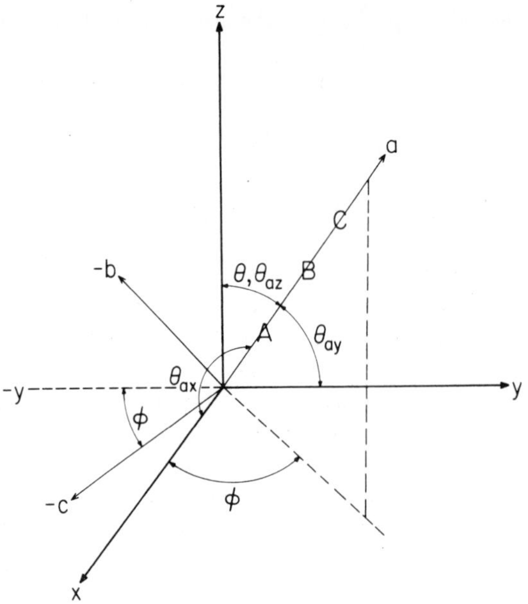

Figure 4-24 Diagram showing the principal inertial axis system for the A–B–C molecule and the angles between the principal inertial axes and the space-fixed axes. The normal spherical polar angles θ and ϕ are shown and the transformation between the xyz and abc axis systems is discussed in the text when for cylindrically symmetric systems we place the a, z, and b axes in the same plane with no loss in generality.

$C_{bz} = -\sin\theta$, and $C_{cz} = 0$, where θ is the spherical polar angle. Substituting these results into Eq.(4-205) and using the average polarizability,

$$\alpha = \tfrac{1}{3}(\alpha_{aa} + \alpha_{bb} + \alpha_{cc}), \tag{4-206}$$

gives

$$\begin{aligned} \mathscr{H}_2 &= -\frac{E_z^2}{6}[(\alpha_{aa} - \alpha_{bb})(3\cos^2\theta - 1)] - \frac{1}{2}E_z^2\alpha \\ &= -\frac{2}{3}\left(\frac{\pi}{5}\right)^{1/2} E_z^2(\alpha_{aa} - \alpha_{bb})Y_{20} - \frac{1}{2}E_z^2\alpha, \end{aligned} \tag{4-207}$$

where we have used Table 4-3 for the spherical harmonics in the last step. Careful examination of Fig. 4-24 gives the entire **C** matrix for a cylindrically symmetric molecule. This matrix, which will be used several other places in the text, is given by

$$\begin{aligned} (x \quad y \quad z) &= (a \quad b \quad c)\tilde{\mathbf{C}} = (a \quad b \quad c)\begin{pmatrix} C_{ax} & C_{ay} & C_{az} \\ C_{bx} & C_{by} & C_{bz} \\ C_{cx} & C_{cy} & C_{cz} \end{pmatrix} \\ &= (a \quad b \quad c)\begin{pmatrix} \sin\theta\cos\phi & \sin\theta\sin\phi & \cos\theta \\ \cos\theta\cos\phi & \cos\theta\sin\phi & -\sin\theta \\ -\sin\phi & \cos\phi & 0 \end{pmatrix}, \end{aligned} \tag{4-208}$$

where we note from this equation, or Fig. 4-24, that $x \longrightarrow b$, $y \longrightarrow c$, and $z \longrightarrow a$ when $\theta = 0$ and $\phi = 0$. The matrix elements of $\mathscr{H}_2$ in the spherical harmonic basis are given by

$$\langle J, M | \mathscr{H}_2 | J', M' \rangle = -\frac{2}{3}\left(\frac{\pi}{5}\right)^{1/2} E_z^2(\alpha_{aa} - \alpha_{bb})\langle J, M | Y_{20} | J', M' \rangle - \frac{E_z^2}{2}\alpha\langle J, M | J', M' \rangle. \tag{4-209}$$

Substituting the functions and integrating shows that $\langle JM | J'M' \rangle = \delta_{JJ'}\delta_{MM'}$ and the Y_{20} matrix elements are nonzero when $\Delta M = M - M' = 0$ and $\Delta J = J - J' = 0$ and ± 2. The first-order or $\Delta J = 0$ correction to the energy will certainly be dominant at low electric fields. Thus, we shall consider only the first-order contribution in the following discussion. The $\langle J, M | Y_{20} | J, M \rangle$ matrix elements are easily computed by direct integration and using the resultant

$$\langle J, M | Y_{20} | J, M \rangle = -\left(\frac{5}{4\pi}\right)^{1/2}\left[\frac{3M^2 - J(J+1)}{(2J-1)(2J+3)}\right]$$

gives the first-order energy correction for a linear molecule:

$$E_{J,M}^{(1)} = \left(\frac{1}{3}\right)E_z^2\left[\frac{3M^2 - J(J+1)}{(2J-1)(2J+3)}\right](\alpha_{aa} - \alpha_{bb}) - \frac{1}{2}E_z^2\alpha. \tag{4-210}$$

The second-order correction involves the $M = M'$, $J' = J \pm 2$ terms in Eq.(4-209) and will lead to E^4-dependence in the final energy expression. We can now combine the energies in Eqs.(4-203) and (4-210) to give the first- and second-order corrections to the energy carried through order E^2 for a linear molecule. The result is

$$E(J, M) = hBJ(J+1) - E_z^2\left[\frac{3M^2 - J(J+1)}{(2J-1)(2J+3)}\right] \times \left[\frac{D_a^2}{2hBJ(J+1)} - \frac{1}{3}(\alpha_{aa} - \alpha_{bb})\right] - \frac{1}{2}E_z^2\alpha. \tag{4-211}$$

There are no linear terms in the electric field. We shall now evaluate the magnitudes of the terms in Eq.(4-211) for a typical molecule and electric field. Choose the OCS molecule where $|D_a| = 0.71 \times 10^{-18}$ SC $\cdot$ cm, $\alpha_{aa} - \alpha_{bb} = 4.7 \times 10^{-24}$ cm^3, and $B = 6081.5$ MHz. Using an electric field of $E = 3000$ V $\cdot$ cm^{-1} $= 10$ SV $\cdot$ cm^{-1} gives (the energy is in cgs units)

$$E(J, M) = J(J+1) \times 4.03 \times 10^{-17} - \left[\frac{3M^2 - J(J+1)}{(2J-1)(2J+3)}\right] \times \left[\frac{6.25 \times 10^{-19}}{J(J+1)} - 1.6 \times 10^{-24}\right] - \frac{1}{2}E_z^2\alpha.$$

The second term involving the electric dipole moment and molecular polarizability is substantially smaller than the zero-order energy. Also, the term involving the polarizability is smaller than the dipole-dependent term. By studying the high J energy

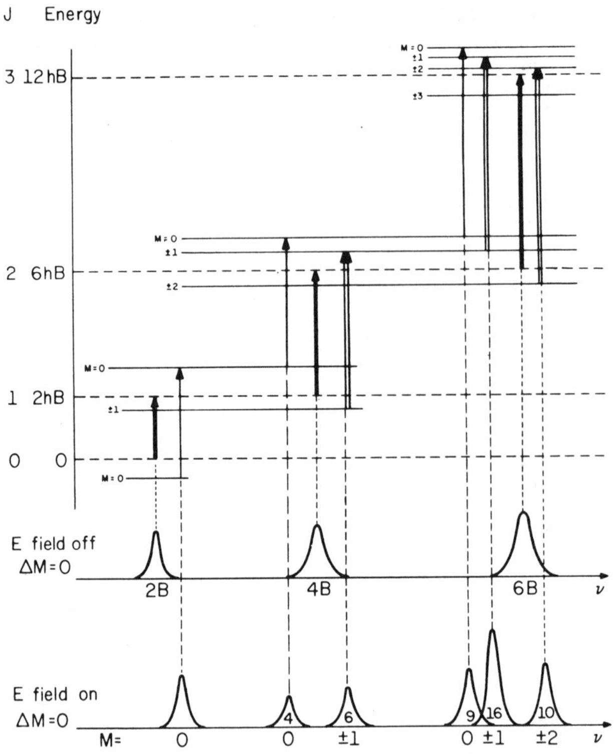

Figure 4-25 The energy levels and electric dipole transitions of a linear molecule with rotational constant B in the presence of an external electric field. The $\Delta M = 0$ selection rules arise from parallel static and electromagnetic electric fields. The energy levels are given in Eq.(4-211) and the relative intensities for the $\Delta M = 0$ transitions are given by squaring the appropriate matrix elements in Eq.(4-212). The relative intensities are indicated by the numbers superimposed on the line shapes.

levels, however, the dipole-dependence decreases due to the inverse $J(J + 1)$ term and by studying the rotational transitions (where the $-\frac{1}{2}E_z^2\alpha$ term cancels), both the polarizability anisotropy, $(\alpha_{aa} - \alpha_{bb})$, and the electric dipole moment, D_a, can be measured.

Rotational transitions in a linear molecule are observed according to the electric dipole moment selection rules and must be the same as in the hydrogen atom as summarized in Table 4-6. The normal radiation source for molecular rotational spectroscopy is an electronic oscillator that generates plane-polarized radiation. When the static field, E_z, is parallel to the radiation field, D_z, $\Delta J = \pm 1$, $\Delta M = 0$ selection rules are appropriate and the nonzero transition moments are given by

$$\langle J, M | D_z | J - 1, M \rangle = D_a \left[\frac{J^2 - M^2}{(2J - 1)(2J + 1)} \right]^{1/2}$$

$$\langle J, M | D_z | J + 1, M \rangle = D_a \left[\frac{(J + 1)^2 - M^2}{(2J + 1)(2J + 3)} \right]^{1/2}. \tag{4-212}$$

These matrix elements are squared and substituted into Eq.(3-176) to give the probability of a transition.

The energy levels and electric dipole $\Delta J = \pm 1$, $\Delta M = 0$ rotational transitions for a linear molecule at zero external field and in the presence of a static external electric field are shown in Fig. 4-25. The plane-polarized radiation field is parallel to the static field. The energy levels are computed from Eq.(4-211) and the relative intensities are obtained by squaring the appropriate matrix elements in Eq.(4-212). Careful measurements by microwave spectroscopy [22] of the frequencies of the transitions sketched in Fig. 4-25 will allow the measurement of the electric dipole moment and electric polarizability anisotropy; several values of D_a and $\alpha_{aa} - \alpha_{bb}$ are listed in Table 4-17. The much weaker $\Delta J = 0$, $\Delta M = \pm 1$ transitions can also be observed by using molecular beam electric resonance methods [23].

We shall now examine the effects of a magnetic field on the rotational energy levels of a linear molecule. Returning to Eqs.(4-191) to (4-194), we note by analogy that the energy of a molecule in a *magnetic field*, $\boldsymbol{H}$, is given by

$$W = W^0 - \boldsymbol{\mu} \cdot \boldsymbol{H} - \tfrac{1}{2}\boldsymbol{H} \cdot \boldsymbol{\chi} \cdot \boldsymbol{H} + \cdots,$$

Magnetic dipole moment

$$\mu_\alpha = -\left(\frac{\partial W}{\partial H_\alpha}\right)_0,$$

Magnetic susceptibility

$$\chi_{\alpha\beta} = -\left(\frac{\partial^2 W}{\partial H_\alpha \, \partial H_\beta}\right)_0, \qquad (4\text{-}213)$$

0

where W is the appropriate Hamiltonian operator.

First, we shall consider the $-\boldsymbol{\mu} \cdot \boldsymbol{H}$ term in Eq.(4-213) for a rotating molecule with zero electronic angular momentum. Previously we have noted that a magnetic moment is proportional to an angular momentum. Thus, in a linear molecule we shall assume that the rotational magnetic moment can be written as

$$\boldsymbol{\mu}_J = g \frac{\mu_0}{\hbar} \boldsymbol{J}, \qquad (4\text{-}214)$$

where $\boldsymbol{\mu}_J$ is the molecular magnetic moment for the Jth rotational state with angular momentum $\boldsymbol{J}$, and g is the proportionality constant referred to as the molecular g-value. μ_0 is the nuclear magneton defined in Eq.(4-155). Equation (4-214) is analogous to Eq.(4-124) for the electron spin magnetic moment ($\boldsymbol{\mu}_S = -g_e(\mu_B/\hbar)\boldsymbol{S}$) where g_e is the value of the intrinsic electronic spin g-value and μ_B is the Bohr magneton. Substituting Eq.(4-214) into the $-\boldsymbol{\mu} \cdot \boldsymbol{H}$ term in Eq.(4-213) for a static magnetic field along the z axis, H_z, gives

$$\mathscr{H}_1 = -(\mu_J)_z H_z = -\frac{\mu_0}{\hbar} g J_z H_z. \qquad (4\text{-}215)$$

Table 4-17 Electric and magnetic parameters in several linear molecules that were determined by measuring the effect on the rotational transitions in the molecules in an electric or magnetic field. The *a* axis is parallel to the internuclear axis.

Molecule	Electric Dipole Moment[a] $\lvert D \rvert$ (10^{-18} SC · cm)	Electric Polarizability Anisotropy,[b] $\alpha_{aa} - \alpha_{bb}$ (10^{-24} cm^3)	Perpendicular Molecular g-Value[c] g_{bb}	Magnetic Susceptibility Anisotropy[c] $\chi_{aa} - \chi_{bb}$ (10^{-29} erg · G^{-2})
H—H	0	0.93	0.88291 ± 0.00007	$+0.092 \pm 0.004$
^{35}ClF	0.88 ± 0.01	—	-0.102 ± 0.035	-3.00 ± 0.30
^{15}N^{15}N	0	2.4	-0.2593 ± 0.0004	-1.42 ± 0.08
^{15}N^{15}N^{16}O	0.17 ± 0.01	3.22 ± 0.05	-0.07606 ± 0.0001	-1.68 ± 0.16
OCO	0	2.0	-0.05508 ± 0.00005	-1.05 ± 0.03
OC^{32}S	0.7152 ± 0.0002	4.67 ± 0.16	-0.028711 ± 0.00004	-1.54 ± 0.03
OC^{80}Se	0.75 ± 0.01	—	-0.01952 ± 0.0002	-1.67 ± 0.03
H—C≡C—F	0.73 ± 0.01	3.3	-0.0077 ± 0.0002	-0.86 ± 0.02

[a] The signs of the electric dipole moments cannot be measured by the methods above and are not given here. We shall discuss the measurement of these signs later. The numerical values and references for the electric dipole moments are given in Ref. [22].

[b] The values for NNO and OCS are from L. H. Scharpen, J. S. Muenter, and V. W. Laurie, *J. Chem. Phys.* **53**, 2513 (1970). The remaining values are estimated from other sources (see Section 6.10).

[c] The magnetic parameters for linear molecules are from W. H. Flygare and R. C. Benson, *Mol. Phys.* **20**, 225 (1971) and W. H. Flygare, *Chem. Rev.* **74**, 653 (1974).

As Y_{JM} is an eigenfunction of the J_z operator, the energy correction is first order and the Hamiltonian matrix remains diagonal:

$$E_{J,M}^{(1)} = -\frac{g\mu_0}{\hbar} H_z \langle J, M | J_z | J, M \rangle = -g\mu_0 H_z M. \qquad (4\text{-}216)$$

Thus, the perturbation removes the M degeneracy. Typical molecular g-values for linear molecules are listed in Table 4-17 and we note that they are usually less than 1.0. Thus, the first-order corrections according to Eq.(4-216) usually lead to molecular M splittings, which are smaller than nuclear M splittings (see Section 4.10).

The second term in Eq.(4-213) involves the effects of the molecular magnetic susceptibility tensor, $\boldsymbol{\chi}$. In the case of a linear molecule, we note that the tensorial forms of $\boldsymbol{\chi}$ and $\boldsymbol{\alpha}$ are identical in symmetry. Thus, we can transfer the arguments from Eqs.(4-204) to (4-211) involving the $\boldsymbol{\alpha}$ tensor to the $\boldsymbol{\chi}$ term in Eq.(4-213) involving the magnetic susceptibility tensor. The resultant energy of a rigid linear molecule in a magnetic field is therefore given by

$$E(J, M) = hBJ(J+1) - g_{bb}\mu_0 H_z M + \frac{H^2}{3}\left[\frac{3M^2 - J(J+1)}{(2J-1)(2J+3)}\right](\chi_{aa} - \chi_{bb}) - \frac{1}{2} H_z^2 \chi.$$

$$(4\text{-}217)$$

a is the symmetry axis and g_{bb} is the perpendicular g-value equal to g in the previous equations (see Problem 4-30 for further explanation). The average magnetic susceptibility is $\chi = \frac{1}{3}(\chi_{aa} + \chi_{bb} + \chi_{cc})$, which is analogous to $\alpha = \frac{1}{3}(\alpha_{aa} + \alpha_{bb} + \alpha_{cc})$.

We shall now consider the relative orders of magnitude in Eq.(4-217) for a typical laboratory magnetic field of $H_z = 10{,}000$ G. We shall again choose the OCS molecule with the molecular g-value of $g_{bb} = -0.0287$ and the magnetic susceptibility anisotropy of $\chi_{aa} - \chi_{bb} = -1.54 \times 10^{-29}$ erg $\cdot$ G^{-2} (see Table 4-17). Substituting these values into Eq.(4-217) gives (in cgs units)

$$E(J, M) = J(J+1) \times 4.03 \times 10^{-17} + M \times 1.45 \times 10^{-21}$$

$$- \left[\frac{3M^2 - J(J+1)}{(2J-1)(2J+1)}\right](5.1 \times 10^{-22}) - \frac{1}{2} H_z^2 \chi. \qquad (4\text{-}218)$$

It is evident from this equation (for low J values) that the magnitude of the quadratic field term is as large as the linear field term for 10,000 G. Comparing Eq.(4-217) for a magnetic field with the corresponding electric field result [Eq.(4-211)] shows that the magnetic susceptibility anisotropy $(\chi_{aa} - \chi_{bb})$ should be easier to measure than the corresponding polarizability anisotropy $(\alpha_{aa} - \alpha_{bb})$. The observed values of g_{bb} and $\chi_{bb} - \chi_{aa}$ are listed in Table 4-17 for several molecules.

The energy levels and electric dipole transitions for both $\Delta J = \pm 1$, $\Delta M = 0$ and $\Delta J = \pm 1$, $\Delta M = \pm 1$ transitions for a linear molecule in the presence of a static external magnetic field are shown in Fig. 4-26. The energy levels are computed from Eq.(4-217). If the static magnetic field is parallel to the electric field of the radiation, the $\Delta J = \pm 1$, $\Delta M = 0$ selection rules are appropriate with relative intensities from

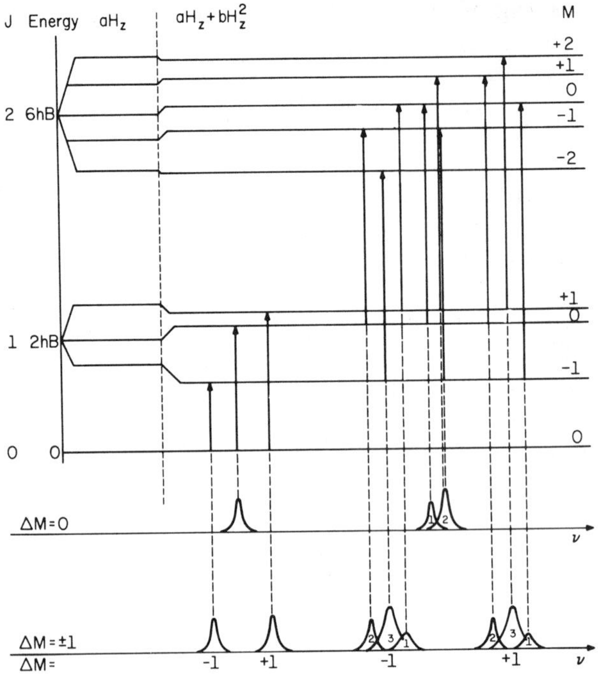

Figure 4-26 The energy levels and electric dipole transitions in a linear molecule with rotational constant B in the presence of an external magnetic field. The diagram shows first the linear field Zeeman effect and then the linear plus quadratic Zeeman effect from the left to right. Negative values of g_{bb} and $\chi_{aa} - \chi_{bb}$ are used for this illustration. The $\Delta M = 0$ electric dipole selection rules arise from a static magnetic field that is parallel to the electric field of the electromagnetic radiation. The $\Delta M = \pm 1$ selection rules arise from perpendicular fields. The relative intensities are indicated by the numbers superimposed on the line profiles.

Eq.(4-212). If the static magnetic field is perpendicular to the electric field of the radiation, the $\Delta J = \pm 1$, $\Delta M = \pm 1$ selection rules are appropriate. Classically we note that if the fields are aligned, the radiation field has no effect on the static field orientation of the molecules and there will be no tendency to change the molecular orientation or M state. If the radiation field is perpendicular to the static field, the radiation field will apply a torque to the molecule tending to change its orientation or corresponding M state.

In the absence of a permanent electric dipole moment in a linear molecule, we can still observe magnetic dipole moment transitions. If the molecule has unpaired electrons (O_2, for instance) or if the electrons in the molecule have a nonzero component of orbital angular momentum, the molecule will possess a magnetic moment on the order of a Bohr magneton. The magnetic dipole selection rules for the rigid rotor are also given in Table 4-6. Further analysis will require the coupling of the electronic angular momentum to the rotational angular momentum, however, and the corresponding coupled case must be considered in detail [16].

In the case of rotational excitation of the electronic angular momentum in a diamagnetic molecule ($L = S = 0$), the magnetic moments will be on the order of a nuclear magneton and the selection rules are obtained again directly from Table 4-6. The $\Delta J = 0$, $\Delta M = \pm 1$ rotational magnetic moment reorientation transitions have been observed in several molecules (see Problem 4-31).

Nonlinear Molecules

We shall now generalize the preceding discussion on the linear molecule to the more general cases of nonlinear polyatomic molecules. Referring back to Fig. 4-7 we can see that if the molecule contained another atom off the diatomic internuclear line, an additional angle, χ, would be required in addition to θ and ϕ to describe all orientations of the molecule in space. Thus, the eigenfunction for a nonlinear rigid rotor is a function of three angles, $f(\theta, \phi, \chi)$. In general, the Hamiltonian will have the form shown in Eq.(2-32), and the Schrödinger equation takes the form

$$\mathscr{H} f(\theta, \phi, \chi) = \left(\frac{J_a^2}{2\mathsf{I}_{aa}} + \frac{J_b^2}{2\mathsf{I}_{bb}} + \frac{J_c^2}{2\mathsf{I}_{cc}}\right) f(\theta, \phi, \chi) = E f(\theta, \phi, \chi), \qquad (4\text{-}219)$$

where the a, b, and c axes represent the coordinates in the *principal* inertial axis system, that is, in the c.m. axis system where the inertial tensor is diagonal. The principal inertial axes are labeled a, b, and c according to the order of $\mathsf{I}_{aa} \le \mathsf{I}_{bb} \le \mathsf{I}_{cc}$. The classification of molecules according to linear, spherical, prolate and oblate symmetric, and asymmetric is discussed in Section 2.1 and several examples of this classification are shown in Fig. 2-2.

We shall now solve Eq.(4-219) by use of the matrix representations of the rotational angular momentum operators. First, we shall find the matrix elements of J_a, J_b, and J_c in some *convenient* basis. Then we can compute the matrix elements of J_a^2, J_b^2, and J_c^2 giving the following matrix equation (we assume that I_{aa}, I_{bb}, and I_{cc} are constants):

$$\frac{1}{2\mathsf{I}_{aa}} \mathbf{J}_a^2 + \frac{1}{2\mathsf{I}_{bb}} \mathbf{J}_b^2 + \frac{1}{2\mathsf{I}_{cc}} \mathbf{J}_c^2 = \mathscr{H}_R. \qquad (4\text{-}220)$$

This matrix can then be diagonalized to give the energy levels of the rigid rotor, $\mathbf{u}^\dagger \mathscr{H}_R \mathbf{u} = \mathbf{E}$.

An appropriate basis is suggested by the fact that J_a, J_b, and J_c are the projections of the total angular momentum vector, $\boldsymbol{J}$, on the principal axes. $\boldsymbol{J}$ can also be referred to the space-fixed system. The square of the total angular momentum is independent of the axis system:

$$J^2 = J_a^2 + J_b^2 + J_c^2 = J_x^2 + J_y^2 + J_z^2.$$

The angular momentum components in the two axis systems are related by the direction cosines according to [see Eq.(4-208)]

$$(J_x \quad J_y \quad J_z) = (J_a \quad J_b \quad J_c)\tilde{\mathbf{C}}, \qquad (J_a \quad J_b \quad J_c) = (J_x \quad J_y \quad J_z)\mathbf{C}. \quad (4\text{-}221)$$

There are three permutations of xyz with respect to abc that lead to different starting points for the axes systems:

CASE I. $a \longrightarrow x, b \longrightarrow y, c \longrightarrow z$ (oblate)

$$\tilde{\mathbf{C}} = \begin{pmatrix} 1 & 0 & 0 \\ 0 & 1 & 0 \\ 0 & 0 & 1 \end{pmatrix} \tag{4-222a}$$

CASE II. $c \longrightarrow x, a \longrightarrow y, b \longrightarrow z$

$$\tilde{\mathbf{C}} = \begin{pmatrix} 0 & 1 & 0 \\ 0 & 0 & 1 \\ 1 & 0 & 0 \end{pmatrix} \tag{4-222b}$$

CASE III. $b \longrightarrow x, c \longrightarrow y, a \longrightarrow z$ (prolate)

$$\tilde{\mathbf{C}} = \begin{pmatrix} 0 & 0 & 1 \\ 1 & 0 & 0 \\ 0 & 1 & 0 \end{pmatrix}. \tag{4-222c}$$

Case III is discussed in Fig. 4-24 and Eq.(4-208) for the cylindrically symmetric system. We shall now examine the commutation properties of **C** with ***J*** for Case I.

Previously (Section 4.5), we examined the commutation properties of J_x, J_y, and J_z [see Eq.(4-72)]. The commutation properties of J_a, J_b, and J_c can also be derived by using Eq.(4-221):

$$\begin{aligned}[J_a, J_b] &= J_aJ_b - J_bJ_a = (J_x\mathrm{C}_{xa} + J_y\mathrm{C}_{ya} + J_z\mathrm{C}_{za})(J_x\mathrm{C}_{xb} + J_y\mathrm{C}_{yb} + J_z\mathrm{C}_{zb}) \\ &\quad - (J_x\mathrm{C}_{xb} + J_y\mathrm{C}_{yb} + J_z\mathrm{C}_{zb})(J_x\mathrm{C}_{xa} + J_y\mathrm{C}_{ya} + J_z\mathrm{C}_{za}).\end{aligned}$$

In order to proceed we need the commutators of the angular momentum operators with the direction cosines. These commutators, which are the same as the commutators of any vector and its components with the angular momentum, are given by [24]

$$\begin{gathered}[J_x, \mathrm{C}_{yg}] = -[J_y, \mathrm{C}_{xg}] = i\hbar\mathrm{C}_{zg}, \qquad [J_x, \mathrm{C}_{xg}] = 0, \\ g = a, b, \text{ or } c, \text{ and cyclic permutations of } x, y, \text{ and } z. \\ [J_a, \mathrm{C}_{fb}] = -[J_b, \mathrm{C}_{fa}] = -i\hbar\mathrm{C}_{fc}, \qquad [J_a, \mathrm{C}_{fa}] = 0, \\ f = x, y, \text{ or } z, \text{ and cyclic permutations of } a, b, \text{ and } c.\end{gathered} \tag{4-223}$$

Substituting these commutators into the expression for $[J_a, J_b]$ above gives $[J_a, J_b] = -i\hbar J_c$ [24], which shows that the commutators of the angular momentum components in the molecular-fixed axis system have the opposite signs of the commutators of the angular momentum components in the space-fixed axis system [Eq.(4-72)]. The components of the angular momentum in any axis system will commute with J^2.

Thus, the commutation properties for the angular momentum can be summarized by

$$
\begin{array}{llll}
[J_x, J_y] = i\hbar J_z & [J_x, J^2] = 0 & [J_a, J_b] = -i\hbar J_c & [J_a, J^2] = 0 \\
[J_y, J_z] = i\hbar J_x & [J_y, J^2] = 0 & [J_b, J_c] = -i\hbar J_a & [J_b, J^2] = 0 \\
[J_z, J_x] = i\hbar J_y & [J_z, J^2] = 0 & [J_c, J_a] = -i\hbar J_b & [J_c, J^2] = 0
\end{array}
$$

$$[J_g, J_\alpha] = 0 \qquad g = x, y, z \quad \text{and} \quad \alpha = a, b, c. \tag{4-224}$$

It is evident from these commutation properties that J^2 and any component of $\boldsymbol{J}$ in the space-fixed and molecular-fixed axes form a set of three mutually commuting operators; choose (J^2, J_z, J_c). We shall now find the eigenfunctions of this set of mutually commuting operators. In the case of the spherical harmonics, we note that $\exp(iM\phi)$ is an eigenfunction of J_z with eigenvalue $\hbar M$. Thus, there must also be an equivalent eigenvalue of J_z in the more general case described by $f(\theta, \phi, \chi)$ in Eq.(4-219). Furthermore, there must also be a corresponding eigenvalue of J_c that is equal to $\hbar K$; this requires the K-dependence to be described by $\exp(iK\chi)$ with the integral K-values having the same limits as the integral M-values. In summary, J^2, J_z, and J_c are three mutually commuting operators that can share the same eigenfunctions labeled by J, K, M with properties given by

$$
\begin{aligned}
J^2 f(\theta, \phi, \chi) &= \hbar^2 J(J+1) f(\theta, \phi, \chi) \\
J_z f(\theta, \phi, \chi) &= \hbar M f(\theta, \phi, \chi) \\
J_c f(\theta, \phi, \chi) &= \hbar K f(\theta, \phi, \chi) \\
M &= -J, -J+1, -J+2, \ldots, 0, 1, 2, \ldots, J \\
K &= -J, -J+1, -J+2, \ldots, 0, 1, 2, \ldots, J.
\end{aligned}
$$

Thus, the two independent axis systems yield the two independent M and K substates. Furthermore, we expect the matrix elements of J_x and J_y in the $f(\theta, \phi, \chi)$ basis to be described by the same equations as obtained in Eq.(4-74) for the spherical harmonics. The matrix elements of J_a and J_b are obtained by analogy with Eq.(4-74) according to the opposite signs for the commutators in Eqs.(4-224). The resulting nonzero matrix elements are

$$
\begin{aligned}
\langle J, K, M | J_x | J, K, M \pm 1 \rangle &= \frac{\hbar}{2} [J(J+1) - M(M \pm 1)]^{1/2} \\
\langle J, K, M | J_y | J, K, M \pm 1 \rangle &= \pm \frac{i\hbar}{2} [J(J+1) - M(M \pm 1)]^{1/2} \\
\langle J, K, M | J_z | J, K, M \rangle &= \hbar M \\
\langle J, K, M | J_a | J, K \pm 1, M \rangle &= \frac{\hbar}{2} [J(J+1) - K(K \pm 1)]^{1/2} \\
\langle J, K, M | J_b | J, K \pm 1, M \rangle &= \mp \frac{i\hbar}{2} [J(J+1) - K(K \pm 1)]^{1/2} \\
\langle J, K, M | J_c | J, K, M \rangle &= \hbar K.
\end{aligned} \tag{4-225}
$$

Note the different sign on the J_b matrix element relative to the J_y matrix element. This sign change is forced by the opposite signs in the commutators in Eqs.(4-224) for the molecular-fixed axes relative to the space-fixed axes.

Returning to our task of writing Eq.(4-220), we shall now compute the matrix elements of J_a^2, J_b^2, and J_c^2 using the results in Eq.(4-225). Remembering that the matrix representations of the operators multiply according to the same order as the operators themselves, we can write

$$\langle J, K, M | J_a^2 | J', K', M' \rangle = \sum_{J'', M'', K''} \langle J, K, M | J_a | J'', K'', M'' \rangle \times \langle J'', K'', M'' | J_a | J', K', M' \rangle.$$

As J_a is diagonal in J and M, $J'' = J = J'$ and $M'' = M' = M$ in this summation. The summing process over K'' is easily carried out with the nonzero terms being given in Eq.(4-225). The nonzero matrix elements of J_a^2, J_b^2, and J_c^2 are given by

Diagonal elements

$$\langle J, K, M | J_a^2 | J, K, M \rangle = \langle J, K, M | J_b^2 | J, K, M \rangle = \frac{\hbar^2}{2} [J(J+1) - K^2]$$

$$\langle J, K, M | J_c^2 | J, K, M \rangle = \hbar^2 K^2$$

Off-diagonal elements

$$\begin{aligned} \langle J, K, M | J_a^2 | J, K \pm 2, M \rangle &= -\langle J, K, M | J_b^2 | J, K \pm 2, M \rangle \\ &= \frac{\hbar^2}{4} [J(J+1) - K(K \pm 1)]^{1/2} \\ &\quad \times [J(J+1) - (K \pm 1)(K \pm 2)]^{1/2}. \end{aligned} \tag{4-226}$$

Substituting Eqs.(4-226) into Eq.(4-220) gives the Hamiltonian matrix for the rotation of any type of molecule. The matrix elements in $\mathscr{H}_R$ are given by

Diagonal

$$\langle J, K, M | \mathscr{H}_R | J, K, M \rangle = \frac{h(A+B)}{2} [J(J+1) - K^2] + hK^2C$$

Off-diagonal

$$\begin{aligned} \langle J, K, M | \mathscr{H}_R | J, K \pm 2, M \rangle &= \frac{h}{4} (A - B)[J(J+1) - K(K \pm 1)]^{1/2} \\ &\quad \times [J(J+1) - (K \pm 1)(K \pm 2)]^{1/2}, \end{aligned} \tag{4-227}$$

where the rotational constants are expressed in hertz according to Eq.(4-190), $A = (\hbar/4\pi \mathrm{I}_{aa})$, etc.

Equation (4-227) is appropriate for an oblate symmetric rotor where $\mathrm{I}_{aa} = \mathrm{I}_{bb} < \mathrm{I}_{cc}$ and the symmetry axis for the symmetric rotor lies along the c axis. For linear and prolate symmetric rotor molecules, however, the a axis is the symmetry axis and it is convenient to switch from Case I to Case III in Eq.(4-220) by permuting the original

a, b, c axes to b, c, a. The result is to replace, in Eq.(4-227), A with B, B with C, and C with A to give

Diagonal

$$\langle J, K, M | \mathscr{H}_R | J, K, M \rangle = \frac{h(B + C)}{2} [J(J + 1) - K^2] + hK^2 A$$

Off-diagonal

$$\langle J, K, M | \mathscr{H}_R | J, K \pm 2, M \rangle = \frac{h}{4}(B - C)[J(J + 1) - K(K \pm 1)]^{1/2}$$
$$\times [J(J + 1) - (K \pm 1)(K \pm 2)]^{1/2}. \quad (4\text{-}228)$$

The permutation of the a, b, and c axes is easily made in Eqs.(4-227) and (4-228) because the changes in phases in the commutation expressions caused by the interchange of the axes will not affect the energy expression.

In the case of a linear molecule, $I_{bb} = I_{cc}$ and thus $B = C$ and the off-diagonal elements in Eq.(4-228) are zero. $I_{aa} = 0$ and $A = \infty$, which also requires K to be zero. Thus, with the a axis along the internuclear line, Eqs.(4-228) reduce to the correct linear molecule result.

In the case of a spherical-type rotor such as CH_4, SF_6, and others, $A = B = C$ and Eqs.(4-227) or (4-228) reduce to the linear molecular form. The three remaining types of molecules (oblate and prolate symmetric and asymmetric) are considered separately below.

PROLATE: a is the symmetry axis, $I_{aa} < I_{bb} = I_{cc}$, $A > B = C$.

The off-diagonal elements in Eqs.(4-228) are zero leading to the following energy:

$$E(J, K, M) = h[BJ(J + 1) + (A - B)K^2]. \quad (4\text{-}229)$$

The energy levels from Eq.(4-229) are shown on the left side in Fig. 4-27. The energy differences between the $\Delta J = +1$, $\Delta K = 0$, $\Delta M = 0$ levels in the prolate symmetric rotor are identical to those in the linear molecule. The moments of inertia and rotational constants for several prolate symmetric rotors are listed in Table 4-18 (see FCH_3 in Fig. 2-2).

It is apparent that most molecules fall under the prolate symmetric or near-prolate symmetric rotor category. This is primarily because of the light hydrogen atoms that are farthest from the c.m. in most molecules. Consider now the oblate symmetric rotor.

OBLATE: c is the symmetry axis, $I_{aa} = I_{bb} < I_{cc}$, $A = B > C$.

When $A = B$ in Eqs.(4-227), the off-diagonal elements are zero, giving

$$E(J, K, M) = h[BJ(J + 1) + (C - B)K^2]. \quad (4\text{-}230)$$

The energy levels for the oblate symmetric rotor are shown on the right side in Fig. 4-27. Again, the oblate symmetric rotor energy level differences ($\Delta J = +1$, $\Delta K = 0$, $\Delta M = 0$) are identical to a linear molecule. The moments of inertia and rotational

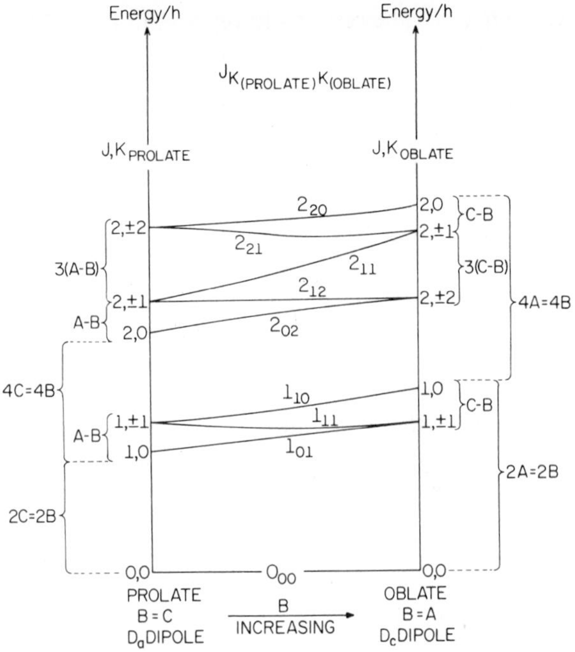

Figure 4-27 The rotational energy levels of prolate and oblate symmetric rotors. The correlation between the prolate and oblate limits contain the energy levels of the asymmetric rotors indicated by J_{ij} where i is the prolate K and j is the oblate K.

constants for several oblate symmetric rotors are listed in Table 4-19 (see HCF_3 in Fig. 2-2).

The rotational selection rules for symmetric rotors can be obtained from the dipole selection rules as applied to the space-fixed moments in Eq.(4-195). When D_a is the only nonzero moment (prolate rotor), the space-fixed components are given by

$$D_x = D_a C_{ax}, \qquad D_y = D_a C_{ay}, \qquad D_z = D_a C_{az}. \tag{4-231}$$

The direction cosine matrix elements in the *symmetric rotor basis* are listed in Table 4-20 [22] in a form that requires multiplying three components together. It is evident

Table 4-18 Moments of inertia and rotational constants of some typical prolate ($I_{aa} < I_{bb} = I_{cc}$) symmetric rotors. The units for the moments are 10^{-40} g · cm² and the units for the rotational constants are megahertz. The conversion factor is $\hbar/4\pi = 8.39215 \times 10^{-35}$ g · cm² · MHz.

	I_{aa}(symmetry axis)	A	$I_{bb} = I_{cc}$	B
FCH_3	5.300	158,319.0	32.864	25,536.12
$ClCH_3$	5.310	158,020.9	63.132	13,292.95
$BrCH_3$	5.320	157,723.8	87.708	9,568.19
ICH_3	5.351	156,839.0	111.876	7,501.31
$CH_3{-}CF_3$	154.424	5,434.5	161.854	5,185.00
$CH_3{-}SiH_3$	10.612	79,084.9	76.508	10,968.96
$CH_3{-}C{\equiv}C{-}H$	5.320	157,723.8	98.202	8,545.84

Table 4-19 Rotational constants and moments of inertia for some typical oblate ($I_{aa} = I_{bb} < I_{cc}$) symmetric rotors. The units and conversion factor are given in Table 4-18.

	$I_{aa} = I_{bb}$	$A = B$	I_{cc} (symmetry axis)	C
HCF_3	81.093	10,348.74	149.122	5,627.70
NF_3	78.571	10,680.96	87.333	9,609.37
PF_3	107.318	7,819.90	111.627	7,518.06

from Eqs.(4-231) and Table 4-20 that the selection rules on K are always $\Delta K = 0$. This is also evident classically as the value of K is not dependent on the orientation of the molecule with respect to the laboratory framework.

In addition to the $\Delta K = 0$ selection rule in symmetric rotors, it is evident from Table 4-20 and Eqs.(4-231) that $\Delta J = 0, \pm 1$, $\Delta M = 0, \pm 1$ selection rules are also possible. The resultant selection rules are summarized here:

$$\left.\begin{array}{lll} & D_z & \Delta J = 0, \pm 1; \quad \Delta M = 0 \\ D_\pm & \begin{cases} D_x & \Delta J = 0, \pm 1; \quad \Delta M = \pm 1 \\ D_y & \Delta J = 0, \pm 1; \quad \Delta M = \pm 1 \end{cases} \end{array}\right\} \Delta K = 0. \tag{4-232}$$

Symmetric rotor transitions can be viewed with the help of Fig. 4-27.

The direction cosine matrix elements in the spherical harmonic basis can also be obtained from Table 4-20 by setting $K = 0$. We can illustrate the use of the matrix elements of the direction cosines in the spherical harmonic basis by returning to Eq.(4-205). Using $C_{az}^2 + C_{bz}^2 + C_{cz}^2 = 1$ and Eq.(4-206) in Eq.(4-205) gives the following diagonal matrix element of $\mathscr{H}_2$ in the spherical harmonic basis:

$$\begin{aligned}\langle J, M|\mathscr{H}_2|J, M\rangle &= -\frac{E_z^2}{2}(\alpha_{aa} - \alpha_{bb})\left(\langle J, M|C_{az}^2|J, M\rangle - \frac{1}{3}\right) - \frac{E_z^2}{2}\alpha \\ &= -\frac{E_z^2}{2}(\alpha_{aa} - \alpha_{bb})\Big(|\langle J, M|C_{az}|J + 1, M\rangle|^2 \\ &\quad + |\langle J, M|C_{az}|J - 1, M\rangle|^2 - \frac{1}{3}\Big) - \frac{E_z^2}{2}\alpha.\end{aligned}$$

According to Table 4-20, when $K = 0$, we can write

$$\langle J, M|C_{az}|J - 1, M\rangle = \left[\frac{J^2 - M^2}{(2J + 1)(2J - 1)}\right]^{1/2},$$

$$\langle J, M|C_{az}|J + 1, M\rangle = \left[\frac{(J + 1)^2 - M^2}{(2J + 1)(2J + 3)}\right]^{1/2}.$$

Substituting these results into the expression for $\langle J, M|\mathscr{H}_2|J, M\rangle$ above gives (after some tedious algebra) the result in Eq.(4-210).

We can also use the direction cosine matrix elements in Table 4-20 to derive the matrix elements that describe the electric field perturbation in the symmetric rotor. For a prolate rotor with the electric dipole along the a axis and resultant perturbation Hamiltonian of $\mathscr{H}_1 = -D_a E_z C_{az}$, we can write the nonzero diagonal matrix elements

Table 4-20 The nonzero direction cosine matrix elements in the symmetric rotor basis, $g = a, b$, or c and $f = x, y$, or z. The matrix elements are appropriate for a prolate rotor [Case III in Eq.(4-222c)]. For an oblate rotor, the bca axes must be permuted to abc [Case I in Eq.(4-222a)]. The direction cosines in the spherical harmonic basis are obtained by setting $K = 0$.

$$\langle J, K, M | C_{gf} | J', K', M' \rangle = \langle J \| C_{gf} \| J' \rangle \langle J, K \| C_{gf} \| J', K' \rangle \langle J, M \| C_{gf} \| J', M' \rangle$$

	$J' = J + 1$	$J' = J$	$J' = J - 1$
$\langle J \| C_{gf} \| J' \rangle$	$\{4(J+1)[(2J+1)(2J+3)]^{1/2}\}^{-1}$	$[4J(J+1)]^{-1}$	$[4J(4J^2-1)^{1/2}]^{-1}$
$\langle J, K \| C_{af} \| J', K \rangle$	$2[(J+1)^2 - K^2]^{1/2}$	$2K$	$-2(J^2 - K^2)^{1/2}$
$\langle J, K \| C_{bf} \| J', K \pm 1 \rangle = \mp i \langle J, K \| C_{cf} \| J', K \pm 1 \rangle$	$\mp[(J \pm K + 1)(J \pm K + 2)]^{1/2}$	$[J(J+1) - K(K \pm 1)]^{1/2}$	$\pm[(J \mp K)(J \mp K - 1)]^{1/2}$
$\langle J, M \| C_{gz} \| J', M \rangle$	$2[(J+1)^2 - M^2]^{1/2}$	$2M$	$-2(J^2 - M^2)^{1/2}$
$\langle J, M \| C_{gx} \| J', M \pm 1 \rangle = \pm i \langle J, M \| C_{gy} \| J', M \pm 1 \rangle$	$\mp[(J \pm M + 1)(J \pm M + 2)]^{1/2}$	$[J(J+1) - (M \pm 1)]^{1/2}$	$\pm[(J \mp M)(J \mp M - 1)]^{1/2}$

of $\mathscr{H}_1$ that lead to first-order corrections to the energy as

$$\langle J, K, M | \mathscr{H}_1 | J, K, M \rangle = -D_a E_z \langle J, K, M | C_{az} | J, K, M \rangle = -\frac{D_a E_z MK}{J(J+1)}. \tag{4-233}$$

Thus, the energy correction appears in first order if $M \neq 0$ and $K \neq 0$, which leads to energy level changes which are linear in the electric field. These general methods can also be used to derive the perturbations of the electric or magnetic fields in symmetric tops beginning with the Hamiltonians in Eqs.(4-194) or (4-213) (see Problem 4-30). The results for asymmetric rotors are obtained by starting in the symmetric rotor basis.

Asymmetric Rotor Molecules

The asymmetric rotor energy levels lie between the prolate and oblate limits shown in Fig. 4-27. The energy levels are obtained directly by diagonalizing the matrix described by Eqs.(4-227) (near-oblate) or Eq.(4-228) (near-prolate) and the designation, $J_{K(\text{prolate})K(\text{oblate})}$, is shown in Fig. 4-27. A large number of asymmetric rotor molecules fall under the near-prolate symmetric rotor category and it is convenient to think in terms of the symmetric rotor limits. A few examples of near-prolate symmetric rotor molecules are listed in Table 4-21.

Equations (4-228) are appropriate for constructing the rigid rotor asymmetric rotor Hamiltonian matrix for the near-prolate molecules in Table 4-21. The nonzero values in this matrix for the lower values of J and K are given by

$$\frac{\mathscr{H}_R}{h} = \begin{array}{cc|c|ccc|ccccc}
 & J & 0 & & 1 & & & & 2 & & \\
 & K & 0 & -1 & 0 & 1 & -2 & -1 & 0 & 1 & 2 \\
J & K & & & & & & & & & \\
\hline
0 & 0 & 0 & 0 & 0 & 0 & 0 & 0 & 0 & 0 & 0 \\
1 & -1 & 0 & \alpha_1 & 0 & \gamma_1 & 0 & 0 & 0 & 0 & 0 \\
1 & 0 & 0 & 0 & \alpha_2 & 0 & 0 & 0 & 0 & 0 & 0 \\
1 & 1 & 0 & \gamma_1 & 0 & \alpha_1 & 0 & 0 & 0 & 0 & 0 \\
2 & -2 & 0 & 0 & 0 & 0 & \alpha_3 & 0 & \gamma_2 & 0 & 0 \\
2 & -1 & 0 & 0 & 0 & 0 & 0 & \alpha_4 & 0 & \gamma_3 & 0 \\
2 & 0 & 0 & 0 & 0 & 0 & \gamma_2 & 0 & \alpha_5 & 0 & \gamma_2 \\
2 & 1 & 0 & 0 & 0 & 0 & 0 & \gamma_3 & 0 & \alpha_4 & 0 \\
2 & 2 & 0 & 0 & 0 & 0 & 0 & 0 & \gamma_2 & 0 & \alpha_3 \\
 & & & & \vdots & & & & & & \ddots
\end{array} \cdots$$

$$\alpha_1 = A + \tfrac{1}{2}(B + C) \qquad \gamma_1 = \tfrac{1}{2}(B - C)$$

$$\alpha_2 = B + C \qquad \gamma_2 = \frac{(6)^{1/2}}{2}(B - C)$$

$$\alpha_3 = 4A + B + C \qquad \gamma_3 = \tfrac{3}{2}(B - C)$$

$$\alpha_4 = A + \tfrac{5}{2}(B + C)$$

$$\alpha_5 = 3(B + C). \tag{4-234}$$

Table 4-21 Near-prolate ($A > B \cong C$) symmetric rotors that have a-axis electric dipole moments. The rotational constants are in units of megahertz.

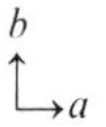	A	B	C
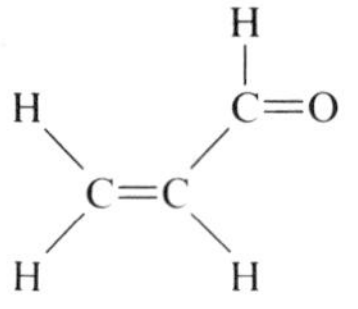	47,353.3	4,659.44	4,242.79
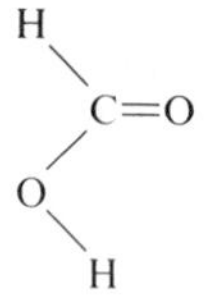	77,510.7	12,055.0	10,416.2
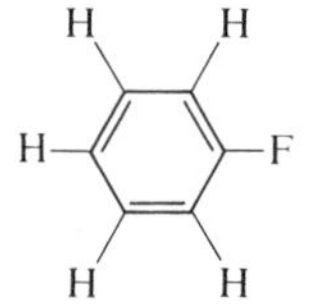	5,663.5	2,570.64	1,767.94
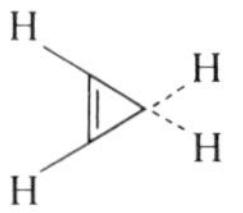	30,063.7	21,825.6	13,795.7
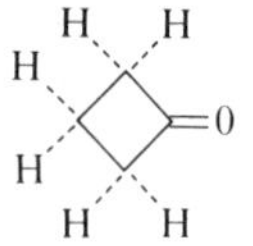	10,785.2	4,806.7	3,558.8

The J blocks are indicated with parentheses. The eigenvalues of Eq.(4-234) are obtained by matrix diagonalization, $\boldsymbol{E}_R = \boldsymbol{u}^\dagger \mathscr{H}_R \boldsymbol{u}$. Analytical solutions to Eq.(4-234) are easily determined for the lower levels of the asymmetric rotor and some results are listed in Table 4-22 [25]. Also listed in Table 4-22 are the prolate rotor limits for the energies obtained by $A > B = C$. The K states for the prolate rotor limit are the first subscript numbers in the $J_{K(\text{prolate})K(\text{oblate})}$ notation in the left-hand column, as also shown in Fig. 4-27. In either the prolate or oblate rotor limit the symmetric rotor selection rules are valid. In an asymmetric rotor, all three principal inertial axis dipole moments may be nonzero. The dipole moment operator in the space-fixed axis is given by Eq.(4-195) where D_a, D_b, and D_c may all be nonzero. The procedure for computing the selection rules is to evaluate the matrix elements of the $D_f(f = x, y, z)$ operators in the symmetric rotor basis using Table 4-20. The $\mathbf{D}_x$, $\mathbf{D}_y$, and $\mathbf{D}_z$

Table 4-22 Rigid rotor energies and transition frequencies for a near-prolate asymmetric rotor in terms of the rotational constants obtained by diagonalizing $\mathscr{H}_R$ in Eq.(4-234). The energies for the prolate rotor limit are also listed. The lower part of the table shows transition frequencies for the low J, $\Delta J = +1$ series of transitions in a near-prolate D_a dipole asymmetric rotor.

Level $J_{K(\text{prolate})K(\text{oblate})}$	*Energy/h*	$A > B = C$ Prolate Rotor
0_{00}	0	0
1_{01}	$B + C$	$2B$
1_{11}	$A + C$	$B + A$
1_{10}	$B + A$	
2_{02}	$2A + 2B + 2C - 2[(B - C)^2 + (A - C)(A - B)]^{1/2}$	$6B$
2_{12}	$A + B + 4C$	$5B + A$
2_{11}	$A + 4B + C$	
2_{21}	$4A + B + C$	$2B + 4A$
2_{20}	$2A + 2B + 2C + 2[(B - C)^2 + (A - C)(A - B)]^{1/2}$	
3_{03}	$2A + 5B + 5C - 2[4(B - C)^2 + (A - B)(A - C)]^{1/2}$	$12B$
3_{13}	$5A + 2B + 5C - 2[4(A - C)^2 - (A - B)(B - C)]^{1/2}$	$11B + A$
3_{12}	$5A + 5B + 2C - 2[4(A - B)^2 + (A - C)(B - C)]^{1/2}$	
3_{22}	$4A + 4B + 4C$	$8B + 4A$
3_{21}	$2A + 5B + 5C + 2[4(B - C)^2 + (A - B)(A - C)]^{1/2}$	
3_{31}	$5A + 2B + 5C + 2[4(A - C)^2 - (A - B)(B - C)]^{1/2}$	$3B + 9A$
3_{30}	$5A + 5B + 2C + 2[4(A - B)^2 + (A - C)(B - C)]^{1/2}$	

Transitions	Frequencies
$0_{00} \longrightarrow 1_{01}$	$B + C$
$1_{11} \longrightarrow 2_{12}$	$B + 3C$
$1_{01} \longrightarrow 2_{02}$	$2A + B + C - 2[(B - C)^2 + (A - C)(A - B)]^{1/2}$
$1_{10} \longrightarrow 2_{11}$	$3B + C$
$2_{12} \longrightarrow 3_{13}$	$4A + B + C - 2[4(A - C)^2 - (A - B)(B - C)]^{1/2}$
$2_{02} \longrightarrow 3_{03}$	$3(B + C) - 2[4(B - C)^2 + (A - B)(A - C)]^{1/2} + 2[(B - C)^2 + (A - B)(A - C)]^{1/2}$
$2_{21} \longrightarrow 3_{22}$	$3(B + C)$
$2_{20} \longrightarrow 3_{21}$	$3(B + C) + 2[4(B - C)^2 + (A - B)(A - C)]^{1/2} - 2[(B - C)^2 + (A - B)(A - C)]^{1/2}$
$2_{11} \longrightarrow 3_{12}$	$4A + B + C - 2[4(A - B)^2 + (A - C)(B - C)]^{1/2}$

transition matrices in the symmetric rotor basis are then transformed to the same basis that diagonalized the Hamiltonian matrix in the symmetric rotor basis. It is evident that the off-diagonal coupling in K in the Hamiltonian matrix will mix K states causing the symmetric top K degeneracy to split as shown in Fig. 4-27. These $\Delta K = 0$ transitions for the near-prolate system are obtained from the energy levels in Table 4-22 by taking $\Delta J = +1$ differences in the $\Delta K(\text{prolate}) = 0$ levels. Some of the transitions and corresponding frequencies are also listed in Table 4-22. The low J near-prolate $\Delta K_{\text{prolate}} = 0$ transitions are illustrated in Fig. 4-28 as compared to the corresponding transitions in linear molecules and prolate symmetric rotors.

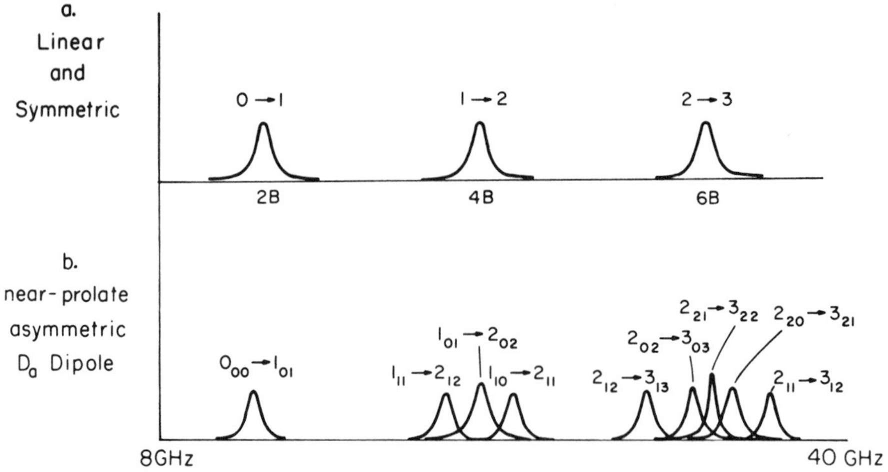

Figure 4-28 a. Typical $J \longrightarrow J + 1$ transitions in linear molecules and prolate symmetric rotors with $B \cong 5000$ MHz. b. $\Delta J = +1$ transitions in a near-prolate asymmetric rotor with $B \cong C \cong 5000$ MHz. See also the energy level diagram in Fig. 4-27 for the symmetric and asymmetric molecules.

Note the dominant $\Delta K_{\text{prolate}} = 0$ selection rules in these transitions. Thus, the symmetric rotor—rigid rotor spectra that overlaps the linear molecule is split (according to Fig. 4-28b) in the slightly asymmetric rotor. Note also that there are several checks and symmetries that appear in this asymmetric rotor spectrum. A few of the obvious symmetries are

$$\nu(0_{00} \longrightarrow 1_{01}) = B + C = \tfrac{1}{3}\nu(2_{21} \longrightarrow 3_{22})$$
$$= \tfrac{1}{4}[\nu(1_{11} \longrightarrow 2_{12}) + \nu(1_{10} \longrightarrow 2_{11})].$$

Also, it is evident that the $2_{02} \longrightarrow 3_{03}$, $2_{21} \longrightarrow 3_{22}$, and $2_{20} \longrightarrow 3_{21}$ transitions form a perfect triplet with separations of

$$\Delta\nu = 2[4(B - C)^2 + (A - B)(A - C)]^{1/2} - 2[(B - C)^2 + (A - B)(A - C)]^{1/2}.$$

As the molecules listed in Table 4-21 are near-prolate rotors ($B \cong C$), the $\Delta J = +1$ spectra will be grouped with separations approximately $2B$ as shown in Fig. 4-28b. Many near-prolate rotors also have a D_b dipole with approximate $\Delta J = +1$ separations equal to $A + C$. Near-oblate rotors ($B \cong A$) also show similar patterns that are easily obtained as shown here for near-prolate rotors.

A more general analysis of the selection rules in an asymmetric rotor can be obtained by the use of group theory (see Appendix E). The eigenfunctions of the asymmetric rotor designated by $J_{K(\text{prolate}),\,K(\text{oblate})}$ generate the irreducible representations of the D_2 group shown in Table 4-23. The representations generated by the electric dipole moments, the eigenfunctions (designated by $J_{\text{even–odd}}$), and the direction cosines are listed. The dipole and direction cosine matrix representations are generated

Table 4-23 The D_2 symmetry group showing the irreducible representations generated by the molecular-fixed D_a, D_b, and D_c dipole operators; the direction cosines ($f = x, y, z$); and the asymmetric rotor eigenfunctions. C_{2a}, C_{2b}, and C_{2c} are rotations through π about the three principal inertial axes. The right-hand column shows the irreducible representations formed from several direct products of interest. The lower part of the table shows the decomposition of the reducible representations formed from the symmetric rotor basis functions.

D_2		E	C_{2a}	C_{2b}	C_{2c}	
J_{ee}	A	1	1	1	1	$(J_{KK'})^2$
J_{eo}, C_{af}, D_a	B_1	1	1	-1	-1	$J_{ee} \times J_{eo}, J_{oo} \times J_{oe}$
J_{oo}, C_{bf}, D_b	B_2	1	-1	1	-1	$J_{ee} \times J_{oo}, J_{eo} \times J_{oe}$
J_{oe}, C_{cf}, D_c	B_3	1	-1	-1	1	$J_{ee} \times J_{oe}, J_{oo} \times J_{eo}$
$J = 0$		1	1	1	1	A
$J = 1$		3	-1	-1	-1	$B_1 + B_2 + B_3$
$J = 2$		5	1	1	1	$2A + B_1 + B_2 + B_3$
$J = 3$		7	-1	-1	-1	$A + 2B_1 + 2B_2 + 2B_3$
⋮		⋮				⋮

according to the techniques developed in Appendix E. The symmetry of the asymmetric rotor functions and the representations generated with these functions require further comment.

It is evident from Eq.(4-219) that the rotational Hamiltonian has a symmetry determined by the principal moments of inertia. It is also evident that the appropriate symmetry group that describes the symmetry of the moments of inertia is D_2 and that $\mathcal{H}_R$ generates the completely symmetric irreducible representation of D_2. It may be convenient to think of the symmetry in terms of a moment of inertia ellipsoid with the dimensions of the ellipsoid being proportional to I_{aa}^{-1}, I_{bb}^{-1}, and I_{cc}^{-1}. This ellipsoid has D_2 symmetry for an asymmetric rotor and $D_{\infty h}$ symmetry for a symmetric rotor. The moment of inertia ellipsoid for an oblate symmetric rotor looks like a pancake where the c axis is perpendicular to the pancake plane and the ellipsoid for a prolate rotor looks like a cigar with the a axis along the long axis of the cigar.

We shall now generate the reducible representations for the symmetry operators in the D_2 group in the symmetric rotor basis. Remembering that the M states represent molecule laboratory projections of $\boldsymbol{J}$, we note that the M states will not be mixed or involved with the symmetry operations. Thus, we shall look at the representations generated by the J, K components of the symmetric rotor rotational wavefunction. In the absence of external perturbations, the rotational energy level is $2K + 1$ degenerate, which arises in the $\exp(i\chi K)$ dependence in the wavefunction [see discussion following Eq.(4-224)] where χ is the angle around the symmetry axis of the symmetric rotor. Consider now the matrix representation of a rotation operator C_α in the $\exp(i\chi K)$ basis where $K = J, J-1, \ldots, 0, -1, -2, \ldots, -J$ and α is an undetermined angle of rotation about the symmetry axis of the asymmetric rotor. Operation on $\exp(i\chi K)$ with C_α gives

$$C_\alpha \exp(i\chi K) = \exp(i\chi K)\exp(-i\alpha K). \tag{4-235}$$

Rewriting this equation in matrix form for a given value of J gives

$$C_\alpha \begin{pmatrix} \exp(iJ\chi) \\ \exp[i(J-1)\chi] \\ \vdots \\ 1 \\ \vdots \\ \exp(-iJ\chi) \end{pmatrix} =$$

$$\begin{pmatrix} \exp(-iJ\alpha) & 0 & & & \\ 0 & \exp[-i(J-1)\alpha] & & & \\ & & \ddots & & \\ & & & 1 & \\ & & & & \ddots \\ & & & & & \exp(iJ\alpha) \end{pmatrix} \begin{pmatrix} \exp(iJ\chi) \\ \exp[i(J-1)\chi] \\ \vdots \\ 1 \\ \vdots \\ \exp(-iJ\chi) \end{pmatrix}. \tag{4-236}$$

The diagonal sum or character of the representation is

$$\begin{aligned} \Gamma(C_\alpha) &= \exp(-iJ\alpha) + \exp[-i(J-1)\alpha] + \exp[-i(J-2)\alpha] \\ &\quad + \cdots + 1 + \cdots + \exp(iJ\alpha) \\ &= \exp(-iJ\alpha)\sum_{k=0}^{2J}[\exp(i\alpha)]^k = \frac{\exp[i(J+\frac{1}{2})\alpha] - \exp[-i(J+\frac{1}{2})\alpha]}{\exp(i\alpha/2) - \exp(-i\alpha/2)} \\ &= \frac{\sin(J+\frac{1}{2})\alpha}{\sin(\alpha/2)}. \end{aligned} \tag{4-237}$$

$\Gamma(C_\alpha)$ is the character of the irreducible representation for C_α under the $D_{\infty h}$ group that is appropriate for the rotational wavefunction for a symmetric rotor. In the case of the lower symmetry asymmetric rotor inertial ellipsoid, the representations formed by C_π about the a, b, and c axes with the symmetric rotor functions can be obtained in reducible form under D_2 from $\Gamma(C_\alpha)$ in Eq.(4-237). The characters for $\alpha = \pi$ are easily determined from Eq.(4-237) to be

$$\Gamma(C_\pi) = (-1)^J. \tag{4-238}$$

Thus, remembering that the character corresponding to E is $(2J+1)$ according to the K degeneracy, we can now decompose the reducible representations under the D_2 character table as shown also in Table 4-23.

The desired symmetry of the J_{ee}, J_{eo}, J_{oo}, and J_{oe} asymmetric rotor levels are obtained by comparing the decomposition above with Fig. 4-27. It is evident from the decomposition of $J = 0$ that J_{ee} transforms according to A. We can use the break-down of $\Gamma(C_\alpha)$ under $D_{\infty h}$ and its correlation with D_2 (along with Fig. 4-27) to show the

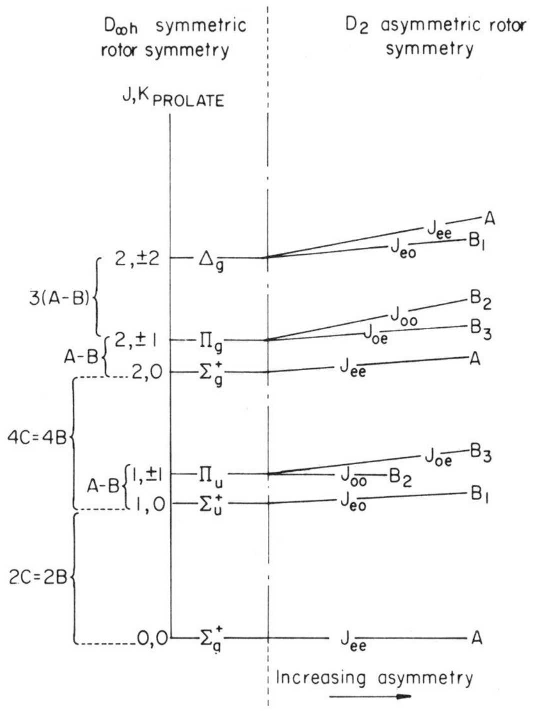

Figure 4-29 Correlation diagram showing the symmetry species of the symmetric rotor functions under $D_{\infty h}$ symmetry being correlated with the D_2 symmetry species of the asymmetric rotor. A direct comparison among this diagram, Table 4-23, and Fig. 4-27 will tie things together.

remaining symmetry species of J_{eo}, J_{oo}, and J_{oe} as shown in Table 4-23. This correlation is shown in Fig. 4-29.

According to the results in Appendix E.5, the product of initial and final rotational state eigenfunctions with the direction cosines in Eq.(4-231) must contain A symmetry under D_2 in order for the transition to be allowed. The irreducible representations formed from the $J_{KK'}$ products are shown in Table 4-23. These products are easily coupled with the direction cosines of the appropriate symmetry to give the selection rules listed in Table 4-24.

Table 4-24 The selection rules for an asymmetric rotor. Even (e) and odd (o) refer to the prolate and oblate K quantum numbers, respectively (see Fig. 4-27). The selection rules on J and M are $\Delta J = 0, \pm 1$ and $\Delta M = 0, \pm 1$ and the selection rules on the K_p and K_o quantum numbers are given here.

D_a dipole	D_b dipole	D_c dipole
$ee \longleftrightarrow eo$	$oo \longleftrightarrow ee$	$ee \longleftrightarrow oe$
$oo \longleftrightarrow oe$	$eo \longleftrightarrow oe$	$eo \longleftrightarrow oo$

4.12 THE HARMONIC OSCILLATOR AND NORMAL COORDINATES

In Section 2.3 we discussed the classical harmonic oscillator that arose from the potential energy obtained from Hooke's law for the restoring force of a particle with mass m connected to a solid wall with a massless spring. The kinetic and potential energies are

$$T = \frac{1}{2} m\dot{s}^2 = \frac{1}{2m}(m\dot{s})^2 = \frac{p_s^2}{2m}, \qquad V = \frac{1}{2} ks^2, \tag{4-239}$$

where s is the displacement coordinate (see Fig. 2-3), k is the force constant, and p_s represents the momentum of the particle. Replacing m with μ, the reduced mass for the diatomic molecule, we write Schrödinger's equation for the diatomic harmonic oscillator:

$$\left(-\frac{\hbar^2}{2\mu}\frac{d^2}{ds^2} + \frac{1}{2}ks^2\right)\psi(s) = E\psi(s), \tag{4-240}$$

where $\psi(s)$ is the eigenfunction in argument s with eigenvalue E. The solution to this differential equation and generation of the Hermite polynomials by the series method is discussed in detail in Appendix C.3. The normalized eigenfunctions are [9]

$$\psi_n(s) = \left[\frac{\gamma}{(\pi)^{1/2}2^n n!}\right]^{1/2} H_n(\gamma s)\exp\left(-\frac{\gamma^2 s^2}{2}\right)$$

$$\gamma = \left[\frac{(\mu k)^{1/2}}{\hbar}\right]^{1/2} \qquad n = 0, 1, 2, 3, \ldots. \tag{4-241}$$

γ is a natural parameter for the harmonic oscillator with inverse length dimensions. $H_n(\gamma s)$ are the Hermite polynomials in argument γs. The first few $\psi_n(s)$ functions are given in Table (4-25).

Table 4-25 Harmonic oscillator wavefunctions, $\psi_n(s)$, for the lower states.

$$\psi_0 = \frac{\gamma^{1/2}}{\pi^{1/4}}\exp\left(-\frac{\gamma^2 s^2}{2}\right)$$

$$\psi_1 = \frac{(2\gamma)^{1/2}}{\pi^{1/4}}\gamma s\exp\left(-\frac{\gamma^2 s^2}{2}\right)$$

$$\psi_2 = \left(\frac{1}{\pi^{1/4}}\right)\left(\frac{\gamma}{2}\right)^{1/2}(-1 + 2\gamma^2 s^2)\exp\left(-\frac{\gamma^2 s^2}{2}\right)$$

$$\psi_3 = \frac{(3\gamma)^{1/2}}{\pi^{1/4}}\left(-\gamma s + \frac{2\gamma^3 s^3}{3}\right)\exp\left(-\frac{\gamma^2 s^2}{2}\right)$$

$$\psi_4 = \left(\frac{1}{\pi^{1/4}}\right)\left(\frac{\gamma}{6}\right)^{1/2}\left(\frac{3}{2} - 6\gamma^2 s^2 + 2\gamma^4 s^4\right)\exp\left(-\frac{\gamma^2 s^2}{2}\right)$$

$$\psi_5 = \left(\frac{1}{\pi^{1/4}}\right)\left(\frac{\gamma}{15}\right)^{1/2}\left(\frac{15}{2}\gamma s - 10\gamma^3 s^3 + 2\gamma^5 s^5\right)\exp\left(-\frac{\gamma^2 s^2}{2}\right)$$

The corresponding eigenvalues in Eq.(4-240) are

$$E_n = \hbar\left(\frac{k}{\mu}\right)^{1/2}\left(n + \frac{1}{2}\right), \tag{4-242}$$

with the Bohr transition frequency given by

$$\nu = \frac{E_{n+1} - E_n}{h} = \frac{\Delta E}{h} = \frac{1}{2\pi}\left(\frac{k}{\mu}\right)^{1/2},$$

which is identical to the classical result in Eq.(2-61). In HCl, $\nu = 8.65 \times 10^{13}$ Hz (Fig. 2-4), which gives

$$k = 4.77 \times 10^5 \text{ dyne} \cdot \text{cm}^{-1}, \qquad \gamma = 9.13 \times 10^8 \text{ cm}^{-1} \tag{4-243}$$

for the harmonic oscillator parameters in HCl. Using these parameters leads to the potential function and wavefunctions shown in Fig. 4-30. Several additional force constants for diatomic molecules [26] are listed in Table 4-26.

The electric dipole transitions in a harmonic oscillator involve the change in an electric dipole moment with the vibration. Noting that the displacement for a molecular vibration in a diatomic molecule is only about 1% of the interatomic distance,

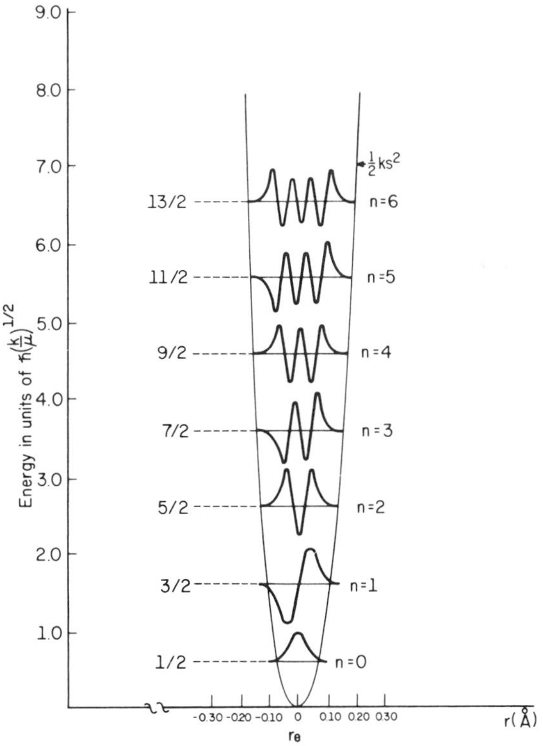

Figure 4-30 Potential function and wavefunctions for a harmonic oscillator modeled after the force constant in the HCl molecule.

Table 4-26 Vibrational energy spacings and the resultant diatomic force constants.

Molecule	$\Delta E/hc$ (cm^{-1})	k (10^5 dyne · cm^{-1})
H_2	4159.2	5.2
HF	3958.4	8.8
HCl	2885.6	4.8
HBr	2559.3	3.8
HI	2230.0	2.9
F_2	892	4.5
Cl_2	556.9	3.2
Br_2	321	2.4
I_2	213.4	1.7
O_2	1556.3	11.4
N_2	2330.7	22.6
CO	2143.3	18.7
NO	1876.0	15.5
Li_2	246.3	1.3
Na_2	157.8	1.7
NaCl	378	1.2
KCl	278	0.8

we expand the electric dipole moment along the internuclear axis in a Taylor series around $s = 0$, giving

$$D = D_0 + \left(\frac{\partial D}{\partial r}\right)_0 s + \frac{1}{2}\left(\frac{\partial^2 D}{\partial r^2}\right)_0 s^2 + \cdots, \tag{4-244}$$

where D_0 is the permanent electric dipole moment at equilibrium with $s = 0$ and $(\partial D/\partial r)_0$ is a constant representing the change in the dipole moment with distance evaluated at the equilibrium internuclear distance. Due to the small displacements in most molecules, Eq.(4-244) is expected to converge rapidly.

First, we consider the selection rules for a diatomic molecule fixed in space (nonrotating) with the internuclear axis parallel to the radiation field axis. The electric dipole transition moment for a harmonic oscillator (Table 4-25) gives the following nonzero contributions:

$$\begin{aligned} D_{nn'} &= \int_0^\infty \psi_n^*(s)\left[D_0 + \left(\frac{dD}{dr}\right)_0 s + \frac{1}{2}\left(\frac{d^2D}{dr^2}\right)_0 s^2 + \cdots\right]\psi_{n'}(s)\, ds \\ &= D_0\delta_{nn'} + \left(\frac{dD}{dr}\right)_0 \langle n|s|n \pm 1\rangle + \frac{1}{2}\left(\frac{d^2D}{dr^2}\right)_0 \langle n|s^2|n \pm 2\rangle + \cdots. \end{aligned} \tag{4-245}$$

The dominant selection rule that allows a change in the vibrational energy is the $\Delta n = \pm 1$ transition that is proportional to $(dD/dr)_0$. This series will usually converge rapidly, which makes the allowed $\Delta n = \pm 2, \pm 3, \ldots$ *overtone* transitions appear with less intensity than the dominant $\Delta n = \pm 1$ transitions. Breakdown in the $\Delta n = \pm 1$ selection rules also occurs due to anharmonic contributions to the potential function.

Reconsider the quadratic $\frac{1}{2}ks^2$ potential function in HCl as shown in Fig. 4-30. There are two observations that show that the vibrational energy levels in HCl deviate from those described by the simple quadratic function. First, the observed $\Delta n = +1, +2, +3, \ldots$ transitions do not occur in exact multiples of $(k/\mu)^{1/2}$ as predicted by using Eq.(4-242) (see Table 4-30). The second observation that discredits the harmonic potential function is that the equilibrium internuclear distance, r_e, increases with increasing vibrational energy as measured by the rotational constants (moments of inertia) in the various vibrational states. The Morse potential. $V(s) = D_e[1 - \exp(-as)]^2$ as discussed in Problem 4-34 (see Fig. 4-39), predicts the experimental observations above better than the quadratic $V(s) = \frac{1}{2}ks^2$ potential. In order to compute the new eigenvalues and eigenfunctions for the Morse oscillator; the standard matrix technique is again appropriate. The Hamiltonian is

$$\mathscr{H} = -\frac{\hbar^2}{2\mu}\frac{d^2}{ds^2} + D_e[1 - \exp(-as)]^2, \tag{4-246}$$

and the matrix representation of the Hamiltonian is easily computed in the harmonic oscillator basis. A sufficiently large matrix can be diagonalized to obtain accurate energies for the lower states (Problem 4-34). The results show that both the equilibrium internuclear distance increases and that the energy spacings between vibrational states decreases with increasing vibrational energy. The selection rules for the Morse oscillator are easily obtained by truncating Eq.(4-245) at the second term and computing the transition dipole matrix in the harmonic oscillator basis. Transforming to the same basis that diagonalized the Hamiltonian matrix, we find $\Delta n = \pm 1, \pm 2, \pm 3, \ldots$ selection rules with the harmonic terms ($\Delta n = \pm 1$) being again dominant. In summary, it is clear that both deviations in harmonic oscillation and the higher-order terms in the dipole moment function in Eq.(4-244) lead to the breakdown of the dominant $\Delta n = \pm 1$ selection rules for a vibrator.

Consider now the rotating diatomic where we assume the vibrational and rotational motions are separable. Writing the total wavefunction as a product of the vibrational, ψ_n, and rotational, Y_{JM}, parts, the vibration-rotation selection rules for the diatomic molecule that is interacting with an electromagnetic electric field along the laboratory z axis are given by

$$\begin{aligned}
&\int Y^*_{J'M'}\,\psi^*_{n'}[D_z]Y_{JM}\,\psi_n \sin\theta\, d\theta\, d\phi\, ds \\
&= \left\langle J', M', n' \left| D_a \cos\theta + \left(\frac{\partial D_a}{\partial r}\right)_0 s\cos\theta + \cdots \right| J, M, n \right\rangle \\
&= D_a\langle J', M', n'|\cos\theta|J, M, n\rangle + \left(\frac{\partial D_a}{\partial r}\right)_0 \langle J', M', n'|s\cos\theta|J, M, n\rangle + \cdots \\
&= D_a\langle J', M'|\cos\theta|J, M\rangle\delta_{n'n} + \left(\frac{\partial D_a}{\partial r}\right)_0 \langle J', M'|\cos\theta|J, M\rangle\langle n'|s|n\rangle + \cdots.
\end{aligned} \tag{4-247}$$

Similar results are obtained for D_x and D_y. Thus, the selection rules for the vibration-rotation transitions are

$$\Delta n = \pm 1, \qquad \Delta J = \pm 1, \qquad \Delta M = 0, \pm 1, \tag{4-248}$$

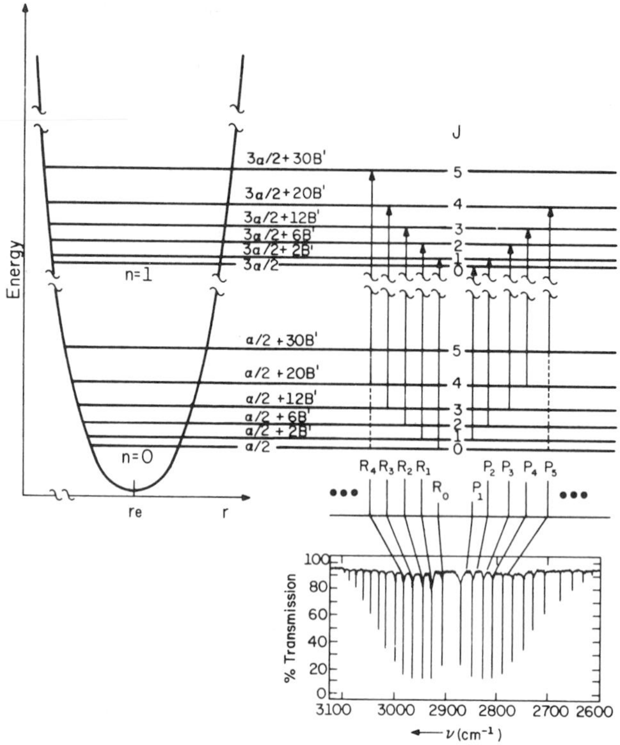

Figure 4-31 Vibration-rotation energy level diagram in HCl showing the $R(J \longrightarrow J+1)$ and $P(J \longrightarrow J-1)$ vibration-rotation transitions. An actual spectrum is also shown where the $H^{35}Cl$ (75%) and $H^{37}Cl$ (25%) are not resolved. $\alpha = (1/2\pi c)(k/\mu)^{1/2} = 2890\ \text{cm}^{-1}$ and $B' = B/c = 10.6\ \text{cm}^{-1}$ in the HCl molecule.

and there are no allowed *pure* vibrational transitions in the rotating diatomic molecule. The diatomic HCl molecule's rotation-vibration transitions and notation are shown in Fig. 4-31. The $\Delta n = \pm 1, \Delta J = +1$ transitions are called R-branch and the $\Delta n = \pm 1,\ \Delta J = -1$ transitions are called P-branch. The $\Delta n = \pm 1,\ \Delta J = 0$ Q-branch transitions are not excited in a diatomic molecule by electromagnetic radiation.

It is evident that $|(dD/dr)_0|^2$ can be determined from the infrared absorption intensities. We can also find additional information about $(dD/dr)_0$ by measuring the molecular electric dipole moment in different vibrational states. As the value of r_e increases with vibrational state, the electric dipole moment should change also. The value of the moment in different vibrational states can be determined by measuring the electric field perturbations in the different vibrational states. The rotational transitions in different vibrational states will be at different frequencies due to the different equilibrium internuclear distances and resultant different moments of inertia as shown in Fig. 4-32. We have assumed that the moment of inertia increases with vibrational excitation, which will certainly be true for diatomic molecules. We have assumed a vibrational $\Delta E/hc = 300\ \text{cm}^{-1}$ and a 300 K temperature for the spectra shown in Fig. 4-32 where the intensities fall off as $\exp(-\Delta E/kT)$ due to the Boltzmann

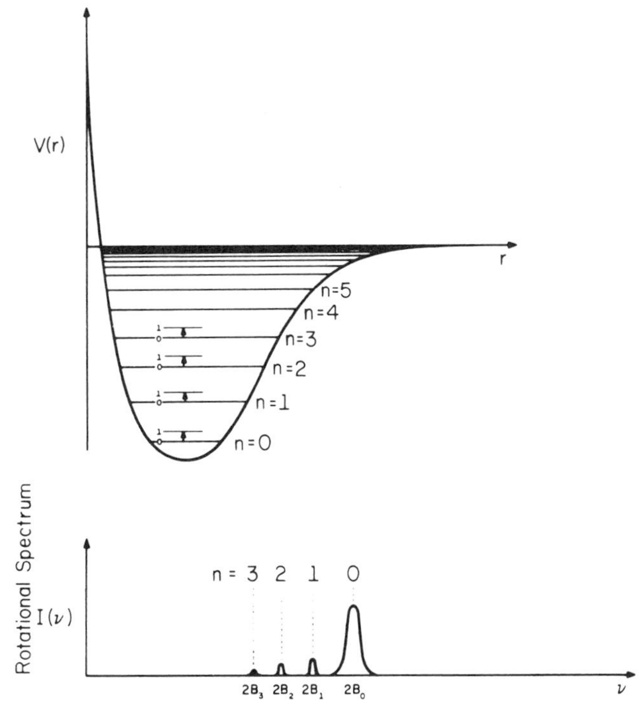

Figure 4-32 Diagram showing the observation of the $J = 0 \longrightarrow 1$ rotational transition in the $n = 0, 1, 2,$ and 3 vibrational states in a diatomic molecule. The rotational transition frequencies decrease with vibrational excitation because of an increase in internuclear distance. The electric dipole moment and rotational Stark effect in each vibrational state will be different. B_n indicates the rotational constant in the nth vibrational state. The intensities in the lower figure fall off with the Boltzmann factor at $T = 300$ K and $\Delta E/hc = 300\ \text{cm}^{-1}$.

factor. The zero-field spectrum in Fig. 4-32 is easily repeated in the presence of an electric field to yield the dipole moment as a function of vibrational state. It has been found experimentally, in the alkali halide molecules for instance, that the electric dipole moment increases successively in the first few vibrational states. A schematic of the possible change in electric moment with vibrational state is shown in Fig. 4-33. r_n represents the equilibrium internuclear distance in the nth vibrational state.

In summary, the absolute value of the slope of the dipole moment function at $r = r_e$ can be obtained by measuring the intensity of a vibrational transition. The

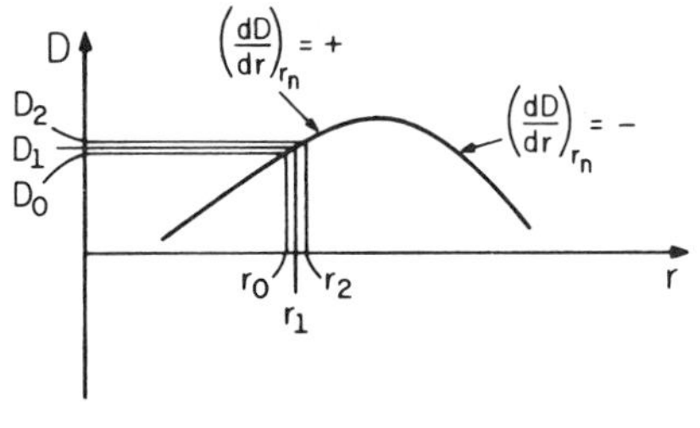

Figure 4-33 Diagram showing the dipole moment of a diatomic molecule as a function of internuclear distance. Values of $(dD/dr)_{r_n}$ for several molecules are listed in Table 4-27. r_n represents the equilibrium internuclear distance in the nth vibrational state.

Table 4-27 Values of bond dipole moments and $(dD/dr)_0$ for several molecules.

Molecule	Bond	Bond Dipole Moment (10^{-18} SC · cm)	$(dD/dr)_0$ (10^{-10} SC)
Cl—H[a]	$^-$Cl—H$^+$	1.085	0.86
Cl—Br[b]	Cl—Br	0.57	0.76
H_2CCl_2[c]	$^-$C—H$^+$	~0.46	|0.25|
	$^-$Cl—C$^+$	~0.95	3.7
H_3CF[d]	$^-$F—C$^+$	~2.0	~4.6
CF_4[d]	$^-$F—C$^+$	~2.0	~4.0
CF_3—CF_3[d]	$^-$F—C$^+$	~1.3	2.3
OCO[e]	$^-$O—C$^+$	~1.5	|5.6|
SCS[e]	$^-$S—C$^+$	~2.0	|5.5|

[a] W. S. Benedict, *et al.*, *J. Chem. Phys.* **26**, 1671 (1957).

[b] W. V. F. Brooks and B. Crawford, *J. Chem. Phys.* **23**, 363 (1955).

[c] J. W. Straley, *J. Chem. Phys.* **23**, 2183 (1955).

[d] D. G. Williams, W. B. Person, and B. Crawford, *J. Chem. Phys.* **23**, 179 (1955).

[e] J. C. D. Brand and J. C. Speakman, *Molecular Structure* (E. Arnold, London, 1960).

permanent moments in the different vibrational states can be measured directly to obtain the points on the curve in Fig. 4-33 and the sign of the slope. The results for several molecules are listed in Table 4-27. It is evident from Table 4-27 that most bond dipole moments increase with increasing vibrational excitation. If $(dD/dr)_0 = 0$, the vibrational transition cannot be excited by electromagnetic radiation. Thus, symmetric diatomic molecules like H_2 and N_2 do not have a dipole-induced vibrational spectrum excited by electromagnetic radiation; however, these transitions can be excited by an electron or neutron scattering process.

Polyatomic Molecules, Normal Coordinates

In Section 2.3 we discussed the vibration of a system of particles in terms of their generalized coordinates, mass-weighted displacement coordinates, and finally the normal coordinates. We shall now show that the normal coordinates conveniently lead to the quantum mechanical solution to the harmonic oscillator problem for a polyatomic molecule. The mass-weighted displacement coordinates are defined in Eq.(2-73); the kinetic and potential energies in terms of the mass-weighted displacement coordinates are in Eqs.(2-74) and (2-75), respectively; and the resultant secular equation for a harmonic oscillator obtained from Lagrange's equations is given in Eq.(2-80). The harmonic oscillator,

$$S_k = a_k \cos(\omega t + \alpha), \tag{4-249}$$

leads to a series of linear equations given by

$$-a_k\omega^2 + \sum_{i=1}^{3n} V_{ki}a_i = 0, \tag{4-250}$$

where a_k is the amplitude of the kth mass-weighted displacement coordinate and ω is the vibrational angular frequency as discussed following Eq.(2-78). Following Eq.(2-79) we set the determinant of the coefficients of a_i in the equation above equal to zero to obtain the secular equation and resultant roots, ω^2, given from

$$\text{Det}\,(\boldsymbol{V} - \omega^2\mathbf{1}) = 0$$

$$\boldsymbol{V} = \begin{pmatrix} V_{11} & V_{12} & V_{13} & \cdots \\ V_{21} & V_{22} & V_{23} & \\ V_{31} & V_{32} & V_{33} & \\ \vdots & & & \ddots \end{pmatrix}, \quad \mathbf{1} = \begin{pmatrix} 1 & 0 & 0 & \cdots \\ 0 & 1 & 0 & \\ 0 & 0 & 1 & \\ \vdots & & & \ddots \end{pmatrix}. \tag{4-251}$$

We now return to Eq.(4-250) and write in matrix form as

$$\begin{pmatrix} V_{11} & V_{12} & V_{13} & \cdots \\ V_{21} & V_{22} & V_{23} & \\ V_{31} & V_{32} & V_{33} & \\ \vdots & & & \ddots \end{pmatrix} \begin{pmatrix} a_{1i} \\ a_{2i} \\ a_{3i} \\ \vdots \end{pmatrix} - \begin{pmatrix} a_{1i}\lambda_i \\ a_{2i}\lambda_i \\ a_{3i}\lambda_i \\ \vdots \end{pmatrix} = 0, \tag{4-252}$$

where $\lambda_i = \omega_i^2$ is the ith root of the secular equation. Now if Eq.(4-252) is valid for each root λ_i, we can write the complete equation as

$$\begin{pmatrix} V_{11} & V_{12} & V_{13} & \cdots \\ V_{21} & V_{22} & V_{23} & \\ V_{31} & V_{32} & V_{33} & \\ \vdots & & & \ddots \end{pmatrix} \begin{pmatrix} a_{11} & a_{12} & a_{13} & \cdots \\ a_{21} & a_{22} & a_{23} & \\ a_{31} & a_{32} & a_{33} & \\ \vdots & & & \ddots \end{pmatrix}$$

$$= \begin{pmatrix} a_{11} & a_{12} & a_{13} & \cdots \\ a_{21} & a_{22} & a_{23} & \\ a_{31} & a_{32} & a_{33} & \\ \vdots & & & \ddots \end{pmatrix} \begin{pmatrix} \lambda_1 & 0 & 0 & \cdots \\ 0 & \lambda_2 & 0 & \\ 0 & 0 & \lambda_3 & \\ \vdots & & & \ddots \end{pmatrix} \tag{4-253}$$

$$\boldsymbol{V}^s\mathbf{a} = \mathbf{a}\boldsymbol{\lambda},$$

where the elements in $\boldsymbol{V}^s$, V_{ij}^s, are defined in terms of the mass-weighted displacement coordinates in Eq.(2-75); $V_{ij}^s = (\partial^2 V/\partial S_i\,\partial S_j)$. $\boldsymbol{\lambda}$ is the eigenvalue matrix with the eigenvalues given by the $i = 3n$ values of ω_i^2 for the normal vibrations. Multiplying Eq.(4-253) by $\mathbf{a}^\dagger$ and requiring $\mathbf{a}$ to diagonalize $\boldsymbol{V}^s$ to give λ yields

$$\mathbf{a}^\dagger\boldsymbol{V}^s\mathbf{a} = \boldsymbol{V}^Q = \boldsymbol{\lambda}. \tag{4-254}$$

As $\boldsymbol{V}^s$ is a symmetric matrix, $\mathbf{a}$ must be unitary; $\mathbf{a}^\dagger\mathbf{a} = \mathbf{1}$. $\boldsymbol{V}^Q$ is the matrix defined by the second derivative with respect to a new set of coordinates, called the *normal coordinates*: $Q_1, Q_2, Q_3, \ldots$. Thus, the $\boldsymbol{V}^Q = \boldsymbol{\lambda}$ matrix has elements

$$V_{ij}^Q = \left(\frac{\partial^2 V}{\partial Q_i\,\partial Q_j}\right)_0 \delta_{ij}, \tag{4-255}$$

leading to the diagonal λ matrix and we can write the classical vibrational frequency, ν, as $\omega_i^2 = (2\pi\nu_i)^2 = (\partial^2 V/\partial Q_i^2)_0$ or $\nu_i = 1/2\pi\ [(\partial^2 V/\partial Q_i^2)_0]^{1/2}$. Furthermore, it is evident from the form of Eq.(4-254) that the mass-weighted displacement coordinates, $\boldsymbol{S}$, are related to the normal coordinates, $\boldsymbol{Q}$, by the unitary transformation,

$$\boldsymbol{Q} = \begin{pmatrix} Q_1 \\ Q_2 \\ Q_3 \\ \vdots \end{pmatrix} = \mathbf{a}\boldsymbol{S} = \begin{pmatrix} \mathrm{a}_{11} & \mathrm{a}_{12} & \mathrm{a}_{13} & \cdots \\ \mathrm{a}_{21} & \mathrm{a}_{22} & \mathrm{a}_{23} & \\ \mathrm{a}_{31} & \mathrm{a}_{32} & \mathrm{a}_{33} & \\ \vdots & & & \ddots \end{pmatrix} \begin{pmatrix} S_1 \\ S_2 \\ S_3 \\ \vdots \end{pmatrix} \tag{4-256}$$

$$\boldsymbol{S} = \mathbf{a}^\dagger \boldsymbol{Q}.$$

We now show that the kinetic and potential energies written in terms of the mass-weighted displacement coordinates in Eqs.(2-74) and (2-76) can be reexpressed by using Eq.(4-256) in terms of the normal coordinates. The kinetic energy from Eq.(2-74) is

$$T = \sum_{i=1}^{3n} \dot{S}_i^2 = \frac{1}{2}\dot{\mathbf{S}}^\dagger \dot{\mathbf{S}}. \tag{4-257}$$

Substituting $\dot{\boldsymbol{S}} = \mathbf{a}^\dagger \dot{\boldsymbol{Q}}$ and $\dot{\mathbf{S}}^\dagger = \dot{\boldsymbol{Q}}^\dagger \mathbf{a}$ from Eq.(4-256) gives

$$T = \frac{1}{2}\dot{\boldsymbol{Q}}^\dagger \mathbf{a}\mathbf{a}^\dagger \dot{\boldsymbol{Q}} = \frac{1}{2}\dot{\boldsymbol{Q}}^\dagger \dot{\boldsymbol{Q}} = \frac{1}{2}\sum_{i=1}^{3n} \dot{Q}_i^2. \tag{4-258}$$

The harmonic potential energy is written similarly from Eq.(2-76) as

$$V = \frac{1}{2}\sum_{i,j} S_i V_{ij}^s S_j = \frac{1}{2}\mathbf{S}^\dagger\, V^s \mathbf{S}. \tag{4-259}$$

Substituting $\boldsymbol{S} = \mathbf{a}^\dagger \boldsymbol{Q}$ and $\boldsymbol{S}^\dagger = \boldsymbol{Q}^\dagger \mathbf{a}$ again gives

$$V = \frac{1}{2}\boldsymbol{Q}^\dagger \mathbf{a} V^s \mathbf{a}^\dagger \boldsymbol{Q} = \frac{1}{2}\boldsymbol{Q}^\dagger V^Q \boldsymbol{Q} = \frac{1}{2}\sum_{i,j} Q_i V_{ij}^Q \delta_{ij} Q_j = \frac{1}{2}\sum_i V_{ii}^Q Q_i^2. \tag{4-260}$$

The normal coordinate representation of $\mathscr{H} = T + V$ also provides a very convenient framework for the quantum mechanical solution to the problem. T and V are converted to operator form by

$$T = \frac{1}{2}\sum_i \dot{Q}_i^2 \longrightarrow -\frac{\hbar^2}{2}\sum_i \frac{\partial^2}{\partial Q_i^2}, \qquad V = \frac{1}{2}\sum_i \omega_i^2 Q_i^2 \longrightarrow \frac{1}{2}\sum_i \omega_i^2 Q_i^2, \tag{4-261}$$

and the resulting Schrödinger equation is

$$\left(-\frac{\hbar^2}{2}\sum_i \frac{\partial^2}{\partial Q_i^2} + \frac{1}{2}\sum_i \omega_i^2 Q_i^2\right)\psi = \psi E. \tag{4-262}$$

As the normal coordinates, Q_i, are orthogonal, the vibrational wavefunction ψ can be expressed as a product of functions in each normal coordinate,

$$\psi(n_1, n_2, \ldots, n_{3n}) = \psi_{n_1}(Q_1)\psi_{n_2}(Q_2)\psi_{n_3}(Q_3)\cdots\psi_{n_{3n}}(Q_{3n}).$$

Table 4-28 Fundamental vibrational $\Delta E/hc$ in units of cm^{-1} for several triatomic molecules as shown in Fig. 2-4. The bending motion in linear molecules is doubly degenerate. ν_1 is the symmetric stretch, ν_2 is the bend, and ν_3 is the asymmetric stretch. The values of ν are listed in wave numbers, cm^{-1}. Multiplying by c, the speed of light, gives the vibrational frequencies in hertz.

Molecule	$(\Delta E/hc)(cm^{-1})$ ν_1	ν_2	ν_3
H_2O	3652	1595	3756
D_2O	2666	1179	2789
H_2S	2611	1290	2684
F_2O	830	490	1110
O_2F	1151	519	1361
O—C—S	833	500	2100
O—C—O	{1286, 1388}	667	2349
S—C—S	657	397	1523
N—N—O	1285	589	2224

Substituting into the Schrödinger equation gives $3n$ separable equations, one in each normal coordinate,

$$\left(-\frac{\hbar^2}{2}\frac{\partial^2}{\partial Q_i^2} + \frac{1}{2}\omega_i^2 Q_i^2\right)\psi(Q_i) = E_{Q_i}\psi(Q_i) \tag{4-263}$$

$$E = E(n_1, n_2, \ldots) = E_{Q_1} + E_{Q_2} + \cdots + E_{Q_{3n}}, \tag{4-264}$$

where $\psi(Q_i)$ are the familiar harmonic oscillator wavefunctions, as discussed previously in the case of a diatomic molecule and $E(n_1, n_2, \ldots)$ is the total energy.

The observed fundamental vibrational frequencies for several triatomic molecules are listed in Table 4-28 and the normal modes of motion in SO_2 are shown in the lower part of Fig. 4-34. Figure 4-34 illustrates the energy levels of the nonlinear triatomic molecule including overtone and combination levels. The normal modes of motion for two four-atom molecules are illustrated in Fig. 4-35 [27].

Consider now the selection rules for polyatomic molecules. The dipole moment along a space-fixed z axis for a general molecule with $3n - 6$ normal modes of vibration ($3n - 5$ for a linear molecule) is given by

$$D_z = \sum_g D_g C_{gz} + \sum_{i=1}^{3n-6} \sum_g \left(\frac{\partial D_g}{\partial Q_i}\right)_0 Q_i C_{gz} + \frac{1}{2}\sum_{i,j=1}^{3n-6}\sum_g \left(\frac{\partial^2 D_g}{\partial Q_i\,\partial Q_j}\right)_0 Q_i Q_j C_{gz} + \cdots. \tag{4-265}$$

The sums over g are over the three principal inertial axes, a, b, and c and the direction cosines represent the transformation of the dipole moment vector from the space-fixed to the molecular-fixed axis system. The rotational transitions arising from the first term in this equation for linear, symmetric, and asymmetric rotors were discussed

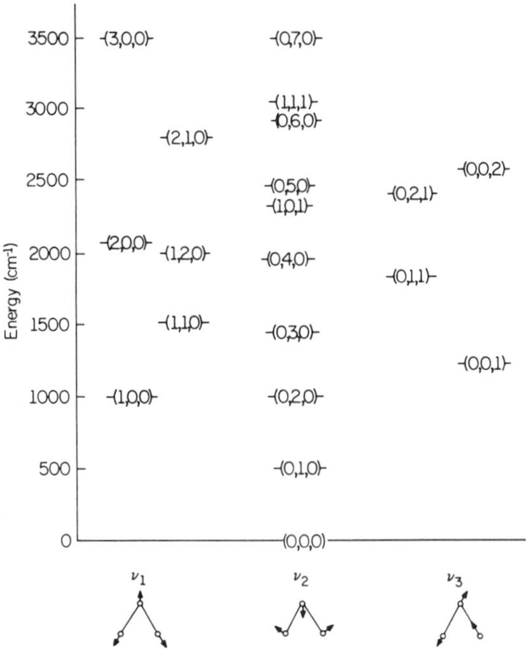

Figure 4-34 Energy level diagram for the nonlinear triatomic molecule with three normal modes of motion where (n_1, n_2, n_3) indicates the energy level with excitation of n_1, n_2, and n_3 in the three normal modes. The energy levels are modeled after SO_2 where the experimental transitions are given in Table 4-29.

in Section 4.11. The second term is the vibration-rotation term. Consider the second term in Eq.(4-265) for a linear triatomic such as HCN, which gives

$$\sum_{i=1}^{4}\sum_{g}\left(\frac{\partial D_g}{\partial Q_i}\right)_0 Q_i C_{gz} = \sum_g C_{gz}\Bigg[\underbrace{\left(\frac{\partial D_g}{\partial Q_1}\right)_0 Q_1}_{\text{symmetric stretch}} + \underbrace{2\left(\frac{\partial D_g}{\partial Q_2}\right)_0 Q_2}_{\text{bend}} + \underbrace{\left(\frac{\partial D_g}{\partial Q_3}\right)_0 Q_3}_{\text{asymmetric stretch}}\Bigg].$$

←H←C—N→   H—C—N (↗ H, ↓ C, ↖ N)   H→←C—N→

(4-266)

The first term in Eq.(4-266) involves a symmetric stretch in which the atoms oscillate *parallel* to the internuclear axis and the vibrational selection rules are the normal $\Delta n_1 = \pm 1$. The rotational selection rules involve the matrix elements of C_{gz} in the linear molecule spherical harmonic basis. These matrix elements are listed in Table 4-20 and lead to $\Delta J = \pm 1$, $\Delta M = 0, \pm 1$ selection rules. Thus, the $\psi(000) \longrightarrow \psi(100)$ vibration-rotation transition in HCN leads to the *P*- and *R*-branch transitions with the *Q*-branch transitions forbidden.

The second term in Eq.(4-266) involves the bending motion in which the atoms oscillate *perpendicular* to the internuclear axis. The vibrational selection rules are $\Delta n_2 = \pm 1$ but the rotational selection rules for a linear molecule are no longer appropriate because the bending vibration distorts the molecule into an asymmetric

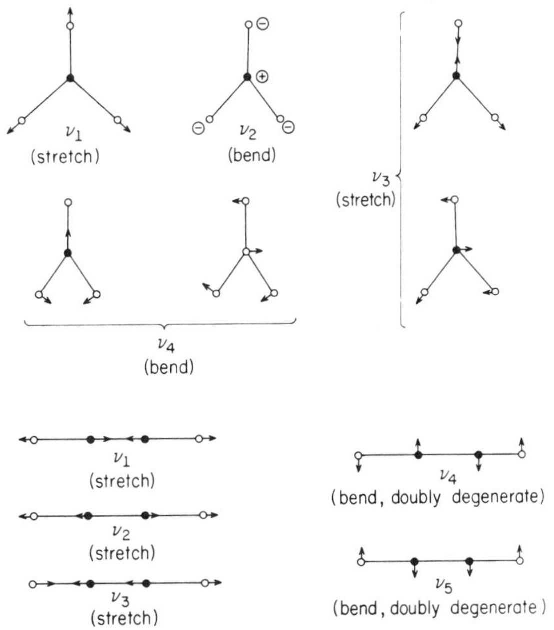

Figure 4-35 Normal modes of motion for the planar symmetric AB_3 and linear A_2B_2 molecules.

rotor. Thus, it is appropriate to compute the rotational selection rules by evaluating the direction cosine matrix elements in the symmetric rotor basis instead of the linear molecule spherical harmonic basis. The matrix elements of C_{gz} from Table 4-20 in the symmetric rotor basis show that the rotational selection rules for the Q_2 bending mode are $\Delta J = 0, \pm 1, \Delta M = 0$. Thus, P-, Q-, and R-branch transitions are all allowed in the Q_2 normal mode. The Q_3 normal mode from the third term in Eq.(4-266) involves a *parallel* vibration where the atoms all vibrate along the line joining the nuclei leading to parallel selection rules as discussed above for Q_1.

In summary, the vibrational modes where the atoms oscillate parallel to the internuclear line in linear molecules are called *parallel modes* and these vibrations lead to P- and R-branches. The vibrational modes where the atoms oscillate perpendicular to the internuclear line are called *perpendicular modes* and these vibrations lead to P-, Q-, and R-branches. The infrared spectrum of both parallel and perpendicular transitions in HCN are shown in Fig. 4-36. The $\psi(000) \longrightarrow \psi(110)$ *perpendicular* combination band is shown in Fig. 4-36a and the $\psi(000) \longrightarrow \psi(101)$ *parallel* combination band is shown in Fig. 4-36b.

Somewhat different results are found for symmetric and asymmetric rotors; that is, the parallel vibrations in a prolate symmetric rotor involve vibrations along the a axis that lead to $\Delta J = 0, \pm 1$ P-, Q-, and R-branch transitions. Most asymmetric rotors are near-symmetric rotors and the characteristic parallel and perpendicular bands are often observed.

The relative intensities for the fundamental and several overtone and combination transitions in SO_2 are listed in Table 4-29 along with the calculated combination and

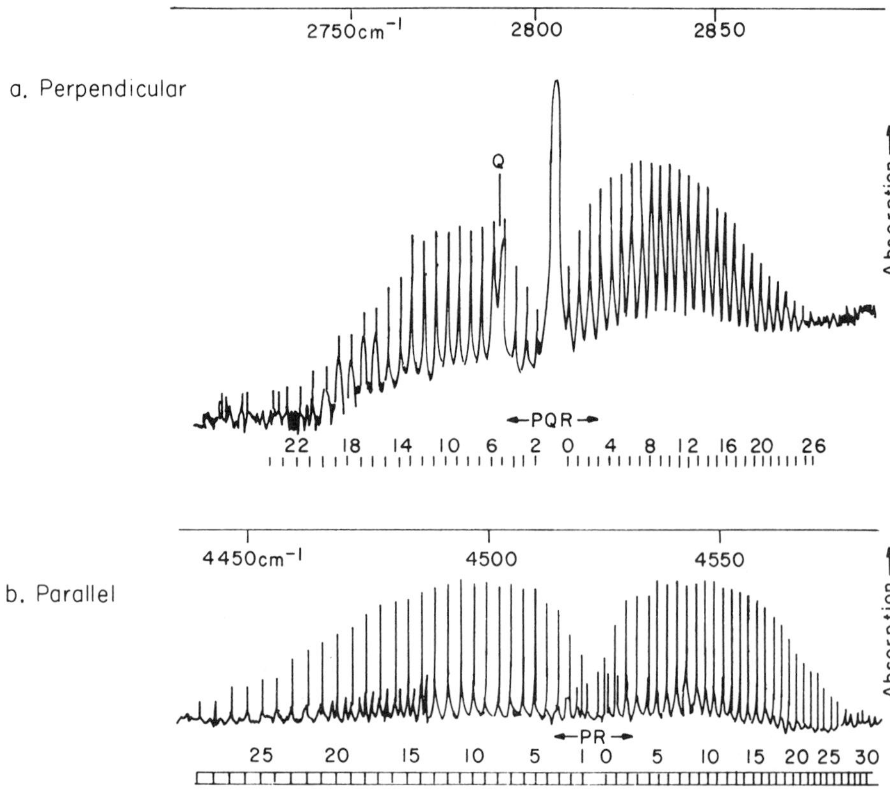

Figure 4-36 The infrared spectra of *perpendicular* and *parallel* bands in HCN. a. The $\psi(000) \longrightarrow \psi(110)$ perpendicular combination band that involves the perpendicular bending mode. The *P*-, *Q*-, and *R*-branches are evident. The spectra is contaminated by a weaker $\psi(110) \longrightarrow \psi(120)$ transition at a slightly lower energy where the *Q*-branch of this transition is indicated with the *Q*. b. The $\psi(000) \longrightarrow \psi(101)$ parallel transition that leads to only *P*- and *R*-branches. This transition is contaminated with the weaker $\psi(010) \longrightarrow \psi(111)$ transition. The spectra and description are adapted from H. C. Allen, E. D. Tidwell, and E. W. Plyler, *J. Chem. Phys.* **25**, 302 (1956) and we note the nonlinear scales.

overtone bands using the three fundamental modes of motion and the pure harmonic approximation. The observed overtone and combination bands all occur at transition energies less than the calculated values. This indicates that the harmonic approximation is not maintaining its integrity at higher vibrational energies. Note the rapid falloff in intensities of the overtones and combinations relative to the fundamentals. The fact that the overtone and combination transitions were observable at all indicates that the harmonic approximation is breaking down as vibrational excitation increases. An adequate explanation of the energy levels in Table 4-29 is obtained by treating the cubic, quartic, and higher terms in Eq.(2-75) as perturbations on the harmonic system according to the usual matrix methods [28].

An extreme example of the higher-order coupling between two normal modes is the Fermi resonance perturbation, which strongly couples two near-degenerate

Table 4-29 Observed transitions for the molecular vibrations in SO_2 from R. D. Shelton, A. H. Nielson, and W. H. Fletcher, *J. Chem. Phys.* **21**, 2178 (1953). See Fig. 4-34 for the form of the normal vibrations and an energy level diagram. The intensities are all relative to the (000) ⟶ (001) transition, which is normalized to 1000.

n_1	n_2	n_3	Observed $(\Delta E/hc)$ (cm^{-1})	Relative Intensities	Calculated $(\Delta E/hc)$ (cm^{-1})[a]
0	0	0	0	—	0
1	0	0	1151.4	565	1151.4
0	1	0	517.7	455	517.7
0	0	1	1361.8	1000	1361.8
0	3	0	1535.0	0.1	1553.1
1	1	0	1665.1	0.1	1669.1
0	1	1	1875.6	6.0	1879.5
2	0	0	2295.9	5.5	2302.8
1	0	1	2499.6	20.0	2513.2
0	0	2	2715.5	0.2	2723.6
2	1	0	2808.3	0.8	2820.5
1	1	1	3011.3	0.02	3030.9
3	0	0	3431.2	0.01	3454.2

[a] Multiples of the fundamental (100), (010), and (001) vibrations.

vibrational energy levels. It is evident from Table 4-28 that the (0, 2, 0) energy in OCO is approximately equal to the (1, 0, 0) energy. The matrix representation that illustrates the perturbation between the (0, 2, 0) and (1, 0, 0) states in CO_2, which is an approximate harmonic triatomic, is shown below in schematic form:

$$\mathscr{H} = \mathscr{H}_0 + \mathscr{H}_1 = \begin{pmatrix} (0,0,0) & & & & & & & & \\ & (1,0,0) & & & & & \langle 1,0,0|\mathscr{H}_1|0,2,0\rangle & & \cdots \\ & & (2,0,0) & & & & & & \\ & & & (3,0,0) & & & & & \\ & & & & \ddots & & & & \\ & & & & (2,1,0) & & & & \\ & & & & & (3,1,0) & & & \\ & & & & & & \ddots & & \\ & \langle 1,0,0|\mathscr{H}_1|0,2,0\rangle & & & & & (0,2,0) & & \\ & & & & & & & (1,2,0) & \\ & & & & & & & & \ddots \end{pmatrix}. \qquad (4\text{-}267)$$

(3, 1, 0), for instance, indicates the harmonic energy of $E(3, 1, 0) = E_{v_1}(n_1 = 3) + E_{v_2}(n_2 = 1) + E_{v_3}(n_3 = 0)$ and an eigenfunction of $\psi(3, 1, 0) = \psi_1(3)\psi_2(1)\psi_3(0)$, where $\psi_1(n_1)$ is the harmonic oscillator for mode 1 with excitation n_1. The off-diagonal element in Eq. (4-267) is given by

$$\langle 1, 0, 0|\mathscr{H}_1|0, 2, 0\rangle = \int_1 \int_2 \int_3 \psi^*(1, 0, 0)\mathscr{H}_1\psi(0, 2, 0)\, dQ_1\, dQ_2\, dQ_3, \qquad (4\text{-}268)$$

where the perturbation is taken from Eq.(2-75) as a sum of cubic and higher terms given by

$$\mathscr{H}_1 = \frac{1}{6}\sum_{i,j,k} V_{ijk}Q_iQ_jQ_k + \frac{1}{24}\sum_{i,j,k,\ell} V_{ijk\ell}Q_iQ_jQ_kQ_\ell + \cdots. \tag{4-269}$$

Cubic $(\frac{1}{6}V_{122}Q_1Q_2^2)$ terms in Eq.(4-269) will lead to a nonzero off-diagonal element in Eq.(4-268) according to the functions in Table 4-25 for the normal coordinates. As the zero-order energies between the (1, 0, 0) and (0, 2, 0) states in CO_2 are nearly degenerate, the second-order perturbation leads to a large correction to the energy. The second-order energy of the two interacting levels is given by (nondegenerate perturbation theory)

$$E^{(2)}_{1,0,0} = \frac{|\langle 1,0,0|\mathscr{H}_1|0,2,0\rangle|^2}{E^0_{1,0,0} - E^0_{0,2,0}}, \qquad E^{(2)}_{0,2,0} = \frac{|\langle 1,0,0|\mathscr{H}_1|0,2,0\rangle|^2}{E^0_{0,2,0} - E^0_{1,0,0}}.$$

The originally near-degenerate (1, 0, 0) and (0, 2, 0) states in CO_2 repel each other; the lower state energy decreases and the upper state increases by an equal amount. Nondegenerate perturbation theory is valid only when $|\langle 1,0,0|\mathscr{H}_1|0,2,0\rangle| \ll E^0_{1,0,0} - E^0_{0,2,0}$. For degenerate systems we can diagonalize the 2×2 coupled block in Eq.(4-267) when $|(1,0,0) - (0,2,0)| \ll \langle(1,0,0)|\mathscr{H}_1|(0,2,0)\rangle$, which gives eigenvalues of

$$E_1 = (1,0,0) + \langle 1,0,0|\mathscr{H}_1|0,2,0\rangle, \qquad E_2 = (0,2,0) - \langle 1,0,0|\mathscr{H}_1|0,2,0\rangle.$$

We now return to the possibility of observing the infrared transitions in OCO. The symmetric stretch, ν_1(←O—C—O→), does not lead to a change in dipole moment with internuclear distance; however, the symmetric bend, ν_2(O—C—O) (with the O atoms displaced up and the C atom down), does lead to a change in dipole moment with vibration. In the absence of Fermi resonance, the ν_1 molecular vibration in OCO does not lead to infrared absorption but the (0, 0, 0) ⟶ (0, 2, 0) bending transition *is* infrared allowed. In the presence of the Fermi resonance discussed above for a near-degeneracy, the ν_1 and $2\nu_2$ vibrations are strongly mixed. The result is that the original ν_1 vibration now has an appreciable amount of ν_2 character leading to an allowed infrared transition. The eigenfunctions or new functions ψ_1 and ψ_2 in the presence of this Fermi resonance are linear combinations of the (1, 0, 0) and (0, 2, 0) zero-order functions given by the transformation that diagonalized the 2×2 coupled block in Eq.(4-267) to give

$$\psi_1 = (\tfrac{1}{2})^{1/2}[\psi(1,0,0) + \psi(0,2,0)], \qquad \psi_2 = (\tfrac{1}{2})^{1/2}[\psi(1,0,0) - \psi(0,2,0)].$$

Thus, the two states are equally mixed and the two transitions from the ground state, (0, 0, 0), will be of equal intensity. These two transitions correspond to the two observed transitions at 1286 and 1388 cm^{-1} in CO_2 listed in Table 4-28. Thus, in many cases of Fermi resonance, a weak or forbidden transition will become observable. This interaction of normal modes through the cubic and higher terms in the potential function appears quite often, especially in large molecules that have large numbers of normal modes and many possible near-degeneracies.

Infrared Selection Rules

The selection rules for infrared vibrational transitions in polyatomic molecules that are correct to all orders of perturbation are obtained directly by symmetry according to group theory. Using the matrix element theory discussed in Appendix E.5 we note that the integrand in the following integral must have a totally symmetric component to be nonzero:

$$\int_1 \int_2 \cdots \int_{3n-6} \psi(n_1', n_2', \ldots, n_{3n-6}') \begin{pmatrix} D_x \\ D_y \\ D_z \end{pmatrix} \psi(n_1, n_2, \ldots, n_{3n-6})\, dQ_1\, dQ_2 \cdots dQ_{3n-6}.$$

(4-270)

$\psi(n_1, n_2, \ldots, n_{3n-6})$ represents a vibrational wavefunction where the primes indicate the excited states. The dipole moment operators D_x, D_y, and D_z for a nonrotating molecule are obtained from Eq.(4-265) by setting the direction cosines equal to unity to give

$$D_z = D_z^0 + \sum_{i=1}^{3n-6} \left(\frac{\partial D_z}{\partial Q_i}\right)_0 Q_i + \frac{1}{2} \sum_{i,j=1}^{3n-6} \left(\frac{\partial^2 D_z}{\partial Q_i \partial Q_j}\right)_{\substack{0\\0}} Q_i Q_j + \cdots.$$

Thus, D_z contains the complete dipole moment functions for the nonrotating molecule and the symmetry classification of the D_x, D_y, and D_z operators will be the same as the x, y, and z functions, which are tabulated in the character tables.

Allowed transitions from the ground vibrational state, $\psi(0, 0, 0, \ldots, 0)$, which generates the totally symmetric representation, are simple to determine. In this case, we merely search for excited state symmetries that generate the same irreducible representation as one of the D_x, D_y, or D_z operators. The square or direct product of two equal irreducible representations always contains the totally symmetric irreducible representation. This technique can be used to obtain the selection rules from the ground states to the molecular levels shown in Fig. E-4 for the series:

$$CH_4 \longleftarrow CH_3D \longleftrightarrow CH_2D_2 \longleftrightarrow CHD_3 \longleftrightarrow CD_4.$$

Allowed infrared transitions from the ground to first excited states from Table E-5.

T_d CH_4, CD_4
(D_x, D_y, D_z) excites the T_2 vibration and the A_1, E, and T_1 vibration are inactive.

C_{3v} CH_3D, CD_3H

D_z excites A_1 and (D_x, D_y) excites E.

C_{2v} CH_2D_2

D_z excites A_1, D_x excites B_1, D_y excites B_2, and the A_2 mode is inactive.

The selection rule for any vibrational transition is easily obtained by examining the direct product from Eq.(4-270) for the totally symmetric representation. We should emphasize, however, that this symmetry analysis only yields the selection rules or whether a transition is allowed or forbidden. The selection rules according to group theory cannot give any information on the *magnitudes* of the transition moments of the allowed transitions. For instance, we can use the C_{2v} group to show that both of the $\psi(0, 0, 0) \longrightarrow \psi(0, 0, 1)$ and $\psi(0, 0, 0) \longrightarrow \psi(0, 1, 1)$ transitions in SO_2 are allowed. The experimental results in Table 4-29 show, however, that the fundamental transition is much stronger than the combination transition. This is due to the dominant harmonic character of the oscillations, which cannot be predicted by group theory.

Raman Selection Rules

The Raman effect is the molecular vibrational (and rotational) modulation of monochromatic light that is being scattered from the molecule [29]. According to the discussion in Section 1.2, the incident electromagnetic radiation induces a molecular dipole moment that scatters the incident light with an intensity proportional to the square of the induced dipole moment. If the induced dipole moment is modulated by internal motion, this information will be transmitted in the scattered light. In general, the induced electric dipole moment is given by

$$\boldsymbol{D}_{\text{ind}} = \boldsymbol{\alpha} \cdot \boldsymbol{E}. \tag{4-271}$$

$\boldsymbol{\alpha}$ is the polarizability tensor and $\boldsymbol{E}$ is the inducing electric field of the electromagnetic radiation. Using Eq.(4-204) relating $\boldsymbol{\alpha}_l$ in the space-fixed axis to $\boldsymbol{\alpha}_m$ in the molecular-fixed axis system gives

$$D_z = E_z[C_{za}^2\alpha_{aa} + \alpha_{bb}(C_{zb}^2 + C_{zc}^2)] = E_z(\alpha_{aa}\cos^2\theta + \alpha_{bb}\sin^2\theta) \tag{4-272}$$

for a linear molecule where D_z is the induced moment along the space-fixed z axis, E_z is the space-fixed electric field, and θ is the spherical polar angle (between the z and a axes). Similar equations are evident for D_x and D_y. Starting with z-polarized radiation propagating along the y axis, we note that D_z leads to polarized scattering and D_y and D_z give depolarized scattering. We shall derive the selection rules for polarized scattering here. Substituting the average polarizability, $\alpha = \frac{1}{3}(\alpha_{aa} + \alpha_{bb} + \alpha_{cc})$, into Eq.(4-272) and consulting Table 4-3 for the spherical harmonics gives

$$D_z = E_z\left[\frac{4}{3}\left(\frac{\pi}{5}\right)^{1/2}(\alpha_{aa} - \alpha_{bb})Y_{20} + \alpha\right].$$

The rotational modulation of the induced moment is given by the Y_{20}-dependence. The vibrational modulation of the induced dipole moment function is obtained by expanding D_z in a displacement series in normal modes around the equilibrium molecular structure as described earlier to give

$$
\begin{aligned}
D_z &= (D_z)_0 + \sum_i \left(\frac{\partial D_z}{\partial Q_i}\right)_0 Q_i + \frac{1}{2}\sum_{ij}\left(\frac{\partial^2 D_z}{\partial Q_i\,\partial Q_j}\right)_0 Q_i Q_j + \cdots \\
&= E_z\left[\frac{4}{3}\left(\frac{\pi}{5}\right)^{1/2}(\alpha_{aa}^0 - \alpha_{bb}^0)Y_{20} + \alpha^0\right] \\
&\quad + E_z \sum_i \left\{\frac{4}{3}\left(\frac{\pi}{5}\right)^{1/2}\left[\left(\frac{\partial \alpha_{aa}}{\partial Q_i}\right)_0 Q_i - \left(\frac{\partial \alpha_{bb}}{\partial Q_i}\right)_0 Q_i\right]Y_{20} + \left(\frac{\partial \alpha}{\partial Q_i}\right)_0 Q_i\right\} \\
&\quad + \frac{E_z}{2}\sum_{ij}\left\{\frac{4}{3}\left(\frac{\pi}{5}\right)^{1/2}\left[\left(\frac{\partial^2 \alpha_{aa}}{\partial Q_i\,\partial Q_j}\right)_0 Q_i Q_j + \left(\frac{\partial^2 \alpha_{bb}}{\partial Q_i\,\partial Q_j}\right)_0 Q_i Q_j\right]Y_{20} \right.\\
&\quad \left. + \left(\frac{\partial^2 \alpha}{\partial Q_i\,\partial Q_j}\right)_0 Q_i Q_j\right\} + \cdots.
\end{aligned}
\tag{4-273}
$$

We now examine the matrix elements of the D_z operator in Eq.(4-273) to determine which transitions will be observed in the scattered light, which is also polarized along the z axis. We shall consider here only the parallel vibrations where the linear molecule rotational functions are appropriate. The zero-order eigenfunctions are a product of rigid rotor spherical harmonics, Y_{JM}, with the product of the harmonic oscillators, $\psi_{n_1}(Q_1)\psi_{n_2}(Q_2)\cdots$. The contributions of the various terms in Eq.(4-273) for parallel vibrations and polarized scattering are summarized below.

$E_z\alpha^0$; $\Delta n = 0$, $\Delta J = 0$, $\Delta M = 0$

The $E_z\alpha^0$ operator is a scalar constant and does not excite any vibrational or rotational transitions. This term gives rise to Rayleigh scattering as discussed in Section 1.2.

$E_z\left(\frac{4}{3}\right)\left(\frac{\pi}{5}\right)^{1/2}(\alpha_{aa}^0 - \alpha_{bb}^0)Y_{20}$; $\Delta n = 0$, $\Delta J = 0, \pm 2$, $\Delta M = 0$

This term leads to the pure rotational Raman effect. There are no vibrational operators giving $\Delta n = 0$ and the Y_{20} rotational operator leads to $\Delta J = 0, \pm 2$, $\Delta M = 0$ rotational selection rules.

$E_z \sum_i \left\{\frac{4}{3}\left(\frac{\pi}{5}\right)^{1/2}\left[\left(\frac{\partial \alpha_{aa}}{\partial Q_i}\right)_0 Q_i - \left(\frac{\partial \alpha_{bb}}{\partial Q_i}\right)_0 Q_i\right]Y_{20}\right\}$; $\Delta n = \pm 1$, $\Delta J = 0, \pm 2$, $\Delta M = 0$

This term leads to a combined vibration-rotation Raman effect by the combined $Q_i Y_{20}$ operators.

$E_z \sum_i \left(\frac{\partial \alpha}{\partial Q_i}\right)_0 Q_i$; $\Delta n = \pm 1$, $\Delta J = 0$, $\Delta M = 0$

This term leads to the $\Delta n = \pm 1$ pure vibrational selection rules through the Q_i operator. There are no rotational operators in this term.

The other terms in Eq.(4-273) lead to the vibrational overtone and combination Raman excitation [30].

We can give a classical description of the Raman process by returning to Eq.(4-273) and substituting the sinusoidal forms of E_z for an electromagnetic wave and Q_i for the ith normal mode,

$$E_z = E_z^0 \cos(\omega_0 t), \qquad Q_i = Q_i^0 \cos(\omega_v t). \tag{4-274}$$

ω_0 is the angular frequency of the radiation and ω_v is the angular frequency of the vibration. Substituting into the first two terms of Eq.(4-273) gives (for a single mode, Q_i)

$$D_z = \gamma E_z + B E_z Q_i, \qquad \gamma = \left[\frac{4}{3}\left(\frac{\pi}{5}\right)^{1/2}(\alpha_{aa}^0 - \alpha_{bb}^0)Y_{20} + \alpha^0\right]$$

$$B = \left\{\frac{4}{3}\left(\frac{\pi}{5}\right)^{1/2}\left[\left(\frac{\partial\alpha_{aa}}{\partial Q_i}\right)_0 - \left(\frac{\partial\alpha_{bb}}{\partial Q_i}\right)_0\right]Y_{20} + \left(\frac{\partial\alpha}{\partial Q_i}\right)_0\right\}$$

$$\begin{aligned} D_z &= \gamma E_z^0 \cos\omega_0 t + B Q_i^0 E_z^0 \cos\omega_0 t \cos\omega_v t \\ &= \underbrace{\gamma E_z^0 \cos\omega_0 t}_{\text{Rayleigh}} + \frac{1}{2} B Q_i^0 E_z^0[\underbrace{\cos(\omega_0 - \omega_v)t}_{\text{Raman Stokes}} + \underbrace{\cos(\omega_0 + \omega_v)t}_{\text{Raman anti-Stokes}}]. \end{aligned} \tag{4-275}$$

According to this equation, the induced dipole, D_z, will oscillate at three frequencies giving rise to Rayleigh scattering at the frequencies of the inducing electric field and Raman scattering at vibrational frequencies below (Stokes) and above (anti-Stokes) the frequency of the inducing radiation.

The discussion above of the Raman effect was limited to the parallel vibrations of a linear molecule where the polarizability tensor in the principal inertial axis system is diagonal with $\alpha_{bb} = \alpha_{cc}$. A more general expression involving the direction cosines is easily obtained by returning to the form of $\boldsymbol{\alpha}_m$. In the case of the $ClCH_3$ symmetric rotor for instance, $\alpha_{ab} = \alpha_{ac} = \alpha_{bc} = 0$, $\alpha_{aa} \neq 0 \neq \alpha_{bb} = \alpha_{cc}$, where the Cl—C bond is the a axis. This leads to

$$\alpha_{zz} = C_{az}^2 \alpha_{aa} + C_{bz}^2 \alpha_{bb} + C_{cz}^2 \alpha_{bb} \tag{4-276}$$

and so on. Little simplification is present in the general asymmetric rotor, which leads to virtually any combination of selection rules on the vibrational and rotational states.

We shall now describe the determination of the Raman selection rules for a nonrotating molecule by group theory. From Eq.(4-271) when $E_x = E_y = 0$, we have

$$D_z \text{ (induced)} = \alpha_{zz} E_z, \qquad D_y \text{ (induced)} = \alpha_{yz} E_z, \qquad D_x \text{ (induced)} = \alpha_{xz} E_z,$$

where x, y, and z are the laboratory-based coordinates. Now, in a manner similar to Eq.(4-270), the following matrix element determines the selection rules:

$$\int_1 \int_2 \cdots \int_{3n-6} \psi(n_1' n_2' \cdots n_{3n-6}') \begin{pmatrix} \alpha_{zz} \\ \alpha_{yz} \\ \alpha_{xz} \end{pmatrix} \psi(n_1 n_2 \cdots n_{3n-6})\, dQ_1\, dQ_2 \cdots dQ_{3n-6}. \tag{4-277}$$

If the integrand contains the totally symmetric representation, the transition is allowed. The polarizabilities are expanded in the normal coordinates according to

$$\alpha_{zz} = \alpha_{zz}^0 + \sum_{i=1}^{3n-6}\left(\frac{\partial \alpha_{zz}}{\partial Q_i}\right)Q_i + \frac{1}{2}\sum_{i,j=1}^{3n-6}\left(\frac{\partial^2 \alpha_{zz}}{\partial Q_i\,\partial Q_j}\right)Q_iQ_j + \cdots.$$

The symmetry classification of the α_{zz}, α_{yz}, and α_{xz} operators will be the same as the z^2, yz, and xz functions that are tabulated in the character tables. Thus, the selection rules are determined by multiplying the irreducible representations formed by the ground-state function, the excited-state function, and the polarizability function. Remembering that the ground state is totally symmetric, we can obtain the Raman selection rules for the fundamental vibrations by finding equivalent polarizability and excited-state symmetries. We shall now return to Fig. E-4 and the

$$CH_4 \longleftrightarrow CH_3D \longleftrightarrow CH_2D_2 \longleftrightarrow CHD_3 \longleftrightarrow CD_3$$

correlation to determine the Raman active ground- to excited-state transitions.

Allowed Raman transitions from the ground to first excited state from Table E-5

T_d CH_4, CD_4

$\alpha_{xx} + \alpha_{yy} + \alpha_{zz}$ excites the A_1 normal mode; $\alpha_{xx} + \alpha_{yy} - 2\alpha_{zz}$ and $\alpha_{xx} - \alpha_{yy}$ excite the E mode; and α_{xy}, α_{xz}, and α_{yz} excite the T_2 mode.

C_{3v} CH_3D, CD_3H

$\alpha_{xx} + \alpha_{yy}$ and α_{zz} excite A_1 and ($\alpha_{xx} - \alpha_{yy}$ and α_{xy}) and (α_{xz} and α_{yz}) excite E.

C_{2v} CH_2D_2

All modes are Raman active.

The selection rule for any Raman vibrational transition is easily obtained by examining the direct product from Eq.(4-277) for the totally symmetric representation.

We shall return to a discussion of the polarization of the Raman scattered light and a discussion of the nature of Rayleigh and Raman scattering in Chapter 8.

PROBLEMS†

1. Draw a potential function describing the barrier to internal rotation about the C—C or C—N bonds in CH_3—CHFCl, CFH_2—CF_3, and COH—NH_2.

† Asterisks indicate problems that require the use of a digital computer.

2. The matrix solution for the energy levels for the hindered internal rotor in the free rotor basis is given in Section 4.2.
   (a) Use perturbation theory and include terms up through second order to calculate the energy of the ground state of a methyl group rotor with $(1/hc)(\hbar^2/2I) = 17.0\ \text{cm}^{-1}$ under the influence of the barrier perturbation with $V_3 = 100\ \text{cm}^{-1}$ (see Fig. 4-4).
   (b) The probability that one of the methyl protons is in the region from $\phi = -\pi/6$ to $\phi = +\pi/6$ is 0.167 in the absence of a barrier. Use perturbation theory and the parameters in part a to compute the same probability in the presence of the barrier.
   (c) A reasonable trial wavefunction for the ground state that will acknowledge the barrier as illustrated in Fig. 4-4 is

$$\psi = \left(\frac{1}{2\pi}\right)^{1/2} + b \cos 3\phi,$$

   with variational parameter, b. Compute the variational energy of the ground state using this trial function, the Hamiltonian in Eq.(4-35), and the parameters in part a.
   *(d) Check by direct matrix diagonalization the results in Figs. 4-4 and 4-5. Use up to a 20×20 or 30×30 matrix.
   *(e) Use the free rotor basis and matrix diagonalization techniques to fit the observed energy differences in CH_3CH_2F as shown in Fig. 4-3.
   *(f) The observed lowest torsional energy spacings in CH_3CH_3 are given by S. Weiss and G. E. Leroi, *J. Chem. Phys.* **48**, 962 (1968). The results are

$n = 0 \longrightarrow n = 1$		$289\ \text{cm}^{-1}$
		$255\ \text{cm}^{-1}$
$n = 1 \longrightarrow n = 2$		
		$258\ \text{cm}^{-1}$

   The $n = 2$ torsional energy level is apparently split into two levels, as shown in Fig. 4-5. Use the free rotor basis and matrix diagonalization to find V_3 and V_6 for ethane from the data above.

3. In some cases, diatomic molecules (and other molecules) are able to rotate in the solid state; the prime example being the nearly free rotation of H_2 in solid hydrogen. Apparently, I_2 also rotates in the pure I_2 solid even at very low temperatures. The apparent form for the potential function resembles a cylindrical box [31] with energy minima occurring when the I_2 internuclear line is aligned along the axis of the cylinder,

$$V(\theta) = \frac{V_2}{2}(1 - \cos 2\theta), \tag{4-278}$$

   where θ is the standard spherical polar angle.
   (a) Set up the matrix representation of the complete Hamiltonian in the spherical harmonic basis including the $J = 0$, $J = 1$, and $J = 2$ states.
   (b) Show schematically by perturbation theory how the original $J = 0$, 1, and 2 states are affected by increasing the value of V_2 from the initial value of $V_2 = 0$.

Consider now the breakdown of the degeneracies by group theory. The matrix representations of the spherical harmonics under the $D_{\infty h}$ group (cylindrical barrier) are reducible.

(c) Find the reducible representations of the spherical harmonics (up to $J = 6$) under the operations of the $D_{\infty h}$ group that is appropriate for the cylindrical symmetry described above.

In the case of a high barrier, the molecule would appear to vibrate in the cylindrical box described above.

(d) Describe how you would modify Eq.(4-278) to describe the vibrational motion.

(e) What is the vibrational energy spacing for I_2 in the cylinder for $V_2 = 3000\ \text{cm}^{-1}$ with an internuclear distance of 2.67×10^{-8} cm?

(f) Generalize the results above to draw a correlation diagram between the free rotor and the vibrator in the cylinder.

(g) Decompose the reducible representations in part c into the irreducible components under $D_{\infty h}$ and label the levels in the correlation diagram above.

*(h) Write down and diagonalize the Hamiltonian matrix for the rigid rotor in the presence of the cylindrical well in Eq.(4-278) to obtain the lower energy levels (up through $J = 3$) for barriers of $V_2 = 10\ \text{cm}^{-1}$, $V_2 = 100\ \text{cm}^{-1}$, and $V_2 = 300\ \text{cm}^{-1}$. Use a matrix that includes terms up to $J = 6$ if possible (49×49 matrix).

*(i) At what value of V_2 does the (ground state) $\longrightarrow$ (lowest excited state) transition frequency in I_2 equal $\frac{1}{2}$ the transition frequency when $V_2 = 0$?

4. Deuterium and hydrogen peroxide are interesting molecules as the equilibrium structure is neither *cis* nor *trans*. The plane of one DOO group in deuterium peroxide is at an equilibrium angle of 110.8° from the plane of the other DOO group as shown in Fig. 4-37 along with the potential function for internal rotation about the O—O bond. It is evident from the potential function and structure of the molecule that there is a dominant V_1 potential due to the deuteron-deuteron repulsions and a smaller V_2 potential that opposes the *trans* conformer as the low energy form. There must also be a V_3 term in order to stabilize the 110.8° angle as shown in the figure.

(a) Use the V_1, V_2, V_3 terms in Eq.(4-42) and the appropriate zero-order basis set in order to construct the Hamiltonian matrix for the internal rotation in D_2O_2.

*(b) Diagonalize the matrix above as a function of V_1, V_2, and V_3 and determine the correct values of the barriers in order to fit the experimental energy spacings in Fig. 4-37. It may be necessary to use a 30×30 matrix (or larger) in order to obtain a good fit of the experimental numbers.

*(c) Construct the transition dipole matrix (electric dipole) in the free rotor basis and transform to the basis that diagonalized the Hamiltonian matrix above. What are the relative intensities for the transitions shown in Fig. 4-37 at $T = 300$ K?

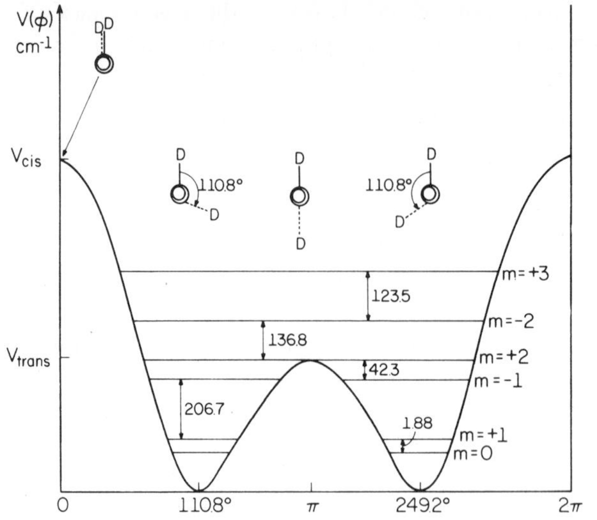

Figure 4-37 The energy levels for internal rotation about the O—O bond in deuterium peroxide as given by R. H. Hunt and R. A. Leacock, *J. Chem. Phys.* **45**, 3141 (1966); see also C. S. Ewig and D. O. Harris, *J. Chem. Phys.* **52**, 6268 (1970). The numbers are in units of cm^{-1}. The structure of D_2O_2 is given here along with $d(O—O) = 1.475$ Å, $d(O—D) = 0.950$ Å, and $<DOD = 94.5°$.

5. The particle-in-a-box is described by a Hamiltonian that constrains the particle to be within the three-dimensional boundaries defined by

$$V_x = V_x^0 \text{ from } x = 0 \longrightarrow \ell_x; \quad V_x = \infty \text{ outside these boundaries,}$$
$$V_y = V_y^0 \text{ from } y = 0 \longrightarrow \ell_y; \quad V_y = \infty \text{ outside these boundaries,}$$
$$V_z = V_z^0 \text{ from } z = 0 \longrightarrow \ell_z; \quad V_z = \infty \text{ outside these boundaries.}$$

Schrödinger's equation is

$$\mathscr{H}\psi(x, y, z) = \left(-\frac{\hbar^2}{2m}\nabla^2 + V_x^0 + V_y^0 + V_z^0\right)\psi(x, y, z) = E\psi(x, y, z),$$

which leads to three separable equations given by

$$-\frac{\hbar^2}{2m}\frac{d^2}{dx^2}\psi(x) = (E_x - V_x^0)\psi(x)$$

$$-\frac{\hbar^2}{2m}\frac{d^2}{dy^2}\psi(y) = (E_y - V_y^0)\psi(y)$$

$$-\frac{\hbar^2}{2m}\frac{d^2}{dz^2}\psi(z) = (E_z - V_z^0)\psi(z), \tag{4-279}$$

where $E = E_x + E_y + E_z$ and $\psi(x, y, z) = \psi(x)\psi(y)\psi(z)$.

(a) Solve the equations above to give

$$E(n_x, n_y, n_z) = \frac{h^2}{8m}\left[\left(\frac{n_x}{\ell_x}\right)^2 + \left(\frac{n_y}{\ell_y}\right)^2 + \left(\frac{n_z}{\ell_z}\right)^2\right] + V_x^0 + V_y^0 + V_z^0$$

$$\psi(n_x, n_y, n_z) = 2\left(\frac{2}{\ell_x \ell_y \ell_z}\right)^{1/2} \sin\left(\frac{n_x \pi}{\ell_x} x\right) \sin\left(\frac{n_y \pi}{\ell_y} y\right) \sin\left(\frac{n_z \pi}{\ell_z} z\right) \quad (4\text{-}280)$$

$$n_x = 1, 2, 3, \ldots; \qquad n_y = 1, 2, 3, \ldots; \qquad n_z = 1, 2, 3, \ldots.$$

(b) Draw an energy level diagram for a particle-in-a-cube where $V_x^0 = V_y^0 = V_z^0$ and $\ell_x = \ell_y = \ell_z = \ell$ showing the degeneracies. Show how the degeneracies of the particle-in-a-cube energy levels are broken as the z face of the cube is extended to give $\ell_x = \ell_y < \ell_z$.

(c) It is possible to trap a hydrogen atom substitutionally in a rare gas lattice. Assuming the particle-in-a-cube model is appropriate, calculate the energy spacings for the translational energy levels of the trapped atom in the Ne, Ar, Kr, and Xe lattices. Where would transitions be observed (microwave, infrared, or optical)?

6. Calculate the quantum mechanical average values of the potential and kinetic energies for the hydrogen atom in the ground state and show that your results obey the virial theorem.

7. Consider the effects of a static electric field along a z axis, E_z, on the electron in the hydrogen atom. The perturbation Hamiltonian is given by

$$\mathscr{H}_1 = E_z er \cos\theta. \quad (4\text{-}281)$$

(a) Write down the 5×5 matrix representation of the complete Hamiltonian [$\mathscr{H}_0 = \mathscr{H}_{in}$ is in Eq.(4-47)] in the $1s$, $2s$, $2p_x$, $2p_y$, and $2p_z$ basis set given in Eq.(4-103) and show that the ground-state energy, corrected to second order, is given by

$$E_{1s} = -\frac{e^2}{2a} + \frac{[(2)^{1/2}(\frac{128}{243})aE_z e]^2}{-(e^2/2a) + (e^2/8a)} = -\frac{e^2}{2a} - (1.480)a^3E_z^2, \quad (4\text{-}282)$$

and that the ground-state wavefunction, corrected to first order, is given by

$$\psi_{1s} = \psi_{1s}^0 - \left[\frac{8(2)^{1/2}}{3}\left(\frac{128}{243}\right)\frac{a^2E_z}{e}\right]\psi_{2p_z}^0 = \psi_{1s}^0 - 1.986\left(\frac{a^2E_z}{e}\right)\psi_{2p_z}^0, \quad (4\text{-}283)$$

where a is defined in Eq.(4-88). Show also that for typical laboratory fields of $E_z = 3000\ \text{V}\cdot\text{cm}^{-1}$ that this function is already normalized to an accuracy of about one part in 10^6.

We shall now use a z-polarized ψ_{1s}^0 orbital and the variation method to calculate the energy of the hydrogen atom in an electric field.

(b) Using the variational function

$$\psi_{1s} = \psi_{1s}^0 + b\left(\frac{z}{a}\right)\psi_{1s}^0 = \psi_{1s}^0\left(1 + \frac{br}{a}\cos\theta\right), \quad (4\text{-}284)$$

where a is in Eq.(4-88) and b is a variational parameter, show that the variational energy is given by

$$E = \frac{\int \psi_{1s}^*(\mathscr{H}_0 + erE_z \cos\theta)\psi_{1s}\,dV}{\int \psi_{1s}\psi_{1s}\,dV} = \frac{-(e^2/2a) + 2beE_z a}{1 + b^2}.$$

It is nontrivial to show that one integral needed above,

$$\int b\left(\frac{r}{a}\cos\theta\right)\psi_{1s}^0 \mathscr{H}_0 b\left(\frac{r}{a}\cos\theta\right)\psi_{1s}^0\,dV,$$

is identically zero.

(c) Show by minimizing the energy with respect to b where $eE_z a \ll e^2/2a$ that

$$b \cong -\frac{2E_z a^2}{e}. \tag{4-285}$$

(d) Substitute b back into E to give the energy. Compare with Eq.(4-282).

(e) Truncate the result in part d at the E_z^2 terms and use Eq.(4-193) to give the scalar polarizability of the hydrogen atom in the ground state.

The results above can also be determined by a straightforward application of the matrix equation given in Eq.(3-125). In the present case with the trial function in Eq.(4-284) we write

$$\psi_{1s} = b_1\psi_{1s}^0 + b_2\left(\frac{z}{a}\right)\psi_{1s}^0 = b_1\chi_1 + b_2\chi_2, \tag{4-286}$$

to identify χ_1 and χ_2 according to the discussion beginning with Eq.(3-121). Using this trial function and the Hamiltonian for the hydrogen atom in an electric field, we can write Eq.(3-125) for this system as

$$\begin{pmatrix} -e^2/2a & eE_z a \\ eE_z a & 0 \end{pmatrix}\begin{pmatrix} b_1 \\ b_2 \end{pmatrix} - \begin{pmatrix} 1 & 0 \\ 0 & 1 \end{pmatrix}\begin{pmatrix} b_1\varepsilon \\ b_2\varepsilon \end{pmatrix} = \begin{pmatrix} 0 \\ 0 \end{pmatrix}. \tag{4-287}$$

We note that the overlap matrix is a unit matrix because χ_1 and χ_2 are orthonormal as defined in Eq.(4-286).

(f) Solve Eq.(4-287) for the ground-state energy, b_1, and b_2, and show that the results are identical to those found earlier in parts c and d.

(g) In order to improve the variational function we now use [see Eq.(4-284)]

$$\psi_{1s} = \psi_{1s}^0\left(1 + \frac{br}{a}\cos\theta + \frac{c}{a}r^2\cos\theta\right), \tag{4-288}$$

where a is in Eq.(4-88) and b and c are variational parameters. In parts b through e we considered the variational energy and function for $c = 0$. Calculate the variational energy, function, and polarizability with Eq. (4-288) with $b = 0$.

(h) Repeat the variational calculation of the energy, function, and polarizability using Eq.(4-288) with nonzero values of both b and c.

8. Use a Gaussian trial function,

$$\psi_{1s} = \exp(-br^2) \tag{4-289}$$

(where b is a variational parameter), and the variation method to compute the ground-state energy for the hydrogen atom. Compare your result to the correct energy.

9. The interaction potential between a neutron and proton is given by

$$V = -A \exp\left(-\frac{r}{a}\right), \tag{4-290}$$

where $A = 32$ MeV and $a = 2.2 \times 10^{-13}$ cm. This equation represents a significantly steeper potential function than the Coulomb potential. Use a trial function of

$$\psi = \exp\left(-\frac{\alpha r}{2a}\right) \tag{4-291}$$

(where α is a variational parameter) to compute by the variation method the energy of the ground state of the proton-neutron system.

10. The raising operator, L_+, and lowering operator, L_-, are defined by

$$L_{\pm} = L_x \pm iL_y, \tag{4-292}$$

where L_x and L_y are the normal angular momentum operators. Consider the raising and lowering operation on the spherical harmonics, Y_{LM}:

$$L_{\pm} Y_{LM} = C(L, M) Y_{LM\pm 1}, \tag{4-293}$$

where $C(L, M)$ is some function of L and M.

(a) Show that the raising and lowering operators obey the following commutator equation:

$$[L_+, L_-] = 2\hbar L_z. \tag{4-294}$$

(b) Show from the $L = 1$ ($M = 0, \pm 1$) matrix representations of L_+, L_-, and L_z in the spherical harmonic basis that

$$\mathbf{L}_+\mathbf{L}_- - \mathbf{L}_-\mathbf{L}_+ = 2\hbar \mathbf{L}_z.$$

$\mathbf{L}_+$ is the matrix representation of L_+ in the spherical harmonic basis.

11. In Section 4.8 we noted that relativistic corrections are needed in addition to the spin-orbit interaction in order to describe the experimental energy levels in Table 4-7. Our previous spin-orbit theory gave only the doublet splittings ($^3P_{1/2}$—$^3P_{3/2}$, $^2D_{3/2}$—$^2D_{5/2}$, etc.). We shall now consider the relativistic corrections to the hydrogen-like atom where the rest mass (m_0) energy of an electron is given by $E_{rm} = m_0 c^2$. The total energy of an electron is given by

$$E_t = mc^2 = \frac{m_0 c^2}{(1 - \beta^2)^{1/2}}, \tag{4-295}$$

where $\beta^2 = v^2/c^2$ and $m = m_0/(1 - \beta^2)^{1/2}$ is the mass of the electron at a velocity v. Thus, the kinetic energy of the electron is

$$T = E_t - E_{rm} = \frac{m_0 c^2}{(1 - \beta^2)^{1/2}} - m_0 c^2 = m_0 c^2\left[\frac{1}{(1 - \beta^2)^{1/2}} - 1\right]. \qquad (4\text{-}296)$$

The square of the linear momentum is given by

$$p^2 = m^2 v^2 = \frac{m_0^2 v^2}{1 - \beta^2} = \frac{m_0^2 c^2 \beta^2}{1 - \beta^2} = m_0^2 c^2\left[\frac{1}{1 - \beta^2} - 1\right].$$

Solving for $1/(1 - \beta^2)$ and substituting this result into Eq.(4-296) gives

$$T = m_0 c^2\left[\left(1 + \frac{p^2}{m_0^2 c^2}\right)^{1/2} - 1\right]. \qquad (4\text{-}297)$$

$p^2/m_0^2 c^2$ will normally be much less than unity, which allows us to rewrite this expression to give

$$T = m_0 c^2\left[\frac{p^2}{2m_0^2 c^2} - \frac{p^4}{8m_0^4 c^4} + \cdots\right]. \qquad (4\text{-}298)$$

Truncating after the p^4 term and adding the electron-proton potential energy gives the Hamiltonian including the relativistic correction,

$$\mathscr{H} = \frac{p^2}{2m_0} - \frac{p^4}{8m_0^3 c^2} - \frac{Ze^2}{r} = \underbrace{\frac{p^2}{2m_0} - \frac{Ze^2}{r}}_{\mathscr{H}_0} \underbrace{- \frac{p^4}{8m_0^3 c^2}}_{\mathscr{H}_1}. \qquad (4\text{-}299)$$

It is easy to show that $\mathscr{H}_1 \ll \mathscr{H}_0$ from classical arguments.

(a) Find the first-order corrections to the energy in the hydrogen atom from $\mathscr{H}_1$ in Eq.(4-299) and show by combining your result with the spin-orbit energy that the $2^2S_{1/2}$ and the $2^2P_{1/2}$ energy levels are now degenerate (review Table 4-7).

(b) Use the Hamiltonian in Eq.(4-299) to derive the quantum mechanical virial expression for the relativistically corrected Hamiltonian beginning with the $(\mathscr{H}, \boldsymbol{r} \cdot \boldsymbol{p})$ commutator in Eq.(3-153). Compare with the results in Eq.(3-153).

12. Compute the low magnetic field perturbation on the Na atom D-line emission ($3^2P_{1/2} \longrightarrow 3^2S_{1/2}$ and $3^2P_{3/2} \longrightarrow 3^2S_{1/2}$) and compare your prediction with the available experimental results.

13. Calculate the rate of spontaneous emission from the $2^2P_{1/2}$, $2^2P_{3/2}$, and $2^2S_{1/2}$ states to the ground state in the hydrogen atom (see also Ref. [14]).

14. Consider the methylene diradical [32]

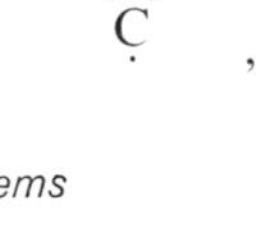

,

where the two unpaired electrons lead to singlet and triplet electronic states. Assuming a large energy separation and no coupling between the singlet and triplet states,

(a) Write the triplet state Hamiltonian describing the methylene radical in the presence of a magnetic field.

(b) Compute the diagonal matrix elements of the Hamiltonian in part a in the uncoupled basis. Make a table of high field energies in order of increasing energy.

(c) What are the coupled basis F states, where F is the total angular momentum?

(d) Draw a correlation diagram between the energy levels determined in the coupled and uncoupled basis sets similar to Figs. 4-12, 4-13, and 4-14.

(e) Show all the allowed high field ESR transitions on your correlation diagram and predict the nature of the ESR spectra.

15. We discussed the ESR spectra of the methyl radical (Fig. 4-14), which has an experimental nuclear spin–electron spin coupling constant of $a = 64.5$ MHz. Consider now the $D_3^{12}C\cdot$ radical.

(a) What is the value of a in $D_3^{12}C\cdot$?

(b) Construct a high field energy level diagram for the $D_3^{12}C\cdot$ radical.

(c) Show the allowed transitions on your energy level diagram. Sketch out the predicted spectra and compare with the $H_3^{12}C\cdot$ spectra.

(d) Predict the high field ESR spectra of $H_2D^{12}C\cdot$ and $H_3^{13}C\cdot$ and find an experimental verification of your predictions.

16. Predict schematically, according to the discussion in Section 4.9 on the nature of the relative magnitudes of electron spin–nuclear spin coupling constants, the high field ESR spectra of the styrene radical.

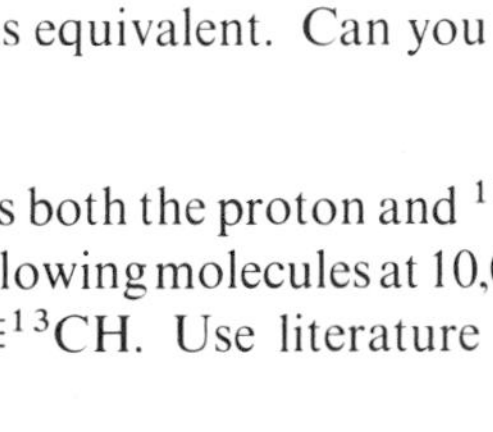

Treat all benzene protons as equivalent. Can you find an experimental verification of your prediction?

17. Predict on a first-order basis both the proton and ^{13}C high field nuclear magnetic resonance spectra for the following molecules at 10,000 G: $^{13}CH_4$, $^{13}CH_2{=}CH_2$, $CH_3{-}^{13}COH$, and $HC{\equiv}^{13}CH$. Use literature values for the coupling constants when appropriate.

18. Predict the ^{19}F and ^{31}P first-order NMR spectra in the following molecules at 10,000 G: $^{31}PF_3$, $H^{31}PF_2$, $H_2^{31}PF$, $H_3^{31}P$, $O^{31}PF_3$, $^{31}PF_5$, $O^{31}PClF_2$, and $O^{31}PCl_2F$. Assume that the Cl atom does not interact with the other magnetic nuclei due to rapid Cl nuclear quadrupole relaxation.

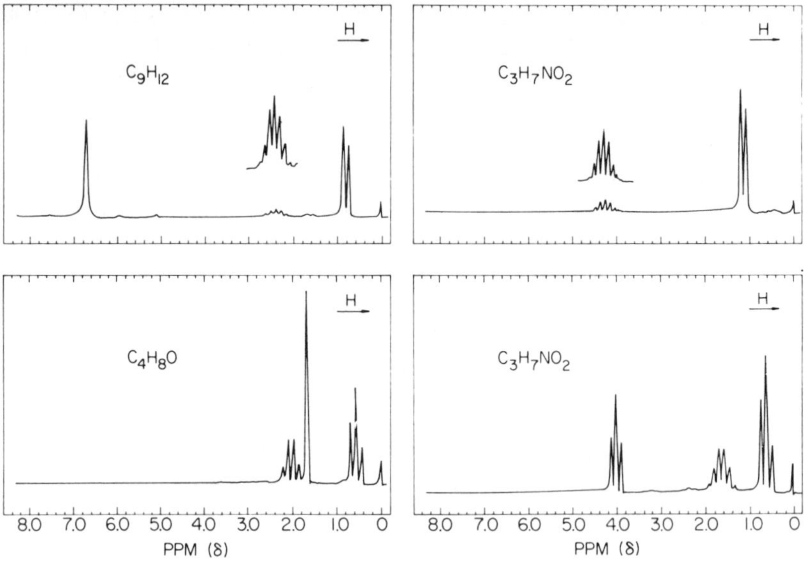

Figure 4-38 Proton NMR spectra where the chemical shifts are relative to $Si(CH_3)_4$ at $\delta = 0.0$. These spectra were recorded by M. Meadows.

19. Figure 4-38 shows a number of proton magnetic resonance spectra near 60.0 MHz. The scales are in ppm chemical shifts relative to $Si(CH_3)_4$ (see Fig. 4-21). Use the chemical shifts and the spin-spin information to deduce the structure of each compound from the empirical formula. Many additional examples are found in J. C. Davis, *Advanced Physical Chemistry* (The Ronald Press Co., New York, 1965) and J. D. Roberts, *An Introduction to the Analysis of Spin-Spin Splitting in High Resolution Nuclear Magnetic Resonance Spectra* (W. A. Benjamin, Inc., New York, 1962).

***20.** Acetaldehyde has a proton magnetic resonance spectra described by the chemical shift and spin-spin coupling constants of $\sigma(CH_3) - \sigma(CHO) = 7.5 \times 10^{-6}$ and $J_{CH_3-CHO} = 2.90$ Hz, respectively. Transform the appropriate four-spin Hamiltonian and transition matrices to predict the proton NMR spectra of CH_3CHO at 1000, 5000 and 10,000 G. Make plots of the spectra.

***21.** ClF_3 has been studied by E. L. Muetterties and W. D. Phillips, *J. Am. Chem. Soc.* **79**, 322 (1957) and L. G. Alexakos and C. D. Cornwell, *J. Chem. Phys.* **39**, 844 (1963). The structure shows that two of the F nuclei in ClF_3 are in equivalent chemical environments.

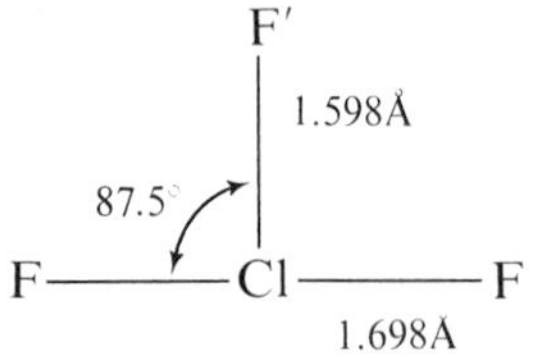

The coupling constant and chemical shift between the two chemically different sets of nuclei are (gas phase) $J_{FF'} = 441$ Hz and $\sigma_{(F')} - \sigma_{(F)} = 125.9 \times 10^{-6}$. Predict the ^{19}F resonance spectra in ClF_3 at 10-, 20-, and 30-MHz frequencies. Compare your results with the published spectra. Is the agreement satisfactory?

22. *n*-Propyl nitrite can exist in two possible isomers:

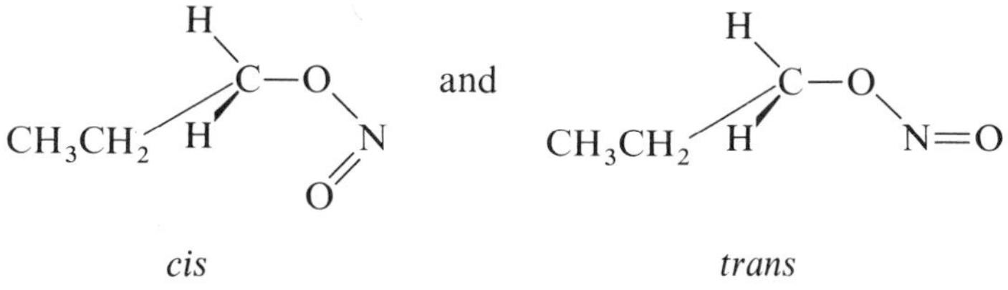

Assume that we know that the barrier about the CO bond axis is low in both forms so that the *n*-propyl group protons nearest the oxygen are equivalent. The proton NMR spectrum at $T = 200$ K shows two lines of relative intensity 2:1 [33]. The stronger line is at a higher frequency at constant field.

(a) Which isomer is at the lowest energy?

(b) What is the approximate energy difference between the *cis* and *trans* forms of *n*-propyl nitrite?

23. Consider the CXF_2CXYZ molecule where X, Y, and Z are magnetically inert nuclei. Predict the low temperature (locked configuration) and high temperature (free internal rotation) ^{19}F NMR spectra.

24. The potential function for $CH_2F—CH_2F$ is shown in Fig. 4-6. Now consider the corresponding $CX_2F—CX_2F$ molecule where X is some magnetically inert group.

(a) Assuming the potential function in Fig. 4-6 is still valid, predict the ^{19}F NMR spectrum with correct relative intensities at $T = 200$ K.

(b) What would the corresponding high temperature spectrum look like? The high temperature spectrum is obtained by assuming that there is free internal rotation about the C—C bond.

25. Consider the use of nuclear magnetic double resonance in the AX HC≡CF system shown in Fig. 4-17 to uncouple the proton and fluorine nuclei. We shall investigate the nature of the proton spectrum under the influence of strong radiation at the fluorine nuclear magnetic resonance as denoted by H{F}. The time-independent Hamiltonian in the rotating coordinate system for the two-spin system can be written as [34]

$$\mathscr{H}_R = -(\omega_H - \omega_F')I_{zH} - (\omega_F - \omega_F')I_{zF} + \frac{hJ_{HF}}{\hbar^2}\mathbf{I}_H \cdot \mathbf{I}_F - (\gamma_F I_{xF} + \gamma_H I_{xH})H_F$$

$$\omega_H = \frac{\mu_0 g_H(1 - \sigma_H)H_z}{\hbar}, \qquad \omega_F = \frac{\mu_0 g_F(1 - \sigma_F)H_z}{\hbar}, \qquad \omega_F' = \frac{\mu_0 g_F(1 - \sigma_F)H_F}{\hbar}. \tag{4-300}$$

H_z is the static magnetic field along the z axis, H_F (along the x axis) is the strong radio-frequency field used at the ^{19}F resonance where $H_z > H_F$. ω'_F is the angular frequency of the rotating coordinate system.

(a) Use the Hamiltonian above to set up the matrix representation in the uncoupled basis.

(b) Show that the proton spectrum during the ^{19}F irradiation with H_F will appear as a doublet when $hJ_{HF} > \mu_0 g_F H_F$. As $hJ_{HJ} \cong \mu_0 g_F H_F$, the doublet spectrum spreads out and another line appears at ν_H. Show further that the single line at ν_H dominates the spectrum as $hJ_{HF} < \mu_0 g_F H_F$.

26. A few years ago there was some question about the molecular structure of a simple linear triatomic in which there was conflicting evidence for either HCP or HPC. However, the microwave spectrum gave the following results:

Transition	HCP or HPC	H^{13}CP or HP^{13}C	DCP or DPC	D^{13}CP or DP^{13}C
$J = 0 \longrightarrow J = 1$	39,951.98	38,278.46	33,968.73	32,855.00
$J = 1 \longrightarrow J = 2$	79,903.28		67,937.07	

(a) Is the molecule HCP or HPC?

(b) What are the bond distances?

27. The molecular structure of the planar formic acid is given here. The bond distances are given in Å units and the angles are in degrees.

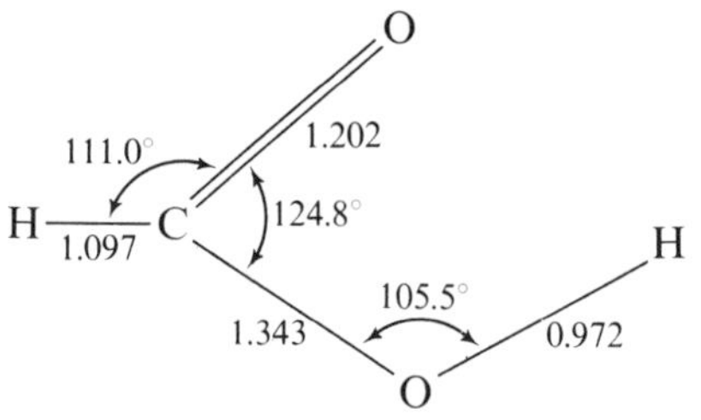

(a) Find the c.m. and compute the principal moments of inertia.

(b) Show that the molecule is a near-prolate symmetric top.

(c) Compute, from the moments of inertia, the frequencies of the near-prolate microwave transitions shown in Fig. 4-28b.

***28.** Use the rotational constants for one of the molecules in Table 4-21 to set up the rotational Hamiltonian matrix in the near-prolate symmetric rotor basis [Eq.(4-228)]. Diagonalize a sufficiently large matrix to predict the lower J rotational transitions and compare with the literature values for the observed transitions.

29. In the absence of any external fields, the probability that a free linear molecule in the ground rotational state is orientated within a cone bounded by $\theta = 0 \longrightarrow \pi/6$ and $\phi = 0 \longrightarrow 2\pi$ is 0.067 or 6.7%. What is the probability that HCN in the ground rotational state is within the same cone in the presence of a field of $E_z = 3000\ \mathrm{V \cdot cm^{-1}}$ (use perturbation theory).

30. Equation (4-214) gives the rotational magnetic moment for a linear molecule as $\boldsymbol{\mu}_J = g(\mu_0/\hbar)\boldsymbol{J}$ where g is the scalar molecular g-value, μ_0 is the nuclear magneton, and $\boldsymbol{J}$ is the rotational angular momentum. In a nonlinear molecule the g-value is a 3×3 Cartesian tensor, $\mathbf{g}$, giving the following rotational magnetic moment:

$$\boldsymbol{\mu}_J = \frac{\mu_0}{\hbar}\mathbf{g} \cdot \boldsymbol{J}. \tag{4-301}$$

The Hamiltonian in Eq.(4-215) is modified for a nonlinear molecule to give

$$\mathscr{H}_1 = -\boldsymbol{H} \cdot \boldsymbol{\mu}_J = -\frac{\mu_0}{\hbar}\boldsymbol{H} \cdot \mathbf{g} \cdot \boldsymbol{J}. \tag{4-302}$$

$\boldsymbol{H}$ is defined in the laboratory axis system and it is convenient to describe $\mathbf{g}$ in the molecular-fixed axis system. This is accomplished by the direction cosine transformation as in Eqs.(4-195) giving

$$\mathscr{H}_1 = -\frac{\mu_0}{\hbar}\boldsymbol{H} \cdot \mathbf{C}\mathbf{g} \cdot \boldsymbol{J}$$

$$= -\frac{\mu_0}{\hbar}(H_x \quad H_y \quad H_z)\begin{pmatrix} C_{xa} & C_{xb} & C_{xc} \\ C_{ya} & C_{yb} & C_{yc} \\ C_{za} & C_{zb} & C_{zc} \end{pmatrix}\begin{pmatrix} \mathrm{g}_{aa} & \mathrm{g}_{ab} & \mathrm{g}_{ac} \\ \mathrm{g}_{ba} & \mathrm{g}_{bb} & \mathrm{g}_{bc} \\ \mathrm{g}_{ca} & \mathrm{g}_{cb} & \mathrm{g}_{cc} \end{pmatrix}\begin{pmatrix} J_a \\ J_b \\ J_c \end{pmatrix}. \tag{4-303}$$

In the case of a linear molecule, the $\mathbf{g}$ tensor is diagonal with $\mathrm{g}_{aa} = 0$ and $\mathrm{g}_{bb} = \mathrm{g}_{cc}$. Substituting these results into Eq.(4-303) gives the result in Eq.(4-215) for a linear molecule (when $H_x = H_y = 0$). The $\mathbf{g}$ tensor for a prolate symmetric rotor with the symmetry axis along the a axis is also diagonal with $\mathrm{g}_{aa} \neq 0$ and $\mathrm{g}_{bb} = \mathrm{g}_{cc} \neq 0$.

(a) Write the Hamiltonian in Eq.(4-303) for a symmetric rotor.

(b) Show by using the direction cosine matrix elements for a symmetric rotor in Table 4-20 that the first-order correction due to the Zeeman effect in a symmetric rotor is given by

$$E(J, M, K) = -\mu_0 M H_z\left[\mathrm{g}_{bb} + (\mathrm{g}_{aa} - \mathrm{g}_{bb})\frac{K^2}{J(J+1)}\right]. \tag{4-304}$$

(c) Now show that the term in Eq.(4-213) involving the magnetic susceptibility tensor, $\mathscr{H}_2 = -\frac{1}{2}\boldsymbol{H} \cdot \boldsymbol{\chi} \cdot \boldsymbol{H}$, leads to the following first-order correction to the rotational energy in a symmetric rotor:

$$E(J, M, K) = -\frac{H^2}{3}\left[\frac{3M^2 - J(J+1)}{(2J-1)(2J+3)}\right]$$

$$\times\left[(\chi_{bb} - \chi_{aa}) - \frac{3K^2}{J(J+1)}(\chi_{bb} - \chi_{aa})\right] - \frac{1}{2}H^2\chi. \tag{4-305}$$

(d) Construct an energy level diagram showing the energy levels according to Eqs.(4-304) and (4-305) in a symmetric rotor in a magnetic field including the $J = 0$, 1, and 2 levels. Draw in the transitions and predict the microwave spectrum for the case of parallel static magnetic and electromagnetic electric fields. Compare your results with Fig. 4-26, which shows the similar situation for a linear molecule.

31. The rotational magnetic moment of the H_2 molecule along the axis perpendicular to the internuclear axis is given in Table 4-17 and Eq.(4-214) to be $\mu_\perp = (0.88291 \pm 0.00005)\mu_0(J/\hbar)$, where J is the magnitude of $\mathbf{J}$. Compute the $\Delta J = 0$, $\Delta M = \pm 1$ transition frequencies that are activated by the rotational reorientation of $\mu_\perp$ in the hydrogen molecule for both the $J = 1$ and $J = 2$ rotational states at 3000 G. Draw the energy level diagram and show the transitions. These types of transitions have been observed in several molecules [35].

32. Show that if s and p are the displacement and linear momentum of the harmonic oscillator, the average values of s^2 and p^2 are

$$\langle s^2 \rangle_n = \frac{E_n}{\mu\omega^2}, \qquad \langle p^2 \rangle_n = \mu E_n,$$

where $n = 0, 1, 2, \ldots$ and μ is the reduced mass of the oscillator. Using the standard $\Delta s = (\langle s^2 \rangle_n - \langle s \rangle_n^2)^{1/2}$ definition of uncertainty, calculate $\Delta s \Delta p$ for the harmonic oscillator. Calculate the quantum mechanical average values of the potential and kinetic energies for the $n = 0$ and $n = 1$ vibrational states in HCl and compare with the corresponding classical results at $T = 300$ K.

33. Assume that the bending mode in ketene has a quartic potential

$$V(s) = ks^4 \qquad (4\text{-}306)$$

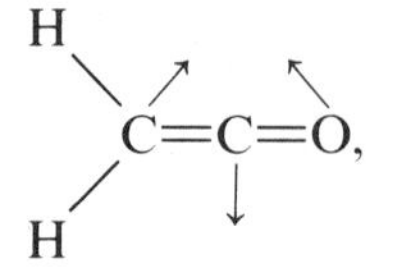

where s is the bending coordinate and reasonable estimates for the quartic force constant and reduced mass for the bending mode are $k = 10^{21}\ \text{erg} \cdot \text{cm}^{-4}$ and $\mu = 4 \times 10^{-23}$ g, respectively. Use the Gaussian function in argument s as a trial wavefunction:

$$\psi = \exp\left(-\frac{\alpha s^2}{2}\right),$$

where α is a variational parameter, and the variation method to compute the ground-state energy of the bending mode in ketene.

(a) Evaluate α by the variation method.

(b) Evaluate the energy of the lowest state of the quartic oscillator.

(c) What additional functional dependence would you include along with the Gaussian to lower the energy of the quartic oscillator?

Table 4-30 The observed and calculated harmonic vibrational fundamental and overtone bands in HCl. The calculated numbers are multiples of the fundamental $0 \longrightarrow 1$ transition. The observed numbers are taken from Ref. [26].

Transition	Observed (cm^{-1})	Calculated (cm^{-1})
$n = 0 \longrightarrow 1$	2885.9	2885.9
$0 \longrightarrow 2$	5668.0	5771.8
$0 \longrightarrow 3$	8346.9	8657.7
$0 \longrightarrow 4$	10,923.1	11,543.6
$0 \longrightarrow 5$	13,396.5	14,429.5

34. It is evident from the observed vibrational transitions in HCl as listed in Table 4-30 that the harmonic oscillator model is breaking down. A more realistic description for the potential energy is the Morse potential given by

$$V(s) = D_e[1 - \exp(-as)]^2, \tag{4-307}$$

as shown in Fig. 4-39. D_e is the dissociation energy from the bottom of the potential well obtained by

$$\lim_{s \to \infty} V(s) = D_e. \tag{4-308}$$

The following questions will involve the determination of D_e and a in the potential function in Eq.(4-307) from the experimental data in Table 4-30.

(a) Expand the Morse potential in a power series about $s = 0$; evaluate the quadratic, cubic, and quartic terms; and write the complete Hamiltonian.

(b) Use cubic and quartic terms in $V(s)$ and perturbation theory through second order to compute the energy levels.

The general form for the vibrational energy levels is given by (see part b)

$$\frac{E_n}{hc} = \frac{v_e}{c}\left(n + \frac{1}{2}\right) - \frac{v_e x_e}{c}\left(n + \frac{1}{2}\right)^2, \tag{4-309}$$

where v_e is the normal fundamental vibrational frequency and x_e is a molecular constant. The Schrödinger equation for the Morse potential function can be solved exactly giving (k is the normal vibrational force constant) [9]

$$\frac{v_e}{c} = \frac{1}{2\pi c}\left(\frac{k}{\mu}\right)^{1/2} = \frac{a}{2\pi c}\left(\frac{2D_e}{\mu}\right)^{1/2}, \qquad \frac{v_e x_e}{c} = \frac{\hbar a^2}{4\pi c \mu}. \tag{4-310}$$

(c) Use Eq.(4-309) and the definitions of v_e and x_e to obtain D_e and a for HCl from the experimental results in Table 4-30.

*(d) Write down the complete Hamiltonian matrix for the Morse oscillator in the lowest 10 states from the work above. Diagonalize this Hamiltonian using a and D_e from the solution in part c and note whether the values E_3, E_4, and E_5 correspond to the experimental data. Remember that $\mathbf{S}^2 = \mathbf{SS}$, $\mathbf{S}^3 = \mathbf{S}^2\mathbf{S} = \mathbf{SSS}$, and so on, where $\mathbf{S}$ is the matrix representation of the displacement coordinate s.

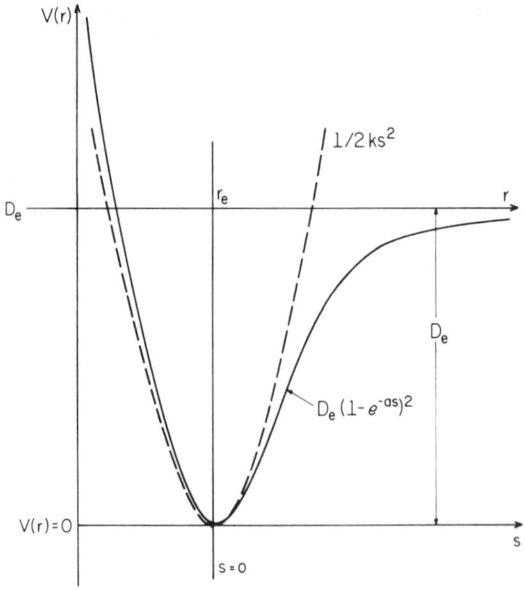

Figure 4-39 The Morse potential function for a diatomic vibrator, $V(s) = D_e[1 - \exp(-as)]^2$. D_e is the dissociation energy and a is an adjustable parameter. The quadratic potential is also shown for comparison.

*(e) Now fit the energy spacings in Table 4-30 by diagonalizing the 10×10 matrix above as a function of D_e and a until a satisfactory fit is obtained.

*(f) Repeat part e for a 20×20 matrix.

(g) Construct the transition dipole moment matrix for HCl in the harmonic oscillator basis. Use only the $(dD/ds)_0\, s$ term in the dipole expansion.

*(h) Transform the harmonic oscillator transition dipole matrix to the final basis obtained in part d or e. Square the final elements in the transition matrix and list the relative intensities for the vibrational Δn transitions shown in Table 4-30 as well as the additional $1 \longrightarrow 2$, $2 \longrightarrow 3$, $3 \longrightarrow 4$, and $2 \longrightarrow 4$ transitions.

Table 4-31 The rotational transitions in $^{12}C^{16}O$ from G. Jones and W. Gordy, *Phys. Rev.* **135**, A295 (1964). The frequencies were observed in the ground vibrational state.

Transition $J \longrightarrow J+1$	ν (MHz)
$0 \longrightarrow 1$	115,271.195 $\pm$ 0.015
$1 \longrightarrow 2$	230,537.974 $\pm$ 0.030
$2 \longrightarrow 3$	345,795.900 $\pm$ 0.090
$3 \longrightarrow 4$	461,040.68 $\pm$ 0.06
$4 \longrightarrow 5$	576,267.75 $\pm$ 0.10
$5 \longrightarrow 6$	691,472.60 $\pm$ 0.60

35. We have discussed the rigid rotor. It is known, however, that as a molecule's angular frequency increases, the bond length will increase. This centrifugal distortion is evident in the $^{12}C^{16}O$ rotational transitions listed in Table 4-31. For instance, six times the observed $J = 0 \longrightarrow J = 1$ frequency is 154.57 MHz higher than the observed $J = 5 \longrightarrow J = 6$ transition. In order to compute the centrifugal distortion effects we balance the centrifugal force, $\mu r \omega^2$ (see Problem 2-3), with Hooke's law force, $k(r - r_0)$,

$$\mu r \omega^2 = k(r - r_0), \qquad (4\text{-}311)$$

where r_0 is the equilibrium bond distance in the ground vibrational state, $r - r_0$ represents the displacement from equilibrium, μ is the diatomic reduced mass, and k is the force constant. Rearranging Eq.(4-311) gives

$$r = \frac{k r_0}{k - \mu \omega^2}. \qquad (4\text{-}312)$$

We can now write the classical Hamiltonian for the rotational motion for the nonrigid rotor:

$$H = \tfrac{1}{2}\mu r^2 \omega^2 + \tfrac{1}{2}k(r - r_0)^2, \qquad (4\text{-}313)$$

where $\frac{1}{2}\mu r^2 \omega^2$ is the rotational kinetic energy and $\frac{1}{2}k(r - r_0)^2$ is the potential energy. Substituting $(r - r_0)^2$ from Eq.(4-311) and remembering that $\mu r^2 \omega = J$, the rotational angular momentum, we can write the rotational Hamiltonian as

$$H = \frac{J^2}{2\mu r^2} + \frac{J^4}{2k\mu^2 r^6}. \qquad (4\text{-}314)$$

The spherical harmonics are eigenfunctions of J^2 and J^4 in Eq.(4-314). The eigenvalues of $\mathscr{H}$ are

$$\langle J, M | \mathscr{H} | J, M \rangle = \frac{\hbar^2 J(J + 1)}{2\mu} \left\langle J, M \left| \frac{1}{r^2} \right| J, M \right\rangle + \frac{\hbar^4 J^2 (J + 1)^2}{2k\mu^2} \left\langle J, M \left| \frac{1}{r^6} \right| J, M \right\rangle, \qquad (4\text{-}315)$$

where we note the $1/r^2$ and $1/r^6$ are still functions of the rotational motion through Eq.(4-312). Returning to Eq.(4-312) we can write

$$\frac{1}{r^2} = \frac{1}{r_0^2}\left(1 - \frac{\mu\omega^2}{k}\right)^2. \qquad (4\text{-}316)$$

Noting that $\mu\omega^2/k \ll 1$ we can expand Eq.(4-316) to give

$$\frac{1}{r^2} = \frac{1}{r_0^2}\left(1 - \frac{2\mu\omega^2}{k}\right). \qquad (4\text{-}317)$$

Substituting Eq.(4-317) into Eq.(4-315) gives

$$\langle J, M|\mathscr{H}|J, M\rangle = \frac{\hbar^2 J(J+1)}{2\mu r_0^2} - \frac{\hbar^2 J(J+1)}{\mu r_0^2 k}\langle J, M|\mu\omega^2|J, M\rangle + \frac{\hbar^4 J^2(J+1)^2}{2k\mu^2}\left\langle J, M\left|\frac{1}{r^6}\right|J, M\right\rangle. \tag{4-318}$$

Now substituting $\mu\omega^2 = (\mu^2 r^4/\mu r^4)\omega^2 = J^2/\mu r^4$ into the second term of Eq.(4-318) gives

$$\langle J, M|\mathscr{H}|J, M\rangle = \frac{\hbar^2 J(J+1)}{2\mu r_0^2} - \frac{\hbar^4 J^2(J+1)^2}{\mu^2 r_0^2 k}\left\langle J, M\left|\frac{1}{r^4}\right|J, M\right\rangle + \frac{\hbar^4 J^2(J+1)^2}{2k\mu^2}\left\langle J, M\left|\frac{1}{r^6}\right|J, M\right\rangle. \tag{4-319}$$

The average values of $1/r^4$ and $1/r^6$ in the last two terms can now be approximated by $1/r_0^4$ and $1/r_0^6$ to give the eigenvalues of

$$\langle J, M|\mathscr{H}|J, M\rangle = \frac{\hbar^2 J(J+1)}{2\mu r_0^2} - \frac{\hbar^4 J^2(J+1)^2}{2k\mu^2 r_0^6}$$

$$\frac{E(J, M)}{h} = BJ(J+1) - DJ^2(J+1)^2, \tag{4-320}$$

where B is the usual rotational constant and $D = \hbar^4/2hk\mu^2 r_0^6$ is called the *centrifugal distortion constant*.

Use Eq.(4-320) and Table 4-31 to evaluate r_0 and the force constant k for CO in the $v = 0$ vibrational state. The least-squares procedure is appropriate (but not necessary for the solution of the problem) for the analysis of the data in Table 4-31 (see Appendix D.4). Compare your value of k with the result in Table 4-26.

Table 4-32 Vibration and rotation constants that are used along with Eq.(4-322) to describe the vibration-rotation energy levels of diatomic molecules. All numbers are in units of cm^{-1} and are taken from Refs. [22] and [26].

Molecule	ν_e/c	$(\nu_e/c)x_e$	B_e/c	α_e/c	$(D/c) \times 10^{10}$
HF	4138.52	90.069	20.939	0.770	—
$H^{35}Cl$	2989.74	52.05	10.5909	0.3019	5.313
HBr	2649.67	45.21	8.473	0.226	—
HI	2309.5	39.73	6.551	0.183	—
$^{35}Cl^{19}F$	793.2	9.9	0.516508	0.004358	0.008753
$^{35}Cl_2$	564.9	4.0	0.2438	0.0017	—
H_2	4395.2	117.99	60.800	2.993	—
$^{14}N_2$	2359.61	14.456	2.010	0.0187	—
$^{16}O_2$	1580.361	12.0730	1.44566	0.01579	—

We can now combine the results in Eqs.(4-320) and (4-309) to describe the combined vibration-rotation energy levels in a diatomic molecule,

$$\frac{E(n, J)}{hc} = \frac{\nu_e}{c}(n + \tfrac{1}{2}) - \frac{\nu_e x_e}{c}(n + \tfrac{1}{2})^2 + \frac{B}{c}J(J + 1) - \frac{D}{c}J^2(J + 1)^2, \quad (4\text{-}321)$$

where n and J are the vibrational and rotational quantum numbers, respectively. We shall make one more refinement in this expression by recognizing that the rotational constant B will be different in different vibrational states. Thus, we rewrite B as

$$\frac{B}{c} = \frac{B_e}{c} - \frac{\alpha_e}{c}(n + \tfrac{1}{2}),$$

where α_e/c is the vibration-rotation coupling constant and B_e is also a constant. Substituting gives the final vibration-rotation energy:

$$\frac{E(n, J)}{hc} = \frac{\nu_e}{c}(n + \tfrac{1}{2}) - \frac{\nu_e x_e}{c}(n + \tfrac{1}{2})^2 + \frac{B_e}{c}J(J + 1)$$
$$- \frac{\alpha_e}{c}(n + \tfrac{1}{2})J(J + 1) - \frac{D}{c}J^2(J + 1)^2. \quad (4\text{-}322)$$

This equation is used to describe the rotational-vibrational energy levels in diatomic molecules [36] and some typical results are shown in Table 4-32.

36. The bending (buckling) motion of ring compounds can be described by a double minimum potential function as shown in Fig. 4-40 for trimethylene sulfide. The kinetic energy is $T = \frac{1}{2}\dot{Q}^2$ as shown in Eq.(4-258) where Q is the normal coordinate describing the buckling motion. The potential energy must be constructed to approximate the potential function shown in Fig. 4-40. One possibility is to sum a quadratic function to a Gaussian bump written as

$$V(Q) = \frac{k}{2}Q^2 + k' \exp(-\beta Q^2). \quad (4\text{-}323)$$

k, k' and β are all parameters to be fit to the experimental energy level spacings shown in Fig. 4-40.

*(a) Set up the Hamiltonian matrix for the trimethylene sulfide buckling motion in the harmonic oscillator basis. Diagonalize the matrix and fit the observed energy level spacings with an appropriate choice of k, k', and β in Eq.(4-323). Use at least a 20×20 matrix to fit the lower levels.

*(b) Now investigate the possible improvement in the results above by the addition of a quartic term that will tend to flatten the bottom of the potential well and increase the steepness of the walls. The total potential function is now given by

$$V(Q) = \frac{k}{2}Q^2 + k' \exp(-\beta Q^2) + k''Q^4. \quad (4\text{-}324)$$

Add the $k''Q^4$ matrix elements to the original matrix in part a and redetermine a best fit using the four parameters, k, k', β, and k''.

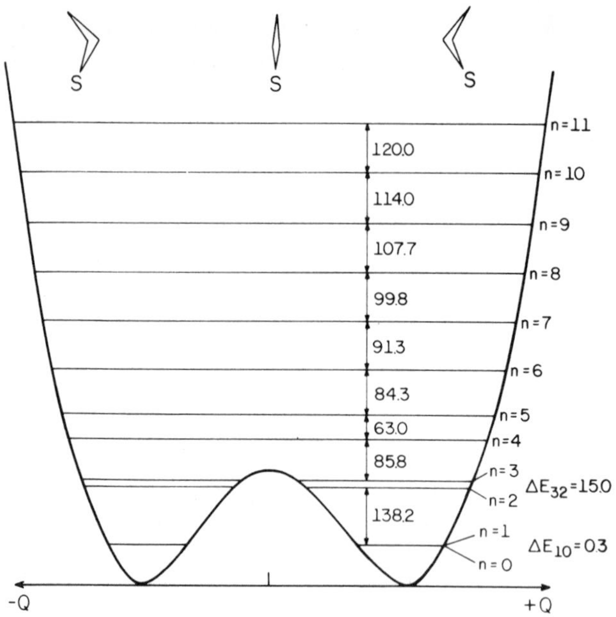

Figure 4-40 The potential function for the ring-buckling motion in trimethylene sulfide. Q is the normal coordinate. All energy differences are in cm^{-1} and these experimental numbers are reported in D. O. Harris, *et al.*, *J. Chem. Phys.* **44**, 3467 (1966) and H. Harrington, *ibid.*, **44**, 3481 (1966).

*(c) Now try to fit the experimental data with a simpler potential function given by

$$V(Q) = k''Q^4 - kQ^2, \tag{4-325}$$

which is a quartic and an inverted harmonic (quadratic) potential which again leads to a double minimum.

37. The low barrier approach to the double minimum potential problem is illustrated in Problem 4-36 where the doubling of the energy levels below the barrier height is evident (see Fig. 4-40). We shall now consider in more detail the high barrier approach to the double minimum problem by application of the inversion problem in NH_3 as shown in Fig. 4-41. We shall start with the potential function about the $s = 0$ point in Fig. 4-41 given by

$$V(s) = ks^4 - k's^2, \tag{4-326}$$

where s represents the displacement coordinate.

(a) Show that the points of minimum potential energy are at

$$s_m = \pm\left(\frac{k'}{2k}\right)^{1/2}$$

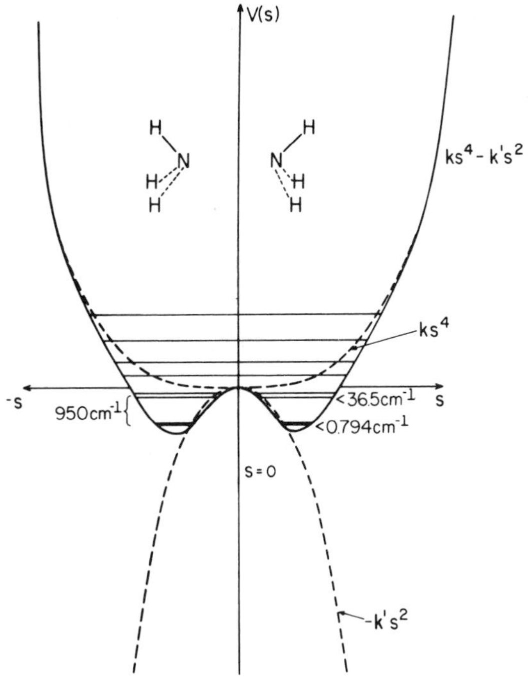

Figure 4-41 Potential function for the inversion motion in NH_3. The potential function is a sum of a positive quartic potential and a negative quadratic potential (dashed lines). The energy spacings in the lower levels are also shown. s_m, the minimum points on the potential function, are at $\pm 0.3 \times 10^{-8}$ cm.

and that the energy difference between the top of the bump and the minimum energy is given by

$$V = V(s = 0) - V(s = s_m) = \frac{1}{4}\frac{(k')^2}{k}. \tag{4-327}$$

(b) First we shall consider an infinitely high barrier V where we can expand the potential function in Eq.(4-326) in a Taylor series around $s_m = \pm(k'/2k)^{1/2}$. Show that, by truncating above $(s - s_m)^2$ terms that the potentials for the left- and right-hand sides of the barrier are given by

$$V_\ell = 2k'\left[s + \left(\frac{k'}{2k}\right)^{1/2}\right]^2 - V, \qquad V_r = 2k'\left[s - \left(\frac{k'}{2k}\right)^{1/2}\right]^2 - V \tag{4-328}$$

and the corresponding high barrier energy levels are given by

$$E_\ell(n') = \varepsilon_V(n' + \tfrac{1}{2}) - V; \qquad n' = 0, 1, 2, \ldots$$

$$E_r(n) = \varepsilon_V(n + \tfrac{1}{2}) - V; \qquad n = 0, 1, 2, \ldots$$

$$\varepsilon_V = 2\hbar\left(\frac{k'}{\mu}\right)^{1/2}. \tag{4-329}$$

μ is the reduced mass for the vibration described by s. According to Eq.(4-329), the energy spacings between vibrational levels on either side of the barrier are given in multiples of ε_V.

In the case of an infinite barrier, the wavefunctions for either the left or right sides do not overlap. According to Table 4-25 the ground-state functions are given by

$$\psi = \frac{\gamma^{1/2}}{\pi^{1/4}} \exp\left[-\frac{\gamma^2}{2}(s - s_m)^2\right]$$

$$\gamma = \left(\frac{(4\mu k')^{1/2}}{\hbar}\right)^{1/2}, \tag{4-330}$$

where $s_m = -(k'/2k)^{1/2}$ for the left, ψ_ℓ, and $s_m = (k'/2k)^{1/2}$ for the right, ψ_r. The energy levels for $E_\ell(n')$ and $E_r(n)$ are degenerate and the total ground-state wavefunction for the system is

$$\boldsymbol{\psi} = (\psi_\ell \quad \psi_r). \tag{4-331}$$

As the barrier becomes finite, however, the ψ_ℓ and ψ_r functions overlap causing a splitting of the doubly degenerate barrier levels.

The matrix representation of the high barrier Hamiltonian

$$-\infty < s < 0; \qquad \mathcal{H}_\ell = -\frac{\hbar^2}{2\mu}\frac{d^2}{ds^2} + 2k'\left[s + \left(\frac{k'}{2k}\right)^{1/2}\right]^2 - V$$

$$0 < s < +\infty; \qquad \mathcal{H}_r = -\frac{\hbar^2}{2\mu}\frac{d^2}{ds^2} + 2k'\left[s - \left(\frac{k'}{2k}\right)^{1/2}\right]^2 - V, \tag{4-332}$$

is given schematically by

$$\boldsymbol{\mathcal{H}} = \begin{pmatrix} (\ell\text{-}\ell) & (\ell\text{-}r) \\ (r\text{-}\ell) & (r\text{-}r) \end{pmatrix} = \begin{pmatrix} \alpha & \beta \\ \beta & \alpha \end{pmatrix}, \tag{4-333}$$

where ℓ-ℓ indicate left-left matrix elements with similar definitions for ℓ-r, r-ℓ, and r-r. In the case of an infinite value of V, the (ℓ-r) block will be zero. In the presence of a finite barrier, however, the ℓ wavefunctions will extend into the regions above $s = 0$ and the r functions will extend below $s = 0$. This will lead to nonzero matrix elements in the ℓ-r off-diagonal blocks in Eq.(4-333). The overlap between the ℓ and r functions will also be nonzero, which means that the orthonormality of the functions is dependent on the barrier height. In this case the secular equation given in Eq.(3-126) is the most convenient form for solving for the eigenvalues for a finite barrier:

$$\begin{vmatrix} \alpha - \lambda & \beta - \lambda S \\ \beta - \lambda S & \alpha - \lambda \end{vmatrix} = 0. \tag{4-334}$$

Expanding the determinant and solving for λ gives

$$\lambda = \frac{\alpha - \beta S \mp (\beta - S\alpha)}{1 - S\alpha}, \tag{4-335}$$

where the α, β, and S integrals are given by

$$S = \int_{-\infty}^{+\infty} \psi_\ell^* \psi_r \, ds$$

$$\alpha = \int_{-\infty}^{0} \psi_\ell^* \mathscr{H}_\ell \psi_\ell \, ds + \int_{0}^{\infty} \psi_\ell^* \mathscr{H}_r \psi_\ell \, ds = \int_{-\infty}^{0} \psi_r^* \mathscr{H}_\ell \psi_r \, ds + \int_{0}^{\infty} \psi_r^* \mathscr{H}_r \psi_r \, ds$$

$$= \frac{\varepsilon_V}{2} - V + \int_{0}^{\infty} \psi_\ell^* \mathscr{H}_r \psi_\ell \, ds \qquad (4\text{-}336)$$

$$\beta = \int_{-\infty}^{0} \psi_\ell^* \mathscr{H}_\ell \psi_r \, ds + \int_{0}^{\infty} \psi_\ell^* \mathscr{H}_r \psi_r \, ds = \int_{-\infty}^{0} \psi_r^* \mathscr{H}_\ell \psi_\ell \, ds + \int_{0}^{\infty} \psi_r^* \mathscr{H}_r \psi_\ell \, ds$$

$$= S(\varepsilon_V - 2V). \qquad (4\text{-}337)$$

(c) Complete the S and α integrals above and express the results as functions of ε_V and V. Substitute the results into Eq.(4-335) to give the energy splitting in the ground state as a function of ε_V and V. The experimental result from Fig. 4-41 is $\Delta E_0/hc = 0.794\ \text{cm}^{-1}$ and the tunneling frequency is $\Delta\nu = \Delta E_0/h$.

(d) Evaluate an approximate value of V in NH_3 with the information in Fig. 4-41 and the theory above.

(e) Use the value of V obtained in part d to estimate $\Delta E_0/hc$ in ND_3.

(f) Derive the tunneling frequency in the first excited vibrational state (high barrier limit), $\Delta\nu_1 = \Delta E_1/h$ in NH_3. The experimental result is given from Fig. 4-41 where $\Delta E_1/hc = 36.5\ \text{cm}^{-1}$.

38. We shall now consider the effect of nuclear spin statistics on rotational and rotational-vibrational energy levels and the resultant transitions. We begin with the Pauli principle, which states that (see Section 5.2)

1. The total molecular wavefunction must be antisymmetric with respect to the interchange of identical half-integral spin nuclei.
2. The total molecular wavefunction must be symmetric with respect to the interchange of identical integral spin nuclei.

The total wavefunction for a molecule is a product of the electronic, ψ_E, vibrational, ψ_V, rotational, ψ_R, and nuclear, ψ_I, wavefunctions given by

$$\psi_{\text{total}} = \psi_E \psi_V \psi_R \psi_I .$$

In a homonuclear diatomic molecule the vibrational functions are symmetric to the exchange of the nuclei. In polyatomic molecules the symmetry depends upon the vibration. In the case of a linear molecule the even J functions are symmetric and the odd J functions are antisymmetric. Thus, the even J rotational states will combine with the antisymmetric nuclear spin states of half-integral spins and the odd J rotational states will combine with symmetric nuclear spin states. In the case of H_2, H—^{12}C≡^{12}C—H, F_2, and other similar $I_1 = I_2 = \frac{1}{2}$ systems,

$$\boldsymbol{I} = \boldsymbol{I}_1 + \boldsymbol{I}_2 \longrightarrow 0, 1$$

$I = 0$,	spin degeneracy = 1:	antisymmetric
$I = 1$,	spin degeneracy = 3,	symmetric.

There are two rules that determine the number and symmetry of the nuclear spin states obtained by combining two identical nuclei:

Rule 1. The total I states are obtained from the maximum and minimum vector sums of I_1 and I_2 and all integral values in between.

Rule 2. The lowest I state is antisymmetric when I_1 and I_2 are half integral and the lowest I state is symmetric if I_1 and I_2 are integral [37]. The remaining symmetries of the I states alternate according to integral changes.

Two examples of combining identical nuclear spins are given here.

$$I_1 = I_2 = \tfrac{3}{2}\ (^{35}Cl_2)$$

$I = I_1 + I_2 \longrightarrow$	0,	1,	2,	3
degeneracy $\longrightarrow$	1	3	5	7

antisymmetric (1 + 5) — symmetric (3 + 7)

(total nuclear degeneracy) antisymmetric (6) symmetric (10)

$$I_1 = I_2 = 1\ (^{14}N_2, D_2)$$

$\boldsymbol{I} = \boldsymbol{I}_1 + \boldsymbol{I}_2 \longrightarrow$	0,	1,	2
degeneracy $\longrightarrow$	1	3	5

symmetric (1 + 5) — antisymmetric (3)

(total nuclear degeneracy) symmetric (6) antisymmetric (3)

Consider first the observation of the pure rotational Raman effect in $^{14}N_2$ in the ground electronic-vibrational state with $\Delta J = \pm 2$, $\Delta M = 0$ rotational selection rules. The $\psi_E \psi_V \psi_R$ wavefunction is symmetric for even J and antisymmetric for odd J. Thus, the symmetric $^{14}N(I = 1)$ nuclear spin states in $^{14}N_2$ combine with the symmetric rotational states (even J) and the antisymmetric ^{14}N nuclear spin states in $^{14}N_2$ combine with antisymmetric rotational states (odd J). In conclusion, the $\Delta J = \pm 2$, $J_{\text{even}} \leftrightarrow J_{\text{even}}$ transitions will be more intense than the $J_{\text{odd}} \leftrightarrow J_{\text{odd}}$ transitions by the 6:3 nuclear spin factor. The predicted pure rotational Raman effect in the ground vibrational state in $^{14}N_2$ is shown in Fig. 4-42 and this prediction has been verified experimentally.

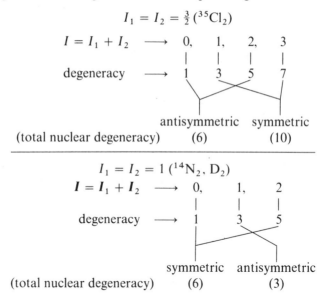

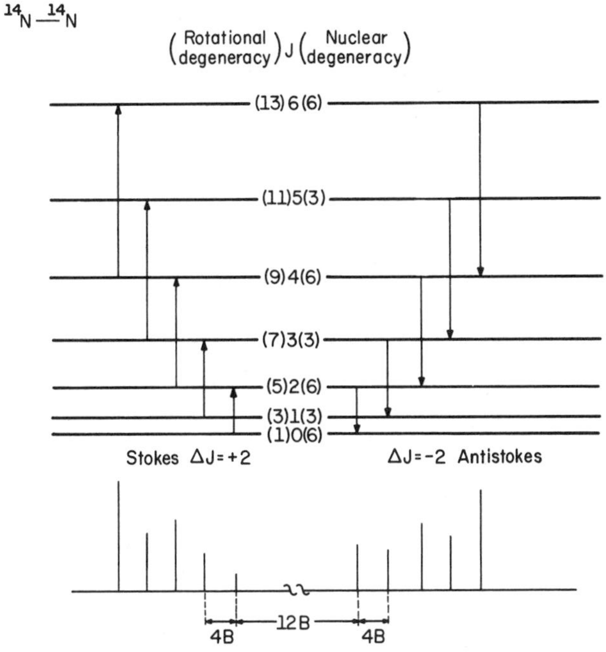

Figure 4-42 The predicted pure rotational Raman effect in $^{14}N_2$ at $T = 300$ K. The rotational spacings are $hBJ(J + 1)$. The relative intensities include the Boltzmann factor, the rotational degeneracy, and the nuclear spin degeneracy. The alternation in intensities is due to the nuclear spin statistics. The spectrum has been verified experimentally as shown in Ref. [26] and A. Weber and J. J. Barrett, *J. Opt. Soc. Am.* **57**, 19 (1967) as shown in Fig. 8-2.

In the case of a parallel vibrational transition from an asymmetric stretch in H—^{12}C≡^{12}C—H, the *IR* selection rules are $\Delta n = \pm 1$, $\Delta J = \pm 1$, $\Delta M = 0, \pm 1$. The ground vibrational state is symmetric with respect to proton interchange and the first excited vibrational state is antisymmetric with respect to interchange of the protons. Thus, according to our previous discussion we have the following nuclear degeneracies for H—^{12}C≡^{12}C—H.

$n = 0$ (*ground vibrational state*)

Rotational State	Nuclear Degeneracy
J_{even}	1
J_{odd}	3

$n = 1$ (*first excited asymmetric stretch*)

Rotational State	Nuclear Degeneracy
J_{even}	3
J_{odd}	1

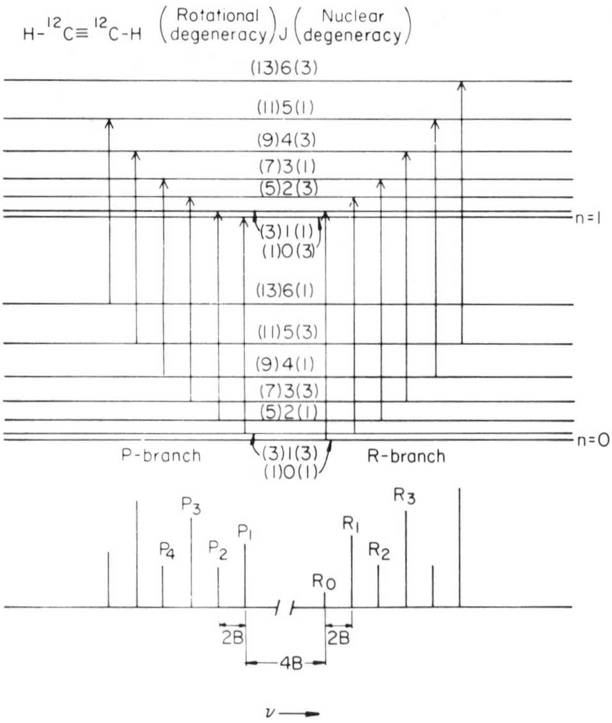

Figure 4-43 The predicted parallel band vibration-rotation spectrum in H—^{12}C≡^{12}C—H at 300 K. The rotational spacings are $hBJ(J+1)$ and the relative intensities include the Boltzmann factor, the rotational degeneracy, and the nuclear spin degeneracy. The alternation in intensities is due to the nuclear spin statistics.

Thus, the $\Delta n = +1$, $\Delta J = \pm 1$ ($J_{\text{odd}} \longrightarrow J_{\text{even}}$) transitions will have a nuclear degeneracy of 3 relative to a degeneracy of 1 for the corresponding $\Delta n = +1$, $\Delta J = \pm 1$ ($J_{\text{even}} \longrightarrow J_{\text{odd}}$) transitions. The former transitions will be three times as intense as the latter. These results are illustrated in Fig. 4-43.

(a) Predict the pure rotational Raman lines (relative intensities in the $\Delta J = \pm 2$, $\Delta M = 0$ transitions only) for $^{15}N_2$, $^{127}I_2$, HCl, OCO, and H—^{12}C≡^{12}C—H in the ground vibrational states. Check your predictions with the literature, if possible.

(b) Predict the rotational-vibrational Raman lines (relative intensities only) for $^{127}I_2$, $^{15}N_2$, and D_2. Check again with the literature, if possible.

(c) Predict the asymmetric stretch infrared rotational-vibrational transitions and relative intensities in H—^{13}C≡^{13}C—H and D—^{12}C≡^{12}C—D to compare with Fig. 4-43.

(d) The (001)⟶(100) parallel rotation-vibration transition in CO_2 is the transition that supports laser oscillation. Draw a diagram and show the P- and R-branch transitions. Use the appropriate nuclear spin statistics for

$^{16}O^{12}C^{16}O$ to show that only even P- and R-branch transitions appear $(\ldots P_6, P_4, P_2, R_0, R_2, \ldots)$. The CO_2 laser can oscillate on any of the available P- and R-branch transitions that range in frequency (in $4hB$ increments) around the hypothetical pure vibrational transition at $960\ \text{cm}^{-1}$ ($\lambda = 10.6\ \mu\text{m}$). The NNO molecule has energy levels very similar to OCO and the (001)⟶(100) transition in NNO is also a quite effective molecular laser. In N_2O, however, there are no nuclear spin statistics and all the P- and R-branch transitions are available (spacing of $2hB$).

39. We shall now discuss the direct dipole–dipole interactions (electron–electron and nuclear–nuclear) that were ignored in discussing the $\boldsymbol{I}_i \cdot \boldsymbol{S}$ and $\boldsymbol{I}_i \cdot \boldsymbol{I}_j$ coupling in Sections 4.9 and 4.10. Consider the magnetic dipole moments $\boldsymbol{\mu}_1$ and $\boldsymbol{\mu}_2$ as shown in Fig. 4-44.

According to Eq.(1-56) the potential at 2 due to $\boldsymbol{\mu}_1$ is (cgs units)

$$\varphi_2 = \frac{\boldsymbol{\mu}_1 \cdot \boldsymbol{R}}{R^3}. \tag{4-338}$$

Thus, the energy of interaction of $\boldsymbol{\mu}_1$ and $\boldsymbol{\mu}_2$ is equal to the vector dot product between $\boldsymbol{\mu}_2$ and the field at 2 due to 1 ($\boldsymbol{E}_2 = -\nabla\varphi_2$),

$$\mathcal{H} = -\boldsymbol{\mu}_2 \cdot \boldsymbol{E}_2 = -\boldsymbol{\mu}_2 \cdot (-\nabla\varphi_2) = \boldsymbol{\mu}_2 \cdot \nabla\left(\frac{\boldsymbol{\mu}_1 \cdot \boldsymbol{R}}{R^3}\right)$$

$$= \frac{\boldsymbol{\mu}_2 \cdot \boldsymbol{\mu}_1}{R^3} - 3\frac{(\boldsymbol{\mu}_2 \cdot \boldsymbol{R})(\boldsymbol{\mu}_1 \cdot \boldsymbol{R})}{R^5} = -\boldsymbol{\mu}_2 \cdot \left[\frac{3\boldsymbol{R}(\boldsymbol{\mu}_1 \cdot \boldsymbol{R})}{R^5} - \frac{\boldsymbol{\mu}_1}{R^3}\right]. \tag{4-339}$$

Consider now the effect of a strong magnetic field (H_z) on the system above. If $\boldsymbol{\mu}_1$ and $\boldsymbol{\mu}_2$ are nuclear magnetic moments, the external magnetic field will orientate both moments along the laboratory-fixed z axis and $\theta_1 = \theta_2 = \theta$ now represents the angle between the space-fixed axis z and the molecular-fixed axis $\boldsymbol{R}$. The energy under these circumstances is given by

$$\mathcal{H} = \frac{\mu_2\mu_1}{R^3}(1 - 3\cos^2\theta). \tag{4-340}$$

The magnetic moment of a proton is (see Table 4-8) $\mu_1 = g_1\mu_0/2 = 1.41 \times 10^{-23}\ \text{erg} \cdot \text{G}^{-1}$, and the energy of interaction between two protons that are 3×10^{-8} cm apart is $E = 7.36 \times 10^{-24} \times (1 - 3\cos^2\theta)\ \text{erg} \cong 3.7 \times 10^{-8} \times (1 - 3\cos^2\theta)\ \text{cm}^{-1}$. These direct dipole–dipole interactions are considerably larger than the electron-coupled dipole–dipole interactions treated in Section 4.10. Thus, in a solid where θ in Eq.(4-340) is fixed, appreciable direct dipole–dipole interactions are observable. If μ_1 and μ_2 are known, Eqs.(4-339) and

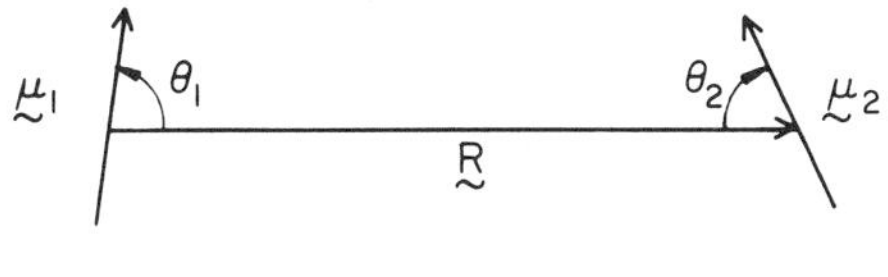

Figure 4-44 Diagram showing two magnetic dipole moments and their relative orientation.

(4-340) can be used to determine R, the distance between the nuclei. Proton–proton distances have been measured in the solid state and detailed molecular structures have been obtained in benzene, cyclopropane, and other molecules by application of these equations for molecules isolated and orientated in the solid state and in liquid crystals [38].

In the case of electron spin–electron spin interactions, the energies are on the order of 10^6 larger than the nuclear spin–nuclear spin interactions. In the case of the spin–spin interaction between two unpaired electrons ($R = 3 \times 10^{-8}$ cm) we have [in a magnetic field, from Eq.(4-340)] an energy of $E = 3.18 \times 10^{-18} \times (1 - 3\cos^2\theta)$ erg.

In the gas phase, the effects of the Hamiltonians described above must be averaged over the rotational wavefunctions. In the liquid state, the rotational angular momentum is quenched ($J = 0$) and the effects described in Eqs.(4-339) and (4-340) average to zero.

(a) Calculate the effect of Eq.(4-340) on the rotational $J = 1$, $M_J = 1, 0, -1$ states in $H^{12}C^{15}N$.

(b) Show that if the rotational angular momentum is quenched ($J = 0$) in the liquid state that the dipole–dipole interaction in Eq.(4-339) goes to zero. The remaining $\boldsymbol{I} \cdot \boldsymbol{S}$ and $\boldsymbol{I}_i \cdot \boldsymbol{I}_j$ terms in Eq.(4-156) are the angular independent contact terms.

We shall now discuss the electron spin–electron spin interaction in a molecule that has two unpaired electrons, $\boldsymbol{S}_1 + \boldsymbol{S}_2 = \boldsymbol{S} \longrightarrow 0, 1$. The $S = 0$ is the singlet and the $S = 1$ is the triplet. Normally in zero magnetic field we would expect the triplet to be composed of three degenerate levels. In the presence of the spin–spin interaction, however, the threefold degeneracy is broken. The coupling between the magnetic moments $\boldsymbol{\mu}_1 = -(g_e\mu_B\boldsymbol{S}_1/\hbar)$ and $\boldsymbol{\mu}_2 = -(g_e\mu_B\boldsymbol{S}_2/\hbar)$ is given by substituting into Eq.(4-339) to give

$$\mathscr{H} = \frac{g_e^2\mu_B^2}{\hbar^2}\left[\frac{\boldsymbol{S}_1 \cdot \boldsymbol{S}_2}{R^3} - \frac{3(\boldsymbol{S}_1 \cdot \boldsymbol{R})(\boldsymbol{S}_2 \cdot \boldsymbol{R})}{R^5}\right]. \tag{4-341}$$

Expanding Eq.(4-341) in its Cartesian components and rearranging gives

$$\begin{aligned}\mathscr{H} = \frac{g_e^2\mu_B^2}{\hbar^2}\Bigg[&S_{1x}S_{2x}\left(\frac{R^2 - 3x^2}{R^5}\right) + S_{1y}S_{2y}\left(\frac{R^2 - 3y^2}{R^5}\right) + S_{1z}S_{2z}\left(\frac{R^2 - 3z^2}{R^5}\right)\\ &- (S_{1x}S_{2y} + S_{1y}S_{2x})\frac{3xy}{R^5} - (S_{1y}S_{2z} + S_{1z}S_{2y})\frac{3yz}{R^5}\\ &- (S_{1z}S_{2x} + S_{1x}S_{2z})\frac{3zx}{R^5}\Bigg] = \frac{1}{\hbar^2}\boldsymbol{S}_1 \cdot \mathbf{D} \cdot \boldsymbol{S}_2,\end{aligned} \tag{4-342}$$

where $\mathbf{D}$ is a second-rank Cartesian tensor with elements given by

$$\mathsf{D}_{xx} = g_e^2\mu_B^2\left(\frac{R^2 - 3x^2}{R^5}\right), \qquad \mathsf{D}_{xy} = -g_e^2\mu_B^2\frac{3xy}{R^5}, \tag{4-343}$$

and cyclic permutations. **D** is sometimes called the zero-field splitting tensor. We can now transform **D** to its principal axis where there are no off-diagonal values giving

$$\mathcal{H} = \frac{1}{\hbar^2}(S_{1X} \quad S_{1Y} \quad S_{1Z})\begin{pmatrix} \mathsf{D}_{XX} & 0 & 0 \\ 0 & \mathsf{D}_{YY} & 0 \\ 0 & 0 & \mathsf{D}_{ZZ} \end{pmatrix}\begin{pmatrix} S_{2X} \\ S_{2Y} \\ S_{2Z} \end{pmatrix}$$

$$= \frac{g_e^2\mu_B^2}{\hbar^2}\left[S_{1X}S_{2X}\left(\frac{R^2 - 3X^2}{R^5}\right) + S_{1Y}S_{2Y}\left(\frac{R^2 - 3Y^2}{R^5}\right) + S_{1Z}S_{2Z}\left(\frac{R^2 - 3Z^2}{R^5}\right)\right]$$

$$= \frac{1}{\hbar^2}\{S_{1X}S_{2X}\mathsf{D}_{XX} + S_{1Y}S_{2Y}\mathsf{D}_{YY} + S_{1Z}S_{2Z}\mathsf{D}_{ZZ}\}. \tag{4-344}$$

(c) Calculate the matrix representation of the Hamiltonian (zero field) in Eq.(4-344) in the uncoupled basis.

(d) Diagonalize the 4×4 matrix in part c and show that the singlet ($S = 0$) has zero energy and the triplet ($S = 1$) is split (at zero field) into three levels with energies $-(\mathsf{D}_{XX}/2)$, $-(\mathsf{D}_{YY}/2)$, and $-(\mathsf{D}_{ZZ}/2)$. The experimental values of the diagonal elements in the **D** tensor have been measured in the triplet state of naphthalene. The results are {C. A. Hutchison and B. W. Mangum, *J. Chem. Phys.* **34**, 908 (1961) and Ref. [19]}.

$$\frac{\mathsf{D}_{XX}}{2hc} = 0.0478 \pm 0.0006 \text{ cm}^{-1}, \qquad \frac{\mathsf{D}_{YY}}{2hc} = 0.0196 \pm 0.0006 \text{ cm}^{-1},$$

$$\frac{\mathsf{D}_{ZZ}}{2hc} = -0.0675 \pm 0.0006 \text{ cm}^{-1}. \tag{4-345}$$

These numbers agree with our earlier classical estimates for the magnitude of the spin–spin interaction.

(e) Use the results above to draw a diagram to show the effects of a magnetic field along the z axis. Set up the appropriate matrix representation in the coupled representation to demonstrate the effects of small magnetic fields.

40. Consider a planar AB_4 molecule with equal A—B bond lengths. The molecule has $D_{4h} = i \times D_4$ symmetry.

(a) Obtain the D_{4h} group by the $i \times D_4$ direct product.

(b) Determine the subgroups of the D_{4h} group.

(c) Use displacement coordinates to obtain the symmetry species of the normal modes of vibration.

(d) Use symmetry coordinates to determine which of the normal modes of vibration are stretching modes, in-plane bending modes, and out-of-plane bending modes.

(e) Use part b and any information you can find in Ref. [27] to sketch all the normal modes of motion in AB_4.

(f) Use the conventional scheme to designate the vibrational wavefunction

$$\psi(n_1 n_2 n_3 n_4 n_5 n_6 n_7),$$

where $n_1 \longrightarrow n_7$ are in successive order in the character table. Determine the symmetry classification of the following vibrational functions:

$$\psi(0010100), \qquad \psi(0321001), \qquad \psi(0110101), \qquad \psi(0113012).$$

41. Consider the planar BF_3 molecule with equal B—F bond distances.

(a) What symmetry group describes the symmetry operations appropriate to this group?

(b) Generate the characters of the reducible representation formed from the displacement coordinates in BF_3.

(c) Decompose the reducible representation into its irreducible components under the symmetry group and identify the translational, rotational, and vibrational symmetry species.

(d) Work out the correlation diagram or breakdown of representations between the BF_3 molecule above and the planar BF_2Cl molecule where the BF bond distances are equal.

(e) Determine the infrared and Raman activity for the transitions from the ground to the first excited states for all the normal modes in BF_3 and BF_2Cl.

42. In Fig. 4-35 we illustrated the normal modes of vibrational motion in a planar

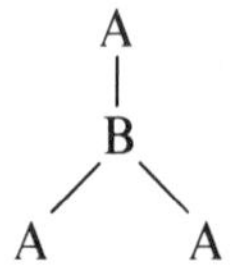

molecule that has D_{3h} symmetry. There were two singly and two doubly degenerate modes. Formaldehyde is a similar molecule which has six singly degenerate normal modes which are all infrared active. Formaldehyde is a near-symmetric top with the near-symmetry axis lying along the C=O bond. The observed fundamental vibrations in formaldehyde are listed in Table 4-33.

(a) Use the C_{2v} group character table and Fig. 4-35 to identify and assign the fundamental transitions above.

(b) Assign each fundamental mode with a motion as shown in Fig. 4-35.

(c) Determine by group theory which of the vibrations in Table 4-33 are infrared and Raman active.

Table 4-33 The observed fundamental vibrational frequencies in formaldehyde from Ref. [27]. The parallel (∥) and perpendicular (⊥) transitions of the near symmetric rotor are identified.

Assignment	Type of Band	Fundamental Frequencies (cm^{-1})
	⊥	1167
	⊥	1280
	∥	1503
	∥	1744
	∥	2780
	⊥	2874

REFERENCES

[1] N. S. BAYLISS, *Quart. Rev.* **6**, 319 (1952); see also Section 6.7.

[2] C. P. SLICHTER, *Principles of Magnetic Resonance* (Harper & Row, New York, 1963).

[3] Other examples of quantum mechanical tunneling in chemistry are found in M. D. HARMONY, *Chem. Soc. Rev.* **1**, 211 (1972). See also problems at the end of this chapter.

[4] A review of conformational studies in small molecules is in E. B. WILSON, *Chem. Soc. Rev.* **1**, 293 (1972).

[5] See also V. LAURIE, *Accts. Chem. Res.* **3**, 331 (1970) and O. BASTIANSEN, H. M. SEIP, and J. E. BOGGS, *Perspectives in Structural Chemistry*, Ed. by J. D. Dunitz and J. A. Ibers, **4**, 60 (John Wiley & Sons, New York, 1971).

[6] H. MARGENAU and G. M. MURPHY, *The Mathematics of Physics and Chemistry* (D. Van Nostrand Co., New York, 1956).

[7] L. I. SCHIFF, *Quantum Mechanics*, 3rd ed. (McGraw-Hill Book Co., New York, 1968).

[8] P. A. M. DIRAC, *The Principles of Quantum Mechanics* (Clarendon Press, Oxford, 1958).

[9] L. PAULING and E. B. WILSON, Jr., *Introduction to Quantum Mechanics* (McGraw-Hill Book Co., New York, 1935).

[10] L. PAULING, *The Chemical Bond* (Cornell University Press, Ithaca, N.Y., 1967).

[11] M. E. ROSE, *Elementary Theory of Angular Momentum* (John Wiley & Sons, Inc., New York, 1957). A. R. EDMUNDS, *Angular Momentum in Quantum Mechanics* (Princeton University Press, Princeton, N.J., 1957).

[12] D. BOHM, *Quantum Theory* (Prentice Hall, Inc., Englewood Cliffs, N.J., 1951).

[13] J. D. JACKSON, *Classical Electrodynamics* (John Wiley & Sons, Inc., New York, 1962).

[14] H. A. BETHE, *Intermediate Quantum Mechanics* (W. A. Benjamin, Inc., New York, 1964).

[15] E. U. CONDON and G. H. SHORTLEY, *The Theory of Atomic Spectra* (Cambridge University Press, 1935).

[16] Molecules with nonzero $\boldsymbol{L}$ and $\boldsymbol{S}$ in the gas phase where $\boldsymbol{L}$ and $\boldsymbol{S}$ couple with the rotational angular momentum are discussed in A. CARRINGTON, D. H. LEVY, and T. A. MILLER, *Adv. Chem. Phys.*, Ed. by I. Prigogine and S. A. Rice, **18**, 149 (John Wiley & Sons, 1970). A. CARRINGTON, *Microwave Spectroscopy of Radicals* (Academic Press, New York, 1975).

[17] R. BERINGER and M. A. HEALD, *Phys. Rev.* **95**, 1474 (1954); P. KUSCH, *Phys. Rev.* **100**, 1188 (1955).

[18] R. W. FESSENDEN and R. H. SCHULER, *J. Chem. Phys.* **39**, 2147 (1963).

[19] Additional examples and more details on ESR are found in J. E. WERTZ and J. R. BOLTON, *Electron Spin Resonance, Elementary Theory and Practical Applications* (McGraw-Hill Book Co., New York, 1972).

[20] A. CARRINGTON and A. D. MCLACHLIN, *Introduction to Magnetic Resonance* (Harper & Row, New York, 1967).

[21] We shall not develop a detailed electron theory of J_{ij} here (see, however, Problem 6-15). The reader should review the perturbation techniques in Sections 6.8 and 6.9 and then refer to M. BARFIELD and D. M. GRANT, *Adv. Mag. Res.* Ed. by J. S. Waugh, **1**, 149 (Academic Press, New York, 1965). See also J. N. MURRELL, *Prog. NMR Spec.* Ed. by J. W. Emsley, J. Feeney, and L. H. Sutcliffe, **6**, 1 (Pergamon Press, New York, 1971).

[22] C. H. TOWNES and A. L. SCHAWLOW, *Microwave Spectroscopy* (McGraw-Hill Book Co., New York, 1955).

[23] J. C. ZORN and T. C. ENGLISH, *Adv. Atomic Mol. Phys.*, Ed. by D. R. Bates and I. Estermann, **9**, 244 (Academic Press, New York, 1973).

[24] J. H. VAN VLECK, *Rev. Mod. Phys.* **23**, 213 (1951).

[25] W. GORDY and R. L. COOK, *Microwave Molecular Spectra* (Interscience, New York, 1970).

[26] G. HERZBERG, *Spectra of Diatomic Molecules*, I. (D. Van Nostrand Co., New York, 1950).

[27] G. HERZBERG, *Infrared and Raman Spectra* (D. Van Nostrand Co., New York, 1945).

[28] Additional details in the determination of the molecular force field are given in T. SHIMAROUCHI, *Physical Chemistry, An Advanced Treatise*, Ed. by D. Henderson, **4**, 233 (Academic Press, New York, 1970).

[29] M. M. SUSHCHINSKII, *Raman Spectra of Molecules and Crystals* (Keter, Inc., New York, 1972). See also detailed development in Chapter 8.

[30] Additional details on the quantum mechanical Raman effect and selection rules can be found in G. PLACZEK, *Rayleigh and Raman Scattering* (National Technical Information Service, Trans. 5266, Washington, D.C., 1962). See also E. B. WILSON, J. C. DECIUS, and P. C. CROSS, *Molecular Vibrations* (McGraw-Hill Book Co., New York, 1955). Experimental techniques and applications of Raman scattering to examine intramolecular motions like torsional and ring buckling motions are found in J. R. DURIG and W. C. HARRIS, "Raman Spectroscopy" *Physical Methods of Chemistry* Vol. I, Part IIIB, Ed. by A. Weissberger and B. Rossiter (Wiley-Interscience, New York, 1972).

[31] L. PAULING, *Phys. Rev.* **36**, 430 (1930).

[32] R. A. BERNHEIM, *et al.*, *J. Chem. Phys.* **64**, 2747 (1976) and references cited therein.

[33] W. D. PHILLIPS, C. E. LOONEY, and C. P. SPAETH, *J. Mol. Spectr.* **1**, 35 (1957).

[34] J. D. BALDESCHWIELER and E. W. RANDALL, *Chem. Rev.* **63**, 81 (1963).

[35] J. W. CEDERBERG, C. H. ANDERSON, and N. F. RAMSEY, *Phys. Rev.* **136**, 906 (1964).

[36] For additional details and extension to polyatomic molecules, see H. C. ALLEN and P. C. CROSS, *Molecular Vib-Rotors* (John Wiley & Sons, Inc., New York, 1963).

[37] J. E. MAYER and M. G. MAYER, *Statistical Mechanics* (John Wiley & Sons, Inc., New York, 1940).

[38] S. MEIBOOM and L. C. SNYDER, *Accts. Chem. Res.* **4**, 81 (1971).

5

atomic theory and atoms in solids and molecules

5.1 ATOMIC ORBITALS WITH APPLICATION TO THE HELIUM ATOM

Following our earlier work on the hydrogen-like atom, we shall now examine the helium atom. A careful study of this system will lead to a model for more complex atoms. The Hamiltonian for the helium-like atom with an infinitely heavy nucleus is given by

$$\mathscr{H} = \underbrace{-\frac{\hbar^2}{2m}\nabla_1^2 - \frac{Ze^2}{r_1}}_{\mathscr{H}_1} \underbrace{- \frac{\hbar^2}{2m}\nabla_2^2 - \frac{Ze^2}{r_2}}_{\mathscr{H}_2} + \underbrace{\frac{e^2}{r_{12}}}_{\mathscr{H}_3}, \tag{5-1}$$

where Z is the nuclear charge; m is the electron mass; r_1 and r_2 are the distances from the nucleus to electrons (1) and (2), respectively; and r_{12} is the interelectronic distance. It is clear from Eq.(5-1) that the first and second terms, $\mathscr{H}_1 + \mathscr{H}_2$, in the Hamiltonian are dependent only on the coordinates of electrons (1) and (2), respectively, and the last term, $\mathcal{H}_3$, couples the two electrons through their mutual repulsion. The Schrödinger equation is

$$\mathcal{H}\psi(\boldsymbol{r}_1, \boldsymbol{r}_2) = E\psi(\boldsymbol{r}_1, \boldsymbol{r}_2). \tag{5-2}$$

We now seek an approximate solution to Eq.(5-2) that is easily transferable to more complex atoms and finally to molecules.

In our search for a many-electron atomic wavefunction, we shall consider several stages of approximation. In all cases, however, the primary concept is the formulation of a many-electron function as a product of one-electron orbitals. The atomic orbital and subsequent product of atomic orbitals for a many-electron function is a very powerful concept and extends over into our approximate description of molecular electronic structure.

The concept of a product of one-electron atomic orbitals is illustrated in the case of the He atom with the Hamiltonian in Eq.(5-1). If $\mathcal{H}_3$ (describing the interelectronic repulsion) is much less important than $\mathcal{H}_1 + \mathcal{H}_2$, the Hamiltonian is approximately separable into independent operators for each electron and the two-electron function, $\psi(1, 2)$, can be written as a simple product of one-electron functions,

$$\psi(1, 2) = \Phi_1(1)\Phi_1(2), \tag{5-3}$$

where $\Phi_i(j)$ denotes the ith spatial orbital containing the jth electron.

The variational energy for He is given from Eqs.(5-1) and (5-3),

$$E_0 \leq W_0 = \frac{\int_1 \int_2 \psi^*(1, 2)(\mathcal{H})\psi(1, 2)\, dV_1\, dV_2}{\int_1 \int_2 \psi^*(1, 2)\psi(1, 2)\, dV_1\, dV_2}, \tag{5-4}$$

where E_0 is the true energy and the integration is over the coordinates of electrons (1) and (2) with volume elements dV_1 and dV_2, respectively. Substituting the function in Eq.(5-3) gives

$$W_0 = \frac{\int_1 \int_2 \Phi_1^*(1)\Phi_1^*(2)[-(\hbar^2/2m)\nabla_1^2 - (Ze^2/r_1) - (\hbar^2/2m)\nabla_2^2 - (Ze^2/r_2) + (e^2/r_{12})]\Phi_1(1)\Phi_1(2)\, dV_1\, dV_2}{\int_1 \Phi_1^*(1)\Phi_1(1)\, dV_1 \int_2 \Phi_1^*(2)\Phi_1(2)\, dV_2}. \tag{5-5}$$

The spatial integral and volume element is given by

$$\int_1 dV_1 = \int_0^{2\pi}\int_0^{\pi}\int_0^{\infty} r_1^2 \sin\theta_1\, dr_1\, d\theta_1\, d\phi_1. \tag{5-6}$$

We start by using the ground state 1s hydrogen-like orbitals for $\Phi(1)$ and $\Phi(2)$:

$$\Phi_1(i) = \left(\frac{1}{\pi}\right)^{1/2}\left(\frac{Z}{a_0}\right)^{3/2} \exp\left(-\frac{Zr_i}{a_0}\right), \tag{5-7}$$

where i indicates either electron (1) or (2).

Substituting the normalized atomic orbitals from Eq.(5-7) into Eq.(5-5) and remembering that

$$\left(-\frac{\hbar^2}{2m}\nabla_1^2 - \frac{Ze^2}{r_1}\right)\Phi_1(1) = E_1\Phi_1(1), \tag{5-8}$$

with $E_1 = -Z^2e^2/2a_0$ for the hydrogen-like atom, gives

$$\begin{aligned} W_0 &= E_1(1) + E_1(2) + \int_1\int_2 \Phi_1^*(1)\Phi_1^*(2)\left(\frac{e^2}{r_{12}}\right)\Phi_1(1)\Phi_1(2)\,dV_1\,dV_2 \\ &= 2E_1 + \frac{1}{\pi^2}\left(\frac{Z}{a_0}\right)^6\int_1\int_2 \exp\left[-\left(\frac{2Z}{a_0}\right)(r_1 + r_2)\right]\left(\frac{e^2}{r_{12}}\right)dV_1\,dV_2. \end{aligned} \tag{5-9}$$

The difficult integral involving $1/r_{12}$ in Eq.(5-9) is evaluated in several books [1]†, and the result is equal to $(\frac{5}{8})Ze^2/a_0$. The resultant energy is

$$\begin{aligned} W_0 &= 2E_1 + \frac{5}{8}Z\frac{e^2}{a_0} = -\frac{Z^2e^2}{a_0} + \frac{5}{8}Z\frac{e^2}{a_0} = \frac{e^2}{a_0}\left(-Z^2 + \frac{5}{8}Z\right) \\ &= (27.2116)\left(-Z^2 + \frac{5}{8}Z\right)\text{ eV}, \end{aligned} \tag{5-10}$$

where $e^2/a_0 = 27.2116$ eV from Tables A-3 and A-4. e^2/a_0 is the *atomic unit* (a.u.) of energy, sometimes referred to as the *Hartree*. Atomic units are often used in atomic and molecular electronic calculations and they are obtained by setting $e = a_0 = \hbar = m = 1.0$. Using $Z = 2$ in the one-electron trial functions in Eq.(5-7) for the He-like atom described above by Eq.(5-10) gives the energy of

$$W_0 = -\frac{e^2}{a_0}(2.7500) = -74.832\text{ eV}. \tag{5-11}$$

The experimental energy necessary to remove the two electrons from the He^{2+} nucleus is $E = 79.014$ eV. Thus, we see from Eq.(5-11) that our simple orbital picture has yielded 95% of the true energy of the atom and we still have not used any variational parameters.

We now return to the variational expression and let Z in the trial function be a variational parameter. It is reasonable to suggest that one electron may shield the other electron from the full influence of the nuclear charge, Ze. Thus, we shall replace Z in Eq.(5-7) with the variational parameter, ζ. The Z in the Hamiltonian remains the same, however. Equation (5-5) now becomes

$$\begin{aligned} W_0 = \frac{1}{\pi^2}\left(\frac{\zeta}{a_0}\right)^6\int_1\int_2 \exp\left[-\left(\frac{\zeta}{a_0}\right)(r_1 + r_2)\right]\Bigg[\left(-\frac{\hbar^2}{2m}\nabla_1^2 - \frac{\zeta e^2}{r_1}\right) + \left(-\frac{\hbar^2}{2m}\nabla_2^2 - \frac{\zeta e^2}{r_2}\right) \\ + \frac{e^2}{r_{12}} - (Z - \zeta)\left(\frac{e^2}{r_1} + \frac{e^2}{r_2}\right)\Bigg]\exp\left[-\left(\frac{\zeta}{a_0}\right)(r_1 + r_2)\right]dV_1\,dV_2, \end{aligned} \tag{5-12}$$

† Numbers in square brackets correspond to the numbered sources found in the Reference Section on p. 328 at the end of the chapter.

where $(-\zeta e^2/r_1) - (\zeta e^2/r_2)$ is added and subtracted from the original Hamiltonian in Eq.(5-1). The first two bracketed operators in Eq.(5-12) lead to the hydrogen-like energies, $-\zeta^2 e^2/2a_0$. The integral over the e^2/r_{12} operator is equal to $(\frac{5}{8})\zeta(e^2/a_0)$ from our previous discussion following Eq.(5-9). The $(Z - \zeta)e^2/r_1$ and $(Z - \zeta)e^2/r_2$ integrals are easily obtained leading, in summary, to

$$W_0 = -\frac{\zeta^2 e^2}{a_0} + \frac{5}{8}\zeta\frac{e^2}{a_0} + 2(\zeta - Z)\zeta\frac{e^2}{a_0} = \frac{e^2}{a_0}\left[\zeta^2 + \zeta\left(\frac{5}{8} - 2Z\right)\right]. \tag{5-13}$$

Minimizing W_0 with respect to ζ gives $\zeta = Z - \frac{5}{16}$. In He, where $Z = 2$, we obtain

$$\zeta = 2 - \tfrac{5}{16} = \tfrac{27}{16} = 1.6875. \tag{5-14}$$

Substituting back into Eq.(5-13) gives the energy,

$$W_0 = -\left(\frac{e^2}{a_0}\right)(2.8477) = -77.490 \text{ eV}, \tag{5-15}$$

which is a considerable improvement over the result in Eq.(5-11). The value above of $\zeta = \frac{27}{16}$ instead of 2.0 indicates an appreciable shielding of the nuclear charge of 2.0 by the other electron in the two-electron system.

There are several possible approaches to improve the results above by adding additional parameters in a way that best reflects the true atomic wavefunction. If we look at the radial functions in Eq.(4-87), Table 4-4, and the value of the function for $n = 1$ in Eq.(5-7) we can see that it might be advisable to make n variable as well as the shielding. It is difficult to apply the concept of a variable n to the radial functions defined in Eq.(4-87). We note from Fig. 4-9, however, that the overall radial shape of the $\eta_{nl}(r)$ functions in Table 4-4 are relatively accurately reproduced by the nodeless $l = n - 1$ functions. The electronic properties near the nucleus are likely to be affected by the inner nodal structure of the functions (as shown in Fig. 4-9), but the overall outer electron shape will be described by shifting the nodeless function along the r axis by a variable shielding parameter. Using this concept, we define the *Slater-type orbitals* [2] (STOs) as the product of the nodeless functions, $\eta_{n,n-1}(r)$ in Table 4-4 with the spherical harmonics in Table 4-3,

$$S_{nlm}(r, \theta, \phi) = f_{nl}(r)Y_{lm}(\theta, \phi) = (2\zeta)^{n+1/2}[(2n)!]^{-1/2}r^{n-1}\exp(-\zeta r)Y_{lm}(\theta, \phi). \tag{5-16}$$

The value of r in this equation is in atomic units, that is, in units of length equal to the Bohr radius, $a_0 = \hbar^2/me^2 = 0.529177 \times 10^{-8}$ cm. Thus, all average values computed with this orbital will be in Bohr radii units of length. The orbital exponent zeta is defined by

$$\zeta = \frac{Z - S}{n}. \tag{5-17}$$

S is the shielding factor, Z is the atomic number, and n is now a variable not necessarily integral in value. If $\zeta = Z/n$ in Eq.(5-17), the radial functions, $f_{nl}(r)$, reduce to the $\eta_{n,n-1}(r)$ nodeless function of the hydrogen-like atom as shown in Table 4-4 and Fig. 4-9. It is important to note that the STOs are not an orthogonal set.

Table 5-1 Calculated nonrelativistic electronic energies in He in atomic units and electron volts (1 a.u. = e^2/a_0 = 27.2116 eV). The true energy is from the 1028-term variational function in C. L. Pekeris, *Phys. Rev.* **115**, 1216 (1959) or from the ionization energy given in Ref. [4].

	Simple Hydrogen-like, No Parameters	Simple Hydrogen-like, One Parameter (ζ)	STO Two Parameters (n, ζ)	Best Numerical Radial Function	True Energy
a.u.	−2.7500	−2.8477	−2.8542	−2.8617	−2.9037
eV	−74.832	−77.490	−77.667	−77.871	−79.014

Using both ζ and n as variables in the Slater atomic orbital in Eq.(5-16) with the variation method as before, gives [3]

$$\zeta = 1.6116, \qquad n = 0.955, \qquad W_0 = -2.8542 \text{ a.u.} = -2.8542\left(\frac{e^2}{a_0}\right) = -77.667 \text{ eV}, \tag{5-18}$$

where we also indicate atomic units (a.u.) of energy (in units of e^2/a_0). The energy in Eq.(5-18) is better than that obtained with the one-parameter function in Eq.(5-15). Collecting numbers, we obtain the sequence of energies in the use of atomic orbitals as shown in Table 5-1 [4]. By adding more parameters and flexibility to our single spatial orbital, we can successively decrease the energy until we ultimately reach the limit of the best numerical radial function. Before examining two different spatial orbitals for the two electrons in He, we shall now define, develop, and solve the atomic Hartree–Fock equations for a general many-electron atom.

5.2 THE HARTREE–FOCK SELF-CONSISTENT FIELD (SCF) ATOMIC ORBITALS AND ANALYTICAL APPROXIMATIONS

In order to extend the concept of the many-electron function being written as a product of one-electron orbitals to atoms with more than two electrons, we must acknowledge the indistinguishability of electrons. The indistinguishability of electrons arises from the uncertainty principle, which, as discussed in Section 3.2, states that the product of the uncertainties in position and velocity (momentum) is on the order of $\hbar$. Thus, if we localize an electron in position, its velocity will be very uncertain. The consequence is that if the electron's velocity is uncertain, we cannot predict its future position with certainty. As a result, in a many-electron system we cannot predict which electron will be in any particular position at any particular time; the electrons are indistinguishable. The same consequences arise for any system of identical particles.

The many-particle wavefunction for n indistinguishable particles is written as $\psi(\boldsymbol{r}_1, \ldots, \boldsymbol{r}_i, \boldsymbol{r}_j, \ldots, \boldsymbol{r}_n)$ where $\boldsymbol{r}_i$ indicates the coordinate variables of the ith particle.

We require also an orthonormal many-particle function,

$$\int_1 \int_2 \cdots \int_n \psi^*(\boldsymbol{r}_1, \boldsymbol{r}_2, \ldots, \boldsymbol{r}_n)\psi(\boldsymbol{r}_1, \boldsymbol{r}_2, \ldots, \boldsymbol{r}_n)\, dV_1\, dV_2 \cdots dV_n = 1.0, \qquad (5\text{-}19)$$

where the volume elements are defined in Eq.(5-6).

Consider now the exchange of two indistinguishable particles in the many-particle wavefunction. Let P_{ij} be the permutation operator for the ith and jth particles, which gives

$$P_{ij}\psi(\boldsymbol{r}_1, \ldots, \boldsymbol{r}_i, \boldsymbol{r}_j, \ldots, \boldsymbol{r}_n) = \psi(\boldsymbol{r}_1, \ldots, \boldsymbol{r}_j, \boldsymbol{r}_i, \ldots, \boldsymbol{r}_n). \qquad (5\text{-}20)$$

Now, due to the normalization requirement, $\psi(\boldsymbol{r}_1, \ldots, \boldsymbol{r}_i, \boldsymbol{r}_j, \ldots, \boldsymbol{r}_n)$ and $\psi(\boldsymbol{r}_1, \ldots, \boldsymbol{r}_j, \boldsymbol{r}_i, \ldots, \boldsymbol{r}_n)$ can vary only by a phase factor, $\exp(i\alpha)$, giving

$$\psi(\boldsymbol{r}_1, \ldots, \boldsymbol{r}_i, \boldsymbol{r}_j, \ldots, \boldsymbol{r}_n) = \exp(i\alpha)\psi(\boldsymbol{r}_1, \ldots, \boldsymbol{r}_j, \boldsymbol{r}_i, \ldots, \boldsymbol{r}_n). \qquad (5\text{-}21)$$

Repeating the P_{ij} operation on this equation gives

$$\begin{aligned} P_{ij}\psi(\boldsymbol{r}_1, \ldots, \boldsymbol{r}_i, \boldsymbol{r}_j, \ldots, \boldsymbol{r}_n) &= \psi(\boldsymbol{r}_1, \ldots, \boldsymbol{r}_j, \boldsymbol{r}_i, \ldots, \boldsymbol{r}_n) \\ &= \exp(i\alpha)P_{ij}\psi(\boldsymbol{r}_1, \ldots, \boldsymbol{r}_j, \boldsymbol{r}_i, \ldots, \boldsymbol{r}_n) \\ &= \exp(i\alpha)\psi(\boldsymbol{r}_1, \ldots, \boldsymbol{r}_i, \boldsymbol{r}_j, \ldots, \boldsymbol{r}_n) \\ &= \exp(i2\alpha)\psi(\boldsymbol{r}_1, \ldots, \boldsymbol{r}_j, \boldsymbol{r}_i, \ldots, \boldsymbol{r}_n). \end{aligned}$$

Thus, $\exp(i\alpha) = \pm 1$ giving $\exp(i2\alpha) = 1$. The result of this exercise is to show that a many-particle wavefunction is either symmetric (use $+1$) or antisymmetric (use -1) to the interchange of any pair of particles. Identical particles with integral spin angular momentum values have symmetric many-particle functions [5]. Identical particles with half-integral spin angular momentum values have antisymmetric many-particle wavefunctions [5] (see also Problem 4-38). Thus, a many-electron wavefunction must be antisymmetric with respect to the exchange of any pair of electrons. The antisymmetrized product of one-electron orbitals, as an approximation to a many-electron wavefunction, is conveniently written as a determinant [6],

$$\psi(1, 2, 3, \ldots, n) = \left(\frac{1}{n!}\right)^{1/2} \begin{vmatrix} \psi_1(1) & \psi_1(2) & \psi_1(3) & \cdots & \psi_1(n) \\ \psi_2(1) & \psi_2(2) & \psi_2(3) & & \\ \psi_3(1) & \psi_3(2) & \psi_3(3) & & \\ \vdots & & & \ddots & \\ \psi_n(1) & & & & \psi_n(n) \end{vmatrix},$$

where $\psi_i(j)$ indicates the ith one-electron orbital containing the jth electron. It is clear that interchanging two electrons (interchange two columns in the determinant) changes the sign of the function. An even number of permutations gives a positive sign and an odd number of permutations gives a negative sign.

The Pauli principle also follows from the determinantal form for the wavefunction [7]. The Pauli principle states that no two electrons can be in the same state.

If two of the rows in the determinant are identical, the determinant vanishes. If the many-electron Hamiltonian contains no electron spin operators, we can write the one-electron orbitals, $\psi_i(j)$ in the determinant above, as a product of spatial, $\Phi_i(j)$, and spin [$\alpha(j)$ or $\beta(j)$] components as shown in Eq.(4-128) to give, for a closed shell configuration of paired electrons,

$$\psi(1, 2, 3, \ldots, n) = \left(\frac{1}{n!}\right)^{1/2} \begin{vmatrix} \Phi_1(1)\alpha(1) & \Phi_1(2)\alpha(2) & \Phi_1(3)\alpha(3) & \ldots & \Phi_1(n)\alpha(n) \\ \Phi_1(1)\beta(1) & \Phi_1(2)\beta(2) & & \ldots & \Phi_1(n)\beta(n) \\ \Phi_2(1)\alpha(1) & \Phi_2(2)\alpha(2) & & & \vdots \\ \Phi_2(1)\beta(1) & \Phi_2(2)\beta(2) & & & \\ \Phi_3(1)\alpha(1) & \Phi_3(2)\alpha(2) & & & \\ \vdots & \vdots & \ddots & & \\ \Phi_{n/2}(1)\beta(1) & & & & \Phi_{n/2}(n)\beta(n) \end{vmatrix}, \tag{5-22}$$

where $\Phi_i(j)$ denotes the ith spatial orbital containing the jth electron, and $\beta(j)$ and $\alpha(j)$ represent the electron spin functions for the jth electron, respectively. In the two-electron system discussed in Section 5.1, the spin and spatial parts of the function separate. As there are no spin operators in the two-electron Hamiltonian for He [Eq.(5-1)], the resultant function is identical to the result used earlier in Eq.(5-3). The spatial and spin parts of the function in Eq.(5-22) do not separate for three and higher numbers of electrons, however.

The complete Hamiltonian for a many-electron atom with an infinitely heavy nucleus is given by

$$\mathscr{H} = -\sum_i \frac{\hbar^2}{2m} \nabla_i^2 - \sum_i \frac{Ze^2}{r_i} + \sum_{i>j} \frac{e^2}{r_{ij}} = \sum_i \mathscr{H}^0(i) + \sum_{i>j} \frac{e^2}{r_{ij}}, \tag{5-23}$$

where $\mathscr{H}^0(i) = -[(\hbar^2/2m)\nabla_i^2] - (Ze^2/r_i)$ is the ith one-electron hydrogen-like operator usually referred to as the *core Hamiltonian.* Substituting the normalized determinant in Eq.(5-22) into the expression for the variational energy and remembering that the spin functions are orthonormal,

$$\int_a \alpha^*(a)\alpha(a)\, dV_a(\text{spin}) = \int_a \beta^*(a)\beta(a)\, dV_a(\text{spin}) = 1.0$$

$$\int_a \alpha^*(a)\beta(a)\, dV_a(\text{spin}) = 0,$$

gives [8]

$$W_0 = \int_1 \int_2 \cdots \int_n \psi^*(1, 2, \ldots, n)\mathscr{H}\psi(1, 2, \ldots, n)\, dV_1\, dV_2 \cdots dV_n$$
$$= \sum_{i=1}^{n/2} \left[2\mathscr{H}_{ii}^0 + \sum_{j=1}^{n/2} (2\langle ij|ij\rangle - \langle ij|ji\rangle)\right]. \tag{5-24}$$

The sums over i and j are over the $n/2$ spatial atomic orbitals in the atom where each spatial orbital accommodates two electrons. The result in Eq.(5-24) is easily checked with a 2×2 example. The terms in Eq.(5-24) are defined below.

1. *The core integral*

$$\mathscr{H}_{ii}^{0} = \int_a \Phi_i^*(a)\mathscr{H}^0(a)\Phi_i(a)\,dV_a = \int_a \Phi_i^*(a)\left(-\frac{\hbar^2}{2m}\nabla_a^2 - \frac{Ze^2}{r_a}\right)\Phi_i(a)\,dV_a \quad (5\text{-}25)$$

2. *The Coulomb integral*

$$\langle ij|ij\rangle = \int_a\int_b \Phi_i^*(a)\Phi_j^*(b)\left(\frac{e^2}{r_{ab}}\right)\Phi_i(a)\Phi_j(b)\,dV_a\,dV_b$$

$$= \int_a \Phi_i^*(a)\left[\int_b \Phi_j^*(b)\left(\frac{e^2}{r_{ab}}\right)\Phi_j(b)\,dV_b\right]\Phi_i(a)\,dV_a = \int_a \Phi_i^*(a)J_j(a)\Phi_i(a)\,dV_a \quad (5\text{-}26)$$

$J_j(a)$ is the Coulomb operator.

3. *The exchange integral*

$$\langle ij|ji\rangle = \int_a\int_b \Phi_i^*(a)\Phi_j^*(b)\left(\frac{e^2}{r_{ab}}\right)\Phi_j(a)\Phi_i(b)\,dV_a\,dV_b$$

$$= \int_a \Phi_i^*(a)\left[\int_b \Phi_j^*(b)\left(\frac{e^2}{r_{ab}}\right)\Phi_i(b)\,dV_b\right]\Phi_j(a)\,dV_a = \int_a \Phi_i^*(a)K_j(a)\Phi_i(a)\,dV_a \quad (5\text{-}27)$$

$K_j(a)$ is the exchange operator and the operator has meaning only when operating on one of the $\Phi_i(a)$ functions. Note that $K_j(a)$ exchanges the a and b electrons in the i and j orbitals.

The next step is to minimize the energy in Eq.(5-24) with respect to the functions under the constraint that the atomic orbitals remain orthonormal. The constraint is handled with a Lagrangian multiplier, ε_{ji}, having the units of energy. Thus, we seek to apply the variation treatment to W' given by

$$W' = W_0 - 2\sum_{i,j}^{n/2}\varepsilon_{ji}S_{ij} = 2\sum_{i}^{n/2}\mathscr{H}_{ii}^{0} + \sum_{i,j}^{n/2}(2\langle ij|ij\rangle - \langle ij|ji\rangle) - 2\sum_{i,j}^{n/2}\varepsilon_{ji}S_{ij}, \quad (5\text{-}28)$$

where

$$S_{ij} = \int \Phi_i^*(a)\Phi_j(a)\,dV_a$$

is the overlap integral. The variation gives

$$\delta W' = 0 = 2\sum_i \delta\mathscr{H}_{ii}^{0} + \sum_{i,j}(2\delta\langle ij|ij\rangle - \delta\langle ij|ji\rangle - 2\varepsilon_{ji}\delta S_{ij}). \quad (5\text{-}29)$$

Thus, we seek the best functions that leave the energy unchanged ($\delta W' = 0$) with respect to the first-order changes in the functions under the orthonormality constraint [9]. The variation in the core integral in Eq.(5-29) is given by

$$\delta\mathscr{H}_{ii} = \delta\int_a \Phi_i^*(a)\mathscr{H}^0(a)\Phi_i(a)\,dV_a$$

$$= \int_a \delta\Phi_i^*(a)\mathscr{H}^0(a)\Phi_i(a)\,dV_a + \int_a \Phi_i(a)^*\mathscr{H}^0(a)\delta\Phi_i(a)\,dV_a$$

$$= 2\int_a \delta\Phi_i^*(a)\mathscr{H}^0(a)\Phi_i(a)\,dV_a, \quad (5\text{-}30)$$

where the one-electron Hamiltonian for the ath electron, $\mathscr{H}^0(a)$, is defined in Eq.(5-25). The last step acknowledges the fact that $\mathscr{H}^0(a)$ is a Hermitian operator. The variation of the Coulomb integral is given by

$$\begin{aligned}\delta\langle ij|ij\rangle &= 2\int_a\int_b \delta\Phi_i^*(a)\Phi_j^*(b)\frac{e^2}{r_{ab}}\Phi_i(a)\Phi_j(b)\,dV_a dV_b \\ &\quad + 2\int_a\int_b \Phi_i^*(a)\delta\Phi_j^*(b)\frac{e^2}{r_{ab}}\Phi_i(a)\Phi_j(b)\,dV_a\,dV_b \\ &= 2\int_a \delta\Phi_i^*(a)J_j(a)\Phi_i(a)\,dV_a + 2\int_b \delta\Phi_j^*(b)J_i(b)\Phi_j(b)\,dV_b. \end{aligned} \tag{5-31}$$

Similarly, the variation of the exchange integral is given by

$$\delta\langle ij|ji\rangle = 2\int_a \delta\Phi_i^*(a)K_j(a)\Phi_i(a)\,dV_a + 2\int_b \delta\Phi_j^*(b)K_i(b)\Phi_j(b)\,dV_b. \tag{5-32}$$

The last term in Eq.(5-29), which is the variation of the overlap, is given by

$$\delta S_{ij} = \delta\int_a \Phi_i^*(a)\Phi_j(a)\,dV_a = \int_a \delta\Phi_i^*(a)\Phi_j(a)\,dV_a + \int_a \Phi_i^*(a)\delta\Phi_j(a)\,dV_a. \tag{5-33}$$

Substituting Eqs.(5-33), (5-32), (5-31) and (5-30) into Eq.(5-29) gives

$$\begin{aligned}0 = \sum_i\Bigg[4\int \delta\Phi_i^*\mathscr{H}^0\Phi_i\,dV &+ \sum_j\Bigg(4\int\delta\Phi_i^*J_j\Phi_i\,dV + 4\int\delta\Phi_j^*J_i\Phi_j dV \\ &-2\int\delta\Phi_i^*K_j\Phi_i\,dV - 2\int\delta\Phi_j^*K_i\Phi_j\,dV - 2\varepsilon_{ji}\int\delta\Phi_i^*\Phi_j\,dV - 2\varepsilon_{ji}\int\delta\Phi_j^*\Phi_i\,dV\Bigg)\Bigg],\end{aligned} \tag{5-34}$$

where we have now dropped the unnecessary index identifying the electrons. The order of summation over i and j inside the $\sum_j$ summation is arbitrary and can be interchanged. Rearranging reduces Eq.(5-34) to

$$0 = \sum_i \int \delta\Phi_i^*\Big[\mathscr{H}^0\Phi_i + \sum_j (2J_j\Phi_i - K_j\Phi_i - \varepsilon_{ji}\Phi_j)\Big]\,dV. \tag{5-35}$$

Thus, as the magnitude of $\delta\Phi_i$ in Eq.(5-35) is arbitrary, the term in brackets must be zero, giving

$$\Big[\mathscr{H}^0 + \sum_j (2J_j - K_j)\Big]\Phi_i = F\Phi_i = \sum_j \Phi_j\varepsilon_{ji}. \tag{5-36}$$

Equation (5-36) is the celebrated *Hartree–Fock equation* [10] for atoms where F is called the *Hartree–Fock operator*. It is interesting to note that the earliest calculation of the electronic energies in atoms by Hartree and co-workers included only simple product functions that were not antisymmetrized; these functions give a resulting eigenvalue equation similar to Eq.(5-36) with the exchange operator missing. The exchange operator is a result of using the antisymmetrized product function that leads to Eq.(5-36).

For a closed shell system with n electrons as considered here, there are $n/2$ spatial orbitals, Φ_i, and the elements ε_{ji} form a square symmetric $(n/2) \times (n/2)$ matrix. Thus, we can write Eq.(5-36) in matrix form as

$$F\mathbf{\Phi} = \mathbf{\Phi}\boldsymbol{\varepsilon}, \qquad \int \mathbf{\Phi}^{\dagger} F \mathbf{\Phi}\, dV = \mathbf{F}^{\Phi} = \boldsymbol{\varepsilon}, \tag{5-37}$$

where $\boldsymbol{\varepsilon}$ is not necessarily a diagonal matrix. As the F operator also contains the $\mathbf{\Phi}$ basis, Eq.(5-36) is a coupled integral-differential equation. The solutions, $\mathbf{\Phi}$ and $\boldsymbol{\varepsilon}$ in Eq.(5-36) or (5-37), can be obtained by using numerical methods [11]. It is usual to transform the $\mathbf{\Phi}$ to a basis that diagonalizes $\boldsymbol{\varepsilon}$ by a unitary transformation, $\mathbf{u}$, to give the Hartree–Fock orbital energies, E_i, given by

$$\mathbf{u}^{\dagger}\mathbf{F}^{\Phi}\mathbf{u} = \mathbf{u}^{\dagger}\boldsymbol{\varepsilon}\mathbf{u} = \mathbf{E}, \qquad \boldsymbol{\psi} = \mathbf{\Phi}\mathbf{u}, \tag{5-38}$$

where $\boldsymbol{\psi}$ contains the final Hartree–Fock orbitals. In other words, if we replace ε_{ji} with $\varepsilon_{ji}\delta_{ij}$ in Eq.(5-36) and solve the integral-differential equations, the resultant $\mathbf{\Phi}$ will equal $\boldsymbol{\psi}$ in Eq.(5-38). In solving Eq.(5-36) with $\varepsilon_{ji}\delta_{ij}$, we note that the $\mathbf{\Phi}$ must be known in order to compute the Hartree–Fock operator, F. Thus, a set $\mathbf{\Phi}$ is chosen to compute F and then the new set $\mathbf{\Phi}$ is computed with Eq.(5-36). The new set is then used to recompute F and a new solution is again obtained. The process is repeated until the functions used to compute F are equal to the final solution—the *self-consistent-field* (SCF) limit.

The Hartree–Fock orbital energy for the ith orbital is given from Eq.(5-36) for the final set $\mathbf{\Phi}$, where $\varepsilon_{ij} = \varepsilon_{ij}\delta_{ij}$, by

$$\varepsilon_i = \mathscr{H}^0_{ii} + \sum_j (2\langle ij|ij\rangle - \langle ij|ji\rangle),$$

where we use the single subscript on ε_i to indicate the final eigenvalues of Eq.(5-36) or (5-37).

Comparing the orbital energy above with Eq.(5-24) for the total energy in the Hartree–Fock picture leads to

$$W_0 = \sum_{i=1}^{n/2} W_{ii} = \sum_{i=1}^{n/2} (\varepsilon_i + \mathscr{H}^0_{ii}) = \sum_{i=1}^{n} \left[\mathscr{H}^0_{ii} + \sum_j \left(\langle ij|ij\rangle - \frac{1}{2}\langle ij|ji\rangle\right)\right]. \tag{5-39}$$

In solving Eq.(5-36) numerically for the atomic orbitals Φ_i and eigenvalues ε_i, we note that the atomic centrosymmetric system leads to a separation in the angular and radial variables. Normally, the spherical harmonics are used for the angular part of Φ_i and only the radial functions are determined numerically. Thus, in the case of the $1s^2$ configuration in He, only the radial orbital is determined numerically. The calculated Hartree–Fock energy for the He atom is equal to -2.8617 a.u. as shown in Table 5-1. The step between a single determinant two-parameter and an infinite-parameter (Hartree–Fock) single determinant function is only 0.204 eV. Thus, we might expect the two- or even the one-parameter function to be fairly revealing with respect to the shape of atomic orbitals.

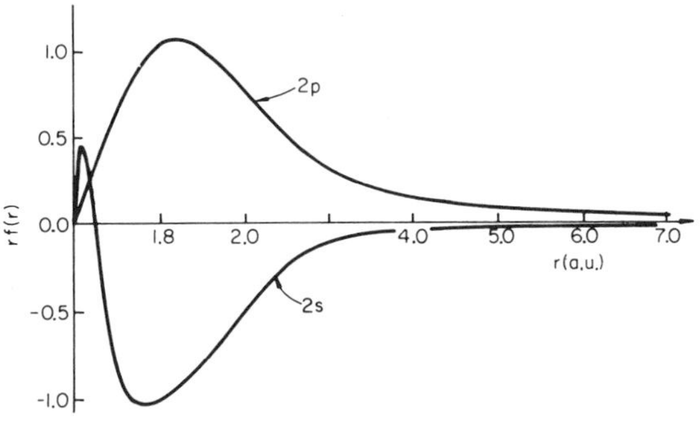

Figure 5-1 The numerical Hartree–Fock SCF radial amplitude functions for the $2s$ and $2p$ atomic orbitals in the F^- ion from C. Froese, *Proc. Cambridge Phil. Soc.* **53**, 206 (1957).

In the case of the closed shell F^- ion with 10 electrons, the atomic orbitals (AOs) are given by

$$\Phi_1 = f_{1s}(r)Y_{00}, \qquad \Phi_2 = f_{2s}(r)Y_{00}, \qquad \Phi_3 = f_{2p}(r)Y_{10},$$
$$\Phi_4 = f_{2p}(r)Y_{1+1}, \qquad \Phi_5 = f_{2p}(r)Y_{1-1}, \tag{5-40}$$

in terms of the spherical harmonics and corresponding radial functions. The SCF radial amplitude or $R_i(r) = rf_i(r)$ functions for the outer shell in F^- are shown in Fig. 5-1. It is clear by comparing the SCF radial amplitude functions in Fig. 5-1 with the hydrogen-like radial amplitude functions given in Table 4-4 that the nodal characteristics in both sets of functions are identical.

The development above where an atom has all electrons paired (each spatial orbital is doubly occupied) is called the *closed shell restricted Hartree–Fock* (RHF) method. In some cases, such as in Na, the outer electron is unpaired. Under these conditions, the *open shell restricted Hartree–Fock* method is used, which requires a modification to the treatment we have given here [12]. Finally, in the case of an atom that has unpaired electrons in different orbitals, a slightly different treatment must be used called the *unrestricted Hartree–Fock* (UHF) method.

The Hartree–Fock energies for several atoms in the first row of the periodic table are listed in Table 5-2 [12,13]. Only light atoms are shown to represent nonrelativistic systems. Relativistic corrections to the energy become evident in heavier atoms due to a Z^4-dependence [14]. The *correlation error*, or the difference between the best single determinant energy (Hartree–Fock) and the true nonrelativistic energy, is also listed in Table 5-2. The correlation error, which is approximately 1.1 eV for each electron pair, arises from the use of the averaged interelectron interaction instead of the true instantaneous interelectronic interaction.

The most important result of the many-electron theory in the atomic orbital scheme up to this point is the concept of a nucleus being shielded by an average SCF field. The concept of one electron acting under the influence of a reduced nuclear charge due to shielding by the average distribution of the other electrons appears to be useful in thinking about atoms. A body of experience has accumulated on the

Table 5-2 The Hartree–Fock SCF energies for a number of atoms in the first row of the periodic table. The experimental results are from Ref. [4]. The correlation errors are also listed.

Z	Atom	Hartree–Fock SCF (a.u.)	Experiment (a.u.)	Correlation Error (eV)
1	H		−0.5000	—
2	He	−2.8617	−2.9037	1.14
3	Li	−7.4327	−7.4768	1.20
4	Be	−14.5730	−14.6660	2.53
5	B	−24.5291	−24.6550	3.43
6	C	−37.6886	−37.8490	4.36

shielding parameters for atoms that are obtained by fitting the Hartree–Fock numerical functions with analytical functions.

Before discussing the analytical approximations to the Hartree–Fock orbitals, we shall discuss the many-electron building-up principle where the electrons are paired in atomic orbitals. The ground-state configuration of electrons in atoms is obtained from experience on the observable atomic properties and the electrons fill the orbitals approximately as shown on Fig. 5-2. The notation for the shell structure is given in Section 4.7 (see Table 4-5), and the electrons fill 2 into the s groups, 6 into the p groups, 10 into the d groups, and so on. A *configuration* of electrons is the assignment or placement of electrons within the shell structure.

The ground-state configurations of atoms are obtained by the building-up principle using Fig. 5-2 as a guide to the orderly filling of electrons into atomic orbitals. For instance, Na has 11 electrons and a ground-state configuration of $1s^2 2s^2 2p^6 3s = [\mathrm{Ne}]3s$ where we use the rare gas atom configuration for the inner closed shell. Excited-state configurations are obtained by promotion of electrons into higher energy configurations. For instance, Si has a ground-state configuration of $[\mathrm{Ne}]3s^2 3p^2$ and an excited state configuration of $[\mathrm{Ne}]3s^2 3p 3d$. Further examples can be obtained from the periodic table in Fig. A-5 (in Appendix A).

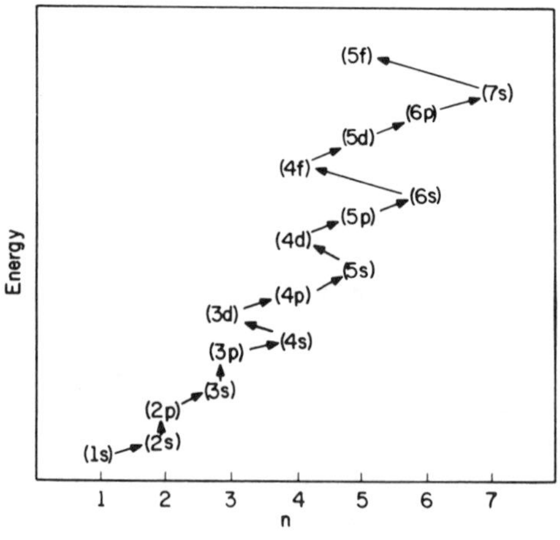

Figure 5-2 Building-up principle for many-electron atoms. The electrons fill the shells in an order approximately as shown by the arrows.

Linear Combination of Atomic Orbitals, LCAOs

We shall now discuss a method of taking linear combinations of analytical functions to approximate the SCF Hartree–Fock numerical orbitals. Let $\boldsymbol{\chi}$ be a set of analytical orbitals such as the STOs [Eq.(5-16)] or hydrogen-like orbitals. We then expand $\boldsymbol{\Phi}$ in Eq.(5-22) in terms of $\boldsymbol{\chi}$,

$$\boldsymbol{\Phi} = \boldsymbol{\chi}\mathbf{a}. \tag{5-41}$$

Equation (5-41) is called a *linear combination of atomic orbitals* (LCAOs). We now return to the development beginning with Eq.(5-24) and use Eq.(5-41) for the atomic orbitals Φ_i. Proceeding to Eq.(5-29) we apply the variation on Φ_i to the linear coefficients in the **a** matrix, keeping $\boldsymbol{\chi}$ fixed. Substituting

$$\Phi_i = \boldsymbol{\chi}\mathbf{a}_i, \qquad \Phi_i^* = \mathbf{a}_i^\dagger\boldsymbol{\chi}^\dagger = (\boldsymbol{\chi}\mathbf{a}_i)^\dagger,$$

where $\mathbf{a}_i$ is a column matrix, into Eq.(5-30) and dropping the unnecessary index for the ath electron gives the variation in the core Hamiltonian;

$$\delta\mathscr{H}_{ii}^0 = \delta\int \mathbf{a}_i^\dagger\boldsymbol{\chi}^\dagger\mathscr{H}^0\boldsymbol{\chi}\mathbf{a}_i\,dV = \delta(\mathbf{a}_i^\dagger\mathscr{H}^\chi\mathbf{a}_i)$$

$$= (\delta\mathbf{a}_i^\dagger)\mathscr{H}^\chi\mathbf{a}_i + \mathbf{a}_i^\dagger\mathscr{H}^\chi(\delta\mathbf{a}_i) = 2(\delta\mathbf{a}_i^\dagger)\mathscr{H}^\chi\mathbf{a}_i, \tag{5-42}$$

where $\mathscr{H}^0(a)$ is a Hermitian operator. The Coulomb integral is given by [see Eq.(5-31)]

$$\delta\langle ij|ij\rangle = \delta\int_a\int_b \mathbf{a}_i^\dagger(a)\boldsymbol{\chi}^\dagger(a)\mathbf{a}_j^\dagger(b)\boldsymbol{\chi}^\dagger(b)\left(\frac{e^2}{r_{ab}}\right)\boldsymbol{\chi}(a)\mathbf{a}_i(a)\boldsymbol{\chi}(b)\mathbf{a}_j(b)\,dV_a\,dV_b$$

$$= 2\int_a [\delta\mathbf{a}_i^\dagger(a)]\boldsymbol{\chi}^\dagger(a)J_j(a)\boldsymbol{\chi}(a)\mathbf{a}_i(a)\,dV_a + 2\int_b [\delta\mathbf{a}_j^\dagger(b)]\boldsymbol{\chi}^\dagger(b)J_i(b)\boldsymbol{\chi}(b)\mathbf{a}_j(b)\,dV_b$$

$$= 2(\delta\mathbf{a}_i)^\dagger\mathbf{J}_j^\chi\mathbf{a}_i + 2(\delta\mathbf{a}_j)^\dagger\mathbf{J}_i^\chi\mathbf{a}_j, \tag{5-43}$$

where the unnecessary index on the electron has been dropped. The Coulomb operator, $J_j(a)$, is given by

$$J_j(a)\Phi_i(a) = J_j(a)\boldsymbol{\chi}(a)\mathbf{a}_i(a) = \left[\int_b \Phi_j^*(b)\left(\frac{e^2}{r_{ab}}\right)\Phi_j(b)\,dV_b\right]\Phi_i(a)$$

$$= \left[\int_b \mathbf{a}_j^\dagger(b)\boldsymbol{\chi}^\dagger(b)\left(\frac{e^2}{r_{ab}}\right)\boldsymbol{\chi}(b)\mathbf{a}_j(b)\,dV_b\right]\boldsymbol{\chi}(a)\mathbf{a}_i(a). \tag{5-44}$$

The matrix representation of the Coulomb operator in the $\boldsymbol{\chi}$ basis is given by

$$\mathbf{J}_j^\chi = \int_a \boldsymbol{\chi}^\dagger(a)J_j(a)\boldsymbol{\chi}(a)\,dV_a. \tag{5-45}$$

The variation in the exchange integral is given by [see Eq.(5-32)]

$$\delta\langle ij|ji\rangle = 2(\delta\mathbf{a}_i)^\dagger\mathbf{K}_j^\chi\mathbf{a}_i - 2(\delta\mathbf{a}_j)^\dagger\mathbf{K}_i^\chi\mathbf{a}_j. \tag{5-46}$$

The corresponding exchange operator and matrix representation in the χ basis are given by analogy with Eqs.(5-44) and (5-45) for the Coulomb operator. The variation in the overlap integral is given by

$$\delta S_{ij} = \delta \int_a \Phi_i^*(a)\Phi_j(a)\, dV_a = \delta \int_a \mathbf{a}_i^\dagger(a)\chi^\dagger(a)\chi(a)\mathbf{a}_j(a)\, dV_a$$

$$= \delta(\mathbf{a}_i^\dagger \mathbf{S}^\chi \mathbf{a}_j) = (\delta\mathbf{a}_i)^\dagger \mathbf{S}^\chi \mathbf{a}_j + (\delta\mathbf{a}_j)^\dagger \mathbf{S}^\chi \mathbf{a}_i, \tag{5-47}$$

where $\mathbf{S}^\chi$ is the overlap matrix in the χ basis (remember that χ need not be orthonormal). Substituting Eqs.(5-47), (5-46), (5-43), and (5-42) into Eq.(5-29) gives

$$0 = \sum_i \left\{(\delta\mathbf{a}_i)^\dagger[\mathscr{H}^\chi\mathbf{a}_i + \sum_j (2\mathbf{J}_j^\chi\mathbf{a}_i - \mathbf{K}_j^\chi\mathbf{a}_i - \varepsilon_{ji}\mathbf{S}^\chi\mathbf{a}_j)]\right\}. \tag{5-48}$$

As the variation in $\mathbf{a}_i$, $(\delta\mathbf{a}_i)$, is arbitrary, the term in square brackets must be zero, giving N equations of the following form (N is the order of expansion basis, χ):

$$\left\{\int \chi^\dagger\left[\mathscr{H}^0 + \sum_j (2J_j - K_j)\right]\chi\, dV\right\}\mathbf{a}_i = \sum_j \varepsilon_{ji}\mathbf{S}^\chi\mathbf{a}_j$$

$$\left[\int \chi^\dagger F \chi\, dV\right]\mathbf{a}_i = \sum_j \mathbf{S}^\chi\mathbf{a}_j\varepsilon_{ji}$$

$$\mathbf{F}^\chi\mathbf{a}_i = \sum_j \mathbf{S}^\chi\mathbf{a}_j\varepsilon_{ji}. \tag{5-49}$$

The usual way to proceed from here is to write a secular equation (as described in Section 3.5) from Eq.(5-49). This is done by replacing ε_{ji} in Eq.(5-49) by $\delta_{ij}\varepsilon_{ji}$ as suggested in the discussion following Eq.(5-38). Doing this gives

$$\mathbf{F}^\chi\mathbf{a}_i = \sum_j \mathbf{S}^\chi\mathbf{a}_j\varepsilon_{ji}\delta_{ij} = \mathbf{S}^\chi\mathbf{a}_i\varepsilon_{ii}, \qquad (\mathbf{F}^\chi - \mathbf{S}^\chi\varepsilon_{ii})\mathbf{a}_i = 0. \tag{5-50}$$

Now, setting the determinant of the coefficients of $\mathbf{a}_i$ to zero gives the secular equation:

$$\text{Det}\,(\mathbf{F}^\chi - \mathbf{S}^\chi\varepsilon_{ii}) = 0. \tag{5-51}$$

The number of roots in this secular equation equals the order of the χ basis in Eq.(5-41).

As the **a** coefficients in Eq.(5-41) must be known in order to calculate the Hartree–Fock operator, F, and the resulting matrix representation, $\mathbf{F}^\chi$ in Eq.(5-51), an iterative method must be used to solve these equations. The method proceeds as follows:

1. Choose a set χ and $\mathbf{\Phi} = \chi\mathbf{a}$ in Eq.(5-41).
2. Compute F in Eq.(5-49) and then the matrix representation of F in the χ basis given by $\mathbf{F}^\chi$.
3. Solve Eq.(5-51) for the orbital energies and use these in Eq.(5-50) to compute the linear coefficients $\mathbf{a}_i$. If these new $\mathbf{a}_i$ are identical with the columns in the original **a** in statement 1 above, then the first choice was excellent. If the solutions **a** from solving Eq.(5-51) differ from the original **a**, then the new **a** is used in Eq.(5-41) to return to statement 2 above to recompute F.

The process is repeated until self-consistency is achieved to the desired accuracy in the eigenvalues. This limit is again called the self-consistent field (SCF) limit. The eigenstates with the lowest $n/2$ final eigenvalues are occupied by the electrons in the atom. Thus, the total electronic energy is obtained by adding the SCF energy to the core Hamiltonian energy for the lowest (occupied) $n/2$ levels [see again Eqs.(5-24) and (5-39)],

$$W_0 = \sum_{i=1}^{n/2} (\mathscr{H}_{ii}^0 + \varepsilon_i), \tag{5-52}$$

where $\mathscr{H}_{ii} = \int \Phi_i^*(a)\mathscr{H}^0(a)\Phi_i(a)\, dV_a$, $\mathscr{H}^0(a)$ is the Hamiltonian for ath electron, Φ_i is the ith final SCF function from Eq.(5-41), and ε_i is from the lowest $n/2$ energy levels from Eq.(5-51).

Atomic Orbital Basis Sets

We shall now describe some of the commonly used atomic orbital basis sets used in the LCAO method described above beginning with Eq.(5-41). STOs as defined in Eq.(5-16), are commonly used. The orbital exponents, ζ, shown in Eq.(5-17) are now fairly well set by experience in AO calculations. One fairly successful set of rules in determining ζ and n is given here [1].

1. The value of n in the orbital in Eq.(5-16) is assigned as follows in reference to the hydrogen-like orbital assignment:

n (hydrogen-like) =	1	2	3	4	5	6
n (Slater) =	1	2	3	3.7	4.0	4.2

2. The screening constant S in the $\zeta = (Z-S)/n$ equation is found by arranging the orbitals in groups as follows:

   $(1s), (2s, 2p), (3s, 3p), (3d), (4s, 4p), (4d), (4f), (5s, 5p), \ldots$

   The screening constant, S, for any electron in one of the groups above is a sum of
   a. nothing from any electron outside and higher in energy than the group,
   b. a contribution of 0.35 from any electron in the same group (except that 0.30 is used for the $1s$ group),
   c. a contribution of 0.85 from each electron from groups with $n - 1$ below an (ns, np) group and a contribution of 1.0 for all electrons in groups lower in energy, and
   d. a contribution of 1.0 for each electron below an (nd) or (nf) group.

As an example, consider the sodium atom.

Na, $1s^2 2s^2 2p^6 3s$

$1s$ electron	$S = 0.30$	$= 0.30$;	$\zeta = 10.70$
$2s$ or $2p$ electron	$S = 2.0(0.85) + 7.0(0.35)$	$= 4.15$;	$\zeta = 6.85/2 = 3.43$
$3s$ electron	$S = 2.0 + 8.0(0.85)$	$= 8.80$;	$\zeta = 2.20/3 = 0.73$

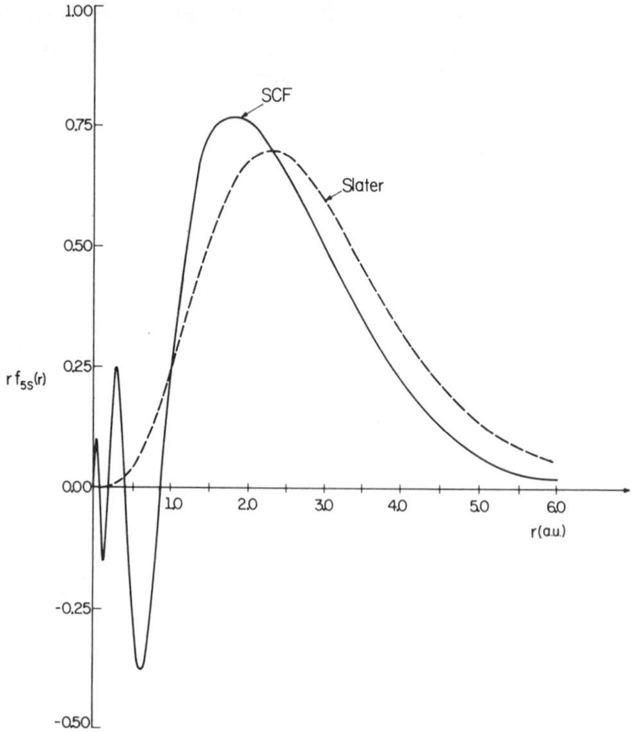

Figure 5-3 Comparison of the Slater radial amplitude function in Eq.(5-53) (dotted line) with an SCF radial amplitude function (solid line) for the 5*s* electron in I^-. The SCF function was obtained from D. F. Mayers.

The values of ζ and n for the 5*s* outer electron in the I^- ion as determined by the preceding set of empirical rules are $n = 4.0$ and $\zeta = 1.81$. From Eq.(5-16) the Slater radial amplitude function is given by (in a.u. of length)

$$rf_{5s}(r) = \frac{(3.62)^{4.5}}{(8!)^{1/2}} r^4 \exp(-1.81r). \tag{5-53}$$

The maximum in this function is equal to 0.71 at $r = 2.21$ a.u. This number compares with the maximum in the SCF 5*s* orbital as shown in Fig. 5-3, which is 0.77 at $r =$ 1.90 a.u. The entire Slater orbital is plotted in Fig. 5-3 to compare with the corresponding 5*s* SCF result in I^-. Figure 5-3 shows that the Slater radial atomic orbital agrees reasonably well with the SCF function in the outer regions of electron density, but near the nucleus considerable differences are apparent. By taking linear combinations of STOs as described above beginning with Eq.(5-41), however, the correct nodal structure near the nucleus can be obtained. Extensive work has been completed on using STOs as a starting basis for atomic orbitals in the atomic SCF procedure and a summary of these results is in order because much of this development in atoms can also be carried over into a study of molecules.

A *minimal basis set* of STOs for an atom includes one function for each SCF occupied orbital with different n and l quantum numbers in Eq.(5-16). The minimal

Table 5-3 Slater-type orbital SCF optimized orbital exponent calculations of the ground-state energies (in atomic units) of some light atoms. Also listed are the Hartree–Fock SCF energies.

	Minimal Basis[a]	Double zeta[b]	Extended Basis[c]	Hartree–Fock SCF[a]
Li	−7.4185	−7.4327	−7.4327	−7.4327
Be	−14.5567	−14.5724	−14.5730	−14.5730
B	−24.4984	−24.5279	−24.5291	−24.5291
C	−37.6224	−37.6868	−37.6886	−37.6886
N	−54.2689	−54.3980	−54.4009	−54.4091
O	−74.5404	−74.8043	−74.8094	−74.8094
F	−98.9421	−99.4013	−99.4093	−99.4093

[a] Reference [13].

[b] S. Huzinaga and C. Arnau, *J. Chem. Phys.* **53**, 451 (1970).

[c] C. Roetti and E. Clementi, *J. Chem. Phys.* **60**, 4725 (1974).

basis set for F is $1s$, $2s$, and $2p$ and for Cl the minimal basis includes $1s$, $2s$, $2p$, $3s$, and $3p$ functions. Using the minimal basis sets of STOs for the first row of atoms in the periodic table and then optimizing the orbital exponents, ζ, in each STO [see Eq.(5-17)] according to the SCF procedure gives the energies in the first column in Table 5-3. The optimized orbital exponents are approximately equal to the results obtained with the rules preceding Eq.(5-53). For instance, the optimized exponents for F are $1s_F$ ($\zeta = 8.6501$), $2s_F$ ($\zeta = 2.5638$), and $2p_F$ ($\zeta = 2.5500$) compared to the rules above that give $1s_F$ ($\zeta = 8.7000$), $2s_F$ ($\zeta = 2.6000$), and $2p_F$ ($\zeta = 2.6000$). The calculated minimal basis Slater orbital exponents, ζ, are given from the screening constants, S, in Fig. 5-4 and Eq.(5-17) where integral n values are employed.

The next order of approximation (and complexity) is to double the number of STOs over the minimal basis discussed above; this is called the *double zeta basis set*. The double zeta set of AOs for F includes $1s$, $1s'$, $2s$, $2s'$, $2p$, and $2p'$. The optimized orbital exponents in the double zeta set leads to considerably better energies as shown in Table 5-3. The optimized double zeta orbital exponents for F (see references in Table 5-3) are $1s_F$ ($\zeta = 10.5136$ and 7.7159), $2s_F$ ($\zeta = 3.1202$ and 1.9333) and $2p_F$ ($\zeta = 4.1746$ and 1.8470).

In order to improve over the double zeta basis set, more functions must be added. Extended basis sets usually refer to adding more functions of the same symmetry to the double zeta basis set. Continuing to add functions of new symmetries to an atom is called the extended basis with polarization. The terminology sometimes becomes clouded however, and it's fair to call basis expansion beyond double zeta sets the extended basis sets. Functions can be added until the Hartree–Fock limit of energy is reached, as shown in Table 5-3. It is evident from Table 5-3, however, that the double zeta set of atomic orbital functions make up the major share of the difference between the minimal basis and the Hartree–Fock limit. The double zeta basis with exponents optimized by the SCF procedure has been given for all atoms from He through Xe and the calculated energies are very close to the Hartree–Fock limits [15].

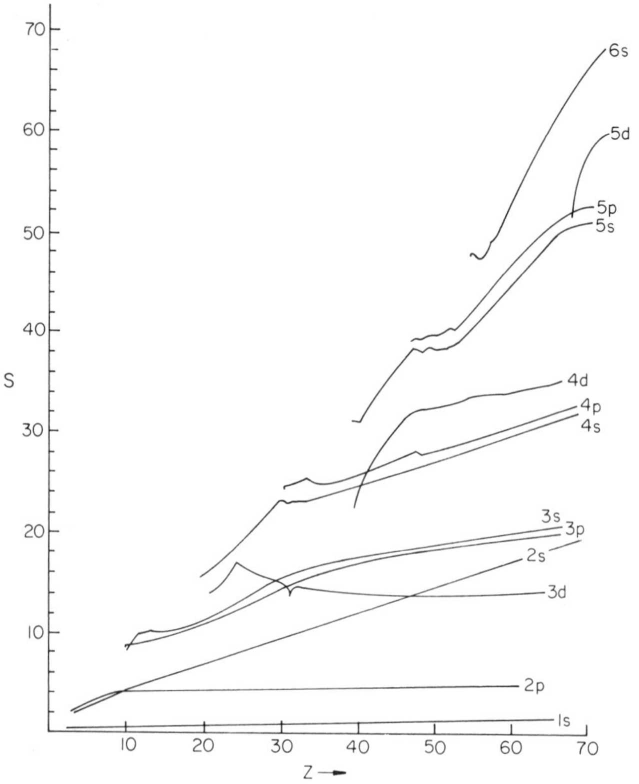

Figure 5-4 Screening constants, S [see Eq.(5-17)], as a function of atomic number for the Slater-type orbitals. These results are from Ref. [13].

Gaussian orbitals have also been used to calculate the SCF energy in atoms and molecules. Gaussian orbitals can be written by analogy with Eq.(5-16) as

$$G_{nlm}(r, \theta, \phi) = Nr^n \exp(-\alpha r^2) Y_{lm}(\theta, \phi), \tag{5-54}$$

where N is the normalization constant, α is the Gaussian orbital exponent, and n is the analog of n in the Slater-type orbital in Eq.(5-16). In practice, it is convenient to use a real Gaussian orbital defined by

$$G_{nlm}(x, y, z) = N' x^n y^l z^m \exp(-\alpha r^2), \tag{5-55}$$

where N' is again a normalization constant and n, l, and m are all integers. The single center matrix elements for atoms in the Hartree–Fock equation [Eq.(5-50)] are no easier to compute using the Gaussian-type orbitals (GTOs) than with STOs. In molecules, however, multicenter integrals occur and considerable simplification is gained in computing these integrals with GTOs. Unfortunately, considerably larger bases sets of GTOs are needed to compute an energy equivalent to the minimal basis STOs. For instance, for O and F, 5 s-type and 3 p-type GTOs yield an energy below the $1s$, $2s$, $2p$ minimal basis STO set. To match the double zeta $1s$, $1s'$, $2s$, $2s'$, $2p$, $2p'$ STO basis sets in O and F, however, 10 s-type and 6 p-type Gaussian functions must be used [16]. In the case of the hydrogen atom, Gaussian sets from 1 through 6 yield energies of -0.42441 (see Problem 4-8), -0.48581, -0.49698, -0.49928, -0.49981, and -0.49994 a.u., respectively, relative to the true value of -0.5 a.u. (see Table 5-2). Apparently GTOs require larger basis sets, requiring longer computational times for solution of the matrix equations and longer times in the subsequent iterations to reach self-consistency [17]. This problem has been largely circumvented with the use of contracted Gaussians that are linear combinations of GTOs with fixed coefficients. Thus, only the coefficients in each SCF orbital of the contracted GTO are determined by the SCF iterative procedure [16]. The obvious limit in this procedure would be to contract the Gaussians to the numerical atomic SCF Hartree–Fock orbitals, however, these contracted Gaussians lack the flexibility needed in molecular calculations.

In summary, the Hartree–Fock atomic orbitals are defined by the SCF single determinant of atomic orbitals. Considerable success has been achieved in using a LCAO scheme for the Hartree–Fock orbitals. The results indicate that the one-electron atomic orbital with screening is a valid and meaningful concept. We shall also use these methods in describing the electronic theories of molecules in Chapter 6.

5.3 CONFIGURATION INTERACTION (CI)

The error remaining in our attempts to compute the true nonrelativistic electron energy with the best single determinant function (Hartree–Fock) is called the *correlation error* [18]. The correlation error arises from replacing the true instantaneous $1/r_{ij}$ interaction by a self-consistent average value of $1/r_{ij}$. Thus, each electron interacts with the average field of the other electrons. The true many-electron wavefunction should have the actual or instantaneous interelectronic repulsion included. Due to difficulties in the direct incorporation of r_{ij}-dependence into the many-electron wavefunction in an atom, an expansion method is often used instead. The method starts with the ground-state configuration defined by the Hartree–Fock determinant and then proceeds by adding additional determinants representing additional configurations of the electrons in the molecule. The new many-electron function is written as

$$\psi(1, 2, \ldots, n) = \sum_I C_I \Delta_I, \qquad (5\text{-}56)$$

where the C_I are variation coefficients and Δ_I is the Ith determinental wavefunction as in Eq.(5-22). The concept of adding other configurations of electrons is referred to as *configuration interaction* (CI). The coefficients in Eq.(5-56) are determined in a standard manner by the variation method leading to the secular equation:

$$(\mathscr{H} - E\mathbf{S})\mathbf{C} = 0, \qquad \text{Det}\,(\mathscr{H} - E\mathbf{S}) = 0,$$

$$\mathscr{H}_{JI} = \int_1 \int_2 \cdots \int_n \Delta_J^* \mathscr{H} \Delta_I \, dV_1 \, dV_2 \cdots dV_n, \quad S_{JI} = \int_1 \int_2 \cdots \int_n \Delta_J^* \Delta_I \, dV_1 \, dV_2 \cdots dV_n. \qquad (5\text{-}57)$$

$\mathbf{C}$ is a column matrix containing the C_I in Eq.(5-56) and E represents the roots of the secular equation. Considerable effort has been expended in devising methods of choosing which configurations to include in Eq.(5-56) [16]. Normally, high energy configurations will not be important. Furthermore, some help in writing the $\mathscr{H}$ matrix is obtained from Brillouin's theorem [19], which states that $\mathscr{H}_{OI} = 0$ (the subscript O indicates the ground-state RHF determinant and the subscript I indicates an excited state configuration) when Δ_0 differs by Δ_I by only a single spin orbital. In He for instance, the radial correlation is improved by expanding in s orbitals according to

$$\psi(1, 2) = C_1\Delta(1s^2) + C_2\Delta(1s2s) + C_3\Delta(2s^2) + \cdots$$

This expansion in s-orbitals will correct the Hartree–Fock $\Delta(1s^2)$ determinant for the radial correlation because each succeeding configuration allows the electrons to be separated further, thereby relieving the interelectronic repulsion.

Another way of writing a sum of determinants in helium is to place the two electrons in different spatial orbitals:

$$\psi(1, 2) = (\tfrac{1}{2})\left\{\begin{vmatrix} \Phi_1(1)\alpha(1) & \Phi_1(2)\alpha(2) \\ \Phi_2(1)\beta(1) & \Phi_2(2)\beta(2) \end{vmatrix} + \begin{vmatrix} \Phi_2(1)\alpha(1) & \Phi_2(2)\alpha(2) \\ \Phi_1(1)\beta(1) & \Phi_1(2)\beta(2) \end{vmatrix}\right\}$$

$$= (\tfrac{1}{2})^{1/2}[\Phi_1(1)\Phi_2(2) + \Phi_2(1)\Phi_1(2)]\{(\tfrac{1}{2})^{1/2}[\alpha(1)\beta(2) - \alpha(2)\beta(1)]\}. \quad (5\text{-}58)$$

We now choose a simple two-parameter function, one parameter for each spatial $1s$ orbital. From Eq.(5-16) we can write

$$\Phi_1 = \frac{(2\zeta)^{3/2}}{2}\left(\frac{1}{2\pi}\right)^{1/2} \exp(-\zeta r_1), \qquad \Phi_2 = \frac{(2\zeta')^{3/2}}{2}\left(\frac{1}{2\pi}\right)^{1/2} \exp(-\zeta' r_2), \quad (5\text{-}59)$$

where ζ and ζ' are the two parameters representing different one-electron $1s$-type atomic orbitals. Substituting Eq.(5-59) into Eq.(5-58) and using the variation method yields the variational energy as a function of ζ and ζ'. Minimizing the variational energy with respect to ζ and ζ' gives [3] $\zeta = 1.1885$, $\zeta' = 2.1732$, and $W_0 = -2.8757$ a.u. $= -78.252$ eV. Comparing with Table 5-1, we find that Eqs.(5-58) and (5-59) yield a considerably better energy than the best single determinant RHF SCF function. In fact, Eq.(5-58) is an example of the UHF method as described following Eq.(5-40). Extending the linear combination of determinants to include large numbers of s-type functions gives the correction to the correlation error in the radial coordinate. The complete calculation includes the radial and angular correlations by including appropriately coupled nonspherically symmetric functions such as Y_{10} in Eq.(5-16). Extensive calculations of LCAO-SCF-CI energies in many atoms [20] have been completed and the agreement with experiment is excellent leaving no doubt about the validity of the method [16]. The process is laborious, however, and a large number of Slater determinants must be employed. We shall discuss CI techniques further in Section 6.3.

Configurations of electrons with net nonzero spin angular momentum can also be handled by a linear combination of determinants. For instance, in the case of helium where we no longer have two electrons for each spatial orbital, we can write the functions for unpaired electrons as

$$\psi(1,2) = (\tfrac{1}{2})^{1/2}\begin{vmatrix}\Phi_1(1)\alpha(1) & \Phi_1(2)\alpha(2)\\ \Phi_2(1)\alpha(1) & \Phi_2(2)\alpha(2)\end{vmatrix} = (\tfrac{1}{2})^{1/2}[\Phi_1(1)\Phi_2(2) - \Phi_1(2)\Phi_2(1)]\alpha(1)\alpha(2)$$

$$\psi(1,2) = (\tfrac{1}{2})^{1/2}\begin{vmatrix}\Phi_1(1)\beta(1) & \Phi_1(2)\beta(2)\\ \Phi_2(1)\beta(1) & \Phi_2(2)\beta(2)\end{vmatrix} = (\tfrac{1}{2})^{1/2}[\Phi_1(1)\Phi_2(2) - \Phi_1(2)\Phi_2(1)]\beta(1)\beta(2)$$

$$\psi(1,2) = \tfrac{1}{2}\left\{\begin{vmatrix}\Phi_1(1)\alpha(1) & \Phi_1(2)\alpha(2)\\ \Phi_2(1)\beta(1) & \Phi_2(2)\beta(2)\end{vmatrix} - \begin{vmatrix}\Phi_2(1)\alpha(1) & \Phi_2(2)\alpha(2)\\ \Phi_1(1)\beta(1) & \Phi_1(2)\beta(2)\end{vmatrix}\right\}$$

$$= (\tfrac{1}{2})^{1/2}[\Phi_1(1)\Phi_2(2) - \Phi_2(1)\Phi_1(2)](\tfrac{1}{2})^{1/2}[\alpha(1)\beta(2) + \alpha(2)\beta(1)]. \tag{5-60}$$

Equation (5-60) represents the components of the triplet state and Eq.(5-58) represents a single state in helium.

5.4 ANGULAR MOMENTUM IN ATOMS

In this section we shall make some general comments on the coupling of angular momentum in atoms. There are several possible coupling schemes, as we shall show. We begin our discussion with what is called L-S coupling or Russell–Saunders coupling.

In a many-electron atom we can write the total orbital angular momentum as a sum of the one-electron orbital angular momenta,

$$\boldsymbol{L} = \boldsymbol{l}_1 + \boldsymbol{l}_2 + \boldsymbol{l}_3 + \cdots \boldsymbol{l}_n.$$

A similar assumption is used for the intrinsic electron spin angular momentum,

$$\boldsymbol{S} = \boldsymbol{s}_1 + \boldsymbol{s}_2 + \boldsymbol{s}_3 + \cdots \boldsymbol{s}_n.$$

The total angular momentum is a sum of $\boldsymbol{L}$ and $\boldsymbol{S}$,

$$\boldsymbol{J} = \boldsymbol{L} + \boldsymbol{S}. \tag{5-61}$$

As there is no net orbital or spin angular momentum in closed shell configurations, ns^2, np^6, nd^{10}, nf^{14}, ..., we need only consider the unfilled shells in our determination of $\boldsymbol{L}$, $\boldsymbol{S}$, and $\boldsymbol{J}$. The same spectroscopic term symbol as discussed earlier in the hydrogen atom is used to denote the atomic state,

$$^{2S+1}L_J \qquad \begin{matrix} L = 0, & 1, & 2, & 3, & 4, & 5, & \ldots \\ S, & P, & D, & F, & G, & H, & \end{matrix}$$

where the J states are at different energies due to the spin-orbit interaction.

Consider two electrons in a [closed shell] $npn'p$ configuration. In this case, the Pauli principle is satisfied for all combinations of spins because the electrons are in different spatial orbitals. From the vector sum $\boldsymbol{l}_1 + \boldsymbol{l}_2$, the values of the magnitude of L will be $|(\boldsymbol{l}_1 + \boldsymbol{l}_2)|_{\max}$, $|(\boldsymbol{l}_1 - \boldsymbol{l}_2)|_{\min}$, and all integral values in between. Thus, for the configuration above, $|\boldsymbol{L}| = L = 2, 1, 0$ in units of $\hbar$. The values of the magnitude of S are similarly determined to be $S = 1, 0$, also in units of $\hbar$. The values of the

Table 5-4 The values of the spin, $\boldsymbol{S}$, orbital, $\boldsymbol{L}$, and total, $\boldsymbol{J}$, angular momenta (in units of $\hbar$) for two nonequivalent p electrons. The corresponding terms are also listed.

L	S	J	Terms
2	1	3, 2, 1	$^3D_3$, $^3D_2$, $^3D_1$
2	0	2	$^1D_2$
1	1	2, 1, 0	$^3P_2$, $^3P_1$, $^3P_0$
1	0	1	$^1P_1$
0	1	1	$^3S_1$
0	0	0	$^1S_0$

magnitude of $\boldsymbol{J}$ also range from $|\boldsymbol{L} + \boldsymbol{S}|$ to $|\boldsymbol{L} - \boldsymbol{S}|$ in integral steps and the results are given in Table 5-4. The appropriate term symbols are also listed. It is evident that both triplet and singlet states exist for each value of L. This vector coupling procedure can be repeated for any combination of unpaired electrons in different shells and a summary of results for two-electron systems is given in Table 5-5. The allowed terms for nonequivalent electrons in the middle column of Table 5-5 are obtained from the simple vector summations and combinations of S and L.

The analysis above was limited to the case where the two electrons were in different spatial orbitals. Under these circumstances, there will always be a singlet state [Eq.(5-58)] and a triplet state [Eq.(5-60)] for each value of L. The spin functions guarantee an antisymmetric function and there are no special symmetry requirements upon the spatial functions. For two equivalent electrons in the same spatial orbital, however, not all the states shown in Tables 5-4 and 5-5 are present. The additional symmetry requirement is inversion symmetry of the spatial orbital associated with the $^{2S+1}L_J$ state. Interchanging electrons is equivalent to inverting the coordinates in the spatial orbital. Thus, the total wavefunction associated with the $^{2S+1}L_J$ state that is a space-spin product must be antisymmetric. The spatial functions alternate

Table 5-5 Terms for several electronic configurations. The J states are not listed and are easily derived.

Configuration for Nonequivalent Electrons	Allowed Terms for Nonequivalent Electrons	Allowed Terms for Equivalent Electrons when Appropriate
ns	^2S	
$nsn's$	^1S, ^3S	^1S
$nsn'sn''s$	^2S, ^4S	
$nsn'p$	^1P, ^3P	
$nsn'd$	^1D, ^3D	
$nsn'f$	^1F, ^3F	
$npn'p$	^1S, ^3S, ^1P, ^3P, ^1D, ^3D	^1S, ^3P, ^1D
$npn'd$	^1P, ^3P, ^1D, ^3D, ^1F, ^3F	
$npn'f$	^1D, ^3D, ^1F, ^3F, ^1G, ^3G	
$ndn'd$	^1S, ^3S, ^1P, ^3P, ^1D, ^3D, ^1F, ^3F, ^1G, ^3G	^1S, ^3P, ^1D, ^3F, ^1G

in symmetry with respect to inversion according to the symmetry of the corresponding spherical harmonics as given by

$$L = \quad 0, \qquad 1, \qquad 2, \qquad 3, \qquad \cdots$$
$$\quad \text{sym} \quad \text{antisym} \quad \text{sym} \quad \text{antisym}$$

The spin functions are sums of $s_1 + s_2 + \cdots$ and the lowest spin state function is always antisymmetric with respect to interchange of the electrons.

$$S = \quad \tfrac{1}{2}, \qquad \tfrac{3}{2}, \qquad \tfrac{5}{2}, \qquad \tfrac{7}{2}, \qquad \cdots$$
$$\quad \text{antisym} \quad \text{sym} \quad \text{antisym} \quad \text{sym}$$

$$S = \quad 0, \qquad 1, \qquad 2, \qquad 3, \qquad \cdots$$
$$\quad \text{antisym} \quad \text{sym} \quad \text{antisym} \quad \text{sym}$$

We can now combine these results in products to obtain the appropriate total antisymmetric function. For instance, in the np^2 configuration we have

$$S = 0 \ (\text{antisym}), \qquad 1 \ (\text{sym})$$

$$L = 0 \ (\text{sym}), \qquad 1 \ (\text{antisym}), \qquad 2 \ (\text{sym}),$$

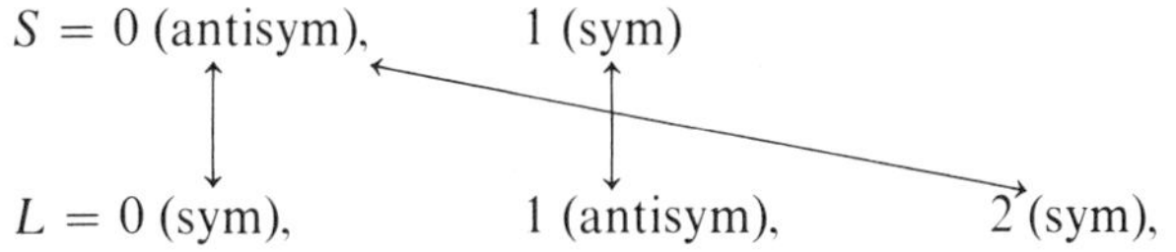

where the arrows show the products that give antisymmetric functions. In the case of nd^2, we have

$$S = \qquad 0 \ (\text{antisym}) \qquad\qquad\qquad\qquad\qquad 1 \ (\text{sym})$$

$$L = 0 \ (\text{sym}), \quad 1 \ (\text{antisym}), \quad 2 \ (\text{sym}), \quad 3 \ (\text{antisym}), \quad 4 \ (\text{sym}),$$

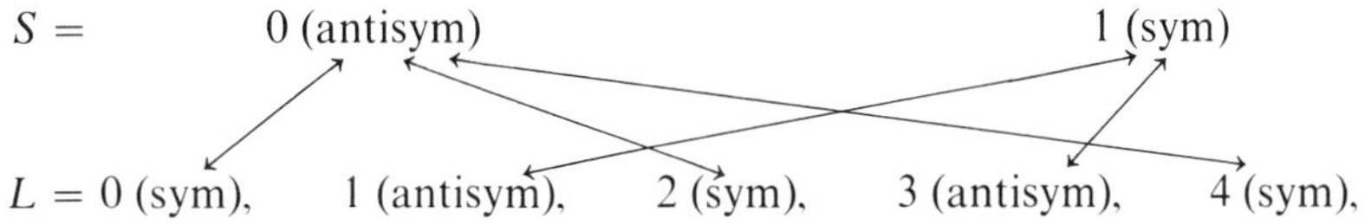

where again the arrows show the allowed total antisymmetric product states. These results and others are shown in the right-hand column in Table 5-5. In summary, we have developed the methods to determine the terms and therefore the symmetry species of the eigenstates of any configuration of two electrons. We should clearly distinguish a *configuration* of electrons that is the placement of electrons within the shell structure (see Fig. 5-2) from the *term states*, or possible energy levels, within a configuration. Thus, an np^2 *configuration* leads to the ^1S, ^3P, and ^1D *terms* or *states* within the configuration.

In the case of more than two electrons, the procedure for determining the terms for a given configuration is similar. Consider three nonequivalent p electrons that have a configuration of $npn'pn''p$. The coupling scheme is

$$l_1 + l_2 = l' = 0, 1, 2$$

$$s_1 + s_2 = s' = 0, 1$$

$$l' + l_3 = L = 1; 0, 1, 2; 1, 2, 3$$

$$s' + s_3 = S = \tfrac{1}{2}; \tfrac{1}{2}, \tfrac{3}{2}.$$

As all possible combinations are valid for nonequivalent electrons, we have the terms shown in Table 5-6.

Table 5-6 Term states for three nonequivalent p electrons.

$S = \frac{1}{2}$ (two states)	^2P, ^2S, ^2P, ^2D, ^2P, ^2D, ^2F
$S = \frac{3}{2}$	^4P, ^4S, ^4P, ^4D, ^4P, ^4D, ^4F

Simple symmetry arguments are more difficult to apply in the case of equivalent electrons in an np^3 configuration. Combining the M_{l_i} and M_{s_i} individual states gives ^2P, ^2D, and ^4S as the only remaining terms for equivalent electrons [21].

It is also important to point out that the term states arising from an np configuration are identical to an np^5 configuration. Similar pairs are $np^2{:}np^4$, $nd{:}nd^9$, $nd^2{:}nd^8$, and so on.

The relative energies of the states obtained from an electron configuration can be determined by experiment and we have a body of experience to guide us in making predictions. Hund's empirical rules [21] can give a rough idea of the relative ordering that places the lowest energy states at

1. Maximum S consistent with the Pauli principle.
2. Maximum L consistent with the Pauli principle.
3. $J = J_{\text{min}} = |\mathbf{L} - \mathbf{S}|$ if the shell is less than half full, and $J = J_{\text{max}} = |\mathbf{L} + \mathbf{S}|$ if the shell is more than half full.

It is instructive to review some of the work above on configurations, terms, and states by looking at the energy levels for a typical atom. The observed energy levels and the corresponding configurations and term states for He are listed in Table 5-7.

Table 5-7 A few of the electronic configurations, term states, and energies for the He atom from Ref. [4]. E_p is the ionization energy for the first electron.

Configuration	Term State	Energy (cm^{-1})
$1s$	^2S (He^+)	198,305 (E_p = 24.587 eV)
$1s3p$	$^1P_1$	186,203.62
$1s3d$	$^1D_2$	186,099.22
$1s3d$	$^3D_{1,2,3}$	186,095.90
	$^3P_0$	185,559.277
$1s3p$	$^3P_1$	185,559.085
	$^3P_2$	185,558.92
$1s3s$	$^1S_0$	184,859.06
$1s3s$	$^3S_1$	183,231.08
$1s2p$	$^1P_1$	171,129.148
	$^3P_0$	169,082.185
$1s2p$	$^3P_1$	169,081.189
	$^3P_2$	169,081.111
$1s2s$	$^1S_0$	166,271.70
$1s2s$	$^3S_1$	159,850.318
$1s^2$	$^1S_0$	0

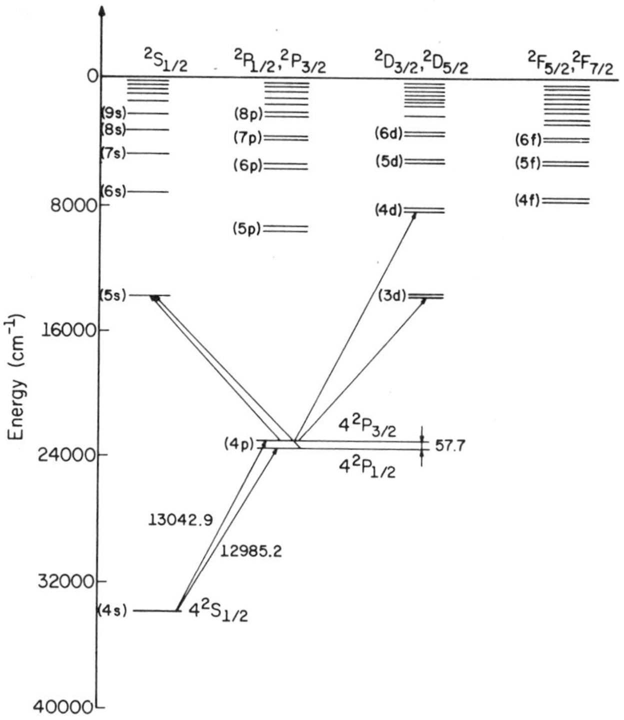

Figure 5-5 Potassium atom energy level diagram for the outer electron excitation. The ground-state electronic configuration is $1s^2 2s^2 2p^6 3s^2 3p^6 4s$. The notation in the diagram of 4s, 4p, 5s, and so on indicates the outer electron, which is in the 4s configuration in the ground state. The difference between the $^2S_{1/2} \longrightarrow {}^2P_{1/2}$ and $^2S_{1/2} \longrightarrow {}^2P_{3/2}$ transitions of 57.7 cm^{-1} is the K atom doublet. This low energy doublet splitting is characteristic of all alkali atoms (see Table 5-8). The numerical values are from Ref. [4].

He is composed of singlet and triplet states as shown in Section 5.3. Note that the triplet state of a given configuration is always at a lower energy, which agrees with Hund's rules. Note also the ordering of the spin-orbit states.

The selection rules for magnetic and electric dipole transitions in a many-electron atom are derived much in the same manner as with the hydrogen-like atom. In complex atoms we have the selection rules given by $\Delta l = \pm 1$; $\Delta L = 0, \pm 1$; and $\Delta M_L = 0, \pm 1$. A few of these transitions are shown in Fig. 5-5 for the potassium atom.

The angular momentum of an atom or ion can be determined by measuring the paramagnetic susceptibility of the system. The magnetic moment of a system in state $^{2S+1}L_J$ is given by

$$\boldsymbol{\mu}_J = g_L \frac{\mu_B}{\hbar} \mathbf{J},$$

where g_L is the Landé g-value defined in Eq.(4-152), μ_B is the Bohr magneton, and $\mathbf{J} = \mathbf{S} + \mathbf{L}$ is the total angular momentum. The average value of the magnetic moment of the atom in state ${}^{2S+1}L_J$ along the direction of an external field H_z is given by

$$\mu_{av} = \frac{\sum_{M=-J}^{J} g_L \mu_B M \exp(g_L \mu_B M H_z/kT)}{\sum_{M=-J}^{J} \exp(g_L \mu_B M H_z/kT)},$$

where $\langle J, M | J_z | J, M \rangle = \hbar M$ and the sums over M include all substates in the ground electronic state. Now, at $T \approx 300$ K, $g_L \mu_B M H_z/kT \ll 1$ and we can expand the exponentials giving

$$\mu_{av} = \frac{g_L \mu_B \sum_M M[1 + g_L \mu_B M H_z/kT)]}{\sum_M [1 + (g_L \mu_B M H_z/kT)]}.$$

Using

$$\sum_M 1 = (2J+1), \qquad \sum_M M = 0, \quad \text{and} \quad \sum_M M^2 = \frac{J(J+1)(2J+1)}{3}$$

gives

$$\mu_{av} = \frac{J(J+1) g_L^2 \mu_B^2}{3kT} H_z,$$

which leads then to the paramagnetic susceptibility given by

$$\frac{\partial \mu_{av}}{\partial H_z} = \chi = \frac{J(J+1) g_L^2 \mu_B^2}{3kT}.$$

The rare earth ions give measured temperature-dependent paramagnetic susceptibilities that agree well with the analysis above [22,23]. The transition metal ions in solutions or complexes give a temperature-dependent susceptibility, however, that indicates that the orbital angular momentum has been quenched and the paramagnetic susceptibility is due entirely to the spin-angular momentum. Thus

$$\lim_{L \to 0} \chi = \frac{S(S+1) g_e^2 \mu_B^2}{3kT}$$

will describe the transition metal systems [24].

5.5 THE SPIN-ORBIT INTERACTION IN ALKALI ATOMS, j–j COUPLING, AND PAIR COUPLING

The spin-orbit interaction in many-electron atoms is treated according to the methods developed for the hydrogen-like atom in Section 4.8. The Hamiltonian for the spin-orbit interaction in a many-electron atom is [see Eq.(4-133)]

$$\mathscr{H}_1 = \frac{g_e \mu_B^2 Z}{\hbar^2} \sum_i \left(\frac{1}{r_i^3}\right) \mathbf{s}_i \cdot \mathbf{l}_i, \tag{5-62}$$

where $\mu_B = e\hbar/2mc$ is the Bohr magneton and g_e is the free electronic g-value. It is clear that the spin-orbit interaction shows a direct Z-dependence and will, therefore, become increasingly important as the higher atomic number atoms are considered.

The alkali metal atom series, Li, Na, K, Rb, and Cs, each contain a single unpaired electron, and we can write Eq.(5-62) to give

$$\mathscr{H}_1 = \frac{g_e \mu_B^2 Z}{\hbar^2}\left(\frac{1}{r^3}\right)\boldsymbol{S}\cdot\boldsymbol{L},$$

for the unpaired outer electron in each atom. The spin-orbit energy is easily obtained for the strong coupling case giving (see Section 4.8)

$$E(S, L, J) = \frac{g_e \mu_B^2 Z}{2}\left(\frac{1}{r^3}\right)_{av}[J(J+1) - S(S+1) - L(L+1)], \tag{5-63}$$

where $(1/r^3)_{av}$ is the average value of $1/r^3$ over the radial function of the single outer electron.

As a typical example, consider the energy spacing between the $3^2P_{3/2}$ and $3^2P_{1/2}$ levels in the sodium atom which gives rise (in a transition from the $3^2S_{1/2}$ ground state) to the sodium doublet which is analogous to the K atom $4^2P_{3/2}$-$4^2P_{1/2}$ doublet shown in Fig. 5-5. The $n^2P_{3/2}$-$n^2P_{1/2}$ doublets for all the alkali atoms (including hydrogen) are listed in Table 5-8. The value of $(1/r^3)_{av}$ can be extracted from the data and Eq.(5-63) where the $^2P_{3/2}$-$^2P_{1/2}$ energy differences are given by

$$\Delta E = E\left(\frac{1}{2}, L, L+\frac{1}{2}\right) - E\left(\frac{1}{2}, L, L-\frac{1}{2}\right) = \frac{(2L+1)g_e \mu_B^2 Z}{2}\left(\frac{1}{r^3}\right)_{av}. \tag{5-64}$$

We shall now compute the average values of $1/r^3$ for the outer electrons in the series shown in Table 5-8. The one-electron average value of $(1/r^3)$ with the nodeless Slater orbitals is easily determined in general. The result using Eq.(5-16) is given by

$$\left(\frac{1}{r^3}\right)_{av}^S = \frac{(2\zeta)^{2n+1}}{(2n)!}\int_0^\infty r^{2n-2}\exp(-2\zeta r)\left(\frac{1}{r^3}\right)r^2\,dr = \frac{(2n-3)!}{(2n)!}(2\zeta)^3, \tag{5-65}$$

in atomic units where the superscript S indicates Slater orbital. In the case of Na where $\zeta = 0.73$ [see discussion preceding Eq.(5-53)] and $n = 3$, we have $(1/r^3)_{av}^S = 2.59 \times 10^{-2}$ a.u. $= 1.75 \times 10^{23}$ cm^{-3}, which is an order of magnitude less than the experimental result for Na shown in Table 5-8. Similar poor results are obtained with all the alkali atoms using the Slater orbitals.

Table 5-8 Experimental spin-orbit splitting in the lowest $n^2P_{1/2}$ and $n^2P_{3/2}$ states in the alkali metal atoms and the hydrogen atom from Ref. [4]. The resultant values of $(1/r^3)_{av}$ obtained from Eq.(5-64) are also shown.

Atom	n	$^2P_{3/2} - {}^2P_{1/2}$ (cm^{-1})	Z	$(1/r^3)_{av}$ (10^{24} cm^{-3})
H	1	0.365	1	0.281
Li	2	0.337	3	0.086
Na	3	17.19	11	1.20
K	4	57.7	19	2.34
Rb	5	237.6	37	4.94
Cs	6	554.1	55	7.75

We shall now repeat the calculation of $(1/r^3)_{av}$ with the $3p$ hydrogen-like orbital taken from Table 4-4. The hydrogen-like $n = 3$ radial orbital in atomic units is given by (let $a = 1$ and $Z/3 = \zeta$)

$$\eta_{3p}(r) = \frac{4}{81(6)^{1/2}}(3\zeta)^{3/2}[6r(3\zeta) - r^2(3\zeta)^2] \exp(-\zeta r), \tag{5-66}$$

where the same definition for $\zeta = (Z - S)/3$ is valid. The average value of $1/r^3$ is given by (the superscript H indicates hydrogen-like)

$$\left(\frac{1}{r^3}\right)_{av}^{H} = \int_0^\infty \eta_{3p}(r)\left(\frac{1}{r^3}\right)\eta_{3p}(r)r^2\,dr = \frac{\zeta^3}{3}, \tag{5-67}$$

compared to Eq.(5-65) for the Slater orbitals. Using $\zeta = 0.73$ for Na gives $(1/r^3)_{av}^H =$ 0.13 a.u. $= 8.77 \times 10^{23}$ cm^{-3}, which is considerably closer to the experimental value of 1.20×10^{24} cm^{-3} than the corresponding Slater value of 0.18×10^{24} cm^{-3}. Similar results are obtained by calculating $(1/r^3)_{av}$ for the other alkali atoms listed in Table 5-8. These results of the calculation of $(1/r^3)_{av}$ with the Slater and hydrogen-like

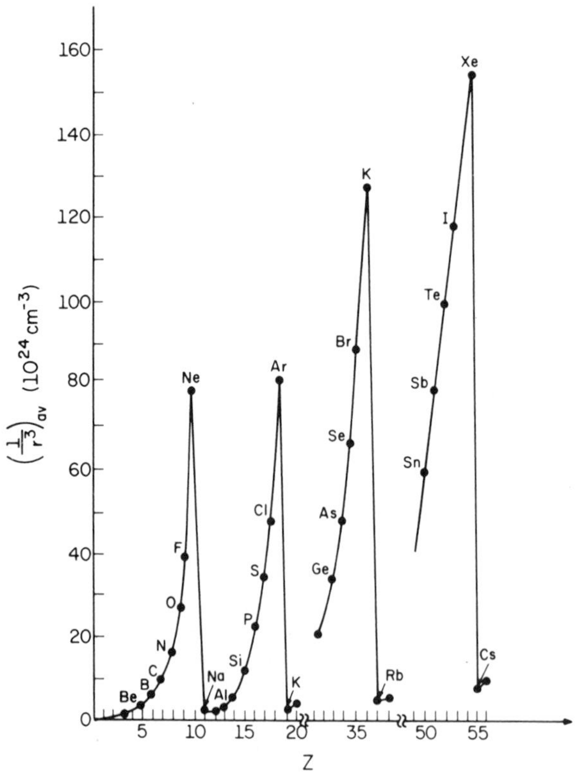

Figure 5-6 Diagram showing the average values of $1/r^3$ for p-type electrons for a number of atoms. The increasing values of $(1/r^3)_{av}$ within any row or column of the periodic table are evident. The numbers are largely determined from atomic spin-orbit data and are from R. G. Barnes and W. V. Smith, *Phys. Rev.* **93**, 95 (1954).

radial orbitals again points out that the hydrogen-like nodal radial orbitals (similar to the SCF orbitals) give a better overall view of the atom's electron distribution than the nodeless Slater orbitals, especially on properties such as $1/r^3$ that are large at the nucleus.

Spin-orbit splittings have been observed by spectroscopy in a large number of atoms in addition to the alkali metals listed in Table 5-8 and the resultant average values of $1/r^3$ for p-type electrons are listed in Fig. 5-6. $(1/r^3)_{av}$ also appears in a number of other observable properties that we shall discuss later. According to Fig. 5-6 and our previous spin-orbit discussion, we note that the halogen atom spin-orbit interactions are much larger than the corresponding alkali atoms in the same row.

An interesting interplay of the ideas above occurs in the observation of photoelectron spectra in the rare gas atoms. In this experiment, He gas is excited in an electric discharge. The electronically excited He atom (see Table 5-7) emits a photon from the excited $^1P_1$ state at 171, 129.148 cm^{-1} = 21.217 eV, which corresponds to a photon wavelength of $\lambda = 584$ Å. This photon collides with a rare gas atom and ejects an electron from the atom to produce a rare gas ion. In the case of Ar, we can write

$$\text{Ar} + h\nu = \text{Ar}^+ + e^- .$$

The photoelectron is then analyzed energetically and some typical data are shown in Fig. 5-7. There are two maxima in the data for each atom that arise from the ground state $^2P_{1/2}$ and $^2P_{3/2}$ [degeneracies of $(2J + 1)$ or 2 and 4, respectively] levels in the rare gas ions. Thus, the splittings in Fig. 5-7 give a direct measurement of the $^2P_{3/2}$-$^2P_{1/2}$ doublet splittings in Ar^+, Kr^+, and Xe^+. The rare gas ions have ground-state electron configurations identical to the isoelectronic neutral halogen atoms that are adjacent in the periodic table. The splittings shown in Fig. 5-7 lead to $(1/r^3)_{av}$ values larger than any shown in Fig. 5-6, which is the expected result.

Our previous discussion on the coupling of angular momentum of electrons in a particular configuration was based on the equations preceding Eq.(5-61) where it was

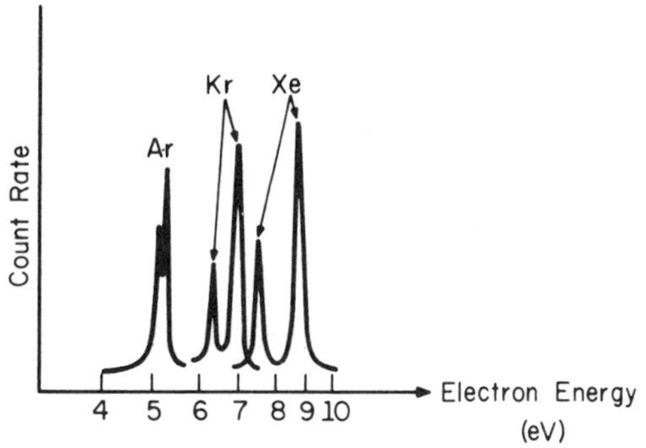

Figure 5-7 The $\lambda = 584$ Å (He) photoelectron spectra of the rare gas atoms. The electron count rate is plotted as a function of the energy of the ejected electron. The doublet splittings arise from the $n^2P_{3/2}$-$n^2P_{1/2}$ splittings in the ground state of the rare gas ions. The data are adapted from D. W. Turner and D. P. May, *J. Chem. Phys.* **45**, 471 (1966).

assumed that the spin-orbit coupling ($\mathbf{s}_i \cdot \mathbf{l}_i$) for an individual electron was small relative to the electrostatic repulsion between electrons. This case is sometimes referred to as the Russell–Saunders weakly coupled limit.

j-j Coupling

An alternate coupling scheme is where the $\mathbf{s}_i \cdot \mathbf{l}_i$ interaction is larger than the electrostatic repulsion. In this case, the coupling is given by

$$\mathbf{s}_i + \mathbf{l}_i = \mathbf{j}_i, \qquad \sum_i \mathbf{j}_i = \mathbf{J}. \tag{5-68}$$

This is called j–j coupling and this scheme may be correlated with the Russell–Saunders coupling scheme discussed above. The states arising from the j–j coupling scheme for an $npn'p$ configuration for nonequivalent electrons are given by

$$s_1 + l_1 = j_1 = \tfrac{1}{2}, \tfrac{3}{2}, \qquad s_2 + l_2 = j_2 = \tfrac{1}{2}, \tfrac{3}{2}, \qquad J = 0, 1; 2, 1; 2, 1; 3, 2, 1, 0. \tag{5-69}$$

These final J states must correlate with the corresponding Russell–Saunders coupling results as shown, for instance, in Table 5-5 [21]. Several of these states are excluded by the Pauli principle in the case of equivalent electrons. The remaining J states must agree (and correlate) with the states obtained by Russell–Saunders coupling. Normally the atoms in the left-hand columns in the periodic table will obey Russell–Saunders coupling and the atoms in the right-hand columns (or near closed shell atoms) will be described better with j–j coupling [25].

Pair Coupling

A third type of coupling, which is an intermediate combination of L–S and j–j coupling, is called *pair coupling*. This coupling scheme is most conveniently used to describe the energy levels arising from various excited-state configurations of the rare gas atoms. In the case of a single electron excited from the original closed shell of the rare gas atom, the coupling scheme is best described by

$$\mathbf{l} + \mathbf{J}_c = \mathbf{K}, \qquad \mathbf{K} + \mathbf{S} = \mathbf{J}. \tag{5-70}$$

$\mathbf{l}$ is the orbital angular momentum of the outer electron that has been excited out of the closed shell, $\mathbf{J}_c$ is the total angular momentum of all the electrons remaining in the core, $\mathbf{S}$ is the spin angular momentum of the outer excited electron, and $\mathbf{J}$ is the final total angular momentum. The electronic configurations, the $nl[K]_J$ term notation, and the experimental energy levels of the Ne atom are shown in Table 5-9. As an example, consider the $2p^5(^2P_{3/2})3p$ configuration (the $^2P_{3/2}$ indicates the inner core term):

$$\begin{aligned} &2p^5; \qquad J_c = \tfrac{3}{2}; \qquad (3p) l = 1 \\ &\mathbf{l} + \mathbf{J}_c = \mathbf{K} \longrightarrow \tfrac{5}{2}, \quad \tfrac{3}{2}, \quad \tfrac{1}{2}, \\ &\qquad\qquad\qquad\quad \wedge \quad \wedge \quad \wedge \\ &\mathbf{K} + \mathbf{S} = \mathbf{J} \longrightarrow 3, 2; 2, 1; 1, 0. \end{aligned}$$

Table 5-9 A few of the lower energy electronic configurations, term states, and energies for the Ne atom from Ref. [4].

Configuration	K	J	$nl[K]_J$	Energy in cm^{-1}
$2p^5(^2P_{1/2})4s'$	$\frac{1}{2}$	1	$4s'[\frac{1}{2}]_1$	159,536.57
		0	$4s'[\frac{1}{2}]_0$	159,381.94
$2p^5(^2P_{3/2})4s$	$\frac{3}{2}$	1	$4s[\frac{3}{2}]_1$	158,797.954
		2	$4s[\frac{3}{2}]_2$	158,603.070
$2p^5(^2P_{1/2})3p'$	$\frac{1}{2}$	0	$3p'[\frac{1}{2}]_0$	152,972.697
		1	$3p'[\frac{1}{2}]_1$	151,040,413
	$\frac{3}{2}$	2	$3p'[\frac{3}{2}]_2$	150,860.468
		1	$3p'[\frac{3}{2}]_1$	150,774.072
$2p^5(^2P_{3/2})3p$	$\frac{1}{2}$	0	$3p[\frac{1}{2}]_0$	150,919.391
	$\frac{3}{2}$	2	$3p[\frac{3}{2}]_2$	150,317.821
		1	$3p[\frac{3}{2}]_1$	150,123.551
	$\frac{5}{2}$	2	$3p[\frac{5}{2}]_2$	149,826.181
		3	$3p[\frac{5}{2}]_3$	149,659.000
	$\frac{1}{2}$	1	$3p[\frac{1}{2}]_1$	148,259.746
$2p^5(^2P_{1/2})3s'$	$\frac{1}{2}$	1	$3s'[\frac{1}{2}]_1$	135,890.670
		0	$3s'[\frac{1}{2}]_0$	134,820.591
$2p^5(^2P_{3/2})3s$	$\frac{3}{2}$	1	$3s[\frac{3}{2}]_1$	134,461.237
		2	$3s[\frac{3}{2}]_2$	134,043.790
$1s^22s^22p^6$	$^1S_0$			0

We shall now examine the atomic electronic states involved in the common He-Ne laser system. First we note from Tables 5-7 and 5-9 that the lowest excited state of the He atom ($^3S_1$, 159,850.32 cm^{-1}) is only 313.75 cm^{-1} higher in energy than the $4s'[\frac{1}{2}]_1$ level (159,536.57 cm^{-1}) of the Ne atom. The $^3S_1$ state in He, which can be excited by electrons in an electrical discharge, is quite long-lived (relative to an allowed electric dipole transition) because of the weak magnetic transition dipole to the ground state. In any event, we might expect an efficient collisional energy transfer between He($^3S_1$) and Ne($^1S_0$) to give He($^1S_0$) and Ne($4s'[\frac{1}{2}]_1$),

$$\text{He}(^3S_1) + \text{Ne}(^1S_0) \longrightarrow \text{He}(^1S_0) + \text{Ne}(4s'[\tfrac{1}{2}]_1) + 313.75\ \text{cm}^{-1}. \qquad (5\text{-}71)$$

The excess 314 cm^{-1}, which is on the order of kT at the discharge temperature, appears as kinetic energy in the atoms. The process above produces excited Ne($4s'[\frac{1}{2}]_1$) atoms that then exchange energy rapidly via Ne-Ne collisions to give near-equal

Table 5-10 Some of the stronger ^{20}Ne laser $(4s, 4s') \longrightarrow (3p, 3p')$ transitions. The energy differences are from Table 5-9. The experimentally observed laser wavelengths are very close to the values in this table, as shown in G. Birnbaum, *Optical Masers* (Academic Press, New York, 1964).

Ne Transition	ΔE (cm^{-1})	λ (μm or 10^4 Å)
$4s[\frac{3}{2}]_2 - 3p[\frac{5}{2}]_3$	8944.07	1.1181
$4s'[\frac{1}{2}]_1 - 3p'[\frac{3}{2}]_2$	8676.10	1.1526
$4s'[\frac{1}{2}]_0 - 3p'[\frac{3}{2}]_1$	8607.87	1.1617
$4s'[\frac{1}{2}]_1 - 3p'[\frac{1}{2}]_1$	8496.16	1.1770
$4s[\frac{3}{2}]_2 - 3p[\frac{3}{2}]_2$	8285.25	1.2070
$4s'[\frac{1}{2}]_1 - 3p'[\frac{1}{2}]_0$	6563.87	1.5235

populations in the $4s$ and $4s'$ states grouped from 158,603 to 159,537 cm^{-1} (see Table 5-9). The Ne atom in any of these four excited states then fluoresces to one of the ten $3p'$ and $3p$ lower energy states ranging from 148,260 to 152,973 cm^{-1}. If the emitted radiation is contained in an optical cavity, additional transitions will be stimulated. Coherent laser oscillation will be possible if the gain in the optical cavity exceeds the losses in the system (see Fig. 7-19 and associated discussion). A continuous oscillation is possible if a continuous supply of excited $4s$ and $4s'$ Ne states is provided by the efficient Ne-Ne energy transfer. Also a rapid relaxation from the lower $3p$ and $3p'$ Ne states is necessary to sustain an oscillation. The strongest observed laser transitions as described above are listed in Table 5-10.

The Doppler half width of the 1.1526-μm $4s'[\frac{1}{2}]_1 \longrightarrow 3p'[\frac{3}{2}]_2$ Ne atom emission at a temperature of 1000 K (in the discharge) is on the order of [Eq.(7-39)] $\Delta\nu \cong 1000$ MHz. The spectra emitting from the Fabry–Perot cavity with a 1-m length will be decomposed into about 15–20 lines under the Doppler profile (see Fig. 7-19) separated by 150 MHz with widths of about 15 MHz (cavity finesse of 10). Finally, if enough gain can be established, oscillation will occur on one of the longitudinal modes that narrows the emitting frequency to a width of a few hertz.

The popular 6328 Å Ne laser transition in the red part of the visible spectrum arises from a $5s'$ Ne excitation followed by an optical emission to a $3p'$ level $\{5s'[\frac{1}{2}]_1 \longrightarrow 3p'[\frac{3}{2}]_2, 166{,}658.5 - 150{,}860.5 = 15{,}798.0 \text{ cm}^{-1}, \lambda = 6328 \text{ Å}\}$. The $5s'[\frac{1}{2}]_1$ (166,658.5 cm^{-1}) state in Ne is excited by a collision with a $^1S_0$ (166,271.7 cm^{-1}) He atom in an electronic discharge.

5.6 NUCLEAR-ELECTRONIC MAGNETIC INTERACTIONS

In Section 4.9 we examined nuclear-electronic angular momentum coupling under certain conditions. Applying these concepts to a many-electron atom with spin angular momentum, $\boldsymbol{S}$, and orbital angular momentum, $\boldsymbol{L}$, the spin-orbit coupling gives rise to a total electronic angular momentum of

$$\boldsymbol{J} = \boldsymbol{S} + \boldsymbol{L}. \tag{5-72}$$

In the presence of further coupling with the nuclear angular momentum, $\boldsymbol{I}$, we obtain the total angular momentum, $\boldsymbol{F}$,

$$\boldsymbol{F} = \boldsymbol{I} + \boldsymbol{J}. \tag{5-73}$$

The Hamiltonian describing these couplings as well as the Zeeman effect of a magnetic field along the z axis is given by

$$\mathscr{H} = \frac{C}{\hbar^2}\boldsymbol{S}\cdot\boldsymbol{L} + h\left(\frac{a}{\hbar^2}\right)\boldsymbol{I}\cdot\boldsymbol{J} + \frac{\mu_B}{\hbar}(g_e S_z + L_z)H_z - \frac{\mu_0}{\hbar}g_i(1-\sigma)I_z H_z, \tag{5-74}$$

where C is the spin-orbit coupling constant and all other terms are defined in Sections 4.9 and 4.10. There are several interesting limits to Eq.(5-74) and the appropriate first-order energies are examined in Problem 5-8.

In the case of the hydrogen and alkali metal atoms in the ns lowest energy configuration, $L = 0$, and the zero-field Hamiltonian and energies for the hydrogen atom are discussed in Section 4.9 with the energy being given in Eq.(4-169). The observed $^2S_{1/2}$ $\boldsymbol{I} + \boldsymbol{S} = \boldsymbol{F}$ doublet splittings in the alkali metal atom series are shown in Table 5-11 where F_u is the F-value of the upper energy level of the $^2S_{1/2}$ doublets (maximum F-values) and a is the coupling constant in Eq.(4-169). We shall now attempt to deduce the nature of the a constants in the alkali metal series (as well as the hydrogen atom) shown in Table 5-11. The nuclear-electronic interaction arises from an $\mathscr{H} = -\boldsymbol{\mu}\cdot\boldsymbol{H}_e$ type term where $\boldsymbol{\mu}$ is the nuclear magnetic moment that is proportional to $g_i\mu_0$ and $\boldsymbol{H}_e$ is the magnetic field at the nucleus due to the electrons. According to the discussion in Section 4-8, the magnetic field (due to the intrinsic electron magnetic moment) at the nucleus is proportional to the magnetic dipole moment of the electron, which is $g_e\mu_B$. Thus, the nuclear spin electron–spin coupling constant, a, should be proportional to $g_i\mu_0 g_e\mu_B$. According to the arguments above, we have multiplied the

Table 5-11 Nuclear magnetic splitting of the ground electronic $^2S_{1/2}$ states of the alkali atoms. I is the nuclear angular momentum in units of $\hbar$, g_i is the nuclear g-value, Z is the atomic number, $\Delta\nu$ are the experimental splittings, and a is the coupling constant where $F_u = (1/\hbar)|\boldsymbol{S} + \boldsymbol{I}|$ is the F-value for the upper energy level in the $^2S_{1/2}$ nuclear coupled doublet. The values of γ are described in the text.

Atom	I	g_i	Z	$\Delta\nu$ (MHz)	a (MHz) $= \Delta\nu/F_u$	$ah/g_i\mu_0 g_e\mu_B = \gamma$ (10^{24} cm^{-3})
^{7}Li	$\frac{3}{2}$	2.1707	3	803.5	402.8	13.1
^{23}Na	$\frac{3}{2}$	1.4774	11	1,771.6	885.8	42.4
^{39}K	$\frac{3}{2}$	0.2606	19	461.7	230.9	62.6
^{85}Rb	$\frac{5}{2}$	0.5393	37	3,035.7	1,011.9	132.6
^{87}Rb	$\frac{3}{2}$	1.8276	37	6,834.7	3,417.8	132.1
^{133}Cs	$\frac{7}{2}$	0.7326	55	9,192.6[a]	2,298.1	221.6
^{1}H	$\frac{1}{2}$	5.5854	1	1,420.4	1,420.4	17.97

[a] We note that this transition can be measured very accurately and is the current standard of time. The second is the duration of 9,192,631,770 periods in the $F = 1 \longrightarrow F = 0$ ($^2S_{1/2}$) transitions in ^{133}Cs (see Appendix A).

values of a in Table 5-11 by Planck's constant and divided by $g_i\mu_0 g_e\mu_B$ and the results are listed as γ, which has cm^{-3} units. The values of γ in Table 5-11 cannot arise from a $(1/r^3)_{av}$ type of term in an $^2S_{1/2}$ state because $(1/r^3)_{av}$ diverges for a spherically symmetric state. We can relate γ in Table 5-11 to $\lim_{r\to 0}|\psi|^2$, however, where ψ is the eigenfunction for the outer electron in the atoms in Table 5-11. In the case of the hydrogen atom in the ground state, we write (see eigenfunctions in the ground state from Tables 4-4 and 4-5)

$$\lim_{r\to 0}|\psi_{1s}(r,\theta,\phi)|^2 = \frac{1}{\pi a_0^3} = \lim_{r\to 0}\frac{1}{4\pi}|\eta(r)|^2 = \frac{1}{4\pi}|\eta(0)|^2,$$

where $\eta(r)$ is the radial function and $\psi(r, \theta, \phi)$ contains also the angular normalization constant. Combining the above we can show that

$$ha(\text{hydrogen}, {}^2S_{1/2}) = \frac{2}{3} g_i\mu_0 g_e\mu_B|\eta(0)|^2$$

$$= \frac{8\pi}{3} g_i\mu_0 g_e\mu_B|\psi(0)|^2 = \left(\frac{8}{3}\right) g_i\mu_0 g_e\mu_B\left(\frac{1}{a_0^3}\right). \qquad (5\text{-}75)$$

The result in Eq.(5-75) agrees numerically to about one part in 10^3 with the result in Table 5-11 for the hydrogen atom. In general, we write (for $^2S_{1/2}$ states)

$$ha(n^2S_{1/2}) = \frac{8\pi}{3} g_i\mu_0 g_e\mu_B|\psi_{ns}(0)|^2. \qquad (5\text{-}76)$$

$|\psi_{ns}(0)|^2$ is sometimes called the Fermi contact term [26]. This contact term, $|\psi_{ns}(0)|^2$, for s electrons at the nucleus also appears in the observation of the Mössbauer effect. The isomer shift observed in nuclear emission from the same nucleus in different chemical environments is due to different amounts of s electron density at the nucleus (see Problem 5-10).

In calculating $|\psi_{ns}(0)|^2$ for the heavier atoms in Table 5-11, we note that the Slater orbitals [Eq.(5-16)] predict $|\psi_{ns}(0)|^2 = 0$ for $n > 1$. Using the hydrogen-like $3s$ radial orbital from Table 4-4,

$$\eta_{3s}(r) = \tfrac{2}{3}\zeta^{3/2}[3 - 6\zeta r + 2(\zeta r)^2]\exp(-\zeta r),$$

where $\zeta = (Z - S)/3$, as in Eq.(5-17), leads to

$$|\eta_{3s}(0)|^2 = 4\zeta^3, \qquad \gamma = \tfrac{8}{3}\zeta^3 = 1.04 \text{ a.u.} = 7.0 \times 10^{24}\ \text{cm}^{-3},$$

where we have used $\zeta = 0.73$ for the $3s$ orbital in the Na atom. This result is somewhat smaller than the experimental result in Table 5-11, which indicates that the hydrogen-like functions applied to Na need a smaller shielding constant, S, and a larger value of ζ. For instance, if S was reduced 25%, ζ would be doubled and the calculation would agree with experiment. The simple hydrogenic functions with screening do give a respectable understanding of the magnetic hyperfine coupling in the $^2S_{1/2}$ state of ^{23}Na. The more refined atomic Hartree–Fock radial orbitals give accurate values for $|\psi_{ns}(0)|^2 = (1/4\pi)|\eta_n(0)|^2$ in Eq.(5-76) [16].

The nuclear electronic magnetic hyperfine coupling in the $^2P_{1/2}$ and $^2P_{3/2}$ atomic states is quite different than in the $^2S_{1/2}$ state. The P states have nodes at the nucleus and therefore the electron cannot penetrate the nucleus; however, the average value of $1/r^3$ for a P state does not diverge. a for non-S electrons (one-electron model) is given by

$$ha = g_e \mu_B g_i \mu_0 \left(\frac{1}{r^3}\right)_{av} \left[\frac{L(L+1)}{J(J+1)}\right]. \tag{5-77}$$

In the sodium atom, the experimental coupling constants are $a(^2P_{1/2}) = a_{1/2} =$ 94.45 MHz, and $a(^2P_{3/2}) = a_{3/2} =$ 19.06 MHz. Using Eq.(5-77) gives the predicted coupling constants of

$$a(^2P_{1/2}) = \frac{g_e \mu_B g_i \mu_0}{h}\left(\frac{8}{3}\right)\left(\frac{1}{r^3}\right)_{av} \quad \text{and} \quad a(^2P_{3/2}) = \frac{g_e \mu_B g_i \mu_0}{h}\left(\frac{8}{15}\right)\left(\frac{1}{r^3}\right)_{av}.$$

Thus, we predict $a_{3/2} \cong \frac{1}{5}a_{1/2}$, which is verified by the experimental numbers. Using the information above gives $(1/r^3)_{av}$ for a P state of $(1/r^3)_{av} = 1.68 \times 10^{24}$ cm^{-3}. This number is in fair agreement with the same value in Table 5-8 that was obtained from the spin-orbit interaction. Thus, the one-electron model will give a fair understanding of both the spin-orbit and the nuclear magnetic hyperfine interactions. It is evident, however, that the magnetic hyperfine interaction is larger for $^2S_{1/2}$ states than the corresponding $^2P_{1/2}$ and $^2P_{3/2}$ states in the same atom.

In conclusion, we note that the nuclear electronic coupling constant, a, in non-S states will be proportional to $(1/r^3)_{av}$. Thus, both this a and the spin-orbit interaction is proportional to the $(1/r^3)_{av}$ shown in Fig. 5-6; however, the nuclear electronic coupling will be less than the spin-orbit coupling by the ratio of the nuclear to Bohr magneton, μ_0/μ_B. The nuclear electronic coupling constant in S states is proportional to the contact term, $|\psi(0)|^2$, or the probability density for the electron at the nucleus [27].

5.7 ATOMS AND CRYSTAL FIELD THEORY

Having obtained some understanding of free unperturbed many-electron atoms or ions and the effects of magnetic fields, we shall now subject the free atom to an electric field in the form of a laboratory static field or a crystal field. A crystal field is the average static electric field experienced by an ion, molecule, or atom due to all the other surrounding atoms, molecules, or ions. The crystal field approximation is obtained by considering the lattice charge distributions as point charges. Crystal field effects are observed in the first row transition elements and ions that have the $3d$ configuration of electrons as shown in Table 5-12.

A wide range of complexing agents and a number of geometries are found in the transition metal complexes. The octahedral complex is quite common with the central transition metal ion being positioned at the center of a cube and surrounded by six nearest neighbors occupying the faces of the cube. Other typical geometries are tetrahedral, tetragonal, and square planar. Some of these arrangements are illustrated

Table 5-12 Atomic and ionic configurations and terms in the first transition metal atom series. The term states from a given configuration are obtained from *LS* coupling and the lowest energy levels are obtained from Hund's rules (Section 5.4).

Atom	Outer Electron Configuration	Ion	Ion Configuration	Ground-State Term of Ion
Sc	$3d4s^2$	Sc^{3+}	$3d^0$	^1S
Ti	d^2s^2	Ti^{2+}, Ti^{3+}	$3d^{2,1}$	^3F, ^2D
V	d^3s^2	V^{2+}, V^{4+}	$3d^{3,1}$	^4F, ^2D
Cr	d^5s	Cr^+, Cr^{2+}, Cr^{3+}	$3d^{5,4,3}$	^6S, ^5D, ^4F
Mn	d^5s^2	Mn^{2+}	$3d^5$	^6S
Fe	d^6s^2	Fe^{2+}, Fe^{3+}	$3d^{6,5}$	^5D, ^6S
Co	d^7s^2	Co^{2+}, Co^{3+}	$3d^{7,6}$	^4F, ^5D
Ni	d^8s^2	Ni^{2+}	$3d^8$	^3F
Cu	$d^{10}s$	Cu^{2+}	$3d^9$	^2D
Zn	$d^{10}s^2$	Zn^{2+}	$3d^{10}$	^1S

in Fig. 5-8. Common complexing agents are given in Table 5-13 in order of increasing crystal fields obtained with these groups.

The electric fields within a crystal (crystal fields) are normally much higher than laboratory electric fields. For instance, the electric field 2 Å from a point electronic charge has a magnitude of $3.6 \times 10^8\ V \cdot cm^{-1}$. Static laboratory electric fields above $10^5\ V \cdot cm^{-1}$ are difficult to obtain; however, large and variable crystal fields can be achieved by using different groups in the spectrochemical series (Table 5-13) or by

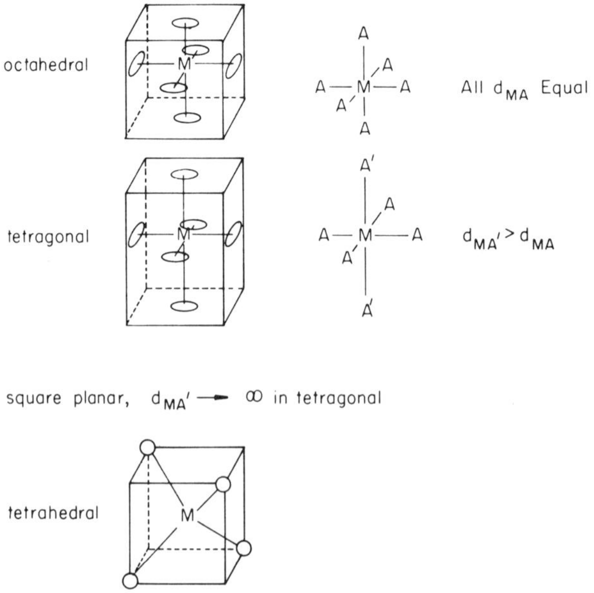

Figure 5-8 Typical geometries for transition metal complexes.

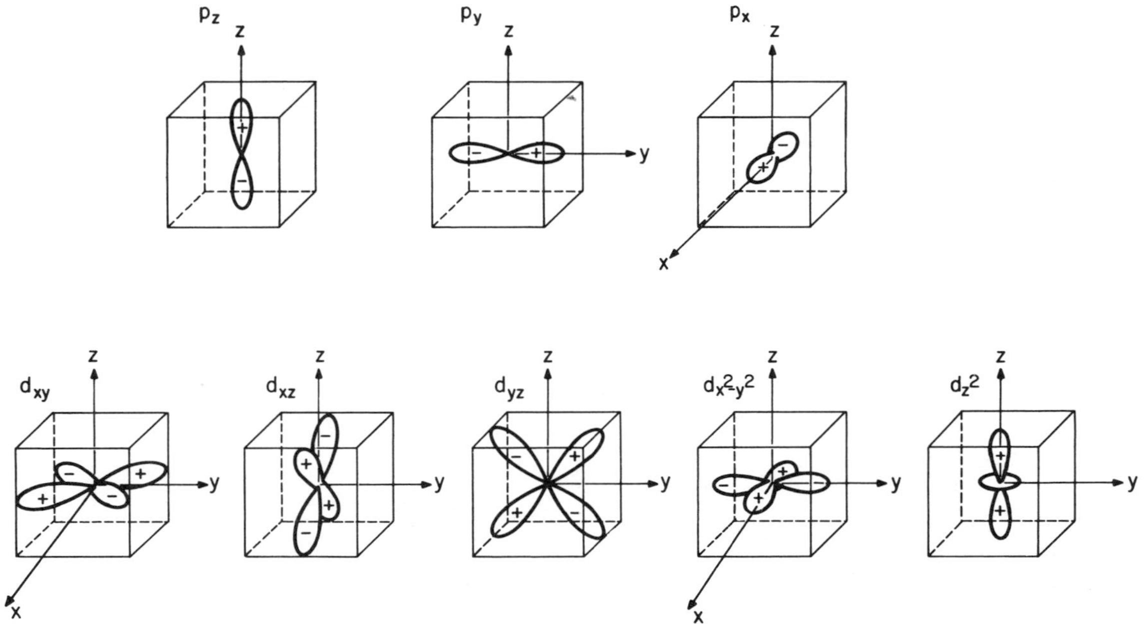

Figure 5-9 The spatial p and d functions for the hydrogen-like orbitals. See Eqs.(4-101) and (4-102) for the analytical forms for these orbitals.

Table 5-13 The spectrochemical series arranged in order of increasing crystal field effects from left to right.

$I^- < Br^- < Cl^- < OH^- < F^- < H_2O < C_5H_5N < NH_3 < NO_2^- < CN^-$

applying high enough pressure on a crystalline sample to change bond distances between the transition metal and the surrounding groups [28].

Consider now an octahedral complex with negative ion ligands in the faces of the cubes shown in Fig. 5-9. If the central ion has a single outer electron in an *np* configuration, it is clear that each of the three p_x, p_y, and p_z orbitals are at equal energy. If the single electron is in an *nd* configuration, however, there would be a difference in energy depending on whether the electron were in the $d_{x^2-y^2}$ or d_{z^2} orbitals or the corresponding d_{xy}, d_{xz}, and d_{yz} group of orbitals. The $d_{x^2-y^2}$ and d_{z^2} orbitals would force the single *nd* electron to be closer to the negative ion ligands than the d_{xy}, d_{xz}, or d_{yz} orbitals. The energy levels are shown in Fig. 5-10. Returning to Fig. 5-9 and the discussion above, we note that the ordering of the (d_{xy}, d_{xz}, d_{yz}) and (d_{z^2}, $d_{x^2-y^2}$) groups in Fig. 5-10 would be reversed if the negative ligands occupied the alternate corners of the cube to give a tetrahedral complex. This reversal in energy levels is also shown in Fig. 5-10. The tetragonal distortion and the resultant perturbation is easily

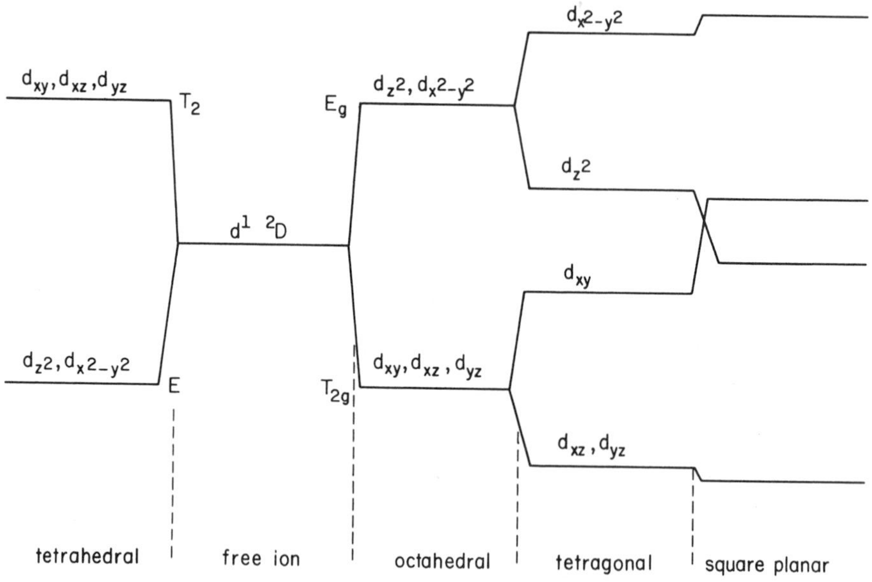

Figure 5-10 The energy levels of an *nd* electron under the influence of octahedral, tetrahedral, tetragonal, and square planar negative ion complexes. The energies are ordered according to the reasoning given in the text with the use of the spatial functions in Fig. 5-9. The symmetry species under the octahedral and tetrahedral group are also shown.

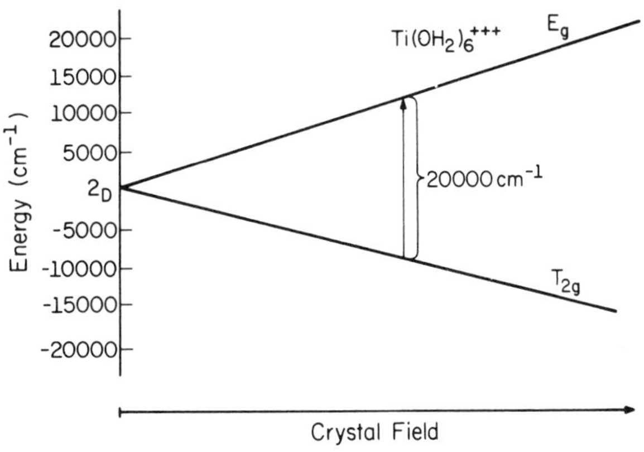

Figure 5-11 The crystal field octahedral OH_2 complex of the Ti^{3+} ion that has a *d* outer electron configuration. The other elements in the spectrochemical series in Table 5-13 will give transitions at different energies along the horizontal axis (crystal field). The higher elements in the spectrochemical series (CN^- for instance) will lead to larger energy splittings.

viewed using Figs. 5-8 and 5-9 with the results in Fig. 5-10. Removing the ligands along the *z* axis gives the square planar configuration as indicated. The energy levels in Fig. 5-10 would also be appropriate for an nd^9 electron configuration, which is equivalent to an *nd* configuration with a reversal in the signs of the energies.

A variable crystal field can be achieved by using different bonding groups in Table 5-13 to produce the crystal field. Figure 5-11 shows the splitting of 20,000 cm^{-1} arising from a ^2D term in the Ti^{3+} ion in the $Ti(OH_2)_6^{3+}$ complex. As H_2O is near the center of effectiveness in producing crystal fields, we would expect various energies along the horizontal axis in Fig. 5-11 to be observable with different groups in the spectrochemical series. The effects are quite visual. $Ti(OH_2)_6^{3+}$ absorbs at 20,000 cm^{-1} or at a wavelength of about 5000 Å and reflects longer wavelength radiation, appearing orange or red. If weaker bonding groups are used, the lower optical frequencies are absorbed and the solution appears blue or green. Other transition metal ions give similar results beginning with a much more complex energy level structure [29].

Multipole Expansion Formula

We shall now consider the form of the potential energy of interaction between a central atom and its surroundings as shown in the coordinate system in Fig. 5-12. The potential energy of interaction of the *i* charges at the center with the peripheral charges *j* is

$$V = \sum_{i,j} \frac{e_i e_j}{r_{ij}}. \tag{5-78}$$

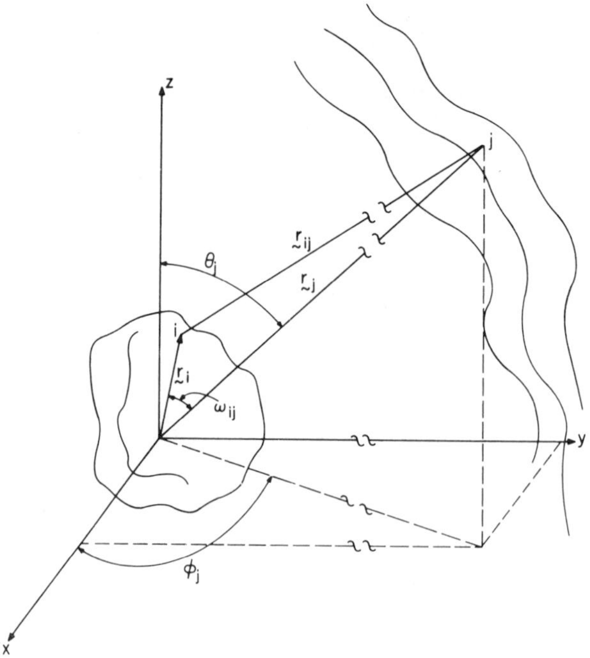

Figure 5-12 The coordinate system used to describe the interaction of a central ion charge distribution (i) with a surrounding charge distribution (j). The crystal field approximation assumes that the j charges are point charges and that j does not penetrate into the region of the central ion charge distribution.

r_{ij} can be expressed as a function of r_i, r_j, and ω_{ij}, where ω_{ij} is the angle between $\mathbf{r}_i$ and $\mathbf{r}_j$ according to Fig. 5-12 to give

$$\mathbf{r}_{ij} = \mathbf{r}_j - \mathbf{r}_i, \qquad \mathbf{r}_{ij} \cdot \mathbf{r}_{ij} = r_{ij}^2 = r_j^2 + r_i^2 - 2\mathbf{r}_i \cdot \mathbf{r}_j = r_j^2 + r_i^2 - 2r_i r_j \cos \omega_{ij}.$$

Substituting into Eq.(5-78) and rearranging gives

$$V = \sum_{i,j} \frac{e_i e_j}{r_j\{1 + [(r_i/r_j)^2 - 2(r_i/r_j)\cos \omega_{ij}]\}^{1/2}}. \tag{5-79}$$

In addition to assuming that the j peripheral charges are point charges, we shall also assume that $r_i \ll r_j$, which allows us to write

$$1 \gg \left[\left(\frac{r_i}{r_j}\right)^2 - 2\left(\frac{r_i}{r_j}\right)\cos \omega_{ij}\right] = x.$$

Using this expression in Eq.(5-79) and using the binomial expansion formula,

$$\frac{1}{(1+x)^{1/2}} = (1+x)^{-1/2} = 1 - \frac{1}{2}x + \frac{1 \cdot 3}{2 \cdot 4}x^2 - \frac{1 \cdot 3 \cdot 5}{2 \cdot 4 \cdot 6}x^3 + \cdots$$

leads to

$$V = \sum_{i,j} \frac{e_i e_j}{r_j}\left(1 - \frac{1}{2}x + \frac{3}{8}x^2 - \frac{15}{48}x^3 + \cdots\right)$$

$$= \sum_{i,j} \frac{e_i e_j}{r_j}\left[1 + \frac{r_i}{r_j}\cos\omega_{ij} + \left(\frac{r_i}{r_j}\right)^2\left(-\frac{1}{2} + \frac{3}{2}\cos^2\omega_{ij}\right) + \cdots\right]$$

$$= \sum_{i,j} \frac{e_i e_j}{r_j} \sum_{L=0}^{\infty} \left(\frac{r_i}{r_j}\right)^L P_L(\cos\omega_{ij}), \tag{5-80}$$

where $P_L(\cos\omega_{ij})$ are the Legendre polynomials in argument $\cos\omega_{ij}$ (see Appendix C.1). The next step in the derivation is to express $P_L(\cos\omega_{ij})$ in Eq.(5-80) as a product of separable common origin functions, each in the specific coordinates of the i central and j peripheral regions of charge. This transformation is realized with the spherical harmonic addition formula [30],

$$P_L(\cos\omega_{ij}) = \left(\frac{4\pi}{2L+1}\right) \sum_{M=-L}^{L} Y^*_{LM}(\theta_i, \phi_i) Y_{LM}(\theta_j, \phi_j). \tag{5-81}$$

Substituting into Eq.(5-80) and rearranging again gives

$$V = \sum_{L=0}^{\infty} \sum_{M=-L}^{L} \left[\left(\frac{4\pi}{2L+1}\right)^{1/2} \sum_i e_i (r_i)^L Y^*_{LM}(\theta_i, \phi_i)\right]$$

$$\times \left[\left(\frac{4\pi}{2L+1}\right)^{1/2} \sum_j \frac{e_j}{(r_j)^{L+1}} Y_{LM}(\theta_j, \phi_j)\right] = \sum_{L=0}^{\infty} \sum_{M=-L}^{L} \mathsf{T}^*_{LM}(i)\mathsf{T}_{LM}(j). \tag{5-82}$$

$\mathsf{T}^*_{LM}(i)$ is the operator representing the central ion (multipole moments) and $\mathsf{T}^*_{LM}(j)$ is the operator representing the surrounding charges. Equation (5-82) is the multipole expansion formula. The successive terms on the central charge distribution are the monopole ($L = 0$), dipole ($L = 1$), quadrupole ($L = 2$), octapole ($L = 3$), ... The successive terms on the surrounding charge distribution are the electric potential ($L = 0$), electric field ($L = 1$), electric field gradient ($L = 2$), ... It is relatively easy to see the nature of these terms for a cylindrically symmetric central charge distribution. In this case $V(\phi_i) = V(\phi_i + \alpha)$ where α is an arbitrary angle. Thus, in order to satisfy this requirement on the potential energy, all terms with $M \neq 0$ must go to zero leading to

$$V_{L=0} = \sum_{i,j} \frac{e_i e_j}{r_j}$$

$$V_{L=1} = \left(\sum_i e_i z_i\right)\left(\sum_j \frac{z_j}{r_j^3}\right)$$

$$V_{L=2} = \left[\frac{1}{2}\sum_i e_i(3z_i^2 - r_i^2)\right]\left[\frac{1}{2}\sum_j \frac{e_j}{r_j^5}(3z_j^2 - r_j^2)\right].$$

$$\vdots$$

The $L = 0$ term is the product of the monopole of the central charge distribution with the electric potential caused by the surrounding charges. The $L = 1$ term is the product of the z-axis dipole moment of the cylindrically symmetric charge distribution with the gradient of the potential (electric field) due to the surrounding charges. The $L = 2$ term is a product of the central charge z-axis quadrupole moment with the gradient of the electric field caused by the surrounding charge distribution.

The Use of Group Theory

The analysis above can be extended to more complex crystal fields with a variety of symmetries. The powerful concepts of group theory can also be applied to these crystal field problems to reduce the number of elements in the sum in Eq.(5-82). The primary use of symmetry in crystal field theory is to catalog the splittings of the atomic states; $S, P, D, F, \ldots$, under the influence of the crystal fields. The radial parts of the electronic eigenfunctions are spherically symmetric. Thus, only the angular parts of the functions, described by the spherical harmonics, will generate the irreducible representations of the free atomic or free atomic ion species. The spherical harmonics will generate the covering operators of a sphere and the diagonal sum or character of the matrix representation of these rotational operations generate the characters of the group describing a sphere.

This rotational character is discussed in Section 4.11. The characters of the full rotation group are given in Eq.(4-237) where α represents any angle of rotation and j represents the order of spherical harmonic. We now write expressions for the characters in Eq.(4-237) for rotations about $\alpha = 0$ (E operation), $\alpha = \pi$ (C_2 operation), $\alpha = 2\pi/3$ (C_3 operation), and $\alpha = \pi/2$ (C_4 operation).

$$\Gamma(E) = 2j + 1$$

$$\Gamma(C_2) = \frac{\sin\,(j + \frac{1}{2})\pi}{\sin\,(\pi/2)}$$

$$\Gamma(C_3) = \frac{\sin\,(j + \frac{1}{2})(2\pi/3)}{\sin\,(\pi/3)}$$

$$\Gamma(C_4) = \frac{\sin\,(j + \frac{1}{2})(\pi/2)}{\sin\,(\pi/4)}$$

We can now evaluate these characters for the various values of $j = 0, 1, 2, \ldots$ and decompose the irreducible representations of the full rotation group above under the lower symmetry crystal fields. Consider the octahedral crystal field as shown in Fig. 5-10. The characters shown above for the full rotation group under the angles given are now reducible representations under the octahedral group and the decomposition of the reducible representations into the irreducible representations under O_h symmetry proceeds as discussed in Appendix E. The characters and decompositions are shown in Table 5-14 [31]. The breakdown of the spherical harmonics under other symmetry groups (T_d, D_{4h}, D_3, and so on) is easily obtained using the standard decomposition techniques that are illustrated in Table 5-14 for an octahedral field.

Another example of weak and strong crystal fields is provided by the Ni^{2+} complexes where Ni^{2+} has a free ion d^8 configuration with terms given by ^1S, ^3P, ^1D, ^3F, and ^1G, as shown in Table 5-5. The energy level diagram of the d^8 configuration under the influence of interelectronic electrostatic and crystal fields is shown in

Table 5-14 The symmetry operations of the O group and the decomposition of the spherical harmonics under the O group. As $O_h = i \times O$, we shall only list the symmetry operations of O here and we shall remember that the even spherical harmonics ($j = 0, 2, 4, \ldots$) generate the even irreducible representations (g) under O_h and the odd spherical harmonics ($j = 1, 3, 5, \ldots$) generate the odd irreducible representations (u) under O_h.

Reducible Representations							
j	State	E	$8C_3$	$3C_2 = 3C_4^2$	$6C_2$	$6C_4$	Irreducible Components Under O_h
0	S	1	1	1	1	1	A_{1g}
1	P	3	0	-1	-1	1	T_{1u}
2	D	5	-1	1	1	-1	$E_g + T_{2g}$
3	F	7	1	-1	-1	-1	$A_{2u} + T_{1u} + T_{2u}$
4	G	9	0	1	1	1	$A_{1g} + E_g + T_{1g} + T_{2g}$
5	H	11	-1	-1	-1	1	$E_u + 2T_{1u} + T_{2u}$
6	I	13	1	1	1	-1	$A_{1g} + A_{2g} + E_g + T_{1g} + 2T_{2g}$
⋮	⋮	⋮	⋮	⋮	⋮	⋮	⋮

Fig. 5-13. The two limits are first, negligible crystal field splittings and large interelectronic repulsions; and second, negligible interelectronic repulsions relative to the larger crystal field effects. The results in the weak crystal field are from Table 5-14. The symmetry species of the strong crystal field limit on the right side of Fig. 5-13 are determined by breaking down the direct product representations into their irreducible representations under the O_h symmetry group. The results of this determination gives

$$t_{2g} \times t_{2g} \longrightarrow A_{1g}, E_g, T_{1g}, T_{2g}$$

$$t_{2g} \times e_g \longrightarrow T_{1g}, T_{2g}$$

$$e_g \times e_g \longrightarrow A_{2g}, E_g, A_{1g}.$$

We now must determine the spin states of these various states under O_h. Consider the $(e_g)^2$ configuration as shown on the right side of Fig. 5-13. It is convenient to start with a lower symmetry group to determine the spin multiplicity and then extend the arguments back to the O_h group. Consider the breakdown of the e_g representation under D_{4h}:

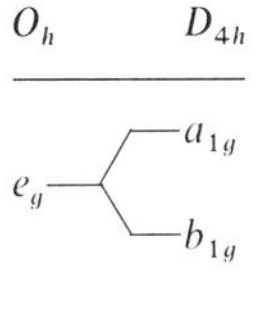

There are four ways of putting two electrons in these levels. The final states are obtained by forming the direct products and decomposing under D_{4h}.

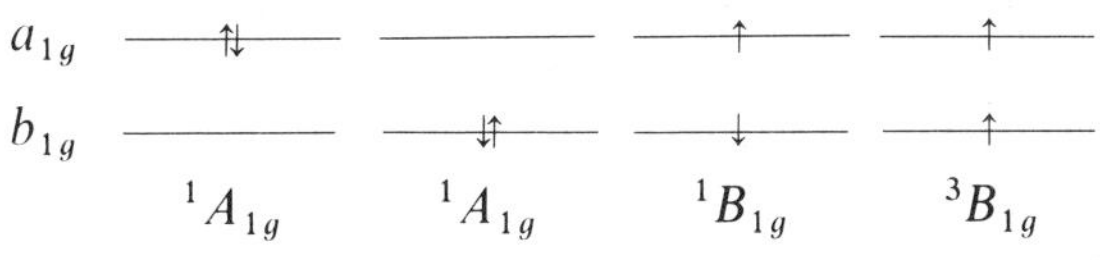

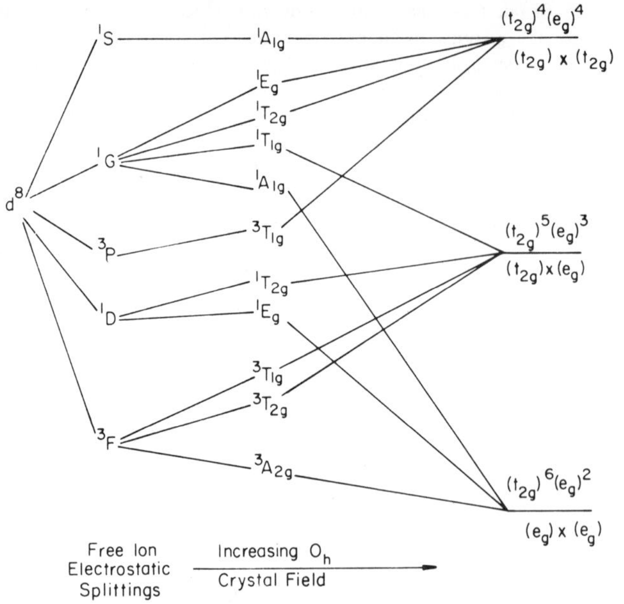

Figure 5-13 The crystal field energy level diagram for the Ni^{2+} octahedral complex under conditions of the free ion (left) and large crystal field (right) for two electrons. The spin-orbit interaction is neglected. On the right, the strong crystal field electronic configurations are shown above the line and the direct products are below the line.

The back correlation to the O_h group is easily obtained by noting that the different symmetries cannot change spin multiplicity giving

D_{4h}		O_h
$^1A_{1g}$	$\longrightarrow$	$^1A_{1g}$
$^3B_{1g}$	$\longrightarrow$	$^3A_{2g}$
$^1A_{1g}$, $^1B_{1g}$	$\longrightarrow$	^1E_g

The final results under O_h ($^1A_{1g}$, $^3A_{2g}$, and ^1E_g) appear in Fig. 5-13. Similar arguments are evident for the $(t_{2g})(e_g)$ and $(t_{2g})^2$ configurations.

We have ignored the effects of spin-orbit coupling up to this time. If the spin-orbit perturbation is less important than the crystal field, the smaller effect can be treated as a perturbation on the crystal field levels. On the other hand, if the spin-orbit coupling is large, the crystal field can be treated as a perturbation on the spin-orbit levels. To illustrate these two limits we shall consider the Cr^{2+} ion surrounded by an octahedral crystal field caused by one of the bonding groups in the electrochemical series in Table 5-13. The halogen ligands on the left will cause a relatively small crystal field relative to the spin-orbit interaction and the opposite order will be experienced with ligands on the stronger side of the electrochemical series such as the CN^- ion. The free Cr^{2+} ion has an outer electron d^4 configuration with the ^5D term being the lowest energy electronic state (see Table 5-12). Thus, the ^5D term will be split into ^5E_g and $^5T_{2g}$ sublevels under the influence of the octahedral crystal field, as shown in Table

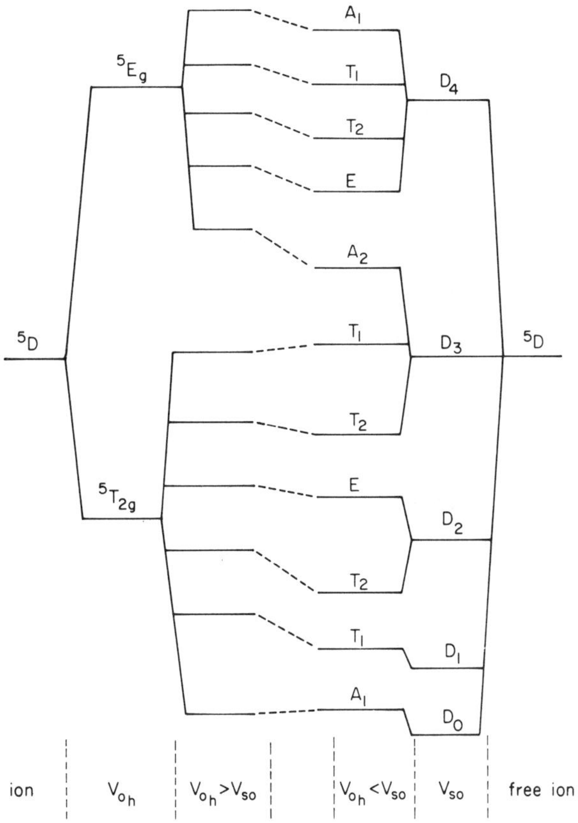

Figure 5-14 An energy level correlation diagram showing the effects of an O_h crystal field and spin-orbit coupling on a ^5D free atom (ion) term. V_{O_h} is the crystal field energy and V_{SO} is the spin-orbit energy. The strong crystal field interaction is shown on the left and the strong spin-orbit interaction is shown on the right. There are 25 total states throughout the correlation.

5-14 for instance. After the crystal field splitting, the spin-orbit coupling is added as a smaller perturbation where V_{O_h} (crystal field) > V_{SO} (spin-orbit). The spin-orbit splitting is obtained from the $L = 2, S = 2$ states of the ^5D term by the direct product $T_2 \times (E + T_2) = A_1 + E + 2T_1 + 2T_2$ and $E \times (E + T_2) = A_1 + A_2 + E + T_1 + T_2$ decompositions under the O group.

The case of zero crystal field showing only the spin-orbit splitting in the ^5D term is shown on the right side of Fig. 5-14. Then, proceeding from right to left, the smaller crystal field effects are added. The D_J spin-orbit levels are decomposed for various J in Table 5-14 [31]. The correlation between the two limiting cases is evident; however, quantitative predictions of the splittings between the energy levels for a particular system are very complicated [32,33].

Returning to the concept of a pure crystal field we note that the distortion of an octahedral field can be handled by a decomposition of the irreducible representations in O_h under the lower symmetry species. This breakdown and correlation is similar to the corresponding analysis of molecular vibrations. For instance, the normal modes of motion under the octahedral ReF_6 molecule will be broken into lower symmetry types if the O_h symmetry is broken by replacing one of the F atoms with a Cl atom (see Problems 5-14 and 5-15). The breakdown of the normal vibrational symmetries is discussed in Appendix E. The corresponding analysis for the influence of the O_h crystal field levels under a tetragonal distortion is treated by identical procedures and several examples are given in the problems at the end of this chapter.

Simple crystal field theory will break down if the charge distributions deviate from the point charge approximation or if there is appreciable chemical bonding between the central atom or ion and its surrounding ligands. *Ligand field theory* employs molecular orbital concepts to explain the nature of the interactions in the transition metal complexes [34].

5.8 NUCLEAR QUADRUPOLE INTERACTIONS

The nuclear quadrupole interaction is the interaction between the electric quadrupole moment of a nucleus with the electric field gradient at the nucleus due to all electronic (and other nuclear) charges in the atom (or molecule). The development from Eq.(5-78) to the result in Eq.(5-82) can be used here as the $r_i \ll r_j$ approximation illustrated in Fig. 5-12 is quite valid for this application. We first note that if the nucleus has a nonzero angular momentum, it will be spinning rapidly about an internal axis (z) and the nuclear shape will be essentially cylindrically symmetric. It is easy to show that a typical nucleus with angular momentum $I = \hbar$ is spinning about its internal axis with an angular velocity that is 10^5 larger than the angular velocity of an electron in an atom with orbital angular momentum $L = \hbar$. As a result, the perturbation in Eq.(5-82) reduces to the single cylindrically symmetric component (z is the symmetry axis):

$$\mathscr{H}_1 = \frac{1}{2}\,\mathsf{Q}_{zz}\mathsf{q}_{zz} = \frac{1}{2}\underbrace{\left[\sum_i e_i(z_i^2 - y_i^2)\right]}_{\substack{\text{cylindrically}\\ \text{symmetric}\\ \text{quadrupole}}}\underbrace{\left[\sum_j e_j\left(\frac{3z_j^2 - r_j^2}{r_j^5}\right)\right]}_{\substack{\text{field}\\ \text{gradient}}}. \tag{5-83}$$

We shall now consider the energy obtained from $\mathscr{H}_1$ in Eq.(5-83) for a nucleus (with a quadrupole moment) in a crystal lattice fixed in the laboratory. The q_{zz} field gradient in Eq.(5-83) is defined in the nuclear-fixed coordinate system. The Cartesian field gradient tensor has the same transformation properties as the polarizability and magnetic susceptibility tensors that were discussed in Section 4.11. Thus, we can relate the field gradient tensor in the nuclear-fixed axis system, $\mathbf{q}_n$, to the field gradient tensor in the laboratory-fixed axis system, $\mathbf{q}_l$, by the direction cosine transformation, $\mathbf{q}_n = \tilde{\mathbf{C}}\mathbf{q}_l\mathbf{C}$ [see Eqs.(B-35)]. If we choose the laboratory framework to coincide with the *principal* field gradient axis system, $\mathbf{q}_I$ is diagonal with elements q_{XX}, q_{YY}, and q_{ZZ} giving

$$q_{zz} = q_{XX}C_{Xz}^2 + q_{YY}C_{Yz}^2 + q_{ZZ}C_{Zz}^2,$$

where again z is along the nuclear axis of spin; X, Y, and Z are in the laboratory framework, and C_{Xz}^2, C_{Yz}^2, and C_{Zz}^2 are the squared direction cosines between the indicated axes. Substituting into Eq.(5-83) gives

$$\mathscr{H}_1 = \tfrac{1}{2}Q_{zz}q_{zz} = \tfrac{1}{2}Q_{zz}(q_{XX}C_{Xz}^2 + q_{YY}C_{Yz}^2 + q_{ZZ}C_{Zz}^2). \tag{5-84}$$

We shall now evaluate the first-order correction due to this perturbation. The zero-order nuclear wavefunction will be a product of the function describing the intrinsic nuclear properties times the nuclear rotational function, $\psi_{I,M}$,

$$\psi_N = \psi_{I,M}\psi_{(\text{intrinsic})} = \psi_{I,M}\psi(r_i, \theta_i, \phi_i).$$

Thus, the model is that of a cylindrical nucleus with a distribution of mass and charges to yield electric and magnetic moments. The intrinsic structure of the nucleus is described by an intrinsic nuclear wavefunction $\psi(r_i, \theta_i, \phi_i)$. The spin axis of the nucleus is allowed to rotate (similar to a linear molecule) with respect to the laboratory frame of reference and the nuclear wavefunction will be a product of the intrinsic and orientation functions as shown. The field gradient is produced by electrons and other nuclei that can be evaluated by the wavefunction Φ. The total wavefunction is

$$\psi = \psi_N\Phi = \psi_{I,M}\psi(r_i, \theta_i, \phi_i)\Phi. \tag{5-85}$$

The first-order energy due to $\mathscr{H}_1$ in Eq.(5-84) is given by

$$E(I, M) = \tfrac{1}{2}(Q_{zz})_{av}\{(q_{XX})_{av}\langle I, M|C_{Xz}^2|I, M\rangle + (q_{YY})_{av}\langle I, M|C_{Yz}^2|I, M\rangle + (q_{ZZ})_{av}\langle I, M|C_{Zz}^2|I, M\rangle\}, \tag{5-86}$$

where

$$(Q_{zz})_{av} = \int\psi^*(r_i, \theta_i, \phi_i)Q_{zz}\psi(r_i, \theta_i, \phi_i)\,dV_i, \qquad (q_{ZZ})_{av} = \int \Phi^* q_{ZZ}\Phi\,dV, \tag{5-87}$$

with similar expressions for q_{XX} and q_{YY}. Now we can evaluate the direction cosine matrix elements by expanding the squares and using Table 4-20 with $K = 0$. Using $(q_{ZZ})_{av} + (q_{YY})_{av} + (q_{XX})_{av} = 0$ from Eq.(1-15) and the evaluation above of the direction cosine matrix elements, we can show that the first-order energy reduces to

$$E(I, M) = \frac{(Q_{zz})_{av}(q_{ZZ})_{av}}{2}\left[\frac{3M^2 - I(I+1)}{(2I-1)(2I+3)}\right]. \tag{5-88}$$

There are also off-diagonal elements, in $\langle I, M|\mathscr{H}_1|I', M'\rangle$, in both I and M; however, the separation between different nuclear I states is in the megavolt range. Thus, the second-order perturbations between I states will be negligible. The off-diagonal elements in M will be proportional to $(q_{XX})_{av} - (q_{YY})_{av}$. Thus, the off-diagonal elements are zero for cylindrical electronic symmetry where $(q_{XX})_{av} = (q_{YY})_{av}$.

The energy expression in Eq.(5-88) differs from the usual expression for the pure nuclear quadrupole energy. Normally, the energy is referenced to the value with $M = I$, which is

$$E(I, M = I) = \frac{(\mathsf{Q}_{zz})_{av}(\mathsf{q}_{ZZ})_{av} I}{2(2I + 3)} = \frac{q_{av}}{2} Q$$

$$Q = \frac{I}{(2I + 3)} (\mathsf{Q}_{zz})_{av}$$

$$= \frac{I}{(2I + 3)} \int \psi^*(r_i, \theta_i, \phi_i) \left[\frac{1}{2} \sum_i e_i r_i^2 (3 \cos^2 \theta_i - 1) \right] \psi(r_i, \theta_i, \phi_i) \, dV_i. \qquad (5\text{-}89)$$

Substituting this result into Eq.(5-88) gives

$$E(I, M) = \frac{q_{av} Q[3M^2 - I(I + 1)]}{2I(2I - 1)}. \qquad (5\text{-}90)$$

The only remaining problem is the matter of convention in the definition of $(\mathsf{Q}_{zz})_{av}$ in Eq.(5-87) and the subsequent results in Eqs.(5-88) and (5-90). Physicists have many times preferred to use twice the value of $(\mathsf{Q}_{zz})_{av}$ used here. The values of $Q_{av} = (\mathsf{Q}_{zz})_{av}$ and Q as defined in Eqs(5-87) and (5-89) are listed for several nuclei in Table 5-15. A plot of the nuclear energy levels according to Eq.(5-90) is given in Fig. 5-15 for $I = \frac{3}{2}$ and $I = \frac{5}{2}$ systems [note that $E(I, M)$ is zero when $I = \frac{1}{2}$].

The model of nuclear quadrupole coupling presented in the last few paragraphs is that of a nucleus with a local cylindrical charge distribution having preferred orientational states in a field gradient caused by a peripheral charge distribution. If a given nucleus is rigidly constructed and the nucleons retain their relative locations in different nuclear I states, the quadrupole interaction model is similar to a linear molecule in a field gradient. Thus, the different nuclear I states will possess nearly the same nuclear quadrupole moment. These states are sometimes referred to as the nuclear rotational states in analogy to the rotational states of a diatomic molecule [35].

Table 5-15 Nuclear quadrupole moments in several nuclei. $Q_{av} = (\mathsf{Q}_{zz})_{av}$ is defined in Eqs.(5-87) and (5-83). $Q = IQ_{av}/(2I + 3)$ from Eq.(5-89) is also listed. The data are from V. S. Shirley, *Table of Nuclear Moments, Appendix C, Hyperfine Structure and Nuclear Radiations* (North Holland Publishing Co., Amsterdam, 1968).

Nucleus	I	Q_{av} (10^{-34} SC · cm^2)	Q (10^{-34} SC · cm^2)
D	1	0.0335	0.0067
^{14}N	1	0.190	0.038
^{17}O	$\frac{5}{2}$	−0.204	−0.064
^{35}Cl	$\frac{3}{2}$	−0.768	−0.192
^{37}Cl	$\frac{3}{2}$	−0.600	−0.15
^{79}Br	$\frac{3}{2}$	3.16	0.79
^{81}Br	$\frac{3}{2}$	2.68	0.67
^{127}I	$\frac{5}{2}$	−5.98	−1.87
^{23}Na	$\frac{3}{2}$	1.40	0.35
^{85}Rb	$\frac{5}{2}$	2.21	0.69
^{133}Cs	$\frac{7}{2}$	−0.026	−0.009

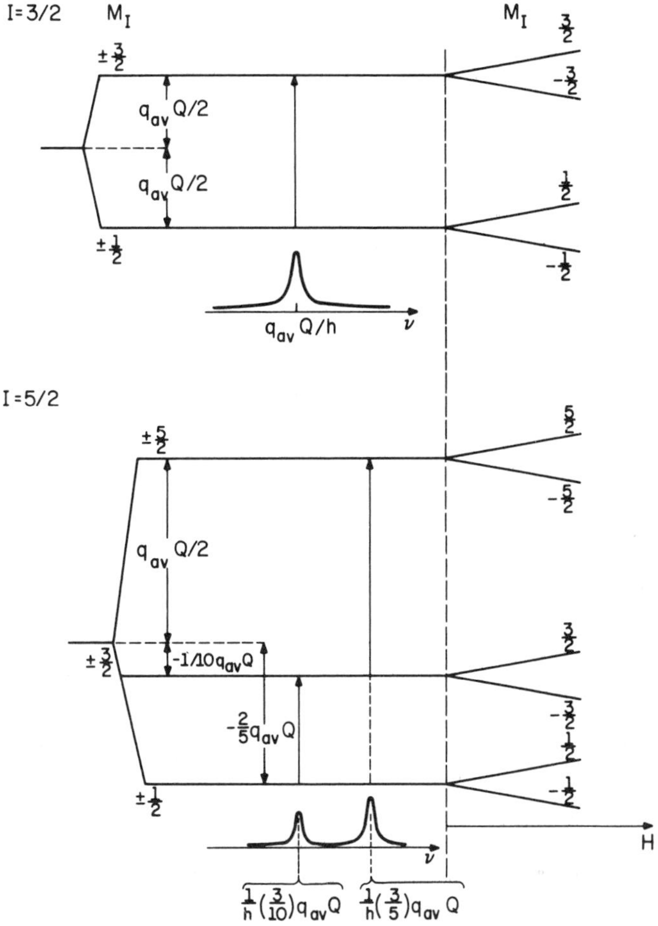

Figure 5-15 The energy levels of nuclei with $I = \frac{5}{2}$ and $I = \frac{3}{2}$ spins according to Eq. (5-90) for zero magnetic fields and positive values of $q_{av}Q$. Typical values of $q_{av}Q$ are listed in Table 5-16. The right side of the diagram shows the effects of external magnetic fields as described in Eq.(5-91). The relative intensities in the lower $I = \frac{5}{2}$ spectra can be obtained from Table 4-6.

On the other hand, there are other nuclear states that may lead to a quite different nuclear structure and resultant quadrupole moment. These latter states differ from the ground (or reference) state by a change in the distribution of nucleons, which is analogous to the excited electronic states in a linear molecule which might lead to a nonlinear configuration. In conclusion, we shall interpret the nuclear quadrupole moment in Eqs.(5-87) and (5-89) for the nucleus as being potentially different in different I states. If the quadrupole moments are essentially identical in two different nuclear I states, the nuclear states are related by a change in angular momentum that is analogous to a change in rotational angular momentum in a linear molecule. If the nuclear quadrupole moments are substantially different in two

Table 5-16 The results from pure nuclear quadrupole resonances in several molecules. Only the absolute value of $q_{av}Q$ is measured. The values of the field gradients are also listed and the ionic character of the halogen-carbon bonds are listed according to Eq.(5-95).

Molecule	Nucleus	$(q_{av}Q)/h$ (MHz)	q_{av} (10^{16} SC · cm^{-3})	$-q_{av}/e$ (10^{24} cm^{-3})	Percent Ionic Character
H_3CCl	^{35}Cl	34.2	−1.08	23.0	39
F_3CCl	^{35}Cl	38.8	−1.23	25.6	33
H_3CBr	^{79}Br	265.0	−2.22	46.2	34
F_3CBr	^{79}Br	302.0	−2.53	52.7	24
H_3CI	^{127}I	883.0	−3.98	83.0	13
F_3CI	^{127}I	1035.0	−4.12	86.0	10

different nuclear I states, the two nuclear states are different in their intrinsic arrangement of nucleons.

The effects of an external magnetic field can also be added to Eq.(5-90). The Zeeman correction for a single nucleus is given according to Eq.(4-172). The resultant energy is

$$E(I, M) = \frac{q_{av}Q[3M^2 - I(I+1)]}{2I(2I-1)} - g_i \mu_0 H_z M(1-\sigma). \tag{5-91}$$

g_i is the nuclear g-value, μ_0 is the nuclear magneton, and σ is the nuclear magnetic shielding. The magnetic field effect is also shown in Fig. 5-15.

The values of $(q_{ZZ})_{av} = q_{av}$ extracted from the observed zero-field pure nuclear quadrupole resonances in a number of nuclei, in various molecules in the solid state, are listed in Table 5-16. The values of q_{av} are obtained by using the values of Q in Table 5-15. Dividing q_{av} by $-e$ gives [see Eqs.(5-87) and (5-83)]

$$-\frac{q_{av}}{e} = \left[\sum_i \frac{1}{r_i^3}(3\cos^2\theta_i - 1)\right]_{av}. \tag{5-92}$$

θ_i is referenced to the C_3 symmetry axis in the molecules in Table 5-16. The sum over i is over all charges in the molecule excluding the interacting nucleus. The value of q_{av} for most molecules is a local atomic quantity due to the $1/r^3$ dependence in the operator. If the local charge distribution in the halogen atoms in Table 5-16 were completely ionic, the charge distribution would be spherically symmetric and $q_{av} = 0$. If the atomic configuration were that of a free atom with five np electrons, the field gradient would be identical to that calculated for a single p electron. Thus, in a free halogen atom we would expect a field gradient of approximately

$$-\frac{(q_{ZZ})_{av}}{e} = \int_0^{2\pi}\int_0^{\pi}\int_0^{\infty} \psi_{np_z}^* \left[\frac{1}{r^3}(3\cos^2\theta - 1)\right]\psi_{np_z} r^2\, dr \sin\theta\, d\theta\, d\phi. \tag{5-93}$$

Using $\psi_{np_z} = \eta_{np}(r)Y_{10}(\theta, \phi)$, where $\eta_{np}(r)$ is the radial function, we obtain

$$-\frac{(q_{ZZ})_{av}}{e} = \frac{4}{5}\int_0^{\infty} \eta_{np}(r)\left[\frac{1}{r^3}\right]\eta_{np}(r) r^2\, dr = \frac{4}{5}\left(\frac{1}{r^3}\right)_{av}^{np}. \tag{5-94}$$

The radial function average of $(1/r^3)$ for a p electron can be taken from the experimental results in Fig. 5-6 obtained from the spin-orbit interaction. In the case of the halogen atoms in Table 5-16, Fig. 5-6 gives $-q/e = 38 \times 10^{24}$ cm^{-3} for Cl, $-q/e = 70 \times 10^{24}$ cm^{-3} for Br, and $-q/e = 95 \times 10^{24}$ cm^{-3} for I. These values of $-q/e$ for a single unfilled p orbital in the Cl, Br, and I series are all larger than the experimental values in Table 5-16. This indicates that the electron distributions around the halogen nuclei in the methyl halides in Table 5-16 are intermediate between the ionic closed shell configuration ($q = 0$) and the free atom unfilled p-orbital values obtained above from Fig. 5-6. It is possible to define an approximate ionic character, or deviation from $q = 0$, from the results above:

$$\text{Percent ionic character} = \left[\frac{q_{\text{free atom}} - q_{\text{molecule}}}{q_{\text{free atom}}}\right] \times 100. \tag{5-95}$$

The resultant ionic characters are also listed in Table 5-16. Thus, it appears that the lighter halogens can achieve a higher ionic character, or more nearly closed shell configurations, than the heavier halogens.

The mechanism of interaction with the electromagnetic field to cause a nuclear transition in pure nuclear quadrupole resonance spectroscopy (NQR) is the magnetic dipole mechanism. This is the same mechanism discussed in Section 4.10 in NMR. Additional details as well as a discussion of the use of pure quadrupole resonance spectroscopy for chemical analysis can be found eslewhere [36].

5.9 ATOMS IN THE SOLID STATE

We shall now give a short qualitative discussion of the solid state. There are four basic ways of combining atoms to form crystalline solids. These are

1. Ionic lattices, NaCl,
2. Van der Waals lattices, rare gases, CH_4, Ar,
3. Metallic lattices, Na, Fe,
4. Covalent lattices, diamond.

Ionic Lattices

The alkali halide ionic crystals and other ionic lattices are made up of positive and negative ions. The energy of the lattice is a combination of the positive repulsive terms due to the interaction of like ions and negative attractive contributions arising from unlike ions.

The alkali and halide ions all have spherically symmetric closed shell (rare gas) electronic configurations (for instance, Na^+ $1s^22s^22p^6$ and Cl^- $1s^22s^22p^63s^23p^6$). The dominant stabilization energy of the alkali halide lattices is the point charge electrostatic energy. In the case of singly charged ions, the total energy is the sum over all point charges, j, from any origin, i, which is given by

$$E_i = -\sum_j \frac{e^2}{R_{ij}} = -\frac{e^2}{R}\sum_j \frac{\pm}{p_{ij}} = -\frac{e^2\alpha}{R}. \tag{5-96}$$

The energy is computed at the ith atom as the origin and the sum starts at the nearest neighbors and runs over all other atoms in the lattice. R is the nearest-neighbor distance and p_{ij} in $R_{ij} = Rp_{ij}$ is a numerical multiplier. $\sum_j \pm/p_{ij} = \alpha$ in Eq.(5-96) is called the *Madelung constant*, which can be evaluated for a given crystal structure by the indicated summation. Consider α in NaCl with Cl^- as the reference location. There are 6 positive ions at $p = 1$ giving a contribution of 6 to α; there are 12 next nearest-neighbor negative ions at $p = (2)^{1/2}$; 8 next-next nearest-neighbor positive ions at $p = (3)^{1/2}$; 6 negative ions at $p = 2.0$, and so on. The value of α for the NaCl structure is [37].

$$\alpha = 6 - \frac{12}{(2)^{1/2}} + \frac{8}{(3)^{1/2}} - \frac{6}{2} + \cdots = 1.74756. \tag{5-97}$$

Substituting this value of α into Eq.(5-96) with $R = 2.820$ Å in NaCl gives

$$E = -14.3 \times 10^{-12}\ \text{erg} = -205.6\ \text{kcal}\cdot\text{mol}^{-1}. \tag{5-98}$$

The experimental cohesive energy in NaCl is $E = -182.6\ \text{kcal}\cdot\text{mol}^{-1}$. Thus, the simple electrostatic picture gives an energy within 10% of the experimental result. The energy can be raised to the experimental result by including a repulsive energy between the nearest-neighbor ions in the alkali halide lattice. The repulsive interaction arises from the repulsion of the electron clouds of the closed shell nearest-neighbor ions [38]. As both the positive and negative ions have closed shells, the overlap of one ion by another will be opposed by the Pauli principle. Therefore, the repulsive energy, E_{rep}, between the closed shells of atoms A and B may be approximated as being proportional to the overlap integral between the atomic electronic functions, S_{AB},

$$E_{\text{rep}} = CS_{\text{AB}}, \tag{5-99}$$

where C is a proportionality constant. S_{AB} has been computed for the outer shell electrons (using ionic Hartree–Fock orbitals) in all the alkali halide ionic pairs [39]. These results show that the overlap is described by an exponential function of the internuclear spacing,

$$S_{\text{AB}} = A \exp\left(-\frac{R}{\rho}\right), \tag{5-100}$$

where A and ρ are parameters depending on the ions. Substituting Eq.(5-100) into Eq.(5-99) and adding this result to Eq.(5-96) give the total energy of

$$E = -\frac{e^2\alpha}{R} + nCA \exp\left(-\frac{R}{\rho}\right), \tag{5-101}$$

Table 5-17 The experimental and theoretical cohesive energies of the alkali halide lattices that have the NaCl structure. The calculated results are from Eq.(5-102) where ρ is from Ref. [39]. The internuclear spacings and experimental binding energies are from Ref. [23].

			Energy (kcal · mol^{-1})	
Lattice	R_{ab} (Å)	ρ (Å)	Eq.(5-102)	Experimental
LiF	2.014	0.293	−246.3	−242.3
LiCl	2.570	0.342	−195.8	−198.9
LiBr	2.751	0.354	−184.2	−189.8
LiI	3.000	0.380	−168.9	−177.7
NaF	2.317	0.301	−217.9	−214.4
NaCl	2.820	0.345	−180.5	−182.6
NaBr	2.989	0.357	−171.0	−173.6
NaI	3.237	0.382	−158.1	−163.2
KF	2.674	0.340	−189.4	−189.8
KCl	3.147	0.372	−162.4	−165.8
KBr	3.298	0.387	−155.3	−158.5
KI	3.533	0.409	−145.2	−149.9
RbF	2.815	0.357	−180.0	−181.4
RbCl	3.291	0.390	−155.4	−159.3
RbBr	3.445	0.401	−148.8	−152.6
RbI	3.671	0.423	−140.0	−144.9

where n is the number of nearest neighbors. At equilibrium ($R = R_{AB}$), $dE/dR = 0$, which gives $nCA \exp(-R_{ab}/\rho) = e^2\alpha\rho/R_{ab}^2$. Substituting into Eq.(5-101) gives the energy at the equilibrium nearest-neighbor distance, R_{ab},

$$E(R_{ab}) = -\frac{e^2\alpha}{R_{ab}}\left(1 - \frac{\rho}{R_{ab}}\right). \tag{5-102}$$

Substituting $\rho = 0.345$ Å from Table 5-17, $R_{ab} = 2.820$ Å, and $\alpha = 1.74756$ into Eq.(5-102) gives E (NaCl lattice) $= -180.5$ kcal · mol^{-1}, which is much closer to the experimental result than obtained in Eq.(5-98). In summary, the cohesive energy is largely due to point charge electrostatic interactions. The electrostatic stabilization is opposed by a small amount of repulsive energy arising from the overlap of the closed shell nearest-neighbor ions in the lattice. The results of this calculation on the alkali halide lattices with the NaCl structure are listed in Table 5-17. The calculated stabilization energies are normally still less than the experimental results. Improvement can be obtained by adding in the induced dipole-induced dipole attractive interactions as described below. Alternatively, better agreement with experiment can be obtained by assigning the values of ρ in Eq.(5-101) by using the experimental bulk modulus [40]. However, the treatment above demonstrating the role of the overlap in the repulsive energy may be a more lucid picture of the energetics involved.

An electron gas-type model for the electronic structure of ions has also been employed to calculate the cohesive energies of the alkali halides [41].

Rare Gas Lattices

The rare gas lattices provide an arrangement of spherically symmetric atoms in a lattice where the point charge attractive interactions are absent. According to the last few paragraphs, the only remaining energy for the spherically symmetric atoms is the respulsive term in Eq.(5-99) and the attractive induced dipole-induced dipole interaction as mentioned above. The induced dipole-induced dipole interaction is considered in detail in Section 6.10 and the result is given in Eq.(6-178) [42],

$$E \cong -\frac{3\alpha_1\alpha_2 E_1 E_2}{2R^6(E_1 + E_2)}, \qquad (5\text{-}103)$$

where α_1 and α_2 are the average polarizabilities $[\alpha = \frac{1}{3}(\alpha_{xx} + \alpha_{yy} + \alpha_{zz})]$ of two different atoms (or molecules). E_1 and E_2 are the first ionization energies and R is the distance between the centers of mass in the two atoms (or molecules). If the interacting pair are identical, Eq.(5-103) reduces to

$$E = -\frac{3\alpha^2 E_1}{4R^6}. \qquad (5\text{-}104)$$

As these attractive energies are of moderately short range, we shall include only the nearest-neighbor interactions and each pair interaction is assumed to be independent of the others. Thus, the total cohesive attractive energy for 12 nearest neighbors in the rare gas lattice is given by

$$E = -\frac{9\alpha^2 E_1}{R^6}. \qquad (5\text{-}105)$$

The cohesive energies computed with Eq.(5-105) are all too large, as shown in Table 5-18. This indicates the necessity of adding the nearest-neighbor repulsive energies, as in Eq.(5-101), giving the total energy of (12 nearest neighbors)

$$E = -\frac{9\alpha^2 E_1}{R^6} + 12B\exp\left(-\frac{R}{\rho}\right). \qquad (5\text{-}106)$$

B is a constant and ρ is an interaction length parameter. At equilibrium where $(dE/dR)_{R=R_{ab}} = 0$, the total energy is found as before to give

$$E(R_{ab}) = -\frac{9\alpha^2 E_1}{R_{ab}^6}\left(1 - \frac{6\rho}{R_{ab}}\right). \qquad (5\text{-}107)$$

ρ-Values from overlap integrals are not available for the rare gases. Therefore, ρ in Eq.(5-107) was fit to the experimental energy with the results listed in Table 5-18. The values of ρ listed in Table 5-18 are very similar to the corresponding overlap parameters (see discussion on alkali halide lattices) for Na^+, K^+, Rb^+, and Cs^+, respectively. In summary, rare gas crystals and other molecular crystals can be described by the cohesive energy expression in Eq.(5-107), which includes an attractive induced dipole-induced dipole term and a repulsive overlap term.

Table 5-18 Atomic and crystalline properties of the rare gas atoms. The experimental data are from H. Margenau, *Rev. Mod. Phys.* **11**, 1 (1939), E. R. Dobbs and G. O. Jones, *Rept. Progr. Phys.* **20**, 560 (1957), and Ref. [23].

Rare Gas	Ne	Ar	Kr	Xe
Ionization energy E(eV)	21.5	15.76	14.00	12.13
α (10^{-24} cm^3)	0.39	1.63	2.46	4.00
Crystalline R_{ab} (Å) Nearest neighbor (12)	3.08	3.75	4.02	4.31
Experimental cohesive energy (kcal · mol^{-1})	0.5	1.9	2.7	3.8
$9\alpha^2 E/R_{ab}^6$ (kcal · mole^{-1})	0.79	3.12	4.16	6.28
ρ from Eq.(5-107) to fit experimental energy	0.19	0.24	0.24	0.28

Another popular alternate to Eq.(5-106) is the Lennard–Jones potential, which is a sum of a stabilizing $1/R^6$ attractive term and a steep $1/R^{12}$ repulsive term. We shall discuss this potential function in more detail in Section 9.4.

Metals

Metal crystals have weak cohesive energies relative to the ionic crystals. Cu metal is described by relatively rigid positive Cu^+ ions distributed in a sea of free electrons that are obtained from the 4*s* shell of the Cu atom. The interatomic (nearest-neighbor) distances in the alkali metals are larger than the corresponding alkali halide distances in the ionic crystals due to the weaker bonding in the former system. Some data on metal crystals are given in Table 5-19.

The free conduction electrons in a metal can be treated by the particle-in-a-box model described in Problem 4-5. The energy levels in a three-dimensional cube are given by

$$E = \frac{h^2}{8m\ell^2}(n_x^2 + n_y^2 + n_z^2), \tag{5-108}$$

where $n_x = 1, 2, 3, \ldots$; $n_y = 1, 2, 3, \ldots$; and $n_z = 1, 2, 3, \ldots$ and ℓ is the dimension of the cube. The conduction electrons fill the energy levels in the cube with two electrons for each state, according to the Pauli principle. The value of ℓ is large, being on the order of centimeters and the particle-in-a-box energy levels are very close together forming a near continuum of levels.

The values of n_x, n_y, and n_z can be treated like the components of a vector, $\boldsymbol{n}$, where $\boldsymbol{n} \cdot \boldsymbol{n} = n^2 = n_x^2 + n_y^2 + n_z^2$ giving

$$E = \frac{h^2}{8m\ell^2} n^2 = \frac{\hbar^2}{2m} k^2, \qquad k^2 = \left(\frac{2\pi}{\lambda}\right) = \left(\frac{\pi n}{\ell}\right)^2, \tag{5-109}$$

Table 5-19 T = 300 K data on some conducting metals with body-centered-cubic (bcc), face-centered-cubic (fcc), or hexagonal-closest-packed (hcp) lattices. E_F and v_F are the Fermi energy and velocity, respectively, ρ_0 is the electron density, and σ is the electronic conductivity. The data are from Ref. [23].

Atom	Configuration	Free Electrons Per Atom	Lattice Type	Nearest-Neighbor Distance (Å)	E_F (eV)	v_F (cm · s^{-1})	ρ_0 (cm^{-3})	σ (ohm^{-1} · cm^{-1})
Li	[He]$2s$	1	bcc	3.03	4.72	1.29×10^8	4.70×10^{22}	1.07×10^5
Na	[Ne]$3s$	1	bcc	3.71	3.23	1.07	2.65	2.11
K	[Ar]$4s$	1	bcc	4.62	2.12	0.86	1.40	1.39
Rb	[Kr]$5s$	1	bcc	4.87	1.85	0.81	1.15	0.80
Cs	[Xe]$6s$	1	bcc	5.24	1.58	0.75	0.91	0.50
Cu	[Ar]$3d^{10}4s$	1	fcc	2.55	7.00	1.57	8.45	5.88
Ag	[Kr]$4d^{10}5s$	1	fcc	2.88	5.48	1.39	5.85	6.21
Au	[Xe]$4f^{14}5d^{10}6s$	1	fcc	2.88	5.51	1.39	5.90	4.55
Be	[He]$2s^2$	2	hcp	2.22	14.14	2.23	2.42	3.08
Mg	[Ne]$3s^2$	2	hcp	3.19	7.13	1.58	8.60	2.33
Ca	[Ar]$4s^2$	2	fcc	3.93	4.68	1.28	4.60	2.78
Ba	[Xe]$5s^2$	2	bcc	4.34	3.95	1.18	3.20	0.26
Zn	[Ar]$3d^{10}4s^2$	2	hcp	2.66	9.39	1.82	13.10	1.69
Cd	[Kr]$4d^{10}5s^2$	2	hcp	2.97	7.46	1.62	9.28	1.38
Al	[Ne]$3s^23p$	3	fcc	2.86	11.63	2.02	18.06	3.65
Ga	[Ar]$3d^{10}4s^24p$	3	hcp	3.55	10.35	1.91	15.30	0.67
Pb	[Xe]$4f^{14}5d^{10}6s^26p^2$	4	fcc	3.49	9.37	1.82	13.30	0.48

where $\boldsymbol{k}$ is the wave vector describing the free electron in quantum state n and we note that there are an integral number of half wavelengths between 0 and ℓ; $\lambda n = 2\ell$. The energy of the highest occupied energy level, E_F, is the Fermi energy and the associated wave vector is k_F. The total number of states contained within the Fermi surface described by k_F is given by the sphere of volume $4\pi k_F^3/3$ divided by the volume element in k space of $k_0^3 = (2\pi/\ell)^3$. The number of states (or number of electrons), N_0, is (the factor of 2 allows two electrons per state)

$$N_0 = 2\frac{4\pi k_F^3 \ell^3}{3(2\pi)^3} = \frac{k_F^3 \ell^3}{3\pi^2} = \frac{V k_F^3}{3\pi^2},$$

where $V = \ell^3$ is the volume of the cube. Rearranging gives $k_F^3 = (3\pi^2 N_0/V)$. Substituting into Eq.(5-109) gives the Fermi energy of

$$E_F = \frac{\hbar^2}{2m}\left(\frac{3\pi^2 N_0}{V}\right)^{2/3}, \tag{5-110}$$

which is dependent on the density of electrons (N_0/V), not the total number of electrons. In the case of Cu (see Table 5-19), the mass density is 8.96 $\text{g}\cdot\text{cm}^{-3}$, which leads to an atom density of 8.45×10^{22} $\text{atom}\cdot\text{cm}^{-3}$. Thus, if each copper atom contributes its $4s$ electron to the conduction band, there are 8.45×10^{22} electrons per cubic centimeter, and the Fermi energy is $E_F = 7.0$ eV. The momentum of the electron at the Fermi surface is $p_F = \hbar k_F$ and the resultant velocity is $v_F = \hbar k_F/m$. In the case of copper, $v_F = 1.57 \times 10^8$ $\text{cm}\cdot\text{s}^{-1}$, which is about the velocity of a thermal electron. The Fermi energies and Fermi velocities of electrons in several other metals are also listed in Table 5-19. We can also review the concept of electronic conductivity from Eq.(2-42), $\sigma = \rho_0 e^2\tau/m$, where ρ_0 is the electron number density, e is the charge, m is the mass, and τ is the free electron relaxation or collision time. The value of σ in Table 5-19 is $\sigma = 5.88 \times 10^5$ $\text{ohm}^{-1}\cdot\text{cm}^{-1}$. The units of ohm according to Table A-1 is ohm = volt $\cdot$ amp^{-1}. Thus, if we write σ in units of $\text{ohm}\cdot\text{m}^{-1}$, the units of σ are inverse time or seconds in either cgs or SI units. Hence, if we write σ in units of $\text{ohm}^{-1}\cdot\text{m}^{-1}$, we can use $\sigma = \rho_0 e^2\tau/m$ in either cgs or SI units. In the case of Cu, $\sigma = 5.88 \times 10^7$ $\text{ohm}^{-1}\cdot\text{m}^{-1} = 5.88 \times 10^7$ s $= \rho_0 e^2\tau/m$. Using ρ_0 from Table 5-19 and using either cgs or SI units for ρ_0, e, and m gives $\tau = 2.47 \times 10^{-14}$ s. Using the Fermi velocity from Table 5-19 gives the mean length between electron collisions or relaxations. In the case of Cu, $\ell = v_F\tau = 3.90 \times 10^{-8}$ m. Thus, a free electron path before a collision is in excess of 100 Cu-Cu interatomic distances in the lattice. The primary relaxation mechanism is the electron interacting with the lattice vibrations (phonons). As the temperature is lowered, the lattice vibrations are cooled and the electron-phonon interaction decreases. At $T = 4$ K, the electron conductivity of Cu is 10^6 times larger than the $T = 300$ K value and the corresponding free electron path before a collision is on the order of 1 cm.

The eigenfunction for the free electron with the eigenvalue in Eq.(5-109) is the plane wave for the electron given by

$$\psi = \left(\frac{1}{V}\right)^{1/2} \exp\,(i\boldsymbol{k}\cdot\boldsymbol{r}).$$

In the presence of periodic positive atom potentials in the lattice, standing waves are obtained or the traveling wave is blocked at certain k-values. Energy gaps result that may exceed kT [43]. Many of the interesting properties of semiconductors concern the distribution of energy levels between the allowed bands or electron zones in crystals [23].

Covalent Lattices

Covalent lattices have properties similar to molecules. The many-electron theory of molecules is given in Chapter 6.

PROBLEMS

1. Substitute Eq.(5-22) into the first of Eqs.(5-24) to derive the final part of Eqs.(5-24) for a 4×4 example (four electrons in two doubly occupied spatial orbitals).

2. Consider a world where electrons possess zero spin. Under these conditions, where the electronic function is symmetric under exchange of electrons, there would be no limit to the number of electrons that could group into each spatial orbital. The ground state of the lithium atom would then have a configuration of $1s^3$. Compute the total electronic energy of the lithium atom with a $1s^3$ electronic configuration. Use Slater orbitals with the simple Slater rules to obtain $n = 1$ and the effective nuclear charge and then calculate the energy. Compare your result with the Hartree–Fock SCF electronic energy of the lithium atom of -7.4327 a.u.

3. Use Eqs.(5-3) and (5-7) for the He atom and calculate separately the electronic kinetic and potential energies. Do the functions defined in Eqs.(5-3) and (5-7) satisfy the virial theorem for the He atom?

4. The energy levels of the outer electron in the K atom are shown in Fig. 5-5. We shall attempt here to compute the $s - p$ electron static energy spacings for the outer electron in an alkali atom. Consider the Li atom. Assume the inner two electrons are at $r = a_1 = a_0/3$, where a_0 is the Bohr radius. Therefore, the potential for the outer electron can be approximated by $V = -e^2/r$ outside a_1 and $V = -3e^2/r$ inside a_1. This is equivalent to complete shielding of two charges on the Li nucleus by the inner two electrons. The Hamiltonian for the outer single electron is, therefore,

$$\mathscr{H} = -\frac{\hbar^2}{2m}\nabla^2 - \frac{e^2}{r}(r > a_1) - \frac{3e^2}{r}(0 \le r \le a_1).$$

This can be rearranged to give

$$\mathscr{H} = \underbrace{-\frac{\hbar^2}{2m}\nabla^2 - \frac{e^2}{r}}_{\mathscr{H}_0} \underbrace{- \frac{2e^2}{r}(0 \le r \le a_1)}_{\mathscr{H}_1}. \tag{5-111}$$

We can now treat the solution of the Li atom outer electron Schrödinger equation with this Hamiltonian and perturbation theory.

(a) What are the eigenfunctions and eigenvalues obtained from $\mathscr{H}_0\Phi_n = E_n\Phi_n$?

(b) The outer electron necessarily resides in the $n = 2$ level in the ground state of the Li atom. Using the four degenerate $n = 2$ eigenfunctions obtained in part a as a starting set of functions $\mathbf{\Phi}$, calculate the Hamiltonian matrix in the $\mathbf{\Phi}$ representation. (Use the complete Hamiltonian.)

(c) Diagonalize the $\mathscr{H}^{\Phi}$ matrix above to give the new eigenvalues and eigenfunctions for the complete Hamiltonian.

(d) Draw an energy level diagram starting with the four degenerate states for the outer electron in the Li atom and then sketch the split levels due to the electrostatic perturbation, $\mathscr{H}_1$.

(e) The experimental energy difference between the $2s$ and $2p$ levels is 14,906 cm^{-1}. Compare this to your calculated splitting due to the Coulombic perturbation. State two ways that your result could be improved.

(f) Repeat parts b through d where a small electric field effect is added to $\mathscr{H}_1$ giving

$$\mathscr{H}_1 = -\frac{2e^2}{r}(0 \le r \le a_1) - eEr\cos\theta.$$

5. Show that the average value of a one-electron operator, R_i, summed over all electrons in the atom ($R = \sum_i R_i$) for a doubly occupied single determinant function is given by

$$\langle R\rangle_{av} = \int_1\int_2\cdots\int_n \psi^*(1, 2, \ldots, n)\left[\sum_i^n R_i\right]\psi(1, 2, \ldots, n)\, dV_1\, dV_2\cdots dV_n$$

$$= 2\sum_{j=1}^{n/2}\int \Phi_j^*(a)R_a\Phi_j(a)\, dV_a, \qquad (5\text{-}112)$$

where $\Phi_j(a)$ is the jth AO containing the ath electron and the sum over j is over the $n/2$ occupied AOs.

6. The diamagnetic susceptibility is defined in Section 4.1. Starting with Eq.(4-28) we write the average diamagnetic susceptibility in an atom as

$$\chi_{av}^d = \frac{1}{3}(\chi_{xx}^d + \chi_{yy}^d + \chi_{zz}^d) = -\frac{e^2}{6mc^2}\left\langle 0\left|\sum_i r_i^2\right|0\right\rangle,$$

where the sum over i is over all electrons in the atom, and $\langle 0|0\rangle$ represents the ground electronic state. Consider now the following atoms and ions: F^+, F, F^-, Na^+, Na, and Na^-.

(a) Use Slater rules and Slater functions to write the one-electron orbitals for all electrons in each of the atoms and ions above.

(b) Use the Slater atomic orbitals above to calculate the values of χ_{av}^d for each of the atoms and ions above. Compare your results with the SCF results in G. Malli and S. Fraga, *Theor. Chim. Acta* (Berlin) **5**, 284 (1966).

7. Use the Slater orbitals for F and Na determined in Problem 5-6 to compute the nuclear diamagnetic shielding in both atoms given by

$$\sigma^d = \frac{e^2}{3mc^2}\left\langle 0 \left| \sum_i \frac{1}{r_i} \right| 0 \right\rangle.$$

Check your result with the results in R. A. Bonham and T. G. Strand, *J. Chem. Phys.* **40**, 3447 (1964).

8. Referring to Eq.(5-74), evaluate the appropriate basis and first-order energies for each of the following situations:
   (a) strong coupling; $C > ha >$ Zeeman splittings (draw an energy level diagram);
   (b) strong spin-orbit, weak nuclear coupling; $C >$ Zeeman splittings $> ha$; and
   (c) high field; Zeeman splittings $> C > ha$.

9. Compute $|\psi_{ns}(0)|^2$ for the outer electron in K, Rb, and Cs (see Section 5.6) using hydrogen-like orbitals for the outer electron and Slater rules for the appropriate shielding constants. Compare your results with the experimental values determined with Table 5-11 and Eq.(5-76).

10. The isomer shift in a number of iodine compounds has been studied by the Mössbauer effect by observing the 27.7-keV decay of ^{129}I (populated by the β decay of ^{129}Te) [44]. The isomer shift is the shift in nuclear energy levels between the gamma emitter and absorber. The ^{129}I Mössbauer isomer shifts are proportional to the 5*s*-electron density at the nucleus. Some data for isomer shifts relative to the ZnTe sources are listed in Table 5-20. Give a reasonable interpretation of these isomer shifts based on electron configurations.

11. Optical pumping experiments [45] have been employed in a static alkali metal gas to cause population differences in the lower levels. The decay of the non-Boltzmann population distribution gives the collisional relaxation times and information about the mechanisms of collision. The reorientation times or relaxation times between the Zeeman energy levels of electronic ground state Rb ($^2S_{1/2}$) atoms have been measured in Ne, Ar, Kr, and Xe. The effective collision sizes or cross sections for the Rb atom relaxation in the various rare gases are Ne (5.2×10^{-23} cm^2), Ar (3.7×10^{-22} cm^2), Kr (5.9×10^{-21} cm^2), and

Table 5-20 Isomer shift data in a number of ^{129}I compounds relative to ZnTe given by D. W. Hafemeister, *Adv. Chem. Series* **68**, 126 (1967).

Compound	Isomer Shift, 10^{-6} eV
$KICl_4 \cdot H_2O$	0.35
IO_3 compounds	0.12
I_2	0.08
I^- halide crystals	−0.04
KIO_4	−0.21
$Na_3H_2IO_6$	−0.31

Xe (1.3×10^{-20} cm^2). The cross sections have approximately a Z^3-dependence. Thus, the perturbation causing the relaxation must have a Z^3-dependence. Suggest a model that gives a Z^3-dependence for the relaxation times.

12. We noted in Section 5.5 that the 632.8-nm Ne transition in He-Ne laser arises from a $5s'[\frac{1}{2}]_1 \longrightarrow 3p'[\frac{3}{2}]_2$ transition at 15,798.0 cm^{-1}. The Ar^+ ion laser has a strong transition at $\lambda = 514.5$ nm. What transition in the Ar^+ atom gives rise to this laser oscillation?

13. Consider the rotation of the OH^- ion that is present substitutionally as an impurity in the place of Cl^- in the KCl lattice. If we assume nearly free rotation with the OH^- center of mass at the lattice site, we can use the techniques described in Section 5.7 to show the nature of the perturbation of the original free rotor levels as a function of the crystalline electric field.

(a) Draw a diagram showing the nature of the crystal field perturbation from zero to very high crystal fields for the OH^-:KCl system above for the $J = 0, 1, 2,$ and 3 states. Use group theory to label the levels.

(b) Repeat part a for HF substituted for one of the Ar atoms in the Ar lattice.

14. The optical spectra of OsF_6 has been observed by W. Moffitt, G. L. Goodman, M. Fred, and B. Weinstock, *Mol. Phys.* **2**, 109 (1959). The electron configuration of the Os atom is [Xe] $4f^{14}5d^6 6s^2$ and the ground-state term for Os^{6+} is ^3F where we remember that the F-state atomic orbitals have odd symmetry.

(a) Predict the crystal field splitting of the ^3F atomic state of Os in the octahedrally symmetric OsF_6 molecule. Draw a diagram with the lowest degeneracies lowest in energy.

(b) Assuming that the spin-orbit interaction is a lower-order perturbation than the crystal field, draw a diagram to show the spin-orbit splittings.

(c) Sketch in all allowed electric dipole transitions from the ground state to all excited states.

(d) Sketch in all allowed magnetic dipole transitions from the ground state to all allowed excited states.

(e) Determine the symmetry species of the normal modes of vibration and show which modes are stretching and which are bending modes.

(f) List which normal modes of vibration could be observed in infrared and Raman experiments.

(g) What are the combinations of atomic orbitals that can form symmetry allowed hybrid orbitals centered at the Os and directed toward the F atoms?

15. Consider an ReF_5Cl molecule that has C_{4v} symmetry. The Re atom has an electronic configuration given by [Xe] $4f^{14}5d^5 6s^2$.

(a) Obtain the ground-state term symbol of the central (free) ion.

(b) Sketch out the crystal field splitting by first ignoring the difference between Cl and F and then show the effects due to the difference between the Cl and F atoms.

(c) What are the allowed hybrid atomic orbitals on the central atom?

(d) Determine the normal modes of vibration for this molecule and show which are stretching modes and which are bending modes. Show also which modes are Raman and infrared active.

16. Consider the $CrSO_4 \cdot 5H_2O$ salt. The Cr^{2+} ion is at the center of a square of four water molecules with two oxygen ions located above and below the square. The six oxygen atoms form approximately a regular octahedran.

The central field seen by the Cr^{2+} ion is predominantly octahedral (O_h) with a smaller component of tetragonal symmetry (D_{4h}). In addition, a distortion of the square of water molecules gives a still smaller rhombic (D_{2h}) crystal field component.

The chromous ion, Cr^{2+}, has a closed shell $3d^4$ electron configuration with a ground-state term of ^5D.

(a) Using the irreducible representations of the full rotation group, obtain the splitting of the ^5D state in the octahedral configuration above.

(b) Next, obtain the further splitting due to the tetragonal distortion.

(c) Next, obtain the splitting (which is small) from the rhombic distortion.

17. Consider a sandwich compound (similar to ferrocene) by placing V^{4+} between two cyclobutane skeletons. Consider the crystal field spectrum of the V^{4+} ion. Ignore spin-orbit interactions.

(a) What is the ground-state term of the V^{4+} ion?

(b) What are the possible configurations (symmetries) with which the two high temperature limit planar cyclobutane molecules could surround the spherically symmetric V^{4+} atom?

(c) Assuming we are starting with O_h symmetry from above, what are the possible distortions from O_h as the low temperature nonplanar cyclobutane molecule is obtained? Show the splittings under O_h in each of the possible cases by a correlation diagram with temperature on the horizontal axis.

18. Consider the motion of the impurity Li^+ ion trapped substitutionally in low concentration in place of K^+ atoms in the KCl lattice. The Li^+ is known to be at its lowest energy in one of the eight off-center points that make up the corners of an internal cube. Thus, we start with an eightfold degenerate ground state that describes the eight equivalent minima positions. Consider now the motion between the various minima positions that include (1) tunneling along the corner of the cube, (2) tunneling along the face diagonal of the cube, and (3) tunneling through the body diagonal of the cube.

(a) What will be the effect of tunneling in (1) on the original eight degenerate levels?

(b) Add the effects of tunneling in (2) to the results in part a. Draw an energy level diagram by assuming that the barrier in (2) is higher than in (1).

(c) Consider now the higher barrier tunneling in (3) and show the effect on the result in part b.

19. Our previous work involving ESR and the nuclear electronic Zeeman effect in molecules was appropriate only for liquids where all orientational information averages to simple scalar parameters. In general, the Hamiltonian is written in tensor form for a single nucleus and a single electron spin:

$$\mathscr{H} = \mu_B \boldsymbol{H} \cdot \mathbf{Cg} \cdot \boldsymbol{S} + \boldsymbol{S} \cdot \mathbf{A} \cdot \boldsymbol{I}. \tag{5-113}$$

The nuclear Zeeman effect is omitted for simplicity. $\boldsymbol{H}$ is the laboratory magnetic field, $\mathbf{g}$ is the molecular electronic g tensor; $\boldsymbol{S}$ and $\boldsymbol{I}$ are the electron spin and nuclear spin angular momenta, respectively; and $\mathbf{A}$ is the nuclear-electronic spin coupling constant. Equation (5-113) simplifies considerably if the $\mathbf{g}$ and $\mathbf{A}$ tensors are diagonal and if the system is cylindrically symmetric, which is common for situations where the free electron is in the region of a cylindrically symmetric bond or orbital. Under these circumstances where a is the cylindrical axis of symmetry, we use $\mathsf{A}_{aa} = \mathsf{A}_{\|}$, $\mathsf{A}_{bb} = \mathsf{A}_{cc} = \mathsf{A}_{\perp}$, $\mathsf{g}_{aa} = \mathsf{g}_{\|}$, $\mathsf{g}_{bb} = \mathsf{g}_{cc} = \mathsf{g}_{\perp}$, and $C_{az} = \cos\theta$ and $C_{bz} = \sin\theta$ for the direction cosines to give

$$\mathscr{H} = \mu_B H_z(\mathsf{g}_{\|} S_a \cos\theta + \mathsf{g}_{\perp} S_b \sin\theta) + \mathsf{A}_{\|} I_a S_a + \mathsf{A}_{\perp}(I_b S_b + I_c S_c). \quad (5\text{-}114)$$

We shall assume that the radical is fixed in space and that θ is a constant.

(a) Compute the high field first-order energy with the Hamiltonian in Eq. (5-114).

(b) Work out the complete matrix representation of $\mathscr{H}$ in Eq.(5-114) in the uncoupled basis for an $I = \frac{1}{2}$, $S = \frac{1}{2}$ system.

20. An F center in the KCl alkali halide lattice is a substitutional replacement (in small concentrations) of Cl^- ions with electrons. Predict the nature of the ESR spectrum of an F center electron at 10,000 G (^{39}K, $I = \frac{3}{2}$; ^{35}Cl, $I = \frac{3}{2}$).

21. The nuclear quadrupole coupling constants of a number of free atoms are listed in Table 5-21.

(a) Complete Table 5-21 by evaluating the atomic field gradients.

(b) Use Slater orbitals to calculate q_{av} for all the atoms in Table 5-21.

22. Consider the nuclear quadrupole coupling in a system without cylindrical symmetry.

(a) Evaluate the nonzero off-diagonal matrix elements in M of the Hamiltonian in Eq.(5-84) in the spherical harmonic basis and show that the off-diagonal elements depend directly on the deviation from electronic cylindrical symmetry through a $(\mathsf{q}_{XX})_{av} - (\mathsf{q}_{YY})_{av}$ dependence.

(b) Set up the complete matrix representation of $\mathscr{H}_1$ in Eq.(5-84) from the results above and in Eq.(5-90) for both the $I = 1$ and $I = \frac{3}{2}$ cases.

(c) Diagonalize the $I = 1$ and $I = \frac{3}{2}$ matrices above and compare the eigenvalues with the results in the literature [46].

(d) Draw a diagram similar to Fig. 5-15 for $I = \frac{3}{2}$ and show the effects of a nonzero value of $(\mathsf{q}_{XX})_{av} - (\mathsf{q}_{YY})_{av}$ on the energy levels with and without the external magnetic field.

Table 5-21 The nuclear quadrupole coupling constants of several free atoms from Ref. [36].

Atom	Electron Giving Rise to q_{av}	$2q_{av}Q/h$ (MHz)	q_{av} (SC · cm^{-3})
^{35}Cl	$3p$	+109.75	
^{79}Br	$4p$	−769.76	
^{127}I	$5p$	+2292.71	

23. Calculate the cohesive energy in CsCl using the methods worked out in Section 5.9. Remember that the CsCl crystal structure is different than NaCl.

24. Calculate the additional attractive energy in the NaCl lattice due to the induced dipole-induced dipole interaction. Ionic alkali and halide ionization potentials and polarizabilities can be found in J. R. Tessman, A. H. Kahn, and W. Shockley, *Phys. Rev.* **92**, 890 (1953).

REFERENCES

[1] H. Eyring, J. Walter, and G. E. Kimball, *Quantum Chemistry* (John Wiley & Sons, Inc., New York, 1944).

[2] J. C. Slater, *Phys. Rev.* **36**, 57 (1930).

[3] R. G. Parr, *The Quantum Theory of Molecular Electronic Structure* (W. A. Benjamin, Inc., New York, 1963).

[4] The experimental energies for atoms are in C. E. Moore, *Atomic Energy Levels*, Nat. Bur. Std. (U.S.) Circ. 467 (1949).

[5] L. D. Landau and E. M. Lifshitz, *Quantum Mechanics* (Pergamon Press, Oxford, 1974).

[6] J. C. Slater, *Phys. Rev.* **34**, 1293 (1929).

[7] A. S. Davydov, *Quantum Mechanics*, 2nd ed. (Pergamon Press, Oxford, 1965).

[8] The details in this manipulation can be found in Ref. [3] and in F. L. Pilar, *Elementary Quantum Chemistry* (McGraw-Hill Book Co., New York, 1968).

[9] Minimization procedures are discussed in D. Garten and B. T. Sutcliffe, *Theoretical Chemistry*, Vol. 1 (The Chemical Society, London, 1974).

[10] V. Fock, *Z. Physik* **61**, 126 (1930) and J. C. Slater, *Phys. Rev.* **35**, 210 (1930).

[11] C. F. Fischer, *The Hartree–Fock Method for Atoms* (Wiley Interscience, New York, 1977).

[12] D. R. Hartree, *The Calculation of Atomic Structures* (John Wiley & Sons, Inc., New York, 1957).

[13] E. Clementi and D. L. Raimondi, *J. Chem. Phys.* **38**, 2686 (1963) and E. Clementi, D. L. Raimondi, and W. P. Reinhart, *J. Chem. Phys.* **47**, 1300 (1967).

[14] A. Fröman, *Rev. Mod. Phys.* **32**, 317 (1960); H. T. Doyle, *Adv. in Atomic and Mol. Phys.*, Ed. by D. R. Barter and I. Esterman, **5**, 337 (Academic Press, New York, 1970).

[15] E. Clementi and C. Roetti, *At. Data Nucl. Data Tables* **14**, 177 (1974).

[16] H. F. Schaefer, *The Electronic Structure of Atoms and Molecules* (Addison-Wesley Pub. Co., Reading, Mass., 1972).

[17] J. L. Whitten, *J. Chem. Phys.* **44**, 359 (1966) and E. Clementi and D. R. Davis, *J. Comput. Phys.* **2**, 223 (1967).

[18] *Adv. Chem. Phys.*, Ed. by R. Lefebvre and C. Moser, **14** (John Wiley & Sons, New York, 1969).

[19] R. McWeeny and B. T. Sutcliffe, *Methods of Molecular Quantum Mechanics* (Academic Press, New York, 1969).

[20] A. Veillard and E. Clementi, *J. Chem. Phys.* **49**, 2415 (1968). S. Frage and G. Malli, *Many-Electron Systems: Properties and Interactions* (W. B. Saunders, Philadelphia, 1968).

[21] G. Herzberg, *Atomic Spectra and Atomic Structure* (Dover, New York, 1944).

[22] L. E. ORGEL, *An Introduction to Transition Metal Chemistry* (John Wiley & Sons, Inc., New York, 1960).

[23] C. KITTEL, *Introduction to Solid State Physics*, 4th ed. (John Wiley & Sons, Inc., New York, 1971).

[24] More details are available in R. J. MYERS, *Molecular Magnetism and Magnetic Resonance Spectroscopy* (Prentice-Hall, Inc., Englewood Cliffs, N.J., 1973).

[25] E. U. CONDON and G. H. SHORTLEY, *The Theory of Atomic Spectra* (Cambridge University Press, London, 1963).

[26] E. FERMI, *Z. Physik* **60**, 320 (1973). See also J. E. WERTZ and J. R. BOLTON, *Electron Spin Resonance* (McGraw-Hill Book Co., New York, 1972). A semiclassical derivation of Eq.(5-76) is given in C. P. SLICHTER, *Principles of Magnetic Resonance* (Harper & Row, New York, 1963) and see also R. A. FENELL, *Am. J. Phys.* **28**, 484 (1960).

[27] We shall not develop a detailed electron theory for the a coupling constant in molecules as determined from ESR spectra (see Section 4.9). Details of this theory can be found in R. MCWEENY, *Spins in Chemistry* (Academic Press, New York, 1970) and J. D. MEMORY, *Quantum Theory of Magnetic Resonance Parameters* (McGraw Hill Book Co., New York, 1968). These references use techniques similar to the perturbation methods developed here in Sections 6.8 and 6.9.

[28] H. G. DRICKAMER and C. W. FRANK, *Electronic Transitions and the High Pressure Chemistry and Physics of Solids* (Chapman and Hall, London, also Halsted-Wiley, New York, 1973).

[29] A summary of the energy levels of the transition metal ions in complexes with groups in the electrochemical series (see Table 5-13) is given by C. K. JØRGENSEN, *Absorption Spectra and Chemical Bonding in Complexes* (Pergamon Press, Oxford, 1962).

[30] H. MARGENAU and G. M. MURPHY, *The Mathematics of Physics and Chemistry* (D. Van Nostrand Co., New York, 1956).

[31] Only integral values of J are decomposed according to Table 5-14. Half-integral values require the concept of the crystal double group as described in M. TINKHAM, *Group Theory and Quantum Mechanics* (McGraw-Hill Book Co., New York, 1964).

[32] B. R. JUDD, *Operator Techniques in Atomic Spectroscopy* (McGraw-Hill Book Co., New York, 1963).

[33] J. S. GRIFFITH, *The Theory of Transition Metal Ions* (Cambridge University Press, Cambridge, 1961).

[34] C. J. BALLHAUSEN, *Ligand Field Theory* (McGraw-Hill Book Co., New York, 1962). J. N. MURRELL, S. F. A. KETTLE, and J. M. TEDDER, *Valence Theory* (John Wiley & Sons, Inc., New York, 1965). T. M. DUNN *Physical Chemistry, An Advanced Treatise*, Ed. by H. Eyring **5**, 205 (Academic Press, New York, 1970).

[35] J. O. RASMUSSEN, *Accts. Chem. Res.* **3**, 166 (1970).

[36] E. A. C. LUCKEN, *Nuclear Quadrupole Coupling Constants* (Academic Press, New York, 1969). E. SCHAMPP and P. J. BRAY, *Physical Chemistry, An Advanced Treatise*, Ed. by D. Henderson **4**, 522 (Academic Press, New York, 1970).

[37] A convenient list of Madelung constants is given in Q. C. JOHNSON and D. H. TEMPLETON, *J. Chem. Phys.* **34**, 2004 (1961).

[38] H. MARGENAU and N. R. KESTNER, *The Theory of Intermolecular Forces*, 2nd ed. (Pergamon Press, Inc. New York, 1971).

[39] D. W. HAFEMEISTER and W. H. FLYGARE, *J. Chem. Phys.* **43**, 795 (1965).

[40] See comparison of ρ-values and assignment of ρ and C in Eq.(5-101) in M. Born and K. Huang, *Dynamical Theory of Crystal Lattices* (Oxford Press, London, 1956).

[41] Y. S. Kim and R. G. Gordon, *Phys. Rev.* **B9**, 3548 (1974).

[42] See also F. London, *Trans. Faraday Soc.* **33**, 8 (1937).

[43] For more details on bonding in metals, see W. A. Harrison, *Physical Chemistry, An Advanced Treatise*, Ed. by H. Eyring **5**, 526 (Academic Press, New York, 1970).

[44] R. H. Herber, *Adv. Chem. Series* **68**, 1 (1967). G. K. Wertheim, *Mössbauer Effect* (Academic Press, New York, 1964).

[45] R. Bernheim, *Optical Pumping* (W. A. Benjamin, Inc., New York, 1965). W. Hopper, *Rev. Mod. Phys.* **44**, 169 (1972).

[46] T. P. Das and E. L. Hahn, *Nuclear Quadrupole Resonance Spectroscopy* (Academic Press, New York, 1958).

6

the electronic structure of molecules

6.1 THE GENERAL HAMILTONIAN, SEPARATION OF COORDINATES, AND THE FORCE THEOREMS

We shall now consider in detail the total energy of a molecule. We shall show the limitations in the separation of rotational, vibrational, and electronic energies and we shall examine the force concept in molecules.

The Total Energy

We start with the kinetic energy. Referring to Fig. 2-1 and the discussion of laboratory-based and c.m.-based coordinate systems, we can write the velocity of the kth particle (both electrons and nuclei) in a molecule with respect to a fixed-laboratory framework as

$$\boldsymbol{v}_k^0 = \dot{\boldsymbol{R}} + \boldsymbol{\omega} \times \boldsymbol{r}_k + \boldsymbol{v}_k, \tag{6-1}$$

where the superscript zero represents the origin of the coordinate system, which is fixed in the x, y, and z laboratory frame. $\boldsymbol{R}$ is the vector from the arbitrary origin in the fixed-laboratory frame to the center of mass (c.m.) of the system of particles, $\dot{\boldsymbol{R}}$ is the corresponding velocity vector of the c.m., and $\boldsymbol{\omega}$ is the angular frequency of the rotating coordinate system attached to the molecule (normally the principal inertial axis system) with respect to the fixed-laboratory frame. $\boldsymbol{r}_k$ is the vector from the center of mass to the kth particle, and $\boldsymbol{v}_k$ is the velocity of the kth particle (with respect to the c.m.) in the rotating coordinate system. The kinetic energy of the system of particles with mass, m_k, is

$$T = \frac{1}{2}\sum_k m_k \boldsymbol{v}_k^0 \cdot \boldsymbol{v}_k^0 = \frac{M}{2}\dot{R}^2 + \frac{1}{2}\sum_k m_k(\boldsymbol{\omega} \times \boldsymbol{r}_k)\cdot(\boldsymbol{\omega} \times \boldsymbol{r}_k) + \frac{1}{2}\sum_k m_k v_k^2$$

$$+ \dot{\boldsymbol{R}}\cdot\boldsymbol{\omega} \times \left(\sum_k m_k \boldsymbol{r}_k\right) + \dot{\boldsymbol{R}}\cdot\left(\sum_k m_k \boldsymbol{v}_k\right) + \boldsymbol{\omega}\cdot\left(\sum_k m_k \boldsymbol{r}_k \times \boldsymbol{v}_k\right), \tag{6-2}$$

where we have also used $\sum_k m_k = M$, the total molecular mass. Now, from the definition of the c.m., $\sum_k m_k \boldsymbol{r}_k = 0$, and noting that the c.m. remains constant for all internal motions of the molecule, we can also imply that $\sum_k m_k \boldsymbol{v}_k = 0$. Substituting into Eq.(6-2) gives

$$T = \frac{M}{2}\dot{R}^2 + \frac{1}{2}\sum_k m_k(\boldsymbol{\omega} \times \boldsymbol{r}_k)\cdot(\boldsymbol{\omega} \times \boldsymbol{r}_k) + \frac{1}{2}\sum_k m_k v_k^2 + \boldsymbol{\omega}\cdot\left(\sum_k m_k \boldsymbol{r}_k \times \boldsymbol{v}_k\right). \tag{6-3}$$

We shall now write the third and fourth terms of Eq.(6-3) in terms of separate sums over electrons, i, and nuclei, α, given by

$$\frac{1}{2}\sum_k m_k v_k^2 = \frac{1}{2}\sum_i m_i v_i^2 + \frac{1}{2}\sum_\alpha m_\alpha v_\alpha^2 \tag{6-4}$$

$$\boldsymbol{\omega}\cdot\left(\sum_k m_k \boldsymbol{r}_k \times \boldsymbol{v}_k\right) = \boldsymbol{\omega}\cdot m\left(\sum_i \boldsymbol{r}_i \times \boldsymbol{v}_i\right) + \boldsymbol{\omega}\cdot\left(\sum_\alpha m_\alpha \boldsymbol{r}_\alpha \times \boldsymbol{v}_\alpha\right). \tag{6-5}$$

m_α is the nuclear mass and m is the electron mass. Defining the displacement vector of the αth nucleus, $\boldsymbol{s}_\alpha$, relative to an equilibrium position in the rotating coordinate system, $\boldsymbol{a}_\alpha$, by $\boldsymbol{s}_\alpha = \boldsymbol{r}_\alpha - \boldsymbol{a}_\alpha$; substituting this into the last term of Eq.(6-5); assuming small displacements; and

$$\boldsymbol{\omega}\cdot\left(\sum_\alpha m_\alpha \boldsymbol{a}_\alpha \times \boldsymbol{v}_\alpha\right) \cong 0$$

in the rotating coordinate system [1]† gives

$$\boldsymbol{\omega}\cdot\left(\sum_\alpha m_\alpha \boldsymbol{r}_\alpha \times \boldsymbol{v}_\alpha\right) \cong \boldsymbol{\omega}\cdot\left(\sum_\alpha m_\alpha \boldsymbol{s}_\alpha \times \boldsymbol{v}_\alpha\right). \tag{6-6}$$

† Numbers in square brackets correspond to the numbered sources found in the Reference section on p. 418 at the end of the chapter.

Substituting this into Eq.(6-5) and substituting the result and Eq.(6-4) into Eq.(6-3) gives (dropping the separable translational term, $M\dot{R}^2/2$)

$$T = T_r + T_e + T_n + T_{re} + T_{rn} = \frac{1}{2}\sum_k m_k(\boldsymbol{\omega} \times \boldsymbol{r}_k)\cdot(\boldsymbol{\omega} \times \boldsymbol{r}_k) + \frac{1}{2}\sum_i m_i v_i^2$$

$$+ \frac{1}{2}\sum_\alpha m_\alpha v_\alpha^2 + \boldsymbol{\omega}\cdot m\left(\sum_i \boldsymbol{r}_i \times \boldsymbol{v}_i\right) + \boldsymbol{\omega}\cdot\left(\sum_\alpha m_\alpha \boldsymbol{s}_\alpha \times \boldsymbol{v}_\alpha\right). \tag{6-7}$$

T_r is the classical kinetic energy for rotational motion (Section 2.1); T_e and T_n are the electronic and nuclear kinetic energies, respectively; T_{re} is the electronic-rotational coupling, and T_{rn} is the rotational-vibrational coupling term sometimes called the *coriolis interaction* [2].

The potential energy of an isolated molecule is given by

$$V = \sum_{i>j}\frac{e^2}{r_{ij}} + \sum_{\alpha>\beta}\frac{Z_\alpha Z_\beta e^2}{r_{\alpha\beta}} - \sum_{i,\alpha}\frac{Z_\alpha e^2}{r_{i\alpha}} = V_{ee} + V_{nn} + V_{en}. \tag{6-8}$$

V_{ee} and V_{nn} are the electron–electron and nuclear–nuclear potential energies and V_{en} is the electron–nuclear potential energy. The sum over i and j is over electrons and the sum over α and β is over nuclei with atomic numbers Z_α and Z_β, respectively. Adding the kinetic and potential energies gives the complete Hamiltonian:

$$\begin{aligned}\mathscr{H} &= (T_e + V_{ee} + V_{en}) + (T_n + V_{nn}) + T_r + T_{re} + T_{rn} \\ &= \mathscr{H}_e + \mathscr{H}_n + \mathscr{H}_r + \mathscr{H}_{re} + \mathscr{H}_{rn}\end{aligned} \tag{6-9}$$

with terms defined in Eqs.(6-7) and (6-8). $\mathscr{H}_r$ is the rotational term that was discussed in Sections 2.1 and 4.11. In our previous work on

$$\mathscr{H}_r = \frac{1}{2}\sum_k m_k(\boldsymbol{\omega} \times \boldsymbol{r}_k)\cdot(\boldsymbol{\omega} \times \boldsymbol{r}_k),$$

we summed k over atoms as though both the mass of the nuclei and the free atom electrons were located at the nuclear points defined by $\boldsymbol{r}_k$. In Section 6.8 we shall consider the $\mathscr{H}_r + \mathscr{H}_{re}$ parts of the Hamiltonian above in detail. We shall break $\mathscr{H}_r$ down into nuclear and electronic contributions and show that the sum over atoms (as noted above) is a good approximation. Furthermore, we shall examine in detail the rotational-electronic coupling in $\mathscr{H}_{re}$. The last term in Eq.(6-9), $\mathscr{H}_{rn}$, the vibration-rotation interaction term, is zero for a rigid rotor. In a vibrating system $\mathscr{H}_{rn}$ will affect both the vibrational and the rotational energy levels as discussed in detail in Problem 4-35.

Separation of Electronic and Nuclear Coordinates

We shall now discuss the conditions of separability for the nuclear and electronic coordinates through the solution of Schrödinger's equation for the $\mathscr{H}_e + \mathscr{H}_n$ Hamiltonian operator:

$$(\mathscr{H}_e + \mathscr{H}_n)\psi(r_i, r_\alpha) = E\psi(r_i, r_\alpha), \tag{6-10}$$

where $\psi(r_i, r_\alpha)$ indicates the state of the system that depends on both the electronic, r_i, and nuclear, r_α, coordinates. Substituting from Eq.(6-9) gives

$$(T_e + V_{ee} + V_{en} + T_n + V_{nn})\psi(r_i, r_\alpha) = E\psi(r_i, r_\alpha).$$

We seek a solution to this equation in the form of a product of an electronic function, $\psi_e(r_i, r_\alpha)$, and a nuclear function, $\psi_n(r_\alpha)$, given by

$$\psi(r_i, r_\alpha) = \psi_e(r_i, r_\alpha)\psi_n(r_\alpha), \qquad \mathscr{H}_e\psi_e(r_i, r_\alpha) = \varepsilon_e(r_\alpha)\psi_e(r_i, r_\alpha),$$
$$\mathscr{H}_n\psi_n(r_\alpha) = \varepsilon_n\psi_n(r_\alpha), \tag{6-11}$$

where both r_i and r_α dependence must be included in $\psi_e(r_i, r_\alpha)$ because of the V_{en} in $\mathscr{H}_e$ that depends on the coordinates of the nuclei. The electronic energy, $\varepsilon_e(r_\alpha)$, also depends on the nuclear coordinates for the same reason. We now seek the conditions under which Eq.(6-10) gives separate electronic and nuclear Schrödinger equations as shown in Eq.(6-11).

Substituting $\psi(r_i, r_\alpha) = \psi_e(r_i, r_\alpha)\psi_n(r_\alpha)$ into Eq.(6-10) gives

$$\begin{aligned}\mathscr{H}\psi_e(r_i, r_\alpha)\psi_n(r_\alpha) &= E\psi_e(r_i, r_\alpha)\psi_n(r_\alpha) = (\mathscr{H}_e + \mathscr{H}_n)\psi_e(r_i, r_\alpha)\psi_n(r_\alpha) \\ &= \psi_n(r_\alpha)\mathscr{H}_e\psi_e(r_i, r_\alpha) + \psi_e(r_i, r_\alpha)V_{nn}\psi_n(r_\alpha) + T_n\psi_e(r_i, r_\alpha)\psi_n(r_\alpha).\end{aligned} \tag{6-12}$$

Now if $T_n\psi_e(r_i, r_\alpha)\psi_n(r_\alpha) = \psi_e(r_i, r_\alpha)T_n\psi_n(r_\alpha)$ in this equation, the separate equations in Eqs.(6-11) follow directly. Expanding the $T_n\psi_n(r_i, r_\alpha)\psi_n(r_\alpha)$ term above and differentiating by parts gives

$$\begin{aligned}T_n\psi_e(r_i, r_\alpha)\psi_n(r_\alpha) &= -\frac{\hbar^2}{2}\sum_\alpha \frac{\nabla_\alpha^2}{m_\alpha}\psi_e(r_i, r_\alpha)\psi_n(r_\alpha) \\ &= -\frac{\hbar^2}{2}\sum_\alpha \frac{1}{m_\alpha}[\psi_e(r_i, r_n)\nabla_\alpha^2\psi_n(r_\alpha) + \psi_n(r_\alpha)\nabla_\alpha^2\psi_e(r_i, r_\alpha) \\ &\quad + 2(\nabla_\alpha\psi_e(r_i, r_\alpha))\nabla_\alpha\psi_n(r_\alpha)].\end{aligned} \tag{6-13}$$

Thus, separability to Eqs.(6-11) is reduced to dropping the last two terms in the brackets in Eq.(6-13). We can justify dropping these terms by noting that

$$\nabla_\alpha\psi_n(r_\alpha) \gg \nabla_\alpha\psi_e(r_i, r_\alpha). \tag{6-14}$$

This inequality is evident by realizing that the nuclei are highly localized in the molecule and the electrons are distributed throughout the molecule. Roughly the $\nabla_\alpha\psi_n(r_\alpha)/\nabla_\alpha\psi_e(r_i, r_\alpha)$ ratio from Eq.(6-14) will be on the order of a typical nuclear displacement (during a normal mode of vibration) to a molecular bond length. These ratios are on the order of 10^{-2} (see Fig. 4-30 for an estimate of 10^{-10} cm as a displacement in HCl). As a result of these discussions, we drop the last two terms in the brackets in Eq.(6-13) and substitute the result into Eq.(6-12) to give [3]

$$\mathscr{H}\psi_e(r_i, r_\alpha)\psi_n(r_\alpha) = E\psi_e(r_i, r_\alpha)\psi_n(r_\alpha) \cong \psi_n(r_\alpha)\mathscr{H}_e\psi_e(r_i, r_\alpha) + \psi_e(r_i, r_\alpha)\mathscr{H}_n\psi_n(r_\alpha). \tag{6-15}$$

We now fix the nuclear coordinates in the molecule and solve the many-electron Schrödinger equation given by

$$\mathscr{H}_e \psi_e(r_i, r_\alpha) = \varepsilon_e(r_\alpha)\psi_e(r_i, r_\alpha). \tag{6-16}$$

Substituting Eq.(6-16) into Eq.(6-15) gives

$$\psi_n(r_\alpha)\varepsilon_e(r_\alpha)\psi_e(r_i, r_\alpha) + \psi_e(r_i, r_\alpha)\mathscr{H}_n\psi_n(r_\alpha) \cong E\psi_e(r_i, r_\alpha)\psi_n(r_\alpha). \tag{6-17}$$

Canceling $\psi_e(r_i, r_\alpha)$ from both sides of this equation gives

$$\mathscr{H}_n\psi_n(r_\alpha) + \varepsilon_e(r_\alpha)\psi_n(r_\alpha) \cong E\psi_n(r_\alpha), \qquad [T_n + V_{nn} + \varepsilon_e(r_\alpha)]\psi_n(r_\alpha) \cong E\psi_n(r_\alpha), \tag{6-18}$$

where E in this equation is called the Born–Oppenheimer energy (see Problem 6-1). $V_{nn} + \varepsilon_e(r_\alpha)$ is the total effective nuclear potential energy in the presence of the electrons that we have approximated previously with a harmonic potential energy.

In summary, the many-electron problem in molecules is treated by solving Eq.(6-16) parametrically as a function of the nuclear coordinates. The energy, $\varepsilon_e(r_\alpha)$, should go through a minimum at the equilibrium molecular structure. $\varepsilon_e(r_\alpha)$ then provides the electronic contribution to the potential energy function for nuclear displacement; the total potential energy function for nuclear displacement is $V_{nn} + \varepsilon_e(r_\alpha)$ and the resulting solution of Eq.(6-18) gives the total energy including the nuclear vibrations. Another way of viewing the Born–Oppenheimer approximation is suggested here. Due to the very fast motion of the electrons in the molecule, the electrons respond rapidly to a nuclear displacement, providing at all displacements a nuclear potential energy given by $\varepsilon_e(r_\alpha)$ in the solution of Eq.(6-16). In the rest of this chapter we shall concentrate on the solution of Eq. (6-16) for the electronic energy for specific molecular structures. First, however, we examine some molecular force theorems.

Force Theorems

At the equilibrium molecular structure, the net force on the nuclei, which is the negative gradient of the potential energy with respect to a displacement s, is zero. Returning to Eq.(6-18) for the potential energy, $V_{nn} + \varepsilon_e(r)$, we write

$$-\frac{\partial}{\partial s}[V_{nn} + \varepsilon_e(r)] = 0$$

$$-\frac{\partial V_{nn}}{\partial s} = \frac{\partial}{\partial s}\varepsilon_e(r) = \frac{\partial}{\partial s}\int \psi_e^*(T_e + V_{ee} + V_{en})\psi_e\,dV$$

$$= \int \frac{\partial \psi_e^*}{\partial s}(\mathscr{H}_e)\psi_e\,dV + \int \psi_e^*\left(\frac{\partial \mathscr{H}_e}{\partial s}\right)\psi_e\,dV + \int \psi_e^*(\mathscr{H}_e)\frac{\partial \psi_e}{\partial s}\,dV, \tag{6-19}$$

where we have also used $\varepsilon_e(r_\alpha) = \int \psi_e^*(\mathscr{H}_e)\psi_e\,dV$ from Eq.(6-16). Remembering that $\mathscr{H}_e$ is Hermitian, we can use

$$\int \psi_e^*(\mathscr{H}_e)\frac{\partial \psi_e}{\partial s}\,dV = \int \frac{\partial \psi_e}{\partial s}(\mathscr{H}_e)\psi_e^*\,dV$$

to write

$$\int \frac{\partial \psi_e^*}{\partial s} (\mathscr{H}_e)\psi_e \, dV + \int \psi_e^*(\mathscr{H}_e) \frac{\partial \psi_e}{\partial s} \, dV = \varepsilon_e(r)\left(\int \frac{\partial \psi_e^*}{\partial s} \psi_e \, dV + \int \psi_e^* \frac{\partial \psi_e}{\partial s} \, dV \right)$$

$$= \varepsilon_e(r) \frac{\partial}{\partial s} \int \psi_e^* \psi_e \, dV = 0,$$

which gives (using $\partial \mathscr{H}_e / \partial s = \partial V_{en} / \partial s$)

$$-\frac{\partial V_{nn}}{\partial s} = \int \psi_e^* \left(\frac{\partial V_{en}}{\partial s} \right) \psi_e \, dV_e. \tag{6-20}$$

Equation (6-20) states that the force on a nucleus in a molecule due to all the other nuclei is just canceled by the average value of the force on that nucleus due to all the electrons in the molecule. This is one result of the more general Hellmann–Feynman theorem [4]. The result in Eq.(6-20) is a useful theorem for testing approximate molecular electronic wavefunctions (see summary of the electronic and nuclear forces in OCS in Table 6-19).

Another development related to the discussion above is the calculation of energy differences in different molecular conformations (or other isoelectronic energy differences) by the *integral* Hellmann–Feynman theorem. Consider the ground electronic states for the staggered, ψ_s, and eclipsed, ψ_{ec}, forms of the ethane molecule (see Fig. 4-3 for a typical potential function). ψ_s and ψ_{ec} represent the electronic eigenfunctions for the electronic Hamiltonians ($T_e + V_{ee} + V_{en}$) in the two forms. The Schrödinger equations are

$$\mathscr{H}_s \psi_s = E_s \psi_s \tag{6-21}$$

$$\mathscr{H}_{ec} \psi_{ec} = E_{ec} \psi_{ec}. \tag{6-22}$$

Multiplying Eq.(6-21) by ψ_{ec}^* and Eq.(6-22) by ψ_s^*, integrating, and rearranging gives

$$E_s = \frac{\int \psi_{ec}^* \mathscr{H}_s \psi_s \, dV}{\int \psi_{ec}^* \psi_s \, dV}, \qquad E_{ec} = \frac{\int \psi_s^* \mathscr{H}_{ec} \psi_{ec} \, dV}{\int \psi_s^* \psi_{ec} \, dV}. \tag{6-23}$$

Remembering that $\mathscr{H}_s$ and $\mathscr{H}_e$ are real Hermitian operators, we can write the difference in electronic energies between the conformers as

$$E_{ec} - E_s = \frac{(\int \psi_s^* \mathscr{H}_{ec} \psi_{ec} \, dV)^* - \int \psi_{ec}^* \mathscr{H}_s \psi_s \, dV}{\int \psi_{ec}^* \psi_s \, dV} = \frac{\int \psi_{ec}^* (\mathscr{H}_{ec} - \mathscr{H}_s) \psi_s \, dV}{\int \psi_{ec}^* \psi_s \, dV}. \tag{6-24}$$

Furthermore, it is clear that T_e and V_{ee} are identical for the two conformers and therefore the difference in $\mathscr{H}_e - \mathscr{H}_s$ must arise from differences in V_{en}, call them ΔV_{en}. This gives the electronic energy difference of

$$E_{ec} - E_s = \frac{\int \psi_{ec}^* (\Delta V_{en}) \psi_s \, dV}{\int \psi_{ec}^* \psi_s \, dV}. \tag{6-25}$$

Subtracting this electronic energy difference from the nuclear–nuclear potential energy difference between the two conformers, ΔV_{nn}, gives the total energy difference between the conformers;

$$\Delta E = \Delta V_{nn} - \frac{\int \psi_{ec}^* (\Delta V_{en}) \psi_s \, dV}{\int \psi_{ec}^* \psi_s \, dV}. \tag{6-26}$$

We can easily compute ΔV_{nn} in ethane, which is just the difference in the proton–proton repulsions in the staggered and eclipsed forms. The result is (use tetrahedral angles, $d_{CC} = 1.543$ Å and $d_{CH} = 1.100$ Å)

$$\Delta V_{nn} = (V_{nn})_s - (V_{nn})_{ec} = 5068 \text{ cal} \cdot \text{mol}^{-1}.$$

Thus, ΔV_{nn} predicts that the eclipsed form is at a higher energy by 5068 cal · mol^{-1} than the corresponding staggered form. Experimentally, it is found that the eclipsed form is higher than the staggered form by only 2930 cal · mol^{-1} (see Table 4-2). Thus, the second term in Eq.(6-26) must be equal to 2138 cal · mol^{-1}. Conceptually it is convenient to think that the origin of the barrier in ethane arises from the proton–proton repulsions; the proton repulsions are shielded by the presence of the electrons, which lead to a smaller energy difference than the pure proton–proton repulsions would predict [5a]. This concept of repulsions of shielded nuclei can also be extended to heavier nuclei. The nature of the origin to the barrier to internal rotation has also been examined in terms of changes in the kinetic energy [5b]; however, this analysis should be equivalent to the potential energy analysis discussed above, according to the virial theorem.

Returning to Eq.(6-26), we note that in order to calculate energy differences between conformers we need the exact many-electron functions for the conformers because Eq.(6-26) is not protected by the variational method.

According to the variational theorem, the electronic energy difference between conformers can be computed by

$$\Delta E = \frac{\int \psi_s^* \mathscr{H}_s \psi_s \, dV}{\int \psi_s^* \psi_s \, dV} - \frac{\int \psi_{ec}^* \mathscr{H}_{ec} \psi_{ec} \, dV}{\int \psi_{ec}^* \psi_{ec}^* \, dV}. \tag{6-27}$$

ψ_{ec} and ψ_s are now the best variational functions. The work involved in computing the energies in Eq.(6-27) is considerably larger than in Eq.(6-26). Even though Eq.(6-26) appears to involve less work, however, one is still bound to the variational theorem (or perturbation theory) to obtain a good wavefunction.

The general expression for the Hellmann–Feynman theorem is obtained by straightforward extensions of the arguments beginning in Eq.(6-19). Let λ be some variable in the Hamiltonian: $\mathscr{H}(\lambda)$. In Eq.(6-19) the variable λ was the displacement s. Now, the change in the energy with respect to the variable λ is given by

$$\begin{aligned}
\frac{\partial E}{\partial \lambda} &= \frac{\partial}{\partial \lambda} \int \psi^* \mathscr{H}(\lambda) \psi \, dV \\
&= \int \frac{\partial \psi^*}{\partial \lambda} \mathscr{H} \psi \, dV + \int \psi^* \frac{\partial \mathscr{H}}{\partial \lambda} \psi \, dV + \int \psi^* \mathscr{H} \frac{\partial \psi}{\partial \lambda} dV \\
&= E \int \frac{\partial \psi^*}{\partial \lambda} \psi \, dV + \int \psi^* \frac{\partial \mathscr{H}}{\partial \lambda} \psi \, dV + E \int \psi^* \frac{\partial \psi}{\partial \lambda} dV \\
&= E \frac{\partial}{\partial \lambda} \int \psi^* \psi \, dV + \int \psi^* \frac{\partial \mathscr{H}}{\partial \lambda} \psi \, dV \\
&= \int \psi^* \frac{\partial \mathscr{H}(\lambda)}{\partial \lambda} \psi \, dV,
\end{aligned} \tag{6-28}$$

where we have used the Hermitian properties of $\mathscr{H}(\lambda)$. Equation (6-28) is the more general statement of the Hellmann–Feynman theorem.

The force theorems (Hellmann–Feynman and the related virial theorems) have been used in interpreting the nature of chemical bonding [6] and in interpreting experimentally determined vibrational potential constants [7]. Further discussion of virial theorems and force theorems and their relation to perturbation and variation theories, is also available [8].

In the next six sections we shall examine models for the solution to Eq.(6-16) for a general many-electron molecule with fixed nuclear positions.

6.2 LCAO-MO IN DIATOMIC MOLECULES

In studies of electronic structure in atoms, the one-electron orbital proved to be a very useful concept. We shall also apply this concept on one-electron *molecular orbitals* (MOs) in molecules by using the H_2^+ one-electron system as a model case for diatomic molecules. The Schrödinger equation for H_2^+ can be solved directly to give the electronic energy as a function of internuclear distance [9]. The other H_2^+ molecular properties can be computed from the molecular wavefunction. We begin here, however, by developing the technique of using *linear combinations of atomic orbitals* (LCAOs) to give the MOs. This one-electron orbital concept can be easily extended to a many-electron system in a manner similar to the development on atoms in Chapter 5.

The H_2^+ Molecule Ion

The H_2^+ molecule ion has historically occupied a very important position in molecular orbital theory as the molecular orbitals developed in this case have served as a model for a diatomic molecular building-up principle similar to the role the hydrogen atom has played in atomic structure [10]. The Hamiltonian for the H_2^+ molecule is

$$\mathscr{H} = -\frac{\hbar^2}{2m}\nabla^2 - \frac{e^2}{r_a} - \frac{e^2}{r_b} + \frac{e^2}{R}, \tag{6-29}$$

where r_a and r_b are the distances from nuclei a and b to the electron and R is the internuclear distance.

The molecular orbitals for H_2^+ must reflect equal probability of the electron being on either atom. Assuming that molecules have electronic shapes that reflect the atoms in the molecule, it is reasonable to write the MO as a LCAO on the two atomic centers in the molecule. Thus, we write the normalized molecular orbital Φ for H_2^+ as a linear combination of atomic orbitals on centers a and b, χ_a and χ_b,

$$\Phi_\pm = \frac{\chi_a \pm \chi_b}{(2 \pm 2S_{ab}^{\chi})^{1/2}}, \tag{6-30}$$

where χ_a and χ_b are normalized, but not necessarily orthogonal. We shall find later that the plus sign leads to molecular *bonding* and the minus sign leads to molecular *antibonding.* $(2 \pm 2S_{ab})$, the normalization, is easily verified where

$$S_{ab} = \int \chi_a^* \chi_b \, dV$$

is the *overlap integral.* The energy of the ground state of H_2^+ at a fixed value of R is determined from the variation theorem:

$$E_{\text{True}} \leq W = \int \Phi^* \mathscr{H} \Phi \, dV$$

$$W_{\pm} = \frac{\int (\chi_a \pm \chi_b)^* [-(\hbar^2/2m)\nabla^2 - e^2/r_a - e^2/r_b + e^2/R](\chi_a \pm \chi_b) \, dV}{2(1 \pm S_{ab})}$$

$$= \frac{\mathscr{H}_{aa} + \mathscr{H}_{bb} \pm 2\mathscr{H}_{ab}}{2(1 + S_{ab})}. \tag{6-31}$$

We now must make our choice of χ_a and χ_b. Certainly the most reasonable starting point will be to use free hydrogen atom $1s$ atomic orbitals, which allows analytic evaluation of the integrals:

$$\mathscr{H}_{aa} = \int \chi_a^* \left(-\frac{\hbar^2}{2m} \nabla^2 - \frac{e^2}{r_a} \right) \chi_a \, dV - e^2 \int \frac{\chi_a^* \chi_a}{r_b} \, dV + \frac{e^2}{R} \int \chi_a^* \chi_a \, dV$$

$$= -\frac{e^2}{2a_0} - e^2 \int \frac{\chi_a^* \chi_a}{r_b} \, dV + \frac{e^2}{R}. \tag{6-32}$$

The matrix element of the $[-(\hbar^2/2m)\nabla^2 - (e^2/r_a)]$ operator in the χ_a basis is from the hydrogen atom results in Chapter 4. $\int (\chi_a^* \chi_a / r_b) \, dV$ is a two-center integral (functions at atom a and operator at atom b) that can be evaluated by transforming to elliptical coordinates to give [11]

$$\int \frac{\chi_a^* \chi_a}{r_b} \, dV = -\left(\frac{1}{a_0} + \frac{1}{R} \right) \exp\left(-\frac{2R}{a_0} \right) + \frac{1}{R}. \tag{6-33}$$

Substituting Eq.(6-33) into Eq.(6-32) gives

$$\mathscr{H}_{aa} = -\frac{e^2}{2a_0} + e^2 \left(\frac{1}{a_0} + \frac{1}{R} \right) \exp\left(-\frac{2R}{a_0} \right). \tag{6-34}$$

The other integrals are derived by similar methods [11]:

$$\mathscr{H}_{bb} = \mathscr{H}_{aa}$$

$$\mathscr{H}_{ab} = S_{ab} \left(-\frac{e^2}{2a_0} + \frac{e^2}{R} \right) - \frac{e^2}{a_0} \left(\frac{R}{a_0} + 1 \right) \exp\left(-\frac{R}{a_0} \right)$$

$$S_{ab} = \int \chi_a \chi_b \, dV = \left[\frac{R^2}{3(a_0)^2} + \frac{R}{a_0} + 1 \right] \exp\left(-\frac{R}{a_0} \right). \tag{6-35}$$

Substituting these equations into Eq.(6-31), rearranging, and setting $R/a_0 = r$ for atomic units of length gives

$$W_{\pm} = \frac{-(e^2/2a_0) + (e^2/a_0)[1 + (1/r)]\exp(-2r) \pm (e^2/a_0)\{[(r^2/3) + r + 1][-\tfrac{1}{2} + (1/r)] - (r + 1)\}\exp(-r)}{1 \pm [(r^2/3) + r + 1]\exp(-r)}. \tag{6-36}$$

The limit of Eq.(6-36) as r approaches infinity is $-e^2/2a_0$, which is the electronic energy of a hydrogen atom and a free proton.

Equation (6-36) is plotted as a function of r in Fig. 6-1 where $e^2/a_0 = 27.2116$ eV. W_+ has a minimum at $r = 2.5$ a.u. or $R = 1.32$ Å and the calculated dissociation energy, D, is 1.76 eV compared to the experimental values of $R = 1.058$ Å and $D = 2.79$ eV. W_+ gives a qualitative picture of the ground state of the hydrogen molecule ion. W_-, the *antibonding* state, does not have a minimum in energy. It is now clear that the $\pm$ signs in Eq.(6-30) refer to *bonding* and *antibonding* orbitals. In Problem 6-2 we examine the bonding values of the kinetic and potential energies, T_+ and V_+, respectively, where $T_+ + V_+ = W_+$. We find a minimum in T_+ but not in V_+. Thus, the bonding in H_2^+ apparently arises from a stabilization of the kinetic energy and not a Coulomb stabilization in the potential energy (see Ref. [9] for further discussion).

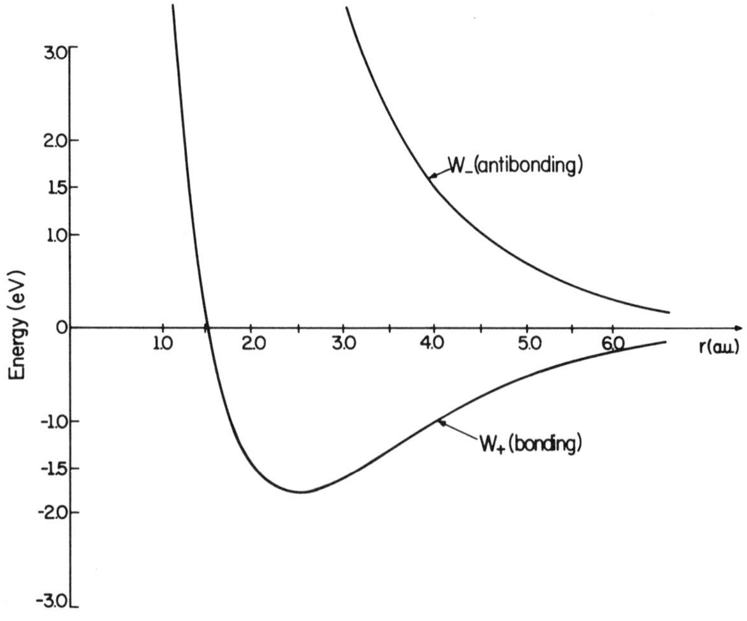

Figure 6-1 The bonding and antibonding energies in H_2^+ as a function of internuclear distance from Eq.(6-36). $W_{\pm} + e^2/2a_0$ is plotted here.

Table 6-1 LCAO-MOs for the H_2^+ diatomic. The z axis contains the nuclei.

Function	Function Shape		$H + H^+$ Energy	Eigenvalues of Operators L_ϕ	i	σ_d	σ_v	Symbol for Electron in the Configuration	MO
$1s_a + 1s_b$	⊕	⊕	$-e^2/2a_0$	0	1	1	1	$\sigma_g 1s$	$\Phi_{\sigma_g 1s}$
$1s_a - 1s_b$	⊕	⊖	$-e^2/2a_0$	0	-1	-1	1	$\sigma_u^* 1s$	$\Phi_{\sigma_u^* 1s}$
$2s_a + 2s_b$	⊕	⊕	$-e^2/8a_0$	0	1	1	1	$\sigma_g 2s$	$\Phi_{\sigma_g 2s}$
$2s_a - 2s_b$	⊕	⊖	$-e^2/8a_0$	0	-1	-1	1	$\sigma_u^* 2s$	$\Phi_{\sigma_u^* 2s}$
$2p_{xa} + 2p_{xb}$ or $2p_{ya} + 2p_{yb}$	⊕ over ⊖	⊕ over ⊖	$-e^2/8a_0$	1	-1	1	-1	$\pi_u 2p$	$\Phi_{\pi_u 2p}$
$2p_{xa} - 2p_{xb}$ or $2p_{ya} - 2p_{yb}$	⊕ over ⊖	⊖ over ⊕	$-e^2/8a_0$	1	1	-1	-1	$\pi_g^* 2p$	$\Phi_{\pi_g^* 2p}$
$2p_{za} + 2p_{zb}$	⊖⊕	⊖⊕	$-e^2/8a_0$	0	-1	-1	1	$\sigma_u^* 2p$	$\Phi_{\sigma_u^* 2p}$
$2p_{za} - 2p_{zb}$	⊖⊕	⊕⊖	$-e^2/8a_0$	0	1	1	1	$\sigma_g 2p$	$\Phi_{\sigma_g 2p}$
	a ——	b							

The concept of the separated atoms and the united atoms along the r coordinate in Fig. 6-1 is very useful where He^+ is the $r = 0$ result,

$$He^+ \xrightarrow{\text{molecule}} H + H^+. \tag{6-37}$$

In writing LCAOs to give the MOs for diatomic molecules, it is helpful to designate the eigenvalues of the symmetry and angular momentum operators to identify the functions (use the H_2^+ functions as a model). The operators are L_ϕ, the angular momentum operator about the axis joining the two nuclei; i, the inversion of the coordinates through the midpoint between the two nuclei; σ_d, the reflection of the electronic coordinates in a plane perpendicular to the line containing the two nuclei; and σ_v, the reflection in a plane containing the two nuclei. A number of the separated atom functions and their symmetry eigenvalues are shown in Table 6-1. The notation under the configuration column follows standard rules. The lowercase Greek letters are used to indicate the eigenvalues of the L_ϕ operator (in units of $\hbar$):

$$\begin{aligned} L_\phi \text{ eigenvalue} &= 0, 1, 2, \ldots \\ \text{Greek designate} &= \sigma, \pi, \delta, \ldots \end{aligned} \tag{6-38}$$

The g and u subscripts are used to denote the eigenvalues for i of $+1$ and -1, respectively. The asterisk is used for -1 eigenvalues of the σ_d operator. The configurations containing the asterisk lead to antibonding and the other configurations lead to bonding.

The ground-state configuration in H_2^+ is $\sigma_g 1s$ as discussed above and the symmetry properties are described in Table 6.1. The spectroscopic state is designated by

$$^{2S+1}(L_\phi)^{\pm}_{u \text{ or } g},$$

where L_ϕ is the capital Greek letter for eigenvalue of the L_ϕ operator, $L_\phi = 0(\Sigma)$, $1(\Pi)$, $2(\Delta), \ldots, \pm$ and u or g are from Table 6.1. Thus, the ground state for H_2^+ is $^2\Sigma_g^+$. Excitation from the $(\sigma_g 1s)^2\Sigma_g^+$ state in H_2^+ occurs in increasing energy increments according to the energies shown in Table 6.1 as summarized in the following ordering scheme:

$$\sigma_g 1s < \sigma_u^* 1s < \sigma_g 2s < \sigma_u^* 2s < \sigma_g 2p < \pi_u 2p < \pi_g^* 2p \lesssim \sigma_u^* 2p.$$

This scheme is the homonuclear diatomic molecule analog of the hydrogen atom energy level ordering shown in Fig. 5-2. Similar ordering diagrams are evident for the 3*s*, 3*p*, and 3*d* systems.

The Electronic Structure of the H_2 *Molecule*

We can now apply the concepts above in LCAO-MO theory on the H_2^+ molecule to the H_2 molecule. Consider the lowest energy configuration of electrons in H_2 where the two electrons would both be in the $\sigma_g 1s$ configuration (see Table 6-1) and the electrons would be paired to satisfy the Pauli principle. The appropriate Slater determinant for two electrons is given by (see Section 5.2)

$$\begin{aligned}
\psi(1, 2) &= \left(\frac{1}{2}\right)^{1/2} \begin{vmatrix} \Phi_{\sigma_g 1s}(1)\alpha(1) & \Phi_{\sigma_g 1s}(2)\alpha(2) \\ \Phi_{\sigma_g 1s}(1)\beta(1) & \Phi_{\sigma_g 1s}(2)\beta(2) \end{vmatrix} \\
&= \left(\frac{1}{2}\right)^{1/2} [\Phi_{\sigma_g 1s}(1)\Phi_{\sigma_g 1s}(2)][\alpha(1)\beta(2) - \alpha(2)\beta(1)] \\
&= \left(\frac{1}{2}\right)^{1/2} \left\{\frac{[\chi_{1s_a}(1) + \chi_{1s_b}(1)][\chi_{1s_a}(2) + \chi_{1s_b}(2)]}{2}\right\} [\alpha(1)\beta(2) - \alpha(2)\beta(1)] \\
&= \frac{1}{2(2)^{1/2}} [\chi_{1s_a}(1)\chi_{1s_a}(2) + \chi_{1s_a}(1)\chi_{1s_b}(2) + \chi_{1s_b}(1)\chi_{1s_a}(2) \\
&\quad + \chi_{1s_b}(1)\chi_{1s_b}(2)][\alpha(1)\beta(2) - \alpha(2)\beta(1)]. \qquad (6\text{-}39)
\end{aligned}$$

Equation (6-39) shows the separability of the spatial and spin components as determined previously for atoms in two-electron systems.

The electronic Hamiltonian for the H_2 molecule is given by (we also add the nuclear e^2/R term)

$$H = -\frac{\hbar^2}{2m}\nabla_1^2 - \frac{\hbar^2}{2m}\nabla_2^2 - \frac{e^2}{r_{1a}} - \frac{e^2}{r_{1b}} - \frac{e^2}{r_{2a}} - \frac{e^2}{r_{2b}} + \frac{e^2}{r_{12}} + \frac{e^2}{R}. \qquad (6\text{-}40)$$

Using the wavefunction in Eq.(6-39) with the free hydrogen atom AOs for χ_{1s_a} and χ_{1s_b} as in our previous work on H_2^+, we obtain the dissociation energy in H_2 of $D = 2.681$ eV and an equilibrium separation of $R = 0.850$ Å compared to the experimental results of $D = 4.75$ eV and $R = 0.740$ Å. Apparently the electronic repulsion energy is canceled by the increase in stability due to more electron density between the two positive nuclei.

We can now go through the same steps on improving the orbitals in the single determinant function as described in Section 5.1 for atoms. The first step is to use a

Table 6-2 Dissociation energies in H_2 calculated for single determinant LCAO-MOs from Ref. [12].

	Simple Hydrogen AOs		Hydrogen Atom AOs with Shielding		Hartree–Fock MOs		Experiment
D (eV)	2.68	$\longrightarrow$	3.49	$\longrightarrow$	3.62	$\longrightarrow$	4.75
R (Å)	0.850	$\longrightarrow$	0.732	$\longrightarrow$	0.740	$\longrightarrow$	0.740

shielding variational parameter in the AOs in the LCAO-MOs. Using the shielding parameter in H_2 leads to $D = 3.49$ eV and $R = 0.732$ Å, which is an impressive improvement over the previous result with the simple hydrogen AOs. The best orbital exponent in the AO in the MO is somewhat larger than the hydrogen atom value of 1.0.

The best single determinant function or numerical restricted Hartree–Fock molecular orbitals have also been determined for H_2 giving $D = 3.62$ eV using the experimental internuclear distance of $R = 0.74$ Å. These results are summarized in Table 6-2 [12] and should be compared to the corresponding table for He (Table 5-1).

We note that the correlation error is $4.75 - 3.62 = 1.13$ eV for the pair of electrons in H_2. This is nearly identical to the correlation error in the He pair of electrons.

We can now construct other states for the H_2 molecule by forming other molecular electronic configurations. The $(\sigma_g 1s)^2$ configuration leads to a singlet spin state as the two electrons are required to be paired by the Pauli principle. A possible configuration for the first excited state would be $(\sigma_g 1s)(\sigma_g 2s)$. The Slater determinants for this configuration are easily formed and the result leads to both a singlet and triplet state. A completely analogous situation was discussed previously in the case of the He atom with the triplet determinants in Eq.(5-60). Successive excitation in the H_2 molecule is evident from Table 6-1:

$$H_2(\sigma_g 1s)^2 \begin{cases} \xrightarrow{\text{singlet}} (\sigma_g 1s)(\sigma_u^* 1s) \longrightarrow (\sigma_g 1s)(\sigma_g 2s) \cdots \\ \xrightarrow{\text{triplet}} (\sigma_g 1s)(\sigma_u^* 1s) \longrightarrow (\sigma_g 1s)(\sigma_g 2s) \cdots \end{cases} \longrightarrow H_2^+ \quad (6\text{-}41)$$

Triplet states of a given configuration usually lie lower in energy than the corresponding states. Thus, the experience we have gained with atoms (Hund's rules) is still of use in interpreting the energy levels of molecules. A highly simplified diagram showing the progression in energy between H_2 and H_2^+ in the singlet state manifold is shown in Fig. 6-2 (many states are omitted in this figure for visual clarity).

Other Diatomic Molecules

The development above on H_2^+ and H_2 can be extended to the heavier diatomics by filling the electrons into a configuration (filling the lowest energy states first), according to the states shown in Table 6-1 as indicated in Eq.(6-39), for instance.

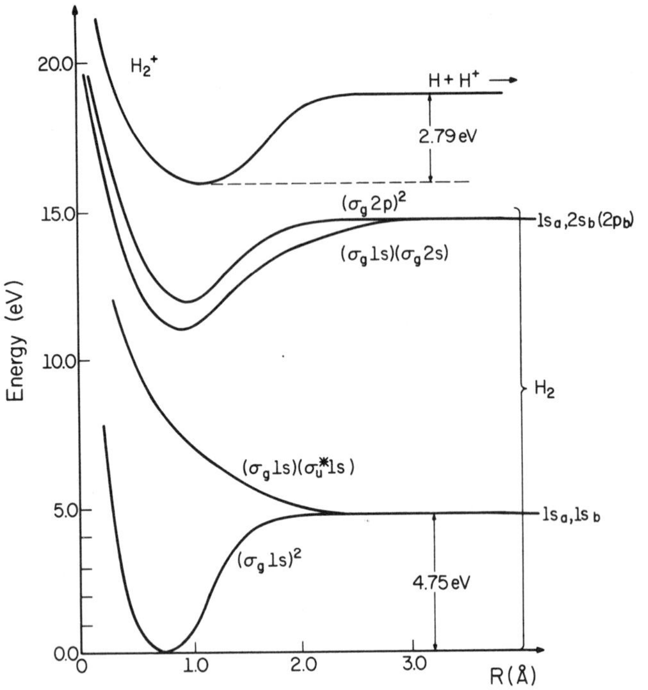

Figure 6-2 Energy level diagram showing some of the electronic states in H_2 and the lowest state of H_2^+ obtained by ionizing H_2. Only the singlet state manifold is shown. A more detailed diagram of the $H_2^- \longrightarrow H_2 \longrightarrow H_2^+$ system is given by T. Sharp, *Atomic Data* **2**, 119 (Academic Press, New York, 1971).

The ground-state electronic configurations, state designation, and other relevant information on the homonuclear diatomics in the first row of the periodic table are shown in Table 6-3. The strength of the bonds are reflected in the vibrational force constants and the force constants correlate well with the predicted bonding by the molecular orbital configurations. A more complete potential energy diagram for the $N_2 \longrightarrow N_2^+$ system is given in Fig. 6-3.

Heteronuclear molecules are treated in a manner similar to the treatment above on homonuclear molecules. Further details can be found elsewhere [10,13,14].

Electronic Transitions

Electronic transitions in diatomic molecules by the electric dipole mechanism are determined by the matrix element of the dipolar operator as given by

$$D_{fi} = \int \psi_f^* D \psi_i \, dV, \tag{6-42}$$

Table 6-3 Ground-state electronic configurations and states for homonuclear diatomic molecules in the first row of the periodic table as adapted from Ref. [14]. Also shown are the bonding electrons, N_b, the antibonding electrons, N_a, and the number of bonds defined by $\frac{1}{2}(N_b - N_a)$. The last column lists the vibrational force constants, which reflect the bond strength.

Molecule	Electronic Configuration	State	N_b	N_a	$\frac{1}{2}(N_b - N_a)$	k (10^5 dyne · cm^{-1})
H_2^+	$(\sigma_g 1s)$	$^2\Sigma_g^+$	1	0	$\frac{1}{2}$	1.56
H_2	$(\sigma_g 1s)^2$	$^1\Sigma_g^+$	2	0	1	5.60
He_2^+	$(\sigma_g 1s)^2(\sigma_u^* 1s)$	$^2\Sigma_g^+$	2	1	$\frac{1}{2}$	3.13
He_2	$(\sigma_g 1s)^2(\sigma_u^* 1s)^2$	$^1\Sigma_g^+$	2	2	0	—
Li_2	$[He_2]\,(\sigma_g 2s)^2$	$^1\Sigma_g^+$	2	0	1	0.25
Be_2	$[He_2](\sigma_g 2s)^2(\sigma_u^* 2s)^2$	$^1\Sigma_g^+$	2	2	0	—
B_2	$[Be_2](\pi_u 2p)^2$	$^3\Sigma_g^-$	4	2	1	3.60
C_2	$[Be_2](\pi_u 2p)^3(\sigma_g 2p)$	$^3\Pi_u$	6	2	2	9.55
N_2^+	$[Be_2](\pi_u 2p)^4(\sigma_g 2p)$	$^2\Sigma_g^+$	7	2	$\frac{5}{2}$	20.1
N_2	$[Be_2](\pi_u 2p)^4(\sigma_g 2p)^2$	$^1\Sigma_g^+$	8	2	3	23.1
O_2^+	$[N_2](\pi_g^* 2p)$	$^2\Pi_g$	8	3	$\frac{5}{2}$	16.6
O_2	$[N_2](\pi_g^* 2p)^2$	$^3\Sigma_g^-$	8	4	2	11.8
F_2	$[N_2](\pi_g^* 2p)^4$	$^1\Sigma_g^+$	8	6	1	4.45
Ne_2	$[N_2](\pi_g^* 2p)^4(\sigma_u^* 2p)^2$	$^1\Sigma_g^+$	8	8	0	—

where the f and i subscripts indicate the final and initial states, respectively. For illustration, we assume plane-polarized incident electric fields in the radiation and D is the electric dipole moment projected along the field axis given by

$$D = D_a \cos\theta, \tag{6-43}$$

where D_a is the internuclear axis in the diatomic molecule and θ is the angle between the radiation electric field axis and the molecular dipole axis. Now, as D_a is a sum over nuclear and electronic charges, we write

$$D_a = e\left(\sum_\alpha Z_\alpha a_\alpha - \sum_i a_i\right), \tag{6-44}$$

where e is the usual electronic charge, Z_α is the atomic number of the αth nucleus, a_α is the c.m. position along the internuclear axis of the αth nucleus, and a_i is the c.m. position of the ith electron in the molecule. The total wavefunction for the initial and final states is given by

$$\psi_i = \psi_e'(r_\alpha, r_i)\psi_n'(r_\alpha)Y_{J'M'}, \qquad \psi_f = \psi_e(r_\alpha, r_i)\psi_n(r_\alpha)Y_{JM}, \tag{6-45}$$

where $\psi_e(r_\alpha, r_i)$ and $\psi_n(r_\alpha)$ are the electronic and nuclear vibrational functions, respectively, as described in Section 6.1 [see Eq.(6-15) for instance] and Y_{JM} is the spherical harmonic rotational function for the linear molecule (diatomic) as discussed in Section 4.11. Substituting Eqs.(6-45), (6-44), and (6-43) into Eq.(6-42) gives

$$D_{fi} = \langle JM|\cos\theta|J'M'\rangle \int_i \int_\alpha \psi_e^*\psi_n^*(e)\left(\sum_\alpha Z_\alpha a_\alpha - \sum_i a_i\right)\psi_e'\psi_n'\,dV_i\,dV_\alpha, \tag{6-46}$$

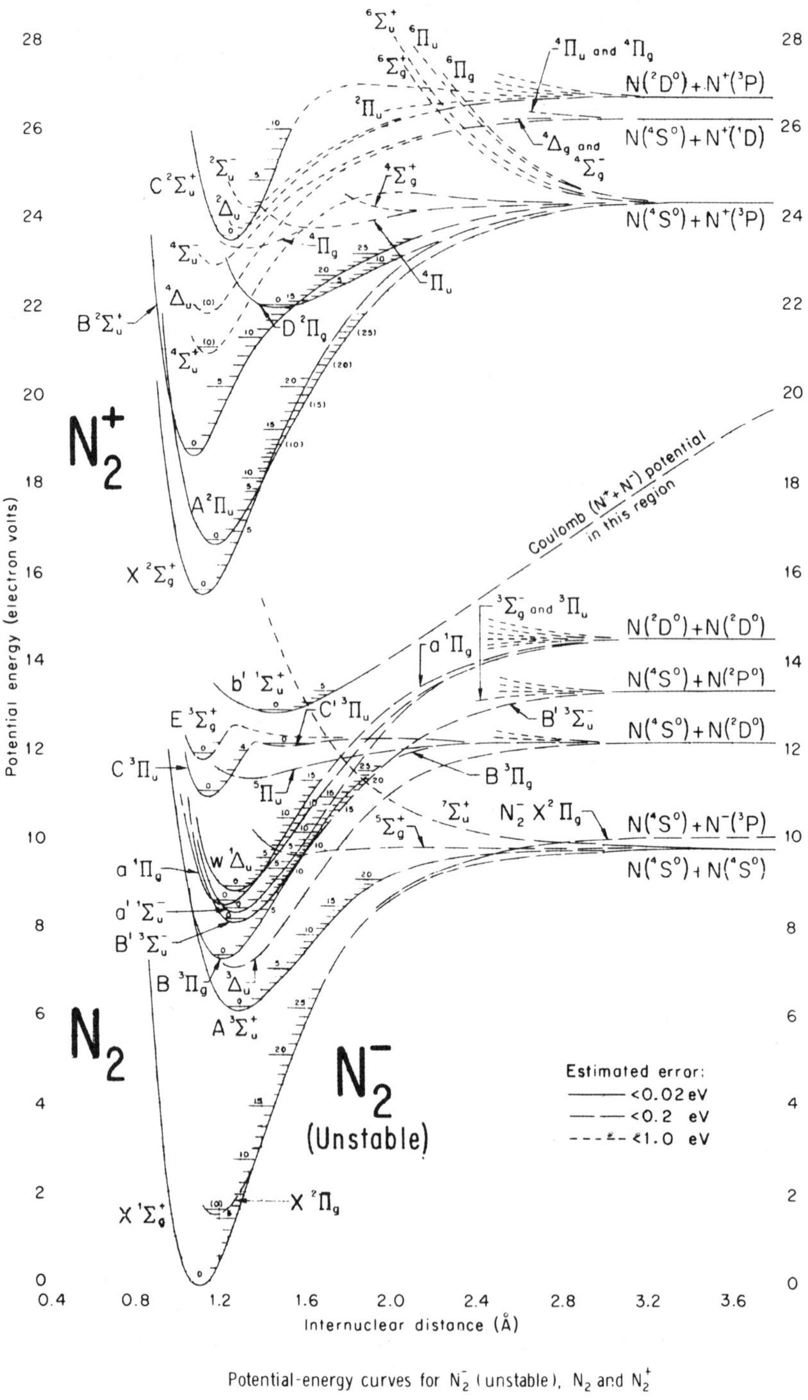

Figure 6-3 Potential energy level diagram for the $N_2 \longrightarrow N_2^+ + e^-$ system from F. R. Gilmore, *J. Quant. Spectrosc. Radiat. Transfer* **5**, 369 (1965).

where the integrals indicate integration over the electronic (i) and nuclear coordinates (α). The $\langle JM|\cos\theta|J'M'\rangle$ matrix elements were examined in detail in Section 4.11 and lead to $\Delta J = \pm 1$, $\Delta M = 0, \pm 1$ selection rules for the rotational transitions. Rewriting the remaining integral in Eq.(6-46) gives

$$\begin{aligned}\text{Int} &= e \int_i \int_\alpha \psi_e^* \psi_n^* \left(\sum_\alpha Z_\alpha a_\alpha - \sum_i a_i \right) \psi_e' \psi_n' \, dV_i \, dV_\alpha \\ &= e \int_\alpha \int_i \psi_e^* \psi_n^* \sum_\alpha Z_\alpha a_\alpha \psi_e' \psi_n' \, dV_i \, dV_\alpha - e \int_\alpha \int_i \psi_e^* \psi_n^* \sum_i a_i \psi_e' \psi_n \, dV_i \, dV_\alpha. \end{aligned} \quad (6\text{-}47)$$

Now, if the Born–Oppenheimer approximation is valid, we remember that ψ_e and ψ_n are separable and we can rewrite the equation above giving

$$\begin{aligned}\text{Int} = {} & e \int_i \psi_e^*(r_i, r_\alpha) \psi_e'(r_i, r_\alpha) \, dV_i \int_\alpha \psi_n^*(r_\alpha) \left(\sum_\alpha Z_\alpha a_\alpha \right) \psi_n'(r_\alpha) \, dV_\alpha \\ & - e \int_\alpha \psi_n^*(r_\alpha) \psi_n'(r_\alpha) \, dV_\alpha \int_i \psi_e^*(r_i, r_\alpha) \sum_i a_i \psi_e'(r_i, r_\alpha) \, dV_i. \end{aligned} \quad (6\text{-}48)$$

The first term in this equation is zero because $\psi_e(r_i, r_\alpha)$ and $\psi_e'(r_i, r_\alpha)$ represent two different electronic states that have orthogonal eigenfunctions because of the Hermitian nature of $\mathscr{H}_e$ in Eq.(6-16). Thus, we now rewrite Eq.(6-48) to give

$$\text{Int} = D_e \int_\alpha \psi_n^*(r_\alpha) \psi_n'(r_\alpha) \, dV_\alpha, \quad (6\text{-}49)$$

where D_e is the electronic dipole integral in Eq.(6-48) and $\psi_n(r_\alpha)$ and $\psi_n'(r_\alpha)$ are the vibrational functions for the excited and ground electronic states, respectively, at fixed r_α. Thus, the electronic transition intensity depends on the overlap in the vibrational functions in the ground and excited electronic states, which is referred to as the Franck–Condon factor [15]. Combining the results above gives the transition selection rules and intensities for the electronic transitions in a diatomic molecule. Further details on electronic spectroscopy are given elsewhere [16]. We shall now develop a more general many-electron theory for molecules.

6.3 *Ab initio* CALCULATIONS OF THE ELECTRONIC STRUCTURE OF MOLECULES; LCAO-MO-SCF AND LCAO-MO-SCF-CI

Ab initio methods described here and in Section 6.4 involve molecular calculations where a basis set of functions is chosen to describe the system and then the many-electron Schrödinger equation in Eq.(6-16) is solved exactly; all integrals are evaluated with no approximation. In the semiempirical methods described in Sections 6.5, 6.6, and 6.7, certain integrals are neglected or are estimated from previous calculations or experimental data. Of course, there are many models for a many-electron system that can be the starting point for an *ab initio* or semiempirical calculation. We shall now examine some of the models used to describe the many-electron structure of molecules.

The atomic orbital SCF method is given from Eqs.(5-23) through (5-39). We can carry that development over here for molecules by replacing the atomic orbitals with molecular orbitals where the core Hamiltonian $\mathscr{H}^0(i)$ [Eq.(5-23)] now contains an additional sum over the α nuclei in a molecule,

$$\mathscr{H} = -\sum_i \frac{\hbar^2}{2m}\nabla_i^2 - \sum_{\alpha,i}\frac{Z_\alpha e^2}{r_{\alpha i}} + \sum_{i>j}\frac{e^2}{r_{ij}} = \sum_i \mathscr{H}^0(i) + \sum_{i>j}\frac{e^2}{r_{ij}}. \tag{6-50}$$

The subscripts i and α indicate electrons and nuclei, respectively. All the language from Eq.(5-23) [replace Eq.(6-50)] to Eq.(5-39) can be carried over to molecules where now we recognize Φ_i as the ith spatial molecular orbital (MO). The numerical SCF calculation in Eq.(5-36) [replace $\mathscr{H}^0$ in Eq.(6-50) for $\mathscr{H}^0$ in Eq.(5-23)] is much more difficult, however, due to the loss of centrosymmetry in molecules. We noted in Chapter 5 that in atoms the variation in atomic orbitals to achieve the best set (Hartree–Fock) is usually executed only on the radial variable. This was a convenient simplification in atoms as the radial and angular components in the integrals are separable. We also discussed the concept of angular and radial correlation error. In molecules, this separation into radial and angular components is not possible and the molecular numerical SCF orbitals must necessarily be calculated and tabulated in an r, ϕ, and θ (or x, y, and z) grid (from the c.m.). Such a calculation would be very time-consuming and the output data would be voluminous.

An alternative and more tractable solution to the molecular SCF equations is to replace the molecular orbitals (MOs) with a *linear combination of atomic orbitals* (LCAOs) [17]. The atomic orbitals originate at the atomic centers in the molecule [this equation is analogous to Eq.(5-41) for atoms]:

$$\mathbf{\Phi} = \chi\mathbf{a} = (\Phi_1\Phi_2\Phi_3\cdots). \tag{6-51}$$

χ are the atomic orbitals and **a** contains the linear coefficients. The functions in χ are centered on the different nuclei in the molecule and therefore **a** is not necessarily a unitary matrix as χ are not necessarily orthonormal. In the case of butadiene ($CH_2{=}CH{=}CH{=}CH_2$) for instance, there are 30 electrons and 15 occupied molecular orbitals. A minimal basis of AOs contains all the AOs that are occupied or partially occupied in the separated free atoms. Therefore, the minimal basis in butadiene contains 26 AOs; six $1s$ orbitals from each of the six hydrogen atoms, and $1s$, $2s$, $2p_x$, $2p_y$, and $2p_z$ orbitals from each of the four carbon atoms. The analytical form for the AOs can be fixed by intuition and the experience we have gained with atoms and atomic structure as discussed in Chapter 5. Thus, we shall pick a fixed set of parameters (or adjust them later) for the AOs and perform the variation in $\mathbf{\Phi}$ by adjusting **a** to minimize the energy. The variation in constants, as described here, was also considered in Section 5.2 from Eq.(5-41) to the results in Eqs.(5-50) and (5-51). The iterative SCF scheme is then summarized in the three items following Eq.(5-51). All the language in atoms can again be transferred to the MOs considered here. For doubly occupied MOs, the total energy is given by the sum over occupied orbitals as given in Eq.(5-52). Before investigating the nature of the AOs in the LCAO-MO-SCF analysis, some appreciation for the nature of the integrals involved will be given.

For an arbitrarily complex polyatomic molecule, the integrals in the molecular Hartree–Fock matrix in Eq.(5-50) are given in terms of from one- to four-center integrals. A one-center (OC) integral is where the operator and both AOs are on the same center. A two-center integral (TC) is where any two of a combination of AOs and the operator are at different atomic centers in the molecule. The three-center (TrC) and four-center (FC) integrals are also defined in a similar manner.

One-Electron Integrals

1. Overlap; OC, TC

$$\int \chi_i(a)\chi_j(a)\, dV_a$$

OC if i and j are on the same atom.
TC if i and j are on different atoms.

2. Kinetic Energy; OC, TC

$$\int \chi_i(a)\left(\frac{\hbar^2}{2m}\nabla_a^2\right)\chi_j(a)\, dV_a$$

OC if i and j are on the same atom.
TC if i and j are on different atoms.

3. Nuclear Attraction; OC, TC, TrC

$$\int \chi_i(a)\left(\frac{Z_\alpha e^2}{r_{\alpha a}}\right)\chi_j(a)\, dV_a$$

OC if i and j are at α.
TC if i and j are on $\beta \neq \alpha$ or i is on β and j is on α.
TrC if i is on γ and j is on β and $\alpha \neq \beta \neq \gamma$.

Two-Electron Integrals $(1/r_{ab})$

1. Coulomb and Exchange Integrals

$$\int_a \int_b \chi_i(a)\chi_j(b)\left(\frac{e^2}{r_{ab}}\right)\chi_l(a)\chi_n(b)\, dV_a\, dV_b.$$

Various combinations of i, j, l, and m in this equation lead to OC, TC, TrC, and FC integrals [18].

We shall now discuss some of the progress that has been made in *ab initio* calculations of various degrees of sophistication. The term *ab initio* means solving the Schrödinger equation with no approximations in any integrals; however, the theory may still be limited. For instance, a minimal basis and an extended basis calculation on the same molecule are both *ab initio* calculations if no approximations are made in carrying out the formal theory but quite different results may arise from the two basis sets. First we shall examine some of the *ab initio* work within the single determinant LCAO-MO-SCF limitation. Then we shall examine some improvements obtained with configuration interaction (CI) methods.

After reviewing the AO basis set discussion for atoms in Section 5.2, we begin here with a discussion of the HF molecule, which has a minimal basis set of STOs of $1s_{\mathrm{H}}$, $1s_{\mathrm{F}}$, $2s_{\mathrm{F}}$, $2p_{x\mathrm{F}}$, $2p_{y\mathrm{F}}$, and $2p_{z\mathrm{F}}$. The orbital exponents can be chosen on the basis of the rules as derived for atoms. This procedure fixes $\boldsymbol{\chi}$ and allows a determination of the SCF energy and the coefficients in the LCAO-MO functions [$\boldsymbol{\Phi}$ in Eq.(6-51)]. However, the best orbital exponents as determined for atoms may not be appropriate

Table 6-4 LCAO-MO-SCF energies for HF from Ref. [12]. The Hartree–Fock limit is estimated.

STO Minimal Basis (6), Optimized orbital Exponents	Best Atom Hartree–Fock Atomic Orbitals	STO Extended Basis, Optimized Orbital Exponents	Hartree–Fock MOs
−99.49 (a.u.)	−99.96	−100.06	(−100.07)

for molecules. Thus, a better method would be to vary the orbital exponents systematically with the LCAO-MO-SCF procedure to minimize the energy. The best orbital exponents in the Slater atomic orbitals above for HF are $1s_H$ ($\zeta = 1.3163$), $1s_F$ ($\zeta = 8.6533$), $2s_F$ ($\zeta = 2.5551$), $2p_{zF}$ ($\zeta = 2.6693$), and $2p_{yF}$ and $2p_{xF}$ ($\zeta = 2.4965$) [19]. The best ζ values for the free F atom were given in Section 5.2 [$1s_F$ ($\zeta = 8.6501$), $2s_F$ ($\zeta = 2.5638$), and $2p_F$ ($\zeta = 2.5500$)]. The values for $1s_F$ are almost identical. The ζ-value for $2s_F$ is slightly larger for the atom, which means that the atomic orbital in the molecule is more effectively shielded. The $2p$ orbital along the internuclear axis (z) in the molecule is shielded less than the corresponding free atom and the $2p$ orbitals perpendicular to the HF bond are more effectively shielded than the free atom. The changes are small, however, and Slater orbitals apparently retain their shape in molecules. If the best atom Hartree–Fock SCF atomic orbitals are used instead of the Slater orbitals, a considerable improvement is achieved in the energy, the result being $E = -99.96$ a.u. compared to $E = -99.4914$ a.u. for the best Slater orbitals (see Table 6-4) [12]. It appears from this result that Hartree–Fock numerical AOs give a better description of the molecule (at least the energy) than the minimal basis of Slater orbitals with optimized orbital exponents. A discussion of the difference in shape between a Slater and SCF radial function is given in Section 5.2 (see Fig. 5-3).

A further improvement is achieved by going to a higher-order expansion in the Slater atomic orbital basis. Adding $2p_z$ on H and $3s$, $3p_x$, $3p_y$, $3p_z$, and $3d_{z^2}$ on F to the minimal basis in HF above gives an energy of $E = -100.06$ a.u. The resultant sequence of energies are shown in Table 6-4 (compare Tables 5-1 and 5-3 for atoms).

An interesting example of a minimal basis LCAO-MO-SCF calculation involves formaldehyde (see molecular structure and axis system in Fig. 6-4) using the unnormalized minimal basis AOs shown below:

Oxygen	*Carbon*	*Hydrogen*
$\chi_1(1s_O) = \exp(-\zeta_O r_O)$	$\chi_6(1s_C) = \exp(-\zeta_C r_C)$	$\chi_{11}(1s_{H_1}) = \exp(-\zeta_H r_{H_1})$
$\chi_2(2s_O) = r_O \exp[-\zeta'_O r_O]$	$\chi_7(2s_C) = r_C \exp(-\zeta'_C r_C)$	$\chi_{12}(1s_{H_2}) = \exp(-\zeta_H r_{H_2})$
$\chi_3(2p_{xO}) = x_O \exp(-\zeta'_O r_O)$	$\chi_8(2p_{x_C}) = x_C \exp(-\zeta'_C r_C)$	
$\chi_4(2p_{yO}) = y_O \exp(-\zeta'_O r_O)$	$\chi_9(2p_{y_C}) = y_C \exp(-\zeta'_C r_C)$	
$\chi_5(2p_{zO}) = z_O \exp(-\zeta'_O r_O)$	$\chi_{10}(2p_{z_C}) = z_C \exp(-\zeta'_C r_C)$	(6-52)

The orbital exponents are $\zeta_O = 7.7$, $\zeta'_O = 2.275$, $\zeta_C = 5.7$, $\zeta'_C = 1.625$, and $\zeta_H = 1.2$.

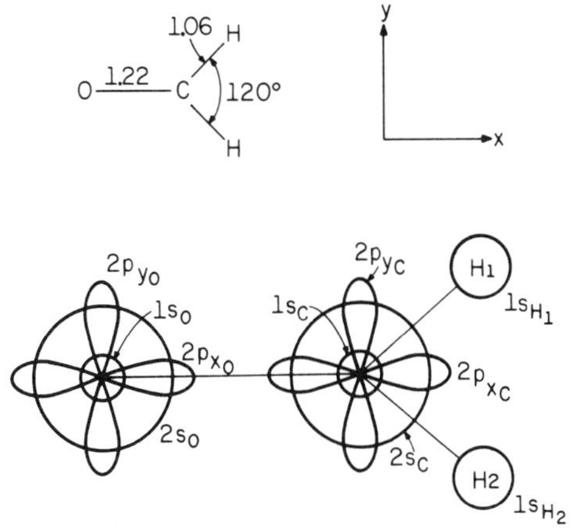

Figure 6-4 Molecular structure and axis system used in the LCAO-MO-SCF calculation in formaldehyde. The bond lengths are in angstrom units. The minimal basis is shown where $2p_{z_O}$ and $2p_{z_C}$ are out of the plane.

The LCAO-MO-SCF results for formaldehyde using the structure in Fig. 6-4 and the orbital exponents listed above are listed in Table 6-5 according to the AOs in Eq.(6-52). Only the lowest eight molecular orbital levels are filled with the 16 electrons in formaldehyde. As mentioned earlier in the study of atoms, the total SCF energy (sum over occupied orbitals) is invariant to a unitary transformation that will result in linear combinations of the orbitals, Φ_i, in Table 6-5. Formal methods of finding linear combinations of Φ_i, which tend to localize the electron distribution in each Φ_i into specific bonding regions, have been devised [20]. We can interpret the Φ_i in Table 6-5 directly, realizing that this choice is arbitrary. The lowest energy molecular orbital, Φ_1, is essentially a pure $1s_O$ atomic orbital as shown by the large coefficient on χ_1 (0.996). Φ_2 is essentially a pure $1s_C$ atomic orbital. It is evident from Φ_1 and Φ_2 that the $1s$ atomic orbitals on both the carbon and oxygen atoms do not mix appreciably with the other AOs in the molecule. Thus, the outer electron or valence shells appear to give rise to the mixing and chemical bonding as is evident from the remaining MOs in Table 6-5. Φ_3 and Φ_4 are predominantly $2s$ (O and C), $2p_x$ (O and C), and hydrogen orbitals. Φ_5 contains the $2p_y$ (O and C) and the hydrogen orbitals. Φ_6 appears to be the SCF version of a carbon–oxygen σ bond and Φ_7 appears to be the carbon-oxygen π bond. Φ_8 is a higher energy version of Φ_5. The coefficients in the LCAO-MOs can also be obtained by starting with hybrid orbitals (at C and O) and then fitting the coefficients by a least-squares method to several one-electron observables (electric dipole moment, magnetic shielding, magnetic susceptibility, and others) [21]. We shall compare the minimal basis LCAO-MO-SCF results obtained by several approximate methods in later sections. The shape of the formaldehyde molecule is shown in Fig. 6-5 from electron density maps determined from a GTO extended basis LCAO-MO-SCF function (experimental molecular structure) that was very near the Hartree–Fock limit.

The LCAO-MO-SCF calculation described above on formaldehyde was for a fixed structure and fixed orbital exponents. Of course, it would be better to minimize the energy with respect to variation in both the orbital exponents and the molecular structure by the SCF procedure. We now describe the calculation of molecular structure by optimizing both the orbital exponents and the molecular structure.

Table 6-5 The LCAO-MO-SCF orbitals and energies for formaldehyde (see Fig. 6-4 for the structure) from M. D. Newton and W. E. Palke, *J. Chem. Phys.* **45**, 2329 (1966). The coefficients listed are from the final SCF **a** matrix in Eq.(6-51), $\mathbf{\Phi} = \boldsymbol{\chi}\mathbf{a}$ where the χ functions are the normalized version of the functions in Eq.(6-52).

χ \ Φ		Occupied								Unoccupied			
		1	2	3	4	5	6	7	8	9	10	11	12
$1s_O$	1	−0.996	0	0.212	0.098	0	0.867	0	0	0	−0.020	0	−0.111
$2s_O$	2	−0.019	0.005	−0.759	−0.445	0	−0.500	0	0	0	0.114	0	0.914
$2p_{x_O}$	3	−0.005	0.001	−0.170	0.172	0	0.692	0	0	0	0.175	0	0.926
$2p_{y_O}$	4	0	0	0	0	−0.426	0	0	−0.880	0	0	0.325	0
$2p_{z_O}$	5	0	0	0	0	0	0	−0.629	0	0.808	0	0	0
$1s_C$	6	0	−0.996	0.111	−0.168	0	−0.025	0	0	0	0.163	0	0.088
$2s_C$	7	0.006	−0.024	−0.284	0.586	0	0.097	0	0	0	−1.404	0	−0.720
$2p_{x_C}$	8	−0.005	0	0.158	0.229	0	−0.460	0	0	0	−0.540	0	1.156
$2p_{y_C}$	9	0	0	0	0	−0.556	0	0	0.184	0	0	−1.255	0
$2p_{z_C}$	10	0	0	0	0	0	0	−0.653	0	−0.789	0	0	0
$1s_{H_1}$	11	0	0.006	−0.026	0.243	−0.274	−0.153	0	0.348	0	0.992	0.954	−0.087
$1s_{H_2}$	12	0	0.006	−0.026	0.243	0.274	−0.153	0	−0.348	0	0.992	−0.954	−0.087
Energy a.u.		−20.590	−11.357	−1.369	−0.837	−0.675	−0.571	−0.470	−0.385	0.247	0.615	0.746	0.821

Electronic energy [Eq.(5-52)] = −144.901 a.u.
Nuclear repulsive energy V_{nn} = 31.451
Total Energy = −113.450

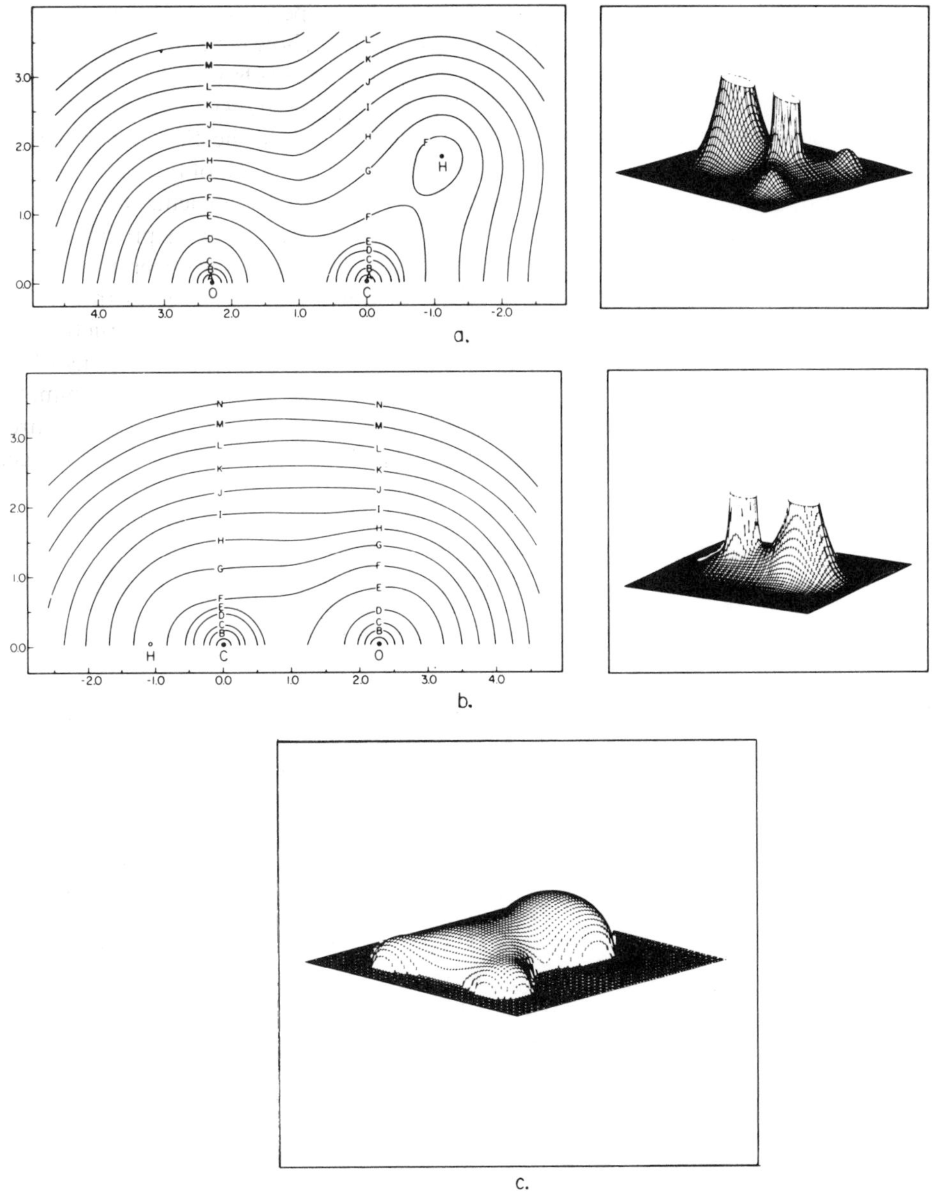

Figure 6-5 Electron contour and three-dimensional perspective plots of the electron densities in formaldehyde from T. H. Dunning and N. W. Winter, *J. Chem. Phys.* **55**, 3360 (1971). a. The contour and density perspective through the molecular plane (xy plane in Fig. 6-4 where $z = 0$). b. The contour and density perspective through the plane perpendicular to the molecular plane (xz plane in Fig. 6-3 where $y = 0$). The values of the electron density for the contours are $A = 0.43455$, $B = 0.21727$, $C = 0.10864$, $D = 0.05432$, $E = 0.02716$, $F = 0.01359$, $G = 0.00697$, $H = 0.00339$, $I = 0.00170$, $J = 0.00085$, $K = 0.00042$, $L = 0.00021$, $M = 0.00011$, and $N = 0.00006$ all in atomic units of electron per unit a_0^3. c. The shape of the formaldehyde molecule taken at the 0.125-a.u. density surface. The base plane is the molecular plane.

Often STOs are replaced with Gaussian-type orbitals (GTOs) to speed up the computations. The first column in Table 6-6 shows the calculated structures for several hydrocarbons obtained by starting with STOs with fixed orbital exponents [$1s_C$ ($\zeta = 5.67$), $2s_C$, $2p_C$ ($\zeta = 1.72$), and $1s_H$ ($\zeta = 1.24$)] and then minimizing the energy with respect to the molecular geometry by the SCF procedure. The first column is labeled STO-3G where 3G indicates that each of the STOs were expanded into a linear combination of 3 GTOs and both the coefficients in the LCGTOs and the orbital exponents were fit to the STOs by a least-squares procedure resulting in STO-3Gs that are very nearly equal to the true STOs. However, the GTOs simplify the calculation of the two-, three-, and four-center integrals in the Hartree–Fock equations. The results of the STO-3G calculations of molecular structure show good agreement with the experimental structures. Additional calculations of molecular structures and other one-electron molecular properties by the STO minimal basis and STO-3G minimal basis discussed above are readily found in the literature [22,23]. The general conclusion, with a few exceptions, is that the minimum energy geometries

Table 6-6 Calculated molecular structures by the LCAO-MO-SCF procedure by using the STO-3G and floating spherical Gaussian basis sets.

Molecule	STO-3G[a,b]	Floating Spherical Gaussian[c]	Experimental
CH_4	$r_{CH} = 1.083$	1.115	1.085[d]
C_2H_2	$r_{CC} = 1.168$	1.214	1.203[e]
	$r_{CH} = 1.065$	1.079	1.061
C_2H_4	$r_{CC} = 1.306$	1.351	1.330[f]
	$r_{CH} = 1.082$	1.101	1.076
	$<HCH = 115.6^\circ$	118.7°	116.6°
C_2H_6	$r_{CC} = 1.538$	1.501	1.531[g]
	$r_{CH} = 1.086$	1.120	1.096
	$<HCH = 108.2^\circ$	108.2°	107.8°
$CH_3{-}C{\equiv}C{-}H$	$r_{C\equiv C} = 1.170$		1.206[h]
	$r_{C-C} = 1.484$		1.459
	$r_{methylCH} = 1.088$		1.105
	$r_{ethylCH} = 1.064$		1.056
	$<HCH = 108.4^\circ$		108.7°
$CH_2{=}C{=}CH_2$	$r_{C=C} = 1.288$		1.308[i]
	$r_{CH} = 1.083$		1.087
	$<HCH = 116.2^\circ$		118.2

[a] W. A. Lathan, W. J. Hehre, and J. A. Pople, *J. Am. Chem. Soc.* **93**, 808 (1971).
[b] L. Radom, *et al.*, *J. Am. Chem. Soc.* **93**, 5339 (1971).
[c] A. A. Frost and R. A. Rouse, *J. Am. Chem. Soc.* **90**, 1965 (1968).
[d] L. S. Bartell, K. Kuchitsu, and R. J. DeNui, *J. Chem. Phys.* **35**, 1211 (1961).
[e] W. J. Lafferty and R. J. Thibault, *J. Mol. Spect.* **14**, 79 (1964).
[f] K. Kuchitsu, *J. Chem. Phys.* **44**, 906 (1966).
[g] D. E. Shaw, D. W. Lepard, and H. L. Welsh, *J. Chem. Phys.* **42**, 3736 (1965).
[h] C. C. Costain, *J. Chem. Phys.* **29**, 864 (1958).
[i] A. G. Maki and R. A. Toth, *J. Mol. Spect.* **17**, 136 (1965).

are given fairly accurately (average bond distance errors are on the order of 0.03 Å and average bond angle errors are on the order of a few degrees).

Molecular structures obtained by the floating spherical Gaussian method are also shown in Table 6-6. For closed shell molecules, this method uses $n/2$ (n is the number of electrons) $1s$ GTOs [Eqs.(5-54) and (5-55)] where the energy is minimized both with respect to the GTO exponent and its position in the molecule. Quite good structural parameters are also obtained with this minimal basis set as shown in Table 6-6.

We now follow the route outlined in Section 5.2 on atoms in calculating better structures and properties of molecules by the LCAO-MO-SCF method. The next step is to use the double zeta set of STOs, which doubles the minimal basis set as explained in Section 5.2. Following this, we add more STOs including functions with higher angular momenta (called *polarization*) to obtain the extended basis sets; however, the results do not always improve with increased effort. In general, the minimal basis sets give structures that are normally larger than indicated by experiment. The double zeta basis set gives structures that come into much better alignment with experiment. Going then to the extended basis sets usually leads to structures that are smaller than indicated by experiment. Thus, if information on equilibrium structure is the goal, it appears best to stop at the double zeta basis set for the most reliable result within the limitations of the single determinant LCAO-MO-SCF scheme [23,24]. The reason for the difficulty in the extended basis STO single determinant scheme is the electron correlation problem, which can be alleviated by using configuration interaction (CI) techniques. Before examining CI methods we shall examine a few more successful calculations using the single determinant approximation.

An interesting comparison has been made of the difference in electron densities between the first-row hydride diatomic molecules and the separated free atoms placed at the equilibrium molecular interatomic distance given by

$$\Delta\rho = [\psi^*(\boldsymbol{r})\psi(\boldsymbol{r})]_{\text{molecule}} - \left[\sum_{\alpha} \psi_{\alpha}^*(\boldsymbol{r})\psi_{\alpha}(\boldsymbol{r})\right]_{\text{atoms}} . \tag{6-53}$$

$\Delta\rho$ is the density difference, $\psi(\boldsymbol{r})$ is the molecular electronic wavefunction, and $\psi_{\alpha}(\boldsymbol{r})$ is the free atom electronic wavefunction and the sum over α is over all atoms in the molecule. Figure 6-6 shows a plot of $\Delta\rho$ for the first-row diatomic hydrides where $\psi(\boldsymbol{r})$ is derived from an extended STO basis that is very close to the Hartree–Fock limit of the best single determinant function for the molecule. $\psi_{\alpha}(\boldsymbol{r})$ are the atomic Hartree–Fock functions. The difference density in LiH indicates a transfer of charge from the Li region to the H region. In general, the electron density near the heavy nucleus decreases and the density builds up on either side of the heavy nucleus. The buildup of charge density on the bonding side of the heavy nucleus is due to bonding effects; the buildup on the other side is due to polarization of the $2s$ nonbonding orbitals out of the bonding region. In all cases, the electron density difference between molecules and separated free atoms is quite small and we note that the shapes of molecules are largely determined by the shapes of the free atoms. For instance, the maximum $\Delta\rho$ in OH is about 0.25 electron $\cdot\, a_0^{-3}$, which is equal to only 0.037

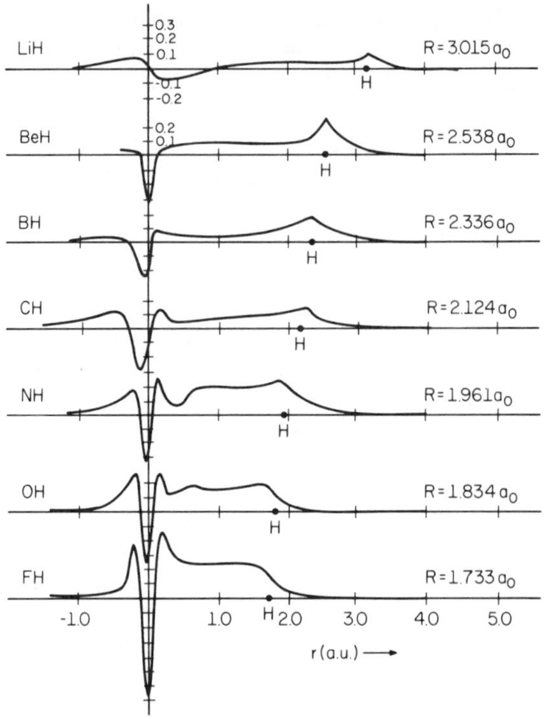

Figure 6-6 Electron density differences between molecules and the corresponding free atoms from Eq.(6-53) along the internuclear axis in the first-row hydrides in atomic units of electrons per unit Bohr volume element. The results are from R. F. W. Bader, I. Keavery, and P. E. Cade, *J. Chem. Phys.* **47**, 3381 (1968).

electron · cm^{-3} as the electron density difference. Plots of $\Delta\rho$ are also available for formaldehyde (see reference in Fig. 6-5).

Another interesting and successful application of the single determinant LCAO-MO-SCF method is the calculation of barriers to internal rotation (see Section 4.2). A minimal basis STO-LCAO-MO-SCF calculation [25] of the barrier to internal rotation in ethane using the experimental structure (staggered form) gave -78.99115 and -78.98593 a.u. for the staggered and eclipsed forms, respectively, with a resultant barrier of 3.3 kcal · mol^{-1} compared to the experimental result of 2.93 ± 0.03 kcal · mol^{-1} (see Problem 4-2f for reference). Later calculations that used minimal basis STOs and minimized the energy with respect to the structure of both the eclipsed and staggered forms confirmed the earlier calculations. Extended basis calculations also gave barriers near 3.0 kcal · mol^{-1}. Notable in these efforts is an extensive calculation using an extended basis of contracted Gaussians [26] that gave energies of -79.2377 and -79.2319 a.u. for the staggered and eclipsed terms giving 3.65 kcal · mol^{-1} for the barrier and another extended basis calculation [27] giving -79.25875 and -79.25364 a.u. with 3.2 kcal · mol^{-1} for the barrier. Even though the energies of either the eclipsed or staggered forms of the molecules vary over 30 kcal · mol^{-1} in these various calculations, the differences in energies or the barrier heights remain nearly constant at 3 kcal · mol^{-1}. Thus, it appears that methyl group internal rotation barriers can be quite accurately estimated at several different levels of *ab initio* sophistication.

Considerably more effort must be expended to compute the barrier to inversion in NH_3 (see Fig. 4-41) and it appears that the calculated energy must be near the Hartree–Fock limit before yielding good results for the barrier. This is quite different than the ethane barrier discussed above where differences in total energy of up to 30 kcal · mol^{-1} between minimal basis and extended basis calculations still all gave correct barriers in the region of 3 kcal · mol^{-1}. Extended basis contracted Gaussian functions and optimized structures have been used [28] to compute the energy in NH_3 of -56.2219 a.u. and the inversion barrier of 5.1 kcal · mol^{-1}, which is in good agreement with experiment. The inclusion of the *d*-type functions in the contracted Gaussians were essential in computing the correct inversion barrier. An extended basis of STOs gives nearly equal results to the contracted Gaussian calculation above [29].

Another barrier problem that we have discussed before is in hydrogen peroxide (H_2O_2) as shown in Problem 4-4 (see Fig. 4-37). This case is apparently more stringent than either the CH_3—CH_3 internal rotation calculation or the NH_3 inversion barrier problem as discussed above. An optimally contracted extended Gaussian basis [30] with full optimization of all geometric parameters was necessary to obtain the experimental potential functions for rotation about the O—O bond in H_2O_2. Apparently the energy must be very near the Hartree–Fock limit to yield an accurate description of the barrier. In summary, the single determinant LCAO-MO-SCF method appears to be adequate to describe relatively complex internal energy changes in molecules as well as predicting fairly well the experimental geometries and several one-electron properties [31].

In spite of the successes outlined above in the LCAO-MO-SCF scheme in the Hartree–Fock limit, there are still significant limitations. For instance, molecular interatomic distances calculated with the extended bases or Hartree–Fock wavefunction (single determinant limit) are usually too small and the resultant vibrational force constants obtained from these functions are usually too large. This overcalculation of the atomic binding in molecules leads to potential functions that are too steep and as a result dissociation products are not usually predicted accurately. The Born–Oppenheimer approximation is normally used in these calculations of the potential energy surface as a function of internuclear separation [$\varepsilon_e(r) + V_{nn}$ in Eq.(6-18), for instance] [32]. Apparently, if the reactants and products conserve the pairing of electrons, the Hartree–Fock limit can predict the heats of reaction [33]. If there is considerable rearrangement of the electron pairs during a dissociation of molecules into atoms or a molecular rearrangement, however, electron correlation is expected to be important.

Configuration Interaction (CI)

The single determinant approximation replaces the true $1/r_{ij}$ interaction by a self-consistent average value of $1/r_{ij}$. Thus, each electron interacts with the average field of the other electrons. The true many-electron wavefunction should have the actual or instantaneous interelectronic repulsion included. Due to difficulties in the direct incorporation of r_{ij} dependence into the many-electron wavefunction in a molecule, the method of configuration interaction (CI) is often used instead. Remembering our previous definition of a configuration (Sections 5.2 and 6.2) we note that CI starts with the ground-state configuration defined by the Hartree–Fock determinant and then proceeds by adding additional determinants representing additional configurations of the electrons in the molecule. The closed shell Hartree–Fock scheme starts with a set of N AOs represented by χ in Eq.(6-51) and then $n/2$ (n is the number of electrons) doubly occupied MOs are formed that are represented by Φ_i in Eq.(6-51) and the determinant in Eq.(5-22). N solutions or eigenvalues to the Hartree–Fock equations are generated from Eq.(5-50) and the lower $n/2$ of these solutions are the occupied orbitals. There are $N - n/2$ unoccupied orbitals. In a CI calculation, the electrons in the highest occupied orbitals are systematically promoted into the unoccupied orbitals and a new wavefunction is formed from a linear combination of the original determinant plus all excited-state configurations according to

$$\psi(1, 2, 3, \ldots, n) = \sum_I \mathrm{C}_I \Delta_I, \tag{6-54}$$

where C_I are variation coefficients and Δ_I is the Ith determinant as described above. Several types of CI are discussed in Section 5.3 for atoms. The coefficients in Eq.(6-54) are determined in a standard manner by the variation method leading to the secular equation as shown in Eq.(5-57). According to the discussion above, the number of configurations that can be formed from the $n/2$ occupied and $N - n/2$ unoccupied orbitals and N basis functions is of the order n^N. Thus, some method of truncating the series in Eq.(6-54) and the resultant order of the matrices in Eq.(5-57) is desirable. Following Eq.(5-57) we noted, by perturbation theory, that the more important configurations in Eq.(6-54) will involve low energy states. We also cited Brillouin's theorem [34] for closed shell systems that excludes single excited electron configurations with respect to the ground-state configuration. In addition, most CI work includes only the doubly excited configurations. Using these concepts, several systematic methods of CI have been developed.

One CI technique that leads to fast convergence is the multiconfigurational self-consistent field (MCSCF) method. In this method, both the C_I in Eq.(6-54) and the **a** matrix in Eq.(6-51) are simultaneously optimized by a double iteration to yield the MCSCF energy. The method of optimized valence configurations (OVC) has also been employed to choose systematically which types of configurations in Eq.(6-54) should be included [35]. The criteria for choosing which Δ_I are to be included in the final set include those Δ_I that are necessary for the dissociation of the molecule into correct Hartree–Fock atoms, those Δ_I that exist in the molecule but not in the dissociated atoms, and finally those Δ_I that describe mainly atomic correlation but change when forming a molecule. Considerable success has been obtained in CI calculations by using these OVC concepts. Another systematic method of truncating the Δ_I in the CI method is by the use of natural orbitals [36]. A survey of *ab initio* calculations on small molecules, which includes a discussion of CI, is available [37]. Schaeffer has given a concise summary of the LCAO-MO-SCF-CI techniques and an

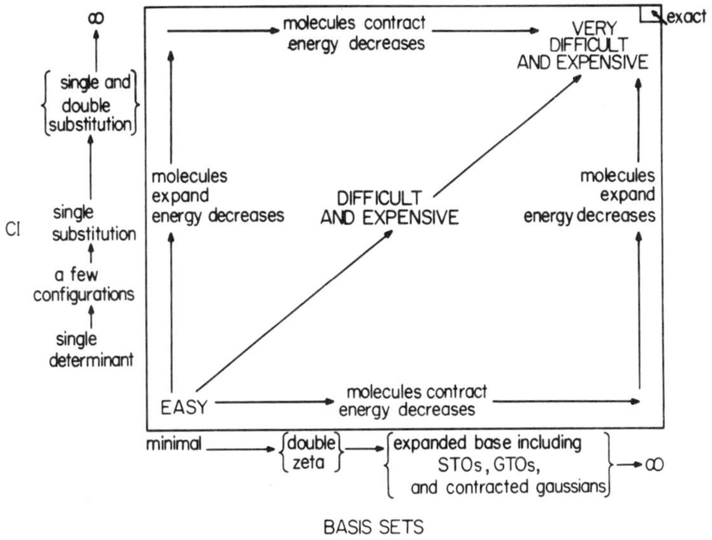

Figure 6-7 A "difficulty plot" summarizing the various steps of increased sophistication and difficulty in the LCAO-MO-SCF and CI methods in molecules.

analysis of the reliability of the results at various stages of sophistication from the minimal basis LCAO-MO-SCF to the MCSCF-CI technique that in principle can ultimately give the exact result [38]. These techniques seem also to give satisfactory results for excited states [39] and potential energy surfaces dissociating to correct atomic states thus leading to the possibility of predicting chemical reactions [40]. There are considerable problems in basis set balance, however, in determining the relative energies of levels of differing symmetry.

At the risk of oversimplification, a "difficulty plot" is shown in Fig. 6-7 that summarizes some of the discussion given in this section.

6.4 VALENCE BOND THEORY

Returning to H_2 and the LCAO-MO complete two-electron function in Eq.(6-39), we note that the first spatial component, $\chi_{1s_a}(1)\chi_{1s_a}(2)$, has both electrons near the a atom, and the last term, $\chi_{1s_b}(1)\chi_{1s_b}(2)$, has both electrons on atom b. Both of these terms are unrealistic as they give too much weight to the $H_a^-{-}H_b^+$ and $H_a^+{-}H_b^-$ ionic configurations of electrons. Thus, it might be reasonable to drop these two terms in Eq.(6-39), giving the following (unnormalized) function for two electrons:

$$\psi(1, 2) = [\chi_{1s_a}(1)\chi_{1s_b}(2) + \chi_{1s_a}(2)\chi_{1s_b}(1)][\alpha(1)\beta(2) - \alpha(2)\beta(1)]. \qquad (6\text{-}55)$$

This is the Heitler–London *valence bond* function [10] for the two-electron molecule that actually preceded the molecular orbital theory by several years. It is clear that the concept of the LCAO-MO is absent as it is not possible to construct $\psi(1, 2)$ in Eq.(6-55) from a single Slater determinant. One can construct Eq.(6-55) from a sum of

Slater determinants, however, that can be represented by a single Slater determinant LCAO-MO, plus some configuration interaction (see Sections 6.3 and 5.3). By dropping the $\chi_{1s_a}(1)\chi_{1s_a}(2)$ and $\chi_{1s_b}(1)\chi_{1s_b}(2)$ terms in the LCAO-MO determinantal function, we have improved the correlation of electrons by removing the functional component that localizes both electrons in the same orbital.

Using the valence bond function in Eq.(6-55) and the variational theorem, we can compute the molecular energy. Using the hydrogen atom atomic orbitals in H_2, the dissociation energy of $D = 3.14$ eV and the internuclear distance of 0.869 Å are obtained. This result is considerably better than that obtained with the simple hydrogen atom LCAO-MO, but it is not as good as the LCAO-MO with shielding in the AOs (see comparison in Table 6-2). Adding the shielding parameters in the AOs in the valence bond function in Eq.(6-55) gives $D = 3.78$ eV and $R = 0.743$ Å, which is now intermediate between the Hartree–Fock MO result and the true energy in Table 6-2.

The next stage of sophistication in the valence bond method is to add some polarization to the MOs. Thus, in place of the simple AOs in Eq.(6-55) or even the screened AOs, we write

$$\chi_a = \lambda_1 \exp(-\zeta r_a) + \lambda_2 z_a \exp(-\zeta r_a) \qquad \chi_b = \lambda_1 \exp(-\zeta r_b) + \lambda_2 z_b \exp(-\zeta r_b), \tag{6-56}$$

where λ_1 and λ_2 are parameters in addition to the screening parameter, ζ. Using these orbitals in the valence bond scheme in Eq.(6-55) is equivalent to expanding the basis in our previous LCAO-MO single determinant scheme. The expanded basis result gives $D = 4.04$ eV and $R = 0.74$ Å.

The next most obvious step is to add back the ionic terms (present in the MO theory) that we ignored in the valence bond expression in Eq.(6-55). Adding in small amounts of $\chi_{1s_a}(1)\chi_{1s_a}(2)$ and $\chi_{1s_b}(1)\chi_{1s_b}(2)$ gives another parameter that, when combined with the screening and polarization (extended basis) above, gives $D = 4.12$ eV and $R = 0.75$ Å. The energy has improved but the internuclear distance has increased slightly from the previous result.

In the case of H_2, it is possible to write an electronic wavefunction that includes the interelectronic coordinate directly. Using this concept (the James–Coolidge expansion [10] in r_{12}) and carrying enough terms, the true energy and equilibrium internuclear distance is obtained. This sequence of results is shown in Table 6-7 (see also Table 6-2).

The distinction between the MO and VB theories is arbitrary and we must choose a point of differentiation. Usually the MO theory refers to the single determinant many-electron function with improvements by employing more determinants (configuration interaction). In VB theory one makes a more intuitive jump to the many-electron function by writing down combinations of MOs (two-electron bond functions) that reflect the writer's picture of the molecule [41]. In some cases the VB function will be better than a minimal basis LCAO-MO function as we found in H_2. The LCAO-MO theory has been more popular due to the ease of computation (using the matrix formulation) with high-speed digital computers. The valence bond method can also be generalized for use on digital computers, however, and in many cases simple orbital pictures emerge [42].

Table 6-7 Calculated dissociation energy in H_2 with several molecular wavefunctions. VB corresponds to valence bond and MO to molecular orbital (see also Table 6-2). The numbers are adapted from Ref. [10].

Function	Dissociation Energy (eV)
MO, hydrogen atom AOs	2.681
VB, hydrogen atom AOs	3.14
MO, with shielding	3.49
MO, Hartree–Fock	3.62
VB, with shielding	3.78
VB, expanded basis (polarization)	4.04
VB, r_{12} expansion	4.72
Experimental	4.75

Generalized Valence Bond (GVB) Method

In the generalized valence bond (GVB) method, we start with Eq.(6-55) and then systematically (by the variational method) find the best functions χ_a and χ_b in the covalent *ab* bond that will minimize the energy. χ_a and χ_b can be expanded at atom *a* and atom *b* into a minimal basis, a double zeta basis, or an extended basis where contracted Gaussians can be used. Thus, the GVB method is related to the valence bond method in the same way that the approaches to the Hartree–Fock limit are related to the LCAO-MO single determinant method; both methods are SCF generalizations of simpler intuitive wavefunctions.

An interesting example of GVB method is the application to formaldehyde, which we have considered in detail in Section 6.3 by the LCAO-MO-SCF method. In one particular calculation [43] the 16 electrons in formaldehyde were distributed as follows:

ψ_1 $1s_O$ paired electrons
ψ_2 $1s_C$ paired electrons
ψ_3 $1s_O$ paired electrons
ψ_4 $2p_O$ paired electrons (in-plane lone pair)

$$\left.\begin{array}{l}\left.\begin{array}{l}\psi_5\\ \psi_6\end{array}\right\}\text{VB CO, }\sigma\text{ bond}\\ \left.\begin{array}{l}\psi_7\\ \psi_8\end{array}\right\}\text{VB CO, }\pi\text{ bond}\\ \left.\begin{array}{l}\psi_9\\ \psi_{10}\end{array}\right\}\text{VB CH, }\sigma\text{ bond}\\ \left.\begin{array}{l}\psi_{11}\\ \psi_{12}\end{array}\right\}\text{VB CH, }\sigma\text{ bond}\end{array}\right\}\begin{array}{l}\text{each orbital}\\ \text{singly occupied}\end{array}$$

The wavefunction has the form

$$\psi(1, 2, 3, \ldots, 16) = N[\psi_1^2\psi_2^2\psi_3^2\psi_4^2(\psi_5\psi_6)(\psi_7\psi_8)(\psi_9\psi_{10})(\psi_{11}\psi_{12})], \quad (6\text{-}57)$$

where N is the normalization constant and the brackets indicate the appropriate spin symmetrization or Slater determinant. The AOs at each center were composed of a double zeta basis set. These AOs were combined in the VB functions described above and the energy was minimized with the many-electron wavefunction given above. The resulting orbital amplitudes of the valence orbitals for the ground state are shown in Fig. 6-8. The electron densities are obtained from the square of the

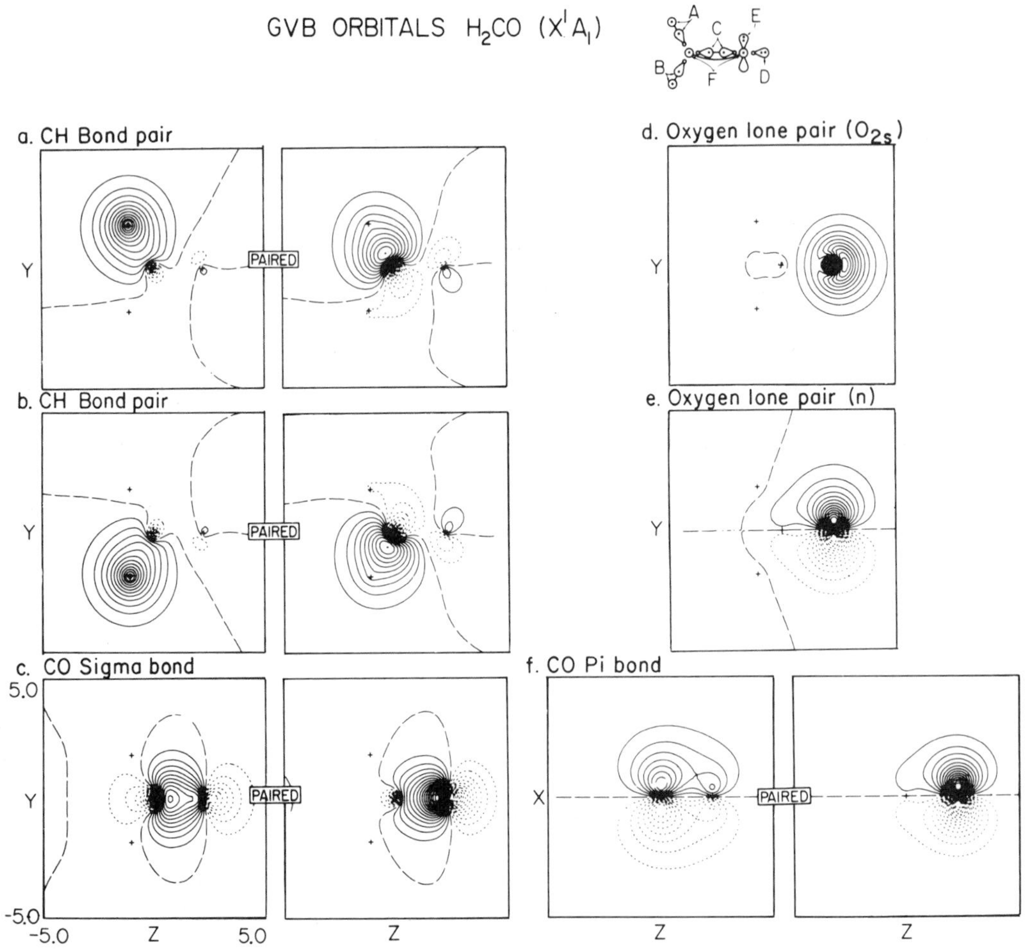

Figure 6-8 The outer GVB orbitals in CH_2O (the 1*s* C and O orbitals are not shown) from Ref. [43]. The spacing between contours is 0.05 a.u. and the long dashes indicate lines of zero amplitude. The *z* axis contains the CO bond and the *y* axis is also in the molecular plane.

orbital amplitudes. It is evident from Fig. 6-8 that the GVB orbitals are localized to rather specific regions of the molecule as expected in the VB approach. The CH bonding pairs are each composed of one essentially hydrogen orbital and a second orbital that is primarily localized on the carbon atom but hybridized toward the hydrogen atom. The CO σ bonding pair shows the expected localization between the C and O atoms. The two oxygen lone pairs are also localized as expected. The CO π bonding pair indicates that one orbital is localized near the oxygen atom but the second orbital is more diffuse, being spread toward the carbon atom. It is interesting to compare these GVB orbitals with the LCAO-MO-SCF results in Table 6-5.

In summary, the GVB method shows that the orbitals are localized into the intuitively expected forms in spite of there being no restrictions made upon the shape or localization of these orbitals; they are unique and provide a quantum mechanical justification for our original intuitive thoughts on the valence bonds in formaldehyde. The energies calculated with the GVB method are lower than the Hartree–Fock limit in the LCAO-MO-SCF scheme. This is expected from the arguments made at the beginning of this section on the VB method and the demonstration that the VB function for H_2 could not be composed from a single determinant of MOs. The calculation of excited state energies and other properties is easily extended by the GVB procedure and several results are available [42,43].

6.5 APPROXIMATE LCAO-MO-SCF THEORIES

From the very beginning of molecular quantum mechanics of electronic structure there has been a trade-off between trying to do *ab initio* calculations from a well-developed theory (Hartree–Fock, for instance) on the one hand and trying to gain an understanding of electronic properties by making clever approximations on the other hand. Large amounts of computer time must be available in order to conduct an effective program in using *ab initio* techniques to study the electronic structure of molecules and only a few laboratories have these expensive facilities. Approximate methods, however, can be tailored to some extent to the resources available by examining a single series of properties (such as electric moments or bond strengths) which are fairly well predicted by one specific approximate method as opposed to other approximate methods which may be useful for other properties. We shall now examine a number of approximate LCAO-MO-SCF methods and try to gain some understanding of how these methods are obtained from the general SCF methods described in Section 6.4.

Working within the single determinant limitation and the LCAO-MO approximation, we review Eqs.(6-51) and (5-41) and the discussion that follows Eq.(5-41) through to the discussion of the atomic and molecular SCF method. Most approximate LCAO-MO theories make assumptions regarding the elements in the $\mathbf{F}^{\chi}$ matrix and the corresponding $\mathbf{S}^{\chi}$ matrix in Eqs.(5-50) and (5-51). The general approach is to use semiempirical methods of estimating the elements in $\mathbf{F}^{\chi}$ and $\mathbf{S}^{\chi}$. The most difficult and time-consuming integrals involving the F operator are the interelectronic repulsion integrals. The matrix elements in $\mathbf{F}^\chi$ are given by [see Eq.(5-49)]

$$\mathscr{H}^\chi_{\alpha\beta} = \int_a \chi_\alpha(a)\mathscr{H}(a)\chi_\beta(a)\,dV_a$$

$$(J_j)^\chi_{\alpha\beta} = \int_a \chi_\alpha(a)^*\left[\int_b \mathbf{a}_j^\dagger(b)\boldsymbol{\chi}^\dagger(b)\frac{e^2}{r_{ab}}\boldsymbol{\chi}(b)\mathbf{a}_j(b)\,dV_b\right]\chi_\beta(a)\,dV_a$$

$$= \sum_{\lambda,\nu} \mathrm{a}^*_{j\lambda}\mathrm{a}_{j\nu}\int_a \chi^*_\alpha(a)\left[\int_b \chi^*_\lambda(b)\left(\frac{e^2}{r_{ab}}\right)\chi_\nu(b)\,dV_b\right]\chi_\beta(a)\,dV_a$$

$$= \sum_{\lambda,\nu} \mathrm{a}^*_{j\lambda}\mathrm{a}_{j\nu}\langle\alpha\lambda|\nu\beta\rangle$$

$$(K_j)^\chi_{\alpha\beta} = \sum_{\lambda,\nu} \mathrm{a}^*_{j\lambda}\mathrm{a}_{j\nu}\langle\alpha\lambda|\beta\nu\rangle, \qquad (6\text{-}58)$$

where $\langle\alpha\lambda|\nu\beta\rangle$ is the Coulomb integral and $\langle\alpha\lambda|\beta\nu\rangle$ is the exchange integral. The two-electron Coulomb and exchange integrals are very difficult to calculate; however, many are zero by symmetry and others are small. Specifically, the values of the Coulomb $\langle\alpha\lambda|\nu\beta\rangle$ and exchange $\langle\alpha\lambda|\beta\nu\rangle$ integrals are small if α and β are on different centers or if λ and ν are on different centers. Ignoring these types of integrals is called the *zero-differential overlap approximation* [12]. The zero-differential overlap approximation to the Coulomb and exchange integrals above leads naturally into a set of approximations called the *complete neglect of differential overlap* (CNDO) [44]. This method has been used extensively to compute energies and other molecular properties.

The basic features of the LCAO-MO-SCF method including the iterative procedure are retained in the CNDO method, which includes the larger, more important interelectronic integrals. The CNDO method (like many approximate MO theories) treats only the valence electrons or outer shell electrons in each atom. This approximation is evidently justified according to the SCF results in Table 6-5 where the $1s_O$ and $1s_C$ orbitals are essentially localized and do not mix with the rest of the molecule. In the case of formaldehyde, the 16-electron problem is reduced to a 12-electron problem using the frozen core approximation. The approximations to the LCAO-MO-SCF procedure in the CNDO method are

1. $\mathbf{S}^\chi = \mathbf{1}$ in Eqs.(5-50).
2. The zero-differential overlap approximation is used in all two-electron integrals.

Table 6-8 Average ionization energies (electron volts) used in the CNDO approximation to the LCAO-MO-SCF equations (see Ref. [44]).

Atom	H	Li	Be	B	C	N	O	F
$1s$	−13.06							
$2s$		−5.39	−9.32	−14.05	−19.44	−25.58	−32.38	−40.20
$2p$		−3.54	−5.96	−8.30	−10.67	−13.19	−15.85	−18.66

Table 6-9 Values of β_A and β_B (electron volts) in Eq.(6-59) used in the CNDO approximation to the LCAO-MO-SCF equations (see Ref. [44]).

Atom	H	Li	Be	B	C	N	O	F
$-\beta_A$	9	9	13	17	21	25	31	39

3. The remaining two-electron integrals are reduced to one value per atom pair. $\langle \alpha\nu | \nu\alpha \rangle = \gamma_{AB}$; χ_α on atom A, χ_ν on atom B. γ_{AB} is the average electrostatic repulsion between any electron on atom A with any electron on atom B that reduces to e^2/R_{AB} for large R_{AB}.
4. The diagonal core Hamiltonian matrix elements are approximated by average ionization energies as shown in Table 6-8.
5. The two-center off-diagonal core matrix elements are proportional to the corresponding overlap integrals,

$$\mathscr{H}^{\chi}_{\mu\nu} = \beta_{AB} S^{\chi}_{\mu\nu} = \tfrac{1}{2}(\beta_A + \beta_B) S^{\chi}_{\mu\nu}, \tag{6-59}$$

where β_i are also semiempirical parameters assigned as shown in Table 6-9.

Consider now the CNDO-LCAO-MO-SCF calculation in formaldehyde where the valence atomic orbitals and molecular structure are shown in Fig. 6-4. The unnormalized AOs for the valence electrons are defined here:

Oxygen	*Carbon*	*Hydrogen*
$\chi_1(2s_O) = r_O \exp(-\zeta_O r_O)$	$\chi_5(2s_C) = r_C \exp(-\zeta_C r_C)$	$\chi_9(1s_{H_1}) = \exp(-\zeta_H r_{H_1})$
$\chi_2(2p_{x_O}) = x_O \exp(-\zeta_O r_O)$	$\chi_6(2p_{x_C}) = x_C \exp(-\zeta_C r_C)$	$\chi_{10}(1s_{H_2}) = \exp(-\zeta_H r_{H_2})$.
$\chi_3(2p_{y_O}) = y_O \exp(-\zeta_O r_O)$	$\chi_7(2p_{y_C}) = y_C \exp(-\zeta_C r_C)$	
$\chi_4(2p_{z_O}) = z_O \exp(-\zeta_O r_O)$	$\chi_8(2p_{z_C}) = z_C \exp(-\zeta_C r_C)$	(6-60)

Table 6-10 The overlap matrix in formaldehyde with the normalized form of the AOs in Eq.(6-60). The coordinate system is shown in Fig. 6-4.

	χ	1	2	3	4	5	6	7	8	9	10
$2s_O$	1	1.0	0	0	0	0.380	−0.460	0	0	0.086	0.086
$2p_{x_O}$	2	0	1.0	0	0	0.310	−0.310	0	0	0.078	0.078
$2p_{y_O}$	3			1.0	0	0	0	0.220	0	0.040	−0.040
$2p_{z_O}$	4				1.0	0	0	0	0.220	0	0
$2s_C$	5					1.0	0	0	0	0.537	0.537
$2p_{x_C}$	6						1.0	0	0	0.262	0.262
$2p_{y_C}$	7			symmetric				1.0	0	0.424	−0.424
$2p_{z_C}$	8								1.0	0	0
$1s_{H_1}$	9									1.0	0.180
$1s_{H_2}$	10										1.0

Table 6-11 CNDO electronic energies and coefficients in the LCAO-MO functions for formaldehyde starting with χ in Eq.(6-60). The coordinate system is shown in Fig. 6-4. The coefficients listed here are from the final SCF **a** matrix in Eq.(6-51), $\mathbf{\Phi} = \chi\mathbf{a}$, where the χ functions are the normalized version of the functions in Eq.(6-60) and the CNDO approximations are used in the LCAO-MO-SCF scheme.

χ \ Φ		Occupied						Unoccupied			
		1	2	3	4	5	6	7	8	9	10
$2s_O$	1	0.770	−0.424	0	−0.299	0	0	0	−0.076	0	−0.361
$2p_{xO}$	2	−0.171	−0.250	0	−0.771	0	0	0	0.171	0	0.532
$2p_{yO}$	3	0	0	0.599	0	0	−0.776	0	0	−0.193	0
$2p_{zO}$	4	0	0	0	0	−0.761	0	0.648	0	0	0
$2s_C$	5	0.494	0.536	0	0.036	0	0	0	0.631	0	0.261
$2p_{xC}$	6	0.288	−0.375	0	0.489	0	0	0	−0.224	0	0.697
$2p_{yC}$	7	0	0	0.619	0	0	0.297	0	0	0.726	0
$2p_{zC}$	8	0	0	0	0	−0.648	0	−0.761	0	0	0
$1s_{H_1}$	9	0.157	0.405	0.358	−0.192	0	0.392	0	−0.508	−0.466	0.125
$1s_{H_2}$	10	0.157	0.405	−0.358	−0.192	0	−0.392	0	−0.508	0.466	0.125
energies (a.u.)		−1.624	−1.083	−0.929	−0.749	−0.682	−0.538	+0.157	+0.259	+0.338	+0.480

$$\text{Orbital energy} = 2\sum_{i=1}^{6} \varepsilon_i = -13.210 \text{ a.u.}$$

$$\text{Total electronic energy} = -44.837$$

$$V_{nn} = 18.015$$

$$\text{Total energy} = -26.822$$

The orbital exponents are $\zeta_O = 1.725$, $\zeta_C = 1.350$, and $\zeta_H = 1.0$. Note the comparison between the unnormalized valence AOs listed here with the original minimal basis in Eq.(6-52). Only the $1s_O$ and $1s_C$ are missing here in the frozen core approximation. The overlap matrix with the normalized form of the AOs in Eq.(6-60) is listed in Table 6-10. The CNDO electronic energies and associated functions are listed in Table 6-11. The overlap integrals and the CNDO results given here were obtained by using a Fortran program from QPCE (address appears in the Preface) according to the methods given here. The correspondence between the SCF orbitals in Table 6-5 and the CNDO orbitals in Table 6-11 is quite evident. The identification of the CNDO orbitals to localized regions in the molecule follows our previous discussion due to the similarity between the SCF and CNDO orbital coefficients.

There are several modifications to the CNDO techniques described here that can be made. The method of intermediate neglect of differential overlap, INDO, restores some of the one-center electron repulsion integrals that were omitted in the CNDO method. This change leads to improvements in properties associated with spin densities [45]. Other types of modified INDO (MINDO) calculations have also been extensively employed [46].

There are many other approximate LCAO-MO-SCF theories that combine various degrees of semiempirical input with judicious omission of certain unimportant or hard to calculate integrals. The various merits of each method should be studied before starting a calculation. Some methods may be better at predicting certain properties than others [47]. Some of these methods include the multicenter zero-differential overlap (MCZDO) method [48], the partial retention of diatomic differential overlap (PRDDO) method [49], the X-α method [50], and others [51]. In addition to these semiempirical methods above, techniques of transferring results in small molecules as molecular fragments to larger molecules have been developed. These include the nonempirical molecular orbital theory (NEMO) method [52] and the simulated *ab initio* molecular orbital (SAMO) method [53]. These methods of transferability of the matrix elements in the Hartree–Fock matrix appear to work fairly well if structural symmetry is preserved in the transfer (the central carbon in propane can be transferred to the central carbon atoms in butane) [54]. Molecular fragment approaches have also been applied to calculations on large molecules by using the floating spherical Gaussian basis set [55]. In these techniques the locations of the floating spherical Gaussian functions in large molecules are fixed according to experience gained in minimizing the energy of smaller molecular fragments to find the optimum locations.

6.6 THE EXTENDED HÜCKEL THEORY AND MOLECULAR CHARGE DISTRIBUTION

The extended Hückel method loses the feature of the self-consistent field because the final $\mathbf{F}^\chi$ matrix is constructed directly by semiempirical rules. The methods described in Section 6.5 made approximations in the matrix elements involving only the χ basis functions, an initial $\mathbf{a}$ was guessed to give $\mathbf{F}^\chi$, and a final set of coefficients was obtained by iteration (SCF). If the elements in $\mathbf{F}^\chi$ are chosen directly by semi-empirical methods, however, $\mathbf{a}$ is also fixed and only a single step of diagonalization and transformation is needed.

The assumptions in the extended Hückel LCAO-MO theory are

1. Only the valence atomic orbitals (outer electrons) participate in the molecular bonding.
2. The integrals in the $\mathbf{F}^\chi$ matrix are fixed as follows:

$$F_{ij}^\chi = \alpha_{ij} \qquad \text{if } i \text{ and } j \text{ are on the same atom.}$$

$$F_{ij}^\chi = \beta_{ij} = \frac{k}{2}(\alpha_{ii} + \alpha_{jj})S_{ij} \qquad \text{if } i \text{ and } j \text{ are adjacent atoms.}$$

$$F_{ij}^\chi = 0 \qquad \text{if } j \text{ and } i \text{ are on nonadjacent atoms.} \tag{6-61}$$

   α_{ii} are obtained from molecular spectra or atomic spectra data and are usually set equal to the (negative value of the) valence state ionization energy. The valence state of an atom is intermediate between the free atom and the molecular state. For instance, for carbon atoms the valence state ionization energy is approximately the energy necessary to remove an electron from the hybrid orbital in the atom. Valence state ionization energies of several common atoms are given in Table 6-12 where the similarities with Table 6-8 are clear. k [in Eqs.(6-61)] is an adjustable parameter and acceptable final results have been obtained for k near 1.7–2.0. [56].
3. The total energy is a sum over the orbital energies.

We can show some validity from the SCF formalism [57] for the seemingly *ad hoc* approximations made above. F_{ij}^χ, in the LCAO-MO $\mathbf{F}^\chi$ matrix in Eq.(5-49),

Table 6-12 Values of valence state ionization energies for several common atoms. The energy is necessary to remove the electron from the specified electron shell and the numbers are from the electroaffinities of atoms from H. O. Pritchard and H. A. Skinner, *Chem. Rev.* **55**, 745 (1955). These numbers are representative of the input data in the Hückel theories of molecular electronic structure.

Atom and Shell	Energy (eV)
H(1*s*)	−13.6
C(2*s*)	−21.4
C(2*p*)	−11.4
N(2*s*)	−27.4
N(2*p*)	−14.5
O(2*s*)	−35.3
O(2*p*)	−17.8
F(2*p*)	−21.0
Cl(3*p*)	−15.1
Br(4*p*)	−13.7
I(5*p*)	−12.6

will be dominated by the first term, $\mathscr{H}_{ij}^{\chi}$, the core Hamiltonian matrix element given for molecules by

$$\mathscr{H}_{ij}^{\chi} = \int_a \chi_i^*(a)\left(-\frac{\hbar^2}{2m}\nabla_a^2 - \sum_\alpha \frac{Z_\alpha e^2}{r_{a\alpha}}\right)\chi_j(a)\, dV_a. \tag{6-62}$$

When χ_i and χ_j are on the same center and $i = j$, the dominant contribution will be given by the energy to remove the ath electron from the shielded nucleus (the valence state ionization energies in Table 6-12). When i and j are on the same center and $i \neq j$, the matrix elements above [Eq.(6-62)] are zero due to the orthogonality of $\boldsymbol{\chi}$. When χ_i and χ_j are on different centers, the dominant terms in the sum over α in Eq.(6-62) will be the ith and the jth, which gives

$$\begin{aligned}
\mathscr{H}_{ij}^{\chi} = \beta_{ij} &= \int_a \chi_i^*(a)\left(-\frac{\hbar^2}{2m}\nabla_a^2 - \frac{Z_i e^2}{r_{ai}} - \frac{Z_j e^2}{r_{aj}} - \cdots\right)\chi_j(a)\, dV_a \\
&= \int_a \chi_i^*(a)\left[\left(-\frac{\hbar^2}{2m}\nabla_a^2 - \frac{Z_i e^2}{r_{ai}}\right) + \left(-\frac{\hbar^2}{2m}\nabla_a^2 - \frac{Z_j e^2}{r_{aj}}\right)\right]\chi_j(a)\, dV_a \\
&\quad - \int_a \chi_i^*(a)\left(-\frac{\hbar^2}{2m}\nabla_a^2\right)\chi_j(a)\, dV_a + \cdots \\
&= (\varepsilon_{ii} + \varepsilon_{jj})\int \chi_i^*(a)\chi_j(a)\, dV_a - \int_a \chi_i^*(a)\left(-\frac{\hbar^2}{2m}\nabla_a^2\right)\chi_j(a)\, dV_a + \cdots \\
&= (\varepsilon_{ii} + \varepsilon_{jj})\mathrm{S}_{ij}^{\chi} - \mathrm{T}_{ij}^{\chi} + \cdots.
\end{aligned} \tag{6-63}$$

T_{ij}^{χ} is the ijth matrix element of the kinetic energy operator. The other terms in the summation in Eq.(6-63) will lead to three-center integrals and smaller contributions. As T_{ij}^{χ} is positive, Eq.(6-63) indicates that the magnitude of $(\varepsilon_{ii} + \varepsilon_{jj})\mathrm{S}_{ij}^{\chi}$ will be reduced by the additional terms. This reduction includes shielding, which is handled empirically in Eqs.(6-61) by the $k/2$ term where k is a variable depending on the system.

Using these methods above, the extended Hückel $\mathbf{F}^{\chi}$ matrix is easily constructed and the energy eigenvalues and LCAO-MO electronic functions are obtained from the secular equation. However, we require the overlap matrix, $\mathbf{S}^{\chi}$, in the original AO basis $\boldsymbol{\chi}$. Once the choice of the $\boldsymbol{\chi}$ basis is made, the $\mathbf{S}^{\chi}$ matrix elements can be obtained numerically or analytically quite easily by a digital computer or by the use of published tables in some cases [58].

Consider formaldehyde as shown in Fig. 6-4. The valence AOs are listed in Eq.(6-60) and the overlap matrix is given in Table 6-10. The resultant $\mathbf{F}^{\chi}$ matrix is shown in Table 6-13. Tables 6-10 and 6-13 are now used along with Eqs.(5-51) to give the results in Table 6-14 where $\boldsymbol{\Phi} = \mathbf{a}\boldsymbol{\chi}$ with $\boldsymbol{\chi}$ being the normalized version of the AOs in Eq.(6-60). Only the lowest six energy levels are filled with the 12 valence electrons. The MOs in Table 6-14 obtained by the extended Hückel method are comparable on a one-to-one basis with the corresponding CNDO results in Table 6-11, except that the energies for the ψ_4-ψ_5 and ψ_8-ψ_9 pairs are interchanged.

An impression of the distribution of electrons in a molecule may be obtained (in formaldehyde) by the coefficients in Tables 6-5, 6-11, or 6-14, where $\boldsymbol{\Phi} = \boldsymbol{\chi}\mathbf{a}$ and

Table 6-13 Elements in the $\mathbf{F}^{\chi}$ matrix in units of electron volts for formaldehyde. See Eq.(6-60) for definition of functions and Fig. 6-4 for the coordinate system. The matrix elements are from Eqs.(6-61) with $k = 2$ and the values of α_{ii} given in Table 6-12 (the $\mathbf{S}^{\chi}$ matrix is in Table 6-10).

χ	1	2	3	4	5	6	7	8	9	10
1	−35.3	0	0	0	−21.6	21.5	0	0	0	0
2		−17.8	0	0	−12.2	9.1	0	0	0	0
3			−17.8	0	0	0	−6.4	0	0	0
4				−17.8	0	0	0	−6.4	0	0
5					−21.4	0	0	0	−18.8	−18.8
6						−11.4	0	0	−6.6	−6.6
7							−11.4	0	−10.6	10.6
8								−11.4	0	0
9		symmetric							−13.6	0
10										−13.6

χ are the normalized AOs. The probability density for the electrons in the jth MO is given by [59]

$$P_j = n_j \psi_j^* \psi_j = n_j \left(\sum_m \chi_m^* a_{mj}^* \right) \left(\sum_n \chi_n a_{nj} \right) = n_j \sum_{m,n} a_{mj}^* a_{nj} \chi_m^* \chi_n, \tag{6-64}$$

where $n_j = 0$, 1, or 2 (the number of electrons in the jth MO) and the electronic charge at each atom (a) from the jth MO is *defined* by

$$q_a^j = e n_j \sum_m |a_{mj}|^2, \tag{6-65}$$

where the sum over m is over only those AOs on center a in the jth MO. Equation (6-65) is obtained from Eq.(6-64) by setting $n = m$ and including only the AOs on center a. Summing q_a^j over the occupied MOs gives the total electronic charge on atomic center a,

$$q_a^e = \sum_j q_a^j, \tag{6-66}$$

and the net charge at center a is obtained by summing the electronic and nuclear contributions. The nuclear charge contributions are the atomic numbers (Z) minus the inner electrons that have not participated in the approximate LCAO-MO scheme. For instance, the nuclear charge at C in the CNDO and extended Hückel methods is $4e$ as the atomic number is 6, and there are two inner $1s$ carbon electrons that approximately cancel two nuclear charges. Thus, the net electronic charge at center a is

$$q_a = e(Z - n_{\text{inner}}) - q_a^e. \tag{6-67}$$

In formaldehyde, we find from this equation and Table 6-14 that $q_{\text{H}} = -e\,(0.053)$, $q_{\text{C}} = e(0.642)$, and $q_{\text{O}} = -e(0.539)$. This charge distribution apparently correctly reflects the sign of the electric dipole moment of $-\text{OCH}_2+$.

Table 6-14 Extended Hückel LCAO-MOs and energies in formaldehyde starting with the normalized χ in Eq.(6-60). The $\mathbf{S}^{\chi}$ and $\mathbf{F}^{\chi}$ matrices are given in Tables 6-10 and 6-13, respectively, and the coordinate system is shown in Fig. 6-4. These results are obtained from Eq.(5-50) or (5-51), where $\mathbf{\Phi} = \chi\mathbf{a}$ and the **a** matrix is listed here.

χ \ Φ		Occupied						Unoccupied			
		1	2	3	4	5	6	7	8	9	10
$2s_O$	1	0.678	0.415	0	0	0.352	0	0	0	0.334	0.363
$2p_{x_O}$	2	0.215	−0.025	0	0	−0.903	0	0	0	0.229	0.292
$2p_{y_O}$	3	0	0	0.374	0	0	0.901	0	0.220	0	0
$2p_{z_O}$	4	0	0	0	0.851	0	0	0.526	0	0	0
$2s_C$	5	0.589	−0.426	0	0	−0.039	0	0	0	0.109	−0.676
$2p_{x_C}$	6	−0.274	−0.512	0	0	0.205	0	0	0	0.763	0.195
$2p_{y_C}$	7	0	0	0.671	0	0	−0.099	0	−0.735	0	0
$2p_{z_C}$	8	0	0	0	0.526	0	0	−0.851	0	0	0
$1s_{H_1}$	9	0.189	−0.437	0.453	0	0.092	−0.299	0	0.453	−0.348	0.380
$1s_{H_2}$	10	0.189	−0.437	−0.453	0	0.092	0.289	0	−0.453	−0.348	0.380
energies (a.u.)		−2.306	−1.457	−1.076	−0.795	−0.749	−0.628	−0.274	0.132	0.248	0.608

6.7 THE SIMPLE π-ELECTRON HÜCKEL THEORY

By analogy with the LCAO-MO in diatomics in Section 6.2, we now consider the π-electron theory in conjugated organic-type molecules. The π-electron MOs are formed from linear combinations of p-type atomic orbitals that are perpendicular to a molecular plane. We assume that the π-electron orbitals lead to energy levels that are higher than and separated from the σ-bonding electrons in the molecule; that is, we assume no nonzero matrix elements between the π and σ AOs [60]. Finally, we now use Eqs.(6-61) to set up $\mathbf{F}^{\chi}$ where we also estimate $\mathbf{S}^{\chi} = \mathbf{1}$. Using these assumptions, we first consider the planar linear polyene chain;

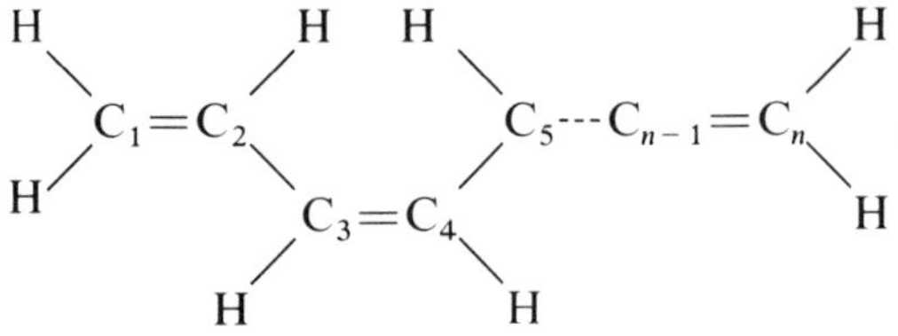

,

which is an alternate single bond–double bond system. Using the scheme above to describe the out-of-plane LCAO-MOs with out-of-plane $2p$ functions as AOs, we can write the secular equation of order n to give

$$D_n = \begin{array}{c|cccccc|} & C_1 & C_2 & C_3 & C_4 & \cdots & C_n \\ \hline C_1 & \alpha - E & \beta & 0 & 0 & \cdots & 0 \\ C_2 & \beta & \alpha - E & \beta & 0 & & 0 \\ C_3 & 0 & \beta & \alpha - E & \beta & & 0 \\ \vdots & \vdots & & & & \ddots & \\ C_n & 0 & 0 & 0 & & \cdots & \alpha - E \end{array} = 0. \tag{6-68}$$

Dividing by β^n and substituting $(\alpha - E)/\beta = -x$ gives

$$D_n = \begin{vmatrix} -x & 1 & 0 & 0 & \cdots \\ 1 & -x & 1 & 0 & \\ 0 & 1 & -x & 1 & \cdots \\ & & & \ddots & \\ & & & & -x \end{vmatrix} = 0. \tag{6-69}$$

Expanding for any order n gives

$$D_1 = -x$$

$$D_2 = x^2 - 1$$

$$D_3 = -x^3 + 2x$$

$$D_4 = x^4 - 3x^2 + 1$$

$$\vdots$$

$$D_n = -xD_{n-1} - D_{n-2}. \tag{6-70}$$

The polynomials in the progression in Eq.(6-70) are similar to the Tschebyscheff polynomials in argument θ that are given by

$$C_n(\theta) = \frac{\sin(n+1)\theta}{\sin\theta} = 2\cos\theta C_{n-1}(\theta) - C_{n-2}(\theta)$$

$$C_1(\theta) = \frac{\sin 2\theta}{\sin\theta} = -2\cos\theta$$

$$C_2(\theta) = \frac{\sin 3\theta}{\sin\theta} = 4\cos^2\theta - 1. \tag{6-71}$$

$$\vdots$$

It is evident by comparing Eqs.(6-71) and (6-70) that if $x = 2\cos\theta$, we can write $D_n = 0 = [\sin(n+1)\theta]/\sin\theta$, which is satisfied if

$$(n+1)\theta = j\pi; \qquad j = 1, 2, 3, \ldots, n$$

$$\theta = \frac{j\pi}{n+1},$$

which gives $x_j = 2\cos\theta = 2\cos[j\pi/(n+1)] = (E-\alpha)/\beta$. Rearranging gives

$$E_j = \alpha + 2\beta\cos\left(\frac{j\pi}{n+1}\right). \tag{6-72}$$

n is the order of the polyene or the number of atoms in the chain and each j from 1 to n indicates one of the solutions or roots in the secular equation. A few of the solutions to Eq.(6-72) are shown below. We remember from Section 6.5 that β and α are negative numbers.

$n = 2; j = 1, 2;$ *ethylene*

$$j = 1; \qquad E_1 = \alpha + 2\beta\cos\frac{\pi}{3} = \alpha + \beta$$

$$j = 2; \qquad E_2 = \alpha + 2\beta\cos\frac{2\pi}{3} = \alpha - \beta \tag{6-73}$$

$n = 3; j = 1, 2, 3;$ *allene radical*

$$j = 1; \qquad E_1 = \alpha + (2)^{1/2}\beta$$

$$j = 2; \qquad E_2 = \alpha$$

$$j = 3; \qquad E_3 = \alpha - (2)^{1/2}\beta \tag{6-74}$$

$n = 4; j = 1, 2, 3, 4;$ *butadiene*

$$j = 1; \qquad E_1 = \alpha + 1.618\beta$$

$$j = 2; \qquad E_2 = \alpha + 0.618\beta$$

$$j = 3; \qquad E_3 = \alpha - 0.618\beta$$

$$j = 4; \qquad E_4 = \alpha - 1.618\beta. \tag{6-75}$$

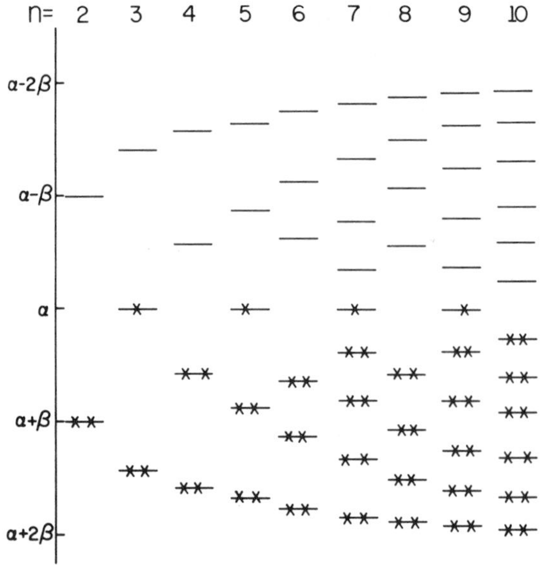

Figure 6-9 Diagram showing the π-electron Hückel energy levels as a function of the length of the polyene chain.

Figure 6-9 demonstrates the energy levels in the electrically neutral polyenes as n increases where the electrons are filled in according to the Pauli principle. Note that the energy for the transition from the highest occupied to the lowest unoccupied levels decreases as n increases. This decrease in the $\pi \longrightarrow \pi^*$ transition energy with increasing molecular size is observed experimentally. As $n \longrightarrow \infty$, a band of energy levels will occur.

The eigenfunctions associated with the solutions in Eqs.(6-73) through (6-75) are obtained from Eq.(6-68) and the secular equation, $(\mathbf{F}^\chi - \mathbf{S}^\chi E_j)\mathbf{a}_j = \mathbf{0}$ for $\mathbf{S}^\chi = \mathbf{1}$ and $\boldsymbol{\Phi} = \boldsymbol{\chi}\mathbf{a}$. For instance, when $n = 3$, we write

$$a_1(\alpha - E) + a_2\beta + 0 = 0, \qquad a_1\beta + a_2(\alpha - E) + a_3\beta = 0,$$
$$0 + a_2\beta + a_3(\alpha - E) = 0.$$

Substituting E_1, E_2, and E_3 from Eq.(6-74), and remembering that ψ_i must be normalized ($a_1^2 + a_2^2 + a_3^2 = 1.0$), gives

Lowest level; $E_1 = \alpha + (2)^{1/2}\beta;\ \psi_1$

$$a_1[-(2)^{1/2}\beta] + a_2\beta = 0$$
$$a_1\beta - a_2(2)^{1/2}\beta + a_3\beta = 0$$
$$a_2\beta - a_3(2)^{1/2}\beta = 0$$
$$a_1^2 + a_2^2 + a_3^2 = 1.0.$$

The solution of these equations is $a_1 = 1/2$, $a_2 = 1/(2)^{1/2}$, and $a_3 = 1/2$, which gives

$$\psi_1 = \frac{1}{2}\chi_1 + \frac{1}{(2)^{1/2}}\chi_2 + \frac{1}{2}\chi_3. \tag{6-76}$$

The solutions for the other levels are also easily obtained:

Middle level; $E_2 = \alpha$; ψ_2

$$\psi_2 = \frac{1}{(2)^{1/2}}\chi_1 - \frac{1}{(2)^{1/2}}\chi_3 \tag{6-77}$$

Highest level; $E_3 = \alpha - (2)^{1/2}\beta$; ψ_3

$$\psi_3 = \frac{1}{2}\chi_1 - \frac{1}{(2)^{1/2}}\chi_2 + \frac{1}{2}\chi_3. \tag{6-78}$$

Equations (6-76) through (6-78) show an increasing number of nodes in the functions as the energy is increased, which is suggestive of the free electron particle-in-a-box system as mentioned in Problem 4-5. The energy levels are given by $E_j = j^2h^2/8mL^2$ with $j = 1, 2, 3, \ldots, n$, where j is an integer indicating the energy levels up to n (the number of atoms in the chain), L is the length of the chain, and m is the mass of the electron. The particle-in-a-box eigenfunctions are $\psi_j = (2/L)^{1/2}\sin(j\pi x/L)$. The lower energy levels coalesce as L increases and the nodal structure is identical to the Hückel molecular orbitals. The particle-in-a-box free electron model has been applied extensively to the linear polyene chains and the cyclic systems [61]. Both the free electron and the Hückel models give similar intuitive feelings for the π electrons in a molecule.

Returning to the Hückel theory, it is interesting to consider the calculated energies and the concept of relative molecular stability. The total π-electron energy of a system is a sum over all occupied levels given by

$$E_T = \sum_{j=1}^{n} n_j E_j(\text{occupied}), \tag{6-79}$$

where n_j is the number of electrons in the jth level. The localized energy for the system is equal to α for each heavy atom and β for each atom participating in a localized double bond,

$$E_l = n\alpha + 2n'\beta, \tag{6-80}$$

where n is the number of heavy atoms and n' is the number of localized double bonds. The additional stabilization of the molecule or delocalization energy is defined as

$$E_d = E_T - E_l, \tag{6-81}$$

which is apparently the energy gained in allowing the electrons, which were originally at each center, to move around throughout the whole molecule.

In the previous examples of $n = 2$, 3, and 4 in Eqs.(6-73), (6-74), and (6-75), the delocalization energies are $E_d = 0, 0.828\beta$, and 0.472β, respectively. Thus, as the chain length increases, less delocalization energy is available for stabilization.

Charge densities and bond orders in the simple Hückel theory are also easily obtained. Charge densities are defined in Eq.(6-65) and bond orders are defined in the Hückel theory by

$$P_{ab} = \sum_j n_j \sum_{m,l} a_{mj}^* a_{lj}, \tag{6-82}$$

where m and l are on adjacent a and b atoms and n_j is the number of electrons in the jth occupied orbital. Bond orders and charges in the allene radical and butadiene are given below. The transformation matrix for the allene radical is given in Eqs.(6-76) to (6-78):

$$\mathbf{\Phi} = \chi \mathbf{a} = (\chi_1 \quad \chi_2 \quad \chi_3)\begin{pmatrix} \frac{1}{2} & \frac{1}{(2)^{1/2}} & \frac{1}{2} \\ \frac{1}{(2)^{1/2}} & 0 & -\frac{1}{(2)^{1/2}} \\ \frac{1}{2} & -\frac{1}{(2)^{1/2}} & \frac{1}{2} \end{pmatrix} \tag{6-83}$$

$$q_1 = 1.0 \qquad q_2 = 1.0 \qquad q_3 = 1.0$$

$$C_1 \text{———} C_2 \text{———} C_3$$

$$P_{12} = \frac{1}{(2)^{1/2}} \qquad P_{23} = \frac{1}{(2)^{1/2}},$$

which indicates equal charges and bond distances. In the case of butadiene the results are

$$\mathbf{\Phi} = \chi \mathbf{a} = (\chi_1 \quad \chi_2 \quad \chi_3 \quad \chi_4)\begin{pmatrix} 0.37 & 0.60 & 0.60 & 0.37 \\ 0.60 & 0.37 & -0.37 & -0.60 \\ 0.60 & -0.37 & -0.37 & 0.60 \\ 0.37 & -0.60 & 0.60 & -0.37 \end{pmatrix}$$

$$q_1 = 1.0 \qquad q_2 = 1.0 \qquad q_3 = 1.0 \qquad q_4 = 1.0$$

$$C_1 \text{———} C_2 \text{———} C_3 \text{———} C_4 \tag{6-84}$$

$$P_{12} = 0.88 \qquad P_{23} = 0.44 \qquad P_{34} = 0.88,$$

which indicates equal charges but unequal bond distances. Larger bond orders are found at the end of the molecule, which predicts correctly the shorter $C_1—C_2$ and $C_3—C_4$ bonds relative to the center $C_2—C_3$ bond.

It is evident that several conformations exist for many of the linear polyene molecules. For instance, butadiene has *cis* and *trans* forms with all atoms in the plane and angles of approximately 120°. The simple Hückel method does not distinguish energetically between the two conformers; however, the extended Hückel theory discussed in Section 6.6 correctly predicts the *trans* form to be more stable than the *cis* form.

Cyclic Conjugated Systems

Cyclic conjugated alternate single-double bond hydrocarbons can also be treated in a straightforward manner by the simple Hückel method. Cyclic hydrocarbon polyenes are similar to the linear polyenes shown earlier, but in the cyclic case the C_1 and C_n atoms are bonded together. The Hamiltonian matrix is easily constructed and the secular equation is identical to the result in Eq.(6-68) except that in the cyclic case a β appears in place of zero for the C_1-C_n matrix element. Following steps given earlier for the chain molecules, we find the following energy expression for the cyclic polyenes:

$$E_j = \alpha + 2\beta \cos \frac{2\pi j}{n}, \qquad j = 1, 2, 3, 4, \ldots, n. \tag{6-85}$$

The results for a few interesting cyclic systems are listed below.

$n = 3; j = 1, 2, 3;$ *cyclopropene radical*

$$\left.\begin{aligned} j = 1; \quad & E_1 = \alpha - \beta \\ j = 2; \quad & E_2 = \alpha - \beta \\ j = 3; \quad & E_3 = \alpha + 2\beta \\ E_d = E_T - E_l = \beta & \end{aligned}\right\} \tag{6-86}$$

$n = 4; j = 1, 2, 3, 4;$ *cyclobutadiene*

$$\left.\begin{aligned} j = 1; \quad & E_1 = \alpha \\ j = 2; \quad & E_2 = \alpha - 2\beta \\ j = 3; \quad & E_3 = \alpha \\ j = 4; \quad & E_4 = \alpha + 2\beta \\ E_d = 0 & \end{aligned}\right\} \tag{6-87}$$

$n = 6; j = 1, 2, 3, 4, 5, 6;$ *benzene*

$$\left.\begin{aligned} j = 1; \quad & E_1 = \alpha + \beta \\ j = 2; \quad & E_2 = \alpha - \beta \\ j = 3; \quad & E_3 = \alpha - 2\beta \\ j = 4; \quad & E_4 = \alpha - \beta \\ j = 5; \quad & E_5 = \alpha + \beta \\ j = 6; \quad & E_6 = \alpha + 2\beta \\ E_d = 2\beta & \end{aligned}\right\} \tag{6-88}$$

According to the results above, the cyclopropene radical and benzene are stabilized by the π electrons, however; cyclobutadiene is not stabilized. The coefficients giving the eigenfunctions in benzene are easily obtained and they show nodal characteristics similar to the corresponding real particle-in-a-ring functions that are discussed in Section 4.1 as given by $\Phi_m = (1/\pi)^{1/2} \cos m\phi;\ m = 0, \pm 1, \pm 2, \pm 3, \ldots.$ A direct comparison is shown here. The degeneracies and nodal structure of the eigenfunctions of both the Hückel and particle-in-a-ring theories are evident from these equations.

	Hückel Orbitals	*Particle-in-a-Ring Orbitals*
$j = 3;$	$\psi_3 = \left(\frac{1}{6}\right)^{1/2}(\chi_1 - \chi_2 + \chi_3 - \chi_4 + \chi_5 - \chi_6);$	$\left(\frac{1}{\pi}\right)^{1/2} \cos(3\phi)$
$j = 2, 4;$	$\left.\begin{aligned} \psi_2 &= \left(\frac{1}{4}\right)^{1/2}(\chi_1 - \chi_2 + \chi_3 - \chi_4) \\ \psi_4 &= \left(\frac{1}{4}\right)^{1/2}(\chi_2 - \chi_3 + \chi_5 - \chi_6) \end{aligned}\right\};$	$\left(\frac{1}{\pi}\right)^{1/2} \cos(\pm 2\phi)$
$j = 1, 5;$	$\left.\begin{aligned} \psi_5 &= \left(\frac{1}{4}\right)^{1/2}(\chi_1 + \chi_2 - \chi_4 - \chi_5) \\ \psi_1 &= \left(\frac{1}{4}\right)^{1/2}(\chi_2 + \chi_3 - \chi_5 - \chi_6) \end{aligned}\right\};$	$\left(\frac{1}{\pi}\right)^{1/2} \cos(\pm \phi)$
$j = 6;$	$\psi_6 = \left(\frac{1}{6}\right)^{1/2}(\chi_1 + \chi_2 + \chi_3 + \chi_4 + \chi_5 + \chi_6);$	$\left(\frac{1}{\pi}\right)^{1/2}$

(6-89)

The Hückel molecular orbitals for any system can be set up in a straightforward manner. Only the simple linear polyenes [Eq.(6-72)] and cyclic systems [Eq.(6-85)] have simple closed solutions. Consider methylene cyclopropene,

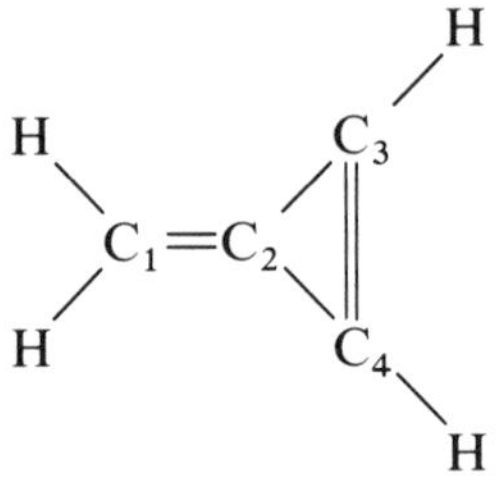

,

where all atoms are again in the plane. The secular equation is

$$\begin{array}{c} \\ C_1 \\ C_2 \\ C_3 \\ C_4 \end{array} \begin{array}{c} \begin{array}{cccc} C_1 & C_2 & C_3 & C_4 \end{array} \\ \begin{vmatrix} \alpha - E & \beta & 0 & 0 \\ \beta & \alpha - E & \beta & \beta \\ 0 & \beta & \alpha - E & \beta \\ 0 & \beta & \beta & \alpha - E \end{vmatrix} \end{array} = 0,$$

and a straightforward solution to this equation can be obtained to give

$$\begin{aligned} E_1 &= \alpha - 1.481\beta, \qquad & E_2 &= \alpha - \beta, \\ E_3 &= \alpha + 0.311\beta, \qquad & E_4 &= \alpha + 2.170\beta. \end{aligned} \tag{6-90}$$

Thus, $E_d = 0.962\beta$ and methylene cyclopropene is predicted to be stable according to the calculated delocalization energy. The electron densities and bond orders are all easily obtained [62].

Atoms other than carbon that may participate in the Hückel molecular orbitals are easily incorporated into the theory. For instance, cyclopropeneone is quite similar to methylene cyclopropene.

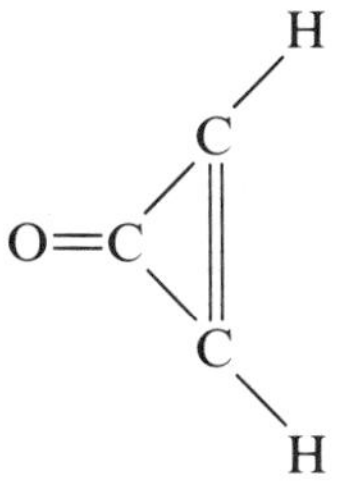

$$\begin{vmatrix} \alpha_O - E & \beta_{CO} & 0 & 0 \\ \beta_{CO} & \alpha_C - E & \beta_{CC} & \beta_{CC} \\ 0 & \beta_{CC} & \alpha_C - E & \beta_{CC} \\ 0 & \beta_{CC} & \beta_{CC} & \alpha_C - E \end{vmatrix} = 0$$

β_{CO} and α_O can be treated parametrically in terms of β_{CC} and α_C according to

$$\alpha_O = k\alpha_C, \qquad \beta_{OC} = k'\beta_{CC}, \tag{6-91}$$

where k and k' are empirical parameters. The values of k and k' for oxygen as well as for any other atom could be obtained from the valence state ionization energies in Table 6-12. For instance, reasonable values for k and k' might be

$$k = \frac{E_O}{E_C}, \qquad k' = \frac{(E_O + E_C)}{E_O}. \tag{6-92}$$

E_O and E_C are the valence state energies for $2p$ electrons in oxygen and carbon (see Table 6-12). These parameters merely indicate that $\alpha_O < \alpha_C$ and $\beta_{CO} < \beta_{CC}$; that is, the oxygen atom provides a deeper well for an electron than the carbon atom. Similar considerations are possible for other atoms. Several books are available that treat the π-electron Hückel theory in more detail [63].

Orbital Symmetry and Chemical Reactions

The concepts of orbitals and energy levels developed here can also be used to predict products of certain reactions that conserve orbital symmetry [64]. The butadiene energy levels and Hückel molecular orbitals are given in Eqs.(6-75) and (6-84), respectively. Comparing these with the σ and π orbitals of cyclobutene, we indicate the wavefunctions in the upper part of Fig. 6-10. We now examine the *conrotatory* and *disrotatory* breaking of the long C—C bond in cyclobutene, as shown in the lower part of Fig. 6-10, where the molecules on either side of the ring compound indicate transition intermediates. Now returning to the upper part of the diagram we see that a *conrotatory* motion will correlate σ with Φ_2, π with Φ_1, π^* with Φ_4, and σ^* with Φ_3. On the other hand, a *disrotatory* motion will correlate

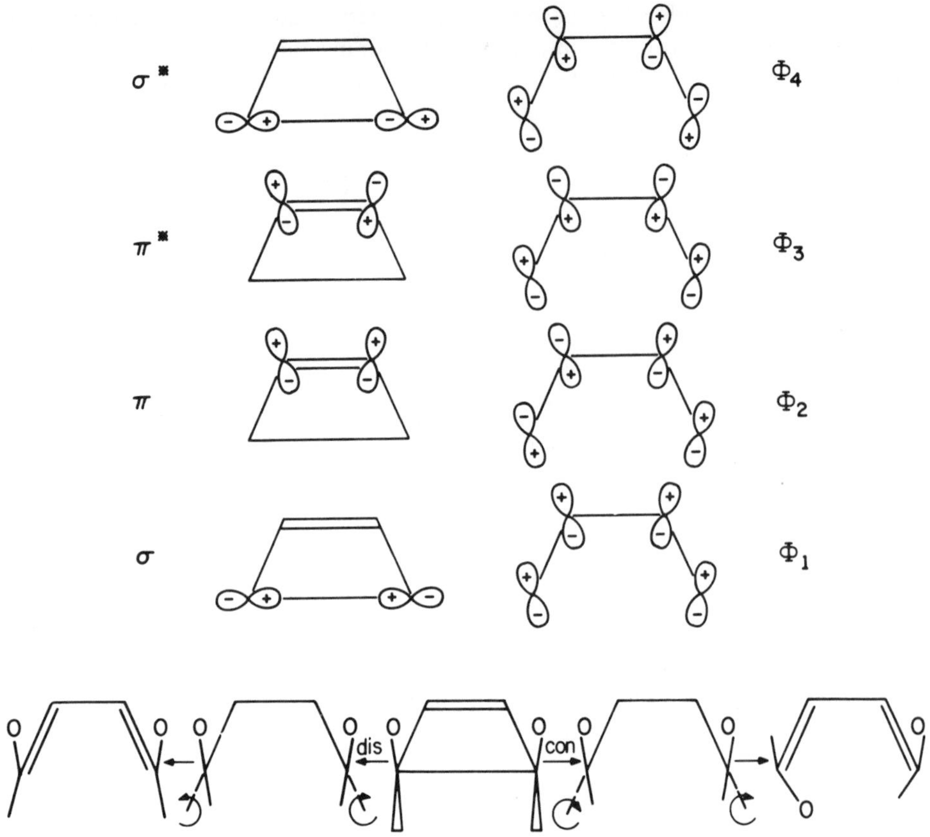

Figure 6-10 Diagram showing the molecular orbitals that are important through symmetry in determining the reactions between butadiene and cyclobutene. The upper part of the diagram shows the four sets of molecular orbitals and the lower part shows the *conrotatory* and *disrotatory* breaking of the long C—C bond in cyclobutene.

σ with Φ_1, π with Φ_3, π^* with Φ_2, and σ^* with Φ_4. Thus, the *disrotatory* motion requires a larger excitation (Φ_1 and Φ_3) to execute the transition than the corresponding *conrotatory* motion (Φ_1 and Φ_2). Thus, the *conrotatory* motion will lead to the facile reaction. A large number of simple organic reactions can be predicted by applying these concepts in orbital symmetry [65].

6.8 MAGNETIC INTERACTIONS IN MOLECULES IN THE ABSENCE OF NUCLEAR SPINS

In this section we shall examine both the effects of magnetic field interactions and the effects of the intramolecular electronic-rotational interactions represented by $T_{re} = m\boldsymbol{\omega}\cdot\sum_i \boldsymbol{r}_i \times \boldsymbol{v}_i$ in Eq.(6-7). We shall limit the discussion in this section to diamagnetic molecules that have zero electronic spin and orbital angular momentum in the ground electronic state. This limitation is not severe as a great majority of molecules satisfy this criterion. We also ignore the effects of molecular vibrations. Rewriting Eq.(6-7) for a rigid molecule ($v_\alpha = 0$) gives

$$T = \frac{m}{2}\sum_i \boldsymbol{v}_i \cdot \boldsymbol{v}_i + \frac{1}{2}\boldsymbol{\omega}\cdot\mathbf{I}_n\cdot\boldsymbol{\omega} + \frac{1}{2}\boldsymbol{\omega}\cdot\mathbf{I}_e\cdot\boldsymbol{\omega} + m\boldsymbol{\omega}\cdot\sum_i \boldsymbol{r}_i \times \boldsymbol{v}_i, \tag{6-93}$$

where we have also separated the rotational term into a pure nuclear term, $\mathbf{I}_n = \sum_\alpha M_\alpha(r_\alpha^2\mathbf{1} - \boldsymbol{r}_\alpha\boldsymbol{r}_\alpha)$, from the nuclear c.m. and a pure electronic term,

$$\mathbf{I}_e = m\sum_i (r_i^2\mathbf{1} - \boldsymbol{r}_i\boldsymbol{r}_i),$$

from the electronic c.m. We also assume that the nuclear c.m. coincides with the electronic c.m. The errors made by this approximation are negligible. The total angular momentum, $\boldsymbol{J}$, is obtained by Lagrangian mechanics by the first derivative of the kinetic energy (the potential energy is zero) with respect to the angular velocity [see Section 2.3, specifically Eq.(2-67)],

$$\boldsymbol{J} = \frac{\partial T}{\partial \boldsymbol{\omega}} = \boldsymbol{\omega}\cdot\mathbf{I}_n + \boldsymbol{\omega}\cdot\mathbf{I}_e + m\sum_i \boldsymbol{r}_i \times \boldsymbol{v}_i. \tag{6-94}$$

We can now rewrite Eq.(6-93) for the kinetic energy to give

$$T = \frac{1}{2}\boldsymbol{\omega}\cdot\boldsymbol{J} + \frac{1}{2}\sum_i \boldsymbol{p}_i\cdot\boldsymbol{v}_i, \tag{6-95}$$

where $\boldsymbol{J}$ is given in Eq.(6-94). The linear momentum of the ith electron, $\boldsymbol{p}_i$, is also obtained from Eq.(6-93) by the first derivative of T with respect to the velocity, giving [remember Eq.(2-66) and $\boldsymbol{\omega}\cdot\boldsymbol{r}_i \times \boldsymbol{v}_i = \boldsymbol{v}_i\cdot\boldsymbol{\omega}\times\boldsymbol{r}_i$]

$$\boldsymbol{p}_i = \frac{\partial T}{\partial \boldsymbol{v}_i} = m\boldsymbol{v}_i + m\boldsymbol{\omega}\times\boldsymbol{r}_i. \tag{6-96}$$

Substituting $\boldsymbol{v}_i$ from this equation into Eq.(6-95) gives an expression in terms of the linear momentum:

$$T = \frac{1}{2}\boldsymbol{\omega}\cdot\boldsymbol{J} + \frac{1}{2m}\sum_i p_i^2 - \frac{1}{2}\sum_i \boldsymbol{p}_i\cdot\boldsymbol{\omega}\times\boldsymbol{r}_i = \frac{1}{2}\boldsymbol{\omega}\cdot\boldsymbol{J} - \frac{1}{2}\boldsymbol{\omega}\cdot\sum_i \boldsymbol{r}_i\times\boldsymbol{p}_i + \frac{1}{2m}\sum_i p_i^2$$

$$= \frac{1}{2}\boldsymbol{\omega}\cdot\boldsymbol{J} - \frac{1}{2}\boldsymbol{\omega}\cdot\boldsymbol{L} + \frac{1}{2m}\sum_i p_i^2 = \frac{1}{2}\boldsymbol{\omega}\cdot(\boldsymbol{J} - \boldsymbol{L}) + \frac{1}{2m}\sum_i p_i^2. \tag{6-97}$$

We have also defined

$$\sum_i \boldsymbol{r}_i \times \boldsymbol{p}_i = \boldsymbol{L} \tag{6-98}$$

as the intrinsic electronic angular momentum or the electronic angular momentum in the rotating frame. The last term in Eq.(6-97), $(1/2m)\sum_i p_i^2 = T_e$, is included in the electronic Hamiltonian to determine the electronic wavefunctions for the non-rotating molecule as shown in Section 6.1, giving rise to a set of zero-order electronic states,

$$\psi_0, \psi_1, \psi_2, \ldots, \psi_k, \tag{6-99}$$

where ψ_0 is the ground electronic state and ψ_k is the kth excited electronic state. We are assuming that the ground electronic state, ψ_0, does not possess any electronic angular momentum, however, the excited states, ψ_k, may possess electronic angular momentum.

We now compare Eqs.(6-97) and (6-93). Substituting Eq.(6-96) into Eq.(6-97) and comparing with Eq.(6-93) shows that $\boldsymbol{\omega}\cdot(\boldsymbol{J}-\boldsymbol{L}) = \boldsymbol{\omega}\cdot\mathbf{I}_n\cdot\boldsymbol{\omega}$. Evidently, $(\boldsymbol{J}-\boldsymbol{L})$ can be expressed as

$$(\boldsymbol{J}-\boldsymbol{L}) = \mathbf{I}_n\cdot\boldsymbol{\omega}, \tag{6-100}$$

where $\mathbf{I}_n$ is the moment of inertia for the point mass nuclei as described before. In the *principal* inertial (nuclei only) axis system, $\mathbf{I}_n$ is diagonal and we can write $(\boldsymbol{J}-\boldsymbol{L})_x = (\mathbf{I}_n)_{xx}\omega_x$ and cyclic permutations. We then define the inverse principal inertial tensor, $\mathbf{I}_n^{-1}$, with diagonal elements, $1/(\mathbf{I}_n)_{xx}$, and so on, where we require $\mathbf{I}_n^{-1}\cdot\mathbf{I}_n = \mathbf{I}_n\cdot\mathbf{I}_n^{-1} = \mathbf{1}$. Using this we can rewrite Eq.(6-100) to give

$$\boldsymbol{\omega} = (\boldsymbol{J}-\boldsymbol{L})\cdot\mathbf{I}_n^{-1}. \tag{6-101}$$

Substituting this result into Eq.(6-97), dropping the $(1/2m)\sum_i p_i^2$ term, and expanding gives

$$T = \frac{1}{2}(\boldsymbol{J}-\boldsymbol{L})\cdot\mathbf{I}_n^{-1}\cdot(\boldsymbol{J}-\boldsymbol{L}) = \frac{1}{2}\boldsymbol{J}\cdot\mathbf{I}_n^{-1}\cdot\boldsymbol{J} - \boldsymbol{J}\cdot\mathbf{I}_n^{-1}\cdot\boldsymbol{L} + \frac{1}{2}\boldsymbol{L}\cdot\mathbf{I}_n^{-1}\cdot\boldsymbol{L}$$

$$= \frac{1}{2}\sum_g \frac{J_g^2}{(\mathbf{I}_n)_{gg}} - \sum_g \frac{J_g L_g}{(\mathbf{I}_n)_{gg}} + \frac{1}{2}\sum_g \frac{L_g^2}{(\mathbf{I}_n)_{gg}}. \tag{6-102}$$

The sums over g are over the three principal inertial axes. The first term leads to the rotational state eigenvalues for the square of the total angular momentum. The second term is the rotational-electronic coupling term, which contributes to the rotational magnetic moment and the spin rotation interaction. The last and smallest term can be dropped in the limit of interest here where $\boldsymbol{L} = 0$ in the ground electronic state.

External Magnetic Field Effects

Next we consider the effects of a magnetic field on a nonvibrating molecule. We start with a discussion of the electrons in a molecule. Using an obvious generalization of the one-electron Hamiltonian in Eq.(4-21), we write the field-dependent terms for the electrons as

$$\mathscr{H} = \frac{e}{2mc}\boldsymbol{H}\cdot\boldsymbol{L} + \frac{e^2}{8mc^2}\boldsymbol{H}\cdot\sum_i (r_i^2\mathbf{1} - \boldsymbol{r}_i\boldsymbol{r}_i)\cdot\boldsymbol{H}, \tag{6-103}$$

where $\boldsymbol{L}$ is the total intrinsic electronic angular momentum [Eq.(6-98)] and i indicates electrons. The corresponding terms that depend on the field interaction with the rigid nuclei are written by analogy. The first term analogous to $(e/2mc)\boldsymbol{H}\cdot\boldsymbol{L}$ is written as

$$-\frac{e}{2c}\boldsymbol{H}\cdot\sum_{\alpha} Z_{\alpha}\boldsymbol{r}_{\alpha}\times\boldsymbol{v}_{\alpha} = -\frac{e}{2c}\boldsymbol{H}\cdot\sum_{\alpha} Z_{\alpha}\boldsymbol{r}_{\alpha}\times\boldsymbol{\omega}_{\alpha}\times\boldsymbol{r}_{\alpha}$$

$$= -\boldsymbol{H}\cdot\left[\frac{e}{2c}\sum_{\alpha} Z_{\alpha}(r_{\alpha}^{2}\mathbf{1} - \boldsymbol{r}_{\alpha}\boldsymbol{r}_{\alpha})\right]\cdot\boldsymbol{\omega},$$

where α indicates the nuclei. The nuclear analog to the last term in Eq.(6-103) is written as

$$\frac{e^2}{8c^2}\boldsymbol{H}\cdot\sum_{\alpha}\frac{Z_{\alpha}^{2}}{M_{\alpha}}(r_{\alpha}^{2}\mathbf{1} - \boldsymbol{r}_{\alpha}\boldsymbol{r}_{\alpha})\cdot\boldsymbol{H}.$$

Adding these nuclear terms to Eq.(6-103) and adding this result to Eq.(6-102) gives the complete perturbation Hamiltonian in the presence of the field:

$$\mathscr{H}' = \frac{1}{2}\boldsymbol{J}\cdot\mathbf{I}_{n}^{-1}\cdot\boldsymbol{J} - \boldsymbol{L}\cdot\overline{\boldsymbol{\omega}} + \frac{1}{2}\boldsymbol{L}\cdot\mathbf{I}_{n}^{-1}\cdot\boldsymbol{L} + \frac{e^2}{8c^2 m}\boldsymbol{H}\cdot\left[\sum_{i}(r_{i}^{2}\mathbf{1} - \boldsymbol{r}_{i}\boldsymbol{r}_{i})\right.$$

$$\left. + m\sum_{\alpha}\frac{Z_{\alpha}^{2}}{M_{\alpha}}(r_{\alpha}^{2}\mathbf{1} - \boldsymbol{r}_{\alpha}\boldsymbol{r}_{\alpha})\right]\cdot\boldsymbol{H} - \frac{e}{2c}\sum_{\alpha} Z_{\alpha}\boldsymbol{H}\cdot(r_{\alpha}^{2}\mathbf{1} - \boldsymbol{r}_{\alpha}\boldsymbol{r}_{\alpha})\cdot\mathbf{I}_{n}^{-1}\cdot\boldsymbol{J}$$

$$\overline{\boldsymbol{\omega}} = \mathbf{I}_{n}^{-1}\cdot\boldsymbol{J} - \frac{e}{2mc}\boldsymbol{H} - \frac{e}{2c}\mathbf{I}_{n}^{-1}\cdot\sum_{\alpha} Z_{\alpha}(r_{\alpha}^{2}\mathbf{1} - \boldsymbol{r}_{\alpha}\boldsymbol{r}_{\alpha})\cdot\boldsymbol{H}. \qquad (6\text{-}104)$$

The $(e/2mc)\boldsymbol{H}$ contribution to $\overline{\boldsymbol{\omega}}$ is the well-known Larmour frequency [see Eq.(4-30) and associated discussion].

We now use standard perturbation theory to obtain the effect of the Hamiltonian in Eq.(6-104) on the zero-order electronic and rotational states for the rigid molecule. First we correct for the electronic effects by using Eq.(6-99) as the zero-order electronic wavefunctions. According to perturbation theory (Section 3.4) the corrections are easily obtained:

1. *First order*

$$E^{(1)} = \frac{1}{2}\boldsymbol{J}\cdot\mathbf{I}_{n}^{-1}\cdot\boldsymbol{J} + \frac{1}{2}\boldsymbol{H}\cdot\left[\left\langle 0\left|\frac{e^2}{4c^2 m}\sum_{i}(r_{i}^{2}\mathbf{1} - \boldsymbol{r}_{i}\boldsymbol{r}_{i})\right|0\right\rangle + \frac{e^2}{4c^2}\sum_{\alpha}\frac{Z_{\alpha}^{2}}{M_{\alpha}}(r_{\alpha}^{2}\mathbf{1} - \boldsymbol{r}_{\alpha}\boldsymbol{r}_{\alpha})\right]\cdot\boldsymbol{H}$$

$$-\boldsymbol{H}\cdot\left[\frac{e}{2c}\sum_{\alpha} Z_{\alpha}(r_{\alpha}^{2}\mathbf{1} - \boldsymbol{r}_{\alpha}\boldsymbol{r}_{\alpha})\cdot\mathbf{I}_{n}^{-1}\cdot\boldsymbol{J}\right] + \frac{1}{2}\sum_{k>0}\langle 0|\boldsymbol{L}|k\rangle\cdot\mathbf{I}_{n}^{-1}\cdot\langle k|\boldsymbol{L}|0\rangle \qquad (6\text{-}105)$$

2. *Second order*

$$E^{(2)} = \sum_{k>0} \frac{\langle 0|\boldsymbol{L}\cdot\boldsymbol{\varpi}|k\rangle\langle k|\boldsymbol{L}\cdot\boldsymbol{\varpi}|0\rangle}{E_0 - E_k} = \boldsymbol{J}\cdot\mathbf{I}_n^{-1}\cdot\mathbf{A}\cdot\mathbf{I}_n^{-1}\cdot\boldsymbol{J} + \frac{1}{2}\boldsymbol{H}\cdot\left(\frac{e^2}{2m^2c^2}\mathbf{A}\right)\cdot\boldsymbol{H}$$

$$+ \frac{e^2}{4c^2}\boldsymbol{H}\cdot\mathbf{I}_n^{-1}\cdot\sum_\alpha Z_\alpha(r_\alpha^2\mathbf{1} - \boldsymbol{r}_\alpha\boldsymbol{r}_\alpha)\cdot\mathbf{A}\cdot\sum_\alpha Z_\alpha(r_\alpha^2\mathbf{1} - \boldsymbol{r}_\alpha\boldsymbol{r}_\alpha)\cdot\mathbf{I}_n^{-1}\cdot\boldsymbol{H}$$

$$- \frac{e}{mc}\boldsymbol{H}\cdot\mathbf{A}\cdot\mathbf{I}_n^{-1}\cdot\boldsymbol{J} - \boldsymbol{H}\cdot\left[\frac{e}{c}\sum_\alpha Z_\alpha(r_\alpha^2\mathbf{1} - \boldsymbol{r}_\alpha\boldsymbol{r}_\alpha)\cdot\mathbf{I}_n^{-1}\cdot\mathbf{A}\cdot\mathbf{I}_n^{-1}\cdot\boldsymbol{J}\right]$$

$$+ \frac{e^2}{2c^2m}\boldsymbol{H}\cdot\mathbf{I}_n^{-1}\cdot\mathbf{A}\cdot\sum_\alpha Z_\alpha(r_\alpha^2\mathbf{1} - \boldsymbol{r}_\alpha\boldsymbol{r}_\alpha)\cdot\boldsymbol{H}$$

$$\mathbf{A} = \sum_{k>0}\frac{\langle 0|\boldsymbol{L}|k\rangle\langle k|\boldsymbol{L}|0\rangle}{E_0 - E_k}. \tag{6-106}$$

Before examining the third-order corrections, we shall combine the first terms of Eqs.(6-105) and (6-106) to give the zero-field or "pure" rotational contributions, which we designate as the rigid rotor Hamiltonian, $\mathscr{H}_{rr}$,

$$\mathscr{H}_{rr} = \tfrac{1}{2}\boldsymbol{J}\cdot\mathbf{I}_n^{-1}\cdot\boldsymbol{J} + \boldsymbol{J}\cdot\mathbf{I}_n^{-1}\cdot\mathbf{A}\cdot\mathbf{I}_n^{-1}\cdot\boldsymbol{J}$$
$$= \tfrac{1}{2}\boldsymbol{J}\cdot[\mathbf{I}_n^{-1}\cdot(\mathbf{1} + 2\mathbf{A}\cdot\mathbf{I}_n^{-1})]\cdot\boldsymbol{J} = \tfrac{1}{2}\boldsymbol{J}\cdot\mathbf{I}_{\text{eff}}^{-1}\cdot\boldsymbol{J}, \tag{6-107}$$

where $\mathbf{I}_{\text{eff}}^{-1}$ is the measured inverse moment of inertia for the molecule at zero magnetic field. Of course, it is evident that $\mathbf{1} \gg 2\mathbf{A}\cdot\mathbf{I}_n^{-1}$, where $\mathbf{A}\cdot\mathbf{I}_n^{-1}$ is on the order of the electron-proton mass ratio. We now sum the terms in Eqs.(6-105) and (6-106) to give the rotational Hamiltonian up to second order in electronic correction. We use $\mathbf{I}_{\text{eff}}^{-1}$ whenever possible and we drop the small last term in Eq.(6-105) to give

$$\mathscr{H}' = \frac{1}{2}\boldsymbol{J}\cdot\mathbf{I}_{\text{eff}}^{-1}\cdot\boldsymbol{J} - \boldsymbol{H}\cdot\left[\frac{e}{2c}\sum_\alpha Z_\alpha(r_\alpha^2\mathbf{1} - \boldsymbol{r}_\alpha\boldsymbol{r}_\alpha)\cdot\mathbf{I}_{\text{eff}}^{-1}\right]\cdot\boldsymbol{J} - \boldsymbol{H}\cdot\left(\frac{e}{mc}\mathbf{A}\cdot\mathbf{I}_n^{-1}\right)\cdot\boldsymbol{J}$$

$$- \frac{1}{2}\boldsymbol{H}\cdot\boldsymbol{\chi}^d\cdot\boldsymbol{H} - \frac{1}{2}\boldsymbol{H}\cdot\boldsymbol{\chi}^p\cdot\boldsymbol{H} + \frac{1}{2}\boldsymbol{H}\cdot\boldsymbol{\gamma}\cdot\boldsymbol{H}$$

$$\boldsymbol{\gamma} = \frac{e^2}{2c^2}\mathbf{I}_n^{-1}\cdot\sum_\alpha Z_\alpha(r_\alpha^2\mathbf{1} - \boldsymbol{r}_\alpha\boldsymbol{r}_\alpha)\cdot\mathbf{A}\cdot\sum_\alpha Z_\alpha(r_\alpha^2\mathbf{1} - \boldsymbol{r}_\alpha\boldsymbol{r}_\alpha)\cdot\mathbf{I}_n^{-1}$$

$$+ \frac{e^2}{c^2m}\mathbf{I}_n^{-1}\cdot\mathbf{A}\cdot\sum_\alpha Z_\alpha(r_\alpha^2\mathbf{1} - \boldsymbol{r}_\alpha\boldsymbol{r}_\alpha) + \frac{e^2}{4c^2}\sum_\alpha\frac{Z_\alpha^2}{M_\alpha}(r_\alpha^2\mathbf{1} - \boldsymbol{r}_\alpha\boldsymbol{r}_\alpha)$$

$$\boldsymbol{\chi}^d = -\frac{e^2}{4c^2m}\left\langle 0\left|\sum_i(r_i^2\mathbf{1} - \boldsymbol{r}_i\boldsymbol{r}_i)\right|0\right\rangle$$

$$\boldsymbol{\chi}^p = -\frac{e^2}{2m^2c^2}\mathbf{A} = -\frac{e^2}{2m^2c^2}\sum_{k>0}\frac{\langle 0|\boldsymbol{L}|k\rangle\langle k|\boldsymbol{L}|0\rangle}{E_0 - E_k}$$

$$\boldsymbol{\chi} = \boldsymbol{\chi}^d + \boldsymbol{\chi}^p. \tag{6-108}$$

χ^d, χ^p, and χ are the diamagnetic, paramagnetic, and total magnetic susceptibilities, respectively [66]. Careful examination of γ will show that the elements in γ are smaller than the elements in χ^d and χ^p by the electron-proton mass ratio.

We now return to the third-order terms and note that the cross terms between $\boldsymbol{L}\cdot\boldsymbol{\bar{\omega}}$ and $\frac{1}{2}\boldsymbol{L}\cdot\mathbf{I}_n^{-1}\cdot\boldsymbol{L}$ produce terms that are similar to γ in Eq.(6-108), as well as converting $\mathbf{I}_n^{-1}$ in Eq.(6-108) to $\mathbf{I}_{\text{eff}}^{-1}$. Noting this and dropping the terms in γ gives

$$\mathscr{H} = \frac{1}{2}\boldsymbol{J}\cdot\mathbf{I}_{\text{eff}}^{-1}\cdot\boldsymbol{J} - \frac{\mu_0}{\hbar}\boldsymbol{H}\cdot\mathbf{g}\cdot\boldsymbol{J} - \frac{1}{2}\boldsymbol{H}\cdot\boldsymbol{\chi}\cdot\boldsymbol{H}$$

$$\mathbf{g} = \mathbf{g}_n + \mathbf{g}_e,$$

$$\mathbf{g}_n = m_p \sum_\alpha Z_\alpha(r_\alpha^2\mathbf{1} - \boldsymbol{r}_\alpha\boldsymbol{r}_\alpha)\cdot\mathbf{I}_{\text{eff}}^{-1},$$

$$\mathbf{g}_e = \frac{2m_p}{m}\mathbf{I}_{\text{eff}}^{-1}\cdot\sum_{k>0}\frac{\langle 0|\boldsymbol{L}|k\rangle\langle k|\boldsymbol{L}|0\rangle}{E_0 - E_k}. \tag{6-109}$$

$\mu_0 = (\hbar e/2m_p c)$ is the nuclear magneton and m and m_p are the electron and proton masses. This equation can now be compared to Eq.(4-213) where $\frac{1}{2}\boldsymbol{J}\cdot\mathbf{I}_{\text{eff}}^{-1}\cdot\boldsymbol{J}$ is the rigid rotor term, $\boldsymbol{\mu}_J = (\mu_0/\hbar)\mathbf{g}\cdot\boldsymbol{J}$ is the rotational magnetic moment, and $\boldsymbol{\chi}$ is the magnetic susceptibility tensor. $\mathbf{g}$ is the molecular g-value tensor [67] with elements given by

$$\mathsf{g}_{xx} = \frac{m_p}{\mathsf{I}_{xx}}\sum_\alpha Z_\alpha(r_\alpha^2 - x_\alpha^2) + \frac{2m_p}{m\mathsf{I}_{xx}}\sum_{k>0}\frac{|\langle 0|L_x|k\rangle|^2}{E_0 - E_k} \tag{6-110}$$

and cyclic permutations for g_{yy} and g_{zz}. It is evident that the rotational magnetic moment, $(\mu_0/\hbar)\mathbf{g}\cdot\boldsymbol{J}$, is a sum of positive nuclear and negative $(E_0 < E_k)$ electronic terms. We can also give a phenomenological description of a molecular magnetic moment that is suggested by the equations above. In the absence of rotation $(J = 0)$ there is no rotational magnetic moment. In a rotating molecule the nuclear current contribution leads to a magnetic moment given by $(\mu_0/\hbar)\mathbf{g}_n\cdot\boldsymbol{J}$. The electrons in the molecule, which are localized to a single atom, will rotate in a direction opposite to the nuclear motion like the seats on a ferris wheel, which maintain their orientation with respect to the laboratory framework as the wheel (overall nuclear framework) rotates. This electronic counterrotation cancels the nuclear contribution to the moment. If the electrons on one atom are coupled to another atom in a molecular bond, however, the electrons will rotate with the molecular framework destroying the cancellation effect described above for uncoupled electrons. The result is a nonzero total magnetic moment.

In summary, according to the results in Eq.(6-109) as developed following Eq.(4-213), the values of the molecular g-values and the magnetic susceptibility anisotropies can be measured in molecules and several results are listed in Table 4-17. The g-value for OCS is listed at $g_{bb} = -0.0287$. The nuclear contribution to g_{bb} in OCS is easily determined from the first term in Eq.(6-110) and the molecular structure in Table 4-16 to give $g_{bb}^n = (m_p/\mathsf{I}_{bb})\sum_\alpha Z_\alpha a_\alpha^2 = 0.504$ where all coordinates are from the molecular c.m. According to these results, g_{bb} is a near cancellation of the nuclear and electronic contributions with the electronic contribution being larger.

The individual diagonal elements in the magnetic susceptibility tensor are also given from Eq.(6-108) by

$$\chi_{xx} = \chi_{xx}^{d} + \chi_{xx}^{p}$$

$$\chi_{xx} = -\frac{e^2}{4c^2 m}\left\langle 0\left|\sum_i (y_i^2 + z_i^2)\right|0\right\rangle - \frac{e^2}{2c^2 m^2}\sum_{k>0}\frac{|\langle 0|L_x|k\rangle|^2}{E_0 - E_k} \tag{6-111}$$

and cyclic permutations for χ_{yy} and χ_{zz}. χ_{xx}^{d} and χ_{xx}^{p} are the diamagnetic and paramagnetic components of the susceptibility, respectively. χ_{xx}^{d} is negative and depends only on the ground-state distribution of electrons. χ_{xx}^{p} is positive as $E_k > E_0$ and depends on a sum over all excited electronic states. These results show that the magnetic field-induced electronic moment, $\frac{1}{2}\boldsymbol{H}\cdot\boldsymbol{\chi}$, is a sum of negative and positive terms. The negative terms arise through the diamagnetic response of the molecule's electrons which gives a moment which opposes the field. The second and positive contribution to the electronic magnetic moment arises through a paramagnetic response of the molecule's electrons; the paramagnetic moment complements the applied magnetic field.

It is clear that the second terms in g_{xx} [Eq.(6-110)] and χ_{xx} [Eq.(6-111)] have the same dependence on the sum over all excited molecular electronic states. Therefore, if the total g_{xx} can be measured and if the nuclear component of g_{xx} can be computed from the known molecular structure, the numerical value for the paramagnetic dependence in the susceptibility can be obtained. Substituting the

$$\sum_{k>0}\frac{|\langle 0|L_x|k\rangle|^2}{E_0 - E_k}$$

dependence in g_{xx} into χ_{xx} gives

$$\chi_{xx} = \chi_{xx}^{d} + \chi_{xx}^{p} = -\frac{e^2}{4c^2 m}\left\langle 0\left|\sum_i (y_i^2 + z_i^2)\right|0\right\rangle - \frac{e^2}{4c^2 m}\left[\frac{g_{xx} I_{xx}}{m_p} - \sum_\alpha Z_\alpha (y_\alpha^2 + z_\alpha^2)\right] \tag{6-112}$$

and cyclic permutations for χ_{yy} and χ_{zz}.

Molecular Quadrupole Moments

It is also evident from these results that the diagonal elements in the molecular quadrupole moment tensor, $\mathbf{Q}$, are also related to $\boldsymbol{\chi}$, $\mathbf{g}$, and $\mathbf{I}$. The value of Q_{zz} is easily written in terms of the magnetic susceptibility anisotropy $(\chi_{xx} + \chi_{yy} - 2\chi_{xx})$, g-values, and moments of inertia. The appropriate relation is

$$\begin{aligned} Q_{zz} &= \frac{e}{2}\sum_\alpha Z_\alpha(3z_\alpha^2 - r_\alpha^2) - \frac{e}{2}\left\langle 0\left|\sum_i (3z_i^2 - r_i^2)\right|0\right\rangle \\ &= \frac{2mc^2}{e}(\chi_{xx} + \chi_{yy} - 2\chi_{zz}) + \frac{e}{2m_p}(g_{xx} I_{xx} + g_{yy} I_{yy} - 2g_{zz} I_{zz}) \end{aligned} \tag{6-113}$$

and cyclic permutations for Q_{yy} and Q_{xx}. If molar χ_{xx} values are used in Eq.(6-113), they must be divided by N_A, Avogadro's number. It is easy to check Eq.(6-113) by substituting from Eqs.(6-110), (6-111), and (6-112). The molecular quadrupole moments of several molecules have been determined by measuring the diagonal elements in the $\mathbf{g}$ and $\mathbf{I}$ tensors and the magnetic susceptibility anisotropies

$(\chi_{xx} + \chi_{yy} - 2\chi_{zz})$ and some results are shown in Table 6-15. Equation (6-113) simplifies for a linear molecule to

$$\mathsf{Q}_{zz} = e\sum_\alpha Z_\alpha z_\alpha^2 - e\left\langle 0\left|\sum_i (z_i^2 - x_i^2)\right|0\right\rangle = \frac{e}{m_p}\mathsf{g}_{xx}\mathsf{I}_{xx} + \frac{4mc^2}{e}(\chi_{xx} - \chi_{zz}). \quad (6\text{-}114)$$

A similar equation is evident for symmetric tops,

$$\mathsf{Q}_{zz} = \frac{e}{m_p}(\mathsf{g}_{zz}\mathsf{I}_{zz} - \mathsf{g}_{xx}\mathsf{I}_{xx}) + \frac{4mc^2}{e}(\chi_{xx} - \chi_{zz}), \quad (6\text{-}115)$$

where z is the symmetry axis for both Eqs.(6-114) and (6-115).

We now note that only the value of the first nonzero electric multipole moment is independent of the origin. All higher-order moments depend on the origin. Consider a molecular quadrupole moment along the internuclear z axis in a linear molecule with respect to the c.m. at two different origins (xyz and $x'y'z'$) along the z axis separated by Z. According to Eq.(6-114), we can write the moment in the primed axis system by

$$\mathsf{Q}'_{zz} = e\sum_\alpha Z_\alpha(z'_\alpha)^2 - e\left\langle 0\left|\sum_i [(z'_i)^2 - (x'_i)^2]\right|0\right\rangle. \quad (6\text{-}116)$$

Now, substituting $z' = z - Z$ and $x' = x$ for the linear molecule gives

$$\begin{aligned}\mathsf{Q}'_{zz} &= e\sum_\alpha Z_\alpha(z_\alpha - Z)^2 - e\left\langle 0\left|\sum_i [(z_i - Z)^2 - x_i^2]\right|0\right\rangle \\ &= e\sum_\alpha Z_\alpha z_\alpha^2 - e\left\langle 0\left|\sum_i (z_i^2 - x_i^2)\right|0\right\rangle - 2Z\left[e\sum_\alpha Z_\alpha z_\alpha - e\left\langle 0\left|\sum_i z_i\right|0\right\rangle\right] \\ &\quad + Z^2\left(-ne + e\sum_\alpha Z_\alpha\right) = \mathsf{Q}_{zz} - 2ZD_z + Z^2 M_0, \end{aligned} \quad (6\text{-}117)$$

where n is the number of electrons, D_z is the electric dipole moment, and M_0 is the molecular monopole moment. If $M_0 = 0$ (neutral molecule) and $D_z = 0$, $\mathsf{Q}_{zz} = \mathsf{Q}'_{zz}$. If the molecule is neutral and if the molecule has a nonzero electric dipole moment,

$$\mathsf{Q}'_{zz} - \mathsf{Q}_{zz} = -2ZD_z. \quad (6\text{-}118)$$

Substituting the functions for Q'_{zz} and Q_{zz} from Eq.(6-114) in terms of $\mathsf{g}_{xx}\mathsf{I}_{xx}$ and $\chi_{xx} - \chi_{zz}$, where we note that the magnetic susceptibility anisotropy $(\chi_{xx} - \chi_{zz})$ is independent of the origin of the coordinate system, gives

$$\frac{e}{2m_p}(\mathsf{g}'_{xx}\mathsf{I}'_{xx} - \mathsf{g}_{xx}\mathsf{I}_{xx}) = -ZD_z. \quad (6\text{-}119)$$

These results are easily generalized to three dimensions, giving

$$\frac{e}{2m_p}(\mathsf{g}'_{xx}\mathsf{I}'_{xx} - \mathsf{g}_{xx}\mathsf{I}_{xx}) = -ZD_z - YD_y. \quad (6\text{-}120)$$

Molecular electric dipole moment signs determined by these methods are listed in Table 6-15.

The anisotropies in the second moment of the electronic charge distributions are also available from the information above, giving

$$\begin{aligned}\langle y^2\rangle - \langle x^2\rangle &= \left\langle 0\left|\sum_i y_i^2\right|0\right\rangle - \left\langle 0\left|\sum_i x_i^2\right|0\right\rangle \\ &= \sum_n Z_n(y_n^2 - x_n^2) + \frac{1}{m_p}(\mathsf{g}_{yy}\mathsf{I}_{yy} - \mathsf{g}_{xx}\mathsf{I}_{xx}) + \frac{4mc^2}{e^2}(\chi_{yy} - \chi_{xx}). \end{aligned} \quad (6\text{-}121)$$

Table 6-15 Moments of inertia, molecular g-values, magnetic susceptibility anisotropies, and molecular quadrupole moments in some typical molecules from W. H. Flygare and R. C. Benson, *Mol. Phys.* **20**, 225 (1971). The dipole moment signs are from Eq.(6-120).

Molecule	Principal Inertial Axis System	I_{aa}, I_{bb}, I_{cc} $\left\{\begin{matrix}10^{-40}\\ g \cdot cm^2\end{matrix}\right\}$	g_{aa}, g_{bb}, g_{cc}	$2\chi_{aa} - \chi_{bb} - \chi_{cc}$, $2\chi_{bb} - \chi_{cc} - \chi_{aa}$ $\left\{\begin{matrix}10^{-6}\\ cm^3 \cdot mol^{-1}\end{matrix}\right\}$	Q_{aa}, Q_{bb}, Q_{cc} $\left\{\begin{matrix}10^{-26}\\ SC \cdot cm^2\end{matrix}\right\}$
$^{15}N—^{15}N—^{16}O$	$\longrightarrow a$	0 73.295 73.295	0 -0.07606 ± 0.0001 -0.07606 ± 0.0001	-20.30 ± 0.30 10.15 ± 0.15	-3.65 ± 0.25 1.82 ± 0.15 1.82 ± 0.15
$F—C{\equiv}C—H$	$\longrightarrow a$	0 86.449 86.449	0 -0.0077 ± 0.0002 -0.0077 ± 0.0002	-10.38 ± 0.30 5.19 ± 0.15	3.96 ± 0.14 -1.98 ± 0.1 -1.98 ± 0.1
$^{-}CH_3—C{\equiv}C—H^{+}$	$\longrightarrow a$	5.300 98.187 98.187	$+0.312 \pm 0.002$ $+0.00350 \pm 0.00015$ $+0.00350 \pm 0.00015$	-15.4 ± 0.28 7.7 ± 0.14	4.82 ± 0.23 -2.41 ± 0.15 -2.41 ± 0.15
FCH_3	$\longrightarrow a$	5.300 32.915 32.915	$+0.310$ -0.0612 ± 0.002 -0.0612 ± 0.002	-16.4 ± 1.6 8.2 ± 0.8	-1.4 ± 1.1 $+0.7 \pm 0.6$ $+0.7 \pm 0.6$

Table 6-15 *continued*

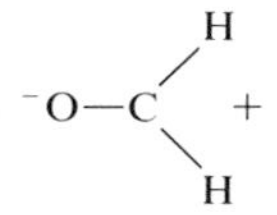	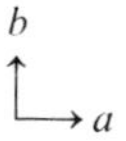	2.975	−2.9017 ± 0.0008	25.5 ± 0.5	−0.1 ± 0.3
		21.606	−0.2243 ± 0.0001	−3.9 ± 0.3	0.2 ± 0.2
		24.677	−0.0994 ± 0.0001		−0.1 ± 0.5
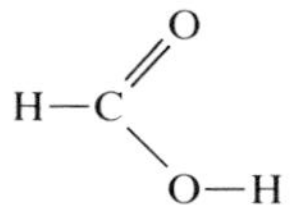	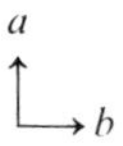	10.825	−0.2797 ± 0.006	3.4 ± 0.5	−5.3 ± 0.6
		69.605	−0.0903 ± 0.0006	9.4 ± 0.3	5.2 ± 0.6
		80.556	−0.0270 ± 0.0006		0.1 ± 0.6
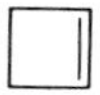	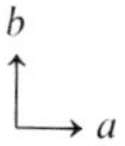	65.082	−0.0516 ± 0.0007	−0.9 ± 0.5	−0.3 ± 0.6
		68.631	−0.0663 ± 0.0007	5.0 ± 0.7	1.6 ± 0.7
		123.101	−0.0219 ± 0.0006		−1.3 ± 1.0
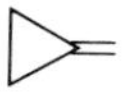	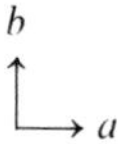	43.213	−0.0672 ± 0.0007	18.3 ± 0.5	−0.7 ± 0.5
		122.012	−0.0231 ± 0.0004	14.9 ± 0.6	0.9 ± 0.6
		154.099	+0.0244 ± 0.0004		−0.2 ± 0.9
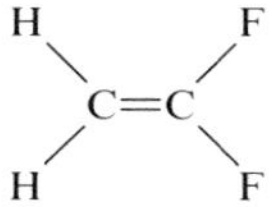	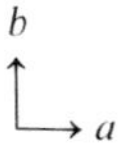	76.264	−0.0421 ± 0.0005	−2.3 ± 0.6	2.4 ± 0.5
		80.458	−0.0466 ± 0.0004	7.7 ± 0.5	−0.9 ± 0.4
		156.975	−0.0119 ± 0.0004		−1.5 ± 0.8

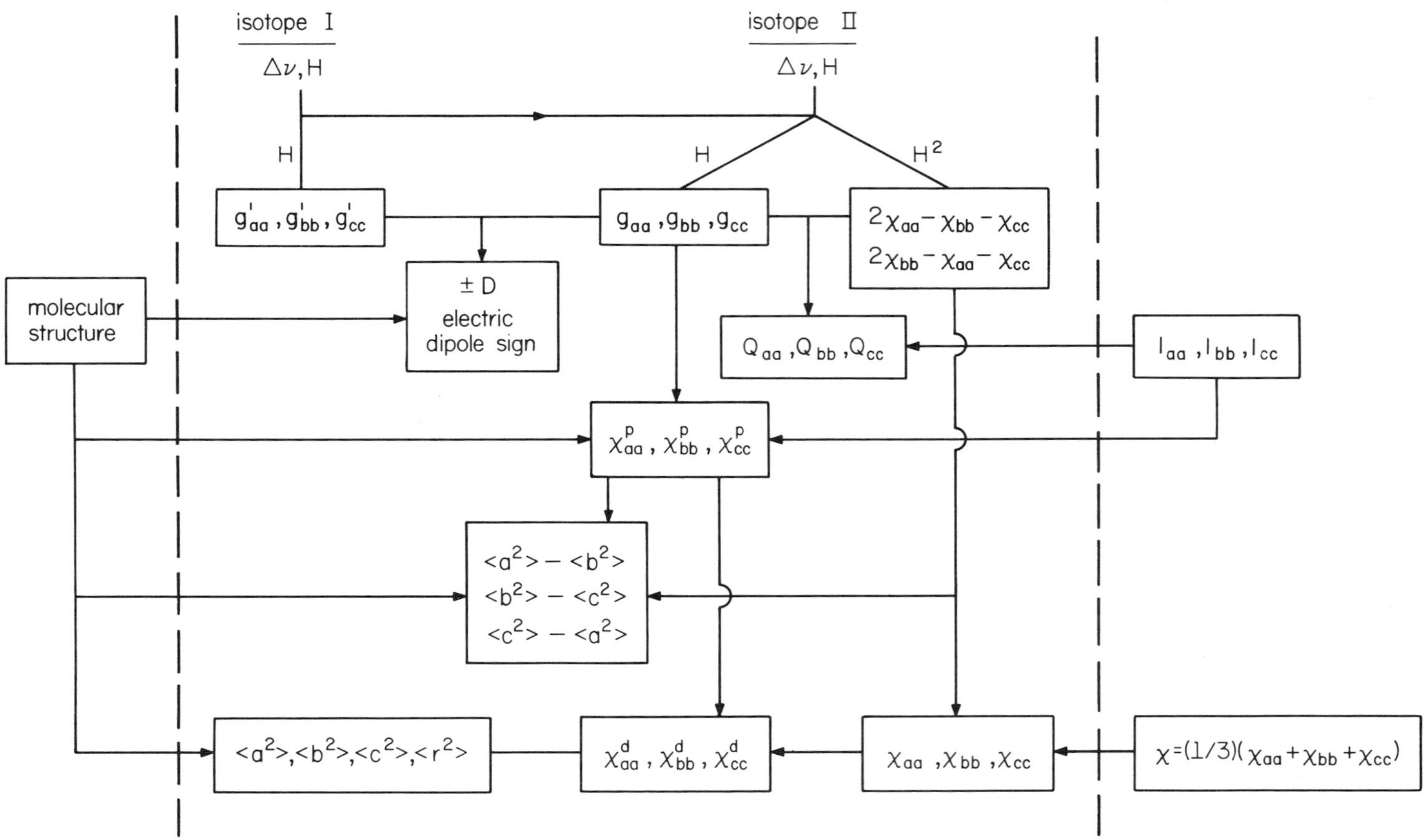

Figure 6-11 Summary of data available from the measurement of the magnetic field (H) giving rise to the frequency shifts ($\Delta\nu$) in the molecular rotational Zeeman effect. The information between the dotted lines must be added to the non-Zeeman information outside the dotted lines when indicated.

Finally, if the bulk magnetic susceptibility is known, the individual tensor elements in $\boldsymbol{\chi}^d$ and $\boldsymbol{\chi}$ can be determined. The bulk or average magnetic susceptibility is given by

$$\chi = \tfrac{1}{3}(\chi_{xx} + \chi_{yy} + \chi_{zz}). \tag{6-122}$$

The individual second moments of the electronic charge distributions can also be determined by

$$\langle x^2 \rangle = -\frac{2mc^2}{e^2}(\chi_{yy}^d + \chi_{zz}^d - \chi_{xx}^d) = -\frac{2mc^2}{e^2}[(\chi_{yy} + \chi_{zz} - \chi_{xx}) - (\chi_{yy}^p + \chi_{zz}^p - \chi_{xx}^p)]. \tag{6-123}$$

A summary of the interconnections between these magnetic parameters is shown in Fig. 6-11. The top line of parameters is measured directly by the molecular Zeeman effect as outlined in Section 4.11. A summary of the parameters above for three molecules is given in Table 6-16.

Table 6-16 The molecular Zeeman parameters in pyridine, ethylene oxide, and formic acid, which include the molecular quadrupole moments, second moment of the charge distributions, and the magnetic susceptibility. χ-values are in units of 10^{-6} cm$^3 \cdot$ mol^{-1}. Q-values are in units of 10^{-26} SC $\cdot$ cm^2 and $\langle a^2 \rangle$ values are in units of 10^{-16} cm^2.

	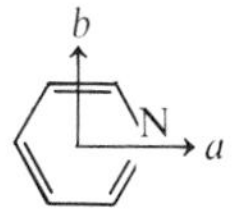	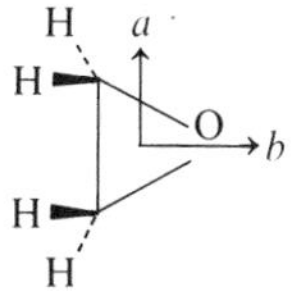	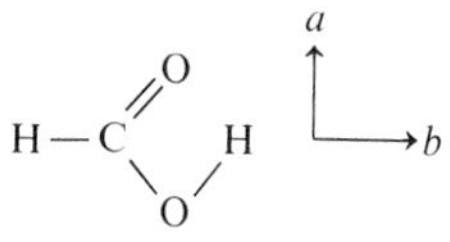
g_{aa}	-0.077 ± 0.0005	-0.0946 ± 0.0003	-0.2797 ± 0.006
g_{bb}	-0.1010 ± 0.0008	0.0189 ± 0.0004	-0.0903 ± 0.0006
g_{cc}	0.0428 ± 0.0004	0.0318 ± 0.0006	-0.0270 ± 0.0006
$2\chi_{aa} - \chi_{bb} - \chi_{cc}$	54.3 ± 0.6	18.06 ± 0.57	3.4 ± 0.5
$2\chi_{bb} - \chi_{aa} - \chi_{cc}$	60.5 ± 0.8	0.78 ± 0.97	9.4 ± 0.3
Q_{aa}	-3.5 ± 0.9	2.5 ± 0.4	-5.3 ± 0.4
Q_{bb}	9.7 ± 1.1	-4.3 ± 0.5	5.2 ± 0.4
Q_{cc}	-6.2 ± 1.5	1.8 ± 0.8	0.1 ± 0.4
χ_{aa}^p	241.5 ± 1.5	60.7 ± 0.8	28.8 ± 0.1
χ_{bb}^p	247.4 ± 2.0	67.3 ± 1.5	106.5 ± 0.1
χ_{cc}^p	393.9 ± 2.0	88.4 ± 2.0	117.2 ± 0.1
$\langle a^2 \rangle - \langle b^2 \rangle$	0.92 ± 0.80	7.6 ± 0.3	17.8 ± 0.3
$\langle b^2 \rangle - \langle c^2 \rangle$	48.28 ± 0.60	15.5 ± 0.3	-4.3 ± 0.3
$\langle c^2 \rangle - \langle a^2 \rangle$	-49.19 ± 0.60	-23.1 ± 0.3	-22.1 ± 0.3
$\chi = \frac{1}{3}(\chi_{aa} + \chi_{bb} + \chi_{cc})$	-48.4 ± 0.1	-30.7	-19.9 ± 0.3
χ_{aa}	-30.4 ± 0.5	-24.7 ± 0.7	-18.8 ± 0.8
χ_{bb}	-28.3 ± 0.6	-30.4 ± 0.7	-16.8 ± 0.8
χ_{cc}	-86.8 ± 0.8	-37.0 ± 0.8	-24.2 ± 0.8
χ_{aa}^d	-217.9 ± 1.6	-85.4 ± 0.9	-47.6 ± 0.8
χ_{bb}^d	-275.7 ± 2.0	-27.7 ± 1.1	-123.3 ± 0.8
χ_{cc}^d	-480.6 ± 2.2	-125.4 ± 2.0	-141.4 ± 0.8
$\langle a^2 \rangle$	57.1 ± 0.8	16.3 ± 0.4	25.6 ± 0.3
$\langle b^2 \rangle$	56.2 ± 0.8	13.3 ± 0.4	7.7 ± 0.3
$\langle c^2 \rangle$	7.9 ± 0.8	6.8 ± 0.4	3.5 ± 0.3

Considerable progress has been accomplished in computing the average molecular magnetic susceptibility as sums of localized atomic contributions plus neighbor group and bond-type corrections [68]. These localized rules can be used to compute fairly accurately the average magnetic susceptibilities of virtually any organic molecule. Progress has also been made in estimating atomic values for the principal diagonal tensor elements in the $\boldsymbol{\chi}$ matrix and these values can be used to estimate the magnetic susceptibility anisotropies in nonaromatic-type molecules [69]. Considerable progress has also been made in calculating magnetic susceptibilities in molecules by *ab initio* and semiempirical approximate methods [70].

6.9 NUCLEAR MAGNETIC SHIELDING AND NUCLEAR SPIN-ROTATION INTERACTIONS

We shall now add to Eq.(6-104) the additional effects in a molecule due to the presence of magnetic nuclei with magnetic moments $\boldsymbol{\mu}_i = (g_i/\hbar)\mu_0 \boldsymbol{I}_i$ for the ith nucleus (where g_i is the nuclear g-value and μ_0 is the nuclear magneton). The spin-rotation interaction arises, in the absence of any external fields, from the interaction of a nuclear magnetic moment with the intramolecular magnetic field at the nucleus, due to the intrinsic electronic and rotational motion of the molecule. The internal field at the kth nucleus, $\boldsymbol{H}^k$, is given by a sum of electronic, $\boldsymbol{H}_e^k$, and nuclear, $\boldsymbol{H}_n^k$, terms given by

$$\boldsymbol{H}^k = \boldsymbol{H}_e^k + \boldsymbol{H}_n^k = -\frac{e}{c}\sum_i \frac{\boldsymbol{r}_{ki} \times \boldsymbol{v}_{ki}}{r_{ki}^3} + \frac{e}{c}\sum_\alpha{}' Z_\alpha \frac{\boldsymbol{r}_{k\alpha} \times \boldsymbol{v}_{k\alpha}}{r_{k\alpha}^3}, \tag{6-124}$$

where all vectors originate at the kth nucleus, which is indicated by $\boldsymbol{r}_{ki}$ and $\boldsymbol{r}_{k\alpha}$ (see Fig. 6-12); the sum over i is over all electrons in the molecule; and the sum over α is over all nuclei (where the prime on the summation indicates exclusion of the $k = \alpha$ term) with atomic number Z_α. The pure nuclear term in Eq.(6-124) can be rewritten by remembering that $\boldsymbol{v}_{k\alpha}$ and $\boldsymbol{r}_{k\alpha}$ depend only on the nuclear coordinates in a nonvibrating molecule where $\boldsymbol{v}_{k\alpha} = \boldsymbol{\omega} \times \boldsymbol{r}_{k\alpha}$. Substituting into $\boldsymbol{H}_n^k$ in Eq.(6-124) gives

$$\boldsymbol{H}_n^k = \frac{e}{c}\sum_\alpha{}' \frac{Z_\alpha}{r_{k\alpha}^3}(r_{k\alpha}^2\mathbf{1} - \boldsymbol{r}_{k\alpha}\boldsymbol{r}_{k\alpha})\cdot\boldsymbol{\omega} = \frac{e}{c}\sum_\alpha{}' \frac{Z_\alpha}{r_{k\alpha}^3}(r_{k\alpha}^2\mathbf{1} - \boldsymbol{r}_{k\alpha}\boldsymbol{r}_{k\alpha})\cdot\mathbf{I}_n^{-1}\cdot(\boldsymbol{J} - \boldsymbol{L}), \tag{6-125}$$

where $\boldsymbol{\omega} = \mathbf{I}_n^{-1}\cdot(\boldsymbol{J} - \boldsymbol{L})$ from Eq.(6-101) has also been used.

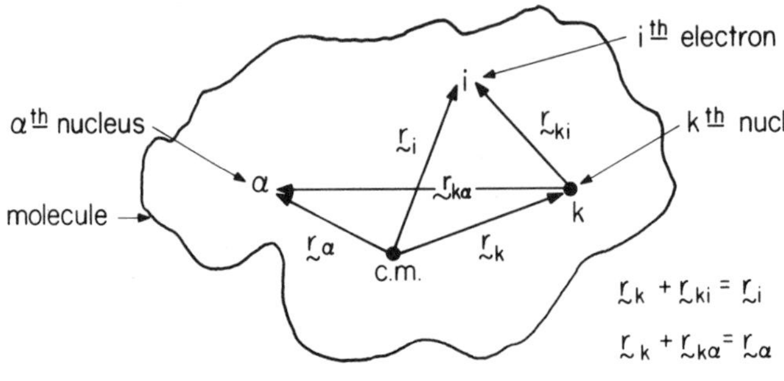

Figure 6-12 The molecular coordinate system showing the vectors from the c.m. to the kth and αth nuclei and ith electron.

External Magnetic Field Effects

We now add an external magnetic field, $\boldsymbol{H}$, and recognize from Eq.(2-94) that the velocity of an electron will be modified in a magnetic field according to $m\boldsymbol{v}_{ki} = \boldsymbol{p}_{ki} + (e/c)\boldsymbol{A}_{ki}$, where $\boldsymbol{A}_{ki} = \frac{1}{2}\boldsymbol{H} \times \boldsymbol{r}_{ki}$ is the vector potential arising from the uniform, static, external magnetic field. However, we now choose a new gauge (origin) at the kth nucleus. Our previous choice of origin in Section 6.8 was at the molecular c.m. The current choice of the kth nucleus requires us to redefine the intrinsic electronic angular momentum in Eq.(6-98) and the following equations with a new origin at the kth nucleus. Substituting these results above into $\boldsymbol{H}_e^k$ in Eq.(6-124) gives

$$\boldsymbol{H}_e^k = -\frac{e}{mc}\sum_i \frac{\boldsymbol{l}_{ki}}{r_{ki}^3} - \frac{e^2}{2mc^2}\sum_i \frac{(r_{ki}^2\mathbf{1} - \boldsymbol{r}_{ki}\boldsymbol{r}_{ki})\cdot\boldsymbol{H}}{r_{ki}^3}, \tag{6-126}$$

where $\boldsymbol{l}_{ki} = \boldsymbol{r}_{ki} \times \boldsymbol{p}_{ki}$. Combining Eqs.(6-125) and (6-126) and the external field, $\boldsymbol{H}$, gives the total field at the nucleus;

$$\begin{aligned}\boldsymbol{H}^k = \boldsymbol{H} &- \frac{e}{mc}\sum_i \frac{\boldsymbol{l}_{ki}}{r_{ki}^3} - \frac{e^2}{2mc^2}\sum_i \left(\frac{r_{ki}^2\mathbf{1} - \boldsymbol{r}_{ki}\boldsymbol{r}_{ki}}{r_{ki}^3}\right)\cdot\boldsymbol{H} \\ &+ \frac{e}{c}\sum_\alpha{}' \frac{Z_\alpha}{r_{k\alpha}^3}(r_{k\alpha}^2\mathbf{1} - \boldsymbol{r}_{k\alpha}\boldsymbol{r}_{k\alpha})\cdot\mathbf{I}_n^{-1}\cdot(\boldsymbol{J} - \boldsymbol{L}).\end{aligned} \tag{6-127}$$

Finally, the Hamiltonian describing the interaction of the magnetic dipole moment of the kth nucleus, $\boldsymbol{\mu}^k$, with the field at the kth nucleus, $\boldsymbol{H}^k$ in Eq.(6-127), is given by

$$\begin{aligned}\mathcal{H}_n &= -\boldsymbol{\mu}^k\cdot\boldsymbol{H}^k = -\gamma_k\boldsymbol{I}_k\cdot\boldsymbol{H}^k = -\frac{\mu_0}{\hbar}g_k\boldsymbol{I}_k\cdot\boldsymbol{H}^k \\ &= -\gamma_k\boldsymbol{I}_k\cdot\boldsymbol{H} + \frac{e}{mc}\gamma_k\boldsymbol{I}_k\cdot\sum_i\frac{\boldsymbol{l}_{ki}}{r_{ki}^3} + \frac{e^2}{2mc^2}\gamma_k\boldsymbol{I}_k\cdot\sum_i\left(\frac{r_{ki}^2\mathbf{1} - \boldsymbol{r}_{ki}\boldsymbol{r}_{ki}}{r_{ki}^3}\right)\cdot\boldsymbol{H} \\ &\quad - \frac{e}{c}\gamma_k\boldsymbol{I}_k\cdot\sum_\alpha{}'\frac{Z_\alpha}{r_{k\alpha}^3}(r_{k\alpha}^2\mathbf{1} - \boldsymbol{r}_{k\alpha}\boldsymbol{r}_{k\alpha})\cdot\mathbf{I}_n^{-1}\cdot(\boldsymbol{J} - \boldsymbol{L}).\end{aligned} \tag{6-128}$$

$\gamma_k = (\mu_0/\hbar)g_k$ is the magnetogyric ratio of the kth nucleus. We now combine the results in Eqs.(6-104) and (6-128) to give the complete Hamiltonian, including the nuclear terms:

$$\begin{aligned}\mathcal{H} = \mathcal{H}' + \mathcal{H}_n &= \frac{1}{2}\boldsymbol{J}\cdot\mathbf{I}_n^{-1}\cdot\boldsymbol{J} - \boldsymbol{L}_k\cdot\boldsymbol{\varpi} + \frac{1}{2}\boldsymbol{L}\cdot\mathbf{I}_n^{-1}\cdot\boldsymbol{L} \\ &+ \frac{e^2}{8c^2m}\boldsymbol{H}\cdot\left[\sum_i(r_i^2\mathbf{1} - \boldsymbol{r}_i\boldsymbol{r}_i) + m\sum_\alpha\frac{Z_\alpha^2}{M_\alpha}(r_\alpha^2\mathbf{1} - \boldsymbol{r}_\alpha\boldsymbol{r}_\alpha)\right]\cdot\boldsymbol{H} \\ &- \frac{e}{2c}\sum_\alpha Z_\alpha\boldsymbol{H}\cdot(r_\alpha^2\mathbf{1} - \boldsymbol{r}_\alpha\boldsymbol{r}_\alpha)\cdot\mathbf{I}_n^{-1}\cdot\boldsymbol{J} - \gamma_k\boldsymbol{I}_k\cdot\boldsymbol{H} + \frac{e}{mc}\gamma_k\boldsymbol{I}_k\cdot\sum_i\frac{\boldsymbol{l}_{ki}}{r_{ki}^3} \\ &+ \frac{e^2}{2mc^2}\gamma_k\boldsymbol{I}_k\cdot\sum_i\left(\frac{r_{ki}^2\mathbf{1} - \boldsymbol{r}_{ki}\boldsymbol{r}_{ki}}{r_{ki}^3}\right)\cdot\boldsymbol{H} \\ &- \frac{e}{c}\gamma_k\boldsymbol{I}_k\cdot\sum_\alpha{}'\frac{Z_\alpha}{r_{k\alpha}^3}(r_{k\alpha}^2\mathbf{1} - \boldsymbol{r}_{k\alpha}\boldsymbol{r}_{k\alpha})\cdot\mathbf{I}_n^{-1}\cdot(\boldsymbol{J} - \boldsymbol{L}),\end{aligned} \tag{6-129}$$

where $\bar{\omega}$ is defined in Eq.(6-104). We note that $\boldsymbol{L}_k$ in the second term now refers to the total electronic angular momentum with origin at the kth nucleus. This change is required because we have used the nuclear origin in our expression involving the vector potential. We now use perturbation theory to examine the contributions to the energy. We have examined in detail the contribution due to $\mathscr{H}'$ in Section 6.8. We now examine only the *additional* effects due to the addition of $\mathscr{H}_n$ in Eq.(6-129) above to give the complete Hamiltonian. Using the zero-order electronic basis in Eq.(6-99) we have the following *additional* contributions to the results considered previously in Section 6.8:

1. *First order*

$$E^{(1)} = -\gamma_k \boldsymbol{I}_k \cdot \boldsymbol{H} + \frac{e^2}{2mc^2}\gamma_k \boldsymbol{I}_k \cdot \left\langle 0 \left| \sum_i \frac{r_{ki}^2 \mathbf{1} - \boldsymbol{r}_{ki}\boldsymbol{r}_{ki}}{r_{ki}^3} \right| 0 \right\rangle \cdot \boldsymbol{H}$$

$$-\frac{e}{c}\gamma_k \boldsymbol{I}_k \cdot \sum_\alpha{}' \frac{Z_\alpha}{r_{k\alpha}^3}(r_{k\alpha}^2 \mathbf{1} - \boldsymbol{r}_{k\alpha}\boldsymbol{r}_{k\alpha})\mathbf{I}_n^{-1} \cdot \boldsymbol{J}$$

2. *Second order*

$$E^{(2)} = \frac{e^2}{2m^2c^2}\gamma_k \boldsymbol{I}_k \cdot \left\{ \sum_{k>0} \frac{\langle 0 | \sum_i \boldsymbol{l}_{ki}/r_{ki}^3 | k \rangle \langle k | \boldsymbol{L}_k | 0 \rangle + \text{c.c.}}{E_0 - E_k} \right\} \cdot \boldsymbol{H}$$

$$-2\frac{e}{c}\gamma_k \boldsymbol{I}_k \cdot \sum_\alpha{}' \frac{Z_\alpha}{r_{k\alpha}^3}(r_{k\alpha}^2 \mathbf{1} - \boldsymbol{r}_{k\alpha}\boldsymbol{r}_{k\alpha}) \cdot \mathbf{I}_n^{-1}\mathbf{A} \cdot \mathbf{I}_n^{-1} \cdot \boldsymbol{J}$$

$$-\frac{e}{mc}\gamma_k \boldsymbol{I}_k \left\{ \sum_{k>0} \frac{\langle 0 | \sum_i \boldsymbol{l}_{ki}/r_{ki}^3 | k \rangle \langle k | \boldsymbol{L}_k | 0 \rangle + \text{c.c.}}{E_0 - E_k} \right\} \cdot \mathbf{I}_n^{-1} \cdot \boldsymbol{J} + \text{higher-order terms.}$$

(6-130)

$\boldsymbol{L}_k = \sum_i \boldsymbol{l}_{ki}$ as discussed above and c.c. means the complex conjugate of the preceding term. Combining the first- and second-order terms, adding third-order terms to convert $\mathbf{I}_n^{-1}$ to $\mathbf{I}_{\text{eff}}^{-1}$ as discussed in Section 6.8, and dropping the higher-order terms gives

$$\mathscr{H} = -\gamma_k \boldsymbol{I}_k \cdot (\mathbf{1} - \boldsymbol{\sigma}) \cdot \boldsymbol{H} - \frac{1}{\hbar^2} \boldsymbol{I}_k \cdot \mathbf{M} \cdot \boldsymbol{J}, \qquad \boldsymbol{\sigma} = \boldsymbol{\sigma}^d + \boldsymbol{\sigma}^p,$$

$$\boldsymbol{\sigma}^d = \frac{e^2}{2mc^2} \left\langle 0 \left| \sum_i \frac{r_{ki}^2 \mathbf{1} - \boldsymbol{r}_{ki}\boldsymbol{r}_{ki}}{r_{ki}^3} \right| 0 \right\rangle,$$

$$\boldsymbol{\sigma}^p = \frac{e^2}{2m^2c^2} \sum_{k>0} \frac{\langle 0 | \sum_i \boldsymbol{l}_{ki}/r_{ki}^3 | k \rangle \langle k | \boldsymbol{L}_k | 0 \rangle + \text{c.c.}}{E_0 - E_k},$$

$$\mathbf{M} = \mathbf{M}_n + \mathbf{M}_e, \qquad \mathbf{M}_n = \frac{e}{c}\gamma_k \hbar^2 \sum_\alpha{}' \frac{Z_\alpha}{r_{k\alpha}^3}(r_{k\alpha}^2 \mathbf{1} - \boldsymbol{r}_{k\alpha}\boldsymbol{r}_{k\alpha}) \cdot \mathbf{I}_{\text{eff}}^{-1},$$

$$\mathbf{M}_e = 2\frac{e}{mc}\gamma_k \hbar^2 \left(\sum_{k>0} \frac{\langle 0 | \sum_i \boldsymbol{l}_{ki}/r_{ki}^3 | k \rangle \langle k | \boldsymbol{L}_k | 0 \rangle}{E_0 - E_k} \right) \cdot \mathbf{I}_{\text{eff}}^{-1}. \tag{6-131}$$

$\mathbf{I}_{\text{eff}}^{-1}$ is defined in Eq.(6-107). $\boldsymbol{\sigma}^d$ and $\boldsymbol{\sigma}^p$ are the diamagnetic and paramagnetic nuclear magnetic shielding tensors [71], respectively, with individual diagonal elements given by

$$\sigma_{xx} = \sigma_{xx}^d + \sigma_{xx}^p = \frac{e^2}{2mc^2}\left\langle 0 \left| \sum_i \frac{(y_{ki}^2 + z_{ki}^2)}{r_{ki}^3} \right| 0 \right\rangle$$

$$+ \frac{e^2}{2m^2c^2} \sum_{k>0} \left[\frac{\langle 0 | \sum_i (\boldsymbol{l}_{ki})_x / r_{ki}^3 | k \rangle \langle k | (\boldsymbol{L}_k)_x | 0 \rangle + \langle 0 | (\boldsymbol{L}_k)_x | k \rangle \langle k | \sum_i (\boldsymbol{l}_{ki})_x / r_{ki}^3 | 0 \rangle}{E_0 - E_k} \right], \tag{6-132}$$

and cyclic permutations for σ_{yy} and σ_{zz}.

It is evident that σ_{xx}^d is the diamagnetic and σ_{xx}^p is the paramagnetic shielding. The diamagnetic shielding is always positive and decreases the net field at the nucleus. σ_{xx}^p is normally negative ($E_0 < E_k$), leading to an enhanced magnetic field at the nucleus. Therefore, we see from Eq.(6-132) that the average field at the nucleus is equal to the laboratory field plus a paramagnetic and minus a diamagnetic field, both of these later effects (fields) being caused by the electrons. In atoms, the paramagnetic term goes to zero as $\langle k | \boldsymbol{L} | 0 \rangle = (\boldsymbol{L})_{av} \langle k | 0 \rangle = 0$ where $(\boldsymbol{L})_{av}$ is one of the matrix elements in Eq.(4-74). $\mathbf{M}$ is the spin-rotation constant with individual diagonal elements given by

$$\mathbf{M}_{xx}^k = \frac{\hbar e g_k \mu_0}{c \mathbf{I}_{xx}} \sum_\alpha{}' \frac{Z_\alpha}{r_{k\alpha}^3} (r_{k\alpha}^2 - x_{k\alpha}^2)$$

$$+ \frac{\hbar e g_k \mu_0}{mc \mathbf{I}_{xx}} \sum_{k>0} \left[\frac{\langle 0 | \sum_i (\boldsymbol{l}_{ki})_x / r_{ik}^3 | k \rangle \langle k | (\boldsymbol{L}_k)_x | 0 \rangle + \langle 0 | (\boldsymbol{L}_k)_x | k \rangle \langle k | \sum_i (\boldsymbol{l}_{ki})_x / r_{ik}^3 | 0 \rangle}{E_0 - E_k} \right] \tag{6-133}$$

and cyclic permutations for M_{yy}^k and M_{zz}^k.

Realizing that the second term in the Hamiltonian in Eq.(6-131) represents the interaction of a nuclear magnetic moment, $\boldsymbol{\mu}_k = (g_k/\hbar)\mu_0 \boldsymbol{I}_k$ with an internal magnetic field, we can rewrite this term as

$$-\frac{1}{\hbar^2} \boldsymbol{I}_k \cdot \mathbf{M} \cdot \boldsymbol{J} = -\frac{1}{\hbar} \boldsymbol{\mu}_k \cdot \mathbf{R} \cdot \boldsymbol{J}, \qquad \mathbf{R} = \frac{1}{g_k \mu_0} \mathbf{M}, \tag{6-134}$$

where the $\mathbf{R}$ tensor has units of magnetic field and the $\boldsymbol{J}/\hbar$ multiplier shows that the magnitude of the internal field is weighted by the rotational angular momentum. The spin-rotation tensor constants, given in Eq.(6-133), are a sum of a positive pure nuclear term and a negative ($E_0 < E_k$) electronic term. We note the similarities in concept between the rotationally induced molecular magnetic moments [g-values in Eqs.(6-109) and (6-110)] and the corresponding rotationally induced magnetic field at the kth nucleus in Eqs.(6-131) and (6-134). In both cases, the bare nuclei contribute and the ground-state electronic distribution of electrons do not contribute. In both cases, the electronic states that possess electronic orbital angular momentum contribute to the electronic term. The ferris wheel discussion following Eq.(6-110) in the c.m. framework is also appropriate here with the kth nucleus as a framework of reference.

Before discussing the connection between the $\boldsymbol{\sigma}$ and M tensors, we shall note that the diagonal elements in M and the anisotropy in the diagonal elements in the $\boldsymbol{\sigma}$ tensor can be measured by rotational spectroscopy in a manner similar to the measurement of $\mathbf{g}$ and χ discussed in Sections 6.8 and 4.11. Consider the spin-rotation interaction from Eq.(6-131) in a linear molecule where $\mathsf{M}_{aa} = 0$ and $\mathsf{M}_{bb} = \mathsf{M}_{cc}$ (a is the symmetry axis), which gives (in the absence of the external field)

$$\mathscr{H} = -\frac{\mathsf{M}_{bb}}{\hbar^2}\boldsymbol{J}\cdot\boldsymbol{I}. \tag{6-135}$$

The coupled basis $\boldsymbol{F} = \boldsymbol{J} + \boldsymbol{I}$ is appropriate for the matrix elements, giving

$$E(I, J, F) = -\frac{\mathsf{M}_{bb}}{2}[F(F+1) - I(I+1) - J(J+1)]. \tag{6-136}$$

Adding this to the rigid rotor energy of $hBJ(J+1)$ gives the total energy. The measured spin-rotation constants in several molecules are listed in Table 6-17.

We can now relate either the values of R_{xx} or M_{xx} to the magnetic shielding, σ_{xx}, as all three constants have identical dependence on the sum over the excited electronic states. We shall relate the magnetic field-induced shielding to the rotationally induced field given by R_{xx}. Comparing Eqs.(6-132), (6-133), and (6-134) gives

$$\sigma_{xx} = \sigma_{xx}^d + \sigma_{xx}^p = \frac{e^2}{2mc^2}\left\langle 0\left|\sum_i \frac{(y_i^2 + z_i^2)}{r_i^3}\right|0\right\rangle + \frac{e^2}{2mc^2}\left[\frac{\mathsf{R}_{xx}\mathsf{I}_{xx}c}{e\hbar} - \sum_\alpha{}' \frac{Z_\alpha}{r_\alpha^3}(y_\alpha^2 + z_\alpha^2)\right] \tag{6-137}$$

and cyclic permutations for σ_{yy} and σ_{zz}. All distances are from the nucleus in question, and the sum over α omits this nucleus. We have mentioned previously that normally only the average magnetic shielding is measured by nuclear magnetic resonance experiments [72]. The average shielding is given by

$$\sigma_{av} = \sigma_{av}^d + \sigma_{av}^p = \frac{1}{3}(\sigma_{xx} + \sigma_{yy} + \sigma_{zz}) = \frac{e^2}{3mc^2}\left\langle 0\left|\sum_i \frac{1}{r_i}\right|0\right\rangle$$

$$+ \frac{e^2}{6mc^2}\left(\frac{\mathsf{R}_{xx}\mathsf{I}_{xx}c}{e\hbar} + \frac{\mathsf{R}_{yy}\mathsf{I}_{yy}c}{e\hbar} + \frac{\mathsf{R}_{zz}\mathsf{I}_{zz}c}{e\hbar} - 2\sum_\alpha{}' \frac{Z_\alpha}{r_\alpha}\right). \tag{6-138}$$

This expression further simplifies for high symmetry molecules. The value of σ_{av} for diatomic molecules is given by

$$\sigma_{av}(\text{diatomic}) = \sigma_{av}^d + \sigma_{av}^p = \frac{e^2}{3mc^2}\left\langle 0\left|\sum_i \frac{1}{r_i}\right|0\right\rangle + \frac{e^2}{3mc^2}\left(\frac{\mathsf{R}_\perp \mathsf{I} c}{e\hbar} - \frac{Z}{r}\right). \tag{6-139}$$

$\mathsf{R}_\perp$ is the perpendicular-induced magnetic field constant, Z is the atomic number of the other nucleus in the diatomic, r is the internuclear distance, and I is the moment of inertia. The paramagnetic shielding [σ_{xx}^p in Eq.(6-132) or σ_{av}^p in Eq.(6-138)] can be computed directly from the molecular spin-rotation constants (or magnetic field

Table 6-17 Spin-rotation, $M_\perp$, and the rotational field, $R_\perp$, constants for several linear molecules. g_k is the nuclear g-value, and $I_\perp$ is the moment of inertia. σ^d_{av} for H_2 is an accurate calculation. σ^d_{av} for the remaining molecules is estimated from Eq.(6-140) and Table 6-18. σ^p_{av} is computed with $I_\perp$, r, and Eq.(6-139). $\sigma_{av} = \sigma^d_{av} + \sigma^p_{av}$.

Molecule	Nucleus	g_k	$M_\perp/h$ (kHz)	$M_\perp/\mu_0 g_k = R_\perp$ (gauss)	r (10^{-8} cm)	$I_\perp$ (10^{-40} g·cm)	σ^p_{av} (10^{-6}) [Eq.(6-139)]	σ^d_{av} (10^{-6}) [Eq.(6-140)]	σ_{av} (10^{-6})
H—H	H	5.854	+112.734[a]	25.2	0.7416	0.467	−6	32	26
H—^{19}F	H	5.854	+71[a]	16.1	0.917	1.34	−80	110	30
H—^{19}F	^{19}F	5.2546	−284[b]	−71.0	0.917	1.34	−63	483	420(410)[b]
^{6}Li—^{19}F	^{19}F	5.2546	−37.3[c]	−9.3	1.525	10.61	−73	489	416
^{7}Li—^{19}F	^{19}F	5.2546	−32.9[c]	−8.2	1.525	11.91	−73	489	416
^{19}F—^{19}F	^{19}F	5.2546	−157[d]	−39.0	1.418	32.0	−750	530	−220
^{15}N − ^{15}N	^{15}N	−0.5660	+22[e]	−51.0	1.094	15.0	−485	386	−99
^{13}C—^{16}O	^{13}C	1.4042	−32.6[f]	−30.5	1.128	15.16	−323	327	+4
^{12}C—^{17}O	^{17}O	−0.7572	+29[g]	−50.3	1.128	14.9	−460	445	−15
H—^{12}C=^{12}C—H	H	5.854	3.58×10^3[a]	0.80	$d_{HC} = 1.060$ $d_{CC} = 1.207$	23.8	−70	98	28

[a] N. F. Ramsey, *Am. Scientist* **49**, 509 (1961).
[b] Corrected for vibrational effects; see D. K. Hinderman and C. D. Cornwell, *J. Chem. Phys.* **48**, 4148 (1968).
[c] See C. H. Townes and A. L. Schawlow, *Microwave Spectroscopy* (McGraw-Hill Book Co., New York, 1955).
[d] M. R. Baker, C. H. Anderson, and N. F. Ramsey, *Phys. Rev.* A **133**, 1533 (1964).
[e] C. W. Kern and W. N. Lipscomb, *J. Chem. Phys.* **37**, 260 (1962).
[f] I. Ozier, L. M. Crapo, and N. F. Ramsey, *J. Chem. Phys.* **49**, 2315 (1968).
[g] W. H. Flygare and V. W. Weiss, *J. Chem. Phys.* **45**, 2785 (1966).

constants) and the molecular structure. The results for several diatomics are listed in Table 6-17 under σ^p_{av}.

The *average* magnetic shielding at a nucleus cannot be measured directly by nuclear magnetic resonance. Only the *chemical shift*, which is the difference in the magnetic shielding for a nucleus in two different chemical environments, is measured by nuclear magnetic resonance methods (see Section 4.10):

$$\Delta\sigma_{av} = \sigma_{av}(A) - \sigma_{av}(B) = \sigma^p_{av}(A) + \sigma^d_{av}(A) - \sigma^p_{av}(B) - \sigma^d_{av}(B).$$

Now, in principle, $\sigma^p_{av}(A)$ and $\sigma^p_{av}(B)$ can be determined from Eq.(6-138) by using the spin-rotation constants and the molecular structure. If we can determine $\sigma^d_{av}(A)$ for the nucleus in molecule A by an independent method, then $\sigma^d_{av}(B)$ can be extracted from the equation above and the measured values of $\Delta\sigma_{av}$, $\sigma^p_{av}(B)$, and $\sigma^p_{av}(A)$. The value of $\sigma^d_{av}(A)$ need only be determined once for each nucleus in any given molecule and other $\sigma^d(B)$ can be extracted from the chemical shifts. Values of $\sigma^d_{av}(A)$ can be determined from *ab initio* calculations with the ground-state electronic wavefunctions in simple molecules that are convenient for chemical shift measurements. For instance, the diamagnetic shielding at the proton in H_2 has been calculated to high accuracy to give

$$\frac{e^2}{3mc^2}\left\langle 0\left|\sum_i \frac{1}{r_i}\right|0\right\rangle_{\text{H in H}_2} = 32.0 \times 10^{-6}.$$

Calculated σ^d_{av} are available for some other molecules. As σ^d_{av} is a local atomic property that weighs most heavily the inner electrons, however, we can approximate σ^d_{av} in molecules by a sum over atomic properties given by [73]

$$\sigma^d_{av}(\text{nucleus } k \text{ in a molecule}) = \sigma^d_{av}(\text{free } k \text{ atom}) + \frac{e^2}{3mc^2}\sum_\alpha{}' \frac{Z_\alpha}{r_\alpha}, \qquad (6\text{-}140)$$

where the prime on the summation excludes the kth term. The free atom diamagnetic shieldings for several atoms are given in Table 6-18. Equation (6-140) implies that other atom corrections may be added by assuming the other atom's charges are distributed at the nuclear points throughout the molecule. The values of σ^d_{av} from Eq.(6-140) in several molecules are listed in Table 6-17. The molecular values of σ^d_{av} in Table 6-17, which were calculated with Eq.(6-140), are all within 1% of the values calculated with the best available molecule wavefunctions. Thus, Eq.(6-140) is a good approximation. If Eq.(6-140) is valid, we can further simplify Eq.(6-138) to give

$$\sigma^d_{av} = \sigma^d_{av}(\text{atom}) + \frac{e}{6m\hbar c}(\mathsf{R}_{xx}\mathsf{I}_{xx} + \mathsf{R}_{yy}\mathsf{I}_{yy} + \mathsf{R}_{zz}\mathsf{I}_{zz}), \qquad (6\text{-}141)$$

where $\sigma^d_{av}(\text{atom})$ is the free atom diamagnetic shielding as in Table 6-18. In a linear molecule, Eq.(6-141) reduces to

$$\sigma_{av}(\text{linear}) = \sigma^d_{av}(\text{atom}) + \left(\frac{e}{3m\hbar c}\right)\mathsf{R}_\perp \mathsf{I}_\perp.$$

Returning to Table 6-17, we note that the net shielding at O in CO, N in N_2, and F in F_2 are all negative (paramagnetic). Thus, the actual fields in these nuclei are larger

Table 6-18 Nuclear diamagnetic shielding in atoms. The values of

$$\sigma_{av}^{d} = \frac{e^2}{3mc^2}\left\langle 0 \left| \sum_i \left(\frac{1}{r_i}\right) \right| 0 \right\rangle$$

are calculated with the atomic Hartree–Fock functions and the resultant values are from G. Malli and C. Froese, *Int. J. Quantum Chem.* **15**, 95 (1967). The value for hydrogen is easily computed.

Atom	$\sigma_{av}^{d}(10^{-6})$
H	17.7
He	59.9
Li	101.5
Be	149.3
B	202.0
C	260.7
N	325.5
O	395.1
F	470.7
Ne	552.7
Na	628.9
Mg	705.6
Al	789.9
Si	874.1
P	961.1
S	1050.5
Cl	1142.6
Ar	1237.6

than the external fields. The other examples in Table 6-17 indicate a net diamagnetic shielding at the nuclei.

In summary, we note again that the average shielding, σ_{av}, for a nucleus in a specific molecule provides a standard of comparison for the shielding at this nucleus in other molecules. The average magnetic shielding cannot be measured directly by nuclear magnetic resonance. Only the chemical shift that is the *difference* in the magnetic shielding is obtained. Thus, in order to obtain the average shielding in any molecule, the shifts must be related to the actual shielding for a specific case as in Table 6-17. The standard shielding is obtained by the link with the spin-rotation interaction and the calculated diamagnetic shielding. A summary of the interconnections between the spin-rotation and shielding tensor elements is shown in Fig. 6-13. This figure is similar in design to Fig. 6-11 showing the relations between molecular g-values and magnetic susceptibilities.

We can now use the principles above to assign an absolute nuclear magnetic shielding scale for the nuclei involved. The magnetic shielding for protons in a number of molecules is shown in Fig. 4-21. The zero shielding case is the bare proton nucleus and the reference is the hydrogen molecule, which shows that the proton shielding is positive in all molecules. Thus, the field at the protons is always reduced due to the molecule's electrons. By combining M_H to give σ_{av}^{p}, and the calculated

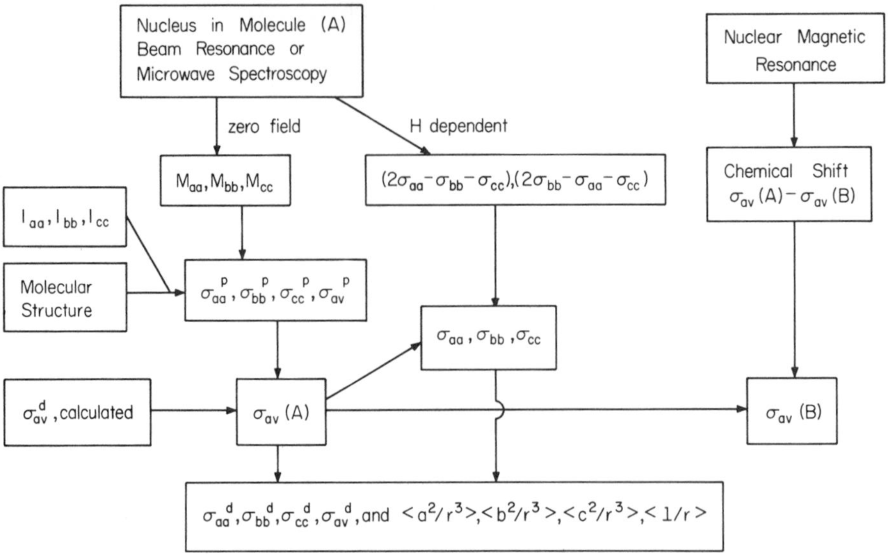

Figure 6-13 Diagram showing the relation between the diagonal elements in the spin-rotation and magnetic shielding tensor elements. $\sigma_{av}^{d} = \frac{1}{3}(\sigma_{aa}^{d} + \sigma_{bb}^{d} + \sigma_{cc}^{d})$, $\sigma_{av}^{p} = \frac{1}{3}(\sigma_{aa}^{p} + \sigma_{bb}^{p} + \sigma_{cc}^{p})$, and $\langle a^2/r^3\rangle = \langle 0|\sum_i a_i/r_i^3|0\rangle$.

σ_{av}^{d}, we have the absolute shielding, which provides the reference for the other molecules listed in Fig. 4-21.

In Fig. 6-14 we show the shielding scales for the ^{19}F, ^{17}O, ^{14}N, and ^{13}C nuclei [74]. The standards of reference are from the results in Table 6-17. Both shielding and antishielding are evident for these nuclei in various molecules.

A great deal of interpretation has been applied to the observed chemical shifts in molecules. Many of these theories assume that the chemical shifts arise primarily from the change in the paramagnetic terms. It is evident from Table 6-17 and Eq.(6-140), however, that the diamagnetic shielding can also vary considerably at the same nucleus in two different molecules.

Summary of One-Electron Properties in Molecules

We have examined the electronic ground-state average values of several one-electron operators here and in Chapter 5 [see Eq. (5-112)]. These various one-electron properties give information about the electron density in various parts of the molecule. The c.m. molecular electric quadrupole moment [Eq.(6-113), for instance] gives an r^2-weighted average value and thus samples the electron density in the outer regions of the molecule. Of course, the diamagnetic susceptibility is intimately related to the molecular quadrupole moment and also gives a c.m. r^2-type dependence. The molecular electric dipole moment given by [see Eq.(6-117)]

$$D_z = e\sum_{\alpha} Z_{\alpha} z_{\alpha} - e\left\langle 0\left|\sum_i z_i\right|0\right\rangle, \tag{6-142}$$

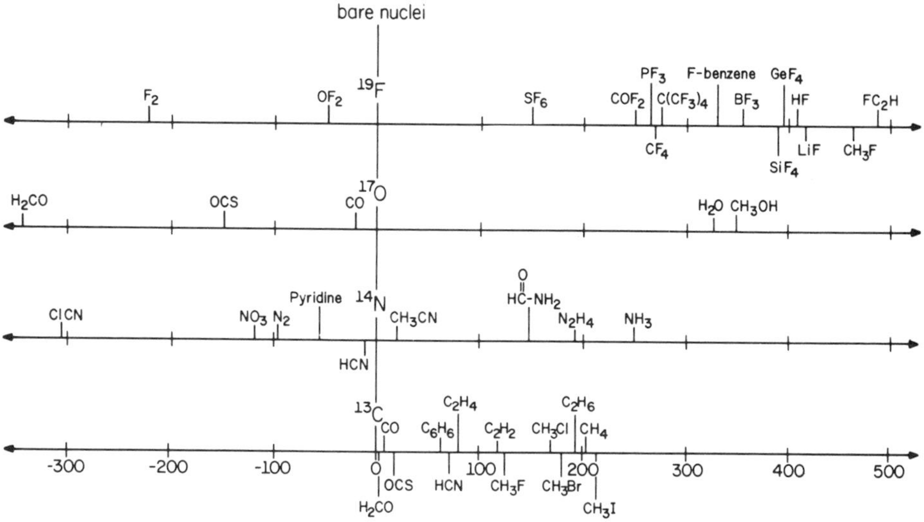

Figure 6-14 Summary of the nuclear magnetic shieldings in ^{19}F, ^{17}O, ^{14}N, and ^{13}C nuclei in different chemical environments relative to the bare nuclei. The scale along the abcissa is the nuclear magnetic shielding, σ, in units of 10^{-6} where $\sigma = 0$ indicates the bare nuclei.

and cyclic permutations, gives a c.m. r-type dependence. We have also examined three properties with origin at a nucleus in a molecule. These are the nuclear diamagnetic shielding [Eq.(6-132)] which gives a nuclear origin $(1/r)$-type average value, the forces at the nuclei which give a nuclear origin $(1/r^2)$-type average value, and the nuclear-based electronic field gradient which gives a $(1/r^3)$-type average value. According to Eq.(6-19) we can express the force at a nucleus along a z axis according to

$$-\frac{\partial V_{nn}}{\partial z} = F_z = 0 = e \sum_{\alpha}' \frac{Z_\alpha z_\alpha}{r_\alpha^3} - e\left\langle 0 \left| \sum_i \left(\frac{z_i}{r_i^3}\right) \right| 0 \right\rangle, \qquad (6\text{-}143)$$

and cyclic permutations, where the sum over α is over all nuclei except the nucleus where the force is being evaluated (the origin for the z_i/r_i^3 operator). The ground electronic state average, $\langle 0|0\rangle$ in Eq.(6-143), shows a nuclear origin $(1/r^2)$-type dependence in the average value. The electric field gradient at a nucleus is given in Section 5.8 to be

$$q_{zz} = \frac{\partial^2 V}{\partial z^2} = \frac{\partial E}{\partial z} = e \sum_{\alpha}' Z_\alpha \left(\frac{3z_\alpha^2 - r_\alpha^2}{r_\alpha^5}\right) - e\left\langle 0 \left| \sum_i \left(\frac{3z_i^2 - r_i^2}{r_i^5}\right) \right| 0 \right\rangle, \qquad (6\text{-}144)$$

and cyclic permutations, where again we have the sum over near-canceling nuclear and electronic contributions where the sum over α excludes the origin nucleus.

In summary, we have discussed five one-electron quantities that depend on the electronic wavefunction for a single electronic state. These include the electric dipole and quadrupole moments that sample the c.m. r and r^2 electronic dependence in a molecule. Also noted are nuclear-based properties including the diamagnetic shielding, the force on a nucleus, and the nuclear field gradients that sample the electron density near the nucleus with $1/r$-, $1/r^2$-, and $1/r^3$-dependence, respectively. These quantities are shown in more detail for the OCS molecule in Table 6-19.

Table 6-19 An analysis of the experimental, nuclear, and electronic values for the dipole moment [Eq.(6-142)], quadrupole moment [Eq.(6-114)], forces [Eq. (6-143)], and field gradients [Eq.(6-144)] in the OCS molecule. The nuclear contributions are computed from the known structure and the electronic contributions are obtained from the experimental values and the calculated nuclear values. The average diamagnetic shielding at the ^{13}C nucleus is also shown [$\sigma^d_{av} = (e^2/3mc^2)\langle 0|\sum_i 1/r_i|0\rangle$ from Eq.(6-138)].

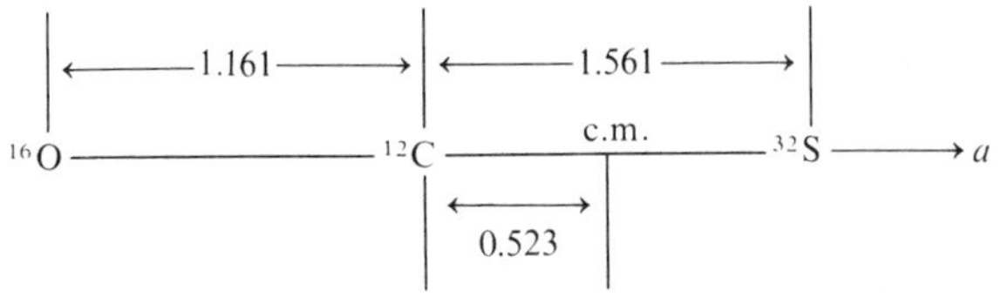

Property	Experimental	Nuclear Contribution	Electronic Contribution	r Dependence
Electric quadrupole moment (10^{-26} SC · cm^2)	$Q_{aa} = -0.88$	199.44	−200.32	r^2
Electric dipole moment (10^{-18} SC · cm)	$D_a = +0.710$	−0.009	0.719	r
Diamagnetic shielding (10^{-6})				
carbon	$\sigma^d_{av} = 420$	0	420	$1/r$
Nuclear forces (10^6 SC · cm^{-2})				
oxygen	$F_z = 0$	31.72	−31.72	$1/r^2$
carbon	$F_z = 0$	3.03	−3.03	
sulfur	$F_z = 0$	−16.99	16.99	
Field gradient (10^{14} SC · cm^{-3})				
oxygen	$q_{av} = 7.58$	44.42	−36.84	$1/r^3$
sulfur	$q_{av} = 60.00$	18.95	41.05	

6.10 ELECTRIC FIELD-DEPENDENT ELECTRONIC INTERACTIONS AND MOLECULAR POLARIZABILITY

The perturbation Hamiltonian for a distribution of electronic and nuclear charges in a molecule in the presence of a static electric field, $\boldsymbol{E}$, is given by

$$\mathscr{H}_1 = -E \cdot \left(e \sum_\alpha Z_\alpha \boldsymbol{r}_\alpha - e \sum_i \boldsymbol{r}_i \right) = -\boldsymbol{E} \cdot \boldsymbol{D}. \tag{6-145}$$

See the $L = 1$ term in Eq.(5-82) or the $e\varphi$ term in Eq.(2-95) for a single charge, where $\varphi = -\boldsymbol{r} \cdot \boldsymbol{E}$ is the potential experienced by an electron at a distance r from the origin in an electric field, $\boldsymbol{E}$. $\boldsymbol{D}$ is the molecular dipole moment operator where α and i are sums over the nuclei and electrons, respectively, and all vectors originate at the molecular c.m. We now compute the first- and second-order corrections to the ground-state energy using the ground- and excited-state electronic functions in Eq.(6-99):

1. *First order*

$$E^{(1)} = -\langle 0|\boldsymbol{D}|0\rangle \cdot \boldsymbol{E} = \boldsymbol{D}_{00} \cdot \boldsymbol{E}$$

$$\boldsymbol{D}_{00} = e\sum_{\alpha} Z_\alpha r_\alpha - \left\langle 0\left| e\sum_i \boldsymbol{r}_i \right| 0\right\rangle, \tag{6-146}$$

where $\boldsymbol{D}_{00}$ is the ground-state permanent electric dipole moment.

2. *Second order*

$$E^{(2)} = \boldsymbol{E}\cdot\left(\sum_{k>0} \frac{\langle 0|\boldsymbol{D}|k\rangle\langle k|\boldsymbol{D}|0\rangle}{E_0 - E_k}\right)\cdot\boldsymbol{E} = \boldsymbol{E}\cdot\left(e^2\sum_{k>0} \frac{\langle 0|\sum_i \boldsymbol{r}_i|k\rangle\langle k|\sum_i \boldsymbol{r}_i|0\rangle}{E_0 - E_k}\right)\cdot\boldsymbol{E}$$

$$= -\frac{1}{2}\boldsymbol{E}\cdot\boldsymbol{\alpha}\cdot\boldsymbol{E}$$

$$\boldsymbol{\alpha} = -2\sum_{k>0} \frac{\langle 0|\boldsymbol{D}|k\rangle\langle k|\boldsymbol{D}|0\rangle}{E_0 - E_k}$$

$$\alpha_{xx} = -2e^2\sum_{k>0} \frac{|\langle 0|\sum_i x_i|k\rangle|^2}{E_0 - E_k} = -2\sum_{k>0} \frac{|\langle 0|D_x|k\rangle|^2}{(E_0 - E_k)}. \tag{6-147}$$

The polarizability tensor is also defined according to the discussion in Section 4.11 starting with Eq.(4-193). The polarizability tensor defined here is similar to the χ tensor in Section 6.8; both the elements in χ and $\boldsymbol{\alpha}$ have cgs units of volume.

The treatment above is appropriate for a static field. In the presence of a time-dependent field, the situation is more complicated. Consider the more general sinusoidal electric field given by

$$\boldsymbol{E} = \boldsymbol{E}_0 \cos(\boldsymbol{k}\cdot\boldsymbol{r} - \omega t) = \frac{E_0}{2}\{\exp[i(\boldsymbol{k}\cdot\boldsymbol{r} - \omega t)] + \exp[-i(\boldsymbol{k}\cdot\boldsymbol{r} - \omega t)]\} \tag{6-148}$$

in Eq.(6-145). Now, if the system is initially in the ground electronic state, $\psi_0 \exp(-iE_0 t/\hbar)$, and if the perburbation due to the field is small, the new state will be given by a linear combination of the functions in Eq.(6-99) according to Eqs.(3-162),

$$\Psi(\boldsymbol{r}, t) = \psi_0 \exp(-iE_0 t/\hbar) + \sum_{k>0}{}' c_k(t)\psi_k \exp(-iE_k t/\hbar). \tag{6-149}$$

ψ_0 and ψ_k in Eq.(6-149) represent the total electronic ground- and excited-state functions from Eq.(6-99) where the prime on the summation excludes the ground state. Using the perturbation function in Eq.(6-149) to calculate the average value of the dipole moment gives

$$\langle \boldsymbol{D}\rangle = \frac{\int \Psi^*(\boldsymbol{r}, t)\boldsymbol{D}\Psi(\boldsymbol{r}, t)\,dV}{\int \Psi^*(\boldsymbol{r}, t)\Psi(\boldsymbol{r}, t)\,dV}$$

$$= \boldsymbol{D}_{00} + \sum_{k>0}{}' c_k^*(t)\exp(i\omega_{k0}t)\boldsymbol{D}_{k0} + \sum_{k>0}{}' c_k(t)\exp(-i\omega_{k0}t)\boldsymbol{D}_{0k}, \tag{6-150}$$

where the smaller $c_k^*(t)c_k(t)$ terms have been dropped and the denominator is unity to first order in the field. $\boldsymbol{D}$ is the dipole moment operator in Eq.(6-145) and $\boldsymbol{D}_{00}$ is defined in Eq.(6-146). The $c_k(t)$ coefficient for the kth state in Eq.(6-150) is given in Eq.(3-163) for small perturbations,

$$\frac{dc_k(t)}{dt} \cong \frac{1}{i\hbar}\langle k|\mathscr{H}_1|0\rangle \exp(i\omega_{k0}t). \tag{6-151}$$

$\mathscr{H}_1$ is given in Eq.(6-145) with the field in Eq.(6-148) and $\omega_{k0} = (E_k - E_0)/\hbar$. Substituting Eq.(6-148) into Eq.(6-151) and using the $\exp(\pm i\boldsymbol{k}\cdot\boldsymbol{r}) \cong 1.0$ approximation as discussed in Eq.(3-170) gives

$$\frac{dc_k(t)}{dt} \cong -\frac{\langle k|\boldsymbol{E}_0\cdot\boldsymbol{D}|0\rangle}{2i\hbar}\{\exp[i(\omega_{k0}-\omega)t] + \exp[i(\omega_{k0}+\omega)t]\}.$$

In order to integrate over time, we start in the distant past ($t = -\infty$) where the value of $c_k(-\infty) = 0$ as $\mathscr{H}_1(t = -\infty) = 0$ and integrate to t. This gradual "turning on" of the perturbations allows this equation to be written as an indefinite integral, giving (to within a constant)

$$c_k(t) = \frac{\langle k|\boldsymbol{E}_0\cdot\boldsymbol{D}|0\rangle}{2\hbar}\left\{\frac{\exp[i(\omega_{k0}-\omega)t]}{\omega_{k0}-\omega} + \frac{\exp[i(\omega_{k0}+\omega)t]}{\omega_{k0}+\omega}\right\}. \tag{6-152}$$

Substituting into Eq.(6-150) and dropping the smaller $c_k(t)c_k^*(t)$ terms gives

$$\begin{aligned}\langle\boldsymbol{D}\rangle &= \boldsymbol{D}_{00} + \boldsymbol{E}_0\cdot\sum_{k>0}\frac{\boldsymbol{D}_{k0}^*\boldsymbol{D}_{k0}}{2\hbar}\left[\frac{\exp(i\omega t)}{\omega_{k0}-\omega} + \frac{\exp(-i\omega t)}{\omega_{k0}+\omega} + \frac{\exp(-i\omega t)}{\omega_{k0}-\omega} + \frac{\exp(i\omega t)}{\omega_{k0}+\omega}\right] \\ &= \boldsymbol{D}_{00} + \boldsymbol{E}_0\cdot\sum_{k>0}\frac{|D_{k0}|^2[2\omega_{k0}\cos\omega t]}{\hbar(\omega_{k0}^2-\omega^2)}.\end{aligned}$$

According to Eq.(6-148), the induced moment is given by $\langle\boldsymbol{D}\rangle_{\text{ind}} = \boldsymbol{\alpha}\cdot\boldsymbol{E} = \boldsymbol{\alpha}\cdot\boldsymbol{E}_0\cos\omega t$, and the polarizability is given then by

$$\boldsymbol{\alpha}(\omega) = \frac{2}{\hbar}\sum_{k>0}\frac{\omega_{k0}\boldsymbol{D}_{k0}^*\boldsymbol{D}_{k0}}{\omega_{k0}^2-\omega^2}. \tag{6-153}$$

In the limit where $\omega_{k0} \gg \omega$, Eq.(6-153) reduces to the result in Eq.(6-147). Of course, all these results could be generalized to the jth state instead of the ground state by replacing all 0 with j and replacing $k > 0$ in the summation with $k \neq j$. We are normally concerned, however, with the polarizability in the ground state of the molecule or atom.

It is interesting to compare the quantum mechanical polarizability in Eq.(6-153) with the classical result from Section 2.2. The classical value of α_{xx} can be obtained by substituting the displacement $x(t)$ in Eq.(2-47) where $\tau \longrightarrow \infty$ into Eq.(2-50) to give

$$\alpha_{xx} = -\frac{ex}{E_x} = \frac{(e^2/m)}{\omega_0^2-\omega^2}, \tag{6-154}$$

where $\omega_0 = (k/m)^{1/2}$ is the harmonic vibrational angular frequency. The ratio of the quantum mechanical to the classical polarizabilities for the $k \longrightarrow j$ transition is called the *oscillator strength*, f_{jk}, for the transition. Consider the α_{xx} component in Eq.(6-153) for the $k \longrightarrow j$ transition where $\omega_{jk} = \omega_{k0}$. Comparing Eq.(6-153) with Eq.(6-154) gives the oscillator strength for the x dipole transition of

$$f^x_{jk} = \frac{2m\omega_{jk}|D^x_{kj}|^2}{\hbar e^2} = \frac{8\pi^2 m v_{jk}|D^x_{kj}|^2}{he^2}. \tag{6-155}$$

We can also define an oscillator strength in terms of the average polarizability of an atom or molecule, $\alpha = \frac{1}{3}(\alpha_{xx} + \alpha_{yy} + \alpha_{zz})$, where for an isotropic system $\alpha_{xx} = \alpha_{yy} = \alpha_{zz} = \frac{1}{3}\alpha$. Under these circumstances we write the average oscillator strength in the form

$$f_{jk} = \frac{8\pi^2 m v_{jk}|D_{kj}|^2}{3he^2}. \tag{6-156}$$

The average polarizability of a molecule can be measured by dielectric techniques where the dielectric constant is related to the polarizability in a dilute gas according to Eq.(1-43). If a permanent electric moment is present, however, the bulk induced electric moment will be temperature dependent due to an additional alignment of the permanent moments in the electric field. Consider a diatomic molecule with a ground-state electric moment along the internuclear axis of D_a. The energy of the dipolar molecule in the presence of the field, correct to second order in perturbation theory, is given in Eq.(4-203). The average dipole moment along the field axis for a molecule in the J, M state, $D(J, M)$, is given by taking the first derivative of the energy in Eq.(4-203) as shown, for instance, in Eq.(4-192),

$$D(J, M) = -\frac{dE^{(2)}}{dE} = -\frac{D_a^2 E}{hBJ(J+1)}\left[\frac{J(J+1) - 3M^2}{(2J-1)(2J+3)}\right],$$

where B is the rotational constant. Now, the average value of the dipole moment in the state J is obtained by summing this expression over all M states within the J state. Doing this, we find $\sum_{J=-M}^{M} D(J, M) = 0$ if $J \neq 0$. Thus, the only state that gives an aligned dipole moment is $J = 0$, $M = 0$; using the energy in Eq.(4-201) and the derivative above gives

$$D(0, 0) = \frac{D_a^2 E}{3hB}. \tag{6-157}$$

The conclusion that only the $J = 0$ state is aligned in the field can be understood by reviewing Fig. 4-25. The $J = 0$ energy level is at a lower energy in the field. All $J \neq 0$ states show M splittings both to higher and lower energy, however, which when summed over M lead to no net energy of stabilization in the field.

The total fraction of molecules in the $J = 0$ state, N_0/ρ_0, where N_0 is the number density of the $J = 0$ state and ρ_0 is the molecular number density, is given by

$$\frac{N_0}{\rho_0} = \frac{\exp(-E_{00}/kT)}{\sum_{J=0}(2J+1)\exp(-E_J/kT)} \cong \frac{1}{\int_0^\infty (2J+1)\exp[-J(J+1)hB/kT]\,dJ}$$

$$\cong \frac{1}{2\int_0^\infty J\exp(-J^2 hB/kT)\,dJ} = \frac{hB}{kT}, \tag{6-158}$$

where the $(2J + 1) \cong 2J$ approximations are evident. Thus, the average induced moment due to the aligned linear molecule is given by the product of the results in Eqs.(6-157) and (6-158). The final observed polarizability per dipolar molecule is given then by

$$\alpha_{\text{observed}} = \alpha + \frac{D_a^2}{3kT}. \tag{6-159}$$

Substituting into Eq.(1-43) gives the dielectric constant of the dilute gas:

$$\text{cgs} \qquad\qquad \text{SI}$$

$$\varepsilon = 1 + 4\pi\rho_0\left(\alpha + \frac{D_a^2}{3kT}\right) \qquad \varepsilon = 1 + \frac{\rho_0}{\varepsilon_0}\left(\alpha + \frac{D_a^2}{3kT}\right). \tag{6-160}$$

It is now evident that in the gas phase only the ground rotational state ($J = 0$) contributes to the dipolar contribution in the dielectric constant. In summary, a measurement of the temperature-dependence of ε in a dilute gas can give both a measure of D_a^2 and $\alpha = \frac{1}{3}(\alpha_{aa} + 2\alpha_{bb})$ for the cylindrically symmetric molecule. In a liquid or dense gas, the equations above must be modified due to the polarization of the medium surrounding the molecule. First we use Eq.(6-160) for a dense medium to give

$$\text{cgs} \qquad\qquad \text{SI}$$

$$\frac{(\varepsilon - 1)}{4\pi}\boldsymbol{E} = \rho_0 \sum_i \boldsymbol{E}_i \alpha_i \qquad \varepsilon_0(\varepsilon - 1)\boldsymbol{E} = \rho_0 \sum_i \boldsymbol{E}_i \alpha_i, \tag{6-161}$$

where $\boldsymbol{E}$ is the electric field in the sample and $\boldsymbol{E}_i$ is the internal "local" field at the ith molecule where $\alpha_i = (\alpha + D_i^2/3kT)$ [75]. Now, the simplest form for the local field correction is the spherical cavity Lorentz model, which gives [76]

$$\boldsymbol{E}_i = \left(\frac{\varepsilon + 2}{3}\right)\boldsymbol{E}. \tag{6-162}$$

Substituting into Eq.(6-161) gives

$$\text{cgs} \qquad\qquad \text{SI}$$

$$\frac{(\varepsilon - 1)}{(\varepsilon + 2)} = \left(\frac{4\pi}{3}\right)\rho_0\left(\alpha + \frac{D_a^2}{3kT}\right) \qquad \frac{(\varepsilon - 1)}{(\varepsilon + 2)} = \frac{\rho_0}{3\varepsilon_0}\left(\alpha + \frac{D_a^2}{3kT}\right), \tag{6-163}$$

which is the well-known Clausius–Mossotti equation [77].

The Kerr Effect and Polarizability Anisotropies

The anisotropies in the polarizabilities (the difference in the diagonal elements of the principal polarizability tensor) can be measured by the electric field perturbations on rotation states (Section 4.11) and by the Kerr effect, which is the optical measurement of a static field-induced anisotropy in the refractive index of the medium.

Consider a static field, E_z^0, along the z laboratory-fixed axis, which will cause a polarization along the z axis due to both the induced moment and the aligned permanent moment. The refractive indices parallel and perpendicular to the static field in a dilute gas are given by Eq.(1-60):

$$\begin{array}{ll} \textit{cgs} & \textit{SI} \\ n_{\parallel} = 1 + 2\pi\rho_0[\alpha + (\alpha_{\parallel})_{av}] & n_{\parallel} = 1 + \dfrac{\rho_0}{\varepsilon_0}[\alpha + (\alpha_{\parallel})_{av}] \\ n_{\perp} = 1 + 2\pi\rho_0[\alpha + (\alpha_{\perp})_{av}] & n_{\perp} = 1 + \dfrac{\rho_0}{\varepsilon_0}[\alpha + (\alpha_{\perp})_{av}], \end{array} \quad (6\text{-}164)$$

where $(\alpha_{\parallel})_{av}$ and $(\alpha_{\perp})_{av}$ are the average static field-dependent polarizabilities along axes parallel and perpendicular to the static field and α is the average polarizability.

We now return to Section 4.11 to examine the effects of static and optical fields on a cylindrically symmetric molecule with dipole moment D_a and principal polarizability tensor elements α_{aa} and $\alpha_{bb} = \alpha_{cc}$ where a is the symmetry axis. The energy of interaction between this molecule and a static E_z^0 field is given in Eqs.(4-197) and (4-205) by

$$\begin{aligned} \mathscr{H} &= -D_a E_z^0 \cos\theta_{az} - \frac{(E_z^0)^2}{2}[\alpha_{aa}\cos^2\theta_{az} + \alpha_{bb}(\cos^2\theta_{bz} + \cos^2\theta_{cz})] \\ &= -D_a E_z^0 \cos\theta - \frac{(E_z^0)^2}{2}(\alpha_{aa}\cos^2\theta + \alpha_{bb}\sin^2\theta), \end{aligned} \quad (6\text{-}165)$$

where the last step relates the direction cosines to the spherical polar angles from Eq.(4-208). The interaction of the molecule with optical fields along the z (parallel) and x (perpendicular) axes are given by similar methods, using Eq.(4-208), to give

$$\begin{aligned} \mathscr{H}_{\parallel} &= -\frac{E_z^2}{2}[\alpha'_{aa}\cos^2\theta + \alpha'_{bb}\sin^2\theta] \\ \mathscr{H}_{\perp} &= -\frac{E_z^2}{2}[\alpha'_{aa}\sin^2\theta\cos^2\phi + \alpha'_{bb}(\sin^2\phi + \cos^2\phi\cos^2\theta)], \end{aligned} \quad (6\text{-}166)$$

where we have assumed that the optical radiation frequency is fast relative to the rotational motion, thereby dropping the $D_a E_z \cos\theta$ term. The primes on α'_{aa} and α'_{bb} indicate the polarizabilities induced by the optical frequencies, which need not be the same as the corresponding static field values of α_{aa} and α_{bb} in Eq.(6-165). The induced moments and corresponding polarizabilities along the parallel and perpendicular axes are given by the negative first derivatives of Eqs.(6-166) with respect to E_z, giving

$$\begin{aligned} \alpha_{\parallel} &= -\frac{1}{E_z}\left(\frac{\partial \mathscr{H}_{\parallel}}{\partial E_z}\right) = \alpha'_{aa}\cos^2\theta + \alpha'_{bb}\sin^2\theta \\ \alpha_{\perp} &= -\frac{1}{E_z}\left(\frac{\partial \mathscr{H}_{\perp}}{\partial E_z}\right) = \alpha'_{aa}\sin^2\theta\cos^2\phi + \alpha'_{bb}(\sin^2\phi + \cos^2\phi\cos^2\theta). \end{aligned} \quad (6\text{-}167)$$

Now, the average value of $\alpha_{\parallel}$ and $\alpha_{\perp}$ in the presence of the static field, E_z^0, is given by the appropriate field-weighted average as follows:

$$(\alpha_\gamma)_{av} = \frac{\int_0^{2\pi} \int_0^{\pi} \alpha_\gamma \exp(-\mathscr{H}/kT) \sin\theta \, d\theta \, d\phi}{\int_0^{2\pi} \int_0^{\pi} \exp(-\mathscr{H}/kT) \sin\theta \, d\theta \, d\phi}, \tag{6-168}$$

where $\gamma = \parallel$ or $\perp$ for $\alpha_{\parallel}$ and $\alpha_{\perp}$ from Eq.(6-167) and $\mathscr{H}$ is in Eq.(6-165). Substituting $\alpha_{\parallel}$ or $\alpha_{\perp}$ from Eqs.(6-167) and $\mathscr{H}$ from Eq.(6-165) into Eq.(6-168), expanding the $\exp(-\mathscr{H}/kT)$ to include all terms through $(E_z^0)^2$ dependence in both the numerator and denominator (remember that $\mathscr{H}/kT \ll 1$ for reasonably low fields), doing all the integrals over the spherical polar angles θ and ϕ, and then keeping only terms through $(E_z^0)^2$-dependence gives

$$(\alpha_{\parallel})_{av} = \frac{2(E_z^0)^2}{45}\left[\frac{1}{kT}(\alpha'_{aa} - \alpha'_{bb})(\alpha_{aa} - \alpha_{bb}) + \frac{D_a^2}{(kT)^2}(\alpha'_{aa} - \alpha'_{bb})\right] + \alpha'$$

$$(\alpha_{\perp})_{av} = -\frac{1}{2}(\alpha_{\parallel})_{av} + \alpha', \tag{6-169}$$

where $\alpha' = \frac{1}{3}(\alpha'_{aa} + \alpha'_{bb} + \alpha'_{cc})$ is the average optical polarizability. Substituting these results into Eqs.(6-164) and taking the $n_{\parallel} - n_{\perp}$ difference gives the static field-induced optical anisotropy in the refractive index of the dilute gas that is proportional to the Kerr constant [78]. The result is (cgs)

$$n_{\parallel} - n_{\perp} = \frac{2\pi\rho_0(E_z^0)^2}{15}\left[\frac{1}{kT}(\alpha_{aa} - \alpha_{bb})(\alpha'_{aa} - \alpha'_{bb}) + \frac{D_a^2}{(kT)^2}(\alpha'_{aa} - \alpha'_{bb})\right]. \tag{6-170}$$

The result in SI units is obtained by multiplying the right-hand side by $(4\pi\varepsilon_0)^{-1}$.

Table 6-20 Molecular optical ($\lambda \cong 5500$ Å) polarizabilities (in units of 10^{-24} cm^3) for several cylindrically symmetric molecules. The data are obtained from Landolt–Börnstein (Springer-Verlag, Berlin, 1951), Vol. I, Part 3, p. 510; N. J. Brige and A. D. Buckingham, *Proc. Roy. Soc.* **295A**, 334 (1966), and Ref. [80].

Molecule	α	$\alpha_{\parallel} - \alpha_{\perp}$	$\alpha_{\parallel}$	$\alpha_{\perp}$
H_2	0.79	0.31	1.00	0.69
N_2	1.77	0.71	2.24	1.53
O_2	1.60	1.10	2.83	1.23
Cl_2	4.61	2.98	6.60	3.62
HF	2.46	0.72	2.88	2.16
HCl	2.63	0.74	3.13	2.39
N_2O	3.01	3.03	5.03	2.00
CO_2	2.64	2.13	4.06	1.94
OCS	5.22	4.18	8.01	3.83
CS_2	8.84	10.02	15.52	5.50
CH_3Cl	4.62	1.58	5.66	4.08
$CHCl_3$	8.56	−2.72	6.74	9.46
HCCH	3.50	1.86	4.74	2.88
C_6H_6	10.50	−5.68	6.68	12.36
NH_3	2.26	0.23	2.41	2.18
CH_3CH_3	4.47	1.51	5.48	3.97

Table 6-21 Empirical parallel and perpendicular bond polarizabilities (in units of 10^{-24} cm^3) adapted from K. G. Denbigh, *Trans. Faraday Soc.* **36**, 936 (1940).

Bond	$\alpha_{\parallel}$	$\alpha_{\perp}$
C—C (aliphatic)	1.9	0.0
C—C (aromatic)	2.3	0.5
C=C	2.9	1.1
C≡C	3.5	1.3
C—H	0.8	0.6
C—Cl	3.7	2.1
C—Br	5.0	2.9
C=O	2.0	0.8
C=S	7.6	2.8
C≡N	3.1	1.4
N—H	0.6	0.8
S—H	2.3	1.7

A large number of polarizability anisotropies have been measured in molecules by Kerr effect measurements where usually the anisotropies are reported after ignoring the small differences between the optical and static values [79]. Furthermore, most of these measurements are in the liquid state and local field corrections must be made to extract single molecule values of $\alpha_{aa} - \alpha_{bb}$.

The intensity of depolarized light scattered from anisotropic molecules in a dilute gas can also be used to extract values of optical polarizabilities in molecules. We shall discuss this technique in Chapter 8. Some representative numbers for optical polarizabilities are given in Table 6-20. The optical value of $\alpha_{\parallel} - \alpha_{\perp} = 4.18$ in OCS can be compared to the static value of 4.7 from Table 4-17 [80].

Attempts have been made to construct a set of empirical bond polarizabilities that are capable, by addition, of fitting the polarizability of a large number of molecules. One set of bond parameters is shown in Table 6-21. The calculated values of the polarizabilities for several molecules are listed below where all values are in units of 10^{-24} cm^3.

1. *Acetylene*, HC≡CH

$$\alpha_{\parallel} = 2\alpha_{\parallel}(\mathrm{C{-}H}) + \alpha_{\parallel}(\mathrm{C{\equiv}C}) = 5.1$$

$$\alpha_{\perp} = 2\alpha_{\perp}(\mathrm{C{-}H}) + \alpha_{\perp}(\mathrm{C{\equiv}C}) = 2.5$$

2. *Carbon dioxide*, OCO

$$\alpha_{\parallel} = (2.0 + 2.0) = 4.0$$

$$\alpha_{\perp} = (0.8 + 0.8) = 1.6$$

3. *Carbonyl sulfide*, OCS

$$\alpha_{\parallel} = (2.0 + 7.6) = 9.6$$

$$\alpha_{\perp} = (0.8 + 2.8) = 3.6$$

$$\alpha_{\parallel} - \alpha_{\perp} = 6.0$$

4. *Formaldehyde*, H_2CO (*see Fig. 6-4 for structure and coordinates*)

$$\alpha_{zz} = [0.8 + 2(0.6)] = 2.0$$

$$\alpha_{yy} = [0.8 + 2(0.8)\cos^2 32^\circ + 2(0.6)\sin^2 32^\circ] = 2.4$$

$$\alpha_{xx} = [2.0 + 2(0.8)\cos^2 58^\circ + 2(0.6)\sin^2 58^\circ] = 3.32$$

In the case of formaldehyde, the direction cosine transformation is used from the CH bond axis system to the principal inertial axis system where the z axis is out of the molecular plane and the x axis is the dipole moment axis as shown in Fig. 6-4. The results for HC≡CH, CO_2, and OCS above agree reasonably well with the results in Table 6-20.

Induced Dipole–Induced Dipole Attractive Energies

We shall now consider an application of the polarizability concepts to the problem of the attraction of two spherical, but polarizable, systems such as rare gas atoms. Consider the interaction between two rare gas atoms at an internuclear distance R. The total Hamiltonian describing the energy is

$$\mathscr{H} = \underbrace{\mathscr{H}_a + \mathscr{H}_b}_{} + \underbrace{V_{ab}}_{}$$

$$\mathscr{H} = \mathscr{H}_0 + \mathscr{H}_1. \tag{6-171}$$

$\mathscr{H}_a$ and $\mathscr{H}_b$ are the independent Hamiltonians for the individual two atoms [see Eq.(5-23)] and V_{ab} is the interaction potential between atoms given by

$$V_{ab} = \frac{e^2 Z_a Z_b}{R_{ab}} - e^2 \sum_j \frac{Z_a}{r_{aj}} - e^2 \sum_i \frac{Z_b}{r_{bi}} + e^2 \sum_{i,j} \frac{1}{r_{ij}}. \tag{6-172}$$

The sums over i and j are over the i electrons on atom a and the j electrons on atom b and Z_a and Z_b are the atomic numbers of atoms a and b separated by R_{ab} as shown in Fig. 6-15. We can expand the last three terms in Eq.(6-172) about R_{ab} by the

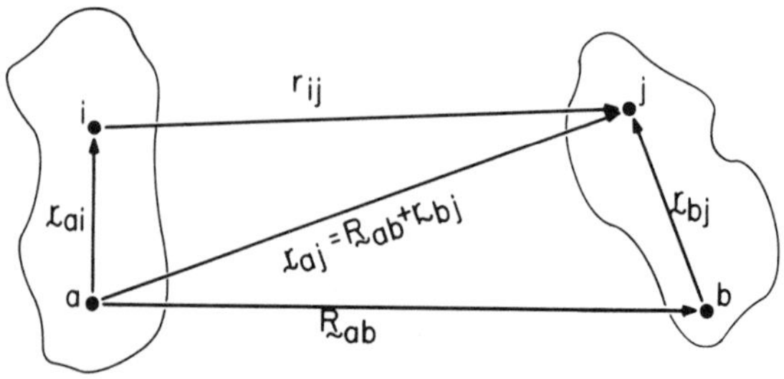

Figure 6-15 Coordinate system showing the interaction of atom a with atom b.

binomial expansion if $R_{ab} > r_{bj}$ and $R_{ab} > r_{ai}$. First, we place the nuclei of the atoms along the z axis and write the following identities according to Fig. 6-15:

$$\frac{1}{r_{aj}} = \frac{1}{[(R_{ab} + z_{bj})^2 + y_{bj}^2 + x_{bj}^2]^{1/2}} = \frac{1}{R_{ab}}\left[1 + 2\left(\frac{z_{bj}}{R_{ab}}\right) + \left(\frac{r_{bj}}{R_{ab}}\right)^2\right]^{-1/2}$$

$$\frac{1}{r_{bi}} = \frac{1}{R_{ab}}\left[1 - 2\left(\frac{z_{ai}}{R_{ab}}\right) + \left(\frac{r_{ai}}{R_{ab}}\right)^2\right]^{-1/2}$$

$$\frac{1}{r_{ij}} = \frac{1}{R_{ab}}\left[1 + \frac{2(z_{bj} - z_{ai})}{R_{ab}} + \frac{(r_{bj} - r_{ai})^2}{R_{ab}^2}\right]^{-1/2}. \tag{6-173}$$

Each of the terms in Eqs.(6-173) can be expanded in a power series when $1 \gg r/R_{ab}$ giving

$$\frac{1}{r_{aj}} = \frac{1}{R_{ab}}\left\{1 - \frac{z_{bj}}{R_{ab}} - \frac{r_{bj}^2}{2R_{ab}^2} + \frac{3}{8}\left[2\frac{z_{bj}}{R_{ab}} + \left(\frac{r_{bj}}{R_{ab}}\right)^2\right]^2 + \cdots\right\}$$

$$= \frac{1}{R_{ab}}\left(1 - \frac{z_{bj}}{R_{ab}} - \frac{r_{bj}^2}{2R_{ab}^2} + \frac{3z_{bj}^2}{2R_{ab}^2} + \cdots\right)$$

$$\frac{1}{r_{bi}} = \frac{1}{R_{ab}}\left(1 + \frac{z_{ai}}{R_{ab}} - \frac{r_{ai}^2}{R_{ab}^2} + \frac{3z_{ai}^2}{2R_{ab}^2} + \cdots\right)$$

$$\frac{1}{r_{ij}} = \frac{1}{R_{ab}}\left[1 - \frac{z_{bj}}{R_{ab}} + \frac{z_{ai}}{R_{ab}} - \frac{(r_{bj} - r_{ai})^2}{2R_{ab}^2} + \frac{3z_{bj}^2}{2R_{ab}^2} - \frac{3z_{ai}z_{bj}}{R_{ab}^2} + \frac{3z_{ai}^2}{2R_{ab}^2} + \cdots\right],$$

where only terms up to $1/R_{ab}^3$ have been given. Substituting into V_{ab} in Eq.(6-172) and using the following sum rules,

$$\sum_j Z_a\left(\frac{1}{R_{ab}}\right) = \frac{Z_aZ_b}{R_{ab}}; \qquad \sum_j r_{bi} = Z_b r_{bi}$$

$$\sum_i Z_b\left(\frac{1}{R_{ab}}\right) = \frac{Z_aZ_b}{R_{ab}}; \qquad \sum_i r_{aj} = Z_a r_{aj}$$

$$\frac{1}{2}\sum_{i,j}\frac{1}{R_{ab}} = \frac{Z_aZ_b}{R_{ab}},$$

gives

$$V_{ab} = e^2\sum_{i,j}\frac{1}{R_{ab}^3}(x_{aj}x_{bi} + y_{aj}y_{bi} - 2z_{aj}z_{bi}) + \cdots. \tag{6-174}$$

The first or $1/R_{ab}^3$ term is the dipole-dipole term or the dipole of one atom in the field produced by the dipole of the second atom. Higher-order terms give dipole-quadrupole and others. We can now treat the potential energy in Eq.(6-174) as a perturbation on the free atom system.

We shall assume that the atoms are sufficiently far apart so their electron densities do not overlap. Thus, the zero-order wavefunction for the system described by $\mathscr{H}_0$ in Eq.(6-171) is given by

$$\boldsymbol{\psi} = \boldsymbol{\psi}_a \times \boldsymbol{\psi}_b; \qquad \psi_{ij}(ab) = \sum_{j=0}^{\infty} \sum_{i=0}^{\infty} \psi_a^i \psi_b^j$$

$$\boldsymbol{\psi}_a = (\psi_a^0 \psi_a^1 \psi_a^2 \cdots), \qquad \boldsymbol{\psi}_b = (\psi_b^0 \psi_b^1 \psi_b^2 \cdots), \tag{6-175}$$

where $\psi_a^0, \psi_a^1, \ldots$ are the ground- and excited-state electronic functions for atom a as indicated in Eq.(6-99), for instance. Using the zero-order function, $\psi_{ij}(ab)$, above for the a-b system gives the following second-order correction to the ground-state energy of the two atoms (the first-order correction is zero):

$$E^{(2)} = \sum_{k>0}^{b} \sum_{m>0}^{a} \frac{\langle 0_a 0_b | V_{ab} | k_a m_b \rangle \langle k_a m_b | V_{ab} | 0_a 0_b \rangle}{E_b^0 - E_b^k + E_a^0 - E_a^m}. \tag{6-176}$$

0_a and k_a represent the ground- and excited-state electronic functions for atom a and 0_b and m_b represent the functions on atom b. Substituting V_{ab} from Eq.(6-174) into Eq.(6-176) and noting that $\langle 0|x|k\rangle = \langle 0|y|k\rangle = \langle 0|z|k\rangle$ for a spherically symmetric atom gives the dipole-dipole interaction energy:

$$E(a-b) = \frac{6e^4}{R_{ab}^6} \sum_{k>0} \sum_{m>0} \left[\frac{(|\langle 0|\sum_j x_{aj}|m\rangle|^2)(|\langle 0|\sum_i x_{bi}|k\rangle|^2)}{E_b^0 - E_b^k + E_a^0 - E_a^m} \right]. \tag{6-177}$$

We can write the dipole-dipole interaction energy in a more familiar form by estimating

$$E(a-b) \cong -\frac{3}{2R_{ab}^6} \frac{\alpha_a E_a \alpha_b E_b}{(E_a + E_b)}, \tag{6-178}$$

where α_a and α_b are the polarizabilities of atoms a and b, and E_a and E_b are the corresponding ionization energies. This result was used previously to describe the dipole-dipole interaction between two atoms [see Eq.(5-103) and associated discussion].

Of course, this treatment above is quite approximate, especially in the last step to give Eq.(6-178). The results using Eq.(6-178) for spherical systems are quite good, however (see Section 5.9). In some cases for close interactions between molecules the expansions following Eq.(6-173) cannot be safely truncated. A more general theory of interaction between nonoverlapping charge distributions as well as reviews of other work in this area can be found elsewhere [81].

PROBLEMS

1. The molecular energy, after using the Born–Oppenheimer approximation, is given in Eq.(6-17). We call this energy the Born–Oppenheimer energy, E_{BO}. Prove that if the true energy is E^0, that $E_{BO} < E^0$ or that E_{BO} is a lower bound for the energy [see S. T. Epstein, *J. Chem. Phys.* **44**, 836 (1966)] [82].

2. Return to an examination of the chemical bonding in H_2^+ as described by Φ_+ and W_+ in Eqs.(6-30) and (6-36), respectively, as shown in Fig. 6-1. Calculate the potential and kinetic energies separately with Φ_+ in Eq.(6-30) and plot these values of T_+ and V_+, respectively, along with $W_+ = T_+ + V_+$ as shown in Fig. 6-1. The results should show T_+ has a minimum in energy and V_+ does not. Thus, according to these results the chemical bond in H_2^+ arises from a minimization of the kinetic energy rather than the Coulomb stabilization in the potential energy.

*3. Use the extended Hückel, CNDO, or other approximate LCAO-MO theory to describe the following molecular properties. If possible, do the problem with two or more methods and compare the results.
   (a) Calculate the vibrational potential function for the O—H bond in formic acid.
   (b) Calculate the vibrational potential function for the proton in the acetic acid–formic acid dimer.

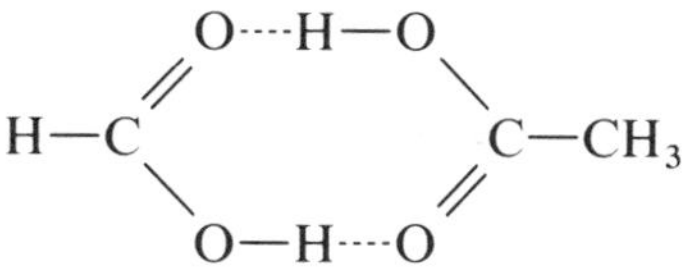

   The hydrogen bonding should lead to a double minimum potential function.
   (c) Predict the relative stabilities of

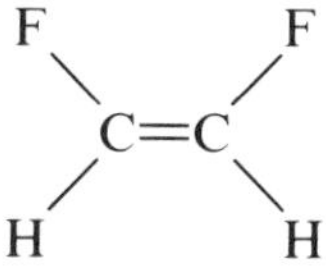 and 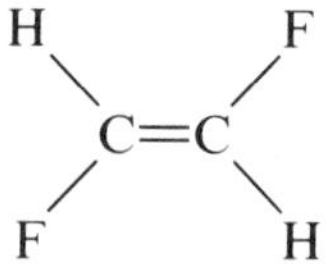

   How does your result agree with the literature?
   (d) Predict the relative stabilities of

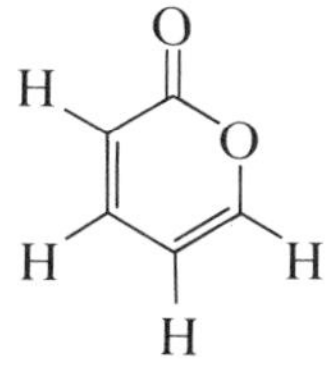 and 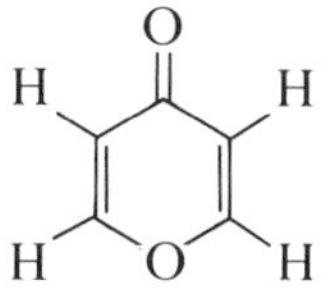

   How does your result agree with the literature?
   (e) Predict the relative stabilities of the chair and boat forms of cyclohexane and compare with the known relative stabilities.

 or 

(f) Predict the relative stabilities of the following two forms of methyl nitrite and compare with the known results.

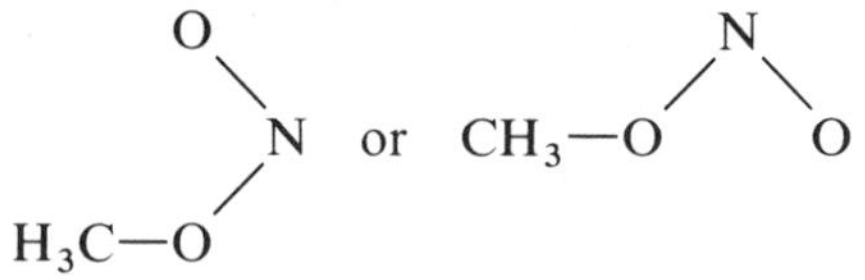

(g) Calculate the vibrational potential function for NH_3 nitrogen inversion and compare your result with the literature (see Problem 4-37).

(h) Calculate the relative stabilities of the following two molecules and compare with the known results.

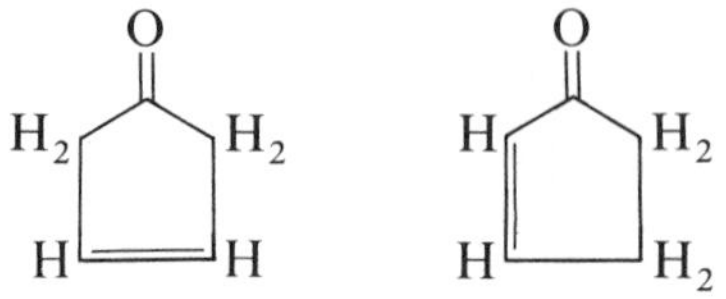

***4.** Calculate the one-electron probability density for the H_2 molecule with both the MO and VB functions. Make an electron density plot.

5. In Section 6.6 we evaluated the effective point charges in H_2CO by use of the extended Hückel LCAO-MO method. The molecular monopole, dipole, and quadrupole moments of formaldehyde are given below:

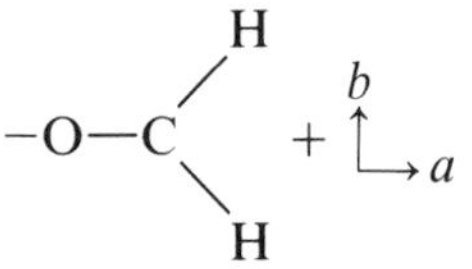

$$M = 0$$

$$D_a = 2.34 \times 10^{-18} \text{ SC cm}$$

$$Q_{aa} = -0.1 \times 10^{-26} \text{ SC cm}^2$$

$$Q_{bb} = +0.2 \times 10^{-26} \text{ SC cm}^2$$

Use a point charge model at the atomic centers and the moments above to evaluate the charges on the three different types of atoms in formaldehyde and compare your results with the results obtained by the extended Hückel method given in the text.

6. Calculate the electron densities and bond orders of

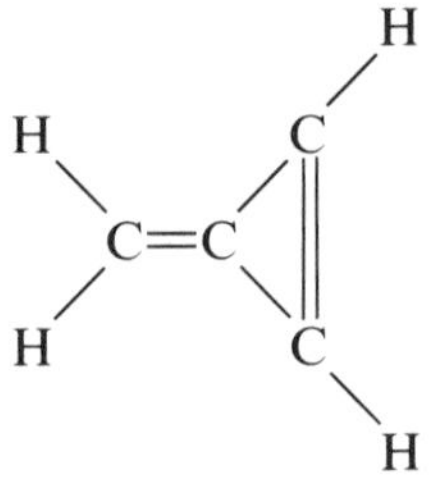

by both Hückel and extended Hückel methods.

7. Calculate the principal moments of inertia in H_2CO using the structure given in Fig. 6-3 and atomic masses. I_{aa}, which is the moment about the CO bond axis, is computed with the hydrogen masses only. The object of this problem is to estimate the ignored contribution to I_{aa} due to the electrons near the oxygen and carbon nuclei. The diamagnetic susceptibility about the a axis in formaldehyde is

$$\chi_{aa}^{d} = -30.9 \times 10^{-6} = -\frac{Ne^2}{4c^2m}\left\langle 0\left|\sum_i (b_i^2 + c_i^2)\right|0\right\rangle.$$

Use this value of χ_{aa}^{d} to estimate the contribution to I_{aa} due to the electrons around the C and O atoms. Show your logic and approximations.

8. Consider the effect of a near-atom point charge on the magnetic shielding at a proton in a molecule.

(a) Start by computing the average nuclear shielding for the free H atom.

Consider now the effect of a point electronic charge some distance R from the original hydrogen atom. We are interested in knowing whether the field at the proton will be larger or smaller when the unit point charge is near. For instance, we might want to know the effect of the —F group on the magnetic field at the proton in COFH relative to the corresponding field at the proton in COH_2.

(b) Assuming that H in COFH is described approximately by the free atom wavefunction, compute a corrected wavefunction for the ground state of the hydrogen atom (by perturbation theory) due to the perturbation of a charge of q that is a distance R from the proton.

(c) Compute the magnetic shielding, σ, with this perturbed wavefunction and find the change in the proton field due to the presence of the charge q at a distance R from the proton. Use $R = 2.0$ Å and $q = -3.0 \times 10^{-10}$ SC.

This problem is discussed in T. W. Marshall and J. A. Pople, *Mol. Phys.* **1**, 199 (1958) and A. D. Buckingham, *Can. J. Chem.* **38**, 300 (1960).

9. Consider the effect on the nuclear Zeeman effect of the kth nucleus due to an induced moment in a neighbor group, $\boldsymbol{\mu}^G$. The induced moment arises from the magnetic susceptibility tensor of the neighbor group, $\boldsymbol{\chi}^G$, which gives $\boldsymbol{\mu}^G = \boldsymbol{\chi}^G \cdot \boldsymbol{H}$, where $\boldsymbol{H}$ is the external field [83]. The energy of interaction between the induced magnetic moment of the neighbor group, $\boldsymbol{\mu}^G$, and the magnetic moment of the kth nucleus, $\boldsymbol{\mu}^k$, is given according to Eq.(4-339) by

$$\mathscr{H} = \frac{\boldsymbol{\mu}^k \cdot \boldsymbol{\mu}^G}{R^3} - 3\frac{(\boldsymbol{\mu}^k \cdot \boldsymbol{R})(\boldsymbol{\mu}^G \cdot \boldsymbol{R})}{R^5}, \qquad (6\text{-}179)$$

where $\boldsymbol{R}$ is the vector from the kth nucleus to the center of the neighbor group. Substituting $\boldsymbol{\mu}^G = \boldsymbol{\chi}^G \cdot \boldsymbol{H}$ and adding this Hamiltonian to the $-\boldsymbol{\mu}^k \cdot (\boldsymbol{1} - \boldsymbol{\sigma}) \cdot \boldsymbol{H}$ term in Eq.(6-131) shows that the neighbor group gives rise to an effective shielding on the kth nucleus equal to

$$\boldsymbol{\sigma}_K^G = \frac{\boldsymbol{\chi}^G}{R^3} - 3\frac{\boldsymbol{\chi}^G \cdot \boldsymbol{RR}}{R^5}. \qquad (6\text{-}180)$$

Show that if the neighbor group is cylindrically symmetric with principal magnetic susceptibility anisotropy $\chi_{\parallel}^{G} - \chi_{\perp}^{G}$ that the average shielding at the kth nucleus is given from Eq.(6-180) by

$$\sigma_k^G = \frac{1}{3}(\sigma_{xx} + \sigma_{yy} + \sigma_{zz})_k^G = \frac{(\chi_{\parallel}^G - \chi_{\perp}^G)}{3R^3}(1 - 3\cos^2\theta), \qquad (6\text{-}181)$$

where θ is the angle between the $\boldsymbol{R}$ vector and the axis of symmetry of the cylindrically symmetric neighbor group.

10. Prove that only the first nonzero electric multipole moment is independent of the origin along the internuclear axis in a linear molecule.

11. Consider the $^{16}O^{12}C^{16}O$ molecule. The molecular g-value, $g_{\perp} = -0.05508 \pm 0.00005$, was measured by J. W. Cederberg, C. H. Anderson, and N. F. Ramsey, *Phys. Rev.* **136**, A960 (1964) and the value of $Q_{\parallel} = -4.3 \pm 0.2 \times 10^{-26}\ \mathrm{SC \cdot cm^2}$ was given by A. D. Buckingham, R. L. Disch, and D. A. Dunmur, *J. Am. Chem. Soc.* **90**, 3104 (1968).

(a) Use $d_{CO} = 1.162$ Å along with the data above to calculate $\chi_{\perp} - \chi_{\parallel}$ for OCO.

(b) Use the information above to estimate the point atomic charge distribution in CO_2.

(c) Define the next two higher electric moments (octapole and hexadecapole) along the internuclear axis of a linear molecule and calculate these two moments using the point charges derived in part b.

12. Show that one of the magnetic susceptibility tensor elements (χ_{zz}, for instance) is independent of the origin. χ_{zz}^{d} and χ_{zz}^{p} are both origin-dependent, but the sum is origin-independent.

13. The molecular Zeeman parameters for ethyleneoxide are

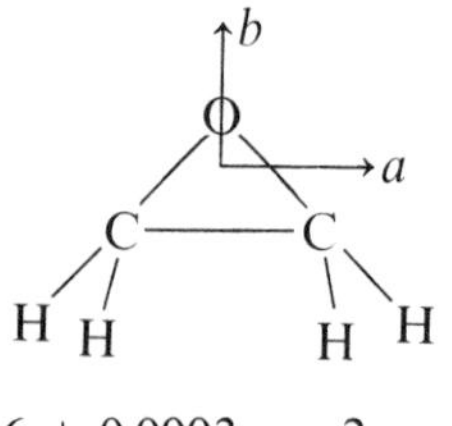

$A = 25{,}483.7$ MHz

$B = 22{,}120.9$ MHz

$C = 14{,}098.0$ MHz

$g_{aa} = -0.0946 \pm 0.0003$

$g_{bb} = +0.0189 \pm 0.0004$

$g_{cc} = +0.0318 \pm 0.0006$

$2\chi_{aa} - \chi_{bb} - \chi_{cc} = 18.1 \pm 0.6 \times 10^{-6}\ \mathrm{erg \cdot G^{-2} \cdot mol^{-1}}$

$-\chi_{aa} + 2\chi_{bb} - \chi_{cc} = 0.8 \pm 1.0 \times 10^{-6}\ \mathrm{erg \cdot G^{-2} \cdot mol^{-1}}$

(a) Calculate the molecular quadrupole moments and show that

$$Q_{aa} + Q_{bb} + Q_{cc} = 0.$$

(b) Calculate χ_{aa}^{p}, χ_{bb}^{p}, and χ_{cc}^{p}.

(c) Use $\chi = -30.7 \times 10^{-6}\ \mathrm{erg \cdot G^{-1} \cdot mol^{-1}}$ to calculate χ_{aa}, χ_{bb}, and χ_{cc}.

(d) Use the information above to calculate $\langle a^2 \rangle$, $\langle b^2 \rangle$, and $\langle c^2 \rangle$.

14. Return to the development from Eq.(6-171) through Eq.(6-174) to extract the dipole-quadrupole term.

15. Consider the nature of the nuclear spin–nuclear spin coupling constant, J_{AB}, discussed in Section 4.10 where the subscript AB indicates coupling of nucleus A with nucleus B. In Problem 4-39 we showed that the direct nuclear spin–nuclear spin coupling averages to zero for randomly orientated molecules (liquid or gas phase). In fact, Ramsey [84] has shown that the dominant interaction, which gives rise to J_{AB}, originates with the Fermi contact term given by [see Eqs.(5-76) and (4-172) when $\boldsymbol{L} = 0$ and $\boldsymbol{J} = \boldsymbol{S}$]

$$\mathscr{H} = \sum_i \left(\frac{8\pi}{3}\right) \frac{g_i \mu_0 g_e \mu_B}{\hbar^2} |\psi_i(0)|^2 \boldsymbol{I}_i \cdot \boldsymbol{S}, \qquad (6\text{-}182)$$

where the sum over i is over all nuclei with nonzero nuclear g-values, g_i, in the molecule that includes, of course, nucleus A and B. $|\psi_i(0)|^2$ is the s-electron density at the ith nucleus as described in Section 5.6.

Consider now the first- and second-order corrections to the electronic ground state due to the Hamiltonian in Eq.(6-182). Now, if we consider only diamagnetic molecules with no electronic angular momentum in the ground state, the only nonzero matrix elements of the electronic $\boldsymbol{S}$ operator will be between the ground and excited electronic states. Thus, the first-order correction is zero.

(a) Evaluate an expression for the second-order correction to the energy of the ground electronic state using the electronic basis in Eq.(6-99) and show how the result reduces to an expression given by the $\mathscr{H} = h(J_{AB}/\hbar^2)\boldsymbol{I}_A \cdot \boldsymbol{I}_B$ term in Eq.(4-172). In conclusion, we see that J_{AB} represents an electron-coupled nuclear spin–nuclear spin coupling.

(b) Use the derived form of J_{AB} above to interpret the observed trend of $J_{CH}(H{-}^{13}C)$ in $CH_4(J_{CH} = 125\text{ Hz})$, $CH_2{=}CH_2(J_{CH} = 158\text{ Hz})$, and $HC{\equiv}CH(J_{CH} = 250\text{ Hz})$.

16. Equation (6-159) gives the observed polarizability of a dipolar linear molecule in the dilute gas. In the dense gas or liquid the rotational states are quenched. The induced moment of a molecule in a liquid is given by

$$\boldsymbol{\mu}_{\text{ind}} = \alpha \boldsymbol{E} + D_a \boldsymbol{E} \langle \cos\theta \rangle,$$

where $\boldsymbol{D}_a \cdot \boldsymbol{E} = D_a E \cos\theta$ and $\langle \cos\theta \rangle$ is the average value of the $\cos\theta$ in the liquid. Calculate $\langle \cos\theta \rangle$ for the linear molecule in the liquid where $kT > D_a E \cos\theta$ and show that the result is equivalent to the result in Eq.(6-159).

17. Consider the planar ClF_3 molecule. Assume that the three F are bonded only to the Cl and you know only that the system is planar.

(a) What are the possible planar symmetry groups that could include ClF_3?

(b) Determine the possible hybrid orbitals on Cl for the various symmetry groups.

(c) Can you predict the structure by considering the energetics of hybridization?

(d) Compare your most probable structure to the actual one.

18. Work out the combination of *s*, *p*, and *d* atomic functions that form appropriate hybrid orbitals from the center atom in the octahedral ReF_6 molecule.

19. Consider an arbitrary basis set of $\mathbf{\Phi} = (\Phi_a \Phi_b \Phi_c \Phi_d \Phi_e)$ representing the five spherically symmetric functions centered at each atom in a tetrahedron like CH_4. Generate the symmetry orbitals from the original set.

REFERENCES

[1] H. C. ALLEN and P. C. CROSS, *Molecular Vib-Rotors* (John Wiley & Sons, Inc., New York, 1963).

[2] G. HERZBERG, *Infrared and Raman Spectra* (D. Van Nostrand Co., New York, 1945).

[3] This is the Born–Oppenheimer approximation; M. BORN and J. R. OPPENHEIMER, *Ann. Physik* **84**, 457 (1927); see also M. BORN and K. HUANG, *Dynamical Theory of Crystal Lattices* (Oxford University Press, New York, 1954).

[4] H. HELLMANN, *Einführung in die Quantenchemie* (Franz Denticke, Leipzig, Germany, 1937), p. 285; and R. P. FEYNMAN, *Phys. Rev.* **56**, 340 (1939). See also a comment by J. I. MUSHER, *Am. J. Phys.* **34**, 267 (1966).

[5] (a) R. G. PARR, *J. Chem. Phys.* **40**, 3726 (1964); R. W. WYATT and R. G. PARR, *J. Chem. Phys.* **41**, 3262 (1964); and B. KIRTMAN, *J. Chem. Phys.* **41**, 775 (1965). See review in R. G. PARR, *Modern Quantum Chemistry*, Ed. by O. Sinanoglu (Academic Press, New York, 1965).
(b) W. H. FINK and L. C. ALLEN, *J. Chem. Phys.* **46**, 2261 (1967).

[6] I. N. LEVINE, *Quantum Chemistry* (Allyn and Bacon, Inc., Boston, 1974) and B. M. DEB, *Mod. Phys.* **45**, 22 (1973).

[7] R. H. SCHWENDEMAN, *J. Chem. Phys.* **44**, 556 (1966) and C. J. H. SCHUTTE, *Struct. and Bond.* **9**, 213 (1970).

[8] S. T. EPSTEIN, *The Variation Method in Quantum Chemistry* (Academic Press, New York, 1974).

[9] J. C. SLATER, *Quantum Theory of Molecules and Solids*, Vol. I (McGraw-Hill Book Co., New York, 1963).

[10] C. A. COULSON, *Valence* (Oxford University Press, London, 1961).

[11] H. EYRING, J. WALTER, and G. F. KIMBALL, *Quantum Chemistry* (John Wiley & Sons, Inc., New York, 1944).

[12] R. G. PARR, *The Quantum Theory of Molecular Electronic Structure* (W. A. Benjamin, Inc., New York, 1963).

[13] G. HERZBERG, *Electronic Spectra of Diatomic Molecules* (D. Van Nostrand Co., New York, 1945).

[14] W. KAUZMANN, *Quantum Chemistry* (Academic Press, New York, 1956). See also J. I. STEINFELD, *Molecules and Radiation* (Harper & Row, New York, 1974).

[15] See, for instance, A. S. DAVYDOV, *Quantum Mechanics*, 2nd ed. (Pergamon Press, New York, 1965).

[16] We shall not discuss further the details of electronic state spectroscopy. More information on diatomic molecules can be obtained in Ref. [13]. Information on the electronic spectra

of polyatomic molecules is given in G. HERZBERG, *Electronic Spectra and Electronic Structure of Polyatomic Molecules* (D. Van Nostrand Co., New York, 1966); G. W. KING, *Spectroscopy and Molecular Structure* (Holt, Rinehart, and Winston, New York, 1964); H. H. JAFFE and M. ORCHIN, *Theory and Applications of Ultraviolet Spectroscopy* (John Wiley & Sons, Inc., New York, 1962); and F. GRUM, "Visible and Ultraviolet Spectrophotometry," *Physical Methods in Chemistry*, Ed. by A. Weissberger and B. Rossiter (Wiley-Interscience, New York, 1972). High resolution spectroscopy is discussed in I. G. ROSS, *Adv. Chem. Phys.*, Ed. by I. Prigogine and S. A. Rice **20**, 341 (Wiley-Interscience, New York, 1971).

[17] C. C. J. ROOTHAAN, *Rev. Mod. Phys.* **23**, 69 (1951). The treatment for closed shells or completely paired electrons is given in the text. The LCAO-MO-SCF matrix equations given here must be modified for open shell configurations as shown in C. C. J. ROOTHAAN, *Rev. Mod. Phys.* **32**, 179 (1960).

[18] More details are given in *Methods of Computational Physics*, Vol. 2, Ed. by B. J. Alder, S. Fernbach, and A. Rotenberg (Academic Press, New York, 1963).

[19] B. J. RANSIL, *Rev. Mod. Phys.* **32**, 245 (1960).

[20] H. WEINSTEIN, R. PAUNZ, and M. COHEN, *Adv. Atom. and Mol. Phys.*, Ed. by D. R. Bates and I. Estermann **7**, 97 (Academic Press, New York, 1971) and K. RUEDENBERG, *Modern Quantum Chemistry*, Ed. by O. Sinanoglu (Academic Press, New York, 1965).

[21] W. HÜTTNER, M. K. LO, and W. H. FLYGARE, *J. Chem. Phys.* **48**, 1206 (1968).

[22] See, for instance, M. D. NEWTON, *et al.*, *J. Chem. Phys.* **52**, 4064 (1970).

[23] L. C. SNYDER and H. BASCH, *Molecular Wave Functions and Properties* (John Wiley & Sons, Inc., New York, 1972).

[24] H. F. SCHAEFER, *The Electronic Structure of Atoms and Molecules* (Addison-Wesley Pub. Co., Reading, Mass., 1972).

[25] R. M. PITZER and W. N. LIPSCOMB, *J. Chem. Phys.* **39**, 1995 (1963).

[26] A. VEILLARD, *Chem. Phys. Lett.* **3**, 128 (1969).

[27] E. CLEMENTI and H. POPKIE, *J. Chem. Phys.* **57**, 4870 (1972).

[28] See A. RAUK, L. C. ALLEN, and E. CLEMENTI, *J. Chem. Phys.* **52**, 4133 (1970) and earlier references cited therein.

[29] R. M. STEVENS, *J. Chem. Phys.* **55**, 1725 (1971).

[30] T. H. DUNNING and N. W. WINTER, *J. Chem. Phys.* **63**, 1847 (1975).

[31] A detailed comparison of calculations on H_2O that range from minimal basis to a number of near Hartree–Fock results is summarized in C. W. KERN and M. KARPLUS, *Water—A Comprehensive Treatise*, Vol. 1, Ed. by F. Franks (Plenum Press, New York, 1974). See also I. SHAVITT, B. S. ROSENBERG, and S. PALALIKIS, *Int. J. Quant. Chem.* **510**, 33 (1976).

[32] The calculation of the potential energy functions $[V_{nn} + \varepsilon_e(r)]$ for diatomic molecules is discussed in detail by J. GOODISMAN, *Diatomic Potential Interaction Theory*, Vol. I and II (Academic Press, New York, 1973). A review of methods of computing $V_{nn} + \varepsilon_e(r)$ and a discussion of the limitations in Hartree–Fock theory and the effects of configuration interaction (CI) is given in R. F. W. BADER and R. A. GANGI, *Theoretical Chemistry*, Vol. 2 (The Chemical Society, London, 1975), p. 1.

[33] L. C. SNYDER and H. BASCH, *J. Am. Chem. Soc.* **91**, 2189 (1969). R. DITCHFIELD, *et al.*, *J. Am. Chem. Soc.*, **92**, 4796 (1970).

[34] R. MCWEENY and B. T. SUTCLIFFE, *Methods of Molecular Quantum Mechanics* (Academic Press, New York, 1969).

[35] A. C. WAHL and G. DAS, *Adv. in Quant. Chem.* **5**, 261 (1970), and *J. Chem. Phys.* **56**, 3532 (1972). See also R. P. HOSTENY, A. R. HINDS, and A. C. WAHL, *Chem. Phys. Lett*, **23**, 9 (1973).

[36] E. R. DAVIDSON, *Reduced Density Matrices in Quantum Mechanics* (Academic Press, New York, 1976).

[37] C. THOMSON, *Theoretical Chemistry*, Vol. 2 (The Chemical Society, London, 1975), p. 83.

[38] H. F. SCHAEFFER, *Critical Evaluation of Chemical and Physical Structural Information*, Ed. by D. R. Lide and M. A. Paul (National Academy of Sciences, Washington, D.C., 1974).

[39] S. D. PEYERINSHOFF and R. J. BUENKER, *Adv. Quant. Chem.* **9**, 69 (1975).

[40] R. F. W. BADER and R. A. GANGI, *Theoretical Chemistry*, Vol. 2 (The Chemical Society, London, 1975), p. 1.

[41] Systematic methods are given in J. GARRAT, *Theoretical Chemistry*, Vol. 1 (The Chemical Society, London, 1974).

[42] W. A. GODDARD, *et al.*, *Accts. Chem. Res.* **6**, 368 (1973). Some interesting MOs are shown in W. L. JORGENSEN and L. SALEM, *The Organic Chemist's Book of Orbitals* (Academic Press, New York, 1973).

[43] L. B. HARDING and W. A. GODDARD III, *J. Am. Chem. Soc.* **97**, 6293 (1975).

[44] J. A. POPLE and D. J. BEVERIDGE, *Approximate Molecular Orbital Theory* (McGraw-Hill Book Co., New York, 1970). J. A. Pople, *Accts. Chem. Res.* **3**, 217 (1970).

[45] R. MCWEENY, *Spins in Chemistry* (Academic Press, New York, 1970). See also Section 5.6.

[46] M. J. S. DEWAR, *Science* **187**, 1037 (1975).

[47] M. J. DEWAR, *The Molecular Orbital Theory of Organic Chemistry* (McGraw-Hill Book Co., New York, 1969).

[48] R. D. BROWN and K. R. ROBY, *Theor. Chim. Acta* **16**, 175, 194, 278, 291 (1970).

[49] T. A. HALGREN and W. N. LIPSCOMB. *J. Chem. Phys.* **58**, 1569 (1973).

[50] J. C. SLATER, *Adv. Quant. Chem.* **6**, 1 (1972). K. H. JOHNSON, *Adv. Quant. Chem.* **7**, 143 (1973).

[51] J. N. MURREL and A. J. HARGET, *Semi-empirical Self-consistent Field Molecular Orbital Theory of Molecules* (Wiley-Interscience, Inc., New York, 1972).

[52] M. D. NEWTON, F. P. BOER, and W. N. LIPSCOMB, *J. Am. Chem. Soc.* **88**, 2353, 2361, 2367 (1966).

[53] J. E. EILERS, and D. R. WHITMAN, *J. Am. Chem. Soc.* **95**, 2067 (1973).

[54] B. O'LEARY, B. L. DUKE, and J. E. EILERS, *Adv. Quant. Chem.* **9**, 1 (1955).

[55] R. E. CHRISTOFFERSEN, *Adv. Quant. Chem.* **6**, 333 (1972) and R. E. CHRISTOFFERSEN, *et al.*, *J. Am. Chem. Soc.* **95**, 8526 (1973).

[56] R. HOFFMAN, *J. Chem. Phys.* **39**, 1397 (1963).

[57] G. BLYHOLDER and C. A. COULSON, *Adv. in Chem. Phys.*, Ed. by J. O. Hirschfelder and D. Henderson, **21**, 251 (John Wiley & Sons, Inc., New York, 1971).

[58] R. S. MULLIKEN, *et al.*, *J. Chem. Phys.* **17**, 1248 (1949).

[59] For example, see L. SALEM, *The Molecular Orbital Theory of Conjugated Systems* (W. A. Benjamin, Inc., New York. 1966).

[60] An interesting survey and historical perspective on σ and π bonds as well as a thorough discussion of the Hückel method is given in C. A. COULSON, *Physical Chemistry, An Advanced Treatise*, Ed. by H. Eyring **5**, 288 and 370 (Academic Press, New York, 1970).

[61] J. R. PLATT in *Handbuch der Physik* (S. Flügge, Ed.), Vol. XXXVII/2 (Springer-Verlag, Berlin, 1961). See also Section 4.1 and Problem 4-5.

[62] Several more examples are found in J. D. ROBERTS, A. STREITWEISER, and C. M. REGAN, *J. Am. Chem. Soc.* **74**, 4579 (1952).

[63] J. D. ROBERTS, *Molecular Orbital Calculations* (W. A. Benjamin, Inc., New York, 1961); A. STREITWEISER, *Molecular Orbital Theory* (John Wiley & Sons, Inc., New York, 1961); and L. SALEM, *The Molecular Orbital Theory of Conjugated Systems* (W. A. Benjamin, Inc., New York, 1966).

[64] R. HOFFMAN and R. B. WOODWARD, *Accts. Chem. Res.* **1**, 17 (1968) and R. B. WOODWARD and R. HOFFMAN, *The Conservation of Orbital Symmetry* (Academic Press, New York, 1970).

[65] See also symmetry arguments applied to inorganic reactions in R. G. PEARSON, IUPAC, XIII International Conference on Coordination Chemistry (Butterworths, London, 1971), p. 145.

[66] J. H. VAN VLECK, *The Theory of Electric and Magnetic Susceptibilities* (Oxford University Press, London, 1932); see also W. WELTNER, *J. Chem. Phys.* **29**, 477 (1958).

[67] J. R. ESHBACH and M. W. P. STRANDBERG, *Phys. Rev.* **35**, 24 (1952).

[68] P. W. SELWOOD, *Magnetochemistry* (Interscience Publishers, Inc., New York, 1943) and A. K. BURNHAM, *et al.*, *J. Am. Chem. Soc.* **99**, 1836 (1977).

[69] W. H. FLYGARE, *Chem. Rev.* **74**, 653 (1974).

[70] R. DITCHFIELD, *MTP International Review of Science, Physical Chemistry*, Series One, Vol. 2 (Butterworths, London, 1972), p. 91.

[71] N. F. RAMSEY, *Phys. Rev.* **78**, 699 (1950).

[72] For a review of experimental and theoretical magnetic shielding and magnetic susceptibility anisotropies, see B. R. APPLEMAN and B. P. DAILEY, *Adv. Mag. Res.* **7**, 231, Ed. by J. S. Waugh (Academic Press, New York, 1974).

[73] W. H. FLYGARE and J. GOODISMAN, *J. Chem. Phys.* **49**, 3122 (1968).

[74] More details including the evaluation of the individual elements in the $\boldsymbol{\sigma}$ tensor can be found in Ref. [72] and T. D. GIERKE and W. H. FLYGARE, *J. Am. Chem. Soc.* **94**, 7277 (1972).

[75] See Problem 6-16 for a recalculation of Eq.(6-159) for a molecule in a liquid.

[76] For a discussion of the spherical Lorentz model as well as more sophisticated models, see C. J. F. BÖTTCHER, *The Theory of Electric Polarization* (Elsevier Pub. Co., New York, 1973).

[77] P. DEBYE, *Z. Physik* **13**, 97 (1912). Additional details and a discussion of the orientation pair correlations and their effect on these measurements can be found in J. M. DEUTCH, *Ann. Rev. Phys. Chem.* **24**, 301 (1973). Discussions of the measurements of dielectric constants and molecular dipole moments by these methods is found in W. E. VAUGHAN, C. P. SMYTH, and J. G. POWLES, "Determination of Dielectric Constant and Loss" and C. P. SMYTH, "Determination of Dipole Moments" both in *Physical Methods of Chemistry*, Ed. by A. Weissberger and B. Rossiter (Wiley-Interscience, Inc., New York, 1976). See also *Dielectric and Related Molecular Processes* (The Chemical Society, London).

[78] A. D. BUCKINGHAM and J. A. POPLE, *Proc. Phys. Soc.* **68A**, 905 (1955). This paper also includes an analysis of the effects on $n_{\parallel} - n_{\perp}$ due to the hyperpolarizabilities arising from the next two terms in Eqs.(4-191) and (4-194).

[79] C. G. LEFEVRE and R. J. W. LEFEVRE, "The Kerr Effect," *Physical Methods of Chemistry*, Ed. by A. Weissberger and B. W. Rossiter (Wiley-Interscience, Inc., New York, 1972).

[80] The method of calculating these differences between the optical and static $\alpha_{\parallel} - \alpha_{\perp}$ is found in G. R. ALMS, A. K. BURNHAM, and W. H. FLYGARE, *J. Chem. Phys.* **63**, 3321 (1975) and references cited therein.

[81] B. LINDER and D. A. RABENOLD, *Adv. Quant. Chem.* **6**, 203 (1972). More details on intermolecular forces are also found in *Adv. Chem. Phys.* **12**, Ed. by J. Hirschfelder (Interscience Publishers, New York, 1967). See also G. C. MAITLAND and E. B. SMITH, *Chem. Soc. Rev.* **2**, 181 (1973) and H. MARGENAU and N. R. KESTNER, *Theory of Intermolecular Forces*, 2nd ed. (Pergamon Press, Oxford, 1971).

[82] A more general discussion of upper and lower bounds is found in F. WEINHOLD, *Adv. in Quant. Chem.* **6**, 299 (1972).

[83] H. M. MCCONNELL, *J. Chem. Phys.* **27**, 226 (1957).

[84] N. F. RAMSEY, *Phys. Rev.* **91**, 303 (1953). See also H. F. HAMEKA, *Advanced Quantum Chemistry* (Addison Wesley Pub. Co., Inc., Reading, Mass., 1965) and D. W. DAVIES, *The Theory of the Electric and Magnetic Properties of Molecules* (John Wiley & Sons, New York, 1967).

7

molecular spectroscopy

7.1 MOLECULAR ABSORPTION SPECTROSCOPY

In this section we shall continue our development of absorption spectroscopy that was introduced in Section 3.7. We shall derive the absorption coefficient for the $k \longrightarrow j$ transition and examine first-order relaxation effects. We have considered three excitation mechanisms in decreasing order of importance: D_{jk}, the electric dipole; μ_{jk}, the magnetic dipole; and Q_{jk}, the electric quadrupole. The following discussion will concern only the electric dipole case with the understanding that the magnetic dipole and electric quadrupole expressions are similar and easily obtained.

Most spectroscopy experiments (see Chapter 4 for examples) involve the measurement of the absorption of electromagnetic energy. Consider the radiation traveling along the y axis, as in Fig. 1-1, and entering an absorbing sample at $y = 0$. The absorption of the radiation intensity along y will be proportional to the intensity,

$$\frac{dI}{dy} = -I\gamma, \tag{7-1}$$

where γ is the absorption coefficient. The solution to this equation is

$$I(y) = I(0) \exp(-\gamma y),$$

as shown, for instance, in Eq.(2-55). $I(0)$ is the incident intensity at $y = 0$ and $I(y)$ is the intensity at point y in the sample. For very weak absorption ($\gamma y \ll 1$), we can expand the exponential and truncate at the second term to give

$$\Delta I = I(0) - I(y) \cong \gamma y I(0), \tag{7-2}$$

where ΔI is the radiation intensity absorbed by the sample. We shall now derive an expression for ΔI that depends on the $k \longrightarrow j$ transition probability in an atomic or molecular absorption process.

Beginning with a monochromatic source of electromagnetic radiation, the number of molecules per unit volume making the $k \longrightarrow j$ and $j \longrightarrow k$ transitions ($E_k < E_j$) will be equal to $N_k P_{jk}(\omega, t)$ and $N_j P_{kj}(\omega, t)$, respectively, where N_k and N_j are the number of molecules per unit volume in the kth and jth states and $P_{jk}(\omega, t)$ is the probability of the $k \longrightarrow j$ transition [see Eq.(3-176)]. We assume low power radiation where $N_k - N_j$ is not appreciably changed in the presence of the radiation. According to these arguments, the radiation energy density being absorbed by the molecules is equal to the net number of molecules (per unit volume) making the transition multiplied by the energy of the transition photon, $\hbar\omega_{jk}$. Thus, the total energy absorbed is given by

$$\Delta E = yA\hbar\omega_{jk}(N_k - N_j)P_{jk}(\omega, t),$$

where A is the radiation beam cross section traveling in the y direction and yA is the sample volume being illuminated by the radiation. The intensity being absorbed is given by the time rate of energy absorption divided by the radiation beam area, A,

$$\Delta I = \left(\frac{1}{A}\right)\frac{d}{dt}\Delta E = y\hbar\omega_{jk}(N_k - N_j)\dot{P}_{jk}(\omega, t),$$

where $\dot{P}_{jk}(\omega, t)$, the probability per unit time that a translation takes place, is the time derivative of $P_{jk}(\omega, t)$. Comparing this equation with Eq.(7-2) gives the absorption coefficient:

$$\gamma_{jk}(\omega, t) = \frac{\hbar\omega_{jk}(N_k - N_j)}{I(0)}\dot{P}_{jk}(\omega, t). \tag{7-3}$$

Substituting $I(0) = c\rho$ (ρ is the radiation energy density), $\dot{P}_{jk}(\omega, t)$ from Eq.(3-176),

$$\dot{P}_{jk}(\omega, t) = \frac{\omega_{jk}^2[(\boldsymbol{A}_0 \cdot \boldsymbol{D})_{jk}]^2}{2\hbar^2 c^2}\frac{\sin(\omega_{jk} - \omega)t}{(\omega_{jk} - \omega)}, \tag{7-4}$$

and (we now use cgs units)

$$[(\boldsymbol{A}_0 \cdot \boldsymbol{D})_{jk}]^2 = D_{jk}^2\left(\frac{8\pi c^2 \rho}{\omega^2}\right) \tag{7-5}$$

for plane-polarized radiation, where D_{jk} contains the angular dependence between $\boldsymbol{A}_0$ and $\boldsymbol{D}$ as described from Eq.(3-178) and associated discussion, gives

$$\gamma_{jk}(\omega, t) = \frac{4\pi D_{jk}^2(N_k - N_j)\omega_{jk}^3 t}{\hbar\omega^2 c} \frac{\sin(\omega_{jk} - \omega)t}{(\omega_{jk} - \omega)t}. \tag{7-6}$$

Thus, near resonance the absorption coefficient increases in time. We know from a good deal of experience, however, that the absorption coefficient in most experiments is independent of time. The reason for this is that the molecule-radiation interaction is continually interrupted by first-order processes. In order to understand this relaxation process, consider the system originally in the kth state at $t = 0$ in the absence of the radiation field. As $t > 0$, the radiation field is on and the system will be in a superposition state of the kth and jth states; $\Psi(t) = c_j(t)\psi_j + c_k(t)\psi_k$. Strong collisions will cause an annihilation of the superposition state. After a collision, the system will have a certain probability (depending on details of the collision) of returning to the kth state after which the radiation will again cause the growth of the superposition state. If n is the number density of systems in the superposition state, the time rate of change of n is given by

$$\frac{dn}{dt} = -\left(\frac{1}{\tau}\right)n,$$

where τ is the relaxation time. The solution to this first-order rate equation is

$$n(t) = n(0) \exp\left(-\frac{t}{\tau}\right),$$

where $n(0)$ is the initial number of molecules in the superposition state. Thus, we have competitive processes of creation of the superposition state by the radiation and interruption or relaxation by a first-order process. As a result, we must average $\gamma_{jk}(\omega, t)$ in Eq.(7-6) over the probability of the interruption or relaxation. This average is given by

$$\gamma_{jk}(\omega) = \frac{\int_0^\infty \gamma_{jk}(\omega, t) \exp(-t/\tau)\, dt}{\int_0^\infty \exp(-t/\tau)\, dt}, \tag{7-7}$$

where τ is the relaxation time for the process.

Substituting from Eq.(7-6) gives (in cgs units)

$$\gamma_{jk}(\omega) = \frac{4\pi^2 D_{jk}^2(N_k - N_j)\omega_{jk}^3}{\hbar\omega^2 c} \mathscr{L}(\omega_{jk} - \omega). \tag{7-8}$$

The absorption coefficient can be obtained in SI units by dividing the result in Eq.(7-8) by $4\pi\varepsilon_0$. $\mathscr{L}(\omega_{jk} - \omega)$ is the normalized Lorentzian function (with maximum amplitude of τ/π) that is observed in many spectroscopy experiments (see Chapter 4):

$$\mathscr{L}(\omega_{jk} - \omega) = \frac{1}{\pi}\left[\frac{1/\tau}{(1/\tau)^2 + (\omega_{jk} - \omega)^2}\right], \qquad \int_{-\infty}^{+\infty} \mathscr{L}(\omega_{jk} - \omega)\, d\omega = 1.0. \tag{7-9}$$

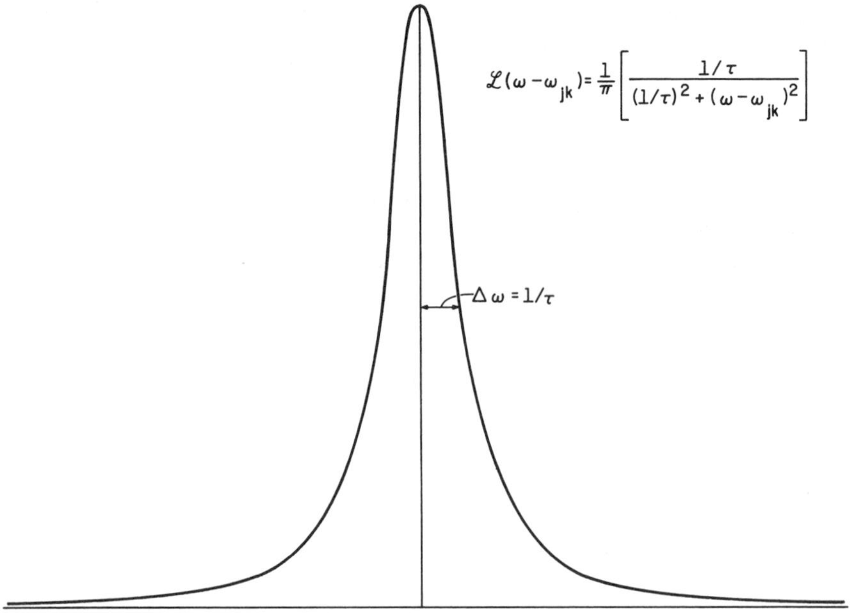

Figure 7-1 The normalized Lorentz absorption curve from Eq.(7-9). $\Delta\omega = 1/\tau$ is the half width at half height where τ is the relaxation time and the maximum in the curve is at $\mathscr{L}(0) = \tau/\pi$.

The Lorentzian curve is shown in Fig. 7-1 where the half width at half height, $\Delta\omega$, is given from

$$\frac{1}{2}\mathscr{L}(0) = \frac{\tau}{2\pi} = \frac{1}{\pi}\left[\frac{1/\tau}{(1/\tau)^2 + (\Delta\omega)^2}\right],$$

which gives $\Delta\omega = 1/\tau$.

Equation (7-8) gives the absorption coefficient for electric dipole transitions in the limit where the line shape is determined by first-order relaxation processes. For magnetic dipole transitions, replace D_{jk} with μ_{jk}; for electric quadrupole transitions, replace D_{jk} with $(\omega_{jk}/c)Q_{jk}$.

We shall now calculate the peak absorption coefficients for rotational, vibrational, and electronic transitions using electric dipole selection rules. We assume that the partition function can be written as separated products of the rotational, vibrational, and electronic partition functions. Furthermore, we assume the thermal equilibrium existence of only a single electronic state (the ground state). At thermal equilibrium, the number density of molecules in the kth vibration-rotation state is given by (ε_k is the energy and g_k is the degeneracy)

$$N_k = \frac{\rho_0 g_k \exp(-\varepsilon_k/kT)}{Q}, \qquad Q = \sum_k g_k \exp\left(-\frac{\varepsilon_k}{kT}\right) \cong Q_R Q_v, \qquad (7\text{-}10)$$

where ρ_0 is the total molecular number density and Q_R and Q_v are the rotational and vibrational partition functions (cgs units are used throughout the rest of this chapter).

Rotational Energy Levels

Using Eq.(7-10) and assuming equal degeneracies for the kth and jth states we write

$$N_k - N_j = N_k\left\{1 - \exp\left[-\frac{(E_j - E_k)}{kT}\right]\right\} \cong N_k\frac{(E_j - E_k)}{kT} = N_k\frac{\hbar\omega_{jk}}{kT}, \quad (7\text{-}11)$$

where $E_j - E_k \ll kT$ has been used in the last step.

Substituting Eq.(7-11) into Eq.(7-8) for plane-polarized radiation gives the absorption coefficient for the $k \longrightarrow j$ transition.

$$\gamma(\omega) = \frac{4\pi^2 D_{jk}^2 N_k \omega_{jk}^4}{\omega^2 ckT}\mathscr{L}(\omega_{jk} - \omega). \quad (7\text{-}12)$$

The peak absorption coefficient is obtained by letting $\omega \longrightarrow \omega_{jk}$, giving

$$\gamma(\omega = \omega_{jk}) = \frac{4\pi D_{jk}^2 N_k}{ckT}\frac{\omega_{jk}^2}{\Delta\omega} = \frac{8\pi^2 D_{jk}^2 N_k}{ckT}\frac{\nu_{jk}^2}{\Delta\nu}. \quad (7\text{-}13)$$

We can now use Eq.(7-10) to compute N_k (for the ground electronic state). In the case of a diatomic molecule, the partition functions are [1]†

$$Q_R = \frac{kT}{hB}, \qquad Q_V = \frac{1}{1 - \exp(-h\nu_V/kT)}, \quad (7\text{-}14)$$

where B is the rotational constant (in units of hertz) and ν_V is the fundamental vibrational frequency. Thus, (the ground vibrational state is the reference point)

$$N_k = N_{Jn} = \frac{(2J + 1)\exp[-hBJ(J + 1)/kT]\exp(-nh\nu_V/kT)}{\dfrac{kT}{hB}\left[\dfrac{1}{1 - \exp(-h\nu_V/kT)}\right]}\rho_0, \quad (7\text{-}15)$$

where n indicates the vibrational state ($n = 0, 1, 2, \ldots$) and J represents the rotational state ($J = 0, 1, 2, \ldots$). ρ_0 is the molecular number density.

N_{Jn} can be estimated by remembering that $[hBJ(J + 1)] \ll kT$, giving

$$N_{Jn} \cong \frac{(2J + 1)hB}{kT}\exp\left(-\frac{nh\nu_V}{kT}\right)\left[1 - \exp\left(-\frac{h\nu_V}{kT}\right)\right]\rho_0.$$

If $h\nu_V > kT$, $[1 - \exp(-h\nu/kT)] \cong 1.0$, so if we choose a rotational transition in the ground vibrational state ($n = 0$), N_{J0} in the equation above reduces to

$$N_{J0} \cong \frac{(2J + 1)hB}{kT}\rho_0.$$

Substituting Eq.(4-189) into this expression gives

$$N_{J0} \cong \frac{h\nu_{J+1,J}}{kT}\rho_0.$$

† Numbers in square brackets correspond to the numbered sources found in the Reference section on p. 493 at the end of the chapter.

Substituting into Eq.(7-13) gives the rotational absorption coefficient at $\omega = \omega_{jk}$,

$$\gamma_R \cong \frac{8\pi^2 D_{jk}^2 h\nu_{jk}^3 \rho_0}{c(kT)^2 \Delta\nu}, \tag{7-16}$$

where $\nu_{jk} = \nu_{J+1,J}$. The number density, ρ_0, and the half width at half height, $\Delta\nu$, are both proportional to pressure and therefore γ_R is independent of pressure. However, the total absorption or the frequency integrated value of $\gamma(\omega)$ is proportional to the number density. For a typical molecule with $D_{jk} = 10^{-18}$ SC · cm, $\nu_{jk} = 10^{10}$ Hz, $\Delta\nu = 10^7$ Hz ($p = 1$ Torr), and $T = 300$ K, we obtain $\gamma_R \cong 10^{-4}$ cm^{-1} for a typical absorption coefficient for diatomic molecules in the microwave region of the electromagnetic spectrum. This result shows that for the conditions stated above the sample cell must be 100 m long in order to absorb $1/e$ of the microwave power.

Vibrational Energy Levels

We now return to Eq.(7-8) for an infrared transition in a molecular system where $\Delta E > kT$ and we assume that $N_k - N_j \cong N_k$, which gives

$$\gamma_V(\omega_{jk}) = \gamma_V = \frac{4\pi D_{jk}^2 N_k}{3\hbar c}\left(\frac{\nu_{jk}}{\Delta\nu}\right). \tag{7-17}$$

The value of N_k can be taken from the work following Eq.(7-15) for the ground rotation-vibration state ($J = 0$, $n = 0$) to give $N_k = (hB/kT)\rho_0$, which when substituted into Eq.(7-17) gives the peak height of

$$\gamma_V = \frac{8\pi^2 D_{jk}^2 B\nu_{jk}\rho_0}{3ckT\Delta\nu}. \tag{7-18}$$

Substituting $\Delta\nu = 1/2\pi\tau = 10^7$ Hz, $D_{jk} = 10^{-18}$ SC · cm, $\nu_{jk} = 10^{14}$ Hz for the vibrational transition, $B \cong 60{,}000$ MHz for a molecule like CO, $T = 300$ K, and $\rho_0 = 3 \times 10^{16}$ molecules · cm^{-3} at 1 Torr pressure gives $\gamma_V \cong 4 \times 10^2$ cm^{-1}. We use $\Delta\nu = 10^7$ Hz because at 1 Torr pressure the half width at half height is expected to be dominated by the rotational relaxation. It is evident by comparing the results above that $\gamma_R \ll \gamma_V$ for equivalent pressures. In the presence of Doppler broadening, different results are obtained for both γ_V and γ_R (see Section 7.2 and Problem 7-4).

Electronic Energy Levels

Equation (7-17) is also appropriate for electronic absorption where nearly all the molecules are in the ground state, N_k. Again we note that for widely separated vibrational and electronic states $\Delta E_E \gg kT$ and $\Delta E_V \gg kT$, giving $Q_R Q_V Q_E \cong Q_R$. If the line width for the electronic transition is dominated by first-order relaxation processes, we would expect the electronic absorption coefficient to be related to the vibrational absorption coefficient by

$$\gamma_E \cong \frac{\nu_E}{\nu_V}\gamma_V.$$

We now return to the Hamiltonian in Eq.(3-167) and the associated theory to develop the absorption coefficient for two-photon processes. Assuming again that the system begins in the kth state, a two-photon absorption could involve the $k \longrightarrow j$ excitation described above by a combination of radiation frequencies, $\omega_1 \pm \omega_2$, which were resonant with $\omega_{jk}(\omega_1 \pm \omega_2 \cong \omega_{jk})$. We shall limit our discussion to the normally stronger electric dipole effects that arise from the $\exp(i\boldsymbol{k} \cdot \boldsymbol{r}) \cong 1.0$ truncation in Eq.(3-170).

We must also examine the possible effects of $\mathscr{H}_2$ in Eq.(3-167). Using $\mathscr{H}_2$ in place of $\mathscr{H}_1$ in Eq.(3-163), using the electric dipole approximation in Eq.(3-170), $\exp(i\boldsymbol{k} \cdot \boldsymbol{r}) = 1.0$, shows that no $k \longrightarrow j$ transitions are excited by $\mathscr{H}_2$ as there are no remaining atomic or molecular operators that survive. The resulting $\int \psi_j^* \psi_k \, dV = \delta_{jk}$ requirement assures that the system remains in the initial or kth state.

We can generate two-photon effects in a three-level system using $\mathscr{H}_1$ in Eq.(3-167) and the theory as described in Eqs.(3-163) and (3-164). We shall consider only absorption processes here and we start with a three-level system, k, j, and m, where $E_k < E_j < E_m$ (the ordering is for visual clarity). If the system begins in the lowest or kth state, we can write the differential equations for the upper state j and m coefficients, $c_j(t)$ and $c_m(t)$, according to Eqs.(3-163) and (3-164):

$$\frac{dc_j(t)}{dt} = \frac{1}{i\hbar}(\mathscr{H}_1)_{jk} \exp(i\omega_{jk}t), \qquad \hbar\omega_{jk} = E_j - E_k. \tag{7-19}$$

$$\frac{dc_m(t)}{dt} = \frac{1}{i\hbar}(\mathscr{H}_1)_{mk} \exp(i\omega_{mk}t) + \frac{1}{i\hbar}(\mathscr{H}_1)_{mj} c_j(t) \exp(i\omega_{mj}t),$$

$$\hbar\omega_{mk} = E_m - E_k, \qquad \hbar\omega_{mj} = E_m - E_j. \tag{7-20}$$

The $c_j(t)$ in Eq.(7-19) will lead to the probability of the first-order $k \longrightarrow j$ transition according to the absorption part of Eq.(3-175), for instance. $dc_m(t)/dt$ is a sum of two terms, the first being the first-order effects of the direct $k \longrightarrow m$ transition and the second being the second-order $j \longrightarrow m$ transition that depends on the $c_j(t)$ determined from Eq.(7-19). We have already considered the direct $k \longrightarrow j$ and $k \longrightarrow m$ effects above. Assuming the direct $k \longrightarrow m$ transition is not allowed ($D_{mk} = 0$), we are left with the second (second-order) term in Eq.(7-20). Substituting the absorption part of Eq. (3-175) into the second term in Eq.(7-20) gives

$$\frac{dc_m(t)}{dt} = \frac{\omega_{jk}}{2i\hbar^2\omega}(\boldsymbol{E}_0 \cdot \boldsymbol{D})_{jk}\left\{\frac{\exp[i(\omega_{jk} - \omega)t] - 1}{\omega_{jk} - \omega}\right\}(\mathscr{H}_1)_{mj} \exp(i\omega_{mj}t). \tag{7-21}$$

Now, according to Eqs.(3-163) and (3-173), we can write the part of $(\mathscr{H}_1)_{mj}$ that will be important for absorption according to

$$\frac{1}{i\hbar}(\mathscr{H}_1)_{mj} = -\frac{\omega_{mj}}{2\hbar c}\exp(-i\omega t)(\boldsymbol{A}_0 \cdot \boldsymbol{D})_{mj}.$$

Using $\boldsymbol{E}_0 = (i\omega/c)\boldsymbol{A}_0$ according to Eq.(1-29) and the discussion following Eq.(3-174) gives

$$(\mathscr{H}_1)_{mj} = -\frac{\omega_{mj}}{2\omega}(\boldsymbol{E}_0 \cdot \boldsymbol{D})_{mj} \exp(-i\omega t).$$

Substituting into Eq.(7-21) gives

$$\frac{dc_m(t)}{dt} = -\frac{\omega_{mj}\omega_{jk}}{4i\omega^2\hbar^2}(\boldsymbol{E}_0 \cdot \boldsymbol{D})_{mj}(\boldsymbol{E}_0 \cdot \boldsymbol{D})_{jk} \times \left\{\frac{\exp[i(\omega_{mj} + \omega_{jk} - 2\omega)t] - \exp[i(\omega_{mj} - \omega)t]}{(\omega_{jk} - \omega)}\right\}.$$

Integrating from $0 \longrightarrow t$ where $c_m(0) = 0$ and remembering that $\omega_{mj} + \omega_{jk} = \omega_{mk}$ gives

$$c_m(t) = \frac{\omega_{mj}\omega_{jk}(\boldsymbol{E}_0 \cdot \boldsymbol{D})_{mj}(\boldsymbol{E}_0 \cdot \boldsymbol{D})_{jk}}{4\omega^2\hbar^2(\omega_{jk} - \omega)} \times \left\{\frac{\exp[i(\omega_{mk} - 2\omega)t] - 1}{(\omega_{mk} - 2\omega)} - \frac{\exp[i(\omega_{mj} - \omega)t] - 1}{(\omega_{mj} - \omega)}\right\}. \tag{7-22}$$

We can calculate the absorption coefficient for the $k \longrightarrow m$ transition according to Eq.(7-3) where we need $\dot{P}_{mk}(\omega, t)$. Using $c_m(t)$ above gives

$$\dot{P}_{mk}(\omega, t) = \frac{d}{dt}|c_m(t)|^2 = \frac{\omega_{mj}^2\omega_{jk}^2 E_0^4 D_{mj}^2 D_{jk}^2}{4\omega^4\hbar^4}\left\{\frac{\sin(\omega_{mj} - \omega)t}{(\omega_{jk} - \omega)(\omega_{mk} - 2\omega)(\omega_{mj} - \omega)} - \frac{\sin(\omega_{mk} - 2\omega)t}{(\omega_{jk} - \omega)(\omega_{mj} - \omega)(\omega_{mk} - 2\omega)} + \frac{\sin(\omega_{jk} - \omega)t}{(\omega_{mk} - 2\omega)(\omega_{mj} - \omega)(\omega_{jk} - \omega)}\right\},$$

where D_{mj} and D_{jk} include the orientational dependence between $\boldsymbol{E}_0$ and $\boldsymbol{D}$. Substituting this result into Eq.(7-3) and taking the long-time average as illustrated previously gives [we also use $I(0) = c\rho = (c/8\pi)E_0^2$]

$$\gamma(\omega) = \frac{4\pi^2\omega_{km}\omega_{jk}^2\omega_{mj}^2(N_k - N_m)E_0^2 D_{mj}^2 D_{jk}^2}{2\omega^4 c\hbar^3} \times \left[\frac{\mathscr{L}(\omega_{mj} - \omega)}{(\omega_{jk} - \omega)(\omega_{mk} - 2\omega)} - \frac{\mathscr{L}(\omega_{mk} - 2\omega)}{(\omega_{jk} - \omega)(\omega_{mj} - \omega)} + \frac{\mathscr{L}(\omega_{jk} - \omega)}{(\omega_{mk} - 2\omega)(\omega_{mj} - \omega)}\right]. \tag{7-23}$$

Now, the valid range for interpreting this second-order absorption curve for the forbidden $k \longrightarrow m$ transition is far off resonance in the $\omega_{jk} - \omega$ and $\omega_{mj} - \omega$ curves but near resonance in $\omega_{mk} - 2\omega$. In the limit where $(1/\tau)^2 \ll (\omega_{mj} - \omega)^2$ and $(1/\tau)^2 \ll (\omega_{jk} - \omega)^2$, we can rewrite Eq.(7-23), after considerable algebra, to give

$$\lim_{\substack{(1/\tau)^2 \ll (\omega_{jk} - \omega)^2 \\ (1/\tau)^2 \ll (\omega_{mj} - \omega)^2}} \gamma(\omega) = C\left\{\frac{1}{(\omega_{mj} - \omega)^2(\omega_{jk} - \omega)^2} \times \left[\frac{1}{\pi\tau} - (\omega_{jk} - \omega)(\omega_{mj} - \omega)\mathscr{L}(\omega_{mk} - 2\omega)\right]\right\}$$

$$\cong -C\frac{\mathscr{L}(\omega_{mk} - 2\omega)}{(\omega_{mj} - \omega)(\omega_{jk} - \omega)},$$

where C is the premultiplier in Eq.(7-23) and $(\omega_{jk} - \omega)(\omega_{mj} - \omega)$ is always negative for any ordering of the k, j, and m levels. $\gamma(\omega)$ is positive in this limit leading to absorption as $(\omega_{mk} - 2\omega)$ approaches resonance. Thus, the second-order resonance curve at $(\omega_{mk} - 2\omega)$ shows a direct observation of the ω_{mk} resonance even in the case where $D_{mk} = 0$. It should also be pointed out that the second-order absorption coefficient in Eq.(7-23) is proportional to the radiation power through the E_0^2-dependence. It is also evident that Eq.(7-23) is valid for any relative ordering of the energy levels, E_k, E_j, E_l, as long as the definitions in Eqs.(7-19) and (7-20) are maintained; for instance, if $E_k < E_m < E_j$, we can still observe an electric dipole forbidden $k \longrightarrow m$ transition according to Eq.(7-23) even when there is no intermediate level between the kth and mth states.

We now return to Eqs.(7-19) and (7-20) to generalize our relations to two different electromagnetic frequencies. We now require an electromagnetic field with amplitude $E_0^{(1)}$ and frequency ω_1 in the $(\mathscr{H}_1)_{jk}$ perturbation in Eq.(7-19) and another electromagnetic wave with amplitude $E_0^{(2)}$ and frequency ω_2 in the $(\mathscr{H}_1)_{mj}$ perturbation in Eq.(7-20). The resultant second-order value of $c_m(t)$ is

$$c_m(t) = \frac{\omega_{jm}\omega_{jk}(\boldsymbol{E}_0^{(1)} \cdot \boldsymbol{D})_{jk}(\boldsymbol{E}_0^{(2)} \cdot \boldsymbol{D})_{mj}}{4\omega_2\omega_1\hbar^2(\omega_{jk} - \omega_1)} \times \left(\frac{\exp\{i[\omega_{mk} - 2(\omega_1 + \omega_2)]\} - 1}{\omega_{mk} - 2(\omega_1 + \omega_2)} - \frac{\exp[i(\omega_{mj} - \omega_2)t] - 1}{(\omega_{mj} - \omega_2)}\right).$$

We now follow through the standard steps to give the absorption coefficient; however, we have to choose a sweep frequency to identify $I(0)$ with the radiation power of the sweep frequency. We choose ω_2 fixed and sweep ω_1 giving $I(0) = c\rho = c/8\pi(E_0^{(1)})^2$. The resultant second-order absorption coefficient in the variable frequency ω_1 is given by

$$\gamma(\omega_1) = \frac{4\pi^2\omega_{mj}^2\omega_{jk}^2\omega_{mk}(N_k - N_m)[E_0^{(2)}]^2 D_{mj}^2 D_{jk}^2}{2\omega_2^2\omega_1^2 ch^3}\left\{\frac{\mathscr{L}(\omega_{mj} - \omega_2)}{(\omega_{jk} - \omega_1)[\omega_{mk} - (\omega_1 + \omega_2)]} - \frac{\mathscr{L}[\omega_{mk} - (\omega_1 + \omega_2)]}{(\omega_{jk} - \omega_1)(\omega_{mj} - \omega_2)} + \frac{\mathscr{L}(\omega_{jk} - \omega_1)}{[\omega_{mk} - (\omega_1 + \omega_2)](\omega_{mj} - \omega_2)}\right\}. \quad (7\text{-}24)$$

We again note that the $\mathscr{L}(\omega_{mj} - \omega_2)$ and $\mathscr{L}(\omega_{jk} - \omega_1)$ second-order resonance curves given here are of negligible amplitude relative to the corresponding first-order terms from Eq.(7-8). We now find the $\mathscr{L}(\omega_{mk} - \omega_1 - \omega_2)$ resonant term that allows a direct measurement of the ω_{mk} resonance when $\omega_1 + \omega_2 = \omega_{mk}$, even in the absence of an electric dipole moment transition between the mth and kth states. There are a number of interesting combinations of the relative energies E_k, E_j, and E_m and frequencies ω_1 and ω_2 that will lead to the $\mathscr{L}(\omega_{mk} - \omega_1 - \omega_2)$ resonance as shown in Eq.(7-24).

We can now extend the arguments above to the third-order terms leading to three-photon events. Consider adding a fourth higher level to the sequence discussed preceding Eqs.(7-19) and (7-20). The four levels now have energies $E_k < E_j < E_m < E_n$. If the system begins in the lowest or kth state, Eq.(7-19) shows the growth of the jth state, Eq.(7-20) shows the growth of the mth state, and the subsequent growth of the nth state is given by extending this sequence to

$$\frac{dc_n(t)}{dt} = \frac{1}{i\hbar}(\mathscr{H}_1)_{nk} \exp(i\omega_{nk} t) + \frac{1}{i\hbar}(\mathscr{H}_1)_{nj} c_j(t) \exp(i\omega_{nj} t) + \frac{1}{i\hbar}(\mathscr{H}_1)_{nm} c_m(t) \exp(i\omega_{nm} t), \tag{7-25}$$

$$\hbar\omega_{nk} = E_n - E_k, \qquad \hbar\omega_{nj} = E_n - E_j, \qquad \hbar\omega_{nm} = E_n - E_m.$$

The first term is first order involving the nk electric dipole term; the second term is the second-order term considered above. The third term in Eq.(7-25) contains both second-order and third-order contributions. If we use $c_m(t)$ from the first-order development in Eq.(3-175), for instance, then the result is a second-order correction as described from Eqs.(7-20) through (7-22). If we use $c_m(t)$ from our second-order development in Eq.(7-22) in the last term in Eq.(7-25), however, the result will be a third-order term and three frequencies can be involved in the absorption process. We shall leave this continuing development to the reader; however, we point out that a considerable experimental activity is evident in multiphoton spectroscopy [2].

7.2 THE DOPPLER EFFECT AND CONVOLUTIONS

In Section 7.1 we examined the frequency-dependence of the near-resonant absorption or emission of electromagnetic energy by a quantum mechanical system. In the first-order relaxation limit, the line shape is Lorentzian with half width at half height of $\Delta\nu = 1/2\pi\tau$ where τ is the relaxation time (see Fig. 7-1). We shall now examine the effects on the line shape that arise from the relative translational motion of the molecules interacting with the electromagnetic radiation: the Doppler effect.

The Doppler effect in sonics is often experienced. The higher-frequency whistle on an approaching train or the lower-frequency whistle on a departing train relative to a stationary observer is an example of the Doppler effect. If ν represents a frequency being absorbed on resonance in a stationary molecule, the frequency must be decreased if the molecule is translating toward the source of the radiation and the frequency must be increased if the molecule is translating away from the source of the radiation in order to maintain the resonant interaction in the molecule. A dramatic example of the Doppler effect is given in Fig. 7-2, which shows the $1_{11} \longrightarrow 1_{10}$ absorption spectrum of interstellar formaldehyde ($H_2^{12}C^{16}O$) at 4830 MHz as seen against the 10 K thermal background radiation in the region near the Sagittarius A constellation. The radiation source is near the galactic center about 30,000 light-years from the Earth and the interstellar clouds, which contain the absorbing formaldehyde molecules, are somewhere between the radio source and Earth. The rest frequency of the $1_{11} \longrightarrow 1_{10}$ transition as observed in the laboratory on Earth is 4829.649 ± 0.001 MHz, which is shown in the figure at a zero radial velocity. The

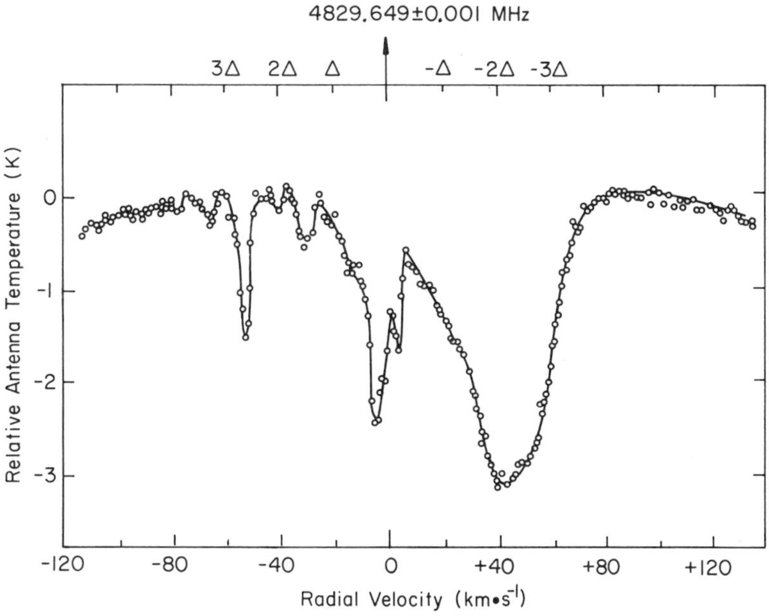

Figure 7-2 The $1_{11} \longrightarrow 1_{10}$ absorption spectra of interstellar formaldehyde against a galactic center radio source in the region of the Sagittarius A constellation from L. E. Snyder and D. Buhl, *Sky and Telescope* **40**, 7 (1970). The absorption is indicated on the ordinate as a lowering of the antenna temperature at the resonance frequencies. An increase in antenna temperature would indicate emission. Radial velocities with respect to the local standard of rest (LSR) are plotted on the abscissa where a radial velocity of $+20\ \text{km} \cdot \text{s}^{-1}$ leads to a Doppler shift of $\Delta = -0.322$ MHz. The several absorption features in the figure are discussed in the text.

different features in the figure apparently all arise from the single $1_{11} \longrightarrow 1_{10}$ formaldehyde absorption in different clouds that have different velocities and corresponding Doppler shifts relative to the rest frequency of 4829.649 MHz.

In radio astronomy, the standard convention is to measure radial velocities with respect to a framework called the local standard of rest (LSR) [3]. The LSR is the centroid of motion of stars near the sun (or local region of the galaxy). By referring radial velocities to the LSR, radio astronomers eliminate the need to describe their observations with respect to the earth's orbit about the sun and the sun's drift with respect to the local region of the galaxy.

Assuming that the radiation source and the LSR are not moving with respect to each other, a $+20\text{-km} \cdot \text{s}^{-1}$ radial cloud velocity means that the cloud is moving away from the LSR and toward the radiation source with a velocity of $20\ \text{km} \cdot \text{s}^{-1}$, which corresponds to a frequency shift for absorption to a lower frequency of $\Delta\nu = (v/c)\nu_0$ $= 0.322$ MHz. Thus, the broad absorption with a maximum at $44\ \text{km} \cdot \text{s}^{-1}$ arises from absorption in a cloud that is traveling away from the LSR at a speed of 44 $\text{km} \cdot \text{s}^{-1}$ and the peak absorption is observed on Earth at a frequency 0.71 MHz below the rest frequency. The narrow feature at $-53\ \text{km} \cdot \text{s}^{-1}$ is observed at a frequency 0.85 MHz above the rest frequency and arises from formaldehyde absorption in a different cloud that is moving toward the LSR at a velocity of 53 km $\cdot$ s^{-1}. The broadness of the feature at +44 km $\cdot$ s^{-1} relative to the other transitions probably arises from a rotating cloud where molecules in different parts of the cloud are moving with different relative velocities. Narrower beam diameters of the observing radio telescope could possibly allow the observation of a single formaldehyde cloud.

Consider now the absorption in gaseous molecules in a fixed container. According to our previous discussion, the resonance frequency of a molecule translating with velocity v along the axis parallel and in the same direction as the radiation is shifted in frequency by ν' according to

$$\nu\left(1 + \frac{v}{c}\right) = \nu + \frac{v}{c}\nu = \nu + \nu', \tag{7-26}$$

where c is the speed of the radiation and normally $v/c \ll 1.0$. The distribution of the velocities along the x axis is given by the normalized Maxwell–Boltzmann distribution [4]:

$$P(v_x) = \left(\frac{M}{2\pi kT}\right)^{1/2} \exp\left(-\frac{Mv_x^2}{2kT}\right). \tag{7-27}$$

$P(v_x)$ is the probability per unit velocity that the molecule has velocity v_x along the x axis, M is the mass of the molecule, and k is Boltzmann's constant. The probability that the molecule has a velocity between v_a and v_b along the x axis is given by

$$P(v_a \longrightarrow v_b) = \left(\frac{M}{2\pi kT}\right)^{1/2} \int_{v_a}^{v_b} \exp\left(-\frac{Mv_x^2}{2kT}\right) dv_x.$$

Integrating through limits from $v_a = -\infty$ to $v_b = +\infty$ gives unity.

The radiation absorption coefficient for a two-level ($k \longrightarrow j$) system (atom or molecule) moving with velocity v along the axis of radiation propagation for a fixed radiation source and detector, $\gamma_v(\nu)$, is given by replacing ν in the Lorentzian in Eq.(7-8) by $\nu[1 + (v/c)]$,

$$\gamma_v(\nu) = \frac{2D_{jk}^2(N_k - N_j)\nu_{jk}}{\hbar c}\left(\frac{1/\tau}{(1/2\pi\tau)^2 + \{\nu_{jk} - \nu[1 + (v/c)]\}^2}\right), \tag{7-28}$$

where we have also used $\omega_{jk}^3/\omega^2 = \omega_{jk} = 2\pi\nu_{jk}$ near resonance and we remember that the angular dependence in the $\boldsymbol{A}_0 \cdot \boldsymbol{D}$ operator is contained in the $\boldsymbol{D}_{ij}$ matrix element. The absorption coefficient at frequency ν in a large number of molecules with the velocity distribution in Eq.(7-27) is given by

$$\begin{aligned}\gamma(\nu) &= \int_{-\infty}^{+\infty} P(v)\gamma_v(\nu)\, dv \\ &= \frac{2D_{jk}^2(N_k - N_j)\nu_{jk}}{\hbar c\tau}\left(\frac{M}{2\pi kT}\right)^{1/2} \int_{-\infty}^{+\infty} \frac{\exp(-Mv^2/2kT)\, dv}{(1/2\pi\tau)^2 + \{\nu_{jk} - \nu[1 + (v/c)]\}^2}.\end{aligned}$$

This integral can be converted to an integral over frequencies by using Eq.(7-26),

$$v = \frac{c}{\nu}\nu', \qquad dv = \frac{c}{\nu}d\nu'. \tag{7-29}$$

The result is

$$\gamma(\nu) = \frac{2D_{jk}^2(N_k - N_j)\nu_{jk}}{\hbar\tau\nu}\left(\frac{M}{2\pi kT}\right)^{1/2}\int_{-\infty}^{+\infty}\frac{\exp\left[-(Mc^2/2kT\nu^2)(\nu')^2\right]d\nu'}{(1/2\pi\tau)^2 + (\nu_{jk} - \nu - \nu')^2}. \quad (7\text{-}30)$$

Equation (7-30) is the standard form for the convolution of a Gaussian function with a Lorentzian function. We can see this by first writing the normalized Gaussian and Lorentzian functions and their half widths at half heights:

$$G(\nu') = \frac{c}{\nu}\left(\frac{M}{2\pi kT}\right)^{1/2}\exp\left[-\frac{M}{2kT}\frac{c^2}{\nu^2}(\nu')^2\right]$$

$$= \frac{1}{\Delta\nu_G}\left(\frac{\ln 2}{\pi}\right)^{1/2}\exp\left[-\frac{\ln 2}{(\Delta\nu_G)^2}(\nu')^2\right]$$

$$\Delta\nu_G = \frac{\nu}{c}\left(\frac{2kT\ln 2}{M}\right)^{1/2} \quad (7\text{-}31)$$

$$\mathscr{L}(\nu_{jk} - \nu - \nu') = \frac{1}{\pi}\left[\frac{(1/2\pi\tau)}{(1/2\pi\tau)^2 + (\nu_{jk} - \nu - \nu')^2}\right], \qquad \Delta\nu_L = \frac{1}{2\pi\tau}. \quad (7\text{-}32)$$

Substituting into Eq.(7-30) gives the convolution integral,

$$\gamma(\nu) = \frac{4\pi^2 D_{jk}^2(N_k - N_j)\nu_{jk}}{c\hbar}\int_{-\infty}^{+\infty} G(\nu')\mathscr{L}(\nu_{jk} - \nu - \nu')\,d\nu'. \quad (7\text{-}33)$$

This Voight function or convolution integral of a Gaussian and Lorentzian fails to give a simple analytical result; however, numerical evaluation with a computer is straightforward [5].

Before interpreting Eq.(7-33) we note a useful theorem in Fourier transforms that states that the Fourier transform of the convolution of two functions is equal to the product of the Fourier transforms. Applying this theorem to the convolution gives [see Eqs.(B-59) and (B-63) and associated discussion]

$$\mathscr{F}\int_{-\infty}^{+\infty} G(\nu')\mathscr{L}(\nu_{jk} - \nu - \nu')\,d\nu' = 2\pi(\mathscr{F}G)(\mathscr{F}\mathscr{L}), \quad (7\text{-}34)$$

where $\mathscr{F}$ indicates the Fourier transform. According to Eq.(7-34) we can interchange the arguments in the Gaussian and Lorentzian functions to give

$$\int_{-\infty}^{+\infty} G(\nu')\mathscr{L}(\nu_{jk} - \nu - \nu')\,d\nu' = \int_{-\infty}^{+\infty} G(\nu_{jk} - \nu - \nu')\mathscr{L}(\nu')\,d\nu'. \quad (7\text{-}35)$$

We now wish to interpret Eq.(7-33) in the Dirac delta function limits where either $\Delta\nu_L$ or $\Delta\nu_G$ go to zero (see Appendix B.3 for a discussion of Dirac delta functions).

The delta function form of the Gaussian function is given by [see Eqs.(7-31) and (B-72)]

$$\lim_{\Delta\nu_G \to 0} G(\nu') = \lim_{\Delta\nu_G \to 0}\left\{\frac{1}{\Delta\nu_G}\left(\frac{\ln 2}{\pi}\right)^{1/2}\exp\left[-\frac{\ln 2}{(\Delta\nu_G)^2}(\nu')^2\right]\right\} = \delta(\nu'). \quad (7\text{-}36)$$

Substituting $\delta(\nu') = G(\nu')$ into Eq.(7-33) for the $\Delta\nu_G \ll \Delta\nu_L$ limit and integrating over ν' converts $\mathscr{L}(\nu_{jk} - \nu - \nu')$ to $\mathscr{L}(\nu_{jk} - \nu)$ and the resultant $\gamma(\nu)$ is the Lorentzian function centered at ν_{jk} with half width at half height of $\Delta\nu_L$. This result is equivalent to Eq.(7-8), which is the zero velocity limit of Eq.(7-28).

Using Eq.(7-35) we can also examine the other limit where $\Delta\nu_G \gg \Delta\nu_L$. The Lorentzian also goes to a delta function in the limit of zero half width (see again Appendix B.3),

$$\lim_{\Delta\nu_L \to 0} \mathscr{L}(\nu') = \lim_{\Delta\nu_L \to 0} \left[\left(\frac{1}{\pi}\right) \frac{\Delta\nu_L}{\Delta\nu_L^2 + (\nu')^2}\right] = \delta(\nu'). \tag{7-37}$$

Thus, by first applying Eq.(7-35) to Eq.(7-33), substituting Eq.(7-37) into the result, and integrating over ν', we obtain

$$\begin{aligned} \gamma(\nu) &= \frac{4\pi^2 D_{jk}^2 (N_k - N_j)\nu_{jk}}{c\hbar} \int_{-\infty}^{+\infty} G(\nu_{jk} - \nu - \nu')\delta(\nu')\, d\nu' \\ &= \frac{4\pi^2 D_{jk}^2 (N_k - N_j)\nu_{jk}}{c\hbar} G(\nu_{jk} - \nu). \end{aligned} \tag{7-38}$$

This function is the Gaussian centered at ν_{jk} with half width given in Eq.(7-31),

$$\Delta\nu_G = \nu_{jk}\left(\frac{2kT \ln 2}{Mc^2}\right)^{1/2} = \nu_{jk}\left(\frac{T}{M'}\right)^{1/2}(3.58115 \times 10^{-7}). \tag{7-39}$$

M is the weight of the molecule and M' is the molecular weight in atomic mass units (u). Consider a molecule with $M' = 30$ u that would correspond to the CO molecule that we discussed earlier (1 u $= 1.660566 \times 10^{-24}$ g). Assume also that T $= 300$ K, which gives $\Delta\nu = \nu_{jk} \times 10^{-6}$ Hz. The Doppler width for electronic, vibrational, and rotational energy spacings are then given by

Electronic ($\nu_{jk} = \omega_{jk}/2\pi \cong 1.6 \times 10^{15}$ Hz)

$$\Delta\nu_E = 1.6 \times 10^9 \text{ Hz} = 1.6 \times 10^3 \text{ MHz}$$

Vibrational ($\nu_{jk} \cong 6.0 \times 10^{13}$ Hz)

$$\Delta\nu_V = 60 \text{ MHz}$$

Rotational [$\nu_{jk}(J = 0 \longrightarrow J = 1) = 10^{11}$ Hz]

$$\Delta\nu_R \cong 0.1 \text{ MHz}.$$

It is interesting to compare these widths with those predicted from relaxation processes and the estimated spontaneous emission rate constants given in Section 3.7. In the cases above, the Doppler line widths exceed the natural line widths from spontaneous emission. However, the widths obtained from collisions (rotational relaxation) in either vibrational or rotational spectroscopy will exceed the Doppler widths at pressures above about 10^{-2} and 10^{-5} atm, respectively (see end of Section 7.4 for further discussion of molecular relaxation).

7.3 HEISENBERG'S REPRESENTATION OF THE ABSORPTION OF ELECTROMAGNETIC ENERGY AND CORRELATION FUNCTIONS

All our previous discussions on relaxation and the spectral line shape were based on the assumption of a first-order relaxation mechanism for the superposition state that leads to an exponential decay, exp $(-t/\tau)$, and the corresponding Lorentz line shape with a half width at half height of $\Delta\nu_L = (1/2\pi\tau)$. In the gas phase at low pressure, the Gaussian Doppler broadening will probably dominate the spectra. As the pressure increases, the relaxation width will increase leading to broader rotational and vibrational-rotational widths. In the liquid state, "collisions" occur several times before the molecule is able to rotate, thereby quenching the rotational motion. The width of a vibrational transition will be a complex function of the rotational motion and the vibrational relaxation processes. Under these circumstances first-order exponential decay may not be observed. In this section we shall investigate a more general expression for the absorption coefficient in which we need not assume exponential decay. In fact, we can use the more general methods to determine the mechanism of decay by examining the line shape of the absorption or emission spectrum. We shall attempt to show that the Heisenberg representation of a time-dependent process is a more convenient framework to view the actual motion of a molecule over very short times. The aim and result of this work is to write the absorption coefficient in terms of the autocorrelation function of the electric dipole moment of the molecule [6].

We shall begin our discussion of the near-resonant absorption in an isolated two-level system with $\dot{P}_{jk}(\omega, t)$ from Eq.(7-4) where we note that

$$\left(\frac{1}{\pi}\right)\frac{\sin(\omega_{jk} - \omega)t}{\omega_{jk} - \omega}$$

has a sharp peak at $\omega = \omega_{jk}$ and its area is unity; therefore this function can represent the delta function ($t > 0$) (see Appendix B.3)

$$\lim_{t\to\infty}\left(\frac{1}{\pi}\right)\frac{\sin(\omega_{jk} - \omega)t}{\omega_{jk} - \omega} = \delta(\omega_{jk} - \omega). \tag{7-40}$$

The long-time limit is appropriate here as we are interested in the steady-state absorption coefficient where the observing time is long relative to any relaxation times in the system. Substituting Eq.(7-40) into $\dot{P}_{jk}(\omega)$ from Eq.(7-4) gives

$$\lim_{t\to\infty}\dot{P}_{jk}(\omega, t) = \frac{\omega_{jk}^2[(\boldsymbol{A}_0\cdot\boldsymbol{D})_{jk}]^2}{2\hbar^2c^2}\pi\delta(\omega_{jk} - \omega) = \dot{P}_{jk}(\omega).$$

We now substitute the Fourier representation for the delta function from Eq. (B-67) to give

$$\dot{P}_{jk}(\omega) = \frac{\omega_{jk}^2}{4\hbar^2c^2}\int_{-\infty}^{+\infty}[(\boldsymbol{A}_0\cdot\boldsymbol{D})_{jk}]^2\exp[i(\omega_{jk} - \omega)t]\,dt. \tag{7-41}$$

We can also rewrite the $[(\boldsymbol{A}_0\cdot\boldsymbol{D})_{jk}]^2$ for a two-level system as

$$[(\boldsymbol{A}_0\cdot\boldsymbol{D})_{jk}]^2 = (\boldsymbol{A}_0\cdot\boldsymbol{D})_{jk}(\boldsymbol{A}_0\cdot\boldsymbol{D})_{jk} = (\boldsymbol{A}_0\cdot\boldsymbol{D})_{kj}(\boldsymbol{A}_0\cdot\boldsymbol{D})_{jk}. \tag{7-42}$$

Substituting Eq.(7-42) into Eq.(7-41) and remembering that $\omega_{jk} = (E_j - E_k)/\hbar$ gives

$$\dot{P}_{jk}(\omega) = \frac{\omega_{jk}^2}{4\hbar^2 c^2} \int_{-\infty}^{+\infty} (\boldsymbol{A}_0 \cdot \boldsymbol{D})_{kj} (\boldsymbol{A}_0 \cdot \boldsymbol{D})_{jk} \exp \left\{ i \left[\frac{(E_j - E_k)}{\hbar} - \omega \right] t \right\} dt$$

$$= \frac{\omega_{jk}^2}{4\hbar^2 c^2} \int_{-\infty}^{+\infty} (\boldsymbol{A}_0 \cdot \boldsymbol{D})_{kj} \left\{ \boldsymbol{A}_0 \cdot \left[\exp\left(\frac{iE_j t}{\hbar}\right) \boldsymbol{D} \exp\left(-\frac{iE_k t}{\hbar}\right) \right] \right\}_{jk} \exp(-i\omega t)\, dt.$$

Using

$$\exp\left(-\frac{iE_k t}{\hbar}\right)\psi_k(\boldsymbol{r}) = \exp\left(-\frac{i\mathcal{H}_0 t}{\hbar}\right)\psi_k(\boldsymbol{r})$$
$$\exp\left(\frac{iE_j t}{\hbar}\right)\psi_j^*(\boldsymbol{r}) = \exp\left(\frac{i\mathcal{H}_0 t}{\hbar}\right)\psi_j^*(\boldsymbol{r}) \tag{7-43}$$

gives

$$\dot{P}_{jk}(\omega) = \frac{\omega_{jk}^2}{4\hbar^2 c^2} \int_{-\infty}^{+\infty} (\boldsymbol{A}_0 \cdot \boldsymbol{D})_{kj} \left\{ \boldsymbol{A}_0 \cdot \left[\exp\left(\frac{i\mathcal{H}_0 t}{\hbar}\right) \boldsymbol{D} \exp\left(-\frac{i\mathcal{H}_0 t}{\hbar}\right) \right] \right\}_{jk} \exp(-i\omega t)\, dt. \tag{7-44}$$

Now, according to Eq.(3-148), in the absence of the radiation-molecule interaction where $\mathcal{H} = \mathcal{H}_0$, we can write

$$\boldsymbol{D}(t) = \exp\left(\frac{i\mathcal{H}_0 t}{\hbar}\right) \boldsymbol{D}(0) \exp\left(-\frac{i\mathcal{H}_0 t}{\hbar}\right). \tag{7-45}$$

$\boldsymbol{D}(t)$ is the value of the $\boldsymbol{D}$ operator at time t and $\boldsymbol{D}(0)$ is the value of the operator at $t = 0$ (in the absence of the radiation perturbation). Substituting gives

$$\dot{P}_{jk}(\omega) = \frac{\omega_{jk}^2}{4\hbar^2 c^2} \int_{-\infty}^{+\infty} [\boldsymbol{A}_0 \cdot \boldsymbol{D}(0)]_{kj} [\boldsymbol{A}_0 \cdot \boldsymbol{D}(t)]_{jk} \exp(-i\omega t)\, dt. \tag{7-46}$$

We can now obtain a closure over j for the two-level system by writing

$$[\boldsymbol{A}_0 \cdot \boldsymbol{D}(0)]_{kj} [\boldsymbol{A}_0 \cdot \boldsymbol{D}(t)]_{jk} = \sum_j [\boldsymbol{A}_0 \cdot \boldsymbol{D}(0)]_{kj} [\boldsymbol{A}_0 \cdot \boldsymbol{D}(t)]_{jk} = [\boldsymbol{A}_0 \cdot \boldsymbol{D}(0) \boldsymbol{A}_0 \cdot \boldsymbol{D}(t)]_{kk}, \tag{7-47}$$

where $[\boldsymbol{A}_0 \cdot \boldsymbol{D}(0)]_{jj} = [\boldsymbol{A}_0 \cdot \boldsymbol{D}(0)]_{kk} = 0$. Substituting gives

$$\dot{P}_{jk}(\omega) = \frac{\omega_{jk}^2}{4\hbar^2 c^2} \int_{-\infty}^{+\infty} [\boldsymbol{A}_0 \cdot \boldsymbol{D}(0) \boldsymbol{A}_0 \cdot \boldsymbol{D}(t)]_{kk} \exp(-i\omega t)\, dt. \tag{7-48}$$

Expanding the term in brackets we write

$$[A_x D_x(0) + A_y D_y(0) + A_z D_z(0)][A_x D_x(t) + A_y D_y(t) + A_z D_z(t)].$$

Choosing plane polarized radiation with the electric field polarized along the z axis gives

$$\boldsymbol{A}_0 \cdot \boldsymbol{D}(0) \boldsymbol{A}_0 \cdot \boldsymbol{D}(t) = A_z^2 D_z(0) D_z(t); \tag{7-49}$$

however, we can also write

$$\boldsymbol{D}(0)\cdot\boldsymbol{D}(t) = D_x(0)D_x(t) + D_y(0)D_y(t) + D_z(0)D_z(t). \tag{7-50}$$

For an isotropic sample $D_x(0)D_x(t) = D_y(0)D_y(t) = D_z(0)D_z(t)$ in the laboratory-fixed axis system that is appropriate for a liquid. Thus, we can rewrite Eq.(7-49) for an isotropic sample and plane-polarized radiation to give

$$\boldsymbol{A}_0\cdot\boldsymbol{D}(0)\boldsymbol{A}_0\cdot\boldsymbol{D}(t) = \tfrac{1}{3}A_0^2\boldsymbol{D}(0)\cdot\boldsymbol{D}(t). \tag{7-51}$$

These arguments are similar to those made following Eq.(3-177). Substituting into Eq.(7-48) gives

$$\dot{P}_{jk}(\omega) = \frac{\omega_{jk}^2 A_0^2}{12\hbar^2 c^2}\int_{-\infty}^{+\infty}[\boldsymbol{D}(0)\cdot\boldsymbol{D}(t)]_{kk}\exp(-i\omega t)\,dt. \tag{7-52}$$

Substituting this result into Eq.(7-3) gives the absorption coefficient $[I(0) = c\rho]$:

$$\gamma_{jk}(\omega) = \frac{\omega_{jk}^3 A_0^2}{12c^3\hbar\rho}(N_k - N_j)\int_{-\infty}^{+\infty}[\boldsymbol{D}(0)\cdot\boldsymbol{D}(t)]_{kk}\exp(-i\omega t)\,dt.$$

Now, substituting $(N_k - N_j) = N_k[1 - \exp(-\hbar\omega_{jk}/kT)]$ from Eq.(7-11) and $A_z^2 = (8\pi c^2/\omega^2)\rho$ from Eq.(3-181) gives

$$\begin{aligned}\gamma_{jk}(\omega) &= \frac{2\pi\omega_{jk}^3[1 - \exp(-\hbar\omega_{jk}/kT)]}{3c\hbar\omega^2}\int_{-\infty}^{+\infty}N_k[\boldsymbol{D}(0)\cdot\boldsymbol{D}(t)]_{kk}\exp(-i\omega t)\,dt \\ &= \frac{A}{2\pi}\int_{-\infty}^{+\infty}N_k[\boldsymbol{D}(0)\cdot\boldsymbol{D}(t)]_{kk}\exp(-i\omega t)\,dt.\end{aligned} \tag{7-53}$$

$N_k[\boldsymbol{D}(0)\cdot\boldsymbol{D}(t)]_{kk}$ is equivalent to an ensemble average of $\boldsymbol{D}(0)\cdot\boldsymbol{D}(t)$ over the equilibrium distribution of initial states k, which we denote by $\langle\boldsymbol{D}(0)\cdot\boldsymbol{D}(t)\rangle$, called the *electric dipole moment autocorrelation function.* Thus, we rewrite Eq.(7-53) as

$$\begin{aligned}\gamma_{jk}(\omega) &= \frac{A}{2\pi}\int_{-\infty}^{+\infty}\langle\boldsymbol{D}(0)\cdot\boldsymbol{D}(t)\rangle\exp(-i\omega t)\,dt \\ &= \frac{A}{\pi}\,\mathrm{Re}\int_0^{\infty}\langle\boldsymbol{D}(0)\cdot\boldsymbol{D}(t)\rangle\exp(-i\omega t)\,dt.\end{aligned} \tag{7-54}$$

In the last step, Re indicates only the real part of the integral. This important result relates the real spectrum, $\gamma_{jk}(\omega)$, to the real part of the $0 \longrightarrow \infty$ time integral of the autocorrelation function and is proved in Eqs.(8-15) through (8-16). The dipole moment autocorrelation function describes the memory of the system or how long (after $t = 0$) one can predict the orientation of the dipole moment before correlation or the predictive ability is destroyed by the microscopic motion of the molecules in the system.

The correlation function can be obtained from Eq.(7-54) by the Fourier transform:

$$\int_{-\infty}^{+\infty} \exp(i\omega t')\gamma_{jk}(\omega)\, d\omega = \frac{A}{2\pi}\int_{-\infty}^{+\infty}\int_{-\infty}^{+\infty} \langle \mathbf{D}(0)\cdot\mathbf{D}(t)\rangle_{jk} \exp(-i\omega t)\exp(i\omega t')\, dt\, d\omega$$

$$= A\int_{-\infty}^{+\infty} \langle \mathbf{D}(0)\cdot\mathbf{D}(t)\rangle_{jk}\,\delta(t-t')\, dt = A\langle \mathbf{D}(0)\cdot\mathbf{D}(t')\rangle, \tag{7-55}$$

where A is defined in Eq.(7-53).

We shall now consider Eqs.(7-54) and (7-55) for the case of a vibrational transition in a freely rotating diatomic molecule. Under these circumstances the time-dependence is simply described by [use Eqs.(7-45) and (7-43)]

$$\langle \mathbf{D}(0)\cdot\mathbf{D}(t)\rangle = N_k D_{jk}^2 \exp(i\omega_{jk} t), \tag{7-56}$$

where the time-dependence in the autocorrelation function is contained in the $\exp(i\omega_{jk}t)$ term with rotational angular frequency ω_{jk}. Substituting this result into Eq.(7-54) gives the absorption coefficient for the transition:

$$\gamma_{jk}(\omega) = \frac{AN_k D_{jk}^2}{2\pi}\int_{-\infty}^{+\infty} \exp(i\omega_{jk}t)\exp(-i\omega t)\, dt = \frac{4\pi^2 D_{jk}^2\omega_{jk}^3}{3\hbar c\omega^2}(N_k - N_j)\delta(\omega_{jk}-\omega). \tag{7-57}$$

This result is equivalent to Eq.(7-8) in the limit of randomly orientated dipoles [see Eq.(3-178)] and where $1/\tau \longrightarrow 0$ and the Lorentzian becomes a delta function. Thus, the absorption coefficient for the vibrational transition in a freely rotating molecule is described by a constant times a delta function. Of course, collisions are usually present that destroy the free rotation and give rise to a finite width. In the presence of collisions in the gas phase, for instance, the pure rotational term in Eq.(7-56) is better described by an exponentially damped sinusoidal function. Thus, substituting $\langle \mathbf{D}(0)\cdot\mathbf{D}(t)\rangle = N_k D_{jk}^2 \exp(i\omega_{jk}t)\exp(-t/\tau)$ into the $0 \longrightarrow \infty$ integral in Eq.(7-54) gives the result in Eq.(7-8).

In summary, the result in Eq.(7-54) gives an alternative view of the absorption coefficient. The absorption coefficient is a constant times the Fourier transform of the dipole moment autocorrelation function. If the correlation function contains an exponential decay in time, the absorption coefficient (in the frequency domain) is a pure Lorentzian. On the other hand, a more complicated relaxation process would lead to a non-Lorentzian curve.

According to Eq.(7-55) the correlation function can be obtained from the measured absorption coefficient, $\gamma(\omega)$, by the Fourier transform. Consider the vibration band in the infrared region of the spectrum. The dipole correlation function for the $k \longrightarrow j$ transition is obtained from Eq.(7-55) (we assume that $\omega_{jk}^2 \cong \omega^2$) by integrating over the band of frequency in the absorption curve,

$$\langle \mathbf{D}(0)\cdot\mathbf{D}(t)\rangle = \frac{3\hbar c}{4\pi^2\omega_{jk}[1-\exp(-\hbar\omega_{jk}/kT)]}\int_{\text{band}} \gamma(\omega)\exp(i\omega t)\, d\omega. \tag{7-58}$$

It is convenient to normalize the correlation function to its initial value at $t = 0$, which gives

$$\langle D^2(0)\rangle = \frac{3\hbar c}{4\pi^2\omega_{jk}[1 - \exp(-\hbar\omega_{jk}/kT)]}\int_{\text{band}} \gamma(\omega)\, d\omega. \tag{7-59}$$

Therefore, the normalized correlation function is given by

$$\langle \boldsymbol{U}(0)\cdot \boldsymbol{U}(t)\rangle = \frac{\langle \boldsymbol{D}(0)\cdot \boldsymbol{D}(t)\rangle}{\langle D^2(0)\rangle} = \frac{\int_{\text{band}} \gamma(\omega)\exp(i\omega t)\, d\omega}{\int_{\text{band}} \gamma(\omega)\, d\omega}, \tag{7-60}$$

where $\boldsymbol{U}$ are unit vectors.

Consider now the experimental absorption curve for the $\nu_2 + \nu_8$ vibration combination band in CH_2Cl_2 in Figs. 7-3(a) and (b) showing the power absorbed in the cell of length ℓ. According to Eq.(7-2), which was derived by assuming weak absorption, the radiation intensity absorbed in the cell with length ℓ by the molecules is given by

$$\Delta I = \gamma \ell I(0), \tag{7-61}$$

where γ is the absorption coefficient and $I(0)$ is the incident intensity. Measuring ΔI for a vibrational band and using Eq.(7-61) in Eq.(7-60) for the normalized correlation function shows that both ℓ and $I(0)$ cancel.

The inertial axes in CH_2Cl_2 are shown in Fig. 2-2 where the a and b axes are in the Cl—C—Cl plane and the b axis bisects the Cl—C—Cl angle. The $\nu_2 + \nu_8$ mode of vibrational motion in CH_2Cl_2 leads to the dipole moment correlation function along the a principal inertial axis. Thus, the rotational motion is about the b and c principal inertial axes. The normalized dipole moment correlation function for the $\nu_2 + \nu_8$ vibrational absorption function can be obtained from the experimental curve in Fig. 7-3(b) and Eqs.(7-61) and (7-60). Of course, we have to assume that the width of the curve is due solely to rotational motion. The resulting D_a dipole moment correlation function is shown in Fig. 7-3(c) where we note that the curve is not exponential at very short times. After about 0.5×10^{-12} s, the correlation function is essentially exponential. The short time nonexponential behavior for the correlation function arises from the failure of the absorption curve to go to zero at frequencies far from resonance.

The correlation function for very short times appears to be similar to that expected from a free rotor. The free rotor correlation function for the D_a dipole is given from Eq.(7-56) by

$$\langle \boldsymbol{U}_a(0)\cdot \boldsymbol{U}_a(t)\rangle = \exp(i\omega_{jk}t), \tag{7-62}$$

where ω_{jk} is the usual resonant rotational angular frequency. If we consider Eq.(7-62) for short times, relative to the period for one molecular rotation, we can expand in a power series giving

$$\langle \boldsymbol{U}_a(0)\cdot \boldsymbol{U}_a(t)\rangle = 1 - \frac{(\omega t)^2}{2} + \cdots, \tag{7-63}$$

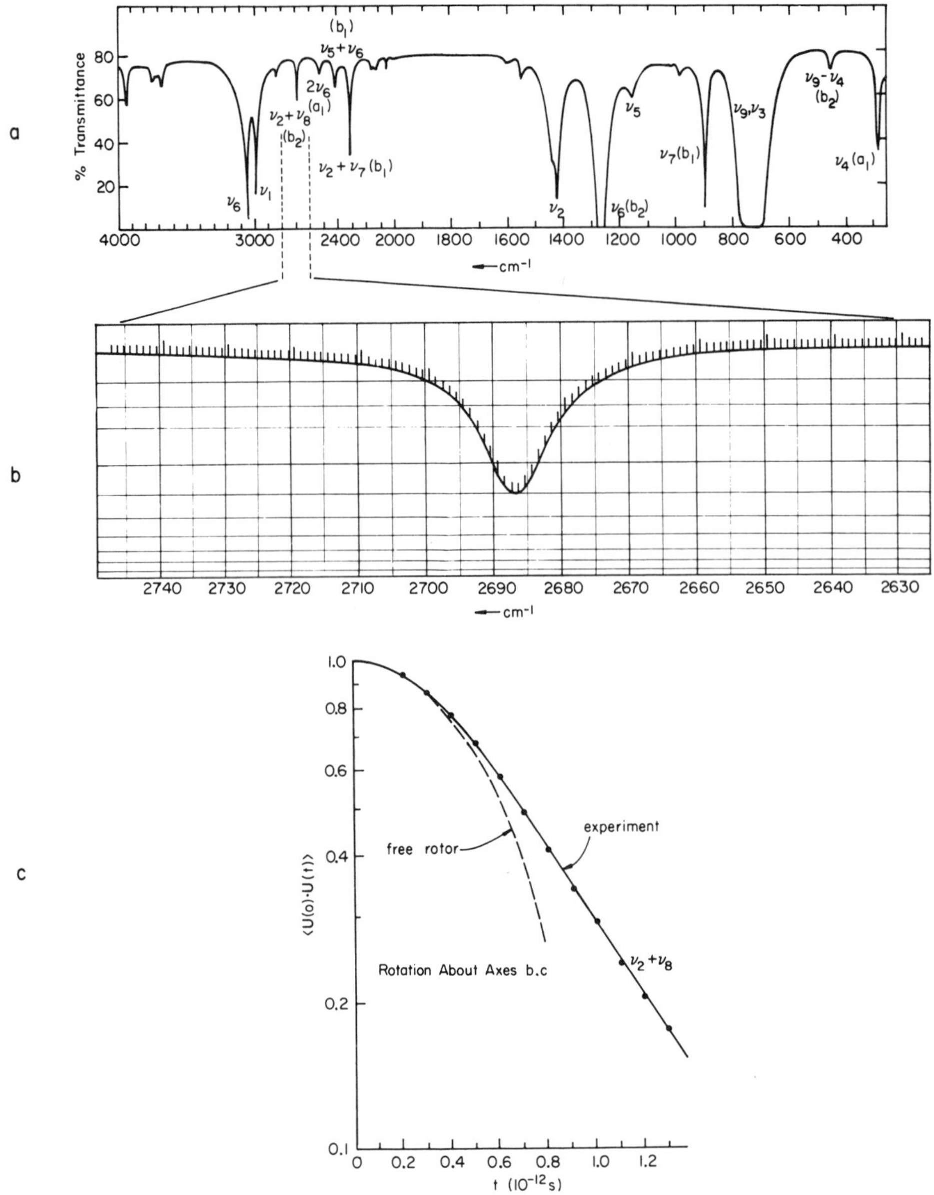

Figure 7-3 a. The infrared spectrum of liquid CH_2Cl_2 at 300 K. The a_1, a_2, b_1, and b_2 designation indicates symmetry species for vibration. These data are from W. G. Rothschild, *J. Chem. Phys.* **53**, 990 (1970). b. The expanded scale spectrum of the b_2 symmetry $\nu_2 + \nu_8$ combination band in CH_2Cl_2. The markers are every cm^{-1}. Assuming that the line width is due to rotational motion leads to the dipole moment autocorrelation function in (c). c. The dipole moment autocorrelation function for rotation of the molecule about the b and c inertial axes. The solid line is obtained from the Fourier transform of the experimental spectrum in (b). The dotted line is for a free rotor as described in Eq.(7-65).

where the $i\omega_{jk}t$ term has been dropped because $\langle \boldsymbol{U}_a(0) \cdot \boldsymbol{U}_a(t) \rangle$ must be a symmetric function in time. For rotation about the a inertial axis we can write the kinetic energy as ($kT = \omega^2 I/2$)

$$\omega^2 = \frac{2kT}{\mathsf{I}} = kT\left(\frac{1}{\mathsf{I}_b} + \frac{1}{\mathsf{I}_c}\right), \tag{7-64}$$

giving for the free rotor correlation function (at short times)

$$\langle \boldsymbol{U}_a(0) \cdot \boldsymbol{U}_a(t) \rangle = 1 - \frac{kT}{2}\left(\frac{1}{\mathsf{I}_b} + \frac{1}{\mathsf{I}_c}\right)t^2 + \cdots. \tag{7-65}$$

The first two terms (at short times) of this free rotor correlation function are also given in Fig. 7-3(c) where $\mathsf{I}_b = 254.51 \times 10^{-40}$ g · cm^2 and $\mathsf{I}_c = 275.71 \times 10^{-40}$ g · cm^2 as shown in Fig. 2-2. The short-time behavior shown in Fig. 7-3(c) can be interpreted in terms of molecular free rotation between collisions. According to Eq.(7-64), $\nu \cong 2.8 \times 10^{11}$ Hz for $T = 300$ K and the average period of rotation for the thermal free CH_2Cl_2 rotor is $\tau = 1/\nu = 3.6 \times 10^{-12}$. Thus, according to this number and Fig. 7-3(c), the molecule apparently makes approximately one-tenth of a full rotation (36°) as a free rotor before it experiences the environment leading to the exponential decay. Larger molecules are expected to be much more restricted in their free rotation. This interpretation may be naïve, however, as molecules are actually in close contact in the liquid phase and we wouldn't expect such definite free rotor-type behavior [7].

A more careful analysis of the data in Fig. 7-3 would include the extraction of the rotationally dependent line shape by deconvoluting the experimental line shape with the instrument profile. Some allowance must also be found for the possible contribution to the infrared line width due to vibrational relaxation.

The instrumental line shape, $S(\omega)$, is the response of the infrared spectrograph to a near monochromatic infrared signal. $S(\omega)$ is due to the optics, limitations, and imperfections in the monochromator, and possible broadening due to the electronic analysis equipment. In the absence of any vibrational relaxation broadening, the experimental infrared line shape $IR(\omega)$ will be a convolution of the molecular absorption coefficient in Eq.(7-54), $\gamma(\omega)$, with the instrumental line shape, $S(\omega)$,

$$IR(\omega') = \int_{-\infty}^{+\infty} S(\omega' - \omega)\gamma(\omega)\, d\omega. \tag{7-66}$$

$\gamma(\omega)$ is not necessarily Lorentzian as discussed above.

Remembering that the Fourier transform of the convolution of two functions is equal to the product of the Fourier transforms of the individual functions [see Eq. (7-34)], we can write

$$\mathscr{F} IR(\omega) = 2\pi[\mathscr{F} S(\omega)][\mathscr{F} \gamma(\omega)], \tag{7-67}$$

where $\mathscr{F}$ indicates the Fourier transform as before. Thus, we can measure the instrumental line width and compute $\mathscr{F} S(\omega)$. Using Eq.(7-67) we can extract the true absorption function $\gamma(\omega)$ from the experimental curve, $IR(\omega)$, by

$$\gamma(\omega) = \mathscr{F}^{-1}\left\{\frac{\mathscr{F}[IR(\omega)]}{\mathscr{F}[S(\omega)]}\right\}, \tag{7-68}$$

where the $\mathscr{F}[IR(\omega)]/\mathscr{F}[S(\omega)]$ ratio is taken before the inverse Fourier transform, $\mathscr{F}^{-1}$.

In some cases, the vibrational relaxation will also contribute to the width of the spectrum in a liquid. According to the discussion at the end of Section 7-4, vibrational relaxation times in the gas phase at 1 atm pressure are on the order of 10^{-5}–10^{-8} s (these relaxation times arise from approximately 10^9 collisions per second). In a liquid, molecules suffer many more collisions (average distance of 2 Å at $v = 4 \times 10^4$ cm · s^{-1} leads to approximately 2×10^{12} collisions · s^{-1}) than in a gas at 1 atm. Vibrational relaxation times in liquids (as measured by fluorescence, for instance) are on the order of 10^{-8}–10^{-12} s, which is a reasonable result on the basis of 2×10^{12} collisions · s^{-1}. Thus, the vibrational relaxation time in a liquid will contribute to the width where $\tau_V = 10^{-11}$ s leads to a Lorentz half width at half height of 0.5 cm^{-1} (see Fig. 7-3 for typical liquid line widths). If the vibrational relaxation is exponential, the line width due to this relaxation will be Lorentzian. We shall specify the line shape arising from the pure vibrational relaxation time by $V(\omega)$. In the presence of $V(\omega)$, the molecular line shape, $M(\omega)$, will be a convolution of $\gamma(\omega)$ with $V(\omega)$,

$$M(\omega) = \int_{-\infty}^{+\infty} V(\omega - \omega')\gamma(\omega')\, d\omega'. \tag{7-69}$$

The observed line shape $\underline{IR}(\omega)$ will be a convolution of $M(\omega)$ with $S(\omega)$. Thus, the Fourier transform gives

$$\mathscr{F}\underline{IR}(\omega) = 2\pi[\mathscr{F}S(\omega)][\mathscr{F}M(\omega)] = 4\pi^2[\mathscr{F}S(\omega)][\mathscr{F}V(\omega)][\mathscr{F}\gamma(\omega)]. \tag{7-70}$$

We can still measure $S(\omega)$ but we cannot unravel the $[\mathscr{F}V(\omega)][\mathscr{F}\gamma(\omega)]$ product by using only infrared absorption methods. If we assume exponential decay in the vibration and if we can obtain τ_V from another experiment, we can use $\mathscr{F}V(\omega) = A \exp(-t/\tau_V)$ (where A is a constant) to obtain $\mathscr{F}\gamma(\omega)$ from the equation above. We shall discuss later the separation and measurement of $V(\omega)$ and $\gamma(\omega)$ in a liquid by Raman scattering (Chapter 8).

7.4 THE BLOCH EQUATIONS, STEADY-STATE SOLUTIONS, POWER SATURATION, AND MOLECULAR RELAXATION

In previous sections we outlined the theory necessary to describe the observations of the long-time or steady-state resonant interaction of radiation and matter in the limit of low radiation power where thermal equilibrium is adequate to describe the steady-state populations and first-order relaxation theory is used to describe the relaxation of radiation-induced superposition states. We now extend the treatment to a more complete theory which includes radiation power levels which can both cause rapidly oscillating superposition states and appreciably change the thermal equilibrium populations in stationary states. This general theory will also describe transient effects that arise immediately after a sudden radiation-induced change in the condition of the system. Power saturation is examined by the time-independent steady-state solution to the Bloch equations here and in Section 7.5. In Section 7.6 we shall examine the time-dependent solution to the Bloch equations to describe transient experiments.

The Bloch equations were first derived to describe the time-dependence of the magnetism in a system of $I = \frac{1}{2}$ nuclear spins under the influence of static H_0 and time-dependent H_1 magnetic fields [8]. The equations can be derived from a classical interpretation of the effects of magnetic fields on a system of nuclear spins. A magnetic field $\boldsymbol{H}$ will produce a torque on a magnetic moment, $\boldsymbol{\mu}$, equal to $\boldsymbol{\mu} \times \boldsymbol{H}$. This torque will tend to align the magnetic moment along the field axis. If the system with the magnetic moment also has an angular momentum, $\boldsymbol{I}$, the effect of the field will be to cause a precession of the angular momentum vector about the axis of the field. The equation of motion for the magnetic moment is obtained by equating the torque to the time rate of change of the angular momentum as shown in Eq.(2-3),

$$\frac{d\boldsymbol{I}}{dt} = \boldsymbol{\mu} \times \boldsymbol{H}.$$

Substituting $\boldsymbol{\mu} = \gamma \boldsymbol{I}$ from Eq.(4-155) where γ is the magnetogyric ratio gives

$$\frac{d\boldsymbol{\mu}}{dt} = \gamma\boldsymbol{\mu} \times \boldsymbol{H}. \tag{7-71}$$

We now assume that we can replace the individual nuclear moments by a macroscopic moment density, $\boldsymbol{M}$, by multiplying Eq.(7-71) by ρ_0, the moment number density [see Eq.(1-46) in cgs units], giving

$$\frac{d\boldsymbol{M}}{dt} = \gamma\boldsymbol{M} \times \boldsymbol{H}. \tag{7-72}$$

Expanding the vector cross product gives three differential equations:

$$\frac{1}{\gamma}\frac{dM_x}{dt} = (\boldsymbol{M} \times \boldsymbol{H})_x = M_y H_z - H_y M_z$$

$$\frac{1}{\gamma}\frac{dM_y}{dt} = (\boldsymbol{M} \times \boldsymbol{H})_y = M_z H_x - H_z M_x$$

$$\frac{1}{\gamma}\frac{dM_z}{dt} = (\boldsymbol{M} \times \boldsymbol{H})_z = M_x H_y - H_x M_y. \tag{7-73}$$

We now define the values of the magnetic field to complete the equations. Normally a large static field is aligned along the z axis, $H_z = H_0$, which leads, in the case of an $I = \frac{1}{2}$ nuclear system, to a quantum mechanical two-level configuration as discussed with the density matrix formulation in the latter part of this section. After applying the $H_0 = H_z$ static field, we then probe the system with weaker time-dependent fields along the x and y axes. Keeping this orientation in mind, we note that any relaxation processes will reduce the transverse M_x and M_y to zero and the longitudinal M_z to an equilibrium value of M_0. Thus, we add the first-order relaxation

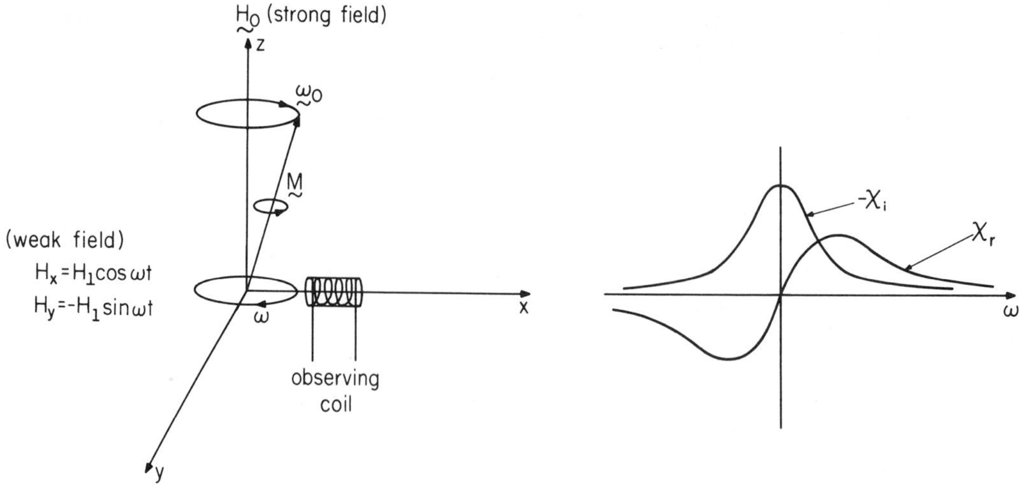

Figure 7-4 Diagram showing the precession of a positive nuclear (macroscopic) magnetic moment $\boldsymbol{M}$ at angular frequency $\omega_0 = \gamma H_0$ about the strong field axis. The weak field components H_x and H_y give a circularly rotating field with angular frequency $-\omega$. An observing coil is placed along the x axis to observe the linear magnetization along the x axis. The resultant real and imaginary components of the magnetic susceptibility as observed along the x axis are also plotted in the right part of the figure.

of M_x, M_y, and M_z to Eq.(7-73) by

$$\begin{aligned}\frac{dM_x}{dt} &= \gamma(M_y H_0 - H_y M_z) - \frac{M_x}{T_2} \\ \frac{dM_y}{dt} &= \gamma(M_z H_x - H_0 M_x) - \frac{M_y}{T_2} \\ \frac{dM_z}{dt} &= \gamma(M_x H_y - H_x M_y) - \frac{M_z - M_0}{T_1},\end{aligned} \tag{7-74}$$

where we assume that the *transverse relaxation time* T_2 (spin-spin) can be different from the *longitudinal relaxation time* T_1 (spin-lattice). Now we note for a positive nuclear magnetic moment that H_0 produces a clockwise precession at frequency ω_0 about the z axis as shown in Fig. 7-4. We can see this from Eq.(7-71) if $\boldsymbol{H} = \boldsymbol{H}_0$ and $d\boldsymbol{\mu}/dt = \boldsymbol{\mu} \times \gamma \boldsymbol{H}_0 = \boldsymbol{\mu} \times \boldsymbol{\omega}_0$, where the angular frequency is $\boldsymbol{\omega}_0 = \gamma \boldsymbol{H}_0$. This precession frequency is related to the quantum mechanical energy spacings of the $I = \frac{1}{2}$ nuclear energy levels in the H_0 field by the Bohr frequency rule $\omega_0 = \omega_{jk} = \Delta E/\hbar = (E_j - E_k)/\hbar$.

In order to probe the two-level system we apply a clockwise rotating magnetic field in the xy plane defined by the components [see Eq.(1-33)]

$$H_x = H_1 \cos \omega t, \qquad H_y = -H_1 \sin \omega t.$$

We also find it convenient to transform the x and y space-fixed values of the macroscopic moments, M_x and M_y, into a set of values u and v in a rotating coordinate system that is rotating about the z axis with angular frequency $-\omega$. This transformation is given by (where $M_x \longrightarrow u$ and $M_y \longrightarrow -v$ when $\omega t \longrightarrow 0$) [9]

$$M_x = u \cos \omega t - v \sin \omega t, \qquad M_y = -(u \sin \omega t + v \cos \omega t). \tag{7-75}$$

Substituting this transformation and H_y into the first of Eqs.(7-74) gives

$$\left(\dot{u} + \Delta\omega v + \frac{u}{T_2}\right) \cos \omega t = \left(\dot{v} - \Delta\omega u + \gamma H_1 M_z + \frac{v}{T_2}\right) \sin \omega t$$

where $\Delta\omega = \omega_0 - \omega$. Substituting the transformation and H_x into the second of Eqs.(7-74) gives

$$-\left(\dot{u} + \Delta\omega v + \frac{u}{T_2}\right) \sin \omega t = \left(\dot{v} - \Delta\omega u + \gamma H_1 M_z + \frac{v}{T_2}\right) \cos \omega t.$$

Simple algebra then shows that the expressions in parentheses are both zero giving equations in $\dot{u}$ and $\dot{v}$. Substituting the transformation and H_x and H_y from Eq.(7-75) into the third of Eqs.(7-74) gives the equation in $\dot{M}_z$. The results give the three coupled equations in $\dot{u}$, $\dot{v}$, and $\dot{M}_z$:

$$\begin{aligned}
\dot{u}(t) &= -\Delta\omega v(t) - \frac{u(t)}{T_2} \\
\dot{v}(t) &= \Delta\omega u(t) - \gamma H_1 M_z(t) - \frac{v(t)}{T_2} \\
\dot{M}_z(t) &= \gamma H_1 v(t) - \frac{M_z(t) - M_0}{T_1} \\
\Delta\omega &= \omega_0 - \omega.
\end{aligned} \tag{7-76}$$

The steady-state solutions to these equations are obtained by setting $\dot{u}(t) = \dot{v}(t) = \dot{M}_z(t) = 0$, which gives the following time-independent solutions:

$$\begin{aligned}
u &= \frac{\gamma H_1 M_0 \Delta\omega}{(\Delta\omega)^2 + (1/T_2)^2 + (T_1/T_2)(\gamma H_1)^2} \\
v &= -\frac{\gamma H_1 M_0/T_2}{(\Delta\omega)^2 + (1/T_2)^2 + (T_1/T_2)(\gamma H_1)^2} \\
M_z &= \frac{M_0[(\Delta\omega)^2 + (1/T_2)^2]}{(\Delta\omega)^2 + (1/T_2)^2 + (T_1/T_2)(\gamma H_1)^2}.
\end{aligned} \tag{7-77}$$

We shall discuss saturation effects arising from the $(T_1/T_2)(\gamma H_1)^2$ terms following the derivation of the corresponding electric dipole analog of the Bloch equations.

The steady-state values of u and v can be measured by observing the magnetization along the x axis as depicted in Fig. 7-4. We define the real part of the magnetic susceptibility, χ_r, as the component that is in phase with the stimulating field along the

x axis, $H_x = H_1 \cos \omega t$, and we define the imaginary part of the magnetic susceptibility, χ_i, as the component along the x axis that is $\pi/2$ out of phase with ωt obtained from

$$\cos\left(\omega t - \frac{\pi}{2}\right) = -\sin \omega t\,;$$

this gives

$$M_x = H_1\chi_r \cos \omega t - H_1\chi_i \sin \omega t \tag{7-78}$$

for the observed magnetization [see the definitions in Eq.(1-48) and the associated discussion]. Comparing M_x in Eq.(7-78) with M_x in Eq.(7-75) gives expressions relating u and v to χ_r and χ_i,

$$H_1\chi_r = u, \qquad H_1\chi_i = v. \tag{7-79}$$

Substituting u and v from Eqs.(7-77) and $M_0 = H_0\chi_0$, where χ_0 is the bulk or average magnetic susceptibility, gives ($\gamma H_0 = \omega_0$)

$$\begin{aligned}\chi_r(\omega) &= \chi_0\omega_0\left[\frac{\Delta\omega}{(\Delta\omega)^2 + (1/T_2)^2 + (T_1/T_2)(\gamma H_1)^2}\right]\\ \chi_i(\omega) &= -\chi_0\omega_0\left[\frac{(1/T_2)}{(\Delta\omega)^2 + (1/T_2)^2 + (T_1/T_2)(\gamma H_1)^2}\right].\end{aligned} \tag{7-80}$$

$\chi_i(\omega)$ and $\chi_r(\omega)$ are also plotted in Fig. 7-4. The $\chi_r(\omega)$ and $\chi_i(\omega)$ curves have the same near-resonance shapes as the corresponding real and imaginary dielectric constants, polarizabilities, or refractive indices as discussed before in Section 2.2.

Electric Field Analog of the Bloch Equations

We shall now use a density matrix method and rederive the Bloch-type equations for the electric dipole–electric field interaction for a two-level system. The complete Hamiltonian, including the interaction of the electric field of the radiation, $\boldsymbol{E}$, with the dipole moment in the molecular system, $\boldsymbol{D}$, can be written as

$$\mathscr{H} = \mathscr{H}_0 - \boldsymbol{D}\cdot\boldsymbol{E}. \tag{7-81}$$

$\mathscr{H}_0$ is the internal time-independent Hamiltonian leading, through Schrödinger's equation, to the stationary states of the system; the ith stationary state, ψ_i, is determined from

$$\mathscr{H}_0\psi_i = E_i\,\psi_i. \tag{7-82}$$

Limiting the dipole components to a single inertial axis D_a (as in a linear molecule) and limiting $\boldsymbol{E}$ to $E_z(t) = E_0 \cos(ky - \omega t)$, a traveling wave (linear polarized in the yz plane) propagating along the y axis with angular frequency ω gives

$$\mathscr{H} = \mathscr{H}_0 - D_a C_{az} E_z = \mathscr{H}_0 - D_a C_{az} E_0 \cos(ky - \omega t), \tag{7-83}$$

where C_{az} is the cosine between the a and z axes. The time-dependent Schrödinger equation for the complete Hamiltonian in Eq.(7-83) can be obtained by expanding $\Psi(\boldsymbol{r}, t)$ according to Eq.(3-157) in terms of the stationary state eigenfunctions, $\boldsymbol{\psi}$ given in Eq.(7-82), where all time-dependence is contained in the $b_j(t)$ coefficients. The most convenient method of proceeding is to form the density matrix as defined in Eq.(3-70) with the $b_j(t)$ coefficients $[\rho_{ij} = b_i(t)^* b_j(t)]$ and then solve Eq.(3-146) for the density matrix,

$$\frac{d}{dt}\boldsymbol{\rho} = \dot{\boldsymbol{\rho}} = \frac{i}{\hbar}[\mathcal{H}, \boldsymbol{\rho}]. \tag{7-84}$$

$\mathcal{H}$ is the matrix representation of the complete Hamiltonian in Eq.(7-83) in the $\boldsymbol{\psi}$ basis set of stationary states defined in Eq.(7-82). At thermal equilibrium in the absence of the time-dependent perturbation, the density matrix is diagonal (the system is in a distribution of stationary states) and the diagonal elements in $\boldsymbol{\rho}$ are related to the Boltzmann relation,

$$\frac{\rho_{jj}}{\rho_{kk}} = \exp\left[-\frac{(E_j - E_k)}{kT}\right], \tag{7-85}$$

for the jth and kth stationary states with equal degeneracies. The number of molecules per unit volume in the jth and kth states is given by

$$N_j = \rho_0 \rho_{jj}, \qquad N_k = \rho_0 \rho_{kk}, \tag{7-86}$$

where ρ_0 is the total molecular number density. We consider only a two-state system here where $\boldsymbol{\psi} = (\psi_k \quad \psi_j)$ and the Hamiltonian matrix in this basis is given by

$$\mathcal{H}(t) = \begin{pmatrix} \mathcal{H}_{jj}(t) & \mathcal{H}_{jk}(t) \\ \mathcal{H}_{kj}(t) & \mathcal{H}_{kk}(t) \end{pmatrix} = \begin{pmatrix} E_j & -D_{kj}E_0 \cos(ky - \omega t) \\ -D_{kj}E_0 \cos(ky - \omega t) & E_k \end{pmatrix}$$

$$D_{kj} = \int \psi_k^* D_a C_{az} \psi_j \, dV, \qquad D_{kk} = D_{jj} = 0, \tag{7-87}$$

where E_j and E_k are the stationary state eigenvalues in Eq.(7-82) and the condition where $D_{kk} = D_{jj} = 0$ includes a large number of interesting systems in electronic, vibrational, rotational, and magnetic resonance (replace D_{jk} with μ_{jk}) spectroscopy.

Substituting Eq.(7-87) into Eq.(7-84) and expanding gives the differential equations for the density matrix elements. Substituting

$$\cos(ky - \omega t) = \tfrac{1}{2}\{\exp[i(ky - \omega t)] + \exp[-i(ky - \omega t)]\},$$

using $(E_j - E_k)/\hbar = \omega_{jk}$, and defining $\Delta = \rho_{kk} - \rho_{jj}$ leads to

$$\dot{\rho}_{jj} = -\frac{i}{2\hbar}E_0\{\exp[i(ky - \omega t)] + \exp[-i(ky - \omega t)]\}(D_{jk}\rho_{kj} - \rho_{jk}D_{kj})$$

$$\dot{\rho}_{jk} = i\omega_{jk}\rho_{jk} - \frac{i}{2\hbar}D_{jk}\Delta E_0\{\exp[i(ky - \omega t)] + \exp[-i(ky - \omega t)]\}$$

$$\dot{\rho}_{kj} = -i\omega_{jk}\rho_{jk} + \frac{i}{2\hbar}D_{kj}\Delta E_0\{\exp[i(ky - \omega t)] + \exp[-i(ky - \omega t)]\}$$

$$\dot{\rho}_{kk} = -\frac{i}{2\hbar}E_0\{\exp[i(ky - \omega t)] + \exp[-i(ky - \omega t)]\}(D_{kj}\rho_{jk} - \rho_{kj}D_{jk}). \tag{7-88}$$

We shall now relate the elements in the density matrix to the electric polarizability of a molecule and to the resultant macroscopic polarization of a large number of molecules. The average value of the dipole moment, D_{av}, of a molecule described by $\Psi(\boldsymbol{r}, t)$, which is an eigenfunction of the complete Hamiltonian in Eq.(7-83), is obtained in terms of the density matrix as shown in Eq.(3-70),

$$D_{av} = \text{Tr}\, \boldsymbol{\rho} \boldsymbol{D}, \tag{7-89}$$

where Tr indicates the diagonal sum, $\boldsymbol{\rho}$ is obtained from the $b_j(t)$ coefficients in Eq. (3-157) as described above, and $\boldsymbol{D}$ is computed in the $\boldsymbol{\psi}$ basis from Eq.(7-82). Expanding for the two-level system gives

$$D_{av} = \rho_{jk} D_{kj} + \rho_{kj} D_{jk}, \tag{7-90}$$

where we remember that $D_{kk} = D_{jj} = 0$. Now, in a manner analogous to the magnetic case [see Eq.(7-78) and associated discussion], we shall write the average dipole moment observed along the field axis (z) in terms of real and imaginary components. The field-induced dipole moment is proportional to the polarizability as discussed in Sections 1.1 and 6.10. We define the real part of the electrical polarizability, α_r, as the component that is in phase with the time-dependence of the stimulating field, $E_z = E_0 \cos(ky - \omega t)$ at any point on the y axis, and we define the imaginary part of the electric polarizability, α_i, as the component at any point on the y axis that is $\pi/2$ out of phase with ωt obtained from

$$\cos\left[ky - \left(\omega t + \frac{\pi}{2}\right)\right] = \cos\left(ky - \omega t - \frac{\pi}{2}\right) = -\sin(ky - \omega t);$$

this gives

$$\begin{aligned} D_{av} &= E_0 \alpha_r \cos(ky - \omega t) - E_0 \alpha_i \sin(ky - \omega t) \\ &= E_0 \alpha_r \cos(\omega t - ky) + E_0 \alpha_i \sin(\omega t - ky) \\ &= \tfrac{1}{2}(\alpha_r - i\alpha_i) E_0 \exp[i(\omega t - ky)] + \tfrac{1}{2}(\alpha_r + i\alpha_i) E_0 \exp[-i(\omega t - ky)]. \end{aligned} \tag{7-91}$$

Next we define the macroscopic dipole moment or *polarization*, P, as the product of the number density with the average value of the dipole moment of a single molecule [see Eqs.(1-41) and associated discussion]

$$\begin{aligned} \rho_0 D_{av} = P &= \frac{\rho_0}{2}(\alpha_r - i\alpha_i) E_0 \exp[i(\omega t - ky)] + \frac{\rho_0}{2}(\alpha_r + i\alpha_i) E_0 \exp[-i(\omega t - ky)] \\ &= \tfrac{1}{2}(P_r + iP_i) \exp[i(\omega t - ky)] + \tfrac{1}{2}(P_r - iP_i) \exp[-i(\omega t - ky)], \end{aligned} \tag{7-92}$$

where P_r and P_i are the real and imaginary parts of the macroscopic polarization of the sample. $\boldsymbol{P}$ is the electric analog of the magnetic $\boldsymbol{M}$ discussed earlier in this section (see also Section 1.1). We also note that we have defined $P_r = \rho_0 \alpha_r E_0$ and $P_i = -\rho_0 \alpha_i E_0$ with the opposite sign on the imaginary components by analogy to the magnetic case (the choice is arbitrary).

The absorption coefficient in Eq.(2-55) can also be written in terms of the imaginary component of polarization, $P_i(\omega, t)$, by relating $P_i(\omega, t)$ to the imaginary component of the refractive index, $n_i(\omega, t)$. Using Eq.(1-60) for a dilute gas and using $P_r = \rho_0 \alpha_r E_0$ and $P_i = -\rho_0 \alpha_i E_0$ from Eq.(7-92) gives

$$n_r + in_i = 1 + 2\pi\rho_0(\alpha_r + i\alpha_i) = 1 + 2\pi \frac{P_r}{E_0} - i2\pi \frac{P_i}{E_0}. \tag{7-93}$$

Substituting $n_i(\omega, t) = -2\pi[P_i(\omega, t)/E_0]$ from Eq.(7-93) into Eq.(2-55) gives the absorption coefficient in terms of the imaginary part of the polarization,

$$\gamma(\omega, t) = -\frac{4\pi\omega_{jk} P_i(\omega, t)}{cE_0}. \tag{7-94}$$

Returning to Eq.(7-90) and multiplying by ρ_0 gives

$$P = \rho_0 \rho_{jk} D_{kj} + \rho_0 \rho_{kj} D_{jk}.$$

Comparing this equation with Eq.(7-92) leads to

$$\begin{aligned} \rho_0 \rho_{jk} D_{kj} &= \tfrac{1}{2}(P_r + iP_i) \exp[-i(ky - \omega t)], \\ \rho_0 \rho_{kj} D_{jk} &= \tfrac{1}{2}(P_r - iP_i) \exp[i(ky - \omega t)], \end{aligned} \tag{7-95}$$

which relates the real and imaginary components of the polarization to the dipole moment matrix elements and density matrix elements. We shall now develop the equations of motion for P_r and P_i. Taking the time derivative of $(P_r + iP_i)$ from Eq.(7-95) gives

$$\begin{aligned} \frac{d}{dt}(P_r + iP_i) &= \dot{P}_r + i\dot{P}_i \\ &= 2\rho_0\{-i\omega\rho_{jk} D_{kj} \exp[i(ky - \omega t)] + \dot{\rho}_{jk} D_{kj} \exp[i(ky - \omega t)]\}, \end{aligned}$$

where $\dot{\mathrm{D}}_{kj} = 0$. Substituting $\rho_0 \rho_{jk} \mathrm{D}_{kj}$ from Eq.(7-95) and $\rho_0 \dot{\rho}_{jk}$ from Eq.(7-88) into the right-hand side of this equation gives

$$\begin{aligned} \dot{P}_r + i\dot{P}_i = &-i\omega(P_r + iP_i) + 2i\omega_{jk} \rho_0 D_{kj} \rho_{jk} \exp[i(ky - \omega t)] \\ &- \frac{i}{\hbar} D_{jk}^2 E_0 \rho_0 \Delta\{1 + \exp[2i(ky - \omega t)]\}. \end{aligned} \tag{7-96}$$

We now assume that P_r and P_i have much slower time-dependence than the sinusoidal $\exp[2i(ky - \omega t)]$ term in this expression. Dropping the $\exp[2i(ky - \omega t)]$ term and substituting again $\rho_0 D_{kj} \rho_{jk} \exp[i(ky - \omega t)] = \frac{1}{2}(P_r + iP_i)$ from Eq.(7-95) gives

$$\begin{aligned} \dot{P}_r + i\dot{P}_i &= i(P_r + iP_i)(\omega_{jk} - \omega) - \frac{i}{\hbar} D_{jk}^2 E_0 \Delta N, \\ \Delta N &= N_k - N_j = \rho_0 \Delta = \rho_0(\rho_{kk} - \rho_{jj}). \end{aligned} \tag{7-97}$$

The real and imaginary components of Eq.(7-97) give two coupled differential equations in P_r, P_i, and ΔN. We now seek a third differential equation involving these variables by starting with

$$\frac{d\Delta N}{dt} = \rho_0(\dot{\rho}_{kk} - \dot{\rho}_{jj}). \tag{7-98}$$

Substituting from Eq.(7-88) gives

$$\frac{d\Delta N}{dt} = -\frac{i\rho_0}{\hbar} E_0\{\exp[i(ky - \omega t)] + \exp[-i(ky - \omega t)]\}(D_{kj}\rho_{jk} - \rho_{kj}D_{jk}).$$

Substituting from Eq.(7-95) gives

$$\frac{d\Delta N}{dt} = \frac{i}{2\hbar} E_0\{(P_r - iP_i)\exp[2i(ky - \omega t)] - 2iP_i - (P_r + iP_r)\exp[-2i(ky - \omega t)]\}. \quad (7\text{-}99)$$

We also assume that the time-dependence in ΔN is slow relative to

$$\exp[\mp 2i(ky - \omega t)].$$

Dropping the $\exp[\pm 2i(ky - \omega t)]$ terms in Eq.(7-99) gives the result. This result along with the real and imaginary components in Eq.(7-97) gives the three coupled differential equations in $P_r(t)$, $P_i(t)$, and $\Delta N(t)$:

$$\frac{dP_r(t)}{dt} = -\Delta\omega P_i(t)$$

$$\frac{dP_i(t)}{dt} = \Delta\omega P_r(t) - \frac{D_{jk}^2 E_0}{\hbar}\Delta N(t)$$

$$\frac{d\Delta N(t)}{dt} = \frac{E_0}{\hbar} P_i(t)$$

$$\Delta\omega = \omega_{jk} - \omega. \quad (7\text{-}100)$$

We can now add the effects of first-order relaxations into these equations. We define here two different relaxation times. T_1 is associated with the relaxation of $\Delta N(t)$ involving the diagonal elements of the density matrix. T_2 is associated with the relaxation of the polarization components, $P_i(t)$ and $P_r(t)$, involving the off-diagonal elements of the density matrix. Adding the first-order relaxations gives the final coupled differential equations:

$$\dot{P}_r(t) = -\Delta\omega P_i(t) - \frac{P_r(t)}{T_2},$$

$$\dot{P}_i(t) = \Delta\omega P_r(t) - \frac{D_{jk}^2 E_0}{\hbar}\Delta N(t) - \frac{P_i(t)}{T_2}$$

$$\Delta\dot{N}(t) = \frac{E_0}{\hbar} P_i(t) - \frac{\Delta N(t) - \Delta N_0}{T_1}, \quad (7\text{-}101)$$

where ΔN_0 is the equilibrium value of ΔN. We can now solve Eqs.(7-101) for $P_i(t)$, $P_r(t)$, and $\Delta N(t)$ under the desired experimental conditions. We are normally interested in $P_i(\omega, t)$, which is needed for the absorption coefficient as shown in Eq.(7-94).

First we examine the steady-state solution to Eqs.(7-101) where $\dot{P}_r(t) = \dot{P}_i(t) = \Delta\dot{N}(t) = 0$. The resultant solutions for P_i, P_r, and ΔN give the following time-independent results:

$$P_r(\omega) = \frac{(1/\hbar)(D_{jk}^2 E_0)\Delta N_0 \Delta\omega}{(\Delta\omega)^2 + (1/T_2)^2 + (T_1/T_2)\xi^2}$$

$$P_i(\omega) = -\frac{(1/\hbar)(D_{jk}^2 E_0)(\Delta N_0/T_2)}{(\Delta\omega)^2 + (1/T_2)^2 + (T_1/T_2)\xi^2}$$

$$\Delta N(\omega) = \frac{\Delta N_0[(\Delta\omega)^2 + (1/T_2)^2]}{(\Delta\omega)^2 + (1/T_2)^2 + (T_1/T_2)\xi^2}$$

$$\xi = \frac{E_0 D_{jk}}{\hbar}. \qquad (7\text{-}102)$$

Before proceeding, we draw attention to the similarities between the electric dipole equations given here and the corresponding magnetic dipole case in Eqs.(7-76) and (7-77) [10]. Equations (7-76) and any of the solutions to Eqs.(7-76) can be obtained from Eqs.(7-101) by making the following substitutions:

$$P_r \longrightarrow u, \qquad P_i \longrightarrow v, \qquad \Delta N \longrightarrow \frac{M_z}{D_{jk}}, \qquad \Delta N_0 \longrightarrow \frac{M_0}{D_{jk}},$$

$$\frac{D_{jk} E_0}{\hbar} \longrightarrow \gamma H_1. \qquad (7\text{-}103)$$

We shall keep the electric dipole–magnetic dipole analogy above in mind in developing both the steady-state and transient equations for the two-level system. Any result in P_r, P_i, and ΔN can be easily transferred to u, v, and M_0 by the relations above; however, we note that, normally, different methods are used in observing the radiation-system interaction. In the magnetic dipole case observed in NMR and ESR the values of χ_i or χ_r are usually measured as described following Eq.(7-78). χ_i and χ_r are related to u and v in Eq.(7-79). In the electric dipole case observed in rotational, vibrational, and optical spectroscopy, the absorption coefficient is normally measured, which is related to P_i by Eq.(7-94). For instance, substituting $P_i(\omega)$ in Eqs.(7-102) into Eq.(7-94) gives the steady-state absorption coefficient in the presence of power saturation [the $(T_1/T_2)\xi^2$ term] where $N_k - N_j = \Delta N_0$ is the equilibrium value:

$$\gamma(\omega) = \frac{4\pi(N_k - N_j)\omega_{jk} D_{jk}^2}{c\hbar}\left[\frac{1/T_2}{(\omega - \omega_{jk})^2 + (1/T_2)^2 + (T_1/T_2)\xi^2}\right]. \qquad (7\text{-}104)$$

In the limit of low radiation power where $(1/T_2)^2 \gg (T_1/T_2)\xi^2$, this expression is identical to Eq.(7-8) near resonance where $\omega/\omega_{jk} \cong 1.0$. The angular dependence between $\boldsymbol{A}_0$ and $\boldsymbol{D}$ is contained in the $\boldsymbol{D}_{jk}$ matrix element as described previously. Furthermore, it is evident by comparing Eqs.(7-104) and (7-8) that the relaxation time measured from the low-power line shapes is equivalent to T_2, the relaxation time for the polarization in Eqs.(7-101). T_1, the relaxation of the population difference

Table 7-1 Linewidth data for low J rotational transitions in several systems. The line widths are half widths at half height per unit Torr pressure at $T \cong 300$ K. τ_R are computed at 1 atm by assuming $\Delta\nu$ is due to rotational relaxation, $\tau_R = 1/2\pi\Delta\nu$.

Molecule	Broadened by	$\Delta\nu/p$ (MHz · Torr^{-1})	τ_R (10^{-11} s)
HCN	HCN	25 ± 8	1.0
OCS	OCS	6 ± 1	3.2
OCS	He	2.0 ± 0.5	10.0
OCS	Ar	3.0 ± 1.0	8.0
OCS	NNO	4.5 ± 0.5	5.0
NNO	NNO	5.0 ± 1.0	4.0

in the two-level system, can be measured when $(1/T_2)^2 \cong (T_1/T_2)\xi^2$ from the line shape in Eq.(7-104), which is still a Lorentzian with half width at half height given by

$$\Delta\omega = \left[\left(\frac{1}{T_2}\right)^2 + \frac{T_1}{T_2}\xi^2\right]^{1/2}.$$

Thus, as the radiation power increases, the peak height decreases and the half width at half height increases. Assuming $T_1 \cong T_2$, and using $P = (cA/8\pi)(E_0)^2$ for the incident radiation power (A is the radiation cross section entering the sample) leads to the conditions under which power saturation can be observed. If $\tau_R = T_1 = T_2 \cong 10^{-11}$ s at 1 atm (see Table 7-1), power saturation effects will occur in the rotational transition at an intensity of approximately 10^6 W · cm^{-2}. At a lower pressure of 10 m Torr, however, where most microwave spectroscopy is done, $\tau_R = T_1 = T_2 = 10^{-6}$ s and power saturation effects will occur at an intensity level of only 10^{-4} W · cm^{-2}. Thus, power saturation is easily observed in microwave spectroscopy. The problem of saturation in electron and nuclear magnetic resonance is examined in Problems 7-5 and 7-6. Normally, it is relatively easy to saturate ESR transitions and even lower powers will saturate NMR transitions where the relaxation times are very long ($T_2 \cong 1$ s in liquids). Conversely, it is more difficult to saturate infrared and optical-type transitions that normally have much faster relaxation times. In summary, T_2 can be measured in steady state from the line width at low power and T_1 can be measured by observing the changes in the Lorentz line shape when power saturation is important. The challenge in this field now is to relate T_1 and T_2 to the microscopic details of collisions.

In the presence of the Doppler effect, the absorption coefficient in Eq.(7-104) must be convoluted with the Gaussian. Using steps as outlined in Section 7.2, the result in Eq.(7-104) will be modified in the presence of the Doppler effect to give

$$\gamma(\nu) = \frac{2D_{jk}^2 \nu_{jk}(N_k - N_j)}{c\hbar T_2} \int_{-\infty}^{+\infty} \frac{G(\nu')\,d\nu'}{(\Delta\nu_L)^2 + (\nu_{jk} - \nu - \nu')^2 + (T_1/T_2)(\xi/2\pi)^2}, \qquad (7\text{-}105)$$

where $G(\nu')$ is the Gaussian defined in Eq.(7-31) and $\Delta\nu_L$ is the Lorentz half width at half height defined in Eq.(7-32). The limit of this expression when $G(\nu') \longrightarrow \delta(\nu')$ gives an expression that is equivalent to Eq.(7-104). Now using Eq.(7-35) in Eq.(7-105) and finding the limit as the Lorentzian goes to a Dirac delta function (relative to the Gaussian) gives

$$\lim_{[(\Delta\nu_L)^2+(T_1/T_2)(\xi/2\pi)^2]\ll\Delta\nu_G} \gamma(\nu) = \frac{4\pi^2 D_{jk}^2 \nu_{jk}(N_k - N_j)G(\nu_{jk} - \nu)}{c\hbar(2\pi T_2)[(1/2\pi T_2)^2 + (T_1/T_2)(\xi/2\pi)^2]^{1/2}},$$

which reduces to Eq.(7-38) in the low-power limit. The broadening due to T_1 and T_2 processes in these equations is called *homogeneous* because all molecules in the sample have identical T_1 and T_2 processes. The Doppler broadening or other types of broadening not related to relaxation effects are called *inhomogeneous* because different molecules have different velocities and different Doppler shifts. Other inhomogeneous effects would arise from inhomogeneities in the static magnetic field in NMR or ESR and inhomogeneities in the electric field on an experiment observing the Stark effect in a molecule.

We emphasize here that the relaxation times T_1 and T_2 were added phenomenologically to Eqs.(7-101) and (7-76) in the electric and magnetic dipole cases, respectively. Thus, a microscopic theory of interparticle collisions (interactions) must be developed to relate the experimentally determined T_1 and T_2 to intermolecular potential functions. Considerable success has been achieved in this task in the magnetic resonance case [11] and a start has also been made in the rotational case [12]. A good deal is also known about T_1 processes among electronic, vibrational, rotational, and translational states and we shall summarize some of these findings here.

Molecular Relaxation (T_1)

In the following discussion we shall distinguish among translational, rotational, vibrational, and electronic states by the use of T, R, V, and E, respectively. It is conceivable and indeed often observed that the temperatures within any of the systems above (T, R, V, and E) may be different from the others [the temperature of the rotational states (R) may be different than the vibrational (V) states]. If thermal equilibrium is achieved, however, then the temperature in each system must be the same as the surrounding environment.

Molecular relaxation processes (T_1) are important in studies of gas kinetics and transport properties, laser applications specifically in producing population inversions for amplifying electromagnetic energy, chemical kinetics where each binary molecular collision is an incipient chemical reaction, and chemical kinetics involving highly excited (vibrational) reaction products. In the latter case, it is important to know whether the molecule can relax vibrationally to produce a stable product before it crosses over to an unstable potential surface causing decomposition. We shall confine this discussion to the gas phase (normally at standard condition of 1 atm pressure and $T = 300$ K) where the dominant relaxation processes are caused by binary collisions and we shall discuss the rate of equilibration or relaxation between each of the internal degrees of freedom, $E \longrightarrow E$, $V \longrightarrow V$, $R \longrightarrow R$, and $T \longrightarrow T$, and equilibration between two or more internal degrees of freedom, $E \longrightarrow V$, $V \longrightarrow R$, $V \longrightarrow T$, $R \longrightarrow T$ and other possible combinations.

We begin by writing the rate constant for a relaxation process according to an Arrhenius-type equation (based on experience),

$$k = \frac{1}{\tau} = \eta \exp\left(-\frac{\Delta E}{kT}\right).$$

τ is the relaxation time for the first-order process where $1/\tau$ is the relaxation rate constant, ΔE is the energy change or activation energy necessary to cause the relaxation, and η is the attempt frequency that is approximately the frequency of kinetic hard sphere collisions (the number of collisions per unit time). The number of hard sphere (binary) collisions per second is given by [4]

$$\eta = \frac{\bar{v}}{\lambda} = (2)^{1/2}\pi\rho_0\, d^2\bar{v},$$

where λ is the mean free path, d is the molecular diameter for a kinetic hard sphere collision, and $\bar{v}$ is the average thermal velocity. For typical molecules like CO or CF_4 at $T = 300$ K, $\eta = 10^9$–10^{10} collisions per second.

As the translational energy levels form a near continuum, $\exp(-\Delta E/kT)$ in the Arrhenius equation will be unity and each binary collision will lead to a change in translational energy. Thus, $T \longrightarrow T$ transfer will occur very fast, on the order of 10^{-9} to 10^{-10} s at standard conditions.

Rotational energy spacings are also normally less than kT at 300 K and $R \longrightarrow R$ relaxation is also expected to be at least as fast as $T \longrightarrow T$ relaxation. In some cases, however, there is reason to expect that $R \longrightarrow R$ may be faster than $T \longrightarrow T$ relaxation due to dipole electric fields. The electric fields 10 Å from a molecule with a dipole moment of 10^{-18} SC · cm can be estimated from Eq.(1-57) to be $E \cong D/R^3 = 10^3$ SV · cm^{-1} = 3×10^5 V · cm^{-1}. This large electric field will cause a large Stark effect on the molecule that in many cases will be as large as the rotational energy spacing. Thus, for molecules that have permanent electric dipole moments, the rotational collision diameters may be larger than the kinetic translational collision diameters. The result of this discussion is to expect that $R \longrightarrow R$ relaxation times may be as fast or faster than $T \longrightarrow T$ relaxation times. $R \longrightarrow R$ relaxation times obtained from line width data (assuming $T_1 = T_2$) for rotational transitions are given in Table 7-1. The HCN-HCN line width is by far the largest due to the large HCN electric dipole moment (see Table 4-16). Calculating $\eta \cong 4.7 \times 10^{-10}$ s for HCN shows that $R \longleftrightarrow R$ relaxation is considerably faster than $T \longleftrightarrow T$ relaxation in HCN. This will be generally true for molecules having large dipole moments. Of course, we have not considered the selection rules for collisional rotational energy transfer, which, in some cases, can deviate from the normally allowed electric dipole transitions [13].

We expect $R \longrightarrow T$ relaxation to be somewhat slower than $T \longrightarrow T$ relaxation due to geometric factors. For instance, two diatomic molecules colliding with each other on an axis parallel to the rotational angular momentum vector will not lead to any $R \longrightarrow T$ exchange. This collision will lead to $T \longrightarrow T$ exchange.

The $V \longrightarrow T$ and $V \longrightarrow R$ vibrational relaxation times are normally slower than the $R \longrightarrow R$, $R \longrightarrow T$, and $T \longrightarrow T$ times. Some vibrational relaxation times are listed in Table 7-2. A physical picture of vibrational $V \longrightarrow T$ relaxation can be obtained by

Table 7-2 $V \longrightarrow T$ relaxation times from the lowest energy excited state to the ground state at 1 atm pressure and 300 K. The τ_V values are from T. L. Cottrell and J. C. McCoubrey, *Molecular Energy Transfer in Gases* (Butterworths, London, 1961).

Molecule	τ_V (10^{-6} s)
Cl_2	4.90
Br_2	0.80
OCO	6.00
NO	1.00
SCS	0.50
SO_2	0.06
CH_4	1.70
FCH_3	0.20

considering the collision of an atom with an oscillator (see Problem 2-7). The duration of the collision must be of the same order of magnitude as the period of oscillation in order to have a nonadiabatic collision. Molecular velocities at 300 K are about $v = 3 \times 10^4$ cm · s^{-1} and vibrational periods for $[(E_1 - E_0)/hc] \cong 1000$ cm^{-1} are $\tau_V \cong 3.3 \times 10^{-14}$ s. Thus, in order for the period of the collision, τ_0, to be 3.3×10^{-14} s, the interaction length, ℓ, must be very short. $\ell/v = \tau_0 = 3.3 \times 10^{-14}$ s; $\ell = 10^{-9}$ cm. This short length would require an extremely steep repulsive potential between the atom and molecule. Exponential repulsion due to overlap between closed shell orbitals normally leads to softer potentials with $\ell \cong 0.5$–0.8 Å (see Section 5.9). Thus, only the molecules and atoms in the high velocity part of the Maxwell velocity distribution curve will have the velocity necessary to effect the $V \longrightarrow T$ transfer of energy.

Enhancement of $V \longrightarrow T$ relaxation times by the $V \longrightarrow R$ process may be obtained in collision with molecules with small moments of inertia (normally hydrogen-containing molecules). A considerable increase in relative velocity is obtained by a light rotor leading to $V \longrightarrow R$ exchange of energy. The angular velocity, ω, is approximately related to the rotational angular momentum J by $J \cong \omega I$. The component of linear velocity is $v = r\omega$ where r is an approximate radius of the rotor and $v = r\omega = Jr/I \cong 10^5$ cm · s^{-1} for $J = \hbar$ in a free methyl rotor where $r \cong 10^{-8}$ cm and $I = 5.3 \times 10^{-40}$ g · cm^2. Thus, according to our previous discussion on nonadiabatic collisions, the rotational contribution to the relative velocity ought to contribute substantially to the rate of vibrational relaxation and we expect $V \longrightarrow R$ transfer of energy to occur at a rate equal or larger than the $V \longrightarrow T$ rate for light molecules.

$V \longrightarrow V$ relaxation is normally faster than either $V \longrightarrow R$ or $V \longrightarrow T$ rates due to the smaller ΔE necessary for many $V \longrightarrow V$ transfers of energy [14]. Referring to Fig. 4-34 for the vibrational energy spacings of a nonlinear triatomic molecule, we note that transfer of energy within any mode should be very efficient because of the small amount of excess kinetic energy produced on the exchange. For instance, for the A-B-C molecule,

$$\text{ABC}(0, 2, 0) + \text{ABC}(0, 0, 0) \xrightarrow{k} 2\text{ABC}(0, 1, 0) + KE,$$

where KE represents the excess kinetic energy, which is zero for a pure harmonic oscillator. The vibrational exchange between different modes is usually not so fast because of larger values of the excess KE that must be liberated or absorbed in rotation or translation to satisfy the conservation of energy. Several other equations listing the energy exchanges of various states of CO_2 (ν_1, ν_2, ν_3) with itself and with N_2, H_2O, and He are listed here:

$$CO_2(0, 0, 1) + CO_2(0, 0, 0) \xrightarrow{k_a} CO_2(0, 0, 0) + CO_2(1, 1, 0) + 340 \text{ cm}^{-1} \quad (7\text{-}106a)$$

$$CO_2(0, 0, 1) + N_2(0) \xrightarrow{k_b} CO_2(0, 0, 0) + N_2(1) + 19 \text{ cm}^{-1} \quad (7\text{-}106b)$$

$$CO_2(0, 0, 1) + CO_2(0, 0, 0) \xrightarrow{k_c} CO_2(0, 1, 0) + CO_2(1, 0, 0) + 340 \text{ cm}^{-1} \quad (7\text{-}106c)$$

$$CO_2(0, 0, 1) + H_2O(0, 0, 0) \xrightarrow{k_d} CO_2(0, 1, 0) + H_2O(0, 1, 0) + 87 \text{ cm}^{-1} \quad (7\text{-}106d)$$

$$CO_2(0, 0, 1) + \text{He} \xrightarrow{k_e} CO_2(0, 4, 0) + \text{He} - 319 \text{ cm}^{-1}. \quad (7\text{-}106e)$$

The vibrational energy spacings are from Tables 4-26 and 4-28 where we estimate 1342 cm^{-1} for ν_1 in CO_2. According to the discussion above, we would expect $k_b > k_d > k_e > k_a \cong k_c$. This trend is indeed found experimentally where the relaxation of $CO_2(0, 0, 1)$ in CO_2 is $\tau_a(V \longrightarrow V) \cong 3.0 \times 10^{-6}$ s and τ_b is known to be much shorter at 7×10^{-8} s [15].

We can summarize our knowledge of V, T, and R relaxation times by considering a fairly large polyatomic molecule with several modes of vibrational motion. The approximate relaxation times in a gas at standard conditions are shown in Fig. 7-5. The number of collisions for each transfer process can be obtained relative to the single collision in the $T \longrightarrow T$ process. It should be emphasized that any of the τ in Fig. 7-5 may overlap with any other τ and many exceptions can be found to this simple scheme.

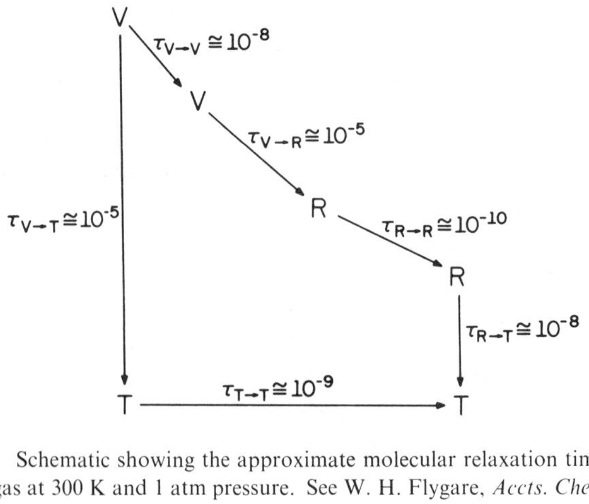

Figure 7-5 Schematic showing the approximate molecular relaxation times in seconds in a gas at 300 K and 1 atm pressure. See W. H. Flygare, *Accts. Chem. Res.* **1**, 121 (1968) for further details.

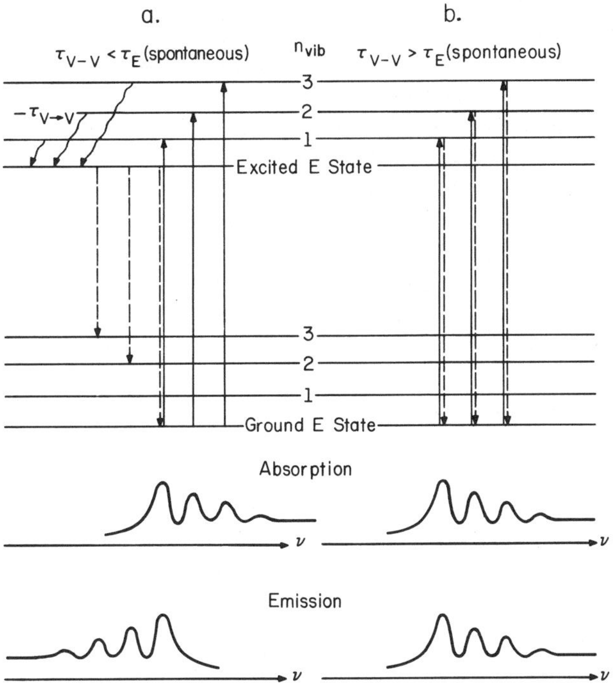

Figure 7-6 Diagram showing the effects of τ_{V-V} in the excited electronic state by observing the absorption (solid lines) and fluorescence (dashed lines) spectra as a function of density (pressure) in the gas phase. The higher density spectra are shown in (a) and the low-pressure limit is shown in (b). In the higher-pressure result in (a), the vibrational relaxation takes place before fluorescence and in (b) the system fluoresces before suffering any vibrational relaxation. $\tau_E = 1/A_E$ where A_E is the rate of spontaneous emission written for electric dipole transitions in Eq.(3-191).

Vibrational relaxation times can be measured by the observation of the decay of vibrational *fluorescence* (spontaneous emission) from an excited vibrational state. Fluorescence is the spontaneous emission from an excited state before the system can relax by nonradiative processes. Vibrational heating can be produced by direct excitation with a laser [14] or by an adiabatic expansion of a gas where the translational and rotational levels reach lower temperatures faster ($\tau_R \cong \tau_T < \tau_V$) than the vibrational levels. A variety of other experiments have been used to measure the vibrational relaxation times that include the frequency dependence of the speed of sound (see Problem 1-10) and the optical-acoustic effect (see Problem 7-3).

Fluorescence is easier to observe in optical (electronic) transitions because the rate of spontaneous emission is proportional to ν^3 as shown in Section 3.7 [τ_E (spontaneous) $\cong 2 \times 10^{-8}$ s and τ_V (spontaneous) $\cong 2 \times 10^{-2}$ s]. Some typical absorption and reemission fluorescence spectra are shown in Fig. 7-6 [16]. Figure 7-6(a) shows a typical liquid fluorescence spectrum where $V \longrightarrow T$ relaxation is very fast indicating that there is enough time after excitation and before fluorescence for the system to relax vibrationally in the excited electronic state. Figure 7-6(b) shows a possible gas phase spectrum where the intermolecular collision rate has decreased, which decreases the $V \longrightarrow T$ process. The result is that the system can now fluoresce before vibrational relaxation occurs, giving a fluorescence spectrum that matches the absorption spectrum. The polarization of the reemitted fluorescent spectra (relative to the incident absorbed radiation) can also be studied to obtain information on the rotational motion of large molecules in solution [17].

The intensity of the fluorescent radiation reemitted in the low-pressure limit [Fig. 7-6(b)] is normally considerably less than the intensity of the exciting radiation. For instance, only about 30% of the absorbed photons are reemitted (in the low-pressure limit) in benzene (2600 Å benzene emission spectra excited by a mercury atom lamp at 2537 Å) [18]. A good deal of the remaining energy is observed at a lower energy reemission (3000–4000 Å) with a smaller rate constant. This delayed fluorescence with longer decay times for spontaneous emission [τ_E (spontaneous) $\cong 10^{-3}$s] is called *phosphorescence* [19]. Phosphorescence normally occurs from a triplet state where the triplet-singlet transition probability is three to four orders of magnitude less than the electric dipole allowed singlet-singlet transitions. The energy levels involved are shown in Fig. 7-7. ${}^1k_E$ is the fluorescence rate of reemission and ${}^3k_E$ is the phosphorescence rate. k_{IS} is the *intersystem crossing rate.*

Decay of the excited state by radiationless intersystem crossing processes can also be first order. The model used to explain intersystem radiationless relaxations involves the excited state, which is near-degenerate with a near continuum of states [20]. First-order time-dependent perturbation theory then leads to a first-order radiationless decay from the single state into the continuum in a manner analogous to spontaneous emission. The nature of the continuum of states in a simple molecule is not clear but they probably involve complex vibrational motions and possibly a

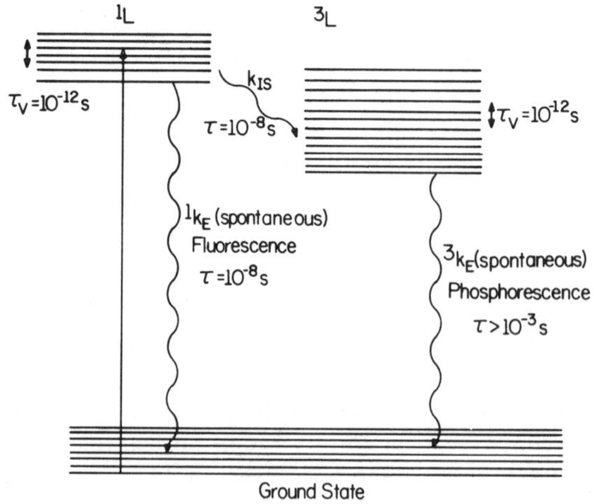

Figure 7-7 Diagram showing fluorescence, phosphorescence, and intersystem crossing in electronic states ${}^1L$ (singlet) and ${}^3L$ (triplet). k_{IS} is the rate of intersystem crossing. k_E is the rate of relaxation due to fluorescence from the singlet state with a similar definition for ${}^3k_E$ in the triplet state. Typical liquid state inverse rates or relaxation times in aromatic molecules are shown here from J. B. Birks, *Photophysics of Aromatic Molecules* (Wiley-Interscience, New York, 1970).

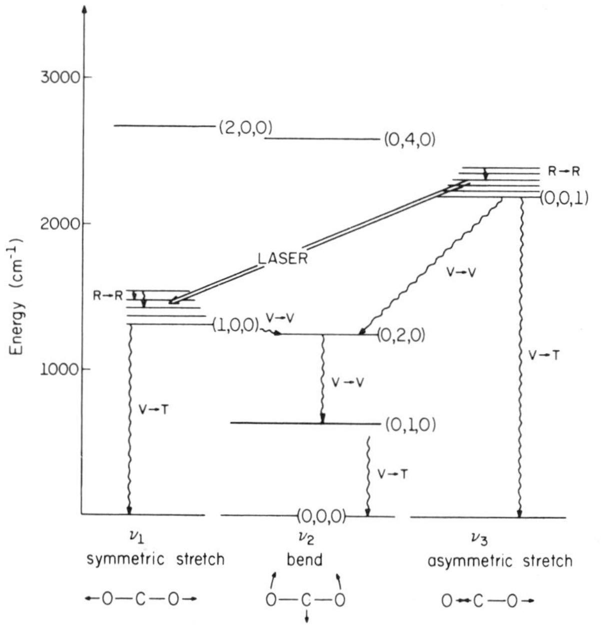

Figure 7-8 The vibrational energy levels of the $^{16}OC^{16}O$ molecule showing various paths of relaxation and the vibrational $(0, 0, 1) \longrightarrow (1, 0, 0)$ transitions that give rise to sustained laser oscillation. Only the even *J*-, *P*-, and *R*-branch rotation-vibration transitions are allowed according to nuclear spin statistics (see Problem 4-38d).

breakdown of the separation of electronic and nuclear motion (the Born–Oppenheimer approximation as described in Section 6.1). Similar first-order radiationless relaxation processes are observed for molecules in condensed phases. In this case, the continuum of states is provided by the lattice or condensed-phase vibrational-electronic states.

One of the more dramatic applications of the interplay of collisional relaxation processes, spontaneous emission, and induced emission is the operation of the maser or laser [21]. Figure 7-8 shows the vibrational energy levels of CO_2 and various paths of molecular relaxation. Each vibrational state has a series of associated rotational states as shown schematically in the (1, 0, 0) and (0, 0, 1) levels (see Problem 4-38d). The $R \longrightarrow R$ relaxation time is very much faster than the various $V \longrightarrow V$ or $V \longrightarrow T$ processes. In the CO_2 vibrational energy system, laser oscillations are observed between the $(0, 0, 1) \longrightarrow (1, 0, 0)$ vibration-rotation energy levels by producing an excess of population in the upper levels. The laser operates by subjecting a mixture of gaseous N_2 and CO_2 to a continuous electrical discharge. The electrons in the discharge excite N_2 to the first excited vibrational state and the excited N_2 molecules can then collide with CO_2 in the (0, 0, 0) vibrational state producing CO_2 in the (0, 0, 1) excited vibrational state as shown by the reverse of Eq.(7-106b). If the rate of N_2 $(v = 1)$ excitation of CO_2 to the (0, 0, 1) state is faster than the (0, 0, 1) state can relax by collisions, an excess population will occur. The (0, 0, 1) state in CO_2 is relatively isolated as shown in Fig. 7-8 and relaxation out of this state is slow. If a single photon is emitted by spontaneous emission from (0, 0, 1) to (1, 0, 0), further (0, 0, 1) $\longrightarrow$ (1, 0, 0) transitions will be stimulated giving rise to a gain of electromagnetic energy provided that the number of molecules in the (0, 0, 1) state exceeds the number in the (1, 0, 0) state. If the electromagnetic energy is contained in an optical cavity (mirrors separated by an integral wavelength of the radiation), the electromagnetic waves with a single wavelength will be selected with an efficiency related to the quality of the reflectors (mirrors) in the optical cavity (see Section 7-7). If sufficient gain via stimulated emission is achieved to offset the losses at the mirrors, laser or coherent oscillation will occur. Fast relaxation out of the (1, 0, 0) state will maintain the unbalance of population and increase the gain in power during the oscillation. The (1, 0, 0) state probably relaxes according to

$$\mathrm{OCO}(1,0,0) + \mathrm{OCO}(0,0,0) \xrightarrow{k_1} \mathrm{OCO}(0,2,0) + \mathrm{OCO}(0,0,0) + KE$$

$$\xrightarrow{k_2} 2\mathrm{OCO}(0,1,0) + KE$$

$$\xrightarrow{k_3} 2\mathrm{OCO}(0,0,0) + KE, \qquad (7\text{-}107)$$

where k_3 is the rate-determining step. The rates in this relaxation process can apparently be increased by adding He. Thus, we see the several processes in interplay. A metastable overpopulation of the (0, 0, 1) state is produced by selective excitation. Relaxation from the (0, 0, 1) state is relatively slow. Spontaneous emission starts the process of cascading stimulated emission of the (0, 0, 1) $\longrightarrow$ (1, 0, 0) rotational-vibrational transitions and relaxation from the (1, 0, 0) state is relatively fast. The optical cavity selects a single wavelength and therefore produces a coherent oscillation at one of the vibration-rotation components of the (0, 0, 1) $\longrightarrow$ (1, 0, 0) transition.

The energy levels of all triatomic molecules are similar to the diagram in Fig. 7-8 (see Fig. 4-34). Thus, any of the molecules listed in Table 4-28 might be appropriate for laser oscillation [22]. CO_2 is apparently one of the most powerful infrared lasers. The failure of other triatomics to outperform CO_2 could be for any or all of the following reasons:

1. The inability to produce an overpopulation of an upper state j relative to a lower state k.
   (a) The jth state may relax too rapidly.
   (b) The kth state may relax too slowly.
2. Too small probability of spontaneous or induced transitions between the upper and lower states (P_{jk} and A_{jk} too small).

The CO_2 laser can be used effectively to excite the vibrational energy levels of other molecules. After excitation, the molecule will vibrationally fluoresce and the observation of the decay in the vibrational fluorescence can lead to a measurement of the vibrational relaxation times. The vibrational excitation and fluorescence of the SF_6 molecule is shown in Fig. 7-9. Both the $\nu_3/c = 940\ \text{cm}^{-1}$ and $\nu_6/c = 363\ \text{cm}^{-1}$ vibrational modes are infrared active in SF_6. The $CO_2(1,0,0) \longrightarrow (0,0,1)$, P_{18}-P_{22} (see Problem 4-38d) transitions will excite the ν_3 mode in SF_6 as shown in Fig. 7-9. It is convenient to pulse the CO_2 laser so the excitation can be cut off, and then

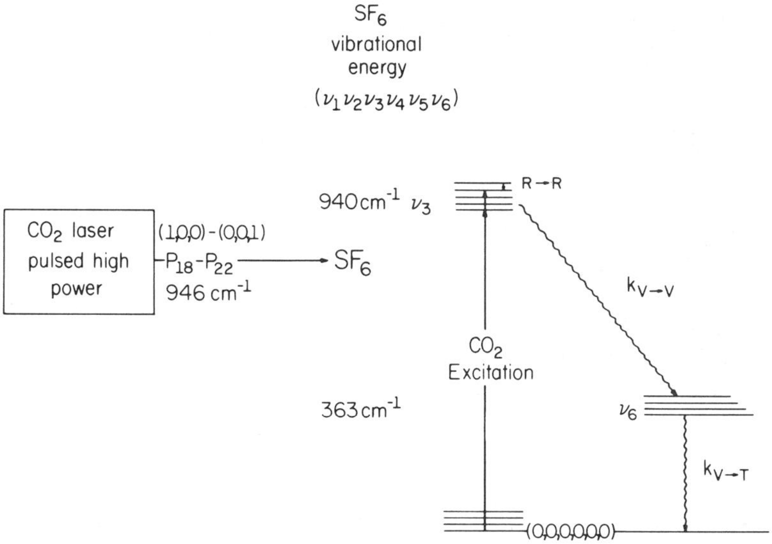

Figure 7-9 Diagram showing the $CO_2[(1, 0, 0) \longrightarrow (0, 0, 1)]$, P_{18}-P_{22} laser excitation of the ν_3 vibrational mode in SF_6. Vibrational fluorescence in SF_6 can be observed at both the ν_3 and ν_6 transitions. By pulsing the CO_2 laser, short periods of high excitation energy are used to excite SF_6. The subsequent time-dependence of the ν_3 and ν_6 fluorescence gives information on the k_{R-R}, k_{V-V}, and k_{V-T} rates. Details can be found in J. L. Steinfeld, *et al.*, *J. Chem. Phys.* **52**, 5421 (1970).

vibrational fluorescence in SF_6 can be observed. The time-dependence of the intensity of vibrational fluorescence from ν_3 in SF_6 will give information about $V \longrightarrow T$ and $V \longrightarrow V$ from that level. Vibrational fluorescence can also be observed from ν_6, which indicates an appreciable $V \longrightarrow V$ transfer rate between ν_3 and ν_6. The other modes of motion ($\nu_1 = 775\ \text{cm}^{-1}$, $\nu_2 = 644\ \text{cm}^{-1}$, $\nu_4 = 615\ \text{cm}^{-1}$, and $\nu_5 = 524\ \text{cm}^{-1}$) in combinations and overtones may provide many alternate routes to $\nu_6 \longleftrightarrow \nu_3$, $V \longrightarrow V$ relaxation. Subsequent monitoring of the time-dependence of the ν_6 fluorescence gives information on the $V \longrightarrow V$, $\nu_3 \longrightarrow \nu_6$ rate as well as $V \longrightarrow T$ relaxation from ν_6.

Another interesting example is the CO_2 laser excitation of the ν_3 C—F stretching vibrational mode in CH_3F near 1000 cm^{-1} and then the subsequent observation of vibrational fluorescence at a higher energy near 3000 cm^{-1} [23]. Apparently the CO_2 laser excites the $n_3 = 1$ state of ν_3 and then very fast $V \longleftrightarrow V$ relaxation within the ν_3 mode of vibration leads to an increased population of $n_3 = 3$, which is near-resonant with the $n = 1$ states of the ν_1 and ν_4 modes in CH_3F. Rapid $n_3 = 3$ exchange with $n_1 = 0$ and $n_4 = 0$ lead to $n_1 = 1$ or $n_4 = 1$ excitation and subsequent fluorescence from these final states.

Under the influence of very high laser power, multiphoton infrared absorption occurs that leads to chemical dissociation of the molecule [24]. Apparently 20–30 infrared photons can be absorbed by SF_6 (see Fig. 7-9) leading to the dissociation of the S—F bond in the molecule. The possibilities of laser-induced specific chemical decomposition as applied to isotope separation and chemical synthesis are evident.

7.5 DOUBLE-RESONANCE SATURATION EFFECTS

We shall now examine some double-resonance techniques where a low-power probe frequency is used to observe the frequency dependence of an absorption coefficient for a transition that is under the influence of fixed frequency high-power radiation (on or near resonance). The most interesting example for these effects is in the case of a Doppler or other inhomogeneously broadened transition. Consider first a two-level $k \longrightarrow j$ transition probed by low-power radiation at frequency ν and radiation electric field E_ν. If the two-level transition is also under the influence of high-power radiation at fixed frequency ν' with radiation electric field $E_{\nu'}$ and if $E_{\nu'} \gg E_\nu$, then the absorption coefficient is written [Eq.(7-94)] by

$$\gamma(\nu, \nu')_v = -\frac{8\pi^2 \nu_{jk}}{cE_\nu} P_i(\nu, \nu')_v, \tag{7-108}$$

where the subscript v indicates molecules moving with velocity v. Using the steady-state value of $P_i(\nu, \nu')_v$ from Eqs.(7-102) when $T_1 = T_2 = \tau$, we write

$$P_i(\nu, \nu')_v = -\left(\frac{E_\nu D_{jk}^2}{4\pi^2\hbar\tau}\right)\frac{\Delta N(\nu_v')}{f_v}, \qquad f_v = \left[\nu_{jk} - \nu\left(1 + \frac{v}{c}\right)\right]^2 + \left(\frac{1}{2\pi\tau}\right)^2,$$

$$\Delta N(\nu_v') = \frac{\Delta N_0 f_v'}{f_v' + B^2}, \qquad B = \frac{\xi}{2\pi} = \frac{E_{\nu'} D_{jk}}{h}, \qquad f_v' = \left[\nu_{jk} - \nu'\left(1 + \frac{v}{c}\right)\right]^2 + \left(\frac{1}{2\pi\tau}\right)^2, \tag{7-109}$$

where $\Delta N(\nu_v')$ is the population difference under the influence of the saturating (high-power) radiation at frequency ν'. Substituting gives

$$\gamma(\nu, \nu')_v = \frac{2\nu_{jk} D_{jk}^2 \Delta N_0}{c\hbar\tau}\left[\frac{f_v'}{f_v(f_v' + B^2)}\right] = \frac{2\nu_{jk} D_{jk}^2 \Delta N_0}{c\hbar\tau}\left[\frac{1}{f_v} - \frac{B^2}{f_v(f_v' + B^2)}\right]. \tag{7-110}$$

We now add the velocity averaging to the absorption coefficient in Eq.(7-110) where we use ν'' as the variable of integration that leads to $v = (c/\nu)\nu''$ and $dv = (c/\nu)\, d\nu''$ from Eqs.(7-26) and (7-29). We also need $\nu'(v/c)$ in f_v', which is easily found to be $\nu'(v/c) = \nu'\nu''/\nu$. The absorption coefficient is given by

$$\gamma(\nu, \nu') = A \int_{-\infty}^{+\infty} \exp\left[-\frac{\ln 2}{(\Delta\nu_G)^2}(\nu'')^2\right]\left\{\frac{1}{\Delta\nu_L^2 + (\nu_{jk} - \nu - \nu'')^2}\right.$$
$$\left. - \frac{B^2}{[\Delta\nu_L^2 + (\nu_{jk} - \nu - \nu'')^2][\Delta\nu_L^2 + (\nu_{jk} - \nu' - \nu'\nu''/\nu)^2 + B^2]}\right\} d\nu''$$

$$A = \frac{2D_{jk}^2}{\hbar\tau}\left(\frac{M}{2\pi kT}\right)^{1/2}. \tag{7-111}$$

This result is a Gaussian convoluted Lorentzian minus a Gaussian convoluted product of Lorentzians that leads to a dip in the absorption curve as $\nu \longrightarrow \nu'$. Several examples are shown in Fig. 7-10.

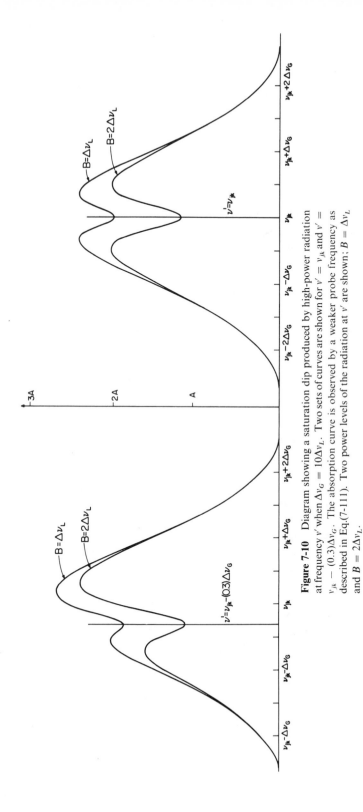

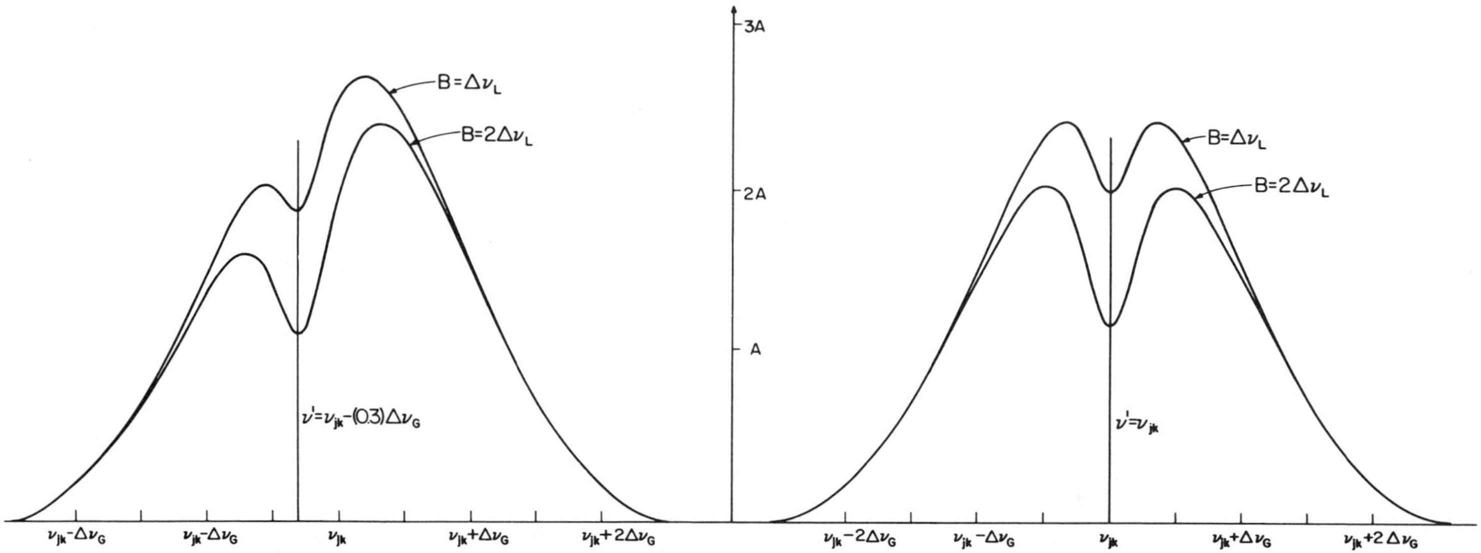

Figure 7-10 Diagram showing a saturation dip produced by high-power radiation at frequency ν' when $\Delta\nu_G = 10\Delta\nu_L$. Two sets of curves are shown for $\nu' = \nu_{jk}$ and $\nu' = \nu_{jk} - (0.3)\Delta\nu_G$. The absorption curve is observed by a weaker probe frequency as described in Eq.(7-111). Two power levels of the radiation at ν' are shown; $B = \Delta\nu_L$ and $B = 2\Delta\nu_L$.

A somewhat different result is obtained by using a single radiation source and reflecting the transmitted radiation back again through the sample, thereby setting up a standing wave in the absorption region. As the radiation approaches saturation levels, saturation dips can be observed at the resonant frequency in a Doppler broadened curve. Consider the absorption of the electromagnetic energy at ν_a, which is slightly off-resonance. As the radiation passes down an absorption cell, all molecules with velocity v_a moving in the same direction as the radiation will absorb the radiation. If the radiation is now reflected and returned in the opposite direction, the molecules that have a velocity component $-v_a$ with respect to the original direction of the incident radiation will now absorb the radiation on the return path. Thus, two *different* sets of molecules ($+v_a$ and $-v_a$) will absorb the radiation during a single reflected passage through the cell. At exact resonance, however, the *same* set of molecules with velocity component $v = 0$ will absorb the radiation on both the forward and reflected passage of the radiation. Under high-power conditions, this double saturation at exact resonance ($v = 0$) leads to a dip in the Gaussian dominated absorption curve at each resonant frequency. This dip will have the line shape characteristics of the molecular relaxation time and in the case where $\Delta\nu_G \gg \Delta\nu_L$, this technique will lead to considerably higher resolution and accuracy by enabling the measurement of ν_0 with a narrower line shape. These concepts have been effectively applied to reduce the Doppler line widths in rotational, vibrational, and electronic transitions to their corresponding narrower relaxation widths. Gas lasers have been very effective in providing the necessary power for these saturation or Lamb dip [25] experiments in the infrared and visible spectral regions. The general technique of setting up standing wave boundary conditions in an atomic or molecular laser system or in a resonant system during saturation is very important in understanding laser systems, multiphoton nonlinear effects, and self-induced transparency effects [21].

In summary, the *traveling wave* double-resonance experiment described in Fig. 7-10 led to a saturation dip (in the Doppler line shape) determined by the frequency of the high-power radiation. In the single frequency *standing wave* reflection experiment, however, a saturation dip (Lamb dip) is observed at the center of the inhomogeneously broadened curve. The case of several transitions within the Doppler line width (or other inhomogeneously broadened line width) cannot be analyzed with the simple theories given here as a sum of two-level effects because of nonlinear mixing of the frequencies between the various transitions. Further details are given elsewhere [21].

Consider now some nuclear magnetic double-resonance effects in the liquid or solid state where there is no Doppler effect. The Overhauser effect [26] results from a selective relaxation mechanism that enables the electron spin Boltzmann distribution to be transferred to the nuclear magnetic resonance energy levels. Thus, the Overhauser effect greatly enhances the intensity of the NMR transition. Figure 7-11 shows the energy levels in Fig. 4-12 for an electron spin ($S = \frac{1}{2}$) and a nuclear spin ($I = \frac{1}{2}$) system when the electron spin–nuclear spin coupling constant is zero ($a = 0$). The coupling constant would be zero if the free electron was rapidly exchanging in solution. Thus, we are considering a system where a is time-dependent on a scale that averages to zero over the period of our experimental observation. In the energy level diagram in Fig. 7-11, it is clear that the ESR transition frequencies are equal, $\nu_{24} = \nu_{13}$, and the NMR transition frequencies are also equal, $\nu_{43} = \nu_{21}$. We now saturate the

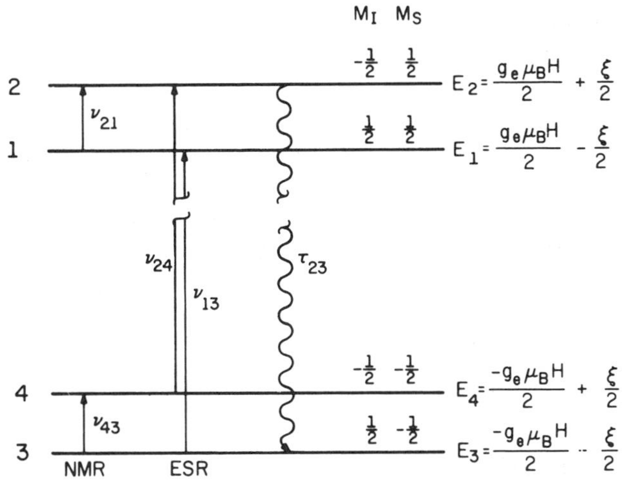

Figure 7-11 Energy level diagram showing the nuclear energy level population difference increase due to the Overhauser effect. The energy levels and notation are from Fig. 4-12 except that in the present case the electron spin–nuclear spin coupling constant, a, is zero. Note that E_1 and E_2 have reversed order in the two diagrams due to the difference in a. $\nu_{24} = \nu_{13}$ (ESR) is saturated leading through τ_{23} to enhanced $\nu_{21} = \nu_{43}$ (NMR) intensities.

ESR transitions with high power that leads to equal populations in the 1, 3 and 2, 4 pairs of levels. Now, in the presence of the time-dependent nuclear spin–electron spin coupling, there will be a relaxation between the 2 and 3 energy levels indicated by the wavy line in Fig. 7-11. The simultaneous $\Delta M_S = \mp 1$, $\Delta M_I = \pm 1$ selection rules are evident from the development in Sections 4.9 and 4.10. Thus, Boltzmann equilibrium at the ambient temperature is established between levels 2 and 3 in Fig. 7-11 given by

$$\frac{N_2}{N_3} = \exp\left\{-\frac{[g_e\mu_B H_z + g_i\mu_0(1-\sigma)H_z]}{kT}\right\}. \tag{7-112}$$

However, the saturating ESR radiation is equalizing the populations of the $M_I = -\frac{1}{2}$ and $M_I = +\frac{1}{2}$ pair of levels. Thus, according to the discussion above, the τ_{23} relaxation drives the $M_I = -\frac{1}{2}$ levels and $M_I = +\frac{1}{2}$ levels apart in population by the same factor as shown above in Eq. (7-112):

$$\frac{N_{M_I=-1/2}}{N_{M_I=+1/2}} = \exp\left\{-\frac{[g_e\mu_B H_z + g_i\mu_0(1-\sigma)H_z]}{kT}\right\}. \tag{7-113}$$

The result shows that the nuclear Zeeman energy level population differences have been enhanced by the electronic Zeeman factor and the NMR intensities are increased by a factor on the order of μ_B/μ_0. Many elegant experiments have arisen from these basic concepts including nuclear ($I_1 = \frac{1}{2}, I_2 = \frac{1}{2}$) Overhauser effects.

ENDOR (electron-nuclear-double-resonance) is another double-resonance technique that employs either steady-state or pulse circumstances. Under certain conditions involving electron and nuclear relaxation where there is no Overhauser effect, changes in the strongly irradiated ESR transition absorption can be observed as the NMR transition is swept through resonance leading to enhanced sensitivity and resolution in observing the NMR transitions [11].

Another important steady-state double-resonance technique used in nuclear magnetic resonance is *spin decoupling*. This technique is a great aid in unraveling complex spin-spin coupled spectra. By strongly irradiating the nuclear magnetic resonance of nucleus 1, ν_1^0, the spin-spin interactions observed in the NMR spectra of nucleus 2, ν_2^0, can be completely removed (see Problem 4-25).

7.6 TRANSIENT EFFECTS, FOURIER TRANSFORM SPECTROSCOPY, AND MULTIPLE PULSE EXPERIMENTS

In the previous sections we have outlined the theory necessary to describe the observations of the steady-state resonant interaction of classical electromagnetic radiation with the stationary states in nuclear, atomic, and molecular systems. The steady-state limit implies an observation time that is long relative to any relaxation times in the system being probed. We shall now examine transient behavior where we observe the absorption (or emission) of electromagnetic energy immediately after a sudden radiation-induced change in the equilibrium condition of the system. Transient phenomena are those time-dependent changes in the system which occur as the system relaxes to the new equilibrium condition which need not be the normal Boltzmann condition. It is evident that transient phenomena must be observed in periods short relative to the relaxation times of the systems involved. The steps in observing the transient behavior are first to prepare the system in an equilibrium condition, second to change the equilibrium condition in a period much shorter than the relaxation time, and third to observe the system relax to the new equilibrium (or steady-state) condition. Of course, for the experiments described here we require coherent radiation sources that can be tuned to atomic, molecular, or magnetic resonance frequencies and have sufficient power to drive a two-level system into a new equilibrium condition.

In nuclear magnetic resonance (NMR) in condensed phases, relaxation times are normally on the order of seconds. Therefore, it is a relatively simple experimental problem to observe transient effects on a time scale of less than one second. In rotational and vibrational spectroscopy, relaxation effects are normally dominated by rotational relaxation times, which are typically on the order of 10^{-6} s · Torr^{-1}. As sensitivity requirements force infrared and microwave observations to be made at pressures near 1 Torr, the corresponding transient effects must be observed in times short relative to 1 μs. This is a considerable experimental challenge and it is quite understandable why the observation of transient effects in the optical, infrared, and microwave spectra of gases has lagged so far behind the NMR development. In spite of the difference of 10^6 in time scales, many of the equations governing NMR relaxation processes can be carried over directly into the discussion of vibrational and rotational relaxation processes. We also keep in mind Eq.(7-103) showing the relation between the electric and magnetic dipole Bloch equations.

We now return to the solution of the coupled differential equations in $P_r(t)$, $P_i(t)$, and $\Delta N(t)$ as given in Eq.(7-101). Consider first the case of an experiment where

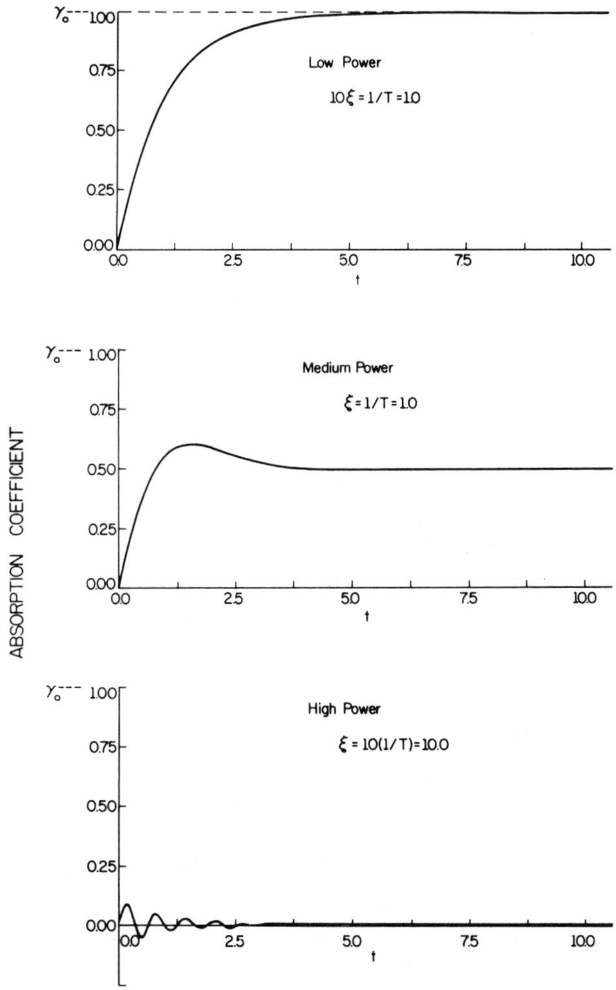

Figure 7-12 Schematic showing the transient effects in the absorption of radiation when radiation, which is initially far off-resonance, is suddenly brought into resonance with a two-level system. The absorption coefficient is from Eq.(7-126) when $T = T_2 = T_1 = 1$. γ_0 is the long-time limit of Eq.(7-126). Note the lower steady-state absorption coefficient ($t \longrightarrow \infty$) with increasing radiation power.

electromagnetic radiation that is initially far off-resonance is instantly (at $t = 0$) brought into resonance with a two-level system. The observed absorption coefficient is shown in Fig. 7-12 as $t > 0$ for three different values of the power factor, ξ. Ignoring the Doppler effect or other inhomogeneous processes for simplicity, we see that at low power the absorption coefficient approaches exponentially the steady-state value of γ_0. At higher powers, oscillations occur in the curve and γ finally approaches, in steady state, values considerably below γ_0.

The time-dependence in the absorption coefficients in these diagrams (Fig. 7-12) are described by the solution of $P_i(t)$ in Eqs.(7-101) and subsequent substitution into the corresponding absorption coefficient in Eq.(7-94). We solve Eqs.(7-101) by the method of Laplace transforms. Taking the Laplace transforms of Eqs.(7-101) we have

$$p\bar{P}_r - P_r(0) = -\Delta\omega\bar{P}_i - \frac{\bar{P}_r}{T_2}$$

$$p\bar{P}_i - P_i(0) = \Delta\omega\bar{P}_r - \frac{D_{jk}^2 E_0}{\hbar}\overline{\Delta N} - \frac{\bar{P}_i}{T_2}$$

$$p\overline{\Delta N} - \Delta N(0) = \frac{E_0}{\hbar}\bar{P}_i - \frac{\overline{\Delta N}}{T_1} + \frac{\Delta N_0}{pT_1}, \tag{7-114}$$

where p is the Laplace transform variable and $\bar{P}_r$, $\bar{P}_i$, and $\overline{\Delta N}$ are the Laplace transforms of $P_r(t)$, $P_i(t)$, and $\Delta N(t)$, respectively. $P_r(0)$, $P_i(0)$, and $\Delta N(0)$ are the initial values at $t = 0$, and ΔN_0 is the thermal equilibrium value. In the experiments described above and in Fig. 7-12, the molecular energy levels are initially far off-resonance for a period of time long relative to the relaxation times and, therefore, according to Eqs.(7-102), we can write $P_r(0) = P_i(0) = 0$ and $\Delta N(0) = \Delta N_0$. Substituting these values into Eq.(7-114) and solving for $\bar{P}_i$ gives

$$\bar{P}_i = \frac{-E_0 D_{jk}^2 \dfrac{\Delta N_0}{\hbar p}}{\left\{\left[p + \dfrac{1}{T_2}\right] + (\Delta\omega)^2\left[\dfrac{1}{p + (1/T_2)}\right] + \xi^2\left[\dfrac{1}{p + (1/T_1)}\right]\right\}}, \tag{7-115}$$

where $\xi = E_0 D_{jk}/\hbar$ is defined in Eqs.(7-102). This equation for $\bar{P}_i$ does not have a closed form for the reverse Laplace transform. If $T_1 = T_2 = \tau$ and $\Delta\omega \neq 0$ or if $T_1 \neq T_2$ and $\Delta\omega = 0$, however, we can find a convenient reverse Laplace transform. The $T_1 = T_2 = \tau$, $\Delta\omega \neq 0$ condition is very appropriate for many situations of physical interest. Using this condition gives

$$\bar{P}_i = -\frac{E_0 D_{jk}^2 \Delta N_0}{\hbar p}\left\{\frac{p + (1/\tau)}{[p + (1/\tau)]^2 + \Omega^2}\right\}, \qquad \Omega^2 = (\Delta\omega)^2 + \xi^2, \tag{7-116}$$

which can be rewritten to give

$$\bar{P}_i = -\frac{E_0 D_{jk}^2 \Delta N_0}{\hbar\tau}\left[\frac{1}{(1/\tau)^2 + \Omega^2}\right]\left\{\frac{1}{p} - \frac{p + (1/\tau)}{[p + (1/\tau)]^2 + \Omega^2} + \frac{\Omega^2\tau}{[p + (1/\tau)]^2 + \Omega^2}\right\}. \tag{7-117}$$

This expression has the following reverse Laplace transform:

$$P_i(t) = -\frac{E_0 D_{jk}^2 \Delta N_0}{\hbar\tau}\left[\frac{1}{(1/\tau)^2 + \Omega^2}\right]\left[1 - \exp\left(-\frac{t}{\tau}\right)(\cos\Omega t - \Omega\tau\sin\Omega t)\right]. \tag{7-118}$$

Substituting into Eq.(7-94) gives the absorption coefficient, which is an exponentially damped sinusoid. The long-time limit of Eq.(7-118) reduces to the steady-state result given in Eqs.(7-102) when $T_1 = T_2 = \tau$.

The general method of solving Eqs.(7-114) is to express $\bar{P}_r(p)$, $\bar{P}_i(p)$, and $\overline{\Delta N}(p)$ in terms of constants and polynomials in the Laplace transform variable p giving [27,28]

$$\bar{f}(p) = \frac{g(p)}{ph(p)}, \tag{7-119}$$

where $\bar{f}(p)$ is $\bar{P}_r(p)$, $\bar{P}_i(p)$, or $\overline{\Delta N}(p)$. $g(p)$ is a cubic in p that is different for each of the $\bar{P}_r$, $\bar{P}_i$, and $\overline{\Delta N}$ functions. $h(p)$ has at least one real negative root. Calling this root a, we can write

$$h(p) = \left(p + \frac{1}{a}\right)\left[\left(p + \frac{1}{b}\right)^2 + \Omega^2\right]. \tag{7-120}$$

Substituting this expression into $\bar{f}(p)$ above and expanding in partial fractions gives

$$\bar{f}(p) = \frac{A}{p + (1/a)} + \frac{B[p + (1/b)]}{[p + (1/b)]^2 + \Omega^2} + \frac{C}{[p + (1/b)]^2 + \Omega^2} + \frac{D}{p}, \tag{7-121}$$

and the corresponding reverse Laplace transform gives

$$f(t) = A\exp\left(-\frac{t}{a}\right) + B\cos\Omega t\exp\left(-\frac{t}{b}\right) + \frac{C}{\Omega}\sin\Omega t\exp\left(-\frac{t}{b}\right) + D,$$

$$\Omega^2 = (\Delta\omega)^2 + \xi^2. \tag{7-122}$$

We followed a similar procedure previously leading to the specific result in Eq.(7-118). The A, B, C, and D coefficients in Eq.(7-122) are found for the particular experimental initial and working conditions. For instance, if $T_1 = T_2 = \tau$, $a = \tau = b$, the result in Eq.(7-118) is easily obtained for the appropriate initial conditions. If $T_1 \neq T_2$, then at resonance ($\Delta\omega = 0$) an exact solution can also be obtained for $P_r(t)$, $P_i(t)$, and $\Delta N(t)$ by using Eq.(7-122). The solutions are

$$P_r(t) = P_r(t_i)\exp\left[-\frac{(t-t_i)}{T_2}\right]$$

$$P_i(t) = \frac{D_{jk}\xi\Delta N_0}{\Omega_p^2}\left(\exp\left[-\frac{(t-t_i)}{T}\right]\left\{\left(\frac{1}{T_2}\right)\cos\Omega_0(t-t_i)\right.\right.$$
$$\left.\left. - \left[\left(\frac{T_1}{T_2}\right)\frac{\xi^2}{\Omega_0} + \left(\frac{1}{T_2}\right)\left(\frac{\varphi}{\Omega_0}\right)\right]\sin\Omega_0(t-t_i)\right\} - \frac{1}{T_2}\right)$$
$$+ \exp\left[-\frac{(t-t_i)}{T}\right]\left\{P_i(t_i)\left[\cos\Omega_0(t-t_i) - \frac{\varphi}{\Omega_0}\sin\Omega_0(t-t_i)\right]\right.$$
$$\left. - \frac{D_{jk}\xi}{\Omega_0}[\Delta N(t_i) - \Delta N_0]\sin\Omega_0(t-t_i)\right\}$$

$$\Delta N(t) = \frac{\Delta N_0}{\Omega_p^2}\left\{\left(\frac{T_1}{T_2}\right)\xi^2\exp\left[-\frac{(t-t_i)}{T}\right]\right.$$
$$\left.\times\left[\cos\Omega_0(t-t_i) + \left(\frac{1}{\Omega_0 T}\right)\sin\Omega_0(t-t_i) + \left(\frac{1}{T_2}\right)^2\right]\right\}$$
$$+ \exp\left[-\frac{(t-t_i)}{T}\right]\left\{P_i(t_i)\frac{\xi}{D_{jk}\Omega_0}\sin\Omega_0(t-t_i)\right.$$
$$\left. + [\Delta N(t_i) - \Delta N_0]\left[\cos\Omega_0(t-t_i) + \frac{\varphi}{\Omega_0}\sin\Omega_0(t-t_i)\right]\right\}$$

$$\frac{1}{T} = \frac{1}{2}\left(\frac{1}{T_1} + \frac{1}{T_2}\right), \qquad \varphi = \frac{1}{2}\left(\frac{1}{T_2} - \frac{1}{T_1}\right),$$

$$\Omega_0 = (\xi^2 - \varphi^2)^{1/2}, \qquad \Omega_p = \left[\left(\frac{1}{T_2}\right)^2 + \frac{T_1}{T_2}\xi^2\right]^{1/2}, \tag{7-123}$$

where $P_r(t_i)$, $P_i(t_i)$, and $\Delta N(t_i)$ are the initial conditions at t_i and $\xi = E_0 D_{jk}/\hbar$ is defined in Eq.(7-102). Now, if the initial conditions are created far off-resonance, $P_i(t_i) = P_r(t_i) = 0$ and $\Delta N(t_i) = \Delta N_0$. Using this and the condition

$$\frac{1}{2}\left(\frac{1}{T_2} - \frac{1}{T_1}\right) \ll \xi, \tag{7-124}$$

which is easy to achieve as $1/T_1 \cong 1/T_2$ in most cases, leads to (let $t_i = 0$)

$$P_r(t) = 0$$

$$P_i(t) = \frac{\hbar\xi^2\Delta N_0}{E_0}$$
$$\times\left(\frac{\exp(-t/T)\{(1/T_2)\cos\xi t - [(T_1/T_2)\xi + (1/T_2)(\varphi/\xi)]\sin\xi t\} - 1/T_2}{\Omega_p^2}\right)$$

$$\Delta N(t) = \Delta N_0\left\{\frac{(T_1/T_2)\xi^2\exp(-t/T)[\cos\xi t + (1/\xi T)\sin\xi t] + (1/T_2)^2}{\Omega_p^2}\right\}. \tag{7-125}$$

Substituting $P_i(t)$ from this equation into Eq.(7-94) gives the absorption coefficient, which describes the on-resonant experiment in Fig. 7-12:

$$\gamma(\omega, t) = -\frac{4\pi\omega_{jk}(N_k - N_j)D_{jk}^2}{c\hbar} \times \left(\frac{\exp(-t/T)\{(1/T_2)\cos\xi t - [(T_1/T_2)\xi + (1/T_2)(\varphi/\xi)]\sin\xi t\} - 1/T_2}{\Omega_p^2}\right). \tag{7-126}$$

The long-time limit of this expression agrees with the result in Eq.(7-104) and defines γ_0 in Fig. 7-12. The low-power result predicts an exponential rise in γ as shown in the $\xi < (1/T)$ response curve in Fig. 7-12. As the power of the radiation increases, the sinusoidal effects occur according to Eq.(7-126). The phenomena depicted in Fig. 7-12 and Eq.(7-126), which are sometimes called *transient nutation* or *transient absorption*, have been observed in NMR, ESR, microwave, infrared, and optical spectroscopy. In the absence of the Doppler effect, Eq.(7-126) can be fit to the experimental data in Fig. 7-12 to extract the relaxation times T_1 and T_2.

We now return to Eqs.(7-101) to examine $P_r(t)$, $P_i(t)$, and $\Delta N(t)$ under short-pulse conditions where $t \ll T_1$ and $t \ll T_2$. The on-resonance ($\Delta\omega = 0$) solutions to Eqs.(7-101) when the terms involving T_1 and T_2 are dropped ($t \ll T_1, t \ll T_2$) can be obtained from Eqs.(7-123) when $T_1 = T_2 = \infty$ giving $1/\mathrm{T} = 0$, $\varphi = 0$, $\Omega_0 = \xi$, and $\Omega_p = \xi$, which leads to

$$P_r(t) = P_r(t_i)$$

$$P_i(t) = P_i(t_i)\cos\xi(t - t_i) - D_{jk}\,\Delta N(t_i)\sin\xi(t - t_i)$$

$$\Delta N(t) = D_{jk}\,P_i(t_i)\sin\xi(t - t_i) + \Delta N(t_i)\cos\xi(t - t_i). \tag{7-127}$$

The conditions for a $\pi/2$-pulse are obtained when

$$\xi(t - t_i) = \frac{\pi}{2}, \tag{7-128}$$

giving

$\pi/2$-pulse	$P_r(t_i) = P_i(t_i) = 0, \Delta N(t_i) = \Delta N_0$
$P_r(t) = P_r(t_i)$	$P_r(t) = 0$
$P_i(t) = -D_{jk}\,\Delta N(t_i)$	$P_i(t) = -D_{jk}\Delta N_0$
$\Delta N(t) = D_{jk}\,P_i(t_i)$	$\Delta N(t) = 0.$

(7-129)

At a later time we have the conditions for the π-pulse when

$$\xi(t - t_i) = \pi \tag{7-130}$$

giving

π-pulse	$P_r(t_i) = P_i(t_i) = 0,\ \Delta N(t_i) = \Delta N_0$
$P_r(t) = P_r(t_i)$	$P_r(t_i) = 0$
$P_i(t) = -P_i(t_i)$	$P_i(t) = 0$
$\Delta N(t) = -\Delta N(t_i)$	$\Delta N(t) = -\Delta N_0.$

(7-131)

The results above in Eqs.(7-129) and (7-131) for the $P_r(t_i) = P_i(t_i) = 0$ and $\Delta N(t_i) = \Delta N_0$ are appropriate for far off-resonance initial conditions. The results in Eq.(7-131) show that a π-pulse leaves P_r unchanged, changes the sign of P_i, and also changes the sign of $\Delta N(t)$.

We are now prepared to discuss another set of experiments, which are shown in Fig. 7-13. In these experiments the initial conditions are created on-resonance and the radiation is switched to off-resonance at $t = 0$. The system is polarized on-resonance at time $t < 0$ to final values of $P_i(0)$ and $P_r(0)$ at time $t = 0$ when the system is then switched to a far off-resonance condition where the radiation field no longer interacts with the two-level system. Assuming also that the field arising from the polarized molecules at $t > 0$ is weak and does not further interact appreciably with the molecules, all terms in Eq.(7-101) containing E_0 can be dropped, which leads to the following solutions for $P_i(t)$ and $P_r(t)$:

$$P_r(t) = \exp\left(-\frac{t}{T_2}\right)[P_r(0)\cos\Delta\omega t - P_i(0)\sin\Delta\omega t]$$

$$P_i(t) = \exp\left(-\frac{t}{T_2}\right)[P_i(0)\cos\Delta\omega t + P_r(0)\sin\Delta\omega t]$$

$$\Delta\omega = \omega_0 - \omega_{jk}, \tag{7-132}$$

where ω_0 is the off-resonant oscillator frequency and ω_{jk} is the resonant frequency. In order to predict the response at the detector we shall evaluate $P_i(t)$ and use the absorption coefficient in Eq.(7-94). We will now choose initial conditions given by Eqs.(7-125). Under conditions of high-power radiation where $\xi T_1 \gg 1$ and $\xi T_2 \gg 1$, Eqs.(7-125) reduce to

$$P_r(t) = 0$$

$$P_i(t) = -D_{jk}\,\Delta N_0 \exp\left(-\frac{t}{T}\right)\sin\xi t$$

$$\Delta N(t) = \Delta N_0 \exp\left(-\frac{t}{T}\right)\cos\xi t. \tag{7-133}$$

We note, therefore, that $P_r(0) = 0$ in Eq.(7-132) giving the following result for $P_i(t)$:

$$\lim_{\substack{\xi T_1 \gg 1 \\ \xi T_2 \gg 1}} P_i(t) = P_i(0)\exp\left(-\frac{t}{T_2}\right)\cos\Delta\omega t. \tag{7-134}$$

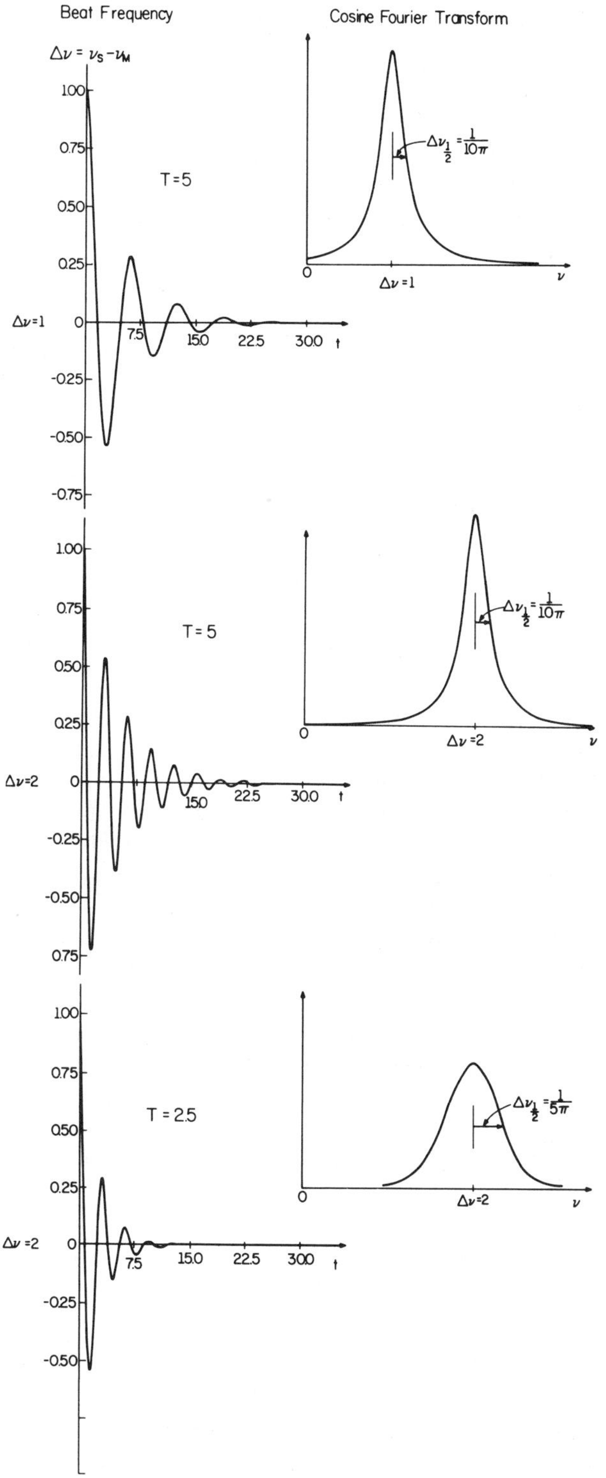

Figure 7-13 Schematic showing the transient effects observed during spontaneous coherent emission of radiation after an on-resonant interaction is instantly changed (at $t = 0$) to an off-resonant condition. The molecules undergo coherent emission at $t > 0$. The response curves are beats or the difference frequencies, $\Delta\nu$, between the emitting frequency and the frequency of the off-resonant oscillator. The response curves are described by Eq.(7-135). The corresponding cosine Fourier transforms are also shown.

Substituting this into Eq.(7-94) gives a function that is proportional to the measured time-dependent signal,

$$g(t) = -\frac{4\pi\omega_{jk}P_i(0)}{cE_0}\exp\left(-\frac{t}{T_2}\right)\cos\Delta\omega t. \tag{7-135}$$

The cosine Fourier transform of this function gives the Lorentzian spectrum of the transition centered at frequency $\Delta\omega = \omega_0 - \omega_{jk}$ (Fig. 7-13) with half width at half height given by $\Delta\omega_{1/2} = 1/T_2$,

$$\begin{aligned} f(\omega) &= \int_{-\infty}^{+\infty} g(t)\cos\omega t\, dt \\ &= -\frac{4\pi^2\omega_{jk}P_i(0)}{cE_0}\int_0^{\infty}\exp\left(-\frac{t}{T_2}\right)\cos\Delta\omega t\cos\omega t\, dt \\ &= -\frac{2\pi^2\omega_{jk}P_i(0)}{cE_0}\left[\left(\frac{1}{\pi}\right)\frac{(1/T_2)}{(1/T_2)^2 + (\Delta\omega - \omega)^2} + \left(\frac{1}{\pi}\right)\frac{(1/T_2)}{(1/T_2)^2 + (\Delta\omega + \omega)^2}\right]. \end{aligned} \tag{7-136}$$

If $\Delta\omega - \omega \cong 0$, the second term can be dropped giving the final result:

$$f(\omega) = -\frac{2\pi^2\omega_{jk}P_i(0)}{cE_0}\mathscr{L}(\Delta\omega - \omega), \qquad \Delta\omega = \omega_0 - \omega_{jk}. \tag{7-137}$$

Both the time-dependent responses and the Fourier transforms are shown in Fig. 7-13 for nonzero values of $P_i(0)$. Of course, the maximum values of $g(t)$ in Eq.(7-135) and $f(\omega)$ in Eq.(7-137) are obtained after a $\pi/2$-pulse as shown in Eqs.(7-129).

The experiment described here and in Fig. 7-13 can also be explained by observing that a polarized sample will commence at $t = 0$ to emit radiation coherently at the resonant frequency [28]. This coherent emission decays by collisional relaxation. The transient response is observed as a beat between the emitted radiation with the reference oscillator in a detector that responds to the difference frequency as shown in Fig. 7-13.

In the case of several possible near-resonant transitions that have all been polarized at $t < 0$, we can write the response after the polarizing radiation has been taken out of resonance by summing terms of the form shown in Eq.(7-135). This result and the Fourier transform gives

$$g(t) = -\sum_j \frac{4\pi\omega_j P_i(0)_j}{cE_z^0}\exp\left(-\frac{t}{T_j}\right)\cos\Delta\omega_j t$$

$$\begin{aligned} f(\omega) &= -\sum_j \frac{2\pi^2\omega_j P_i(0)_j}{cE_0}\left\{\frac{1}{\pi}\left[\frac{1/T_j}{(1/T_j)^2 + (\Delta\omega_j - \omega)^2}\right]\right\} \\ &= -\sum_j \frac{2\pi^2\omega_j P_i(0)_j}{cE_0}\mathscr{L}_j(\Delta\omega_j - \omega), \end{aligned} \tag{7-138}$$

where ω_j, $P_i(0)_j$, $\Delta\omega_j$, and T_j are the resonant frequency, imaginary polarization, frequency difference between the resonant frequency and reference oscillator, and

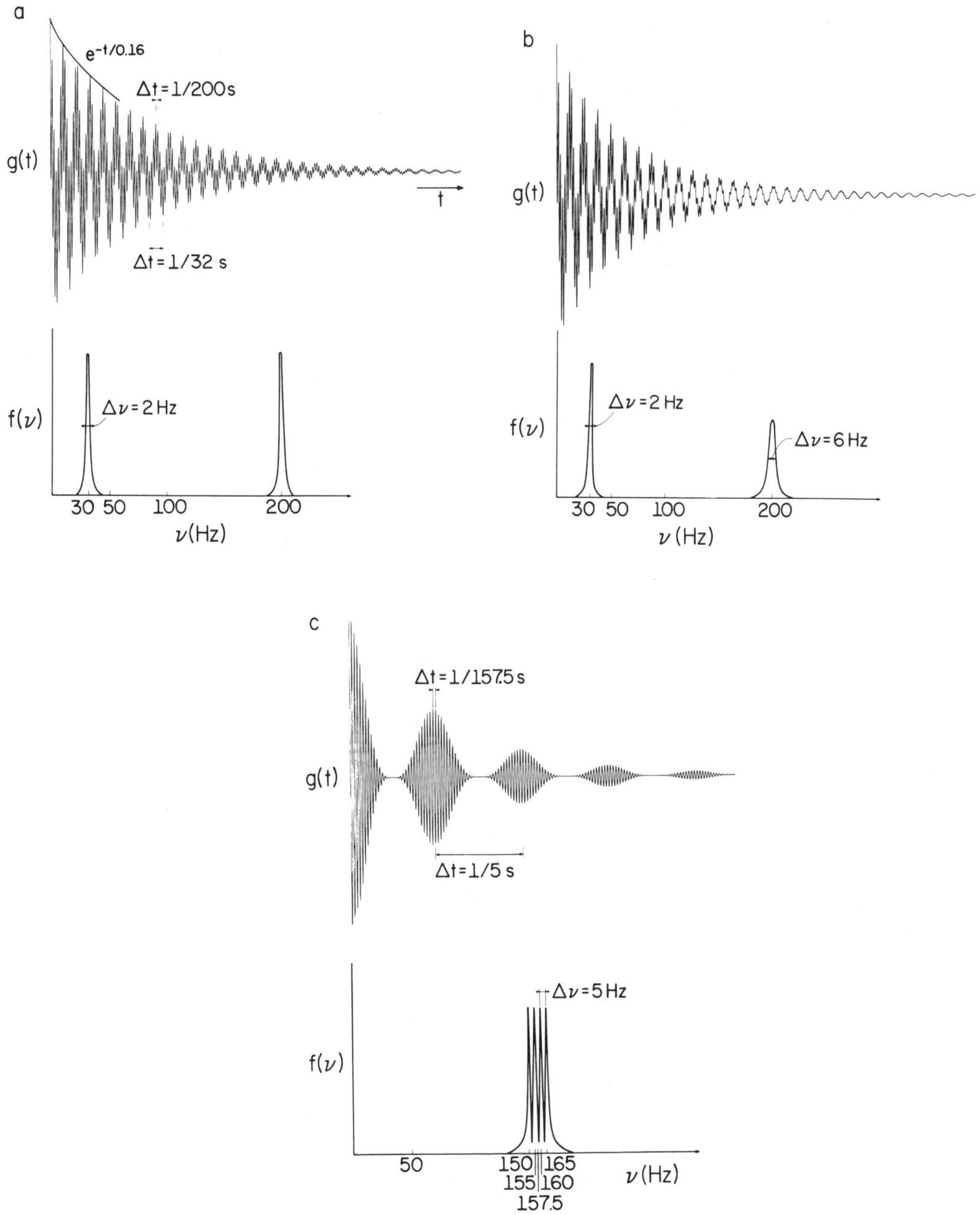

Figure 7-14 Three diagrams showing spontaneous coherent transient emission from two and four transitions, their beats with a reference oscillator, $g(t)$, and their cosine Fourier transforms, $f(\nu)$. a. $g(t)$ shows two signals as described by Eq.(7-138) where the relaxation times for both emitting transitions are the same. b. The same as (a) except that the higher-frequency signal has a shorter relaxation time, which is evident in $g(t)$ and the resulting $f(\nu)$ line widths. c. $g(t)$ and $f(\nu)$ signals for four closely spaced equal intensity, equal line width signals.

T_2 for the jth resonant transition, respectively. Several illustrations of multiple signals are shown in Fig. 7-14.

Substituting the maximum polarization $P_i(0)$ from the $\pi/2$-pulse condition from Eqs.(7-129) into Eq.(7-138) shows that equal $\pi/2$-pulse conditions for several transitions require equal values for the transition moment, D_{jk}. Thus, the spectrum obtained from the Fourier transform method will have amplitudes governed by $P_i(0)$ from Eq.(7-129), the nature of the pulse times, and relative transition moments.

All the equations above were derived by assuming negligible Doppler effects due to translational motion. If appreciable translational motion is present, the derivation above can be repeated for a single velocity, and this result must be averaged over the velocity distribution as shown in detail in Sections 7.2 and 7.5.

The $\pi/2$-pulse experiments described above are normally performed using a repetitive pulse train with pulses every three to five relaxation times. Pulse trains and pulse experiments are used extensively in nuclear magnetic resonance [29]. A pulse train of monochromatic radiation at frequency ν_0 is shown in Fig. 7-15 where t_p is the pulse length and p is the pulse repetition time. The Fourier transform of the pulse train into the frequency domain is also shown in Fig. 7-15 where it is now clear that maximum polarization is achieved when t_p is shorter than the relaxation times (leading to a broad spectral excitation) and p is long relative to the relaxation times.

The expressions above were given for the electric field–electric dipole interaction. The corresponding expressions for the magnetic field–magnetic dipole interactions are obtained by making the replacements given in Eq.(7-103). The actual form of the premultipliers in Eqs.(7-138) will be somewhat different for the two cases depending on the different geometry and detection system used in the magnetic resonance experiment relative to the normal absorption experiment discussed in detail here.

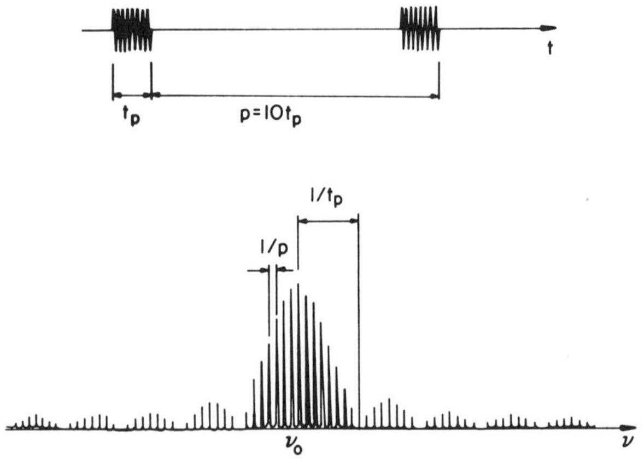

Figure 7-15 Schematic showing a pulse train in the time domain and the corresponding Fourier transform to the frequency domain. The pulse train represents the on-off switching of an oscillator with frequency ν_0. The spacings between the frequency components centered around ν_0 in the lower curve are spaced every $1/p$ where p is the time between the pulses. This curve is adapted from L. W. Anderson and W. W. Beeman, *Electric Circuits and Modern Electronics* (Holt, Rinehart and Winston, Inc., New York, 1973).

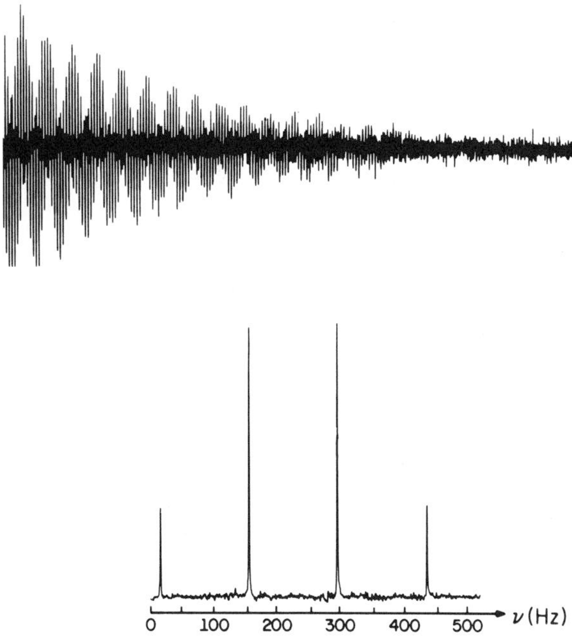

Figure 7-16 The free induction decay (FID) and corresponding Fourier transform to give the spectrum of the natural abundance (1 %) ^{13}C spectrum of CH_3OH. The FID taken in the upper curve covers a time of 1.47 s and includes a time average of 200 scans taken between 19 μs pulses at a reference of 25.16 MHz. The period between the pulses, during which time the 1.47 s FID shown here was stored in a computer for averaging, was 3 s. The pulse train can be viewed with the help of Fig. 7-15. The cosine Fourier transform of the FID gives the expected ^{13}C quartet that arises from the ^{13}C—H spin-spin interaction with the three equivalent H atoms in methanol. The FID and Fourier transform were taken by R. L. Thrift.

The free induction decay and corresponding Fourier transform is illustrated in Fig. 7-16 for the ^{13}C nuclear magnetic resonance spectra of CH_3OH (methyl alcohol). The quartet is due to the three equivalent methyl group protons coupling with ^{13}C (see Section 4.10). Working in the time domain offers a distinct advantage in sensitivity in the case of a relatively large spectrum of narrow lines. If the spectrum were observed by normal resonant absorption methods, most of the time would be spent in the spectral regions between the transitions. In the time domain, only the transitions and their decays are observed. Ernst and Anderson [30] have shown that the sensitivity in the time domain can exceed the sensitivity in the frequency domain by the square root of the ratio of the band of frequencies that are initially polarized to the half width of the transitions in the spectra.

There are several multiple pulse techniques that have been developed. We shall describe here the echo phenomenon that is observed in a $\pi/2$-τ-π-τ (echo) sequence on an inhomogeneously broadened system where τ is an experimentally adjusted delay time. The echo appears only in systems where the polarization decays by faster mechanisms than the normal T_2 processes (inhomogeneous processes). The Doppler effect for a gas phase transition is a case in point. For instance, consider

$P_i(t)$ and $P_r(t)$ in Eq.(7-132) for a single transition and initial conditions where $T_1\xi \gg 1$, $T_2\xi \gg 1$, and $P_r = 0$, which gives

$$\lim_{\substack{\xi T_1 \gg 1 \\ \xi T_2 \gg 1}} \begin{cases} P_i(t) = P_i(0) \exp\left(-\dfrac{t}{T_2}\right) \cos \Delta\omega t \\ P_r(t) = -P_i(0) \exp\left(-\dfrac{t}{T_2}\right) \sin \Delta\omega t, \end{cases} \tag{7-139}$$

for the time evolution of P_i and P_r following the initial polarization $P_i(0)$ and $P_r(0)$. These equations are valid, however, only for a stationary molecule. The value of $P_i(t)$ for a molecule with velocity v traveling in the same direction as the initial radiation is given by

$$P_i(t)_v = P_i(0) \exp\left(-\frac{t}{T_2}\right) \cos 2\pi\left(\Delta\nu + \frac{v}{c}\nu_{jk}\right)t. \tag{7-140}$$

This polarization must now be averaged over a Maxwell–Boltzmann velocity distribution as shown in detail in Section 7.2. Using $\nu' = (v/c)\nu_{jk}$ and $dv = (c/\nu_{jk})\, d\nu'$ leads to

$$\begin{aligned} P_i(t) &= P_i(0)\left(\frac{1}{\Delta\nu_G}\right)\left(\frac{\ln 2}{\pi}\right)^{1/2} \exp\left(-\frac{t}{T_2}\right) \\ &\quad \times \int_{-\infty}^{\infty} \exp\left[-\frac{\ln 2}{(\Delta\nu_G)^2}(\nu')^2\right] \cos 2\pi(\Delta\nu + \nu')t\, d\nu' \\ &= P_i(0) \exp\left(-\frac{t}{T_2}\right) \exp\left[-\frac{(2\pi\,\Delta\nu_G)^2}{4 \ln 2} t^2\right] \cos (2\pi\,\Delta\nu t), \end{aligned} \tag{7-141}$$

where $\Delta\nu_G$ is the Doppler half width at half height. It is evident from this expression that if $\Delta\nu_G \gg \Delta\nu_L = 1/2\pi T_2$, $P_i(t)$ will decay with an $\exp(-at^2)$ dependence and T_2 cannot be measured by observing $\gamma(\omega, t)$ in Eq.(7-94). Assuming that $\xi/(2\pi\,\Delta\nu_G) \gg 1.0$ where $\xi = E_o D_{jk}/\hbar$ in Eq.(7-102), however, we can substitute Eq.(7-141) into Eq.(7-94) and take the cosine Fourier transform to give the correct Doppler broadened spectrum. We shall now consider the $\pi/2$-τ-π-τ (echo) sequence experiment.

Using the $\pi/2$ initial condition from Eq.(7-129) in Eq.(7-139) gives

$$P_i(t) = -D_{jk}\,\Delta N_0 \exp\left(-\frac{t}{T_2}\right) \cos \Delta\omega t, \qquad P_r(t) = D_{jk}\,\Delta N_0 \exp\left(-\frac{t}{T_2}\right) \sin \Delta\omega t. \tag{7-142}$$

After a period τ, we have

$$P_i(\tau) = -D_{jk}\,\Delta N_0 \exp\left(-\frac{\tau}{T_2}\right) \cos \Delta\omega\tau, \qquad P_r(\tau) = D_{jk}\,\Delta N_0 \exp\left(-\frac{\tau}{T_2}\right) \sin \Delta\omega\tau. \tag{7-143}$$

Now, in a period of time short relative to τ we apply a π-pulse to the system that reverses the sign of $P_i(\tau)$ and leaves $P_r(\tau)$ unchanged [see Eq.(7-131)]. Substituting these initial conditions in place of $P_i(0)$ and $P_r(0)$ in Eq.(7-132) gives the evolution of $P_i(t)$ after the π-pulse:

$$P_i(t) = D_{jk}\,\Delta N_0 \exp\left[-\frac{(t+\tau)}{T_2}\right](\cos\Delta\omega\tau\cos\Delta\omega t + \sin\Delta\omega\tau\sin\Delta\omega t)$$

$$= D_{jk}\,\Delta N_0 \exp\left[-\frac{(t+\tau)}{T_2}\right]\cos\Delta\omega(\tau - t), \tag{7-144}$$

where now $t = 0$ at the end of the π-pulse.

For a molecule moving with velocity v we must write

$$P_i(t)_v = D_{jk}\,\Delta N_0 \exp\left[\frac{(t+\tau)}{T_2}\right]\cos 2\pi(\Delta\nu + \nu')(\tau - t),$$

where $\nu' = \nu_{jk}(v/c)$. Using the normal velocity average over the Maxwell–Boltzmann distribution as described following Eq.(7-27) gives

$$P_i(t) = D_{jk}\,\Delta N_0 \exp\left[-\frac{(t+\tau)}{T_2}\right]\exp\left[-\frac{(2\pi\,\Delta\nu_G)^2}{4\ln 2}(\tau - t)^2\right]\cos 2\pi\,\Delta\nu(\tau - t). \tag{7-145}$$

Substituting into Eq.(7-94) for the absorption coefficient gives

$$\gamma(t) = -\frac{4\pi\omega_{jk}D_{jk}\,\Delta N_0}{cE_0}\exp\left[-\frac{(t+\tau)}{T_2}\right]$$

$$\times \exp\left[-\frac{(2\pi\,\Delta\nu_G)^2}{4\ln 2}(\tau - t)^2\right]\cos 2\pi\,\Delta\nu(\tau - t). \tag{7-146}$$

This amazing result shows that after the π-pulse ($t = 0$), the absorption coefficient decreases to a maximum negative value at ($t = \tau$); this is called the echo and the maximum value is given by

$$\gamma(t = \tau) = -\frac{4\pi\omega_{jk}D_{jk}\,\Delta N_0}{cE_0}\exp\left(-\frac{2\tau}{T_2}\right). \tag{7-147}$$

The half width at half height of the Gaussian echo in the time domain is given from Eq.(7-146) by $\Delta t_{1/2} = \ln 2/(\pi\,\Delta\nu_G)$. In summary, a $\pi/2$-pulse is applied to the system followed by fast decay of the polarization due to the Doppler effect (or other inhomogeneous effects). After a period τ, a π-pulse is applied that leads to an echo at a time τ after the π-pulse. Thus, we have the $\pi/2$-τ-π-τ (echo) sequence. By measuring $\gamma(\tau)$ according to Eq.(7-147), the value of T_2 can be extracted from a transition whose line shape is dominated by the Doppler broadening or other inhomogeneous effect.

Another pulse sequence of interest involves the measurement of T_1 by a π-τ-$\pi/2$ sequence. The values of $P_i(0) = P_r(0) = 0$ and $\Delta N(0) = -\Delta N_0$ following a π-pulse are given in Eqs.(7-131). Following the π-pulse, P_i and P_r will remain at zero, but $\Delta N(t)$ will evolve according to

$$\Delta N(t) = \Delta N_0\left[1 - 2\exp\left(-\frac{t}{T_1}\right)\right] \tag{7-148}$$

back to the equilibrium value of $\Delta N(\infty) = \Delta N_0$. Now after a delay τ, a $\pi/2$-pulse is applied to the system where again we assume that the time necessary to produce a $\pi/2$-pulse is short relative to τ. The $\pi/2$-pulse at $t = \tau$ produces a maximum P_i given by

$$P_i(\tau) = -D_{jk}\,\Delta N_0\left[1 - 2\exp\left(-\frac{\tau}{T_1}\right)\right]. \tag{7-149}$$

Thus, a measurement of the amplitude of P_i as a function of the delay time τ leads to an exponential curve in T_1 and a resultant measurement of T_1. Equation (7-149) is also valid in the limit where the Doppler effect dominates the spectrum ($\Delta\nu_L \ll \Delta\nu_G$) as long as $\xi/(2\pi\Delta\nu_G) \gg 1.0$ where $\xi = E_0 D_{jk}/\hbar$.

Other pulse sequences that are combinations or extensions of basic π-τ-$\pi/2$ or $\pi/2$-τ-π sequences given here are discussed elsewhere [29].

We can also combine the pulse methods described here to multiple level systems. Remembering from Eq.(7-131) that a π-pulse inverts the population difference in a two-level system, we can use these principles on the NMR-ESR system shown in Fig. 4-12. Referring to Fig. 4-12, we note that a π-pulse at the $4 \longrightarrow 2$ transition will transfer the electron spin population difference to the nuclear levels, thus enhancing any transient signals observed in the $3 \longrightarrow 4$ and $2 \longrightarrow 1$ transitions. For instance, a $\pi/2$-pulse in the $3 \longrightarrow 4$ and $1 \longrightarrow 2$ transitions immediately following the $4 \longrightarrow 2$ π-pulse will lead to $3 \longrightarrow 4$ and $1 \longrightarrow 2$ spontaneous coherent emission with intensities stronger by μ_B/μ_0 than in the absence of the π-pulse.

In summary, we have developed the equations of motion to describe the strong resonant interaction of electromagnetic radiation with a two-level quantum mechanical system through either the electric or magnetic dipole interaction. The development, results, and conclusions are transferable from one type of spectroscopy to another, which includes nuclear magnetic resonance, electron spin resonance, microwave rotational spectroscopy, infrared vibrational spectroscopy, and electronic optical spectroscopy. Transient experiments have been developed extensively in all of these areas [31]. The strong interaction between radiation and a quantum system described here is also essential to an understanding of laser oscillators [21].

7.7 INTERFEROMETERS, LASERS, PICOSECOND PULSES, AND MORE FOURIER TRANSFORM SPECTROSCOPY

In this section we shall discuss applications of Fabry-Perot and Michelson interferometers. Fabry-Perot interferometers are used as spectrum analyzers, they are used in laser cavities, and they are used with mode-locking for the generation of electromagnetic radiation pulses. Michelson interferometers are used in Fourier transform spectroscopy.

A Fabry-Perot interferometer is shown in Fig. 7-17 where two highly reflecting parallel plates are placed at a distance d from each other (this device is called an

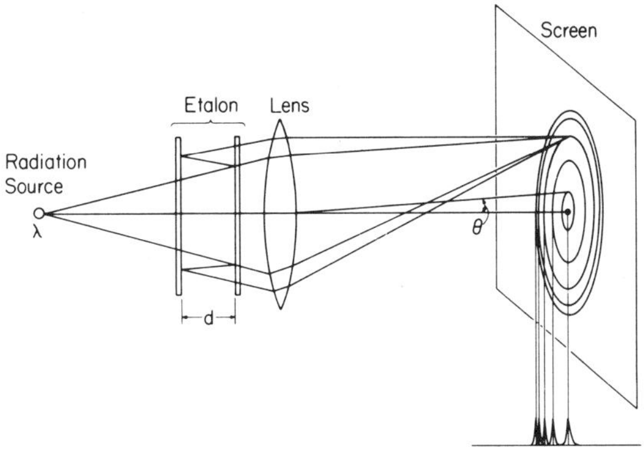

Figure 7-17 The Fabry-Perot interferometer with the transmitted intensity as a function of d, θ, and λ as given in Eq.(7-150). The characteristic ring pattern is shown on the screen with the intensity distribution given below the screen.

etalon). The wavefronts from the source of radiation are reflected back and forth between the plates (mirrors) before passing to the screen. When the waves interfere destructively, a dark region is observed on the screen; when the waves constructively interfere, a bright point is observed on the screen, as shown in the figure. The intensity incident on the screen is given by the Airy function [32], which is a function of θ, d, and λ,

$$A(\theta, \lambda, d) = \frac{I(0)}{1 + [4r/(1-r)^2]\sin^2 \delta/2}, \qquad \delta = \frac{2\pi}{\lambda}(2d\cos\theta), \tag{7-150}$$

where $I(0)$ is the incident intensity of the radiation and r is the reflectance of the etalon plates. The θ-dependence is shown schematically in the diagram below the screen in Fig. 7-17 where the intensity at $\theta = 0$ is given by

$$A(0, \lambda, d) = \frac{I(0)}{1 + [4r/(1-r)^2]\sin^2 (2\pi d/\lambda)}. \tag{7-151}$$

Consider the intensity of a white light source transmitted through the etalon according to Eq.(7-151). $A(\lambda, d)$ in Eq.(7-151) is maximized according to the well-known resonant condition for longitudinal modes or standing waves when the argument of the sine function is some multiple of π ($\sin(2\pi d/\lambda) = \sin n\pi$), giving

$$n\lambda = 2d, \qquad \lambda = \frac{2d}{n}. \tag{7-152}$$

The white light response according to Eqs.(7-151) and (7-152) is shown in Fig. 7-18. According to the equations above, the free spectral range for a fixed cavity length d is given by ($\nu = c/\lambda$)

$$\Delta\nu_{\Delta n = 1} = c\left(\frac{1}{\lambda_{n+1}} - \frac{1}{\lambda_n}\right) = \frac{c}{2d}. \tag{7-153}$$

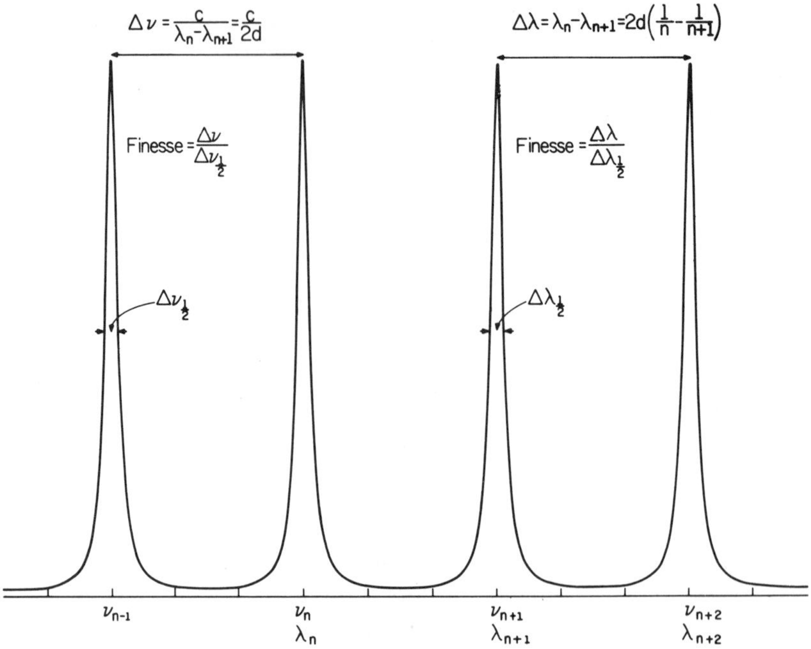

Figure 7-18 The wavelength or frequency response of a Fabry-Perot interferometer at $\theta = 0$ (see Fig. 7-17). A resonant-type curve appears every λ_n and the free spectral range is the spacing (in length, $\Delta\lambda$, or frequency, $\Delta\nu$) between the resonant-type curves. The resonant-type curve shown here is described by Eq.(7-151) when $r = 0.8$ and $d = 2$ m giving $\Delta\nu = 75$ MHz. The resultant finesse is about 15.

The quality factor or finesse is defined as the ratio of the free spectral range to the full width at half height of the Fabry-Perot line (see Fig. 7-18). $\Delta\nu$ is narrower and the finesse is larger if the reflectance, r, is larger. In optical Fabry-Perot interferometers, the finesse is decreased by lack of flatness of the reflecting surfaces of the etalon where surfaces smooth to $\lambda_0/100$ (about the best technically possible) lead to a finesse up to 100. In the infrared and microwave region, however, the wavelengths are longer and the relative flatness is much higher, leading to higher finesse. Finesse values of 50,000 are not unusual in the high-frequency microwave region ($\nu = 100$ GHz).

It is evident from Fig. 7-18 and the discussion above that the Fabry-Perot interferometer can be used as a spectrum analyzer. The Fabry-Perot interferometer is very useful for studying the line shape of Rayleigh scattered light as described in Chapter 8, where a monochromatic laser is used as a source. The subsequent spectrum, $S(\nu)$, which is scattered from a sample, is then measured with the Fabry-Perot interferometric spectrum analyzer with a shape function given by $A(\nu)$. The observed signal will be a convolution of $A(\nu)$ and $S(\nu)$,

$$I(\nu) = \int_{-\infty}^{+\infty} A(\nu')S(\nu - \nu')\, d\nu'. \tag{7-154}$$

$A(\nu)$ can be measured by analyzing the monochromatic laser source of radiation at frequency ν_0. Thus, the desired $S(\nu)$ spectrum of the scattered light centered at ν_0

can be obtained by deconvolution. Taking the Fourier transform of both sides of the equation above gives [see Eq.(B-63)]

$$\mathscr{F}I(\nu) = 2\pi\mathscr{F}A(\nu)\mathscr{F}S(\nu).$$

Dividing by $2\pi\mathscr{F}A(\nu)$ and reverse Fourier transforming gives $S(\nu)$,

$$S(\nu) = \mathscr{F}^{-1}\left[\frac{\mathscr{F}I(\nu)}{2\pi\mathscr{F}A(\nu)}\right].$$

We are now prepared to discuss the use of the Fabry-Perot interferometer as a laser cavity. A typical CO_2 laser operates at a combined CO_2, N_2, and He pressure of about 30 Torr. At this pressure the relaxation half widths of the vibration-rotation transitions in the $(0, 0, 1) \longrightarrow (1, 0, 0)$ CO_2 transitions near 1000 cm^{-1} (see Fig. 7-8) are approximately the same as the Doppler half widths. We estimate the half widths for a CO_2 vibration-rotation transition as $\Delta\nu \cong 35$ MHz. The spacing between the vibration-rotation transitions are in excess of 30,000 MHz. The Fabry-Perot resonant frequencies for a 2-m cavity are every $\Delta\nu = c/2d = 75$ MHz; for a typical finesse of 100, the line width response of the interferometer is 0.75 MHz, which is considerably less than the Doppler width. Thus, the CO_2 laser interferometric cavity is tuned to overlap with one of the vibration-rotation transitions, leading to feedback and coherent oscillations.

Another example of the Fabry-Perot interferometer is shown in Fig. 7-19 for the $\lambda_0 = 514.5$-nm resonance in the Ar^+ laser system. Under the conditions of electronic discharge, Ar^+ fluoresces with a Doppler half width of 1000 MHz. The 2-m cavity response every 75 MHz is shown in Fig. 7-19 and laser oscillation can occur at any of these Fabry-Perot responses. The addition of a second 2-cm etalon in the cavity provides an additional selectivity by overlap of the two-etalon-frequency bandpasses leading to more stable oscillation.

We can now use Fig. 7-19 to describe mode-locking and the generation of subnanosecond (10^{-9} s) or picosecond (10^{-12} s) pulses that can be used to examine kinetic processes in molecules at these very short times. Consider the Ar^+ laser cavity in the absence of the second 2-cm etalon. The Ar^+ ion laser can oscillate at any single one of the Fabry-Perot frequency responses shown in the Doppler line width. We have previously considered the cavity with fixed reflector distance d. An externally applied sinusoidal vibration with frequency ν_e on one of the reflectors in the cavity will produce sidebands $\pm\nu_e$ around the laser frequency. Now, if the sidebands, $\pm\nu_e$, reach the frequency of the Fabry-Perot spacing (75 MHz for the 2-m cavity), all frequencies separated by 75 MHz in the cavity will oscillate simultaneously. The result is a series of monochromatic frequencies distributed at each Fabry-Perot resonance that appears approximately as shown in the lower part of Fig. 7-15. The Fourier transform of this distribution of frequencies into the time domain gives the train of pulses shown in the upper part of Fig. 7-15. Thus, the pulse width of the mode-locked laser is approximately equal to the $t_p = 1/(2\pi\,\Delta\nu_G)$ where $\Delta\nu_G$ is the Doppler line half width of the original Ar^+ ion resonance. In this example for the Ar^+ laser, $\Delta\nu_G = 10^9$ Hz giving $t_p \cong 100$ ps $= 10^{-10}$ s. In order to achieve shorter time pulses, a laser with a larger intrinsic line width must be employed. A number of ions imbedded in glasses have intrinsic line widths that allow the generation of picosecond pulses. In fact, the Nd^{3+}-glass laser is capable of producing pulses shorter than 10^{-12} s [33].

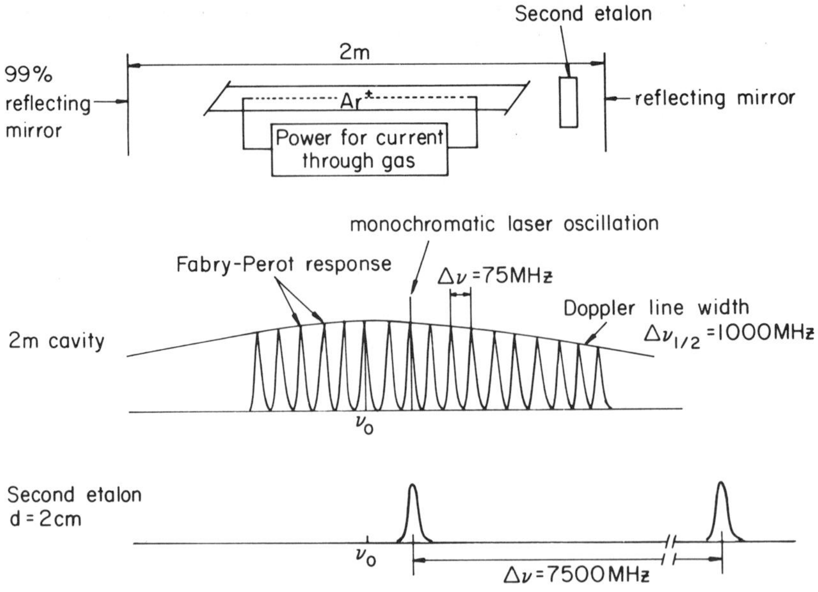

Figure 7-19 Schematic of an Ar^+ laser with fluorescent Doppler half width of approximately 1000 MHz. The Fabry-Perot response is shown in the lower diagram. The feedback in the cavity leads to the coherent oscillation. A second 2-cm etalon is also shown, which along with the 2-m cavity leads to a double-bandpass filter and a higher-frequency selectivity for the laser oscillation. Further details can be found in B. A. Lengyel, *Introduction to Laser Physics*, 2nd ed. (Wiley-Interscience, Inc., New York, 1971).

Short pulses, subnanosecond to picosecond, can be used to measure relaxation and other energy transfer processes on the time scale of the pulse width [34]. Referring to Fig. 7-20, we shall describe an experiment using picosecond pulses to measure the vibrational relaxation time in an excited electronic state in azulene [34]. The general energy level scheme, the measurement of fluorescence and phosphorescence, and the concept of vibrational and intersystem relaxation times, τ_V and τ_{IS}, are illustrated in Fig. 7-7. The experiment to measure τ_V as shown in Fig. 7-20 is to first pulse-excite one of the vibrational states in S_1 with the second harmonic of the Nd^{3+}-glass laser at $2\nu_0/c = 18{,}862\ \text{cm}^{-1}$. The S_1 state is then excited to S_2 with a delayed pulse from the Nd^{3+}-glass laser at $\nu_0/c = 9431\ \text{cm}^{-1}$. The azulene molecule then fluoresces from the S_2 state. The S_2 fluorescence intensity depends on the population of the initial state in S_1 that is excited by the second delayed pulse. Thus, a longer delay before the second pulse will give the S_1 excited vibrational state longer to relax, which leads to a lower S_2 fluorescent intensity. Measuring the S_2 fluorescence as a function of the delay between the $2\nu_0/c$ and ν_0/c pulses gives an excited vibrational S_1 state relaxation time of $\tau = 7.5 \times 10^{-12}$ s. This relaxation time is probably a vibrational relaxation time. The intersystem relaxation time, τ_{IS}, can also be measured by applying picosecond pulses [35].

We shall now discuss the Michelson interferometer shown in Fig. 7-21, which can also be used as a spectrum analyzer; however, the Fourier transform of the spectrum

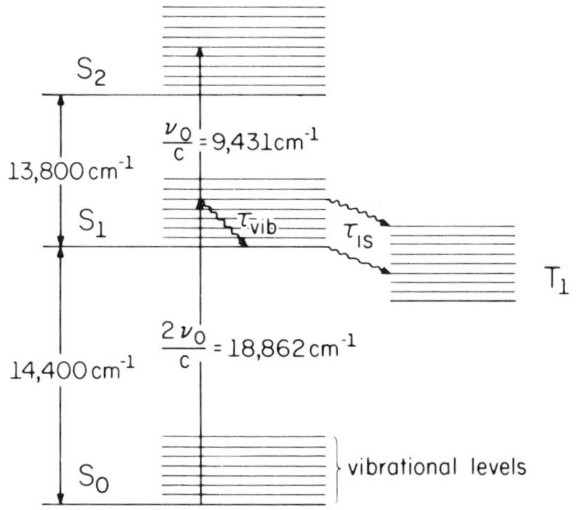

Figure 7-20 Schematic showing the energy level diagram and relaxation processes in azulene. S_0, S_1, and S_2 are the singlet states and T_1 is the triplet state. τ_V is the vibrational relaxation time and τ_{IS} is the intersystem singlet-triplet relaxation time. One of the vibrational states in the S_1 state of azulene is excited by a picosecond pulse from the doubled frequency ($2\nu_0/c = 18{,}862\ \text{cm}^{-1}$) of a Nd^{3+}-glass laser. A second picosecond pulse, which is delayed slightly from the first, now excites S_1 to S_2 ($\nu_0/c = 9431\ \text{cm}^{-1}$). Azulene then fluoresces from the S_2 state and the magnitude of the fluorescence depends on the population of the level being excited from the S_1 state with ν_0 from the laser. Measuring the fluorescence intensity as a function of the pulse delay between $2\nu_0$ and ν_0 gives the relaxation time of the state excited by $2\nu_0$.

is generated by the Michelson interferometer. The radiation from a monochromatic source with frequency ν can be described by the plane wave complex electric field of

$$E = E_0 \exp\left[i(kz - \omega t)\right],$$

where E_0 is the amplitude, $k = 2\pi/\lambda = \omega/c$, and z is the distance from the detector to the point of measurement. Consider now the *S-R-F-R-D* path in Fig. 7-21 where d_f is the total distance traveled to the detector. The field at the detector for this path is (using $k = \omega/c$)

$$E_f = E_0 \exp\left[i\left(\frac{\omega}{c} d_f - \omega t\right)\right]. \tag{7-155}$$

The corresponding *S-R-M-R-D* path, where d_m is the total distance traveled, has a field component given by

$$E_m = E_0 \exp\left[i\left(\frac{\omega}{c} d_m - \omega t\right)\right]. \tag{7-156}$$

The compensator assures that the phases of E_m and E_f will be identical when the distances *F-R* and *M-R* are equal.

The total field at the detector for the monochromatic radiation is given by a sum of Eqs.(7-155) and (7-156) giving

$$E = E_f + E_m = E_0 \exp(-i\omega t)\left[\exp\left(i\frac{\omega}{c} d_m\right) + \exp\left(i\frac{\omega}{c} d_f\right)\right]. \tag{7-157}$$

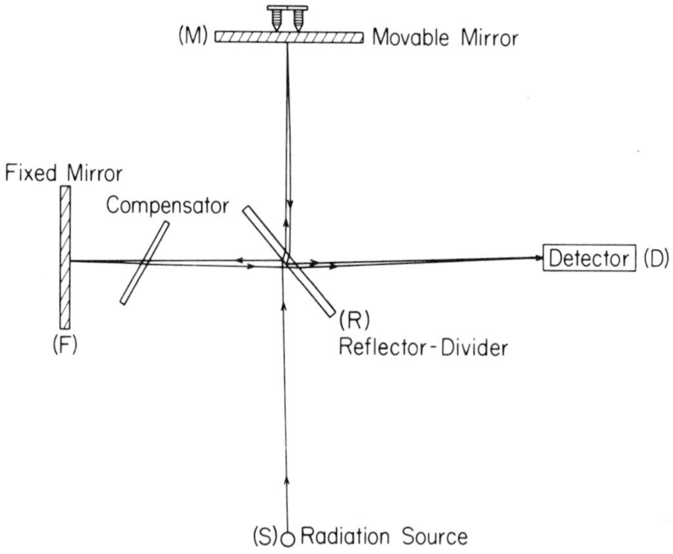

Figure 7-21 Schematic of a Michelson interferometer. The radiation from the source is split at the reflector-divider; half of the radiation goes to the fixed mirror and returns and the other half goes to the movable mirror and returns. The path length difference is varied with the movable mirror leading to the interference effects at the detector.

The intensity of the radiation at the detector can be calculated from the total field in Eq.(7-157) as a function of $d_m - d_f$, giving

$$I(d_m - d_f) = \left(\frac{c}{4\pi}\right)E^*E = \frac{cE_0^2}{2\pi}\left[1 + \cos\frac{\omega}{c}(d_m - d_f)\right],$$

$$I(\ell) = \frac{cE_0^2}{2\pi}\left(1 + \cos\frac{\omega}{c}\ell\right), \qquad \ell = d_m - d_f. \tag{7-158}$$

Consider now the total intensity at the detector for a radiation source with a spectral density distribution function, $I(\omega)$, in units of intensity per unit frequency. The intensity of the radiation centered at ω_j with band $\Delta\omega$ is given by

$$I_j = I(\omega_j)\,\Delta\omega = \frac{c(E_j^0)^2}{2\pi},$$

where E_j^0 is the field amplitude of the jth component of the field with width $\Delta\omega$. Thus, $I(\ell)$ in Eq.(7-158) for a source distribution function, $I(\omega)$, is given by

$$I(\ell) = c\sum_j \frac{(E_j^0)^2}{4\pi}\left(2 + 2\cos\frac{\omega_j \ell}{c}\right) = \sum_j I(\omega_j)\left(2 + 2\cos\frac{\omega_j \ell}{c}\right)\Delta\omega$$

$$= \int_0^{+\infty} I(\omega)\left(2 + 2\cos\omega\frac{\ell}{c}\right)d\omega, \tag{7-159}$$

where the last step converts the summation to an integral for small intervals $\Delta\omega$. According to Eq.(7-159),

$$I(\ell = 0) = 4 \int_0^\infty I(\omega)\, d\omega,$$

which we now use to rewrite Eq.(7-159) giving

$$I(\ell) = 2 \int_0^\infty I(\omega)\, d\omega + 2 \int_0^\infty I(\omega) \cos \omega \frac{\ell}{c}\, d\omega = \frac{1}{2} I(0) + 2 \int_0^\infty I(\omega) \cos \omega \frac{\ell}{c}\, d\omega. \tag{7-160}$$

It is now evident that $I(\omega)$ and $I(\ell) - [I(0)/2] = J(\ell)$ are cosine Fourier transform pairs where $I(\omega)$ is an even function (see Appendix B.2),

$$J(\ell) = I(\ell) - \frac{I(0)}{2} = 2 \int_0^{+\infty} I(\omega) \cos \omega \frac{\ell}{c}\, d\omega, \qquad I(\omega) = \frac{1}{\pi} \int_0^{+\infty} J(\ell) \cos \omega \frac{\ell}{c}\, d\ell. \tag{7-161}$$

$I(\ell) - [I(0)/2] = J(\ell)$ is the interferogram and $I(\omega)$ is the corresponding spectrum. It is evident from the nature of the Fourier transform that the interferogram will have spatial periods equal to the wavelength of the radiation being detected.

The shape or form of the interferogram is obtained by returning to Fig. 7-13 and replacing $g(t)$ with $J(\ell)$ in the first column for all the diagrams. Of course, we also replace the time axis t with a distance axis ℓ. The spacings between the maxima of the interferograms for a single-resonance spectrum are given along the distance axis as

$$\Delta\ell = \frac{c}{\nu_0} = \lambda_0, \tag{7-162}$$

where ν_0 is the resonance frequency, c is the speed of light, and λ_0 is the wavelength at resonance. Thus, the reference frequency in Fig. 7-13 goes to zero for the Fourier transform of the interferogram. The damping of the interferogram for a single transition (beginning at $\ell = 0$) is an exponential with damping distance constant

$$\ell_e = \frac{c}{2\pi\, \Delta\nu}, \tag{7-163}$$

where $\Delta\nu$ is the half width at half height and ℓ_e is the e^{-1} point in the exponential decay in the interferogram.

In summary, the interferogram of a single transition at frequency ν_0 with half width $\Delta\nu$ is a damped cosine with damping constant given in Eq.(7-163) and oscillation period given in Eq.(7-162). Figure 7-13 can be used directly to understand interferograms by making the following replacements: $g(t) \longrightarrow J(\ell)$, $t \longrightarrow \ell$, and $\nu_0 \longrightarrow 0$.

It is evident from the discussion above that in order to generate an interferogram that can be Fourier transformed to give a true spectrum, we must be able to sweep the interferometric distance from $\ell = 0$ to $\ell \gg c/(2\pi\, \Delta\nu)$, where $\Delta\nu$ is the half width of the transition. In addition, we must have recording resolution along the ℓ axis in excess of λ_0, the wavelength at resonance. Mechanically it is convenient to use Michelson interferometry in the infrared and far infrared regions [36]. According to Eq.(7-163), if we require an interferometric sweep of $\ell_{\max} \cong 10\ell_e = 1$ cm, we are limited to line

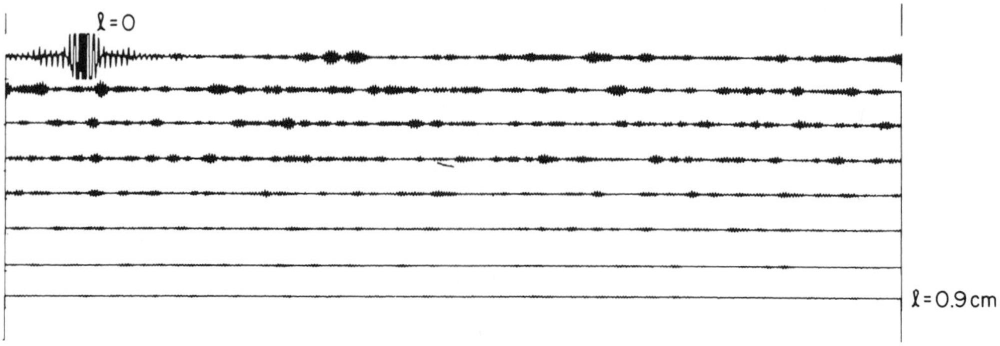

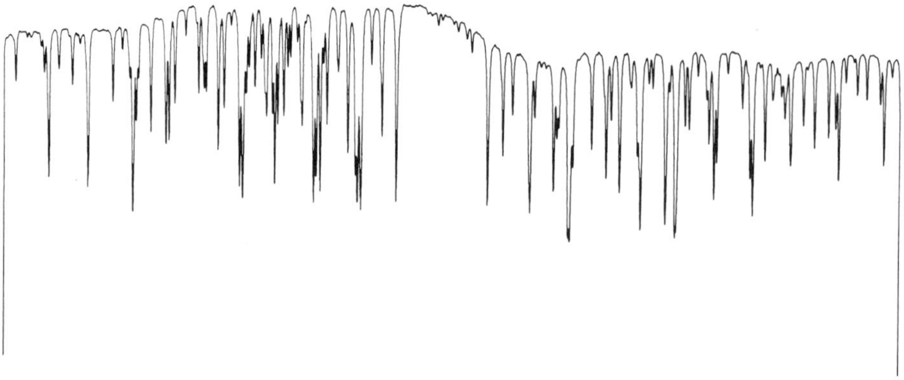

Figure 7-22 The Michelson interferogram and corresponding Fourier transform giving the absorption spectrum of water vapor in the atmosphere at approximately 40% relative humidity at 1 atm pressure. The top curve shows the interferogram starting at $\ell = 0$ and continuing until $\ell = 0.9$ cm at the end of the eighth line. The Fourier transform of the interferogram according to Eqs.(7-161) gives the spectrum. Only the small part of the spectrum from 1400 to 1800 cm^{-1} is shown here where the resolution or narrowest line width is 0.5 cm^{-1}. The Michelson interferogram and spectrum were provided by J. D. McDonald.

widths in excess of $\Delta\nu = 10c/2\pi \cong 10^{10}$ Hz $\cong 0.3$ cm^{-1}. Infrared transitions with lines narrower than 0.3 cm^{-1} will not be faithfully reproduced with the Fourier transform of an $\ell = 0$ to 1 cm Michelson interferogram. The phenomenon of beats in an interferogram can also be understood according to Fig. 7-14 and the considerations above.

A Michelson interferogram from an $\ell = 0$ to 0.9 cm sweep from the absorption spectrum of water vapor in the atmosphere is shown in Fig. 7-22. According to the discussion above, the Fourier transform of $J(\ell)$ (the interferogram) gives the entire spectrum from $\nu = 0$ to ∞. Only the spectral region from 1400 to 1800 cm^{-1} is shown in Fig. 7-22.

The increase in sensitivity in observing the interferogram, $J(\ell)$, relative to observing the spectrum directly by sweeping the frequency is proportional to the square root of the ratio of the total band of frequencies to the half width of a single transition [37].

PROBLEMS

1. Calculate the electric dipole transition moments in the $1s \longrightarrow 2s$ and $1s \longrightarrow 2p$ transitions in the hydrogen atom and the $J = 0 \longrightarrow J = 1$ and $J = 0 \longrightarrow J = 2$ rotational transitions in OCS.

2. The vibrational radiative lifetime of the first excited vibrational state of CO is 0.033 s [38]. What is the value of the transition dipole moment in CO?

3. Vibrational relaxation times can be measured with a spectrophone [39], which measures the internal energy–translational energy exchange rates. Experimentally, the optical-acoustical excitation is initiated by stimulating a vibrational transition with electromagnetic radiation. The vibrational temperature increases as the infrared energy is absorbed. Molecular collisions relax the vibrational states, however, which results in an increased translational temperature.

   As the translational temperature increases, the pressure of the gas increases ($PV = nRT$) and the pressure change can be detected by a variety of techniques. The spectrophone detects a pressure change that is stimulated through $V \leftrightarrow T$ relaxation following a vibrational excitation. The time lag in the increase in the translational temperature (pressure) after the infrared excitation is the vibrational-translational relaxation time.

   Experiments by T. L. Cottrell, *et al.*, *Trans. Far. Soc.* **62**, 2655 (1966), have yielded the vibrational relaxation times by exciting both ν_3 and ν_4 in methane. The relaxation times are $\tau_{\nu_3} = 1.6 \pm 0.4\ \mu s$ and $\tau_{\nu_4} = 1.6 \pm 0.4\ \mu s$ at 1 atm pressure. The normal modes in methane, $E(\nu_1 \nu_2 \nu_3 \nu_4)$, are ν_1 (symmetric stretch) $= 2914.2\ cm^{-1}$, ν_2 (symmetric bend) $= 1526.0\ cm^{-1}$, ν_3 (asymmetric stretch) $= 3020.3\ cm^{-1}$, and ν_4 (asymmetric bend) $= 1306.2\ cm^{-1}$ as shown in Fig. E-4. Use Fig. E-4 and explain the mechanism of $V \longleftrightarrow T$ relaxation from (0, 0, 0, 1) and (0, 0, 1, 0) to the ground state in methane.

4. The peak absorption coefficient for rotational levels in a diatomic molecule in the limit where the Lorentzian relaxation width exceeds the Doppler width is shown in Eq.(7-16) to be independent of pressure and proportional to the cube of the resonant frequency, ν_{jk}^3. Derive the corresponding peak absorption coefficient in the Doppler broadened limit. Compare the pressure and ν_{jk}-dependence in γ_R in both cases.

5. A typical width of a lower-power ESR transition of an organic free radical in solution is 3.0 MHz at about 7000 G. At what microwave power would you expect to experience the effects of power saturation? Compare your answer with the result in Problem 7-6 for NMR.

6. Consider the ^{19}F nuclear magnetic resonance at 15,000 G and 60 MHz in a molecule in a liquid at room temperature. At very low power the line width has a half width at half height of 0.5 Hz.
   (a) At what radio-frequency power level would you expect to see the effects of power saturation?

(b) Using the result in part a, what is the magnitude of the radio-frequency magnetic field (G) needed to excite the nuclear magnetic dipole transition in the ^{19}F nucleus to a point just approaching saturation?

***7.** Use Eq. (7-105) and a digital computer to show the effects of saturation and the Doppler effect. Choose several examples. Show also the validity of the convolution theorem in Eq.(7-35). Finally, reproduce the plots in Fig. 7-10.

8. In the absence of Doppler broadening, calculate the loss in total integrated intensity for absorption in the presence of power saturation from Eq.(7-104).

9. In some cases, the observation of a line width can give information about the rate constant in a chemical reaction. Consider the following reaction:

$$\mathrm{A + B \longrightarrow C.}$$

The rate of the reaction is given by

$$\text{rate} = -\frac{d[\mathrm{A}]}{dt} = -\frac{d[\mathrm{B}]}{dt} = \frac{d[\mathrm{C}]}{dt} = k[\mathrm{A}][\mathrm{B}],$$

where [A], [B], and [C] are the concentrations of A, B, and C, respectively, and k is the rate of the reaction. We can rewrite this equation as

$$\frac{d[\mathrm{A}]}{[\mathrm{A}]} = -k[\mathrm{B}]\,dt.$$

Integrating for a constant value of [B] gives

$$[\mathrm{A}]_t = [\mathrm{A}]_0 \exp\left(-k[\mathrm{B}]t\right), \tag{7-164}$$

where $[\mathrm{A}]_t$ and $[\mathrm{A}]_0$ are the concentrations of A at $t = t$ and $t = 0$, respectively. According to Eq.(7-164), the lifetime of [A], or the time after which $[\mathrm{A}]_0$ has been reduced to $[\mathrm{A}]_0/e$, is given by $\tau_{\mathrm{A}} = 1/[\mathrm{B}]k$. Thus, if we were observing any spectral (electronic, vibrational, rotational, ESR, NMR, etc.) characteristic of species A in the presence of a constant concentration of B during the reaction, we would expect the line width to be affected by the lifetime caused by the reaction. According to our analysis above, the relaxation is first order and the line shape contribution due to the reaction is expected to be Lorentzian; call it $\mathscr{L}_R(\nu)$. Thus, the total observed line width will be given by the convolution of the reaction limited width, $\mathscr{L}_R(\nu)$, with the width due to other causes, $N(\nu)$. In the case where the width of $\mathscr{L}_R(\nu)$ due to the reaction far exceeds other sources of width, the observed shape will be Lorentzian due to $\mathscr{L}_R(\nu)$ with a half width at half height of

$$\Delta\nu_{1/2} = \frac{[B]k}{2\pi}. \tag{7-165}$$

S. Weissman and R. L. Ward, *J. Am. Chem. Soc.* **79**, 2086 (1957), have measured the rate of reaction between naphthalene ($C_{10}H_8$) and the naphthalenide radical

($C_{10}H_8^-$) by observing the width of the electron spin resonance (ESR) transition in the $C_{10}H_8^-$ radical:

$$C_{10}H_8 + C_{10}H_8^- \longrightarrow C_{10}H_8^- + C_{10}H_8.$$

In the case of naphthalene ($C_{10}H_8$) dissolved in a tetrahydrofuran solution of potassium naphthalenide, they report an increase in the width of the ESR line of the $C_{10}H_8^-$ radical with increasing $C_{10}H_8$ concentration, which leads to a rate constant of $k = (5.7 \pm 1) \times 10^7$ liter · mol^{-1} · s^{-1}.

(a) According to these data, what is the electron transfer rate between $C_{10}H_8$ and $C_{10}H_8^-$ when the $C_{10}H_8$ concentration is 5 mol · liter^{-1}?

(b) Estimate the $C_{10}H_8$–$C_{10}H_8^-$ collision frequency at 5 mol · liter^{-1} $C_{10}H_8$ in a dilute solution of $C_{10}H_8^-$ ions and compare with the calculated electron transfer frequency. Does an electron transfer take place during each $C_{10}H_8$–$C_{10}H_8^-$ collision?

10. Fourier transform spectroscopy in nuclear magnetic resonance is accomplished with radio-frequency pulses as described in Section 7.6. Fourier transform spectroscopy in the infrared is accomplished with the Michelson interferometer as described in Section 7.7. In the microwave range from 10–100 GHz, however, it is difficult to apply either the radiation pulse method or the Michelson interferometric method to design a Fourier transform microwave or electron spin resonance spectrograph. Show the limitations by examining the pulse times and radiation power needed in a pulse system and the necessary distance variations in an interferometric system to record the Fourier transform of a microwave spectrum from 10–20 GHz (see Fig. 4-28, for instance).

11. Consider a solution of HCl in a CCl_4 solvent at a concentration of 0.01 mol HCl per liter of solvent. The infrared absorption of HCl appears as a single transition (rotational motion is quenched) at about 2800 cm^{-1} with a half width at half height of 10 cm^{-1} and a peak absorption coefficient of $\gamma = 0.001$ cm^{-1}. What is the approximate value of the electric dipole matrix element, D_{jk}, for the HCl vibrational transition?

***12.** Use a digital computer to plot $\gamma(\omega)$ in Eqs.(7-24) for a variety of conditions to show the nature of the two-photon absorption processes.

REFERENCES

[1] N. Davidson, *Statistical Mechanics* (McGraw-Hill Book Co., New York, 1962). D. A. McQuarrie, *Statistical Thermodynamics* (Harper & Row, New York, 1973).

[2] W. M. McClain, *Accts. Chem. Res.* **7**, 129 (1974). W. L. Peticolas, *Ann. Rev. Phys. Chem.* **18**, 233 (1967). Peticolas has given a summary of multiphoton absorption and scattering processes involving both $\mathscr{H}_1$ and $\mathscr{H}_2$ in Eq.(3-167) according to the Feynman time-ordered diagrams.

[3] J. D. Kraus, *Radio Astronomy* (McGraw-Hill Book Co., New York, 1966), p. 45.

[4] S. M. BLINDER, *Advanced Physical Chemistry* (The Macmillan Co., Toronto, Canada, 1969).

[5] J. T. DAVIES and J. M. VAUGHAN, *Astrophys. J.* **137**, 1302 (1963).

[6] R. G. GORDON, *Advances in Magnetic Resonance*, Ed. by J. Waugh **3**, 1 (Academic Press, New York, 1968) .

[7] D. CHANDLER, *Accts. Chem. Res.* **7**, 246 (1974).

[8] F. BLOCH, *Phys. Rev.* **70**, 460 (1946).

[9] See, for instance, J. A. POPLE, W. G. SCHNEIDER, and H. J. BERNSTEIN, *High-Resolution Nuclear Magnetic Resonance* (McGraw-Hill Book Co., New York, 1959).

[10] The equivalence between Eqs.(7-76) and (7-101) along with an interesting geometric interpretation is discussed in R. P. FEYNMAN, R. W. HELLWARTH, and F. L. VERNON, *J. Appl. Phys.* **28**, 49 (1957).

[11] A. ABRAGAM, *The Principles of Nuclear Magnetism* (Clarendon Press, Oxford, 1971). A. G. REDFIELD, *Adv. Magn. Res.* Ed. by J. S. Waugh **1**, 1 (Academic Press, New York, 1965).

[12] See review in W. H. FLYGARE and T. G. SCHMALZ, *Accts. Chem. Res.* **9**, 385 (1976).

[13] T. OKA, *Advances in Atomic and Molecular Physics*, Ed. by D. R. Bates and I. Esterman, **9**, 127 (Academic Press, New York, 1973).

[14] C. B. MOORE, *Adv. Chem. Phys.*, Ed. by I. Prigogine and S. A. Rice **23**, 41 (John Wiley & Sons, Inc., New York, 1973). C. B. MOORE, *Accts. Chem. Res.* **2**, 103 (1969).

[15] R. C. AMME, *Adv. Chem. Phys.*, Ed. by I. Prigogine and S. A. Rice **28**, 171 (Wiley Interscience, New York, 1975).

[16] Additional details on fluorescence quenching in small molecules can be found in J. I. STEINFELD, *Accts. Chem. Res.* **3**, 313 (1970) and a more general discussion of luminescence (fluorescence and phosphorescence) can be found in D. M. HERCULES, *Fluorescence and Phosphorescence Analysis* (Interscience Publishers, John Wiley & Sons, Inc., New York, 1965). Effects of interference in fluorescence and the effects of electric and magnetic fields to produce level crossings are given in R. N. ZARE, *Accts. Chem. Res.* **11**, 361 (1971).

[17] G. WEBER, *Adv. Protein Chem.* **8**, 415 (1953).

[18] C. S. PARMENTER, *Adv. Chem. Phys.*, Ed. by I. Prigogine and S. A. Rice **22**, 365 (John Wiley & Sons, Inc., New York, 1972) .

[19] N. WOTHERSPOON, G. A. OSTER, and G. OSTER, "The Determination of Fluorescence and Phosphorescence," *Physical Methods of Chemistry*, Ed. by A. Weissberger and B. Rossiter (Wiley-Interscience, New York, 1972).

[20] G. W. ROBINSON and R. P. FROSCH, *J. Chem. Phys.* **38**, 1187 (1963).

[21] M. SARGENT, M. O. SCULLY, and W. E. LAMB, *Laser Physics* (Addison-Wesley Publishing Co., London, 1974).

[22] R. H. BALLIS, *Adv. Chem. Phys.*, Ed. by I. Prigogine and S. A. Rice, **28**, 423 (Wiley-Interscience, New York, 1975).

[23] E. WERTZ, G. FLYNN, and A. W. RONN, *J. Chem. Phys.* **56**, 6060 (1972).

[24] See papers in *Tunable Lasers and Applications*, Ed. by A. Mooradian, T. Jaeger, and P. Stokseth (Springer, New York, 1976) and R. V. AMBARTZUMIAN and V. S. LETOKOV, *Accts. Chem. Res.* **10**, 61 (1977).

[25] R. A. MCFARLANE, W. R. BENNETT, and W. E. LAMB, *Appl. Phys. Lett.* **2**, 189 (1963) and W. E. LAMB, *Phys. Rev.* **A134**, 1429 (1964).

[26] C. P. SLICHTER, *Principles of Magnetic Resonance* (Harper & Row, New York, 1963).

[27] H. C. Torrey, *Phys. Rev.* **76**, 1059 (1949).

[28] J. C. McGurk, T. G. Schmalz, and W. H. Flygare, *Adv. Chem. Phys.*, Ed. by I. Prigogine and S. A. Rice **25**, 1 (John Wiley & Sons, Inc., New York, 1974).

[29] See, for example, T. C. Farrar and E. D. Becker, *Pulse and Fourier Transform NMR* (Academic Press, New York, 1971).

[30] R. R. Ernst and W. A. Anderson, *Rev. Sci. Inst.* **37**, 93 (1966). See also R. R. Ernst, *Adv. Mag. Res.*, Ed. by J. S. Waugh **2**, 1 (Academic Press, New York, 1966).

[31] See an interesting historical summary in J. D. Macomber, *The Dynamics of Spectroscopic Transitions* (John Wiley & Sons, Inc., New York, 1976).

[32] J. F. James and R. S. Sternberg, *The Design of Optical Spectrometers* (Chapman and Hall, Ltd., London, 1969).

[33] A. J. DeMario, W. H. Glenn, and M. E. Mack, *Phys. Today*, July 19, 1972.

[34] P. M. Rentzepis, *Science* **169**, 239 (1970).

[35] See details in P. M. Rentzepis, *Adv. Chem. Phys.*, Ed. by I. Prigogine and S. A. Rice, **23**, 189 (John Wiley & Sons, Inc., New York, 1973).

[36] R. J. Bell, *Introductory Fourier Transform Spectroscopy* (Academic Press, New York, 1972).

[37] Comparisons between different Fourier transform methods can be found in E. D. Becker and T. C. Farrar, *Science* **178**, 361 (1972).

[38] R. C. Millikan, *J. Chem. Phys.* **43**, 1439 (1965).

[39] See an interesting historical account in W. R. Harshbarger and M. B. Robin, *Accts. Chem. Res.* **6**, 329 (1973).

8

electromagnetic scattering

8.1 INTRODUCTION

We start by writing some general relations between measurements of electromagnetic radiation in the time and frequency domains. The *total intensity* of light (electromagnetic radiation) is written as a long-time average of the instantaneous intensity, $I(t)$,

$$I = I\left(\frac{\text{energy}}{\text{area} \cdot t}\right) = \lim_{T \to \infty}\left[\frac{1}{2T}\int_{-T}^{T} I(t)\,dt\right] = \langle I(t)\rangle, \tag{8-1}$$

where the angular brackets indicate the long-time average. The instantaneous intensity is given in terms of the electric fields of the electromagnetic radiation in a vacuum according to Eq.(1-51), where $E = H$ in cgs units, as used throughout this chapter,

$$I(t) = \frac{c}{4\pi} E^*(t)E(t). \tag{8-2}$$

We use the complex form for the instantaneous electric field, $E(t)$. We also define the spectral distribution function, $I(\omega)$, which has units of energy/(area $\cdot t \cdot \omega$). Thus, the total intensity, which is equal to I in Eq.(8-1), is given by integrating $I(\omega)$ over all frequencies:

$$I = \int_{-\infty}^{+\infty} I(\omega)\, d\omega = \lim_{T\to\infty} \left[\frac{1}{2T}\int_{-T}^{T} I(t)\, dt\right]. \tag{8-3}$$

Substituting Eq.(8-2) into Eq.(8-3) gives

$$I = \int_{-\infty}^{+\infty} I(\omega)\, d\omega = \langle I(t)\rangle = \frac{c}{4\pi}\langle E^*(t)E(t)\rangle, \tag{8-4}$$

where the angular brackets again indicate the long-time average.

We now define the *correlation function* for the electric field [1],†

$$C(\tau) = \lim_{T\to\infty} \frac{c}{4\pi}\left[\frac{1}{T}\int_0^T E^*(t)E(t+\tau)\, dt\right] = \frac{c}{4\pi}\langle E^*(t)E(t+\tau)\rangle, \tag{8-5}$$

where we note from Eqs.(8-4) and (8-5) that

$$C(0) = C(\tau = 0) = \frac{c}{4\pi}\langle E^*(t)E(t+0)\rangle = I. \tag{8-6}$$

We shall now show that $C(\tau)$ and $I(\omega)$ are Fourier transform pairs. In order to make this connection, we assume that we can expand $E(t)$ in a complex Fourier series given by

$$E(t) = \sum_j A_j \exp(-i\omega_j t). \tag{8-7}$$

As mentioned previously, we are using the complex form for the electric field. As long as we are consistent in using complex incident and scattered fields, the resultant scattered intensity (which is proportional to the incident radiation intensity) will be correct. Of course, the fields are actually real and we can use the real components throughout and show the same proportionality between the incident and scattered field intensities (see Problem 8-1). The arithmetic is less burdensome, however, if we use complex fields as in Eq.(8-7). Substituting Eq.(8-7) into Eq.(8-5) gives

$$C(\tau) = \frac{c}{4\pi}\langle E^*(t)E(t+\tau)\rangle = \frac{c}{4\pi}\left\langle \sum_{j,l} A_j^* A_l \exp[-i(\omega_l - \omega_j)t]\exp(-i\omega_l\tau)\right\rangle. \tag{8-8}$$

We now factor out the time independent $A_j^* A_l \exp(-i\omega_l\tau)$ parts and note that the long-time average of $\exp[-i(\omega_l - \omega_j)t]$ yields the Kronecker delta (see Appendix B.3),

$$\langle \exp[-i(\omega_l - \omega_j)t]\rangle = \delta_{lj}. \tag{8-9}$$

† Numbers in square brackets correspond to the numbered sources found in the Reference section on p. 569 at the end of the chapter.

Excluding all $j \neq l$ terms in Eq.(8-8), we rewrite to give

$$C(\tau) = \frac{c}{4\pi} \sum_l |A_l|^2 \exp(-i\omega_l \tau). \tag{8-10}$$

Returning to Eq.(8-4), converting the integral in $d\omega$ into a sum, and substituting $E(t)$ from Eq.(8-7) gives

$$\int_{-\infty}^{+\infty} I(\omega)\, d\omega = \sum_l I(\omega_l)\, \Delta\omega = \frac{c}{4\pi} \left\langle \sum_{j,l} A_j^* A_l \exp[-i(\omega_j - \omega_l)t] \right\rangle = \frac{c}{4\pi} \sum_l |A_l|^2.$$

Of course, the last term in this equation is just equal to the initial condition on $C(\tau)$ in Eq.(8-8):

$$C(0) = I = \frac{c}{4\pi} \sum_l |A_l|^2.$$

Thus, term by term, it is evident that

$$I(\omega_l)\, \Delta\omega = \frac{c}{4\pi} |A_l|^2. \tag{8-11}$$

Substituting $|A_l|^2$ from Eq.(8-11) into Eq.(8-10) gives

$$C(\tau) = \sum_l I(\omega_l) \exp(-i\omega_l \tau)\, \Delta\omega = \int_{-\infty}^{+\infty} I(\omega) \exp(-i\omega\tau)\, d\omega, \tag{8-12}$$

where the last step changes the summation to an integral. We can use the reverse Fourier transform to give $I(\omega)$. Applying $\exp(+i\omega'\tau)$ to both sides of Eq.(8-12) and integrating first over τ followed by integration over ω gives

$$\begin{aligned} \int_{-\infty}^{+\infty} C(\tau) \exp(i\omega'\tau)\, d\tau &= \int_{-\infty}^{+\infty} \int_{-\infty}^{+\infty} I(\omega) \exp[-i(\omega - \omega')\tau]\, d\omega\, d\tau \\ &= 2\pi \int_{-\infty}^{+\infty} I(\omega)\delta(\omega - \omega')\, d\omega = 2\pi I(\omega'), \end{aligned} \tag{8-13}$$

where we have recognized the Fourier representation of the delta function [Eq.(B-67)]. The final result is

$$I(\omega) = \frac{1}{2\pi} \int_{-\infty}^{+\infty} C(\tau) \exp(i\omega\tau)\, d\tau, \tag{8-14}$$

which shows that $C(\tau)$ and $I(\omega)$ are *Fourier transform pairs.* It is now evident that if we can measure either $C(\tau)$ or $I(\omega)$, we can use the Fourier transform to obtain the other member of the pair.

The definitions and equations above are general and can be used to analyze the spectrum of a given distribution of electromagnetic fields. We shall now introduce the more useful ensemble average over all positions and momenta that is essential to evaluating the correlation function for a system of scattering atoms or molecules. First, we define a *stationary system*, which requires that the time average in Eq.(8-5) leading to the correlation function is independent of the origin in time. Thus, for a stationary system we can write

$$C(\tau) = \frac{c}{4\pi} \langle E^*(t)E(t + \tau)\rangle = \frac{c}{4\pi} \langle E^*(0)E(\tau)\rangle. \tag{8-15}$$

Now for a stationary system, the *ergodic hypothesis* states that each scattering system in the ensemble of particles will pass through all values accessible to it, given a sufficiently long time. Thus, the time average is essentially the same for all systems of the ensemble. The result is that for a stationary ergodic system the time average is equivalent to the ensemble average [2]. When the brackets in Eq.(8-5) indicate an ensemble average, the relation between $C(\tau)$ and $I(\omega)$ discussed above, which is based on the expansion in Eq.(8-7), is called the Wiener–Khintchine theorem [3].

We shall find it useful to examine more carefully the properties of the correlation functions and spectra as related to a scattering experiment. We first note that the measured scattered spectrum $I(\omega)$ is a real quantity, $I(\omega) = I^*(\omega)$, and we can take the complex conjugate of Eq.(8-14) to give

$$I^*(\omega) = \frac{1}{2\pi}\int_{-\infty}^{+\infty} C^*(\tau) \exp(-i\omega\tau)\, d\tau = \frac{1}{2\pi}\int_{-\infty}^{+\infty} C^*(-\tau) \exp(i\omega\tau)\, d\tau,$$

where the last step is made by interchanging the integration variable τ to $-\tau$. Comparing $I(\omega) = I^*(\omega)$ from the equations above shows that $C(\tau) = C^*(-\tau)$. We can now rewrite the expression in Eq.(8-14) according to

$$\begin{aligned} I(\omega) &= \frac{1}{2\pi}\int_{-\infty}^{+\infty} C(\tau) \exp(i\omega\tau)\, d\tau \\ &= \frac{1}{2\pi}\int_{-\infty}^{0} C(\tau) \exp(i\omega\tau)\, d\tau + \frac{1}{2\pi}\int_{0}^{\infty} C(\tau) \exp(i\omega\tau)\, d\tau. \end{aligned}$$

Changing the variable of integration in the $-\infty \longrightarrow 0$ term from τ to $-\tau$ and using $C(\tau) = C^*(-\tau)$ and $C^*(\tau) = C(-\tau)$ gives

$$\begin{aligned} I(\omega) &= \frac{1}{2\pi}\left[\int_{0}^{\infty} C(-\tau) \exp(-i\omega\tau)\, d\tau + \int_{0}^{\infty} C(\tau) \exp(i\omega\tau)\, d\tau\right] \\ &= \frac{1}{2\pi}\left[\int_{0}^{\infty} C^*(\tau) \exp(-i\omega\tau)\, d\tau + \int_{0}^{\infty} C(\tau) \exp(i\omega\tau)\, d\tau\right]. \end{aligned}$$

We now note that the first integral is the complex conjugate of the second and we can write

$$\begin{aligned} I(\omega) &= \frac{1}{2\pi}\left\{\left[\int_{0}^{\infty} C(\tau) \exp(i\omega\tau)\, d\tau\right]^* + \int_{0}^{\infty} C(\tau) \exp(i\omega\tau)\, d\tau\right\} \\ &= \frac{1}{\pi}\,\mathrm{Re}\int_{0}^{\infty} C(\tau) \exp(i\omega\tau)\, d\tau, \end{aligned} \tag{8-16}$$

where the last step indicates only the real part of the $0 \longrightarrow \infty$ integral as the imaginary part vanishes. In summary, we use Eqs.(8-16) and (8-12) to obtain the real spectrum and the real correlation function from their corresponding Fourier transforms. As a final point, we now integrate $I(\omega)$ over all frequencies to give

$$I = \int_{-\infty}^{+\infty} I(\omega)\, d\omega = \frac{1}{2\pi}\int_{-\infty}^{+\infty} C(\tau) \int_{-\infty}^{+\infty} \exp(i\omega\tau)\, d\omega\, d\tau = \int_{-\infty}^{+\infty} C(\tau)\delta(\tau)\, d\tau = C(0), \tag{8-17}$$

where $C(0)$ is the static or $\tau = 0$ correlation function. Thus, the time or ensemble averaged total intensity is equal to the correlation function at $\tau = 0$ as shown previously in Eq.(8-6).

These equations show the basic relations between measurements in the time and frequency domains. We shall use these equations in describing various aspects of light scattering where it is most convenient to begin with the description of the correlation function. To describe a light scattering experiment, we begin with an incident complex monochromatic electromagnetic field, $E(t)$, with intensity $I_0 = (c/4\pi)|E(t)|^2$. The incident field interacts with the system and gives rise to a scattered field that we shall also describe as a complex field. The scattered light intensity is proportional to the incident light intensity, I_0.

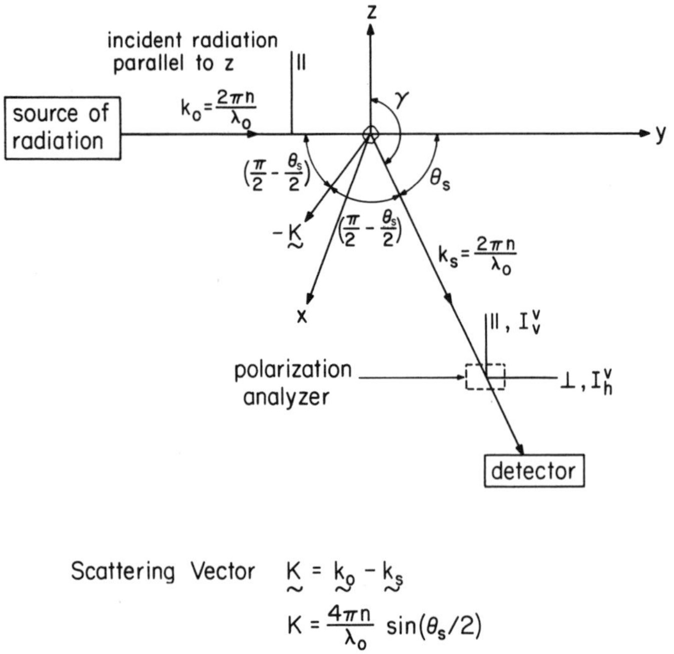

Figure 8-1 Basic scattering diagram showing plane-polarized z axis (||) incident radiation that is scattered along the line shown to the detector at angles γ and θ_s. Choosing $\gamma = \pi/2$, we note either polarized (z axis, ||) or depolarized (xy plane, ⊥) scattered light designated by I_v^v and I_h^v, respectively. $\boldsymbol{k}_0$ and $\boldsymbol{k}_s$ are the incident and scattered wave vectors, n is the refractive index in the scattering medium, λ_0 is the vacuum wavelength of the radiation, and $\boldsymbol{K}$ is the scattering vector.

Starting with zy-polarized radiation traveling along the y axis, as shown in Fig. 1-2, we can write the complex field scattered from the jth scatterer into a detector at a distance R from the origin of the scattering system according to [see Eq.(1-65)]

$$\boldsymbol{E}_j(t) = \frac{\omega_0^2 \sin \gamma}{Rc^2} \exp(-i\omega_0 t) E_0 \hat{\boldsymbol{\gamma}} \cdot \boldsymbol{\alpha}_j(t) \exp[i\boldsymbol{K} \cdot \boldsymbol{r}_j(t)], \tag{8-18}$$

where $\hat{\boldsymbol{\gamma}}$ is the unit vector along the polar angle γ. ω_0 and E_0 are the frequency and amplitude of the incident radiation, respectively; $\boldsymbol{\alpha}_j(t)$ is the polarizability tensor of the jth scatterer in the laboratory-fixed axis system, which is time-dependent due to the rotation of a nonspherical scatterer; $\boldsymbol{K}$ is the scattering vector, which bisects the angle between the incident and scattered radiation as shown in Fig. 1-4; and $\boldsymbol{r}_j(t)$ is the center of mass (c.m.) position of the jth scatterer from an arbitrary origin in the scattering system. $\boldsymbol{r}_j(t)$ is, of course, time-dependent if the scatterer has translational freedom (liquid or gas). Figure 8-1 shows again the basic scattering diagram where the incident light, traveling along the y axis, is plane-polarized in the vertical z direction ($E_x^0 = E_y^0 = 0$) and the scattered light is observed along a line that is at an angle of γ with respect to the z axis. The scattering angle θ_s is between the y axis and the line of observation. Of course, we could also start with x-polarized incident radiation. Of the large number of incident and scattered polarizations and values γ and θ_s, we shall usually choose a set of values which corresponds to those which are most often used for experimental studies. These correspond to zy-polarized incident radiation and observation in the xy plane ($\gamma = \pi/2$) of either parallel ($\|$) or perpendicular ($\perp$) polarized scattered radiation. These choices are shown schematically in Fig. 8-1. A polarization analyzer selects the scattered radiation that is polarized parallel to either the $z(\|)$ or $y(\perp)$ axes. These fields observed in the xy plane are given from Eq.(8-18):

$$\|: \quad E_{zj}(t) = \frac{\omega_0^2}{Rc^2} E_z^0 \alpha_{zz}^j(t) \exp(-i\omega_0 t) \exp[i\boldsymbol{K} \cdot \boldsymbol{r}_j(t)]$$

$$\perp: \quad E_{yj}(t) = \frac{\omega_0^2}{Rc^2} E_z^0 \alpha_{zy}^j(t) \exp(-i\omega_0 t) \exp[i\boldsymbol{K} \cdot \boldsymbol{r}_j(t)]. \tag{8-19}$$

We can now write the total fields scattered in phase into the detector by summing over all j scatterers. Substituting this sum into Eq.(8-15) gives the correlation functions:

$$\|: \quad C_z(\tau) = \frac{c}{4\pi} \langle E_z^*(0) E_z(\tau) \rangle$$

$$= A \left\langle \sum_{i,j} \alpha_{zz}^i(0) \alpha_{zz}^j(\tau) \exp(-i\omega_0 \tau) \exp\{-i\boldsymbol{K} \cdot [\boldsymbol{r}_i(0) - \boldsymbol{r}_j(\tau)]\} \right\rangle \tag{8-20}$$

$$\perp: \quad C_y(\tau) = \frac{c}{4\pi} \langle E_y^*(0) E_y(\tau) \rangle$$

$$= A \left\langle \sum_{i,j} \alpha_{zy}^i(0) \alpha_{zy}^j(\tau) \exp(-i\omega_0 \tau) \exp\{-i\boldsymbol{K} \cdot [\boldsymbol{r}_i(0) - \boldsymbol{r}_j(\tau)]\} \right\rangle \tag{8-21}$$

$$A = \frac{\omega_0^4}{R^2 c^4} (E_z^0)^2 \frac{c}{4\pi} = \frac{\omega_0^4}{R^2 c^4} I_0 = \frac{k_0^4}{R^2} I_0,$$

where the sums over i and j are independent with all terms in i at $t = 0$ and all terms in j at $t = \tau$ being included. I_0 is the incident intensity for the plane-polarized radiation. If the incident radiation is unpolarized, we write

$$A = \frac{\omega_0^4}{c^4 R^2}\left(\frac{1}{2}\right)(1 + \cos^2 \theta_s) I_0, \tag{8-22}$$

as shown in Problem 1-5. $C_y(\tau)$, which indicates the correlation function for y-polarized scattered light, is valid at any angle θ_s in the xy plane ($\gamma = \pi/2$). In this expression we use $C_y(\tau) = (c/4\pi)\langle E_y(t)E_y^*(t + \tau)\rangle$. However,

$$C_y(\tau) = \frac{c}{4\pi}\langle E_x(t)E_x^*(t + \tau)\rangle$$

is equally valid. We can write the scattered field at angle θ_s in Fig. 8-1 as a linear combination of $E_x(t)$ and $E_y(t)$, $E_{\theta_s} = E_x \sin \theta_s + E_y \cos \theta_s$, where $\langle E_{\theta_s}(t)E_{\theta_s}^*(t + \tau\rangle = \sin^2 \theta_s \langle E_x(t)E_x^*(t + \tau)\rangle + \cos^2 \theta_s \langle E_y(t)E_y^*(t + \tau)\rangle$, where $\langle E_x(t)E_y(t + \tau)\rangle = 0$. Now, it is easy to show that

$$\langle E_x(t)E_x^*(t + \tau)\rangle = \langle E_y(t)E_y^*(t + \tau)\rangle$$

and, therefore,

$$\langle E_{\theta_s}(t)E_{\theta_s}^*(t + \tau)\rangle = \langle E_x(t)E_x^*(t + \tau)\rangle = \langle E_y(t)E_y^*(t + \tau)\rangle.$$

The correlation functions in Eqs.(8-20) and (8-21) are written in terms of the initial $t = 0$ and later $t = \tau$ positions and orientations and the angular brackets indicate the time average or the equivalent ensemble average according to the ergodic hypothesis.

The orientational correlation is contained in the $\alpha_{zz}^i(0)\alpha_{zz}^j(\tau)$ and $\alpha_{zy}^i(0)\alpha_{zy}^j(\tau)$ terms and the translational correlation is contained in the $\exp\{-i\boldsymbol{K} \cdot [\boldsymbol{r}_i(0) - \boldsymbol{r}_j(\tau)]\}$ phase factors. The $i = j$ terms are called the *self terms* and the $i \neq j$ terms are called the *distinct terms*. In gases, it is reasonable to assume that the dominant contributions will arise from the self terms. Of course, in liquid crystal-like systems [4] we would expect significant $i \neq j$ contributions and in solids we expect even larger $i \neq j$ contributions.

We shall now examine the nature of the polarizability tensor elements that enter the correlation functions in Eqs.(8-20) and (8-21). We review Section 6.10 and remember that in the presence of an electromagnetic field with frequency ω the system (molecule) is described by the time-dependent function $\Psi(\boldsymbol{r}, t)$ in Eq.(6-149) in terms of the stationary states, $\psi_i(\boldsymbol{r})$, and time-dependent coefficients, $c_i(t)$. The $c_i(t)$ coefficients are given in Eq.(6-152) for the dominant interaction that involves the electric dipole moment operator defined in Eq.(6-145). Using these results, we write the time-dependent wavefunction for the kth state as

$$\Psi(\boldsymbol{r}, t) = \psi_k(\boldsymbol{r}) \exp\left(-\frac{iE_k t}{\hbar}\right) + \sum_i{}' c_i(t) \exp\left(-\frac{iE_i t}{\hbar}\right)\psi_i(\boldsymbol{r})$$

$$c_i(t) = \frac{1}{2\hbar}(\boldsymbol{E}_0 \cdot \boldsymbol{D})_{ik}\left\{\frac{\exp[i(\omega_{ik} - \omega_0)t]}{\omega_{ik} - \omega_0} + \frac{\exp[i(\omega_{ik} + \omega_0)t]}{\omega_{ik} + \omega_0}\right\}, \tag{8-23}$$

where the prime on the summation in $\Psi(\boldsymbol{r}, t)$ excludes the $i = k$ term, ω_0 is the incident radiation frequency, and $\omega_{ik} = (E_i - E_k)/\hbar$ [the development following Eq.(6-148) is similar to that used here]. We note from above that the system is initially in the kth state. We can think of the kth state as a vibration-rotation state in the ground electronic state and the sum over i is over the excited electronic (vibration-rotation) states. We now examine the transition dipole operator matrix element between the k' and k states for the time-dependent system:

$$(\boldsymbol{D})_{k'k} = \frac{\int \Psi_{k'}^*(\boldsymbol{r}, t)(\boldsymbol{D})\Psi_k(\boldsymbol{r}, t)\, dV}{\int \Psi_{k'}^*(\boldsymbol{r}, t)\Psi_k(\boldsymbol{r}, t)\, dV} = \int \left[\psi_{k'}^* \exp\left(\frac{iE_{k'}t}{\hbar}\right) + \sum_i{}' c_i^*(t) \exp\left(\frac{iE_i t}{\hbar}\right)\psi_i^*\right](\boldsymbol{D})$$

$$\times \left[\psi_k \exp\left(-\frac{iE_k t}{\hbar}\right) + \sum_i{}' c_i(t) \exp\left(-\frac{iE_i t}{\hbar}\right)\psi_i\right] dV$$

$$= D_{k'k} \exp(i\omega_{k'k}t) + \sum_i{}' \exp(i\omega_{ik}t)c_i^*(t)\boldsymbol{D}_{ik} + \sum_i{}' \exp(i\omega_{k'i}t)c_i(t)\boldsymbol{D}_{k'i},$$

where the denominator is unity to first order in the coefficients, $c_i(t)$, and the smaller $c_i^*(t)c_i(t)$ terms have been dropped. We note from this expression that k and k' can represent two different vibration-rotation states in the ground electronic state and the primes on the summation signs omit the $i = k$ and $i = k'$ terms, respectively. Substituting $c_i(t)$ from Eq.(8-23) into the equation above for $(\boldsymbol{D})_{k'k}$ gives

$$(\boldsymbol{D})_{k'k} = \exp(i\omega_{k'k}t)\boldsymbol{D}_{kk'} + \frac{\boldsymbol{E}_0}{2\hbar}\cdot\sum_i{}' \boldsymbol{D}_{ki}\boldsymbol{D}_{ik'}$$

$$\times \left\{\frac{\exp[i(\omega_{ik} - \omega_{ik'} + \omega_0)t]}{\omega_{ik'} - \omega_0} + \frac{\exp[i(\omega_{ik} - \omega_{ik'} - \omega_0)t]}{\omega_{ik'} + \omega_0}\right\}$$

$$+ \frac{\boldsymbol{E}_0}{2\hbar}\cdot\sum_i{}' \boldsymbol{D}_{k'i}\boldsymbol{D}_{ik}\left\{\frac{\exp[i(\omega_{ik} + \omega_{k'i} - \omega_0)t]}{\omega_{ik} - \omega_0} + \frac{\exp[i(\omega_{ik} + \omega_{k'i} + \omega_0)t]}{\omega_{ik} + \omega_0}\right\}, \tag{8-24}$$

where we have ignored the small terms of order $|c_i(t)|^2$ as mentioned above. $\boldsymbol{D}_{ki}\boldsymbol{D}_{ik'}$ is a dyadic product. Remembering that

$$\omega_{ik} - \omega_{ik'} = \omega_{ik} + \omega_{k'i} = \frac{1}{\hbar}(E_i - E_k - E_i + E_{k'}) = \frac{1}{\hbar}(E_{k'} - E_k) = \omega_{k'k} = -\omega_{kk'},$$

where ω_{ik} is an electronic transition frequency and $\omega_{kk'}$ is a vibration-rotation or rotational transition frequency in the ground electronic state, we can rewrite this equation to give

$$(\boldsymbol{D})_{k'k} = \exp(i\omega_{k'k}t)\left\{\boldsymbol{D}_{kk'} + \frac{\boldsymbol{E}_0}{2\hbar}\cdot\sum_i{}' \boldsymbol{D}_{ki}\boldsymbol{D}_{ik'}\left[\frac{\exp(i\omega_0 t)}{\omega_{ik'} - \omega_0} + \frac{\exp(-i\omega_0 t)}{\omega_{ik'} + \omega_0}\right]\right.$$

$$\left. + \frac{\boldsymbol{E}_0}{2\hbar}\cdot\sum_i{}' \boldsymbol{D}_{k'i}\boldsymbol{D}_{ik}\left[\frac{\exp(-i\omega_0 t)}{\omega_{ik} - \omega_0} + \frac{\exp(i\omega_0 t)}{\omega_{ik} + \omega_0}\right]\right\}. \tag{8-25}$$

If $k = k'$, $\omega_{kk'} = 0$, and if we let the $k = k'$ state be the ground state, Eq.(8-25) reduces to $\langle \boldsymbol{D} \rangle$ following Eq.(6-152), leading to the polarizability in Eq.(6-153). Returning to Eq.(8-25), we note that normally $\omega_{ik} \gg \omega_{k'k}$ because ω_{ik} is the angular frequency of an electronic transition and $\omega_{k'k}$ is the angular frequency of a vibrational or rotational transition. Thus, it is reasonable to assume that $\omega_{ik} \cong \omega_{ik'}$ for Raman and Rayleigh scattering.

If $\omega_{ik} \cong \omega_{ik'}$ in Eq.(8-25), we can write

$$\lim_{\omega_{ik} \cong \omega_{ik'}} (\boldsymbol{D})_{k'k} = \exp(i\omega_{k'k}t)\left[\boldsymbol{D}_{kk'} + \frac{2}{\hbar}\boldsymbol{E}_0 \cdot \sum_i{}' \frac{\boldsymbol{D}_{ki}\boldsymbol{D}_{ik'}\omega_{ik} \cos \omega_0 t}{(\omega_{ik}^2 - \omega_0^2)}\right]. \tag{8-26}$$

Thus, the polarizability is given from $\boldsymbol{D}_{\text{ind}} = \boldsymbol{\alpha} \cdot \boldsymbol{E}$ to be

$$\boldsymbol{\alpha} = 2 \exp(i\omega_{k'k}t) \sum_i{}' \frac{\boldsymbol{D}_{ki}\boldsymbol{D}_{ik'}\omega_{ik}}{\hbar(\omega_{ik}^2 - \omega_0^2)}. \tag{8-27}$$

If $k' = k$, this value of $\boldsymbol{\alpha}$ reduces to the result in Eq.(6-153). The general expression for the polarizability tensor in Eq.(8-27) shows that the polarizability and resultant intensity of the scattering will increase as $\omega_0 \longrightarrow \omega_{ik}$. This condition of resonant Rayleigh ($k = k'$) [5] and resonant Raman ($k \neq k'$) [6] scattering is useful in enhancing the sensitivity of the scattering if a suitable light source is available with a frequency, ω_0, near one of the electronic resonant frequencies, ω_{ik}. Resonant vibrational Raman scattering has been particularly useful in identifying specific local vibrations on a site in a large macromolecule [7].

We can relate the space-fixed polarizability tensor in Eq.(8-27), $\boldsymbol{\alpha}(xyz)$, to the corresponding values in the molecular-fixed axes, $\boldsymbol{\alpha}(abc)$, by the direction cosine transformation in Eq.(B-35), which gives $\boldsymbol{\alpha}(xyz) = \mathbf{C}\boldsymbol{\alpha}(abc)\tilde{\mathbf{C}}$. According to Eqs.(8-20) and (8-21), we need

$$\begin{aligned} \alpha_{zz} &= C_{za}\alpha_{aa}C_{az} + C_{zb}\alpha_{bb}C_{bz} + C_{zc}\alpha_{cc}C_{cz} \\ \alpha_{zy} &= C_{za}\alpha_{aa}C_{ay} + C_{zb}\alpha_{bb}C_{by} + C_{zc}\alpha_{cc}C_{cy}. \end{aligned} \tag{8-28}$$

Using the direction cosine transformation in Eq.(4-208) for a cylindrically symmetric molecule where $\alpha_{aa} \neq \alpha_{bb} = \alpha_{cc}$ (a is the symmetry axis) gives

$$\alpha_{zz} = \alpha + \tfrac{2}{3}(\alpha_{aa} - \alpha_{bb})P_2(\cos\theta), \qquad \alpha_{zy} = (\alpha_{aa} - \alpha_{bb}) \cos\theta \sin\theta \sin\phi, \tag{8-29}$$

where $P_2(\cos\theta) = \frac{1}{2}(3\cos^2\theta - 1)$ is the Legendre polynomial of order $l = 2$ (see Appendix C.1).

If the scatterer (molecule) is vibrating, we must also include the vibrational dependence in the polarizability tensor elements by expanding each tensor element in the molecular-fixed axis system about the small amplitude molecular vibrations [see Eq.(4-273)],

$$\alpha_{aa} = \alpha_{aa}^0 + \sum_i \left(\frac{\partial \alpha_{aa}}{\partial Q_i}\right)_0 Q_i + \cdots, \tag{8-30}$$

where α_{aa}^0 is the equilibrium structure polarizability. The change in polarizability with normal coordinate, $(\partial \alpha_{aa}/\partial Q_i)$, is evaluated at equilibrium indicated by the zero subscript.

In summary, we have developed expressions to write the correlation functions in Eqs.(8-20) and (8-21) for the general vibrating-rotating molecule.

8.2 RAYLEIGH AND RAMAN SCATTERING IN GASES, KINETIC THEORY, AND ROTATIONAL QUENCHING

We shall now examine the spectrum of light scattered from a cylindrically symmetric molecule in the dilute gas limit where the mean free path of the scatterer (molecule) is long relative to the radiation wavelength. We start by assuming, in this dilute gas limit, that only the self $(i = j)$ terms in Eqs.(8-20) and (8-21) contribute. Furthermore, we assume separation of translational, rotational, and vibrational motion of the molecule. The translational phase factor in Eqs.(8-20) and (8-21) for the molecule moving with velocity $\boldsymbol{v}$ is given by $(i = j)$

$$\exp\{-i\boldsymbol{K}\cdot[\boldsymbol{r}_i(0) - \boldsymbol{r}_i(\tau)]\} = \exp(i\boldsymbol{K}\cdot\boldsymbol{v}\tau), \tag{8-31}$$

where $\boldsymbol{r}_i(\tau) - \boldsymbol{r}_i(0) = \boldsymbol{v}\tau$ has been used. Using the independent particle assumption, substituting Eq.(8-30) into Eqs.(8-29), substituting these results along with Eqs.(8-31) and (8-27) into Eqs.(8-20) and (8-21), and using the ergodic hypothesis gives the correlation functions for the $k \longrightarrow k'$ transition (all cross terms vanish for an isotropic gaseous sample):

$$\begin{aligned}
C_z(t) = {} & NA \exp(-i\omega_0 t) \exp(i\boldsymbol{K}\cdot\boldsymbol{v}t) P_k \exp(i\omega_{k'k} t) \\
& \times \Bigg(\left(\int \psi_{k'}^* \alpha^0 \psi_k \, dV \right)^2 + \left\{ \int \psi_{k'}^* \left[\frac{2}{3} (\alpha_{aa}^0 - \alpha_{bb}^0) P_2(\cos\theta) \right] \psi_k \, dV \right\}^2 \\
& + \left(\int \psi_{k'}^* \left(\frac{\partial \alpha}{\partial Q_j} \right)_0 Q_j \psi_k \, dV \right)^2 \\
& + \left\{ \int \psi_{k'}^* \left[\left(\frac{\partial \alpha_{aa}}{\partial Q_j} \right)_0 - \left(\frac{\partial \alpha_{bb}}{\partial Q_j} \right)_0 \right] Q_j P_2(\cos\theta) \psi_k \, dV \right\}^2 \Bigg)
\end{aligned}$$

$$\begin{aligned}
C_y(t) = {} & NA \exp(-i\omega_0 t) \exp(i\boldsymbol{K}\cdot\boldsymbol{v}t) P_k \exp(i\omega_{k'k} t) \\
& \times \Bigg(\left[\int \psi_{k'}^* (\alpha_{aa}^0 - \alpha_{bb}^0) \cos\theta \sin\theta \sin\phi \psi_k \, dV \right]^2 \\
& + \left\{ \int \psi_{k'}^* \left[\left(\frac{\partial \alpha_{aa}}{\partial Q_j} \right)_0 - \left(\frac{\partial \alpha_{bb}}{\partial Q_j} \right)_0 \right] Q_j \cos\theta \sin\theta \sin\phi \psi_k \, dV \right\}^2 \Bigg)
\end{aligned}$$

$$P_k = \frac{N_k}{N}. \tag{8-32}$$

N is the total number of scatterers equal to $\rho_0 V_s$, the number density times the illuminated volume focused onto the detector. P_k is the probability that the system is in the initial state k where N_k is the number of molecules in the kth state. ψ_k represents the product of the rotational and vibrational functions in the kth state where the averages over the electronic wavefunctions are now included in the α_{aa}^0, α_{bb}^0, $(\partial\alpha_{aa}/\partial Q_j)_0$, and $(\partial\alpha_{bb}/\partial Q_j)_0$ terms. The $\psi_{k'}\psi_k$ matrix elements in $C_z(t)$, where the primes indicate the final state, determine the selection rules for the scattering processes.

We shall now specialize our discussion to $C_z(t)$ in Eq.(8-32) and include only parallel molecular vibrations. Substituting $\psi_k = Y_{JM}(\theta, \phi)\psi_{n_j}(Q_j)$, where $Y_{JM}(\theta, \phi)$ is the rotational eigenfunction for the linear molecule and $\psi_{n_j}(Q_j)$ is the harmonic oscillator function for the jth parallel normal mode (where n_j is the vibration state), gives the following expression for $C_z(t)$:

$$
\begin{aligned}
C_z(t) = {} & A \exp(-i\omega_0 t) \exp(i\boldsymbol{K}\cdot\boldsymbol{v}t)\Bigg(N(\alpha^0)^2 + N_{JM} \exp(i\omega_R t)\left[\frac{4}{3}\left(\frac{\pi}{5}\right)^{1/2}(\alpha_{aa}^0 - \alpha_{bb}^0)\right. \\
& \times \left.\int_0^{2\pi}\int_0^{\pi} Y_{JM}^* Y_{20} Y_{J'M'} \sin\theta\, d\theta\, d\phi\right]^2 + N_{n_j}\exp(i\omega_j t)\left[\left(\frac{\partial\alpha}{\partial Q_j}\right)_0 \langle n_j|Q_j|n_j'\rangle\right]^2 \\
& + N_{n_jJM}\exp(i\omega_{jR}t)\left\{\frac{4}{3}\left(\frac{\pi}{5}\right)^{1/2}\left[\frac{\partial(\alpha_{aa}-\alpha_{bb})}{\partial Q_j}\right]_0 \langle n_j|Q_j|n_j'\rangle\right. \\
& \times \left.\int_0^{2\pi}\int_0^{\pi} Y_{JM}^* Y_{20} Y_{J'M'} \sin\theta\, d\theta\, d\phi\right\}^2\Bigg)
\end{aligned}
$$

$$
\omega_R = \frac{(E_{J'M'} - E_{JM})}{\hbar}, \qquad \omega_j = \frac{(E_{n_j'} - E_{n_j})}{\hbar}, \qquad \omega_{jR} = \frac{(E_{n_j'} + E_{J'M'} - E_{n_j} - E_{JM})}{\hbar}. \tag{8-33}
$$

$P_2(\cos\theta) = 2(\pi/5)^{1/2} Y_{20}$ has also been used in this equation. ω_R is the pure rotational frequency, ω_j is the pure vibrational frequency, and ω_{jR} is the rotational-vibrational frequency. N is the total number of scatterers, N_{JM} is the number of molecules in the JM rotational state, N_{n_j} is the number of molecules in the nth state of the jth vibrational mode, and N_{n_jJM} is the number of molecules in the n_j, J, M vibrational-rotational state. The selection rules for this case of rotational, parallel vibration, and parallel vibration-rotation Raman scattering were considered following Eq.(4-273). The first term in Eq.(8-33) is the Rayleigh scattering weighted by $(\alpha^0)^2$. The second term is the pure rotational Raman scattering term with selection rules $\Delta J = 0, \pm 2, \Delta M = 0$ and we abbreviate the rotational-dependent amplitude factor by

$$
f_{JM} = N_{JM}\left[\frac{4}{3}\left(\frac{\pi}{5}\right)^{1/2}(\alpha_{aa}^0 - \alpha_{bb}^0)\int_0^{2\pi}\int_0^{\pi} Y_{JM}^* Y_{20} Y_{J'M'} \sin\theta\, d\theta\, d\phi\right]^2. \tag{8-34}
$$

The third term is the pure vibrational Raman term with selection rules $\Delta n = +1$ for the harmonic oscillator and we abbreviate the vibrational scattering amplitude for the jth parallel normal mode by

$$
f_j = N_{n_j}\left[\left(\frac{\partial\alpha}{\partial Q_j}\right)_0 \langle n_j|Q_j|n_j'\rangle\right]^2. \tag{8-35}
$$

The fourth term in Eq.(8-33) is the vibration-rotation Raman term with selection rules $\Delta n = +1, \Delta J = 0, \pm 2, \Delta M = 0$ and the vibration-rotation scattering amplitude is abbreviated by

$$f_{JMj} = N_{n_jJM}\left\{\frac{4}{3}\left(\frac{\pi}{5}\right)^{1/2}\left[\left(\frac{\partial \alpha_{aa}}{\partial Q_j}\right)_0 - \left(\frac{\partial \alpha_{bb}}{\partial Q_j}\right)_0\right]\langle n_j|Q_j|n_j'\rangle\right.$$

$$\left.\times \int_0^{2\pi}\int_0^{\pi} Y_{JM}^* Y_{20} Y_{J'M'} \sin\theta \, d\theta \, d\phi\right\}^2. \tag{8-36}$$

Substituting Eqs.(8-34), (8-35), and (8-36) into Eq.(8-33) gives

$$C_z(t) = A \exp(-i\omega_0 t)\exp(i\boldsymbol{K}\cdot\boldsymbol{v}t)[N(\alpha_0)^2 + f_{JM}\exp(i\omega_R t)$$

$$+ f_j \exp(i\omega_j t) + f_{JMj}\exp(i\omega_{jR} t)]. \tag{8-37}$$

Now, we note that Eq.(8-37) was derived for a single molecular velocity, $\boldsymbol{v}$, and independent particles with no collisional relaxations. We shall now add the average over the Boltzmann velocity distribution as well as include both rotational and vibrational relaxation properties.

First, we examine the velocity average of the exp $(i\boldsymbol{K}\cdot\boldsymbol{v}t)$ phase factor in Eq.(8-37). The average of $\exp(i\boldsymbol{K}\cdot\boldsymbol{v}t)$ over all velocities in three dimensions is obtained by generalizing Eq.(7-27) for the velocity probability according to

$$[\exp(i\boldsymbol{K}\cdot\boldsymbol{v}t)]_{av} = \left(\frac{M}{2\pi kT}\right)^{3/2}\int_{-\infty}^{+\infty}\int_{-\infty}^{+\infty}\int_{-\infty}^{+\infty}\exp(i\boldsymbol{K}\cdot\boldsymbol{v}t)$$

$$\times \exp\left[-\frac{M}{2kT}(v_x^2 + v_y^2 + v_z^2)\right] dv_x\, dv_y\, dv_z$$

$$= \exp\left[-\left(\frac{kT}{2M}\right)K^2t^2\right]. \tag{8-38}$$

The half width at half height of this Gaussian (in time) is

$$\Delta t = \frac{1}{K}\left(\frac{2M \ln 2}{kT}\right)^{1/2}. \tag{8-39}$$

Finally, we add the effects of first-order molecular relaxation of the rotational, τ_R, and vibrational, τ_V, states where we expect that $\tau_V \gg \tau_R$. Adding the appropriate first-order relaxations [see discussion following Eq.(7-6)] to Eq.(8-37) and substituting Eq.(8-38) for the velocity average gives

$$C_z(t) = A\exp(-i\omega_0 t)\exp\left[-\left(\frac{kT}{2M}\right)K^2t^2\right]\left[N(\alpha^0)^2 + f_{JM}\exp(i\omega_R t)\exp\left(-\frac{t}{\tau_R}\right)\right.$$

$$\left. + f_j\exp(i\omega_j t)\exp\left(-\frac{t}{\tau_j}\right) + f_{JMj}\exp(i\omega_{jR}t)\exp\left(-\frac{t}{\tau_R}\right)\exp\left(-\frac{t}{\tau_j}\right)\right]. \tag{8-40}$$

N is the number of scatterers; A is defined in Eqs.(8-21) and (8-22); and f_{JM}, f_j, and f_{JMj} are defined in Eqs.(8-34), (8-35), and (8-36), respectively. τ_R is the rotational relaxation time of the JMth rotational state and τ_j is the vibrational relaxation time of the jth vibrational state.

There are two interesting limits to Eq.(8-40); the low-pressure limit where the $\exp[-(kT/2M)K^2t^2]$ damping terms are dominant relative to the $\exp(-t/\tau_R)$ and $\exp(-t/\tau_j)$ relaxation terms and the high-pressure limit where the relaxation damping exceeds the Gaussian $\exp[-(kT/2M)K^2t^2]$ damping. We shall now examine the spectra obtained in these two limits. Substituting Eq.(8-40) into Eq.(8-16) gives the intensity per unit frequency for the N scatterers [Δt is defined in Eq.(8-39)]:

Low pressure: $\Delta t \ll \tau_R$, $\Delta t \ll \tau_j$

$$I_v^v(\omega) = \frac{1}{\pi}\,\mathrm{Re}\int_0^\infty C_z(t)\exp(i\omega t)\,dt$$
$$= \frac{A}{K}\left(\frac{M}{2\pi kT}\right)^{1/2}\left\{N(\alpha^0)^2\exp\left[-\left(\frac{M}{2K^2kT}\right)(\omega-\omega_0)^2\right]\right.$$
$$+ f_{JM}\exp\left[-\left(\frac{M}{2K^2kT}\right)(\omega+\omega_R-\omega_0)^2\right]$$
$$+ f_j\exp\left[-\left(\frac{M}{2K^2kT}\right)(\omega+\omega_j-\omega_0)^2\right]$$
$$\left.+ f_{JMj}\exp\left[-\left(\frac{M}{2K^2kT}\right)(\omega+\omega_{jR}-\omega_0)^2\right]\right\} \tag{8-41}$$

High pressure: $\Delta t \gg \tau_R$, $\Delta t \gg \tau_j$

$$I_v^v(\omega) = \frac{NA}{K}\left(\frac{M}{2\pi kT}\right)^{1/2}(\alpha^0)^2\exp\left[-\left(\frac{M}{2K^2kT}\right)(\omega-\omega_0)^2\right]$$
$$+\frac{A}{\pi}f_{JM}\left[\frac{(1/\tau_R)}{(1/\tau_R)^2+(\omega+\omega_R-\omega_0)^2}\right]+\frac{A}{2\pi}f_j\left[\frac{(1/\tau_j)}{(1/\tau_j)^2+(\omega+\omega_j-\omega_0)^2}\right]$$
$$+\frac{A}{2\pi}f_{JMj}\left[\frac{(1/\tau_j+1/\tau_R)}{(1/\tau_j+1/\tau_R)^2+(\omega+\omega_{jR}-\omega_0)^2}\right]. \tag{8-42}$$

The first terms in Eqs.(8-41) and (8-42) are the Rayleigh scattering terms, the second terms with center frequencies $\omega_0 - \omega_R$ represent the pure rotational Raman scattering, the third terms with center frequencies $\omega_0 - \omega_j$ are the pure vibrational Raman terms, and the fourth terms with frequencies $\omega_0 - \omega_{jR}$ are the rotation-vibration Raman terms [see definitions of ω_R, ω_j, and ω_{jR} in Eqs.(8-33)]. All four terms in Eq.(8-41) for the Doppler broadened low-pressure limit are Gaussian curves with half widths at half heights of $\Delta\omega = K(2kT\ln 2/M)^{1/2}$, where K is the magnitude of the scattering vector. Thus, we find the interesting result that the half width is proportional to the scattering vector, $K = (4\pi/\lambda)\sin(\theta_s/2)$, and we conclude that, in this low-pressure limit, very high resolution is possible at low scattering angles (see Fig. 8-1) [8].

Equation (8-42) in the high-pressure limit shows a Gaussian for the Rayleigh scattering (centered at ω_0) and Lorentzian line shapes for all the Raman scattering. The half widths at half heights for the remaining terms are $\Delta\omega = 1/\tau_R$ for the pure rotational term, $\Delta\omega = 1/\tau_j$ for the pure vibrational term, and $\Delta\omega = (1/\tau_j + 1/\tau_R)$ for the vibration-rotation term.

In the intermediate case where both Doppler effects and relaxation processes are present, Eq.(8-40) can be Fourier transformed directly to the spectra, which will be a convolution of the Gaussian and Lorentzian curves as shown in detail in Section 7.2.

We also note from the definition of the f coefficients in Eqs.(8-34), (8-35), and (8-36) and the definition of the N_k factor of Eqs.(8-32), that when $E(n_jJM) > E(n_j'J'M')$ with any combination of quantum numbers n_j, J, M, n_j', J' and M', where the primes indicate the final state, then $\omega_{n_j'J'M',n_jJM}$ is negative leading to increased Raman frequencies over the incident frequency ω_0. These transitions are called *anti-Stokes transitions.* The *Stokes transitions* give rise to observed frequencies lower than ω_0 due to the $E(n_jJM) < E(n_j'J'M')$ condition leading to positive values of $\omega_{n_j'J'M',n_jJM}$. Of course, it also follows from the definition of N_k in Eqs.(8-32) that the Stokes transitions $[E(n_jJM) < E(n_j'J'M')]$ will be more intense than the anti-Stokes transitions $[E(n_jJM) > E(n_j'J'M')]$.

The experimental polarized $I_v^v(v)$ vibration and vibration-rotation Raman spectra for N_2 and O_2 and the Rayleigh and pure rotational Raman spectra of CO_2 are shown in Fig. 8-2. Remembering the $P\,(\Delta J = -1)$, $Q\,(\Delta J = 0)$, $R\,(\Delta J = +1)$ notation as shown in Fig. 4-31, we now include $O\,(\Delta J = -2)$ and $S\,(\Delta J = +2)$ notation. Using these, we note that the S-branch $(\Delta J = +2)$ lower-frequency rotational Raman–Stokes transitions are more intense than the corresponding O-branch $(\Delta J = -2)$ anti-Stokes transitions. The pure vibrational Stokes $(\Delta n = +1)$ and the Stokes S-branch $(\Delta J = +2)$ and anti-Stokes O-branch $(\Delta J = -2)$ rotational-vibrational spectra are shown in N_2 and O_2.

We now return to $C_y(t)$ in Eqs.(8-32). Substituting

$$\cos\theta \sin\theta \sin\phi = i\left(\frac{2\pi}{15}\right)^{1/2}(Y_{2-1} - Y_{21})$$

and going through the steps described following Eqs.(8-32) gives an expression for $C_y(t)$ similar to Eq.(8-40):

$$C_y(t) = A \exp(-i\omega_0 t) \exp\left[-\left(\frac{kT}{M}\right)K^2t^2\right]\left[f'_{JM} \exp(-i\omega_R t) \exp\left(-\frac{t}{\tau_R}\right)\right.$$
$$\left. + f'_{JMj} \exp(-i\omega_{jR} t) \exp\left(-\frac{t}{\tau_R}\right) \exp\left(-\frac{t}{\tau_j}\right)\right]$$
$$f'_{JM} = N_{JM}\left(\frac{2\pi}{15}\right)\left[(\alpha_{aa}^0 - \alpha_{bb}^0)\int_0^{2\pi}\int_0^{\pi} Y_{JM}^*(Y_{2-1} - Y_{21})Y_{J'M'} \sin\theta\, d\theta\, d\phi\right]^2$$
$$f'_{JMj} = N_{n_jJM}\left(\frac{2\pi}{15}\right)\left\{\left[\left(\frac{\partial\alpha_{aa}}{\partial Q_j}\right)_0 - \left(\frac{\partial\alpha_{bb}}{\partial Q_j}\right)_0\right]\langle n_j'|Q_j|n_j\rangle\right.$$
$$\left.\times \int_0^{2\pi}\int_0^{\pi} Y_{JM}^*(Y_{2-1} - Y_{21})Y_{J'M'} \sin\theta\, d\theta\, d\phi\right\}^2. \tag{8-43}$$

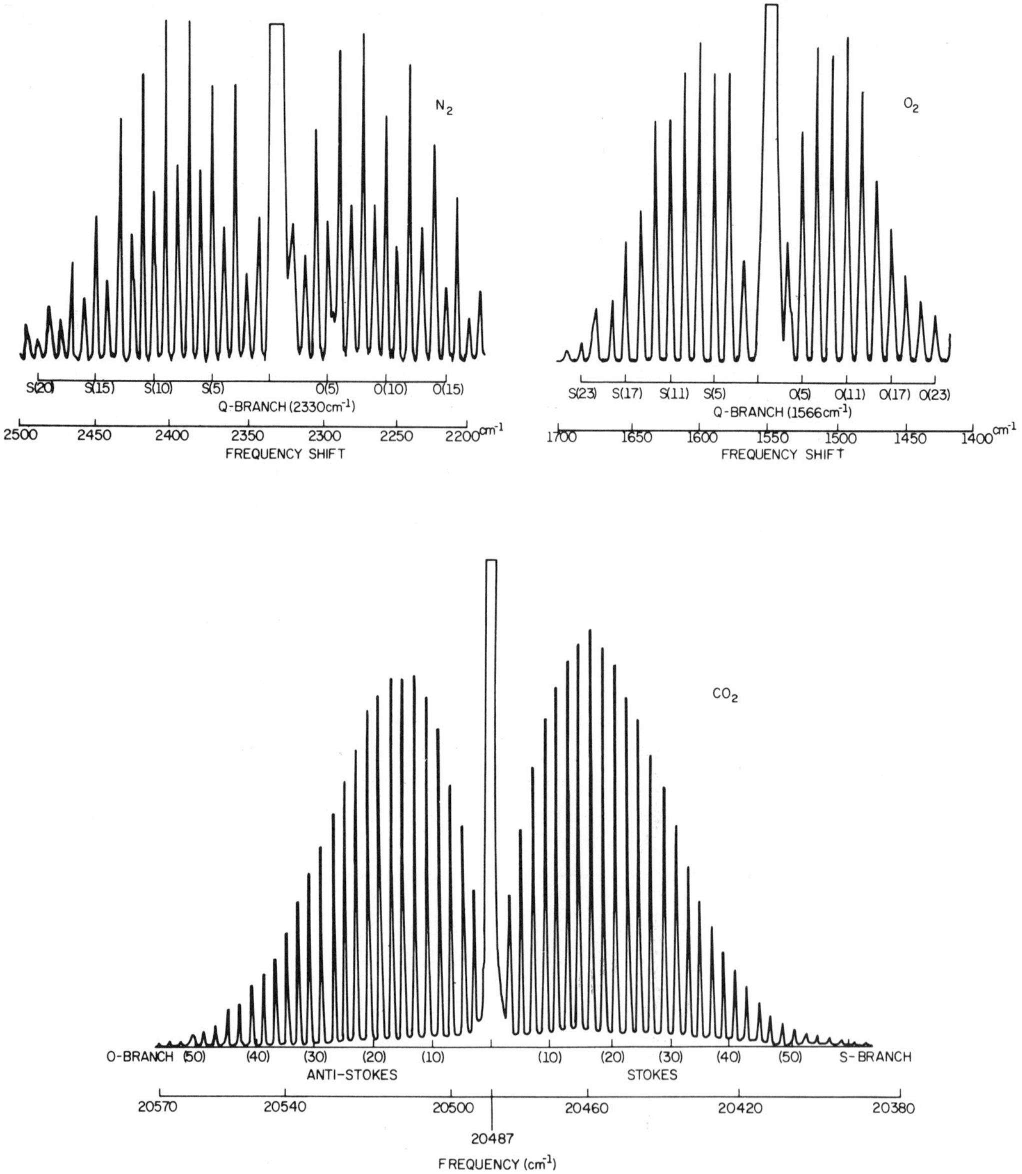

Figure 8-2 Polarized, $I_v^v(v)$, vibration-rotation Raman spectra in N_2 and O_2 and polarized Rayleigh and Raman rotational spectra in CO_2 all taken at pressures near 1 atm. The center line in the O_2 and N_2 spectra is a pure vibrational transition and the center line in the CO_2 spectra is the Rayleigh line. The exciting light is the $\lambda_0 =$ 4881 Å (or 20,487 cm^{-1}) Ar^+ ion laser line. The alternation in intensities in the N_2 spectrum is discussed in Problem 4-38. These spectra were obtained from J. J. Barrett [see J. J. Barrett and N. I. Adams, *J. Opt. Soc. Am.* **58**, 311 (1969)].

Thus, the $C_y(t)$ correlation function, which leads to the depolarized spectra, has a pure rotational term with selection rules $\Delta J = 0, \pm 2$, $\Delta M = \pm 1$, and a vibration-rotation term with selection rules for parallel vibrations of $\Delta n_i = \pm 1$, $\Delta J = 0, \pm 2$, $\Delta M = \pm 1$. The time-frequency Fourier transform of $C_y(t)$ in Eq.(8-43) to give the $I_h^v(\omega)$ spectrum follows the methods discussed previously to obtain $I_v^v(\omega)$ for both the high- and low-pressure limits. These straightforward steps are left to the reader [9].

We shall now return to Eqs.(8-20) and (8-21) to examine the correlation functions at very high pressures or in the condensed phase (liquid) where the time between collisions is short relative to a molecular rotation period. In this limit of rotational state quenching, the concept of a rotational state is no longer meaningful and we must use an alternate development for the correlation functions leading to the spectra. To simplify our discussion, we shall initially assume a rigid nonvibrating rotor and substitute Eqs.(8-29) into Eqs.(8-20) and (8-21) to give

$$\parallel:\quad C_z(t) = \frac{c}{4\pi}\langle E_z(0)E_z^*(t)\rangle$$
$$= A\left\langle \sum_{i,j}\left[\alpha + \frac{2}{3}(\alpha_{aa} - \alpha_{bb})P_2(\cos\theta)\right]_{oi}\left[\alpha + \frac{2}{3}(\alpha_{aa} - \alpha_{bb})P_2(\cos\theta)\right]_j \right.$$
$$\left. \times \exp(-i\omega_0 t)\exp\{-i\boldsymbol{K}\cdot[\boldsymbol{r}_i(0) - \boldsymbol{r}_j(t)]\}\right\rangle. \tag{8-44}$$

$$\perp:\quad C_y(t) = \frac{c}{4\pi}\langle E_y(0)E_y^*(t)\rangle$$
$$= A\left\langle \sum_{i,j}(\cos\theta\sin\theta\sin\phi)_{oi}(\cos\theta\sin\theta\sin\phi)_j(\alpha_{aa} - \alpha_{bb})_{oi}(\alpha_{aa} - \alpha_{bb})_j \right.$$
$$\left. \times \exp(-i\omega_0 t)\exp\{-i\boldsymbol{K}\cdot[\boldsymbol{r}_i(0) - \boldsymbol{r}_j(t)]\}\right\rangle, \tag{8-45}$$

where θ and ϕ are time-dependent due to molecular rotation. The subscript 0 indicates the $t = 0$ values in the correlation functions. If the scatterers are randomly orientated in space (isotropic distribution), the averages of the $\alpha P_2(\cos\theta)_j$ and $\alpha P_2(\cos\theta)_i$ cross terms in Eq.(8-44) will vanish. We also assume that all nonvibrating molecules or scatterers in the gas, liquid, or solid are equivalent and that α and $(\alpha_{aa} - \alpha_{bb})$ are time-independent. Acknowledging the statements above, we rewrite Eqs.(8-44) and (8-45) to give

$$C_z(t) = A\exp(-i\omega_0 t)\alpha^2\left\langle \sum_{i,j}\exp\{-i\boldsymbol{K}\cdot[\boldsymbol{r}_i(0) - \boldsymbol{r}_j(t)]\}\right\rangle$$
$$+ \frac{4}{9}A\exp(-i\omega_0 t)(\alpha_{aa} - \alpha_{bb})^2\left\langle \sum_{i,j}P_2(\cos\theta)_{oi}P_2(\cos\theta)_j \right.$$
$$\left. \times \exp\{-i\boldsymbol{K}\cdot[\boldsymbol{r}_i(0) - \boldsymbol{r}_j(t)]\}\right\rangle \tag{8-46}$$

$$C_y(t) = A\exp(-i\omega_0 t)(\alpha_{aa} - \alpha_{bb})^2\left\langle \sum_{i,j}(\cos\theta\sin\theta\sin\phi)_{oi}(\cos\theta\sin\theta\sin\phi)_j \right.$$
$$\left. \times \exp\{-i\boldsymbol{K}\cdot[\boldsymbol{r}_i(0) - \boldsymbol{r}_j(t)]\}\right\rangle. \tag{8-47}$$

As mentioned following Eq.(8-22), the $i = j$ terms are called the *self terms* and the $i \neq j$ terms are called the *distinct terms*. Using the ergodic hypothesis, the correlation functions for identical scatterers are given by the following averages over position and orientation:

$$C_z(t) = A \exp(-i\omega_0 t)N\alpha^2 \int_{V_s} \exp(i\boldsymbol{K}\cdot\boldsymbol{R})P(\boldsymbol{R}, t)\, dV_s$$

$$+ \frac{4}{9} A \exp(-i\omega_0 t) \frac{N(\alpha_{aa} - \alpha_{bb})^2}{4\pi} \int_0^{2\pi}\int_0^{2\pi}\int_{-1}^{1}\int_{-1}^{1}\int_{V_s}$$

$$\times \exp(i\boldsymbol{K}\cdot\boldsymbol{R})P_2(\cos\theta_0)P_2(\cos\theta)P(\boldsymbol{R}, \theta, \phi, t)\, dV_s\, d\cos\theta_0\, d\cos\theta\, d\phi_0\, d\phi \tag{8-48}$$

$$C_y(t) = A \exp(-i\omega_0 t) \frac{N(\alpha_{aa} - \alpha_{bb})^2}{4\pi} \int_0^{2\pi}\int_0^{2\pi}\int_{-1}^{1}\int_{-1}^{1}\int_{V_s}$$

$$\times \exp(i\boldsymbol{K}\cdot\boldsymbol{R})(\cos\theta\sin\theta\sin\phi)_0(\cos\theta\sin\theta\sin\phi)$$

$$\times P(\boldsymbol{R}, \theta, \phi, t)\, dV_s\, d\cos\theta_0\, d\cos\theta\, d\phi_0\, d\phi, \tag{8-49}$$

where we have replaced $\int_0^{\pi} \sin\theta\, d\theta$ with $\int_{-1}^{1} d\cos\theta$ for convenience. N is the number of scatterers within the scattering volume V_s and dV_s indicates the volume element and corresponding integral over the scattering volume (the illuminated volume that is focused onto the detector). The $P(a)$ functions in these expressions are the space-time and space-time-orientation correlation functions. $P(\boldsymbol{R}, t)$ is the probability per unit volume that if a scattering center is at position $\boldsymbol{R} = 0$ at $t = 0$, there will also be a scattering center at $\boldsymbol{R}$ at t. Thus, it is evident that $P(\boldsymbol{R}, t)$ contains both the self and distinct terms in both space and time. $P(\boldsymbol{R}, \theta, \phi, t)$ is the space-time-orientation correlation function, which is the probability per unit volume that if a particle's center of interaction is at $\boldsymbol{R} = 0$ with orientation θ_0 and ϕ_0 at $t = 0$, there will also be a particle at $\boldsymbol{R}$ with orientation θ and ϕ at t. The additional factor of $(1/4\pi)$ associated with the $P(\boldsymbol{R}, \theta, \phi, t)$ probabilities assumes normalization of the probabilities to the $t = 0$ initial conditions. When $P(\boldsymbol{R}, t) = \delta(\boldsymbol{R})$, the three-dimensional delta function at $t = 0$, we write

$$\int_{V_s} \exp(i\boldsymbol{K}\cdot\boldsymbol{R})P(\boldsymbol{R}, 0)\, dV_s = \int_{V_s} \exp(i\boldsymbol{K}\cdot\boldsymbol{R})\delta(\boldsymbol{R})\, dV_s = 1.0, \tag{8-50}$$

and when $P(\boldsymbol{R}, \theta, \phi, t) = \delta(\boldsymbol{R})\delta(\cos\theta - \cos\theta_0)\delta(\phi - \phi_0)$, we can write

$$\int_0^{2\pi}\int_0^{2\pi}\int_{-1}^{1}\int_{-1}^{1}\int_{V_s} \exp(i\boldsymbol{K}\cdot\boldsymbol{R})P(\boldsymbol{R}, \theta, \phi, 0)\, dV_s\, d\cos\theta_0\, d\cos\theta\, d\phi_0\, d\phi$$

$$= \int_0^{2\pi}\int_0^{2\pi}\int_{-1}^{1}\int_{-1}^{1}\int_{V_s} \exp(i\boldsymbol{K}\cdot\boldsymbol{R})\delta(\boldsymbol{R})\delta(\cos\theta - \cos\theta_0)$$

$$\times \delta(\phi - \phi_0)\, dV_s\, d\cos\theta_0\, d\cos\theta\, d\phi_0\, d\phi = \int_0^{2\pi}\int_{-1}^{1} d\cos\theta_0\, d\phi_0 = 4\pi. \tag{8-51}$$

The results given in Eqs.(8-50) and (8-51) use the important approximation that the integral over the scattering volume, V_s, can be extended to an integral over a scattering volume where $R \longrightarrow \infty$. This approximation is good if the size of the scatterer is small relative to the actual scattering volume. Using this approximation leads to the Fourier transform of the three-dimensional delta function $\delta(\boldsymbol{R})$ in Eqs.(8-50) and (8-51) leading to unity. More generally, we write

$$\int_{V_s} \exp(i\boldsymbol{K}\cdot\boldsymbol{R})P(\boldsymbol{R}, t)\, dV_s \cong \bar{P}(K, t). \tag{8-52}$$

Thus, $\bar{P}(K, t)$ is the three-dimensional spatial Fourier transform of $P(\boldsymbol{R}, t)$.

Equations (8-48) and (8-49) will be used at various points in the remaining parts of this chapter to describe the scattering from rotationally quenched systems. We shall use hydrodynamic theories to obtain expressions for $P(\boldsymbol{R}, t)$ and $P(\boldsymbol{R}, \theta, \phi, t)$. For instance, $P(\boldsymbol{R}, t)$ in the first term of Eq.(8-48) is obtained for a pure liquid from a solution of three coupled equations: the continuity equation, the Navier–Stokes equation, and the energy transfer equation, which leads to Rayleigh–Brillouin scattering.

The total intensities from $C_z(t = 0)$ and $C_y(t = 0)$ are obtained by substituting the appropriate initial or static conditions into Eqs.(8-48) and (8-49). Using $P(\boldsymbol{R}, 0)$ and $P(\boldsymbol{R}, \theta, \phi, 0) = P(\boldsymbol{R}, 0)\delta(\cos\theta - \cos\theta_0)\delta(\phi - \phi_0)$ in Eqs.(8-48) and (8-49) gives the intensities [we also use Eq.(C-14)]:

$$I_v^v = C_z(0) = AN\bar{P}(K, 0)[\alpha^2 + \tfrac{4}{45}(\alpha_{aa} - \alpha_{bb})^2] \tag{8-53}$$

$$I_h^v = C_y(0) = AN\bar{P}(K, 0)(\tfrac{1}{15})(\alpha_{aa} - \alpha_{bb})^2. \tag{8-54}$$

Thus, if z-polarized incident radiation is used and observations are made in the xy plane, the ratio of perpendicular- to parallel-polarized light intensities is

$$\frac{I_h^v}{I_v^v} = \frac{C_y(0)}{C_z(0)} = \frac{\frac{1}{15}(\alpha_{aa} - \alpha_{bb})^2}{\alpha^2 + \frac{4}{45}(\alpha_{aa} - \alpha_{bb})^2} = \frac{3(\alpha_{aa} - \alpha_{bb})^2}{45\alpha^2 + 4(\alpha_{aa} - \alpha_{bb})^2}. \tag{8-55}$$

This depolarization ratio is a well-known result. Wilson, Decius, and Cross [10] have derived a number of depolarization ratios by using similar methods. Equations (8-53) and (8-54) are general for any type of molecule by replacing $(\alpha_{aa} - \alpha_{bb})^2$ with $\frac{1}{2}[(\alpha_{aa} - \alpha_{bb})^2 + (\alpha_{bb} - \alpha_{cc})^2 + (\alpha_{cc} - \alpha_{aa})^2]$. Experimental depolarization ratios have been measured for several cylindrically symmetric molecules and several representative polarizabilities are listed in Table 6-20. Using the results in Table 6-20 and Eq.(8-55) gives $(I_h^v/I_v^v) \times 100$ values of 1.02 for N_2, 2.07 for Cl_2, 4.03 for CO_2, and 5.96 for N_2O.

Returning now to Eqs.(8-53) and (8-54), we write

$$I_v^v = I_{\text{iso}} + \tfrac{4}{3}I_{\text{anis}}, \qquad I_h^v = I_v^h = I_h^h = I_{\text{anis}}. \tag{8-56}$$

The I_v^v and I_h^v notation is shown in Fig. 8-1 and the other types of $I_{\text{observed}}^{\text{incident}}$ are also evident from the geometry in Fig. 8-1. The I_h^h expression is easily derived giving a correlation function equal to the result in Eqs.(8-47) or (8-49) with the $\cos\theta \sin\theta \sin\phi$ dependence being replaced by $\sin^2\theta \sin\phi \cos\phi$. It is then easy to show that $I_h^h = I_h^v$ as indicated in Eqs.(8-56).

If unpolarized incident radiation is used, the ratio of perpendicular to parallel observed intensities is given by ($I_b^a = I_{\text{observed}}^{\text{incident}}$)

$$\frac{I_h^{\text{unpolarized}}}{I_v^{\text{unpolarized}}} = \frac{I_h^v + I_h^h}{I_v^v + I_v^h} = \frac{2I_h^v}{I_v^v + I_v^h} = \frac{6(\alpha_{aa} - \alpha_{bb})^2}{45\alpha^2 + 7(\alpha_{aa} - \alpha_{bb})^2}. \tag{8-57}$$

We now return again to Eqs.(8-20) and (8-21) and repeat the analysis described from Eqs.(8-44) and (8-45) to Eqs.(8-48) and (8-49) including the effects of parallel vibrations in the linear molecule. Substituting Eqs.(8-29) and (8-30) into Eqs.(8-20) and (8-21) and repeating the quenched rotational state analysis described above gives [use also Eq.(8-52)]

$$C_z(t) = A \exp(-i\omega_0 t) N(\alpha^0)^2 \bar{P}(K, t) + \frac{4A}{9} \exp(-i\omega_0 t) \frac{N(\alpha_{aa}^0 - \alpha_{bb}^0)^2}{4\pi}$$

$$\times \bar{P}(K, \theta, \phi, t) + A \exp(-i\omega_0 t) \exp(i\omega_{n_j', n_j} t) \bar{G}(K, t) f_j \exp\left(-\frac{t}{\tau_j}\right)$$

$$+ \frac{4A}{9} \exp(-i\omega_0 t) \exp(i\omega_{n_j', n_j} t) \left(\frac{N}{4\pi}\right) \bar{G}(K, \theta, \phi, t) h_j \exp\left(-\frac{t}{\tau_j}\right)$$

$$C_y(t) = A \exp(-i\omega_0 t) \frac{N(\alpha_{aa}^0 - \alpha_{bb}^0)^2}{4\pi} \bar{\bar{P}}(K, \theta, \phi, t) + A \exp(-i\omega_0 t)$$

$$\times \exp(i\omega_{n_j', n_j} t) \left(\frac{N}{4\pi}\right) \bar{\bar{G}}(K, \theta, \phi, t) h_j \exp\left(-\frac{t}{\tau_j}\right)$$

$$\bar{P}(K, t) = \int_{V_s} \exp(i\boldsymbol{K} \cdot \boldsymbol{R}) P(\boldsymbol{R}, t)\, dV_s$$

$$\bar{P}(K, \theta, \phi, t) = \int_0^{2\pi} \int_0^{2\pi} \int_{-1}^{1} \int_{-1}^{1} \int_{V_s} \exp(i\boldsymbol{K} \cdot \boldsymbol{R}) P_2(\cos\theta_0) P_2(\cos\theta)$$

$$\times P(\boldsymbol{R}, \theta, \phi, t)\, dV_s\, d\cos\theta_0\, d\cos\theta\, d\phi_0\, d\phi$$

$$\bar{G}(K, t) = \int_{V_s} \exp(i\boldsymbol{K} \cdot \boldsymbol{R}) G(\boldsymbol{R}, t)\, dV_s$$

$$\bar{G}(K, \theta, \phi, t) = \int_0^{2\pi} \int_0^{2\pi} \int_{-1}^{1} \int_{-1}^{1} \int_{V_s} \exp(i\boldsymbol{K} \cdot \boldsymbol{R}) P_2(\cos\theta_0) P_2(\cos\theta)$$

$$\times G(\boldsymbol{R}, \theta, \phi, t)\, dV_s\, d\cos\theta_0\, d\cos\theta\, d\phi_0\, d\phi$$

$$\bar{\bar{P}}(K, \theta, \phi, t) = \int_0^{2\pi} \int_0^{2\pi} \int_{-1}^{1} \int_{-1}^{1} \int_{V_s} \exp(i\boldsymbol{K} \cdot \boldsymbol{R}) (\cos\theta \sin\theta \sin\phi)_0 (\cos\theta \sin\theta \sin\phi)$$

$$\times P(\boldsymbol{R}, \theta, \phi, t)\, dV_s\, d\cos\theta_0\, d\cos\theta\, d\phi_0\, d\phi$$

$$\bar{\bar{G}}(K, \theta, \phi, t) = \int_0^{2\pi} \int_0^{2\pi} \int_{-1}^{1} \int_{-1}^{1} \int_{V_s} \exp(i\boldsymbol{K} \cdot \boldsymbol{R}) (\cos\theta \sin\theta \sin\phi)_0 (\cos\theta \sin\theta \sin\phi)$$

$$\times G(\boldsymbol{R}, \theta, \phi, t)\, dV_s\, d\cos\theta_0\, d\cos\theta\, d\phi_0\, d\phi$$

$$f_j = \frac{1}{2} N_{n_j} \left[\left(\frac{\partial\alpha}{\partial Q_j}\right)_0 \langle n_j | Q_j | n_j' \rangle\right]^2$$

$$h_j = \frac{1}{2} N_{n_j} \left[\left(\frac{\partial(\alpha_{aa} - \alpha_{bb})}{\partial Q_j}\right)_0 \langle n_j | Q_j | n_j' \rangle\right]^2 \qquad \omega_{n_j', n_j} = \frac{E_{n_j'} - E_{n_j}}{\hbar}. \tag{8-58}$$

N_{n_j} is the number of molecules in the n_j state. The first two terms in $C_z(t)$ are identical to the result in Eq.(8-48) and the first term in $C_y(t)$ is identical to the result in Eq.(8-49). The remaining terms in $C_z(t)$ and $C_y(t)$ are due to parallel molecular vibrations in the cylindrically symmetric molecule ($\alpha_{aa} \neq \alpha_{bb} = \alpha_{cc}$). The $P(\mathbf{R}, t)$ and $P(\mathbf{R}, \theta, \phi, t)$ probability functions and their spatial Fourier transforms (which includes the orientational averaging) have been discussed previously. These terms contain both self and distinct terms. $G(\mathbf{R}, t)$ and $G(\mathbf{R}, \theta, \phi, t)$ differ from $P(\mathbf{R}, t)$ and $P(\mathbf{R}, \theta, \phi, t)$, respectively, in that the G functions contain only the self terms. This is because the distinct terms in the G functions involve the molecular vibrations of pairs of different molecules that will have random phases with respect to each other. Thus, the distinct terms, involving the sums over pairs of vibrating molecules, are expected to vanish. Of course, if we are examining a system where distinct terms are in general negligible, the G and P functions will be identical. We have also added the exponential vibrational relaxation process to the correlation functions in Eqs.(8-58) where τ_j is the vibrational relaxation time for the jth normal mode of vibration. f_j, h_j, ω_{n_j',n_j}, and N_{n_j',n_j} are similar in definition to the corresponding parameters in Eqs.(8-40) and (8-43).

Returning to Eq.(8-56), we can generalize to include the Raman terms:

$$I_v^v(\omega) = I_{\text{isot}}^{\text{RAY}}(\omega) + \tfrac{4}{3}I_{\text{anis}}^{\text{RAY}}(\omega) + I_{\text{isot}}^{\text{RAM}}(\omega) + \tfrac{4}{3}I_{\text{anis}}^{\text{RAM}}(\omega)$$

$$I_h^v(\omega) = I_v^h(\omega) = I_h^h(\omega) = I_{\text{anis}}^{\text{RAY}}(\omega) + I_{\text{anis}}^{\text{RAM}}(\omega), \tag{8-59}$$

where the superscripts RAY and RAM indicate Rayleigh and Raman scattering, respectively.

In the next sections we shall examine in detail the nature of $\bar{P}(K, t)$, $\bar{P}(K, \theta, \phi, t)$, $\bar{G}(K, t)$, $\bar{G}(K, \theta, \phi, t)$, $\bar{\bar{P}}(K, \theta, \phi, t)$, and $\bar{\bar{G}}(K, \theta, \phi, t)$ for a variety of condensed phase scattering systems.

8.3 ISOTROPIC RAYLEIGH AND BRILLOUIN SCATTERING IN DENSE GASES AND PURE LIQUIDS

We now examine the spectra arising from the isotropic first term in $C_z(t)$ in Eqs.(8-58),

$$C_z(t) = A \exp(-i\omega_0 t) N (\alpha^0)^2 \bar{P}(K, t), \tag{8-60}$$

where $\bar{P}(K, t)$ is also defined in Eqs.(8-58) as the Fourier transform of $P(\mathbf{R}, t)$, the space-time correlation function. Of course, both the self and distinct correlations are contained in $P(\mathbf{R}, t)$. The contribution made by this $P(\mathbf{R}, t)$ term to the $I_v^v(\omega)$ spectrum in a pure liquid is most easily observed in systems where $\alpha_{aa} - \alpha_{bb} = 0$ and the remaining terms in $C_z(t)$ in Eqs.(8-58) go to zero [11].

In Section 8.2 we used a kinetic model to describe $\bar{P}(K, t)$ that led to a dilute gas Gaussian spectrum for the isotropic Rayleigh scattered light. This model breaks down

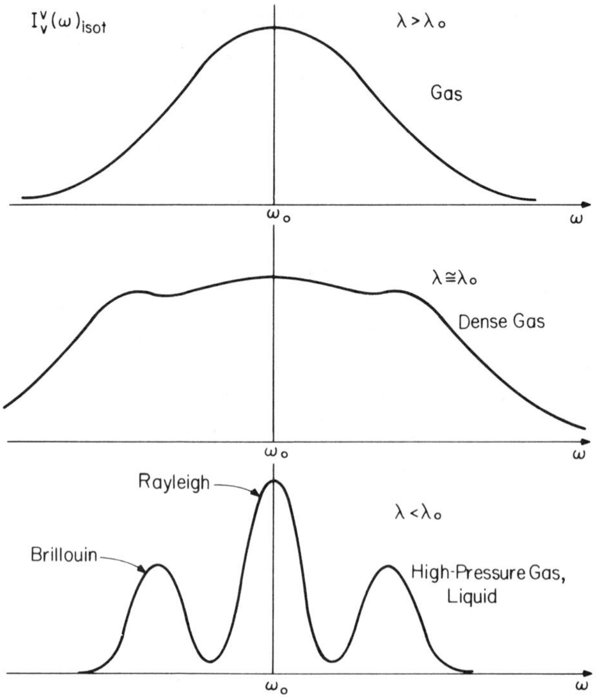

Figure 8-3 Schematic describing the progression from dilute gas to a liquid for the isotropic Rayleigh scattering (see the center line in the CO_2 spectra in Fig. 8-2) where λ is the mean free path and λ_0 is the radiation wavelength. The gas phase Gaussian spectrum is described in Eqs.(8-41) and (8-42). In the dense gas, the Brillouin side bands due to scattering from longitudinal sound waves start to appear. In a high-pressure gas, liquid, or solid, the Brillouin side-band spectrum is clearly evident.

at pressures where $1/K$ starts to exceed the molecular mean free path and cooperative propagation of longitudinal sound waves becomes apparent, giving rise to periodic density variations that scatter the radiation. In the case of a small molecule like CO at $T = 300$ K and a pressure of approximately $P = 10^{-3}$ atm, the mean free path equals the radiation wavelength of $\lambda_0 = 5145$ Å. Thus, at pressures considerably less than 1 atm, we would expect the spectra derived by considering only self correlations from the kinetic theory to show deviations from the Gaussian line shape [see the first term in $I_v^v(\omega)$ from Eqs.(8-41) and (8-42)].

In Fig. 8-3, we illustrate the effects of increasing pressure in the isotropic Rayleigh spectra beginning with the isotropic terms in Eqs.(8-41) and (8-42) leading finally to the liquid state that gives rise to the Rayleigh–Brillouin spectra described in this section.

Mountain [12] and Pecora [13] have discussed the evaluation of $\bar{P}(K, t)$ in a dense gas or liquid in terms of the density-density space-time autocorrelation function. This is equivalent to evaluating $\bar{P}(K, t)$ directly from the following three coupled differential equations [14] in $P(\boldsymbol{R}, t)$, which is the reverse spatial Fourier transform of $\bar{P}(K, t)$ needed in Eq.(8-60):

1. *The continuity equation*

$$\frac{\partial P(\boldsymbol{R}, t)}{\partial t} = -\nabla \cdot \boldsymbol{I}. \tag{8-61}$$

2. *The Navier–Stokes equation*

$$\frac{\partial \boldsymbol{I}}{\partial t} + \frac{C_v v_s^2}{C_p} \nabla P(\boldsymbol{R}, t) + \frac{C_v v_s^2 \xi \rho_0}{C_p} \nabla T(\boldsymbol{R}, t) - \left(\frac{\frac{4}{3}\eta_s + \eta_B}{\rho}\right) \nabla^2 \boldsymbol{I} = 0. \tag{8-62}$$

3. *The energy transport equation*

$$\rho_0 C_v \frac{\partial T(\boldsymbol{R}, t)}{\partial t} - \frac{(C_p - C_v)}{\xi} \frac{\partial P(\boldsymbol{R}, t)}{\partial t} - N_A \chi \nabla^2 T(\boldsymbol{R}, t) = 0. \tag{8-63}$$

$P(\boldsymbol{R}, t)$ is the probability per unit volume that a scatterer is at $\boldsymbol{R}$ at t and $\boldsymbol{I}$ is the probability current or flux (the number of particles passing through a unit cross section per unit time). $T(\boldsymbol{R}, t)$ is the temperature at $\boldsymbol{R}$ at time t; v_s is the *velocity of sound* in the medium; C_p and C_v are the *heat capacities* at constant pressure and volume, respectively; ξ is the *thermal expansion coefficient*, η_s and η_B are the *shear and bulk viscosities*, respectively; χ is the *thermal conductivity*; and ρ is the equilibrium *mass density* of the medium. The mass density denoted here by ρ is related to the number density, denoted by ρ_0, by $\rho = (M/N_A)\rho_0$, where M is the molecular weight and N_A is Avogadro's number.

The use of the linearized equations above will be valid in the hydrodynamic realm with small excursions from equilibrium. Only the longitudinal coupling of the velocity to the density is included. This simplification, where angular correlations between molecules is unimportant, will limit the final results to polarized spectra, I_v^v. We also note that density, or probability per unit volume, and temperature are used as the independent variables. Equations (8-61), (8-62), and (8-63) can be solved for $\bar{P}(K, t)$ [the Fourier transform of $P(\boldsymbol{R}, t)$ as shown in Eqs.(8-58)] by using Laplace and Fourier transform methods to give

$$\bar{P}(K, t) \cong \bar{P}(K, 0)\left[\left(1 - \frac{C_v}{C_p}\right) \exp(-\kappa K^2 t) + \frac{C_v}{C_p} \exp(-\Gamma K^2 t) \cos v_s K t\right]$$

$$\kappa = \frac{\chi N_A}{\rho_0 C_p}, \qquad \Gamma = \frac{1}{2}\left[\frac{\frac{4}{3}\eta_s + \eta_B}{\rho} + \frac{\chi N_A}{\rho_0 C_v}\left(1 - \frac{C_v}{C_p}\right)\right]. \tag{8-64}$$

$\bar{P}(K, 0)$ is the static correlation, which we shall discuss later in this section. κ is recognized as the *thermal diffusion coefficient* and Γ is the *effective mass diffusion coefficient* for sound waves in the medium. Equation (6-84) is only valid in the limit where $v_s K \gg \kappa K^2$.

The first part of the $\bar{P}(K, t)$ in Eq.(8-64) arises from the fluctuations in entropy at constant pressure. The decay of these fluctuations has a time constant of $\tau = 1/\kappa K^2$. The second part of $\bar{P}(K, t)$ in Eq.(8-64) arises from fluctuations in pressure at constant entropy, which leads to a propagating sound wave with velocity v_s and sonic decay time constant of $\tau_s = 1/\Gamma K^2$ [15].

We now proceed to evaluate the spectra of the scattered light. Substituting Eq.(8-64) into Eq.(8-60) gives the correlation function for the isotropic scattering

$$C_z(K, t) = AN\alpha^2\bar{P}(K, 0)\left(\left(1 - \frac{C_v}{C_p}\right)\exp(-i\omega_0 t - \kappa K^2 t)\right.$$

$$\left. + \frac{C_v}{2C_p}\left\{\exp[-i(\omega_0 - v_s K)t - \Gamma K^2 t] + \exp[-i(\omega_0 + v_s K)t - \Gamma K^2 t]\right\}\right), \tag{8-65}$$

where we have expanded the $\cos v_s Kt$ term into its complex components. The real Fourier transform of $C_z(K, t)$ gives the isotropic spectrum as a sum of normalized Lorentzian functions, $\mathscr{L}_\kappa$ and $\mathscr{L}_\Gamma$:

$$I_v^v(K, \omega)_{\text{isot}} = AN\alpha^2\bar{P}(K, 0)\left\{\left(1 - \frac{C_v}{C_p}\right)\mathscr{L}_\kappa(\omega_0 - \omega)\right.$$

$$\left. + \frac{C_v}{2C_p}[\mathscr{L}_\Gamma(\omega_0 - v_s K - \omega) + \mathscr{L}_\Gamma(\omega_0 + v_s K - \omega)]\right\}$$

$$\mathscr{L}_\kappa(\omega_0 - \omega) = \frac{1}{\pi}\left[\frac{\kappa K^2}{(\omega_0 - \omega)^2 + (\kappa K^2)^2}\right],$$

$$\mathscr{L}_\Gamma(\omega_0 \pm v_s K - \omega) = \frac{1}{\pi}\left[\frac{\Gamma K^2}{(\omega_0 \pm v_s K - \omega)^2 + (\Gamma K^2)^2}\right]. \tag{8-66}$$

The isotropic spectrum predicted in Eq.(8-66) is composed of a Rayleigh line centered around the incident radiation frequency, ω_0, with half width at half height of $\Delta\omega = \kappa K^2$. In addition to the central line at ω_0, there are two Brillouin lines shifted symmetrically from ω_0 by $\pm v_s K$ on each side of the central line with half widths given by $\Delta\omega = \Gamma K^2$. Typical isotropic Rayleigh–Brillouin spectra from CCl_4 are shown in Fig. 8-4 for these scattering angles [16]. The experimental half widths and Brillouin frequency shifts can be used to extract the Γ, κ, and v_s parameters by using Eq.(8-66). In addition, the ratio of integrated intensities for the Rayleigh and Brillouin curves gives the value of C_p/C_v from Eq.(8-66):

$$\frac{I_\kappa}{2I_\Gamma} = \frac{C_p}{C_v} - 1, \qquad I_\kappa = AN\alpha^2\bar{P}(K, 0)\left(1 - \frac{C_v}{C_p}\right)\int_{-\infty}^{+\infty}\mathscr{L}_\kappa(\omega_0 - \omega)\,d\omega,$$

$$2I_\Gamma = AN\alpha^2\bar{P}(K, 0)\left(\frac{C_v}{C_p}\right)\int_{-\infty}^{+\infty}\mathscr{L}_\Gamma(\omega_0 + v_s K - \omega)\,d\omega. \tag{8-67}$$

This result is the well-known Landau–Placzek ratio. It is evident from Eqs.(8-66) and (8-67) that if $C_p = C_v$, the intensity of the Rayleigh line vanishes.

Returning to a discussion of Fig. 8-4 for CCl_4, we note that the experimental $I_v^v(\omega)^{\text{RAY}}$ spectra are shown. According to Eq.(8-59),

$$I_v^v(\nu)^{\text{RAY}} = I_{\text{isot}}^{\text{RAY}}(\nu) + \tfrac{4}{3}I_{\text{anis}}^{\text{RAY}}(\nu). \tag{8-68}$$

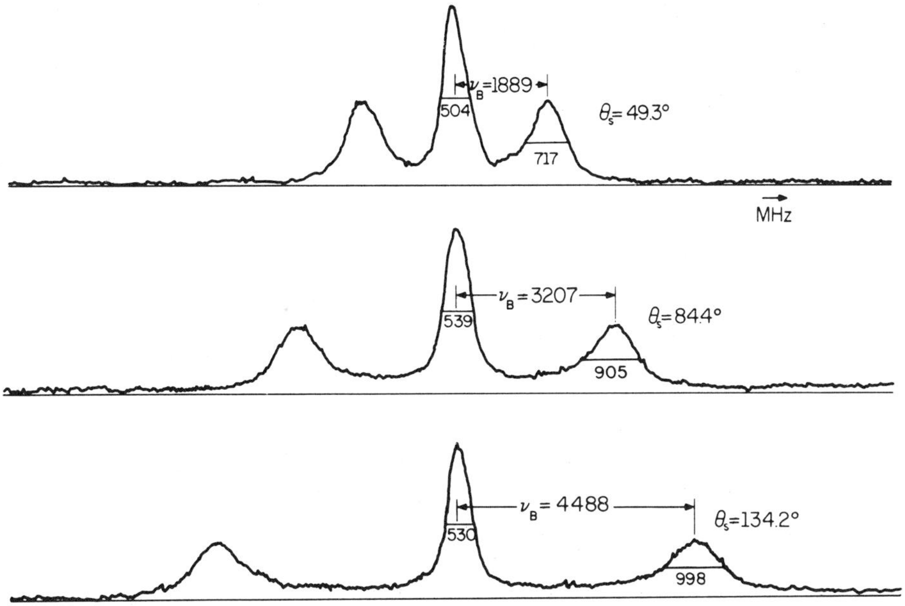

Figure 8-4 Isotropic Rayleigh–Brillouin spectra, $I_v^v(\omega)_{\text{isot}}^{\text{RAY}}$, scattered from liquid CCl_4 ($T = 293$ K) at three different scattering angles with $\lambda_0 = 6328$ Å for the He-Ne laser. The $I_v^v(v)_{\text{anis}}^{\text{RAY}}$ contribution to these spectra $[I_v^v(v)^{\text{RAY}} = I_{\text{isot}}^{\text{RAY}}(v) + \frac{4}{3}I_{\text{anis}}^{\text{RAY}}(v)]$ is expected to have zero or very low intensity because $(\alpha_{ii}^0 - \alpha_{jj}^0)$ in CCl_4 is expected to be zero. The shifts between the central frequency and the Brillouin lines are shown as ν_B (all frequencies are in megahertz); after deconvolution of the instrumental line shapes, the C_v/C_p, Γ, κ, and v_s parameters can be extracted by fitting the spectra with Eq.(8-66).

As the intensity of $I_{\text{anis}}^{\text{RAY}}(v)$ is proportional to the polarizability anisotropy, we would expect this term to vanish in CCl_4, which agrees with the data in Fig. 8-4. We shall discuss $I_{\text{anis}}^{\text{RAY}}(v)$ in Section 8.4.

Returning again to Fig. 8-4, we note that the instrumental line width of the Fabry–Perot interferometer used to record the spectra is larger than the width of the central Rayleigh line. This is easily determined by looking up the value of the thermal diffusivity, κ, given in Eq.(8-64). The values of the thermal conductivity, χ, mass density, ρ, and specific heat at constant pressure are listed in Table 8-1 for CCl_4. The result for κ is $\kappa = N_A\chi/\rho_0 C_p = (\chi/\rho C_p)M = 7.8 \times 10^{-4}\ \text{cm}^2 \cdot \text{s}^{-1}$, where M is the molecular weight of CCl_4. This result leads to a half width at half height for the largest angle shown in Fig. 8-4 (134.2°) of $[K = (2\pi n/\lambda_0)\sin(\theta_s/2) = 2.669 \times 10^5\ \text{cm}^{-1}$ for the $\lambda_0 = 6328$ Å He–Ne laser] $\Delta\nu_R = (1/2\pi)(\kappa K^2) = 8.8$ MHz, which is considerably smaller than the widths of the central Rayleigh lines shown in Fig. 8-4 as observed with a Fabry–Perot interferometer. Thus, the instrumental half width is on the order of $\Delta v = 260$ MHz, according to Fig. 8-4 and the reasoning above, and considerably higher resolution is necessary in order to measure κ. The values of χ, ρ, C_p, and κ for several other molecules are listed in Table 8-2. All the κ values are similar to the result for CCl_4 above predicting very narrow central Rayleigh lines [17].

Table 8-1 Physical constants for CCl_4 at $T = 293$ K from Ref. [16]. K^{-1} indicates units of inverse temperature (Kelvin).

Molecular weight, M	153.8 a.u.
Density (mass), ρ	1.595 $(\text{g} \cdot \text{cm}^{-3})$
Refractive index, n	1.459
Thermal conductivity, χ	2.5×10^{-4} $(\text{cal} \cdot \text{s}^{-1} \cdot \text{cm}^{-1} \cdot \text{K}^{-1})$
Specific heat, C_p	30.8 $(\text{cal} \cdot \text{mol}^{-1} \cdot \text{K}^{-1})$
C_p/C_v	1.47
Low-frequency sound velocity, v_{so}	9.38×10^4 $\text{cm} \cdot \text{s}^{-1}$
Low-frequency mass diffusion constant, Γ_0	0.11 $\text{cm}^2 \cdot \text{s}^{-1}$
Low-frequency sound absorption coefficient divided by the square of the sonic frequency (see Fig. 8-5) (γ_0/v_s^2)	52×10^{-16} $\text{cm}^{-1} \cdot \text{s}^2$

We shall now attempt a superficial interpretation of the Brillouin spectra in Fig. 8-4 using Eq.(8-66), which requires the line widths to be much smaller than the spacings between the central Rayleigh and the side-band Brillouin spectra. We also ignore the weak broad background spectra that arises from a coupling of the propagating sound wave with the intramolecular relaxation processes. Data relevant for interpreting Fig. 8-4 are listed in Table 8-3.

First, we examine the integrated intensities from the data in Fig. 8-4. Simple integration of the curves and using Eq.(8-67) gives the Landau–Placzek ratio and resultant value of C_p/C_v for CCl_4, which is in good agreement with the value in Table 8-1.

The values of $K = (2\pi n/\lambda_0) \sin(\theta_s/2)$ for the three angles displayed in Fig. 8-4, $\lambda_0 = 6328$ Å for the He–Ne laser, and $n = 1.459$ for CCl_4 are listed in the first column in Table 8-3. The second column lists the observed Brillouin splittings, ν_B, from Fig. 8-4. The third column lists the velocities of sound calculated from the scattering vector and ν_B values and the fourth column lists the wavelength of the sound waves in the fluid from

$$\lambda_s = \frac{v_s}{\nu_B} = \frac{\lambda_0}{2n \sin(\theta_s/2)} = \frac{2\pi}{K}. \tag{8-69}$$

It is evident from these results that forward scattered light will observe long λ_s, $\lambda_s = \infty$, and backward-scattered light can observe λ_s as short as $\lambda_s = \lambda_0/2n$, where λ_0 is the radiation wavelength. It is also evident from the results in Table 8-3 that the velocity of sound depends on the frequency of the sound wave. The low-frequency limit for the velocity of sound, v_{so}, can be obtained by extrapolating the values in Table 8-3 and the result is given in Table 8-1 as 9.38×10^4 $\text{cm} \cdot \text{s}^{-1}$. The variation of the velocity of sound with frequency, called *dispersion* by analogy to dispersion in optics, is quite apparent from the results in Table 8-3. The simplified theory used here in Eq.(8-66) for the spectral interpretation must be modified to include dispersion. We shall not pursue this extension here, but instead we now interpret the widths in the Brillouin lines.

The half widths of the Brillouin lines are listed under $\Delta\nu_B$ in the parentheses in Table 8-3. The value of the resulting effective mass diffusion coefficient is given under

Table 8-2 Physical constants at $T = 297$ K for several molecules. These results allow a calculation of the half width of the central Rayleigh line in the isotropic spectra from $\Delta\omega = \kappa K^2$ as shown in Eq.(8-66). The numbers are from Ref. [15]. $\kappa = \chi/\rho C_p$ using the units in this table.

	ρ (g · cm^{-3})	Thermal Conductivity χ (10^{-4} cal · cm^{-1} · s^{-1} · K^{-1})	Heat Capacity at Constant Pressure C_p (cal · g^{-1} · K^{-1})	Thermal Diffusion Coefficient κ (10^{-4} cm^2 · s^{-1})
Benzene	0.879	3.3	0.42	8.9
Carbon disulfide	1.262	3.8	0.24	12.6
Carbon tetrachloride	1.595	2.5	0.20	7.8
Acetic acid	1.049	4.3	0.48	8.5
Acetone	0.792	4.2	0.51	10.5
Toluene	0.866	3.5	0.40	9.7

Table 8-3 Data taken from Fig. 8-4 for the isotropic Rayleigh–Brillouin scattering in CCl_4 for three different angles. A He-Ne laser was used with vacuum wavelength $\lambda_0 = 6328$ Å. The first numbers under $\Delta\nu_B$ are from the data in Fig. 8-4. The numbers in parentheses are the estimated Brillouin half widths obtained by deconvoluting the experimental curves with the instrumental line width.

Scattering Angle	Scattering Vector $K = (2\pi n/\lambda_0)\sin(\theta_s/2)$ ($n = 1.459$)	Shift from Center to Brillouin Lines ν_B	Velocity of Sound $v_s = 2\pi\nu_B/K$	Wavelength of Sound $\lambda_s = v_s/\nu_B$	Half Width of Brillouin Line $\Delta\nu_B$	Effective Mass Diffusion Coefficient $\Gamma = 2\pi\Delta\nu_B/K^2$	Absorption Coefficient $\gamma = \Gamma K^2/v_s$	Sonic Relaxation Time $\tau_s = 1/\Gamma K^2$	Attenuation Length $(1/\gamma)$
49.3°	1.208×10^5 cm^{-1}	1889 MHz	9.8×10^4 cm · s^{-1}	5150 Å	359(200) MHz	8.6×10^{-2} cm^2 · s^{-1}	1.28×10^4 cm^{-1}	7.97×10^{-10} s	7813 Å
84.4°	1.946×10^5	3207	10.4×10^4	3250	453(253)	4.2×10^{-2}	1.53×10^4	6.29×10^{-10}	6536
134.2°	2.669×10^5	4488	10.5×10^4	2355	499(270)	2.4×10^{-2}	1.63×10^4	5.85×10^{-10}	6142

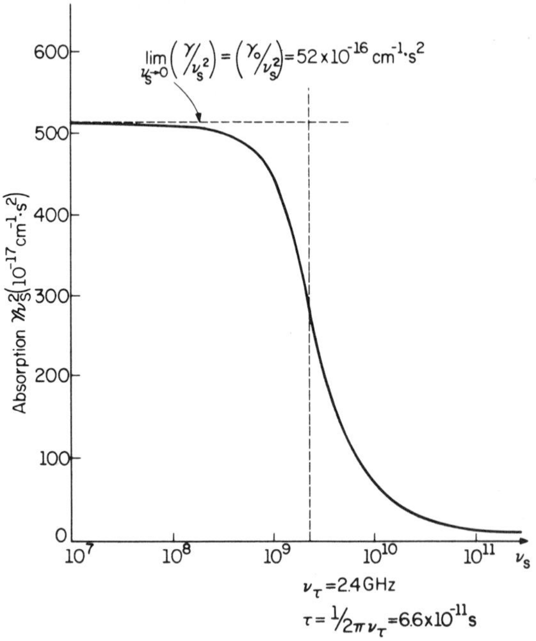

Figure 8-5 Diagram showing the results of the measurement of the absorption of sound in CCl_4 as a function of the sonic frequency, ν_s. γ (cm^{-1}) is the absorption coefficient and γ/ν_s^2 is plotted here as a function of ν_s. The low-frequency limit of $\gamma/\nu_s^2 \longrightarrow \gamma_0/\nu_s^2$ is directly comparable to Γ_0, the low-frequency limit of the mass diffusion coefficient as shown in Eq.(8-72). The inflection point, ν_τ, on the γ/ν_s^2 versus ν_s curve is the internal relaxation frequency in CCl_4.

Γ and we note a strong dependence on the scattering angle or sonic frequency. We shall estimate the low-frequency limit, Γ_0, by extrapolating the values in Table 8-3 to $\nu_B \longrightarrow 0$ to give $\Gamma_0 \cong 10.8 \times 10^{-2}\ \text{cm}^2 \cdot \text{s}^{-1}$. This number can be compared to the low-frequency limit of the absorption coefficient, γ_0, for sound by standard sonic absorption methods. In the sound absorption experiments, the value of γ/ν_s^2 is obtained as a function of ν_s where ν_s is the sonic frequency and γ is the absorption coefficient for the absorption of sound. A plot of γ/ν_s^2 as a function of ν_s for CCl_4 is shown in Fig. 8-5 where we note the leveling off at low frequency. The low-frequency limit of γ/ν_s^2 for CCl_4 is taken directly from Fig. 8-5 to be

$$\lim_{\nu_s \to 0} \frac{\gamma}{\nu_s^2} = \frac{\gamma_0}{\nu_s^2} = 52 \times 10^{-16}\ \text{cm}^{-1} \cdot \text{s}^2. \tag{8-70}$$

Thus, at frequencies below 300 MHz (see Fig. 8-5), the absorption coefficient in CCl_4 is given by the value in Eq.(8-70) times the square of the sonic frequency. The low-frequency limit of γ/ν_s^2 from the sonic absorption experiment and the low-frequency value of Γ obtained from the light scattering experiment are obviously related. Returning to the Brillouin term of Eq.(8-65) or Eq.(8-66), we note that the low-frequency relaxation time of the sound wave is $\tau_s = (\Gamma_0 K^2)^{-1}$. Thus, the corresponding low-frequency absorption coefficient, γ_0, is given by

$$\gamma_0 = \frac{1}{\tau_s v_{so}} = \frac{\Gamma_0 K^2}{v_{so}}, \tag{8-71}$$

where v_{so} is the low-frequency sonic velocity. Remembering that $K = 2\pi/\lambda_s$ from Eq.(8-69) leads to $\gamma_0 = (2\pi)^2\Gamma_0 v_s^2/v_{so}^3$. Thus, it is evident that Γ_0 and γ_0/v_s^2 are related by

$$\frac{\gamma_0}{v_s^2} = \frac{(2\pi)^2\Gamma_0}{v_{so}^3}. \tag{8-72}$$

Using Γ_0, v_{so}, and γ_0/v_s^2 from Table 8-1 for CCl_4 shows that Eq.(8-72) checks. Equations (8-71) and (8-72) are also valid at any frequency and we write

$$\frac{\gamma}{v_s^2} = \frac{(2\pi)^2\Gamma}{v_s^3}, \qquad \gamma = \frac{1}{\tau_s v_s} = \frac{\Gamma K^2}{v_s} = \frac{2\pi\Delta\nu_B}{v_s}. \tag{8-73}$$

The absorption (attenuation) coefficients, γ, relaxation times, τ_s, and the attenuation lengths, $1/\gamma$, are also listed in Table 8-3. These latter equations are valid for all frequencies in comparing the light scattering and ultrasonic absorption experiments.

We also note that the inflection point of the γ/v_s^2 versus ν_s curve in Fig. 8-5 represents an internal relaxation frequency. Above ν_τ the attenuation of sound is considerably less than below ν_τ. The reason for the increased absorption at lower frequencies is because an internal mode of vibrational motion can be excited that gives rise to an additional trap for the kinetic energy of the propagating longitudinal sound wave. At frequencies above ν_τ, the alternative compression and decompression of the sound wave occurs too fast to allow the kinetic translational energy to be converted into the potential energy of a molecular vibration. At the lower frequencies of sound, the kinetic energy can be converted into the potential energy and subsequently dissipated. In summary, the inflection point in Fig. 8-5 represents the relaxation frequency and the relaxation time for the internal conversion is $\tau = 1/(2\pi\nu_\tau)$. In the case of liquid CCl_4, $\tau = 6.6 \times 10^{-11}$ s. We mentioned earlier that the CCl_4 spectra in Fig. 8-4 are a superposition of the isotropic Rayleigh–Brillouin spectra described by Eq.(8-66) with a broad background line that arises from a coupling of the ultrasonic transfer of kinetic energy with the internal relaxation processes in the molecule [18]. We now note that the broad background is due to the vibrational relaxation processes as observed in ultrasonic techniques and the half width of the broad background will be equal to ν_τ from Fig. 8-5. In CCl_4, ν_τ is on the order of the Brillouin shifts as shown in Fig. 8-4 and corrections in the Rayleigh–Brillouin spectra should be made to account for this broad background. In many liquids, the internal relaxation frequency is less than the inverse half width of the central Rayleigh line and no problems are encountered in interpreting the Brillouin spectra.

Ultrasonic techniques have been used extensively to measure molecular relaxation processes [14]. The tie-in by analyzing the Brillouin half widths has been a more recent development. In principle, according to the simple first-order theory given here, classical sound absorption experiments which measure γ/v_s^2 and v_s as a function of v_s give the same information through Eq.(8-73) as the Brillouin shifts, v_B, and widths, Δv_B, which give the velocity and mass diffusion coefficient, respectively, as a function of frequency. In the simple first-order theory, Eq.(8-73) is valid at all frequencies.

In order to evaluate the static correlation, $\bar{P}(K, 0)$ in Eq.(8-64) and the equations following Eq.(8-64) in this section, we return to our earlier discussion of the first terms in Eqs.(8-46) and (8-48) leading to the isotropic term in $I_v^v(\omega)$ as finally written in Eq.(8-65). Rewriting the $N\bar{P}(K, 0)$ part of $C_z(K, t = 0)_{\text{isot}}$ from Eqs.(8-46) and (8-48) gives

$$C(K, t = 0)_{\text{isot}} = A\alpha^2 N \int \exp(i\boldsymbol{K} \cdot \boldsymbol{R}) P(\boldsymbol{R}, 0)\, dV_s = A\alpha^2 N\bar{P}(K, 0)$$

$$\begin{aligned} N\bar{P}(K, 0) &= N \int \exp(i\boldsymbol{K} \cdot \boldsymbol{R}) P(\boldsymbol{R}, 0)\, dV_s = \left\langle \sum_{i,j} \exp\{-i\boldsymbol{K} \cdot [\boldsymbol{r}_i(0) - \boldsymbol{r}_j(0)]\} \right\rangle \\ &= \left\langle \sum_{i=j} \exp\{-i\boldsymbol{K} \cdot [\boldsymbol{r}_i(0) - \boldsymbol{r}_i(0)]\} \right\rangle \\ &\quad + \left\langle \sum_{i \neq j} \exp\{-i\boldsymbol{K} \cdot [\boldsymbol{r}_i(0) - \boldsymbol{r}_j(0)]\} \right\rangle, \end{aligned} \tag{8-74}$$

where $N = \rho_0 V_s$ is the total number of scatterers in the scattering volume, V_s. We have rewritten the bracketed term as a sum of self $(i = j)$ and distinct $(i \neq j)$ terms. Accordingly, we can write $P(\boldsymbol{R}, 0)$ in terms of a self and distinct part. The self part of $P(\boldsymbol{R}, 0)$ is clearly a delta function in $\boldsymbol{R}$ and the second (distinct) term can be written in terms of a two-body radial distribution function, $g(\boldsymbol{R})$, giving [$g(\boldsymbol{R})$ is dimensionless]

$$P(\boldsymbol{R}, 0) = \delta(\boldsymbol{R}) + \rho_0 g(\boldsymbol{R}). \tag{8-75}$$

$g(\boldsymbol{R})$ is the probability of finding a particle at $\boldsymbol{R}$ if there is another particle at the origin. $g(\boldsymbol{R})$ is normalized to unity at large distances,

$$\lim_{R \to \infty} g(\boldsymbol{R}) = 1.0. \tag{8-76}$$

A typical radial pair distribution function is shown in Section 8.5 (Fig. 8-14) where it is evident that for simple liquids the structure in $g(\boldsymbol{R})$ extends out only a few molecular diameters. Thus, at large $\boldsymbol{R}$, $P(\boldsymbol{R}, 0)$ reduces to the average particle density, ρ_0:

$$\lim_{R \to \infty} P(\boldsymbol{R}, 0) = \rho_0. \tag{8-77}$$

Substituting $P(\boldsymbol{R}, 0)$ in Eq.(8-75) into Eq.(8-74) and using

$$\int \exp(i\boldsymbol{K} \cdot \boldsymbol{R})\, dV_s = \delta(K) \quad \text{and} \quad \int \exp(i\boldsymbol{K} \cdot \boldsymbol{R}) \delta(\boldsymbol{R})\, dV_s = 1,$$

we can write

$$C_z(K, 0)_{\text{isot}} = A\alpha^2 N\bar{P}(K, 0) = A\alpha^2 N[S(K) + \rho_0 \delta(K)]$$

$$S(K) = \int \exp(i\boldsymbol{K} \cdot \boldsymbol{R})\{\delta(\boldsymbol{R}) + \rho_0[g(\boldsymbol{R}) - 1]\}\, dV_s. \tag{8-78}$$

$S(K)$ is called the *structure factor* for the liquid [19] and the $\delta(K)$ term leads to the forward scattered light, which will be indistinguishable from the forward traveling incident light. Thus, the $S(K)$ term is the only measurable K-dependent term in the scattered light intensity. The integral in Eq.(8-78) can be simplified considerably if we are dealing with optical radiation where the distances probed by the radiation are considerably larger than the distances from $R = 0$ to the first few molecular diameters or periodic variations in $g(\mathbf{R})$. Under these circumstances, $\exp(-i\mathbf{K}\cdot\mathbf{R}) = 1 - i\mathbf{K}\cdot\mathbf{R} + \cdots \cong 1$ and we can write

$$\rho_0 \int \exp(i\mathbf{K}\cdot\mathbf{R})[g(\mathbf{R}) - 1]\, dV_s \cong \rho_0 \int [g(\mathbf{R}) - 1]\, dV_s. \tag{8-79}$$

This final integral can be evaluated by statistical mechanics [20] to give

$$\rho_0 \int [g(\mathbf{R}) - 1]\, dV_s = \rho_0 kT\left(-\frac{1}{V}\frac{\partial V}{\partial p}\right)_T - 1 = \rho_0 kT\beta_T - 1. \tag{8-80}$$

$\beta_T = [-(1/V)(\partial V/\partial p)]_T$ is the gas or fluid isothermal compressibility at temperature T and k is Boltzmann's constant (β_T has units of inverse pressure). This final result is independent of K. Of course, if static fluctuations extend out to a distance λ_0 (radiation) or if shorter wavelength radiation were used, the $\exp(i\mathbf{K}\cdot\mathbf{R})$ part of the integrand of Eq.(8-79) must be included leading to a K-dependence in the final result. This discussion will be resumed in Section 8.5 when we discuss X-ray scattering.

Substituting Eqs.(8-80) and (8-79) into Eq.(8-78) gives the low K limit structure factor for optical light scattering in small molecule liquids:

$$\lim_{K\to 0} S(K) = kT\rho_0\left(-\frac{1}{V}\frac{\partial V}{\partial p}\right)_T = kT\rho_0\beta_T. \tag{8-81}$$

Returning to Eq.(8-78) we note that in the low K limit where $1/K$ is large relative to the mean free path in a gas or where $1/K$ is large relative to the scatterer–scatterer distance in a liquid, the intensity of the isotropic scattered light (which excludes the forward scattered light) is proportional to the compressibility of the scattering medium,

$$\bar{P}(K, 0) = \rho_0 kT\beta_T. \tag{8-82}$$

Substituting this result into Eqs.(8-65) through (8-67) gives the complete result.

In the case of a perfect gas ($pV = nRT$ for n moles of gas) it is easy to show that $\beta_T = 1/p$ and $\bar{P}(K, 0) = kT\rho_0\beta_T = 1$. Substituting this into Eq.(8-65) and setting $t = 0$ gives the total intensities from the N scatterers [$C_y(K, 0)$ is also given]

$$C_z(K, 0) = I_v^v = \int_{-\infty}^{+\infty} I_v^v(\omega)\, d\omega = I_{\text{isot}} + \frac{4}{3} I_{\text{anis}} = \frac{\omega_0^4}{R^2c^4} I_0 N\left[\alpha^2 + \frac{4}{45}(\alpha_{aa} - \alpha_{bb})^2\right]$$

$$C_y(K, 0) = I_h^v = \int_{-\infty}^{+\infty} I_h^v(\omega)\, d\omega = I_{\text{anis}} = \frac{\omega_0^4}{R^2c^4}\left(\frac{I_0 N}{15}\right)(\alpha_{aa} - \alpha_{bb})^2, \tag{8-83}$$

Table 8-4 The isothermal compressibilities, β_T, particle densities, ρ_0 and $\bar{P}(K,0) = \rho_0 kT\beta_T$ factors of several common liquids from *The Handbook of Chemistry and Physics* (The Chemical Rubber Co., Cleveland, Ohio, 1964). The total Rayleigh–Brillouin intensities are proportional to the dimensionless number $\rho_0 kT\beta_T$, where k is Boltzmann's constant and T is the temperature. $\rho_0^2 kT\beta_T$ times the scattering volume, V_s, gives the total molecules scattering the light.

Liquid	T (K)	ρ_0 (molecules $\cdot$ cm^{-3})	β_T (cm$^2 \cdot$ dyne^{-1})	$\bar{P}(K,0) = \rho_0 kT\beta_T$ (dimensionless)	$\rho_0 \bar{P}(K,0) = \rho_0^2 kT\beta_T$ (molecules $\cdot$ cm^{-3})
Perfect gas					
(P = 1 atm)		2.69×10^{19}	9.8×10^{-7}	1.0	2.69×10^{19}
Acetone	293	1.04×10^{22}	12.7×10^{-11}	5.34×10^{-2}	5.55×10^{20}
Benzene	273	6.9×10^{21}	8.1×10^{-11}	2.25×10^{-2}	1.55×10^{20}
CCl_4	293	1.04×10^{21}	10.7×10^{-11}	2.68×10^{-2}	0.28×10^{20}

where I_0 is the incident intensity. The first term in Eq.(8-83) is identical to the result in Eq.(1-59) for N scatterers when $\gamma = \pi/2$. In fact, Eqs.(8-83) and the gas phase depolarization ratios discussed from Eqs.(8-53) to (8-57) are often used to measure the elements of the electric polarizability tensor in molecules. Several results are listed in Table 6-20.

In the case of dense gases or liquids, the isothermal compressibilities are needed in order to evaluate the intensities. For gases, the nonideality at high pressures can be handled by a dense gas equation of state such as the van der Waals gas equation or the virial equation. The compressibilities can be calculated with the equations of state. In the case of liquids, the isothermal compressibilities are readily available in the literature. A few typical compressibilities in some common liquids are listed in Table 8-4. The isothermal compressibility of a perfect gas at 1 atm pressure is $\beta_T = 1/p = 0.98 \times 10^{-6}$ cm$^2 \cdot$ dyne^{-1}. Also listed in Table 8-4 are the number densities, ρ_0, and the dimensionless scattering factors, $\bar{P}(K,0) = \rho_0 kT\beta_T$. The last two columns in Table 8-4 suggest an interpretation for the loss in scattering intensity due to destructive interference in a dense gas or liquid. The last column listing $\rho_0(\rho_0 kT\beta_T)$ can be interpreted as the effective number of molecules per unit volume that scatter light. For a perfect gas, all molecules in the unit volume scatter the light, $\rho_0 kT\beta_T = 1.0$. In the dense fluid, however, the destructive interference of the correlated, closely spaced particles reduces the value of ρ_0 by the factor $\rho_0 kT\beta_T$. It is also evident from the last column in Table 8-4 that the total intensities per unit scattering volume in the pure liquids have the following order: I(acetone) > I(benzene) > $I(CCl_4)$. Similar information on other molecules is readily available by finding the known densities, isothermal compressibilities, and their respective pressure and temperature dependences.

The previous results show that there is no K-dependence in the intensity of the scattered light if $g(\boldsymbol{R})$ approaches 1.0 in a distance short relative to $1/K$. There are certain circumstances, however, when $g(\boldsymbol{R})$ has considerable structure well past $1/K$ for optical radiation. We shall discuss this point in more detail in Section 8.5 where we develop the scattering theory for short wavelength X-ray radiation ($\lambda_0 \cong 1$ Å) that can probe structural variations on an atomic scale of 1 Å. According to our previous discussion, the shortest distance that can be probed by backscattered ($\theta = \pi$) radiation is $r = \lambda_0/2n$, where n is the refractive index and λ_0 is the radiation wavelength. Thus, in order for a K-dependence to be observable in a dense gas or liquid with optical radiation ($\lambda \cong 5000$ Å), spatial correlations must be present which extend over 3000–5000 Å. This is highly unlikely in a normal small molecule liquid and no K-dependence in the total I_v^v intensity is observed. Near the critical point, however, long-range fluctuations occur and the time-independent spatial correlation function apparently takes the form

$$g(R) - 1 = \frac{A \exp(-R/a)}{R}, \tag{8-84}$$

where a is a two-body correlation length and A is a constant. The resultant structure factor according to Eq.(8-78) is given by

$$S(K) = \int \exp(i\mathbf{K} \cdot \mathbf{R}) \left[\delta(\mathbf{R}) + \rho_0 \frac{A \exp(-R/a)}{R} \right] dV_s = 1 + \frac{\rho_0 A(1/a)}{K^2 + (1/a)^2}, \tag{8-85}$$

which shows a K-dependence [1]. Of course, static fluctuations may exist over a large distance in amorphous solids, polymers (solids), and glasses. If the static fluctuations or inhomogeneities are correlated over distances on the order of $1/K$, a K-dependence will be observed in the intensity of the scattered light as shown above (see end of Section 8.5). The static inhomogeneities in the two-phase systems can also be studied by these methods.

8.4 ANISOTROPIC RAYLEIGH AND RAMAN AND ISOTROPIC RAMAN SCATTERING IN LIQUIDS; TRANSLATIONAL AND ROTATIONAL DIFFUSION

We start by reviewing Eqs.(8-58) and (8-59). In this section, we shall examine $\bar{P}(K, \theta, \phi, t)$ leading to $I_{\text{anis}}^{\text{RAY}}(\omega)$, $\bar{G}(K, \theta, \phi, t)$ leading to $I_{\text{anis}}^{\text{RAM}}(\omega)$, and $\bar{G}(K, t)$ leading to $I_{\text{isot}}^{\text{RAM}}(\omega)$. In Fig. 8-6, the $I_v^v(\nu)^{\text{RAY}} = I_{\text{isot}}^{\text{RAY}}(\nu) + \frac{4}{3} I_{\text{anis}}^{\text{RAY}}(\nu)$ and $I_h^v(\nu)^{\text{RAY}} = I_{\text{anis}}^{\text{RAY}}(\nu)$ scattered spectra of nitrobenzene are shown. The depolarized spectrum in the right-hand curve is pure anisotropic and the left-hand curve is a combination of the isotropic and anisotropic spectra. The $I_{\text{isot}}^{\text{RAY}}(\nu)$ triplet arises from density fluctuations as described in Section 8.3. Several depolarized or anisotropic Rayleigh and Raman spectra are shown in Fig. 8-7. We note that the half widths at half height for both $I_{\text{anis}}^{\text{RAY}}(\nu)$ and $I_{\text{anis}}^{\text{RAM}}(\nu)$ in CS_2 and benzene are considerably larger than in nitrobenzene (Fig. 8-6). Typical spectral line widths at $\theta_s = \pi/2$ (see Fig. 8-1) are on the order of $0.1 \longrightarrow 10\ \text{cm}^{-1}$ or $\Delta\nu = 3 \times 10^9 \longrightarrow 3 \times 10^{11}$ Hz for these small molecules.

First, we examine $\bar{G}(K, \theta, \phi, t)$ or $G(\mathbf{R}, \theta, \phi, t)$ in Eqs.(8-58), which leads to $I_{\text{anis}}^{\text{RAM}}(\nu)$ as shown, for instance, in Fig. 8-7. We remember that $G(\mathbf{R}, \theta, \phi, t)$ is composed entirely of self terms, the distinct terms being zero due to the random phases of vibration. Thus, $G(\mathbf{R}, \theta, \phi, t)$ contains only single particle contributions. We shall now develop the hydrodynamic Debye model for $G(\mathbf{R}, \theta, \phi, t)$ that describes both the

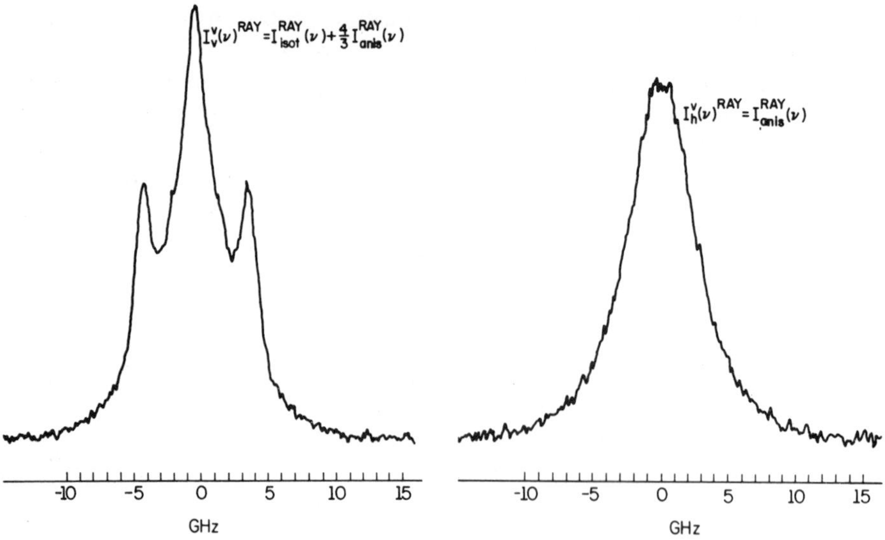

Figure 8-6 The $I_v^v(\nu)^{RAY}$ and $I_h^v(\nu)^{RAY}$ $(\theta_s = \pi/2)$ spectra of nitrobenzene recorded with a Fabry–Perot interferometer at $T = 297$ K by A. K. Burnham and S. J. Bertucci with an Ar^+ ion laser with $\lambda_0 = 5145$ Å [see also A. Szöke, E. Courtens, and A. Ben-Reuven, *Chem. Phys. Lett.* **1**, 87 (1967)]. The $I_v^v(\nu)^{RAY}$ spectrum is a combination of the isotropic and anisotropic parts and the $I_h^v(\nu)^{RAY}$ spectrum is due only to the anisotropic component.

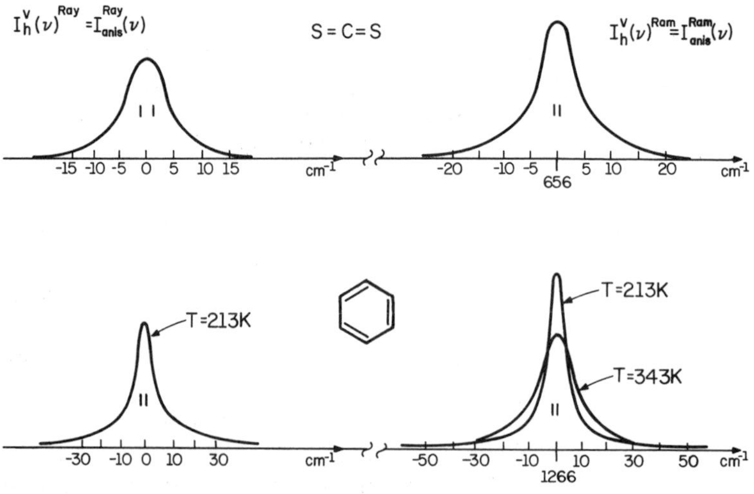

Figure 8-7 Depolarized $I_h^v(\nu) = I_{anis}(\nu)$ Rayleigh and Raman spectra for CS_2 and benzene $(\theta_s = \pi/2)$. The vertical bars indicate the approximate instrumental widths. Polarized lasers were used as radiation sources and the Rayleigh spectra on the left were taken with a Fabry–Perot interferometer and the Raman spectra on the right were taken with a grating optical spectrograph where we also note that 1 cm^{-1} = 30 GHz. The data are adapted from S. L. Shapiro and H. P. Broida, *Phys. Rev.* **154**, 129 (1967) and F. J. Bartoli and T. A. Litovitz, *J. Chem. Phys.* **56**, 404 (1972) and the unpublished work of A. K. Burnham and S. J. Bertucci.

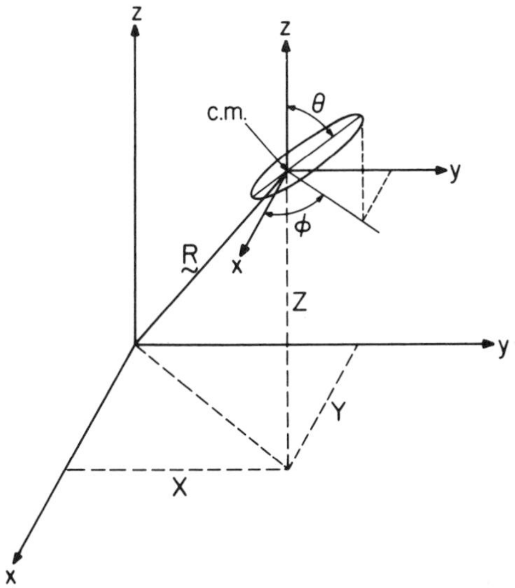

Figure 8-8 Diagram showing the c.m. position and orientation of the anisotropic cylindrically symmetric molecule that is scattering light from a single scattering origin in the molecule. We are interested in the time evolution of $\boldsymbol{R}$, θ, and ϕ as described in the text.

center of mass (c.m.) position (translational diffusion) and orientation (rotational diffusion) of the particle. The Debye model assumes that many collisions are required to reorientate the molecule. Beginning with the $\bar{G}(K, \theta, \phi, t)$ in Eqs.(8-58), we average over the initial angles to give

$$\begin{aligned}\bar{G}(K, \theta, \phi, t) &= \int_0^{2\pi}\int_0^{2\pi}\int_{-1}^{1}\int_{-1}^{1}\int_{V_s} \exp(i\boldsymbol{K}\cdot\boldsymbol{R})P_2(\cos\theta_0)P_2(\cos\theta) \\ &\quad \times G(\boldsymbol{R}, \theta, \phi, t)\,dV_s\,d\cos\theta_0\,d\cos\theta\,d\phi_0\,d\phi \\ &= \int_0^{2\pi}\int_{-1}^{1}\int_{V_s} \exp(i\boldsymbol{K}\cdot\boldsymbol{R})P_2(\cos\theta)\mathscr{G}(\boldsymbol{R}, \theta, \phi, t)\,dV_s\,d\cos\theta\,d\phi \\ &= \int_0^{2\pi}\int_{-1}^{1} P_2(\cos\theta)\bar{\mathscr{G}}(K, \theta, \phi, t)\,d\cos\theta\,d\phi, \end{aligned} \tag{8-86}$$

where $\mathscr{G}(\boldsymbol{R}, \theta, \phi, t)$ is the probability per unit volume of finding the molecular c.m. at $\boldsymbol{R}$ with orientation θ and ϕ at time t. The positional and orientational coordinates are shown in Fig. 8-8. In order to evaluate $\bar{G}(K, \theta, \phi, t)$ for a single molecule in a fluid will develop the hydrodynamic models of translational and rotational diffusion.

We start with a discussion of one-dimensional translational diffusion of a sphere from a planar delta function in number density. The initial delta function in concentration in the xy plane is shown schematically in Fig. 8-9. The flux (number of particles per unit area per unit time), $J(y)$, away from this plane of high concentration is proportional to the gradient of the number density, $N(y, t)$, along the y axis. The proportionality constant is the *diffusion coefficient*, D, in units of $\text{cm}^2 \cdot \text{s}^{-1}$,

$$J(y) = -D\frac{dN(y, t)}{dy}. \tag{8-87}$$

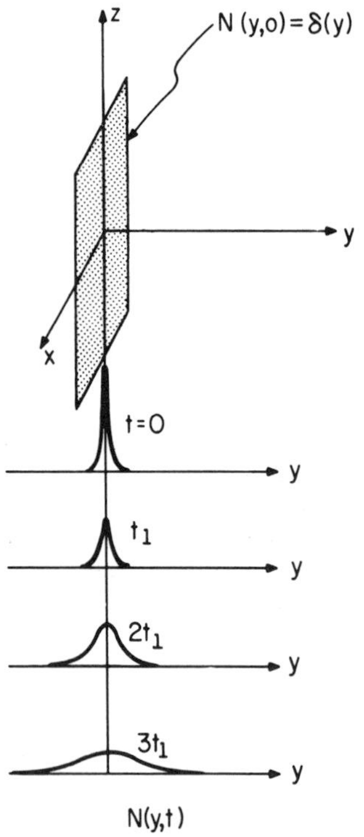

Figure 8-9 A concentration or number density fluctuation of spheres in the xz plane that will cause a flux along the y axis given by $J(y) = -D[dN(y, t)/dy]$, where D is the translational diffusion coefficient. The time-dependence of $N(y, t)$ starting with an xz planar (an infinite plane) delta function is shown in the lower part of the diagram. These $N(y, t)$ distributions are from Eq.(8-93).

We can combine this equation with the y axis mass continuity equation, $dJ(y)/dy = -[dN(y, t)/dt]$, to give the diffusion equation,

$$\frac{dN(y, t)}{dt} = D\frac{d^2N(y, t)}{dy^2}. \tag{8-88}$$

In order to solve this equation, we take the spatial Fourier transform to give

$$\int_{-\infty}^{+\infty} \exp(-iK_y y)\left[\frac{d}{dt}N(y, t)\right] dy = D\int_{-\infty}^{+\infty} \exp(-iK_y y)\left[\frac{d^2N(y, t)}{dy^2}\right] dy$$

$$\frac{d}{dt}\bar{N}(K_y, t) = -K_y^2 D\bar{N}(K_y, t), \tag{8-89}$$

which has a solution of

$$\bar{N}(K_y, t) = \bar{N}(K_y, 0)\exp(-K_y^2 Dt), \tag{8-90}$$

where $\bar{N}(K_y, t)$ is the Fourier transform of $N(y, t)$. The initial condition in K_y space, $\bar{N}(K_y, 0)$, can be obtained from the initial condition in real space. Assuming $N(y, 0) = \delta(y)$, which is the planar delta function in number density in the infinite xz plane as shown in Fig. 8-9, and taking the Fourier transform gives

$$\bar{N}(K_y, 0) = \int_{-\infty}^{+\infty} \exp(-iK_y y)N(y, 0)\,dy = \int_{-\infty}^{+\infty} \exp(-iK_y y)\delta(y)\,dy = 1.0. \tag{8-91}$$

Using this result in Eq.(8-90) gives

$$\overline{N}(K_y, t) = \exp(-K_y^2 Dt). \tag{8-92}$$

We can now use the reverse Fourier transform to give $N(y, t)$,

$$N(y, t) = \frac{1}{2\pi}\int_{-\infty}^{+\infty} \exp(-K_y^2 Dt)\exp(iK_y y)\,dK_y = \left(\frac{1}{4\pi Dt}\right)^{1/2}\exp\left(-\frac{y^2}{4Dt}\right). \tag{8-93}$$

The time-dependence in this Gaussian function is shown in the lower part of Fig. 8-9. We can also write the diffusion equation in three dimensions for a point concentration fluctuation, $N(\boldsymbol{R}, 0) = \delta(\boldsymbol{R})$, for spherical particles. The results are

$$\frac{d}{dt}N(\boldsymbol{R}, t) = D\nabla^2 N(\boldsymbol{R}, t), \qquad \overline{N}(K, t) = \exp(-K^2 Dt),$$

$$N(R, t) = \left(\frac{1}{4\pi Dt}\right)^{3/2}\exp\left(-\frac{R^2}{4Dt}\right), \qquad R^2 = x^2 + y^2 + z^2, \qquad D_x = D_y = D_z = D. \tag{8-94}$$

The mean squared distance traveled along the x axis by the spherical diffusing particle, which is initially at the origin ($x = 0$), is obtained from the standard statistical average given by

$$(\overline{x^2}) = \int_{-\infty}^{+\infty} x^2 N(x, t)\,dx = \left(\frac{1}{4\pi Dt}\right)^{1/2}\int_{-\infty}^{+\infty} x^2 \exp\left(-\frac{x^2}{4Dt}\right)dx = 2Dt. \tag{8-95}$$

We can now return to the discussion in Section 2.2 to relate the diffusion coefficient, D, to the friction coefficient, f. We begin with Eq.(2-36) for a neutral particle or a charged particle in the absence of the external field, E_x. In the absence of the qE_x force, the particle experiences random Brownian motion which can be described by replacing the external field force (qE_x) with a fluctuating force, $F(t)$, which arises from the continuous collisional motion experienced by a particle in a medium of other particles in motion to give

$$F(t) = M\ddot{x} + f\dot{x}. \tag{8-96}$$

This is Langevin's equation and the fluctuating force will lead to Brownian motion of the particle [2,21]. We shall now compute the mean squared distance traveled along the x axis due to the random fluctuating force as described in Eq.(8-96). We can then compare our result with Eq.(8-95) to obtain a relation between D and f. We can find $(\overline{x^2})$ from the fluctuating force $F(t)$ in Eq.(8-96) by first multiplying Eq.(8-96) by x to give $xF(t) = xM\ddot{x} + xf\dot{x}$ and noting that $d^2x/dt^2 = 2\dot{x}^2 + 2x\ddot{x}$ to give

$$xF(t) = \frac{M}{2}\frac{d^2}{dt^2}(x^2) - M(\dot{x})^2 + \frac{1}{2}f\frac{d}{dt}(x^2), \tag{8-97}$$

where we have also used $fx\dot{x} = \frac{1}{2}f(d/dt)x^2$. We now take the long-time average of this equation indicated by brackets, define $g = \langle d/dt(x^2)\rangle$, and note that x and $F(t)$ are uncorrelated in time, giving $\langle xF(t)\rangle = 0$ and

$$\frac{M}{2}\frac{d}{dt}g = \langle M(\dot{x})^2\rangle - \frac{1}{2}fg. \tag{8-98}$$

Substituting $\langle M(\dot{x})^2\rangle = kT$ for the equipartition of translational kinetic energy and rearranging gives ($\tau = M/f$)

$$\frac{d}{dt}g + \frac{1}{\tau}g = \frac{2kT}{M}. \tag{8-99}$$

This equation is similar to Eq.(2-37) with solution

$$g(t) = g(t = \infty)\left[1 - \exp\left(-\frac{t}{\tau}\right)\right] = \frac{2kT\tau}{M}\left[1 - \exp\left(-\frac{t}{\tau}\right)\right]. \tag{8-100}$$

Thus, the long-time limit solution where $t \gg \tau$ is $g(\infty) = 2kT\tau/M$ and the average value of $\langle x^2\rangle$, over a time where $t \gg \tau$ is satisfied, is given according to the previous definition of $g = \langle dx^2/dt\rangle$ by

$$\int_0^t d\langle x^2\rangle = \int_0^t \frac{2kT\tau}{M}\,dt. \tag{8-101}$$

If $\langle x^2\rangle = 0$ at $t = 0$, we obtain

$$(\bar{x}^2) = \langle x^2\rangle = \frac{2kT\tau t}{M} = \frac{2kT}{f}t \tag{8-102}$$

for the mean squared average displacement. Equating Eqs.(8-102) and (8-95) gives an expression for the diffusion coefficient in terms of the frictional force constant, f,

$$D = \frac{kT}{f}. \tag{8-103}$$

Furthermore, it is evident from this equation and Eq.(2-41) that the ratio of the mobility, μ, and the diffusion coefficient, D, is given by

$$\frac{\mu}{D} = \frac{q}{kT}, \tag{8-104}$$

where q is the charge on the ion. Finally, the conductivity, σ, is also related to the diffusion coefficient according to the results above and Eq.(2-41) according to

$$\sigma = \rho_0 q^2\frac{\tau}{M} = \frac{\rho_0 q^2}{f} = \rho_0 q\mu = \frac{\rho_0 q^2}{kT}D. \tag{8-105}$$

Returning now to a discussion of diffusion, we note that Eqs.(8-87) through (8-94) take on a different form for anisotropic translational diffusion [22]. In a cylindrically symmetric molecule, the translational diffusion constant along the cylindrical axis, D_{aa}, may be different from the perpendicular translational diffusion constants, $\mathsf{D}_{bb} = \mathsf{D}_{cc}$. A more general expression for Eq.(8-87) in the space-fixed axis system is

$$J(z) = -\mathsf{D}_{zz} \frac{dN(z, t)}{dz}, \tag{8-106}$$

where D_{zz} is the z laboratory axis translational diffusion coefficient. In the absence of external or internal orientating fields, the fluid will be isotropic and $\mathsf{D}_{xx} = \mathsf{D}_{yy} = \mathsf{D}_{zz}$. However, $\mathsf{D}_{aa} \neq \mathsf{D}_{bb} = \mathsf{D}_{cc}$ for a cylindrically symmetric molecule. Using the mass continuity equation leads to the diffusion equation,

$$\frac{dN(z, t)}{dt} = \mathsf{D}_{zz} \frac{d^2}{dz^2} N(z, t). \tag{8-107}$$

$\mathbf{D}(xyz)$ can be written in terms of the molecular-fixed axis (abc) according to Eq.(B-35), for instance, $\mathbf{D}(xyz) = \mathbf{C}\mathbf{D}(abc)\tilde{\mathbf{C}}$, where $\mathbf{C}$ contains the direction cosines. Using arguments similar to those preceding Eq.(8-29), we find

$$\mathsf{D}_{zz} = D + \tfrac{2}{3}(\mathsf{D}_{aa} - \mathsf{D}_{bb})P_2(\cos\theta),$$

where $D = \tfrac{1}{3}(\mathsf{D}_{aa} + \mathsf{D}_{bb} + \mathsf{D}_{cc})$. Substituting this result into Eq.(8-107) gives

$$\frac{dN(z, t)}{dt} = \left[D + \frac{2}{3}(\mathsf{D}_{aa} - \mathsf{D}_{bb})P_2(\cos\theta)\right] \frac{d^2}{dz^2} N(z, t), \tag{8-108}$$

where θ is the spherical polar angle as before. For an isotropic fluid, translation along the laboratory x, y, and z axes are equivalent and we generalize this expression to give

$$\frac{dN(\boldsymbol{R}, t)}{dt} = \left[D + \frac{2}{3}(\mathsf{D}_{aa} - \mathsf{D}_{bb})P_2(\cos\theta')\right] \nabla^2 N(\boldsymbol{R}, t), \tag{8-109}$$

where θ' is now the angle between the internuclear molecular axis of the linear molecule and the $\boldsymbol{R}$ vector. In the absence of any rotational motion, the $P_2(\cos\theta')$ term in Eq.(8-109) vanishes over an isotropic distribution of molecules. If the molecule is rotating as well as translating, however, the $\mathsf{D}_{aa} - \mathsf{D}_{bb}$ term above will lead to a coupling between rotation and translation. For isotropic molecular diffusion, $\mathsf{D}_{aa} = \mathsf{D}_{bb} = \mathsf{D}_{cc}$, Eq.(8-109) reduces to the first of Eqs.(8-94).

Using the developments above, we now write a general equation for $\mathscr{G}(\boldsymbol{R}, \theta, \phi, \tau)$ as needed in Eq.(8-86). First, we note that the Laplacian operator, ∇^2, for a cylindrically symmetric molecule can be expressed as a sum of two terms; the first term is the c.m. Cartesian Laplacian and the second term is the internal coordinate part written in spherical polar coordinates [see Eq.(4-46)]:

$$\nabla^2 = \nabla^2_{\text{c.m.}} + \sum_i \left[\frac{1}{r_i^2}\frac{\partial}{\partial r_i}\left(r_i^2 \frac{\partial}{\partial r_i}\right) + \frac{1}{r_i^2}\frac{1}{\sin\theta}\frac{\partial}{\partial\theta}\left(\sin\theta\frac{\partial}{\partial\theta}\right) + \frac{1}{r_i^2\sin^2\theta}\frac{\partial^2}{\partial\phi^2}\right]$$

$$\nabla^2_{\text{c.m.}} = \frac{\partial^2}{\partial X^2} + \frac{\partial^2}{\partial Y^2} + \frac{\partial^2}{\partial Z^2}, \tag{8-110}$$

where r_i is the distance from the c.m. to the ith atom. We now substitute this Laplacian into Eq.(8-109) and review the previous development to write an equation in

$\mathscr{G}(\boldsymbol{R}, \theta, \phi, t)$ in Eq.(8-86). First, we note that there is no r_i-dependence in $\mathscr{G}(\boldsymbol{R}, \theta, \phi, t)$, which allows us to omit the first term in brackets in ∇^2. We shall also drop the $\mathsf{D}_{aa} - \mathsf{D}_{bb}$ terms here as $\mathsf{D}_{aa} \cong \mathsf{D}_{bb}$ in near spherical molecules, but we note that the effects can be obtained by using perturbation techniques. We are left with an equation in D and $\Theta = D/\sum_i r_i^2$, the rotational diffusion coefficient (with units of s^{-1}):

$$\frac{\partial \mathscr{G}(\boldsymbol{R}, \theta, \phi, t)}{\partial t} = D\nabla^2_{\text{c.m.}} \mathscr{G}(\boldsymbol{R}, \theta, \phi, t)$$

$$+ \Theta\left[\frac{1}{\sin\theta}\frac{\partial}{\partial\theta}\left(\sin\theta\frac{\partial}{\partial\theta}\right) + \frac{1}{\sin^2\theta}\frac{\partial^2}{\partial\phi^2}\right]\mathscr{G}(\boldsymbol{R}, \theta, \phi, t). \quad (8\text{-}111)$$

Taking the spatial Fourier transform gives

$$\frac{\partial \overline{\mathscr{G}}(K, \theta, \phi, t)}{\partial t} = \left\{-K^2 D + \Theta\left[\frac{1}{\sin\theta}\frac{\partial}{\partial\theta}\left(\sin\theta\frac{\partial}{\partial\theta}\right) + \frac{1}{\sin^2\theta}\frac{\partial^2}{\partial\phi^2}\right]\right\}\overline{\mathscr{G}}(K, \theta, \phi, t). \quad (8\text{-}112)$$

We can now write $\overline{\mathscr{G}}(K, \theta, \phi, t)$ in terms of separated variables,

$$\overline{\mathscr{G}}(K, \theta, \phi, t) = \overline{\mathscr{G}}(K, t)\mathscr{P}(\theta, \phi),$$

which gives

$$\frac{1}{\Theta}\left(\frac{1}{\overline{\mathscr{G}}}\frac{\partial \overline{\mathscr{G}}}{\partial t} + K^2 D\right) = \frac{1}{\mathscr{P}}\left[\frac{1}{\sin\theta}\frac{\partial}{\partial\theta}\left(\sin\theta\frac{\partial}{\partial\theta}\right) + \frac{1}{\sin^2\theta}\frac{\partial^2}{\partial\phi^2}\right]\mathscr{P}. \quad (8\text{-}113)$$

Both sides of this equation are equal for all values of K, θ, and ϕ only if both sides are equal to a constant, call it $-\lambda$, which gives two differential equations:

$$\left[\frac{1}{\sin\theta}\frac{\partial}{\partial\theta}\left(\sin\theta\frac{\partial}{\partial\theta}\right) + \frac{1}{\sin^2\theta}\frac{\partial^2}{\partial\phi^2}\right]\mathscr{P}(\theta, \phi) = -\lambda\mathscr{P}(\theta, \phi) \quad (8\text{-}114)$$

$$\frac{\partial \overline{\mathscr{G}}(K, t)}{\partial t} = (-\lambda\Theta - K^2 D)\overline{\mathscr{G}}(K, t). \quad (8\text{-}115)$$

If $\lambda = l(l + 1)$, Eq.(8-114) has a solution $\mathscr{P}(\theta, \phi) = Y_{lm}(\theta, \phi)$ where $Y_{lm}(\theta, \phi)$ are the spherical harmonics [see Section 4.4, Eq.(4-58), and Table 4-3]. Furthermore, Eq.(8-115) is easily solved to give

$$\overline{\mathscr{G}}(K, t) = \overline{\mathscr{G}}(K, 0)\exp\{-[l(l + 1)\Theta + K^2 D]t\}. \quad (8\text{-}116)$$

Remembering that $\mathscr{G}(\boldsymbol{R}, t)$ is composed entirely of self terms, we can write $\mathscr{G}(\boldsymbol{R}, 0) = \delta(\boldsymbol{R})$, which gives $\overline{\mathscr{G}}(K, 0) = 1$. Substituting into the equation above gives

$$\overline{\mathscr{G}}(K, t) = \exp\{-[l(l + 1)\Theta + K^2 D]t\}.$$

The most general solution to Eq.(8-112) will be given by a linear combination of the solutions above in l, given by

$$\overline{\mathscr{G}}(K, \theta, \phi, t) = \sum_l A_l Y_{lm}(\theta, \phi)\exp\{-[l(l + 1)\Theta + K^2 D]t\}. \quad (8\text{-}117)$$

We find the appropriate values of A_l from the $t = 0$ initial conditions of Eqs.(8-117) and (8-86). Using the $G(\boldsymbol{R}, \theta, \phi, 0) = \delta(\boldsymbol{R})\delta(\cos\theta - \cos\theta_0)\delta(\phi - \phi_0)$ initial condition in Eq.(8-86) gives

$$\bar{\mathscr{G}}(K, \theta, \phi, 0) = \int_0^{2\pi}\int_{-1}^{1} P_2(\cos\theta_0)\delta(\cos\theta - \cos\theta_0)\delta(\phi - \phi_0)\, d\cos\theta_0\, d\phi_0$$

$$= P_2(\cos\theta).$$

Comparing with the $t = 0$ condition on Eq.(8-117), $\bar{\mathscr{G}}(K, \theta, \phi, 0) = \sum_l A_l\, Y_{lm}(\theta_0, \phi_0)$ restricts the sum over l to $l = 2$ and requires that $m = 0$, and $A_2 = 1$. In summary, we find

$$\bar{\mathscr{G}}(K, \theta, \phi, t) = P_2(\cos\theta)\exp\left[-(6\Theta + K^2D)t\right].$$

Substituting this result into Eq.(8-86) and integrating gives

$$\bar{G}(K, \theta, \phi, t) = \frac{4\pi}{5}\exp\left[-(6\Theta + K^2D)t\right], \tag{8-118}$$

where $\tau_{or} = 1/6\Theta$ is the single-particle reorientation relaxation time. It is also easy to show by the methods above that $\bar{\bar{G}}(K, \theta, \phi, t)$ in Eqs.(8-58) is also equal to $\frac{1}{3}\bar{G}(K, \theta, \phi, t)$ in Eq.(8-118) for the cylindrically symmetric scatterers considered here. Furthermore it is evident that $\bar{G}(K, t)$ in Eqs.(8-58) for single-particle scatterers is given by

$$\bar{G}(K, t) = \exp(-K^2Dt). \tag{8-119}$$

Substituting these results into the appropriate parts of the correlation functions in Eqs.(8-58) taking the real Fourier transform, and remembering Eqs.(8-59), gives the Raman spectra in the rotationally quenched limit:

$$I^{\text{RAM}}_{\text{isot}}(\omega) = ANf_j\,\mathscr{L}(\omega - \omega_0 + \omega_{n_j', n_j})_{\text{isot}}$$

$$I^{\text{RAM}}_{\text{anis}}(\omega) = \left(\frac{A}{15}\right)Nh_j\,\mathscr{L}(\omega - \omega_0 + \omega_{n_j', n_j})_{\text{anis}}$$

$$\mathscr{L}(\omega - \omega_0 + \omega_{n_j', n_j})_{\text{isot}} = \frac{1}{\pi}\left\{\frac{DK^2 + (1/\tau_j)}{(\omega - \omega_0 + \omega_{n_j', n_j})^2 + [(1/\tau_j) + DK^2]^2}\right\}$$

$$\mathscr{L}(\omega - \omega_0 + \omega_{n_j', n_j})_{\text{anis}} = \frac{1}{\pi}\left\{\frac{DK^2 + 6\Theta + (1/\tau_j)}{(\omega - \omega_0 + \omega_{n_j', n_j})^2 + [(1/\tau_j) + DK^2 + 6\Theta]^2}\right\}. \tag{8-120}$$

Normally only the Raman–Stokes transitions are observed $[E_{n_j} < E_{n_j'}$ in $\omega_{n_j', n_j} = (E_{n_j'} - E_{n_j})/\hbar]$ due to the probability factors N_{n_j}, f_j, and h_j [Eqs.(8-58)].

The $\mathscr{L}(\omega)_{\text{isot}}$ Raman spectrum has a half width at half height of $\Delta\omega = 1/\tau_j + DK^2$ where τ_j is the vibrational relaxation time for the jth normal mode of vibration.

Table 8-5 Vibrational relaxation times, τ_{vib}, and rotational orientation times $\tau_{or} = 1/6\Theta$ for several molecular liquids from the Raman spectra.

Molecule	Vibrational Transition (cm^{-1})	τ_{vib} (10^{-12} s)	$1/6\Theta = \tau_{or}$ (10^{-12} s)
Carbon disulfide	656	10.6[a]	1.5[a]
Acetonitrile	2943	3.2[b]	0.9[b]
Methyl iodide	525	2.0[a]	1.5[a]
	1245	2.0[a]	1.4[a]
Chloroform	667	2.0[a]	1.5[a]
	3019	1.1[a]	1.5[a]
Bromoform	222	4.1[a]	5.3[a]
	3019	1.2[c]	4.4[c]
Benzene	992	4.7[d]	2.8[d]
Hexafluorobenzene	558	2.2[a]	6.6[a]

[a] F. J. Bartoli and T. A. Litovitz, *J. Chem. Phys.* **56**, 404 (1972).
[b] J. E. Griffiths, *J. Chem. Phys.* **59**, 751 (1973).
[c] G. D. Patterson and J. E. Griffiths, *J. Chem. Phys.* **63**, 2406 (1975).
[d] K. T. Gillen and J. E. Griffiths, *Chem. Phys. Lett.* **17**, 359 (1972).

Noting that $D \cong (10^{-4}\text{–}10^{-5})\ \text{cm}^2 \cdot \text{s}^{-1}$ for most liquids and $\tau_j \cong 10^{-12}$ s for most molecular vibrations in liquids, we can safely write

$$\frac{1}{\tau_j} \gg K^2 D \tag{8-121}$$

for optical radiation and any scattering angle. Thus, a measure of the Raman $\mathscr{L}(\omega)_{\text{isot}}$ gives, from the half width at half height, a direct measurement of the vibrational relaxation times. Several values of τ_{vib} obtained in this way are listed in Table 8-5.

The $\mathscr{L}(\omega)_{\text{anis}}$ Raman spectra have a half width at half height of $\Delta\omega = 1/\tau_j + DK^2 + 6\Theta$. Typical small molecule values of Θ range from $10^9\text{–}10^{12}\ \text{s}^{-1}$ and if we are using optical radiation, we are safe in writing

$$6\Theta \gg DK^2 \tag{8-122}$$

for any scattering angle. Thus, a measure of the Raman $\mathscr{L}(\omega)_{\text{anis}}$ gives, from the half width at half height and a known value of τ_j, a direct measurement of the rotational diffusion coefficient, Θ, or the orientational relaxation time $\tau_{or} = 1/6\Theta$. Several values of τ_{or} obtained in this way are listed in Table 8-5. Of course, $I_v^v(\nu)^{\text{RAM}} = I_{\text{isot}}^{\text{RAM}}(\nu) + \frac{4}{3}I_{\text{anis}}^{\text{RAM}}(\nu)$ and $I_h^v(\nu)^{\text{RAM}} = I_{\text{anis}}^{\text{RAM}}(\nu)$ are measured. $I_{\text{isot}}^{\text{RAM}}(\nu)$ can be extracted from $I_v^v(\nu)^{\text{RAM}}$ by subtracting $\frac{4}{3}I_h^v(\nu)^{\text{RAM}}$.

Keeping in mind our original model of a cylindrically symmetric near-spherical-shaped molecule reorientating about its symmetry axis, we note that the values of Θ and D can be written in terms of the sheer viscosity, η, and partical radius r according to [23,24,25]

$$D = \frac{kT}{f} = \frac{kT}{6\pi\eta r}, \qquad \Theta = \frac{kT}{8\pi\eta r^3} = \frac{kT}{6V^*\eta}, \tag{8-123}$$

where k is Boltzmann's constant, r is the spherical molecule's radius, and $V^* = \frac{4}{3}\pi r^3$ is the effective molecular volume.

We now examine $I_{\text{anis}}^{\text{RAY}}(\nu)$ that arises from the Fourier transform of $\bar{P}(K, \theta, \phi, t)$ in Eqs.(8-58). $\bar{P}(K, \theta, \phi, t)$ is similar to $\bar{G}(K, \theta, \phi, t)$ considered above where $\bar{G}(K, \theta, \phi, t)$ contains only the self terms and $\bar{P}(K, \theta, \phi, t)$ contains both the self terms and the distinct terms. Thus, we expect the difference between $\bar{P}(K, \theta, \phi, t)$ and $\bar{G}(K, \theta, \phi, t)$ or $I_{\text{anis}}^{\text{RAM}}(\nu)$ and $I_{\text{anis}}^{\text{RAY}}(\nu)$, respectively, to reveal the distinct effects or the two-particle orientational pair correlations. A direct comparison of $I_{\text{anis}}^{\text{RAY}}(\nu)$ and $I_{\text{anis}}^{\text{RAM}}(\nu)$ for CS_2 and benzene is shown in Fig. 8-7. A careful analysis of the data shows that the half width at half height of $I_{\text{anis}}^{\text{RAY}}(\nu)$ is slightly smaller than in $I_{\text{anis}}^{\text{RAM}}(\nu)$ in CS_2, thus, reflecting the effects of the orientational pair correlations. In benzene, however, the half widths of $I_{\text{anis}}^{\text{RAY}}(\nu)$ and $I_{\text{anis}}^{\text{RAM}}(\nu)$ are the same, indicating no orientational pair correlation effect. The values of $\tau_{or}^{\text{RAY}} = 1/\Delta\omega$ [where $\Delta\omega$ is the half width at half height in $I_{\text{anis}}^{\text{RAY}}(\nu)$] for several molecules are listed in Table 8-6. Comparing τ_{or}^{RAY} in Table 8-6 with τ_{or} in Table 8-5, shows that $\tau_{or} \leq \tau_{or}^{\text{RAY}}$ or that the effects of orientational pair correlations cause an effectively longer rotational relaxation time. The orientational pair correlations also affect the integrated intensities of the anisotropic Raman and Rayleigh scattered light. According to Eqs.(8-58), Eq.(8-59), the discussion above, and arguments similar to those preceding Eq.(8-74), we can write

$$I_{\text{anis}}^{\text{RAY}} = \int_{-\infty}^{+\infty} I_{\text{anis}}^{\text{RAY}}(\omega)\, d\omega = \left(\frac{AN}{3}\right)\frac{(\alpha_{aa}^0 - \alpha_{bb}^0)^2}{4\pi}\bar{P}(K, \theta, \phi, 0)$$

$$I_{\text{anis}}^{\text{RAM}} = \int_{-\infty}^{+\infty} I_{\text{anis}}^{\text{RAM}}(\omega)\, d\omega = \left(\frac{AN}{3}\right)\left(\frac{h_j}{4\pi}\right)\bar{G}(K, \theta, \phi, 0)$$

$$N\bar{P}(K, \theta, \phi, 0) = N\bar{G}(K, \theta, \phi, 0)$$

$$+ \left\langle \sum_{i \neq j} P_2(\cos\theta)_{oi} P_2(\cos\theta)_{oj} \exp\{-i\boldsymbol{K}\cdot[\boldsymbol{r}_i(0) - \boldsymbol{r}_j(0)]\}\right\rangle, \quad (8\text{-}124)$$

for the jth normal mode in $I_{\text{anis}}^{\text{RAM}}$ where $N = V_s \rho_0$ is the number of scatterers. Substituting from Eq.(8-118) and assuming that rotational and translational motions are separable gives

$$N\bar{P}(K, \theta, \phi, 0) = \frac{4\pi}{5}N + \left\langle \sum_{i \neq j} P_2(\cos\theta)_{oi} P_2(\cos\theta)_j \right\rangle. \quad (8\text{-}125)$$

The remaining independent sum over i and j where $i \neq j$ is over all pairs of molecules within the volume element V_s. Thus, if each of the scattering molecules are identical, all terms in one of the sums will be the same and we can write

$$N\bar{P}(K, \theta, \phi, 0) = \frac{4\pi}{5}N + (N - 1)\left\langle \sum_{j} P_2(\cos\theta)_i P_2(\cos\theta)_j \right\rangle, \quad (8\text{-}126)$$

where the last term includes the long-time average of the $N - 1$ identical i terms that are summed over j ($j \neq i$). We now use the spherical harmonic addition formula [Eq.(5-81)] to write $P_2(\cos\theta)_i P_2(\cos\theta)_j$ in terms of $P_2(\cos\theta_{ij})$, where θ_{ij} is the angle

Table 8-6 Rotational reorientation times, τ_{or}^{RAY}, which are obtained from the half width at half height of the depolarized Rayleigh lines, $\Delta\omega = 1/\tau_{or}^{RAY}$, as shown for instance in Fig. 8-7. The temperatures are near 300 K.

Molecule	τ_{or}^{RAY} $(10^{-12}$ s)
Carbon disulfide	1.8[a]
Acetonitrile	1.7[a]
Methyl iodide	2.3[a]
Chloroform	2.9[b]
Bromoform	10.1[c]
Benzene	2.9[d]
Hexafluorobenzene	14.0[e]

[a] S. J. Bertucci and A. K. Burnham, unpublished data (1976).

[b] G. R. Alms, *et al.*, *J. Chem. Phys.* **59**, 5310 (1973).

[c] G. D. Patterson and J. E. Griffiths, *J. Chem. Phys.* **63**, 2407 (1975).

[d] G. R. Alms, *et al.*, *J. Chem. Phys.* **58**, 5570 (1973).

[e] D. R. Bauer, J. I. Brauman, and R. Pecora, *J. Chem. Phys.* **63**, 53 (1975).

between the cylindrical symmetry axes of the ij-pair of molecules. Using the ergodic hypothesis, we replace the time average with a spatial average. Making these changes, we write

$$N\bar{P}(K,\theta,\phi,0) = \frac{4\pi}{5}N + \frac{4\pi}{5}(N-1)\left\langle \sum_j P_2(\cos\theta_{ij})\right\rangle = \frac{4\pi}{5}Ng_2$$

$$g_2 = 1 + \frac{(N-1)}{N}\left\langle \sum_j P_2(\cos\theta_{ij})\right\rangle \cong 1 + \left\langle \sum_j P_2(\cos\theta_{ij})\right\rangle, \qquad (8\text{-}127)$$

where the last step assumes that $(N-1)/N = 1$, which requires a large number of particles in the scattering volume. $\langle \sum_j P_2 \cos\theta_{ij}\rangle$ is the sum of average values of $P_2(\cos\theta_{ij}) = \frac{1}{2}(3\cos^2\theta_{ij} - 1)$ between the N-1 ij-pairs of molecules in the fluid. Substituting the result in Eq.(8-127) into $I_{\text{anis}}^{\text{RAY}}$ in Eq.(8-124) and substituting Eq.(8-118) into $I_{\text{anis}}^{\text{RAM}}$ in Eq.(8-124) gives the integrated intensities,

$$I_{\text{anis}}^{\text{RAY}} = \frac{1}{15}Ag_2N(\alpha_{aa}^0 - \alpha_{bb}^0)^2, \qquad I_{\text{anis}}^{\text{RAM}} = \frac{A}{15}Nh_j. \qquad (8\text{-}128)$$

Thus, according to these results, if a pure solution of uncorrelated monomeric cylindrically symmetric molecules would suddenly dimerize with symmetry axes aligned and if there were no correlations between different dimers, $\langle \sum_j P_2(\cos\theta_{ij})\rangle = 1$, and $g_2 = 2$, thereby increasing the intensity of $I_{\text{anis}}^{\text{RAY}}$. If the dimerization would occur with symmetry axes perpendicular, $\langle \sum_j P_2(\cos\theta_{ij})\rangle = -\frac{1}{2}$, $g_2 = \frac{1}{2}$, and the intensity of $I_{\text{anis}}^{\text{RAY}}$ would decrease. A diagram of g_2 as a function of density from $I_{\text{anis}}^{\text{RAY}}$ in the isotropic liquid phase in MBBA, a rodlike molecule that forms a liquid crystal phase with the rod axes aligned, is shown in Fig. 8-10. The evidence for increasing alignment with increasing density, as measured from $I_{\text{anis}}^{\text{RAY}}$ and the resultant g_2, is

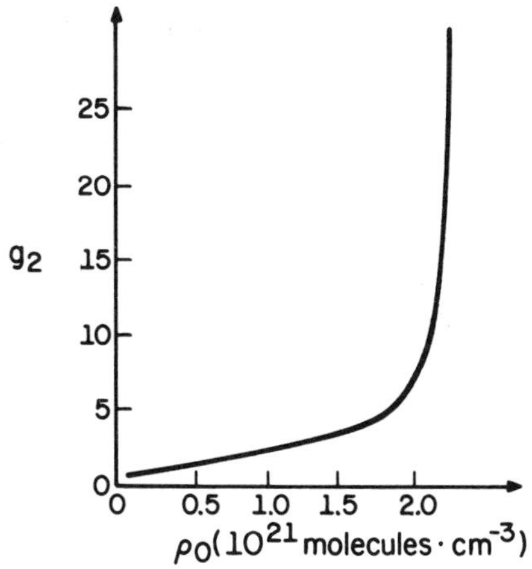

Figure 8-10 Experimental determination of g_2 as a function of density, ρ_0, from $I^{RAY} = \int_{-\infty}^{+\infty} I_{anis}^{RAY}(v)\,dv$ as shown in Eq.(8-128). The system studied is the isotropic liquid *p*-methoxybenzylidene-*n*-butylaniline (MBBA) at 318 K, which is above the transition temperature for the liquid crystal phase. The data are from G. R. Alms, T. D. Gierke, and W. H. Flygare, *J. Chem. Phys.* **61**, 4083 (1974).

quite convincing. Of course, $g_2 \longrightarrow N$ as the system approaches the liquid crystal phase.

In the limit of Debye diffusion, where many collisions are necessary to cause a molecular reorientation, Keyes and Kivelson [26] have shown that the time-dependent part of the correlation function also contains g_2 according to [see Eq.(8-118) where $\tau_{or} = 1/6\Theta$]

$$\bar{P}(K, \theta, \phi, t) = \frac{4\pi}{5} g_2 \exp\left[-\left(\frac{1}{\tau_{or} g_2} + K^2 D\right)t\right], \tag{8-129}$$

where $\tau_{or} g_2 = \tau_{or}^{RAY}$ is defined in Table 8-6 as obtained from the half width at half height of the $I_{anis}^{RAY}(\omega)$ spectrum.

8.5 X-RAY SCATTERING: FREE ATOMS, FREE MOLECULES, CONDENSED PHASES, AND LOW-ANGLE SCATTERING

In this section we describe X-ray scattering from atoms and molecules in the gas and liquid states. We emphasize the determination of the scattering system's electron density by an analysis of the intensity of the scattered X rays as a function of the scattering vector, $K = (4\pi/\lambda_0) \sin(\theta_s/2)$. An analysis of the spectra of the scattered X rays to obtain information about the dynamics of the scattering system is not feasible at this time due to the unavailability of a narrow-band X-ray source of radiation. We are interested here in elastic scattering, the scattering that can be described by a wave-like description of the X-ray electromagnetic radiation. The coherent scattering of X rays is complicated by the presence of inelastic scattering that arises from the Compton effect as described in Section 1.5.

In the previous sections we have discussed the scattering of optical radiation from gases and liquids. We considered only nonresonant scattering where the radiation frequency is considerably less than any electronic transition frequencies of the scattering system. In general, the electromagnetic field-induced dipole moment is proportional to the polarizability given, for a single bound electron, by (see Problem 2-11, where $\tau = \infty$)

$$\alpha_e = \frac{e^2}{m}\left(\frac{1}{\omega_0^2 - \omega^2}\right), \tag{8-130}$$

where ω_0 is the resonant frequency of the bound electron and ω is the frequency of the incident radiation. In the case of X-ray radiation we expect that $\omega_0 \ll \omega$, giving the following expression for the X-ray scattering from an electron:

$$\lim_{\omega \gg \omega_0} \alpha_e = -\frac{e^2}{m\omega^2}. \tag{8-131}$$

The corresponding ratio of the scattered to incident light intensity for X rays scattered from N free electrons is given by substituting Eq.(8-131) into Eq.(1-59), where we replace $\sin^2 \gamma$ with $\frac{1}{2}(1 + \cos^2 \theta_s)$ for unpolarized incident X-ray radiation,

$$\frac{I}{I_0} = \frac{e^4 N}{m^2 c^4 R^2}\left[\tfrac{1}{2}(1 + \cos^2 \theta_s)\right] = \frac{N r_e^2}{2R^2}(1 + \cos^2 \theta_s). \tag{8-132}$$

θ_s is the scattering angle (see Fig. 8-1) and $r_e = e^2/mc^2 = 2.8178 \times 10^{-13}$ cm is the classical electron radius. Thus, I/I_0 for X rays scattered from one mole of free electrons is given by

$$\frac{I}{I_0} \cong \frac{2.4 \times 10^{-2}}{R^2}(1 + \cos^2 \theta_s).$$

The corresponding expression for the Ar^+ ion laser ($\lambda_0 = 5145$ Å) scattering from one mole of Ar atoms ($\alpha = 1.63 \times 10^{-24}$ cm^3) is given in Eq.(1-59) where we use unpolarized incident radiation to give

$$\frac{I}{I_0} = \frac{16\pi^4 \alpha^2 N_A}{R^2 \lambda_0^4}\frac{1}{2}(1 + \cos^2 \theta_s) \cong \frac{1.8 \times 10^{-4}}{R^2}(1 + \cos^2 \theta_s).$$

Comparing these last two results shows that I/I_0 for X rays and a number of free electrons is much larger than the corresponding ratio for optical scattering from an equal number of argon atoms. It is also evident from the inverse squared mass dependence in Eq.(8-132) that the scattering intensity of X rays from electrons exceeds the intensity of X rays scattered from nuclei by at least 10^6.

We shall now examine the nature of the short wavelength ($\lambda_0 \cong 0.3$–2.0 Å) elastic X-ray scattering from atoms and molecules. We assume again a classical scattering model where the incident X-ray radiation is described by a complex electric field, $\boldsymbol{E}(t) = \boldsymbol{E}_0 \exp(-i\omega_0 t)$, and the corresponding incident intensity is given by $I_0 = (c/4\pi)|E_0|^2$. The scattered radiation field for X rays is given by an expression similar to Eq.(8-18). Our previous discussion examined the scattering from a single molecular or atomic scattering center with the electrical asymmetry given by the complete polarizability tensor in the c.m. coordinate system. No intramolecular interference effects were considered. Describing the scattering from a single center in a molecule is appropriate if the size of the molecule is small relative to the radiation

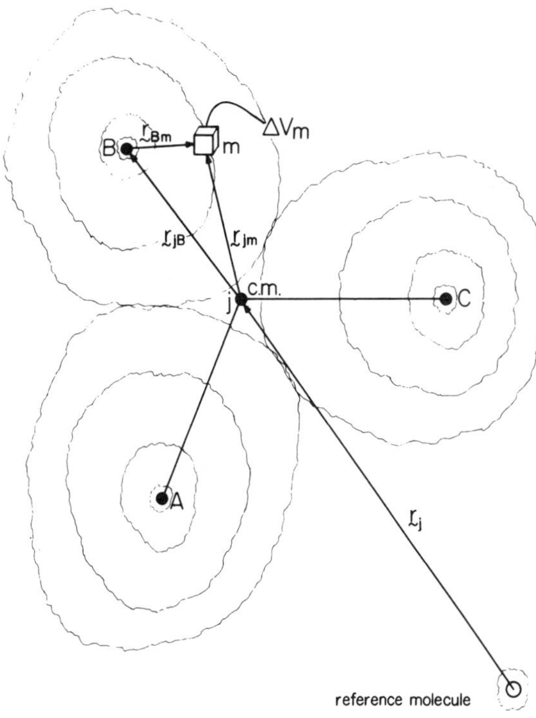

Figure 8-11 Schematic representation of a molecule composed of three atoms, A, B, and C. The wavy lines indicate the electron density, c.m. is the center of mass, and the cube labeled m is the mth scattering center in the jth molecule with volume ΔV_m. $\mathbf{r}_j$ is the vector from the origin or reference molecule to the c.m. of the jth molecule and $\mathbf{r}_{jm}$ is the vector from the jth molecule's c.m. to the mth scattering volume element. We break the sum over molecular electrons down into atomic sums where $\mathbf{r}_{jB}$ is the vector from the c.m. to the Bth atom and $\mathbf{r}_{Bm}$ is the vector from the B atom's nucleus to the mth scattering center associated with the free B atom. In the case of scattering from a free atom, the c.m. or scattering origin is the nucleus.

wavelength. In the case of X-ray scattering, however, the wavelength of the radiation is on the order of the interatomic spacings in molecules and intramolecular interference effects will become apparent, enabling the measurement of interatomic distances in molecules.

The model used to describe the X-ray scattering from a molecule is illustrated in Fig. 8-11. Scattered fields are summed from all points in the molecule where the field is weighted by the probability that an electron is within the scattering volume in question. Starting with a zy-polarized incident field (see Fig. 8-1), the total field scattered from the jth molecule has the following magnitude and phase relative to an arbitrary origin in the scattering system [see again Eq.(8-18)]:

$$\mathbf{E}_j(t) = \frac{\omega_0^2 \alpha_e \sin\gamma}{Rc^2} E_o \hat{\gamma} \exp(-i\omega_0 t) \exp[i\mathbf{K}\cdot\mathbf{r}_j(t)] \sum_m \rho_m \exp[i\mathbf{K}\cdot\mathbf{r}_{jm}(t)] \Delta V_m, \tag{8-133}$$

where $\hat{\gamma}$ is the unit vector along γ, $\alpha_e = -e^2/m\omega^2$ is the X-ray polarizability of a free electron, and ρ_m is the electron density in the volume element identified in Fig. 8-11 by ΔV_m, where the sum over m is over volume elements that include the entire volume of the molecule. $\boldsymbol{r}_j$ and $\boldsymbol{r}_{jm}$ are identified in Fig. 8-11. We now replace the sum over the scattering element volumes in the last part of Eq.(8-133) by an integral over the space occupied by the molecule.

$$\sum_m \rho_m \exp(i\boldsymbol{K}\cdot\boldsymbol{r}_{jm})\,\Delta V_m = \int_0^{2\pi}\int_0^{\pi}\int_0^{\infty} \rho_j(r,\theta,\phi)\exp(i\boldsymbol{K}\cdot\boldsymbol{r})r^2\,dr\sin\theta\,d\theta\,d\phi, \quad (8\text{-}134)$$

where the electron density function for the jth molecule, $\rho_j(r, \theta, \phi)$, with n_j total electrons, is given from the molecular electronic wavefunction, $\psi_j(1, 2, 3, \ldots, n_j)$, by

$$\rho_j(r,\theta,\phi) = |\psi_j|^2. \quad (8\text{-}135)$$

r is the center of mass coordinate and θ and ϕ are the usual spherical polar coordinates. In the low K limit where $\exp(i\boldsymbol{K}\cdot\boldsymbol{r}) \cong 1$, Eq.(8-134) reduces to

$$\lim_{K\to 0}\int_0^{2\pi}\int_0^{\pi}\int_0^{\infty} \rho_j(r,\theta,\phi)r^2\,dr\sin\theta\,d\theta\,d\phi = n_j.$$

Substituting Eq.(8-134) into Eq.(8-133) and summing over j gives the total in-phase fields scattered into the detector:

$$\boldsymbol{E}(t) = \sum_j \boldsymbol{E}_j(t) = \frac{e^2\sin\gamma}{mRc^2}E_o\hat{\gamma}\exp(-i\omega_0 t)\sum_j \exp[i\boldsymbol{K}\cdot\boldsymbol{r}_j(t)]$$
$$\times\int_0^{2\pi}\int_0^{\pi}\int_0^{\infty}\exp(i\boldsymbol{K}\cdot\boldsymbol{r})\rho_j(r,\theta,\phi)\,r^2\,dr\sin\theta\,d\theta\,d\phi. \quad (8\text{-}136)$$

The translational time-dependence is in $r_j(t)$ and the orientational dependence is in the $\boldsymbol{K}\cdot\boldsymbol{r} = Kr\cos\theta'$ term, where θ' is the angle between the space-fixed K vector and molecular-fixed r vector in the electron density function, $\rho(r, \theta, \phi)$. We now use Eq.(8-5) to write the correlation function as a long-time average. Combining fields to obtain unpolarized incident X-ray radiation and using the ergodic hypothesis to write the correlation function as a spatial average, we obtain

$$C(K,t) = I_e\frac{N}{2}\exp(-i\omega_0 t)\int_0^{\pi}\int_0^{\pi}\int_{V_s}\exp(i\boldsymbol{K}\cdot\boldsymbol{R})P(\boldsymbol{R},\theta',t)$$
$$\times\left\{\left[\int_0^{2\pi}\int_0^{\pi}\int_0^{\infty}\exp(iKr\cos\theta_0')\rho(r,\theta,\phi)r^2\,dr\sin\theta\,d\theta\,d\phi\right]\right.$$
$$\times\left.\left[\int_0^{2\pi}\int_0^{\pi}\int_0^{\infty}\exp(iKr\cos\theta')\rho(r,\theta,\phi)r^2\,dr\sin\theta\,d\theta\,d\phi\right]\right\}$$
$$\times\,dV_s\sin\theta_0'\,d\theta_0'\sin\theta'\,d\theta'$$

$$I_e = \frac{e^4 I_0}{m^2R^2c^4}\left[\frac{1}{2}(1+\cos^2\theta_s)\right]. \quad (8\text{-}137)$$

N is the number of scatterers (molecules), I_0 is the incident X-ray intensity, and I_e [according to Eq.(8-132)] is the X-ray scattering intensity for one free electron for unpolarized incident X rays. $\boldsymbol{R}$ is the c.m. coordinate for the scatterer and r is the c.m. to electron coordinate where θ'_0 is the initial angle between $\boldsymbol{K}$ and $\boldsymbol{r}$ and θ' is the angle between $\boldsymbol{K}$ and $\boldsymbol{r}$ at a later time t. $P(\boldsymbol{R}, \theta', t)$ is the probability (per unit volume) that the c.m. position is at $\boldsymbol{R}$ with orientation θ' at t given, of course, the initial conditions at $t = 0$ of $\boldsymbol{R} = 0$ and θ'_0. The factor of $\frac{1}{2}$ in Eq.(8-137) assumes a normalized probability over angles θ'_0 and θ' for the delta function initial condition, as discussed following Eq.(8-49), and dV_s indicates the volume element and corresponding integral over the scattering volume. Equation (8-137) is the general expression for the correlation function for a system of identical scatterers described by the electronic wavefunction $\psi(1, 2, \ldots, n)$ and the space-time–orientation correlation function, $P(\boldsymbol{R}, \theta', \tau)$. In previous sections we developed $P(\boldsymbol{R}, \theta', \tau)$ for gas kinetic and hydrodynamic models. In the case of X-ray scattering, the same principles are applicable. The gas kinetic model will lead to Doppler line widths in the scattered radiation. $K \cong 10^8$ for $\theta_s = \pi/2$ X-ray scattering, however, which leads to significantly larger Doppler line widths than considered previously for optical radiation. It is more risky to apply the hydrodynamic realm to evaluate $P(\boldsymbol{R}, \theta', \tau)$ for X-ray scattering where $K^2 D > 6\Theta$. In principal, self-diffusion could be measured; however, the X-ray polarizability anisotropy is probably near zero, which reduces the depolarized X-ray scattering to zero. These arguments above are currently academic though, because there are apparently no narrow-band continuous X-ray sources with sufficient radiation power for scattering experiments. If the frequency bandwidth of the source greatly exceeds the width of the information in the scattered X rays, it is clear that only the source width is observed. This being the case, we shall now examine the spatial or static correlations as measured by $C(\tau = 0)$ for X rays from Eq.(8-137). Perhaps in another decade, narrow-band X-ray sources will be available, giving rise to an examination of the time correlations as observed in the spectrum of scattered X rays.

Free Atoms

If $C(K, 0)$ in Eq.(8-137) represents the scattering from free atoms, r in the $\exp(irK \cos\theta')$ and $\exp(irK \cos\theta'_0)$ terms has a nuclear origin. In this case all $\cos\theta'$ and $\cos\theta'_0$ for the spherically symmetric atom have equal probability. The static correlation in $\boldsymbol{R}$ or the initial condition on $\boldsymbol{R}$ is described by the delta function. Thus, the initial condition on $P(\boldsymbol{R}, \theta', \tau)$ for a system of free atoms is given by $P(\boldsymbol{R}, \theta', 0) = \frac{1}{2}\delta(\boldsymbol{R})$, where the factor of $\frac{1}{2}$ assumes normalization of the angular part. Substituting this initial condition into Eq.(8-137) gives the total intensity of the coherently scattered X rays from a system of free atoms with a many-electron wavefunction given by $\psi(1, 2, \ldots, n)$:

$$I(K) = C(K, 0) = N \frac{I_e}{4} \int_0^{\pi} \int_0^{\pi} \left\{ \left[\int_0^{2\pi} \int_0^{\pi} \int_0^{\infty} \rho(r, \theta, \phi) \exp(iKr \cos\theta'_0) r^2 \, dr \sin\theta \, d\theta \, d\phi \right] \right.$$

$$\left. \times \left[\int_0^{2\pi} \int_0^{\pi} \int_0^{\infty} \rho(r, \theta, \phi) \exp(iKr \cos\theta') r^2 \, dr \sin\theta \, d\theta \, d\phi \right] \right\} \sin\theta'_0 \, d\theta'_0 \sin\theta' \, d\theta',$$

where we have used

$$\int_{V_s} \exp(i\boldsymbol{K}\cdot\boldsymbol{R})\delta(\boldsymbol{R})\,dV_s = 1.0.$$

The integrals over θ_0' and θ' can be evaluated by using

$$\int_0^{\pi} \exp(\pm iKr\cos\theta')\sin\theta'\,d\theta' = \int_0^{\pi}[\cos(Kr\cos\theta') \pm i\sin(Kr\cos\theta')]\sin\theta'\,d\theta'$$

$$= \frac{2\sin Kr}{Kr}, \tag{8-138}$$

which gives

$$I(K) = I_e N\left[\int_0^{2\pi}\int_0^{\pi}\int_0^{\infty}\rho(r,\theta,\phi)\frac{\sin Kr}{Kr}r^2\,dr\sin\theta\,d\theta\,d\phi\right]^2 = I_e N f(K)^2, \tag{8-139}$$

where $f(K)$ is called the *atomic scattering factor*, which is a function of K. Remembering that $\lim_{x\to 0}(\sin x/x) = 1.0$, we can write the $K \longrightarrow 0$ limit of $f(K)$ as

$$\lim_{K\to 0} f(K) = n, \tag{8-140}$$

where n is the number of electrons in the scattering atom. It is also evident that

$$\lim_{K\to\infty} f(K) = 0. \tag{8-141}$$

According to these expressions we can write the corresponding limits for $I(K)$ in Eq.(8-139) as

$$\lim_{K\to 0} I(K) = NI_e n^2, \qquad \lim_{K\to\infty} I(K) = 0, \tag{8-142}$$

where N is the number of scattering atoms and n is the number of electrons in the scattering atom.

The electronic function $\psi(1, 2, \ldots, n)$ must be available to evaluate $\rho(r, \theta, \phi)$ in Eq.(8-135) and finally $f(K)$ in Eq.(8-139). We have discussed the nature of $\psi(1, 2, \ldots, n)$ for atoms in Chapter 5. It is now clear that coherent X-ray scattering through Eq.(8-139) will give a $(\sin Kr/Kr)$-weighted average of the atomic electron density.

We can illustrate the analytical form for $f(K)$ in Eq.(8-139) with the single-electron hydrogen atom. Using $\psi(r)$ in Eq.(5-7) for the ground state of the hydrogen atom gives the X-ray scattering factor for the atom of

$$f_{\mathrm{H}}(K) = \int_0^{2\pi}\int_0^{\pi}\int_0^{\infty}\frac{1}{\pi}\left(\frac{1}{a_0}\right)^3\exp\left(-\frac{2r}{a_0}\right)\frac{\sin rK}{rK}r^2\,dr\sin\theta\,d\theta\,d\phi = \frac{1}{(1+\frac{1}{4}K^2a_0^2)^2}. \tag{8-143}$$

Substituting Eq.(8-143) into Eq.(8-139) gives the following ratio of the scattering intensity of a free H atom, $I(K)$, over the scattering intensity of a free electron, I_e:

$$\frac{I(K)}{I_e} = [f_{\mathrm{H}}(K)]^2 = \left[\frac{1}{(1 + \frac{1}{4}K^2 a_0^2)^2}\right]^2, \qquad (8\text{-}144)$$

where $[f_{\mathrm{H}}(K)]^2$ is shown in Fig. 8-12.

It is evident from these developments that the forward-scattered X rays ($K = 0$) scatter off atoms with an intensity equal to the free-electron scattering intensity; however, the scattered X-ray intensity falls off with increasing scattering angle. The half-intensity point in Fig. 8-12 for free H atoms and $\lambda_0 = 1.0$ Å X rays is given from Eq.(8-144) to be $\theta_s = 15.0°$.

Atomic scattering factors have been calculated for a wide range of atoms and ions using the best available many-electron atomic functions. A partial listing of f_A values is shown in Table 8-7. The calculated value of f_A^2 for the argon atom, as taken from the numbers in Table 8-7 for $\lambda_0 = 1.54$ Å, is shown in the lower curve

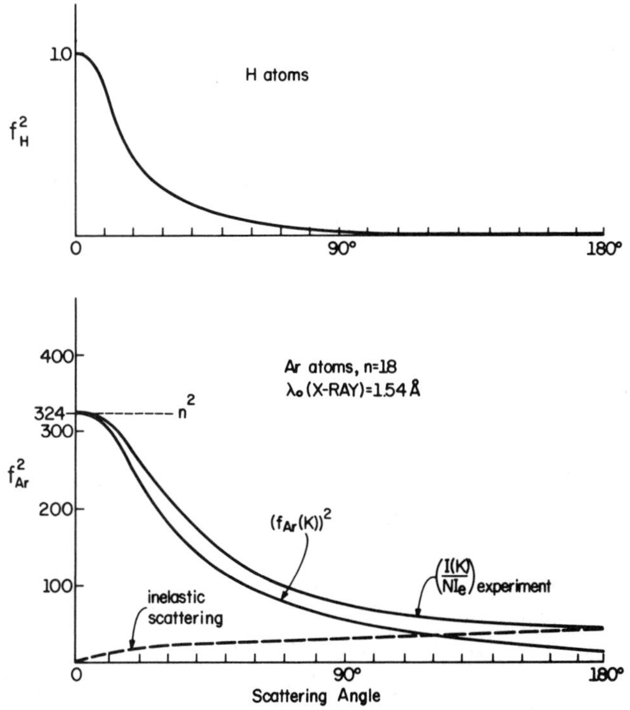

Figure 8-12 X-Ray scattering factors for free hydrogen and argon atoms. The top curve shows f_{H}^2 calculated from Eq.(8-144) for λ_0 (X-ray) = 1.0 Å. The lower curve shows the corresponding calculated value of f_{Ar}^2 for argon atoms from the numbers in Table 8-7 using $\lambda_0 = 1.54$ Å. $n = 18$ is the number of electrons in Ar. Also shown is the experimental value of $I(K)/NI_e$ for argon. The deviation between the calculated f_{Ar}^2 and the experimental $I(K)/NI_e$ arises from inelastic (Compton) scattering shown by the dotted line as calculated with Eq.(8-145) and Table 8-8. The lower curve is adapted from Ref. [27].

Table 8-7 Atomic scattering factors, f [Eq.(8-139)], for several atoms and ions from *International Tables for X-Ray Crystallography* (The Kynoch Press, Birmingham, England, 1962). f-Values are given for $[(1/\lambda)\sin(\theta_s/2)] \times 10^{-8}$ values of 0, 0.2, 0.4, 0.6, 0.8, and 1.0. A useful analytical form for the scattering factors has been given by D. T. Cramer and J. B. Mann, *Acta Cryst.* **A24**, 321 (1968).

Atomic Number	Element	$[(1/\lambda)\sin(\theta_s/2)] \times 10^{-8}$ 0	0.2	0.4	0.6	0.8	1.0
1	H	1.0	0.48	0.13	0.04	0.02	0.01
2	He	2.0	1.45	0.74	0.36	0.18	0.10
3	Li^+	2.0	1.76	1.27	0.82	0.51	0.32
3	Li	3.0	1.74	1.27	0.82	0.51	0.32
4	Be^{2+}	2.0	1.87	1.55	1.18	0.86	0.61
4	Be	4.0	2.07	1.53	1.20	0.88	0.62
5	B	5.0	2.71	1.69	1.41	1.15	0.90
6	C	6.0	3.58	1.95	1.54	1.32	1.11
7	N	7.0	4.60	2.40	1.70	1.44	1.26
8	O	8.0	5.63	3.01	1.94	1.57	1.37
9	F	9.0	6.09	3.76	2.31	1.74	1.48
9	F^-	10.0	7.13	3.79	2.32	1.75	1.49
10	Ne	10.0	7.82	4.62	2.79	1.98	1.61
11	Na^+	10.0	8.39	5.51	3.42	2.31	1.79
11	Na	11.0	8.34	5.47	3.40	2.31	1.78
14	Si	14.0	9.67	7.20	5.31	3.75	2.69
15	P	15.0	10.34	7.54	5.83	4.28	3.11
16	S	16.0	11.21	7.83	6.31	4.82	3.56
17	Cl	17.0	12.00	8.07	6.64	5.27	4.00
17	Cl^-	18.0	12.20	8.03	6.64	5.27	4.00
18	Ar	18.0	12.93	8.54	6.86	5.61	4.43
36	Kr	36.0	28.53	21.34	16.54	12.57	9.66
80	Hg	80.0	67.14	52.65	42.31	34.64	28.59

in Fig. 8-12. The experimental scattering intensity from argon for K_α copper atom X rays ($\lambda_0 = 1.54$ Å) divided by the constant I_e [Eq.(8-139)] times the number of scatterers, N, is also shown in Fig. 8-12. It is evident that the scattered X-ray intensity $I(K)/NI_e$ ratio agrees with the calculated f_{Ar}^2 only at $K = 0$. At all other scattering angles, the experimental scattered intensity exceeds the calculated value. The discrepancy arises from inelastic Compton scattering.

In the case of inelastic scattering, the scattered wave vector, k_s, has a magnitude less than the incident wave vector, $k_0 = 4\pi/\lambda_0$. This inelastic scattering was described in Section 1.5 by assuming a linear momentum for the X ray and the resultant changes in wavelength of the scattered X rays were evaluated as a function of the scattering angle. Of course, the complete theory of radiation-electron scattering must contain both the elastic ($k_s = k_0$) and inelastic ($k_s \neq k_0$) contributions to the scattering intensity. We shall not discuss in any detail the theory leading to the intensity distribution of inelastic scattering. According to the summary by Pirenne [27], the intensity of the inelastic component of the scattered X rays is proportional to I_e and

Table 8-8 Values of K' and $F(K')$ needed in Eqs.(8-145) and (8-146) to evaluate the intensity of the inelastic X-ray scattering in atoms.

K'	$F(K')$
0.05	0.319
0.10	0.486
0.20	0.674
0.30	0.776
0.40	0.839
0.50	0.880
0.60	0.909
0.70	0.929
0.80	0.944
0.90	0.954
1.00	0.963

the inelastic-elastic intensity ratio decreases with increasing number of electrons in the atomic scatterer. According to Pirenne, an adequate semiempirical evaluation of the intensity of the inelastic scattering is given by

$$I(K)_{\text{inelastic}} = \frac{I_e Z F(K')}{[1 + (h/mc\lambda_0)(1 - \cos\theta_s)]^3}, \qquad K' = K\left(\frac{0.176}{Z} \times 10^{-8}\right). \tag{8-145}$$

Z is the atomic number, K' is a reduced scattering vector, and $F(K')$ is a function that is evaluated from a specific model describing the intensity. Several values of K' and $F(K')$ are listed in Table 8-8, which should allow the evaluation of $I(K)_{\text{inelastic}}$ for any atom. The $I(K)_{\text{inelastic}}$ for the argon atom as calculated from Eq.(8-145) is shown in the lower curve in Fig. 8-12. In summary, the total scattering intensity of X rays from free atoms is given by

$$I(K)_{\text{total}} = I(K)_{\text{elastic}} + I(K)_{\text{inelastic}} = NI_e\left[f^2 + \frac{ZF(K')}{[1 + (h/mc\lambda_0)(1 - \cos\theta_s)]^3}\right]. \tag{8-146}$$

In principle, it is possible to separate $I(K)_{\text{elastic}}$ and $I(K)_{\text{inelastic}}$ experimentally because the X rays in $I(K)_{\text{elastic}}$ all have frequencies or energies equal to the incident radiation and the X rays in $I(K)_{\text{inelastic}}$ all have frequencies or energies less than the incident radiation. In any event, the value of $I(K) = I_e f^2$ can be extracted from the experimental result leading through Eqs.(8-135) and (8-139) to a measure of the many-electron wavefunction for the free scattering atom.

Free Molecules

We shall now return to Fig. 8-11 and Eq.(8-137) for a discussion of X-ray scattering from free molecules [28]. First we recall some of our earlier discussion on the nature of atoms in molecules. In Fig. 6-6 and the associated discussion, as well as in Section 6.9, we argued that the electron densities of most molecules are very similar to the electron densities of the corresponding free atoms with the same spatial configurations as in the molecule. This simple atoms-in-molecules approach was very useful in describing a number of one-electron molecular properties. If molecules are made up of atoms that largely retain their overall atomic charge distributions, the integrals over all molecular electrons in Eq.(8-137) can be replaced by a sum over atoms with the corresponding electronic integrals restricted to free atom expressions. Specifically we refer to Fig. 8-11 and we replace $\boldsymbol{r}_{jm}$ in Eq.(8-134) with the vector $\boldsymbol{r}_{j\alpha}$ plus $\boldsymbol{r}_{\alpha m}$ where α indicates the αth nucleus,

$$\boldsymbol{r}_{jm} = \boldsymbol{r}_{j\alpha} + \boldsymbol{r}_{\alpha m}. \tag{8-147}$$

Substituting Eq.(8-147) into Eq.(8-134) for a molecule gives

$$\int_0^{2\pi}\int_0^{\pi}\int_0^{\infty} \rho_j(r, \theta, \phi) \exp(i\boldsymbol{K}\cdot\boldsymbol{r}) r^2\, dr \sin\theta\, d\theta\, d\phi$$

$$= \sum_m \rho_m \exp(i\boldsymbol{K}\cdot\boldsymbol{r}_{jm})\,\Delta V_m = \sum_\alpha \sum_m \rho_{\alpha m} \exp[i\boldsymbol{K}\cdot(\boldsymbol{r}_{j\alpha} + \boldsymbol{r}_{\alpha m})]\,\Delta V_m$$

$$= \sum_\alpha \exp(i\boldsymbol{K}\cdot\boldsymbol{r}_{j\alpha}) \int_0^{2\pi}\int_0^{\pi}\int_0^{\infty} \rho_\alpha(r, \theta, \phi) \exp(i\boldsymbol{K}\cdot\boldsymbol{r}) r^2\, dr \sin\theta\, d\theta\, d\phi. \tag{8-148}$$

$\rho_j(r, \theta, \phi)$ is the molecular electronic density function and $\rho_\alpha(r, \theta, \phi)$ is the αth atomic electron density function in the jth molecule. The sum over α is over all atoms in the free molecule and the sum over m is over all volume elements ΔV_m in the free α atom. Thus, the integral over the molecular electron density is replaced by a sum of integrals over the corresponding free atom densities. If we also assume that the free atoms in the molecules are spherically symmetric, we can average the $\boldsymbol{K}\cdot\boldsymbol{r}$ terms for each α atom over all orientations to give

$$\sum_\alpha \exp(i\boldsymbol{K}\cdot\boldsymbol{r}_\alpha)\left[\int_0^{2\pi}\int_0^{\pi}\int_0^{\infty} \rho(r, \theta, \phi) \exp(i\boldsymbol{K}\cdot\boldsymbol{r}) r^2\, dr \sin\theta\, d\theta\, d\phi\right]_{av}$$

$$= \sum_\alpha \exp(i\boldsymbol{K}\cdot\boldsymbol{r}_\alpha) f_\alpha(K), \tag{8-149}$$

where $f_\alpha(K)$ is the X-ray scattering factor [Eq.(8-139)] for the αth atom. Substituting Eq.(8-149) into Eq.(8-137) gives

$$C(K, t) = I_e \frac{N}{2} \exp(-i\omega_0 t) \sum_{\alpha,\beta} f_\alpha(K) f_\beta(K) \int_0^{\pi}\int_0^{\pi}\int_{V_s} \exp(i\boldsymbol{K}\cdot\boldsymbol{R})$$

$$\times \exp[i(r_\alpha K \cos\theta_0' - r_\beta K \cos\theta')] P(\boldsymbol{R}, \theta', t)\, dV_s \sin\theta_0'\, d\theta_0' \sin\theta'\, d\theta', \tag{8-150}$$

where the summations over α and β are taken independently over all atoms in the molecule. We now write the total intensity as a function of K by setting $t = 0$ in Eq.(8-150) and defining the appropriate initial conditions for $P(\boldsymbol{R}, \theta', 0)$. In the case of a rigid molecule we must write

$$P(\boldsymbol{R}, \theta', 0) = \delta(\boldsymbol{R})\delta(\cos\theta_0' - \cos\theta'). \tag{8-151}$$

Making the changes above and substituting Eq.(8-151) into Eq.(8-150) gives

$$I(K) = I_e \frac{N}{2} \sum_{\alpha,\beta} f_\alpha(K) f_\beta(K) \int_0^\pi \int_0^\pi \exp[iK(r_\alpha \cos\theta_0' - r_\beta \cos\theta')]$$
$$\times\, \delta(\cos\theta_0' - \cos\theta') \sin\theta_0'\, d\theta_0' \sin\theta'\, d\theta'$$
$$= I_e \frac{N}{2} \sum_{\alpha,\beta} f_\alpha(K) f_\beta(K) \int_0^\pi \exp[iK(r_\alpha - r_\beta)\cos\theta'] \sin\theta'\, d\theta'$$
$$= I_e N \sum_{\alpha,\beta} f_\alpha(K) f_\beta(K) \frac{\sin K(r_\alpha - r_\beta)}{K(r_\alpha - r_\beta)}$$
$$= I_e N \left\{ \sum_\alpha [f_\alpha(K)]^2 + \sum_{\alpha \neq \beta} f_\alpha(K) f_\beta(K) \frac{\sin K(r_\alpha - r_\beta)}{K(r_\alpha - r_\beta)} \right\}. \tag{8-152}$$

Thus, the intensity of the coherently scattered X rays from a free molecule is given by a free atom term plus a product of atomic scattering factors with $[\sin K(r_\alpha - r_\beta)]/[K(r_\alpha - r_\beta)]$, which gives a damped sine wave as a function of K with maxima every $K(r_\alpha - r_\beta) = 2\pi$ for *each* interatomic difference $r_\alpha - r_\beta$. If the atomic scattering factors f_α and f_β are fairly well localized to an atom in a molecule and therefore fall off fast in K relative to $[\sin K(r_\alpha - r_\beta)]/[K(r_\alpha - r_\beta)]$, the $[\sin K(r_\alpha - r_\beta)]/[K(r_\alpha - r_\beta)]$ oscillations will be sharp leading to quite distinguishable interatomic spacings. On the other hand, if f_α and f_β extend over distances that overlap neighboring atoms, the periodic behavior in $I(K)$ will be less evident. The two cases described above are demonstrated in Fig. 8-13 by the scattering intensities from N_2, O_2., and Cl_2. The scattering from N_2 shows very little structure. The longer bond distances in O_2 and Cl_2 lead to the beginning of structure in the intensity functions. The $I(K)$ curve for CCl_4 is also shown in Fig. 8-13 where the long Cl—Cl distances and relatively

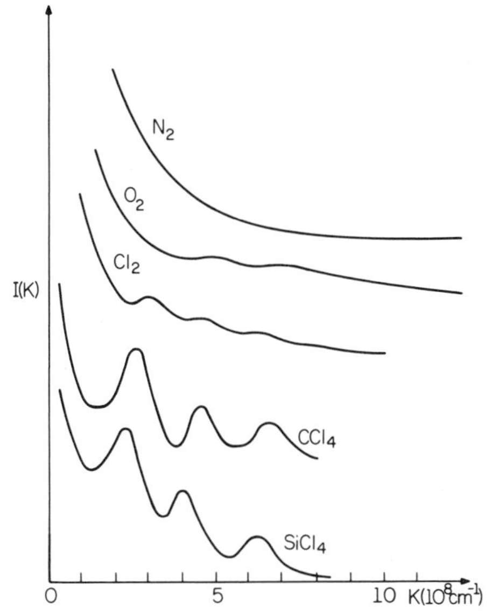

Figure 8-13 Total X-ray scattering intensities [Eq.(8-152)] as a function of K for N_2, O_2, Cl_2, CCl_4, and $SiCl_4$. The X rays are K_α Cu ($\lambda = 1.54$ Å). The curves are adapted from A. H. Compton and S. K. Allison, *X-Rays in Theory and Experiment* (D. Van Nostrand Co., Inc., New York, 1935). The CCl_4 and $SiCl_4$ are adapted from Ref. [27].

localized values of f_{Cl} lead to definite structure in $I(K)$. Also shown in Fig. 8-13 is the $I(K)$ function for $SiCl_4$, which demonstrates that the larger Cl—Cl distances in $SiCl_4$ (3.27 Å) give $I(K)$ maxima which are closer together in K than the corresponding results for CCl_4 with smaller Cl—Cl distances (2.96 Å).

Perhaps the best method of extracting the molecular internuclear distances from Eq.(8-152) is by constructing a trial function for $I(K)$ using a set of assumed values of $r_\alpha - r_\beta$ and the atomic scattering factors in Table 8-7. The calculated $I(K)$ can then be compared to the experimental function and further refinements can cause the calculated and experimental values of $I(K)$ to merge within an uncertainty defined in a systematic way as in the least-squares method. Fourier transform methods are not very useful here as the $f_\alpha(K)$ and $f_\beta(K)$ functions do not fall off fast enough in K. We shall show in Chapter 9 that elastic electron scattering (electron diffraction) has $I(K)$ functions that look similar to Eq.(8-152). For high incident electron energies, however, the electron-atom scattering factors, $f_\alpha(K)$, fall off very fast in K allowing convenient Fourier transforms to obtain the structural information.

Condensed Phases

We discussed static correlations in liquids toward the end of Section 8.3 where we introduced the pair correlation function, $g(R)$, in Eq.(8-75). We shall now apply X-ray scattering to the measurement of $g(R)$ in an atomic liquid. We begin by returning to Eq.(8-137) where we need to write an expression for $P(\mathbf{R}, \theta', t)$ for an atomic liquid. Assuming an isotropic atomic liquid where θ' and θ'_0 have equal probability, we write the static correlation function as

$$P(\mathbf{R}, \theta', 0) = \tfrac{1}{2}P(\mathbf{R}, 0) = \tfrac{1}{2}[\delta(\mathbf{R}) + \rho_0 g(\mathbf{R})], \tag{8-153}$$

where the last step follows by substituting from Eq.(8-75). The factor of $\frac{1}{2}$ assumes normalization of the averages over θ'_0 and θ' as before. The expression in Eq.(8-153) contains both self and distinct terms and it is evident that $\rho_0 \longrightarrow 0$ for a dilute atomic gas. $P(\mathbf{R}, \theta', 0)$ goes to $\frac{1}{2}\delta(\mathbf{R})$ as $\rho_0 \longrightarrow 0$, leading to the results given above for free atoms [see Eq.(8-139)]. Substituting Eq.(8-153) into Eq.(8-137), setting $t = 0$, and integrating over θ'_0 and θ' gives the coherent scattered intensity,

$$I(K) = NI_e[f(K)]^2[S(K) + \rho_0 \delta(K)], \tag{8-154}$$

where $f(K)$ is the atomic X-ray scattering factor (Table 8-7) and the liquid structure factor, $S(K)$, is defined in Eq.(8-78).

Assuming that the experimental $I(K)$ curve can be corrected for the incoherent scattering by methods described previously, we are left with an experimental value of the coherent scattering intensity, $I(K)$, in Eq.(8-154). $S(K)$ can be extracted from $I(K)$ by excluding the forward-scattering term, $\rho_0 \delta(K)$, and using the calculated (from Table 8-7) or experimental values of the free atom atomic scattering factors, $f(K)$. The resulting structure factor, $S(K)$, for X-ray scattering in liquid potassium is shown in Fig. 8-14 for two temperatures.

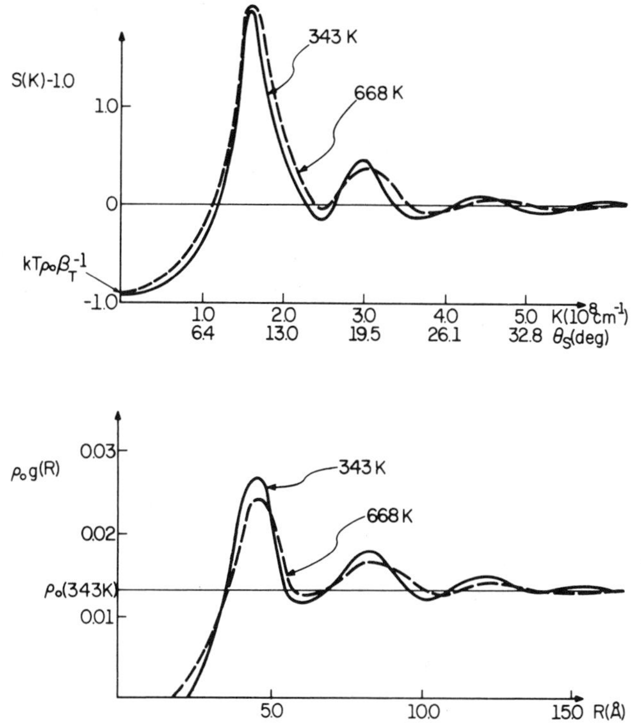

Figure 8-14 The structure factor, $S(K)$, evaluated from the elastic $I(K)$ X-ray scattering from potassium metal at two different temperatures. The X-ray radiation was from K_α Mo where $\lambda_0 = 0.71$ Å. The first maximum in $S(K) - 1.0$ occurs at a scattering angle of $\theta_s = 10.6°$. The lower curve shows the product of the average number density of potassium atoms times the corresponding pair distribution function, $g(R)$, obtained from $S(K)$ by the Fourier transform in Eq.(8-157). According to Eq.(8-81), the $K = 0$ intercepts in the $S(K)$ curves are proportional to the isothermal compressibility, β_T. The data are adapted from C. D. Thomas and N. S. Gingrich, *J. Chem. Phys.* **6**, 411 (1938) and more recent data in A. J. Greenfield, J. Wellendorf, and N. Wiser, *Phys. Rev.* **A4**, 1607 (1971).

In order to extract $g(R)$ from $S(K)$, we use the appropriate Fourier transform. Rewriting Eq.(8-78) for $S(K)$ gives

$$S(K) = 1.0 + \rho_0 \int_0^{2\pi} \int_0^{\pi} \int_0^{\infty} \exp(iKR\cos\theta)[g(R) - 1]R^2\, dR \sin\theta\, d\theta\, d\varphi$$

$$= 1.0 + 4\pi\rho_0 \int_0^{\infty} \frac{R}{K} \sin KR[g(R) - 1]\, dR. \tag{8-155}$$

The large K limit of $S(K)$ equal to 1.0 is evident from this equation. Both the large K and small K limits are evident in Fig. 8-14. The $K \longrightarrow 0$ limit gives a direct measurement of the isothermal compressibility as shown in Fig. 8-14 and Eq.(8-81).

In order to obtain $g(R)$ from Eq.(8-155), we multiply both sides by $K \sin KR'$, rearrange, and integrate over K from 0 to $+\infty$ (approximating the experimental range of K) to give

$$\int_0^\infty K[S(K) - 1] \sin KR' \, dK = 4\pi\rho_0 \int_0^\infty \int_0^\infty [g(R) - 1] R \sin KR \sin KR' \, dK \, dR. \tag{8-156}$$

Comparing the right-hand side of this equation with Eq.(B-55) in Appendix B describing Fourier transform methods shows that the right-hand side is equal to $2\pi^2\rho_0[g(R') - 1]R'$ and we note that $K[S(K) - 1]$ and $2\pi^2\rho_0[g(R) - 1]R$ are Fourier transform pairs. The result for $g(R)$ is

$$g(R) = 1 + \frac{1}{2\pi^2 R\rho_0} \int_0^\infty K[S(K) - 1] \sin KR \, dK. \tag{8-157}$$

The values of the pair correlation functions, $g(R)$, for liquid potassium at two different temperatures are shown in the lower curves in Fig. 8-14. Equations (8-155) and (8-157) show the transformation between $S(K)$ and $g(R)$ and Fig. 8-14 shows the corresponding experimental results for the potassium liquid metal. The average number of first nearest-neighbor potassium atoms in the liquid state at 343 K and 668 K can be obtained directly from $\rho_0 g(R)$ in Fig. 8-14 by integration over a sphere. For instance, the number of nearest neighbors, n', is reasonably well defined by

$$n' = 4\pi\rho_0 \int_0^{R_0} g(R) R^2 \, dR, \tag{8-158}$$

where R_0 is the approximate distance to the first minimum in the $g(R)$ curve; $R_0 \cong 6$ Å from the $\rho_0 g(R)$ curve for potassium at $T = 343$ K. Doing a rough numerical integration on the $T = 343$ K curve in Fig. 8-14 gives $n' \cong 12$. This appears to be a common feature of liquids composed of spherical particles; there are approximately 12 nearest neighbors and then the location of the next nearest and atoms further out becomes much less well defined.

We can also use the theory above to describe X-ray scattering from a dense medium of molecules by starting with Eq.(8-150) and using Eq.(8-153) for $P(\mathbf{R}, \theta', 0)$, where $g(\mathbf{R})$ indicates c.m. correlations, which gives

$$I(K) = I_e N \sum_{\alpha, \beta} f_\alpha(K) f_\beta(K) \frac{\sin K(r_\alpha - r_\beta)}{K(r_\alpha - r_\beta)} [S(K) + \rho_0 \delta(K)], \tag{8-159}$$

where $S(K)$ is the structure factor as defined in Eq.(8-155). The $I(K)$ in this equation has structure in K from $f_\alpha(K)$, $f_\beta(K)$, $[\sin K(r_\alpha - r_\beta)]/[K(r_\alpha - r_\beta)]$, and $S(K)$. The $g(R)$ in the liquid phase can be obtained by comparing the result in the condensed phase in Eq.(8-159) with the free atom result from the gas phase in Eq.(8-152). Considerable progress has been made recently in interpreting $I(K)$ X-ray and neutron scattering data according to Eq.(8-159) to obtain information on the structure of molecular liquids [29]. Orientational information can also be obtained by these methods [29,30].

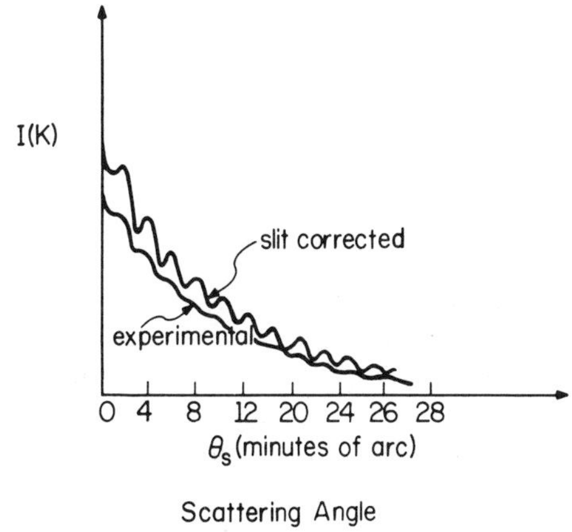

Figure 8-15 Low-angle X-ray scattering from a static emulsion of closely packed latex spheres with a diameter of $d = 2500$ Å. The X rays have $\lambda_0 = 1.54$ Å. The structure is clearly observable where the average peak-to-peak spacing of 2.11′ of arc leads to the distance of 2500 Å between scattering centers. The data are from Ref. [32].

Low-Angle X-Ray Scattering

It is evident that at low angles where K approaches zero, larger and larger distances in the scattering medium are probed. We shall now develop a few applications of low-angle X-ray scattering to examine the extent of some long-range spatial correlations. The low-angle X-ray scattering intensity as a function of K from an emulsion of closely packed uniform polystyrene latex spheres of diameter $d = 2500$ Å is shown in Fig. 8-15. A careful examination of the curve in Fig. 8-15 shows an average peak-to-peak distance of 2.11′ of arc in the scattering angle. Equation (8-152) is appropriate for interpreting the data in Fig. 8-15 where f_α and f_β are the scattering factors for the identical latex spheres, and $r_\alpha - r_\beta$ is the average center-center distance between the spheres. Thus, according to Eq.(8-152) and this interpretation, we require $\sin K(r_\alpha - r_\beta) = \sin 2\pi$ between the maxima. Using this equation, $K = (4\pi/\lambda_0)\sin(\theta_s/2)$, and the appropriate value of $\lambda_0 = 1.54$ Å used in the experiment shown in Fig. 8-15, we obtain $d = r_\alpha - r_\beta \cong 2500$ Å as the diameter of the spheres.

Other applications of low-angle scattering include the study of large-range fluctuations in pure liquids and mixtures of liquids near critical points (see end of Section 8.3), the distances between various layered structures (for instance, liquid crystals), and the determination of the hydrodynamic properties of macromolecules in solutions [31]. Some of these applications were mentioned in earlier sections and we shall examine additional applications in more detail here.

First we note that in the limit of low angles all the equations we have derived in this section must reduce to the corresponding Rayleigh scattering expressions for optical radiation that were developed earlier. For instance, the low K [$Kr \ll 1$ form of Eq.(8-137)] reduces to

$$C(K, t) = I_e N n^2 \exp(-i\omega_0 t) \int_{V_s} \exp(i\boldsymbol{K} \cdot \boldsymbol{R}) P(\boldsymbol{R}, t)\, dV_s, \qquad (8\text{-}160)$$

where n is the number of electrons in the scatterer. The intensity as a function of K is given by

$$I(K) = I_e N n^2 \int_{V_s} \exp(i\boldsymbol{K} \cdot \boldsymbol{R}) P(\boldsymbol{R}, 0)\, dV_s. \qquad (8\text{-}161)$$

In the absence of strains or any other anisotropic perturbation, $P(\mathbf{R}, 0)$ depends only on the absolute value of $\mathbf{R}$ and not on its direction. Using this simplification, noting that $\mathbf{K} \cdot \mathbf{R} = KR \cos \theta$ for this case (aligning R along the z axis), and using the volume element $dV_s = R^2 \, dR \sin \theta \, d\theta \, d\phi$ gives

$$I(K) = 4\pi I_e N n^2 \int_0^\infty R^2 \left(\frac{\sin KR}{KR} \right) P(R, 0) \, dR. \tag{8-162}$$

In the case of an extremely dilute system of scatterers, we can ignore interscatterer correlations and interpret $P(R, 0)$ as an intrascatterer correlation function. This development allows an estimate of the radius of gyration of a large molecule in solution or the corresponding correlation length in an isotropic amorphous solid or liquid crystal. Returning to Eq.(8-162) we can expand the sin KR function if $KR \ll 1$ to give

$$I(K) = I_e N n^2 4\pi \int_0^\infty R^2 \left[1 - \frac{(KR)^2}{6} + \cdots \right] P(R, 0) \, dR, \tag{8-163}$$

where $P(R, 0)$ is the distribution function for a single scatterer. Dividing this expression by $I(K = 0)$ gives

$$\frac{I(K)}{I(K = 0)} = 1 - \frac{(K^2/6) \int_0^\infty R^4 P(R, 0) \, dR}{\int_0^\infty R^2 P(R, 0) \, dR} + \cdots = 1 - \frac{K^2 L^2}{6} + \cdots. \tag{8-164}$$

L is defined as the correlation length according to Eq.(8-164). We now approximate the expansion in Eq.(8-164) with a Gaussian giving

$$\frac{I(K)}{I(K = 0)} = 1 - \frac{1}{6} K^2 L^2 + \cdots \cong \exp\left(- \frac{K^2 L^2}{6} \right). \tag{8-165}$$

Measurements of $I(K)/I(K = 0)$ and interpretation according to Eq.(8-165) have been applied to liquid crystals and other systems [32].

We shall now show the relation between the correlation length defined here in Eq.(8-164) as L and the radius of gyration, R_G, which is often used in the interpretation of the K-dependence of the light intensity scattered from a dilute solution of macromolecules. L^2, in Eq.(8-164), is the average square of the distance between scattering points in the scatterer described by $P(R, 0)$. R_G^2 is defined as the average square of the distance between the center of the scatterer (macromolecule) to the outside periphery of the scatterer described by $P(R, 0)$. Thus, $R_G^2 < L^2$ and it is easy to show by considering a scatterer as a distribution of smaller scatterers that $2R_G^2 = L^2$. Thus, we can write Eq.(8-165) in terms of the radius of gyration as

$$\frac{I(K)}{I(0)} = 1 - \frac{1}{3} K^2 R_G^2 + \cdots \cong \exp\left(- \frac{K^2 R_G^2}{3} \right). \tag{8-166}$$

Equation (8-166) has been used extensively to interpret the low-angle X-ray $I(K)$ function in dilute solutions of macromolecules to measure the radius of gyration, R_G. Considerable caution should be exercised in using the Gaussian forms of Eqs.(8-165) and (8-166) to interpret the $I(K)/I(K = 0)$ data. At the Gaussian half width, KL or KR_G are on the order of unity and the original assumption, $KR \ll 1$, is no longer valid. Thus, the expanded forms of Eqs.(8-165) and (8-166) should be used to interpret only the low K data in order to extract L or R_G.

We can also derive the expression for the radius of gyration by starting with Eq.(8-152) for the scattering from N free macromolecules,

$$I(K) = NI_e \sum_{\alpha,\beta} f_\alpha(K) f_\beta(K) \frac{\sin K(r_\alpha - r_\beta)}{K(r_\alpha - r_\beta)},$$

where $f_\alpha(K)$ and $f_\beta(K)$ are the free atom values in the macromolecule as defined previously. In the case of low-angle scattering where $K(r_\alpha - r_\beta) < 1$, we can expand the sine function in a power series and approximate the low K values of $f_\alpha(K)$ and $f_\beta(K)$ by n_α and n_β, the number of electrons in the α and β atoms, respectively. Taking these steps gives

$$I(K) = NI_e \sum_{\alpha,\beta} n_\alpha n_\beta \left[1 - \frac{K^2(r_\alpha - r_\beta)^2}{6} + \cdots\right]. \tag{8-167}$$

The sum over atoms in this equation can also be expressed by the radius of gyration where we estimate

$$R_G^2 = \frac{1}{2n^2} \sum_{\alpha,\beta} n_\alpha n_\beta (r_\alpha - r_\beta)^2. \tag{8-168}$$

n is the total number of electrons in the molecule and the factor of $\frac{1}{2}$ acknowledges the double counting in the α and β summations over all atoms in the molecule. Substituting this definition of R_G into Eq.(8-167) gives Eq.(8-166). Several values of R_G measured by these techniques are listed in Table 8-9.

The development above was for a system of free particles in the low K region. In the case of a solution of macromolecules, we replace n^2 with $(n - n_0)^2$ where n

Table 8-9 Radii of gyration for several large molecules as measured by low-angle X-ray scattering.

Molecule	R_G (Å)
Myoglobin	15.5[a]
Bovine serum albumin	29.8[b]
Human mercaptalbumin	31.0[b]
Human mercaptalbumin dimer	37.5[b]
Catalase	39.8[c]
Bushy stunt virus	120.0[d]

[a] K. Ibel and H. B. Stuhrmann, *J. Mol. Biol.* **93**, 255 (1975).

[b] J. W. Anderegg, *et al.*, *J. Am. Chem. Soc.* **77**, 2927 (1955).

[c] A. G. Malmon, *Biochim. et Biophys. Acta* **26**, 233 (1957).

[d] B. R. Leonard, *et al.*, *Biochim. et Biophys. Acta* **12**, 499 (1953).

is still the number of electrons in the solute molecule and n_0 is the number of electrons in the solvent molecules comprising the volume occupied by the solute molecule. Thus, $(n - n_0)$ is the difference in the number of electrons in the solute over that of the solvent for a constant volume.

We shall now consider the total low-angle X-ray scattering intensities from noncrystalline solids. Starting with Eq.(8-162) we note that the K-dependence (scattering angle) of $I(K)$ is given by the sine Fourier transform of $RP(R, 0)$ [33]. In the case of an isotropic amorphous (noncrystalline) solid, the correlation function, $P(R, 0)$, is expected to be an exponential [34],

$$P(R, 0) = \exp\left(-\frac{R}{a}\right), \tag{8-169}$$

where a is the correlation length. A scatterer in this case is a region in the amorphous solid that deviates from the average distribution; this deviation could be a static density fluctuation manifested as a void or a high point in density. Substituting Eq.(8-169) into Eq.(8-162) gives

$$I(K) = \frac{4\pi I_e N n^2}{K} \int_0^\infty \mathrm{Re}\left(-\frac{R}{a}\right) \sin RK \, dR = \frac{8\pi I_e N n^2 a^3}{(1 + K^2 a^2)^2}. \tag{8-170}$$

The exponential static correlation function in Eq.(8-169) and the result in Eq.(8-170) have been used extensively to interpret the light scattering intensities in noncrystalline isotropic solids including glasses and polymers. Other forms for the correlation function are given by Ross [35] where he examines and gives explicit transforms from $P(R, 0)$ to $I(K)$ for a wide variety of functional forms for the static correlation function, $P(R, 0)$.

In amorphous solids that contain two components, the angular distribution function of the polarized scattered light, $I(K)$, may require a combination of two correlation functions,

$$P(R, 0) = f P_1(R) + (1 - f) P_2(R). \tag{8-171}$$

In the two-component system, $P_2(R)$ represents the density fluctuations of solute in the solvent and $P_1(R)$ represents the fluctuations within the region of high (or low) solute concentration. f in Eq.(8-171) represents the fractional contribution of each type of static fluctuation. Exponentials may be used for both static correlations in Eq.(8-171). A Gaussian is often used, however, to describe the long-range correlation function (observed at low angles),

$$P_2(R) = \exp\left(-\frac{R^2}{a_2^2}\right), \tag{8-172}$$

with the same meaning for a_2 as in the exponential in Eq.(8-169). Substituting the correlation function in Eq.(8-172) into Eq.(8-162) gives $I(K)$ for the Gaussian correlation:

$$I(K) = 4\pi I_e N n^2 \int^\infty R^2 \frac{\sin RK}{RK} \exp\left(-\frac{R^2}{a_2^2}\right) dR = I_e N n^2 \pi^{3/2} a_2^3 \exp\left(-\frac{K^2 a_2^2}{4}\right). \tag{8-173}$$

Using the Gaussian in Eq.(8-172) to fit the low-angle scattering intensity will lead to a measurement of the correlation length a_2. Equations (8-172) and (8-169) can be combined in Eq.(8-171) to give $I(K)$, which are combinations of Eqs.(8-170) and (8-173). A large amount of work has been done in interpreting low-angle X-ray and light scattering in solid polymer systems using both the exponential and Gaussian distribution functions. In systems requiring two expressions for the correlation functions, the Gaussian appears to be most appropriate for the long-range correlations observed at low angles and the exponential appears to be most appropriate for short-range correlations observed at higher angles.

The equations above, beginning with Eq.(8-162), can also be developed for light scattering if the spatial fluctuations approach or exceed the optical wavelengths. The polarized isotropic light scattering is obtained from Eq.(8-162) by replacing $I_e n^2$ with $A\alpha^2$, giving

$$I_v^v(K) = A\alpha^2 N 4\pi \int_0^\infty R^2 \frac{\sin KR}{KR} P(R, 0)\, dR,$$

where α is the scatterer's polarizability, N is the number of scatterers, and A is defined in Eq.(8-21) or (8-22) for polarized and unpolarized incident light, respectively.

The Debye theory of isotropic scattering in a noncrystalline solid has also been extended to a solid containing static fluctuations in the anisotropy of the polarizability of the scatterer [36]. Thus, the polarized and depolarized scattering can be observed to give depolarization ratios and the radial distribution for both the isotropic and anisotropic scatterers. Normally, the orientation correlation will fall off faster (in distance) than the corresponding isotropic correlations. The isotropic and anisotropic scattering intensities (both the θ_s- and γ-dependence in Fig. 8-1) can be examined to obtain information about morphology and supermolecular structure in polymers, glasses, and other amorphous systems. The work by Borch, *et al.* [37] contains many useful pictures of polymer film textures and their corresponding I_h^v scattering patterns. Similarly, Samuels [38] has examined both I_h^v and I_v^v for a wide variety of morphologies and textures and some interesting scattering pictures from paracrystals in biopolymers and synthetic polymers has also been given [39]. The effects of scattering from particles in the limit where the size of the particle exceeds $2\pi/K$ (Mie scattering) has also been treated in detail [40].

8.6 CONCENTRATION FLUCTUATIONS AND ELECTRIC FIELD EFFECTS

In this section we shall examine a number of principles that are applied to light scattering from solute molecules in solutions. Before discussing scattering from the solute molecules in a dilute solution, we remember that the *solvent* will certainly scatter light, as described in Sections 8.3 and 8.4 where we considered pure liquids. In review, we remember that in pure liquids consisting of molecules that are small relative to λ_0 (radiation) and where we assume a single scattering center in the anisotropic molecule, the polarized scattered light, $I_v^v(\omega)$, is a combination of an isotropic Rayleigh–Brillouin triplet and an anisotropic, normally broader, line as shown in Fig. 8-6 for nitrobenzene. The half width of the center isotropic Rayleigh line is $\Delta\omega = \kappa K^2$ where κ is the thermal diffusivity, listed for several molecules in Table 8-2. The spectrum of the anisotropic component of $I_v^v(\omega)$ is dominated by rotational diffusion where the half widths are given by $\Delta\omega = 1/\tau_{or} = 6\Theta$, when $g_2 = 1$ [Eq.(8-129)], Θ is the rotational diffusion constant, and τ_{or} is the rotational reorientation time, listed for several molecules in Table 8-5. We also noted in the discussion following Eq.(8-120) that the translational diffusion coefficient, D, could not be measured from the anisotropic scattering of optical radiation from small molecules because of the dominant contribution of rotational diffusion to the width of the spectra. As a final point, we remember that the intensity of the light scattered from a pure liquid is proportional to the isothermal compressibility.

In this section we shall show that concentration fluctuations of solute molecules give rise to an additional *isotropic* scattering, which allows the measurement of the translational diffusion coefficient of the solute molecule in the solvent. We shall consider in some detail the intensities of the scattered light due to solute molecule concentration fluctuations relative to the intensities in the scattering due to the density fluctuations in a pure liquid. After discussing small molecule (solute) scattering in solutions we examine some of the features of light scattering from macromolecules in solution including the effects of an electric field and the measurement of mobilities.

In the case of a small dilute solute nonvibrating molecule with cylindrical symmetry ($\alpha_{aa} \neq \alpha_{bb} = \alpha_{cc}$), the distinct terms in the correlation function will be insignificant and the resulting correlation function and spectrum can be given by arguments similar to those leading to Eq.(8-120) for Raman scattering. The results for concentration fluctuations of a solute nonvibrating molecule give the following spectra:

$$I_v^v(\omega)^{\text{RAY}} = I_{\text{isot}}^{\text{RAY}}(\omega) + \frac{4}{3} I_{\text{anis}}^{\text{RAY}}(\omega)$$

$$= A\alpha^2 N \mathscr{L}(\omega - \omega_0)_{\text{isot}} + \frac{4A}{45}(\alpha_{aa} - \alpha_{bb})^2 N \mathscr{L}(\omega - \omega_0)_{\text{anis}}$$

$$I_h^v(\omega)^{\text{RAY}} = \frac{A}{15}(\alpha_{aa} - \alpha_{bb})^2 N \mathscr{L}(\omega - \omega_0)_{\text{anis}}$$

$$\mathscr{L}(\omega - \omega_0)_{\text{isot}} = \frac{1}{\pi}\left[\frac{DK^2}{(\omega - \omega_0)^2 + (DK^2)^2}\right]$$

$$\mathscr{L}(\omega - \omega_0)_{\text{anis}} = \frac{1}{\pi}\left[\frac{DK^2 + 6\Theta}{(\omega - \omega_0)^2 + (DK^2 + 6\Theta)^2}\right], \qquad (8\text{-}174)$$

where A is defined in Eqs.(8-21) and (8-22) and all other terms have been defined previously (Section 8.4). In the case of a very dilute solute, D is the solute self-diffusion coefficient in the solvent. In a binary mixture of A and B at a higher concentration of solute, the measured diffusion coefficient will be the mutual diffusion coefficient, D_{AB}, given to first order (in an ideal A–B solution) by [41]

$$D_{\text{AB}} = D_{\text{A}} X_{\text{B}} + D_{\text{B}} X_{\text{A}}, \qquad (8\text{-}175)$$

where X_{A} is the mole fraction of A in the solution.

We must now relate the single-particle intensities in Eqs.(8-174) to the properties of a solute in a solution. We assume that the fluctuations in concentration, which give rise to the scattered spectra in Eq.(8-174), are independent of the density fluctuations giving rise to the Rayleigh–Brillouin spectra described in Section 8.3. We have treated the case of uncorrelated solute scattering in a dilute solution in Section 1.2. Comparing the discussion following Eq.(1-62) with the total intensity from $I_{\text{isot}}^{\text{RAY}}(\omega)$ in Eq.(8-174) shows that in the case of a dilute solution we can write (for $\gamma = \pi/2$)

$$\alpha^2 = \frac{M^2 n_0^2}{4\pi^2}\left[\left(\frac{\partial n}{\partial C}\right)_0\right]^2, \tag{8-176}$$

where C is the concentration ($C/\rho_0 = M$), M is the mass of the scatterer, n is the refractive index of the solution, and n_0 is the refractive index of the solvent. A more rigorous derivation by statistical mechanics [25] gives a more complete expression including static correlations in concentrated solutions,

$$\alpha^2 = \frac{M^2 n_0^2}{4\pi^2}\left[\left(\frac{\partial n}{\partial C}\right)_0\right]^2\left(\frac{1}{M}\right)\left[\frac{1}{(1/M) + 2B_1 C + 3B_2 C^2 + \cdots}\right], \tag{8-177}$$

which reduces to Eq.(8-176) when $C \to 0$. B_1 and B_2 are the virial coefficients that give rise to a decrease in the isotropic scattering intensity at higher concentrations.

Finally, we can substitute Eq.(8-177) into Eqs.(8-174) and assume no orientational correlations (see Section 8.4) to give the complete spectral function for polarized incident radiation (Fig. 8-1) in a dilute solution of N small solute scatterers.

$$\begin{aligned} I_v^v(\omega)^{\text{RAY}} &= \frac{A}{M}\left[\frac{1}{(1/M) + 2B_1 C + 3B_2 C^2 + \cdots}\right]\left\{\frac{M^2 N n_0^2}{4\pi}\left[\left(\frac{\partial n}{\partial C}\right)_0\right]^2\right\}\mathscr{L}(\omega - \omega_0)_{\text{isot}} \\ &\quad + \frac{4}{45}N(\alpha_{aa} - \alpha_{bb})^2 \mathscr{L}(\omega - \omega_0)_{\text{anis}} \\ I_h^v(\omega)^{\text{RAY}} &= A\left(\frac{N}{15}\right)(\alpha_{aa} - \alpha_{bb})^2 \mathscr{L}(\omega - \omega_0)_{\text{anis}}, \end{aligned} \tag{8-178}$$

where $\mathscr{L}(\omega - \omega_0)_{\text{isot}}$ and $\mathscr{L}(\omega - \omega_0)_{\text{anis}}$ are given in Eq.(8-174). The relative intensities of the isotropic concentration dependent effect in Eq.(8-178) can be compared directly with the total Rayleigh–Brillouin intensity for the isotropic density fluctuation effects considered in Section 8.3. Substituting Eq.(8-82) into $C(K, 0)$ in Eq.(8-65) and comparing with the frequency integrated form of the isotropic term in Eq.(8-178) shows that the relative intensities are given by

Pure liquid	*Solute in dilute solution*
$\alpha^2 N \rho_0 k T \beta_T$	$\frac{M^2 N' n_0^2}{4\pi^2}\left[\left(\frac{\partial n}{\partial C}\right)_0\right]^2,$

where $N = \rho_0 V_s$ is the number of pure liquid scatterers and $N' = \rho_B V_s$ is the number of solute scatterers where V_s is the scattering volume.

The purpose of this detailed comparison between the density fluctuations in a pure liquid and the concentration fluctuations in a dilute solution is to compare the relative intensities. In Table 8-4 and associated discussion, we argued that the $\rho_0 kT\beta_T$ term in pure liquids could be interpreted as decreasing the effective number of scatterers in the dense pure fluid relative to a gas due to the destructive interferences of adjacent closely packed scatterers. We shall now show that the solute molecules in a dilute solution can scatter with a relative intensity that is higher than either the pure solute liquid or the background solvent. The value of $(\partial n/\partial C)_0$ needed in Eq.(8-178) can be measured for the particular system in question. We can estimate $(\partial n/\partial C)_0$ by assuming an ideal solution where the refractive index of the solution can be evaluated by the mole fraction weighted sum of the individual pure-fluid refractive indices. From Eq.(6-163) where $\varepsilon = n^2$, we note that $(n^2 - 1)/(n^2 + 2)$ for the solution is proportional to the number density. Thus, for an ideal solution we can write

$$\frac{(n^2 - 1)}{(n^2 + 2)} = X_a\left(\frac{n_a^2 - 1}{n_a^2 + 2}\right) + X_b\left(\frac{n_b^2 - 1}{n_b^2 + 2}\right) = \frac{n_a^2 - 1}{n_a^2 + 2} + X_b\left(\frac{n_b^2 - 1}{n_b^2 + 2} - \frac{n_a^2 - 1}{n_a^2 + 2}\right). \tag{8-179}$$

n is the refractive index of the solution, n_a and n_b are the refractive indices of liquids A and B, and X_a and X_b are their respective mole fractions. Now, when $n_a - n_b \ll 1.0$, we can write

$$n_a^2 - n_b^2 = (n_a - n_b)(n_a + n_b) \cong 2n_a(n_a - n_b) \quad \text{and} \quad n^2 \cong n_a^2 + 2n_a X_b(n_b - n_a). \tag{8-180}$$

The mole fraction of a dilute solution of solute B in the solvent A is given by

$$X_b = \frac{\text{moles}_b}{\text{moles}_a + \text{moles}_b} \cong \frac{\text{moles}_b}{\text{moles}_a} = \frac{C}{M_b \rho_a}, \tag{8-181}$$

where $C = M_b \rho_b$ is the concentration of the solute B, M_b is the mass of a solute molecule, and ρ_a is the number density of the solvent. Substituting Eq.(8-181) into Eq.(8-180) and differentiating with respect to the concentration of B, gives $(\partial n/\partial C) = (n_b - n_a)/(M_b \rho_a)$. Substituting this result into the expression above under "Solute in dilute solution," gives

$$\frac{M_b^2 \rho_b V_s n_0^2}{4\pi^2}\left[\left(\frac{\partial n}{\partial C}\right)_0\right]^2 \cong \frac{C V_s (n_b - n_a)^2 n_a^2}{4\pi^2 M_b \rho_a^2}. \tag{8-182}$$

It is evident from this equation that the solute (B) scattering is proportional to the square of the difference in refractive indices between the solvent and solute. The refractive indices for several common liquids are listed in Table 8-10.

Consider now the benzene solute concentration fluctuation scattering intensity in a CCl_4 solvent. The $\alpha^2 N\rho_0 kT\beta_T = \alpha^2 V_s \rho_0^2 kT\beta_T$ scattering factor for the CCl_4 solvent at 293 K can be obtained by using $\alpha = 10.5 \times 10^{-24}$ cm^3 for CCl_4 and the value of $\rho_0^2 kT\beta_T$ listed in Table 8-4 for CCl_4 to give $\alpha^2 \rho_0^2 kT\beta_T = 3 \times 10^{-27}$. The benzene solute concentration fluctuation scattering factor in a CCl_4 solvent is obtained from Eq.(8-182), where $\rho_a(CCl_4)$ is from Table 8-4, $M_b = 1.297 \times 10^{-22}$ g,

Table 8-10 The refractive indices in several common liquids from *Handbook of Chemistry and Physics* (The Chemical Rubber Co., Cleveland, Ohio, 1976).

Molecule	n (293 K)
Acetone	1.3588
Benzene	1.5011
CCl_4	1.4601
CH_3Cl	1.4459
n-heptane	1.3878
Toluene	1.4961
Diethyl ether	1.3526
H_2O	1.3330
CS_2	1.6319

and n_a and n_b are from Table 8-10, which gives $n_a^2 C(n_b - n_a)^2/(4\pi^2 M_b \rho_a^2) = 6.5 \times 10^{-25}\, C$. According to these numbers, the total benzene scattering intensity in the CCl_4 solvent will exceed the solvent scattering at concentrations above $C = 0.05$ $\mathrm{g \cdot cm^{-3}}$ (for equal scattering volumes), which is a relatively low concentration. We have also chosen a solute-solvent $(n_a - n_b)$ that is quite small. In conclusion, it is evident that concentration fluctuation scattering can be much more intense than the background solvent Rayleigh scattering even at relatively low concentrations. The relative intensities for any solute-solvent system can be estimated by using the refractive indices and the arguments above.

We now note that the intensity of the anisotropic Lorentzian [Eq.(8-178)] will normally be less than the isotropic term. Of course, in CCl_4, $\alpha_{aa} - \alpha_{bb} = 0$ and the anisotropic term is zero. Assuming that $\alpha_{aa} - \alpha_{bb} = 10^{-24}\,\mathrm{cm}^3$ and remembering that $\rho_0 = C/M$, however, we can write the multiplier of the second Lorentzian in brackets in $I_v^v(\nu)^{\mathrm{RAY}}$ in Eq.(8-178) as $\frac{4}{45}\rho_0(\alpha_{aa} - \alpha_{bb})^2 = \frac{4}{45}(C/M)(\alpha_{aa} - \alpha_{bb})^2 = 6.8 \times 10^{-28} C$, which is considerably less than the multiplier of the first Lorentzian in $I_v^v(\nu)^{\mathrm{RAY}}$ as shown above. Thus, the Lorentzian with $\Delta\omega = K^2 D$ normally dominates the $I_v^v(\omega)^{\mathrm{RAY}}$ spectrum in Eq.(8-178). The $I_v^v(\omega)$ Rayleigh spectra for concentration fluctuations of ethyl ether dissolved in a CS_2 solvent has been observed as a function of K and the spectral half widths as a function of K^2 are plotted in Fig. 8-16. The corresponding half widths versus K^2 in a dilute solution of diethyl diethylene glycol in CS_2 are also shown in Fig. 8-16. The resultant diffusion coefficients are also shown. Comparing the results above on $\mathscr{L}(\omega - \omega_0)_{\mathrm{isot}}$ half widths at half height with the corresponding line widths in $\mathscr{L}(\omega - \omega_0)_{\mathrm{anis}}$ (see Table 8-6, for instance) shows that the DK^2 half widths are always several orders of magnitude less than the 6Θ half widths. Thus, $\mathscr{L}(\omega - \omega_0)_{\mathrm{isot}}$ will appear as a narrow line on top of the broad $\mathscr{L}(\omega - \omega_0)_{\mathrm{anis}}$ spectra in $I_v^v(\omega)^{\mathrm{RAY}}$. It is also evident that the DK^2 half widths considered here will be smaller than the κK^2 half widths considered in Section 8.3 (see κ-values in Table 8-2 and note that $\kappa > D$) [42]. Of course, we have also assumed that there is no coupling between the concentration fluctuations described here and the density fluctuations described in Section 8.3. As the solute particles become larger relative to the solvent molecules, the separation above between the concentration fluctuations (and the measurement of the translational diffusion) and the other

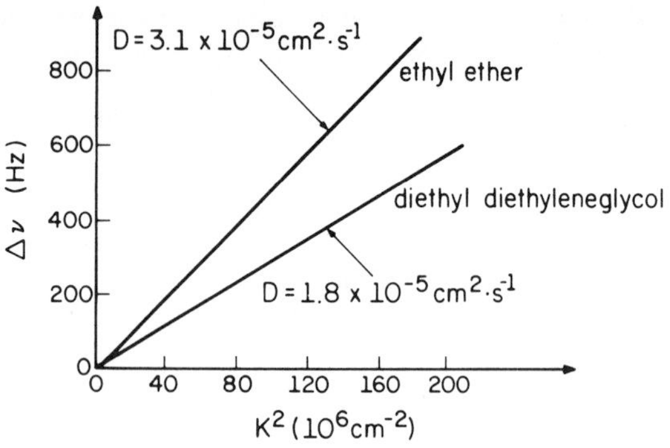

Figure 8-16 Half widths at half height as a function of K^2 for $I_v^v(\nu)$ from ethyl ether and diethyl diethylene glycol, both in a CS_2 solvent. The solute concentration is 10% by weight in both cases. The scattering angles were all below $\theta = 4°$ for the data in this figure from Ref. [42]. The slopes of the $\Delta\nu$ versus K^2 curves give the translational diffusion coefficients from $\Delta\nu = K^2 D/2\pi$.

effects described above become more pronounced. Several diffusion coefficients for macromolecules are listed in Table 8-11.

In some of the large molecules listed in Table 8-11, the size of the scatterer approaches the wavelength of the radiation. Under these conditions, scattering from different parts of the same molecule leads to both static and dynamic (if the molecule is rotating) correlations. The static correlations lead to a K-dependence in the total scattered isotropic intensity leading to a measurement of the radius of gyration (see Section 8.5). The dynamic correlations lead to spectral characteristics that allow the measurement of the rotational diffusion constant, Θ, by examining the spectrum of the isotropically scattered light [43].

Table 8-11 Diffusion coefficients measured at 293 K in water solutions and molecular weights of several molecules of various sizes. The data are from Ref. [25] and S. B. Dubin, J. H. Lunacek, and G. B. Benedek, *Proc. Nat. Acad. Sci. U.S.* **57**, 1164 (1967).

Molecule	D (10^{-7} cm$^2\cdot$s^{-1})	M (a.u.)
Glycine	93.4	75
Sucrose	45.9	342
Ribonuclease	10.7	13,683
Bovine serum albumin	6.1	67,000
Ovalbumin	7.1	45,000
Lyzozyme	11.5	14,100
Tropomyosin	2.20	93,000
Fibrinogen	2.00	330,000
Myosin	1.10	490,000
Tobacco mosaic virus	0.29	40×10^6
Latex spheres (d = 910 Å)	0.52	
Calf thymus DNA	0.2	$\sim 5 \times 10^6$

We shall now examine the principles involved in electrophoretic light scattering, which involves the observation of laser light scattered from a solution of charged macromolecules which are moving with a drift velocity v_d in the presence of an electric field. We shall consider only relatively small molecule scattering here where Eq.(8-178) gives an appropriate description of the spectrum of the light scattered in the absence of any perturbing electric fields.

Most protein macromolecules are charged at a given pH, and in the presence of an electric field the ions in solution will experience a force in the field causing them to translate with a drift velocity, $\boldsymbol{v}_d$, given by [see Eq.(2-39)]

$$\boldsymbol{v}_d = \mu \boldsymbol{E}, \tag{8-183}$$

where $\boldsymbol{E}$ is the electric field vector and μ is the scalar mobility. In the case of spheres, the mobilities and diffusion coefficients of the ionic molecules are related, according to Eq.(8-104), by

$$\frac{\mu}{D} = \frac{eZ}{kT}, \tag{8-184}$$

where e is the electronic charge, Z is the effective number of charges on the translating ion (including the effects of the electrophoretic counter-ions), k is Boltzmann's constant, and T is the temperature.

We shall consider here only the single-particle isotropic scattering. Returning to Fig. 8-1 we start with zy-polarized radiation electric fields and detection of $I_v^v(\nu)$ in the xy plane. Return now to Fig. 8-9 and consider the flux of molecules along the y axis following the concentration fluctuation in the xz plane. In the presence of an electric field component along the y axis, the electrostatic force on the charged (ionic) molecules in solution will give rise to an additional flux along the y axis. The total flux is given by modifying Eq.(8-87) according to

$$J(y) = -D\frac{dP(y,t)}{dy} + v_d P(y,t).$$

Combining this expression with the y-axis mass continuity expression gives the modified one-dimensional Fick's law in $P(y, t)$ in the presence of the electric field according to

$$\frac{dP(y,t)}{dt} = D\frac{d^2P(y,t)}{dy^2} + v_d\frac{dP(y,t)}{dy}. \tag{8-185}$$

The solution to this equation is easily obtained by Fourier transform methods as described following Eq.(8-88) to give

$$\bar{P}(K_y, t) = \bar{P}(K_y, 0)\exp(-K_y^2 Dt - iK_y v_d t), \tag{8-186}$$

where $\bar{P}(K_y, 0) = 1.0$ for uncorrelated scatterers. In three dimensions Eq.(8-186) takes the form

$$\bar{P}(K, t) = \exp(-K^2Dt - i\boldsymbol{K}\cdot\boldsymbol{v}_d t) = \exp(-K^2Dt - i\mu\boldsymbol{K}\cdot\boldsymbol{E}t). \tag{8-187}$$

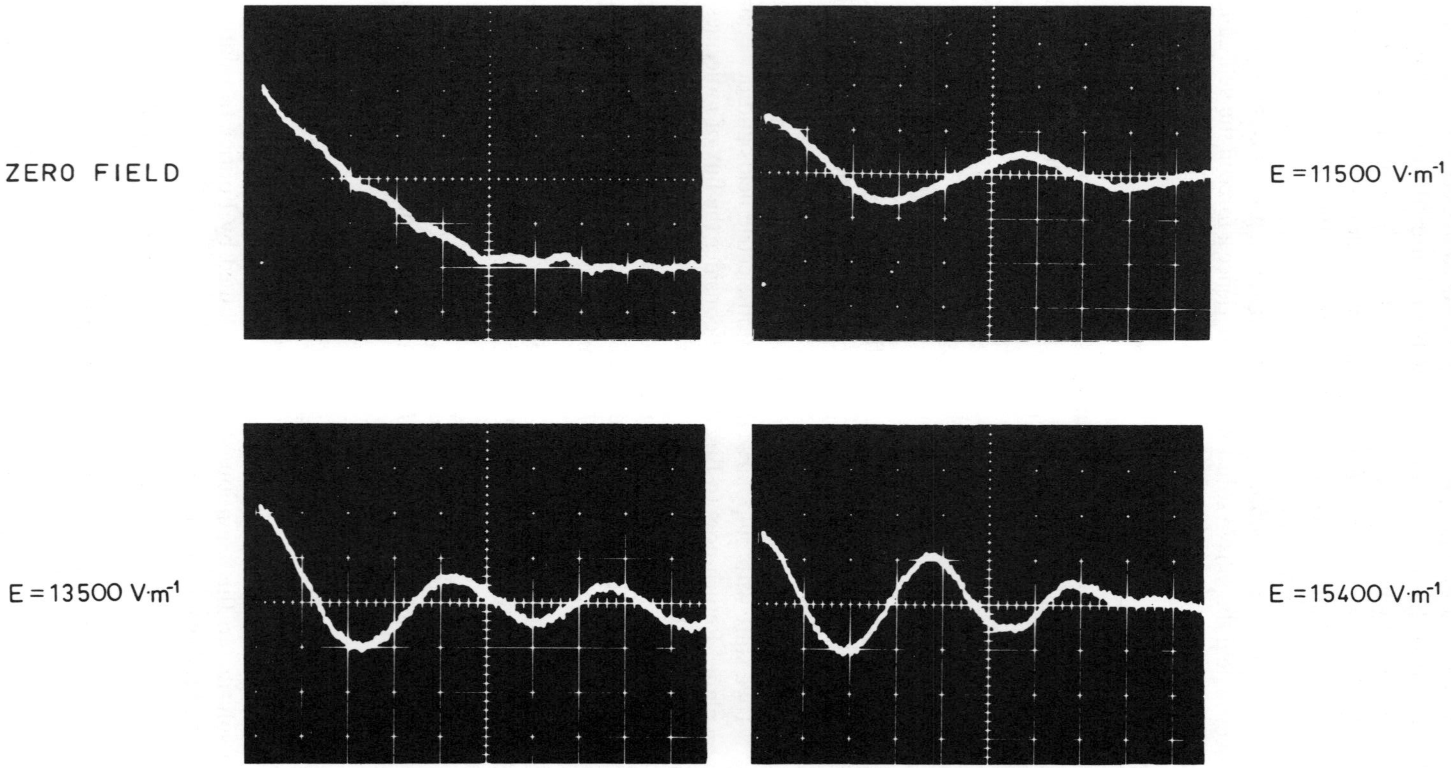

Figure 8-17 The observed real part of the correlation function in Eq.(8-188) for several electric fields from B. R. Ware and W. H. Flygare, *Chem. Phys. Lett.* **12**, 81 (1971).

$\boldsymbol{K} \cdot \boldsymbol{v}_d = \mu \boldsymbol{K} \cdot \boldsymbol{E} = \mu K E \cos \alpha$, where α is the angle between the electric field vector and the K vector. Remember that the K vector always bisects the angle between the direction of the incident and scattered light (see Fig. 8-1). The corresponding reverse Fourier transform back to real space of $\bar{P}(K, t)$ in Eq.(8-187) gives

$$P(\boldsymbol{R}, t) = \left(\frac{1}{4\pi D t}\right)^{3/2} \exp\left[-\frac{(\boldsymbol{R} - \boldsymbol{v}_d t)^2}{4Dt}\right],$$

where $P(\boldsymbol{R}, t)$ is now the three-dimensional probability per unit volume. We have, of course, assumed no coupling between the translational motion imposed by the electric field and the corresponding concentration fluctuations. Substituting Eq.(8-187) into the first term of $C_z(t)$ in Eqs.(8-58) for self terms only gives the correlation function in the presence of the electric field,

$$C_z(t) = AN\alpha^2 \exp(-i\omega_0 t - i\mu \boldsymbol{K} \cdot \boldsymbol{E} t - K^2 D t). \tag{8-188}$$

Using Eq.(8-177) and taking the real Fourier transform [see Eq.(8-16)] gives

$$I_v^v(\omega)_{\text{isot}} = \frac{A}{M}\left[\frac{1}{(1/M) + 2B_1 C + 3B_2 C^2 + \cdots}\right]\left\{\frac{M^2 N n_0^2}{4\pi}\left[\left(\frac{\partial n}{\partial C}\right)_0\right]^2\right\}$$
$$\times \mathscr{L}(\omega_0 - \omega - \mu \boldsymbol{K} \cdot \boldsymbol{E})_{\text{isot}}$$

$$\mathscr{L}(\omega_0 - \omega - \mu \boldsymbol{K} \cdot \boldsymbol{E})_{\text{isot}} = \frac{1}{\pi}\left[\frac{K^2 D}{(\omega_0 - \omega - \mu \boldsymbol{K} \cdot \boldsymbol{E})^2 + (K^2 D)^2}\right]. \tag{8-189}$$

It is quite apparent that the only difference between this expression and the isotropic result in Eq.(8-178) is the translation in ω of $\mu \boldsymbol{K} \cdot \boldsymbol{E}$. By using optical mixing techniques the real part of the correlation function for concentration fluctuations has been observed in solutions of bovine serum albumin (BSA) in the presence of an electric field and the data are shown in Fig. 8-17 for several electric fields. A spectrum showing the field-dependent Doppler shift appears in Fig. 8-18. The experiments illustrated here are all at low angles of scattering and the electric field orientation is perpendicular to the direction of the incident light. This configuration leads to the highest resolution in the electric field effect where the shift in frequency divided by the half width of the line shape defines the resolution, R_θ:

$$R_\theta = \frac{\boldsymbol{v}_d \cdot \boldsymbol{K}}{K^2 D} = \frac{v_d K \cos(\theta_s/2)}{K^2 D} = \frac{\lambda_0 \mu E \cos(\theta_s/2)}{4\pi n D \sin(\theta_s/2)}. \tag{8-190}$$

For very small angles,

$$\lim_{\theta_s \to \text{small}} R_{\theta_s} = \frac{\lambda_0 \mu E}{2\pi n D \theta_s} = \frac{Z e \lambda_0 E}{2\pi n k T \theta_s}, \tag{8-191}$$

where n is the solution refractive index and $\mu/D = eZ/kT$ from Eq.(8-184) is used in the last step in Eq.(8-191).

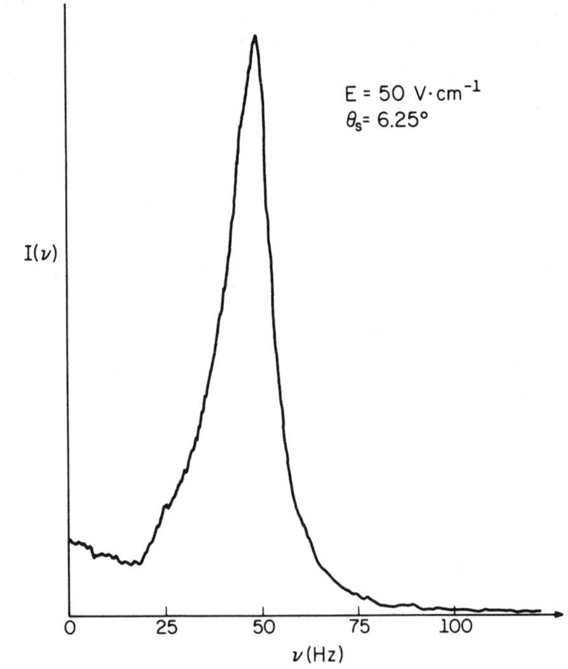

Figure 8-18 The spectrum of light scattered from concentration fluctuations in calf thymus DNA where optical mixing techniques reduce the laser frequency reference to zero. 0.1 mg DNA was dissolved per milliliter of water with an ionic strength of 0.01 at 273 K. The analysis of the shift in frequency of 50 Hz at 50 $V \cdot cm^{-2}$ leads to a mobility of $\mu = 3.5 \times 10^{-4}\ cm^2 \cdot V^{-1} \cdot s^{-1}$. The half width at half height leads to a translational diffusion coefficient of $D = 1.2 \times 10^{-7}\ cm^2 \cdot s^{-1}$. The data are from S. L. Hartford and W. H. Flygare, *Macromolecules* **8**, 80 (1975).

PROBLEMS

1. We have defined the correlation functions and spectra in terms of a complex electric field (see Section 8.1). Rederive Eq.(8-20) starting with the real component of the incident radiation field and real component of the scattered field in Eq.(8-18). Show that the result, when written in terms of the incident intensity, is equivalent to the result in Eq.(8-20).

2. Show that all cross terms mentioned in the parentheses preceding Eq.(8-32) vanish for the case of parallel vibrations in a linear molecule.

3. Consider the rotational Raman scattering in a gas phase sample of linear molecules with a polarizability anisotropy $\alpha_{aa} - \alpha_{bb} = 10^{-24}\ cm^3$, $D = 1 \times 10^{-18}$ SC · cm, and $B = 10$ GHz. Assuming that the minimum detectable photon rate at a detector (collection aperature of 10 mm) placed 10 cm from the scattering molecules is 1 photon per second, what is the minimum number of molecules that will produce detectable scattered light from a single $J = 1 \longrightarrow J = 2$ rotational Raman transition? Use a 1-W Ar^+ laser with $\lambda_0 = 5145$ Å.

4. Construct in some detail a schematic of the $I_v^v(\omega)$ and $I_h^v(\omega)$ spectra as in Fig. 8-2 for (a) the perpendicular vibrations of a linear molecule, and (b) all normal modes of CF_4. Show also the progression from a low-pressure gas to a high-pressure gas to a liquid, as shown in one case in Fig. 8-3. Identify clearly the transitions that are pure vibrational and therefore not affected by the rotational relaxations. Use the methods developed here and the selection rules in Chapter 4 in your work.

5. According to Eq.(8-41) and associated discussion the rotational, vibrational, and rotational-vibrational Raman line widths are all proportional to $K = (4\pi/\lambda_0) \sin(\theta_s/2)$ in a dilute gas where the Doppler width dominates. Thus, the line width goes to zero as the scattering angle goes to zero allowing the possibility of very high resolution Raman spectroscopy. What are the major limitations to putting these principles above into practice?

6. In Eq.(8-109) and the following development we wrote a general expression for the probability distribution of a linear molecule in terms of translational diffusion described by $\mathsf{D}_{aa} \neq \mathsf{D}_{bb} = \mathsf{D}_{cc}$ and rotational diffusion described by Θ. We then assumed $\mathsf{D}_{aa} \cong \mathsf{D}_{bb}$ and solved the separated equations of motion for translation and rotation with the result in the equation preceding Eq.(8-118). Return now to Eq.(8-109) and use the $\mathsf{D}_{aa} = \mathsf{D}_{bb} = \mathsf{D}_{cc}$ zero-order solutions and perturbation theory to develop an expression for $\mathscr{G}(\boldsymbol{R}, \theta, \phi, t)$ that includes the most important term in $\mathsf{D}_{aa} - \mathsf{D}_{bb}$. Carry through to the correlation function and resultant spectra to show the effects of $(\mathsf{D}_{aa} - \mathsf{D}_{bb})$ on the spectra for comparison with the $\mathsf{D}_{aa} = \mathsf{D}_{bb}$ spectra given in Eq.(8-120).

7. Consider the Rayleigh–Brillouin spectra of liquid acetone at 293 K. Describe the effects on the Rayleigh–Brillouin triplet caused by the following changes.
   (a) Increased T, constant p (pressure).
   (b) Increased p, constant T.
   (c) Decreased gravitational field, constant T and p.

8. In Section 8.3 we showed that the total intensity of light scattered by a dense gas or liquid was proportional to its isothermal compressibility,

$$\beta_T = -\left(\frac{1}{V}\frac{\partial V}{\partial p}\right)_T .$$

Consider now a dense gas.
   (a) Use a virial expansion to evaluate β_T for the virial equation of state for a dense gas. Include the first few virial coefficients.
   (b) Look up the first and second virial coefficients (choose the gas) and make an estimate on the pressure dependence for the intensity of the scattered light (constant volume) for the dense gas described by the virial equation.

9. Figure 8-19 shows the sound absorption as a function of frequency for methylene chloride (CH_2Cl_2). Use this information to predict the corresponding Brillouin shifts and half widths for scattering angles of 20°, 50°, and 150° using an Ar^+ laser with $\lambda_0 = 5145$ Å as the incident radiation.

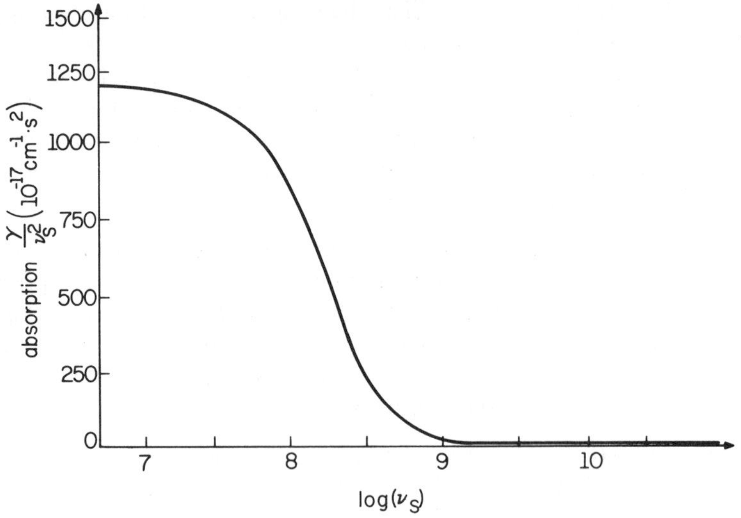

Figure 8-19 The sound absorption curve for CH_2Cl_2 from Ref. [15].

10. Assuming a self-translational diffusion coefficient of $D = 10^{-5}$ $cm^2 \cdot s^{-1}$ and a rotational diffusion coefficient, Θ, from Table 8-5 for CH_3CN, describe the spectrum of the depolarized scattered light at $\theta_s = \pi/2$ from a plane-polarized X-ray source of wavelength $\lambda_0 = 1$ Å. Assume hydrodynamic theory is valid. What is the line width and what angle variation can you suggest to possibly secure the measurement of both D and Θ by this X-ray experiment?

11. Use Fig. 8-14 to estimate the average number of next-nearest neighbors in liquid potassium at 343 K.

***12.** Use the data in Fig. 8-13, the atomic scattering factors in Table 8-7, and Eq.(8-152) to determine the molecular structure of CCl_4. Fit the data in the figure as a function of the structural parameters until a satisfactory fit is obtained.

13. Draw a schematic showing the $I_v^v(\omega)$ and $I_h^v(\omega)$ spectrum scattered at $\theta_s = \pi/2$ from a solution of 10^{-3} $g \cdot cm^{-3}$ of CH_3Cl solute in an acetone solvent at 293 K. Use z-polarized incident radiation from an Ar^+ ($\lambda_0 = 5145$ Å) ion laser. Assume that the translational diffusion coefficient of CH_3Cl in acetone is $D = 2 \times 10^{-5}$ $cm^2 \cdot s^{-1}$. Take into account the $I_v^v(\omega)$ and $I_h^v(\omega)$ solvent spectra arising from density fluctuations as well as the $I_v^v(\omega)$ and $I_h^v(\omega)$ solute spectra arising from concentration fluctuations. Calculate all intensities and spectral parameters and draw the complete spectra for $\theta_s = \pi/2$.

14. Figure 8-20 shows the homodyne spectrum (convolution of the predicted spectra with itself) from a dilute solution of tobacco mosaic virus (TMV). Evaluate the average translational diffusion coefficient, D, from this spectrum.

15. Consider the total isotropic scattering intensity for the light scattered from a *large* spherical particle as described in the $I_v^v(K)$ equation following Eq.(8-173).

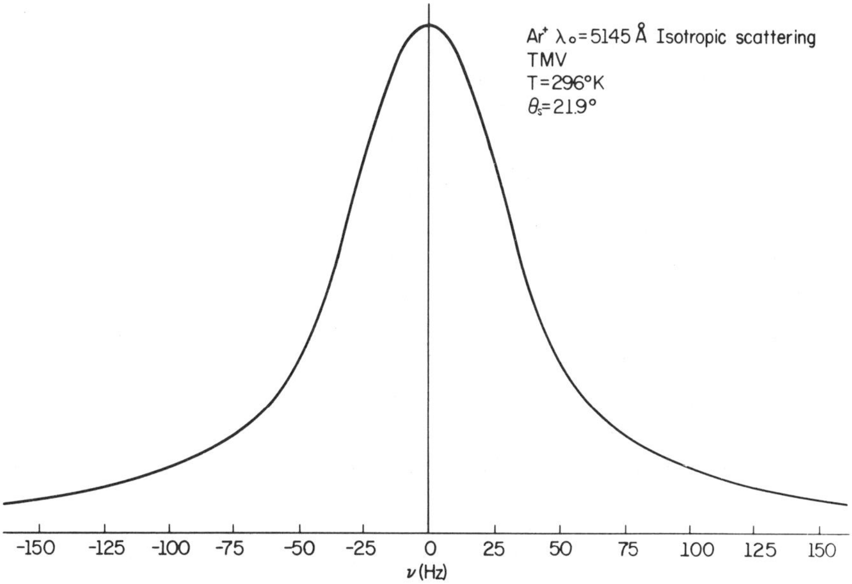

Figure 8-20 The homodyne spectrum scattered from concentration fluctuations in a dilute solution of tobacco mosaic virus (TMV). The source of radiation was an Ar^+ ion laser giving plane-polarized light with wavelength $\lambda_0 = 5145$ Å. The spectrum was taken by S. L. Hartford and T. D. Gierke.

For a dilute distribution of water droplets, aerosol particles, or dust particles, $P(R, 0) = 1$, if the radius of the particle is small relative to the radiation wavelength. Develop a model where $P(R, 0)$ describes the shape of a spherical particle and show that $I_v^v(K)$ decreases with increasing particle radius as the particle radius approaches the radiation wavelength, λ_0.

REFERENCES

[1] B. Chu, *Laser Light Scattering* (Academic Press, New York, 1974). A more general treatment of correlation functions is given in P. C. Martin, *Measurements and Correlation Functions* (Gordon and Breach, New York, 1968).

[2] F. Reif, *Fundamentals of Statistical and Thermal Physics* (McGraw-Hill Book Co., New York, 1965).

[3] C. Kittel, *Elementary Statistical Physics* (John Wiley & Sons, Inc., New York, 1958).

[4] M. J. Stephen and J. P. Straley, *Rev. Mod. Phys.* **46**, 617 (1974).

[5] D. R. Bauer, B. Hudson, and R. Pecora, *J. Chem. Phys.* **63**, 588 (1975).

[6] J. Behringer, *Molecular Spectroscopy*, Vol. 2 (Chemical Society, London, 1975), p. 100; *ibid.*, Vol. 3, p. 163.

[7] T. G. Spiro, *Accts. Chem. Res.* **7**, 339 (1974).

[8] An experimental verification is given in W. R. L. Clements and B. P. Stoicheff, *J. Mol. Spectr.* **33**, 183 (1970).

[9] Additional details about molecular dynamics and light scattering in gases is given in H. F. P. KNAPP and P. LALLEMAND, *Ann. Rev. Phys. Chem.* **26**, 59 (1975).

[10] E. B. WILSON, Jr., J. C. DECIUS, and P. C. CROSS, *Molecular Vibrations* (McGraw-Hill Book Co., New York, 1955).

[11] Even if $\alpha_{aa} - \alpha_{bb} = 0$ for an isolated molecule, a small average $\alpha_{aa} - \alpha_{bb}$ may exist in a dense gas or liquid due to collisional contributions. We exclude these considerations here and refer the reader, for more details, to W. M. GELBART, *Adv. Chem. Phys.* Ed. by I. Prigogine and S. A. Rice **26**, 1 (Wiley-Interscience, Inc., New York, 1974).

[12] R. D. MOUNTAIN, *Rev. Mod. Phys.* **38**, 205 (1966).

[13] R. PECORA, *J. Chem. Phys.* **40**, 1604 (1964).

[14] K. F. HERZFELD and T. A. LITOVITZ, *Absorption and Dispersion of Ultrasonic Waves* (Academic Press, New York, 1959).

[15] I. L. FABILINSKY, *Molecular Scattering of Light* (Plenum Press, New York, 1968).

[16] G. I. A. STEGEMAN, *et al., J. Acoust. Soc. Am.* **49**, 979 (1971).

[17] The width of the central line can be measured by using optical mixing techniques as shown by J. B. LASTOVKA and G. B. BENEDEK, *Phys. Rev. Lett.* **17**, 1039 (1966) who measured κ in toluene by this method.

[18] R. D. MOUNTAIN, *J. Research N.B.S.* **72A**, 95 (1968). Acoustic studies have been used extensively to study molecular conformational changes as described in A. M. NORTH and R. A. PETHRICK, *MPT International Review of Science*, Phys. Chem. Series One, Vol. 2 (Butterworths, London, 1972), p. 159.

[19] P. A. EGELSTAFF, *An Introduction to the Liquid State* (Academic Press, New York, 1967).

[20] T. L. HILL, *Statistical Mechanics* (McGraw-Hill Book Co., New York, 1956).

[21] C. KITTEL, *Elementary Statistical Physics* (John Wiley and Sons, Inc., New York, 1958).

[22] More details and more general expressions can be found in W. A. STEELE, *Adv. Chem. Phys.* Ed. by I. Prigogine and S. A. Rice **34**, 1 (Wiley-Interscience, New York, 1976).

[23] L. D. LANDAU and E. M. LIFSHITZ, *Fluid Mechanics* (Addison Wesley Pub. Co., Reading, Mass., 1959).

[24] J. FRENKEL, *Kinetic Theory of Liquids* (Dover Publications, Inc., New York, 1955).

[25] C. TANFORD, *Physical Chemistry of Macromolecules* (John Wiley & Sons, Inc., New York, 1961).

[26] T. KEYES and D. KIVELSON, *J. Chem. Phys.* **56**, 1057 (1972).

[27] H. M. PIRENNE, *The Diffraction of X-Rays and Electrons by Free Molecules* (Cambridge University Press, Cambridge, 1946).

[28] P. DEBYE, *Proc. Phys. Soc. London* **42**, 340 (1930).

[29] L. J. LOWDEN and D. CHANDLER, *J. Chem. Phys.* **61**, 5228 (1974).

[30] J. F. KARNICKY and C. J. PINGS, *Adv. Chem. Phys.*, Ed. by I. Prigogine and S. A. Rice **34**, 157 (Wiley-Interscience, New York, 1976); L. Blum and A. H. Narten, *ibid.*, p. 203.

[31] A. GUINIER, *et al., Small-Angle Scattering of X-Rays* (John Wiley & Sons, Inc., New York, 1955).

[32] G. W. BRADY, *Accts. Chem. Res.* **4**, 367 (1971).

[33] A. M. BUECHE and P. DEBYE, *J. Appl. Phys.* **20**, 518 (1949).

[34] H. R. ANDERSON, H. BRUMBERGER, and P. DEBYE, *J. Appl. Phys.* **28**, 679 (1957).

[35] G. ROSS. *Opt. Acta* **16**, 95 (1969).

[36] W. H. FLYGARE and T. D. GIERKE, *Ann. Rev. Mat. Science* **4**, 255 (1974).

[37] J. Borch, *et al.*, *J. Appl. Phys.* **42**, 4570 (1971).

[38] R. Samuels, *J. Polymer Sci.* Part A-2 **9**, 2165 (1971).

[39] R. Hosemann, *Endeavour* **117**, 99 (1973).

[40] H. C. van de Hulst, *Light Scattering by Small Particles* (John Wiley & Sons, Inc., New York, 1957).

[41] See S. Bertucci and W. H. Flygare, *J. Chem. Phys.* **63**, 1 (1975) and references cited therein.

[42] P. Berge, *et al.*, *Phys. Rev. Lett.* **24**, 89 (1970).

[43] For details, see B. Berne and R. Pecora, *Dynamic Light Scattering* (John Wiley & Sons, Inc., New York, 1976).

9

the scattering of particles

9.1 CLASSICAL SCATTERING

In this section we shall examine some of the equations that describe the scattering of particles from central field potentials. We start with the definitions of differential and total cross-sections for scattering. The essential features in any scattering experiment are illustrated in Fig. 9-1. The origin of the laboratory-fixed coordinate system contains the source of particles with the beam intensity or flux along the z axis of I_0(number of particles/area · time). The beam of particles (with cross-sectional area A, large with respect to the size of the target particles) hits the target region that has small dimensions relative to the target region-detector distance, R. An expanded view of the target region containing scatterers with number density ρ_0 shows that for very weak scattering the intensity of the incident beam falls off exponentially in the target region with scattering coefficient, γ. The pattern of the scattered particle intensity on the detector screen is also shown in Fig. 9-1 where the hatched area is given in terms of the appropriate spherical polar coordinates about the incident z axis..

Referring to Fig. 9-1, we note that $R^2 \sin\theta\, d\theta\, d\phi$ is the *differential area* on the detecting screen and $d\Omega = \sin\theta\, d\theta\, d\phi$ is called the *solid angle* with $\int_0^{2\pi}\int_0^{\pi} d\Omega = 4\pi$ *steradians*. The number of scattered particles per unit target particle per unit time that strike the small area of $R^2\, d\Omega$ shown on the detector screen is given by

$$N(\theta, \phi) = I(R, \theta, \phi) R^2 \sin\theta\, d\theta\, d\phi, \tag{9-1}$$

where $I(R, \theta, \phi)$ is the intensity (particles/area · time) of scattered particles per unit target particle. Dividing and multiplying the right-hand side of Eq.(9-1) by the incident intensity of particles, I_0, leads to

$$N(\theta, \phi) = I_0\left[\frac{I(R, \theta, \phi)}{I_0} R^2\right] \sin\theta\, d\theta\, d\phi = I_0 \sigma(\theta, \phi) \sin\theta\, d\theta\, d\phi, \tag{9-2}$$

where $\sigma(\theta, \phi)$, which has units of area per unit target particle per steradian, is called the *differential cross-section* for the scattering process.

The total *cross-section*, σ, can be obtained by integrating the differential cross-section, $\sigma(\theta, \phi)$, over the full solid angle of scattering,

$$\sigma = \int_0^{2\pi}\int_0^{\pi} \sigma(\theta, \phi) \sin\theta\, d\theta\, d\phi. \tag{9-3}$$

σ is the effective cross-section for a single scattering particle in the target region. The total cross-section for the hard sphere collision of a spherical incident particle with

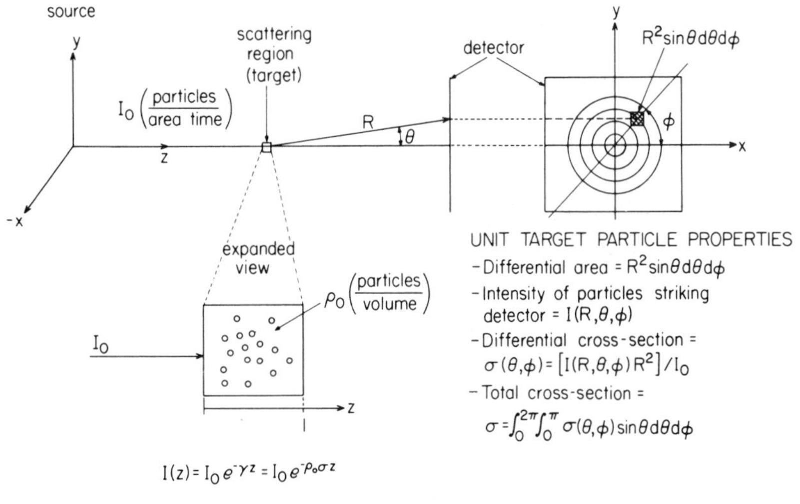

Figure 9-1 Schematic diagram showing the basic elements of a scattering experiment in spherical polar coordinates. The beam of incident particles with intensity I_0 enters the scattering region with target number density ρ_0. The scattering intensity per unit target particle, $I(R, \theta, \phi)$, is measured at the detector screen (as observed looking back at the source). $I(R, \theta, \phi)$ is related to the differential cross-section in Eq.(9-2) and the total cross-section (of a single particle) is given by Eq.(9-3). The scattering coefficient, γ (describing the loss of particles from the incident beam), is given by the product of the total cross-section with the scatterer number density.

radius r_1 and a spherical target particle with radius r_2 is obviously $\sigma = \pi(r_1 + r_2)^2$. According to Eq.(9-3), we can also write

$$\sigma(\theta, \phi) = \frac{d\sigma}{d\Omega}, \qquad d\Omega = \sin\theta \, d\theta \, d\phi. \tag{9-4}$$

The total number of incident particles scattered per unit time through all angles, N_T, is also given from the considerations above (for a single target particle) by

$$N_T = \int_0^{2\pi} \int_0^{\pi} I(R, \theta, \phi) R^2 \sin\theta \, d\theta \, d\phi = I_0 \sigma, \tag{9-5}$$

where I_0 is the incident intensity as before.

In summary, the development above describes the scattering processes for an incident beam of particles scattering from a single stationary particle in the scattering region. The total scattering cross-section, σ, for spherical incident and target particles undergoing hard sphere collisions, should be equivalent to the cross-sectional area of a spherical scatterer, πr_0^2, where r_0 is the sum of radii of the incident and scattering particles. Of course, in most cases, collisions between atoms and molecules are not hard sphere-like and the relative velocity and energies of the two colliding particles are important in determining the final result of the scattering experiment. In some cases, the target will be a gas with a Maxwell–Boltzmann velocity distribution. In this latter case, an effective or velocity averaged scattering cross-section is measured.

Referring to Fig. 9-1, we can also define the loss of incident beam intensity as it passes through the scattering region with scatterer particle number density ρ_0. The loss in intensity or attenuation of the beam along the z axis (see Fig. 9-1) for very weak scattering will be proportional to the intensity,

$$\frac{dI(z)}{dz} = -\gamma I(z), \tag{9-6}$$

where the proportionality constant, γ, is called the scattering coefficient or beam attenuation coefficient. The solution to this equation is

$$I(z) = I(0) \exp(-\gamma z) = I_0 \exp(-\gamma z). \tag{9-7}$$

If γ is small over the length, ℓ, of the scattering region, $\gamma\ell \ll 1.0$, we can expand the exponential giving

$$I(z) = I_0\left[1 - \gamma z + \frac{(\gamma z)^2}{2} - \cdots\right] \cong I_0(1 - \gamma z). \tag{9-8}$$

We now return to Fig. 9-1 and note that $I(z)$ is equal to the incident intensity minus the intensity of particles scattered through all angles after the incident beam has entered a distance z into the scattering region. Using N_T for the total incident particles scattered from a single target particle, as in Eq.(9-5), we can write

$$I(z) = I_0 - \frac{N_T}{A} \rho_0 V_s(z), \tag{9-9}$$

where ρ_0 is the scatterer number density, $V_s(z)$ is the scattering volume or the volume in the scattering region swept out by the incident beam up to a distance z, and A is the area of the incident beam. Substituting Eq.(9-5) into Eq.(9-9) gives

$$I(z) = I_0\left[1 - \rho_0 \sigma \frac{V_s(z)}{A}\right] = I_0(1 - \rho_0 \sigma z), \tag{9-10}$$

where the scattering volume per unit area is just the penetration depth, z, of the incident beam. Comparing Eqs.(9-10) and (9-8) gives the scattering coefficient, γ, in terms of the total cross-section times the number density,

$$\gamma = \rho_0 \sigma, \qquad I(z) = I_0 \exp(-\rho_0 \sigma z). \tag{9-11}$$

Attenuation data for the scattering of a thermal beam of CsCl molecules by low-pressure CH_2F_2 and Ar gases are shown in Fig. 9-2. It is evident that the logarithm of $I(z)/I_0$ is linear in pressure or ρ_0 as indicated in Eq.(9-11). The total cross-sections corresponding to the data in Fig. 9-2 are examined in Problem 9-1.

The equations above concerning the differential, $\sigma(\theta, \phi)$, and total, σ, cross-sections involve the scattering processes for single particles. We shall now evaluate $\sigma(\theta, \phi)$ and σ from the basic kinematics of a collision; however, first we shall apply these scattering concepts to light scattering and several other processes.

Returning to the discussion following Eq.(1-59), consider the total scattering cross-section for the collision between an optical photon with a neon atom. I/I_0 in Eq.(1-59) is an intensity ratio per scattering particle at scattering angles γ and θ_s

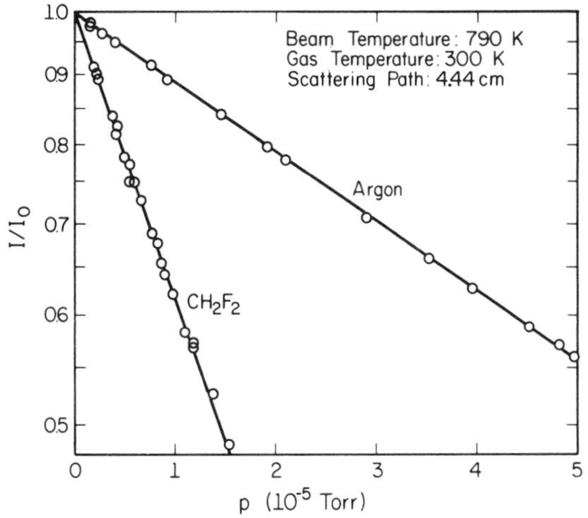

Figure 9-2 Scattering of a thermal CsCl beam from low-pressure CH_2F_2 and Ar gases. The total intensity (divided by the incident intensity) of the CsCl beam after passing through the scattering gas is plotted as a function of target gas pressure as obtained from H. Schumacher, R. B. Bernstein, and E. W. Rothe, *J. Chem. Phys.* **33**, 584 (1960). The loss in intensity with pressure is described by the exponential expression in Eq.(9-11).

and $(I/I_0)R^2$ is equivalent to $\sigma(\theta, \phi)$ defined here in Eqs.(9-1) and (9-2). The corresponding differential scattering cross-section for the photon-atom scattering process is, therefore, given by (compare Figs. 1-2 and 9-1)

$$\sigma(\theta, \phi) = \frac{16\pi^4\alpha^2 \sin^2 \gamma}{\lambda_0^4},$$

for polarized incident radiation where γ is the angle between the axis of polarization and the axis of detection. For unpolarized incident radiation, the differential cross-section is given by

$$\sigma(\theta, \phi) = \frac{16\pi^4\alpha^2}{\lambda_0^4}\frac{1}{2}(\cos^2 \theta + 1),$$

where the intensity of unpolarized incident radiation is defined as one-half the sum of intensities from polarized components along the two perpendicular axes that are both perpendicular to the direction of propagation of the incident radiation. The total scattering cross-section for the photon (with wavelength λ_0) and the atom (with polarizability α) is given by

$$\sigma = \int_0^{2\pi}\int_0^{\pi} \sigma(\theta, \phi) \sin \theta \, d\theta \, d\phi = \frac{16\pi^4\alpha^2}{\lambda_0^4}\int_0^{2\pi}\int_0^{\pi} \sin^2 \gamma \sin \theta \, d\theta \, d\phi$$

$$= \frac{16\pi^4\alpha^2}{\lambda_0^4}\frac{1}{2}\int_0^{2\pi}\int_0^{\pi}(1 + \cos^2 \theta) \sin \theta \, d\theta \, d\phi = \frac{16\pi^4\alpha^2}{\lambda_0^4}\left(\frac{8\pi}{3}\right). \tag{9-12}$$

The equivalence of the expressions for the total cross-sections for polarized and unpolarized incident light is readily seen by comparing Figs. 1-2 and 9-1 and noting that $\sin^2 \gamma = \sin^2 \theta \cos^2 \phi + \cos^2 \theta$. Using an argon ion laser giving polarized radiation with $\lambda_0 = 5145$ Å and $\alpha = 10^{-24}$ cm^3 for neon scatterers gives $\sigma = 1.86 \times 10^{-27}$ cm^2 for the nonresonant photon-neon total cross-section. This is considerably smaller than we would expect for direct particle-particle cross-sections.

It is interesting to compute the cross-section for photon (X ray)-electron scattering processes (see the Compton effect described in Section 1.5). Using $\alpha(\text{X ray}) = -(e^2/m\omega^2)$ for a single electron from Eq.(8-131) where ω is the X-ray frequency gives $(\omega/c = 2\pi/\lambda)$

$$\sigma(\text{X ray} - e^-) = \left(\frac{8\pi}{3}\right)\frac{e^4}{m^2c^4}. \tag{9-13}$$

Note the inverse m^2-dependence in this expression that shows that the X-ray-electron cross-section is 10^6 larger than the X-ray-proton (or neutron) cross-section. Remembering that $\sigma = \pi r_0^2$, where r_0 is the sum of the radii of the spherical incident and scattering particles, and assuming the radius of an X-ray photon is zero, we note from Eq.(9-13) that the classical radius of an electron obtained by this analysis is given by $r = (\frac{8}{3})^{1/2}e^2/mc^2$. The scale factor $(\frac{8}{3})^{1/2}$ depends on the assumption about the radius of the X ray and other scale factors are obtained by other considerations [1].† Therefore, the classical radius of the electron is conventionally defined as

$$r_e = \frac{e^2}{mc^2} = 2.817938 \times 10^{-13}\ \text{cm}. \tag{9-14}$$

Thus, X-ray–electron collisions will yield cross-sections on the order of $\sigma = \pi r_e^2 \cong 10^{-25}\ \text{cm}^2$. Similarly, neutrons have an effective radius on the order of 10^{-13} cm and classical scattering processes between the electron and neutron will also have scattering cross-sections on the order of 10^{-26} to $10^{-25}\ \text{cm}^2$. Neutrons are used commonly as a scattering probe of nuclei. The scattering cross-section for neutron-nuclei scattering processes depends on the nuclear radius, which can be estimated by

$$r_A = 1.4 \times A^{1/3} \times 10^{-13}\ \text{cm},$$

where r_A is the radius of the nucleus with atomic number (number of protons plus neutrons) of A. Thus, neutron-nuclear scattering cross-sections will increase with $A^{2/3}$, starting with $\sigma \cong 10^{-25}\ \text{cm}^2$ with the first row atoms in the periodic table and then increasing as A increases [2].

The intensities of electron-electron or electron-proton scattering processes are described by the Coulomb scattering differential cross-section given in Eq.(9-89) as $\sigma(\theta) = (Z_1 Z_2 e^2/4E_0)^2[1/\sin^4(\theta/2)]$, where Z_1 and Z_2 are the number of electron charges on the incident and target particles, respectively, and E_0 is the relative kinetic energy between the colliding target and incident particles. The corresponding X-ray–electron differential cross-section for unpolarized X rays is given by $\sigma(\theta) = (e^4/2m^2c^4)(1 + \cos^2\theta)$. Comparing these two expressions for backscattering where $\theta = \pi$ gives

$$\frac{\sigma(\text{electron–electron})}{\sigma(\text{X-ray–electron})} = \left(\frac{E_r}{4E_0}\right)^2,$$

where $E_r = mc^2 = 0.5110$ MeV. Thus, unless the value of E_0 approaches the relativistic energy, electron-electron scattering will far exceed X-ray–electron scattering. For typical laboratory electrons with 1-keV energy, electron-electron backscattering cross-sections will exceed X-ray–electron backscattering cross-sections by 10^5.

We shall now examine, in more detail, central field scattering where the interaction force between the target and incident particle leads to a simple central radial field potential energy given by

$$V(r) = Cr^{-n}. \tag{9-15}$$

C is a constant and n describes the order of the interaction ($n = 1$ for a Coulomb force). Using the same axis system as in Fig. 9-1, Fig. 9-3 shows the scattering of an incident particle from a fixed target particle where the interaction force is repulsive. The diagram to the right shows $d\sigma$ [see $d\sigma = \sigma(\theta, \phi)\, d\Omega$ in Eq.(9-4)] in terms of ϕ and the scattering impact parameter, b, which is the perpendicular distance between the center of force and the line traveled by the incident particle. b is the distance of closest approach if no potential is present between the incident and target particle. According

† Numbers in square brackets correspond to the numbered sources found in the Reference section on p. 621 at the end of the chapter.

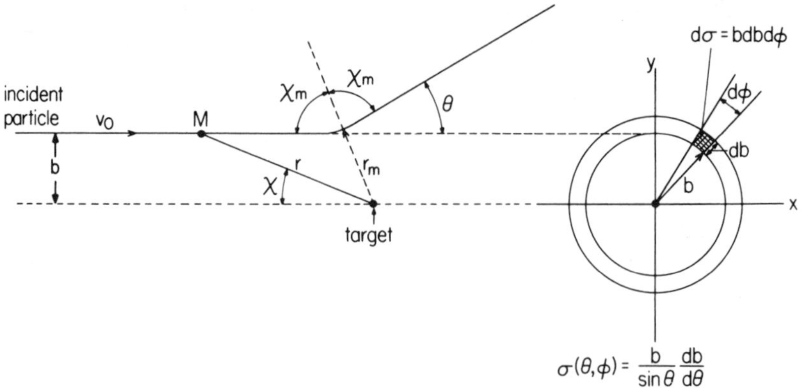

Figure 9-3 Scattering diagram showing the repulsive scattering of a particle with velocity v_0 and mass M from the target. θ is the scattering angle as in Fig. 9-1, r and χ are the variable distance and angle describing the collision, and r_m and χ_m are the distance and corresponding angle of closest approach. b is the impact parameter and the differential cross-section, $\sigma(\theta, \phi)$, which is measured at the detector, is given in terms of the impact parameter and scattering angle [see Eq.(9-16)]. Note that the notation for the angles in this diagram conform to the scattering diagram in Fig. 9-1 and are opposite to the conventional choice.

to Fig. 9-3, the area $d\sigma$ can be written as $d\sigma = b\, db\, d\phi$ and the total cross-section, σ, is obtained by integrating over ϕ and b. In the case of hard spheres, the range of impact parameters is from zero to the distance of closest approach, r_m, which gives

$$\sigma = \int_0^{r_m} b\, db \int_0^{2\pi} d\phi = \pi r_m^2.$$

If $r_m = r_1 + r_2$, where r_1 and r_2 are the spherical radii of the incident and target particles, the result conforms to the discussion following Eq.(9-3). Using Eq.(9-4) we can write the differential cross-section for the scattering in terms of the impact parameter and its variation with θ,

$$\frac{d\sigma}{d\Omega} = \sigma(\theta, \phi) = \frac{b}{\sin\theta}\left|\frac{db}{d\theta}\right|, \tag{9-16}$$

where we use the absolute value of $db/d\theta$.

The mechanics of the central field scattering process is also illustrated in Fig. 9.3. We assume a stationary target and an incident particle of mass M scattered through an angle θ [3]. The interparticle distance is r and the angular momentum of the colliding pair of particles is

$$\boldsymbol{L} = M\boldsymbol{r} \times \dot{\boldsymbol{r}}, \tag{9-17}$$

with magnitude ($\dot{r} = r\dot{\chi}$)

$$L = Mr^2\dot{\chi}. \tag{9-18}$$

If both the incident and target particles are in motion, we replace M with $\mu = MM'/(M + M')$, the reduced mass where M' is the mass of the target particle.

The total kinetic energy is equal to the sum of the radial and angular contributions,

$$T = \frac{1}{2} M\dot{r}^2 + \frac{L^2}{2I} = \frac{1}{2} M(\dot{r}^2 + r^2\dot{\chi}^2), \tag{9-19}$$

where $I = Mr^2$ is the moment of inertia for the two-body system.

We can also relate the angular momentum to the impact parameter. Since b is perpendicular to the initial velocity, $\boldsymbol{v}_0$, we can write the angular momentum as (stationary target particle)

$$L = Mv_0 b. \tag{9-20}$$

The angular momentum is conserved during the collision. Therefore, Eqs.(9-18) and (9-20) are equivalent giving

$$\dot{\chi} = \frac{bv_0}{r^2}. \tag{9-21}$$

Substituting this result into Eq.(9-19) and adding the spherically symmetric potential energy, $V(r)$, gives the total energy of

$$E = \frac{1}{2} M\left(\dot{r}^2 + \frac{b^2 v_0^2}{r^2}\right) + V(r) = \frac{1}{2} M\dot{r}^2 + \frac{b^2 E_0}{r^2} + V(r) = \frac{1}{2} M\dot{r}^2 + \frac{L^2}{2I} + V(r), \tag{9-22}$$

where

$$E_0 = \tfrac{1}{2} M v_0^2 \tag{9-23}$$

is the initial kinetic energy and we have used the r-dependent moment of inertia, I. $E_0 b^2/r^2 = L^2/2I$ in Eq.(9-22) is called the *centrifugal energy*, which is positive and acts as a repulsive barrier when the two colliding particles approach each other. As the colliding particles approach each other, the centrifugal barrier term, $b^2 E_0/r^2$, increases, becoming a maximum at the classical turning point, r_m (see Fig. 9-3), where the radial kinetic energy term, $\frac{1}{2}M\dot{r}^2$, vanishes. We can find this classical turning point using $E = E_0$ (conservation of energy) in Eqs.(9-22) and (9-23) to give an expression for $\dot{r}$,

$$\dot{r} = \pm v_0 \left[1 - \frac{b^2}{r^2} - \frac{V(r)}{E_0}\right]^{1/2}. \tag{9-24}$$

$\dot{r}$ is negative for the incoming trajectory and positive for the outgoing trajectory. Thus, the minimum incident particle–target particle distance, r_m, can be obtained when $\dot{r}$ changes sign. Using $\dot{r} = 0$ in Eq.(9-24) leads to a value for r_m (see Fig. 9-3) of

$$r_m = \frac{b}{\{1 - [V(r_m)/E_0]\}^{1/2}}. \tag{9-25}$$

Of course, the potential function and b must be known before calculating r_m.

Combining Eqs.(9-24) and (9-21) gives a differential equation for the trajectory of the scattered particle,

$$\frac{d\chi}{dr} = -\frac{b}{r^2}\left[1 - \frac{b^2}{r^2} - \frac{V(r)}{E_0}\right]^{-1/2}. \tag{9-26}$$

The negative sign is used for Eq.(9-26) because $d\chi/dr$ is for the incoming particle. We can now calculate χ_m for the minimum distance of approach shown in Fig. 9-3. Using Eq.(9-26) we write

$$\chi_m = \int_{\infty}^{r_m} \left(\frac{d\chi}{dr}\right) dr = -b \int_{\infty}^{r_m} \frac{dr}{r^2[1 - V(r)/E_0 - b^2/r^2]^{1/2}}. \tag{9-27}$$

We now note that the trajectory is symmetric about r_m and according to Fig. 9-3 the scattering or deflection angle is given by

$$\theta = \pi - 2\chi_m = \pi + 2b \int_{\infty}^{r_m} \frac{dr}{r^2[1 - V(r)/E_0 - b^2/r^2]^{1/2}}. \tag{9-28}$$

These equations show that the trajectory and corresponding deflection angle, θ, are dependent on the initial conditions: the energy, E_0, and the impact parameter, b. We shall now evaluate the scattering angle, θ, in Eq.(9-28) for several examples.

Consider first the scattering properties of hard spheres where r_1 and r_2 are radii of the target and incident particles, respectively. The potential for the hard sphere interaction is given by

$$\begin{aligned} V(r) &= 0; \qquad r > r_1 + r_2 \\ V(r) &= \infty; \qquad r \le r_1 + r_2. \end{aligned} \tag{9-29}$$

We can use Eq.(9-28) in this example to obtain an expression for the impact parameter as a function of the scattering angle θ. Using the conditions in Eqs.(9-29) in Eq.(9-28) gives

$$\theta = \pi + 2b \int_{\infty}^{r_1+r_2} \frac{dr}{r(r^2 - b^2)^{1/2}} = \pi + 2\cos^{-1}\left|\frac{b}{r}\right|\Bigg|_{\infty}^{r_1+r_2} = 2\cos^{-1}\left|\frac{b}{r_1 + r_2}\right|$$

and finally

$$b = (r_1 + r_2)\cos\left(\frac{\theta}{2}\right), \tag{9-30}$$

where $\cos^{-1} x = \arccos x$. In the limit of low scattering angles, Eq.(9-30) gives $b = r_1 + r_2$ for this hard sphere model. The differential cross-section for the hard sphere model is evaluated using Eqs.(9-30) and (9-16) to give

$$\sigma(\theta, \phi) = \frac{b}{\sin\theta}\left|\frac{db}{d\theta}\right| = \frac{(r_1 + r_2)^2}{4}. \tag{9-31}$$

The total cross-section is given from Eq.(9-3) and the differential cross-section above as

$$\sigma = \int_0^{2\pi}\int_0^{\pi} \sigma(\theta, \phi) \sin\theta \, d\theta \, d\phi = \pi(r_1 + r_2)^2. \tag{9-32}$$

Equation (9-32) is the expected result; the total cross-section for the scattering of hard spheres is the area of a circle with radius $r_1 + r_2$. The low-angle result of $b = r_1 + r_2$ [Eq.(9-30)] is also expected.

Consider now the scattering angle in Eq.(9-28) and the corresponding cross-sections for the more general central field potential in Eq.(9-15) that gives

$$\theta = \pi + 2b \int_{\infty}^{r_m} \frac{dr}{r^2[1 - (b^2/r^2) - (Cr^{-n}/E_0)]^{1/2}}. \tag{9-33}$$

First we can examine the low-angle or high incident energy limit $[E_0 \gg V(r)]$ in Eq.(9-33), where $1 - b^2/r^2 > Cr^{-n}/E_0$ for all values of r from ∞ to r_m where n is a positive integer. Using these conditions, the denominator of Eq.(9-33) can be expanded giving

$$\begin{aligned}\frac{1}{[1 - (b^2/r^2) - (Cr^{-n}/E_0)]^{1/2}} &= [1 - (b^2/r^2)]^{-1/2}\left\{1 - \frac{Cr^{-n}}{E_0[1 - (b^2/r^2)]}\right\}^{-1/2} \\ &\cong \left(1 - \frac{b^2}{r^2}\right)^{-1/2}\left\{1 + \frac{Cr^{-n}}{2E_0[1 - (b^2/r^2)]}\right. \\ &\quad \left. + \frac{3}{8}\frac{C^2 r^{-2n}}{E_0^2[1 - (b^2/r^2)]^2} + \cdots\right\}.\end{aligned} \tag{9-34}$$

Substituting this result into Eq.(9-33) and using the $r_m \cong b$ estimate for the r_m small angle result gives

$$\theta = \pi + 2b \int_{\infty}^{r_m \cong b} \left[\frac{dr}{r(r^2 - b^2)^{1/2}} + \frac{Cr^{-n+1}\,dr}{2E_0(r^2 - b^2)^{3/2}} + \cdots\right].$$

The first integral cancels the π, leaving, after truncation, the following result:

$$\theta \cong \frac{bC}{E_0} \int_{\infty}^{r_m \cong b} \left[\frac{r^{-n+1}\,dr}{(r^2 - b^2)^{3/2}}\right] = \frac{bC}{E_0} \int_{\infty}^{r_m \cong b} \frac{dr}{r^{n+2}[1 - (b^2/r^2)]^{3/2}}. \tag{9-35}$$

Integrating by parts,

$$\int u\,dv = uv - \int v\,du,$$

where $u = r^{-n}$ and $v = -(r^2 - b^2)^{-1/2}$, gives

$$\theta \cong -\frac{bC}{E_0}\left[\frac{1}{r^n(r^2 - b^2)^{1/2}}\right]\Bigg|_{\infty}^{r_m \cong b} - \frac{nbC}{E_0} \int_{\infty}^{r_m \cong b} \frac{dr}{r^{n+1}(r^2 - b^2)^{1/2}}. \tag{9-36}$$

The first term in Eq.(9-36) can be rewritten as

$$-\frac{bC}{E_0}\left(\frac{1}{r^n}\right)\left(\frac{1}{r^2 - b^2}\right)^{1/2}\Bigg|_{\infty}^{r_m \cong b} = -\frac{bV(r)}{E_0}\left(\frac{1}{r^2 - b^2}\right)^{1/2}\Bigg|_{\infty}^{r_m \cong b}. \tag{9-37}$$

In the limit where $V(r) \ll E_0$ for all r, the expression above goes to zero leaving in Eq.(9-36) [4]

$$\theta \cong -\frac{nbC}{E_0} \int_{\infty}^{r_m \cong b} \frac{dr}{r^{n+2}[1 - (b^2/r^2)]^{1/2}}. \tag{9-38}$$

We now make a change in variables in Eq.(9-38),

$$x = \left(\frac{b}{r}\right)^2, \qquad r^{n+2} = b^{n+2}x^{-(1/2)(n+2)}, \qquad dr = -\tfrac{1}{2}bx^{-3/2}\,dx,$$

which gives a standard definite integral for θ of

$$\theta \cong \frac{nC}{2E_0 b^n}\int_0^1 \frac{x^{(1/2)(n-1)}}{(1-x)^{1/2}}\,dx = \frac{nC}{2E_0\,b^n}\,\frac{\Gamma[(n+1)/2]\Gamma(\frac{1}{2})}{\Gamma[(n/2)+1]}, \tag{9-39}$$

where $\Gamma(a)$ is the gamma function with

$$\Gamma(\tfrac{1}{2}) = (\pi)^{1/2}, \qquad \Gamma(1) = 1.0, \qquad \Gamma(a+1) = a\Gamma(a), \tag{9-40}$$

where $a > 0$. Substituting into Eq. (9-39) gives

$$\theta \cong \frac{C(\pi)^{1/2}\Gamma[(n+1)/2]}{E_0 b^n \Gamma(n/2)}. \tag{9-41}$$

The scattering angle in Eq.(9-41) is now easily evaluated for a given value of n in the potential function $[V(r) = Cr^{-n}$ in Eq.(9-15)]. For instance, for the Coulomb potential where $n = 1$, Eq.(9-41) gives

$$\theta_{n=1} = \frac{C}{E_0 b}. \tag{9-42}$$

The $n = 1$ scattering problem (Rutherford scattering) is considered in more detail in Problem 9-2.

The differential cross-section for the small angle scattering considered here is obtained by beginning with Eq.(9-41) where we write

$$b = \left\{\left(\frac{C(\pi)^{1/2}}{E_0\theta}\right)\frac{\Gamma[(n+1)/2]}{\Gamma(n/2)}\right\}^{1/n} \tag{9-43}$$

$$\left|\frac{db}{d\theta}\right| = \left\{\left(\frac{C(\pi)^{1/2}}{E_0}\right)\frac{\Gamma[(n+1)/2]}{\Gamma(n/2)}\right\}^{1/n}\left(\frac{1}{n}\theta^{-(1/n)-1}\right). \tag{9-44}$$

Substituting into Eq.(9-16) for small angles where $\sin\theta \cong \theta$ gives

$$\sigma(\theta) = \frac{b}{\sin\theta}\left|\frac{db}{d\theta}\right| \cong \frac{b}{\theta}\left|\frac{db}{d\theta}\right| = \frac{1}{n}\left\{\left(\frac{(\pi)^{1/2}C}{E_0}\right)\frac{\Gamma[(n+1)/2]}{\Gamma(n/2)}\right\}^{2/n}\theta^{-(2/n)-2}. \tag{9-45}$$

Thus, in principle, the measurement of the differential cross-section, as shown above, will yield the values of n and C in the potential energy function $[V(r) = Cr^{-n}$ in Eq.(9-15)]. A log $\sigma(\theta)$ versus log θ plot is shown in Fig. 9-4 for the measurement of the differential cross-section of the attractive scattering of a monoenergetic beam of potassium atoms scattering from a mercury atom target. According to Eq.(9-45), the slope of the log-log plot should be equal to the θ exponent of $-(2/n) - 2$. The best fit of the experimental data in Fig. 9-4 gives $-\frac{7}{3} = -(2/n) - 2$, which leads to $n = 6$ for the spherically symmetric attractive potential function, $V(r) = -Cr^{-n}$ in Eq.(9-15). This inverse sixth power attractive potential is normally associated with the attractive part of the Lennard–Jones potential [5] for the interaction between two

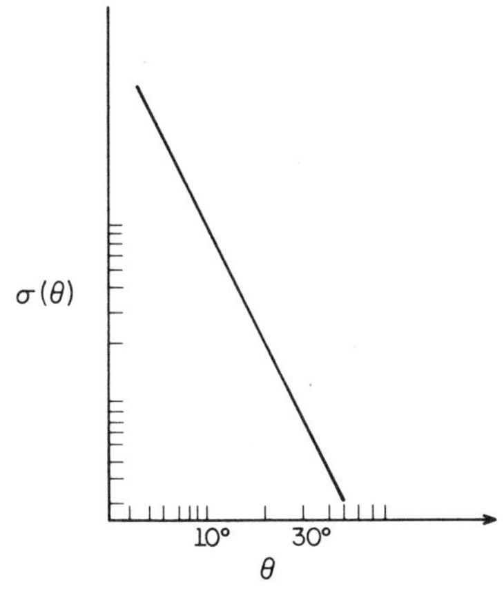

Figure 9-4 Log $\sigma(\theta)$ as a function of log θ for the scattering of neutral potassium atoms from neutral mercury atoms from F. A. Morse and R. B. Bernstein, *J. Chem. Phys.* **37**, 2019 (1962). The solid line is a best smooth line fit of the data. The slope of the curve was found to be independent of the relative incident particle–target particle velocities over a small range from 4×10^2 to $7 \times 10^2\ \mathrm{m \cdot s^{-1}}$. The slope of this curve is equal to $-\frac{7}{3}$, which gives a value of $n = 6$ from Eq.(9-45) for the attractive potential $[V(r) = Cr^{-n}]$.

spherically symmetric nonbonded atoms or molecules [see Eq.(9-141), Table 9-2, and associated discussion].

The total cross-section, σ, can be obtained experimentally, as shown in Figs. 9-1 and 9-2, by measuring the loss in intensity (attenuation) of the incident beam. This experiment requires a carefully collimated monoenergetic incident beam and a high resolution detector. Let θ_0 be the minimum angle of scattering included at the detector, which is placed with its center on the z axis (Fig. 9-1) to measure the attenuation of the incident beam. The total cross-section measured by the attenuation is given from Eq.(9-3) by

$$\sigma_{\theta_0} = \int_0^{2\pi} \int_{\theta_0}^{\pi} \sigma(\theta, \phi) \sin\theta \, d\theta \, d\phi. \tag{9-46}$$

Substituting Eq.(9-16) gives

$$\sigma_{\theta_0} = 2\pi \int_0^{b(\theta_0)} b \, db = \pi[b(\theta_0)]^2. \tag{9-47}$$

$b(\theta_0)$ is obtained from Eq.(9-43) giving the total cross-section as measured with a detector along the z axis with resolution or acceptance angle θ_0:

$$\sigma_{\theta_0} = \pi \left\{ \left(\frac{C(\pi)^{1/2}}{E_0 \theta_0} \right) \frac{\Gamma[(n+1)/2]}{\Gamma(n/2)} \right\}^{2/n}. \tag{9-48}$$

Thus, the measurement of the attenuation of the primary beam as a function of E_0, as described above, will give the total cross-section and values of C and n in Eq.(9-48). A number of these experiments have been reported with angular resolution on the order of $\theta_0 \cong 1°$ giving $n \cong 6$ with a negative deflection angle indicating an attractive potential. Of course, the energy range specifies the distance of nearest approach, or the value of r, for which the potential function $[V(r) = Cr^{-n}]$ is being examined. For high relative energy (or velocity, $E_0 = \frac{1}{2}Mv^2$), $V(r_m) \ll E_0$ in Eq.(9-25) and $r_m = b(\theta_0)$. $r_m = b(\theta_0)$ is obtained from Eq.(9-43) as demonstrated above.

A plot of the total scattering cross-section for a beam of He atoms scattered from a He target in the region of high relative velocity (or energy) is shown in Fig. 9-5. The data in Fig. 9-5(a), which are taken in the laboratory axis system, are transformed into

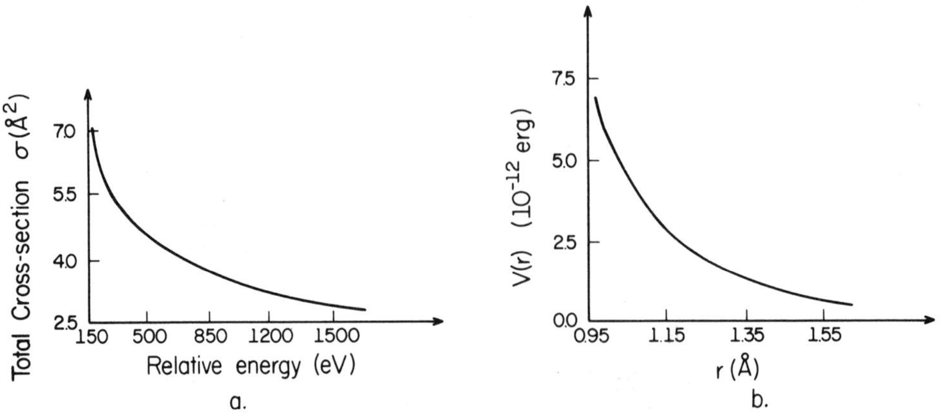

Figure 9-5 a. A plot of the experimental total cross-section as a function of relative energy for the collision of a beam of He atoms with a He gas from I. Amdur, J. E. Jordan, and S. O. Colgate, *J. Chem. Phys.* **34**, 1525 (1961). The total cross-section is measured as shown in Fig. 9-1 by the attenuation of the incident beam. The solid line is the best smooth line fit of the data. After transforming the laboratory based data in (a) to the center of mass frame, the values of n and C are extracted by fitting the data to an expression similar to Eq.(9-47) through the range of $r = b(\theta_0)$ as calculated from Eq.(9-43). b. The potential function, $V(r)$, which fits the data in part (a) [see also Eq.(9-49)].

the center of mass axis system. The results are then fit to an expression similar to Eq.(9-48) to give C and n over a given range of r. The relative energy or relative velocity range of 9×10^4 to 24×10^4 m · s^{-1} leads to an r_m range of 0.97×10^{-10} to 1.48×10^{-8} cm and a repulsive potential function of [6]

$$V(r) = \frac{5.56 \times 10^{-12}}{r^{5.03}} \text{ erg}, \tag{9-49}$$

Table 9-1 Repulsive potentials for several atom-atom, atom-molecule, and molecule-molecule systems from Ref. [6]. More data can be found in E. A. Mason and J. T. Vanderslice, "High Energy Elastic Scattering of Atoms, Molecules, and Ions," *Atomic and Molecular Processes* (Academic Press, New York, 1962), p. 663.

System	$V(r)$ (eV)	Range (Å)
He—He	$(1/r) \exp(1.07/r)$	0.52–0.98
He—He	$3.47 \times r^{-5.03}$	0.97–1.48
Ar—Ar	$3250 \times r^{-9.97}$	1.98–2.41
$Ar—O_2$	$1360 \times r^{-8.34}$	2.01–2.50
$Ar—H_2$	$159 \times r^{-6.28}$	1.81–2.36
Ar—He	$22.2 \times r^{-5.42}$	1.40–1.81
$Ar—N_2$	$567 \times r^{-7.06}$	2.04–2.53
Ar—CO	$551 \times r^{-6.99}$	2.09–2.68
$He—CH_4$	$602 \times r^{-9.43}$	1.92–2.37
$He—CF_4$	$6.18 \times 10^6 \times r^{-17.51}$	2.43–2.74
$CH_4—CH_4$	$5.64 \times 10^6 \times r^{-15.47}$	2.47–3.06
$CF_4—CF_4$	$1.17 \times 10^{22} \times r^{-39.27}$	3.43–3.77

which is illustrated in Fig. 9-5(b). This potential function and data on potential functions for additional systems are listed in Table 9-1.

Considerably more detail in the complete potential function, $V(r)$, can be obtained by a more detailed analysis of the elastic scattering cross-sections [3,7] (see Section 9-4).

9.2 QUANTUM MECHANICAL SCATTERING; WAVE PACKETS

In this section we shall introduce the concepts necessary to understand particle scattering from a quantum mechanical viewpoint. We start by a continuation of our previous discussion in Section 3.2 concerning the simulation of a particle by a wave packet. Equation (3-24) describes a free particle with mass m as a superposition of waves along the z axis for continuous k states and energy values. Following Eq.(3-24), we examined a simple k-pulse form for $f(k)$ and the results are summarized in Eqs. (3-25) and (3-26).

Consider now the nature of the wave packet formed from an initial Gaussian function in k centered around k_0. A Gaussian function is characterized by an amplitude and half width at half height as shown in Fig. 9-6. The normalized Gaussian is given by [see Eq.(7-31)]

$$G(k) = \frac{a}{(\pi)^{1/2}} \exp\left[-a^2(k - k_0)^2\right], \qquad \Delta k = \frac{(\ln 2)^{1/2}}{a}. \tag{9-50}$$

a is a constant and Δk is the half width at half height of $G(k)$. Substituting Eq.(9-50) for $f(k)$ into Eq.(3-24) gives

$$\Psi(z, t) = \frac{a}{(\pi)^{1/2}} \int_{-\infty}^{+\infty} \exp\left[-a^2(k - k_0)^2\right] \exp\left(-\frac{i\hbar t}{2M} k^2\right) \exp(ikz)\, dk. \tag{9-51}$$

This nontrivial integral can be evaluated by multiplying the integrand by $\exp(-ik_0 z)$ $\exp(ik_0 z)$, combining $(k - k_0)$ terms, changing variables to $U = k - k_0$, factoring out all terms not dependent on U, and integrating to give (see Problem 9-4) the normalized results:

$$\Psi(z, t) = \frac{a^{1/2}(1/2\pi)^{1/4}}{[a^2 + (i\hbar t/2M)]^{1/2}} \exp\left\{-\frac{(z - v_0 t)^2}{4(a^2 + (i\hbar t/2M)]}\right\} \exp\left[ik_0\left(z - \frac{v_0}{2} t\right)\right], \tag{9-52}$$

$$\Psi(z, 0) = \left[\frac{1}{a}\left(\frac{1}{2\pi}\right)^{1/2}\right]^{1/2} \exp\left(-\frac{z^2}{4a^2}\right) \exp(ik_0 z), \tag{9-53}$$

where v_0 is the velocity of the particle along the z axis and $k_0 = Mv_0/\hbar$.

The probability per unit length along the z axis is obtained from Eqs.(9-52) and (3-135) to give

$$P(z, t) = |\Psi(z, t)|^2 = \frac{a(1/2\pi)^{1/2}}{[a^4 + (\hbar^2 t^2/4M^2)]^{1/2}} \exp\left\{-\frac{a^2(z - v_0 t)^2}{2[a^4 + (\hbar^2 t^2/4M^2)]}\right\}. \tag{9-54}$$

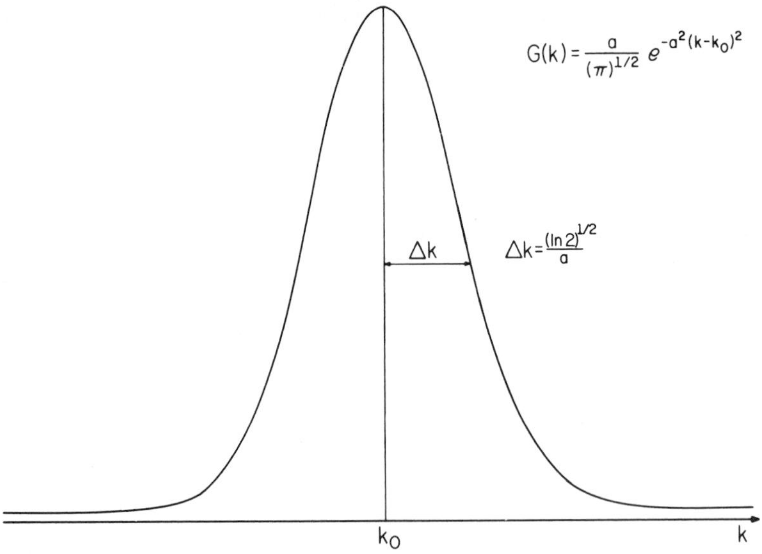

Figure 9-6 The normalized Gaussian function in Eq.(9-50) with half width at half height given by Δk.

The half width at half height of this Gaussian wave packet along the z axis is a function of time given by

$$z_{1/2}(t) = \frac{(2 \ln 2)^{1/2}}{a}\left(a^4 + \frac{\hbar^2 t^2}{4M^2}\right)^{1/2}, \qquad z_{1/2}(t = 0) = a(2 \ln 2)^{1/2} = 1.1774a. \quad (9\text{-}55)$$

It is evident from these equations that the half width of the Gaussian describing the particle spreads in time around the $v_0 t$ centroid of the distribution.

The wave packet has a group velocity that can be calculated at $t = 0$ by

$$v_{\text{group}} = \int_{-\infty}^{+\infty} \Psi^*(z, 0)\left(-\frac{i\hbar}{M}\frac{\partial}{\partial z}\right)\Psi(z, 0)\, dz = \frac{\hbar k_0}{M} = v_0. \quad (9\text{-}56)$$

At $t \longrightarrow \infty, \Psi(z, t \longrightarrow \infty) \longrightarrow 0$ and the group velocity approaches zero. At short times the wave packet is localized and the particle has a classical group velocity of v_0. At long times the packet becomes space delocalized and the group velocity goes to zero. The particle maintains its classical behavior if the packet spreading (or broadening) is small during the course of an observational experiment on the particle.

The particle current density along the z axis for the particle described by Eq.(9-52) is given by Eq.(3-141). For small times ($t \ll a^2 M/\hbar$), we can write

$$I = \frac{\hbar}{2iM}\left[\Psi^* \frac{\partial}{\partial z}\Psi - \left(\frac{\partial}{\partial z}\Psi^*\right)\Psi\right] = \frac{\hbar k_0}{M}|\Psi(z, t)|^2 = \frac{p_0}{M}|\Psi(z, t)|^2. \quad (9\text{-}57)$$

$|\Psi(z, t)|^2$ is obtained from Eq.(9-54). The integral of the current density, I in Eq. (9-57), along the z axis ($-\infty \le z \le \infty$) will yield the group velocity result in Eq. (9-56).

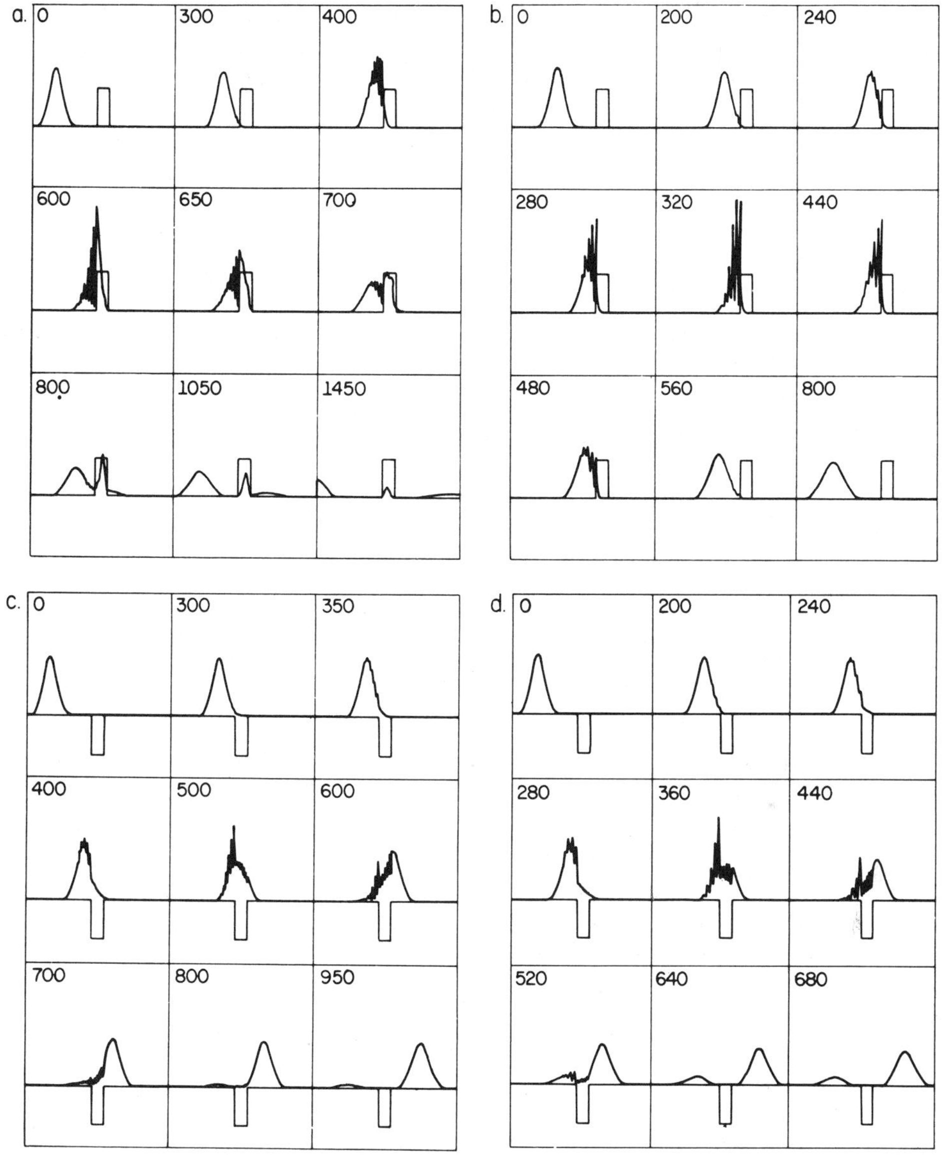

Figure 9-7 The scattering of a Gaussian wave packet moving from left to right into a square barrier and over a square well from A. Goldberg, H. M. Schey, and J. L. Schwartz, *Am. J. Physics* **35**, 177 (1967). The numbers indicate the time lapse in arbitrary units. a. The mean energy of the wave packet is equal to the barrier height. b. The mean energy is one-half the barrier height. c. The mean energy is equal to the well depth. d. The mean energy is one-half the well depth.

The scattering of a nonspreading wave packet traveling from $-z$ to $+z$ through a barrier and over a well has been examined by computer calculation and the results are shown in Fig. 9-7. A small transmission is apparent in (a) where the mean wave packet energy is equal to the barrier height. Transmission is not observed in (b) where the mean energy is one-half the barrier height. Corresponding effects are observed in (c) and (d) when the wave packet passes over a well.

In the case of a classical particle, as discussed in Section 9.1, $\hbar/M$ is very small and this limiting form of Eq. (9-54) gives

$$\lim_{\hbar t/2M \to 0} |\Psi(z, t)|^2 = \frac{1}{a(2\pi)^{1/2}} \exp\left[-\frac{(z - v_0 t)^2}{2a^2}\right],$$

for the probability density. The limit of the result above as $a \longrightarrow 0$ is a Dirac delta function (Appendix B.3) in $(z - v_0 t)$,

$$\lim_{a \to 0}\left[\lim_{\hbar t/2M \to 0} |\Psi(z, t)|^2\right] = \delta(z - v_0 t),$$

which describes a free classical particle, localized in space with velocity v_0, as assumed in Section 9.1.

9.3 QUANTUM MECHANICAL SCATTERING; PLANE WAVES

In order for the classical treatment in Section 9.1 to be valid, both the impact parameter and the angle of deflection must be well defined. We can now use the uncertainty principle to show the onset of quantum behavior in a scattering experiment. Return to Fig. 9-1 and consider a particle described by a quantum mechanical wave packet moving [as described by Eq.(9-52)] from $-z$ to $+z$. The uncertainty in position is defined as the square root of the difference between the average value of the square and the square of the average value,

$$\Delta z = \left[\int_{-\infty}^{+\infty} \psi^* z^2 \psi \, dz - \left(\int_{-\infty}^{+\infty} \psi^* z \psi \, dz\right)^2\right]^{1/2}, \tag{9-58}$$

with a similar definition for Δp_z along the z axis. If we assume no spreading of the wave packet in time, we can use $\Psi(z, k_0, t = 0)$ in Eq.(9-53) to compute Δz and Δp_z to give

$$\Delta z = a, \qquad \Delta p_z = \frac{\hbar}{2a}, \qquad \Delta z \, \Delta p_z = \frac{\hbar}{2}, \tag{9-59}$$

where a is defined by the original Gaussian function in Eq.(9-50) and Fig. 9-6. The consequence in Eq.(9-59), which indicates a limitation in the simultaneous determination of the position and momentum of a particle, is the Heisenberg uncertainty principle [8].

In applying Eqs.(9-59) to Eqs.(9-52) and (9-54), we note that if the uncertainty in the velocity ($\Delta p_z = M\,\Delta v_z$) along the z axis is large, the resultant uncertainty in position can be small and the complete wave function in Eq.(9-52) is necessary to describe the particle during its journey from the source into the target region and finally onto the detector. If the uncertainty in the velocity distribution is small and the resultant uncertainty in position is large, however, the wave functions in Eqs.(9-52) and (9-53) reduce to plane waves. This is evident by separating the real and imaginary parts in the exponent of Eq.(9-52) to give

$$\Psi(z, t) = A \exp\left[ik_0\left(z - \frac{v_0 t}{2}\right)\right] \exp\left(\frac{i\hbar t}{2M}\left\{\frac{(z - v_0 t)^2}{4[a^4 + (\hbar^2 t^2/4M^2)]}\right\}\right)$$

$$\times \exp\left\{-\frac{a^2(z - v_0 t)^2}{4[a^4 + (\hbar^2 t^2/4M^2)]}\right\}, \tag{9-60}$$

where A is the exponential premultiplier in Eq.(9-52). The last two exponentials in this expression reduce to unity as a becomes large, giving

$$\lim_{a \to \text{large}} \Psi(z, t) = A \exp\left(-\frac{ik_0 v_0 t}{2}\right) \exp(ik_0 z). \tag{9-61}$$

The first exponential gives the time-dependence and the second $\exp(ik_0 z)$ term is the spatial plane wave. It is evident that the total and differential cross-sections are both time-independent and the important aspects in scattering theory can be obtained by using the spatial plane wave for the incident-free particle along the z axis, where the velocity is well defined.

Returning again to Fig. 9-1, we note that after the collision the incident particle will be deflected through an angle θ. In order to have a well-defined angle of deflection, the uncertainty in the position of the scattered particle along the transverse axis (x or y axis) must be small relative to the position along that axis,

$$\Delta x \ll x. \tag{9-62}$$

We also note that any component of transverse velocity, Δv_x, in the incident beam will also affect the measurement of the low-angle scattering. The angle of scattering must be large relative to the ratio of the width of the transverse velocity distribution to the incident velocity, v_z (see again Fig. 9-1),

$$\theta \gg \frac{\Delta v_x}{v_z}. \tag{9-63}$$

Using the uncertainty principle along the x axis, $\Delta p_x = M\,\Delta v_x = \hbar/(2\,\Delta x)$, and using the minimum distance along x as $\Delta x = b$ (impact parameter) that is appropriate for low-angle scattering, we can write $\Delta v_x = \hbar/(2\,\Delta x M) = \hbar/2bM$. Substituting into Eq. (9-63) gives

$$\theta \gg \frac{\hbar}{2v_z bM}. \tag{9-64}$$

Using $v_z = 10^3\ \text{m}\cdot\text{s}^{-1}$, $b = 3 \times 10^{-10}$ m, and $M = 4.7 \times 10^{-26}$ kg for a typical N_2 molecular beam gives $\theta \gg 3.7 \times 10^{-3}$ rad $= 0.21°$. According to this estimate and

Fig. 9-1, if the scattering region–detector distance is 1 m, the classical scattering equations are not valid unless detection in the xy plane takes place well outside a ring with a radius of 0.37 cm. Another view of this limitation is seen by examining Eq.(9-41) for central field scattering ($V = Cr^{-n}$). Estimating $\theta \cong C/(E_0 b^n) = 2C/(Mv_z^2 b^n)$ and using Eq.(9-64) leads to

$$\frac{C}{b^{n-1}} \gg \frac{\hbar v_z}{4}. \tag{9-65}$$

Thus, the classical description will break down if the impact parameter is large, especially for potentials where $n > 6$. In summary, we note that the classical description of scattering breaks down for large impact parameters at low scattering angles. We shall now develop a more complete theory of quantum mechanical scattering.

The time-independent Schrödinger equation for a particle with mass M is given by

$$\frac{\hbar^2}{2M} \nabla^2 \psi(\boldsymbol{r}) + [E - V(\boldsymbol{r})]\psi(\boldsymbol{r}) = 0, \tag{9-66}$$

where E is the total beam particle–target particle energy and $V(\boldsymbol{r})$ is the beam particle–target particle potential energy that is a function of the vector distance between the particles, $\boldsymbol{r}$. A solution to this wave equation for a free particle $[V(\boldsymbol{r}) = 0]$ traveling along the z axis with no uncertainties in velocity is given by

$$\psi^0(\boldsymbol{r}) = \exp(i\boldsymbol{k}_0 \cdot \boldsymbol{r}) = \exp(ik_0 z), \tag{9-67}$$

where the incident particle direction is along the z axis, $k_0 = p_0/\hbar$, and p_0 is the linear momentum of the particle. According to our previous discussion of Eq.(9-52) [which follows Eq.(9-59)], Eq.(9-67) represents a plane wave describing a free particle with a definite velocity of $v_0 = \hbar k_0/M$ along the z axis.

We now seek a solution to Eq.(9-66) for the general interaction potential, $V(\boldsymbol{r})$. In most cases, we shall relax the general interaction potential into a spherically symmetric potential where $V(\boldsymbol{r}) = V(r)$, which leads to equal scattering at all angles ϕ (see Fig. 9-1). The solution to the complete Eq.(9-66) must contain the plane wave contribution, $\psi^0(\boldsymbol{r})$, in Eq.(9-67) as well as a scattered wave contribution, $\psi^s(\boldsymbol{r})$, which arises from the target-incident particle interaction, $V(\boldsymbol{r})$. In summary,

$$\psi(\boldsymbol{r}) = \psi^0(\boldsymbol{r}) + \psi^s(\boldsymbol{r}) = \exp(ik_0 z) + \psi^s(\boldsymbol{r}). \tag{9-68}$$

Substituting Eq.(9-68) into Eq.(9-66) gives

$$\left(E + \frac{\hbar^2\nabla^2}{2M}\right)[\exp(i\boldsymbol{k}_0 \cdot \boldsymbol{r}) + \psi^s(\boldsymbol{r})] = V(r)[\exp(i\boldsymbol{k}_0 \cdot \boldsymbol{r}) + \psi^s(\boldsymbol{r})]. \tag{9-69}$$

As this equation is valid for all r, we note that $\lim_{r\to\infty} V(\boldsymbol{r}) = 0$, which requires E to be the incident particle–target particle relative kinetic energy ($E = \hbar^2 k_0^2/2M$). Thus

$$\left(E + \frac{\hbar^2\nabla^2}{2M}\right) \exp(i\boldsymbol{k}_0 \cdot \boldsymbol{r}) = 0.$$

and Eq.(9-69) reduces to

$$\left(E + \frac{\hbar^2\nabla^2}{2M}\right)\psi^s(\boldsymbol{r}) = V(\boldsymbol{r})\psi(\boldsymbol{r}), \tag{9-70}$$

where we remember that $\psi^s(\boldsymbol{r})$ and $\psi(\boldsymbol{r})$ are the scattered and total wave functions, respectively, describing the target particle–incident particle system at any interparticle distance, $\boldsymbol{r}$. Using $E = \hbar^2k_0^2/2M$ and rearranging Eq. (9-70) gives

$$(\nabla^2 + k_0^2)\psi^s(\boldsymbol{r}) = \frac{2M}{\hbar^2} V(\boldsymbol{r})\psi(\boldsymbol{r}). \tag{9-71}$$

Our objective is to write $\psi^s(\boldsymbol{r})$, in terms of the interaction potential energy, $V(\boldsymbol{r})$. In order to solve the differential equation in Eq.(9-71), we use the method of Green's functions [9,10,11]. We use this method by recognizing that

$$(\nabla^2 + k_0^2)\frac{\exp(ik_0|\boldsymbol{r} - \boldsymbol{r}'|)}{|\boldsymbol{r} - \boldsymbol{r}'|} = -4\pi\delta(\boldsymbol{r} - \boldsymbol{r}'). \tag{9-72}$$

The parallel lines around $\boldsymbol{r} - \boldsymbol{r}'$ indicate the magnitude of the $\boldsymbol{r} - \boldsymbol{r}'$ vector and $\delta(\boldsymbol{r} - \boldsymbol{r}')$ is a three-dimensional Dirac delta function in $(\boldsymbol{r} - \boldsymbol{r}')$ [12]. The Dirac delta function is zero if $\boldsymbol{r} \neq \boldsymbol{r}'$ and the integral of $\delta(\boldsymbol{r} - \boldsymbol{r}')$ over the appropriate volume for the system is unity. The Dirac delta function is only defined in terms of its transformation properties on another function $f(\boldsymbol{r})$,

$$f(\boldsymbol{r}') = \int f(\boldsymbol{r})\delta(\boldsymbol{r} - \boldsymbol{r}')\,d\boldsymbol{r}. \tag{9-73}$$

$(|\boldsymbol{r} - \boldsymbol{r}'|)^{-1}\exp(ik_0|\boldsymbol{r} - \boldsymbol{r}'|)$ in Eq.(9-72) is the Green's function that we can use to solve the differential equation in Eq.(9-71). We now combine Eqs.(9-72) and (9-71) to write $\psi^s(\boldsymbol{r})$ for the scattered wave as

$$\psi^s(\boldsymbol{r}) = -\frac{1}{4\pi}\left(\frac{2M}{\hbar^2}\right)\int \frac{\exp(ik_0|\boldsymbol{r} - \boldsymbol{r}'|)}{|\boldsymbol{r} - \boldsymbol{r}'|} V(\boldsymbol{r}')\psi(\boldsymbol{r}')\,d\boldsymbol{r}'. \tag{9-74}$$

Substituting Eq.(9-74) into Eq.(9-71) and using Eqs.(9-72) and (9-73) for the delta function shows that Eq.(9-74) is correct. Thus, Eq.(9-74) gives an expression for the scattered wave as a function of $V(\boldsymbol{r})$.

We shall now examine the asymptotic form of $\psi^s(\boldsymbol{r})$ in Eq. (9-74), which is the appropriate form for the scattered wave at the detector. Remembering that $V(\boldsymbol{r}')$ goes to zero as $\boldsymbol{r}' \longrightarrow \infty$ for typical short-range potentials, we note that the integrand in Eq.(9-74) is important only for small values of $\boldsymbol{r}'$. Thus, we can expand the $(|\boldsymbol{r} - \boldsymbol{r}'|)^{-1}\exp(ik_0|\boldsymbol{r} - \boldsymbol{r}'|)$ part of the integrand in Eq.(9-74) around $|\boldsymbol{r}'|/|\boldsymbol{r}|$. Expanding $|\boldsymbol{r} - \boldsymbol{r}'|$ when $|\boldsymbol{r}'|/|\boldsymbol{r}| \ll 1$ gives

$$|\boldsymbol{r} - \boldsymbol{r}'| = r - \boldsymbol{r}'\cdot\boldsymbol{U}_s + \cdots \cong r - \boldsymbol{r}'\cdot\boldsymbol{U}_s,$$

where $\boldsymbol{U}_s$ is a unit vector along $\boldsymbol{r}$, the direction of the scattered wave. Substituting into the Green's function gives

$$\frac{\exp(ik_0|\boldsymbol{r} - \boldsymbol{r}'|)}{|\boldsymbol{r} - \boldsymbol{r}'|} \cong \frac{\exp[ik_0(r - \boldsymbol{r}'\cdot\boldsymbol{U}_s)]}{r[1 - (\boldsymbol{r}'\cdot\boldsymbol{U}_s/r)]} \cong \frac{\exp[ik_0(r - \boldsymbol{r}'\cdot\boldsymbol{U}_s)]}{r}, \tag{9-75}$$

where the last step again acknowledges that $|\boldsymbol{r}'|/|\boldsymbol{r}| = r'/r \ll 1$ in the denominator. Substituting the result in Eq.(9-75) into Eq.(9-74) gives

$$\psi^s(\boldsymbol{r}) \cong -\frac{M}{2\pi\hbar^2}\frac{\exp(ik_0 r)}{r}\int \exp(-ik_0\boldsymbol{r}'\cdot\boldsymbol{U}_s)V(\boldsymbol{r}')\psi(\boldsymbol{r}')\,d\boldsymbol{r}' = f(\theta,\phi)\frac{\exp(ik_0 r)}{r} \tag{9-76}$$

$$f(\theta,\phi) = -\frac{M}{2\pi\hbar^2}\int \exp(-i\boldsymbol{r}'\cdot\boldsymbol{k}_s)V(\boldsymbol{r}')\psi(\boldsymbol{r}')\,d\boldsymbol{r}'. \tag{9-77}$$

We also use $\boldsymbol{U}_s k_0 = \boldsymbol{k}_s$, where $\boldsymbol{k}_s$ is the wave vector in the direction of the scattered particle (scattered wave vector). $\psi^s(\boldsymbol{r})$ in Eq.(9-77) has the form of a spherical outgoing wave, $(1/r)\exp(ik_0 r)$, multiplied by a geometric amplitude factor, $f(\theta, \phi)$. $f(\theta, \phi)$ is the scattering amplitude for elastic scattering ($k_0 = k_s$) that depends on the form of the potential, $V(\boldsymbol{r})$, where θ and ϕ are the spherical polar angles defined in terms of the scattering geometry in Fig. 9-1. Substituting the result in Eq.(9-76) into Eq.(9-68) gives the asymptotic solution for the wavefunction describing the scattered particle,

$$\psi(\boldsymbol{r}) = \psi^0(\boldsymbol{r}) + \psi^s(\boldsymbol{r}) = \exp(ik_0 z) + f(\theta,\phi)\frac{\exp(ik_0 r)}{r}. \tag{9-78}$$

The incident intensity of particles is obtained from the probability density, $|\psi^0(\boldsymbol{r})|^2$, times the average velocity, v_0,

$$I_0 = v_0|\psi^0(\boldsymbol{r})|^2 = v_0, \tag{9-79}$$

where the last step indicates a single incident particle. The scattered intensity is obtained from the second term of Eq.(9-78) giving

$$I(R,\theta,\phi) = v_0|\psi^s(\boldsymbol{r})|^2 = \frac{v_0}{R^2}|f(\theta,\phi)|^2. \tag{9-80}$$

Substituting Eqs.(9-79) and (9-80) into Eq.(9-2) gives the differential cross-section,

$$\sigma(\theta,\phi) = |f(\theta,\phi)|^2 = \left(\frac{M}{2\pi\hbar^2}\right)^2\left|\int \exp(-i\boldsymbol{r}'\cdot\boldsymbol{k}_s)V(\boldsymbol{r}')\psi(\boldsymbol{r}')\,d\boldsymbol{r}'\right|^2. \tag{9-81}$$

The total cross-section for the scattering process is obtained from Eq.(9-3).

The results in Eqs.(9-76), (9-77), (9-78), and (9-81) give the scattering amplitude for the scattered wave in terms of $V(\boldsymbol{r}')$, the potential energy, and $\psi(\boldsymbol{r}')$, the final total wavefunction for the scattering system. Although this result may not seem initially very helpful, these equations are convenient starting points for approximate methods. For instance, the well-known Born approximation [13] is obtained by replacing the total final wavefunction $\psi(\boldsymbol{r}')$ by the plane wavefunction, $\psi^0(\boldsymbol{r}') = \exp(i\boldsymbol{k}_0\cdot\boldsymbol{r}') = \exp(ik_0 z')$. The first Born approximation assumes weak scattering with negligible effects due to multiple scattering events, which gives the following approximation to the scattering amplitude:

$$f(\theta, \phi) \cong -\frac{M}{2\pi\hbar^2}\int \exp(-i\mathbf{r}'\cdot\mathbf{k}_s)V(\mathbf{r}')\exp(i\mathbf{k}_0\cdot\mathbf{r}')\,d\mathbf{r}'$$

$$= -\frac{M}{2\pi\hbar^2}\int \exp[i\mathbf{r}'\cdot(\mathbf{k}_0 - \mathbf{k}_s)]V(\mathbf{r}')\,d\mathbf{r}'. \tag{9-82}$$

Equation (9-82) now has a form that describes the scattering amplitude as a Fourier transform of the incident particle–target particle interaction potential energy, $V(\mathbf{r})$. The second Born approximation for $f(\theta, \phi)$ is obtained by using the result in Eq.(9-82) to substitute back into $\psi^s(\mathbf{r})$ in Eq.(9-76) to give

$$\psi^s(\mathbf{r}) = f(\theta, \phi)\frac{\exp(ik_0 r)}{r} = -\frac{M}{2\pi\hbar^2}\frac{\exp(ik_0 r)}{r}\int \exp[i\mathbf{r}'\cdot(\mathbf{k}_0 - \mathbf{k}_s)]V(\mathbf{r}')\,d\mathbf{r}'. \tag{9-83}$$

This result is now used as an estimate for $\psi(\mathbf{r}')$ in Eq.(9-77) to recalculate $f(\theta, \phi)$. This process can be repeated to obtain the higher Born approximations. Equation (9-82) for $f(\theta, \phi)$ according to the first Born approximation can be simplified by defining the scattering wave vector, $\mathbf{K}$, as the difference between the incident particle wave vector, $\mathbf{k}_0$, and scattered particle wave vector, $\mathbf{k}_s$, $\mathbf{K} = \mathbf{k}_0 - \mathbf{k}_s$, as shown in Fig. 1-4. If the interaction potential in Eq.(9-82) is a spherically symmetric potential, $V(\mathbf{r}) = V(r)$, we can now write ($\mathbf{r}\cdot\mathbf{K} = rK\cos\theta$)

$$f(\theta) = -\frac{M}{2\pi\hbar^2}\int_0^{2\pi}\int_0^{\pi}\int_0^{\infty} \exp(irK\cos\theta)V(r)r^2\sin\theta\,dr\,d\theta\,d\phi$$

$$= -\frac{2M}{\hbar^2}\int_0^{\infty}\frac{\sin Kr}{Kr}V(r)r^2\,dr. \tag{9-84}$$

There is no ϕ-dependence in the scattering amplitude for spherically symmetric potentials.

Substituting the Coulomb potential,

$$V(r) = -\frac{Z_1 Z_2 e^2}{r}, \tag{9-85}$$

into Eq.(9-84) gives an indeterminant integral. We can evaluate the integral, however, by using the screened Coulomb potential (Yukawa potential) given by

$$V(r) = -\frac{Z_1 Z_2 e^2}{r}\exp(-\lambda r). \tag{9-86}$$

The value of $f(\theta)$ is obtained in the $\lambda \longrightarrow 0$ limit. Substituting gives

$$f(\theta) = \frac{2MZ_1Z_2e^2}{\hbar^2 K}\lim_{\lambda\to 0}\int_0^{\infty}\sin Kr\exp(-\lambda r)\,dr$$

$$= \frac{2MZ_1Z_2e^2}{\hbar^2 K}\lim_{\lambda\to 0}\left(\frac{K}{\lambda^2 + K^2}\right) = \frac{2MZ_1Z_2e^2}{\hbar^2 K^2}. \tag{9-87}$$

For elastic collisions where the magnitude of the incident particle's wave vector is equal to the magnitude of the scattered particle's wave vector ($k_0 = k_s$), we can write $K = 2k_0 \sin(\theta/2)$ as shown in Eq.(1-66) and Fig. 1-4, where θ is the scattering angle.

Substituting $K = 2k_0 \sin(\theta/2)$ into Eq.(9-87) and then substituting this result into Eq.(9-81) for the differential cross-section for Coulomb scattering gives

$$\sigma(\theta) = |f(\theta)|^2 = \frac{1}{4}\left(\frac{Z_1 Z_2 e^2 M}{\hbar^2 k_0^2}\right)^2 \frac{1}{\sin^4(\theta/2)}. \tag{9-88}$$

Remembering that $E_0 = \hbar^2 k_0^2/2M$ is the energy of the incident particle gives

$$\sigma(\theta) = \left(\frac{Z_1 Z_2 e^2}{4E_0}\right)^2 \frac{1}{\sin^4(\theta/2)}, \tag{9-89}$$

which is equivalent to the classical result obtained in Problem 9-2. It is noteworthy that the Born approximation, which equals the classical result, is also the exact result. According to Eq.(9-89), this angle-dependent differential cross-section for Coulomb scattering goes to infinity for forward-scattered particles. Thus, the total cross-section is also infinity. This anomalous and incorrect result at low angles arises because of the infinite range of the Coulomb scattering potential. In spite of these difficulties, the first Born approximation as developed here has been extremely effective in describing both elastic and inelastic processes [14].

9.4 QUANTUM MECHANICAL SCATTERING; PARTIAL WAVES

In Section 9.3 we wrote the solution to the Schrödinger equation in Eq.(9-66) in terms of an incident and scattered wave as shown in Eq. (9-68). In this section we shall develop the method of partial waves, which is most conveniently applied in the realm of low kinetic energy incident particles. In the case of particles with low incident energy, where $kr_0 \ll 1.0$ (r_0 is the effective range of the potential), the scattering tends toward cylindrical symmetry (ϕ-independent, see Fig. 9.1). Low energy incident particles have low momentum and the scattered wave will be predominantly spherically symmetric with zero angular momentum [15]. We shall now show that these low energy scattering processes can be described in terms of a phase shift in the incident plane wave.

Returning to the plane wave solution to the free particle wave equation [$V(\mathbf{r}) = 0$] given in Eq.(9-67), we start by expanding the plane wave in terms of spherical waves,

$$\psi(r, \theta) = \exp(ik_0 z) = \exp(i\mathbf{k}_0 \cdot \mathbf{r}) = \exp(ik_0 r \cos\theta) = \sum_{l=0}^{\infty} A_l u_l(r) P_l(\cos\theta). \tag{9-90}$$

This expression is called the *partial wave expansion* in the function $u_l(r)$ where A_l is a constant and $P_l(\cos\theta)$ is the unnormalized Legendre polynomial (see Appendix

C.1). Substituting Eq.(9-90) into Eq.(9-66) when $V(\mathbf{r}) = 0$ gives

$$\left(\frac{\hbar^2}{2M}\nabla^2 + E\right)\sum_{l=0}^{\infty} A_l\, u_l(r) P_l(\cos\theta) = 0. \tag{9-91}$$

We can now develop a differential equation for $u_l(r)$. Using $E = p_0^2/2M = \hbar^2 k_0^2/2M$ gives

$$(\nabla^2 + k_0^2)\sum_{l=0}^{\infty} A_l\, u_l(r) P_l(\cos\theta) = 0. \tag{9-92}$$

In Chapter 4 we showed in detail that

$$\nabla^2 = \frac{1}{r^2}\left[\frac{\partial}{\partial r}\left(r^2\frac{\partial}{\partial r}\right) - \frac{L^2}{\hbar^2}\right] \quad \text{and} \quad \frac{L^2}{\hbar^2} P_l(\cos\theta) = l(l+1) P_l(\cos\theta)$$

where L^2 is the square of the total angular momentum operator for the incident particle–target particle system and $l = 0, 1, 2, 3, \ldots$. Making these substitutions gives

$$\sum_{l=0}^{\infty} A_l\left[\frac{1}{r^2}\frac{d}{dr}\left(r^2\frac{d}{dr}\right) - \frac{L^2}{\hbar^2 r^2} + k_0^2\right] u_l(r) P_l(\cos\theta)$$

$$= \sum_{l=0}^{\infty} A_l\, P_l(\cos\theta)\left\{\frac{1}{r}\left[r\frac{d^2u_l(r)}{dr^2} + 2\frac{du_l(r)}{dr}\right] - \frac{l(l+1)u_l(r)}{r^2} + k_0^2 u_l(r)\right\}. \tag{9-93}$$

We can simplify this equation by substituting

$$g_l(r) = r u_l(r), \tag{9-94}$$

where

$$\frac{dg_l(r)}{dr} = u_l(r) + r\frac{du_l(r)}{dr}, \qquad \frac{d^2 g_l(r)}{dr^2} = 2\frac{du_l(r)}{dr} + r\frac{d^2u_l(r)}{dr^2}. \tag{9-95}$$

Substituting into Eq.(9-93) gives

$$\sum_{l=0}^{\infty} A_l P_l(\cos\theta)\left[\frac{d^2}{dr^2} - \frac{l(l+1)}{r^2} + k_0^2\right] g_l(r) = 0. \tag{9-96}$$

Multiplying this equation by $P_{l'}(\cos\theta)$ and integrating over θ,

$$\int_0^{\pi} P_l(\cos\theta) P_{l'}(\cos\theta)\sin\theta\, d\theta = \left(\frac{2}{2l+1}\right)\delta_{ll'}$$

from Eq.(C-14) in Appendix C.1, gives a set of differential equations for $g_l(r)$:

$$\left[\frac{d^2}{dr^2} - \frac{l(l+1)}{r^2} + k_0^2\right] g_l(r) = 0. \tag{9-97}$$

This differential equation is one of the forms of Bessel's differential equation having a general solution of [16]

$$g_l(r) = (k_0 r)^{1/2}[a_l J_{l+1/2}(k_0 r) + b_l J_{-l-1/2}(k_0 r)]. \tag{9-98}$$

a_l and b_l are arbitrary constants and J_l are the cylindrical Bessel functions given by

$$J_l(x) = \sum_{m=0}^{\infty} \frac{(-1)^m (x/2)^{l+2m}}{m!\Gamma(m+l+1)}, \tag{9-99}$$

where $\Gamma(m + l + 1)$ is the gamma function with argument $(m + l + 1)$. According to Eqs.(9-90) and (9-94), we require $g_l(0) = 0$ and, therefore, we must force $b_l = 0$ in Eq. (9-98), giving the following satisfactory solution to Eq.(9-97):

$$g_l(r) = (k_0 r)^{1/2} a_l J_{l+1/2}(k_0 r) = a_l \left(\frac{2}{\pi}\right)^{1/2} k_0 r j_l(k_0 r), \tag{9-100}$$

where we have also defined the spherical Bessel functions, $j_l(k_0 r)$, in terms of $J_{l+1/2}(k_0 r)$ [16],

$$j_l(k_0 r) = \left(\frac{\pi}{2k_0 r}\right)^{1/2} J_{l+1/2}(k_0 r), \qquad j_0(k_0 r) = \frac{\sin k_0 r}{k_0 r},$$

$$j_1(k_0 r) = \frac{\sin k_0 r}{(k_0 r)^2} - \frac{\cos k_0 r}{k_0 r}, \qquad j_2(k_0 r) = \frac{3}{k_0 r} j_1(k_0 r) - j_0(k_0 r). \tag{9-101}$$

Substituting Eqs.(9-100) and (9-94) into Eq.(9-90) gives

$$\psi(r, \theta) = \exp(ik_0 r \cos\theta) = \sum_{l=0}^{\infty} (A_l a_l) \left(\frac{2}{\pi}\right)^{1/2} k_0 j_l(k_0 r) P_l(\cos\theta), \tag{9-102}$$

where the $A_l a_l$ constant must still be determined. Multiplying both sides of this equation by $P_{l'}(\cos\theta)$ and integrating over θ gives

$$\frac{2l+1}{2} \int_0^{\pi} \exp(ik_0 r \cos\theta) P_l(\cos\theta) \sin\theta \, d\theta = (A_l a_l) \left(\frac{2}{\pi}\right)^{1/2} k_0 j_l(k_0 r). \tag{9-103}$$

We now integrate the left-hand side for several values of l and use also Eq.(9-101) for the right-hand side to show that the general form for the constant is $(A_l a_l) = (1/k_0)(\pi/2)^{1/2}(2l + 1)i^l$, giving, for Eq.(9-102), the final result:

$$\psi(r, \theta) = \exp(ik_0 r \cos\theta) = \sum_{l=0}^{\infty} (2l + 1) i^l j_l(k_0 r) P_l(\cos\theta). \tag{9-104}$$

This is the commonly used expansion of a plane wave in terms of spherical waves. The $P_l(\cos\theta)$ dependence in Eq.(9-104) represents part of the plane wave with orbital angular momentum $\hbar[l(l + 1)]^{1/2}$ with respect to the center of force. Spherical waves of all l will be present in the plane wave. In analogy with the nomenclature for the hydrogen atom orbitals, we refer to the terms (called *partial waves*) with $l = 0, 1, 2, \ldots$ as *s*, *p*, *d*, ... waves

$l = 0$, *s-wave*

$$\psi_{l=0} = j_0(k_0 r) P_0(\cos\theta) = \frac{\sin k_0 r}{k_0 r} = \frac{1}{2ik_0 r}[\exp(ik_0 r) - \exp(-ik_0 r)]. \tag{9-105}$$

The free particle *s*-wave is a sum of ingoing and outgoing radial waves.

$l = 1$, *p-wave*

$$\psi_{l=1} = j_1(k_0 r)P_1(\cos\theta) = \frac{1}{k_0 r}\left(\frac{\sin k_0 r}{k_0 r} - \cos k_0 r\right)\cos\theta. \tag{9-106}$$

This function has a maximum near $k_0 r \cong 1.0$. Thus, this function describes particles with orbital angular momentum approximately equal to $\hbar\{l = 1$ giving $\hbar[l(l+1)]^{1/2} \cong \hbar\}$, which passes probably no closer than $r_0 = 1/k_0 = \lambda/2\pi$ from the origin. Thus, the $l = 1$ incoming wave avoids the origin and the $l = 0$ incoming wave goes through the origin.

Equation (9-104) will be a useful form in describing the wavefunction for the incident particles. To establish the form of the asymptotic limit where $r \longrightarrow \infty$, we note that the cylindrical Bessel functions in the solution in Eq.(9-98) have asymptotic forms given by [16]

$$\lim_{r\to\infty} J_{l+1/2}(k_0 r) = \left(\frac{2}{\pi k_0 r}\right)^{1/2} \sin\left(k_0 r - \frac{l\pi}{2}\right)$$

$$\lim_{r\to\infty} J_{-(l+1/2)}(k_0 r) = \left(\frac{2}{\pi k_0 r}\right)^{1/2} (-1)^l \cos\left(k_0 r - \frac{l\pi}{2}\right), \tag{9-107}$$

which implies

$$\lim_{r\to\infty} j_l(k_0 r) = \frac{1}{k_0 r}\sin\left(k_0 r - \frac{l\pi}{2}\right). \tag{9-108}$$

Thus, the asymptotic solution for $g_l(r)$ in Eq.(9-98), in the case where $V(\mathbf{r}) = 0$ and $b_l = 0$, is given by

$$\hat{g}_l(r) = \lim_{r\to\infty} g_l(r) = a_l\left(\frac{2}{\pi}\right)^{1/2} \sin\left(k_0 r - \frac{l\pi}{2}\right), \tag{9-109}$$

compared to our earlier result in Eq.(9-100). If we now use this function for $\hat{g}_l(r)$ and Eqs.(9-94) and (9-90) along with the methods given from Eqs.(9-102) to (9-104), we obtain the wavefunction in the asymptotic limit:

$$\lim_{r\to\infty} \psi(r, \theta) = \exp(ik_0 r\cos\theta) = \frac{1}{k_0 r}\sum_{l=0}^{\infty}(2l+1)i^l \sin\left(k_0 r - \frac{l\pi}{2}\right)P_l(\cos\theta). \tag{9-110}$$

Now, we return to the solution of the scattering Schrödinger equation in the presence of a spherically symmetric potential, $V(r)$. Substituting Eqs.(9-90) into Eq.(9-66) and repeating the previous development leads to

$$\left[\frac{d^2}{dr^2} - \frac{l(l+1)}{r^2} + k_0^2 - U(r)\right]f_l(r) = 0, \qquad U(r) = \frac{2M}{\hbar^2}V(r), \tag{9-111}$$

which reduces to Eq.(9-97) when $V(r) = 0$. The solution to this equation will have the same form as Eq.(9-98) with the asymptotic forms in Eqs.(9-107). The resultant new function $f_l(r)$ in the asymptotic limit in the presence of $V(r)$ is

$$\hat{f}_l(r) = \lim_{r\to\infty} f_l(r) = \left(\frac{2}{\pi}\right)^{1/2}\left[a_l\sin\left(k_0 r - \frac{l\pi}{2}\right) + b_l(-1)^l\cos\left(k_0 r - \frac{l\pi}{2}\right)\right], \tag{9-112}$$

which can be written in terms of the simple phase-shifted sinusoidal function to give

$$\hat{f}_l(r) = B_l\left(\frac{2}{\pi}\right)^{1/2} \sin\left(k_0 r - \frac{l\pi}{2} + \eta_l\right), \qquad \tan \eta_l = (-1)^l\left(\frac{b_l}{a_l}\right), \qquad B_l = \frac{a_l}{\cos \eta_l}. \tag{9-113}$$

η_l is the phase shift that, when compared to Eq.(9-109), is seen to result from the presence of the potential energy, $V(r)$. Trigonometric identities show the equivalence of $\hat{f}_l(r)$ in Eqs.(9-112) and (9-113). In the limit of very small effects due to $V(r)$, $\cos \eta_l \cong 1.0$ and Eq. (9-113) for $\hat{f}_l(r)$ reduces to a phase-shifted form of $\hat{g}_l(r)$ in Eq. (9-109). The sum of these arguments leads to an expression for the asymptotic form for the wavefunction in the presence of $V(r)$ that describes the scattering. The result, analogous to Eq.(9-110), is

$$\lim_{r\to\infty} \psi(r, \theta) = \frac{1}{k_0 r}\sum_{l=0}^{\infty} B_l(2l + 1)i^l \sin\left(k_0 r - \frac{l\pi}{2} + \eta_l\right)P_l(\cos \theta), \tag{9-114}$$

where we note that $B_l \longrightarrow 1$ as the phase shift goes to zero [$\cos \eta_l \longrightarrow 1$ in Eq.(9-113)]. We can now compare the exact plane wave expansion for a free particle into spherical waves in Eq.(9-104), the asymptotic form for the free particle in Eq.(9-110), and the asymptotic form for the particle (in the presence of a scattering potential) in Eq. (9-114). Thus, the phase shift has a clear meaning; it is the difference between the asymptotic solutions to Eqs.(9-97), where $V(r) = 0$, and the asymptotic solution to Eq.(9-111), where $V(r) \neq 0$.

We shall now relate the phase shift described here back to the scattered wave analysis in Section 9.3. According to our development above, Eq.(9-114) should be equal to Eq.(9-78) where Eq.(9-110) is used for the unperturbed plane wave giving

$$\begin{aligned}\psi(r, \theta) &= \psi^0(r, \theta) + \psi^s(r, \theta) = \exp(ik_0 r \cos \theta) + f(\theta)\frac{\exp(ik_0 r)}{r} \\ &= \frac{1}{k_0 r}\sum_{l=0}^{\infty} (2l + 1)i^l \sin\left(k_0 r - \frac{l\pi}{2}\right)P_l(\cos \theta) + f(\theta)\frac{\exp(ik_0 r)}{r} \\ &= \frac{1}{k_0 r}\sum_{l=0}^{\infty} B_l(2l + 1)i^l \sin\left(k_0 r - \frac{l\pi}{2} + \eta_l\right)P_l(\cos \theta).\end{aligned} \tag{9-115}$$

From Eq.(9-114), we shall now expand these expressions, collect coefficients, and find an expression for the scattering amplitude $f(\theta)$ in terms of the phase shift, η_l. Rewriting Eq.(9-115) gives

$$\begin{aligned}f(\theta) &= \frac{\exp(-ik_0 r)}{k_0}\sum_{l=0}^{\infty} (2l + 1)i^l P_l(\cos \theta) \\ &\times\left[B_l \sin\left(k_0 r - \frac{l\pi}{2} + \eta_l\right) - \sin\left(k_0 r - \frac{l\pi}{2}\right)\right].\end{aligned} \tag{9-116}$$

Writing the sine functions in exponential form gives

$$f(\theta) = \frac{1}{2ik_0} \sum_{l=0}^{\infty} (2l + 1)i^l P_l(\cos\theta)\left\{B_l\left[\exp\left(-\frac{il\pi}{2} + i\eta_l\right) - \exp\left(-2ik_0 r + \frac{il\pi}{2} - i\eta_l\right)\right] - \left[\exp\left(-\frac{il\pi}{2}\right) - \exp\left(-2ik_0 r + \frac{il\pi}{2}\right)\right]\right\}. \quad (9\text{-}117)$$

As $f(\theta)$ must be independent of r for all θ, the terms containing r in this expression must vanish. Collecting the r-dependent terms gives

$$0 = \sum_{l=0}^{\infty} (2l + 1)i^l P_l(\cos\theta) \exp\left(-2ik_0 r + \frac{il\pi}{2}\right)[B_l \exp(-i\eta_l) - 1]. \quad (9\text{-}118)$$

This expression is satisfied if

$$B_l = \exp(i\eta_l). \quad (9\text{-}119)$$

Substituting this result back into Eq.(9-117) to cancel the r-dependent terms gives

$$f(\theta) = \frac{1}{2ik_0} \sum_{l=0}^{\infty} (2l + 1)i^l P_l(\cos\theta) \exp\left(-\frac{il\pi}{2}\right)[\exp(2i\eta_l) - 1].$$

Using the identity

$$\exp\left(-\frac{il\pi}{2}\right) = \cos\frac{l\pi}{2} - i\sin\frac{l\pi}{2} = (-i)^l \quad (9\text{-}120)$$

gives the final result for the scattering amplitude in terms of the phase shift, η_l:

$$f(\theta) = \frac{1}{2ik_0} \sum_{l=0}^{\infty} (2l + 1)[\exp(2i\eta_l) - 1]P_l(\cos\theta) = \sum_{l=0}^{\infty} f_l(\theta), \quad (9\text{-}121)$$

where we have written the scattering amplitude as a sum over $l = 0 \longrightarrow \infty$ terms. Substituting Eq.(9-119) into Eq.(9-114) gives the scattered wavefunction:

$$\psi(r, \theta) = \frac{1}{k_0 r} \sum_{l=0}^{\infty} (2l + 1)i^l \exp(i\eta_l) \sin\left(k_0 r - \frac{l\pi}{2} + \eta_l\right)P_l(\cos\theta). \quad (9\text{-}122)$$

We are now able to relate the phase shift, η_l, directly to the interaction potential, $V(r)$. First we rewrite Eq.(9-113) using Eq.(9-119) for B_l giving

$$\hat{f}_l(r) = \left(\frac{2}{\pi}\right)^{1/2} \exp(i\eta_l) \sin\left(k_0 r - \frac{l\pi}{2} + \eta_l\right), \quad (9\text{-}123)$$

which is the asymptotic solution to Eq.(9-111). The $\eta_l \longrightarrow 0$ limit of this expression must equal $\hat{g}_l(r)$ in Eq.(9-109), which is the asymptotic solution of Eq.(9-97). Thus, $a_l = 1$ in Eq.(9-109) giving

$$\hat{g}_l(r) = \left(\frac{2}{\pi}\right)^{1/2} \sin\left(k_0 r - \frac{l\pi}{2}\right). \quad (9\text{-}124)$$

In order to relate $V(r)$ to η_l we postmultiply Eq.(9-97) by $f_l(r)$, postmultiply Eq. (9-111) by $g_l(r)$, subtract the first expression from the second, and integrate over r from $0 \longrightarrow \infty$ to give

$$g_l(r)\frac{df_l(r)}{dr}\bigg|_0^\infty - f_l(r)\frac{dg_l(r)}{dr}\bigg|_0^\infty = \int_0^\infty g_l(r)U(r)f_l(r)\,dr. \tag{9-125}$$

Using $g_l(0) = f_l(0) = 0$ from Eqs.(9-97) and (9-111) we now need $g_l(\infty)$ and $f_l(\infty)$ or the asymptotic limits of $g_l(r)$ and $f_l(r)$. Substituting the asymptotic limits of $f_l(r)$ and $g_l(r)$ from Eqs.(9-123) and (9-124) into the left-hand side of Eq.(9-125) gives

$$\sin \eta_l = -\frac{1}{k_0}\int_0^\infty g_l(r)U(r)f_l(r)\,dr, \tag{9-126}$$

which is the direct exact expression relating the phase shift, η_l, to the interaction potential energy. Initially it would appear that this expression would not be very useful as $f_l(r)$ must be known before computing $\sin \eta_l$. We can make what is equivalent to a Born approximation, however, and replace $f_l(r)$ with $g_l(r)$ in this expression giving

$$\sin \eta_l \cong -\frac{1}{k_0}\int_0^\infty g_l^2(r)U(r)\,dr. \tag{9-127}$$

It is now evident that $[U(r) = (2M/\hbar^2)V(r)]$

$$\begin{aligned} &V(r) \text{ repulsive (positive)} \qquad \eta_l < 0 \\ &V(r) \text{ attractive (negative)} \qquad \eta_l > 0. \end{aligned} \tag{9-128}$$

Figure 9-8 demonstrates the repulsive and attractive potential energy phase shifts in the sinusoidal spherical wave.

The phase shifts can be used to express directly the differential and total cross-sections. According to Eq.(9-81), the differential cross-section, for a spherically symmetric potential, is given by

$$\sigma(\theta) = |f(\theta)|^2, \tag{9-129}$$

where $|f(\theta)|^2$ is the sum of the squares of the real and imaginary parts of $f(\theta)$ in Eq.(9-121) given by

$$\begin{aligned} f(\theta) &= \sum_{l=0}^{\infty}(2l+1)P_l(\cos\theta)\left(\frac{\cos 2\eta_l - 1}{2ik_0} + \frac{\sin 2\eta_l}{2k_0}\right) \\ &= \sum_{l=0}^{\infty}(2l+1)P_l(\cos\theta)\left(\frac{\sin 2\eta_l}{2k_0}\right) + i\sum_{l=0}^{\infty}(2l+1)P_l(\cos\theta)\left(\frac{1-\cos 2\eta_l}{2k_0}\right) \\ &= f_r(\theta) + if_i(\theta). \end{aligned} \tag{9-130}$$

$f_r(\theta)$ and $f_i(\theta)$ are the real and imaginary parts of $f(\theta)$, respectively. Thus, the differential cross-section is given by

$$\sigma(\theta) = [f_r(\theta)]^2 + [f_i(\theta)]^2. \tag{9-131}$$

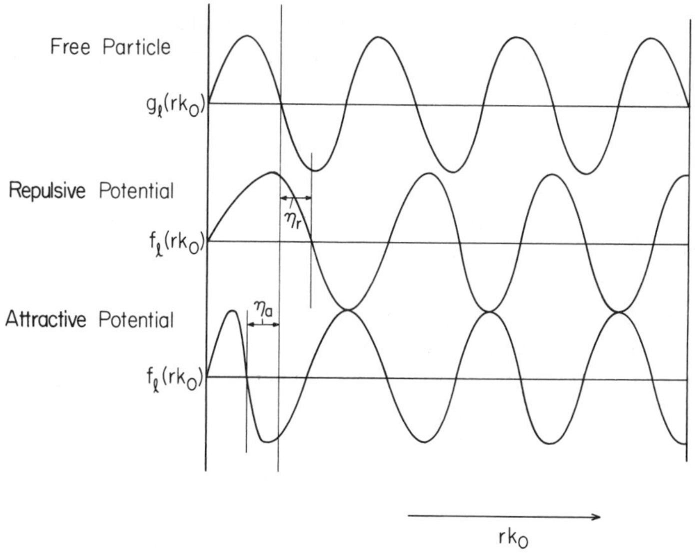

Figure 9-8 Illustration of the phase shift between the wave equation for a free particle and the particle in the presence of a scattering potential, $V(r)$. η_a represents the positive phase shift for an attractive potential and η_r represents the negative phase shift for a repulsive potential. Phase shifts of $\eta_a - \eta_r = 90°$ are shown for convenience in viewing.

The total cross-section is also easily obtained from

$$\sigma = \int_0^{2\pi} \int_0^{\pi} |f(\theta)|^2 \sin\theta \, d\theta \, d\phi. \tag{9-132}$$

Substituting $f(\theta)$ from Eq.(9-121) gives

$$\sigma = \frac{2\pi}{4k_0^2} \sum_{l=0}^{\infty} \sum_{l'=0}^{\infty} (2l+1)(2l'+1)[\exp(2i\eta_l) - 1][\exp(-2i\eta_{l'}) - 1]$$

$$\times \int_0^{\pi} P_l(\cos\theta) P_{l'}(\cos\theta) \sin\theta \, d\theta$$

$$= \frac{2\pi}{k_0^2} \sum_{l=0}^{\infty} (2l+1)(1 - \cos 2\eta_l) = \frac{4\pi}{k_0^2} \sum_{l=0}^{\infty} (2l+1) \sin^2 \eta_l. \tag{9-133}$$

Comparing this expression with Eq.(9-130) shows that the imaginary part of $f(\theta = 0)$ is also related to the total cross-section [$P_l(\cos\theta) = 1$ when $\theta = 0$]. Direct comparison shows that

$$\sigma = \frac{4\pi}{k_0} [f_i(\theta = 0)]. \tag{9-134}$$

The result in Eq.(9-134) is called the *optical theorem*.

It is easy to check the phase shift formalism by the low energy hard sphere collision considered in Section 9.1 beginning with Eqs.(9-29) describing the hard sphere potential,

$$V(r) = 0; \qquad r > a$$
$$V(r) = \infty; \qquad r \leq a, \tag{9-135}$$

where a is the hard sphere distance between the centers of the spherical stationary target particle and incident particle. In the present case, the incident particle is described by a plane wave and is not localized in position (see discussion at the beginning of Section 9.3). This interpretation leads to a slightly different definition for the hard sphere potential. According to Eq.(9-135), $f_l(r)$ must equal zero when $r \leq a$ and $f_l(r)$ is given in Eq.(9-123) when $r > a$. $f_l(r)$ must be continuous through the $r = a$ boundary giving

$$f_l(r \leq a) = 0 = f_l(r = a) = \left(\frac{2}{\pi}\right)^{1/2} \exp(i\eta_l) \sin\left(k_0 a - \frac{l\pi}{2} + \eta_l\right)$$
$$= a_l\left(\frac{2}{\pi}\right)^{1/2}\left[\sin\left(k_0 a - \frac{l\pi}{2}\right) + \tan\eta_l \cos\left(k_0 a - \frac{l\pi}{2}\right)\right], \tag{9-136}$$

where we have also used Eqs.(9-113) and (9-112). Rewriting this equation gives

$$\tan\eta_l = -\frac{\sin[k_0 a - (l\pi/2)]}{\cos[k_0 a - (l\pi/2)]} = -\tan\left(k_0 a - \frac{l\pi}{2}\right). \tag{9-137}$$

Now, for low energy scattering, only s-waves will scatter and $k_0 r \ll 1$. Using this gives

$$\tan\eta_{l=0} = \tan\eta_0 = -\tan k_0 a \cong -\sin k_0 a \cong -k_0 a \cong \sin\eta_0 \cong \eta_0. \tag{9-138}$$

Thus, according to Eq.(9-133), the total scattering cross-section for the s-wave is given by

$$\sigma = \sigma_{l=0} = \frac{4\pi}{k_0^2}\sin^2 k_0 a = 4\pi a^2, \tag{9-139}$$

which is four times the classical result in Eq.(9-32) as $r_1 + r_2 = a$. We can also obtain this same result from Eqs.(9-134) and (9-130). From Eq.(9-130) when $l = 0$, we have (for small phase shifts η_0)

$$f_i(\theta = 0) = \frac{1 - \cos 2\eta_0}{2k_0} = \frac{\sin^2\eta_0}{k_0} \cong \frac{\sin^2 k_0 a}{k_0} \cong a^2 k_0, \tag{9-140}$$

where the last steps use Eq.(9-138). Substituting Eq.(9-140) into Eq.(9-134) gives the results in Eq.(9-139). In this limit where $k_0 \ll a$, the wave appears to scatter off the surface of a sphere with area $4\pi a^2$ as opposed to a classical collision where the scattering is proportional to the area of a circle, πa^2.

We now have the formal theory needed to extract the potential function for the interaction between two particles from the experimental differential scattering cross-section. The method is to solve Eq.(9-111), usually numerically, for a trial potential function, $V(r)$, for a series of l-values giving $f_l(r)$. This $f_l(r)$ can then be compared to

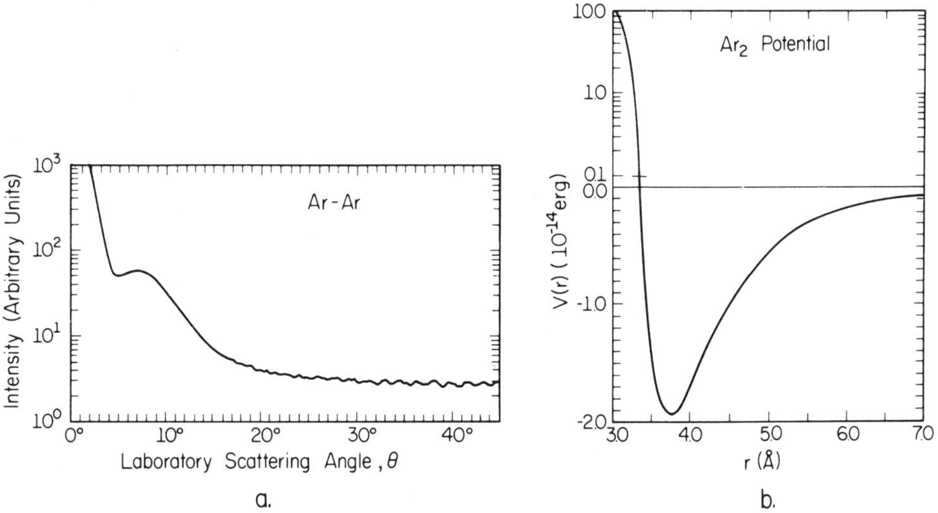

Figure 9-9 The experimental differential scattering cross-section and resulting potential function for the Ar-Ar system adapted from J. M. Parson, P. E. Siska, and Y. T. Lee, *J. Chem. Phys.* **56**, 1511 (1972). a. The scattering cross-section is shown in the laboratory axis system. For like-atom scattering, however, the center of mass results are obtained by multiplying the angle shown here by two. The relative kinetic energy for the two-particle system is $E = 10^{-13}$ erg. b. The potential function for Ar-Ar scattering that reproduces the differential scattering cross-section in (a) by the phase shift method.

$g_l(r)$ to give the phase shifts, η_l. These phase shifts are then used to calculate the differential scattering cross-section in Eqs.(9-131), carrying the l order to the desired accuracy (achieving satisfactory convergence). The form of the potential function is then varied and the calculation repeated until a satisfactory fit with experiment is obtained. An example of this type of analysis is shown in Fig. 9-9 where both the differential scattering cross-section and the resulting potential function is shown for Ar-Ar scattering. The maximum in the differential cross-section at the laboratory scattering angle of about 7° arises from a classical phenomenon very similar to the scattering of light from water droplets to produce a rainbow [17]; hence, the term *rainbow angle* is used to describe this phenomenon in particle scattering [18]. The rainbow angle is related by the optical analogy by the depth of the potential well, shown in Fig. 9-9. There are numerous other examples of fitting an observed differential cross-section to an appropriate potential function [7,19].

The potential functions for the like-atom rare gas atoms are shown in Fig. 9-10. The potential functions are obtained from the experimental differential cross-sections and the phase shift deconvolution procedures described above. One analytical approximation to the experimental potential energy functions shown in Fig. 9-10 is the Lennard–Jones potential, which is a long-range $1/r^6$ attractive term and a steeper $1/r^{12}$ repulsive term [5] given by

$$V(r) = 4\varepsilon\left[\left(\frac{d}{r}\right)^{12} - \left(\frac{d}{r}\right)^{6}\right]. \qquad (9\text{-}141)$$

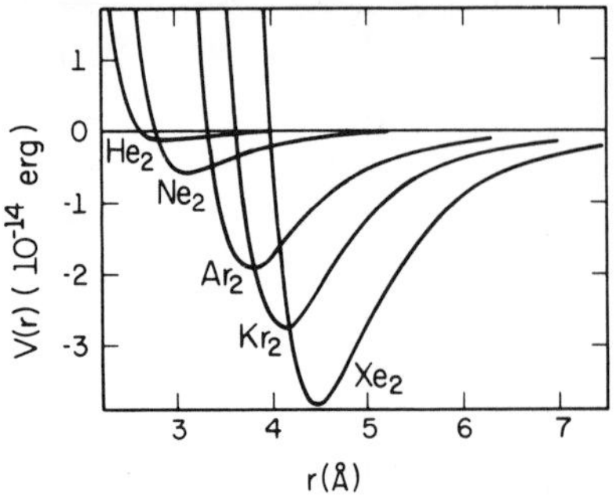

Figure 9-10 Rare gas interatomic potential functions obtained from the experimental differential cross-sections by J. M. Farrar, T. P. Schafer, and Y. T. Lee, *Transport Phenomena*, Ed. by J. Kestin, A. I. P. Conference Proceedings No. 11 (1973) as illustrated in R. C. Levine and R. B. Bernstein, *Molecular Reaction Dynamics* (Oxford University Press, New York, 1974).

The minimum in this potential is at $r_0 = (2)^{1/6}d$ and the well depth at r_0 is ε. d is the $V(r = d) = 0$ crossing point in the function. Potential constants, ε and d, for the Lennard–Jones potential for the interaction between rare gas atoms are listed in Table 9-2.

Using the concepts above, we note that the potential minimum, ε, can be lower than the thermal energy ($kT \cong 4.1 \times 10^{-14}$ erg at $T = 300$ K). Thus, at low temperatures we can expect rare gas diatomics and higher polymers to be stable. These van der Waals molecules are readily observable by adiabatically expanding a gas into a vacuum. Several van der Waals dimers and polymers of rare gases and other molecules have been observed [20].

In summary, we have developed the theory of elastic scattering and given examples of using the experimental total and differential cross-sections to extract potential functions for the interactions between atoms. In the case of atom-molecule or molecule-molecule collisions, the experimental differential cross-sections usually indicate the rainbow maximum and a dominant spherically symmetric potential function. More complicated beam experiments can be used to extract the anisotropies in the interaction potential.

Another realm of scattering, which is important even at thermal energies when molecules are involved, is the inelastic scattering process. In the case of an incident beam of atoms colliding with a molecule, the experimental differential and total cross-sections will include the inelastic processes whereby the target molecule is excited into a different electronic, rotational, or vibrational state in addition to simply being involved in an elastic event [21]. Reactions between the incoming atom and

Table 9-2 Lennard–Jones parameters to fit Eq.(9-141) for the rare gas atoms from N. Bernardes, *Phys. Rev.* **112**, 1534 (1958). See also a more recent summary in Y. S. Kim and R. G. Gordon, *J. Chem. Phys.* **61**, 1 (1974).

Atom Pair	ε (10^{-14} erg)	d (Å)
Ne—Ne	0.50	2.74
Ar—Ar	1.67	3.40
Kr—Kr	2.25	3.65
Xe—Xe	3.20	3.98

the target molecule can also give rise to inelastic scattering processes. Returning to Eq.(9-11), we now write

$$I(z) = I_0 \exp\left[-\rho_0 z(\sigma + \sigma_i)\right] = I_0 \exp\left(-\rho_0 z \sigma_T\right), \tag{9-142}$$

where σ is the previously considered elastic cross-section, σ_i contains the inelastic processes, and σ_T is the total cross-section. As an example of inelastic scattering, consider the reactive scattering for the following reaction:

$$K + HBr \longrightarrow KBr + H. \tag{9-143}$$

The experimental K atom differential cross-section for the collision of perpendicular K and HBr beams is shown in Fig. 9-11. Also shown is the corresponding experimental differential cross-section for the isoelectronic K + Kr collision system, which is, at these low energies, purely elastic. In both experiments, the K atom intensities are measured to give K atom differential cross-sections. Note that both curves in Fig. 9-10 show the characteristic rainbow scattering, which indicates the interaction potential for elastic scattering as shown in Fig. 9-9. The values of $\sigma(\theta)_T \sin\theta$ for the K + HBr collisions, however, fall off faster at high scattering angles than the corresponding K + Kr scattering. This increased falloff in $\sigma(\theta)_T \sin\theta$ in the K + HBr system indicates an increased removal of K atoms from the incident beam, due to the reaction shown in Eq.(9-143). It is also evident from these results that the inelastic reactive scattering cross-section will be considerably smaller than the elastic cross-section.

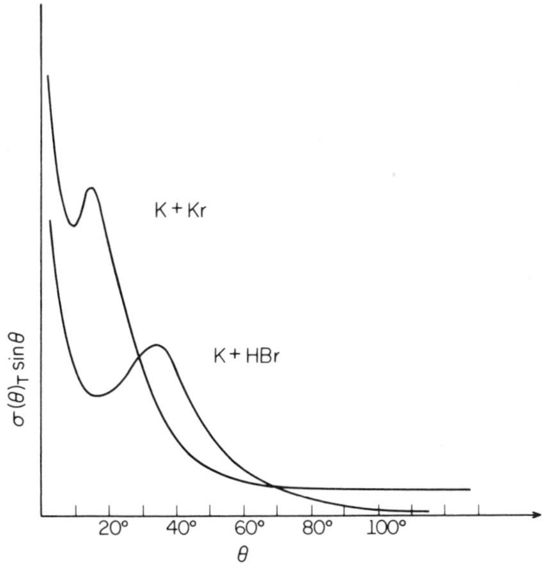

Figure 9-11 Plot of the experimental $\sigma(\theta)_T \sin\theta$ as a function of θ for K atoms for the isoelectronic K + Kr and K + HBr collisions adapted from D. Beck, *J. Chem. Phys.* **37**, 2884 (1962). The relative energies for both processes are near $E_0 = 1.5\,\text{kcal}\cdot\text{mol}^{-1}$. The ordinate is in arbitrary units, which is different for the two curves.

According to our previous discussions, the scattering at low angles will involve large impact parameters (Fig. 9-3) that will probably not excite chemical reactions. Thus, the low-angle data (below 60°) for the K + HBr process can be used to obtain the total elastic cross-section or the elastic part of $\sigma(\theta)_T \sin\theta$ in Fig. 9-11. The total reactive (inelastic) cross-section can then be obtained from the difference between the experimental, $\sigma(\theta)_T$, and elastic, $\sigma(\theta)$, cross-sections according to Eqs.(9-142) and (9-3),

$$\sigma_i = 2\pi \int_0^{\pi} [\sigma(\theta)_T - \sigma(\theta)] \sin\theta \, d\theta. \tag{9-144}$$

Of course, σ_i is still a function of E_0, the relative energy of the collision pair. It is also evident that the inelastic or, in this case, the reactive differential cross-section can be measured directly by observing the angular dependence of the intensity of the reaction product molecules [KBr, see Eq.(9-143)] [22].

The total reactive cross-section, σ_i, is related to the standard rate constant for the chemical reaction, $k(E_0)$, which is a function of the relative kinetic energies, by [23]

$$k(E_0) = v\sigma_i, \tag{9-145}$$

where v is the relative velocity and $E_0 = \frac{1}{2}\mu v^2$ of the collision pair. It is now evident that a great deal can be learned about rate processes in chemical reactions by inelastic molecular beam scattering processes [23,24].

We have now developed the scattering theory needed to extract the potential of interaction between colliding atoms from the measured differential cross-section as a function of scattering angle. In the case of reactive scattering, an additional channel or degree of freedom is open that further complicates the interpretation of the experimental data as discussed above. We shall not develop this theory to any further extent here, but we shall define the scattering matrix that is the starting point for more sophisticated treatments of both elastic and inelastic scattering [25,26,27]. We start with Eq.(9-123), which is the asymptotic solution to Eq.(9-111). Writing the sine function in Eq.(9-123) in exponential form gives

$$\begin{aligned}
\hat{f}_l(r) &= \left(\frac{2}{\pi}\right)^{1/2} \exp(i\eta_l)\left(\frac{1}{2i}\right)\left[\exp\left(ik_r - i\frac{l\pi}{2} + i\eta_l\right) - \exp\left(-ik_r + i\frac{l\pi}{2} - i\eta_l\right)\right] \\
&= \left(\frac{2}{\pi}\right)^{1/2}\left(\frac{1}{2i}\right)\left[\exp\left(ik_r - i\frac{l\pi}{2} + i2\eta_l\right) - \exp\left(-ik_r + i\frac{l\pi}{2}\right)\right] \\
&= \left(\frac{2}{\pi}\right)^{1/2}\left(\frac{1}{2i}\right)\left[S_l \exp\left(ik_r - i\frac{l\pi}{2}\right) - \exp\left(-ik_r + i\frac{l\pi}{2}\right)\right] \\
S_l &= \exp(i2\eta_l).
\end{aligned} \tag{9-146}$$

In this expression, the first term describes an outgoing spherical wave, exp $[ik_r - i(l\pi/2)]$, multiplied by $S_l = \exp(i2\eta_l)$, which is the diagonal element in the scattering matrix for this elastic process. The second term in this expression is an incoming spherical wave that is unperturbed. Thus, the scattering matrix formalism is related to the phase shift development presented earlier in this section. The scattering matrix formalism is much more generally useful, however, for treating both elastic and inelastic scattering processes [28].

9.5 ELASTIC ELECTRON SCATTERING FROM ATOMS

A discussion of the determination of molecular structures from the analysis of elastic electron scattering data arises naturally from the development given in the last two sections. Electron diffraction studies normally employ electrons in the 1- to 100-keV energy range and a static gas of atoms or molecules. This energy range assures minor relativistic corrections ($mc^2 = 511$ keV) in the scattering analysis. A typical electron diffraction apparatus is arranged geometrically, as shown in Fig. 9-1, with the incident electrons along the z axis and the scattering region (target) is produced by an expanding gas through a nozzle directed along the x or y axis. The atomic or molecular velocity of the target molecules along the x or y axis will be thermal or about 10^4–10^5 $\text{cm} \cdot \text{s}^{-1}$ depending on the mass. The nonrelativistic velocity of the 1-keV electron along the z axis is 1.875×10^9 $\text{cm} \cdot \text{s}^{-1}$. Thus, the atomic or molecular target gas is essentially stationary relative to the high velocity electrons and no correction from the laboratory or observed scattering data to the center of mass system is necessary.

The nonrelativistic de Broglie wavelength ($\lambda = h/p$) for 1-keV electrons is $\lambda = 0.388$ Å and the wavelengths are shorter for more energetic electrons. Interference effects will be observable in molecules when these small wavelength electrons are used as incident particles. Many of the interference phenomena discussed in Sections 1.2 and 8.5 for light and X rays also carry over to electrons whenever the $\lambda = h/p$ wavelength of the electron approaches the dimensions of correlated scatterers (such as atoms in molecules).

We shall assume in the following analysis that only single scattering events are important. This requirement forces the use of a relatively low-pressure target gas due to the intense nature of Coulomb scattering. Finally, we shall also assume negligible inelastic scattering in our analysis.

Reviewing again the experimental arrangement for electron scattering shown in Fig. 9-1 and reviewing Eqs.(9-76) through (9-80), we write the intensity of the scattered electrons as

$$I(R, \theta, \phi) = v_0 |\psi^s(\boldsymbol{r})|^2 = \frac{v_0}{R^2} |f(\theta, \phi)|^2 = \frac{v_0}{R^2}\,\sigma(\theta, \phi). \qquad (9\text{-}147)$$

$\psi^s(\boldsymbol{r})$ is the wavefunction for the scattered electrons and v_0 is the velocity of the incident and scattered electrons in the coherent elastic scattering considered here.

First we shall consider atomic scattering where the atomic nucleus is the origin of the coordinate system and r is the nucleus-incident electron distance. R in Eq. (9-147) is the scattering region-detector distance as shown in Fig. 9-1. According to Eq.(9-147) the intensity of scattered electrons on the detector screen (Fig. 9-1) can be predicted from a knowledge of the scattering factor, $f(\theta)$, or its square, the differential cross-section, $\sigma(\theta)$, where we have dropped the ϕ as there is no ϕ-dependence for symmetric scattering about the z axis.

The scattering of electrons by argon atoms has been analyzed in detail by the partial wave analysis described in Section 9.4 and the results are listed in Table 9-3

Table 9-3 Phase shifts calculated for the e^--Ar scattering for 40-keV electrons using the interaction potential in Eq.(9-148). The results are from Ref. [29].

	η_l (rad)	
l	First Born Approximation[a]	Numerical Interaction[b]
0	1.2840	1.2898
1	0.9379	0.9486
2	0.7691	0.7754
3	0.6598	0.6644
4	0.5801	0.582
5	0.5182	0.519
6	0.4681	0.468

[a] Calculated from Eq.(9-127) for small phase shifts where $\sin \eta_l \cong \eta_l$.

[b] Obtained by numerically integrating Eqs.(9-97) and (9-111) and comparing the phases of the numerical results.

[29]. The first column under η_l contains the calculated phase shifts for e^--Ar scattering using the Born approximation in Eq.(9-127) for several orders, l. The potential energy for the e^--Ar interaction, $V(r)$, used in this calculation is given by [30]

$$U(r) = \frac{2m}{\hbar^2} V(r)$$

$$V(r) = -\frac{\hbar^2}{2m}\left(\frac{2Z}{a_0 r}\right)[0.50529 \exp(-2.68764r) + 0.43447 \exp(-9.06392r) + 0.06071 \exp(-46.4985r)]. \tag{9-148}$$

m is the electron mass, Z is the Ar atom atomic number, and a_0 is the Bohr radius. The potential energy function for the e^--Ar interaction was obtained by solving Poisson's equation [Eq.(1-14)] relating the incident electron's electrostatic potential $\varphi(r)$ to the charge density of the atom, $\rho(r, \theta, \phi)$, to give (cgs units)

$$\nabla^2 \varphi(r) = \nabla^2\left[\frac{V(r)}{e}\right] = -4\pi\rho(r, \theta, \phi). \tag{9-149}$$

$\rho(r, \theta, \phi)$, in this case, is obtained from a many-electron wavefunction for the Ar atom, $\rho(r, \theta, \phi) = |\psi|^2$ as shown, for instance, in Eq.(8-135). The second column under η_l in Table 9-3 is obtained by using the potential function in Eq.(9-148) in Eq.(9-111) and numerically integrating to give a numerical result. The phase of this result is then compared with the numerical integration of Eq.(9-97). The numerical phase shift is listed in the second column of Table 9-3.

It is evident from the results in Table 9-3 that the first Born approximation gives quite good results for the phase shifts when compared to the more accurate numerical results. The calculated phase shifts can now be used in Eq.(9-130) to calculate the electron-atomic scattering factor, $f(\theta)$. $f(\theta)$ can then be used with Eq.(9-147) to evaluate the intensity of the electrons scattered from the Ar atom. The values of $|f(\theta)|$ calculated from the phase shifts described in Table 9-3 are listed in Table 9-4 in the first two columns under $|f(\theta)|$. The value of $|f(\theta)|$ calculated with the first Born approximation in Eq.(9-84) using $V(r)$ in Eq.(9-148) is listed in the third column under $|f(\theta)|$. The comparison between the values of $|f(\theta)|$ calculated by these three methods is quite good. These results are also in good agreement with the experimental electron-Ar scattering factors as obtained from the intensities of the scattered electrons [31]. Electron scattering factors, $f(\theta)$, for neutral atoms from $Z = 1$ to $Z = 54$ have been calculated by the phase shift method and the results are tabulated for electron energies of 10, 40, 70, and 100 keV [32].

We shall now use the first Born approximation to write the electron-atom scattering factor, $f(\theta)$, in terms of a sum of electron-nuclear scattering and electron-electron scattering terms. This development will also relate the electron scattering

Table 9-4 Electron-Ar scattering factors, $|f(\theta)|$, for 40-keV electrons. The relationship between the electron-Ar and X-ray–Ar scattering factors as generally expressed in Eq.(9-155) is also demonstrated. The first three columns under $|f(\theta)|$ are from the electron-atom potential in Eq.(9-148) and the results are from Ref. [29]. The last column uses Eq.(9-155), which is a model assuming that there is no electron-atom perturbation during the scattering process.

			$|f(\theta)|$ (10^{-8} cm^{-1})			
θ (deg)	$K = [(4\pi/\lambda)\sin(\theta/2)](10^8\ \text{cm}^{-1})$	f_{Ar}[a]	Born Phase Shifts[b]	Numerical Integration Phase Shifts[c]	First Born Approximation[d]	Eq.(9-155)
0	0	18.0	5.418	5.417	5.521	4.595[e]
3	5.47	8.2	1.238	1.237	1.285	1.237
6	10.93	5.1	0.434	0.434	0.453	0.408
9	16.38	2.0	0.219	0.219	0.227	0.225
12	21.83	1.5	0.131	0.131	0.135	0.131
15	27.26	1.2	0.0871	0.0875	0.0896	0.0870
18	32.67	0.9	0.0621	0.0627	0.0636	0.0630
21	38.06	0.6	0.0466	0.0472	0.0475	0.0472
24	43.42	—	0.0363	0.0370	0.0369	0.0361
27	48.75	—	0.0291	0.0298	0.0295	0.0286
30	54.05	—	0.0239	0.0245	0.0241	0.0233

[a] Argon atom X-ray scattering factors from Table 8-7.

[b] Equation (9-130) using phases as described in Table 9-3 up to order 125.

[c] Equation (9-130) using phases as described in Table 9-3 by numerical integration.

[d] Equation (9-84) and $V(r)$ from Eq.(9-148).

[e] The $\theta = 0$ result is from Eq.(9-158) and $\langle r^2 \rangle = 7.29 \times 10^{-16}\ \text{cm}^2$ from a convenient table of $\langle r^2 \rangle$ for atoms in G. Malli and S. Fraga, *Theor. Chim. Acta* **5**, 284 (1966).

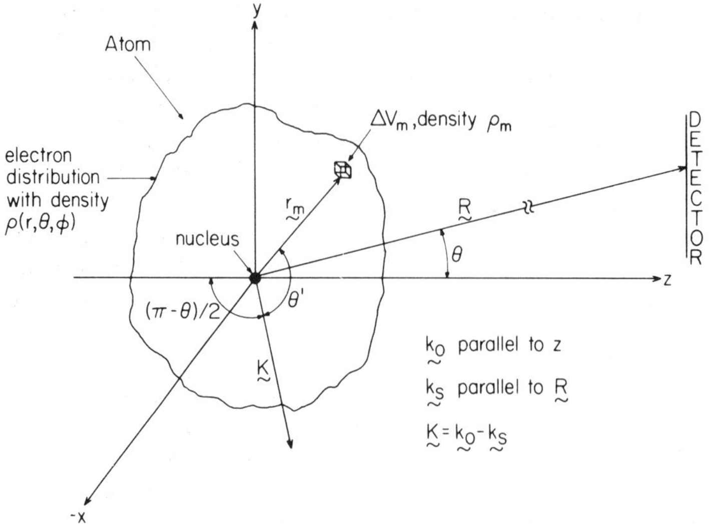

Figure 9-12 Scattering diagram for electrons scattering from an atom. The incident electrons are directed along the positive z axis. This figure has the same geometry as in Fig. 9-1 with θ as the scattering angle and θ' as the angle between the laboratory-fixed $\boldsymbol{K}$ vector and atomic-fixed $\boldsymbol{r}_m$ vector.

factor for atom A, $f_A(\theta)$, to the X-ray scattering factor for atom A, f_A, where f_A is discussed in detail in Section 8.5.

The wavefunction for the scattered electrons is given in Eq.(9-76), for instance, with the Coulomb value for $f(\theta)$, according to the first Born approximation, given in Eq.(9-87). Now, consider the electron-atom (with atomic number Z_A) scattering as a sum of scattering from the nucleus and electrons. We write the wavefunction for the scattered electrons with Eqs.(9-76) and (9-87) and Fig. 9-12 as

$$\psi^s(R) = -\frac{2me^2}{\hbar^2 K^2}\left[Z_A \frac{\exp(ik_0 R)}{R} - \frac{\exp(ik_0 R)}{R}\sum_m \rho_m \exp(-i\boldsymbol{K}\cdot\boldsymbol{r}_m)\,\Delta V_m\right]. \quad (9\text{-}150)$$

R is the constant distance between the detector and the nucleus of the scattering atom and $\boldsymbol{r}_m$ is the vector from the atomic nucleus to the mth scattering volume (see Fig. 9-12) denoted by ΔV_m with electron density, ρ_m. $\boldsymbol{K}$ is the scattering vector. The sum over m includes the entire volume of the atom. The first term is the electron-nuclear scattering and the second term is the electron-electron scattering. The phases of the second terms in Eq.(9-150) are written in light of the discussion leading to and associated with Eq.(1-65) concerning the corresponding phases for the scattered fields in electromagnetic scattering. The $\boldsymbol{K}\cdot\boldsymbol{r}_m$ phase factor in Eq.(9-150) assures that the phase of the scattering electron from the ΔV_m in the electronic part of the atom is equivalent to the phase of the electron scattering from the nucleus. Thus, Eq.(9-150) represents those components of $\psi^s(R)$ that all reach the detector in phase leading, through constructive interference, to a detectable signal. Of course, we have also assumed that $R \gg r_m$ for all m in order to write Eq.(9-150). Replacing the summation over m in Eq.(9-150) with an integral gives

$$\psi^s(R) = -\frac{2me^2}{\hbar^2 K^2}\frac{\exp(ik_0 R)}{R}$$

$$\times \left[Z_A - \int_0^{2\pi}\int_0^{\pi}\int_0^{\infty} \rho_A(r, \theta, \phi) \exp(-i\boldsymbol{K}\cdot\boldsymbol{r}) r^2 \sin\theta \, dr \, d\theta \, d\phi\right]. \quad (9\text{-}151)$$

$\rho_A(r, \theta, \phi)$ is the electron density of the A atom [see discussion following Eq.(8-135)]. The $\boldsymbol{K}\cdot\boldsymbol{r} = Kr\cos\theta'$ term of Eq.(9-151) contains the angle θ', which is the angle between the laboratory-fixed $\boldsymbol{K}$ vector and atomic- or molecular-fixed $\boldsymbol{r}$ vector. For electron scattering from a random gas target, the second term in Eq.(9-151) must be averaged over all orientations of $\boldsymbol{r}$ with respect to $\boldsymbol{K}$ to give

$$\left[\int_0^{2\pi}\int_0^{\pi}\int_0^{\infty} \rho_A(r, \theta, \phi) \exp(-iKr\cos\theta') r^2 \sin\theta \, dr \, d\theta \, d\phi\right]_{\text{ave}}$$

$$= \frac{1}{4\pi}\int_0^{2\pi}\int_0^{\pi}\left[\int_0^{2\pi}\int_0^{\pi}\int_0^{\infty} \rho(r, \theta, \phi) \exp(-iKr\cos\theta') r^2 \, dr \sin\theta \, d\theta \, d\phi\right] \sin\theta' \, d\theta' \, d\phi'$$

$$= \frac{1}{4\pi}\int_0^{2\pi}\int_0^{\pi}\int_0^{\infty} \rho(r, \theta, \phi)\left[\int_0^{2\pi}\int_0^{\pi} \exp(-iKr\cos\theta') \sin\theta' \, d\theta' \, d\phi'\right] r^2 \sin\theta \, dr \, d\theta \, d\phi$$

$$= \int_0^{2\pi}\int_0^{\pi}\int_0^{\infty} \rho(r, \theta, \phi) \frac{\sin Kr}{Kr} r^2 \sin\theta \, dr \, d\theta \, d\phi. \quad (9\text{-}152)$$

Using this average gives the scattered electron wavefunction for a random distribution of scattering atoms,

$$\psi^s(R) = -\frac{2me^2}{\hbar^2 K^2}\frac{\exp(ik_0 R)}{R}\left[Z_A - \int_0^{2\pi}\int_0^{\pi}\int_0^{\infty} \rho_A(r, \theta, \phi) \frac{\sin Kr}{Kr} r^2 \sin\theta \, dr \, d\theta \, d\phi\right]. \quad (9\text{-}153)$$

The second term within the brackets in Eq.(9-153) is identically equal to the X-ray atomic scattering factor for atom A, f_A, given in Eq.(8-139). Atomic X-ray scattering factors are discussed in detail in Section 8.5 and a partial listing for several atoms is given in Table 8-7. Substituting f_A into Eq.(9-153) gives the final result of

$$\psi^s(R) = -\frac{2me^2}{\hbar^2 K^2}\frac{\exp(ik_0 R)}{R}(Z_A - f_A). \quad (9\text{-}154)$$

Comparing this result with Eq.(9-76) shows that the electron-atomic scattering factor for atom A is given by

$$f_A(\theta) = -\frac{2me^2}{\hbar^2 K^2}(Z_A - f_A). \quad (9\text{-}155)$$

At very low scattering angles we can expand the sin Kr function in f_A and obtain, in the limit, the following result:

$$\lim_{K\to 0} f(\theta) = f(\theta = 0) = \lim_{K\to 0} \left\{ -\frac{2me^2}{\hbar^2 K^2} \left[Z_A - \int_0^{2\pi} \int_0^{\pi} \int_0^{\infty} \rho(r, \theta, \phi) \right.\right.$$

$$\left.\left. \times \left[\frac{Kr - (Kr)^3/6 + \cdots}{Kr} \right] r^2 \sin\theta \, dr \, d\theta \, d\phi \right] \right\}$$

$$= -\frac{2me^2}{\hbar^2 K^2} \left[Z_A - \int_0^{2\pi} \int_0^{\pi} \int_0^{\infty} \rho(r, \theta, \phi) r^2 \sin\theta \, dr \, d\theta \, d\phi \right.$$

$$\left. + \frac{K^2}{6} \int_0^{2\pi} \int_0^{\pi} \int_0^{\infty} \rho(r, \theta, \phi) r^4 \sin\theta \, dr \, d\theta \, d\phi + \cdots \right]. \quad (9\text{-}156)$$

First we note that the first integral gives the number of electrons in the atom, n_A [see equation following Eq.(8-135)],

$$\int_0^{2\pi} \int_0^{\pi} \int_0^{\infty} \rho(r, \theta, \phi) r^2 \sin\theta \, dr \, d\theta \, d\phi = n_A, \quad (9\text{-}157)$$

which cancels Z_A, giving as the dominant term in Eq.(9-156) the following result:

$$f(\theta = 0) = -\frac{me^2}{3\hbar^2} \int_0^{2\pi} \int_0^{\pi} \int_0^{\infty} \rho(r, \theta, \phi) r^4 \sin\theta \, dr \, d\theta \, d\phi = -\frac{me^2}{3\hbar^2} \langle r^2 \rangle, \quad (9\text{-}158)$$

where we have indicated the value of the integral as the electronic average value of r^2 for the atom. $\langle r^2 \rangle$ is related to the atomic bulk (diamagnetic), magnetic susceptibility, as shown in Section 6.8. Substituting Eq.(6-111) into Eq.(6-122) and remembering that $\chi_{xx}^p = \chi_{yy}^p = \chi_{zz}^p = 0$ for atoms gives

$$\chi = -\frac{e^2}{6mc^2} \langle r^2 \rangle, \quad (9\text{-}159)$$

which is the average magnetic susceptibility per atom. Substituting this expression for $\langle r^2 \rangle$ into Eq.(9-158) gives

$$f(\theta = 0) = \frac{2m^2c^2}{\hbar^2} \chi, \quad (9\text{-}160)$$

which relates the forward ($\theta = 0$) scattering factor for the electrons to the average magnetic susceptibility of the target atoms. The result in Eq.(9-158) and the convenient relationship to the atomic diamagnetic susceptibility is a consequence of the Born approximation used in Eq.(9-84) with the result in Eq.(9-87) for $f(\theta)$ in the case of Coulomb scattering. We used these relationships to write down our original scattering function in Eq.(9-150) leading finally to Eqs.(9-154) and (9-158). In any event, using Eq.(9-158) and $\langle r^2 \rangle$ for Ar from Table 9-4 (footnote e) gives the resultant $f(\theta = 0)$ in the fourth column under $|f(\theta)|$ in Table 9-4. The number is about 20% lower than the more accurate result in the first three columns. The discrepancy arises in part from the Born approximation as discussed above and secondly we note that

Table 9-5 e^--Ar total cross-sections for 40-keV electrons using the optical theorem in Eq.(9-134) and $f_i(\theta = 0)$ from Eq.(9-130). The phase shifts needed are from the results in Table 9-3 as obtained by both the Born and numerical methods as described in Ref. [29].

Total cross-section	
First Born Approximation[a]	Numerical Integration[a]
0.0761×10^{-16} cm^2	0.0762×10^{-16} cm^2

[a] See Table 9-3.

the treatment beginning with Eq.(9-150) and ending with Eqs.(9-155) and (9-158) assumes that the target atom's electron distribution remains unchanged during the incident electron-atom scattering interaction.

f_A values for Ar are listed in Table 9-4 as obtained from the atomic X-ray data in Table 8-7. Using these values and Eq.(9-154) gives the remaining $\theta \neq 0$ results in the last column under $|f(\theta)|$ in Table 9-4 [33]. As $\theta \longrightarrow \pi$, the X-ray scattering factor goes to zero and we can write $f(\theta)$ in Eq.(9-155) in this limit as

$$\lim_{\theta \to \pi} f(\theta) = f(\theta = \pi) = -\frac{2me^2 Z_A}{\hbar^2 K^2}, \tag{9-161}$$

which is used for the lower two rows in the last column under $|f(\theta)|$ in Table 9-4. The agreement between the four methods of calculating $|f(\theta)|$ shown in Table 9-4 is quite good.

The total cross-section for the 40-keV e^--Ar scattering process can also be evaluated from the information above and the results are listed in Table 9-5.

Comparing the e^--Ar scattering cross-sections in Table 9-5 with the corresponding cross-sections for photon (optical)-Ar, X-ray–Ar, or neutron-Ar scattering processes (see Section 9.1 for discussion) shows the relatively intense nature of electron-atom scattering.

We have presented a simple theory adequate to describe elastic scattering of high energy electrons. At lower energies there are, of course, many other interesting inelastic and resonant [34] processes that require the full electron correlation problem. The references will give a feeling for the progress in this active area of research.

9.6 ELASTIC ELECTRON SCATTERING FROM MOLECULES; ELECTRON DIFFRACTION

We now have the theoretical framework to interpret high energy electron scattering from a random distribution of molecules. We shall now consider scattering from a low-pressure gas where only single scattering events occur and where there is negligible intermolecular correlation (single molecule scattering only). Scattering from a

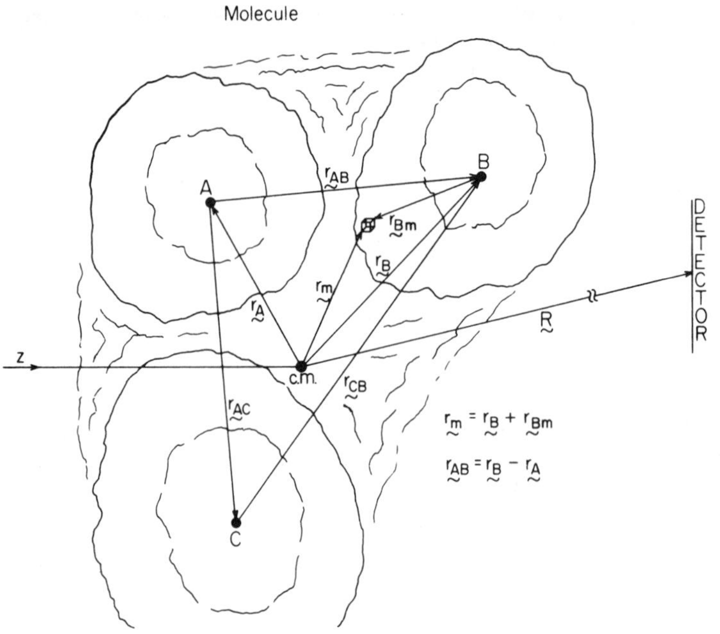

Figure 9-13 Coordinate system showing the electron scattering from a molecule with the c.m. as the origin (see also Fig. 9-12). The small square at $\boldsymbol{r}_m$ is the scattering volume ΔV_m with density ρ_m as indicated in Eq.(9-162).

molecule is shown in Fig. 9-13, which is analogous to Fig. 9-12 describing atoms. The appropriate origin of the coordinate system is the molecular c.m., which is invariant to molecular orientation. The sum of in-phase waves scattered from the molecule is given by

$$\psi^s(R) = -\frac{2me^2}{\hbar^2 K^2}\frac{\exp(ik_0 R)}{R}\left[\sum_\alpha Z_\alpha \exp(-i\boldsymbol{K}\cdot\boldsymbol{r}_\alpha) - \sum_m \rho_m \exp(-i\boldsymbol{K}\cdot\boldsymbol{r}_m)\,\Delta V_m\right], \tag{9-162}$$

where $\boldsymbol{r}_\alpha$ and $\boldsymbol{r}_m$ both originate at the c.m. as indicated in Fig. 9-13. The sum over α is over all nuclei in the molecule with atomic number, Z_α. ΔV_m is the volume element with molecular electron density ρ_m and the sum over volume elements ΔV_m must include the entire molecule. Of course, the molecule-detector distance is large relative to r_m and any of the internuclear distances. Equation (9-162) is broken down into a sum of waves all of which are in phase from all α nuclei plus the contributions from the entire electron distribution of the molecule, which is described by components, ρ_m.

We now assume that molecules are made up of atoms that retain their free electron densities (see discussion under "Free Molecules" on p. 547). According to this assumption we shall break the summation over the molecular volume elements ΔV_m into sums over the individual atoms in the molecule. According to Fig. 9-13,

$$\boldsymbol{r}_m = \boldsymbol{r}_\alpha + \boldsymbol{r}_{\alpha m}, \tag{9-163}$$

for each atom α. Substituting into the second term in Eq.(9-162) gives

$$\psi^s(R) = -\frac{2me^2}{\hbar^2K^2}\frac{\exp(ik_0R)}{R}$$
$$\times\left\{\sum_\alpha Z_\alpha \exp(-i\boldsymbol{K}\cdot\boldsymbol{r}_\alpha) - \sum_\alpha\sum_m \rho_{\alpha m}\exp[-i\boldsymbol{K}\cdot(\boldsymbol{r}_\alpha + \boldsymbol{r}_{\alpha m})]\,\Delta V_{\alpha m}\right\}, \quad (9\text{-}164)$$

where now $\rho_{\alpha m}$ and $\Delta V_{\alpha m}$ represent the electron density and scattering volume element for the αth free atom; the sum over all α will approximate the molecular electron density as a composite of free atom densities. Rewriting Eq.(9-164) and replacing the sum over m in the αth atom with the appropriate integral over the atomic electron density $\rho_\alpha(r, \theta, \phi)$ gives

$$\psi^s(R) = -\frac{2me^2}{\hbar^2K^2}\frac{\exp(ik_0R)}{R}\sum_\alpha \exp(-i\boldsymbol{K}\cdot\boldsymbol{r}_\alpha)$$
$$\times\left[Z_\alpha - \int_0^{2\pi}\int_0^{\pi}\int_0^{\infty}\rho_\alpha(r,\theta,\phi)\exp(i\boldsymbol{K}\cdot\boldsymbol{r})r^2\sin\theta\,dr\,d\theta\,d\phi\right]. \quad (9\text{-}165)$$

The term in brackets in Eq.(9-165) is now identical to the free atom result in Eq.(9-151) summed here over all α atoms in the molecule. If we now average the atomic electronic $\boldsymbol{K}\cdot\boldsymbol{r}$ term over all orientation of the electron distribution with respect to the molecular framework, we obtain

$$\psi^s(R) = -\frac{2me^2}{\hbar^2K^2}\frac{\exp(ik_0R)}{R}\sum_\alpha \exp(-i\boldsymbol{K}\cdot\boldsymbol{r}_\alpha)(Z_\alpha - f_\alpha), \quad (9\text{-}166)$$

where f_α is the X-ray scattering factor for the αth atom as defined in Eq.(8-139). According to Eqs.(9-166) and (9-76) the electron scattering factor for a molecule with a fixed orientation of nuclei is given by

$$f(\theta) = -\frac{2me^2}{\hbar^2K^2}\sum_\alpha \exp(-i\boldsymbol{K}\cdot\boldsymbol{r}_\alpha)(Z_\alpha - f_\alpha). \quad (9\text{-}167)$$

The scattering intensity of electrons is proportional to the differential cross-section given from Eq.(9-81) by

$$\sigma(\theta) = |f(\theta)|^2 = \frac{4m^2e^4}{\hbar^4K^4}\sum_{\alpha,\alpha'}\exp[-i\boldsymbol{K}\cdot(\boldsymbol{r}_\alpha - \boldsymbol{r}_{\alpha'})](Z_\alpha - f_\alpha)(Z_{\alpha'} - f_{\alpha'}). \quad (9\text{-}168)$$

We now average this quantity over all orientations of the molecule relative to $\boldsymbol{K}$ in the space-fixed axis giving [see discussion following Eq.(9-151)]

$$\sigma(\theta) = \frac{4m^2e^4}{\hbar^4K^4}\sum_{\alpha,\alpha'}(Z_\alpha - f_\alpha)(Z_{\alpha'} - f_{\alpha'})\frac{\sin K(r_\alpha - r_{\alpha'})}{K(r_\alpha - r_{\alpha'})}$$
$$= \frac{4m^2e^4}{\hbar^4K^4}\left[\sum_\alpha (Z_\alpha - f_\alpha)^2 + \sum_\alpha\sum_{\substack{\alpha' \\ \alpha\neq\alpha'}}(Z_\alpha - f_\alpha)(Z_{\alpha'} - f_{\alpha'})\frac{\sin K(r_\alpha - r_{\alpha'})}{K(r_\alpha - r_{\alpha'})}\right]$$
$$= \sum_\alpha |f_\alpha(\theta)|^2 + \sum_\alpha\sum_{\substack{\alpha' \\ \alpha\neq\alpha'}} f_\alpha(\theta)f_{\alpha'}(\theta)\frac{\sin K(r_\alpha - r_{\alpha'})}{K(r_\alpha - r_{\alpha'})}, \quad (9\text{-}169)$$

where the last two expressions separate the differential cross-section into a sum over pure atomic terms (as discussed in detail earlier), plus an interference term when $\alpha \neq \alpha'$. We also note the use of Eq.(9-155) in the last step. The interference term is a damped sine wave as a function of K with maxima every $K(r_\alpha - r_{\alpha'}) = 2\pi$ for each interatomic distance $r_\alpha - r_{\alpha'}$.

Experimentally, the intensity [or cross-section, $\sigma(\theta)$] of scattered electrons will also contain an inelastic term. A great deal of successful experimental and theoretical effort has been expended in electron diffraction studies of molecules to correct the observed scattering for the inelastic and pure atomic terms leaving only the interference term for interpreting the molecular structure [35].

Two additional points concerning the experimental determination of the last term in Eq.(9-169) must now be covered. First we note that most electron diffraction experiments are done with electron energies of 40 keV or higher with de Broglie wavelengths of $\lambda = 0.06$ Å or less. At these small wavelengths, the corresponding $K = (4\pi/\lambda)\sin(\theta/2)$ values at any given angle are much larger than the corresponding K-values for the normal $\lambda = 1.54$ Å X rays. Thus, the value of the X-ray scattering factor in Eq.(9-155) goes rapidly to zero as shown by examining Tables 8-7 and 9-4, for instance. Thus, we shall approximate $f_\alpha(\theta)$ in the second term of Eq.(9-169) with the nuclear term only, giving for the molecular scattering term

$$\sigma_M(\theta) = \sum_{\alpha}\sum_{\substack{\alpha' \\ \alpha \neq \alpha'}} f_\alpha(\theta) f_{\alpha'}(\theta) \frac{\sin K(r_\alpha - r_{\alpha'})}{K(r_\alpha - r_{\alpha'})} \cong \frac{4m^2e^4}{\hbar^4 K^4} \sum_{\alpha}\sum_{\substack{\alpha' \\ \alpha \neq \alpha'}} Z_\alpha Z_{\alpha'} \frac{\sin K(r_\alpha - r_{\alpha'})}{K(r_\alpha - r_{\alpha'})}, \tag{9-170}$$

where Z_α and $Z_{\alpha'}$ are the atomic numbers of nuclei α and α', respectively.

The second point is that it is now evident from Eq.(9-170) that the K^4-dependence can also be factored out giving a modified differential cross-section defined by

$$\bar{\sigma}_M(\theta) = K^4\sigma(\theta)_M = \frac{4m^2e^4}{\hbar^4} \sum_{\alpha}\sum_{\substack{\alpha' \\ \alpha \neq \alpha'}} Z_\alpha Z_{\alpha'} \frac{\sin K(r_\alpha - r_{\alpha'})}{K(r_\alpha - r_{\alpha'})}. \tag{9-171}$$

We now rewrite this expression in terms of the probability of there being a unit charge at position R, $P(R)$, giving

$$\bar{\sigma}_M(\theta) = \frac{4m^2e^4}{\hbar^4} \sum_{\alpha}\sum_{\substack{\alpha' \\ \alpha \neq \alpha'}} \int_0^\infty P(R)_{\alpha,\alpha'} \frac{\sin KR}{KR} R^2\, dR. \tag{9-172}$$

Of course, if all nuclei in the molecule are rigidly fixed,

$$P(R)_{\alpha,\alpha'} = Z_\alpha Z_{\alpha'} \delta[R - (r_\alpha - r_{\alpha'})],$$

where $\delta[R - (r_\alpha - r_{\alpha'})]$ is the one-dimensional Dirac delta function and Eq.(9-172) reduces to Eq.(9-171); however, molecular vibrations will cause a slight broadening of the delta function representation. $P(R)_{\alpha,\alpha'}/R$ is the radial correlation function, $D(R)_{\alpha,\alpha'}$, for nuclei α and α' times the product of the atomic numbers Z_α and $Z_{\alpha'}$ for the two nuclei

$$\frac{P(R)_{\alpha,\alpha'}}{R} = Z_\alpha Z_{\alpha'} D(R)_{\alpha,\alpha'}. \tag{9-173}$$

$D(R)_{\alpha, \alpha'}$ is normalized to unity for each α-α' pair,

$$\int_0^\infty D(R)_{\alpha, \alpha'} R^2 \, dR = 1.0. \tag{9-174}$$

We now return to Eq.(9-172) and rewrite the equation in terms of a modified molecular differential scattering cross-section,

$$\bar{\sigma}_M(\theta) = \frac{4m^2e^4}{\hbar^4} \int_0^\infty P(R) \frac{\sin KR}{KR} \, dR, \tag{9-175}$$

where $P(R)$ now contains the sum over all internuclear pairs, α-α', in the molecule. The sine Fourier transform of $K\bar{\sigma}_M(\theta)$ in Eq.(9-175) gives

$$\int_0^\infty K\bar{\sigma}_M(\theta) \sin KR' \, dK = \frac{4m^2e^4}{\hbar^4} \int_0^\infty \int_0^\infty \frac{P(R)}{R} \sin KR \sin KR' \, dR \, dK. \tag{9-176}$$

Comparing with Eq.(B-55) shows that

$$\int_0^\infty K\bar{\sigma}_M(\theta) \sin KR' \, dK = \frac{4m^2e^4}{\hbar^4} \left(\frac{\pi}{2}\right) \left[\frac{P(R')}{R'}\right], \tag{9-177}$$

and we realize that $K\bar{\sigma}_M(\theta)$ and $(4\pi m^2e^4/\hbar^4)[P(R)/R]$ are Fourier transform pairs.

Experimental plots of $\bar{\sigma}_M(\theta)$ as a function of K are shown in Fig. 9-14 for CCl_4 and CO_2. The calculated curves for $\bar{\sigma}_M(\theta)$ shown directly under the experimental

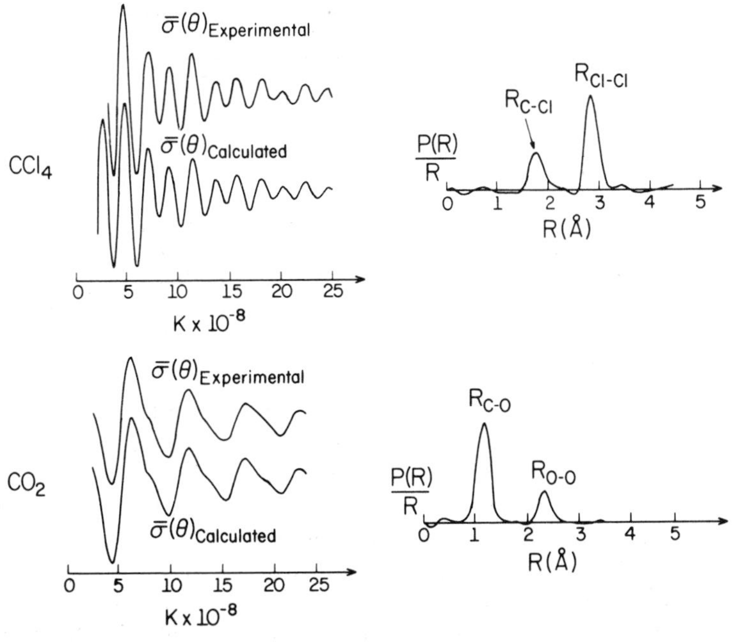

Figure 9-14 Experimental cross-sections for electron scattering and radial distribution functions for CCl_4 and CO_2 from I. L. Karle and J. Karle, *J. Chem. Phys.* **17**, 1052 (1949). The electron energy was 40 keV and the internuclear distances found from the radial distribution curves were $R_{C-Cl} = 1.77 \pm 0.01$ Å and $R_{Cl-Cl} = 2.88 \pm 0.02$ Å in CCl_4 and $R_{C-O} = 1.16 \pm 0.01$ Å and $R_{O-O} = 2.31 \pm 0.02$ Å in CO_2.

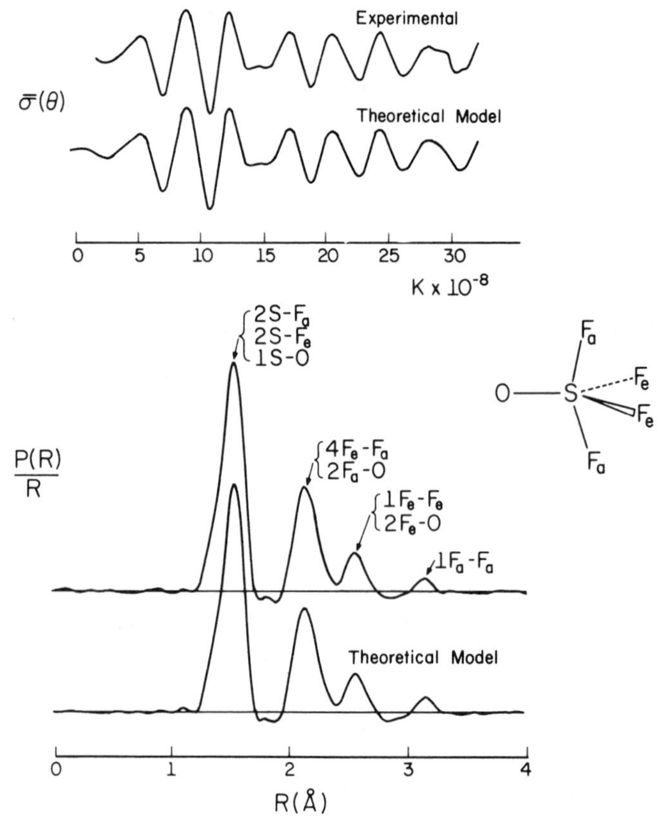

Figure 9-15 Experimental and theoretical molecular differential cross-sections and radial distributions for coherent 40-keV electron scattering from SOF_4 from G. Gunderson and K. Hedberg, *J. Chem. Phys.* **51**, 2500 (1969). The molecular structure is also shown; O, S, and both axial F atoms are in a plane and the equatorial F atoms are in front and behind the plane as shown. The structural parameters used to fit the data are $R_{SO} = 1.403$ Å, $R_{SF_e} = 1.552$ Å, $R_{SF_a} = 1.575$ Å, $R_{F_e-F_e} = 2.545$ Å, $R_{F_a-O} = 2.121$ Å, $R_{F_e-O} = 2.621$ Å, $R_{F_a-F_e} = 2.204$ Å, and $R_{F_a-F_a} = 3.150$ Å. The first peak in $P(R)/R$ is a composite of 2S—F$_a$, 2S—F$_e$, and 1S—O $P(R)/R$ curves. Similar overlaps are found throughout the $P(R)/R$ curve.

curves are obtained using an expression similar to Eq.(9-170) by adjusting the internuclear distances to fit the curve. Also shown in Fig. 9-14 are the corresponding radial distribution functions, $P(R)/R$, obtained by the Fourier transforms as discussed above. In practice, it is evident that the $\bar{\sigma}_M(\theta)$ experimental results are not carried to $K \longrightarrow \infty$ and some approximations must be made to achieve proper convergence of the Fourier transform integrals.

As the size of the molecule examined by electron diffraction increases, considerable overlap in the radial distribution function is obtained. The experimental electron differential scattering cross-section for SOF_4 is shown in Fig. 9-15. The Fourier transform giving the radial distribution function, $P(R)/R$, is also shown. Considerable overlap is found in the radial distribution function where the overlap problem obviously becomes more severe as the number of atoms in the molecule increases. In spite of these problems, a great many molecular structures have been determined by electron diffraction techniques [36].

The molecular structures obtained by diffraction techniques give average distances between correlated centers in all molecules in the scattering gas. Molecular structures obtained by spectroscopic methods must be extracted from the moments of inertia or inverse squared distances by examining transitions between stationary states. It is evident, therefore, that the diffraction structures will involve a different vibrational average of interatomic distances than the corresponding results from the moments of inertia. Detailed comparisons and discussions of the two methods of determining molecular structures are available [37].

In the discussions above, the scattering was described by an atoms-in-molecules approach where the scattered wave in Eq.(9-162) was approximated by a sum over atoms in Eqs.(9-164) to (9-169). We then noted that the electron scattering factor, f_α in Eq.(9-167), for instance, is negligible relative to Z_α for large K leading to the result in Eq.(9-170). At low scattering angles, however, the contribution to the scattering due to f_α cannot be ignored. Thus, low-angle or low-energy electron scattering from molecules can be compared to the scattering factors for atoms to obtain the difference in electron density due to the chemical bonding in molecules [38].

It is evident from our previous discussion that the differential cross-section for Coulomb scattering is inversely proportional to the electron's energy. Thus, low energy electrons will experience larger scattering intensities than higher energy electrons [39]. In fact, low energy electrons have large enough cross-sections to study surface structure as most of the incident electrons will be scattered by surface particle or particles one or two layers into the substrate. This technique of low-energy electron diffraction (LEED) has been used very effectively to study surfaces and the structure and conformation of molecules on a surface [40].

The methods examined here can also be applied to inelastic electron scattering to determine the internal state structure of atoms and molecules. This important field of electron impact spectroscopy is treated elsewhere [41].

Finally, we note that the theory of inelastic electron scattering (electron diffraction) given here should prepare the reader for studies of the determination of molecular structure by neutron diffraction.

PROBLEMS

1. Use Eq.(9-11) and Fig. 9-2 to evaluate the total cross-sections for a thermal beam of CsCl molecules colliding with either CH_2F_2 molecules or Ar atoms. Equation (9-11) assumes a simple hard sphere scattering process. Another point worthy of mention here is that the mean free path, λ, is given in terms of the total cross-section and number density by

$$\lambda = \frac{1}{\rho_0 \sigma}. \tag{9-178}$$

Select target pressures of 10^{-4} Torr and calculate λ for the two scattering events above.

2. Consider the exact solution of Eq.(9-28) for the Coulomb repulsive potential, $V(r) = Z_1 Z_2 e^2/r$, where Z_1 and Z_2 indicate the magnitude and sign of the electrostatic charges on incident particle Z_1 and target particle Z_2.
   (a) Integrate Eq.(9-28) for $V(r) = Z_1 Z_2 e^2/r$ and use r_m in Eq.(9-25) to obtain an exact expression for χ_m.
   (b) Examine the result in (a) and show what approximations (what range of energy and impact parameter) are necessary to obtain the standard differential cross-section for Coulomb scattering given by

$$\sigma(\theta) = \left(\frac{Z_1 Z_2 e^2}{4E_0}\right)^2 \frac{1}{\sin^4(\theta/2)}, \tag{9-179}$$

   which is the Rutherford scattering formula used to describe α-particle scattering from a metal foil [42]. The result of Rutherford's analysis led to a rejection of the Lorentz plum pudding model for atoms and molecules and gave rise to a planetary theory of atomic electronic structure.
   (c) Calculate the total cross-section for Rutherford scattering and the resultant absorption (or attenuation) coefficient for α-particles scattering from a 1-μm thick gold foil. Assume that the detector in the attenuation experiment has an angular resolution of 1°.

3. Consider a beam of N_2 molecules being emitted from a velocity selector with a velocity of $3 \times 10^4 \text{ cm} \cdot \text{s}^{-1}$ with negligible uncertainty. Assume that a single N_2 molecule in this molecular beam can be described by the wavefunction in Eq. (9-53) where $a = 3 \times 10^{-8}$ cm, which is the approximate molecular diameter. The wave packet half width at half height at the exit of the velocity selector is given in Eq.(9-55) as $z_{1/2}(t = 0) = a(2 \ln 2)^{1/2}$. How far along the z axis from the velocity selector origin will the particle have to travel before the initial wave packet half width is doubled?

4. Work the integral in Eq.(9-51) and then normalize the function to give Eq.(9-52).

5. Use the Born approximation in Eq.(9-84) to calculate the scattering amplitude, $f(\theta)$, for the spherically symmetric repulsive Gaussian potential, $V(r) = A \exp(-br^2)$.

6. Follow the directions following Eq.(9-103) for $l = 0$, 1, and 2 to demonstrate that $(A_l a_l) = (1/k_0)(\pi/2)^{1/2}(2l + 1)i^l$.

7. Follow the steps from Eq.(9-125) to prove the result in Eq.(9-126).

8. Consider the production of NO^+ by the collision of an electron beam with an NO gas.

$$e^- + \text{NO} \longrightarrow \text{NO}^+ + 2e^-. \tag{9-180}$$

   Starting with 1 A of electron current directed into a 1-cm long region of NO gas at 10-Torr pressure, calculate the current of NO^+ that can be achieved if the e^--NO total cross-section is 0.1×10^{-16} cm^2.

9. Scattering results from 40-keV electrons from Ar atoms are given in Tables 9-3 and 9-4 and associated discussion. Use relativistic theory [see Eq.(1-83)] to calculate the de Broglie wavelength of a 40-keV electron.

10. Calculate the total cross-sections for
(a) $\lambda = 5145$ Å photon-Ar scattering.
(b) $\lambda = 1.54$ Å X-ray–Ar scattering.
(c) Neutron-Ar scattering and compare with the corresponding e^--Ar scattering cross-sections in Table 9-5.

***11.** Use a digital computer to Fourier transform numerically the $\bar{\sigma}_M(\theta)$ curve in Fig. 9-15 for SOF_4 to give the radial distribution function $P(R)/R$, which is also shown in Fig. 9-15.

12. Evaluate the total cross-section according to Eq.(9-133) for a high energy particle. The low energy result is in Eq.(9-139).

REFERENCES

[1] R. P. FEYNMAN, *Lectures on Physics*, Vol. II, Sec. 28-3 (Addison-Wesley Pub. Co., Reading, Mass., 1964).

[2] A comparison between X-ray and neutron scattering cross-sections is given in G. E. BACON, *MTP International Review of Science, Physical Chemistry*, Series One, Vol. 2 (Butterworths, London, 1972), p. 21.

[3] In the case of a light, fast incident particle and a heavy target particle that is virtually stationary, the laboratory and center of mass coordinate systems are identical and the laboratory measurement leads to the correct cross-sectional parameters. In general, for the scattering of particles of near equal masses at lower relative velocities, the laboratory measurement must be transformed into the center of mass system to obtain the correct scattering parameters; see R. B. BERNSTEIN, *Science* **144**, 141 (1964).

[4] Equation (9-38) can also be derived by using an impulse approximation. See R. E. WESTON and H. A. SCHWARZ, *Chemical Kinetics* (Prentice-Hall, Inc., Englewood Cliffs, N.J., 1972).

[5] J. O. HIRSCHFELDER, C. F. CURTISS, and R. B. BIRD, *Molecular Theory of Gases and Liquids* (John Wiley & Sons, Inc., New York, 1954).

[6] I. AMDUR and J. E. JORDAN, "Molecular Beams," *Adv. Chem. Phys.*, Ed. by J. Ross **10**, 29 (Academic Press, New York, 1966).

[7] R. B. BERNSTEIN, "Molecular Beams," *Adv. Chem. Phys.*, Ed. by J. Ross **10**, 75 (Academic Press, New York, 1966).

[8] W. HEISENBERG, *The Principles of the Quantum Theory* (Dover Pub., New York, 1930).

[9] S. MATHEWS and R. L. WALKER, *Mathematical Methods of Physics* (W. A. Benjamin, Inc., New York, 1965).

[10] D. BOHM, *Quantum Theory* (Prentice-Hall, Inc., Englewood Cliffs, N.J., 1951).

[11] R. COURANT and D. HILBERT, *Methods of Mathematical Physics* (Interscience Pub., New York, 1953).

[12] E. MERZBACHER, *Quantum Mechanics*, 2nd ed. (John Wiley & Sons, Inc., 1970). See also Appendix B.3 for a detailed discussion of Dirac delta functions.

[13] N. F. MOTT and H. S. W. MASSEY, *Theory of Atomic Collisions* (Oxford University Press, London and New York, 1949).

[14] K. L. BELL and A. E. KINGSTON, *Adv. Atom. and Mol. Phys.*, Ed. by D. R. Bates and B. Bederson, **10**, 53 (Academic Press, New York, 1974).

[15] For semiclassical arguments, see D. Parks, *Introduction to Quantum Theory* (McGraw-Hill Book Co., New York, 1974) and Ref. [10].

[16] G. N. WATSON, *Theory of Bessel Functions* (Cambridge University Press, London and New York, 1944).

[17] H. C. VAN DE HULST, *Light Scattering by Small Particles* (John Wiley & Sons, Inc., New York, 1957).

[18] Details in the optical and molecular scattering analogs for rainbow and glory phenomena are found in E. A. MASON, R. J. MUNN, and F. J. SMITH, *Endeavor* **30**, 91 (1971).

[19] U. BUCK, *Rev. Mod. Phys.* **46**, 369 (1974).

[20] G. E. EWING, *Accts. Chem. Res.* **8**, 185 (1975).

[21] A review of rotationally and vibrationally inelastic scattering of molecules is given in J. P. TOENNIES, *Chem. Soc. Rev.* **3**, 407 (1974).

[22] D. BECK, E. F. GREEN, and J. ROSS, *J. Chem. Phys.* **37**, 2895 (1962). G. GRICE, *Adv. Chem. Phys.*, Ed. by I. Prigogine and S. A. Rice, **30**, 247 (Wiley-Interscience, New York, 1975).

[23] R. D. LEVINE and R. B. BERNSTEIN, *Molecular Reactions Dynamics* (Oxford University Press, New York, 1974).

[24] D. R. HERSCHBACH, "Molecular Beams" *Adv. Chem. Phys.*, Ed. by J. Ross **10**, 319 (Interscience, New York, 1966); M. A. D. FLUENDY and K. P. LAWLEY, *Chemical Applications of Molecular Beam Scattering* (Chapman and Hall, London, 1973). Ion-molecule reactions and the relation between the potential surfaces and the electronic states of the reactants and products are discussed by B. H. MAHAN, *Accts. Chem. Res.* **8**, 55 (1975).

[25] E. H. S. BURHOP, "Theory of Collisions" in *Quantum Theory I, Elements*, Ed. by D. R. Bates (Academic Press, New York, 1961).

[26] R. D. LEVINE, *Quantum Mechanics of Molecular Rate Processes* (Clarendon Press, Oxford, 1969).

[27] R. NEWTON, *Scattering Theory of Waves and Particles* (McGraw-Hill Book Co., New York, 1966).

[28] R. A. MARCUS, *Fara. Disc. of Chem. Soc.* **55**, 9 (1973). W. H. MILLER, *Adv. Chem. Phys.*, Ed. by I. Prigogine and S. A Rice **30**, 77 (Wiley-Interscience, New York, 1975).

[29] J. KARLE and R. A. BONHAM, *J. Chem. Phys.* **40**, 1396 (1964).

[30] R. A. BONHAM and T. G. STRAND, *J. Chem. Phys.* **39**, 2200 (1963).

[31] Additional experiments on electron scattering in Ne, Ar, Kr, and Xe along with comparison with theory are given in R. A. BONHAM and H. L. COX, *J. Chem. Phys.* **47**, 3508 (1967).

[32] H. L. COX and R. A. BONHAM, *J. Chem. Phys.* **47**, 2599 (1967).

[33] See experimental work and discussion in L. S. BARTELL and L. O. BROCKWAY, *Phys. Rev.* **90**, 833 (1953).

[34] P. G. BURKE, *Adv. in Atom. and Mol. Phys.*, Ed. by D. R. Bates and I. Estermann, **4**, 173 (Academic Press, New York, 1968); H. S. TAYLOR, *Adv. Chem. Phys.*, Ed. by I. Prigogine and S. A. Rice, **18**, 19 (John Wiley & Sons, Inc., New York, 1970). See also a discussion of

low energy electron-molecule scattering in D. E. GORDON, *et al.*, *Rev. Mod. Phys.* **43**, 642 (1971).

[35] L. S. BARTELL, *Techniques of Chemistry*, Vol. I: *Physical Methods of Chemistry*, Ed. by A. Weissberger and B. W. Rossiter (Wiley-Interscience, New York, 1972), Chap. II. J. KARLE, *Determination of Organic Structures by Physical Methods*, Vol. V, Ed. by F. C. Nachod and J. J. Zuckerman (Academic Press, New York, 1972).

[36] S. H. BAUER, *Physical Chemistry, An Advanced Treatise*, Ed. by D. Henderson **4**, 741, (Academic Press, New York, 1970). K. KUCHITSU, *MTP International Review of Science, Physical Chemistry*, Series I, Vol. 2 (1972), p. 203. See also *Tetrahedron* **17**, 125–191 (1962).

[37] *Critical Evaluation of Chemical and Physical Structural Information*, Ed. by D. R. Lide and M. A. Paul (National Academy of Science, Washington, D.C., 1974).

[38] D. A. KOHL and L. S. BARTELL, *J. Chem. Phys.* **51**, 2891 and 2896 (1969).

[39] D. ANDRICK, *Adv. Atom. and Mol. Phys.*, Ed. by D. R. Bates and I. Estermann, **9**, 207 (Academic Press, New York, 1973).

[40] G. A. SOMORJAI, *Principles of Surface Chemistry* (Prentice-Hall, Inc., Englewood Cliffs, N.J., 1972). See also C. B. DUKE, *Adv. Chem. Phys.*, Ed. by I. Prigogine and S. A. Rice **27**, 1 (John Wiley & Sons, Inc., New York, 1974).

[41] E. H. LASSETTRE and A. SKERBELE, *Methods of Experimental Physics*, 2nd ed. (Academic Press, New York, 1974). See also S. TRAJMAR, J. K. RICE and A. KUPPERMANN, *Adv. Chem. Phys.*, Ed. by I. Prigogine and S. A. Rice, **18**, 15 (John Wiley & Sons, New York, 1970). An interesting account of rotational and vibrational excitation in molecules by slow electron impact can be found in F. LINDER, *Endeavour* **120**, 124 (1974).

[42] H. GEIGER and E. MARSDEN, *Proc. Roy. Soc.* (London) **82**, 495 (1909). E. RUTHERFORD, *Phil. Mag.* **25**, 605 (1911).

appendix

Appendix A UNITS, CONSTANTS, AND CONVERSION FACTORS

We use l, m, and t to denote the base physical quantities of *length*, *mass* and *time*, respectively. The notation for other physical quantities are listed in the left column of Table A-1. Also listed in Table A-1 are the names, symbols, and units for the two basic systems of units. The *centimeter-gram-second* (cgs)-Gaussian combined *electrostatic* unit (esu) and *electromagnetic* unit (emu) system is most convenient in describing basic molecular properties. We hesitate to abandon this system as a great deal of past literature contains data listed in these units and the Gaussian units are convenient in describing Maxwell's equations and electric and magnetic field effects in atoms and molecules. On the other hand, there are a number of reasons to adopt and to continue to use the more practical, laboratory-based *meter-kilogram-second* (mks) International System of Units (SI). It is quite clear that it will be necessary to understand both sets of units in order to read the past, present, and future literature.

Table A-1 Basic physical quantities, derived quantities, and conversions.

Physical Quantity	mks-SI Name	mks-SI Symbol	mks-SI Unit	cgs-Gaussian (esu and emu) Name	cgs-Gaussian Symbol	cgs-Gaussian Unit	$N_{cgs}Q = N_{SI}^{a}$ Q
length, l	meter	m	m	centimeter	cm	cm	(m/cm) × 10^{-2}
mass, m	kilogram	kg	kg	gram	g	g	(kg/g) × 10^{-3}
time, t	second	s	s	second	s	s	
electric current, i	ampere	A	A				
thermodynamic temperature, T	kelvin	K	K				
amount of substance	mole	mol	mol				
frequency, ν	hertz	Hz	s^{-1}				
energy, E	joule	J	$m^2 \cdot kg \cdot s^{-2}$	erg	erg	$cm^2 \cdot g \cdot s^{-2}$	(J/erg) × 10^{-7}
force, F	newton	N	$J \cdot m^{-1}$	dyne	dyne	$erg \cdot cm^{-1}$	(N/dyne) × 10^{-5}
power, P	watt	W	$J \cdot s^{-1}$				
intensity, I			$W \cdot m^{-2}$			$W \cdot cm^{-2}$	$(W \cdot m^{-2}/W \cdot cm^{-2}) \times 10^4$
pressure, p	pascal	Pa	$J \cdot m^{-3}$			$dyne \cdot cm^{-2}$	$(Pa/dyne \cdot cm^{-2}) \times 10^{-1}$
electric charge	coulomb	C	$A \cdot s$	statcoulomb	SC	SC	$(C/SC) \times c^{-1} \times 10$
electric potential difference	volt	V	$J \cdot C^{-1}$	statvolt	SV	$erg \cdot SC^{-1}$	$(V/SV) \times c \times 10^{-8}$
capacitance	farad	F	$C \cdot V^{-1}$				
electric field, E		E	$V \cdot m^{-1}$			$SV \cdot cm^{-1}$	$(V \cdot m^{-1}/SV \cdot cm^{-1}) \times c \times 10^{-6}$
electric resistance	ohm	Ω	$V \cdot A^{-1}$				
magnetic flux	weber	Wb	$V \cdot s$				
inductance	henry	H	$V \cdot s \cdot A^{-1}$				
magnetic field			$T \cdot m \cdot H^{-1}(A \cdot m^{-1})$	gauss(oersted)	G	emu	$(T \cdot m \cdot H^{-1}/G) \times \mu_0^{-1} \times 10^{-4}$
magnetic flux density, B	tesla	T	$Wb \cdot m^{-2}$	gauss	G	emu	$(T/G) \times 10^{-4}$
electric dipole moment, D_x			$C \cdot m$			$SC \cdot cm$	$(C \cdot m/SC \cdot cm) \times c^{-1} \times 10^{-1}$
electric quadrupole moment, Q_{xx}			$C \cdot m^2$			$SC \cdot cm^2$	$(C \cdot m^2/SC \cdot cm^2) \times c^{-1} \times 10^{-3}$
electric polarizability, α_{xx}			$C \cdot m^2 \cdot V^{-1}$			cm^3	$(m^3/cm^3) \times (4\pi\varepsilon_0) \times 10^{-6}$
magnetic dipole moment, μ_x			$J \cdot T^{-1}$			$erg \cdot G^{-1}$	$(J \cdot T^{-1}/erg \cdot G^{-1}) \times 10^{-3}$
magnetic susceptibility, χ_{xx}			$J \cdot T^{-2}$			$erg \cdot G^{-2}(cm^3)$	$(J \cdot T^{-2}/erg \cdot G^{-2}) \times 10$
moment of inertia, I_{xx}			$kg \cdot m^2$			$g \cdot cm^2$	$(kg \cdot m^2/g \cdot cm^2) \times 10^{-7}$

[a] N_{cgs} is a number in cgs units, N_{SI} is the corresponding number in SI units, and Q is the conversion factor. The relationship is $N_{SI} = QN_{cgs}$ or $N_{cgs} = Q^{-1}N_{SI}$. c is the speed of light in a vacuum (see Table A-3) in $cm \cdot s^{-1}$.

Table A-2 Prefixes.

Prefix	Symbol	Meaning; *Move decimal point to left:*	Prefix	Symbol	Meaning; *Move decimal point to right:*
deci	d	1 place 10^{-1}	deca	da	1 place 10^{1}
centi	c	2 places 10^{-2}	hecto	h	2 places 10^{2}
milli	m	3 places 10^{-3}	kilo	k	3 places 10^{3}
micro	μ	6 places 10^{-6}	mega	M	6 places 10^{6}
nano	n	9 places 10^{-9}	giga	G	9 places 10^{9}
pico	p	12 places 10^{-12}	tera	T	12 places 10^{12}
femto	f	15 places 10^{-15}	peta	P	15 places 10^{15}
atto	a	18 places 10^{-18}	exa	E	18 places 10^{18}

Thus, we use both sets of units, making clear as often as necessary their interconversion, as shown in Table A-1.

Most of the symbols for the physical quantities used in the text are listed in Table A-1. Prefixes used in the text are listed in Table A-2. The symbols for Cartesian vector quantities are printed in boldface italic (sloping) type, $\boldsymbol{V}$, and vector components are in italic type, V_x. The symbols for second-order tensors (dyadics) are printed in sans serif type, T (T_{xx}) or lowercase Greek, $\boldsymbol{\chi}$ (χ_{xx}). The various types of vector and tensor products are illustrated in Appendix B.1. The symbol for a matrix is bold-face type, **A**, with components in normal Roman type, A_{ii}. Eigenvalues will carry a single subscript and are written in italic type, A_i.

Some of the entries in Table A-1 deserve special comment. The electric charge in SI units is the *coulomb* (C), which is related to the cgs-Gaussian units by

$$1 \text{ coulomb} = 1 \text{ C} = \frac{c}{10} \text{ statcoulomb} = \frac{c}{10} \text{ SC} = \frac{1}{10} \text{ abcoulomb}, \tag{A-1}$$

where the *statcoulomb* (SC) is the esu unit of charge and the *abcoulomb* is the emu unit of charge. The *ampere* ($\mathrm{C \cdot s^{-1}}$) is the SI unit of current and the corresponding Gaussian units of current are esu ($\mathrm{SC \cdot s^{-1}}$) and emu ($\mathrm{abcoulomb \cdot s^{-1}}$). An electron has a negative charge of

$$e = 1.6021892 \times 10^{-19} \text{ C} = 1.6021892 \times 10^{-20} \text{ abcoulomb}$$
$$= 4.803242 \times 10^{-10} \text{ statcoulomb.}$$

The statcoulomb (SC) is defined as the charge Q that gives a force of 1 $\mathrm{dyne \cdot cm^{-1}}$ when placed 1 cm from an equal charge Q, according to Coulomb's law in cgs Gaussian units:

$$F = \frac{Q^2}{r^2}. \tag{A-2}$$

The coulomb (C) is defined as the charge Q that gives a force of 1 newton ($J \cdot m^{-1}$) when placed 1 m from an equal charge Q, according to Coulomb's law in mks SI units:

$$F = \frac{Q^2}{4\pi\varepsilon_0 r^2}, \tag{A-3}$$

where

$$\varepsilon_0 = \frac{10^7}{4\pi c^2} C^2 \cdot N^{-1} \cdot m^{-2} (F \cdot m^{-1}) = 8.854187818 \times 10^{-12} s^4 \cdot A^2 \cdot kg^{-1} \cdot m^{-3}$$

is the *permutivity of a vacuum* [c is in SI units ($m \cdot s^{-1}$)].

Electromagnetic units (emu) are defined according to Ampere's law for the force between two linear conductors. The absolute coulomb or abcoulomb (the emu unit of charge) is defined by the current, i, in emu (abcoulomb $\cdot\, s^{-1}$), which gives a force of 1 dyne $\cdot\, cm^{-1}$ between two 1-cm long conductors (l) that are separated by 1 cm (d) according to

$$F = \frac{2i^2 l}{d}. \tag{A-4}$$

As the current is charge per unit time, we write $i = q \cdot t^{-1}$ to give

$$F = \frac{2q^2 l}{t^2 d}. \tag{A-5}$$

q in Eq.(A-5) and Q in Eq.(A-2) have different units. Q is in esu and q is in emu and the units are related by c ($cm \cdot s^{-1}$), the speed of light according to $Q = cq$, or, in general,

$$(\text{esu}) \cdot (\text{emu})^{-1} = c\ (cm \cdot s^{-1}). \tag{A-6}$$

Now, Ampere's law in mks SI units is given by

$$F = \frac{2\mu_0 i^2 l}{4\pi d}. \tag{A-7}$$

A force of 1 newton is produced by two 1-m conductors, with currents of 1 A separated by 1 m. $\mu_0 = 4\pi \times 10^{-7}\ kg \cdot m \cdot s^{-2} \cdot A^{-2}$ ($H \cdot m^{-1}$) is the *permeability of a vacuum.* It is evident that

$$\left(\frac{1}{\mu_0 \varepsilon_0}\right)^{1/2} = c. \tag{A-8}$$

The electric *potential difference* in SI units is the *volt* (V), which is related to the cgs-Gaussian-esu unit of potential difference, $SC \cdot cm^{-1} = erg \cdot SC^{-1}$ = *statvolt* (SV) according to

$$1\ V = c^{-1} \times 10^8 \text{ statvolt} = c^{-1} \times 10^8\ SV, \tag{A-9}$$

where c is the speed of light in a vacuum in cgs units ($c = 2.9979258 \times 10^{10}$ cm $\cdot$ s^{-1}). The electric *field intensity* (E) in SI units is V $\cdot$ m^{-1} and E in cgs-Gaussian-esu electrostatic units is SC $\cdot$ cm^{-2} = SV $\cdot$ cm^{-1}. The two are related by

$$\begin{aligned} 1\ \mathrm{V \cdot m^{-1}} &= c^{-1} \times 10^{6}\ \mathrm{SV \cdot cm^{-1}} \\ 1\ \mathrm{V \cdot cm^{-1}} &= c^{-1} \times 10^{8}\ \mathrm{SV \cdot cm^{-1}} \\ &\cong (\tfrac{1}{300})\ \mathrm{SV \cdot cm^{-1}}, \end{aligned} \tag{A-10}$$

where we estimate $(c^{-1} \times 10^{8}) \cong \frac{1}{300}$.

The cgs-Gaussian units of *magnetic induction* are shown in Table A-1 as *gauss* (G), which is an electromagnetic unit (emu). Normally, we shall be considering the vacuum magnetic induction, which is equivalent to the *magnetic field intensity*, $H = B$. Of course, we also see the relation to SI units in Table A-1:

$$1\ \text{gauss} = 10^{-4}\ \text{tesla}, \qquad 1\ \mathrm{G} = 10^{-4}\ \mathrm{T}. \tag{A-11}$$

Two common units of *pressure*, the *atmosphere* and *Torr*, are related to the SI *pascal* (Pa) and cgs units by

$$1\ \text{atm} = 760\ \text{Torr} = 101,325\ \text{Pa} = 1.01325 \times 10^{6}\ \text{dyne} \cdot \text{cm}^{-2}. \tag{A-12}$$

The *angstrom* unit of length (Å) is also used quite often:

$$1\ \text{Å} = 10^{-8}\ \text{cm} = 10^{-10}\ \text{m} = 10\ \text{nm}, \tag{A-13}$$

when referring to atomic and molecular dimensions.

Atomic units (a.u.) are also used occasionally in the text. Atomic units are obtained by setting a_0 (Bohr radius) $= \hbar = e = 1.0$.

Fundamental constants used in this text are listed in Table A-3. The definition of the SI base units are reproduced here [1].†

The kilogram is the unit of mass; it is equal to the mass of the international prototype of the kilogram.

The second is the duration of 9,192,631,770 periods of the radiation corresponding to the transition between the two hyperfine levels of the ground state of the cesium-133 atom (see Table 5-11).

The ampere is that constant current that, if maintained in two straight parallel conductors of infinite length and negligible circular cross section, and placed 1 m apart in vacuum, would produce between these conductors a force equal to 2×10^{-7} newton per meter of length.

The kelvin, unit of thermodynamic temperature, is the fraction $(273.16)^{-1}$ of the thermodynamic temperature of the triple point of water.

The mole is the amount of substance of a system that contains as many elementary entities as there are atoms in 0.012 kg of carbon-12. When the mole is used, the elementary entities must be specified and may be atoms, molecules, ions, electrons, other particles, or specified groups of such particles.

† Numbers in square brackets correspond to the numbered sources found under References on p. 681.

Table A-3 Fundamental constants compiled by E. R. Cohen and B. N. Taylor under the auspices of the CODATA Task Group on Fundamental Constants. This set has been officially adopted by CODATA and is reprinted from *J. Phys. Chem. Ref. DATA*, Vol. 2, no. 4, p. 714 (1973), CODATA Bulletin no. 11 (December 1973), and NBS Dimension, January 1974.

Quantity	Symbol	Value †	Uncert. (ppm)	SI ← Units →	cgs*
Speed of light in vacuum	c	299792458(1.2)	0.004	$m \cdot s^{-1}$	10^{2} $cm \cdot s^{-1}$
Permeability of vacuum	μ_0	4π		10^{-7} $H \cdot m^{-1}$	
		$=12.5663706144$		10^{-7} $H \cdot m^{-1}$	
Permittivity of vacuum, $1/\mu_0 c^2$	ϵ_0	8.854187818(71)	0.008	10^{-12} $F \cdot m^{-1}$	
Fine-structure constant, $[\mu_0 c^2/4\pi](e^2/\hbar c)$	α	7.2973506(60)	0.82	10^{-3}	10^{-3}
	α^{-1}	137.03604(11)	0.82		
Elementary charge	e	1.6021892(46)	2.9	10^{-19} C	10^{-20} emu
		4.803242(14)	2.9		10^{-10} esu
Planck constant	h	6.626176(36)	5.4	10^{-34} $J \cdot s$	10^{-27} $erg \cdot s$
	$\hbar = h/2\pi$	1.0545887(57)	5.4	10^{-34} $J \cdot s$	10^{-27} $erg \cdot s$
Avogadro constant	N_A	6.022045(31)	5.1	10^{23} mol^{-1}	10^{23} mol^{-1}
Atomic mass unit, $10^{-3} kg \cdot mol^{-1} N_A^{-1}$	u	1.6605655(86)	5.1	10^{-27} kg	10^{-24} g
Electron rest mass	m_e	9.109534(47)	5.1	10^{-31} kg	10^{-28} g
		5.4858026(21)	0.38	10^{-4} u	10^{-4} u
Proton rest mass	m_p	1.6726485(86)	5.1	10^{-27} kg	10^{-24} g
		1.007276470(11)	0.011	u	u
Ratio of proton mass to electron mass	m_p/m_e	1836.15152(70)	0.38		
Rydberg constant, $[\mu_0 c^2/4\pi]^2 (m_e e^4/4\pi\hbar^3 c)$	R_∞	1.097373177(83)	0.075	10^{7} m^{-1}	10^{5} cm^{-1}
Bohr radius, $[\mu_0 c^2/4\pi]^{-1}(\hbar^2/m_e e^2) = \alpha/4\pi R_\infty$	a_0	5.2917706(44)	0.82	10^{-11} m	10^{-9} cm
Classical electron radius, $[\mu_0 c^2/4\pi](e^2/m_e c^2) = \alpha^3/4\pi R_\infty$	$r_e = \alpha \lambda\!\!\!^{-}_C$	2.8179380(70)	2.5	10^{-15} m	10^{-13} cm
Thomson cross section, $(8/3)\pi r_e^2$	σ_e	0.6652448(33)	4.9	10^{-28} m^2	10^{-24} cm^2
Free electron g-factor, or electron magnetic moment in Bohr magnetons	$g_e/2 = \mu_e/\mu_B$	1.0011596567(35)	0.0035		
Bohr magneton, $[c](e\hbar/2m_e c)$	μ_B	9.274078(36)	3.9	10^{-24} $J \cdot T^{-1}$	10^{-21} $erg \cdot G^{-1}$
Electron magnetic moment	μ_e	9.284832(36)	3.9	10^{-24} $J \cdot T^{-1}$	10^{-21} $erg \cdot G^{-1}$
Nuclear magneton, $[c](e\hbar/2m_p c)$	μ_N	5.050824(20)	3.9	10^{-27} $J \cdot T^{-1}$	10^{-24} $erg \cdot G^{-1}$
Compton wavelength of the electron, $h/m_e c = \alpha^2/2R_\infty$	λ_C	2.4263089(40)	1.6	10^{-12} m	10^{-10} cm
	$\lambda\!\!\!^{-}_C = \lambda_C/2\pi = \alpha a_0$	3.8615905(64)	1.6	10^{-13} m	10^{-11} cm
Compton wavelength of the proton, $h/m_p c$	$\lambda_{C,p}$	1.3214099(22)	1.7	10^{-15} m	10^{-13} cm
	$\lambda\!\!\!^{-}_{C,p} = \lambda_{C,p}/2\pi$	2.1030892(36)	1.7	10^{-16} m	10^{-14} cm
Compton wavelength of the neutron, $h/m_n c$	$\lambda_{C,n}$	1.3195909(22)	1.7	10^{-15} m	10^{-13} cm
	$\lambda\!\!\!^{-}_{C,n} = \lambda_{C,n}/2\pi$	2.1001941(35)	1.7	10^{-16} m	10^{-14} cm
Molar volume of ideal gas at s.t.p.	V_m	22.41383(70)	31	10^{-3} $m^3 \cdot mol^{-1}$	10^{3} $cm^3 \cdot mol^{-1}$
Molar gas constant, $p_0 V_m/T_0$ ($T_0 \equiv 273.15$ K; $p_0 \equiv 101325$ Pa $\equiv$ 1atm)	R	8.31441(26)	31	$J \cdot mol^{-1} \cdot K^{-1}$	10^{7} $erg \cdot mol^{-1} \cdot K^{-1}$
		8.20568(26)	31	10^{-5} $m^3 \cdot atm \cdot mol^{-1} \cdot K^{-1}$	10 $cm^3 \cdot atm \cdot mol^{-1} \cdot K^{-1}$
Boltzmann constant, R/N_A	k	1.380662(44)	32	10^{-23} $J \cdot K^{-1}$	10^{-16} $erg \cdot K^{-1}$
Stefan-Boltzmann constant, $\pi^2 k^4/60\hbar^3 c^2$	σ	5.67032(71)	125	10^{-8} $W \cdot m^{-2} \cdot K^{-4}$	10^{-5} $erg \cdot s^{-1} \cdot cm^{-2} \cdot K^{-4}$
Gravitational constant	G	6.6720(41)	615	10^{-11} $m^3 \cdot s^{-2} \cdot kg^{-1}$	10^{-8} $cm^3 \cdot s^{-2} \cdot g^{-1}$

† Note that the numbers in parentheses are the one standard-deviation uncertainties in the last digits of the quoted value computed on the basis of internal consistency, that the unified atomic mass scale $^{12}C \equiv 12$ has been used throughout, that u=atomic mass unit, C=coulomb, F=farad, G=gauss, H=henry, Hz=hertz=cycles/s, J=joule, K=kelvin (degrees kelvin), Pa=pascal=$N \cdot m^{-2}$, T=tesla (10^4 G), V=volt, Wb=weber=$T \cdot m^2$, and W=watt. In cases where formulas for constants are given (e.g., R_∞), the relations are written as the product of two factors. The second factor, in parentheses, is the expression to be used when all quantities are expressed in cgs units, with the electron charge in electrostatic units. The first factor, in brackets, is to be included only if all quantities are expressed in SI units. We remind the reader that with the exception of the auxiliary constants which have been taken to be exact, the uncertainties of these constants are correlated, and therefore the general law of error propagation must be used in calculating additional quantities requiring two or more of these constants.

* In order to avoid separate columns for "electromagnetic" and "electrostatic" units, both are given under the single heading "cgs Units." When using these units, the elementary charge e in the second column should be understood to be replaced by e_m or e_s, respectively.

Table A-4 Useful conversion factors. The source is the same as Table A-3 and uncertainties are all about 3 ppm. The last column is based on 1 calorie = 4.1868 J and N_A from Table A-3.

	J	erg	eV	cm^{-1}	$cal \cdot mol^{-1}$
J	1.0	10^7	6.24146×10^{18}	5.03404×10^{22}	1.43834×10^{23}
erg	10^{-7}	1.0	6.24146×10^{11}	5.03404×10^{15}	1.43834×10^{16}
eV	1.60219×10^{-19}	1.60219×10^{-12}	1.0	8065.48	2.30450×10^4
cm^{-1}	1.98648×10^{-23}	1.98648×10^{-16}	1.23985×10^{-4}	1.0	2.85724
$cal \cdot mol^{-1}$	6.95246×10^{-24}	6.95246×10^{-17}	4.33934×10^{-5}	3.49989×10^{-1}	1.0

We now make some additional comments concerning conversion between different units of energy as shown in Table A-4. The *electron volt* (eV) is the energy of one electron in the electric potential of one volt. From Tables A-1 and A-3, we have

$$1 \text{ eV} = 1.6021892 \times 10^{-19} \text{ J} = 1.6021892 \times 10^{-12} \text{ erg}$$
$$1 \text{ erg} = 6.241460 \times 10^{11} \text{ eV}. \qquad \text{(A-14)}$$

Another common unit used in place of energy is the *wavenumber* (cm^{-1}) or the inverse wavelength corresponding to an energy given by $E = h\nu = hc/\lambda$; $1/\lambda = E/hc$, where

$$\frac{1}{hc} = 1.98648 \times 10^{-16} \text{ cm}^{-1} \cdot \text{erg}^{-1} \qquad \text{(A-15)}$$

is the conversion factor. The conversion factors among J, erg, eV, and cm^{-1} are summarized in Table A-4. Occasionally, the *calorie* unit occurs; this can be converted to J by the definition

$$1 \text{ cal} = 4.1868 \text{ J}. \qquad \text{(A-16)}$$

Angles are given in degrees and minutes indicated by superscript zero and prime, respectively:

$$30 \text{ degrees } 20 \text{ minutes} = 30°20', \qquad 1° = 60'. \qquad \text{(A-17)}$$

Degrees are related to radians (rad) by

$$\pi \text{ rad} = 180°. \qquad \text{(A-18)}$$

The atomic weights of the common isotopes of most of the atoms are listed in the periodic table in Table A-5.

In the table headings and figure notation, we adopt the convention of listing the quantity and then the units (often abbreviated) following in parentheses. For instance, the abscissa in Fig. 9-2 is listed as p (10^{-5} Torr), which means that the numbers are in units of 10^{-5} Torr. One of the headings in Table 9-1 is $V(r)$ (eV). which means that the energies, $V(r)$, are in units of electron volts.

Table A-5 The periodic table showing electronic configurations, percent abundance, and atomic weights.

Atomic Number	(Configuration)	Element Symbol
% Abundance		Atomic Weight (a.u.)

n				
1	1 (1s) H 99.985 1.00783 0.015 2.01410			
2	3 (He $2s^1$) Li 7.42 6.01513 92.58 7.01601	4 (He $2s^2$) Be 100.0 9.01219	5 ($2s^2 2p^1$) B 19.6 10.01294 80.4 11.00931	6 ($2s^2 2p^2$) C 98.89 12.00000 1.11 13.00335
3	11 (Ne $3s^1$) Na 100.0 22.98977	12 (Ne $3s^2$) Mg 78.70 23.98504 10.13 24.98584 11.17 25.98259	13 ($3s^2 3p^1$) Al 100.0 26.98153	14 ($3s^2 3p^2$) Si 92.21 27.97693 4.70 28.97649 3.09 29.97376
4	19 (Ar $4s^1$) K 93.10 38.96371 6.88 40.96184	20 (Ar $4s^2$) Ca 96.970 39.96259 0.640 41.95863 2.060 43.95549 and others	31 ($4s^2 4p^1$) Ga 60.4 68.9257 39.6 70.9249	32 ($4s^2 4p^2$) Ge 20.52 69.9243 27.43 71.9217 7.76 72.9234 36.54 73.9212 7.76 75.9214
5	37 (Kr $5s^1$) Rb 72.15 84.9117 27.85 86.9092	38 (Kr $5s^2$) Sr 0.56 83.9134 9.86 85.9094 7.02 86.9089 82.56 87.9056	49 ($5s^2 5p^1$) In 4.28 112.9041 95.72 114.9039	50 ($5s^2 5p^2$) Sn 14.30 115.9021 7.61 116.9031 24.03 117.9018 and others
6	55 (Xe $6s^1$) Cs 100.0 132.9054	56 (Xe $6s^2$) Ba 7.81 135.9044 11.32 136.9056 71.66 137.9055	81 ($6s^2 6p^1$) Tl 29.50 202.9723 70.50 204.9745	82 ($6s^2 6p^2$) Pb 1.5 203.9730 23.6 205.9745 22.6 206.9759 52.3 207.9766
7	87 (Rn $7s^1$) Fr	88 (Rn $7s^2$) Ra	6d & 5f → 89–102 Actinides	

Transition Elements

3d	4d
21(Ar $3d^1 4s^2$)Sc	39(Kr $3d^1 5s^2$)Y
22(Ar $3d^2 4s^2$)Ti	40(Kr $4d^2 5s^2$)Zr
23(Ar $3d^3 4s^2$)V	41(Kr $4d^4 5s^1$)Nb
24(Ar $3d^5 4s^1$)Cr	42(Kr $4d^5 5s^1$)Mo
25(Ar $3d^5 4s^2$)Mn	43(Kr $4d^5 5s^2$)Tc
26(Ar $3d^6 4s^2$)Fe	44(Kr $4d^7 5s^1$)Ru
27(Ar $3d^7 4s^2$)Co	45(Kr $4d^8 5s^1$)Rh
28(Ar $3d^8 4s^2$)Ni	46(Kr $4d^{10} 5s^0$)Pd
29(Ar $3d^{10} 4s^1$)Cu	47(Kr $4d^{10} 5s^1$)Ag
30(Ar $3d^{10} 4s^2$)Zn	48(Kr $4d^{10} 5s^2$)Cd

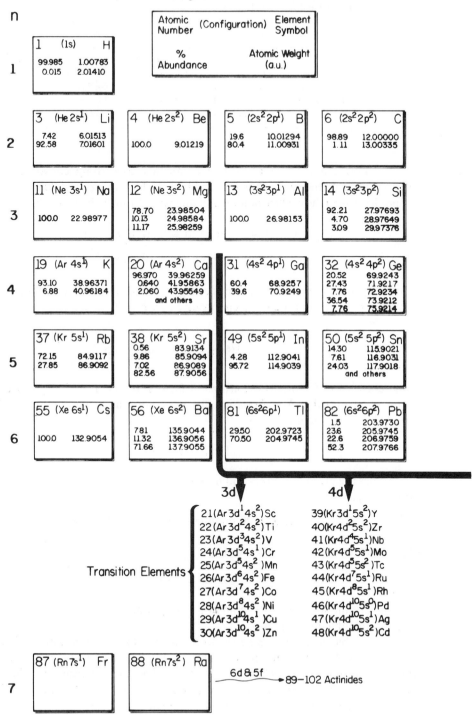

Table A-5 *(continued)*

			2 ($1s^2$) He 100.0 4.00260
7 ($2s^22p^3$) N 99.63 14.00307 0.37 15.00011	8 ($2s^22p^4$) O 99.759 15.99491 0.037 16.99914 0.204 17.99946	9 ($2s^22p^5$) F 100.0 18.99840	10 ($2s^22p^6$) Ne 90.920 19.99244 0.257 20.99395 8.820 21.99138
15 ($3s^2 3p^3$) P 100.0 30.97376	16 ($3s^2 3p^4$) S 95.00 31.97207 0.76 32.97146 4.22 33.96786 0.014 35.96709	17 ($3s^2 3p^5$) Cl 75.53 34.96885 24.47 36.96590	18 ($3s^2 3p^6$) Ar 0.337 35.96755 0.063 37.96272 99.600 39.96238
33 ($4s^24p^3$) As 100.0 74.9216	34 ($4s^24p^4$) Se 9.02 75.9192 7.58 76.9199 23.52 77.9173 49.82 79.9165 9.19 81.9167	35 ($4s^24p^5$) Br 50.54 78.9183 49.46 80.9163	36 ($4s^24p^6$) Kr 2.27 79.9164 11.56 81.9135 11.55 82.9145 56.90 84.9115 17.37 85.9106
51 ($5s^25p^3$) Sb 57.25 120.9038 42.75 122.9042	52 ($5s^25p^4$) Te 18.71 125.9033 31.79 127.9045 34.48 129.9062 and others	53 ($5s^25p^5$) I 100.0 126.90447	54 ($5s^25p^6$) Xe 26.44 128.9048 21.18 130.9051 26.89 131.9042 and others
83 ($6s^26p^3$) Bi 100.0 208.9804	84 ($6s^26p^4$) Po	85 ($6s^26p^5$) At	86 ($6s^26p^6$) Rn

5d

57($Xe5d^16s^2$)La —4f→ 58-71 Rare Earths $4f^1 \rightarrow 4f^{14}$

72($Xe4f^{14}5d^26s^2$)Hf

73($Xe4f^{14}5d^36s^2$)Ta

74($Xe4f^{14}5d^46s^2$)W

75($Xe4f^{14}5d^56s^2$)Re

76($Xe4f^{14}5d^66s^2$)Os

77($Xe4f^{14}5d^76s^2$)Ir

78($Xe4f^{14}5d^96s^1$)Pt

79($Xe4f^{14}5d^{10}6s^1$)Au

80($Xe4f^{14}5d^{10}6s^2$)Hg

Appendix B MATHEMATICAL MISCELLANY

B.1 MATRICES, VECTORS, AND DYADICS

Matrices

A *matrix* is a square or rectangular array of numbers [2,3],

$$\mathbf{A} = \begin{pmatrix} A_{11} & A_{12} & A_{13} & \cdots \\ A_{21} & A_{22} & A_{23} & \\ A_{31} & A_{32} & A_{33} & \\ \vdots & & & \ddots \end{pmatrix}. \tag{B-1}$$

The sum of two matrices **A** and **B** is a new matrix **C** whose elements are the sums of the corresponding elements in **A** and **B** given by

$$\mathbf{A} + \mathbf{B} = \mathbf{C}$$

$$\begin{pmatrix} A_{11} & A_{12} & A_{13} & \cdots \\ A_{21} & A_{22} & A_{23} & \\ A_{31} & A_{32} & A_{33} & \\ \vdots & & & \ddots \end{pmatrix} + \begin{pmatrix} B_{11} & B_{12} & B_{13} & \cdots \\ B_{21} & B_{22} & B_{23} & \\ B_{31} & B_{32} & B_{33} & \\ \vdots & & & \ddots \end{pmatrix} = \begin{pmatrix} C_{11} & C_{12} & C_{13} & \cdots \\ C_{21} & C_{22} & C_{23} & \\ C_{31} & C_{32} & C_{33} & \\ \vdots & & & \ddots \end{pmatrix}$$

$$\mathbf{C} = \begin{pmatrix} A_{11} + B_{11} & A_{12} + B_{12} & A_{13} + B_{13} & \cdots \\ A_{21} + B_{21} & A_{22} + B_{22} & A_{23} + B_{23} & \\ A_{31} + B_{31} & A_{32} + B_{32} & A_{33} + B_{33} & \\ \vdots & & & \ddots \end{pmatrix}, \tag{B-2}$$

where it is evident that the individual elements are related by

$$C_{ij} = A_{ij} + B_{ij}. \tag{B-3}$$

A, **B**, and **C** need not be square matrices but they must have equivalent row and column dimensions. The multiplication of a matrix by a constant, ε, implies multiplication of each element in the matrix by the constant,

$$\varepsilon\mathbf{A} = \begin{pmatrix} \varepsilon A_{11} & \varepsilon A_{12} & \cdots \\ \varepsilon A_{21} & \varepsilon A_{31} & \cdot \\ \vdots & & \ddots \end{pmatrix}. \tag{B-4}$$

Multiplication of matrices is also straightforward:

$$\mathbf{AB} = \mathbf{C}$$

$$\begin{pmatrix} A_{11} & A_{12} & A_{13} & \cdots \\ A_{21} & A_{22} & A_{23} & \\ A_{31} & A_{32} & A_{33} & \\ \vdots & & & \ddots \end{pmatrix} \begin{pmatrix} B_{11} & B_{12} & B_{13} & \cdots \\ B_{21} & B_{22} & B_{23} & \\ B_{31} & B_{32} & B_{33} & \\ \vdots & & & \ddots \end{pmatrix} = \begin{pmatrix} C_{11} & C_{12} & C_{13} & \cdots \\ C_{21} & C_{22} & C_{23} & \\ C_{31} & C_{32} & C_{33} & \\ \vdots & & & \ddots \end{pmatrix}. \tag{B-5}$$

A single element in **C** is obtained by a corresponding row-column multiplication in **A** and **B** (the number of columns in **A** must be equal to the number of rows in **B**) given by

$$C_{ij} = \sum_k A_{ik} B_{kj}. \tag{B-6}$$

The *complex conjugate* (indicated by an asterisk) of a matrix is obtained by taking the complex conjugate [change all $i = (-1)^{1/2}$ to $-i$] of all elements

$$\mathbf{A}^* = \begin{pmatrix} A^*_{11} & A^*_{12} & \cdots \\ A^*_{21} & A^*_{22} & \\ \vdots & & \ddots \end{pmatrix}. \tag{B-7}$$

The *transpose* of a matrix (indicated by a tilde, $\tilde{\mathbf{A}}$) is obtained by interchanging rows and columns in a matrix:

$$\tilde{\mathbf{A}} = \begin{pmatrix} A_{11} & A_{21} & A_{31} & \cdots \\ A_{12} & A_{22} & A_{32} & \\ A_{13} & A_{23} & A_{33} & \\ \vdots & & & \ddots \end{pmatrix}. \tag{B-8}$$

An *adjoint matrix* (indicated with a dagger, $A^\dagger$) is obtained by taking the transposed complex conjugate:

$$\mathbf{A}^\dagger = \tilde{\mathbf{A}}^* = \begin{pmatrix} A^*_{11} & A^*_{21} & A^*_{31} & \cdots \\ A^*_{12} & A^*_{22} & A^*_{32} & \\ A^*_{13} & A^*_{23} & A^*_{33} & \\ \vdots & & & \ddots \end{pmatrix}. \tag{B-9}$$

Inspection will show that the transpose of a product of matrices is equal to the reverse order of the transposed matrices:

$$\begin{aligned} \mathbf{C} &= \mathbf{AB} & \mathbf{C}^* &= \mathbf{A}^*\mathbf{B}^* \\ \tilde{\mathbf{C}} &= \tilde{\mathbf{B}}\tilde{\mathbf{A}} & \mathbf{C}^\dagger &= \mathbf{B}^\dagger\mathbf{A}^\dagger. \end{aligned} \tag{B-10}$$

More complicated products can be constructed from these examples.

A square matrix, **u**, is referred to as a *unitary matrix* if

$$\mathbf{uu}^\dagger = \mathbf{u}^\dagger\mathbf{u} = \mathbf{1} = \begin{pmatrix} 1 & 0 & 0 & \cdots \\ 0 & 1 & 0 & \\ 0 & 0 & 1 & \\ \vdots & & & \ddots \end{pmatrix}. \tag{B-11}$$

1 is the *unit matrix*. Expanding the components in Eq.(B-11) leads to

$$\sum_j u^\dagger_{kj} u_{jl} = \sum_j u^*_{jk} u_{jl} = \delta_{kl}, \tag{B-12}$$

where δ_{kl} is the *Kronecker delta* defined by

$$\delta_{kl}\begin{cases} =1 & \text{if } k = l \\ =0 & \text{if } k \neq l. \end{cases} \qquad \text{(B-13)}$$

It should be evident that

$$\mathbf{1A} = \mathbf{A} \qquad \mathbf{B1} = \mathbf{B}, \qquad \text{(B-14)}$$

where the order of the square **1** matrix must equal the number of columns of **A** and rows of **B**.

The inverse matrix $\mathbf{B}^{-1}$ of a square matrix **B** is defined as

$$\mathbf{BB}^{-1} = \mathbf{B}^{-1}\mathbf{B} = \mathbf{1}, \qquad \text{(B-15)}$$

which requires that **B** have a nonzero determinant. In the case of a unitary matrix where $\mathbf{u}^{\dagger}\mathbf{u} = \mathbf{1}$,

$$\mathbf{u}^{\dagger} = \mathbf{u}^{-1}. \qquad \text{(B-16)}$$

For the special case of all real elements in a unitary matrix, **u**,

$$\mathbf{u}^{\dagger}\mathbf{u} = \mathbf{1} = \tilde{\mathbf{u}}\mathbf{u}; \qquad \tilde{\mathbf{u}} = \mathbf{u}^{-1}. \qquad \text{(B-17)}$$

Thus, the transpose of a real unitary matrix is equal to its inverse.

Another type of matrix multiplication is the direct product, $\mathbf{A} \times \mathbf{B}$, defined by multiplying each element in **A** by each element in **B**:

$$\mathbf{A} \times \mathbf{B} = \begin{pmatrix} A_{11} & A_{12} \\ A_{21} & A_{22} \end{pmatrix} \times \begin{pmatrix} B_{11} & B_{12} \\ B_{21} & B_{22} \end{pmatrix} = \begin{pmatrix} A_{11}B_{11} & A_{11}B_{12} & A_{12}B_{11} & A_{12}B_{12} \\ A_{11}B_{21} & A_{11}B_{22} & A_{12}B_{21} & A_{12}B_{22} \\ A_{21}B_{11} & A_{21}B_{12} & A_{22}B_{11} & A_{22}B_{12} \\ A_{21}B_{21} & A_{21}B_{22} & A_{22}B_{21} & A_{22}B_{22} \end{pmatrix}. \qquad \text{(B-18)}$$

Vectors and Relation to Matrices

A *scalar* is a number. A *vector* is a number that also contains orientation information. Vectors and scalars obey the normal associative and distributive laws of multiplication.

A three-dimensional Cartesian vector is characterized by components along the three Cartesian axes, given by

$$\boldsymbol{A} = \hat{\boldsymbol{\imath}} A_x + \hat{\boldsymbol{\jmath}} A_y + \hat{\boldsymbol{k}} A_z, \qquad \text{(B-19)}$$

where $\hat{\boldsymbol{\imath}}$, $\hat{\boldsymbol{\jmath}}$, and $\hat{\boldsymbol{k}}$ are unit vectors along x, y, and z, respectively, and A_x, A_y, and A_z are the magnitudes of the vector along the Cartesian axes. Another way of writing Eq.(B-19) is as a product of a row and column matrix given by

$$\boldsymbol{A} = (\hat{\boldsymbol{\imath}} \quad \hat{\boldsymbol{\jmath}} \quad \hat{\boldsymbol{k}}) \begin{pmatrix} A_x \\ A_y \\ A_z \end{pmatrix} = \hat{\boldsymbol{\imath}} A_x + \hat{\boldsymbol{\jmath}} A_y + \hat{\boldsymbol{k}} A_z. \qquad \text{(B-20)}$$

It follows that the scalar product between the two Cartesian vectors $\boldsymbol{A}$ and $\boldsymbol{B}$ can be written as

$$\boldsymbol{A}\cdot\boldsymbol{B} = (A_x \quad A_y \quad A_z)\begin{pmatrix}\hat{\boldsymbol{i}}\\ \hat{\boldsymbol{j}}\\ \hat{\boldsymbol{k}}\end{pmatrix}\cdot(\hat{\boldsymbol{i}} \quad \hat{\boldsymbol{j}} \quad \hat{\boldsymbol{k}})\begin{pmatrix}B_x\\ B_y\\ B_z\end{pmatrix}$$

$$= (A_x \quad A_y \quad A_z)\begin{pmatrix}\hat{\boldsymbol{i}}\cdot\hat{\boldsymbol{i}} & \hat{\boldsymbol{i}}\cdot\hat{\boldsymbol{j}} & \hat{\boldsymbol{i}}\cdot\hat{\boldsymbol{k}}\\ \hat{\boldsymbol{j}}\cdot\hat{\boldsymbol{i}} & \hat{\boldsymbol{j}}\cdot\hat{\boldsymbol{j}} & \hat{\boldsymbol{j}}\cdot\hat{\boldsymbol{k}}\\ \hat{\boldsymbol{k}}\cdot\hat{\boldsymbol{i}} & \hat{\boldsymbol{k}}\cdot\hat{\boldsymbol{j}} & \hat{\boldsymbol{k}}\cdot\hat{\boldsymbol{k}}\end{pmatrix}\begin{pmatrix}B_x\\ B_y\\ B_z\end{pmatrix}$$

$$= (A_x \quad A_y \quad A_z)\begin{pmatrix}1 & 0 & 0\\ 0 & 1 & 0\\ 0 & 0 & 1\end{pmatrix}\begin{pmatrix}B_x\\ B_y\\ B_z\end{pmatrix} = (A_x \quad A_y \quad A_z)\begin{pmatrix}B_x\\ B_y\\ B_z\end{pmatrix}$$

$$= A_xB_x + A_yB_y + A_zB_z, \tag{B-21}$$

where we have used the orthonormality of the scalar product (indicated by the dot) of the unit vectors given by

$$\hat{\boldsymbol{i}}\cdot\hat{\boldsymbol{j}} = \delta_{ij}\begin{cases} =1 & \text{if } i=j\\ =0 & \text{if } i\neq j,\end{cases} \tag{B-22}$$

where δ_{ij} is the Kronecker delta. Thus, the dot or scalar product of two vectors $\boldsymbol{A}$ and $\boldsymbol{B}$ is given also by the row-column multiplication of the components of the vector. It is easy to generalize these equations to n-dimensional space. We emphasize this point here because of the strong analogy between vector space and vector scalar products with components of a state vector in quantum mechanics and the corresponding scalar product of state vectors. The kth state vector is defined in Eq.(3-46) by analogy to Eq.(B-20) for the Cartesian vector. Rewriting Eq.(3-46) here gives

$$\psi_k = \mathbf{\Phi}\mathbf{b}_k = (\Phi_1\Phi_2\cdots)\begin{pmatrix}\mathrm{b}_{1k}\\ \mathrm{b}_{2k}\\ \vdots\end{pmatrix} = \sum_j \Phi_j \mathrm{b}_{jk}. \tag{B-23}$$

The scalar product of state vectors ψ_k and ψ_j is given by the integral over the space defined by $\boldsymbol{\psi}$ and $\mathbf{\Phi}$ according to Eq.(B-21) to give (we use the complex conjugate transpose or the adjoint in place of the transpose for normal vectors)

$$\int \psi_k^*\psi_j\, dV = \mathbf{b}_k^\dagger\left(\int \mathbf{\Phi}^\dagger\mathbf{\Phi}\, dV\right)\mathbf{b}_j$$

$$= \mathbf{b}_k^\dagger \mathbf{1}\mathbf{b}_j = \mathbf{b}_k^\dagger\mathbf{b}_j$$

$$= (\mathrm{b}_{1k}^* \quad \mathrm{b}_{2k}^* \quad \mathrm{b}_{3k}^* \quad \cdots)\begin{pmatrix}\mathrm{b}_{1j}\\ \mathrm{b}_{2j}\\ \mathrm{b}_{3j}\\ \vdots\end{pmatrix}$$

$$= \mathrm{b}_{1k}^*\mathrm{b}_{1j} + \mathrm{b}_{2k}^*\mathrm{b}_{2j} + \mathrm{b}_{3k}^*\mathrm{b}_{3j} + \cdots = \sum_i \mathrm{b}_{ik}^*\mathrm{b}_{ij}. \tag{B-24}$$

The analogy between final results in Eqs.(B-21) and (B-24) is evident. Now, normally in quantum mechanics, the state vector is composed of orthonormal components and Eq.(B-24) is rewritten to give

$$\int \psi_k^* \psi_j \, dV = \delta_{kj} = \sum_i \mathrm{b}_{ik}^* \mathrm{b}_{ij}. \tag{B-25}$$

Returning now to the three-dimensional Cartesian vectors in Eq.(B-21), we note that the components of a vector can be transformed in the three-dimensional space according to

$$\boldsymbol{A} = \mathbf{C}\boldsymbol{A}'$$

$$\begin{pmatrix} A_x \\ A_y \\ A_z \end{pmatrix} = \begin{pmatrix} \mathrm{C}_{11} & \mathrm{C}_{12} & \mathrm{C}_{13} \\ \mathrm{C}_{21} & \mathrm{C}_{22} & \mathrm{C}_{23} \\ \mathrm{C}_{31} & \mathrm{C}_{32} & \mathrm{C}_{33} \end{pmatrix} \begin{pmatrix} A'_x \\ A'_y \\ A'_z \end{pmatrix}. \tag{B-26}$$

$\boldsymbol{A}$ and $\boldsymbol{A}'$ are two different vectors in the three-dimensional space that are related by the matrix transformation, $\mathbf{C}$. The use of the direction cosine transformation between the laboratory-fixed (xyz) and molecular-fixed (abc) axes are used extensively in this text. This transformation is given by

$$\begin{pmatrix} x \\ y \\ z \end{pmatrix} = \begin{pmatrix} \mathrm{C}_{xa} & \mathrm{C}_{xb} & \mathrm{C}_{xc} \\ \mathrm{C}_{ya} & \mathrm{C}_{yb} & \mathrm{C}_{yc} \\ \mathrm{C}_{za} & \mathrm{C}_{zb} & \mathrm{C}_{zc} \end{pmatrix} \begin{pmatrix} a \\ b \\ c \end{pmatrix} = \mathbf{C} \begin{pmatrix} a \\ b \\ c \end{pmatrix}$$

$$\begin{pmatrix} a \\ b \\ c \end{pmatrix} = \begin{pmatrix} \mathrm{C}_{ax} & \mathrm{C}_{ay} & \mathrm{C}_{az} \\ \mathrm{C}_{bx} & \mathrm{C}_{by} & \mathrm{C}_{bz} \\ \mathrm{C}_{cx} & \mathrm{C}_{cy} & \mathrm{C}_{cz} \end{pmatrix} \begin{pmatrix} x \\ y \\ z \end{pmatrix} = \tilde{\mathbf{C}} \begin{pmatrix} x \\ y \\ z \end{pmatrix}, \tag{B-27}$$

where C_{xa} is the cosine of the angle between the a and x axes. $\mathbf{C}$ is a real unitary matrix, $\tilde{\mathbf{C}}\mathbf{C} = \mathbf{1}$. A clockwise rotation of a and b about the c axis is given by

$$\begin{pmatrix} x \\ y \\ z \end{pmatrix} = \begin{pmatrix} \cos\theta & -\sin\theta & 0 \\ \sin\theta & \cos\theta & 0 \\ 0 & 0 & 1 \end{pmatrix} \begin{pmatrix} a \\ b \\ c \end{pmatrix}. \tag{B-28}$$

The direction cosine transformation between the laboratory-fixed xyz axes to the molecular fixed abc axes in a cylindrically symmetric molecule is illustrated in Fig. 4-24 with the results given in Eq.(4-208) in terms of the spherical polar coordinates.

The corresponding direction cosine transformation between the laboratory-fixed xyz axes to the molecular-fixed abc axes in a molecule with arbitrary symmetry can be written in terms of the three Euler angles. These results have been given elsewhere [4].

Dyadics

A three-dimensional Cartesian *dyadic* is given by the following product of vectors [5,6]:

$$\mathsf{T} = \begin{pmatrix} A_x \\ A_y \\ A_z \end{pmatrix} (B_x \quad B_y \quad B_z) = \begin{pmatrix} A_x B_x & A_x B_y & A_x B_z \\ A_y B_x & A_y B_y & A_y B_z \\ A_z B_x & A_z B_y & A_z B_z \end{pmatrix}. \qquad \text{(B-29)}$$

Normally, we shall deal with symmetric dyadics given by

$$\mathsf{T} = \begin{pmatrix} A_x A_x & A_x A_y & A_x A_z \\ A_y A_x & A_y A_y & A_y A_z \\ A_z A_x & A_z A_y & A_z A_z \end{pmatrix} = \begin{pmatrix} \mathsf{T}_{xx} & \mathsf{T}_{xy} & \mathsf{T}_{xz} \\ \mathsf{T}_{yx} & \mathsf{T}_{yy} & \mathsf{T}_{yz} \\ \mathsf{T}_{zx} & \mathsf{T}_{zy} & \mathsf{T}_{zz} \end{pmatrix}. \qquad \text{(B-30)}$$

The scalar product of a three-dimensional Cartesian vector, $\boldsymbol{B}$, with a dyadic, T, is given by

$$\boldsymbol{D} = \boldsymbol{B} \cdot \mathsf{T} = \mathsf{T} \cdot \boldsymbol{B}, \qquad \text{(B-31)}$$

where the last step is valid for a symmetric dyadic. Multiplying out the matrices in Eq.(B-31) shows that

$$\begin{aligned} D_x &= B_x \mathsf{T}_{xx} + B_y \mathsf{T}_{yx} + B_z \mathsf{T}_{zx} \\ D_y &= B_x \mathsf{T}_{xy} + B_y \mathsf{T}_{yy} + B_z \mathsf{T}_{zy} \\ D_z &= B_x \mathsf{T}_{xz} + B_y \mathsf{T}_{yz} + B_z \mathsf{T}_{zz}. \end{aligned} \qquad \text{(B-32)}$$

The scalar product of $\boldsymbol{D}$ with another vector $\boldsymbol{F}$ is given by

$$\boldsymbol{F} \cdot \boldsymbol{D} = \boldsymbol{F} \cdot \mathsf{T} \cdot \boldsymbol{B}. \qquad \text{(B-33)}$$

We can transform a dyadic $\mathsf{T}(xyz)$ in the laboratory xyz axes system to $\mathsf{T}(abc)$ in the molecular-fixed abc axes system by the direction cosine transformation. First we define the dyadics according to Eq.(B-29):

$$\mathsf{T}(xyz) = \begin{pmatrix} x \\ y \\ z \end{pmatrix} (x \quad y \quad z) = \begin{pmatrix} xx & xy & xz \\ yx & yy & yz \\ zx & zy & zz \end{pmatrix}$$

$$\mathsf{T}(abc) = \begin{pmatrix} a \\ b \\ c \end{pmatrix} (a \quad b \quad c) = \begin{pmatrix} aa & ab & ac \\ ba & bb & bc \\ ca & cb & cc \end{pmatrix}. \qquad \text{(B-34)}$$

Now, substituting Eqs.(B-27) and (B-28) gives

$$\mathsf{T}(xyz) = \mathbf{C} \begin{pmatrix} a \\ b \\ c \end{pmatrix} (a \quad b \quad c) \tilde{\mathbf{C}} = \mathbf{C}\mathsf{T}(abc)\tilde{\mathbf{C}}$$

$$\mathsf{T}(abc) = \tilde{\mathbf{C}}\mathsf{T}(xyz)\mathbf{C}, \qquad \text{(B-35)}$$

where **C** is the direction cosine transformation. In problems of molecular interest, it is normally convenient to define a *principal axis system* for the $\mathsf{T}(abc)$ tensor; the principal axis system is the axis system that leads to a diagonal $\mathsf{T}(abc)$ tensor. Thus, we require the unitary diıection cosine transformaıion in Eq.(B-35) to diagonalize $\mathsf{T}(xyz)$ to give a diagonal $\mathsf{T}(abc)$. In some cases, $\mathsf{T}(xyz)$, as calculated in the laboratory (or arbitrary) axis system, will be block diagonal in the form

$$\mathsf{T}(xyz) = \begin{pmatrix} \mathsf{T}_{xx} & \mathsf{T}_{xy} & 0 \\ \mathsf{T}_{yx} & \mathsf{T}_{yy} & 0 \\ 0 & 0 & \mathsf{T}_{zz} \end{pmatrix}.$$

Under these circumstances the symmetric $\mathsf{T}(xyz)$ matrix can be diagonalized by a simple counterclockwise rotation given by

$$\mathsf{T}(abc) = \begin{pmatrix} \mathsf{T}_{aa} & 0 & 0 \\ 0 & \mathsf{T}_{bb} & 0 \\ 0 & 0 & \mathsf{T}_{cc} \end{pmatrix}$$

$$= \begin{pmatrix} \cos\theta & \sin\theta & 0 \\ -\sin\theta & \cos\theta & 0 \\ 0 & 0 & 1 \end{pmatrix} \begin{pmatrix} \mathsf{T}_{xx} & \mathsf{T}_{xy} & 0 \\ \mathsf{T}_{yx} & \mathsf{T}_{yy} & 0 \\ 0 & 0 & \mathsf{T}_{zz} \end{pmatrix} \begin{pmatrix} \cos\theta & -\sin\theta & 0 \\ \sin\theta & \cos\theta & 0 \\ 0 & 0 & 1 \end{pmatrix}. \quad \text{(B-36)}$$

Multiplying out the $\tilde{\mathbf{C}}\mathsf{T}(xyz)\mathbf{C}$ matrix and noting that $\mathsf{T}_{xy} = \mathsf{T}_{yx}$ and $\mathsf{T}_{ab} = 0$ gives

$$\tan 2\theta = \frac{2\mathsf{T}_{xy}}{\mathsf{T}_{xx} - \mathsf{T}_{yy}}$$

$$\mathsf{T}_{aa} = \mathsf{T}_{xx}\cos^2\theta + 2\mathsf{T}_{xy}\cos\theta\sin\theta + \mathsf{T}_{yy}\sin^2\theta$$

$$\mathsf{T}_{bb} = \mathsf{T}_{xx}\sin^2\theta - 2\mathsf{T}_{xy}\cos\theta\sin\theta + \mathsf{T}_{yy}\cos^2\theta. \quad \text{(B-37)}$$

Equations (B-31) and (B-32) reduce to a simple form in some circumstances. For instance, in Eq.(1-5) we write

$$\boldsymbol{D} = \boldsymbol{\varepsilon}\cdot\boldsymbol{E} = \begin{pmatrix} D_x \\ D_y \\ D_z \end{pmatrix} = \begin{pmatrix} \varepsilon_{xx} & \varepsilon_{xy} & \varepsilon_{xz} \\ \varepsilon_{yx} & \varepsilon_{yy} & \varepsilon_{yz} \\ \varepsilon_{zx} & \varepsilon_{zy} & \varepsilon_{zz} \end{pmatrix} \begin{pmatrix} E_x \\ E_y \\ E_z \end{pmatrix},$$

which relates the electric displacement, $\boldsymbol{D}$, to the electric field, $\boldsymbol{E}$, through the dielectric constant of the medium represented by the dyadic, $\boldsymbol{\varepsilon}$. $\boldsymbol{E}$ is the externally applied field and $\boldsymbol{D}$ is the vector representing the displacement. In general, $\boldsymbol{D}$ is not parallel to $\boldsymbol{E}$. If the medium is *isotropic*, however, then the response of the medium is equivalent along each of the x, y, and z axes and we must write

$$D_x = \varepsilon E_x, \qquad D_y = \varepsilon E_y \qquad D_z = \varepsilon E_z.$$

Expanding the $\boldsymbol{\varepsilon}\cdot\boldsymbol{E}$ expression above shows that $\varepsilon_{xy} = \varepsilon_{yx} = \varepsilon_{yz} = \varepsilon_{zy} = \varepsilon_{xz} = \varepsilon_{zx} = 0$ and $\varepsilon_{xx} = \varepsilon_{yy} = \varepsilon_{zz} = \varepsilon$ where ε is the scalar dielectric constant and

$$\boldsymbol{D} = \varepsilon\boldsymbol{E},$$

for an isotropic medium.

Vector Manipulations

We end appendix B with a summary of some useful vector manipulations [5,6]. We remember from Eq.(B-21) that the scalar or dot product of two vectors $\boldsymbol{A}$ and $\boldsymbol{B}$ is given by

$$\boldsymbol{A} \cdot \boldsymbol{B} = \boldsymbol{B} \cdot \boldsymbol{A} = AB \cos \theta = A_x B_x + A_y B_y + A_z B_z, \tag{B-38}$$

where A is the magnitude of $\boldsymbol{A}$ and θ is the angle between the vectors $\boldsymbol{A}$ and $\boldsymbol{B}$. The cross product of two vectors $\boldsymbol{A}$ and $\boldsymbol{B}$ is a vector given by

$$\begin{aligned} \boldsymbol{A} \times \boldsymbol{B} &= -\boldsymbol{B} \times \boldsymbol{A} = (\boldsymbol{A} \times \boldsymbol{B})_x + (\boldsymbol{A} \times \boldsymbol{B})_y + (\boldsymbol{A} \times \boldsymbol{B})_z \\ &= \hat{\boldsymbol{i}}(A_y B_z - A_z B_y) + \hat{\boldsymbol{j}}(A_z B_x - A_x B_z) + \hat{\boldsymbol{k}}(A_x B_y - A_y B_x). \end{aligned} \tag{B-39}$$

The magnitude of $\boldsymbol{A} \times \boldsymbol{B}$ is equal to $AB \sin \theta$, which shows that the cross product of any vector with itself is zero. The $\boldsymbol{A} \times \boldsymbol{B}$ vector is perpendicular to both $\boldsymbol{A}$ and $\boldsymbol{B}$ and is conveniently written in the form of a determinant as

$$\boldsymbol{A} \times \boldsymbol{B} = \begin{vmatrix} \hat{\boldsymbol{i}} & \hat{\boldsymbol{j}} & \hat{\boldsymbol{k}} \\ A_x & A_y & A_z \\ B_x & B_y & B_z \end{vmatrix}. \tag{B-40}$$

Consider now the following equation:

$$\boldsymbol{A} \cdot (\boldsymbol{B} \times \boldsymbol{R}) = \text{scalar}.$$

Writing this scalar in determinantal form gives

$$\begin{aligned} \boldsymbol{A} \cdot (\boldsymbol{B} \times \boldsymbol{R}) &= \begin{vmatrix} A_x & A_y & A_z \\ B_x & B_y & B_z \\ R_x & R_y & R_z \end{vmatrix} \\ &= A_x(B_y R_z - B_z R_y) + A_y(B_z R_x - B_x R_z) + A_z(B_x R_y - B_y R_x). \end{aligned} \tag{B-41}$$

Noting that the value of a determinant changes sign for odd permutation of rows or columns and remains unchanged for even permutations, we can write

$$\boldsymbol{A} \cdot (\boldsymbol{B} \times \boldsymbol{R}) = -\boldsymbol{B} \cdot (\boldsymbol{A} \times \boldsymbol{R}) = \boldsymbol{B} \cdot (\boldsymbol{R} \times \boldsymbol{A}) = -\boldsymbol{R} \cdot (\boldsymbol{B} \times \boldsymbol{A}) = \boldsymbol{R} \cdot (\boldsymbol{A} \times \boldsymbol{B}). \tag{B-42}$$

Several additional vector expressions that are used often include

$$\begin{aligned} \boldsymbol{A} \times (\boldsymbol{B} \times \boldsymbol{R}) &= \boldsymbol{B}(\boldsymbol{A} \cdot \boldsymbol{R}) - \boldsymbol{R}(\boldsymbol{A} \cdot \boldsymbol{B}) \\ (\boldsymbol{A} \times \boldsymbol{B}) \cdot (\boldsymbol{R} \times \boldsymbol{S}) &= (\boldsymbol{A} \cdot \boldsymbol{R})(\boldsymbol{B} \cdot \boldsymbol{S}) - (\boldsymbol{B} \cdot \boldsymbol{R})(\boldsymbol{A} \cdot \boldsymbol{S}). \end{aligned} \tag{B-43}$$

Several additional results including expressions involving the differential vector operator, $\boldsymbol{\nabla}$, are given elsewhere [7].

B.2 FOURIER TRANSFORMS

If a square integrable function $F(x)$ exists, then its Fourier transform, $f(k)$, is defined by [8]

$$f(k) = \frac{1}{2\pi} \int_{-\infty}^{+\infty} F(x) \exp(ikx)\, dx, \tag{B-44}$$

and the reverse transform gives

$$F(x) = \int_{-\infty}^{+\infty} f(k) \exp(-ikx)\, dk. \tag{B-45}$$

$F(x)$ and $f(k)$, defined by the transforms above, are called *Fourier transform pairs.* The sine and cosine Fourier transforms for odd and even $f(k)$ and $F(x)$ functions are easily obtained by using

$$\exp(\pm ikx) = \cos kx \pm i \sin kx$$

in Eqs.(B-44) and (B-45), giving

$$f(k) = \frac{1}{\pi}\int_0^{\infty} F(x) \cos kx\, dx; \qquad F(x) = F(-x), \text{ even} \tag{B-46}$$

$$f(k) = \frac{1}{\pi}\int_0^{\infty} F(x) \sin kx\, dx; \qquad F(x) = -F(-x), \text{ odd} \tag{B-47}$$

$$F(x) = 2\int_0^{\infty} f(k) \cos kx\, dk; \qquad f(k), \text{ even} \tag{B-48}$$

$$F(x) = 2\int_0^{\infty} f(k) \sin kx\, dk; \qquad f(k), \text{ odd.} \tag{B-49}$$

Beginning with the definitions in Eqs.(B-44) and (B-45), several other useful transform expressions can be obtained. Starting with Eq.(B-45), let $x = \xi$, giving

$$F(\xi) = \int_{-\infty}^{+\infty} f(k) \exp(-ik\xi)\, dk. \tag{B-50}$$

Substituting $f(k)$ from Eq.(B-44) gives

$$F(\xi) = \frac{1}{2\pi}\int_{-\infty}^{+\infty}\int_{-\infty}^{+\infty} F(x) \exp[ik(x - \xi)]\, dk\, dx. \tag{B-51}$$

Comparing Eq.(B-51) with Eqs.(B-66) shows that the

$$\frac{1}{2\pi}\int_{-\infty}^{+\infty} \exp[ik(x - \xi)]\, dk$$

integral acts like a delta function by converting $F(x)$ to $F(\xi)$ by integrating over x. Equation (B-51) can now be used to derive several other useful Fourier transform formulas and other results. We start with Eq.(B-51) and expand the exponential function giving

$$\exp[ik(x - \xi)] = (\cos kx + i \sin kx)(\cos k\xi - i \sin k\xi). \tag{B-52}$$

Substituting Eq.(B-52) into Eq.(B-51) for even and odd $F(x)$ functions gives the following results:

$F(x) = F(-x)$, *even function*

$$F(\xi) = \frac{1}{2\pi}\int_{-\infty}^{+\infty}\int_{-\infty}^{+\infty} F(x) \cos kx \cos k\xi\, dk\, dx. \tag{B-53}$$

The $F(x) \cos kx$ product gives an even integrand in x and the $\cos kx \cos k\xi$ product gives an even integrand in k. All other sine and cosine products in Eq.(B-52) times $F(x)$ are odd in either x or k and the integrals vanish. As the integrand is even, we can rewrite Eq.(B-53) as

$$F(\xi) = \frac{2}{\pi} \int_0^{\infty} \int_0^{\infty} F(x) \cos kx \cos k\xi \, dk \, dx, \tag{B-54}$$

for the final result. In a similar manner we write the integral for an odd function, $F(x)$ as

$F(x) = -F(-x)$, *odd function*

$$F(\xi) = \frac{2}{\pi} \int_0^{\infty} \int_0^{\infty} F(x) \sin kx \sin k\xi \, dk \, dx. \tag{B-55}$$

Another important result of Fourier theory is the generalized Parseval theorem given by

$$\int_{-\infty}^{+\infty} F(x)^* G(x) \, dx = 2\pi \int_{-\infty}^{+\infty} f(k)^* g(k) \, dk, \tag{B-56}$$

where $F(x)$ and $f(k)$, as well as $G(x)$ and $g(k)$, are both Fourier transform pairs. Of course, if $F(x) = G(x)$, then $f(k) = g(k)$, giving

$$\int_{-\infty}^{+\infty} |F(x)|^2 \, dx = 2\pi \int_{-\infty}^{+\infty} |f(k)|^2 \, dk. \tag{B-57}$$

Both Eqs.(B-56) and (B-57) are easily proved, given the preceding development.

Another useful result of Fourier transform theory involves convolutions. Let $F_1(\xi)$ and $F_2(\xi)$ be two different square integrable functions of ξ with Fourier transforms given by

$$f_1(k) = \frac{1}{2\pi} \int_{-\infty}^{+\infty} F_1(\xi) \exp(ik\xi) \, d\xi, \qquad f_2(k) = \frac{1}{2\pi} \int_{-\infty}^{+\infty} F_2(\xi) \exp(ik\xi) \, d\xi. \tag{B-58}$$

The *convolution* of $F_1(\xi)$ and $F_2(\xi)$ is defined by

$$G(x) = \int_{-\infty}^{+\infty} F_1(\xi) F_2(x - \xi) \, d\xi. \tag{B-59}$$

Of course, if $F_2(x - \xi) = \delta(x - \xi)$, the Dirac delta function (see Appendix B.3), then $F_1(x) = G(x)$ as shown, for instance, by comparing Eq.(B-59) with Eq.(B-66). We now take the Fourier transform of both sides of Eq.(B-59) giving

$$g(k) = \frac{1}{2\pi} \int_{-\infty}^{+\infty} G(x) \exp(ikx) \, dx = \frac{1}{2\pi} \int_{-\infty}^{+\infty} \int_{-\infty}^{+\infty} F_1(\xi) F_2(x - \xi) \exp(ikx) \, d\xi \, dx, \tag{B-60}$$

where $g(k)$ is the Fourier transform of $G(x)$. We now write $F_1(\xi)$ in terms of the Fourier representation of the delta function, as described in detail in Section B.3:

$$F_1(\xi) = \frac{1}{2\pi} \int_{-\infty}^{+\infty} \int_{-\infty}^{+\infty} F_1(x') \exp[ik'(x' - \xi)]\, dk'\, dx' = \int_{-\infty}^{+\infty} F_1(x')\delta(x' - \xi)\, dx'. \tag{B-61}$$

Substituting this equation into Eq.(B-60) gives

$$g(k) = \frac{1}{4\pi^2} \int_{-\infty}^{+\infty} \int_{-\infty}^{+\infty} \int_{-\infty}^{+\infty} \int_{-\infty}^{+\infty} F_1(x') \exp(ik'x') \exp(-ik'\xi) F_2(x - \xi)$$
$$\times \exp(ikx)\, dk'\, dx'\, dx\, d\xi.$$

Integrating first over x' gives the Fourier transform of $F_1(x')$, which is $f_1(k')$:

$$g(k) = \frac{1}{2\pi} f_1(k') \int_{-\infty}^{+\infty} \int_{-\infty}^{+\infty} \int_{-\infty}^{+\infty} \exp(ik'\xi) F_2(x - \xi) \exp(ikx)\, dk'\, dx\, d\xi.$$

Holding $f_1(k')$ constant and integrating first over k' and then over ξ converts $F_2(x - \xi)$ to $F_2(x)$ according to

$$g(k) = f_1(k') \int_{-\infty}^{+\infty} \int_{-\infty}^{+\infty} \delta(\xi) F_2(x - \xi) \exp(ikx)\, dx\, d\xi$$
$$= f_1(k') \int_{-\infty}^{+\infty} F_2(x) \exp(ikx)\, dx = 2\pi f_1(k') f_2(k). \tag{B-62}$$

Comparing Eqs.(B-59) and (B-62), we have shown that the Fourier transform of the convolution of two functions is equal to 2π times the product of the Fourier transforms of the functions. Introducing the symbol $\mathscr{F}$ to indicate the Fourier transform and $\mathscr{F}^{-1}$ to indicate the reverse Fourier transform, we rewrite Eqs.(B-44), (B-45), and (B-59) as

$$f(k) = \mathscr{F} F(x)$$
$$F(x) = \mathscr{F}^{-1} f(k)$$
$$g(k) = \mathscr{F} G(x) = 2\pi[\mathscr{F} F_1(x)][\mathscr{F} F_2(x)], \tag{B-63}$$

respectively.

B.3 DIRAC DELTA FUNCTIONS

The Dirac delta function [9], $\delta(x)$, in linear space has a single sharp maximum as x approaches zero and rapidly attenuates to zero as x becomes nonzero. The integral of $\delta(x)$ over all x is unity:

$$\int_{-\infty}^{+\infty} \delta(x)\, dx = 1.0. \tag{B-64}$$

$\delta(x)$ has dimension of inverse length. The three-dimensional delta function has inverse volume dimensions,

$$\int_0^\infty \int_0^\pi \int_0^{2\pi} \delta(r, \theta, \phi) r^2 \sin\theta \, dr \, d\theta \, d\phi = \int \delta(\mathbf{r}) \, d\mathbf{r} = 1.0. \qquad \text{(B-65)}$$

The Dirac delta function is only defined in terms of its transformation properties on another function according to

$$F(\xi) = \int_{-\infty}^{+\infty} F(x)\delta(x - \xi) \, dx, \qquad F(0) = F(x = 0) = \int_{-\infty}^{+\infty} F(x)\delta(x) \, dx. \quad \text{(B-66)}$$

We shall now list a number of functions that are conveniently employed as Dirac delta functions.

Fourier Representation

The Fourier representation of the Dirac delta function arises directly from the definitions of the Fourier transforms in Eqs.(B-44) and (B-45). Comparing Eq.(B-51) with Eqs.(B-66) shows that the

$$\frac{1}{2\pi} \int_{-\infty}^{+\infty} \exp\left[-ik(x - \xi)\right] dk$$

integral acts like a Dirac delta function by converting $F(x)$ to $F(\xi)$. Thus, the Fourier representation of the Dirac delta function can be written as

$$\lim_{k \to \infty} \left\{ \frac{1}{2\pi} \int_{-k}^{k} \exp\left[ik(x - \xi)\right] dk \right\} = \delta(x - \xi), \qquad \lim_{k \to \infty} \left[\frac{1}{2\pi} \int_{-k}^{k} \exp(ikx) \, dk \right] = \delta(x). \quad \text{(B-67)}$$

If $F(x)$ in Eqs.(B-66) are even or odd functions, the sine and cosine forms for the delta functions can be used [compare Eqs.(B-54) and (B-55) with Eq.(B-66)] giving

$F(x) = F(-x)$ *in Eq.(B-66)*

$$F(x = \xi) = 2 \int_0^\infty F(x)\delta(x - \xi) \, dx, \qquad \delta(x - \xi) = \lim_{k \to \infty} \left(\frac{1}{\pi} \int_0^k \cos kx \cos k\xi \, dk \right) \quad \text{(B-68)}$$

$F(x) = -F(-x)$ *in Eq.(B-66)*

$$F(x = \xi) = 2 \int_0^\infty F(x)\delta(x - \xi) \, dx, \qquad \delta(x - \xi) = \lim_{k \to \infty} \left(\frac{1}{\pi} \int_0^k \sin kx \sin k\xi \, dk \right). \quad \text{(B-69)}$$

Sine Function

$$\lim_{k \to \infty} \left[\frac{1}{\pi} \frac{\sin (x - x')k}{(x - x')} \right] = \delta(x - x'). \qquad \text{(B-70)}$$

Lorentz Function

$$\lim_{\Delta\nu_L \to 0} [\mathscr{L}(\nu - \nu_0)] = \lim_{\Delta\nu_L \to 0} \left[\frac{1}{\pi} \frac{\Delta\nu_L}{(\Delta\nu_L)^2 + (\nu - \nu_0)^2}\right] = \delta(\nu - \nu_0). \tag{B-71}$$

$\Delta\nu_L$ is the half width at half height of the normalized Lorentz function, $\mathscr{L}(\nu - \nu_0)$, which is centered at ν_0.

Gaussian Function

$$\lim_{\Delta\nu_G \to 0} [G(\nu - \nu_0)] = \lim_{\Delta\nu_G \to 0} \left\{\frac{1}{\Delta\nu_G} \left(\frac{\ln 2}{\pi}\right)^{1/2} \exp\left[-\frac{\ln 2}{(\Delta\nu_G)^2}(\nu - \nu_0)^2\right]\right\} = \delta(\nu - \nu_0). \tag{B-72}$$

$\Delta\nu_G$ is the half width at half height of the normalized Gaussian function, $G(\nu - \nu_0)$, which is centered at ν_0.

Green's Function

$$-\frac{1}{4\pi}(\nabla^2 + k^2)\frac{\exp(ik|\boldsymbol{r} - \boldsymbol{r}'|)}{|\boldsymbol{r} - \boldsymbol{r}'|} = \delta(\boldsymbol{r} - \boldsymbol{r}'). \tag{B-73}$$

$\exp(ik|\boldsymbol{r} - \boldsymbol{r}'|)/|\boldsymbol{r} - \boldsymbol{r}'|$ is the Green's function [10] and $\delta(\boldsymbol{r} - \boldsymbol{r}')$ is a three-dimensional delta function [see Eq.(B-65)].

Appendix C DIFFERENTIAL EQUATIONS

C.1 ASSOCIATED LEGENDRE POLYNOMIALS

The associated Legendre polynomials, $\Theta(\theta)$, are given as the solution to the differential equation in Eq.(4-64) according to

$$\sin\theta \frac{d}{d\theta}\left(\sin\theta \frac{d}{d\theta}\right)\Theta(\theta) + \lambda \sin^2\theta\Theta(\theta) = m^2\Theta(\theta). \tag{C-1}$$

Before solving the associated Legendre differential equation in Eq.(C-1), we shall consider the less general Legendre differential equation obtained by letting $m = 0$ in Eq.(C-1), giving

$$\sin\theta \frac{d}{d\theta}\left(\sin\theta \frac{d}{d\theta}\right)\Theta(\theta) + \lambda \sin^2\theta\Theta(\theta) = 0. \tag{C-2}$$

Introducing a new independent variable,

$$z = \cos\theta, \tag{C-3}$$

which varies between the limits $-1 \leq z \leq +1$, replacing $\Theta(\theta)$ by $P(z)$, and noting that

$$\frac{d\Theta}{d\theta} = \frac{dP}{dz}\frac{dz}{d\theta} = -\frac{dP}{dz}\sin\theta, \tag{C-4}$$

gives the following expression in place of Eq.(C-2):

$$\frac{d}{dz}\left[(1 - z^2)\frac{dP(z)}{dz}\right] + \lambda P(z) = 0. \quad \text{(C-5)}$$

We shall now attempt a power series solution to Eq.(C-5) around $z = 0$ written as

$$P(z) = \sum_{n=0}^{\infty} a_n z^n. \quad \text{(C-6)}$$

Substituting into Eq.(C-5) gives

$$\sum_n n(n-1)a_n z^{n-2} - \sum_n n(n+1)a_n z^n + \lambda \sum_n a_n z^n = 0.$$

Expanding the summation over n and rearranging the terms in the equation to collect the coefficients with identical powers of z gives

$$\sum_n z^n[(n+2)(n+1)a_{n+2} - n(n+1)a_n + \lambda a_n] = 0. \quad \text{(C-7)}$$

According to Eq.(C-7), each coefficient of z^n in Eq.(C-7) must be identically zero, giving the following relation between the a_n coefficients:

$$a_{n+2} = \frac{[n(n+1) - \lambda]}{(n+2)(n+1)} a_n. \quad \text{(C-8)}$$

Thus, according to Eq.(C-8), we have a general solution to the differential equation in two constants, a_n and a_{n+2}. We can write all higher even a_n in terms of an arbitrary a_0 and we can write all higher odd a_n in terms of an arbitrary a_1. In the case of even coefficients, we can start with $a_0 = 1$, which gives

$$a_0 = 1, \qquad a_2 = -\frac{\lambda}{2} a_0 = -\frac{\lambda}{2},$$

$$a_4 = \frac{(6-\lambda)}{12} a_2 = -\frac{(6-\lambda)\lambda}{24} a_0, \qquad a_6 = \left(\frac{20-\lambda}{30}\right) a_4 = \cdots, \quad \text{(C-9)}$$

where the a_n coefficients are still a function of λ. Furthermore, at very high values of n, $a_{n+2} \cong a_n$ and the series will not converge at $z = \pm 1$. If we require the series in Eq.(C-6) to converge, that is, if we require $P(z)$ to be well-behaved at $z = \pm 1$, we must limit the expansion of n. According to Eq.(C-8), if we limit the expansion in even coefficients to the nth term, the $n + 2$ term and all higher terms must be zero. This is accomplished by setting $\lambda = n(n + 1)$ for the terminal term in the summation in Eq.(C-6). Thus, if we write l for the terminal value of n, we have

$$\lambda = l(l+1). \quad \text{(C-10)}$$

Now, returning to the $P_l(z)$ functions in Eq.(C-6), we have

$$\begin{aligned} l &= 0 \qquad P_0(z) = a_0 = 1 \\ l &= 2 \qquad P_2(z) = a_0 + a_2 z^2 = \tfrac{1}{2}(-1 + 3z^2) \\ l &= 4 \qquad P_4(z) = a_0 + a_2 z^2 + a_4 z^4 = \tfrac{1}{8}(3 - 30z^2 + 35z^4). \quad \text{(C-11)} \\ &\vdots \qquad\qquad \vdots \end{aligned}$$

In the examples above, a_0 for each l is chosen so that $P_l(z = 1) = 1.0$. Continuing now with the odd l-values, we still require the series to terminate at $n = l$. Starting again with $a_1 = 1$ and using Eqs.(C-6) and (C-8) gives

$$\begin{aligned} l &= 1 \qquad P_1(z) = a_1 z = z \\ l &= 3 \qquad P_3(z) = \tfrac{1}{2}(-3z + 5z^3) \\ l &= 5 \qquad P_5(z) = \tfrac{1}{8}(15z - 70z^3 + 63z^5). \\ &\vdots \qquad\qquad \vdots \end{aligned} \tag{C-12}$$

The Legendre polynomials as listed in Eqs.(C-11) and (C-12) can also be generated by Rodrigues' formula given by

$$P_l(z) = \frac{1}{2^l l\,!}\left(\frac{d}{dz}\right)^l (z^2 - 1)^l. \tag{C-13}$$

This formula gives a convenient method of obtaining the higher-order terms.

We now note that the $P_l(z)$ functions are orthogonal but unnormalized functions. They are normalized over the $-1 \leq z \leq 1$ interval by

$$\int_{-1}^{1} P_l(z)P_{l'}(z)\,dz = \frac{2}{2l + 1}\delta_{ll'}, \tag{C-14}$$

where $\delta_{ll'}$ is the Kronecker delta; $\delta_{ll'} = 0$ if $l \neq l'$ and $\delta_{ll'} = 1$ if $l = l'$. Thus, the complete set of orthonormal Legendre polynomials are given by

$$\left(\frac{2l + 1}{2}\right)^{1/2} P_l(z), \tag{C-15}$$

where $P_l(z)$ is given in Eqs.(C-11), (C-12), and (C-13).

We shall now return to the differential equation for the associated Legendre polynomials in Eq.(C-1). Substituting the change of variables in Eqs.(C-3) and (C-4) into Eq.(C-1), where $\lambda = l(l + 1)$ gives

$$\frac{d}{dz}(1 - z^2)\frac{dP(z)}{dz} + \left[l(l + 1) - \frac{m^2}{(1 - z^2)}\right]P(z) = 0. \tag{C-16}$$

Careful inspection of Eq.(C-16) shows that we can substitute

$$P(z) = [(1 - z^2)^{1/2}]^m G(z) \tag{C-17}$$

into Eq.(C-16) and show that the resultant differential equation is equivalent to the mth derivative of the Legendre differential equation in Eq.(C-5). Substituting Eq.(C-17) into Eq.(C-16) gives

$$(1 - z^2)\frac{d^2G(z)}{dz^2} - 2(m + 1)z\frac{dG(z)}{dz} + [l(l + 1) - m(m + 1)]G(z) = 0.$$

Differentiating Eq.(C-5) m times $[\lambda = l(l + 1)]$ gives

$$(1 - z^2)\frac{d^{m+2}P_l(z)}{dz^{m+2}} - 2(m + 1)z\frac{d^{m+1}P_l(z)}{dz^{m+1}} + [l(l + 1) - m(m + 1)]P_l(z) = 0.$$

According to these last two equations, the Legendre polynomials, $P_l(z)$, are related to the $G(z)$ functions by

$$G(z) = \frac{d^m}{dz^m} P_l(z).$$

Substituting this result into Eq.(C-17) gives

$$P(z) = P_l^m(z) = (1 - z^2)^{m/2} \frac{d^m}{dz^m} P_l(z), \tag{C-18}$$

where $P_l^m(z)$ are the associated Legendre polynomials which are defined in terms of the Legendre polynomials which are given in Eqs.(C-11), (C-12), and (C-13). The associated Legendre polynomials, $P_l^m(z)$, satisfy the more general differential equation in Eq.(C-16). Substituting Eq.(C-3) and normalizing gives the orthonormal associated Legendre polynomials shown in Eq.(4-65). Multiplying the associated Legendre polynomials in Eq.(4-65) by $\Phi_m(\phi) = (1/2\pi)^{1/2} \exp(-i\phi m)$ gives the orthonormal spherical harmonics in Eq.(4-66). The spherical harmonics for the lower values of l and m are listed in Table 4-3.

C.2 ASSOCIATED LAGUERRE POLYNOMIALS

The associated Laguerre polynomials satisfy the differential equation in Eq.(4-84) given by

$$x \frac{d^2 U(x)}{dx^2} + (2l + 2 - x) \frac{dU(x)}{dx} + (n - l - 1)U(x) = 0. \tag{C-19}$$

The radial function for the hydrogen-like atom is related to the $U(x)$ function above through Eq.(4-83) as discussed in Section 4.6. The range of x in Eq.(C-19) is $0 \leq x \leq \infty$ and we shall require $U(x)$ to be well-behaved over this interval. We can solve the differential equation in Eq.(C-19) by a series expansion in x around $x = 0$ given by

$$U(x) = \sum_{q=0}^{\infty} a_q x^q. \tag{C-20}$$

Substituting into Eq.(C-19) gives

$$\sum_q q(q-1)a_q x^{q-1} + (2l+2) \sum_q q a_q x^{q-1} - \sum_q q a_q x^q - (n - l - 1) \sum_q a_q x^q = 0.$$

Expanding the summations in q, rearranging, and collecting coefficients with equal powers of x gives

$$\sum_q x^q [(q+1)(q+2l+2)a_{q+1} - (q + 1 + l - n)a_q] = 0. \tag{C-21}$$

In order for Eq.(C-21) to be satisfied, each coefficient of x^q has to be equal to zero, giving

$$a_{q+1} = \frac{(q + 1 + l - n)}{(q+1)(q+2l+2)} a_q. \tag{C-22}$$

It is evident from Eq.(C-22) that the series in Eq.(C-20) will not go to zero as $x \longrightarrow \infty$ as the a_q coefficients in Eq.(C-20) behave in a manner similar to the coefficients in the e^x expansion around $x = 0$:

$$e^x = \sum_{s=0}^{\infty} \frac{x^s}{s!} = \sum_{s=0}^{\infty} b_s x^s = 1 + \frac{x^2}{2} + \frac{x^3}{6} + \cdots + \frac{x^l}{l!} + \frac{x^{l+1}}{(l+1)!} + \cdots. \quad \text{(C-23)}$$

The ratio of the b_{s+1} to b_s coefficients in Eq.(C-23) is

$$\lim_{s \to \infty} \frac{b_{s+1}}{b_s} = \frac{1}{s}. \quad \text{(C-24)}$$

Now, returning to Eq.(C-22) for the expansion coefficients in Eq.(C-20) for the solution to the differential equation in Eq.(C-19), we have

$$\lim_{q \to \infty} \frac{a_{q+1}}{a_q} = \frac{1}{q}. \quad \text{(C-25)}$$

According to Eqs.(C-24) and (C-25), e^x in Eq.(C-23) and $U(x)$ in Eq.(C-20) have the same behavior at large x; they do not converge. Thus, we must limit the expansion in Eq.(C-20) to finite terms and rely on the exp $(-x/2)$ term in the radial function for the hydrogen atom [Eq.(4-83)] to reduce the function to the proper convergent behavior at large values of x. Accordingly, we shall terminate the expansion in q to the qth term by setting $a_{q+1} = 0$ in Eq.(C-22), giving

$$n = l + 1 + q. \quad \text{(C-26)}$$

Of course, all higher terms in the sum over q will also be zero. According to Eq.(C-26), we note that l can have values 0, 1, 2, ... and q can have values of 0, 1, 2, ... Therefore, n in Eq.(C-26) can have values of 1, 2, 3, ... and n cannot equal zero. For any given value of n, $l = n - 1, n - 2, n - 3, \ldots, 0$ depending on the value of q. In summary, Eqs.(C-22) and (C-20) define the associated Laguerre polynomials in argument x, $U(x) = L_{n+l}^{2l+1}(x)$, which can be defined in terms of the $2l + 1$ derivative of the Laguerre polynomial, $L_{n+l}(x)$,

$$U(x) = L_{n+l}^{2l+1}(x) = \frac{d^{2l+1}}{dx^{2l+1}} L_{n+l}(x) = \frac{d^{2l+1}}{dx^{2l+1}} \left\{ \exp(x) \frac{d^{n+l}}{dx^{n+l}} [x^{n+l} \exp(-x)] \right\}. \quad \text{(C-27)}$$

We can also solve Eq.(C-19) in a manner more analogous to the method used in Section C.1 on the differential equation for the associated Legendre polynomials. For instance, the Laguerre differential equation is obtained from Eq.(C-19) by letting $2l + 1 = 0$, giving

$$\frac{x\, d^2 L_{n+l}(x)}{dx^2} + (1 - x)\frac{dL_{n+l}(x)}{dx} + (n + l)L_{n+l}(x) = 0. \quad \text{(C-28)}$$

Solving this equation by a series expansion around $x = 0$ to generate the Laguerre polynomials,

$$L_s(x) = \sum_{s=0}^{\infty} a_s x^s, \tag{C-29}$$

gives

$$L_s(x) = \exp(x) \frac{d^s}{dx^s}[x^s \exp(-x)]$$

$$L_0(x) = 1$$

$$L_1(x) = 1 - x$$

$$L_2(x) = \frac{1}{2}(2 - 4x + x^2)$$

$$L_3(x) = \frac{1}{6}(6 - 18x + 9x^2 - x^3)$$

$$L_4(x) = \frac{1}{4!}(24 - 96x + 72x^2 - 16x^3 + x^4)$$

$$L_5(x) = \frac{1}{5!}(120 - 600x + 600x^2 - 200x^3 + 25x^4 - x^5) \tag{C-30}$$

$$\vdots \qquad \vdots$$

We can now obtain the differential equation for the associated Laguerre polynomials by differentiating Eq.(C-28) $2l + 1$ times to give Eq.(C-19), where

$$U(x) = L_{l+n}^{2l+1}(x) = \frac{d^{2l+1}}{dx^{2l+1}} L_{l+n}(x), \tag{C-31}$$

as in Eq.(C-27). The associated Laguerre polynomials can be normalized and combined with the $x^l \exp(-x/2)$ factors in Eq.(4-83) to give the normalized radial functions for the hydrogen-like atom in Eq.(4-87) and Table 4-4.

C.3 HERMITE POLYNOMIALS

The differential equation for the linear harmonic oscillator is given in Eq.(4-240) according to

$$-\frac{\hbar^2}{2\mu}\frac{d^2\psi(s)}{ds^2} + \left(\frac{1}{2}ks^2 - E\right)\psi(s) = 0. \tag{C-32}$$

Rearranging gives

$$\frac{d^2\psi(s)}{ds^2} + (\lambda - \gamma^4 s^2)\psi(s) = 0, \tag{C-33}$$

where we have used

$$\lambda = \frac{2\mu E}{\hbar^2}, \qquad \gamma = \left[\frac{(\mu k)^{1/2}}{\hbar}\right]^{1/2}. \tag{C-34}$$

γ has dimensions of inverse length. The asymptotic solution of Eq.(C-33) can be obtained at large s. The limit of Eq.(C-33) as $s \longrightarrow \infty$ is

$$\frac{d^2\psi(s)}{ds^2} - \gamma^4 s^2 \psi(s) = 0, \tag{C-35}$$

which has an approximate solution for large s of

$$\psi(s) = \exp\left(\pm \frac{\gamma^2 s^2}{2}\right). \tag{C-36}$$

The $+\gamma^2 s^2/2$ argument in Eq.(C-36) is excluded because we require the function to go to zero at large s. Thus, the solution to Eq.(C-32) for all s will be of the form

$$\psi(s) = \exp\left(-\frac{\gamma^2 s^2}{2}\right) H(\gamma s), \tag{C-37}$$

where $H(\gamma s)$ is a polynomial that can be generated by a power series expansion about $s = 0$ given by

$$H(\gamma s) = \sum_{j=0}^{\infty} a_j (\gamma s)^j. \tag{C-38}$$

Substituting Eqs.(C-38) and (C-37) into Eq.(C-32) gives an equation in powers of γs. Rearranging and collecting the coefficients for the various powers of $(\gamma s)^j$ gives

$$\sum_{j=0}^{\infty} [(j+2)(j+1)a_{j+2} + (\lambda - 2\gamma^2 j - \gamma^2)a_j](\gamma s)^j = 0. \tag{C-39}$$

This equation can hold only if all coefficients of the $(\gamma s)^j$ are separately zero giving

$$a_{j+2} = \frac{2\gamma^2 j + \gamma^2 - \lambda}{(j+2)(j+1)} a_j. \tag{C-40}$$

Equation (C-40) shows that $a_2, a_4, \ldots, a_{\text{even}}$ are all determined in terms of a_0 and $a_3, a_5, a_7, \ldots, a_{\text{odd}}$ are all determined in terms of a_1. Thus, we have a general solution to the differential equation and we need only investigate its acceptability as a quantum mechanical wavefunction for the system. According to the coefficients in Eq.(C-40) [when substituted into Eqs.(C-38) and (C-37)], $H(\gamma s)$ is continuous and single-valued; however, the series in Eq.(C-38) fails to converge as $s \longrightarrow \infty$. In order to maintain a physically convergent and acceptable solution, we must terminate the expansion in Eq.(C-38) at the jth term. The convergence of the quantum mechanical wavefunction for the terminated polynomial will be assured by the exp $(-\gamma^2 s^2/2)$ factor in Eq.(C-37). If we require the a_{j+2} term in Eq.(C-40) to be zero, we have, for the final $j = n$ term in Eq.(C-39),

$$\lambda = \gamma^2(2n+1). \tag{C-41}$$

Returning to substitute Eqs.(C-34) into Eq.(C-41) gives the energy levels as a function of n for the harmonic oscillator:

$$E_n = \hbar\left(\frac{k}{\mu}\right)^{1/2}\left(n + \frac{1}{2}\right) = h\nu\left(n + \frac{1}{2}\right), \tag{C-42}$$

where $\nu = (1/2\pi)(k/\mu)^{1/2}$. We shall now generate the Hermite polynomials in the truncated series described in Eqs.(C-38) and (C-39). Substituting Eq.(C-41) into Eq.(C-40) gives (n is the highest term in the summation over j)

$$a_{j+2} = \frac{\gamma^2[(2j + 1) - (2n + 1)]}{(j + 2)(j + 1)} a_j. \tag{C-43}$$

It is now possible to generate the Hermite polynomials in Eq.(C-38) from the coefficients in Eq.(C-43). Each n function requires the definition of all higher a_j coefficients in terms of the first term, however (a_0 for the even functions and a_1 for the odd functions). The *Hermite polynomials* are the functions that define the initial coefficients in the different n functions according to the generation formula given by

$$H_n(\gamma s) = (-1)^n \exp(\gamma^2 s^2) \frac{d^n}{d(\gamma s)^n} \exp(-\gamma^2 s^2), \tag{C-44}$$

which gives the following first few Hermite polynomials, which are consistent with the coefficients in Eq.(C-43),

$$H_0(\gamma s) = 1, \qquad H_1(\gamma s) = 2(\gamma s),$$
$$H_2(\gamma s) = 4(\gamma s)^2 - 2, \qquad H_3(\gamma s) = 8(\gamma s)^3 - 12(\gamma s),$$
$$H_4(\gamma s) = 16(\gamma s)^4 - 48(\gamma s)^2 + 12, \qquad H_5(\gamma s) = 32(\gamma s)^5 - 160(\gamma s)^3 + 120(\gamma s). \tag{C-45}$$

The resulting harmonic oscillator wavefunctions in argument γs are obtained by multiplying the Hermite polynomials defined above times $\exp(-\gamma^2 s^2/2)$ as shown in Eq.(C-37). The resulting normalized function is given in Eq.(4-241) and the first few functions are listed in Table 4-25.

Appendix D NUMERICAL METHODS

D.1 ORTHOGONALIZATION METHODS

Orthogonalization methods are necessary in a number of applications in quantum chemistry; for instance, in the solution of a secular equation where the basis set of functions are not orthonormal. The method of solving the secular equations by matrix methods when a nonorthogonal set is used is discussed in Section 3.5 starting with Eq.(3-127),

$$\mathscr{H}^\chi \mathbf{b} = \mathbf{S}^\chi \mathbf{b}\varepsilon, \qquad \mathbf{b}^\dagger \mathscr{H}^\chi \mathbf{b} = \mathbf{b}^\dagger \mathbf{S}^\chi \mathbf{b}\varepsilon. \tag{D-1}$$

$\mathscr{H}^\chi$ is the Hamiltonian matrix in the nonorthonormal χ basis. $\mathbf{b}$ is a nonunitary transformation that diagonalizes the $\mathscr{H}^\chi$ matrix and transforms $\mathbf{S}^\chi$ to $\mathbf{1}$.

Consider a normalized but nonorthogonal basis $\boldsymbol{\chi}$, which we wish to transform to an orthonormal basis, $\boldsymbol{\Phi}$; that is, we seek a transformation

$$\boldsymbol{\Phi} = \boldsymbol{\chi}\mathbf{c}, \tag{D-2}$$

where $\mathbf{c}$ is not necessarily unitary and $\boldsymbol{\Phi}$ is an orthonormal set but is not necessarily an eigenfunction of the Hamiltonian. From Eq.(D-2), we can write

$$\int \boldsymbol{\Phi}^{\dagger}\boldsymbol{\Phi} = \mathbf{1} = \mathbf{c}^{\dagger}\left(\int \boldsymbol{\chi}^{\dagger}\boldsymbol{\chi}\, dV\right)\mathbf{c} = \mathbf{c}^{\dagger}\mathbf{S}^{\chi}\mathbf{c}, \tag{D-3}$$

where $\mathbf{S}^{\chi}$ is the symmetric overlap matrix in the normalized but nonorthogonal $\boldsymbol{\chi}$ basis. The superscript χ will be dropped in the following discussion. A unitary matrix will diagonalize $\mathbf{S}$ but it will not necessarily transform $\mathbf{S}$ into the unit matrix. $\mathbf{c}$, of course, is not a unique matrix and we shall consider two methods of obtaining this matrix. These methods are the symmetric orthogonalization method and the Schmidt orthogonalization method.

D.2 SYMMETRIC ORTHOGONALIZATION

In the symmetric orthogonalization method, $\mathbf{c}$ is given by

$$\mathbf{c} = \mathbf{S}^{-1/2}. \tag{D-4}$$

$\mathbf{S}^{-1/2}$ is obtained by first diagonalizing $\mathbf{S}$ with a unitary transformation, $\mathbf{u}$,

$$\mathbf{u}^{\dagger}\mathbf{S}\mathbf{u} = \boldsymbol{\lambda}, \tag{D-5}$$

where $\boldsymbol{\lambda}$ is a diagonal matrix. The $\boldsymbol{\lambda}^{1/2}$ matrix is obtained by taking the square root of each element in $\boldsymbol{\lambda}$. The $\boldsymbol{\lambda}^{-1/2}$ matrix is obtained by dividing each diagonal element in $\boldsymbol{\lambda}^{1/2}$ into 1. Thus,

$$\boldsymbol{\lambda}^{1/2} = \begin{pmatrix} (\lambda_1)^{1/2} & 0 & 0 & \cdots \\ 0 & (\lambda_2)^{1/2} & 0 & \\ 0 & 0 & (\lambda_3)^{1/2} & \\ \vdots & & & \ddots \end{pmatrix};$$

$$\boldsymbol{\lambda}^{-1/2} = \begin{pmatrix} \left(\dfrac{1}{\lambda_1}\right)^{1/2} & 0 & 0 & \cdots \\ 0 & \left(\dfrac{1}{\lambda_2}\right)^{1/2} & 0 & \\ 0 & 0 & \left(\dfrac{1}{\lambda_3}\right)^{1/2} & \\ \vdots & & & \ddots \end{pmatrix}. \tag{D-6}$$

To obtain $\mathbf{S}^{-1/2}$, the reverse transformation [see Eq.(D-5)] is performed on $\boldsymbol{\lambda}^{-1/2}$,

$$\mathbf{S}^{-1/2} = \mathbf{u}\boldsymbol{\lambda}^{-1/2}\mathbf{u}^{\dagger}. \tag{D-7}$$

The adjoint of $\mathbf{S}^{-1/2}$ is

$$(\mathbf{S}^{-1/2})^{\dagger} = \mathbf{u}(\boldsymbol{\lambda}^{-1/2})^{\dagger}\mathbf{u}^{\dagger} = \mathbf{u}\boldsymbol{\lambda}^{-1/2}\mathbf{u}^{\dagger} = \mathbf{S}^{-1/2}, \tag{D-8}$$

as $\boldsymbol{\lambda}^{-1/2}$ is real and diagonal. It is also easy to show that $\mathbf{S}^{-1/2}\mathbf{S}^{-1/2} = \mathbf{S}^{-1}$:

$$\mathbf{S}^{-1/2}\mathbf{S}^{-1/2} = \mathbf{u}\boldsymbol{\lambda}^{-1/2}\mathbf{u}^{\dagger}\mathbf{u}\boldsymbol{\lambda}^{-1/2}\mathbf{u}^{\dagger} = \mathbf{u}\boldsymbol{\lambda}^{-1/2}\boldsymbol{\lambda}^{-1/2}\mathbf{u}^{\dagger} = \mathbf{u}\boldsymbol{\lambda}^{-1}\mathbf{u}^{\dagger} = \mathbf{S}^{-1}. \tag{D-9}$$

The last step in Eq.(D-9) is obtained from the reverse transformation of $\boldsymbol{\lambda}^{-1}$ from Eq.(D-5). Substituting Eq.(D-4) into Eq.(D-2) gives

$$\boldsymbol{\Phi} = \boldsymbol{\chi}\mathbf{c} = \boldsymbol{\chi}\mathbf{S}^{-1/2}. \tag{D-10}$$

The nonunitary $\mathbf{S}^{-1/2}$ matrix transforms $\boldsymbol{\chi}$ to an orthonormal set, $\boldsymbol{\Phi}$.

Equation (D-10) with a unitary transformation on $\mathscr{H}^{\Phi}$ may now be used to solve the Schrödinger matrix equation instead of the secular equation solution developed in Section 3.5 for nonorthogonal basis sets. Substituting Eq.(D-10) into $\mathbf{a}^{\dagger}\mathscr{H}^{\Phi}\mathbf{a} = \mathbf{E}$ gives

$$\begin{aligned} \mathbf{E} &= \mathscr{H}^{\psi} = \mathbf{a}^{\dagger}\mathscr{H}^{\Phi}\mathbf{a} = \mathbf{a}^{\dagger}(\mathbf{S}^{-1/2})^{\dagger}\mathscr{H}^{\chi}\mathbf{S}^{-1/2}\mathbf{a} \\ \boldsymbol{\psi} &= \boldsymbol{\Phi}\mathbf{a} = \boldsymbol{\chi}\mathbf{S}^{-1/2}\mathbf{a} = \boldsymbol{\chi}\mathbf{b} \\ \mathbf{b} &= \mathbf{S}^{-1/2}\mathbf{a}. \end{aligned} \tag{D-11}$$

Thus, the $\mathbf{S}^{-1/2}$ matrix in Eq.(D-4) transforms $\mathscr{H}^{\chi}$ to $\mathscr{H}^{\Phi}$ and $\mathscr{H}^{\Phi}$ is diagonalized by the unitary **a** transformation to yield the eigenvalues. **b** in Eqs.(D-11) is identical to the **b** in Eqs.(D-1) that was obtained on the basis of the solution of the secular equation. Thus, the symmetric orthogonalization method is the method used in Section 3.5 with the secular equation [11].

D.3 SCHMIDT ORTHOGONALIZATION

Returning to Eq.(D-2), we seek a matrix, **c**, to transform the nonorthogonal basis, $\boldsymbol{\chi}$, to an orthogonal set. In order to generate a new orthogonalized basis $\boldsymbol{\Phi}$ from the nonorthogonal but normalized $\boldsymbol{\chi}$ basis, choose

$$\Phi_1 = \chi_1. \tag{D-12}$$

The Φ_1 function is then normalized and each successive Φ_i is constructed to be orthogonal to all preceding normalized functions; that is,

$$\Phi_2 = \chi_2 + \Phi_1\alpha_{12}, \tag{D-13}$$

where α_{12} is chosen to make

$$\int \Phi_1^*\Phi_2 \, dV = 0$$

over the interval defined by the coordinates of the problem:

$$\int \Phi_1^*\Phi_2 \, dV = \int \chi_1^*(\chi_2 + \chi_1\alpha_{12}) \, dV = S_{12} + \alpha_{12} = 0; \qquad \alpha_{12} = -S_{12}.$$

Normalizing Φ_2 gives

$$\Phi_2 = \frac{\chi_2 - S_{12}\chi_1}{(1 - S_{12}^2)^{1/2}}. \tag{D-14}$$

The third function, Φ_3, is calculated to be orthogonal to Φ_1 and Φ_2,

$$\Phi_3 = \chi_3 + \Phi_2\alpha_{23} + \Phi_1\alpha_{13} = \chi_3 + \chi_2\left[\frac{\alpha_{23}}{(1 - S_{12}^2)^{1/2}}\right] + \chi_1\left[\alpha_{13} - \frac{S_{12}\alpha_{23}}{(1 - S_{12}^2)^{1/2}}\right], \tag{D-15}$$

where α_{23} and α_{13} are determined from the two simultaneous orthogonality relations,

$$\int \Phi_1^*\Phi_3\, dV = 0 \quad \text{and} \quad \int \Phi_2^*\Phi_3\, dV = 0,$$

to be

$$\alpha_{13} = -S_{13} \quad \text{and} \quad \alpha_{23} = \frac{(S_{12}S_{13} - S_{23})(1 + S_{12}^2)}{(1 - S_{12}^2)}.$$

Φ_3 is then normalized and the method continues, to give the **c** matrix in Eq.(D-2).

As an interesting exercise in the use of the Schmidt orthogonalization method, the reader can show that the Legendre polynomials can be generated from the starting basis of normalized 1, z, z^2, ... functions given by

$$\chi = \left(\left(\frac{1}{2}\right)^{1/2} \quad \left(\frac{3}{2}\right)^{1/2} z \quad \left(\frac{5}{2}\right)^{1/2} z^2 \quad \left(\frac{7}{2}\right)^{1/2} z^3 \quad \cdots\right).$$

Using the methods above and Eq.(D-2) will lead to the results in Eqs.(C-11) and (C-12).

D.4 THE METHOD OF LEAST SQUARES

The normal unweighted least-squares method is used to analyze measurable quantities that are related to a set of unknowns (bearing information about molecular constants) through a set of linearly independent equations. Consider a system of m equations that contain n unknown constants (parameters) of interest. Each of the m equations are equal to a measured quantity in a column matrix **M**. There are n unknowns in a **U** column matrix and the coefficients in the linear equations in an **A** matrix. The equation relating the three matrices is

$$\mathbf{AU} = \mathbf{M}$$

$$\begin{pmatrix} A_{11} & A_{12} & A_{13} & \cdots & A_{1n} \\ A_{21} & A_{22} & A_{23} & & \\ A_{31} & A_{32} & A_{33} & & \\ \vdots & & & \ddots & \\ A_{m1} & & & & A_{mn} \end{pmatrix} \begin{pmatrix} U_1 \\ U_2 \\ U_3 \\ \vdots \\ U_n \end{pmatrix} = \begin{pmatrix} M_1 \\ M_2 \\ M_3 \\ \vdots \\ M_m \end{pmatrix}. \tag{D-16}$$

Remembering that there are n unknowns in $\mathbf{U}$ and m measurements in $\mathbf{M}$, the dimensions of $\mathbf{A}$ are $m \times n$ (row $\times$ column) $\tilde{\mathbf{A}}\mathbf{A}$ is a square $n \times n$ matrix. An example of the analysis of data by Eq.(D-16) is found in Problem 4-35 where six rotational transitions in CO were listed in Table 4-31. According to the theory developed in Problem 4-35, the six measured rotational frequencies can be fit with two parameters: r_0, the internuclear distance, and k, the force constant for the molecular vibration. Thus, $\mathbf{U}$ is a 2×1 matrix containing r_0 and k, $\mathbf{M}$ is a 6×1 matrix containing the six observed frequencies, and the 6×2 $\mathbf{A}$ matrix can be generated by Eq.(4-320) for the given J states. The least-squares method is a systematic way of fitting the six measurables in Table 4-31 with the two parameters, r_0 and k.

In the least-squares method, the difference between the $\mathbf{AU}$ and $\mathbf{M}$ matrices is minimized in a systematic way. The deviation matrix, $\boldsymbol{\delta}$, is defined by

$$\boldsymbol{\delta} = \mathbf{AU} - \mathbf{M}. \tag{D-17}$$

In the least-squares procedure, $\tilde{\boldsymbol{\delta}}\boldsymbol{\delta}$ is minimized with respect to the unknowns or numbers in the $\mathbf{U}$ matrix. $\tilde{\boldsymbol{\delta}}\boldsymbol{\delta}$ is a row matrix times a column matrix or a single number. Starting with

$$\begin{aligned}\tilde{\boldsymbol{\delta}}\boldsymbol{\delta} &= (\tilde{\mathbf{U}}\tilde{\mathbf{A}} - \tilde{\mathbf{M}})(\mathbf{AU} - \mathbf{M}) \\ &= \tilde{\mathbf{U}}\tilde{\mathbf{A}}\mathbf{AU} - \tilde{\mathbf{U}}\tilde{\mathbf{A}}\mathbf{M} - \tilde{\mathbf{M}}\mathbf{AU} + \tilde{\mathbf{M}}\mathbf{M},\end{aligned} \tag{D-18}$$

we minimize $\tilde{\boldsymbol{\delta}}\boldsymbol{\delta}$ with respect to the ith element in $\mathbf{U}$, U_i, to give

$$\frac{\partial \tilde{\boldsymbol{\delta}}\boldsymbol{\delta}}{\partial \mathrm{U}_i} = 0 = (\tilde{\mathbf{A}}\mathbf{AU})_i + (\tilde{\mathbf{U}}\tilde{\mathbf{A}}\mathbf{A})_i - (\tilde{\mathbf{A}}\mathbf{M})_i - (\tilde{\mathbf{M}}\mathbf{A})_i, \tag{D-19}$$

where $(\tilde{\mathbf{A}}\mathbf{AU})_i$ and $(\tilde{\mathbf{A}}\mathbf{M})_i$ are the ith rows in the $\tilde{\mathbf{A}}\mathbf{AU}$ and $\tilde{\mathbf{A}}\mathbf{M}$ column matrices, respectively, and $(\tilde{\mathbf{U}}\tilde{\mathbf{A}}\mathbf{A})_i$ and $(\tilde{\mathbf{M}}\mathbf{A})_i$ are the ith columns in the $\tilde{\mathbf{U}}\tilde{\mathbf{A}}\mathbf{A}$ and $\tilde{\mathbf{M}}\mathbf{A}$ row matrices, respectively. Noting that $(\tilde{\mathbf{A}}\mathbf{AU})_i = (\tilde{\mathbf{U}}\tilde{\mathbf{A}}\mathbf{A})_i$ and $(\tilde{\mathbf{A}}\mathbf{M})_i = (\tilde{\mathbf{M}}\mathbf{A})_i$ and repeating the step in Eq.(D-19) for all elements U_i in $\mathbf{U}$ leads to the following equation for column matrices:

$$\tilde{\mathbf{A}}\mathbf{AU} = \tilde{\mathbf{A}}\mathbf{M}. \tag{D-20}$$

Now, remembering that $\tilde{\mathbf{A}}\mathbf{A}$ is a square symmetric $n \times n$ matrix, we find the inverse of the square $\tilde{\mathbf{A}}\mathbf{A}$ matrix and premultiply Eq.(D-20) by this inverse to give the matrix of the unknowns:

$$\mathbf{U} = (\tilde{\mathbf{A}}\mathbf{A})^{-1}\tilde{\mathbf{A}}\mathbf{M}. \tag{D-21}$$

$\tilde{\mathbf{A}}\mathbf{A}$ must, of course, have an inverse for the least-squares method to be valid. Thus, as the $\mathbf{A}$ and $\mathbf{M}$ matrices are known, Eq.(D-21) defines the unknowns according to the least-squares procedure.

The uncertainty in the fitting by the least-squares procedure is determined by the sum of squared deviations, $\gamma = \tilde{\boldsymbol{\delta}}\boldsymbol{\delta}$. The standard deviation is given by

$$\text{standard deviation} = \left(\frac{\gamma}{m - 1}\right)^{1/2}, \tag{D-22}$$

where m is the number of measurements and γ is given by [see Eq.(D-18)]

$$\gamma = \tilde{\boldsymbol{\delta}}\boldsymbol{\delta} = \tilde{\mathbf{U}}(\tilde{\mathbf{A}}\mathbf{AU} - \tilde{\mathbf{A}}\mathbf{M}) - \tilde{\mathbf{M}}\mathbf{AU} + \tilde{\mathbf{M}}\mathbf{M}.$$

Substituting Eq.(D-20) gives

$$\gamma = \tilde{\mathbf{M}}\mathbf{M} - \tilde{\mathbf{M}}\mathbf{AU}. \tag{D-23}$$

The reader may want to extend the analysis above by adding in the uncertainties in the measurements or alternatively by using weighting factors for higher or lower confidence in a certain measurement relative to others.

Appendix E SYMMETRY AND GROUP THEORY

E.1 MOLECULAR SYMMETRY AND GROUPS

Molecular symmetry is classified according to the number of symmetry elements and corresponding symmetry operations that transform a molecule into itself. A symmetry element is a geometric location, line, or plane that acts as a molecular reference for the symmetry operations of a molecule. For instance, ethylene,

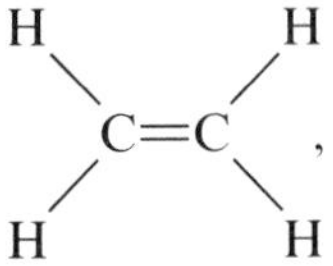

, has a center of inversion symmetry element that lies midway between the carbon atoms. The inversion operator is a symmetry operator that changes the signs on the x, y, and z coordinates for each particle in the molecule. The basic symmetry operators are listed in Table E-1.

The eight basic symmetry operators or transformations that transform the ethylene molecule into itself are illustrated in Fig. E-1. In addition, several successive operations are also illustrated. Of course, the protons and carbon atoms are actually indistinguishable in ethylene and the numbers are used to illustrate the transformations. It is also easy to show that an operator, R, times its own inverse, R^{-1}, is equal to the identity, E.

Table E-1 The basic symmetry elements and symmetry operations.

Symmetry Element	Symbol	Symmetry Operation
Identity	E	No change.
Plane	σ	Reflection through the plane
Center of inversion	i	Inversion of all atoms through the center
Axis of symmetry	C_n	Rotation about the axis by $2\pi/n$
Rotation-reflection axis of symmetry	S_n	Rotation about the axis by $2\pi/n$ followed by a reflection through the plane perpendicular to the axis of rotation

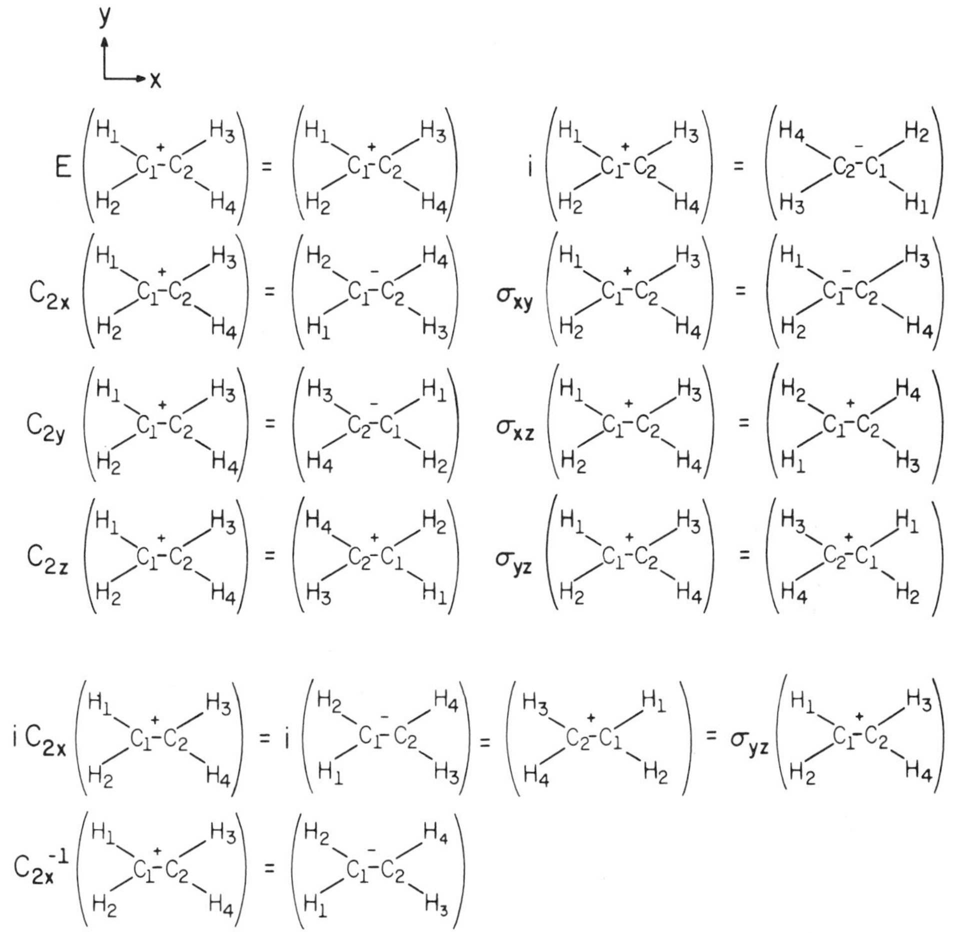

Figure E-1 The eight basic symmetry operations that transform ethylene into itself. A product and an inverse operation are also shown. The + and − indicate the sign of the z coordinate.

A *group* is a collection of operators that are interrelated by the following four rules:

1. The product of any two elements of a group (including the square) must also be a member of the group. The group elements need not, however, commute.
2. Each group must contain the identity operator.
3. The associative law of multiplication must hold for the elements of a group.
4. Each element in the group must have an inverse that is also an element of the group.

The *order* of a group is defined as the number of symmetry elements in the group. A multiplication table for the ethylene symmetry group is shown in Table E-2 where some of the multiplications are shown in Fig. E-1 and the remaining products are

Table E-2 The group multiplication table for the symmetry operators of ethylene (D_{2h}). The order of multiplication is row first followed by the column operator.

	E	C_{2z}	C_{2y}	C_{2x}	i	σ_{xy}	σ_{xz}	σ_{yz}
E	E	C_{2z}	C_{2y}	C_{2x}	i	σ_{xy}	σ_{xz}	σ_{yz}
C_{2z}	C_{2z}	E	C_{2x}	C_{2y}	σ_{xy}	i	σ_{yz}	σ_{xz}
C_{2y}	C_{2y}	C_{2x}	E	C_{2z}	σ_{xz}	σ_{yz}	i	σ_{xy}
C_{2x}	C_{2x}	C_{2y}	C_{2z}	E	σ_{yz}	σ_{xz}	σ_{xy}	i
i	i	σ_{xy}	σ_{xz}	σ_{yz}	E	C_{2z}	C_{2y}	C_{2x}
σ_{xy}	σ_{xy}	i	σ_{yz}	σ_{xz}	C_{2z}	E	C_{2x}	C_{2y}
σ_{xz}	σ_{xz}	σ_{yz}	i	σ_{xy}	C_{2y}	C_{2x}	E	C_{2z}
σ_{yz}	σ_{yz}	σ_{xz}	σ_{xy}	i	C_{2x}	C_{2y}	C_{2z}	E

easily determined. Note that each row and each column contain each of the eight elements of the original group. The elements in the multiplication table indicate two *subgroups* which are smaller groups within the larger group which satisfy the four rules for a group. The two subgroups in the larger group are (E) and (E, C_{2x}, C_{2y}, and C_{2z}).

Another way of separating smaller sets of the larger group is by the *class structure* which is defined by the set of elements which are related to each other by the $R^{-1}AR = B$ relation involving the inverse operations. A and B are in the same class and R represents the other operators in the group. It is easy to show in the example (D_{2h}) above that each element forms a class by itself as each element is equal to its own inverse. The orders of the classes must be integral divisions of the order of the original group.

The more common groups for the interpretation of molecular structure and other relevant applications are listed in Table E-3.

E.2 MATRIX REPRESENTATION AND CHARACTER TABLES

A representation or matrix representation of a symmetry operator can be generated in any known basis set of functions. Let R be a symmetry operator and ψ be an arbitrary basis set. The matrix representation of R in the ψ basis is $\mathbf{R}^{\psi}$ as shown in Section 3.3. Consider the representation generated by the operators in the C_{3v} group by operating on the x, y, and z coordinates. The equilateral triangle-based pyramid (the NH_3 molecule) has the symmetry appropriate to the C_{3v} group with the symmetry operators shown in Fig. E-2. Consider, for instance, the counterclockwise rotation, $C_3\circlearrowleft$, which can be obtained from the rotational transformation in Eq.(B-28) to give

$$C_3\circlearrowleft\begin{pmatrix} x \\ y \\ z \end{pmatrix} = \begin{pmatrix} \cos\theta & -\sin\theta & 0 \\ \sin\theta & \cos\theta & 0 \\ 0 & 0 & 1 \end{pmatrix}\begin{pmatrix} x \\ y \\ z \end{pmatrix}.$$

Table E-3 The common symmetry groups used for the interpretation of atomic and molecular structure and other applications.

Group	Symmetry Elements	System
C_1	E	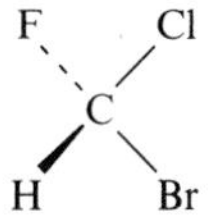
C_2	E, C_2	hydrogen peroxide (see Fig. 4-37)
C_i	E, i	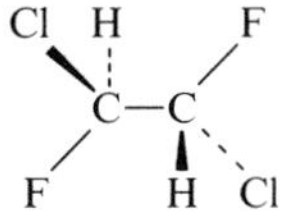
C	E, σ	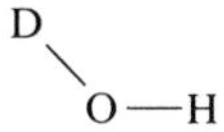
C_{2v}	$E, C_2, 2\sigma$	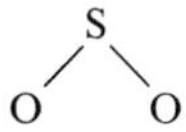
C_{3v}	$E, 2C_3, 3\sigma$	NH_3, FCH_3
$C_{\infty v}$	$E, C_\infty, \infty\,\sigma_\parallel$	OCS, HCN, ClF
C_{2h}	E, C_2, σ, i	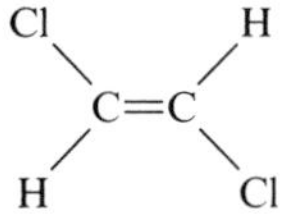
D_2	$E, 3C_2$	asymmetric top rotational wavefunctions
D_{2d}	$E, 3C_2, 2S_4, 2\sigma$	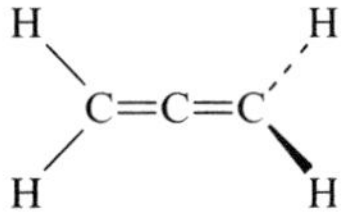
D_{3d}	$E, 2C_3, 3C_2, 2S_6, i, 3\sigma$	cyclohexane
D_{2h}	$E, 3C_2, 3\sigma, i$	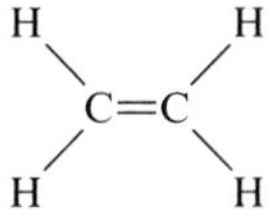
D_{3h}	$E, 2C_3, 3C_2, 4\sigma, 2S_3$	BF_3 (planar)
D_{4h}	$E, 2C_4, 5C_2, 5\sigma, 2S_4, i$	cyclobutane
D_{6h}	$E, 2C_6, 7C_2, 7\sigma, 2C_3, 2S_6, i, 2S_3$	benzene
$D_{\infty h}$	$E, C_\infty, \infty C_2, \infty\sigma_\parallel, \sigma_\perp, i$	OCO, HC≡CH, and symmetric top rotational wavefunctions
T_d	$E, 3C_2, 8C_3, 6\sigma, 6S_4$	CH_4
O_h	$E, 6C_4, 8C_3, 6S_4, 9C_2, 9\sigma, 8S_6, i$	ReF_6

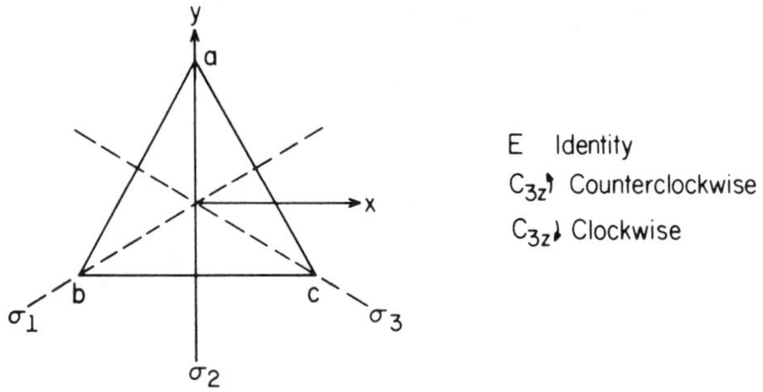

Figure E-2 The symmetry elements of the equilateral triangular-based pyramid, NH_3 or $ClCH_3$. The C_3 operations are about the z axis (axis of highest rotational symmetry).

Substituting $\theta = 120^\circ$ gives the result. Similar transformation matrices can be constructed for the remaining operators in the C_{3v} group giving

$$R\begin{pmatrix} x \\ y \\ z \end{pmatrix} = \mathbf{R}\begin{pmatrix} x \\ y \\ z \end{pmatrix}$$

Class 1:

$$\mathbf{E} = \begin{pmatrix} 1 & 0 & 0 \\ 0 & 1 & 0 \\ 0 & 0 & 1 \end{pmatrix}$$

Class 2:

$$\mathbf{C}_3\circlearrowleft = \begin{pmatrix} -\dfrac{1}{2} & -\dfrac{(3)^{1/2}}{2} & 0 \\ \dfrac{(3)^{1/2}}{2} & -\dfrac{1}{2} & 0 \\ 0 & 0 & 1 \end{pmatrix}, \quad \mathbf{C}_3\circlearrowright = \begin{pmatrix} -\dfrac{1}{2} & \dfrac{(3)^{1/2}}{2} & 0 \\ -\dfrac{(3)^{1/2}}{2} & -\dfrac{1}{2} & 0 \\ 0 & 0 & 1 \end{pmatrix}$$

Class 3:

$$\boldsymbol{\sigma}_1 = \begin{pmatrix} \dfrac{1}{2} & \dfrac{(3)^{1/2}}{2} & 0 \\ \dfrac{(3)^{1/2}}{2} & -\dfrac{1}{2} & 0 \\ 0 & 0 & 1 \end{pmatrix}, \quad \boldsymbol{\sigma}_2 = \begin{pmatrix} -1 & 0 & 0 \\ 0 & 1 & 0 \\ 0 & 0 & 1 \end{pmatrix},$$

$$\boldsymbol{\sigma}_3 = \begin{pmatrix} \dfrac{1}{2} & -\dfrac{(3)^{1/2}}{2} & 0 \\ -\dfrac{(3)^{1/2}}{2} & -\dfrac{1}{2} & 0 \\ 0 & 0 & 1 \end{pmatrix}. \qquad \text{(E-1)}$$

The matrix representations clearly multiply according to the same order as the corresponding operators. Of the many possible matrix representations of the symmetry operators of groups, the *irreducible representation*, which is the lowest-order (dimension) block-diagonal matrix, is of special interest. The $(x \quad y \quad z)$ function generates a threefold (3×3) reducible representation for the C_{3v} group [see Eqs.(E-1)]. The irreducible blocks in the matrices in Eqs.(E-1) include a 1×1 irreducible representation generated by z and a 2×2 irreducible representation generated by $(x \quad y)$.

The irreducible representations of a group are related by the celebrated great orthogonality theorem [12],

$$\sum_R \Gamma_i^*(R)_{mn}\Gamma_j(R)_{m'n'} = \frac{h}{(l_i l_j)^{1/2}}\delta_{ij}\delta_{mm'}\delta_{nn'}, \tag{E-2}$$

where the sum over R represents the sum over all the operations in the group, h is the order of the group, l_i is the dimension of the ith irreducible representation that is given by the $\mathbf{\Gamma}_i(R)$ matrix of which $\Gamma_i(R)_{mn}$ is the mnth matrix element. The remarkable feature of the great orthogonality theorem is that the unifying features of symmetry can be obtained from only the diagonal sum (character) of the matrices forming the irreducible representations. Several rules concerning groups and *characters* can be derived directly from the great orthogonality theorem. These rules are listed here [12]:

1. The number of irreducible representations is equal to the number of classes in a group.
2. The sums of squares of the dimensions of the irreducible representations is equal to the order of the group, h.

$$l_1^2 + l_2^2 + l_3^2 + \cdots = h.$$

3. The sum of the squares of the characters in any irreducible representation is equal to the order of the group, $\sum_R [\chi_i(R)]^2 = h$. $\chi_i(R)$ indicates the character of the ith irreducible representation and the sum is over all the operators in the group.
4. The products of the characters of two different irreducible representations summed over all operations in the group is zero,

$$\sum_R \chi_i(R)\chi_j(R) = h\delta_{ij}. \tag{E-3}$$

5. The characters in any given representation are identical within each class of the group.

We shall now apply these rules to the C_{3v} group and the representations shown in Eqs.(E-1). We find three classes in the C_{3v} group: (E), (C_3^1, C_3^2), and $(\sigma_1, \sigma_2, \sigma_3)$. Thus, according to Rule 1, there are three irreducible representations in the C_{3v} group. The order of the irreducible representations is obtained from Rule 2: $l_1 = 1$, $l_2 = 1$, and $l_3 = 2$ to give $l_1^2 + l_2^2 + l_3^2 = 6$. We noted earlier that the only further reduction of the reducible representation in Eq.(E-1) was into a 1×1 and 2×2 set. Thus, the 1×1 irreducible representation generated by z can account for l_1 and the

2 × 2 irreducible representation can correspond to l_3. We now arrange the l_1 and l_3 irreducible characters in the C_{3v} character table according to symmetry operators:

C_{3v}	E	$C_3^{\circlearrowright}$	$C_3^{\circlearrowleft}$	σ_1	σ_2	σ_3
z	1	1	1	1	1	1
$(x \quad y)$	2	−1	−1	0	0	0
A_2	1	1	1	−1	−1	−1

We note the validity of Rules 3, 4, and 5. We still have an additional one-dimensional irreducible representation (the number of irreducible representations must equal the number of classes) to fill out the C_{3v} character table. This representation is obtained from applying Rules 3, 4, and 5 and the result is listed above under A_2. The character table is usually written for each class because the characters of a given irreducible representation are equal within each class. The final results are given in Table E-4.

The z- and x, y-generating functions, which generate the A_1 and E irreducible representations, are listed in the right column in Table E-4. Other functions that generate irreducible representations include the rotational operators, R_x, R_y, and R_z and the d-type functions [see Eq.(4-102)]; z^2, $x^2 - y^2$, xy, xz, yz. The rotational operators represent clockwise rotations about the indicated axes. Consider R_z, which is a rotation about the z axis, $ER_z = R_z$. $C_3^{\circlearrowright} R_z = R_z$ and $C_3^{\circlearrowleft} R_z = R_z$ as a C_3 rotation does not change the direction of the rotation of R_z; however, $\sigma_1 R_z = -R_z$, $\sigma_2 R_z = -R_z$, and $\sigma_3 R_z = -R_z$ as a reflection will change the sense of the rotation from clockwise to counterclockwise. Thus, it is evident that R_z generates the A_2 irreducible representation of the C_{3v} group. The character tables for the more commonly used groups listed in Table E-3 are given in Table E-5.

We shall now illustrate the decomposition of a *reducible representation* into its *irreducible* components. Remembering that a unitary transformation, which block diagonalizes a reducible representation into its irreducible blocks, does not change the value of the character (see Problem 3-14), we can write the character of the reducible

Table E-4 The C_{3v} character table showing several of the functions that generate the irreducible representations. The breakdown of the cross-product representations into irreducible components is also shown in the lower part of the table.

C_{3v}	E	$2C_3$	3σ	
A_1	1	1	1	$z, x^2 + y^2, z^2$
A_2	1	1	−1	R_z
E	2	−1	0	$(x, y)(R_x, R_y), (x^2 - y^2, xy)(xz, yz)$
$A_1 \times A_1$	1	1	1	$= A_1$
$A_1 \times A_2$	1	1	−1	$= A_2$
$A_2 \times A_2$	1	1	1	$= A_1$
$A_1 \times E$	2	−1	0	$= E$
$E \times E$	4	1	0	$= A_1 + A_2 + E$
$E \times E \times E$	8	−1	0	$= A_1 + A_2 + 3E$

Table E-5 Group character tables most often used in examining molecular structure. The character tables for groups indicated by the direct product are obtained [see Eq.(B-18)] by the products of the characters in the two combining groups. A and B indicate one-dimensional representations and E and T indicate two- and three-dimensional representations, respectively.

C_1	E
A	1

C_2			E	C_2
x^2, y^2, z^2, xy	R_z, z	A	1	1
xz, yz	x, y; R_x, R_y	B	1	-1

C_3			E	C_3	C_3^2	
x^2+y^2, z^2	R_z, z	A	1	1	1	
(xz, yz) (x^2-y^2, xy)	(x, y) (R_x, R_y)	E	1	ω	ω^2	$\omega = \exp(2\pi i/3)$
			1	ω^2	ω	

C_4 (4)			E	C_2	C_4	C_4^3
x^2+y^2, z^2	R_z, z	A	1	1	1	1
x^2-y^2, xy		B	1	1	-1	-1
(xz, yz)	(x, y) (R_x, R_y)	E	1	-1	i	$-i$
			1	-1	$-i$	i

C_5			E	C_5	C_5^2	C_5^3	C_5^4	
x^2+y^2, z^2	R_z, z	A	1	1	1	1	1	
(xz, yz)	(x, y) (R_x, R_y)	E'	1	ω	ω^2	ω^3	ω^4	
			1	ω^4	ω^3	ω^2	ω	$\omega = \exp(2\pi i/5)$
(x^2-y^2, xy)		E''	1	ω^2	ω^4	ω	ω^3	
			1	ω^3	ω	ω^4	ω^2	

C_6			E	C_6	C_3	C_2	C_3^2	C_6^5	
x^2+y^2, z^2	R_z, z	A	1	1	1	1	1	1	
		B	1	-1	1	-1	1	-1	
(xz, yz)	(x, y) (R_x, R_y)	E'	1	ω	ω^2	ω^3	ω^4	ω^5	$\omega = \exp(2\pi i/6)$
			1	ω^5	ω^4	ω^3	ω^2	ω	
(x^2-y^2, xy)		E''	1	ω^2	ω^4	1	ω^2	ω^4	
			1	ω^4	ω^2	1	ω^4	ω^2	

Table E-5 *(continued)*

D_2			E	C_2^z	C_2^y	C_2^x
x^2, y^2, z^2		A_1	1	1	1	1
xy	R_z, z	B_1	1	1	-1	-1
xz	R_y, y	B_2	1	-1	1	-1
yz	R_x, x	B_3	1	-1	-1	1

D_3			E	$2C_3$	$3C_2'$
$x^2 + y^2, z^2$		A_1	1	1	1
	R_z, z	A_2	1	1	-1
$\left.\begin{matrix}(xz, yz)\\(x^2 - y^2, xy)\end{matrix}\right\}$	$\left.\begin{matrix}(x, y)\\(R_x, R_y)\end{matrix}\right\}$	E	2	-1	0

D_4			E	$C_2 = C_4^2$	$2C_4$	$2C_2'$	$2C_2''$
$x^2 + y^2, z^2$		A_1	1	1	1	1	1
	R_z, z	A_2	1	1	1	-1	-1
$x^2 - y^2$		B_1	1	1	-1	1	-1
xy		B_2	1	1	-1	-1	1
(xz, yz)	$\left.\begin{matrix}(x, y)\\(R_x, R_y)\end{matrix}\right\}$	E	2	-2	0	0	0

D_5			E	$2C_5$	$2C_5^2$	$5C_2'$	
$x^2 + y^2, z^2$		A_1	1	1	1	1	
	R_z, z	A_2	1	1	1	-1	
(xz, yz)	$\left.\begin{matrix}(x, y)\\(R_x, R_y)\end{matrix}\right\}$	E_1	2	$2\cos x$	$2\cos 2x$	0	$x = 2\pi/5$
$(x^2 - y^2, xy)$		E_2	2	$2\cos 2x$	$2\cos 4x$	0	

D_6			E	C_2	$2C_3$	$2C_6$	$3C_2'$	$3C_2''$
$x^2 + y^2, z^2$		A_1	1	1	1	1	1	1
	R_z, z	A_2	1	1	1	1	-1	-1
		B_1	1	-1	1	-1	1	-1
		B_2	1	-1	1	-1	-1	1
(xz, yz)	$\left.\begin{matrix}(x, y)\\(R_x, R_y)\end{matrix}\right\}$	E_1	2	-2	-1	1	0	0
$(x^2 - y^2, xy)$		E_2	2	2	-1	-1	0	0

C_{2v}			E	$C_2(z)$	$\sigma_v(xz)$	$\sigma'_v(yz)$
x^2, y^2, z^2	z	A_1	1	1	1	1
xy	R_z	A_2	1	1	-1	-1
xz	R_y, x	B_1	1	-1	1	-1
yz	R_x, y	B_2	1	-1	-1	1

C_{3v}			E	$2C_3$	$3\sigma_v$
$x^2 + y^2, z^2$	z	A_1	1	1	1
	R_z	A_2	1	1	-1
$\left.\begin{matrix}(x^2 - y^2, xy)\\(xz, yz)\end{matrix}\right\}$	$\left.\begin{matrix}(x, y)\\(R_x, R_y)\end{matrix}\right\}$	E	2	-1	0

C_{4v}			E	C_2	$2C_4$	$2\sigma_v$	$2\sigma_d$
$x^2 + y^2, z^2$	z	A_1	1	1	1	1	1
	R_z	A_2	1	1	1	-1	-1
$x^2 - y^2$		B_1	1	1	-1	1	-1
xy		B_2	1	1	-1	-1	1
(xz, yz)	$\left.\begin{matrix}(x, y)\\(R_x, R_y)\end{matrix}\right\}$	E	2	-2	0	0	0

C_{5v}			E	$2C_5$	$2C_5^2$	$5\sigma_v$	
$x^2 + y^2, z^2$	z	A_1	1	1	1	1	
	R_z	A_2	1	1	1	-1	$x = 2\pi/5$
(xz, yz)	$\left.\begin{matrix}(x, y)\\(R_x, R_y)\end{matrix}\right\}$	E_1	2	$2 \cos x$	$2 \cos 2x$	0	
$(x^2 - y^2, xy)$		E_2	2	$2 \cos 2x$	$2 \cos 4x$	0	

C_{6v}			E	C_2	$2C_3$	$2C_6$	$3\sigma_d$	$3\sigma_v$
$x^2 + y^2, z^2$	z	A_1	1	1	1	1	1	1
	R_z	A_2	1	1	1	1	-1	-1
		B_1	1	-1	1	-1	-1	1
		B_2	1	-1	1	-1	1	-1
(xz, yz)	$\left.\begin{matrix}(x, y)\\(R_x, R_y)\end{matrix}\right\}$	E_1	2	-2	-1	1	0	0
$(x^2 - y^2, xy)$		E_2	2	2	-1	-1	0	0

C_{1h}			E	σ_h
x^2, y^2, z^2, xy	R_z, x, y	A'	1	1
xz, yz	R_x, R_y, z	A''	1	-1

Table E-5 *(continued)*

C_{2h}			E	C_2	σ_h	i
x^2, y^2, z^2, xy	R_z	A_g	1	1	1	1
	z	A_u	1	1	-1	-1
xz, yz	R_x, R_y	B_g	1	-1	-1	1
	x, y	B_u	1	-1	1	-1

$C_{3h} = C_3 \times \sigma_h$			E	C_3	C_3^2	σ_h	S_3	$(\sigma_h C_3^2)$
$x^2 + y^2, z^2$	R_z	A'	1	1	1	1	1	1
	z	A''	1	1	1	-1	-1	-1
$(x^2 - y^2, xy)$	(x, y)	E'	$\begin{cases}1\\1\end{cases}$	ω ω^2	ω^2 ω	1 1	ω ω^2	ω^2 ω
(xz, yz)	(R_x, R_y)	E''	$\begin{cases}1\\1\end{cases}$	ω ω^2	ω^2 ω	-1 -1	$-\omega$ $-\omega^2$	$-\omega^2$ $-\omega$

$\omega = \exp(2\pi i/3)$

$$C_{4h} = C_4 \times i$$

$$C_{5h} = C_5 \times \sigma_h$$

$$C_{6h} = C_6 \times i$$

S_2			E	i
$x^2, y^2, z^2, xy, xz, yz$	R_x, R_y, R_z	A_g	1	1
	x, y, z	A_u	1	-1

S_4			E	C_2	S_4	S_4^3
$x^2 + y^2, z^2$	R_z	A	1	1	1	1
	z	B	1	1	-1	-1
(xz, yz) $(x^2 - y^2, xy)$	(x, y) (R_x, R_y)	E	$\begin{cases}1\\1\end{cases}$	-1 -1	i $-i$	$-i$ i

$$S_6 = C_3 \times i$$

D_{2d}			E	C_2	$2S_4$	$2C_2'$	$2\sigma_d$
$x^2 + y^2, z^2$		A_1	1	1	1	1	1
	R_z	A_2	1	1	1	-1	-1
$x^2 - y^2$		B_1	1	1	-1	1	-1
xy	z	B_2	1	1	-1	-1	1
(xz, yz)	(x, y) (R_x, R_y)	E	2	-2	0	0	0

$$D_{3d} = D_3 \times i$$

$$D_{2h} = D_2 \times i$$

$D_{3h} = D_3 \times \sigma_h$			E	σ_h	$2C_3$	$2S_3$	$3C_2'$	$3\sigma_v$
$x^2 + y^2, z^2$		A_1'	1	1	1	1	1	1
	R_z	A_2'	1	1	1	1	−1	−1
		A_1''	1	−1	1	−1	1	−1
	z	A_2''	1	−1	1	−1	−1	1
$(x^2 - y^2, xy)$	(x, y)	E'	2	2	−1	−1	0	0
(xz, yz)	(R_x, R_y)	E''	2	−2	−1	1	0	0

$$D_{4h} = D_4 \times i$$

$$D_{5h} = D_5 \times \sigma_h$$

$$D_{6h} = D_6 \times i$$

T		E	$3C_2$	$4C_3$	$4C_3'$	
	A	1	1	1	1	
	E	{1	1	ω	ω^2	
		{1	1	ω^2	ω	$\omega = \exp(2\pi i/3)$
(R_x, R_y, R_z) (x, y, z)	T	3	−1	0	0	

$$T_h = T \times i$$

O		E	$8C_3$	$3C_2 = 3C_4^2$	$6C_2$	$6C_4$
	A_1	1	1	1	1	1
	A_2	1	1	1	−1	−1
$(x^2 - y^2, 3z^2 - r^2)$	E	2	−1	2	0	0
(R_x, R_y, R_z) (x, y, z)	T_1	3	0	−1	−1	1
(xy, yz, zx)	T_2	3	0	−1	1	−1

$$O_h = O \times i$$

T_d		E	$8C_3$	$3C_2$	$6\sigma_d$	$6S_4$
	A_1	1	1	1	1	1
	A_2	1	1	1	−1	−1
	E	2	−1	2	0	0
(R_x, R_y, R_z)	T_1	3	0	−1	−1	1
(x, y, z)	T_2	3	0	−1	1	−1

$C_{\infty v}$			E	$2C_\phi$	σ_v
$x^2 + y^2, z^2$	z	$A_1(\Sigma^+)$	1	1	1
	R_z	$A_2(\Sigma^-)$	1	1	−1
(xz, yz)	(x, y) (R_x, R_y)	$E_1(\Pi)$	2	$2 \cos \phi$	0
$(x^2 - y^2, xy)$		$E_2(\Delta)$	2	$2 \cos 2\phi$	0
		—		—	

Table E-5 (*continued*)

	$D_{\infty h}$		E	$2C_\phi$	C'_2	i	$2iC_\phi$	iC'_2
$x^2 + y^2, z^2$		$A_{1g}(\Sigma_g^+)$	1	1	1	1	1	1
		$A_{1u}(\Sigma_u^+)$	1	1	1	−1	−1	−1
	R_z	$A_{2g}(\Sigma_g^-)$	1	1	−1	1	1	−1
	z	$A_{2u}(\Sigma_u^-)$	1	1	−1	−1	−1	1
(xz, yz)	(R_x, R_y)	$E_{1g}(\Pi_g)$	2	$2\cos\phi$	0	2	$2\cos\phi$	0
	(x, y)	$E_{1u}(\Pi_u)$	2	$2\cos\phi$	0	−2	$-2\cos\phi$	0
$(x^2 - y^2, xy)$		$E_{2g}(\Delta_g)$	2	$2\cos 2\phi$	0	2	$2\cos 2\phi$	0
		$E_{2u}(\Delta_u)$	2	$2\cos 2\phi$	0	−2	$-2\cos 2\phi$	0
		—				—		

representation, $\chi(R)$, as a linear combination of the characters of the irreducible components, $\chi_j(R)$, according to

$$\chi(R) = \sum_j a_j \chi_j(R). \qquad \text{(E-4)}$$

a_j is the number of times the character of the jth irreducible representation appears in the reducible representation. Multiplying Eq.(E-4) by $\chi_j(R)$, summing over R, and using Eq.(E-3) gives

$$a_j = \frac{1}{h} \sum_R \chi_j(R)\chi(R). \qquad \text{(E-5)}$$

It is easy to show by these techniques that the (3×3) reducible representation in Eqs.(E-1) decomposes into the already evident $A_1 + E$ irreducible representations.

We shall now use Eq.(E-5) to reduce the representation formed from the $3n = 9$ displacement coordinates in a symmetric ABC (the A and C atoms are equal) nonlinear triatomic molecule, which has C_{2v} symmetry, into $3n - 6 = 3$ vibrational components and 3 components each of both translational and rotational motion. Let each atom have small displacements Δx, Δy, and Δz for $3n$ total displacements. Now let the z axis be the C_2 axis of symmetry and let the x axis be perpendicular to the molecular plane (see Table E-6). The matrix representation for the C_{2z} operator in the C_{2v} group in the displacement coordinate basis is given by

$$C_{2z}\begin{pmatrix} \Delta x_a \\ \Delta y_a \\ \Delta z_a \\ \Delta x_b \\ \Delta y_b \\ \Delta z_b \\ \Delta x_c \\ \Delta y_c \\ \Delta z_c \end{pmatrix} = \begin{pmatrix} 0 & 0 & 0 & 0 & 0 & 0 & -1 & 0 & 0 \\ 0 & 0 & 0 & 0 & 0 & 0 & 0 & -1 & 0 \\ 0 & 0 & 0 & 0 & 0 & 0 & 0 & 0 & 1 \\ 0 & 0 & 0 & -1 & 0 & 0 & 0 & 0 & 0 \\ 0 & 0 & 0 & 0 & -1 & 0 & 0 & 0 & 0 \\ 0 & 0 & 0 & 0 & 0 & 1 & 0 & 0 & 0 \\ -1 & 0 & 0 & 0 & 0 & 0 & 0 & 0 & 0 \\ 0 & -1 & 0 & 0 & 0 & 0 & 0 & 0 & 0 \\ 0 & 0 & 1 & 0 & 0 & 0 & 0 & 0 & 0 \end{pmatrix}\begin{pmatrix} \Delta x_a \\ \Delta y_a \\ \Delta z_a \\ \Delta x_b \\ \Delta y_b \\ \Delta z_b \\ \Delta x_c \\ \Delta y_c \\ \Delta z_c \end{pmatrix}. \qquad \text{(E-6)}$$

Table E-6 The C_{2v} character table showing the reducible representation, Γ_{red}, formed from the nine displacement coordinates of a nonlinear symmetric triatomic molecule.

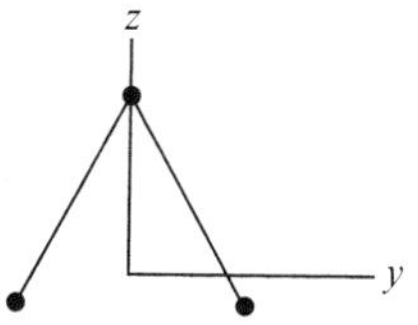

C_{2v}	E	$C_2(z)$	$\sigma_v(xz)$	$\sigma_v'(yz)$	
A_1	1	1	1	1	z, x^2, y^2, z^2
A_2	1	1	−1	−1	R_z, xy
B_1	1	−1	1	−1	x, R_y, xz
B_2	1	−1	−1	1	y, R_x, yz
Γ_{red}	9	−1	1	3	$3A_1 + A_2 + 2B_1 + 3B_2$

The other matrices are also easily worked out for the symmetry elements in C_{2v} giving the characters, $\Gamma_{red}(R)$, for the reducible representation as shown in Table E-6. The reducible representation decomposes into the irreducible components by using Eq.(E-5) to give $\Gamma_{red} = 3A_1 + A_2 + 2B_1 + 3B_2$, as shown also in Table E-6. Subtracting the rotational (R_x, R_y, and R_z) and the translational (x, y, and z) components from Γ_{red} leaves the vibrational degrees of freedom, $\Gamma_{vib} = \Gamma_{red} - \Gamma_{rot} - \Gamma_{trans} = 2A_1 + B_2$. Thus, the normal modes of vibrational motion generate two A_1 and one B_2 irreducible representations. This is quite evident also from the nature of the normal modes as illustrated in Fig. 4-34 where v_1 and v_2 show A_1 symmetry and v_3 shows B_2 symmetry. The symmetry of the normal modes of motion of any complex molecule can be obtained by determining first the reducible representation formed from the displacement coordinates and then the translational and rotational components are subtracted leaving the symmetry species of the vibrational normal modes. We shall discuss further the concept of symmetry in molecular vibrations in Section E.3.

Before proceeding, we shall illustrate the direct product of the irreducible representations and the subsequent decomposition according to Eq.(E-5). The direct product of irreducible or reducible representations is obtained by the direct product of the matrices according to Eq.(B-18),

$$\mathbf{\Gamma}(R) = \mathbf{\Gamma}_i(R) \times \mathbf{\Gamma}_j(R), \qquad \text{(E-7)}$$

where R is an operator in the group and the direct or cross product indicates multiplication of every element in $\mathbf{\Gamma}_i(R)$ with every element in $\mathbf{\Gamma}_j(R)$. Thus, $\mathbf{\Gamma}(R)$ has a dimension equal to the product of the dimensions of $\mathbf{\Gamma}_i(R)$ and $\mathbf{\Gamma}_j(R)$. It is easy to show that the character (diagonal sum) of $\mathbf{\Gamma}(R)$ in Eq.(E-7) is equal to the product of the characters of $\mathbf{\Gamma}_i(R)$ and $\mathbf{\Gamma}_j(R)$,

$$\chi(R) = \chi_i(R)\chi_j(R). \qquad \text{(E-8)}$$

Thus, we can now use Eq.(E-5) to decompose the reducible representations formed from direct products into their irreducible components. Consider the direct products of several combinations of the irreducible representations of the C_{3v} group and the subsequent decomposition into the irreducible representation as illustrated in the lower part of Table E-4. Examples for other groups are easily obtained. One result that is evident in Table E-4 or in the case of any other group is that the direct product of any irreducible representation with itself must contain, in its reduction, the totally symmetric irreducible representation of the group.

E.3 NORMAL COORDINATE SYMMETRY AND THE SYMMETRY OF VIBRATIONAL WAVEFUNCTIONS

We noted in Section E.2 that the *normal coordinates* of a given molecule generated certain of the irreducible representations of the symmetry group appropriate to the molecule. The vibrational harmonic oscillator eigenfunctions for the normal modes also generate certain of the irreducible representations of the corresponding group. We have noted before that the total vibrational eigenfunction is a product of the individual harmonic oscillator functions for each normal mode. For instance, in the case of a triatomic molecule [see discussion following Eq.(4-261)] we write $\psi(n_1 n_2 n_3) = \psi_{n_1}(Q_1)\psi_{n_2}(Q_2)\psi_{n_3}(Q_3)$, where n_1, n_2, and $n_3 = 0, 1, 2, 3, \ldots$ represent the vibrational state with the individual harmonic oscillator functions as listed in Table 4-25. The ground state is given by

$$\psi(0, 0, 0) = \frac{(\gamma_1\gamma_2\gamma_3)^{1/2}}{\pi^{3/4}} \exp\left(-\frac{\gamma_1^2 Q_1^2}{2} - \frac{\gamma_2 Q_2^2}{2} - \frac{\gamma_3 Q_3^2}{2}\right). \tag{E-9}$$

This function clearly generates the totally symmetric representation in any symmetry group. The first excited symmetric stretching mode is given by

$$\begin{aligned}\psi(1, 0, 0) &= \frac{(2\gamma_1\gamma_2\gamma_3)^{1/2}}{\pi^{3/4}} \gamma_1 Q_1 \exp\left(-\frac{\gamma_1^2 Q_1^2}{2} - \frac{\gamma_2^2 Q_2^2}{2} - \frac{\gamma_3 Q_3^2}{2}\right)\\ &= (2)^{1/2}\gamma_1 Q_1 \psi(0, 0, 0).\end{aligned} \tag{E-10}$$

This function transforms according to the same irreducible representation as the normal coordinate Q_1. Similar statements are evident for the $\psi(0, 1, 0)$ and $\psi(0, 0, 1)$ functions. Thus, in general, we note that the first excited vibrational state of the normal coordinate Q_i transforms according to the same symmetry classification as Q_i itself.

Before discussing the symmetry of overtones and combination bands, we shall discuss the nature of the isotope effect in the following progression of molecules:

$$\begin{array}{ccccccccc} CH_4 & \longleftrightarrow & CH_3D & \longleftrightarrow & CH_2D_2 & \longleftrightarrow & CHD_3 & \longleftrightarrow & CD_4 \\ T_d & \longleftrightarrow & C_{3v} & \longleftrightarrow & C_{2v} & \longleftrightarrow & C_{3v} & \longleftrightarrow & T_d \end{array}$$

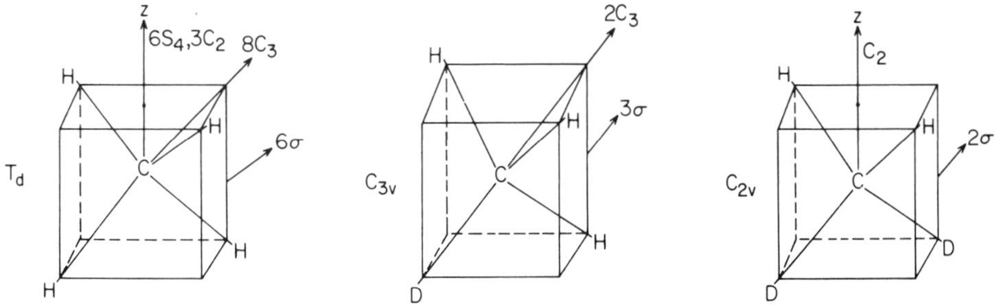

Figure E-3 Symmetry elements in the CH_4, CH_3D, and CH_2D_2 molecules.

Consider first the tetrahedral CH_4 molecule. The molecule is placed in a cube and the symmetry elements are outlined in Fig. E-3. Also outlined in Fig. E-3 are the symmetry elements for CH_3D (C_{3v}) and CH_2D_2 (C_{2v}). It is evident that both C_{3v} and C_{2v} are subgroups of T_d but C_{2v} is not a subgroup of C_{3v}. First we shall generate the reducible representation of the displacement coordinates in CH_4. These results are shown in the T_d character table in Table E-7. We can subtract the translational (T_2) and rotational (T_1) components from Γ_{red} in Table E-7 to give the irreducible representations generated by the normal (vibrational) coordinates; $\Gamma_{vib} = A_1 + E + 2T_2$. Thus, there is a singly degenerate, doubly degenerate, and two triply degenerate vibrational normal modes in a tetrahedral molecule. The vibrational wavefunction is noted by $\psi(n_1 n_2 n_3 n_4)$. $A_1(n_1)$ is a symmetric stretch where all hydrogen atoms breathe (stretch) with respect to the central stationary carbon atom. One T_2 (n_3) is the triply degenerate stretch and the other T_2 (n_4) is a triply degenerate bending motion. The E (n_2) is a doubly degenerate bending motion.

The symmetry species corresponding to the stretching and bending modes can be obtained by using *internal coordinates* that correspond to bond length and bond angles in the molecule. The reducible representation of the four CH vectors, $\boldsymbol{r}_1, \boldsymbol{r}_2, \boldsymbol{r}_3$, and $\boldsymbol{r}_4$, is also shown in Table E-7 leading to the $A_1 + T_2$ irreducible components; these are the stretching normal modes of motion. Comparing with Γ_{vib} shows that the bending modes have E and T_2 symmetry. Single-vibrational mode, first-excited state energies and appropriate symmetries for CH_4 and CD_4 are shown on either side of the correlation diagram in Fig. E-4.

Table E-7 The T_d character table. The reducible representation formed from the 15 displacement coordinates in CH_4 is noted by Γ_{red}.

T_d	E	$8C_3$	$3C_2$	$6\sigma_d$	$6S_4$	
A_1	1	1	1	1	1	$x^2 + y^2 + z^2$
A_2	1	1	1	−1	−1	
E	2	−1	2	0	0	$(x^2 + y^2 - 2z^2, x^2 - y^2)$
T_1	3	0	−1	−1	1	(R_x, R_y, R_z)
T_2	3	0	−1	1	−1	$(T_x, T_y, T_z)(xy, xz, yz), (x, y, z)$
Γ_{red}	15	0	−1	3	−1	$= A_1 + E + T_1 + 3T_2$
$\Gamma_{stretch}(r_1 r_2 r_3 r_4)$	4	1	0	2	0	$= A_1 + T_2$

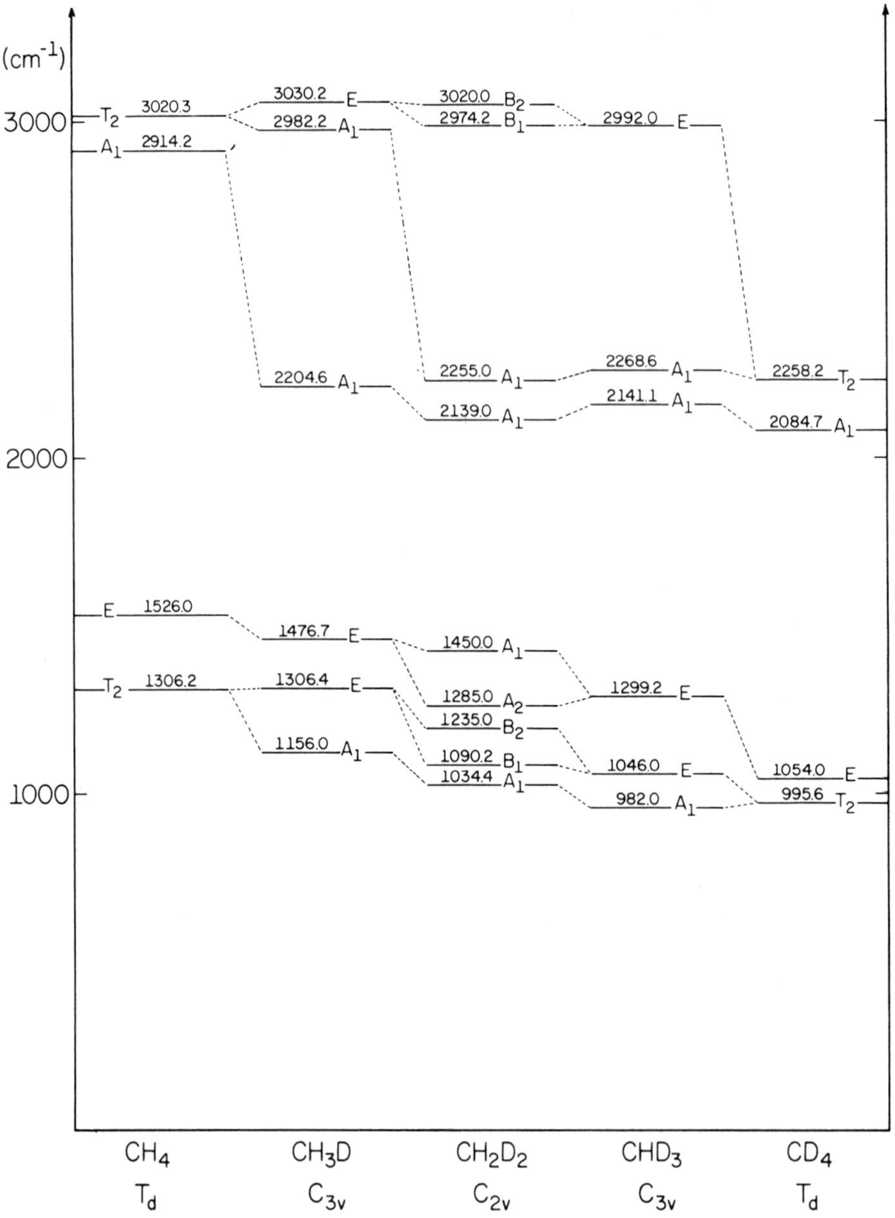

Figure E-4 The correlation in vibrational frequencies starting on the left with CH_4 and then proceeding to the right with increased *D* substitution. The vibrational frequencies are from G. Herzberg, *Infrared and Raman Spectra* (D. Van Nostrand Co., Inc., 1945).

Table E-8 The C_{3v} character table showing the decomposition of the A_1, E, and T_2 representations that originate under the T_d group (see Fig. E-3).

C_{3v}	E	$2C_3$	3σ	
A_1	1	1	1	
A_2	1	1	−1	
E	2	−1	0	
from T_d A_1	1	1	1	$= A_1$
from T_d E	2	−1	0	$= E$
from T_d T_2	3	0	1	$= A_1 + E$

Symmetry reduction and the decomposition of the irreducible representations in T_d under the influence of the lower symmetry under C_{2v} and C_{3v} are also evident in Fig. E-4. Consider first the $CH_4 \longrightarrow CH_3D$ correlation where the symmetry operations of CH_3D in a cube are shown in Fig. E-3. In the $T_d \longrightarrow C_{3v}$ progression, the $6S_4$, $3C_2$, $6C_3$, and 3σ symmetry elements are lost; however, $2C_3$ and 3σ under T_d are also present in C_{3v}. The C_{3v} group is shown again in Table E-8 where we now decompose the A_1, E, and T_2 representations from the T_d group under the lower symmetry C_{3v} group. The characters under the T_d group carry over under the same symmetry operations in the C_{3v} group that is a subgroup of T_d. The A_1 and E representations remain irreducible, but the T_2 irreducible representation under T_d breaks down into a singly degenerate A_1 and a doubly degenerate E irreducible representation under the constraints of lower symmetry in C_{3v}. Physically this means that the T_2 triply degenerate vibrational modes under CH_4 break into two separate vibrations in CDH_3 that have A_1 and E symmetry under the C_{3v} group as illustrated in Fig. E-4.

Further symmetry reduction is experienced with the substitution of two D atoms to give CH_2D_2, as also shown in the correlation diagram in Fig. E-4. As C_{2v} is not a subgroup of C_{3v}, a direct reduction from C_{3v} to C_{2v} is not possible. C_{2v} is a subgroup of T_d, however, so we can decompose the original A_1, E, and $2T_2$ modes in CH_4 under C_{2v} (CH_2D_2) by carrying the characters over according to the common symmetry elements as illustrated in Fig. E-3. The resultant decomposition is shown in Table E-9.

The resultant reduction in symmetry and the symmetry of the new normal modes in CH_2D_2 are also shown in Fig. E-4. An identical correlation is obtained by

Table E-9 The C_{2v} character table showing the decomposition of the A_1, E, and T_2 representations which originate under the T_d group.

C_{2v}	E	$C_2(z)$	$\sigma_v(xz)$	$\sigma_h(yz)$	
A_1	1	1	1	1	
A_2	1	1	−1	−1	
B_1	1	−1	1	−1	
B_2	1	−1	−1	1	
from T_d A_1	1	1	1	1	$= A_1$
from T_d E	2	2	0	0	$= A_1 + A_2$
from T_d T_2	3	−1	1	1	$= A_1 + B_1 + B_2$

starting on the right side of Fig. E-4 with CD_4 by adding protons to give CD_3H and then CH_2D_2.

This concept of the reduction of symmetry and decomposition of higher group symmetry under lower-order groups is very useful. A large number of examples have been tabulated by Wilson, Decius, and Cross [13].

We shall now return to a more complete discussion of the symmetries of the excited vibrational states in a molecule.

First Excited State

Our previous work has shown that the symmetry of the first excited vibrational state is equal to the symmetry of the normal coordinate. For instance, in methane, as shown in Fig. E-4, we have $\psi(1, 0, 0, 0) \longrightarrow A_1$ (stretch), $\psi(0, 1, 0, 0) \longrightarrow E$ (bend), $\psi(0, 0, 1, 0) \longrightarrow T_2$ (stretch), and $\psi(0, 0, 0, 1) \longrightarrow T_2$ (bend).

Combination of Singly Excited Normal Modes

The direct product is used here and can be illustrated with methane by using the T_d group character table in Table E-7 to decompose the direct products to give, for example, $\psi(1, 1, 0, 0) \longrightarrow A_1 \times E = E$, $\psi(1, 1, 1, 0) \longrightarrow A_1 \times E \times T_2 = T_1 + T_2$, $\psi(1, 0, 1, 0) \longrightarrow A_1 \times T_2 = T_2$, and $\psi(1, 1, 1, 1) \longrightarrow A_1 \times E \times T_2 \times T_2 = A_1 + A_2 + 2E + 2T_1 + 2T_2$.

Overtones of One Normal Coordinate

Singly degenerate normal modes

In this case, the simple direct product of the character of the irreducible representation is used. In methane (see Table E-7), we have

$$\psi(4, 0, 0, 0) \quad \longrightarrow \quad A_1 \times A_1 \times A_1 \times A_1 = A_1.$$

In SO_2 (see Table E-9), we have

$$\psi(0, 0, 3) \quad \longrightarrow \quad B_2 \times B_2 \times B_2 = B_2 \quad \text{and} \quad \psi(0, 0, 2) \quad \longrightarrow \quad B_2 \times B_2 = A_1.$$

Doubly degenerate normal modes

In this case, the direct product cannot be used. For instance, consider doubly degenerate modes v_a and v_b that transform according to E. If the doubly degenerate mode is in the second excited state as $\psi(0, 2, 0, 0)$ in methane, the vibrational excitation could be distributed between v_a and v_b according to $\psi(E) = \psi(2) \longrightarrow \psi(v_a v_b) \longrightarrow \psi(2, 0)$, $\psi(1, 1)$, or $\psi(0, 2)$. Thus, $\psi(0, 2, 0, 0)$ in methane is actually triply degenerate and not fourfold degenerate as indicated from the direct product $(E \times E)$. Another case is the third excited state given by $\psi(3) \longrightarrow \psi(v_a v_b) \longrightarrow \psi(3, 0)$, $\psi(2, 1)$, $\psi(1, 2)$, or $\psi(0, 3)$, which is fourfold degerate. In general,

$$\text{degeneracy} = (n_E + 1), \tag{E-11}$$

where $n_E = 0, 1, 2, \ldots$ indicates the excitation of the doubly degenerate E vibrational mode. The symmetry species of the multiply excited E symmetry vibrational modes must be investigated individually. Symmetry tables have been determined for the common groups by Wilson, Decius, and Cross [13] in their Table X-13. The results for methane give $\psi(0, 2, 0, 0) \longrightarrow A_1 + E$, $\psi(0, 3, 0, 0) \longrightarrow A_1 + A_2 + E$, and $\psi(0, 4, 0, 0) \longrightarrow A_1 + 2E$.

Triply degenerate normal modes

The difficulty with triply degenerate modes is similar but more complex than the doubly degenerate symmetry species discussed above. Triply degenerate modes have the following degeneracies for overtones:

$$\text{degeneracy} = \tfrac{1}{2}(n_T + 1)(n_T + 2), \tag{E-12}$$

where $n_T = 0, 1, 2, \ldots$ indicates the excitation of the triply degenerate modes. The symmetries of these functions are also given from Table X-13 in Wilson, Decius, and Cross [13] and the results are given here for a few examples in methane: $\psi(0, 0, 2, 0) \longrightarrow A_1 + E + T_1$, $\psi(0, 0, 0, 2) \longrightarrow A_1 + E + T_2$, and $\psi(0, 0, 3, 0) \longrightarrow A_2 + 2T_1 + T_2$.

Combination Functions

Combinations of overtones from different normal modes are obtained by direct products of our previous results. For instance, in methane we have

$$\psi(1, 0, 2, 0) \quad \longrightarrow \quad A_1 \times (A_1 + E + T_1) = A_1 + E + T_1$$

and

$$\psi(0, 2, 2, 0) \quad \longrightarrow \quad (A_1 + E) \times (A_1 + E + T_1) = 2A_1 + A_2 + 3E + 2T_1 + T_2.$$

E.4 CONSTRUCTION OF HYBRID ORBITALS

The angular dependence in the s, p_x, p_y, p_z, d_{xy}, $d_{x^2-y^2}$, d_{xz}, d_{yz}, and d_{z^2} atomic orbitals are listed in Eqs.(4-100), (4-101), and (4-102). See also Fig. 5-9. We shall now determine which combinations of these spatial atomic orbitals will give the desired molecular geometry. For instance, CH_4 is a tetrahedral molecule and we ask which carbon atomic orbital combinations will form a tetrahedral configuration. The bonds directed between the atoms are called σ (sigma) bonds and bonds that form from atomic orbitals directed perpendicular to the interatomic bonding line are called π (pi) bonds.

First we shall consider the σ-bonding in CH_4, which is represented by bond vectors from the central carbon atom to the bonding hydrogen atoms. The reducible representation formed from the four bond vectors reduces to $A_1 + T_2$ components as shown in Table E-7. Thus, the atomic orbitals which combine to yield the tetrahedral configuration must also generate a reducible representation which decomposes to give $A_1 + T_2$; this limits the atomic orbitals to two combinations according to Table E-7: (s, p_x, p_y, p_z and s, d_{xy}, d_{xz}, d_{yz}). Thus, tetrahedral orbitals are formed from either sp^3 or sd^3 atomic orbitals. The linear combination of atomic orbitals to form σ hybrid orbitals, as well as π orbitals, are easily determined for other symmetry groups [14].

E.5 QUANTUM MECHANICS AND MATRIX ELEMENT THEOREMS

The Hamiltonian, $\mathscr{H}$, for an isolated system must be invariant to the operations corresponding to the symmetry elements of the group that describes the system. Thus, the Hamiltonian must always generate the totally symmetric irreducible representation for the appropriate group.

Let $\boldsymbol{\theta}$ generate the irreducible representations of a group with $\boldsymbol{\theta}^i$ being that component of $\boldsymbol{\theta}$ that generates the ith irreducible representation. Consider the overlap matrix formed from $\boldsymbol{\theta}^i$ and $\boldsymbol{\theta}^j$ given by $\mathbf{S} = \int \boldsymbol{\theta}^{i\dagger}\boldsymbol{\theta}^j \, dV$. We shall now examine the nature of the overlap matrix $\mathbf{S}$ with the help of group theory. We note that $\mathbf{S}$ need not be a square matrix if $\boldsymbol{\theta}^i$ and $\boldsymbol{\theta}^j$ generate irreducible representations with different dimensions. First, apply a symmetry operation R to $\mathbf{S}$ to give $R\mathbf{S} = \mathbf{S}$ as the $\mathbf{S}$ matrix is an array of numbers that are all invariant to the symmetry operations. Continuing, we write

$$\int \boldsymbol{\theta}^{i\dagger}\boldsymbol{\theta}^j \, dV = R\int \boldsymbol{\theta}^{i\dagger}\boldsymbol{\theta}^j \, dV = \int (R\boldsymbol{\theta}^{i\dagger})(R\boldsymbol{\theta}^j)\, dV$$

$$= \tilde{\boldsymbol{\Gamma}}_i(R)\mathbf{S}\boldsymbol{\Gamma}_j(R) = \mathbf{S}, \tag{E-13}$$

where we have used

$$R\boldsymbol{\theta}^j = \boldsymbol{\theta}^j\boldsymbol{\Gamma}_j(R), \qquad R\boldsymbol{\theta}^{i\dagger} = R^*\boldsymbol{\theta}^{i\dagger} = \boldsymbol{\Gamma}_i(R)\boldsymbol{\theta}^{i\dagger} \tag{E-14}$$

for real symmetry operators and real $\boldsymbol{\Gamma}(R)$ matrices where $\boldsymbol{\Gamma}(R)^\dagger = \boldsymbol{\Gamma}(R)$. Consider now a single matrix element in the $\mathbf{S}$ matrix according to Eq.(E-13),

$$\mathrm{S}_{nm} = \sum_{lr} \Gamma_i(R)_{ln}\mathrm{S}_{lr}\Gamma_j(R)_{rm}. \tag{E-15}$$

Summing both sides of this equation over all h symmetry elements R in the group and comparing with Eq.(E-2) gives

$$\sum_R \mathrm{S}_{nm} = h\mathrm{S}_{nm} = \sum_R \sum_{l,r} \Gamma_i(R)_{ln}\mathrm{S}_{lr}\Gamma_j(R)_{rm} = \frac{h}{(l_i l_j)^{1/2}}\delta_{ij}\delta_{nm}\sum_{l,r}\delta_{lr}\mathrm{S}_{lr}. \tag{E-16}$$

According to this result, all the elements in the $\mathbf{S}$ matrix are zero unless $i = j$ or unless both $\boldsymbol{\theta}^i$ and $\boldsymbol{\theta}^j$ generate the same irreducible representation.

Consider now the matrix representation of the Hamiltonian operator over $\boldsymbol{\theta}^i$ and $\boldsymbol{\theta}^j$,

$$\int \boldsymbol{\theta}^{i\dagger}\mathscr{H}\boldsymbol{\theta}^j \, dV = \mathscr{H}. \tag{E-17}$$

Operating on both sides of this equation with a symmetry operator R gives

$$\mathrm{R}\mathscr{H} = \mathscr{H} = \int (R\boldsymbol{\theta}^{i\dagger})(R\mathscr{H})(R\boldsymbol{\theta}^j)\, dV = \int (R\boldsymbol{\theta}^{i\dagger})\mathscr{H}(R\boldsymbol{\theta}^j)\, dV. \tag{E-18}$$

It is now evident, by repeating the steps from Eqs.(E-13) to (E-16), that there are no nonzero matrix elements of the Hamiltonian operator between functions that generate different irreducible representations. In addition, there are no nonzero matrix elements of the Hamiltonian between functions that generate the different rows of the same irreducible representation. We now have a systematic method of reducing the number of off-diagonal matrix elements in the Hamiltonian matrix in an arbitrary basis $\mathbf{\Phi}$. From Eq.(3-74) we write

$$\mathcal{H}^{\psi} = \mathbf{E} = \mathbf{a}^{\dagger}\mathcal{H}^{\Phi}\mathbf{a}, \qquad \boldsymbol{\psi} = \mathbf{\Phi}\mathbf{a}, \tag{E-19}$$

where $\boldsymbol{\psi}$ are the eigenfunctions of $\mathcal{H}$ and $\mathbf{\Phi}$ is a convenient basis for calculating the matrix representation of $\mathcal{H}$. We now seek a basis $\boldsymbol{\theta}$ which are eigenfunctions of the symmetry operators and which will reduce $\mathcal{H}^{\Phi}$ to a more nearly diagonal form, $\mathcal{H}^{\theta}$. $\boldsymbol{\theta}$ and $\mathbf{\Phi}$ are related by a unitary transformation $\mathbf{U}$ according to

$$\boldsymbol{\theta} = \mathbf{\Phi}\mathbf{U}, \qquad \mathcal{H}^{\theta} = \mathbf{U}^{\dagger}\mathcal{H}^{\Phi}\mathbf{U}. \tag{E-20}$$

$\boldsymbol{\theta}^i$ and $\boldsymbol{\theta}^j$ are components in $\boldsymbol{\theta}$ that generate the ith and jth irreducible representations of the symmetry operations. We start by generating the reducible representation, $\mathbf{\Gamma}(R)^{\Phi}$, of the symmetry operators, R, in the $\mathbf{\Phi}$ basis according to

$$R\mathbf{\Phi} = \mathbf{\Phi}\mathbf{\Gamma}(R)^{\Phi}. \tag{E-21}$$

Using $\boldsymbol{\theta} = \mathbf{\Phi}\mathbf{U}$ from Eq.(E-20), we can write

$$\mathbf{\Gamma}(R)^{\theta} = \mathbf{U}^{\dagger}\mathbf{\Gamma}(R)^{\Phi}\mathbf{U}, \qquad \mathbf{\Gamma}(R)^{\Phi} = \mathbf{U}\mathbf{\Gamma}(R)^{\theta}\mathbf{U}^{\dagger}. \tag{E-22}$$

Substituting into Eq.(E-21) gives

$$R\mathbf{\Phi} = \mathbf{\Phi}\mathbf{U}\mathbf{\Gamma}(R)^{\theta}\mathbf{U}^{\dagger} = \boldsymbol{\theta}\mathbf{\Gamma}(R)^{\theta}\mathbf{U}^{\dagger}. \tag{E-23}$$

Now, rewrite Eq.(E-23) for a single component in $\mathbf{\Phi}$ giving

$$R\Phi_m = \boldsymbol{\theta}\mathbf{\Gamma}(R)^{\theta}\mathbf{U}^{\dagger}_m, \tag{E-24}$$

where $\mathbf{U}^{\dagger}_m$ is the mth column in $\mathbf{U}^{\dagger}$. Multiplying both sides of this equation by the character of the jth irreducible representation, $\chi_j(R)$, and summing over R gives

$$\sum_R \chi_j(R)R\Phi_m = \sum_R \chi_j(R)\,\boldsymbol{\theta}\mathbf{\Gamma}(R)^{\theta}\mathbf{U}^{\dagger}_m = \sum_R \chi_j(R)\sum_i \sum_{l,n} \theta^i_l \Gamma_i(R)^{\theta}_{ln}\mathrm{U}_{mn}, \tag{E-25}$$

where $\Gamma_i(R)^{\theta}_{ln}$ is the lnth matrix element in the ith irreducible representation and θ^i_l is the lth component in the symmetrized function $\boldsymbol{\theta}^i$ that generates the ith irreducible representation. In the case of one-dimensional irreducible representations, Eq.(E-25) reduces to

$$\sum_R \chi_j(R)R\Phi_m = \sum_R \chi_j(R)\sum_i \chi_i(R)\theta^i\mathrm{U}_{mi} = h\mathrm{U}_{mj}\theta^j, \tag{E-26}$$

where Eq.(E-2) is used in the last step. $h\mathrm{U}_{mj}$ is the normalizing constant for the symmetrized functions formed from Φ_m.

In the case of irreducible representations of order 2 or larger, the left-hand side of Eq.(E-26) can be used to generate any one of the component functions; the remaining components of the set are constructed from combinations of Φ_m that are orthogonal to the first set.

Table E-10 Generation of symmetry orbitals for the triangular-based pyramid. The initial basis is $\mathbf{\Phi} = (\Phi_a \Phi_b \Phi_c \Phi_d)$ where the indices *a*, *b*, *c*, and *d* indicate the points on the pyramid and each Φ_i is spherically symmetric. The symmetry group and $R\Phi_a$ and $R\Phi_d$ are shown for all symmetry elements in C_{3v} (see Fig. E-2).

C_{3v}	E	C_3 ↺	C_3 ↻	σ_1	σ_2	σ_3
A_1	1	1	1	1	1	1
A_2	1	1	1	−1	−1	−1
E	2	−1	−1	0	0	0
$R\Phi_a$	Φ_a	Φ_b	Φ_c	Φ_b	Φ_a	Φ_c
$R\Phi_d$	Φ_d	Φ_d	Φ_d	Φ_d	Φ_d	Φ_d

Consider the generation of symmetry orbitals from a basis set of spherically symmetric functions at each point in a triangular-based pyramid with C_{3v} symmetry. The original basis set is $\mathbf{\Phi} = (\Phi_a \Phi_b \Phi_c \Phi_d)$ with *a*, *b*, and *c* being identified in Fig.(E-2) at the base points and *d* is located at the apex of the pyramid. The generation of $R\Phi_a$ and $R\Phi_d$ for all operators in the C_{3v} group is given in Table E-10. The results for $R\Phi_a$ and $R\Phi_d$ are now summed over all R for each of the A_1, A_2, and E irreducible representations giving, according to Eq.(E-26), the following results:

	$\sum_R \chi_j(R) R\Phi_m$
Φ_a	
$j = A_1$	$\Phi_a + \Phi_b + \Phi_c$
$j = A_2$	0
$j = E$	$2\Phi_a - \Phi_b - \Phi_c$
Φ_d	
$j = A_1$	Φ_d
$j = A_2$	0
$j = E$	0

Thus, Φ_d and $\Phi_a + \Phi_b + \Phi_c$ form symmetry functions which are totally symmetric and $2\Phi_a - \Phi_b - \Phi_c$ forms one of the pair of functions which have E symmetry. Another possible pair with $2\Phi_a - \Phi_b - \Phi_c$ is the orthogonal $\Phi_b - \Phi_c$ combination. Thus, the unnormalized symmetry orbitals are given by

$$\theta^{A_1} = \Phi_d, \qquad \theta^{A_1} = \Phi_a + \Phi_b + \Phi_c, \qquad \boldsymbol{\theta}^E = (\theta_1^E \, \theta_2^E),$$
$$\theta_1^E = 2\Phi_a - \Phi_b - \Phi_c, \qquad \theta_2^E = \Phi_b - \Phi_c. \tag{E-27}$$

These functions are easily normalized.

We now combine an observation made at the end of Section E.2 with the results in this section; the direct product of any irreducible representation with itself must contain, in its reduction, the totally symmetric representation. It is evident that an integral describing some property of a system (overlap, energy, or other average value) will be zero unless the integrand generates a representation that contains some component of the totally symmetric representation of the group describing the system. This conclusion is of help in determining selection rules for electromagnetic transitions.

Appendix F GENERAL REFERENCES

References to the original literature, review articles, and books are cited sequentially in each chapter. In addition to the standard journals and periodicals, there are several series of books or monographs that are being continuously published. Some of these, which will be of future interest and influence in the study of molecular structure and dynamics, are listed here.

1. *Advances in Chemical Physics* (Wiley-Interscience, New York).
2. *Physical Chemistry, An Advanced Treatise* (Academic Press, New York).
3. *MTP Review of Science, Physical Chemistry, Series One* (Butterworths, London).
4. *Advances in Magnetic Resonance* (Academic Press, New York).
5. *Advances in Atomic and Molecular Physics* (Academic Press, New York).
6. *Advances in Quantum Chemistry* (Academic Press, New York).
7. *Historical Studies in the Physical Sciences* (University of Pennsylvania Press, Philadelphia).
8. *Annual Reviews of Physical Chemistry* (Annual Reviews, Palo Alto, California).
9. *Specialist Periodical Reports* (The Chemical Society, London). These reports commenced publication in 1973 and give an approximate annual review of the various field of study involved in the subgroups noted by
   Molecular Structure by Diffraction Methods
   Molecular Spectroscopy
   Electron Spin Resonance
   Nuclear Magnetic Resonance
   Dielectric and Related Molecular Processes
   Theoretical Chemistry.

REFERENCES IN APPENDIX

[1] M. L. McGlashon, *Ann. Rev. Phys. Chem.* **24**, 51 (1973).

[2] J. D. Jackson, *Mathematics for Quantum Mechanics* (W. A. Benjamin, Inc., New York, 1962).

[3] H. Margenau and G. M. Murphy, *The Mathematics of Physics and Chemistry*, 2nd ed. (D. Van Nostrand Co., Inc., New York, 1956).

[4] H. Goldstein, *Classical Mechanics* (Addison-Wesley Pub. Co., Reading, Mass., 1950).

[5] G. E. Hay, *Vector and Tensor Analysis* (Dover Publications, New York, 1953).

[6] D. D. FITTS, *Vector Analysis in Chemistry* (McGraw-Hill Book Co., New York, 1974).

[7] P. LORRAIN and D. CORSON, *Electromagnetic Fields and Waves* (W. H. Freeman and Co., San Francisco, 1970).

[8] J. MATHEWS and R. L. WALKER, *Mathematical Methods of Physics* (W. A. Benjamin, Inc., New York, 1965).

[9] P. A. M. DIRAC, *The Principles of Quantum Mechanics* (Oxford University Press, Oxford, 1957).

[10] W. H. WYLD, *Mathematical Methods for Physics* (W. A. Benjamin, Inc., New York, 1976).

[11] P.-O. LÖWDIN, *Adv. in Quant. Chem.* **5**, 185 (1970).

[12] M. TINKHAM, *Group Theory and Quantum Mechanics* (McGraw-Hill Book Co., New York, 1964).

[13] E. B. WILSON, Jr., J. C. DECIUS, and P. C. CROSS, *Molecular Vibrations* (McGraw-Hill Book Co., New York, 1955).

[14] More examples of hybrid orbitals, as well as other examples of the use of group theory, are found in F. A. COTTON, *Chemical Applications of Group Theory* (Wiley-Interscience, New York, 1971).

index